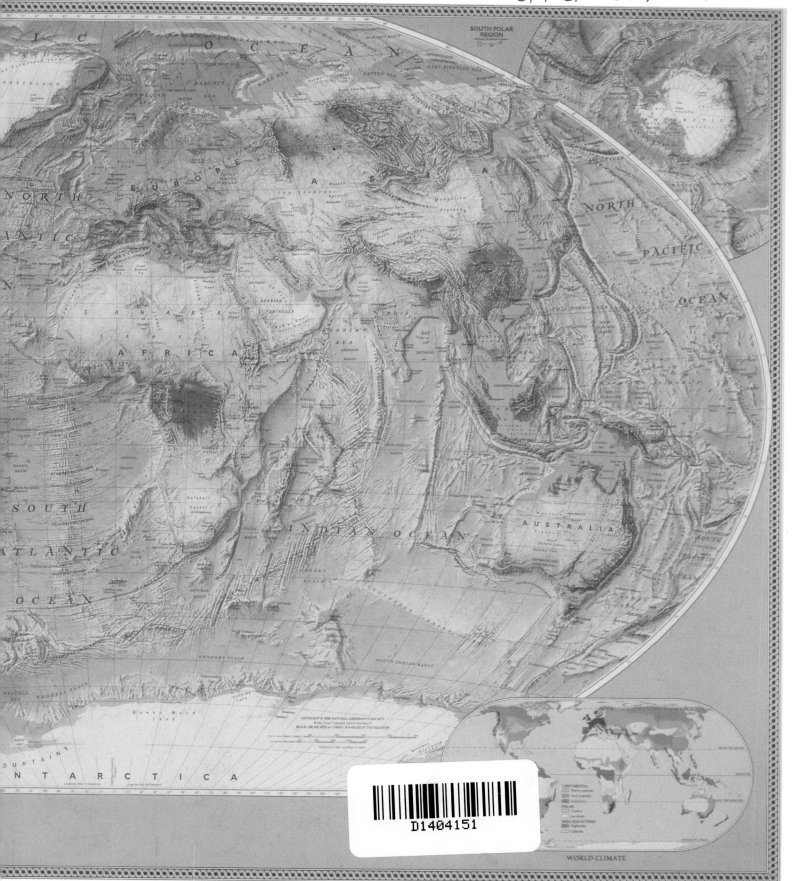

WORLD REGIONAL GEOGRAPHY

SIXTH EDITION

JOSEPH J. HOBBS

WORLD REGIONAL GEOGRAPHY

6TH EDITION

JOSEPH J. HOBBS
University of Missouri

Cartography by Andrew Dolan

BROOKS/COLE
CENGAGE Learning

Australia • Mexico • Singapore • Spain • United Kingdom • United States • Japan • Korea

BROOKS/COLE
CENGAGE Learning·

World Regional Geography
Joseph J. Hobbs

Acquisitions Editor: Peter Adams

Development Editor: Rebecca Heider

Assistant Editor: Alexandra Brady

Technology Project Manager: Melinda Newfarmer

Marketing Manager: Joseph Rogove

Marketing Assistant: Ashley Pickering

Marketing Communications Manager: Belinda Krohmer

Project Manager, Editorial Production: Michelle Cole

Creative Director: Rob Hugel

Art Director: Vernon Boes

Print Buyer: Linda Hsu

Permissions Editor: Bob Kauser

Production Service: Jonathan Peck

Text Designer: Patrick Devine Design

Photo Researcher: Kim Adams

Copy Editor: Bruce Emmer

Cover Designer: Patrick Devine Design

Cover Image: © Keren Su/Getty Images

Compositor: Newgen–Austin

For product information and technology assistance, contact us at
Cengage Learning Customer & Sales Support, 1-800-354-9706.

For permission to use material from this text or product,
submit all requests online at **cengage.com/permissions**.
Further permissions questions can be e-mailed to
permissionrequest@cengage.com

Library of Congress Control Number: 2007939467
ISBN-13: 978-0-495-38950-7
ISBN-10: 0-495-38950-1

Brooks/Cole
10 Davis Drive
Belmont, CA 94002-3098
USA

Cengage Learning is a leading provider of customized learning solutions with office locations around the globe, including Singapore, the United Kingdom, Australia, Mexico, Brazil, and Japan. Locate your local office at **international.cengage.com/region**.

Cengage Learning products are represented in Canada by Nelson Education, Ltd.

For your course and learning solutions, visit academic.cengage.com.
Purchase any of our products at your local college store or at our preferred online store **www.ichapters.com**.

Printed in Canada
1 2 3 4 5 6 7 12 11 10 09 08

brief contents

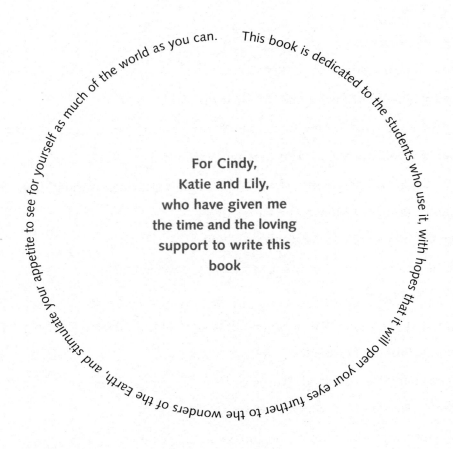

For Cindy,
Katie and Lily,
who have given me
the time and the loving
support to write this
book

This book is dedicated to the students who use it, with hopes that it will open your eyes further to the wonders of the Earth, and stimulate your appetite to see for yourself as much of the world as you can.

contents

maps

preface

World Regional Geography, Sixth Edition, is a comprehensive introduction to the interactions between people and environments around the world. It uses the tried-and-true convention of organizing the human experience on earth principally around regions rather than themes. This approach reflects my training as a geographer, as well as what I have found to be most useful for students in two decades of teaching geography. I introduce geographic themes to students and then immerse them in the interconnected but different realms on earth where these themes are manifested in distinctive and generally predictable ways.

My personal experience with geography underpins my approach to writing this text. I was trained as a regional geographer (focusing on the Middle East) at the University of Texas at Austin. Dr. Ian Manners was my mentor in all things Middle Eastern, while Dr. Robin Doughty helped develop my thematic and practical abilities in ethnographic research, humanistic geography, and natural history; Robin himself had been tutored by the great Berkeley cultural geographer Carl Sauer. I have been able to put the universal "Doughty skills" to work on research in many regions around the world but have only truly mastered the problems at hand where I can also tap into the regional "Manners skills": near-fluency in the Arabic language; intimate knowledge of history, politics, cultures, and environments of the Middle East; and, above all, long-term on-the-ground field-work in that region.

The regional approach is an extremely effective way for students to understand what is happening in the world today. Let us take the example of Iraq. An individual with thematic or systematic expertise in economic or medical geography might be placed on the ground there, or in any number of countries, and quickly come to terms with Iraq's fundamental problems in those areas. He or she might come up with a plan that would logically address or solve the problems, and may even have implemented similar plans successfully elsewhere. But what particular cultural, social, religious, linguistic, political, environmental, or other unique attributes of Iraq (or even within one its particular areas or ethnic groups) would have to be considered for that plan to work effectively in that particular place? Training in the regional context helps ensure the most effective outcomes. If we are to be wise stewards and decision makers in the world, we must have intimate and detailed knowledge about places.

The world is "globalized," but it is not—as the social commentator Thomas Friedman points out—"flat." Its human and physical terrains are extremely rich and varied, and no process will unfold the same way in every location.

This book is the student's "training manual" for the world. It begins with three introductory chapters laying the systematic and thematic foundations of geography, the regional approach, and the overarching human and physical processes at work around the world. It then leads the reader into eight world regions. In each region, the opening chapter lays the systematic and thematic foundations of area and population, physical geography and human adaptations, cultural and historical geographies, economic geography, and geopolitical issues. Subsequent modules give the reader an opportunity to explore the region in greater detail, on a subregional, country-by-country, and topical basis.

This book is comprehensive. It opens the door for the reader to become thoroughly introduced to a region without having to resort to outside readings and other supplements. It allows for flexibility in using the book in a one- or two-semester course (more about this is explained in the section "How to Use This Book").

What Makes This Book Different?

Many professors have lamented that most world regional texts are heavily weighted toward either human or physical geography. This text seeks a balance between the two, offering professors the option of assigning more or less of the cultural or natural material as he or she sees fit.

On the human geography side, readers will notice that I offer exceptional depth in both historical geography and current events. There is nothing contradictory about this. Things that have just happened and things that have yet to occur are seldom unprecedented. I have established the groundwork in both the physical and human realms that will help students picking up today's paper to understand the news in its regional, historical, and cultural contexts. Using the book's foundation, readers should also be able to develop reasonable forecasts of what is to come. Sections dedicated to geopolitical issues are particularly useful for honing those skills.

Underscoring the point that the world is not flat, more attention is paid in this text than in most to the issues of culture, ethnicity, language, religion, and indigenous peoples. Though the world seems to change quickly, leveling the playing field—with more people accessing the Internet, using the English language, and adopting other international cultural attributes—regional and local traditions often endure and even strengthen. As the book points out, the Internet is in many cases helping resuscitate dying languages. And while I do not expect that professors will require students to memorize the names of all the languages spoken in a particular region, I do hope they will ask students to recognize how diverse the world's peoples are by appreciating, in the written text and on the maps, their rich linguistic landscapes.

What's New in This Edition?

For longtime users, the first remarkable feature of the book is that Kit Salter is no longer coauthor. Kit is happily retired and active on other projects. I have missed working with him. For readers, however, the advantage is a more consistent approach, tone, and style throughout the book. It has been a challenge, but a very exciting and informative one, for me to try to keep up with literally everything going on in the world and to distill this into a manageable book for students.

The format is entirely new. The previous exhaustive examination of physical and human processes that operate in each of the various world regions has been cleaved into two chapters, with new attention given to tectonic processes, landforms, and water resources in the physical geography chapter. The book now consists of 11 chapters (3 introductory and 8 regional) rather than 24. There are also 14 more detailed "breakout modules" designed for further exploration or longer courses.

Much of the economic geography in the previous edition had grown long in the tooth. The Sixth Edition still lays the foundation for understanding how regional economies evolved, however the postindustrial characteristics of the more developed countries get the attention they deserve (the final chapter on the United States has been overhauled most extensively). And perhaps most significant of all, the economic "800-pound (and growing) gorilla" of China tramps through every region in this edition.

Andrew Dolan has drafted a number of new maps to reflect these changing economies and other shifts in landscape use and impact. There are also more satellite images to give readers a better feel for natural and cultural landscapes from the extraordinary perspective of space. I have photographed in more than 100 countries and selected images that I think provide a good and accurate illustration of the people and places of the world.

Ancillaries

This text is accompanied by a number of ancillary publications to assist instructors and enhance student learning:

Online Instructor's Manual with Test Bank

Contains chapter outlines, key terms, and themes for general discussion as well as perspectives on selected discussion questions from the text. The Test Bank offers nearly 1,000 multiple-choice, classroom-tested questions and answers keyed to the test.

Places of the World: A Study Guide for Achieving Geographical Literacy

by James W. Lett

Key geographical terms, political and blank maps, review exercises, summary, facts and figures for each region, and much more.

Exam View® Computerized Testing

Create, deliver, and customize tests and study guides (both print and online) in minutes with this easy-to-use assessment and tutorial system.

PowerLecture

This one-stop presentation tool makes it easy to assemble, edit, and present custom lectures using Microsoft™ PowerPoint™. PowerLecture includes animated versions of many figures and maps from the text, allowing you to integrate dynamic visuals into your course in a snap. You'll also find chapter outlines, summaries of key chapter concepts, art and photographs from the text, and JoinIn™ content to use with your Student Response System.

The Comparative World Atlas

This atlas from Hammond features detailed maps for each continent as well as the world so students can see and evaluate regional terrain and political boundaries. Maps, charts, and images help students grasp how the world is physically and culturally interdependent. Also includes a Master Index, Major City Populations Listing, Quick Reference Guide, Global Politics Map, World Statistical Tables, Time Zone Map, and a guide to using the atlas.

Acknowledgments

Many reviewers and advisors helped to bring this sixth edition to completion. Thanks to each and every one of these distinguished geographers and outstanding teachers and students! They include Elizabeth Leppman, Saint Cloud State University; Kefa Otiso, Bowling Green State University; Mark Guizlo, Lakeland Community College; Tarek Joseph, Henry Ford Community College; Joy Adams, Texas State University; Kyong (Keith) Chu, Bergen Community College; Keith Bell, Volunteer State Community College; William Strong, University of North Alabama; Mary Kimsey, James Madison University; Rebecca A. Howze, South Florida Community College; Robert Maguire, Trinity College; Jeffrey Bury, San Francisco State University; J. Duncan Shaeffer, Arizona State University; Kelly Ann Victor, Eastern Michigan University; David Wall, Saint Cloud State University; Norman Carter, California State University—Long Beach; Gail Hobbs, Pierce College; Jim Nemeth, University of Toledo; Fritz Gritzner, South Dakota State University; Wayland Bauer, St. Ambrose University; James Lindberg, University of Iowa; Richard W. Smith, Harford Community College; Ningchuan Xiao, Ohio State University; Audrey Mohan, Texas State University—San Marcos; Thomas Orf, Los Positas College; Alexander Diener, Pepperdine University; L. Edward Hicks, Faulkner University; Thomas William Hainze, Jr., Trinity Valley Community College; Jia Lu, University of Wisconsin—Stevens Point;and at the University of Missouri, using and advising on improvements from the Fourth Edition, Larry Brown, Jennifer Presberry,

Victor Lozano, Jason Jindrich, Nathaniel Albers, Rob Long, Jeff Pickles, Sunny Stevens, Johnny Finn, Trent Holmes, and Paul Hammond.

Deep thanks and appreciation go to the Brooks/Cole editorial, marketing, and production staff who labored so diligently to help us all meet the tight deadlines and the technical challenges of the Sixth Edition, especially Mindy Newfarmer, Peter Adams, Alex Brady, Rebecca Heider, Joe Rogove, Ashley Pickering, Belinda Krohmer, Michelle Cole, Vernon Boes, Linda Hsu, and Bob Kauser. Kim Adams worked quickly to supply a range of stock photo choices.

I want especially to recognize two persons without whom this book would have suffered a dramatic loss of quality. I first met Jon Peck by email as I worked on the book in a hotel room in the United Arab Emirates. He introduced himself as the book's project manager, and quickly assumed more and more responsibilities for seeing that everything would come together to get the book published. He also brought his extensive knowledge of the world to bear, politely correcting my errors and introducing new pieces of information. I am deeply grateful for his dedication to this effort. I have the same depth of gratitude to Andrew Dolan, formally known as the book's cartographer and research assistant. Andy is a perfectionist in the best sense of the word, never compromising on the accuracy and currency of his maps and data updates. He has also been meticulous in bringing new developments to my attention, and helping to ensuring the trustworthiness of my narrative

Joseph J. Hobbs

about the author

Joseph J. Hobbs received his B.A. at the University of California Santa Cruz in 1978 and his M.A. and Ph.D. at the University of Texas Austin in 1980 and 1986 respectively. He is a professor and chairman of the Department of Geography at the University of Missouriand a geographer of the Middle East with many years of field research on biogeography and Bedouin peoples in the deserts of Egypt. Joe's interest in the region grew from a boyhood lived in Saudi Arabia and India. His research in Egypt has been supported by Fulbright fellowships, the American Council of Learned Societies, the American Research Center in Egypt, and the National Geographic Society Committee for Research and Exploration. In the 1990s he served as the team leader of the Bedouin Support Program, a component of the St. Katherine National Park project in Egypt's Sinai Peninsula. In 2007 he led an effort to establish a national plan for environmental management in the United Arab Emirates. His current research interests are indigenous peoples' participation in protected areas in the Middle East, Southeast Asia and Cen-

Katie Hobbs

tral America, and the establishment of joint research and education programs between Vietnam and the United States. He is the author of Bedouin Life in the Egyptian Wilderness and Mount Sinai (both University of Texas Press), co-author of The Birds of Egypt (Oxford University Press), and co-editor of Dangerous Harvest: Drug Plants and the Transformation of Indigenous Landscapes (Oxford). He teaches graduate and undergraduate courses in world regional geography, environmental geography, the geography of the Middle East, the geography of caves, and the geography of global current events, the geographies of drugs and terrorism, and a field course on the ancient Maya geography of Belize. He has received the University of Missouri's highest teaching award, the Kemper Fellowship. During summers he has he led "adventure travel" tours to remote areas in Latin America, Africa, the Indian Ocean, Asia, Europe, and the High Arctic, and done his own research abroad. Joe lives in Missouri with his wife, Cindy, daughters Katie and Lily, and an animal menagerie.

how to use this book

Before you begin reading this Sixth Edition of Essentials of World Regional Geography, there are two important features of the book that you need to know about.

Organization. The text is organized with maximum flexibility in mind. It is suited for both one- and two-term courses. For a one-term course, the three thematic introductory chapters and the eight chapters introducing each region are a good match. These may be complemented by a few more in-depth readings about particular issues or places in the modules. For a two-term course, the three introductory thematic chapters, the eight introductory regional chapters, and as many complete or partial modules as wanted may be used. The regions do not need to be covered in any particular order. For example, although they appear last in the book, the North America chapters may be read first if so desired. For current events that deserve discussion—a vol-canic eruption in Indonesia or Middle East peace talks, for example—the instructor may easily call attention to those parts of the book that provide useful background.

Cross-Referencing. The book is written with global interconnections in mind. "Globalization" is understandable as a concept, but how exactly does it work? The page and figure numbers in the book's margins serve to tie the diverse strands of global issues together. For example, when you read that Europeans generally dislike genetically modified foods, page numbers in the margin next to that discussion lead you to descriptions of how farmers in Africa and Latin America decide to plant crops accordingly. As you read about China's economic growth and appetite for raw materials, you will likewise be directed to places all around the world where these forces come into play.

The bright lights of prosperous North American and Europe are in marked contrast with the darkness of Africa, the least developed continent.

Geographers are interested in the interactions between landscapes, natural systems, and peoples. The relationships have recognizable patterns in different parts of the world. This is a scene on the Perfume River in Vietnam.

Joe Hobbs

CHAPTER 1
OBJECTIVES AND TOOLS OF WORLD REGIONAL GEOGRAPHY

1.1 Welcome to World Regional Geography

In its once-a-decade survey conducted in 2002, the National Geographic Society asked a large group of United States citizens aged 18 to 24 to identify on maps some of the world's most important places. Eleven percent of those surveyed could not locate the United States on a blank map of the world. Forty-nine percent could not find New York City, ground zero for the most spectacular of the terrorist attacks of September 11, 2001. Eighty-three percent did not know where Afghanistan is, despite that country's steady presence in the news. Eighty-seven percent did not know where to situate Iraq, which U.S. forces were preparing to invade. The National Geographic survey also tested the geographic awareness of 18- to 24-year-olds from Canada, France, Germany, Italy, Japan, Mexico, Sweden, and Great Britain. The Americans came in next to last, ahead only of the young people of Mexico. (Sweden, incidentally, ranked number one, followed by Germany and Italy).

chapter outline

chapter objectives

This chapter should enable you to
→ Know what geography is
→ Understand the world regional approach to geography
→ Identify the six essential elements of geography
→ Learn some key concepts in geography
→ Appreciate the book's overall objectives
→ Learn the basic language of maps
→ Understand what GIS and remote sensing are
→ Know what geographers do and what kinds of jobs in geography are available

Most of you reading this book are in that group of 18- to 24-year-old Americans. The dismal findings about your peers a few years ago are not meant to shame you or confront you with what you might not know. Instead, they pose a challenge. They make us think about why it is important to know what is going on in the world and where events are taking place. Does it matter if a lot of us don't know where we are on a map? Does it matter if we don't know where other people and countries are? In short, does geography matter?

Years ago, geography earned a reputation for forcing students to memorize long lists of facts. And the truth is, by itself, knowledge of these facts probably means little. In the context of daily life, however, they have great power to transform our lives and contribute to the welfare of our communities and our country. By the end of this chapter, you will know what geography is, recognize the benefits you might gain from learning world regional geography, and understand the organization and objectives of this book.

What Is Geography?

Geography, a term first used by the Greek scholar Eratosthenes in the 3rd century B.C.E.,[1] literally means "description of the earth" but is probably best summed up as "the study of the earth as the home of humankind." Focusing on interactions between people and the environments in which we live, the modern academic discipline of geography has its roots in the Greek and Roman civilizations and emerged from that classical tradition through the Scientific Revolu-

tion in Europe (●Figure 1.1). Geography today is carried out today by professionals who have inherited that legacy. They are trained mainly in Western scientific techniques and publish online and in print in a host of languages.

There have always been other geographies, too. Arab geographers in places like Baghdad and Damascus, for example, kept the lights of geographic knowledge burning while Europe experienced its "Dark Ages" before a new dawn around 1000 C.E. All the world's other civilizations carried out explorations and scientific inquiry in fields related to what we call geography today. And for thousands of years, there have also been systems of geographic knowledge used by indigenous peoples who have not had written languages. Such **ethnogeographies,** still found among foraging, agricultural, and other rural peoples today, have gone largely unknown and untapped in the Western geographic tradition. Some geographers are actively collecting this kind of indigenous knowledge in the field, hoping to assemble and publish it before it is lost.

In another important exercise, the National Geographic Society commissioned a team of geographers to identify the core features of the discipline of geography. The team came up with what are known as the **six essential elements of geography.** These elements summarize what geographic research and education involve today and what distinguishes geography from other fields. They are as follows:

1. *The World in Spatial Terms.* Geography studies the relationships between people, places, and environments by mapping information about them into a spatial context.

Visual Arts Library (London)/Alamy

● **Figure 1.1** By measuring the lengths of shadows in the Egyptian cities of Alexandria and Aswan (Syene) and applying a simple mathematical formula, the Greek geographer Eratosthenes—the first person known to use the term *geography*—was able in the 3rd century B.C.E. to determine the circumference of the earth as 252,000 stadia—between 24,660 and 28,965 miles (39,690 and 46,620 km; we are not certain of the length of his stadion). The circumference of the earth, measured around the poles, is recognized today as 24,901 miles (40,075 km); Eratosthenes was right "in the ball park"—not bad for 2,000-year-old science!

2. *Places and Regions.* The identities and lives of individuals and peoples are rooted in particular places and in human constructs called regions.

3. *Physical Systems.* Physical processes shape earth's surface and interact with plant and animal life to create, sustain, and modify ecosystems.

4. *Human Systems.* People are central to geography; human activities, settlements, and structures help shape the earth's surface, and humans compete for control of the earth's surface.

5. *Environment and Society.* The physical environment is influenced by the ways in which human societies value and use the earth's physical features and processes.

6. *Uses of Geography.* Knowledge of geography enables people to develop an understanding of the relationships between people, places, and environments over time—that is, of the earth as it was, is, and might be.

While "essential," these six elements are extremely comprehensive. Geography is probably the most all-encompassing of the sciences—a fact of which geographers are duly proud. Broadly, the discipline has two major branches, **physical geography** and **human geography**, each of which has roots and relationships with other disciplines in the social and physical sciences (●Figure 1.2). Geographers very often bridge the social and natural sciences in their research (another fact they are proud of, as it is inappropriate or impossible for most disciplines to do this). Consider, for example, the context of this study, published in 2005 in the *Annals of the Association of American Geographers* (the "flagship" journal for geographers in the United States): "Contemporary Human Impacts on Alpine Ecosystems in the Sagarmatha (Mt. Everest) National Park, Khumbu, Nepal."[2] Alton Byers, the geographer who wrote this article, documented the natural history of one of the world's great national parks, came to understand the human social dynamics bringing about changes in the ecosystems there, and mapped those changes. His work involved every one of geography's six essential elements. It was a classic piece of research in a discipline concerned above all with the theme of **human-environment interaction.**

This interest in the **human agency** (humans' role in changing the face of the earth) has emerged time and again as one of geography's central themes. The great German geographer Alexander von Humboldt (1769–1859) initiated geography's modern era in a series of classic studies on this theme. From field observations in Venezuela, he concluded, "Felling the trees which cover the sides of the mountains provokes in every climate two disasters for future generations: a want of fuel and a scarcity of water."[3] Today, as we observe, for example, the direct connection between deforestation in Nepal's Himalaya Mountains and devastating floods downstream in Bangladesh, we can appreciate the foresight in Humboldt's warnings.

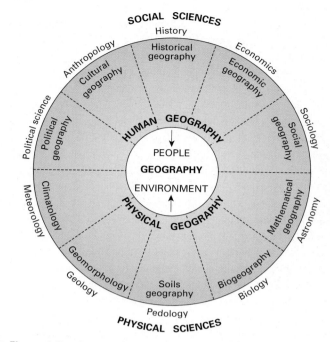

● **Figure 1.2** Selected subfields of geography. These are the main subject areas in human geography and physical geography and their links with the most closely related disciplines in the social and natural sciences.

In Humboldt's wake, other geographers in Europe and the United States wrote about climatic changes resulting from destruction of forest and expansion of farmland. Civilization leads to aridity, they concluded, and they chastised people for violating the harmony and balance that seemed inherent in the natural world. George Perkins Marsh (1801–1882), who served as President Abraham Lincoln's ambassador in Italy, was troubled by the legacy of ancient Greece and Rome on the landscapes of the eastern Mediterranean region. He used the past as a cautionary tale for the future, writing in his book *Man and Nature,* "Man has too long forgotten that the earth was given to him for usufruct alone, not for consumption, still less for profligate waste."[4] *Usufruct* means the nondestructive use of something that belongs to someone else, so Marsh was warning us that we should not be stealing resources and quality of life from one another and from future generations.

A century later, Carl Sauer (1889–1975), another American, wrote:

> We have accustomed ourselves to think of ever expanding productive capacity, of ever fresh spaces of the world to be filled with people, of ever new discoveries of kinds and sources of raw materials, of continuous technical progress operating indefinitely to solve problems of supply. We have lived so long in what we have regarded as an expanding world, that we reject in our contemporary theories of economics and of population the realities that contradict such views. Yet our modern expansion has been effected in large measure at the cost of an actual and permanent impoverishment of the world.[5]

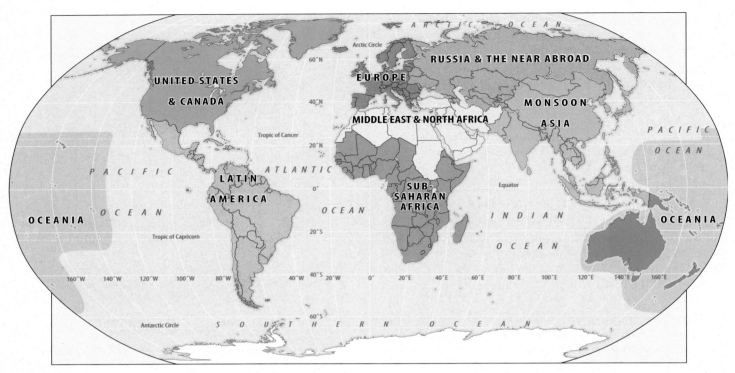

● **Figure 1.3** World regions as identified and used in this book.

These words have a modern ring to them, but Sauer, a geographer at the University of California–Berkeley, wrote them in 1938. Sauer is credited with founding the **landscape perspective** in American geography, based on the method of studying the transformation through time of a **natural landscape** to a **cultural landscape.** People are the agents of that metamorphosis. Sauer defined the geographer as the scientist determined to "grasp in all of its meaning and color the varied terrestrial scene."[6] Geographers continue to be interested in identifying how the forces of nature and culture have been at work in the creation of the **landscape**—the collection of physical and human geographic features on the earth's surface—and in particular the roles that human ideas, activities and cultures play in modifying the landscape. **Culture**—mainly thought of as the realm of anthropology—is an important component in the study of geography and may be defined as a shared, learned, symbolic system of values, beliefs, and attitudes that shapes and influences perception and behavior.[7]

The World Regional Approach to Geography

The **world regional approach** ranges across the human and physical subfields of geography, synthesizing, simplifying, and characterizing the human experiences of earth as home. This text is designed to introduce you in a logical and manageable way to the entire world, first by dividing the earth into eight regions, then by presenting a thematic profile of each region, and finally by examining in greater detail the geographic qualities of countries and other areas within each region.

Because it is impossible to introduce order and logic to something so massive and diverse as our planet without an organized framework built on smaller units, in this book, the world is divided into eight spatial subdivisions, or **regions** (●Figure 1.3 and Table 1.1). Regions are human constructs, not "facts on the ground." People create and draw boundaries around regions in an effort to define distributions of relatively homogeneous characteristics. The Middle East and North Africa, where Europe, Asia, and Africa meet, comprise a region where, very generally, there is a majority Arab population, most of whom are Muslims, and an arid environment. But different analysts draw different boundaries for this region, and those boundaries contain many non-Arab and non-Muslim peoples and non-arid environments. There are often recognizable transition zones between regions; the country of Sudan, for example, is both Middle Eastern and African in its cultures. A region is simply a convenience and a generalization, helping us become acquainted with the world and preparing us for more detailed insight.

Three types of regions are used by geographers and in this book. Each is helpful in its own way in conveying information about different parts of the world:

TABLE 1.1   The Major World Regions: Basic Data

Region	Area (sq mi; thousands)	Area (sq km; thousands)	Estimated Population (millions)	Annual Rate of Increase (%)	Estimated Population Density (sq mi)	Estimated Population Density (sq km)	Human Development Index[a]	Urban Population (%)	Arable Land (%)	GNI PPP Per Capita[b] ($US)
Europe	1,959.2	5,072.2	523.5	0.1	258	100	0.901	74	24	22,611
Russia and the Near Abroad	8,533.2	22,100.8	280.3	−0.1	32	12	0.771	64	8	6,577
Middle East and North Africa	5,916.7	15,324.2	479.0	1.9	160	62	0.690	54	6	5,345
Monsoon Asia	8,013.5	20,755.0	3,504.2	1.2	437	169	0.677	37	20	5,019
Pacific World	3,305.7	8,561.7	32.7	1.0	10	4	0.846	71	5	21,553
Africa South of the Sahara	8,406.5	21,772.5	693.7	2.4	85	33	0.439	31	6	1,767
Latin America	7,946.2	20,580.7	549.4	1.6	69	27	0.773	74	7	7,262
North America	8,403.8	21,765.8	325.6	0.5	38	15	0.939	79	10	37,013
World Regional Totals	52,484.8	135,935.6	6,396.0	1.3	123	47	0.696	48	10	7,590
USA (for comparison)	3,717.8	9,629.1	293.6	0.6	79	30	0.939	79	10	37,800

[a] HDI is a numeric index developed by the United Nations that reflects gross domestic product per capita, literacy rate, level of education on average, and life expectancy. The maximum HDI possible is 1.0, representing the highest level of development.

[b] Per capita gross national income purchasing power parity (per capita GNI PPP) combines gross national income (GNI) and purchasing power parity (PPP) as a method of comparing the real value of output between different countries' economies. This will be fully discussed in Chapter 3.

Source: *World Population Data Sheet*, Population Reference Bureau, 2007; *U.N. Human Development Report*, United Nations, 2007; *World Factbook*, CIA, 2007.

A **formal region** (also called a **uniform** or **homogeneous region**) is one in which all the population shares a defining trait or set of traits. A good example is an administrative political unit such as a county or a state, where the regional boundaries are defined and made explicit on a map.

A **functional region** (also called a **nodal region**) is a spatial unit characterized by a central focus on some activity (often an economic one). At the center of a functional region—a good example is the distribution area for a metropolitan newspaper—the activity is most intense, while toward the edges of the region, the defining activity diminishes in importance.

A **vernacular region** (or **perceptual region**) is a region that exists in the mind of a large number of people and may play an important role in cultural identity but does not necessarily have official or clear-cut borders. Good examples are concepts of the Midwest, the Bible Belt, and the Rust Belt in the United States (•Figure 1.4). These regional terms have economic and cultural connotations, but 10 people might have 10 different definitions of the qualities and boundaries of these regions. Vernacular or perceptual regions, constructed by individuals and cultures, serve as shorthand for place and regional identity, helping us organize, simplify, and make sense of the world around us.

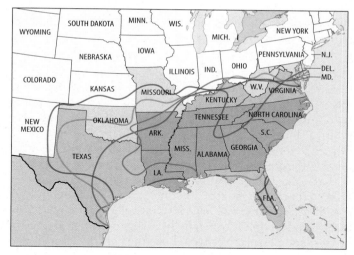

• **Figure 1.4** Definitions of a vernacular region, the American South. Purple shading represents three state-based delineations; colored lines delimit various religious, linguistic, and cultural "Souths." These are just a few of the many different interpretations of the region.

The Objectives of This Book

Mindful of the six essential elements of geography and the comprehensive task of introducing the entire world, this book is written to help you achieve four objectives:

1. *To understand important geographic problems and their potential solutions.* As the home of humankind, the earth requires maintenance; it has some problems that need attention and can be remedied. Many of these problems are environmental, and like geography in general, world regional geography is very concerned with environmental problems and their solutions. Some of these, such as climate change, are global issues that may be addressed with global agreements, like the Kyoto Protocol described in Chapter 2. Others are national and international.

Many regional and international security problems have roots that are environmental. Why can't some countries feed themselves? In some cases, their populations have exceeded their resource bases. What are their prospects then? The least appealing outcome is that famine will lower the population. Another unsavory option is for the country to aggressively seek the resources of others. You will see many examples of countries that have felt pushed into a corner by agricultural failure, debt, or geographic constrictions have lashed out against their neighbors or threatened them (the situation of North Korea comes immediately to mind). Another option is for rich countries to assist the country in need. As the world's sole superpower, what can the United States do to help a country that is in trouble? As a major financial power, the country can help by investing in education, employment, and infrastructure, for example. But should it help? Some authorities (like Garrett Hardin, discussed in Chapter 3) have insisted that Americans should feel no obligation to do so. But if Americans decide to be engaged, there are achievable solutions available. The book will introduce some of these.

2. *To learn to make connections between different kinds of information as a means of understanding the world.* To understand the problems of North Korea, for example, you must consider many variables, including resources, population, economic development, ethnicity, history, and geopolitical interests. Is that too much information for you to take in? Not if you have the right tool with which to filter and synthesize the information. World regional geography is that tool. Geography is all about making connections. Using geography's holistic and integrative approach in a regional framework, you will look at relationships between places and people. You will be using information, techniques, and perspectives from both the natural sciences and the social sciences. You will become familiar with issues in political science, history, economics, anthropology, sociology, geology, atmospheric science, and many other areas. You will pull these issues and perspectives together and find the links between them. In short, you will be doing geography.

This approach will be rewarding for you in both tangible and intangible ways. Your overall university experience should become richer as you become better able to tie together the various courses you are taking. In the future, you will be more likely to interact with the world with new wisdom that can be rewarding both professionally and personally. More complete knowledge of the world—good geography—is good business. American businesses working around the world are doing much better now than in the past in large part because of a greater understanding of local cultures and environments. If you end up working in a business environment, chances are you will be among the best informed of your colleagues when it comes to the cultural and physical settings of a particular place, and that knowledge may offer you new professional opportunities.

3. *To understand current events.* You should become able to pick up a newspaper or turn on the TV news with a greatly enhanced understanding of the issues underlying the world's news. Incidents of violence in the Middle East, earthquakes in the western Pacific, and outbreaks of bird flu are not random, unpredictable events but instead are rooted in consistent, recognizable problems that have geographic dimensions. You should find it very satisfying to feel like something of an expert on a problem that appears in the news.

4. *To develop the ability to interpret places and read landscapes.* Studying geography, you are concerned both with **space**—the precise placement of locations on the face of the earth—and with **place**—the physical and cultural context of a location. Place is much more subjective than location, because it often is defined by the meanings of a particular location; for example, your perceptions of what New York is like may be very different from those of your friend and may be shaped by personal experience in the Big Apple or by photographs or movies you have seen. In this book, there is much discussion of the "sense of place" that individuals and groups have about various locations and regions. Perception of place can have a very strong influence on how we make decisions and interact with others. Perception of place can even have a strong impact on world events. In the Middle East chapters, for example, you will see how Jewish and Muslim perceptions about the sanctity of places within a few meters of each other in Jerusalem play a regular role in both conflict and peacemaking in the Middle East and beyond.

As an example of how you can come to define and identify place, study the photograph in •Figure 1.5, and try to identify the country or region in which this photograph was taken.* What clues in the physical and human geographies of this place help you locate it? As you use this book, you should be increasingly able to look at a photograph of a place or travel to a place with new insight into many of the features that make up its identity: its climate, vegetation, and landforms, for example, and the language, religion, history, and livelihoods of the people who live there.

This book uses photographs that will teach you much about the characteristics of places. There are also aerial and satellite images that will help you identify places on a broader scale and appreciate the world as if you were looking at it from above. (I recommend that you *do* look at the world from above whenever you have the chance. When you fly, ask for a window seat that is not over the airplane's wing.) These images, plus something a book cannot give

Joe Hobbs

• **Figure 1.5** Where is this place?* What clues on the landscape or in the women's appearance might tell you where you are?

place, location, landscape, and especially, maps. As geographers study the relationships between people, places, and environments, they usually (but not always) collect and depict information that can be mapped. In other words, they are interested in the **spatial** context of the things; the world in spatial terms is the first essential element of geography. *Spatial* means "pertaining to space" and in the geographic context refers to the distribution of various phenomena on the earth's surface.

The science of making maps is called **cartography.** Cartographers are the geographers and other skilled technicians who create maps through the research and presentation of geographic information. Using maps, the cartographer shows where things and places are located relative to one another. Cartographers today almost always use computers to interpret and display spatial data, but it was not so long ago that they drafted maps laboriously by hand.

Because maps are among the essential tools in the study of world regional geography, it is important to know how to read them. Major elements of their language (beyond the title, date, and north arrow) are *scale, coordinate systems, projections,* and *symbolization.*

you but you should use whenever you can—a globe—are important tools in your gaining the "whole earth" perspective that geography requires.

1.2 The Language of Maps

We turn now to other important geographers' tools that will help you explore and explain relationships on our planet. Although they may at times seem to act like anything from anthropologists to zoologists, geographers reveal their true identities by their preoccupation with space,

Scale

A map is a reducer; it shrinks an area to the manageable size of a chart, piece of paper, or computer monitor. The amount of reduction appears on the map's **scale,** which shows the actual distance on earth as represented by a given linear unit on the map. A common way of denoting scale is to use a representative fraction or ratio, such as 1:10,000 or 1:10,000,000. In other words, one linear unit on the map (for example, 1 inch or 1 centimeter) represents 10,000 or 10,000,000 such real-world units on the ground. A **large-scale map** is one with a relatively large representative fraction (for example 1:10,000 or even 1:100) that portrays a relatively small area in more detail. A **small-scale map** has a relatively small representative fraction (such as 1:1,000,000 or 1:10,000,000) that, in contrast, portrays a relatively large area in more generalized terms. Compare the two maps in •Figure 1.6. Figure 1.6a is a small-scale map showing San Francisco and surrounding parts of the Bay Area. Figure 1.6b is a large-scale map that "zeroes in" on part of San Francisco. Remember this inverse relationship: a small-scale map shows a large area, and a large-scale map shows a small area.

Coordinate Systems

A principal concern in maps is **location.** In addition to core and peripheral locations (see Insights, page 9) there are two general types of locational information discussed in this book: absolute location and relative location.

Relative location defines a place in relationship to other places. It is one of the most basic reference tools of everyday

*This scene is outside a village in Guangxi, China's southernmost province. Agricultural terraces like these are usually associated with high population densities as large numbers of rural people try to get maximum crop returns from steep landscapes, and terrace building requires a lot of labor. These conditions are true of China but could also describe parts of the South American Andes. The clouds suggest a humid climate, found in southern China and also parts of the Andes. Even the women's dark hair could be Asian, Native American, or Hispanic. Their dress is distinctively that of an ethnic minority in China, however.

• **Figure 1.6** Small-scale (a) and large-scale (b) maps of San Francisco and environs.

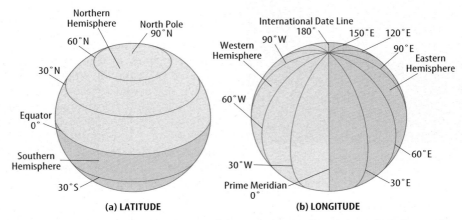

• **Figure 1.7** (a) Earth's lines of latitude (parallels) in increments of 30 degrees, from the equator (0 degrees) to the North Pole (90 degrees north latitude). (b) Earth's lines of longitude (meridians) in increments of 30 degrees.

life; you might say you live south of the city, just west of the shopping mall, or next door to a good friend. Relative location will become part of your basic geographic literacy as you learn, for example, that Sumatra is northwest of Java and that Bolivia is tantalizingly close to, but cut off from, the Pacific Ocean. Understanding the implications of relative location can be quite useful in following world affairs; be watchful for Bolivia's diplomatic efforts to retrieve its lost access to the sea, for example.

Absolute location, also known as **mathematical location,** uses different ways to label places on the earth so that every place has its own unique location, or "address." **Coordinate systems** are used to determine absolute location. These coordinate systems use grids consisting of horizontal and vertical lines covering the entire globe. The intersections of these lines create the addresses in the global coordinate system, giving each location a specific, unique, and mathematical situation.

The most common coordinate system uses **parallels** of **latitude** and **meridians** of **longitude.** The term *latitude* denotes position with respect to the equator and the poles. Latitude is measured in **degrees** (°), **minutes** ('), and **seconds** ("). Each degree of latitude, which is made up of 60 minutes, is about 69 miles (111 km) apart; these distances vary a little because of earth's slightly ellipsoid shape (see •Figure 1.7a). Each minute of latitude, which is made up of 60 sec-onds, is thus roughly a mile apart. The **equator,** which circles the globe east and west midway between the poles, has a

INSIGHTS

Core Location and Peripheral Location

Among the concepts that will prove useful to you in studying world regional geography are **core location** and **peripheral location;** in this book, the region of Europe is subdivided along these lines, for example. Some locales have greater importance in local, regional, or world affairs because they have a central, or core, location relative to others. Other locales are less important because they are situated far from "where the action is." A comparison of two countries, the United Kingdom (U.K.) and New Zealand, provides an example (●Figure 1.A). Both are island countries. Their climates are remarkably similar, although they are in opposite hemispheres and are about as far apart as two places on earth can be. Westerly winds blow off the surrounding seas, bringing abundant rain and moderate temperatures throughout the year to both.

But there are important differences. The U.K. is located in the Northern Hemisphere, which has the bulk of the world's land (and may thus be described as the **land hemisphere**) and most of its principal centers of population and industry; New Zealand is on the other side of the equator, in the Southern Hemisphere, where there is less land and less economic activity.

The U.K. is located near the center of the world's landmasses and is separated by a narrow channel from the densely populated industrial areas of western continental Europe; New Zealand is surrounded by vast expanses of ocean in the **water hemisphere.** The U.K. is located in the western seaboard area of Europe, where many major ocean routes of the world converge; New Zealand is far away from the centers of world commerce. For centuries, the United Kingdom has shared in the development of northwestern Europe as an organizing center for the world's economic and political life; New Zealand, meanwhile, has languished in comparative isolation.

The United Kingdom, in other words, has a core or central location in the modern framework of human activity on earth, whereas New Zealand has a peripheral location. Centrality of location is a very important consideration when it comes to assessing the economic and political geographies of places, countries, and regions.

● **Figure 1.A** In the top map, note how the major landmasses are grouped around the margins of the Atlantic and Arctic Oceans. The British Isles and the northwestern coast of Europe lie in the center of the "land hemisphere," which constitutes 80 percent of the world's total land area and has about 90 percent of the world's population. New Zealand lies near the center of the opposite hemisphere, or "water hemisphere," which has only 20 percent of the land and about 10 percent of the population.

latitude of 0°. All other latitudinal lines are parallel to the equator and to each other, which explains why they are called *parallels*.

Every point on a given parallel has the same latitude. Places north of the equator are in **north latitude.** Places south of the equator are in **south latitude.** The highest latitude a place can have is 90°N (the **North Pole**) or 90°S (the

South Pole). Places near the equator are said to be in **low latitudes;** places near the poles are in **high latitudes.** The **Tropic of Cancer** and the **Tropic of Capricorn,** at 23.44°N and 23.44°S, respectively, and the **Arctic Circle** and **Antarctic Circle,** at 66.56°N and 66.56°S, respectively, form the most commonly recognized boundaries of the low and high latitudes. Places occupying an intermediate position with respect to the poles and the equator are said to be in the **middle latitudes.** Incidentally, there are no universally accepted definitions for the boundaries of the high, middle, and low latitudes. The northern half of the earth between the equator and the North Pole is called the **Northern Hemisphere,** and the southern half between the equator and the South Pole is the **Southern Hemisphere.**

Meridians of longitude are straight lines connecting the poles (see Figure 1.7b). Every meridian is drawn due north and south. All the meridians converge at the poles and are farthest apart at the equator. Longitude, like latitude, is measured in degrees, minutes, and seconds, but because the longitude lines are not equidistant across the globe, their values vary. At the equator, the distance between lines of longitude is about 69.15 statute miles (111.29 km), while at the Arctic Circle, it is only about 27.65 miles (44.50 km). A **statute mile** ("land mile") is the mile you are most likely familiar with, 5,280 feet (1,609 m); a **nautical mile** ("sea mile") is based on one minute of arc of a great circle and equals 6,076 feet (1,852 m). The meridian most used as a base or starting point is the one running through the Royal Astronomical Observatory in Greenwich, England. Known as the **meridian of Greenwich** or the **prime meridian,** it has a longitude of 0°. Places east of the prime meridian are in **east longitude;** places west of it are in **west longitude.**

The meridian of 180°, exactly halfway around the world from the prime meridian, is the basis for the other dividing line between places east and west of Greenwich and is called the **International Date Line.** This line, which has a few zigzags in it for political and practical reasons (see Figure 8.1, page 407), separates two consecutive calendar days. The date west of the line is one day ahead of the date east of the line. All of the earth's surface eastward from the prime meridian to the International Date Line is in the **Eastern Hemisphere,** and all of the earth's surface westward from the prime meridian to the International Date Line is in the **Western Hemisphere.**

Using the simple tools of latitude and longitude notation, you may now establish the absolute location of any given place. Here is a useful exercise to ensure you understand how latitude and longitude work. On the map in •Figure 1.8, the latitude of Madrid, Spain, is approximately 41 degrees north latitude, 4 degrees west longitude (41°N, 4°W). What are the approximate latitude and longitude coordinates of Oslo, Norway? In which hemispheres is Oslo located?

Projections

A **map projection** is a way of depicting the curved surface of the earth—which can be represented accurately only on a globe—on a flat surface such as a piece of paper. There are four metric relationships or properties of objects on a globe: area, shape, distance, and direction. A flat map cannot replicate all of these simultaneously; most projections can preserve only one. As a result, there will inevi-

● **Figure 1.8** What are the approximate latitude and longitude coordinates of Oslo, Norway?*

tably be some distortion on a flat map. On maps depicting very small areas (large-scale maps), the distortion may be minimal enough that it can be disregarded. On small-scale maps, however, the effects of distortion can become very serious. Fortunately, there are thousands of different projections cartographers can choose from to display the geographic area and data they are mapping. There is no one "perfect" projection for any particular map, however.

There are three geometric "developable surfaces" (surfaces that can be flattened without stretching or tearing) that many map projections are based on: the plane, the cylinder, and the cone (●Figure 1.9).

❋ Projections onto a plane are called **azimuthal projections** and are typically used for maps of the polar regions.

❋ **Cylindrical projections** are mostly used for areas around the equator or to depict the entire world; regular cylindrical projections have straight meridians, while pseudocylindrical variants have curved ones.

❋ **Conic projections** are common for middle latitudes; **polyconic projections** can be used for larger areas.

*Using this map, you should determine Oslo's location as about 60 degrees north, 11 degrees east (60°N, 11°E). Oslo is therefore in the Northern Hemisphere, the Eastern Hemisphere, and the land hemisphere.

a An azimuthal projection (North Pole Azimuthal Equidistant)

b A cylindrical projection (Mercator)

c A conic projection (Albers Equal Area Conic)

● **Figure 1.9** Common map projections.

Projections not based on developable surface geometry are called **compromise projections** (or *mathematical projections*). These projections attempt to create a balance of distortion among the four metric properties to create an aesthetically pleasing map. The **Robinson projection,** used for all the world maps in this book, is an example of a compromise projection.

Map projections are also classified by which metric property they preserve (or distort the least):

✳ **Equal-area projections** preserve areas consistently across the entire map; each area on the map is proportional to the area it occupies on the earth's surface. Some equal-area projections are segmented into various "lobes" to preserve area size while attempting to minimize distortion of shapes. The result looks something like the peeled skin of an orange laid flat.

✳ **Equidistant projections** show accurate distances, but only from the center of the projection (in most cases). For example, an equidistant map centered on New York City would accurately show the distance between New York and London and between New York and Los Angeles, but it would not show the correct distance between London and Los Angeles. Flat maps cannot be equidistant and equal-area at the same time.

✳ **Conformal projections** keep the map's scale the same in every direction from any given point, preserving shapes in very small, localized areas. Sizes are usually distorted, especially toward the edges (no flat map can be both conformal and equal-area). The best-known conformal map is the **Mercator projection,** which was designed for navigation. A straight line drawn between any two points on a Mercator map indicates a true direction (a "rhumb line"). However, to maintain the accuracy of those rhumb lines, Mercator maps increasingly distort the north-south dimension away from the equator and distort the east-west dimension near the poles because of the parallel meridians that conic projections employ (on the three dimensional reality of a globe, the meridians converge at the poles). Therefore, the sizes of areas near the poles on Mercator maps are greatly exaggerated, so much so that Greenland and South America appear to be the same size, even though South America is actually about eight times larger than Greenland! This distortion is clearly visible in the Mercator projection in Figure 1.9b.

Symbolization

Maps allow us to extract certain information from the totality of things, to see patterns of distribution, and to compare these patterns with one another. No map is a complete record of an area. It represents instead a selection of certain details, shown by symbols, that a cartographer uses for a particular purpose. Unprocessed data must be classified into categories that symbols can represent. The details selected for **symbolization** may be categories of physical or cultural forms like rivers or roads, aggregates such as 100,000 people or 1 million barrels of crude oil, or averages such as population density (the number of persons per square mile or kilometer within a defined area).

Cartographers often order aggregates or averages into ranked categories in a graded series (for example, population densities of 0 to 49, 50 to 99, and 100 or more people per square mile) to be portrayed on a map. The categories are usually not self-evident. The cartographer must select them, sometimes through elaborate statistical procedures. To represent different phenomena, the cartographer may choose from a wide range of symbols, including lines, dots, circles, squares, shadings, colors, and typography.

There are two major types of maps that employ symbols: reference maps and thematic maps. **Reference maps** (such as Figure 4.1 on page 69) are concerned mainly with the locations of various features on the earth's surface and their spatial relationships with each other. Highway maps are reference maps. **Thematic maps** also show spatial relationships, but they have a more specific purpose: they often show the distribution of just one phenomenon. Thematic maps are sometimes called *statistical maps* because they typically show distributions of numeric data, such as population density in given area. They can also show qualitative (nonmathematical) information. Several types of thematic maps are used throughout this book:

* On a **choropleth map** (the most common thematic map type in the book), each political unit is filled in with a distinguishing color or pattern representing some derived value, such as per capita income or number of college students per 100,000 people (Figure 3.5 on page 43, showing wealth and poverty by country, is a good example).

* **Isarithmic maps** do not use political boundaries with solid colors or patterns but instead use lines to join points of equal value across the mapped area (for example figure 1.4 on page 5). Topographic maps based on contour lines are also isarithmic maps.

* As another alternative to colored political units, **graduated symbol maps** use a simple symbol, such as a circle, square, or bar graph. The symbol is scaled proportionally to the quantity of the data being mapped. On a graduated symbol map of world population, for example, the circle for China would have an area over four times larger than the circle representing the United States. A graduated symbol map that scales the often highly generalized outlines of political units to the data is called a **cartogram.**

Isarithmic and graduated symbol maps can display the same data very differently. Figures 3.13 and 3.15, on pages 56 and 58, clearly illustrate that China and India are the most populous nations on earth. Figure 3.13 is a cartogram: it shows that China and India are the world's demographic giants but not where the populations within those two countries are concentrated. Figure 3.15 is an isarithmic map: you cannot tell from this map how many people live in China or India, but the contrast between areas of dense settlement and virtually uninhabited land in both countries is very clear.

* **Dot maps** use dots to represent a stated amount of some phenomenon within a political unit. Using a scale equating one dot with 1,000 people, 12 dots in one area would represent 12,000 people living in that area.

* **Flow maps** use arrows to detail the movement of people or goods from one area to another (see Figure 3.16 on page 58 for an example).

Maps need to be used judiciously, and you will gradually acquire the skill of evaluating map quality. A well-known book in the field of geography is titled *How to Lie with Maps,* suggesting, like the motto "let the buyer beware," that the fact that something has been mapped does not guarantee that it has been mapped accurately.

Mental Maps

When someone asks you, "Could you draw me a map of how to get there?" you might quickly draw some lines, write down some street names, point out familiar landmarks, and apologize for how crude your map is. Your map would probably end up looking very different from that of another person asked the same question. Our understanding of location is not completely objective. Each of us has a personal sense of space and place and associations with them.

A **mental map,** rather like a vernacular region, is a collection of personal geographic information that each of us uses to organize spatially the images and facts we have about places, both local and distant. Sometimes we use that information to create actual maps, as when a guest asks you to draw a map showing how to get to your house or when a nomad in the Middle East draws lines in the sand diagramming the territories of his and others' tribes (•Figure 1.10). Such maps are not accurate, precise, or scientific, but they portray useful information and tell us much about the individuals and cultures that create them. They are subjective,

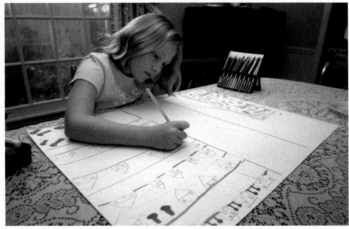

• **Figure 1.10** As we grow, our minds fill with mental maps, which we use consciously or unconsciously to orient ourselves and navigate through daily life. In drawing these maps for others, we often discover the importance of scale, direction, and other map essentials.

just like the various cultural and other regions identified and used in this book.

1.3 New Geographic Technologies and Careers

To close this chapter, we turn to some of the most exciting and innovative developments in geography and consider how you or someone you know might become part of the action in this growing field.

Geographic Information Systems and Remote Sensing

Geographic information systems (GIS) represent the cutting edge of geography today, and are by far the field's leading area of growth and employment. In recent decades GIS has become an essential tool for geographers, and increasingly for people in many other scientific disciplines; it is also widely used by private businesses and governmental and other public agencies. In essence, a GIS is a computerized system designed to help people analyze, manage, and visualize geographic data. By combining many tools including cartography, remote sensing (described below), statistics, and computer science, GIS users can discover relationships between disparate types of spatial information and make informed decisions to solve complex geographic problems. GIS applications are most used for land and resource management, though it has an almost infinite number of other uses (examples include locating the most efficient route for a truck to deliver goods; plotting the locations of various types of crime to detect patterns; determining the best location to open a new store; creating 3-D models of terrain to model the effects of floods; and many more).

GIS data is stored in "layers" (●Figure 1.11), with each layer containing a related set of spatial data. One layer may have a neighborhood's street network, another layer may have the location of every tree in that neighborhood, a third layer might have the routes of electrical wiring or water pipes running below the surface, and yet another layer could contain elevation data. One of the aspects that sets apart each of these layers from a simple map showing the same data are "attributes," or further information, about each item in those layers. For example, a click on a water main on that layer will bring up a box containing information about how long the pipe is and when it was last serviced. The ability to create, select and manipulate all this spatial information has revolutionized both geography itself and all other fields dealing with spatial relationships.

Remote sensing uses various kinds of satellite imagery and photographic coverage to assess land use or other geographic patterns. Many different technologies are associated with remote sensing, and it can be use in an almost countless variety of applications. Aerial photography, one of the earliest geographic uses of remote sensing, allows

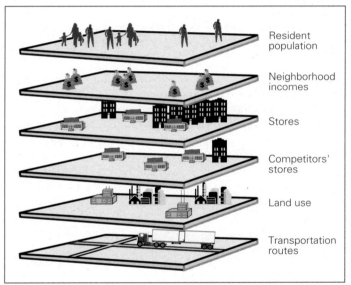

● **Figure 1.11** Geographic information systems (GIS) create and use layers of different types of spatial data. The computer-generated images and models that result have a wide range of academic and practical applications.

for the creation of maps of limited areas entirely from photographs. Radar is used to map elevation and detect rainfall, and a newer technology called LIDAR (laser imaging detection and ranging, similar to radar but using light waves instead of radio waves) is often used to create three-dimensional models, for example, of the remains of the World Trade Center and surrounding buildings after 9/11. And since its introduction in 1960, satellite imagery has been used for an extraordinary range of purposes, from detecting ancient cities covered by the sands of Arabia to tracking the shrinking polar ice cap and monitoring weather systems. Remote sensing is an exceptionally good tool for helping geographers in their mission of understanding how people modify the earth. In the remote-sensing

• **Figure 1.12** A remote-sensing image shows one of the few international political boundaries clearly visible from space. High population densities and intensive cultivation have nearly denuded the lush tropical vegetation of southern Mexico, while north-central Guatemala's forests are nearly intact. Guatemalan populations are clustered in the mountain areas farther south. Compare this image with the political map of the region, Figure 10.1.1 on page XXX: the vegetation and political boundaries correlate perfectly. The black area in the wedge on the right side of the photo is area not included in this satellite photo.

• **Figure 1.13** The most in-demand tool of geography today is geographic information systems (GIS). Here the book's cartographer, Andrew Dolan, is using GIS to create a population density map of the United States.

image in •Figure 1.12, for example, human modification of the landscape is clearly apparent.

To see a fascinating and increasingly popular application of remote-sensing and GIS technologies right on your own computer, try Google Earth (http://earth.google.com). Type in your address, any other address, or a place name like "Giza pyramids." You will enjoy a fly-in from outer space to a mosaic of satellite and aerial photographs (from a variety of sources and vintages) of the location you entered. Google Earth acts as a very simple GIS viewer; using the "Layers" sidebar, you can explore various GIS layers of data about the location.

You can see almost anything on the earth with this program. Want to see Iran's nuclear facility of most concern? No problem: type in "Natanz." You can be sure that many governments around the world are very unhappy about this extraordinary software. A Russian intelligence official said of Google Earth, "Terrorists don't need to reconnoiter their target. Now an American company is working for them."[8]

What Do Geographers Do for a Living?

These days, it is appropriate for this question to follow discussion of GIS and remote sensing. More and more, GIS and geography are becoming synonymous, and people with the ability to work with GIS are finding jobs in a number of fields, often at very attractive starting salaries (•Figure 1.13).

Table 1.2 shows a few of the jobs of recent graduates of geography in just one department, the University of Texas at Austin. Note that even with just undergraduate degrees in geography, students are finding work in the GIS and remote-sensing fields. Many graduates with these technical skills find employment in government agencies at the local, state, regional, and federal levels. Firms working with land use decisions at all scales need employees with these skills. Retail firms find the technical and problem-solving know-how of geographers to be useful in management, survey design and implementation, estimation of public response to innovations, and decision making about location expansion or curtailment.

After some years in decline, geography is making a strong comeback as an academic discipline for holders of the doctoral degree. These researchers typically specialize in fields that are either regional (some are Middle East experts, for example), systematic (they study topics of various physical or human geographic elements), or technical (they rely strongly on GIS and remote sensing). Geographers with **regional specialties** in the United States and Europe are most numerous. The relative shortage of U.S. geographers with professional expertise in non-Western and developing countries is a disadvantage, but it does present opportunities for younger geographers willing to undertake the challenges of foreign-language study and fieldwork. As might be expected, there is particularly high demand among intelligence agencies for trained geographers who are acquainted with regions deemed to be strategic in the national interest, such as the Middle East (ironically, though, very few Middle East geographers are currently in training). However, most geographers with foreign-area expertise have university research and teaching positions.

Most academic geographers—many of whom use GIS, remote sensing, and other technical tools of the discipline—practice a **systematic specialty** rather than a regional one. Experts in *physical geography* study spatial patterns and associations of natural features. Prominent subfields include **geomorphology** (the study of landforms and a field in which geography intersects with geology), **climatology** (cli-

TABLE 1.2   Positions Occupied by Recent Graduates of the Department of Geography, University of Texas at Austin[9]

Recipients of a Bachelor of Arts Degree in Geography

Information system analyst, Transportation Planning and Programming Division, Texas Department of Transportation

CFO and geospatial engineer, Spatial Innovations International

Preserve manager, Nature Conservancy

GIS analyst, Environmental Section, Information Solutions Department, PBS&J, Inc.

Natural science educator, Gore Range Natural Science School, Colorado

GIS administrator, Planning and Zoning Department, City of Longview, Texas

Recipients of a Master of Arts Degree in Geography

Senior conservation scientist, Nature Conservancy, Guatemala

Geographer, Texas Water Science Center, United States Geological Survey

Lead environmental scientist, Aquatic Weed Unit, California Department of Boating and Waterways

CEO, Trinidad and Tobago Insurance Consultants, Ltd., Trinidad and Grenada, West Indies

Vice president for market research, Crescent Real Estate Equities, Ltd.

Recipients of a Ph.D. in Geography

Associate vice president, Community Partnerships, San Diego Foundation

Water resources scientist, Gulf States Natural Resource Center, National Wildlife Federation

Managing principal, SWCA Environmental Consultants, Albuquerque, New Mexico

Executive director, Chihuahuan Desert Research Institute

Supervisor, Geographical Information Sciences, Texas Commission on Environmental Quality

Assistant professor, Department of Geography, University of South Carolina

matic processes and patterns), **biogeography** (the geography of plants and animals), and **soils geography.** Closely related to physical geography is the large interdisciplinary field of **environmental studies,** which is concerned with reciprocal relationships between society and the environment. Another allied specialty is **medical geography,** which focuses on spatial associations between the environment and human health and on locational aspects of disease and health care delivery.

Specialists in **economic geography** study spatial aspects of human livelihood. Major occupations and products are their main concerns, and they have an active theoretical component that intersects with the field of economics. **Marketing geography** is a relatively small but active applied offshoot of economic geography that is of particular utility in commercial planning and zoning. The subfields of **agricultural geography** and **manufacturing geography** are concerned, respectively, with the productive management of soil and water resources and with networks and hierarchies of economic production. **Urban geography** considers the locational associations, internal spatial organization, socioeconomic characteristics, and functions of cities. It

offers an alternative pathway to the applied field of urban planning.

Geographers specializing in **cultural geography** are concerned today, as they were a century ago, with spatial and other aspects of cultural regions, origins, diffusions and interactions, and the cultural forces behind changing landscapes. Closely allied with cultural geography are the fields of **cultural ecology** and **political ecology,** focusing on the relationships between environments and human adaptations to them and on how social groups organize themselves politically to take advantage of environmental and other opportunities. **Political geography** considers topics such as spatial organization of geopolitical units, international power relationships, nationalism, boundary issues, military conflicts, and regional separatism within states. **Social geography** deals with spatial aspects of human social relationships, especially in urban settings. **Population geography** assesses population composition, distribution, migration, and change. **Historical geography** examines the geography of past periods and the evolution of geographic phenomena such as cities, industries, agricultural systems, and rural settlement patterns.

TABLE 1.3   Specialty Groups of the Association of American Geographers

Africa	Disability	Hazards	Recreation, Tourism, and Sport
Applied Geography	Economic Geography	Historical Geography	Regional Development and Planning
Asian Geography	Energy and Environment	History of Geography	Remote Sensing
Bible Geography	Environmental Perception and Behavioral Geography	Human Dimensions of Global Change	Rural Geography
Biogeography		Indigenous Peoples	Russian, Central Eurasian, and East European Geography
Canadian Studies	Ethics, Justice and Human Rights	Latin America	
Cartography	Ethnic Geography	Medical Geography	Sexuality and Space
China	European	Middle East	Socialist and Critical Geography
Climate	Geographic Information Science and Systems	Military Geography	Spatial Analysis and Modeling
Coastal and Marine		Mountain Geography	Study of the American South
Communication Geography	Geographic Perspectives on Women	Paleoenvironmental Change	Transportation Geography
Cryosphere	Geography Education	Political Geography	Urban Geography
Cultural and Political Ecology	Geography of Religions and Belief Systems	Population	Water Resources
Cultural Geography		Qualitative Research	Wine
Developing Areas	Geomorphology		

Source: Association of American Geographers, http://www.aag.org/sg/sg_display.cfm.

So what exactly do geographers do? As you see in Table 1.3, which lists the 55 specialty groups of the approximately 9,000 members of the **Association of American Geographers (AAG),** almost anything may fall in their realms of interest, activities, and employment.

There are other kinds of employment, too, notably in the travel and tourism industry, for geographers trained at the undergraduate and graduate levels. Even if you are not a geography major or you have not yet seen an advertisement proclaiming "Geographer Wanted," chances are that many professional opportunities will arise if you use geography's tools to grow knowledge of the world in which you live. This book will help you do just that.

SUMMARY

→ A recent study showed that U.S. citizens aged 18 to 24 generally have poor knowledge of world geography.

→ There are six essential elements of the national geography standards: the world in spatial terms, places and regions, physical systems, human systems, environment and society, and the uses of geography.

→ Geography means "description of the earth" and is also defined as "the study of the earth as the home of humankind."

→ Four main objectives of the text are for readers (1) to understand important geographic problems and their potential solutions, (2) to become better able to make connections between different kinds of information as a means of understanding the world, (3) to understand current events, and (4) to develop skills in interpreting places and reading landscapes.

→ Maps are the geographers' most basic tools. The basic language of maps includes the concepts and terms of scale, coordinate systems, projection, and symbolization. Maps can depict spatial data in a variety of ways.

→ Individuals and cultures generate their own unique "mental maps." Regions are in effect mental maps that help us make sense of a complex world. Modern geographic thought derives from a long legacy of interest in how people interact with the environment. The dominant approach has been to understand how people have changed the landscape or face of the earth.

→ The discipline of geography may be divided into regional and systematic specialties, with the systematic fields having the most followers. Their concerns overlap many disciplines in the natural and social sciences. Geographers are employed in many private and public capacities. The strongest growth area, with the most jobs, is in geographic information systems (GIS).

KEY TERMS + CONCEPTS

absolute location (p. 8)
agricultural geography (p. 15)
Antarctic Circle (p. 9)
Arctic Circle (p. 9)
Association of American Geographers (AAG) (p. 16)
biogeography (p. 15)
cartogram (p. 12)
cartography (p. 7)
choropleth map (p. 12)
climatology (p. 15)
coordinate systems (p. 8)
core location (p. 9)
cultural ecology (p. 15)
cultural geography (p. 15)
cultural landscape (p. 4)
culture (p. 4)
degrees (p. 8)
dot map (p. 12)
economic geography (p. 15)
environmental studies (p. 15)
equator (p. 8)
ethnogeography (p. 2)
flow map (p. 12)
formal region (p. 5)
functional region (p. 5)
geographic information systems (GIS) (p. 12)
geography (p. 2)
geomorphology (p. 15)
graduated symbol map (p. 12)
hemisphere
    Eastern Hemisphere (p. 10)
    land hemisphere (p. 9)
    Northern Hemisphere (p. 10)
    Southern Hemisphere (p. 10)
    water hemisphere (p. 9)
    Western Hemisphere (p. 10)

historical geography (p. 15)
homogeneous region (p. 5)
human agency (p. 3)
human-environment interaction (p. 3)
human geography (p. 3)
International Date Line (p. 10)
isarithmic map (p. 12)
landscape (p. 4)
landscape perspective (p. 4)
large-scale map (p. 7)
latitude (p. 8)
    high (p. 9)
    low (p. 9)
    middle, (p. 10)
    north (p. 9)
    south (p. 9)
location (p. 7)
longitude (p. 8)
    east (p. 10)
    west (p. 10)
manufacturing geography (p. 15)
map projection (p. 10)
    azimuthal (p. 10)
    compromise (p. 11)
    conformal (p. 11)
    conic (p. 10)
    cylindrical (p. 10)
    equal-area (p. 11)
    equidistant (p. 11)
    Mercator (p. 11)
    polyconic (p. 10)
    Robinson (p. 11)
marketing geography (p. 15)
mathematical location (p. 8)
medical geography (p. 15)
mental map (p. 12)
meridian (p. 8)
meridian of Greenwich (p. 10)

minutes (p. 8)
natural landscape (p. 4)
nautical mile (p. 10)
nodal region (p. 5)
North Pole (p. 9)
parallel (p. 8)
perceptual region (p. 5)
peripheral location (p. 9)
physical geography (p. 3)
place (p. 6)
political ecology (p. 15)
political geography (p. 15)
population geography (p. 15)
prime meridian (p. 10)
reference map (p. 12)
region (p. 4)
regional specialties (p. 14)
relative location (p. 7)
remote sensing (p. 13)
scale (p. 7)
seconds (p. 8)
six essential elements of geography (p. 2)
small-scale map (p. 7)
social geography (p. 15)
soils geography (p. 15)
South Pole (p. 9)
space (p. 6)
spatial (p. 7)
statute mile (p. 10)
symbolization (p. 11)
systematic specialty (p. 14)
thematic map (p. 12)
Tropic of Cancer (p. 9)
Tropic of Capricorn (p. 9)
urban geography (p. 15)
uniform region (p. 5)
vernacular region (p. 5)
world regional approach (p. 4)

REVIEW QUESTIONS

1. What is geography? Is it a natural or a social science? What are some of its characteristic approaches?

2. What are the six essential elements of geography, as defined by the National Geographic Society? What does each element indicate about geography's concern with space, place, or the environment?

3. What does *spatial* mean, and how does geography's interest in space differentiate it from other disciplines?

4. What geographic features make the United Kingdom and New Zealand different?

5. What are the major terms and concepts associated with scale, coordinate systems, projections, and symbolization?

6. Why is a map made with the Mercator projection more suitable for navigation than a map made with the Robinson projection?

7. What is the difference between a dot map and a choropleth map?

8. What is a mental map?

9. What is GIS, and what typically makes it different from old-fashioned manual cartography? What are some applications of remote sensing?

10. What do geographers study, and what do they do for a living?

NOTES

1. The abbreviation B.C.E. stands for "before the Common Era," which is a reference to the dating system invented by European Christians that sets the birth of Jesus Christ as year 1. In Christian cultures, dates before that year are expressed as B.C., meaning "before Christ," and later years are identified as A.D., which stands for *anno Domini* (Latin, "in the year of our Lord"). Religion-neutral dating systems such as the one used in this book employ B.C.E. ("before the Common Era") and C.E. ("Common Era"), respectively, but the years are numbered the same in the two systems.

2. Alton Byers, "Contemporary Human Impacts on Alpine Ecosystems in the Sagarmatha (Mt. Everest) National Park, Khumbu, Nepal." *Annals of the Association of American Geographers*, 95, 2005, pp. 112–140.

3. Quoted in Geoffrey J. Martin and Preston E. James, *All Possible Worlds: A History of Geographical Ideas* (New York: Wiley, 1993), p. 150.

4. Quoted in Ajavit Gupta, *Ecology and Development in the Third World* (New York: Routledge, 1988), p. 2.

5. Quoted in Andrew Goudie, *The Human Impact on the Natural Environment* (Oxford: Blackwell, 1986), p. 6.

6. Quoted in Larry Grossman, "Man-Environment Relations in Anthropology and Geography." *Annals of the Association of American Geographers*, 67, 1977, p. 129.

7. Kathleen A. Dahl, "Culture," http://www2.eou.edu/~kdahl/cultdef.html.

8. Lieutenant General Leonid Sazhin, Federal Security Service, quoted in Katie Hafner and Saritha Rai, "Google Offers a Bird's-Eye View, and Some Governments Tremble." *New York Times*, December 20, 2005, p. 1.

9. Information courtesy of Bill Doolittle, University of Texas at Austin.

On June 30, 1975, as I stood at the edge of the Grand Canyon of the Yellowstone in Yellowstone National Park, Wyoming, an earthquake of magnitude 6.2 struck. In photo a, you can see dust raised by landsliding down the canyon from an initial, small tremor. The 6.2 jolt triggered massive collapse of the canyon walls (photo b) and even made the pine trees sway back and forth. Such events are part of physical processes that change the face of the earth.

Joe Hobbs

Joe Hobbs

CHAPTER 2
PHYSICAL PROCESSES THAT SHAPE WORLD REGIONS

Many issues in world regional geography relate to the interaction of people and the natural environment. This chapter and the next introduce geography's basic vocabulary about how the natural world works and explain how people have interacted with the environment to change the face of the earth.

This chapter deals with physical geography, introducing the three "spheres" that make up the earth's habitable environment:

* The **lithosphere** is the earth's outer "rind" of rock, varying in thickness from about 50 to 125 miles (60 to 200 km); it is subdivided into the categories of continental and oceanic.

* The **hydrosphere** is made up of the world's oceans and other water features, including lakes and rivers.

* The **atmosphere** is the layer of gases (mainly nitrogen and oxygen) surrounding the earth to roughly 60 miles (100 km) out, where space begins.

chapter outline

chapter objectives

This chapter should enable you to

→   Understand the tectonic forces behind some of the world's major landforms and natural hazards

→   Recognize consistent global patterns in the distribution of temperatures, precipitation, and vegetation types

→   Identify the natural areas most threatened by human activity and explain how natural habitat loss may endanger human welfare

→   Appreciate the important roles of the world's oceans in making the earth habitable

→   Describe the potential impacts of global climate change and international efforts to prevent them

Discussion starts with the landforms that set the terrestrial stage for human activities. The earth's lithosphere is a work in progress, and you will see how its changing surfaces provide both opportunities and threats to people. We will consider the climate and vegetation types that play such a large role in human activities and appreciate the rich diversity of wild plant and animal species. We will look briefly at the planet's often overlooked oceans and the resources they hold. Finally, we will examine how the climate may be changing and what this may mean for life on earth.

2.1 Geologic Processes and Landforms

The earth's surface is in motion. Those of you living on the West Coast of the United States have probably had some personal experience of this, the unforgettable sensation of the ground moving beneath you in a mild "temblor" or even a large earthquake. These experiences reflect global processes.

Plate Tectonics

About a century ago (1912), a German geologist named Alfred Wegener came up with an outlandish theory known as **continental drift.** Noting among other things the jigsaw puzzle–like geometry of Africa's west coast and South America's east coast, he proposed that the continents were once joined in a supercontinent (which he named Pangaea) but that they "drifted apart" over time. However, Wegener was not able to explain the forces behind these movements, and he was derided by a scientific community that seemed unwilling to accept anything other than a *terra firma* that was truly firm. Peers pronounced his conclusions "utter damned rot" and "mere geopoetry."[1] Later discoveries in deep-sea science led Wegener's basic proposition to be accepted as fact, and today a good deal is known about how the drift occurs.

The earth's lithosphere is made up of about a dozen giant and several smaller sections called **plates,** and these move in various directions in processes known collectively as **plate tectonics** (●Figure 2.1). On the ocean floors in places such as the Mid-Atlantic Ridge and the East Pacific Rise, new lithosphere is "born" as molten material rises from the earth's mantle and cools into solid rock (●Figure 2.2). Plate tectonics are often explained by the useful analogy of a "conveyor belt" (the convection cell in Figure 2.2) in constant motion. On either side of the long, roughly continuous ridges, the two young plates move away from one another, carrying islands with them; this process is called **seafloor spreading.**

Seafloor spreading has few impacts on people, but when the earth's plates collide, there is cause for great concern: **tectonic forces** are among the planet's greatest **natural hazards.** The **seismic activity** (*seismic* refers to earth vibrations, mainly earthquakes) that causes earthquakes and **tsunamis** (tidal waves) and the **volcanism** (movement of molten earth material) of volcanoes and related features are the most dangerous tectonic forces.

135, 340, 396, 421

Plates collide and converge in different ways and with different consequences. In some parts of the world—offshore from the U.S. Pacific Northwest and the east coast of Japan, for example—one plate "dives" below another in a process known as **subduction** (see Figure 2.2, and an up-close view of the process in Japan in Figure 7.4.A, page 397). The descending lithosphere is melted again as it dives into the earth's mantle along a deep linear feature known as a **trench** (for example, the Mariana Trench off Japan). Subduction is another stage along the "conveyor belt" process that will eventually see this material recycled as newborn lithospheric crust.

This subduction process releases enormous amounts of energy. Periodically, the great stress of one plate pushing beneath another is released in the form of an earthquake. The world's largest recorded earthquakes—registering 9.5 (Chile, 1960), 9.2 (United States, 1964), and 9.1 (Indonesia, 2004), respectively, on the **Richter scale,** which measures the strength of the earthquake at its source—struck along these subduction zones. This sudden displacement of a section of oceanic lithosphere is also what triggers a tsunami and the attendant loss of life and property such a powerful wave can cause (see the feature on the Great Tsunami of 2004 on page 340).

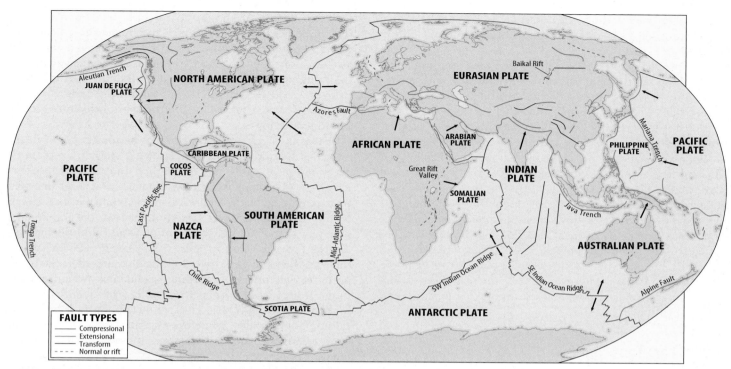

FAULT TYPES
—— Compressional
—— Extensional
—— Transform
- - - Normal or rift

● **Figure 2.1** Major tectonic plates and their general direction of movement. Earthquakes, volcanoes, and other geologic events are concentrated where plates separate, collide, or slide past one another. Where they separate, **rifting** produces very low land elevations (well below sea level at the Dead Sea of Israel and Jordan, for example) or the emergence of new crust on the ocean floor (in the middle of the Atlantic Ocean, for example).

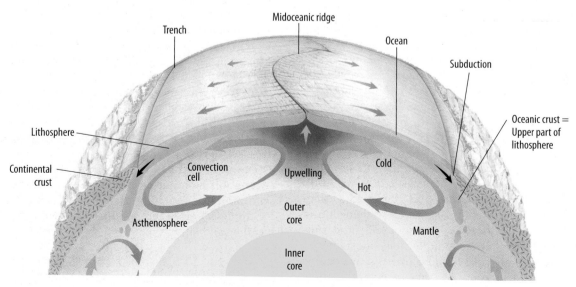

● **Figure 2.2** Generalized cross section of the earth, showing its main concentric layers and the process by which its lithosphere is recycled.

In other places where they meet, the lithospheric plates grind and slide along one another, as in the state of California and in Turkey.

135,
212,
336,
340,
396,
559,
602
The processes of rock crowding together or pulling apart along these fracture lines is known as **faulting,** with movement along various kinds of **faults** causing earthquakes, the emergence of new landforms, and other consequences.

Volcanism generally takes place along and near **subduction zones** (see Figure 2.2) and also in the world's several dozen geologic hot spots where molten material has broken through the crust as a *plume* (as in Yellowstone National Park and in Hawaii—see page 420). Despite their posing a host of natural hazards to people living on their slopes or downwind—pyroclastic flows (fast-moving currents of rock fragments, hot gases, and ash), lava flows, ashfalls, and other dangers—volcanoes have beneficial qualities. Volcanic rock generally breaks down to form fertile soils (as in Ethiopia and the Nile Valley), and the range of climate and vegetation
252,
583,
357
types on their slopes poses opportunities for farming and raising livestock.

Major Landform Types

As a student of world regional geography, one of the most basic questions you will ask or answer about places or countries is "What does the landscape look like?" You will probably want to begin by describing the terrain—is it mountainous, flat, or something in between?—and then move on to the climate, vegetation, human activities, and other characteristics that shape the landscape.

In this book, the landforms of each of the world's eight regions are depicted in the "physical geography" map that introduces each region. The planet has an enormous variety of landform types, but to generalize, there are four. Each poses different opportunities for human activities and for natural vegetation and wildlife.

Hill lands and mountains. These are the high and steep features of the earth's surface. There is no formal definition to distinguish hills from mountains, and usage of these terms is usually rooted in local or regional perceptions. In the shadow of the world's highest mountain range, the Himalayas, the people of India call their southernmost high area the "Palni Hills," although the peaks there rise to 6,500 feet (2,000 m). Some people living in the midwestern United States call the rugged region of southern Missouri and northern Arkansas the "Ozark Mountains," although they rise to only 2,500 feet (760 m). Generally, the so-called mountains are higher—typically at least 2,000 to 3,000 feet (600 to 900 m) above sea level—and more rugged than hills and offer fewer possibilities for human settlement and use. Consequently, whereas many areas of hill land support moderate to dense human populations, most mountain areas are sparsely populated. The planet's mountains tend to be along and near zones of convergence of tectonic plates; not just volcanoes but other mountains (including the Himalayas) have risen as a consequence of tectonic forces.

Plains. Most of the world's people live on plains, so they are a very important landscape element to consider in world regional geography. A **plain** is ordinarily a relatively level area of slight elevation, although this usage varies too; the "High Plains" of the United States, for example, rise to over 5,000 feet (about 1,500 m) above sea level. Some plains are almost table-flat, but most are broken by gentle or moderate slopes.

Plateaus. A **plateau** is an elevated plain. Again, while no universal definition exists, a plateau is typically a plain at an elevation of 2,000 feet (610 m) or higher. By some definitions, a plateau should be terminated on at least one side by a steep edge, called an **escarpment,** that marks a sharp boundary with the lower elevation. In areas with abundant precipitation, water may have cut plateaus into hilly terrain, but the "parent" landscape is still a plateau. Depending on their latitude and elevation, and the climatic and vegetation patterns that prevail accordingly, plateaus present a wide range of livelihood options for people. The Tibetan Plateau, for example, is high, arid, and sparsely populated by pastoral nomadic peoples, while the plateaus of East Africa tend to be relatively lush, well watered, and more densely populated by nomads and farmers alike.

We turn now to the climate and vegetation types that give characteristic appearances and patterns to the world's landforms.

2.2 Patterns of Climate and Vegetation

"Everybody talks about the weather, but nobody does anything about it," Mark Twain reportedly said. The day's weather certainly does influence our conversations and decisions. More important for us as students of geography, the cumulative impacts of weather shape landscapes and influence human activities and settlements in patterns recognizable around the world. Once you know these patterns, you will be prepared to step into any world region and appreciate it better.

As you experience a warm, dry, cloudless summer day or a cold, wet, overcast winter day, you are encountering the **weather**—the atmospheric conditions occurring at a given time and place. **Climate** is the average weather of a place over a long time period. Along with surface conditions such as elevation and soil type, climatic patterns have a strong correlation with patterns of natural vegetation and in turn with human opportunities and activities on the landscape. Precipitation and temperature are the key variables in weather and climate.

Precipitation

Water is essential for life on Earth, and it is very useful to understand the processes of weather and climate that dis-

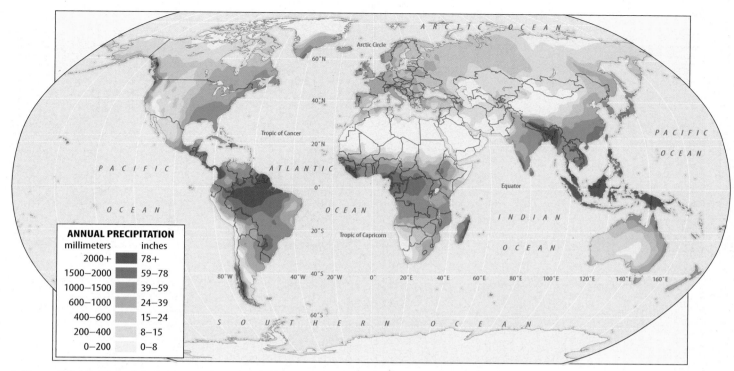

ANNUAL PRECIPITATION

millimeters		inches
2000+		78+
1500–2000		59–78
1000–1500		39–59
600–1000		24–39
400–600		15–24
200–400		8–15
0–200		0–8

● **Figure 2.3** World precipitation map. Some geographers argue that this is the most important of all maps in understanding life on Earth.

tribute this precious resource (●Figure 2.3). Warm air holds more moisture than cool air, and **precipitation**—rain, snow, sleet, and hail—is best understood as the result of processes that cool the air to release moisture. Precipitation results when water vapor in the atmosphere cools to the point of condensation, changing from a gaseous to a liquid or solid form. The amount of cooling needed for this change depends on the original temperature and the amount of water vapor in the air.

For this cooling and precipitation to occur, generally air must rise in one of several ways. In equatorial latitudes or in the high-sun season (summer, when the sun's rays strike the earth's surface more vertically) elsewhere, air heated by intense surface radiation can rise rapidly, cool, and produce a heavy downpour of rain, an event that occurs often in summer over much of the United States (●Figure 2.4). Precipitation that originates in this way and is released from tall cumuloform clouds is called **convectional precipitation** (this and the other types are depicted in ●Figure 2.5).

Orographic (mountain-associated) precipitation results when moving air strikes a topographic barrier and is forced upward. Most of the precipitation falls on the windward side of the barrier, and the leeward (sheltered) side is likely to be very dry.

Cyclonic (frontal) precipitation is generated in traveling low-pressure cells, called **cyclones,** that bring air masses with different characteristics of temperature and moisture into contact (●Figures 2.6 and 2.7). In the atmosphere, air moves from areas of high pressure to areas of low pressure. Air from masses of different temperature and moisture con-

tent is drawn into a cyclone, which has low pressure. One air mass is normally cooler, drier, and more stable than the other. These masses do not mix readily but instead come in contact within a boundary zone 3 to 50 miles (c. 5 to 80 km) wide called a **front**. A front is named after which air mass is

● **Figure 2.4** A "supercell" thunderstorm complex can produce heavy rain, hail, and tornadoes.

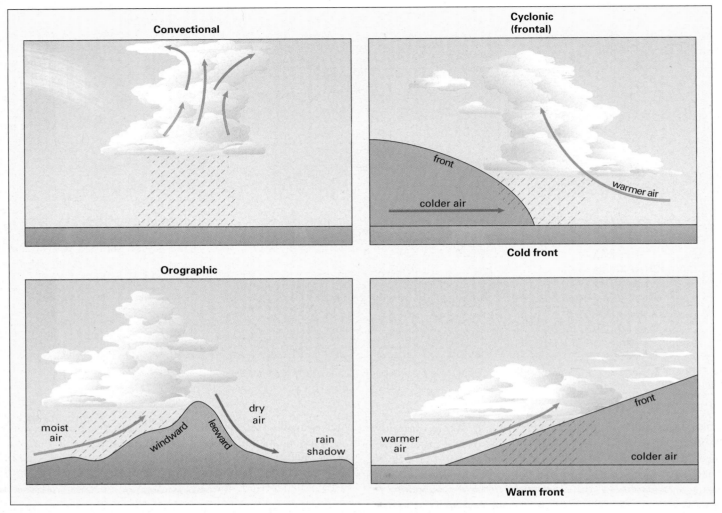

● **Figure 2.5** Diagrams showing the origins of convectional, orographic, and cyclonic precipitation. Note that cyclonic precipitation results from both cold fronts and warm fronts.

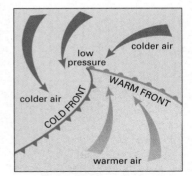

● **Figure 2.6** Diagram of a cyclone in the Northern Hemisphere.

● **Figure 2.7** Satellite image of Hurricane Fran approaching Florida's east coast in September 1996. Also known as cyclones and typhoons, hurricanes are the largest cyclonic features to form in the earth's atmosphere.

advancing to overtake the other. In a **cold front,** the colder air wedges under the warmer air, forcing it upward and back. In a **warm front,** the warmer air rides up over the colder air, gradually pushing it back. Whether from a warm or a cold front, **frontal precipitation** is likely to result because the warmer air mass rises and condensation takes place.

Aridity

The flip side of precipitation is, of course, aridity, and drought or any significant shortages in rainfall are among the natural hazards people around the planet must cope with. There are several causes of low precipitation:

High pressure. Large areas of the earth receive little precipitation because of subsiding air masses having high atmospheric pressure, called **high-pressure cells,** or **anticyclones** (•Figure 2.8). In a high-pressure cell, the air is descending and becoming warmer under the increased pressure (weight) of the air above it. As it warms, its capacity to hold water vapor increases, its relative humidity decreases, and its potential to make precipitation falls. Streams of dry, stable air moving outward from the anticyclones often bring prolonged drought to the areas below their path. Most famous of these air streams are the **trade winds** that originate in semipermanent anticyclones on the margins of the tropics and are attracted equatorward (becoming more moisture-laden as they go) by a semipermanent low-pressure cell, the **equatorial low.** As Figure 2.8 illustrates, high-pressure cells rotate clockwise in the Northern Hemisphere and counterclockwise in the Southern Hemisphere. Conversely, low-pressure systems rotate counterclockwise in the Northern Hemisphere and clockwise in the Southern Hemisphere. This pattern of opposite movements in the hemispheres is caused by the earth's rotation and is known as the **Coriolis effect.**

Atmospheric stability in coastal regions. Cold ocean waters are responsible for coastal deserts in some parts of the world, such as the Atacama Desert in Chile and the Namib Desert in southwestern Africa. Here, air moving onshore over cooler ocean waters is overlain by warm air being blown seaward from the land. Warmer air is lighter than cooler air, and a layer of warm air sitting on top of cooler air creates an **inversion.** This **atmospheric stability** prevents the cooler air from rising and thus precludes cloud formation and precipitation.

507
559

Rain shadow. As noted earlier, the windward side of mountains often receives abundant orographic precipitation. The leeward side, in contrast, is usually very dry. Nevada, which is on the leeward side of the Sierra Nevada in the western United States, is in such a **rain shadow.** Rain shadows are the main cause of arid and semiarid lands in some regions, but other areas with very low precipitation result from a combination of influences. The Sahara of northern Africa, for example, is mainly a product of high atmospheric pressure, but the rain shadow effect of the Atlas Mountains and the presence of cold Atlantic Ocean waters along its western coast also contribute to the Sahara's dryness.

247,
267,
319,
522,
670

Many regions that regularly experience long periods of dry weather are also subject to changes in weather patterns that bring welcome rainfall. These shifts are usually seasonal and are ushered in by changing temperatures.

283,
440,
458

Seasons

In the middle and high latitudes, the most significant factor in determining temperatures is seasonality, which is related to the inclination of the earth's polar axis as the planet orbits the sun over a period of 365 days (•Figure 2.9). On or about June 22, the first day of summer in the Northern Hemisphere, the northern tip of the earth's axis is inclined toward the sun at an angle of 23.44 degrees from a line perpendicular to the plane of the ecliptic (the plane of the planet's orbit around the sun). This is the **summer solstice** in the Northern Hemisphere and the **winter solstice** in the Southern Hemisphere. A larger portion of the Northern Hemisphere than of the Southern Hemisphere remains in daylight, and warmer temperatures prevail. On or about September 23, and again on or about March 20, the earth reaches the **equinox** position. Its axis does not point toward or away from the sun, so days and nights are of equal length at all latitudes on earth. On or about December 22, the first day of winter in the Northern Hemisphere, the southern tip of the earth's axis is inclined toward the sun at an angle of 23.44 degrees from a line perpendicular to the plane of the ecliptic. This is the winter solstice in the Northern Hemisphere and the summer solstice in the Southern Hemisphere. A larger portion of the Southern Hemisphere than of the Northern Hemisphere remains in daylight, and warmer temperatures prevail.

Great regional differences exist in the world's annual and seasonal temperatures. In lowlands near the equator, temperatures remain high throughout the year, while in areas near the poles, temperatures remain low for most of the year. The intermediate (middle) latitudes have well-marked seasonal changes of temperature, with warmer

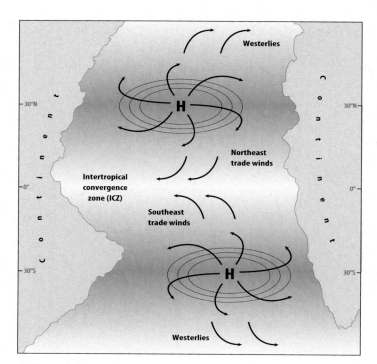

• **Figure 2.8** Idealized wind and pressure systems. Shading indicates wetter areas.

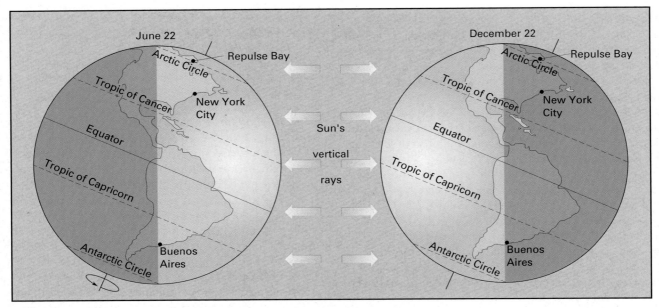

• **Figure 2.9** Geometric relationships between the earth and the sun at the solstices.

temperatures generally in the summer season of high sun and cooler temperatures in the winter season of low sun (when sunlight's angle of impact is more oblique and daylight hours are shorter). Intermittent incursions of polar and tropical air masses increase the variability of temperature in these latitudes, bringing unseasonably cold or warm weather.

Climate and Vegetation Types

The myriad combinations of precipitation, temperature, latitude, and elevation produce a great variety of local climates. Geographers group these local climates into a number of major climate types, each of which occurs in more than one part of the world (•Figure. 2.10) and is associated

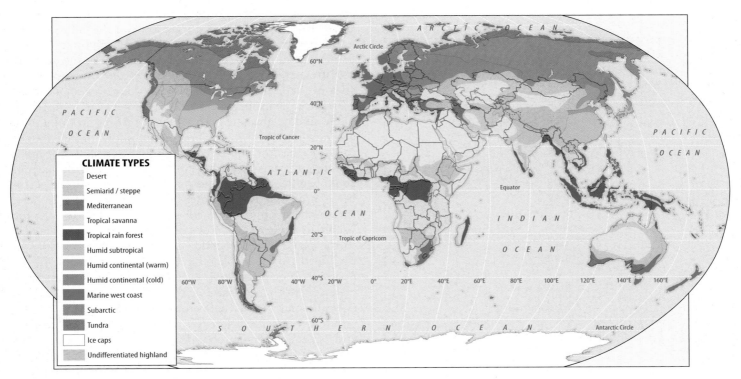

• **Figure 2.10** World climates.

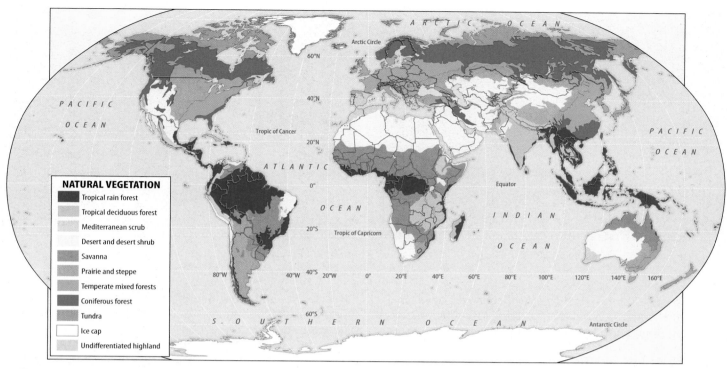

● **Figure 2.11** World biomes (natural vegetation) map.

NATURAL VEGETATION

- Tropical rain forest
- Tropical deciduous forest
- Mediterranean scrub
- Desert and desert shrub
- Savanna
- Prairie and steppe
- Temperate mixed forests
- Coniferous forest
- Tundra
- Ice cap
- Undifferentiated highland

closely with other types of natural features, especially vegetation. Geographers recognize 10 to 20 major types of terrestrial ecosystems, called **biomes,** which are categorized by dominant types of natural vegetation (●Figures 2.11 and 2.12). Climate plays the main role in determining the distribution of biomes, but differing soils and landforms may promote different types of vegetation where the climatic pattern is essentially the same. Vegetation and climate types are sufficiently related that many climate types take their names from vegetation types—for example, the tropical rain forest climate and the tundra climate. You may easily see the geographic links between climate and vegetation by

comparing the maps in Figures 2.10 and 2.11. The spatial distributions of climate and vegetation types do not overlap perfectly, but there is a high degree of correlation.

In the **ice cap, tundra,** and **subarctic** climates, the dominant feature is a long, severely cold winter, making agriculture difficult or impossible. The summer is short and cool. Perpetual ice in the form of glaciers may be found at very high elevations in the lower latitudes (even in equatorial regions) and in the polar realms; the ice cap biome (●Figure 2.12a) is devoid of vegetation, except in those very few spots where enough ice or snow melts in the summer to allow tundra vegetation to grow. Tundra vegetation (●Figure 2.12b)

● **Figure 2.12** (a) Ice cap biome, glacier in British Columbia, Canada

● **Figure 2.12** (b) Tundra, northern Norway

● **Figure 2.12** (c) Coniferous forest, British Columbia, Canada

● **Figure 2.12** (e) Steppe, eastern Turkey

is composed of mosses, lichens, shrubs, dwarfed trees, and some grasses. Needleleaf evergreen **coniferous trees** can stand long periods when the ground is frozen, depriving them of moisture. Thus **coniferous forests**, often called **boreal forests** or **taiga**, their Russian name (●Figure 2.12c), occupy large areas where the climate is subarctic.

In **desert** and **semiarid** or **steppe** climates, the dominant feature is aridity or semiaridity. Deserts and steppes occur in both low and middle latitudes. Agriculture in these areas usually requires irrigation. The earth's largest dry region extends in a broad band across northern Africa and southwestern and central Asia. The deserts of the middle and low latitudes are generally too dry for either trees or grasslands. They have **desert shrub** vegetation (●Figure 2.12d), often only in dry riverbeds, and in many places have no vegetation at all. The bushy desert shrubs are **xerophytic** (from Greek roots meaning "dry plant"), having small leaves, thick bark, large root systems, and other adaptations to absorb and retain moisture. Grasslands dominate in the moister steppe climate, a **transitional zone** between very

arid deserts and more humid areas. The biome composed mainly of short grasses is also called the **steppe**, or **temperate grassland** (●Figure 2.12e). The temperate grassland region of the United States and Canada originally supported both tall-grass and short-grass vegetation types, known in those countries as **prairies.**

The **tropical rain forest** and **tropical savanna** climates are rainy, low-latitude climates. The key difference between them is that the tropical savanna type has a pronounced dry season, which is short or absent in the tropical rain forest climate. Heat and moisture are almost always present in the tropical rain forest biome (●Figure 2.12f), where broadleaf evergreen trees (which lose their leaves, but not seasonally or in tandem) dominate the vegetation. In tropical areas with a dry season that still have enough moisture for tree growth, **tropical deciduous forest** (●Figure 2.12g) replaces the rain forest. Here the broadleaf trees are not green throughout the year; they lose their leaves and are dormant during the dry season and then add foliage and resume their growth during the wet season. The tropical

● **Figure 2.12** (d) Desert shrub, southern Sinai Peninsula, Egypt

● **Figure 2.12** (f) Tropical rain forest, Dominica, West Indies

• **Figure 2.12** (g) Tropical deciduous forest, Gir Forest, western India (note Asiatic lions in the center of the photo)

• **Figure 2.12** (i) Savanna, southern Kenya

deciduous forest approaches the luxuriance of the tropical rain forest in wetter areas but thins out to low, sparse scrub and thorn forest (•Figure 2.12h) in drier areas. **Savanna** vegetation (•Figure 2.12i), which has taller grasses than the steppe, occurs in areas of greater overall rainfall and more pronounced wet and dry seasons.

In the **marine west coast** climate, occupying the western sides of continents in the higher middle latitudes (the Pacific Northwest region of the United States, for example), warm ocean currents moderate the winter temperatures, and summers tend to be cool. Coniferous forest dominates some cool, wet areas of marine west coast climate; a good example is the redwood forest of northern California. These humid middle-latitude regions have mild to hot summers and winters ranging from mild to cold, with several other types of climate interspersed.

The **Mediterranean** climate (named after its most prevalent area of distribution, the lands around the Mediterranean Sea) is typically found between a marine west coast climate and a lower-latitude steppe or desert climate. In the summer high-sun period, it lies under high atmospheric pressure and is rainless. In the winter low-sun period, it lies in a westerly wind belt and receives cyclonic or orographic precipitation. **Mediterranean scrub forest** (•Figure 2.12j), known locally by such names as *maquis* and *chaparral*, characterizes Mediterranean climate areas. Because of the hot, dry summers, the natural vegetation consists primarily of xerophytic shrubs.

The **humid subtropical** climate occupies the eastern portion of continents between approximately 20 and 40 degrees of latitude and is characterized by hot summers, mild to cool winters, and ample precipitation for agriculture. The **humid continental** climate lies poleward of the humid subtropical type; it has cold winters, warm to hot summers, and enough rainfall for agriculture, with the greater part of the precipitation falling in the summer. This climate type is often subdivided into warm and cold subtypes, indicating the greater severity of winter in the zones closer to the poles. In middle-latitude areas with these humid subtropical and humid continental climate types, a **temperate mixed forest** with mostly

• **Figure 2.12** (h) Scrub and thorn forest, northern Zimbabwe

• **Figure 2.12** (j) Mediterranean scrub forest, southern California, United States

• **Figure 2.12** (k) Temperate mixed forest, southern Missouri, United States

broadleaf but also coniferous trees (as in the U.S. Midwest and Northeast; •Figure 2.12k) is found. As cold winter temperatures freeze the water within reach of plant roots, broadleaf trees shed their now-colorful leaves and cease to grow, thus reducing water loss. They then produce new foliage and grow vigorously during the hot, wet summer. Coniferous forests can thrive in some hot and moist locations where porous, sandy soil allows water to escape downward, giving conifers (which can withstand drier soil conditions) an advantage over broadleaf trees. Pine forests on the coastal plains of the southern United States are an example.

Undifferentiated highland climates have a range of conditions according to elevation and exposure to wind and sun. The vegetation in these climates (•Figure 2.12l) differs greatly, depending on elevation, degree and direction of slope, and other factors. These climates are "undifferentiated" in the context of world regional geography in that a small mountainous area may contain numerous biomes, and it would be impossible to map them on a small scale.

• **Figure 2.12** (l) Undifferentiated highland vegetation, San Juan Mountains, Colorado, United States

The world's mountain regions have a complex array of natural conditions and opportunities for human use. Increasing elevation lowers temperatures by a predictable **lapse rate**, on average about 3.6°F (2.0°C) for each increase of 1,000 feet (305 m) in elevation. In climbing from sea level to the summit of a high mountain peak near the equator (for example, in western Ecuador), a person would experience many of the major climate and biome types found in a sea-level walk from the equator to the North Pole!

2.3 Biodiversity

Anyone who has admired the nocturnal wonders of the "barren" desert or appreciated the variety of the "monotonous" Arctic can vouch for an astonishing diversity of life even in the biomes that seem least hospitable to it. But geographers and ecologists recognize the exceptional importance of some biomes because of their **biological diversity (or biodiversity)**—the number of plant and animal species present and the variety of genetic materials these organisms contain.

The most diverse biome is the tropical rain forest. From a single tree in the Peruvian Amazon region, the entomologist Terry Erwin recovered about 10,000 insect species. From another tree several yards away, he counted another 10,000, many of which differed from those of the first tree. Alwyn Gentry of the Missouri Botanical Garden recorded 300 tree species in a 1-hectare (2.47-acre) plot of the Peruvian rain forest. Less than 50 years ago, scientists calculated that there were 4 to 5 million species of plants and animals on earth. Now, however, their estimates are much higher, in the range of 40 to 80 million. This startling revision is based on research, still in its infancy, on species inhabiting the rain forest.

The Importance of Biodiversity

Such diversity is important in its own right, but it also has vital implications for nature's ongoing evolution and for people's lives on earth. Humankind now relies on a handful of crops as staple foods. In our agricultural systems, the trend in recent decades has been to develop high-yield varieties of grains and to plant them as vast **monocultures** (single-crop plantings). This trend, which is the cornerstone of the so-called **Green Revolution,** is controversial. On the one hand, it puts more food on the global table. But on the other, it may render agriculture more vulnerable to pests and diseases and thus pose long-term risks of famine. In evolutionary terms, the Green Revolution has reduced the natural diversity of crop varieties that allows nature and farmer to turn to alternatives when adversity strikes. At the same time, while we remove tropical rain forests and other natural ecosystems to provide ourselves with timber, agriculture, or living space, we may be eliminating the foods, medicines, and raw materials of tomorrow, even before we

299, 321, 377

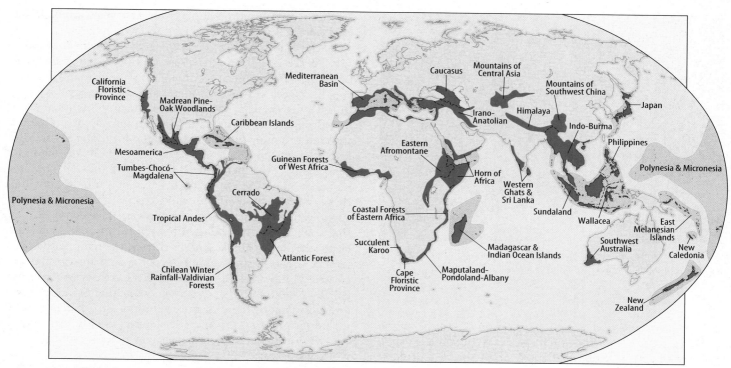

• **Figure 2.13** World biodiversity hot spots.

have collected them and assigned them scientific names. "We are causing the death of birth," laments the biologist Norman Myers.[2]

Regions where human activities are rapidly depleting the rich variety of plant and animal life are known as **biodiversity hot spots,** which scientists believe deserve immediate attention for study and conservation. Thirty-four priority regions identified by Conservation International are depicted in •Figure 2.13. By referring to the map of the earth's biomes in Figure 2.11, you will see that most of these hot spots are in tropical rain forest areas. Many are islands that tend to have high biodiversity because species on them have evolved in isolation to fulfill special roles in these ecosystems and because human pressures on island ecosystems are particularly intense. Efforts are under way in most of these hot spots to establish national parks and other protected areas. Conservationists are also turning their attention increasingly to the state of the world's oceans, which play critical roles in the earth's physical processes.

2.4 The World's Oceans

About 71 percent of the world's surface is comprised of water, but most world regional geography textbooks (including previous editions of this one) devote little, if any, attention to the oceans and their critical resources. Let us not forget them!

Life on earth would be impossible without the roles and resources of the hydrosphere, which includes both oceans and freshwater sources such as lakes and rivers. This book deals often with issues of freshwater, which is appropriate considering the projection by the World Commission on Water that by 2025, half of the world's population will live under conditions of severe water stress.[3] In this section, the focus is on the world's oceans.

Why Should We Care about Oceans?

I live in Missouri, where a day at the beach or a plate of fresh seafood is just a dream. Even those of you living close to the sea may have very little interaction with it. Why bother thinking about the world's oceans, much less worrying about them?

It's a watery world. With so much of the earth's surface covered by seawater, it is vital that we consider how that water shapes life on the planet, including human life. Oceans have the largest role in the **hydrologic cycle,** in which solar radiation evaporates seawater into vapor that is then released onto land in the form of freshwater precipitation. Without the seas, the earth's usable freshwater resources would be extremely limited.

The oceans feed us. Missourians and other landlubbers may not eat fish every day or even every week, but about a billion people, or 15 percent of the world's population, rely mainly on fish for their protein, and even the average American eats 17 pounds (7.7 kg) of fish and shellfish annually (•Figure 2.14). Unfortunately, growing human populations and technological improvements in the fishing industry are putting these resources under unprecedented pressure.[4] In

• **Figure 2.14** The Tsukuji fish market in Tokyo. Humanity's appetite for fish and seafood is growing, but what about the supplies?

2006, scientists issued a stern warning on the future of the world's fisheries: unless dramatic steps were taken quickly, there would be a "global collapse" of all species currently fished by 2050.[5]

Defining "collapse" as a population less than 10 percent of its previous levels, these authorities noted that 30 percent of the fish species currently exploited had already collapsed (•Figure 2.15). Their good news was that this gloomy scenario could be averted by action on several fronts, including reducing the number of unwanted fish caught in nets and reducing overfishing where it is recognized to be a problem. More good news is that **aquaculture** (including **fish farming**) has the potential to catch up with and increasingly substitute for wild-caught fish stocks. There has been explosive growth in aquaculture in China within the last decade, and there is unrealized potential in that country and others. Some of this aquaculture is what may be described as "sustainable

201,
339,
405,
634

seafood," including the unlikely harvest of shrimp raised in well water deep in the Arizona desert.[6]

They provide energy and other raw materials for human use. As terrestrial sources of conventional energy dwindle, our fossil fuel–addicted economies create demand for new supplies. Deep seas are the final frontier for energy exploration. Under the U.S. territorial waters of the Gulf of Mexico, there are as many as 40 billion barrels of petroleum (enough to supply the U.S. oil demand for five years at 2007 levels of consumption). But getting to that oil is going to be a problem: it lies beneath 10,000 feet (3,050 m) of water and then 5 miles (8 km) of rock, salt, and sand. Meanwhile, there is a huge potential to capture unconventional energy supplies from the sea; Scottish engineers have demonstrated that wave power can be used to spin turbines and generate electricity, for example, and in theory, the temperature difference between surface and deeper waters in warm tropical seas can be exploited to produce electricity in a process called *ocean thermal energy conversion* (OTEC). Prospects are increasing for the deep-sea mining of other minerals, including gold, silver, and the copper, cobalt, and nickel held within manganese nodules strewn across much of the world's seafloor.

They play important roles in trade and commerce. The seafaring days of humankind are far from over. A remarkable 90 percent of global trade is seaborne. What about all of that Chinese merchandise finding its way into U.S. retail stores? Some 98 percent of it travels across the Pacific on cargo ships.[7] There is an economic rationale behind this heavy reliance on seaborne trade. Air freight can cost 20 times as much as sea freight. Safeguarding the security of global seaborne commerce is one of the enduring themes of geopolitical strategy making and has all too often been the trigger for regional conflicts and wider wars.

229,
305

2.5 Global Environmental Change

Governments have had great success in recent years in delimiting national parks and other reserves, both on land and at sea, for the protection of biodiversity. Yet even with the most stringent efforts to protect the living things in these wilderness areas, their future is uncertain. Atmospheric changes are occurring that will have profound effects on natural systems and on human uses of the earth. Many of these changes are attributable to human activities, and the final section of this chapter is a prelude to the following chapter on human processes affecting the planet.

Climatic Change

Until about 2000, there was considerable uncertainty in the science of climate change. Most scientists at that time insisted that human activities were responsible for a doc-

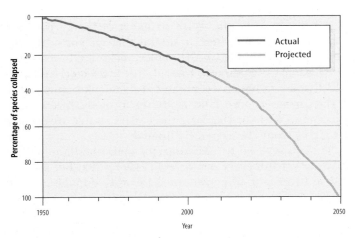

• **Figure 2.15** Percentage of ocean fish species "collapsed" (defined as less than 10 percent remaining).

umented warming of the earth's surface, but a significant minority either insisted that warming was not occurring or, if it was, that a natural climatic cycle was responsible. By 2007, however, most of the uncertainty had evaporated.

In its 2007 report, the atmospheric scientists (representing 154 countries) who make up the United Nations–sponsored **Intergovernmental Panel on Climate Change (IPCC)** concluded that global warming is "unequivocal." The global mean temperate increased by 1.4 degrees Fahrenheit (1.4°F) (about 0.8 degrees Celsius, 0.8°C) since the late 19th century. Human production of **greenhouse gases** such as **carbon dioxide (CO_2)** was, according to the IPCC, "very likely" responsible for this warming (•Figure 2.16). In the scientists' jargon, "very likely" means a more than 90 percent probability; just six years earlier, the IPCC said the human factor was "likely," meaning a 66 to 90 percent probable cause.[8]

The Greenhouse Effect

In 1827, the French mathematician Jean-Baptiste Fourier established the concept of the **greenhouse effect,** noting that the earth's atmosphere acts like the transparent glass cover of a greenhouse (•Figure 2.17)—for modern purposes, think of a car's windshield. Visible sunlight passes through the glass to strike the planet's surface. Ocean and land (the floor of the greenhouse or the car's upholstery) reflect the incoming solar energy as invisible infrared radiation (heat). Acting like the greenhouse glass or car windshield, the earth's atmosphere traps some of that heat.

In the atmosphere, naturally occurring greenhouse gases such as carbon dioxide and water vapor make the earth habitable by trapping heat from sunlight. Concern over global warming focuses on human-derived sources of greenhouse gases, which trap abnormal amounts of heat. Carbon dioxide released into the atmosphere from the burning of coal, oil, and natural gas is the greatest source of concern, but **methane** (from rice paddies and the guts of ruminat-

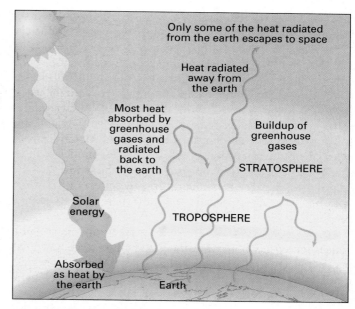

• **Figure 2.17** The greenhouse effect. Some of the solar energy radiated as heat (infrared radiation) from the earth's surface escapes into space, while greenhouse gases trap the rest. Naturally occurring greenhouse gases make the earth habitable, but carbon dioxide and greenhouse gases emitted by human activities accentuate the greenhouse effect, making the planet unnaturally warmer.

ing animals like cattle), **nitrous oxide** (from the breakdown of nitrogen fertilizers), and **chlorofluorocarbons,** or **CFCs** (used as coolants and refrigerants) are also greenhouse gases resulting from human activities. (CFCs also destroy stratospheric ozone, a gas that has the important effect of preventing much of the sun's harmful ultraviolet radiation from reaching the earth's surface.) In the car-as-earth metaphor, continued production of these greenhouse gases has the effect of rolling up the car windows on a sunny day, with the result of increased temperatures and physical overheating of the occupants.

Although formal records of meteorological observations began just over a century ago, past climates left evidence in the form of marine fossils, corals, glacial ice, fossilized pollen, and annual growth rings in trees. These indirect sources, along with formal records dating back to 1861, indicate that the 20th century was the warmest of the past six centuries and that the five warmest years in the past 600 years were 1998, 2002, 2003, 2005, and 2007.

The Effects of Global Warming

Scientists have long predicted that global warming due to human activities will have profound and wide-ranging effects. In 1896, the Swedish scientist Svante Arrhenius feared that Europe's growing industrial pollution would eventually double the amount of carbon dioxide in the earth's atmosphere and as a consequence raise the global mean temperature as much as 9°F (5°C). In its 2007 report, the IPCC

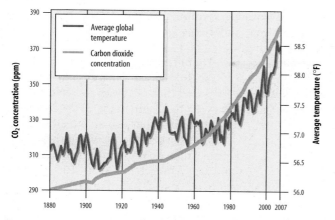

• **Figure 2.16** Industrialization and the burning of tropical forests have produced a steady increase in carbon dioxide emissions. Many scientists believe that these increased emissions explain the corresponding steady increase in the global mean temperature.

reached a remarkably similar conclusion: it calculated the impact of doubling carbon dioxide from the preindustrial 1750 levels as a mean global temperature rise of 3.5°F to 8°F (2°C to 4.4°C). Other climate change models today calculate a doubling of carbon dioxide from year 2000 levels, concluding that the mean global temperatures might warm from 3°F to 11°F (1.7°C to 6.1°C) by 2100. The estimated range and consequences of the increases vary widely because of different assumptions about the little-understood roles of oceans and clouds. Most of these models concur that because of the slowness with which the world's oceans respond to changes in temperature, the effects of this rise will be delayed by decades. Briefly, here are the major predictions:

A warmer climate overall, but . . . Geographers are most concerned with where the anticipated climatic changes will occur and what effects these changes will have. Computer models conclude that there will not be a uniform temperature increase across the entire globe but rather that the increases will vary spatially and seasonally. Differing models sometimes produce contradictory results about the timing and impacts of warming. Models predicting warmer summers generally forecast that crop productivity will decline, while models predicting warmer winters generally forecast an increase in crop yields.

More precipitation overall, but also more pronounced drought. Higher temperatures will mean increased evaporation from the world's oceans. This will result in more precipitation, but it will be distributed irregularly. The 2007 IPCC report forecasts more precipitation in the higher latitudes but less, with intense and longer droughts, in the lower latitudes. What the IPCC calls "heavy precipitation events" will be more common, and the magnitude of hurricanes (whose strength increases with warmer ocean surface waters) will increase.

Pronounced warming in the polar regions. Geographically, the impacts of global warming appear to be greatest at the higher latitudes. Sailing on a Russian nuclear-powered icebreaker, I saw open water at the North Pole in August 1996 (•Figure 2.18), and subsequent years have seen accelerated shrinkage of polar ice. The 2007 IPCC report concluded that the average coverage of Arctic sea ice has shrunk 2.7 percent per decade since 1978, with summertime ice decreasing 7.4 percent per decade. The trend is self-perpetuating, because the darker ocean waters that replace the white ice cover absorb ever more solar radiation. Some quarters are cheering the trend. A further retreat of polar ice would bode well for maritime shipping through the long-icebound Northwest Passage across Arctic Canada and through the northern sea route across the top of Russia, which in the past has been navigable only in summer and only with the aid of icebreakers.

Shifting biomes, with species extinction. There is general agreement that with global warming, the distribution of climatic conditions typical of biomes will shift poleward

• **Figure 2.18** The passengers and crew of the Russian icebreaker *Yamal* found open water at the North Pole on August 16, 1996. Much of the voyage from northeastern Siberia was ice-free. Similar observations since then have contributed to widespread concern about global warming. The ship's global-positioning instrument indicated that the vessel's bow was at 90 degrees north latitude—the North Pole—when this photo was taken.

worldwide and upward in mountainous regions. Many animal species will be able to migrate to keep pace with changing temperatures, but plants, being stationary, will not. Conservationists who have struggled to maintain islands of habitat as protected areas are particularly concerned about rapidly changing climatic conditions. For example, ecologists have documented that two-thirds of Europe's butterfly species have already shifted their habitat ranges northward by 22 to 150 miles (35 to 240 km), coinciding with the continent's warming trend. The World Wildlife Fund recently issued a report warning that due to global warming, as much as 70 percent of the natural habitat could be lost, and 20 percent of the species could become extinct, in the arctic and in subarctic regions of Canada, Scandinavia, and Russia. The report predicted that more than a third of existing habitats in the American states of Maine, New Hampshire, Oregon, Colorado, Wyoming, Idaho, Utah, Arizona, Kansas, Oklahoma, and Texas would be irrevocably altered by global warming.[9]

Rising sea levels. There is also general agreement that if global temperatures rise, so will sea levels as glacial ice melts and as seawater warms and thus occupies more volume. Sea levels rose 6 to 9 inches (15 to 23 cm) in the 20th century. The 2007 IPCC report forecast a further rise of 7 to 24 inches (18 to 61 cm) rise by 2100, but other scientists call these figures too conservative because they do not account enough for the melting of Greenland and Antarctic land ice. Research published in the journal *Science* in 2007 predicted a rise in sea level of 20 to 55 inches (51 to 140 cm) by 2100, and NASA scientist James Hansen forecast even higher waters.

Decision makers in coastal cities throughout the world, in island countries, and in nations with important lowland areas adjacent to the sea are worried about the implications of rising sea levels. The coastal lowland countries of the Netherlands and Bangladesh have for many years been outspoken advocates for reducing greenhouse gases. The president of the Maldives, an Indian Ocean country comprised entirely of islands barely above sea level, pleaded, "We are an endangered nation!"

Geopolitical instability. There is growing evidence that the poorest and most overcrowded nations will be hit hardest by climate change and that political and economic instability will result. The IPCC forecasts that rising sea levels will flood tens of millions of poor people out of their homes each year and that by 2080, possibly 200 to 600 million people will face starvation because of failing crops. By then, 1 to 3 billion people in developing countries will face shortages of freshwater related to growing aridity. Epidemics of malaria and other killer diseases will be more widespread.

Such environmental crises will likely lead to large movements of refugees and to conflict over resources, upsetting the fragile balance of power between countries. In a report published in 2007, U.S. military thinkers concluded that global warming "presents significant national security challenges to the United States," adding, "Projected climate change will seriously exacerbate already marginal living standards in many Asian, African and Middle Eastern nations, causing widespread political instability and the likelihood of failed states. The chaos that results can be an incubator of civil strife, genocide and the growth of terrorism. The U.S. may be drawn more frequently into these situations." Reacting to the report, Congressman Edward Markey of Massachusetts wrote that "global warming's impacts on natural resources and climate systems may create the fiercest battle our world has ever seen. If we don't cut pollution and head off severe global warming at the pass, we could see extreme geopolitical strain over decreased clean water, environmental refugees and other impacts."[10]

There is a growing international political will to take the strong and costly measures that may be necessary to prevent such dire scenarios. Present efforts to reduce the production of greenhouse gases focus on the world's wealthiest nations, but there is a good reason for the growing pressure on China to join these efforts. China and the United States produced about the same amount of carbon emissions in 2007 and together accounted for roughly 40 percent of the entire global emissions output. As recently as 1994, China was predicted to overtake the United States as the world's largest CO_2 producer by 2019. The fact that China achieved this distinction in 2007 instead is testimony to the dizzying pace of that country's development—a theme that echoes throughout this book. It also underscores the seriousness of the threats posed by global warming. Scientists have begun speaking of a **tipping point** in climate change, a yet unknown point at which feedback effects amplify temperature changes. The tipping point would bring a rapid acceleration

in temperature and of effects like more violent storms, dramatic crop losses, and spreading deserts.

What Can Be Done about Global Climate Change?

Some scientists believe that the die has already been cast. No matter how hard we try to reduce greenhouse gases now, the earth's atmosphere will warm dramatically anyway, especially because the oceans are so slow to warm up. Most scientists, however, are more proactive, urging dramatic steps to mitigate global warming. Many policymakers have also been action-oriented, especially in Europe, and even the traditional holdout on climate change action—the United States—has begun to join in the chorus. These are the most commonly prescribed actions:

Negotiate and implement international treaties to reduce greenhouse gas emissions. Most scientists and policymakers agree that the best way to confront global climate change is to implement international treaties to reduce emissions (see Geography of Energy, page 36). There are precedents proving that countries can unite in effective action against global environmental change. Thanks to the **Montreal Protocol** and its amendments signed by 37 countries in the late 1980s, the production of CFCs worldwide has all but ceased; the wealthier countries no longer produce them, and the less developed countries are scheduled to cease production by 2010. The anticipated result is that there will be a marked reduction in the size of stratospheric ozone "holes" that have been observed seasonally over the southern and northern polar regions since about 1985. CFC molecules have very long life spans, but scientists predict that the effects of the Montreal Protocol will be noticeable by 2010, with the ozone layer recovering to pre-1980 levels by 2050. This recovery will also depend on the successful phase-out of another ozone-destroying refrigerant known as HCFC-22 (or R22), which is scheduled to be eliminated worldwide by 2040.

Cut emissions through market-based incentives. Compared with richer nations, the less developed countries other than China and India release relatively little CO_2 into the atmosphere. However, they do contribute to the greenhouse effect, especially because of their heavy reliance on trees and other vegetation for fuel. When burned, trees release the CO_2 that they stored in the process of photosynthesis. At the same time, they cease to exist as organisms that in the process of photosynthesis remove CO_2 from the atmosphere.

Because the less developed countries produce relatively little CO_2, policymakers are considering innovative ways to keep them from doing so and to encourage them to develop clean industrial technologies. One approach is that of a system of **tradable permits** (also called a **cap-and-trade** or **emission-trading system**). In these schemes, already being used by the United Nations, the European Union, and

GEOGRAPHY OF ENERGY

The Kyoto Protocol

In 1997, at a world conference on climate change, 84 countries (of the 160 countries represented at the conference) signed the **Kyoto Protocol,** a landmark international treaty regarding climate change. The agreement requires 38 more developed countries (known as the Annex I countries) to reduce carbon dioxide emissions to 5 percent or more below their 1990 levels by the year 2012. The United States pledged to cut its emissions to 7 percent below 1990 levels by that date. This would be a huge cut; in 2007, the United States produced over 30 percent more carbon dioxide than it did in 1990. The European Union made a promise of 8 percent below 1990 levels, and Japan promised a 6 percent reduction.

Meeting these pledges would require substantial legislative, economic, and behavioral changes in these more affluent countries. Transportation and other technologies that use fossil fuels would have to become more energy-efficient, making these technologies at least temporarily more expensive. Gasoline prices would rise, so consumers would feel the pinch. Advocates of the protocol argue that the initial sacrifices would soon be rewarded by a more efficient and competitive economy powered by cleaner and cheaper sources of energy derived from the sun. Opponents, however, feel that higher fuel prices will be too costly for the United States and other industrialized economies to bear. Their position made it difficult for some signatories of the Kyoto Protocol to ratify the treaty. For the protocols to take effect, at least 55 countries must have ratified them, and the industrialized Annex I countries that ratify must have collectively produced at least 55 percent of the world's total greenhouse gas emissions in 1990.

By mid-2004, a total of 120 countries—including all the European Union members at that time and Japan—had ratified it, and the critical threshold of 55 percent of the 1990 emissions by Annex I countries was near. Ratification by either the United States or Russia would cross that threshold, putting the treaty into effect.

Much to the dismay of the European Union countries, U.S. President George W. Bush rejected the Kyoto Protocol soon after taking office in 2001. The Bush administration had two objections: the potentially high economic cost of implementing the treaty and the fact that China, along with all the world's less developed countries, was not required by the Kyoto Protocol to take any steps to reduce greenhouse gas emissions. China argued then, as it still does, that it is a developing country whose growth should not be hampered by pressure from countries that became rich by burning fossil fuels.

The European Union and Japan expressed outrage that the United States broke ranks with them on global warming. All eyes turned to Russia, whose ratification of the Kyoto Protocol would fulfill the 55 percent of emissions requirement and therefore finally put the treaty into effect in all ratifying countries. Anticipating financial gains to be earned by selling carbon emissions credits, Russia ratified the treaty late in 2004, and it went into force for its signatory nations in 2005. The European countries seized the initiative in cutting emissions, but tremendous hurdles remained for Kyoto's targets to be met. The Kyoto agreement will expire in 2012, and discussions are already under way about what new—and presumably more dramatic—measures will be needed to succeed it.

---

the unregulated "voluntary market," managers require the richer nations to achieve a net reduction in carbon emissions. It is very expensive for the already highly energy-efficient richer countries to cut their greenhouse gas emissions but relatively inexpensive for rich countries to pay a poorer, energy-inefficient country to cut its emissions. These schemes therefore allow the richer countries either to make the cuts at home or to offset their emissions by buying credits from poorer countries that actually cut carbon emissions. The poorer country selling the credits is obliged to use the income to invest in energy-efficient and nonpolluting technologies.

China is making big profits from emissions trading, using European money to reduce emissions of HFC-23, a potent greenhouse gas produced in many Chinese refrigerator factories. Russia is another big advocate of emission-trading schemes like these. Its industries are producing well below their 1990 carbon dioxide emission levels, so the country has much to gain by selling its unused emission rights. One buyer is Denmark, which is paying to convert coal-fired electricity plants to cleaner-burning gas in Siberia. In such a trading scheme, both countries apparently benefit, as does the global atmosphere. However, this practice is still

young, and critics say it ultimately lets the big polluters in the industrialized countries avoid the hard work of cutting emissions further.

Increase carbon sequestration. In the process of photosynthesis, plants absorb carbon dioxide. Collectively, the world's forests, farmlands, and seas (which contain carbon-absorbing phytoplankton) serve as giant **carbon sinks.** Some forest- and farm-rich countries want to receive credit for having these natural buffers against global warming, but the Kyoto Protocol does not allow such so-called **carbon sequestration** to be factored in. The Kyoto treaty negotiators felt that giving credit for carbon sinks would keep big polluters like the United States from attacking the root problem, the greenhouse gases. But not given credit for carbon sinks also means that developing countries lack an important incentive to plant new forests or replace forests that have been cut.

The next chapter discusses many ways in which the world's developed and less developed countries must confront problems like global climate change differently. Like this chapter, it will conclude with some specific ideas about how to deal with some of the most pressing issues of our time.

SUMMARY

→ The earth's three layers of habitable space are the hydrosphere, atmosphere, and lithosphere. The lithosphere is made up of separate plates that are in motion, a process known as plate tectonics. These movements result in mountain building, volcanic activity, earthquakes, and other consequences.

→ The earth's major landform types may be classified as hill lands and mountains, plateaus, and plains.

→ Weather refers to atmospheric conditions prevailing at one time and place. Climate is a typical pattern recognizable in the weather of a region over a long period of time. Climatic patterns have a strong correlation with patterns of vegetation and in turn with human opportunities and activities in the environment.

→ Warm air holds more moisture than cool air, and precipitation is best understood as the result of processes that cool the air to release moisture.

→ Most of the sun's visible short-wave energy that reaches the earth is absorbed, but some of it returns to the atmosphere in the form of infrared long-wave radiation, which generates heat and helps warm the atmosphere. This is the earth's natural greenhouse effect.

→ Geographers group local climates into major climate types, each of which occurs in more than one part of the world and is associated with other natural features, particularly vegetation. Geographers recognize 10 to 20 major types of ecosystems or biomes, which are categorized by the type of natural vegetation. Vegetation and climate types are so sufficiently related that many climate types are named for the vegetation types.

→ Some biomes are particularly important because of their biological diversity—the number of plant and animal species and the variety of genetic materials these organisms contain. Regions where human activities are rapidly depleting a rich variety of plant and animal life are known as biodiversity hot spots, places scientists believe deserve immediate attention for study and conservation.

→ Oceans cover about 71 percent of the earth's surface. They play the key role in the hydrologic cycle, sustain large numbers of people through the protein in fish and seafood, and contain valuable mineral resources.

→ The scientific community is convinced with 90 percent or greater certainty that human activities, particularly the production of carbon dioxide and other greenhouse gases, is responsible for global warming. Computer-based climate change models use the scenario of a doubling of atmospheric carbon dioxide and indicate that the mean global temperature might warm up by an additional 11 degrees Fahrenheit (6.1 degrees Celsius) by the year 2100. The general scientific consensus is that with global warming, the distribution of climatic conditions typical of biomes will shift poleward and upward in elevation, sea levels will rise, and mean global precipitation will increase (but with drought intensified in some areas).

→ The Kyoto Protocol is an international agreement requiring the industrialized countries that ratified it to make substantial cuts in their carbon dioxide emissions to reduce global warming. The United States dropped its support for the treaty. It went into effect for the ratifying countries after Russia ratified it in 2004, and in its present form, it will expire in 2012.

KEY TERMS + CONCEPTS

aquaculture (p. 32)
anticyclone (p. 25)
atmosphere (p. 19)
atmospheric stability (p. 25)
biodiversity (p. 30)
biodiversity hot spots (p. 31)
biological diversity (p. 30)
biomes (p. 27)
   boreal forest (p. 28)
   coniferous forest (p. 28)
   desert shrub (p. 28)
   Mediterranean scrub forest (p. 29)
   prairie (p. 28)
   savanna (p. 29)
   steppe (p. 28)
   taiga (p. 28)
   temperate grassland (p. 28)

   temperate mixed forest (p. 29)
   tropical deciduous forest (p. 28)
cap-and-trade system (p. 35)
carbon sequestration (p. 36)
carbon sink (p. 36)
climate (p. 22)
   desert (p. 28)
   humid continental (p. 29)
   humid subtropical (p. 29)
   ice cap (p. 27)
   marine west coast (p. 29)
   Mediterranean (p. 29)
   semiarid (p. 28)
   subarctic (p. 27)
   tropical rain forest (p. 28)
   tropical savanna (p. 28)
   tundra (p. 27)

undifferentiated highland (p. 30)
coniferous trees (p. 28)
continental drift (p. 20)
Coriolis effect (p. 25)
cyclone (p. 23)
emission-trading system (p. 35)
equatorial low (p. 25)
equinox (p. 25)
escarpment (p. 22)
fault (p. 22)
faulting (p. 22)
fish farming (p. 32)
front (p. 23)
   cold (p. 24)
   warm (p. 24)
Green Revolution (p. 30)
greenhouse effect (p. 33)

greenhouse gases (p. 33)
    carbon dioxide (CO_2) (p. 33)
    chlorofluorocarbons (CFCs) (p. 33)
    methane (p. 33)
    nitrous oxide (p. 33)
high-pressure cell (p. 25)
hydrologic cycle (p. 31)
hydrosphere (p. 19)
Intergovernmental Panel on Climate
    Change (IPCC) (p. 33)
inversion (p. 25)
Kyoto Protocol (p. 36)
lapse rate (p. 30)
lithosphere (p. 19)
modified Mercalli scale (p. xxx)
monoculture (p. 30)

Montreal Protocol (p. 35)
natural hazards (p. 20)
plain (p. 22)
plateau (p. 22)
plates (p. 20)
plate tectonics (p. 20)
precipitation (p. 23)
    convectional (p. 23)
    cyclonic (p. 23)
    frontal (p. 23)
    orographic (mountain-associated)
       (p. 23)
rain shadow (p. 25)
Richter scale (p. 20)
rifting (p. 21)
seafloor spreading (p. 20)

seismic activity (p. 20)
subduction (p. 20)
subduction zone (p. 22)
summer solstice (p. 25)
tectonic forces (p. 20)
tipping point (p. 35)
tradable permits (p. 35)
trade winds (p. 25)
transitional zone (p. 28)
trench (p. 20)
tsunami (p. 20)
volcanism (p. 20)
weather (p. 22)
winter solstice (p. 25)
xerophytic (p. 28)

REVIEW QUESTIONS

1. What are the three "spheres" of habitable life on earth?

2. What is plate tectonics, and what are some of the main consequences of tectonic activity?

3. What is the difference between weather and climate? What are the main forces that produce precipitation and aridity?

4. What are the major climate types and their associated biomes? Where do they tend to occur on earth?

5. Where are the biodiversity hot spots? What kinds of locations and biomes do many of them occur in?

6. What important roles do the world's oceans play, and what are their major resources? In what ways are these resources threatened?

7. What apparent long-term effects on the earth's atmosphere can be attributed to modern technology? What predictions are there for future changes, and what steps are being taken or considered to avert these changes?

NOTES

1. Wolf Roder, "Three Near Misses: Are These Science or Pseudoscience?" *Cincinnati Skeptics Newsletter,* June 1996, http://www.cincinnatiskeptics.org/newsletter/vol5/n5/misses .html. Accessed November 14, 2006.

2. Norman Myers, *Primary Source: Tropical Forests and Our Future* (New York: Norton, 1984), p. x.

3. World Bank, "On World Water Day, World Bank Calls for Investments in Water Infrastructure and Better Governance," press release, March 22, 2007.

4. Review the United Nations' *Annual Report of the State of the World's Fisheries and Aquaculture* at http://www.fao.org/sof/ sofia/index_en.htm.

5. Boris Worm et al., "Impacts of Biodiversity Loss on Ocean Ecosystem Services." *Science,* November 3, 2006, pp. 787– 790.

6. See the Monterey Bay Aquarium's guide to sustainable seafood, "Seafood Watch," at http://www.mbayaq.org/cr/seafood watch.asp.

7. Amy Roach Partridge, "Global Trucking Woes," *Global Logistics,* November 2006, http://www.inboundlogistics.com/ articles/global/global1106.shtml. Accessed September 27, 2007.

8. Intergovernment Panel on Climate Change, "Climate Change 2007," http://www.ipcc.ch. Accessed July 4, 2007.

9. Sarah Lyall, "Global Warming Report Predicts Doom for Many Species." *New York Times,* September 1, 2000.

10. Brad Knickerbocker, "Could Warming Cause War?" *Christian Science Monitor,* April 19, 2007, p. 2.

40°N

N

N

0°

20°S

40°S

140

80°W

40°W

Joe Hobbs

The trail to Mt. Everest, Nepal. The human imprint is almost everywhere on the landscape, even in the most challenging environments.

CHAPTER 3 HUMAN PROCESSES THAT SHAPE WORLD REGIONS

This chapter continues your introduction to geography's basic vocabulary, focusing on how people have interacted with the environment to change the face of the earth, with an emphasis on the human roles. Modern trends in people-land associations are examined as the products of revolutionary changes in the past: the arrival of agriculture and of industrialization. The chapter explains where rich and poor countries are located on the earth's surface and accounts for some of these patterns of prosperity and poverty. It considers where and why populations are increasing and what the implications of that growth are. Finally, it shares ideas about how to solve some of the most important global problems of our time.

chapter outline

chapter objectives

This chapter should enable you to

→   Gain a historical perspective on the capacity of human societies to transform environments and landscapes

→   Understand why some countries are rich and others poor and recognize the geographic distribution of wealth and poverty

→   Explain the simultaneous trends of falling population growth in the richer countries and rapid population growth worldwide

→   Explore the principles of sustainable development

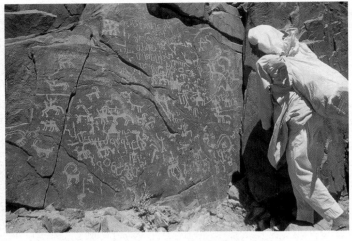

• **Figure 3.1** Until the relatively recent past—just a few thousand years ago—people were exclusively hunters and gatherers. This rock art in Egypt's Sinai Peninsula was created over a long time span, as it depicts Neolithic-period hunting of ibex, later uses of camels and horses, and writing from the Nabatean period (1st century D/F/).

3.1 Two Revolutions That Have Changed the Earth

The geographer's approach to understanding a current landscape—in almost all cases, a cultural landscape that has been fashioned by human activity—is sometimes deeply historical, involving study of its development from the prehuman or early human natural landscape. With a perspective of great historical depth, the current spatial patterns of our relationship with the earth may be seen as products of two "revolutions": the Agricultural Revolution that began in the Middle East about 10,000 years ago and the Industrial Revolution that began in 18th-century Europe. Each of these revolutions transformed humanity's relationship with the natural environment. Each increased substantially our capacity to consume resources, modify landscapes, grow in number, and spread in distribution.

Hunting and Gathering

Until about 10,000 years ago, our ancestors lived by **hunting and gathering** (also known as **foraging**). We were apparently quite good at it—it served us well for more than 100,000 years, until we began experimenting with the revolutionary technologies of agriculture.

Foraging was quite different from the farming and industrial ways of living that succeeded it. Joined in small bands of extended family members, hunters and gatherers were nomads with no villages, homes or other fixed dwellings. They moved to take advantage of changing opportunities on the landscape. These foragers scouted large areas to locate foods such as seeds, tubers, foliage, fish, and game ani-

mals (•Figure 3.1). Moving their small group from place to place, they had a relatively limited impact on the natural environment, especially compared with the impacts left by agricultural and industrial societies.

Hunters and gatherers may have been the **"original affluent society."**[1] Many scholars have praised these preagricultural people for the apparent harmony they maintained with the natural world in both their economies and their spiritual beliefs. After short periods of work to collect the foods they needed, they enjoyed long stretches of leisure time. Studies of those few hunter-gatherer cultures that lingered into modern times, such as the San (Bushmen) of southern Africa and several Amerindian groups of South America, suggest that although their life expectancy was low, they suffered little from the mental illnesses and broken family structures that characterize industrial societies.

Hunters and gatherers did modify their landscapes. These foragers were not always at peace with one another or with the natural world. With upright posture, stereoscopic vision, opposable thumbs, an especially large brain, and no mating season, *Homo sapiens* became after its emergence in southern Africa about 125,000 years ago an **ecologically dominant species**—one that competes more successfully than other organisms for nutrition and other essentials of life or that exerts a greater influence than other species on the environment. Using fire to flush out or create new pastures for the game animals they hunted, preagricultural people shaped the face of the land on a vast scale relative to their small numbers. Many of the world's prairies, savannas, and steppes where grasses now prevail developed as hunters and gatherers repeatedly set fires.

These people also overhunted and in some cases eliminated animal species. The controversial **Pleistocene overkill hypothesis** states that rather than being at harmony with nature, hunters and gatherers of the Pleistocene Era (2 million to 10,000 years ago) hunted many species to extinction, including the elephantlike mastodon of North America.

The Revolutionary Aspects of Farming

Despite these excesses, the environmental changes that hunters and gatherers could cause were limited. Humans' power to modify landscapes took a giant step with **domestication,** the controlled breeding and cultivation of plants and animals. Domestication brought about the **Agricultural Revolution,** also known as the **Neolithic Revolution** or the **Food-Producing Revolution.**

Why people began to produce rather than continue to hunt and gather plant and animal foods—first in the Middle East and later in Asia, Europe, Africa, and the Americas—is uncertain. Although there are many ideas about this process, surprisingly little hard evidence exists to explain it confidently. Two theories are most often put forward. One is that the climate changed. Increasing drought and reduced plant cover may have forced people and wild plants and animals into smaller areas, where people began to tame wild herbivores and sow wild seeds to produce a more dependable food supply.

A more widely accepted theory is that their own growing populations in areas originally rich in wild foods compelled people to find new food sources, so they began sowing cereal grains and breeding animals. The latter process may have begun about 8000 B.C.E. in the Zagros Mountains of what is now Iran. The culture of domestication spread outward from there but also developed independently in several world regions.

Now that humans were producing as well as consuming foods, their landscape uses, cultures, social organizations, and other characteristics changed dramatically. Among other things, in choosing to breed plants and animals, people settled down. Gradually abandoning the nomadic life and **extensive land use** of hunting and gathering, they came to favor the **intensive land use** of agriculture and animal husbandry. They could sow and harvest crops in specific places year after year. With less need to move around, they began living in fixed dwellings, at least on a seasonal basis. These developed into villages, small settlements with fewer than 5,000 inhabitants.

The settlements raised larger and more reliable stocks of food, making it possible to support their growing populations. Through **dry farming,** which involved planting and harvesting according to the seasonal rainfall cycle, population densities could be 10 to 20 times higher than they were in the hunting and gathering mode. By about 4000 B.C.E., people along the Tigris, Euphrates, and Nile Rivers began **irrigation** of crops—bringing water to the land artificially by using levers, channels, and other technologies—an innovation that allowed them to grow crops year round, independent of seasonal rainfall or river flooding (●Figure 3.2). Irrigation allowed even more people to make a living off the

● **Figure 3.2** The Tigris River in southeastern Turkey. The brown areas are rainless in the long, hot summers and are capable of producing only one crop a year through dry (unirrigated) farming. The green areas along the river are irrigated and can produce two or three crops a year. Irrigation was thus a revolutionary technology that greatly increased the number of people the land could support.

land; irrigated farming yields five to six times more food per unit area than dry farming. In ecological terms, the expanding food surpluses of the Agricultural Revolution raised the earth's **carrying capacity,** the size of a species' population (in this case, humans) that an ecosystem can support.

Culture became more complex, and society became more stratified. The steep increase in food production freed more people from the actual work of producing food, and they undertook a wide range of activities unrelated to subsistence needs. Irrigation and the dependable food supplies it provided thus set the stage for the development of **civilization,** the complex culture of urban life characterized by the appearance of writing, economic specialization, social stratification, and high population concentrations. By 3500 B.C.E., for example, 50,000 people lived in the southern Mesopotamian city of Uruk, in what is now Iraq. Other **culture hearths**—regions where civilization followed the domestication of plants and animals—emerged between 8000 and 2500 B.C.E. in China, Southeast Asia, the Indus River Valley, Egypt, West Africa, Mesoamerica, and the Andes.

Human impacts on the natural environment increased. The agriculture-based urban way of life that spread from these culture hearths had larger and more lasting impacts on the natural environment than either hunting and gathering or early agriculture. Acting as agents of humankind, domesticated plants and animals proliferated at the expense of the wild species that people came to regard as pests and competitors. For example, *Bos primigenius*, a wild bovine that was the ancestor of most of the world's domesticated cattle, was hunted to extinction by 1627. Farmers who resented the animals' raid on their crops were probably among the most ardent hunters. Agriculture's permanent and site-specific nature magnified the human imprint on the land, while the pace and distribution of that impact increased with growing numbers of people.

The Industrial Revolution

The human capacity to transform natural landscapes took another giant leap with the **Industrial Revolution,** which began in Europe around 1750 C.E. This new pattern of human-land relations was based on breakthroughs in technology that several factors made possible (•Figure 3.3):

* Western Europe had the economic capital necessary for experimentation, innovation, and risk. Much of this money was derived from the lucrative trade in gold and slaves undertaken initially in the Spanish and Portuguese empires after 1400.

* Significant improvements in agricultural productivity took place in Europe prior to 1500. New tools such as the heavy plow and more intensive and sustainable use of farmland led to increased crop yields. Human populations grew correspondingly.

* Population growth itself was a factor. More people freed from work in the fields could devote their talents and labor to experimentation and innovation. As agricultural innovations spread and industrial productivity improved, more and more of the growing European population was freed from farming, and for the first time in history, a region emerged in which city dwellers outnumbered rural folk. The process of industrialization, having now spread to nearly all parts of the world, continues to promote urbanization today.

Most geographers see population growth today as a drain on resources, but the Industrial Revolution illustrates that given the right conditions, more people do create more

• **Figure 3.3** A tweed mill in Stornoway, on Scotland's Lewis Island. The process of industrialization that began in 18th-century Europe has rapidly transformed the face of the earth.

resources. Innovations such as the steam engine tapped the vast energy of fossil fuels—initially coal and later oil and natural gas. This energy, the photosynthetic product of ancient ecosystems, allowed the earth's carrying capacity for humankind to be raised again, this time into the billions.

Industrialization, Colonization, and Environmental Change

As they began to deplete their local supplies of resources needed for industrial production, Europeans started to look for these materials abroad. As early as their **Age of Discovery,** also known as the **Age of Exploration,** which began in the 15th century, Europeans probed ecosystems across the globe to feed a growing appetite for innovation, economic growth, and political power. The process of European **colonization**—the extension of European countries' political and economic control over foreign areas—was thus linked directly to the Industrial Revolution. Mines and plantations from such faraway places as central Africa and India supplied the copper and cotton that fueled economic growth in colonizing countries such as Belgium and England (•Figure 3.4).

No longer dependent on the foods and raw materials they could procure within their own political and ecosystem boundaries, European vanguards of the Industrial Revolution had an impact on the natural environment that was far more extensive and permanent than that of any other people in history. Among the many measures of the unprecedented changes that the Industrial Revolution and its wake have wrought on the earth's landscapes, these are just a few:

* Between 1750 and the present, the total forested area on earth declined by more than 20 percent.

* During the same period, total cropland grew by nearly 500 percent, with more expansion in the period from 1950 to today than in the century from 1750 to 1850.

* Human use of energy increased more than 100-fold from 1750 to now.

* Today, fully 40 percent of the earth's land-based photosynthetic output is dedicated to human uses, especially in agriculture and forestry. Of particular interest to geographers is the unequal distribution of the costs and benefits of such expansion around the globe.

3.2 The Geography of Development

One of the most striking characteristics of human life on earth is the large disparity between wealthy and poor people, both within and between countries. At a high level of generalization, the world's countries can be divided into "haves" and "have-nots" (•Figure 3.5 and Table 3.1). Writers refer to these distinctions variously as "developed" and

● **Figure 3.4** In the 18th century, European merchant fleets carried goods, slaves, and information all around the world, profoundly transforming cultures and natural environments.

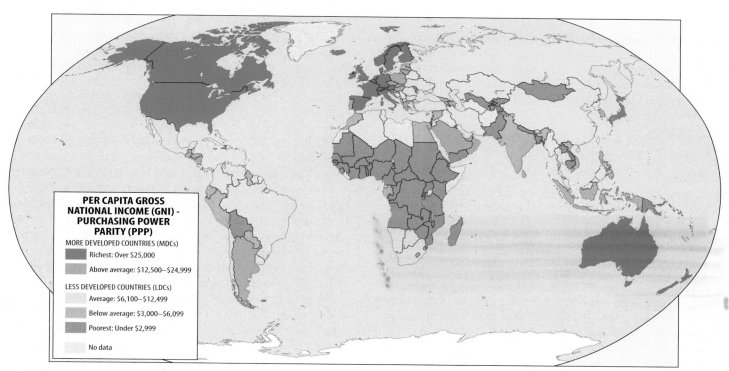

● **Figure 3.5** Wealth and poverty by country. Note the concentration of wealth in the middle latitudes of the Northern Hemisphere.

"underdeveloped," "developed" and "developing," "more developed" and "less developed," "industrialized" and "nonindustrialized," and "North" and "South," based on the concentration of wealthier countries in the middle latitudes of the Northern Hemisphere and the abundance of poorer nations in the Southern Hemisphere. This text uses the terms **more developed countries (MDCs)** and **less developed countries (LDCs).** It must be emphasized that this framework is an introductory tool and cannot account for the tremendous variations and continuous changes in

TABLE 3.1   Characteristics of More Developed Countries (MDCs) and Less Developed Countries (LDCs)

Characteristic	MDC	LDC
Per capita GDP and income	High	Low
Percentage of population in the middle class	High	Low
Percentage of population involved in manufacturing	High	Low
Energy use	High	Low
Percentage of population living in cities	High	Low
Percentage of population living in rural areas	Low	High
Birth rate	Low	High
Death rate	Low	Low[a]
Population growth rate	Low	High
Percentage of population under age 15	Low	High
Percentage of population that is literate	High	Low
Amount of leisure time available	High	Low
Life expectancy	High	Low

[a] Although death rates are high in the LDCs relative to the MDCs, in most countries they are quite low compared to what they once were in the LDCs.

TABLE 3.2   Top and Bottom Countries in Per Capita GNI PPP (U.S. dollars, 2007)[a]

Top Countries	Per Capita GNI PPP	Bottom Countries	Per Capita GNI PPP
Luxembourg	55,970	Madagascar	960
United States	44,260	Congo, Republic	940
Norway	43,920	Yemen	920
Switzerland	40,630	Sierra Leone	850
Denmark	36,110	Guinea-Bissau	830
Iceland	35,980	Niger	830
Netherlands	35,800	Tanzania	740
United Kingdom	35,690	Congo, Dem. Rep.	720
Ireland	35,540	Malawi	720
Austria	35,300	Burundi	710

[a] This table excludes city-states, territories, colonies, and dependencies. Only countries with available data are listed.
Source: 2007 World Population Sheet, Population Reference Bureau

economic and social welfare that characterize the world today. Some countries, including those known as the "Asian Tigers," are best described as **newly industrializing countries (NICs)** because they do not fit the MDC or LDC idealized types. The relevant regional chapters and modules describe these cases.

228, 298

Measuring Development

There is no single, universally acceptable standard for measuring wealth and poverty on the global scale. However, you are likely to encounter these in your university studies:

Annual per capita gross domestic product. **Gross domestic product (GDP)** is the total output of goods and services that a country produces for home use in a year. Divided by the country's population, the resulting figure of **per capita GDP** is one of the most commonly used measure of economic well-being. A closely related measure is per capita **gross national product (GNP)**, which includes foreign output by domestically owned producers.

Annual per capita gross national income purchasing power parity. Although quite a mouthful, **per capita gross national income purchasing power parity (per capita GNI PPP)** is a useful figure in the study of world regional geography. **Gross national income (GNI)** includes gross domestic production plus income from abroad from sources such as rents, profits, and labor. **Purchasing power parity (PPP)** conversion factors consider differences in the relative prices of goods and services, providing a better overall way of com-

paring the real value of output between different countries' economies. GNI PPP is measured in current "international dollars," which indicate the amount of goods and services one could buy in the United States with a given amount of money. Definitions vary, but in this book, an MDC is a county with an annual per capita GNI PPP of $12,500 or more; all others are considered LDCs.

The gulf between the world's richest and poorest countries is startling (Table 3.2). The average per capita GNI PPP in the MDCs is nearly twelve times greater than in the LDCs. In 2007, with a per capita GNI PPP of $55,970, Luxembourg was the world's richest country, whereas Burundi was the poorest with only $710. By another measure, 1.2 billion people, or about 18 percent of the world's population, were "abjectly poor" (by the World Bank's definition), living on less than $1 per day. These raw numbers suggest that economic productivity and income alone characterize **development**, which, according to a common definition, is a process of improvement in the material conditions of people through diffusion of knowledge and technology.

The Human Development Index. Definitions like these, and statistics like per capita GDP and per capita GNI PPP, reveal little about measures of well-being such as income distribution, gender equality, literacy, and life expectancy. Recognizing the shortcomings of strictly economic definitions, the United Nations Development Programme created the **Human Development Index (HDI)**, a scale that considers attributes of quality of life. This book uses HDI in the Basic Data tables of the chapter that introduces each region (for example, Table 4.1 on page 70). On the HDI scale, a measure of 1.0 means "perfect." According to this index, with a rating of 0.965, Norway is "the world's best place to live," although it ranks third in per capita GNI PPP. Following Norway, in descending order, are Iceland, Australia, Ireland, and Sweden. In HDI terms, with a measure of 0. 311, Niger is the world's worst place to live; Sierra Leone, Mali, Burkina Faso, and Guinea-Bissau fare only slightly better. Note that all five of these low-rated countries are in Africa, for reasons discussed in Chapter 9 and Module 9.1.

On the basis of per capita GNI PPP, 1.1 billion, or 16 percent, of the world's people, inhabit the MDCs. Most citizens of these countries, such as the United States, Canada, Japan, Australia, New Zealand, and the nations of Europe, enjoy an affluent lifestyle with freedom from hunger. Employed in industries or services, most of the people live in cities rather than in rural areas. Disposable income, the money that people can spend on goods beyond their subsistence needs, is generally high. There is a large middle class. Population growth is low as a result of low birth rates and low death rates (these demographic terms are explained in Section 3.3 on population). Life expectancy is long, and the literacy rate is high.

Life for the planet's other 84 percent, or about 5.5 billion people, is very different. In the LDCs, including most countries in Latin America, Africa, and Asia, poverty and often hunger prevail. The leading occupation is subsistence agriculture, and the industrial base is small. The middle class tends to be small but growing, with an enormous gulf between the vast majority of poor and a very small wealthy elite, which owns most of the private landholdings. With high birth rates and falling death rates, population growth is high. Life expectancy is short, and the literacy rate is low. However, the fate of LDCs is not predetermined. Wise policies and activities in certain countries or even in certain states or regions of particular countries can make a large difference. For example, while India's overall per capita income is $3,800, its state of Kerala, whose government has long grappled with problems of health and equity, has a per capita income 25 percent higher than the national average, and India's overall life expectancy of 64 years is exceeded in Kerala by 9 years.

With four-fifths of the world's people living in the poorer, less developed countries, it is important to understand the root causes of underdevelopment and to appreciate how wealth and poverty affect the global environment in very different but equally profound ways.

Why Are Some Countries Rich and Others Poor?

Many theories attempt to explain the disparities between MDCs and LDCs. It is important to recognize that there is no single, widely accepted explanation about development in general or in a particular region or country. Probably the best thing a discerning student can do is to weigh the various explanations and see which one, or which combination of them, seem to fit the situation of a given country or region. Here are the main explanations likely to be encountered.

Embraced most strongly in the LDCs, **dependency theory** argues that the worldwide economic pattern established by the Industrial Revolution and the attendant process of colonialism persists today. In his book *Ecological Imperialism,* the historian and geographer Alfred Crosby explains how dependency led to the rich-poor divide, depicting the two very different ways in which European powers used foreign lands during the Industrial Revolution.[2] In the pattern of **settler colonization,** Europeans sought to create new Europes, or **neo-Europes,** in lands much like their own: temperate middle-latitude zones with moderate rainfall and rich soils where they could raise wheat and cattle. Consequently, between 1630 and 1930, more than 50 million Europeans emigrated from their homelands to create European-style settlements in what are now Canada, the United States, Argentina, Uruguay, Brazil, South Africa, Australia, and New Zealand. These lands were destined to become some of the world's wealthier regions and countries.

In contrast to their preference to settle familiar mid-latitude environments, Europeans viewed the world's tropical lands mainly as sources of raw materials and markets for their manufactured goods. The environment was too different from home to make settlement attractive. In establishing a pattern of **mercantile colonialism,** Crosby explains, Europeans were less inhabitants than conquering occupiers of the colonies, overseeing indigenous peoples and resettled slaves in the production of primary or unfinished products: sugar in the Caribbean; rubber in Latin America, West Africa, and Southeast Asia; and gold and copper in southern Africa, for example. Colonialism required huge migrations of people to extract the earth's resources, including 30 million slaves and contracted workers from Africa, India, and China to work mines and plantations around the globe.

338, 470, 501, 616

In the mercantile system, the colony provided raw materials to the ruling country in return for finished goods; this meant that people in India would purchase clothing made in England from the raw cotton they themselves had harvested. The relationship was most advantageous to the colonizer. England, for example, would not allow its colony India to purchase finished goods from any country but England. It prohibited India from producing any raw materials the empire already had in abundance, such as salt (India's Mohandas Gandhi defiantly violated this prohibition in his famous "March to the Sea"). Finished products are **value-added products,** meaning they are worth much more than the raw materials they are made from, so manufacturing in the ruling country concentrated wealth there while limiting industrial and economic development in the colony.

The colony was obliged to contribute to, but was prohibited from competing with, the economy of the ruling country—a relationship that dependency theorists insist continues today. Dependency theory asserts that to participate in the world economy, the former colonies, now independent countries, continue to depend on exports of raw materials to and purchases of finished goods from their former colonizers and other MDCs, and this disadvantageous position keeps them poor. Dependency theorists call this relationship **neocolonialism.** With independence, the former colonies needed revenue. To earn that money, they continued to produce the goods for which markets already existed—generally the same unprocessed primary products they supplied in colonial times. Dependency theorists argue that when the former colonies try to break their dependency by becoming exporters of manufactured goods, the MDCs impose trade barriers and quotas to block that development (see Insights, page 47).

Whether because of neocolonialism or a more complex array of variables, many developing countries continue to rely heavily on income from the export of a handful of raw materials. This makes them vulnerable to the whims of nature and the world economy. The economy of a country heavily dependent on rubber exports, for example, may suffer if an insect pest wipes out the crop or if a foreign laboratory develops a synthetic substitute. When demand for rubber rises, that country may actually harm itself trying to increase its market share by producing more rubber because in doing so, it drives down the price. (Consider oil: members of the OPEC cartel of oil-producing countries drove oil prices to all-time lows in the mid-1980s when they overproduced oil in a bid to earn more revenue.) If the rubber-producing country withholds production to shore up rubber's price, it provides consuming countries with an incentive to look for substitutes and alternative sources. (Again look at oil: after OPEC embargoed shipments of oil to the United States in the 1970s, the United States began developing domestic oil supplies and becoming more energy-efficient, thus temporarily reducing oil prices and OPEC's revenues.) The developing country is in a dependent and disadvantaged position.

Many geographers view dependency theory and neocolonialism as simplistic and politically biased explanations of development. They consider a wider and more complex set of factors, including culture, location, and natural environment, to explain why some countries are wealthy and others are poor. These include the following factors:

Advantageous and disadvantageous location. Does a country's location play a role in how rich or poor it is? In some cases, yes. Location can influence a country's economic fortunes. For example, as discussed in Chapter 1, because it is situated close to a great mainland with which to trade, the island of Great Britain enjoys a core location favorable for economic development. Japan has a similar location relative to the Asian landmass. In contrast, landlocked nations such as Bolivia in South America and numerous nations in Africa have locations unfavorable for trade and economic development, and they have not overcome this disadvantage. But it is important to recognize that geographic location is never the sole decisive factor in development. Like Japan and Britain, Madagascar and Sri Lanka are island nations situated close to large mainlands, but for a variety of (mainly political) reasons discussed in

the relevant chapters and modules, neither has experienced prosperity.

Resource wealth or poverty. Having or lacking a diversity or abundance or **natural resources** plays a significant role in development. Superabundance of an especially valuable resource (for example, oil in the Persian/Arabian Gulf countries) or a diversity of natural resources has helped some countries become more developed than others. The former Soviet Union and the United States achieved superpower status in the 20th century in large part by using the enormous natural resources of both countries.

Cultural and historical factors. In some cases, human industriousness has helped compensate for resource limitations and promoted development. For example, Japan has a rather small territory with few natural resources (including almost no petroleum). Yet in the second half of the 20th century, it became an industrial powerhouse largely because the Japanese people united in common purpose to rebuild from wartime devastation, placing priorities on education, technical training, and seaborne trade from their advantageous island location. Conversely, cultural or political problems like corruption and ethnic factionalism can hinder development in a resource-rich nation, as in the mineral-wealthy Democratic Republic of Congo.

401

480

Environmental Impacts of Underdevelopment

Geographers are very interested in the environmental impacts of relations between MDCs and LDCs, especially on the poorer countries. LDCs generally lack the financial resources needed to build roads, dams, energy grids, and other infrastructure critical to development. They turn to the World Bank, International Monetary Fund (IMF), and other institutions of the MDCs to borrow funds for these projects. Many borrowers are unable to pay even the interest on these loans, which is sometimes huge; debtor nations have been known to spend as much as 40 percent of annual revenues on interest payments to the lenders, or more than they spend on education and health combined.

When lender institutions threaten to cut off assistance, borrowing countries often try to raise cash quickly to avoid this prospect, generally by one or two methods, or both. One method is to dedicate more high-quality land to the production of **cash crops** (also known as **commercial crops**). These are items such as coffee, tea, sugar, coconuts, and bananas exported to the MDCs, where they may be perceived as luxuries or as staple items. Governments or foreign corporations often buy out, force out, or otherwise displace subsistence food farmers in the search for new lands on which to grow these commercial crops. In this process, known as **marginalization,** poor subsistence farmers are pushed onto fragile, inferior, or marginal lands that cannot support crops for long and end up depleted by cultivation. In Brazil's Amazon Basin, for example, peasant migrants arrive from Atlantic coastal regions, where government and wealthy private landowners cultivate the best

INSIGHTS

Globalization: The Process and the Backlash

One of the most remarkable international trends of recent decades has been **globalization**—the spread of free trade, free markets, investments, and ideas across borders and the political and cultural adjustments that accompany this diffusion. There has been much debate and sometimes-violent conflict over the pros and cons of globalization.

Advocates of globalization argue that the newly emerging global economy will bring increased prosperity to the entire world. Innovations in one country will be transferred instantly to another country, productivity will increase, and standards of living will improve. They propose that one obvious solution to the problem of the LDCs' inability to compete in the world economy, and therefore escape their dependency, is to reduce the trade barriers that MDCs have erected against them (some of these barriers are discussed on page 623 in Chapter 11). With free trade, free enterprise will prosper, pumping additional capital into national economies and raising incomes for all. Much of the support for globalization comes from **multinational companies** (also called *transnational companies*, meaning companies with operations outside their home countries) as they increase their investments abroad. Most of the companies are based in the MDCs, but multinational corporations have also grown in LDCs such as China, India, and Mexico.

Opponents of globalization argue that the process will actually increase the gap between rich and poor countries, and even within a country; a selected few developing nations (or people within a country) will prosper from increased foreign investment and resulting industrialization, but the hoped-for "global" wealth will bypass other countries (or people within a country) altogether. And the increasing interdependence of the world economies will make all the players more vulnerable to economic and political instability. The multinational companies will recognize huge profits at the expense of poor wage laborers. In addition, environments will be harmed if environmental regulations are reduced to a lowest common denominator (for example, the high standards of air quality demanded by the U.S. Clean Air Act would be deemed noncompliant with World Trade Organization rules because they make it harder for countries with "dirtier" technologies to compete in the marketplace). Such concerns often lead to massive protests against "corporate-led globalization," especially at meetings of the World Trade Organization, the International Monetary Fund, and the World Bank. Many more confrontations like these can be anticipated. Typical protesters' demands are that working conditions be improved in the foreign "sweatshops," where textiles and other goods are produced at low cost for U.S. corporations, and that corporations like Starbucks should sell only "fair trade" coffee beans bought at a price giving peasant coffee growers a living wage rather than at the "exploitive" price typically paid.

354, 540

The problems of hot money and the digital divide are also cited as drawbacks to globalization. **Hot money** refers to short-term (and often volatile) flows of investment that can cause serious damage to the "emerging market" economies of less developed countries. Individual and corporate investors in the MDCs can invest such money heavily in stock market securities of the LDCs,

reasoning that these developing nations have much greater economic growth potential than the mature economies of the MDCs. This investment can bring rapid wealth to at least some sectors of the economies of LDCs. The problem, however, is that the capital can be withdrawn as quickly as it is pumped in, resulting in huge economic consequences.

344

It is often assumed that the rapid growth and spread of computer and wireless technologies are benefiting all of humankind, but the notion of the **digital divide** challenges this view. The divide is between the handful of countries that are the technology innovators and users and the majority of nations that have little ability to create, purchase, or use new technologies. The United States, most European countries, and Japan are leaders in **information technology (IT),** and the LDCs are well behind. For example, more than 70 percent of U.S. citizens use the World Wide Web, compared with 20 percent of Russians, 4 percent of Indians, and 10 percent of Middle Easterners (●Figure 3.A). Statistically, an American must save a month's salary to buy a computer, but a Bangladeshi must save eight years' wages—a fact that may change rapidly if the nonprofit organization One Laptop per Child is successful in producing, distributing, and selling its $100 Taiwanese-made "XO" laptop (see http://www.laptop.org).

U.S. companies earn a large share of the revenues from the global Internet business. The fear in the technology-lagging countries is that the growth of e-commerce will concentrate the wealth generated by that commerce in the technology leadership countries. This would enable the leaders to make even further advances in technology and realize even bigger economic gains from a so-called **knowledge economy** based on innovation and services, while the technology laggards fail to catch up and simply become poorer. Will globalization prove to be good or bad? It may depend on where and who you are.

● **Figure 3.A** An Internet café in Mérida, Yucatán, Mexico. Information technology—especially computers, the Internet, and cellular telephones—is spreading rapidly around the world. Some people believe that there is now an unfolding Information Revolution comparable to the Agricultural and Industrial Revolutions in its capacity to change humanity's relationship with the earth.

soils for sugarcane and other cash crops. The newcomers to Amazonia slash and burn the rain forest to grow rice and other crops that exhaust the soil's limited fertility in a few years. Then they move on to cultivate new lands, and in their wake come cattle ranchers, whose land use further degrades the soil.

472, 530, 587, 589

Another way for debtor countries to raise cash quickly is to sell off their natural assets. National decision makers often face a difficult choice between using the environment to produce immediate or long-term economic rewards. In most cases, they feel compelled to take short-term profits and, by cash cropping and other strategies, initiate a sequence having sometimes-tragic environmental consequences. This is because most LDCs have resource-based economies that rely not on industrial productivity but on stocks of productive soils, forests, and fisheries. The long-term economic health of these countries could be assured by the perpetuation of these natural assets, but to pay off international debts and meet other needs, the LDCs generally draw on their ecological capital faster than nature can replace it. In ecosystem terms, they exceed **sustainable yield** (also known as the **natural replacement rate**), the highest rate at which a **renewable resource** can be used without decreasing its potential for renewal. In tropical biomes, for example, people cut down 10 trees for every tree they plant. In Africa, that ratio is 29 to 1. In 1950, about 30 percent of Ethiopia's land surface was covered with forest. Today, less than 1 percent is forested.

In other words, these countries find themselves in what might be termed **ecological bankruptcy:** they have exhausted their environmental capital. This environmental poverty plays itself out in several damaging ways. First, there are negative effects on the health and well-being of the people living in these countries. People in the LDCs feel the impacts of environmental degradation more directly than people in the MDCs (•Figure 3.6). Many of the world's poor drink directly from untreated water supplies; an es-

Joe Hobbs

• **Figure 3.6** The Thu Bon River in Vietnam. In the LDCs, it is very common for people to rely on polluted water sources for drinking, bathing, cooking, and washing their clothes and eating utensils. Health effects can be severe; one of the most tragic is infant diarrhea, typically a waterborne illness that kills an estimated 2 million youngsters worldwide every year.

timated 1 billion lack access to clean water. They tend to cook their meals with fuelwood rather than with fossil fuels. They are more dependent on nature's abilities to replenish and cleanse itself, and hence they suffer more when those abilities are diminished (see Insights, below).

In addition, many political and social crises result from this ecological bankruptcy. Revolutions, wars, and refugee migrations in developing nations often have underlying environmental causes. Such problems, as related in the earlier discussion of global warming, are central to the national interests of the United States and other countries far beyond the affected nations. Do you think **deforestation** in Mexico or Haiti is of no concern to you if you are an American? You may want to consider why people from those countries

328, 331, 460

INSIGHTS

The Fuelwood Crisis

The removal of tree cover in excess of sustainable yield (a characteristic problem in Nepal, for example) illustrates the many detrimental effects that a single human activity can have in the LDCs (•Figure 3.B). These effects cause a complex problem, widespread in the world's LDCs, known as the **fuelwood crisis.** As people remove trees to use as fuel, for construction, or to make room for crops, their existing crop fields lose protection against the erosive force of wind. Less water is available for crops because in the absence of tree roots to funnel water downward into the soil, it runs off quickly. Increased salinity (salt content) generally accompanies increased runoff, causing the quality of irrigation and drinking water downstream to decline. Eroded topsoil resulting from reduced

plant cover can choke irrigation channels, reduce water delivery to crops, raise floodplain levels, and increase the chances that floods will destroy fields and settlements. As reservoirs fill with silt, hydroelectric generation, and hence industrial production, is diminished. Upstream, where the problem began, fewer trees are available to use as fuel.

Owing to the depletion of wood, people living in rural areas must now change their behavior (the fuelwood crisis often refers only to this part of the problem). Women, most often the fuelwood collectors, must walk farther to gather fuel. The use of animal dung and crop residues as sources of fuel deprives the soil of the fertilizers these poor people most often use when farming.

INSIGHTS

The Fuelwood Crisis *continued*

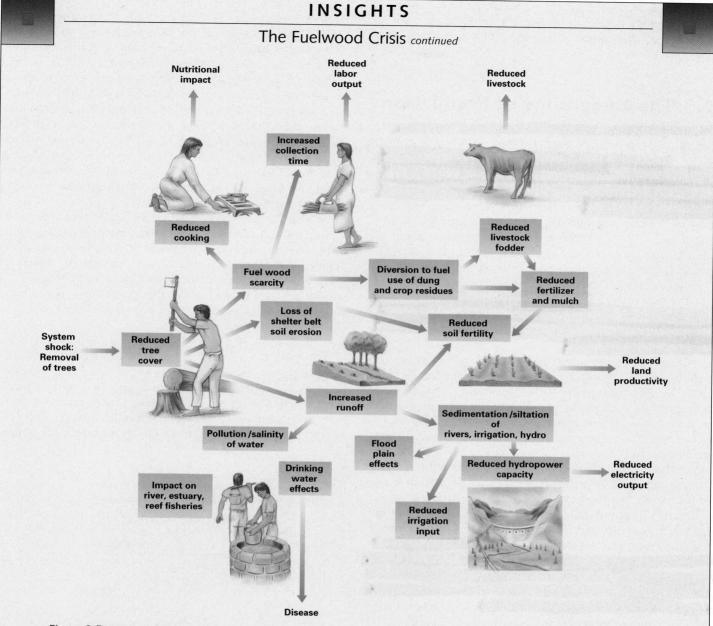

● **Figure 3.B** Impacts of deforestation in the LDCs.

Reduced food output is the result. A family may eventually give up one cooked meal a day or tolerate colder temperatures in their homes because fuel is lacking—steps that have a negative impact on the family's health.

The fuelwood crisis is a problem that governments and societies can confront successfully. A combination of new technologies, government policies, and sound economic growth can halt and even reverse deforestation. The biogas digester, a simple household device usable in warm rural areas, ferments animal manure for the production of methane cooking gas and thus spares fuelwood.

Governments can delimit forest reserves but also assist people in earning livelihoods from **nontimber forest products (NTFPs),** such as handicrafts made from bamboo (see http://www.ntfp.org). The process of economic development can itself lead to less pressure on forest resources; the world's wealthier countries have more forest now than they did in 1990, and forest cover actually increased in 22 of world's 50 most forested countries since that time. A key challenge for the wealthier forested countries is to avoid depleting other forests to meet their needs.

try to move to the United States, legally or otherwise. One of the reasons is that the home environments are so degraded that they simply cannot support farming as they once did.

3.3 The Geography of Population

Population may be the most critical issue in geography. It certainly is one of most important issues for human life on earth. The welfare of humanity and of the planet's other species and natural habitats is tied closely to two related issues: the number of people there are and the rates at which we consume resources. In some quarters, there are fears that the human **population explosion** the world has experienced since 1800 will lead to a precipitous crisis: a massive disease outbreak, famine, or some other kind of catastrophe triggered by too many people living too close together without sufficient food and other resources to sustain them. On the other hand, the human population growth rate is slowing, and our numbers are expected to stabilize—ideally at a population that can be sustained with minimal risk of famine or other suffering.

Meanwhile, large numbers of people move across and within political borders, usually by choice but in some cases by force. Migrants bring new cultures, ideas, and opportunities with them, but their reception is not always warm: tension and violence have characterized relations between majority and migrant minority groups in recent years in Europe and Australia, for example, and in the United States there is a heated debate about "immigration reform" directed at illegal migrants from Mexico. **Migration** is one of the major themes discussed throughout this book.

Interest in population is not confined to geography but is shared by many other fields, including biology, sociology, anthropology, and political science. The study of population is known as **demography.** The field of demography is most concerned with patterns of birth, death, marriage, and related issues in themselves, with less attention to issues of migration and population distributions. What most distinguishes population geography is its focus on spatial variations. This section of the book examines how many people have lived on earth and how many we may be in the future, mindful of who these people are and why they have been so few or so many, and paying especially close attention to the fundamentally geographic issue of where they are.

How Many People Have Ever Lived on Earth?

Around 100,000 years ago, our *Homo sapiens* ancestors came out of Africa across the land bridge of Suez and began to populate Eurasia. By around 10,000 years ago, before plants and animals were domesticated, there were probably about 5.3 million humans in the world—roughly the current number of residents of the city of Chicago or the country of Finland. By 1 C.E., humans numbered between 250 and 300 million, about the population of the United States today. The first billion was reached around 1800. Then a staggering population explosion occurred in the wake of the Industrial Revolution. The second billion came in 1930, the fourth in 1975, and the sixth in 1999 (Table 3.3 and •Figure 3.7). At 6.6 billion, *Homo sapiens* is now by far the most populous large mammal on earth and has succeeded where no other animal has in extending its range to the world's farthest corners. Using a 24-hour time period to represent the 125,000 years of our species' history, we reached our first billion within just the last 3 minutes. The 6.6 billion people alive today represent a remarkably large 5.8 percent of all the people who have ever lived on earth!

Figure 3.7 illustrates very dramatically that right after about 1800, the human population surged. What happened after tens of thousands of years that made our numbers suddenly skyrocket? In addition to the factor, addressed earlier, of greater food surpluses that made it possible to feed more people, we answer that question by considering some of the basic vocabulary of population geography.

How Can We Measure Population Changes?

Excluding the issue of migration for now, two principal variables determine population change in a given village, city, or country or the entire planet: birth rate and death rate. The **birth rate** is the annual number of live births per 1,000 people in a population. The **death rate** is the annual

42

73, 600

TABLE 3.3 How Many People Have Ever Lived on Earth?

Year	Population
50,000 B.C.E.	few
8,000 B.C.E.	5,000,000
1 C.E.	300,000,000
1200	450,000,000
1650	500,000,000
1750	795,000,000
1850	1,265,000,000
1900	1,656,000,000
1950	2,516,000,000
1995	5,760,000,000
2007	6,625,000,000
Number who have ever been born	106.8 billion
Percent of those ever born who were alive in 2007	6.2%

Source: Modified from Population Reference Bureau.

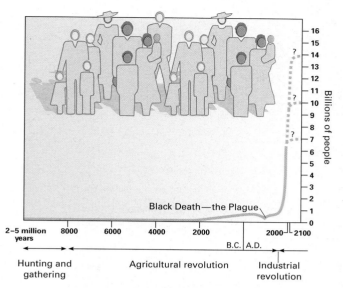

● **Figure 3.7** The human population has exploded since the Industrial Revolution, in a classic J-shaped curve of exponential growth.

number of deaths in that same sample population of 1,000. The **population change rate**—the figure that is often called the "population growth rate" but that may represent either growth or loss—is the birth rate minus the death rate in that population.

We can put these measures to work to appreciate the earth's population today. Supposing a perfect sample of 1,000 people representing the world's population in 2007, the birth rate was 21 per 1,000 and the death rate was 9 per 1,000. This means that by the end of that year, among the 1,000 people, 21 babies had been born and 9 people had died, resulting in a net growth of 12. That figure, 12 per 1,000, or 1.2 percent, represents the 2007 population change rate for the world.

We may now look at some of different combinations of birth rates, death rates, and population change rates and the various forces behind them.

What Determines Family Size?

Many factors affect birth rates, which tend to be much higher in the less developed countries of the world and among the poorer residents of more affluent countries. Some of the motivations and circumstances may seem strange at first glance, but there are good reasons for them:

❋ Better-educated and wealthier people have fewer children. The parents of most of you reading this book probably considered the economic cost of raising you and sending you to college and made decisions about family size in part because of their education and yours.

❋ Conversely, less educated and poorer people generally want and have more children. One reason for this is

economic: poorer parents are often convinced that "extra" children will help bring more income to the family by working in fields or factories and will help care for them in their old age.

❋ People in cities tend to have fewer children than those in rural areas.

❋ Those who marry earlier generally have more children (their reproductive life span is longer).

❋ Couples with access to and understanding of contraception generally have fewer children.

❋ Value systems and cultural norms play very important roles. Even where contraception is available and understood, a couple may decide not to interfere with what they perceive as God's will or may seek the social status associated with a larger family.

What Determines Death Rates?

Death rates are correlated mainly with health factors, particularly the level of nutrition and level of medical care available. Improvements in food production and distribution help reduce death rates. Better sanitation, better hygiene, and cleaner drinking water eliminate fatal diseases such as infant diarrhea, a common cause of infant mortality in the less developed countries. Furthermore, the availability of antibiotics, immunizations, insecticides, and other improvements in medical and public health technologies have a marked correlation with declining death rates.

Human death rates overall have been on a steady trend of decline for decades. But death rates sometimes rise, of course, especially with the outbreak of epidemics such as HIV/AIDS. Large proportions of the world's population have in fact been killed in disease epidemics and natural disasters (Table 3.4). The "Black Death" (caused by bubonic plague) in Europe and Asia killed about 15 percent of the world's population between 1334 and 1349; at least 40 percent of Europe's population died. And we cannot assume that such devastation will never happen again. Estimates for the death toll from the H5NI virus, commonly known as "bird flu," should it become transmissible between humans, range as high as 150 million, or 2.2 percent of the current world population.

Closely related to the measure of death rates is that of *life expectancy,* the number of years a person may expect to live in a given environment (typically defined as a country—see ●Figure 3.8—and differentiated between women, who usually live longer, and men). As death rates fall, life expectancy increases, and the reverse is also true. In the United States in 2007, life expectancy for women was 80 years, and for men, 75 years. In Swaziland, the country hit harder than any other by the HIV/AIDS epidemic, a woman could expect to live 34 years and a man 33 years. But as recently as 1986, before the virus was so widespread in Swaziland, the death rate was much lower, and a woman could look forward to 60 years and a man to 53 years.

TABLE 3.4 The World's Most Costly Epidemics and Disasters

Number of Deaths	Place	Date	Cause
unknown	Sumatra (Indonesia)	c. 74,000 years ago	Toba supervolcano eruption [a]
300 million–500 million	worldwide	20th century	smallpox [b]
c. 100 million	Europe	540–590	plague
55 million–75 million	Europe and Asia	1300s	bubonic plague
25 million–100 million	worldwide	1918–1919	influenza
25 million	worldwide	1981–present	AIDS
10 million	China	1892–1896	bubonic plague
3.7 million	China	1931	river flooding, disease, and famine
3 million	India	1900	drought-related famine
3 million	China	1941	drought-related famine
2 million–3 million	worldwide	annually	malaria
more than 2 million	China	1959	river flooding, disease, and famine
2 million	China	1887	river flooding, disease, and famine
1.5 million	India	1965–1967	drought-related famine
1.1 million	Egypt and Syria	1201	earthquakes
1 million	China	1938–1939	river flooding, disease, and famine
1 million	worldwide	1957	influenza
1 million	worldwide	1968	influenza

[a] This single event and its aftermath killed most of our human ancestors. Based on our genetic heritage from that time, estimates vary, but it is believed that 40 to 500 individual female *Homo sapiens* of childbearing age (suggesting a total human population of just a few hundred to several thousand) survived both Toba's initial supervolcanic cataclysm and the subsequent volcanic winter, which caused temperatures to plummet worldwide. (Several competing hominid species were apparently completely wiped out by the Toba eruption and its effects.)

[b] Historically, this disease is thought to have been responsible, at least in Europe, for as many deaths as the plague. It was a great scourge in whole populations during several periods through history, cutting down about one-third in Rome early in the Common Era and decimating the majority of the members of several American Indian tribes in the 16th through 19th centuries. Until the last case was diagnosed in 1977, it was a greater killer by far, in the 20th century, than all wars combined.

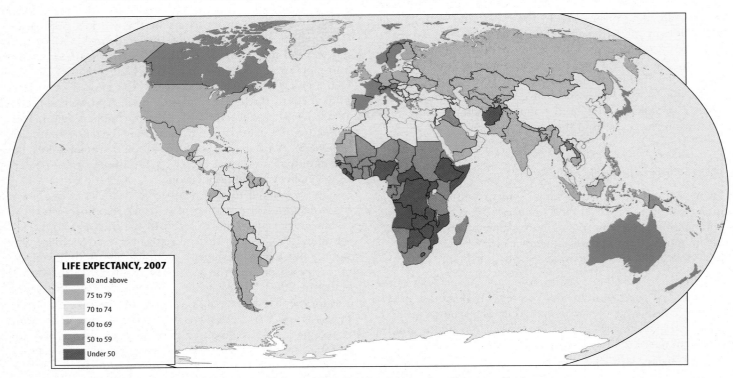

LIFE EXPECTANCY, 2007
- 80 and above
- 75 to 79
- 70 to 74
- 60 to 69
- 50 to 59
- Under 50

● **Figure 3.8** Life expectancy is closely tied to economic well-being; people live longer where they can afford the medicines and other amenities and technologies that prolong life.

What Determines the Population Change Rate?

Throughout history, natural disasters, diseases, and wars have taken huge bites out of our numbers. Overall, however, with birth rates higher than death rates, the trend line has been one of growth —and since 1800, of spectacular growth. In 1968, the rate of population growth hit an all-time high of 2.0 percent. To appreciate how rapid that growth rate was, you can calculate the doubling time. Applying the often-used rule of 70, in which 70 is divided by the growth rate, the doubling time is the number of years required for the human population to double (assuming that the rate of growth would be unchanged over the entire period). In 1968, the human population was growing at a rate that would, if unchanged, have doubled in 35 years. That rate did slow down, however, illustrating the limited usefulness that doubling time has in projecting future growth. At the 2007 population change rate of 1.2 percent, our numbers would double in 58 years.

Doubling time may be only approximate, but it is a good tool for comparing between countries and also illustrates that human populations have the potential for exponential growth; that is, not an incremental or arithmetic increase from 1 to 2 to 3 to 4, and so on, but geometric growth from 2 to 4 to 8 to 16, for example. (Exponential growth in the context of the famous Malthusian scenario is discussed on page 60.) The annual rate of population change worldwide is depicted by country in •Figure 3.9.

Why Has the Human Population "Exploded?"

If both birth rates and death rates are high (as they were when our ancestors were hunters and gatherers and as recently as the dawn of the Industrial Revolution), population growth is minimal: many people are born to a given population in a given period, but many also die in that same period, so they "cancel each other out." Population growth is also negligible if both birth and death rates are low (as they are today in countries like Japan and Italy). But when the birth rate is high and death rates are low, population surges, as seen in •Figure 3.10. This scenario of high birth rates, plunging death rates, and surging growth is exactly what played out for our species beginning around 1800 in western Europe and after about 1950 in the LDCs.

It is vital to appreciate the fact that the explosive growth in world population since the beginning of the Industrial Revolution is the result not of a rise in birth rates but of a dramatic decline in death rates, particularly in the less developed countries. The death rate has fallen as improvements in agricultural and medical technologies have diffused from the richer to the poor countries. Until recently, however, there were no strong incentives for people in the less developed countries to have fewer children. With birth rates remaining high and death rates falling quickly, the population has grown sharply; the LDCs are generally in stage 2 of an important model that demographers call the **demographic transition** (Figure 3.10).

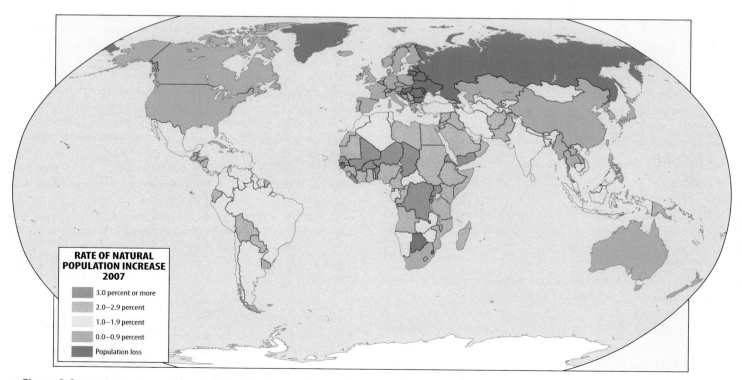

**RATE OF NATURAL
POPULATION INCREASE
2007**

- 3.0 percent or more
- 2.0—2.9 percent
- 1.0—1.9 percent
- 0.0—0.9 percent
- Population loss

• **Figure 3.9** Population change rates are highest in the countries of Africa and other regions of the developing world and lowest in the more affluent countries.

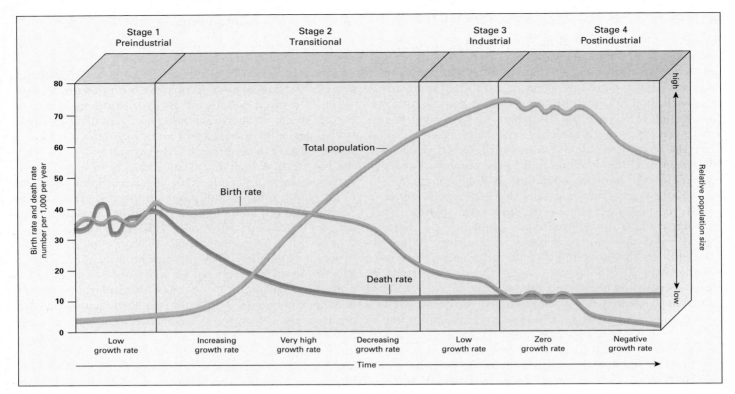

• **Figure 3.10** The demographic transition models population change in the world's wealthier countries. Note how the population surged in the wake of the Industrial Revolution as death rates fell while birth rates remained high but then leveled out and began to decline as economic development advanced.

In its entirety, from stages 1 to 4 as seen in Figure 3.10, the demographic transition model depicts the change from high birth rates and high death rates to low birth rates and low death rates that accompanied economic growth in the more developed countries (for example, western European nations, Japan, and the United States). The first two are the same two stages that humankind as a whole experienced from our earliest days until the present. The latter two have generally been experienced only by people in the wealthier countries. Note how birth rates, death rates, population change rates, and economic development correspond in this model:

☀ In the **first** or **preindustrial stage** (from the earliest humans to about 1800 C.E.), birth rates and death rates were high, and population growth was negligible.

☀ In the **second** or **transitional stage,** birth rates remained high, but death rates dropped sharply after about 1800 due to medical and other innovations of the Industrial Revolution.

☀ In the **third** or **industrial stage,** beginning around 1875, birth rates began to fall as affluence spread.

☀ Finally, after about 1975, some of the industrialized countries entered the **fourth** or **postindustrial stage,** with both low birth rates and low death rates and therefore, once again, as in stage 1, low population growth.

In this model, the United States, with population growth of about 0.6 percent per year, is in the early years of the postindustrial stage. Other industrialized, affluent countries have made their way well into that stage. Some wealthier countries, including Austria, Portugal, and Greece, have in recent years officially registered **zero population growth (ZPG).** Because of the growing desire of women in Japan and Germany to pursue their own careers and postpone marriage, birth rates in those countries have fallen below death rates, meaning that these countries are actually losing population or even experiencing a **population implosion.** In other words, the fertility rate of Japan and some other postindustrial countries is below the **population replacement level,** the number of new births required to keep the population steady (generally calculated as 2.1 children per woman in the MDCs). Other countries—notably Hungary and Latvia—that qualify as MDCs on the basis of per capita GNI PPP are experiencing population losses unfortunately due as much to rising death rates as to falling birth rates. Trends of this sort can be most easily appreciated by looking at a very useful and informative device, the **age structure diagram.**

The Age Structure Diagram

About 9 of every 10 babies born in the world today are in the poorer countries. Both current and projected rates of population growth are distributed quite unevenly between the poorer and richer nations. This phenomenon is apparent in the age-structure diagrams typical of these countries. An

156

age structure diagram (often called a **population pyramid**) classifies a population by gender and by five-year age increments (●Figure 3.11). One important index these profiles show is the percentage of a population under age 15. A country like Niger in Figure 3.11 is typically poor and faces the prospect of increasing poverty because so many new jobs, food, and other resources will have to be created to meet the demands of those children as they mature and have their own children. The bottom-heavy age structure diagram also suggests a continued surge in population as those children grow to enter their reproductive years. A large, youthful population in which competition for jobs, education, and land is intense is a social environment ripe for discord. A recent study by Population Action International found that 80 percent of the civil conflicts of the 1970s, 1980s, and 1990s took place in countries where at least 60 percent of the population was younger than 30.[3]

The bottom-heavy, pyramid-shaped age structure diagram of Niger contrasts markedly with the more chimney-shaped structures of the United States and Germany in Figure 3.11. These wealthier countries have a much more even distribution of population through age groups, with a modest share under age 15. Such profiles suggest that, not considering migration, their population growth will be low in the near future.

Collectively, about 31 percent of the population of the poorer countries is under age 15, while the corresponding figure for the wealthier countries is 17 percent (●Figure 3.12). As a whole, then, the developing world faces the critical challenge of providing for a burgeoning population in the future, even while struggling to meet the needs of the people alive today.

Where Do We Live?

Where do all these people live? The world population cartogram in ●Figure 3.13 shows clearly that China and India are the most populous countries, with 1.3 and 1.1 billion people, respectively; in fact, about 35 percent of all people on earth in 2007 were either Chinese or Indian.

Why are so many people in just two countries? There are several reasons. Both countries are large. China is 20 percent larger than the contiguous 48 U.S. states, and India is about one-third the size of China. Both have been populated since very ancient times, and successful intensive agriculture has been practiced in both for more than 4,000 years. Both have large areas of productive soils and high rainfall, promoting successful farming. Both are developing countries in which birth rates remained high while death rates fell. The resulting population surge prompted China's government to adopt an aggressive population control policy, and India followed suit with less forceful measures—meaning that India will likely overtake China in population by 2030. Overall, Monsoon Asia has about 54 percent of the world's people, and its slice of the population "pie" will continue to grow (●Figure 3.14).

279, 280, 286

The United States and Indonesia rank third and fourth among the world's most populous countries, but different

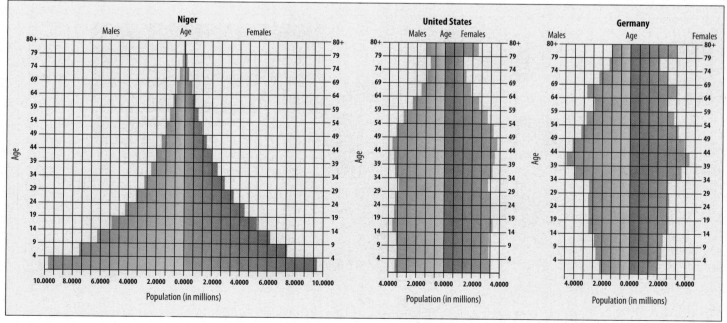

● **Figure 3.11** Population by age and sex. The pyramid-shaped age structure diagram for Niger contrasts remarkably with those of the far more affluent United States and postindustrial Germany, with their chimneylike shapes. A poor country, Niger has a relatively high birth rate, with about 48 percent of the population under age 15. The United States' population is growing slowly, while Germany and some other industrialized nations are losing populations.

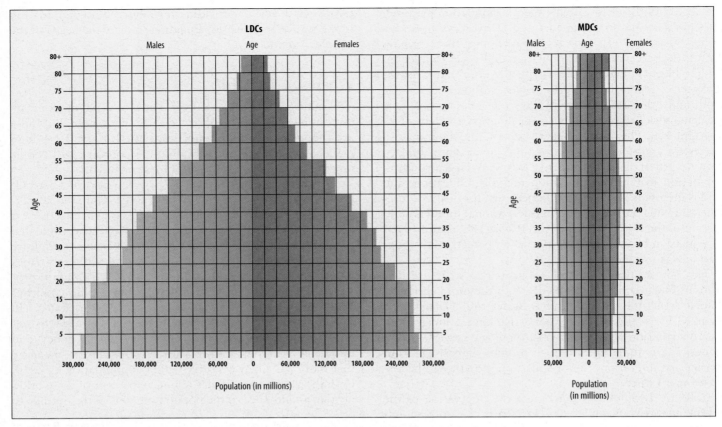

• **Figure 3.12** Population by age and sex. This diagram summarizes some of the most important facts about the human population: the lion's share lives in the poorer countries, and the lion's share of that share is young.

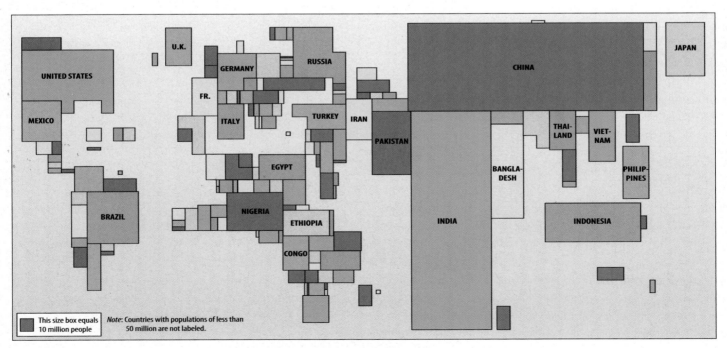

• **Figure 3.13** The demographic heavyweights of China and India stand out in the world population cartogram. The United States and Indonesia, the world's third and fourth most populous countries, are prominent too.

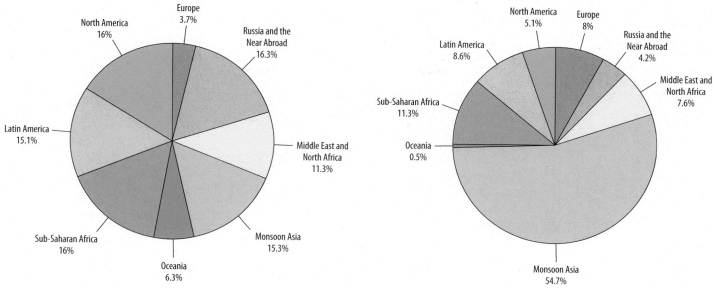

• **Figure 3.14** World regions by land areas (left) and human populations (right).

reasons account for their size. The United States has huge swaths of productive farmland, and so the environment has been able to sustain large numbers of people, but migration in periodic large waves has played a far stronger role than in China, India, or Indonesia in increasing the population. Indonesia, like China and India, is a developing country with an ancient productive agricultural environment and a recent history of high population growth. Its colonial past also played a role in population growth. The Dutch, who colonized the area known as the East Indies in the 19th century, introduced the **culture system,** a scheme to boost the output of valuable food and cash crops by requiring people on the island of Java to contribute their fields and their labors. The agricultural successes of this harsh system contributed to a **positive feedback loop** (in which change in one direction produces more change in that direction) of population growth: more people produced more food, which made it possible to support more people who grew more food. Are there limits to growth of this kind? That question is addressed later in this chapter.

While there are many variables to consider in explaining why large numbers of people are clustered in particular countries, including cultural factors and family planning policies, the natural setting is by far the most important factor. The population densities shown in •Figure 3.15 correlate generally with agricultural and other environmental conditions. The deepest reds showing the highest population densities are in the more humid and fertile regions of both China and India. Conversely, in the very western part of China, including the high Tibetan Plateau, very dry conditions limit agriculture to a few favored areas. The world's highest mountains, the Himalayas, rise just north of the deep red area of high population density in northern India. The moisture-laden winds that bring so much productive

rainfall to India cannot cross that mountain barrier, which has been a divide between densely and sparsely populated regions for thousands of years. Looking at the lightly populated areas of the world, you can recognize similar environmental factors at work: in the Sahara of northern Africa, the Arctic of northern Canada, and the Amazon Basin of South America, conditions are too dry, too cold, or too wet and infertile to support large numbers of people.

The Geography of Migration

Migration refers to the movement of people from one location to another in any setting, whether within a community, within a country, or between countries. Migration is one of the most dynamic and most problematic human processes on earth (•Figure 3.16). A migrant is always both an **emigrant** (one who moves from a place) and an **immigrant** (one who moves to a place). Migration is usually associated with either **push factors,** as when hunger or lack of land "pushes" peasants out of rural areas into cities or warfare pushes people from one place to another, or **pull factors,** when, for example, an educated villager takes advantage of a job opportunity in the city or another country. People responding to push factors are often referred to as **nonselective migrants,** and people reacting to pull factors are called **selective migrants.**

Both push and pull forces are behind the **rural-to-urban migration** pattern that is characteristic of most countries. The growth of cities, known as **urbanization,** is due in part to the natural growth rate of people already living in urban areas, but it is particularly strong in the LDCs because of this internal migration of both selective and nonselective migrants.

Within and between countries today, there are considerable movements of **refugees,** the victims of such severe

358

111,
220,
289,
297,
525

● **Figure 3.15** This isarithmic map of population density shows the approximate distribution of people around the world.

● **Figure 3.16** The global picture of people on the move. The major trends are of migrants in search of work in more affluent countries and of refugees driven by warfare or environmental adversity.

push factors as persecution, political repression, and war. They may be on the move either as illegal immigrants or as people granted **asylum,** or permission to immigrate on the grounds that they would be harmed or persecuted in their country of origin. Immigration and asylum laws and quotas vary widely around the world, depending on host countries' political, economic, and social systems. Among the most disadvantaged of the world's peoples are the **internally displaced persons (IDPs)** who are dislodged and impoverished by strife in their home country but have little prospect of

255

emigrating. The African nation of Sudan has the world's highest number of internally displaced persons, and in the pages of this book you will read many accounts of others moved by forces beyond their control.

Migration is almost never clearly "good" or "bad" but is almost always a controversial mixed blessing. Migration is bad for the country whose most talented people are leaving but good for the country they go to; the Indian doctor down the street benefits your community in the United States but may be sorely missed in India, for example. The doctor is illustrative of the **brain drain,** the emigration of educated and talented people from a place that needs them. The Mexican immigrant who does low-wage labor in the United States may be perceived as an **illegal alien** threatening to overwhelm social services and take jobs away from local people or as a **guest worker** who performs important services that no one else wants to do. Some host-country peoples are more welcoming of new cultures that enrich their ethnic mosaic (Americans are generally perceived as among the most accommodating cultures in the world), while others fear losing their ethnic majorities and privileges.

How Many People Will Live on Earth?

Population geographers are fairly confident in their calculations of how many people have lived on earth at various times in the past (see Table 3.3). Projecting future numbers is another matter, however. There are many uncertainties. Will birth rates fall faster than anticipated in the developing world? Will death rates surge because of HIV/AIDS or some other epidemic? These are some of the wild cards in the population deck.

In asking how many people will live on earth in the future, we are essentially asking, "Will the poorer countries of the world go through the demographic transition?" Will the current, relatively high birth rates in the less developed

countries continue their present slow decline (●Figure 3.17)? In 1970, Kenya had a birth rate of 51 and Bangladesh had a birth rate of 45; by 2007, Kenya's birth rate had dropped to 40 and Bangladesh's to 27. Considering what might have been, given the country's huge population, China's decline may be the most impressive of all, from a rate of 38 in 1965 to 12 in 2007. The Chinese government's generally harsh but successful **one-child policy,** using a combination of incentives and punishments to encourage couples to bear only one offspring, has been the main reason for the decline. In Bangladesh, growing literacy and the slow but steady economic progress of women have brought down the birth rate. Family planning policies and levels of education and economic well-being have played various roles in the poorer countries, but have collectively combined to bring birth rates down since 1968. If the processes of increasing economic and social development continue in the LDCs as a whole, they should make steady progress through the demographic transition, and the earth's population should cease to grow and should stabilize.

But will the poorer countries complete the transition successfully? An unsavory but possible scenario would see death rates rise dramatically (due to the scourge of HIV/ AIDS, for example) in at least some of the countries, in effect pushing them back to stage 1 of the demographic transition, where both birth and death rates are high. Others could remain in stage 2, with high birth rates and falling death rates, long enough to bring unexpectedly high numbers of people into the world. With such different scenarios in mind, the United Nations prefers to use a widely respected model with three projections: high, moderate, and low growth (●Figure 3.18).

In view of declining birth rates worldwide, the United Nations has revised its projection for future population growth downward. The agency predicts 9.2 billion by 2050 (half a billion fewer than it had estimated in earlier

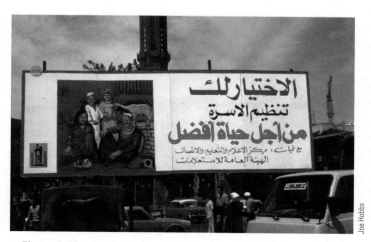

● **Figure 3.17** National family planning programs can have a pronounced impact on birth rates. This sign on the main square in Cairo, Egypt, urges parents to have no more than two children "for the sake of a better life."

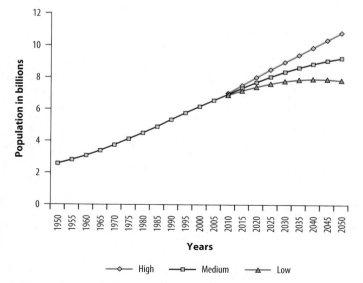

● **Figure 3.18** United Nations projections for population growth.

projections) and stabilization—the maximum number of people that will ever live on earth at one time—at 10.8 billion in 2150 (the earlier projection was 12.6 billion).

This downward revision has prompted a rash of popular and academic articles proclaiming that "the population explosion is over" and even some essays arguing there would soon be too few people on earth, particularly in the richer countries. Other experts have cautioned that it is too soon to declare the population bomb defused. "World population growth turned a little slower," a Population Institute report argues. "The difference, however, is comparable to a tidal wave surging toward one of our coastal cities. Whether the tidal wave is 80 feet or 100 feet high, the impact will be similar."[4] (To put that in more concrete terms, while the world population grew by an all-time high of 88 million people in 1994 and the population growth rate fell after that, it still grew by 70 million in 2007.) Not content with its own projected figure for stabilization, the United Nations has pledged to do its best to help stabilize the global population at no more than 9.8 billion after the year 2050. The organization's plan is to focus on enhancing the education and employment of women in the less developed countries as a means of bringing down birth rates.

The Malthusian Scenario

Even while the birth rate falls, the population increases. Already large and growing numbers pose fundamental questions: Can the earth sustain 9.8 billion or more people? Will we exceed the planet's carrying capacity, and what will happen if we do? Early in the Industrial Revolution, an English clergyman named Thomas Malthus (1766–1834) postulated that human populations, growing geometrically or exponentially, would exceed food supplies, which grow only arithmetically or linearly. He predicted a catastrophic human die-off as a result of this irreconcilable equation (●Figure 3.19). He could not have foreseen that the exploitation of new lands and resources, including tapping the

energy of fossil fuels, would permit food production to keep pace with or even outpace population growth for at least the next two centuries.

This **Malthusian scenario** of the lost race between food supplies and mouths to feed remains a source of constant and important debate today. On one side of the debate are optimists, the so-called **technocentrists** or **cornucopians,** who argue that human history provides insight into the future. Thanks to their technological ingenuity, people have always been able to conquer food shortages and other problems and therefore always will (●Figure 3.20). The late Julian Simon, a University of Maryland economist, argued that far from being a drain on resources, additional people create additional resources. Technocentrists therefore insist that people can raise the earth's carrying capacity indefinitely and that the die-off that Malthus predicted will always be averted. Our more numerous descendants will instead enjoy more prosperity than we do.

In contrast, the **neo-Malthusians** (heirs of the reasoning of Thomas Malthus) argue that although we've been successful so far, we cannot increase the earth's carrying capacity indefinitely. There is an upper limit beyond which growth cannot occur, with calculations ranging from 8 to 40 billion people. Neo-Malthusians insist that the poorer countries cannot remain indefinitely in the second stage of the demographic revolution. Either they must intentionally bring birth rates down further and make it successfully through the demographic transition, or they must unwillingly suffer nature's solution, a catastrophic increase in death rates. Thus by either the **birth rate solution** or the **death rate solution** to the problem, the neo-Malthusians argue, the less developed world must confront its population crisis (see pages 62–64).

Whereas technocentrists view the equation between people and resources passively, insisting that no corrective action is needed, neo-Malthusians tend to be activists who describe terrible scenarios of a death rate solution in order to motivate people to adopt the birth rate solution. The biolo-

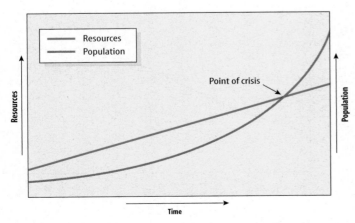

● **Figure 3.19** Malthus envisioned a race between people and resources, in which people lost.

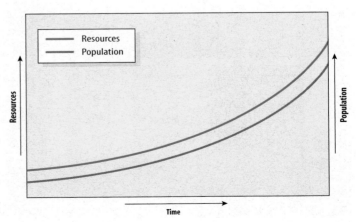

● **Figure 3.20** The technocentrists reason that production of food and other resources will always stay ahead of population growth.

gist Paul Ehrlich thinks that we are increasingly vulnerable to a Malthusian catastrophe, particularly as HIV and other viruses diffuse around the globe with unprecedented speed. "The only big question that remains," Ehrlich writes, "is whether civilization will end with the bang of an all-out nuclear war, or the whimper of famine, pestilence and ecological collapse."[5] Such dire warnings have earned many neo-Malthusians the reputation of being "gloom-and-doom pessimists." The neo-Malthusians insist that there are simply too many people already for the world to support, especially in the LDCs, where there are not enough resources to support them. But what does "too many people" mean? How many are too many, and based on what criteria?

What Is "Overpopulation"?

It is probably most useful to think about two distinct types of overpopulation, one characteristic of the poor countries and the other of the rich (•Figure 3.21). **People overpopulation** is an apparent problem in the poorer countries. The environmental problems characteristic of LDCs are intensified by the relatively high rates of human population growth in those countries. More people cut more trees,

hunt more wildlife, and otherwise use more resources. Many persons, each using a small quantity of natural resources daily to sustain life, have a great collective impact on the environment and may add up to too many people for the local environment to support. Common consequences are malnutrition and even the famine emergencies in which richer countries are called on to provide relief.

Consumption overpopulation is characteristic of the MDCs. In the wealthier countries, there are fewer persons, but each uses a large quantity of natural resources from ecosystems around the world. Their collective impacts also degrade the environment, and even their smaller numbers may be "too many" at such unsustainably high levels of consumption, particularly with their habit of feeding so high on the food chain (Insights, page 62). With less than 5 percent of the world's population, the United States may be regarded as the world's leading overconsumer, accounting, for example, for about one-quarter of the world's annual energy consumption. According to a useful measure known as the **ecological footprint**—the amount of biologically productive land needed to sustain a person's consumption and absorb his or her wastes—for every acre needed to support an average Ethiopian, 69 acres are needed to support an

287,
415,
455,
617

Developing Countries

Population (**P**)   X   Consumption per person (affluence, **A**)   X   Technological impact per unit of consumption (**T**)   =   Environmental impact of population (**I**)

Developed Countries

• **Figure 3.21** Two types of overpopulation, calculated according to this formula: *number of people × number of units of resources used per person × environmental degradation and pollution per unit of resource used × environmental impact.* Circle size shows the relative importance of each factor. People overpopulation is caused mostly by growing numbers of people and is typical of LDCs. Consumption overpopulation is caused mostly by growing affluence and is typical of MDCs.

INSIGHTS

The Second Law of Thermodynamics

It is useful to understand how energy flows through ecosystems because this process is central to the question of how many people the earth can support. **Food chains** (for example, in which an antelope eats grass and a leopard eats an antelope) are short, seldom consisting of more than four **trophic (feeding) levels.** The collective weight (known as **biomass**) and absolute number of organisms decline substantially at each successive trophic level (there are fewer leopards than antelopes). The reason is a fundamental rule of nature known as the **second law of thermodynamics,** which states that the amount of high-quality usable energy diminishes dramatically as the energy passes through an ecosystem. In living and dying, organisms use and lose the high-quality, concentrated energy that green plants produce and that is passed up the food chain. As an organism is consumed in any given link in the food chain, about 90 percent of that organism's energy is lost, in the form of heat and feces, to the environment. The amount of animal biomass that can be supported at each successive level thus declines geometrically, and little energy remains to support the top carnivores. So, for example, there are many more deer than there are mountain lions in a natural system.

The second law of thermodynamics has important consequences and implications for the human use of resources. The higher we feed on the food chain (the more meat we eat), the more energy we use. The question of how many people the earth's resources can support thus depends very much on what we eat. The affluent consumer of meat demands a huge expenditure of food energy because that energy flows from grain (producer) to livestock animal (primary consumer, with a 90 percent energy loss) and then from livestock animal to human consumer (secondary consumer, with another 90 percent energy loss). Each year, the average U.S. citizen eats 100 pounds of beef, 50 pounds of pork, and 45 pounds of poultry. By the time it is slaughtered, a cow has eaten about 10 pounds of grain per pound of its body weight; a pig, 5 pounds; and a chicken, 3 pounds. Thus in a year, a single American consumes the energy captured by two-thirds of a ton of grain (1,330 pounds) in meat products alone. By skipping the meat and eating just the corn that fattens these animals, many people could live on the energy required to sustain just one meat-eater. The point is not to feel guilty about eating meat but to realize that there is no answer to the question "How many people can the world support?" The question needs to be rephrased: "How many people can the world support based on people consuming _____?" (with the blank filled in with a specific number of calories per day or with a general lifestyle).

Development means a more affluent lifestyle, including eating more meat (feeding higher on the food chain). Some analysts fear the burden that the changing dietary habits accompanying development will pose to the planet's food energy supply. Lester Brown of the WorldWatch Institute wrote a provocative essay titled "Who Will Feed China?"[6] Keeping in mind the second law of thermodynamics, he argued that if China continues on its present course of growing affluence, with a move away from reliance on rice as a dietary staple to a diet rich in meats and beer, the entire planet will shudder. How could more than a billion people be sustained so high on the food chain?

average American. The disparity suggests that if Ethiopia is "underdeveloped," the United States is "overdeveloped."

If the vast majority of the world's population were to consume resources at the rate that U.S. citizens do, the environmental results might be ruinous (see Insights, above). Even if consumption levels in the LDCs do not rise substantially, the sheer increase in numbers of people in those countries suggests that degradation of the environment will accelerate in the coming decades.

3.4 Addressing Global Problems

The cornucopian view is comforting: we have nothing to worry about. By keeping up the good, ingenious work as we have in the past, our futures will be secure. If, however, one diverges modestly from this premise or even accepts the most dire neo-Malthusian view, it is appropriate to ask what can be done to prevent or solve some of the world's critical problems involving natural resources and human population numbers.

The Death Rate Solution and Lifeboat Ethics

One option is to "let nature take it course" and allow people imperiled by famine or other catastrophe to perish. The neo-Malthusian ecologist Garrett Hardin introduced **lifeboat ethics**—the question of whether or not the wealthy should rescue the "drowning" poor—in a distressing and challenging essay. "People turn to me," wrote Hardin, "and say, 'My children are starving. It's up to you to keep them alive.' And I say, 'The hell it is. I didn't have those children.'"[7] He described the world not as a single "spaceship *Earth*" or "global village" with a single carrying capacity, as many environmentalists do, but as a number of distinct "lifeboats," each occupied by the citizens of single countries and each having its own carrying capacity. Each rich nation is a lifeboat comfortably seating a few people. The world's poor are in lifeboats so overcrowded that many fall overboard. They swim to the rich lifeboats and beg to be brought aboard. What should the passengers of the rich lifeboat do? The choices pose an ethical dilemma (●Figure 3.22).

Hardin set out the following scenario. There are 50 rich passengers in a boat with a capacity of 60. Around them country diminishes the quality of life for subsequent generations," Hardin concluded. "For the foreseeable future, survival demands that we govern our actions by the ethics of a lifeboat."[8] Incidentally, Hardin committed suicide in 2003.

The Birth Rate Solution and Sustainable Development

In recent decades, new concepts and tools for managing the earth and its resources in an effective, long-term way have emerged. Known collectively as **sustainable development** (or **ecodevelopment**), these ideas and techniques consider what both MDCs and LDCs can do to avert the possible Malthusian dilemma and improve life on the planet. By promoting the birth rate solution and other concrete actions, sustainable development offers an activist agenda without the peril of doom depicted by the neo-Malthusians.

The World Conservation Union defines sustainable development as "improving the quality of human life while living within the carrying capacity of supporting ecosystems."[9] Sustainable development refutes what its proponents perceive as the current pattern of unsustainable development, whereby economic growth is based in large part on excessive resource use. Advocates of sustainable development point out that a country that depletes its resource base for short-term profits gained through deforestation increases its gross domestic product (GDP) and appears to be more "developed" than a country that protects its forests for a long-term harvest of sustainable yield. Deforestation appears to be beneficial to a country because it raises GDP through the production of pulp, paper, furniture, and charcoal. However, GDP growth does not measure the negative impacts of deforestation, such as erosion, flooding, siltation, and malnutrition. These consequences are known as **external costs,** or **externalities,** and they are not taken into account in the prices of goods and services. Advocates of sustainable development argue that these externalities should be added or "internalized" before a good or a service is marketed because the true high costs of producing these goods and services would then be recognized. The lower true costs of less destructive practices would then be evident, providing stronger incentives for individuals, companies, and nations to invest in sustainable practices and technologies.

Sustainable development is a complex assortment of theories and activities, but its proponents call for eight essential changes in the way people perceive and use their environments:

1. People must change their worldviews and value systems, recognizing the finiteness of resources and reducing their expectations to a level more in keeping with the earth's environmental capabilities. Proponents of sustainable development argue that this change in perspective is needed especially in the MDCs, where instead of trying to "keep up with the Joneses," people should try to enjoy life through more social rather than material pursuits.

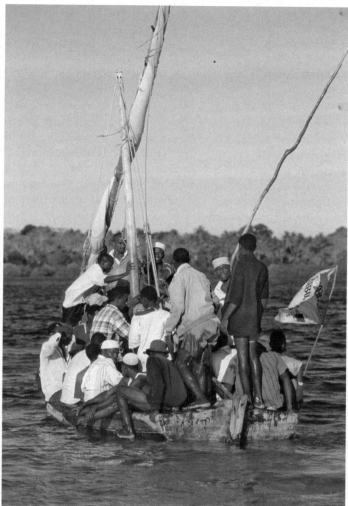

● **Figure 3.22** An overcrowded wooden boat ferries passengers from Mafia Island to mainland Tanzania. What if the boat capsizes?

are 100 poor swimmers who want to come aboard. The rich boaters have three choices. First, they could take in all the swimmers, capsizing the boat with "complete justice, complete catastrophe." Second, as they enjoy an unused excess capacity of 10, they could admit just 10 from the water. But which 10? And what about the margin of comfort that excess capacity allows them? Finally, the rich could prevent any of the doomed from coming aboard, ensuring their own safety, comfort, and survival.

Translating this metaphor into reality, as an occupant of the rich lifeboat *United States,* for example, what would you do for the drowning refugees from lifeboat *Haiti* or lifeboat *Afghanistan*?

Hardin's choice was the third: drowning. To preserve their own standard of living and ensure the planet's safety, the wealthy countries must cease to extend food and other aid to the poor and must close their doors to immigrants from poor countries. "Every life saved this year in a poor

2. People should recognize that development and environmental protection are compatible. Rather than viewing environmental conservation as a drain on economies, we should see it as the best guarantor of future economic well-being. This is especially important in the LDCs, with their resource-based economies.

3. People all over the world should consider the needs of future generations more than we do now. Much of the wealth we generate is in effect borrowed or stolen from our descendants. Our economic system values current environmental benefits and costs far more than future benefits and costs, and so we try to improve our standard of living today without regard to tomorrow.

4. Communities and countries should strive for self-reliance, particularly through the use of appropriate technologies. For example, remote villages could rely increasingly on solar power for electricity rather than be linked into national grids of coal-burning plants.

5. LDCs need to limit population growth as a means of avoiding the destructive impacts of people overpopulation. Advances in the status of women, improvements in education and social services, and effective family planning technologies can help limit population growth.

6. Governments need to practice land reform, particularly in the LDCs. Poverty is often not the result of too many people on too little total land area but of a small, wealthy minority holding a disproportionately high share of quality land. To avoid the environmental and economic consequences of marginalization, a more equitable distribution of land is needed.

7. Economic growth in the MDCs should be slowed to reduce the effects of consumption overpopulation. If economic growth, understood as the result of consumption of natural resources, continues at its present rate in excess of sustainable yield, the earth's "environmental capital" will continue to diminish rapidly.

8. Wealth should be redistributed between the MDCs and LDCs. Because poverty is such a fundamental cause of environmental degradation, the spread of a reasonable level of prosperity and security to the LDCs is essential. Proponents argue that this does not mean that rich countries should give cash outright to poor countries. Instead, the lending institutions of MDCs can forgive some existing debts owed by LDCs or use such innovations as **debt-for-nature swaps,** in which a certain portion of debt is forgiven in return for the borrower's pledge to invest that amount in national parks or other conservation programs. Reducing or eliminating **trade barriers** that MDCs impose against products from LDCs would also help redistribute global wealth.

561

Some geographers and other scientists believe that sustainable development (rather than information technology) will bring about the **"Third Revolution,"** a shift in human ways of interacting with the earth so dramatic that it will be compared with the origins of agriculture and industry. The formidable changes called for in sustainable development are attracting increasing attention, perhaps because at present there are no comprehensive alternative strategies for dealing with some of the most critical issues of our time.

SUMMARY

→ A useful way to begin to appreciate the current spatial patterns of our relationship with the earth is to view these patterns as products of two "revolutions" in the relatively recent past: the Agricultural or Neolithic (New Stone Age) Revolution and the Industrial Revolution. Each transformed humanity's relationship with the natural environment. After the Industrial Revolution, the human impact on earth grew dramatically.

→ The world is markedly divided between the haves and the have-nots, characterized at the largest scale by contrasts between more developed and less developed countries (MDCs and LDCs). Explanations for these disparities include dependency theory, cultural factors, geographic location, and natural resource base.

→ LDC economies often rely on the export of a few raw materials or commercial crops and tend to draw down their natural "capital" quickly, with profound impacts on the environments.

→ There is disagreement on whether globalization is beneficial or detrimental to poorer countries and to the poorer people who live in those countries.

→ Population growth is measured by birth rates, death rates, and migration. Growth rates tend to be higher in the LDCs, due to relatively high but falling birth rates and much lower and falling death rates. The demographic transition, a model of what happened to populations in the MDCs, shows the wealthier countries passing from high birth rates and death rates to low birth rates and death rates.

→ The explosive growth in world population since the beginning of the Industrial Revolution is the result not of a rise in birth rates but of a dramatic decline in death rates, particularly in the LDCs. High rates of human population growth intensify environmental problems characteristic of the LDCs. More people use more resources, a phenomenon that suggests there is a problem of people overpopulation in the poorer countries. Consumption overpopulation is more characteristic of the MDCs, where fewer people use large quantities of natural resources from around the world.

→ Human demands for more food and other photosynthetic products are growing, and it is questionable whether supplies can keep pace. The Malthusian scenario, which has not been realized so far, insists that a catastrophic die-off of people will occur when their numbers exceed food supplies.

→ New concepts and tools for managing the earth and its resources in an effective, long-term way have emerged in the past two decades. Known collectively as sustainable development, or ecodevelopment, these ideas and techniques consider what humanity can do to avert the Malthusian dilemma and improve life on the planet.

KEY TERMS + CONCEPTS

Age of Discovery (p. 42)
Age of Exploration (p. 42)
age structure diagram (p. 54)
Agricultural Revolution (p. 41)
asylum (p. 58)
biomass (p. 62)
birth rate (p. 50)
birth rate solution (p. 60)
brain drain (p. 59)
carrying capacity (p. 41)
cash crops (p. 46)
civilization (p. 41)
commercial crops (p. 46)
consumption overpopulation (p. 61)
cornucopians (p. 60)
culture hearth (p. 41)
culture system (p. 57)
death rate (p. 50)
death rate solution (p. 60)
debt-for-nature swap (p. 64)
deforestation (p. 48)
demographic transition (p. 53)
    first (preindustrial) stage (p. 54)
    second (transitional) stage (p. 54)
    third (industrial) stage (p. 54)
    fourth (postindustrial) stage (p. 54)
demography (p. 50)
dependency theory (p. 45)
development (p. 44)
digital divide (p. 47)
domestication (p. 41)
dry farming (p. 41)
ecodevelopment (p. 63)
ecological bankruptcy (p. 48)
ecological footprint (p. 61)
ecologically dominant species (p. 40)
emigrant (p. 57)
extensive land use (p. 41)
external costs (p. 63)

externalities (p. 63)
feeding levels (p. 62)
food chain (p. 62)
Food-Producing Revolution (p. 41)
foraging (p. 40)
fuelwood crisis (p. 48)
globalization (p. 47)
gross domestic product (GDP) (p. 44)
gross national income (GNI) (p. 44)
gross national product (GNP) (p. 44)
guest worker (p. 59)
hot money (p. 47)
Human Development Index (HDI)
    (p. 44)
hunting and gathering (p. 40)
illegal alien (p. 59)
immigrant (p. 57)
Industrial Revolution (p. 42)
information technology (IT) (p. 47)
intensive land use (p. 41)
internally displaced persons (IDPs) (p. 58)
irrigation (p. 41)
knowledge economy (p. 47)
less developed countries (LDCs) (p. 43)
lifeboat ethics (p. 62)
Malthusian scenario (p. 60)
marginalization (p. 46)
mercantile colonialism (p. 45)
migration (p. 50)
more developed countries (MDCs)
    (p. 43)
multinational companies (p. 47)
natural replacement rate (p. 48)
natural resource (p. 46)
neocolonialism (p. 46)
neo-Europes (p. 45)
Neolithic Revolution (p. 41)
neo-Malthusians (p. 60)

newly industrializing countries (NICs)
    (p. 44)
nontimber forest products (NTFPs)
    (p. 49)
nonselective migrants (p. 57)
"original affluent society" (p. 40)
one-child policy (p. 59)
people overpopulation (p. 61)
per capita GDP (p. 44)
per capita gross national income
    purchasing power parity
    (per capita GNI PPP) (p. 44)
Pleistocene overkill hypothesis (p. 40)
population change rate (p. 51)
population explosion (p. 50)
population implosion (p. 54)
population pyramid (p. 55)
population replacement level (p. 54)
positive feedback loop (p. 57)
pull factors (p. 57)
purchasing power parity (PPP) (p. 44)
push factors (p. 57)
refugees (p. 57)
renewable resource (p. 48)
rural-to-urban migration (p. 57)
second law of thermodynamics (p. 62)
selective migrants (p. 57)
settler colonization (p. 45)
sustainable development (p. 63)
sustainable yield (p. 48)
technocentrists (p. 60)
"Third Revolution" (p. 64)
trade barriers (p. 64)
trophic levels (p. 62)
urbanization (p. 57)
value-added products (p. 45)
zero population growth (ZPG) (p. 54)

REVIEW QUESTIONS

1. What were the Agricultural and Industrial Revolutions? In what ways did they initiate important changes in human-earth relationships?

2. What are the typical differences between MDCs and LDCs?

3. According to dependency theory, what are the causes of disparities between MDCs and LDCs? What other factors may explain global wealth and poverty?

4. What is the fuelwood crisis? What are some of the effects of deforestation on human lives and economies in the LDCs?

5. What are the two types of overpopulation, and how do they differ?

6. What has been the main cause of the world's explosive population growth since the beginning of the Industrial Revolution?

7. What variables distinguish the four stages of the demographic transition, and what explains them?

8. Which world regions have the most and the fewest people, and what factors might account for these differences?

9. How do technocentrists and neo-Malthusians view the balance between people and resources?

10. What do Hardin's lifeboats represent—what is his metaphor?

11. What are the goals and methods of sustainable development?

NOTES

1. Marshall B. Sahlins, *Stone Age Economics* (Chicago: Aldine-Atherton, 1972).

2. Alfred W. Crosby, *Ecological Imperialism: The Biological Expansion of Europe, 900–1900* (New York: Cambridge University Press, 1986).

3. Celia W. Dugger, "Very Young Populations Contribute to Strife, Study Concludes." *New York Times*, April 4, 2007, p. A6.

4. Cited in Steven A. Holmes, "Global Crisis in Population Still Serious, Group Warns." *New York Times*, December 31, 1997, p. A7.

5. Paul R. Ehrlich, "Populations of People and Other Living Things." In *Earth '88: Changing Geographic Perspectives*, ed. Harm De Blij (Washington, D.C.: National Geographic Society, 1988), p. 309.

6. Lester R. Brown, "Who Will Feed China?" *WorldWatch Magazine*, September-October 1994, p. 10.

7. Garrett Hardin, "The Tragedy of the Commons." *Science, 162*, 1968, pp. 1243–1248.

8. Garrett Hardin, "Lifeboat Ethics: The Case against Helping the Poor." *Psychology Today*, September 1974, p. 6; and "Living on a Lifeboat," *BioScience 24*, 1974, p. 568.

9. World Wildlife Fund, *Sustainable Use of Natural Resources: Concepts, Issues and Criteria* (Gland, Switzerland: World Wildlife Fund, 1995), p. 5.

A composite satellite image of Europe.

People have long left their imprint on Europe's landscapes, often with beautiful and legendary monuments. This is Kalmar Castle, in southern Sweden.

Joe Hobbs

CHAPTER 4
A GEOGRAPHIC PROFILE OF EUROPE

Many **culture hearths** have exerted influences on peoples and landscapes around the world, but none more impressively than Europe. For more than 500 years, the experiences of European nation building, scientific and technological developments, and colonization of far-flung lands have reshaped cultures and ecosystems around the world. Mainly as a result of self-inflicted wounds in the form of two world wars, Europe has taken a backseat to the United States in global influence but still today wields enormous economic and political clout. This chapter explores the human and physical geographies of this influential world region and sets the stage for subsequent exploration of its subregions.

chapter objectives

This chapter should enable you to

→ Recognize Europe as a postindustrial region with a wealthy, declining population

→ Become familiar with Europe's immigration issues

→ Learn the region's distinguishing geographic characteristics, landforms, and climates

→ Get acquainted with Europe's major ethnic groups, languages, and religions

→ Understand how Europe rose to global political and economic dominance and then declined

→ Trace Europe's emergence from wartime divisions to supranational unity within the European Union (EU) and know why the EU is important

→ Appreciate important distinctions between Europeans and Americans

4.1 Area and Population

Europe has traditionally been classified as one of the world's seven continents, but one look at the globe reveals what an arbitrary designation that is. Europe is not a distinct landmass such as Australia or Africa. It is rather an appendage or a subcontinent of the world's greatest landmass, Eurasia (a term combining the names Europe and Asia). Nevertheless, the popular and scholarly designation of Europe as a region distinct from the rest of Eurasia is a time-honored and useful organizing device. In this text, Europe is the culture region made up of the countries of Eurasia lying west of Turkey, Russia, and three former constituent republics of the Soviet Union: Belarus, Ukraine, and Moldova. (Incidentally, the traditional *physical* dividing line between Europe and Asia is drawn from the Ural Mountains down to the Caucasus, technically placing Turkey, a good part of Russia, and all three of those republics within Europe.) Under our definition, Europe is a great peninsula, fringed by lesser peninsulas and islands and bounded on its seaward sides by the Arctic and Atlantic Oceans, the Mediterranean Sea, and the Black Sea (•Figure 4.1).

Europe's Subregions

This text classifies Europe into four subregions (Table 4.1).

The **European core** is northwestern and north central Europe. It consists of the United Kingdom, Ireland, France, the **Benelux** countries (Belgium, the Netherlands, and Luxembourg), Switzerland, Austria, and Germany. These countries have the largest populations and play major economic and political roles in contemporary Europe. Within the European core are also the **microstates** of Andorra, Monaco, and Liechtenstein.

Northern Europe is made up of Denmark, Iceland, Norway, Sweden, and Finland.

Southern Europe includes Portugal, Spain, Italy, Greece, Malta, and Cyprus.

Eastern Europe includes Estonia, Latvia, Lithuania, Poland, the Czech Republic, Slovakia, Hungary, Romania, Bulgaria, Albania, Serbia, Montenegro, Bosnia and Herzegovina, Croatia, Macedonia, and Slovenia.

The Europeans

All of Europe has an area only about half as large as that of the 48 contiguous United States. The average European country is some 50,000 square miles (130,000 sq km), or about the size of Arkansas. Four countries—Germany, the United Kingdom, France, and Italy—far outsize the others in Europe in population. Their respective populations range from about 82 million for Germany to approximately 60 million each for France, Italy, and the United Kingdom. Together, the four countries represent about half of Europe's population, which is about 87 percent of the population of the United States.

Europe contains one of the world's great clusters of human population (•Figure 4.2; see also Figures 3.16 and 3.17 and Table 4.1). Europe's population of approximately 532 million people in 2007 was nearly twice that of the United States. One of every 12 people in the world is a European, and they live in a space half the size of the United States. Europe's population density varies widely, from 1,020 persons per square mile (394/sq km) in the intensively developed Netherlands to about 7 per square mile (3/sq km) in Iceland but overall is greater than the world average of 127 per square mile (49/sq km). Comparisons with the United States reveal about the same range, but the European countries are generally more densely populated.

Industrialization and the Development of Europe

The greatest densities in population are in two belts of industrialization and urbanization near historical sources of coal and hydroelectric power; these energy resources were critical in the location and rapid development of many European cities (see Insights, page 71, and Figure 4.1.3, page 98). One belt extends from the United Kingdom to Italy. In addition to large parts of Britain, it includes

● **Figure 4.1** Political geography of Europe. In nearly all European countries, the political capital is also the largest and most important city.

extreme northeastern France, most of Belgium and the Netherlands, Germany's Rhineland, northern Switzerland, and northern Italy (south of the Alps). The second belt of dense population goes from Britain to southern Poland and continues out of the region into Ukraine. It overlaps the first belt as far east as the Rhineland, where it forms a narrow strip eastward across Germany, the western Czech Republic, and Poland.

Although these two belts cover a minor portion of Europe's geography, they contain more large cities and generate a greater value of industrial output than the rest of Europe combined. Only in eastern North America and Japan are there urban-industrial belts of such complexity and importance. All these regions have overwhelmingly urban populations. Europe's population overall is 74 percent urban. In the United Kingdom, Iceland, and Belgium,

395, 600

TABLE 4.1   Europe: Basic Data

Political Unit	Area (thousands; sq mi)	Area (thousands; sq km)	Estimated Population (millions)	Estimated Population Density (sq mi)	Estimated Population Density (sq km)	Annual rate of Natural Increase (%)	Human Development Index	Urban Population (%)	Arable Land (%)	Per Capita GNI PPP ($US)
European Core	**549.5**	**1,422.6**	**253.1**	**461**	**178**	**0.2**	**0.939**	**81**	**28**	**33,560**
Andorra	0.2	0.5	0.1	500	193	0.7	N/A	91	2	N/A
Austria	32.4	83.8	8.3	256	99	0.0	0.944	67	17	35,300
Belgium	11.8	30.5	10.6	898	347	0.1	0.945	97	23	34,460
France	212.9	551.2	61.7	290	112	0.4	0.942	77	33	32,310
Germany	137.8	356.7	82.3	597	231	−0.2	0.932	75	34	31,280
Ireland	27.1	70.1	4.4	162	63	0.9	0.956	60	15	35,540
Liechtenstein	1.0	2.6	0.0	40	15	0.5	N/A	15	25	N/A
Luxembourg	0.1	0.1	0.5	8333	3218	0.4	0.945	83	23	55,970
Monaco	0.001	0.002	0.03	30000	11583	0.9	N/A	100	0	N/A
Netherlands	15.8	40.9	16.4	1038	401	0.3	0.947	62	26	35,800
Switzerland	15.9	41.1	7.5	472	182	0.2	0.947	68	10	40,630
United Kingdom	94.5	244.6	61.2	648	250	0.3	0.940	90	23	35,690
Northern Europe	**485.8**	**1,257.7**	**24.9**	**51**	**20**	**0.2**	**0.951**	**77**	**6**	**36,820**
Denmark	16.6	43.0	5.5	331	128	0.2	0.943	72	54	36,110
Finland	130.6	338.1	5.3	41	16	0.2	0.947	62	7	34,810
Iceland	39.8	103.0	0.3	8	3	0.8	0.960	93	0	35,980
Norway	125.1	323.8	4.7	38	15	0.4	0.965	78	3	43,920
Sweden	173.7	449.7	9.1	52	20	0.2	0.951	84	6	34,780
Southern Europe	**401.9**	**1,040.5**	**128.0**	**318**	**123**	**0.0**	**0.934**	**79**	**25**	**28,020**
Cyprus	3.6	9.3	1.0	278	107	0.5	0.903	62	7	21,480
Greece	51.0	132.0	11.2	220	85	0.0	0.921	59	21	24,570
Italy	116.3	301.1	59.3	510	197	0.0	0.940	68	27	29,840
Malta	0.1	0.2	0.4	4000	1544	0.2	0.875	95	28	18,620
Portugal	35.5	91.9	10.7	301	116	0.0	0.904	55	21	20,850
San Marino	0.02	0.05	0.03	1500	579	0.4	N/A	84	16	N/A
Spain	195.4	505.9	45.3	232	90	0.3	0.938	77	26	28,420
Vatican City	0.1	0.04	0.001	10	4	0.0	N/A	100	0	N/A
Eastern Europe	**522.0**	**1,351.4**	**125.7**	**241**	**93**	**−0.1**	**0.844**	**60**	**37**	**14,180**
Albania	11.1	28.7	3.2	288	111	0.8	0.784	45	21	5,840
Bosnia and Herzegovina	19.7	51.0	3.8	193	74	0.0	0.800	46	13	N/A
Bulgaria	42.8	110.8	7.7	180	69	−0.5	0.816	71	40	10,140
Croatia	21.8	56.4	4.4	202	78	−0.2	0.846	56	26	13,670
Czech Republic	30.4	78.7	10.3	339	131	0.0	0.885	74	40	21,160

TABLE 4.1   Europe: Basic Data *continued*

Political Unit	Area (thousands; sq mi)	Area (thousands; sq km)	Estimated Population (millions)	Estimated Population Density (sq mi)	Estimated Population Density (sq km)	Annual rate of Natural Increase (%)	Human Development Index	Urban Population (%)	Arable Land (%)	Per Capita GNI PPP ($US)
Estonia	17.4	45.0	1.3	75	29	−0.2	0.858	69	16	17,530
Hungary	35.9	92.9	10.1	281	109	20.3	0.869	65	50	17,920
Latvia	24.9	64.4	2.3	92	36	20.5	0.823	68	29	15,340
Lithuania	25.2	65.2	3.4	135	52	20.4	0.857	67	45	14,930
Macedonia	9.9	25.6	2.0	202	78	0.2	0.796	59	22	7,610
Montenegro	5.4	14.0	0.6	111	43	0.2	N/A	64		N/A
Poland	124.8	323.1	38.1	305	118	0.0	0.862	62	46	14,530
Romania	92.0	238.1	21.6	235	91	20.2	0.805	55	40	9,820
Serbia	34.1	88.3	9.5	279	108	0.0	N/A	52	33	N/A
Slovakia	18.9	48.9	5.4	286	110	0.0	0.856	56	30	16,910
Slovenia	7.8	20.2	2.0	256	99	0.0	0.910	49	8	23,960
Summary Total	1,959.2	5,072.2	531.7	271	105	0.1	0.917	74	24	28,160

Sources: World Population Data Sheet, Population Reference Bureau, 2007; Human Development Report, United Nations, 2006; World Factbook, CIA, 2007.

INSIGHTS

Site and Situation

Site and situation are related but distinct dimensions of place and are crucial to understanding the geography of Europe and the world's other regions. **Site** refers to the physical properties of a piece of land on which something is located (or will be located): Rome is often described as "the city of seven hills," for example, while Venice is on flat ground, penetrated and often inundated by the sea. **Situation** is the larger geographical context of the site; the saying "all roads lead to Rome" indicates that among other things, the Eternal City was situated as a transportation hub. In studying Europe, you will see many examples of cities that grew up at the crossroads of transportation routes or where coal, hydropower, and other resources could be obtained nearby; these are examples of how situation plays an important role in urban and economic geography.

90 percent or more of the population is urban; in Belgium, it is 97 percent. The least urban are the countries of Albania, Bosnia, and Slovenia, with urban percentages between 45 and 49 percent.

Why Is Europe's Population Declining?

54   Europe essentially defined the demographic transition, with a trajectory from preindustrial high birth rates and high death rates to postindustrial low birth rates and low death rates (see Figure 3.10, page 54). Over a period of decades, Europe recovered from the demographic setbacks of two world wars (for example, France lost more than 3 percent of its prewar population in World War I, and Poland lost 19 percent in World War II). The population peaked in 1997 and then began a slow, steady decline to what is widely acknowledged as a **"birth dearth."** Today, birth rates are low in this region of relative affluence and high urbanization, where increasing numbers of employed and educated women are saying, "There is no time for motherhood, and children are too expensive anyway."[1]

The region's fertility rate is in fact below population replacement level, with no European country maintaining its population through births. Typical of Europe are the 54

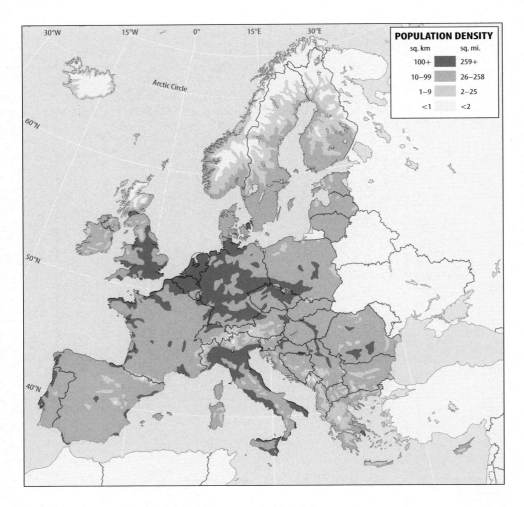

POPULATION DENSITY

sq. km	sq. mi.
100+	259+
10–99	26–258
1–9	2–25
<1	<2

• **Figure 4.2** Population distribution (right) and population cartogram (below) of Europe. The continent's demographic heavyweights are Germany, the United Kingdom, France, and Italy.

United Kingdom and the Netherlands, both with annual population change rates of 0.3 percent; the highest growth rate is in Ireland (0.9 percent) and the lowest in Bulgaria and Latvia (both −0.5 percent). Birth rates in eastern Europe are particularly low because of economic challenges faced following cuts in state subsidies (many of the former communist governments once provided free housing, education, and child care). Perceiving better working opportunities elsewhere in Europe, women or couples from the east are also moving westward in search of education and work, delaying childbirth or deciding not to have children.

Europe's population is aging faster than that of any other world region. The median age in Italy is projected to rise from 42 in 2007 to 53 by 2050. The populations of the Czech Republic and Spain are expected to contract by about 20 percent by 2050, and the Netherlands' may shrink by 10 percent. The Italian government is trying to slow a similar trend by offering parents a **"baby bounty"** equivalent to $1,200 for each child after their first—but only if the mother is Italian or carries a **European Union (EU)** passport. France offers similar cash incentives for third children.

REGIONAL PERSPECTIVE

Europe's Immigration Issues

Read

It is estimated that more than 15 million people of non-European origin live in Europe today. The major groups are Turks, Kurds, Arabs, Indians, Pakistanis, Sri Lankans, and a variety of peoples from Sub-Saharan Africa, the Caribbean, and Latin America. Some of the immigrant communities have been in Europe for a generation or more, including the West Indians who came to Britain to help with worker shortages in the 1950s and the Turks whom Germany invited to come and work during its economic boom years beginning around 1960. There are more recent streams of migrants, too, including Bangladeshis moving to Britain. Some host countries specialize in "importing" people from certain countries outside the region. In Portugal, for example, Brazilians make up 12 percent of resident foreigners, and Moroccans and Algerians constitute about 20 percent of the immigrants in France. There are also internal migrations within Europe, such as Albanians moving to Italy and Poles to Germany.

During periods of more rapid economic growth, prior to a downturn in the late 1990s and early 2000s, the immigrants often provided needed labor resources and were generally welcomed. The guest workers were encouraged to come to industrial and agricultural centers that were not able to meet their labor needs with domestic populations. But recent events, including growing domestic unemployment, terrorist attacks, and tax burdens, have changed native attitudes toward immigrants considerably. Europeans overall are not as familiar with immigration as Americans are, and they are trying to figure out what to do with their increasingly multiethnic mosaic.

Of particular concern to many Europeans are recent and incoming Muslim immigrants (●Figure 4.A). Some fears relate to terrorism: the London subway bombings of 2005 and the London and Glasgow car bombing attempts of 2007 were attributed to Muslim dissidents of Pakistani and Middle Eastern origin, for example. But more than anything, deep-seated cultural issues sometimes lead to violence and recrimination. In 2004, a Dutch citizen of Moroccan descent murdered Theo van Gogh, a Dutch filmmaker (and distant relative of the artist Vincent van Gogh) who had criticized the status of women in Muslim societies. In 2005, after a Danish newspaper published satirical cartoons of the Prophet Muhammad—among other things, wearing a bomb in his turban—a wave of sometimes violent demonstrations against Denmark erupted across the Muslim world. For all Muslims except some Shiite groups, it is heresy to depict the Prophet in any form, so such negative and stereotypical caricatures touched a particularly raw nerve. Many European media fueled the fire by publishing the cartoons in their own outlets, and the cycle of taunting and reaction intensified. Waves of bitterness again swept between the cultures in 2006 after Pope Benedict XVI made unflattering historical references to the Prophet Muhammad. Prejudice and suspicion have

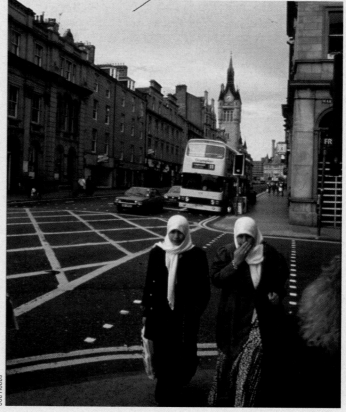

● **Figure 4.A** Muslim women in Aberdeen, Scotland. Europeans are not of one voice in answering questions about their ethnic and religious diversity.

Joe Hobbs

deepened among the majority ethnic Europeans, who increasingly believe that Islam cannot be accommodated within their value systems and societies.

Despite these apparent "culture wars," some governments believe that immigration is good for the economy and continue to encourage and accept more foreigners; Germany is one. Other countries stress that immigration does not always help the receiving country and may harm it. Some countries prohibit students and people seeking asylum from working. Not all employable immigrants are fully employed. Many are either in very low-paying and part-time jobs or are out of work. The unemployment rate for migrants is twice that of citizens in Denmark, three times that of citizens in Finland, and four times that of citizens in the Netherlands. Immigrants often have large families whose members end up on welfare, thus costing the state and its taxpayers rather than generating new tax revenue.

103, 115

Cause background
Ethnic
religion

58 Europe's shrinking, aging population could use a youthful boost from immigration, which is already the main source of population growth in most European countries. Studies have shown what additional levels of immigration would be needed to prevent major population declines in Europe and to maintain the existing ratio of workers aged 15 to 64 to those over 65 needing support. If birth rates remain at their current low level, the EU will have a shortfall of 20 million workers by 2030. The EU countries would need an annual inflow of about 3 million migrants a year (about twice the current rate) to prevent the future support ratio from dropping sharply. But that prospect raises a number of concerns in the individual countries and for the European Union itself (see the Regional Perspective on page 73).

In some countries, notably Denmark and the Netherlands, welfare benefits were extremely generous and in themselves provided a strong magnet for immigration. There is now a backlash. Denmark has passed tough anti-immigration laws and reduced welfare benefits for new arrivals. Increasing numbers of Dutch nationals are expressing weariness with paying the tax burden to support literacy, technical training, and health care costs for immigrants, who now make up 10 percent of the country's population. The country's second-largest city, Rotterdam, has almost as many nonnative (mainly Moroccan and Turkish) as native Dutch inhabitants. The native Dutch fear they may become a minority in their own country, pointing out that Moroccan-born women who reside in the Netherlands have a birth rate four times that of Dutch women. The country is now exploring ways to slow down immigration, even as increasing numbers of the Dutch, citing their fears, emigrate to Australia, the United States, and Canada.

Powerful economic forces tend to trump governmental restrictions on migration. With the rich countries of Europe creating jobs, the underemployed young in the poor countries want to fill them; simple laws of supply and demand keep the flow going. Governments are often reluctant to impose harsh measures that would restrict migration. Borders are porous. Today, about 1.3 million people enter the European Union legally each year, and about 500,000 come illegally. Germany is home to an estimated 1 million "illegals," as Europeans call their **undocumented** **59** **workers**, and France has 300,000. Most of the illegal aliens cross land or sea borders unlawfully, but perhaps a third are visitors who overstay tourist or student visas. Concerned about a new flood of undocumented workers that might inundate the older, richer EU countries, the European Union will not allow citizens of the 12 newest EU countries the freedom of movement between member states now enjoyed by the others.

92 The EU is considering offering African immigrants temporary visas as a move to encourage **"circular migration,"** in which the immigrant would work for no longer than a specified period in a European host country and then return home. The Europeans believe this approach would offer three advantages: African countries would benefit from

remittances (earnings sent home by migrant workers), migrants would have both legal jobs and safe means of reaching them, and Europe would fill some of its labor shortages.

In Europe, especially northern Europe, a large source of legal immigration is people seeking asylum. A 1967 protocol signed by the European countries gives everyone the right to seek asylum in Europe and gives refugees the right not to be sent to a place where they may be tortured or persecuted. At first, few people took advantage of this opportunity, but a tide of asylum seekers rose during the conflicts in the former Yugoslavia in the 1990s. Asylum claims in EU countries have fallen by more than half since the early 1990s but still number about 200,000 people each year. Iranians, Afghans, and others still want to come, claiming asylum from persecution but perhaps more often motivated by poverty. This accounts for why about 80 percent of the applicants are rejected.

4.2 Physical Geography and Human Adaptations

Europe has diverse physical features. Among its most distinctive physical characteristics are its irregular shape, its high latitude, and its temperate climate. One of its most striking characteristics, readily seen on a map, is its jagged coastal outline. The main peninsula of Europe is fringed by numerous smaller peninsulas, notably the Scandinavian, Jutland, Iberian, Italian, and Balkan Peninsulas (•Figure 4.3). Offshore are numerous islands, including Great Britain, Ireland, Iceland, Sicily, Sardinia, Corsica, and Crete. The island of Cyprus lies far in the eastern Mediterranean. Around the indented shores of Europe, arms of the sea penetrate the land in the form of **estuaries** (the tidal mouths of rivers), and harbors offer protection for shipping. This complex mingling of land and water provides many opportunities for maritime activity, and much of Europe's history has focused on seaborne trade, sea fisheries, and sea power.

Another remarkable environmental characteristic of Europe is its northerly location: much of Europe lies north of the 48 conterminous United States (•Figure 4.4). Despite their moderate climate, the British Isles are at the same latitude as Hudson Bay in Canada. Athens, Greece, is only slightly farther south than Saint Louis, Missouri. The high latitudes have predictable implications for people. Northern Europeans revel in the long summer days but pay a price with long, dark nights in winter. Scandinavians in particular celebrate the summer solstice as "midsummer" with great fervor.

Why Is Europe So Warm?

Europe's mild climates, especially in winter, are particularly surprising given its high latitudes. London, for example, has about the same average temperature in January

● **Figure 4.3** Physical geography of Europe.

as Richmond, Virginia, which is 950 miles (c. 1,500 km) farther south. Relatively warm currents of water account for Europe's warmth. The **Gulf Stream** and its appendage the **North Atlantic Drift** (see Figure 4.3), which originate in tropical western parts of the Atlantic Ocean, flow to the north and east and cause the waters around Europe to be much warmer in winter than the latitude would warrant.

With winds blowing mainly from the west (from sea to land) along the Atlantic coast of Europe, the moving air in winter absorbs heat from the ocean and transports it to the land. This makes winter temperatures abnormally mild for this latitude, especially along the coast. In the summer, the climatic roles of water and land are reversed: instead of being warmer than the land, the ocean is now

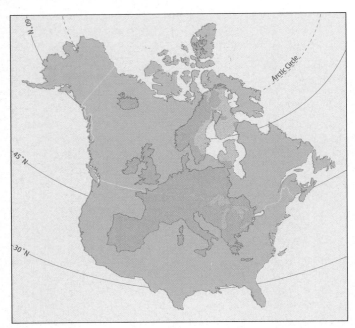

● **Figure 4.4** Europe in terms of latitude and area compared with the United States and Canada.

cooler, so the air brought to the land by **westerly winds** (also known as *westerlies*) in the summer has a cooling effect. These influences are carried as far north as the Scandinavian Peninsula and into the Barents Sea, giving the Russians a warm-water port at Murmansk, even though it lies above the Arctic Circle.

The same winds that bring warmth in winter and coolness in summer also bring abundant moisture. Most of this falls as rain, although the higher mountains and more northerly areas have considerable snow. Abundant, well-distributed, and relatively dependable moisture has always been one of Europe's major assets (see the world precipitation map, Figure 2.3 on page 23). The average precipitation in most places in the European lowlands is 20 to 40 inches (c. 50 to 100 cm) per year. This is enough for a wide range of crops, because mild temperatures and high atmospheric humidity reduce the rate at which plants lose their moisture.

The Relationship of Human Settlement and Production to Europe's Varied Landscapes

Europe's topographic features are extremely diverse, with plains, plateaus, hill lands, mountains, and water bodies all well represented. Enriched by the human associations of an eventful history, Europe offers distinctive and often highly scenic landscapes.

One of the most prominent surface features of Europe is an undulating or rolling plain that extends without a break from near the French-Spanish border, across western and northern France, central and northern Belgium, the Netherlands, Denmark, northern Germany, Poland, and

far into Russia. Known as the **North European Plain** (see Figure 4.3), it has outliers in Great Britain, the southern part of the Scandinavian Peninsula, and southern Finland. The North European Plain contains the greater part of Europe's cultivated land, and it is underlain in some places by deposits of coal, iron ore, potash (used in the production of soaps, glass, and certain compounds), and other minerals that were important in the region's industrial development.

Geographic characteristics of the North European Plain have had important impacts on human settlement. Many of the largest European cities, including London, Paris, and Berlin, developed on the plain. From northeastern France eastward, a band of especially dense population extends along the southern edge of the plain. It coincides with fertile soils formed from deposits of **loess** (a windblown soil of generally high fertility), providing an important natural transportation route skirting the highlands to the south, and a large share of the region's mineral deposits.

South of this northern plain, Europe is mainly mountainous or hilly, with scattered plains, valleys, and plateaus. The hills are geologically old, erosion having reduced their heights over long periods. But many mountains in southern Europe are geologically young and often high and ruggedly spectacular, with jagged peaks and snowcapped summits. They reach their zenith of height and grandeur in the Alps with the summit of Mount Blanc (15,771 ft/4,807 m) on the French-Italian border (●Figure 4.5). Another formidable mountain chain, the Pyrenees, forms a wall up to 10,695 feet (3,404 m) high along the border of Spain and France. Lower but legendary chains of mountains, the Transylvanian Alps and the Carpathian Mountains, meet in Romania.

Glaciation—a geologic and climatologic process in which great ice sheets formed in the Arctic and Antarctic and advanced toward the equator—played a major role in shaping the landscapes of Europe. Continental ice sheets

● **Figure 4.5** Switzerland's Alps, seen from space. Note the glaciers, whose extent has been retreating in recent decades.

formed over the continent, beginning on the Scandinavian Peninsula and Scotland (•Figure 4.6), during the **Ice Age** of the Pleistocene (c. 2 million to 10,000 years ago). Evidence suggests that there were four major periods of widespread glacial coverage. Today, landscapes of **glacial scouring**—the erosive action of ice masses in motion—characterize most of Norway and Finland, much of Sweden, parts of the British Isles, and Iceland. These changes created many favorable sites for hydroelectric installations. **Glacial deposition**—the process of offloading rock and soil in glacial retreat or lateral movement—had a major effect on the present landscape. Glacial deposits were left behind on most of the North European Plain and are productively farmed today.

The North European Plain is bordered by glaciated lowlands and hill lands in Finland and eastern Sweden and by rugged, ice-scoured mountains and fjords in western Sweden and most of Norway. The British Isles also have extensive areas of glacially scoured hill country and low mountains, along with lowlands where glacial deposition occurred.

Diversity of Climate and Vegetation

Despite its relatively small size, Europe also has remarkable climatic and biotic diversity (•Figure 4.7; see also Figures 2.10 and 2.11). A marine west coast climate extends from the coast of Norway to northern Spain and inland to west central Europe. Comparable to the climate found in the U.S. Pacific Northwest, the main characteristics are mild winters, cool summers, and ample rainfall, with many drizzly, cloudy, and foggy days. Throughout the year, changes of weather follow each other in rapid succession as different air masses temporarily dominate or collide with each other along weather fronts. Most precipitation is frontal in origin or results from a combination of frontal and orographic (highland) influences (see Chapter 2, pages 23–24). In lowlands, winter snowfall is light, and the ground is seldom covered for more than a few days at a time. Summer days are longer, brighter, and more pleasant than the short, cloudy days of winter, but even in summer, there are many chilly and overcast days. The frost-free season of 175 to 250 days is long enough for most crops grown in the middle latitudes to mature, although most areas have summers that are too cool for heat-loving crops such as corn (maize) to ripen (•Figure 4.8).

Inland from the coast, in western and central Europe, the marine climate gradually changes. Winters become colder and summers are hotter; cloudiness and annual precipitation decrease. Influences of maritime air masses from the Atlantic diminish and are modified by continental air masses from inner Asia. Farther inland, conditions are different, and two new climate types are apparent: the humid continental short-summer (cold) climate in the north (principally in southern parts of Sweden and Finland, Estonia, Latvia, Lithuania, Poland, Slovakia, eastern Hungary, and northern Romania) and the humid continental long-summer (warm) climate in the warmer south (mainly in southern Romania, Serbia, and northern Bulgaria). The natural vegetation is mostly temperate mixed forest, and soils vary in quality. Among the best soils are those formed from alluvium and loess along the roughly 1,800-mile (3,000-km) valley of the Danube River.

Southernmost Europe has a distinctive climate: the Mediterranean climate, with a pattern of dry summers and wet winters. This pattern results from seasonal shifts in atmospheric belts. In winter, the belt of westerly winds shifts southward, bringing precipitation at a time of relatively low evaporation and helping replenish subsurface waters. In summer, the belt of subsiding high atmospheric pressure over the Sahara shifts northward, bringing desert conditions. Mediterranean summers are warm to hot, and little precipitation occurs during the summer months when temperatures are most advantageous for crop growth. Winters are mild, and frosts are rare. Drought-resistant trees originally covered Mediterranean lands, but little of that forest remains, having been depleted by centuries of human use. It has been replaced by the wild scrub that the French call *maquis* (called *chaparral* in the United States and Spain) or by cultivated fields, orchards, or vineyards (•Figure 4.9). Much of the land consists of rugged, rocky, and eroded slopes where thousands of years of deforestation, overgrazing, and excessive cultivation have taken their toll. Subtropical temperatures allow a great variety of crops, but irrigation is needed in the dry summers. This climate is also a draw for summer travelers, and tourism has immense economic importance in Mediterranean Europe.

• **Figure 4.6** The maximum extent of glaciation in Europe, about 18,000 years ago.

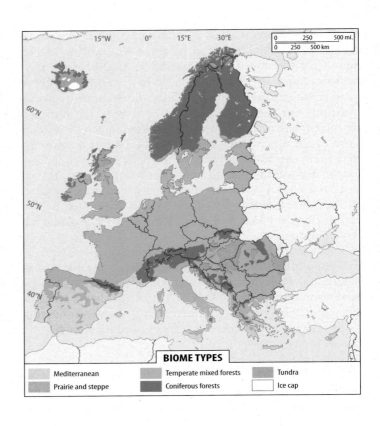

● **Figure 4.7** Climate types (left) and biomes (right) of Europe.

● **Figure 4.8** Land use in Europe.

● **Figure 4.9** Mediterranean landscape in Greece.

Some northerly sections of Europe experience the harsh conditions associated with subarctic and tundra climates. The subarctic climate, characterized by long, severe winters and short, rather cool summers, covers most of Finland, much of Sweden, and parts of Norway. A short frost-free season, coupled with thin, highly leached, acidic soils, makes agriculture difficult. Human settlement is sparse, and forest of needleleaf conifers such as spruce and fir covers most of the land. In the tundra climate of northernmost Norway and much of Iceland, cold winters combine with brief, cool summers and strong winds to create conditions hostile to tree growth. An open, windswept landscape results, covered with lichens, mosses, grass, low bushes, dwarf trees, and wildflowers (a typical Norwegian tundra landscape may be seen in Figure 2.12b on page 27). Human inhabitants are few, and agriculture is largely impracticable. In the taiga region, south of the tundra, a slight increase in moisture and temperature promotes coniferous forest.

The higher mountains of Europe, like high mountains in other parts of the world, have an undifferentiated highland climate varying with elevation and differential exposure to sun, wind, and precipitation. The variety can be startling. The Italian slope of the Alps, for example, ascends from subtropical conditions at the base of the mountains to tundra and ice cap climates at the highest elevations. The ice cap climate experiences temperatures that historically have averaged below freezing every month of the year. At the higher latitudes, these extreme conditions are found even at sea level. Europe's glaciers are following the almost worldwide trend of retreat in the past half century due mainly, climatologists believe, to increasing levels of greenhouse gases in the earth's atmosphere and the accompanying global warming. Probably because of their interior continental location, the Alpine glaciers are melting much faster than the world average.

34

Rivers and Waterways

Europe has many river systems important for transport, water supply, electricity generation, and recreation. Rivers were a critical part of Europe's transport system at least as far back as Roman times, and they still make it possible to move cargo at low cost. The most important rivers for transportation are those of the highly industrialized areas in the core area of west north central Europe.
98
An extensive system of canals connects and supplements the rivers. Some are quite old; the Romans built canals throughout northern Europe and Britain, mainly for military transport.

The Dutch developed the pound (pond) lock for canals in the late 14th century, leading to the expansion of regional connections of waterways through canals and to the use of canals to link farmlands with the coasts. Starting in the late 1700s, the British built canals to move raw materials toward factory locations. By the mid-19th century, the expanding influence of railroads and the steam engine competed with canals. Although canals continue to play a significant transport role, since the 1950s, railroads and highways have been the main means to transport goods throughout Europe.

Important seaports developed along the lower courses of many rivers, and some became major cities. London on the Thames, Antwerp on the Scheldt, Rotterdam in the

● **Figure 4.10** The Burg Katz Castle overlooks the busy Rhine River at St. Goar, Germany.

delta of the Rhine, and Hamburg on the Elbe are outstanding examples. Many river mouths are wide and deep, allowing ocean ships to travel a considerable distance upstream and inland. This is true even of short rivers such as the Thames and Scheldt.

The Rhine and the Danube are particularly important European rivers, touching or crossing the territory of many countries (see Figure 4.3). The Rhine countries include Switzerland, Liechtenstein, Austria, France, Germany, and the Netherlands. Highly scenic for much of its course, the Rhine is Europe's most important inland waterway (●Figure 4.10). Along and near it, an axis of intense urban industrial development and high population density developed. At its North Sea end, the Rhine connects to world commerce at one of the world's most active seaports, Rotterdam, in the Netherlands. The Danube, on its journey from the Black Forest of southwestern Germany to the Romanian shore of the Black Sea, touches or crosses more countries than any other river in the world. Within Europe, the Danube is unusual in its southeasterly course. The river is an important artery for the flow of goods and is also held in deep esteem by the cultures that share its waters. The Danube delta region of eastern Romania is one of the world's most important wetlands for waterfowl, and it draws growing numbers of ecotourists.

4.3 Cultural and Historical Geographies

The apparent crowding of so many countries into the relatively small land area of Europe is an indication of the region's extraordinary cultural diversity. Travelers in Europe experience this richness. A brief train ride, for example, takes the visitor from French-speaking western Switzerland eastward into the country's mainly Germanic region and southward through mountain tunnels into the Italian world. Today, almost all of Europe's peoples coexist harmoniously, but this harmony was a long time in coming.

Linguistic and Ethnic Groups of Europe

Europe emerged from prehistory as the homeland of many different peoples. In ancient and medieval times, some of them expanded vigorously, and their languages and cultures became widely diffused. The first millennium B.C.E. witnessed a great expansion of the **Greek** and **Celtic** (pronounced *kel*'-tik) peoples. In peninsulas and islands bordering the Aegean and Ionian Seas, the early Greeks developed a civilization that reached previously unsurpassed heights of philosophical inquiry and literary and artistic expression. Greek adventurers, traders, and colonists used the Mediterranean Sea as their highway to spread classical Greek civilization and its language along much of the Mediterranean shoreline. Evidence of the geographic range and influence of the Greek language and culture is apparent in the many Greek elements in modern European languages. But over time, the use of Greek in most areas disappeared as new peoples and languages were introduced and expanded. Several language families are represented in Europe today, and language is one of they key components of European ethnicity (•Figure 4.11).

Europe's Celtic languages expanded at roughly the same time as Greek and, like Greek, are represented today only by remnants. Preliterate Celtic-speaking tribes radiated from a culture hearth in what is now southern Germany and Austria, eventually occupying much of continental Europe and the British Isles. Conquest and cultural influence by later arrivals, mainly Romans and Germans, eventually eliminated the Celtic languages except for a few traces that survive today as Gaelic (or Goeidelic) and Brythonic tongues in isolated pockets in the British Isles (notably in Wales, Scotland, and Ireland) and in French Brittany.

In present-day Europe, the overwhelming majority of the people speak Romance, Germanic, and Slavic languages. Like Greek and Celtic, these are **Indo-European languages.** The **Romance languages** evolved from **Latin,** originally the language of ancient Rome and a small district around it. In the few centuries before and just after the beginning of the Common Era, the Romans subdued territories extending through western Europe as far west as Great Britain, and the use of Latin spread to this large empire. It is often said that the Mediterranean Sea became the "Roman Lake." Latin had the greatest impact in the less developed and less populous western parts of the empire. Over a long period, well beyond the collapse of this part of the empire in the 5th century, regional dialects of Latin survived and evolved into **Italian, French, Spanish, Catalan** (still spoken in the northeastern corner of Spain), **Portuguese, Romanian,** and other Romance languages of today.

It is important to point out that language frontiers do not always coincide with political frontiers. For example, the French language extends to western Switzerland and also to southern Belgium, where it is known as **Walloon.** French was Europe's dominant language as recently as the 18th century, when France was the continent's intellectual capital. French is still widely spoken outside the handful of countries in which it is the leading language and is second only to English as a leading tongue in European affairs.

In the middle centuries of the first millennium C.E., the power of Rome declined, and a prolonged expansion by Germanic and Slavic peoples began. Germanic (Teutonic) peoples first appeared in history as groups inhabiting the coasts of Germany and much of Scandinavia. They later expanded southward into Celtic lands east of the Rhine. They repelled Roman attempts at conquest, and the Latin language had little impact in Germany. In the 5th and 6th centuries, Germanic incursions overran the western Roman Empire, but in many areas, the conquerors were eventually absorbed into the culture and language of their Latinized subjects.

However, the **German** language expanded into, and remains, the language of present-day Germany, Austria, Luxembourg, Liechtenstein, the greater part of Switzerland, the previously Latinized part of Germany west of the Rhine, and parts of easternmost France (Alsace and part of Lorraine). The **Germanic languages** of Europe include many tongues other than German itself. In the Netherlands, **Dutch** developed as a language closely related to dialects of northern Germany, and **Flemish**—virtually identical to Dutch—became the language of northern Belgium. **Danish, Norwegian, Swedish,** and **Icelandic** descended from the same ancient Germanic tongue.

English is a Germanic language, but it has many words and expressions derived from French, Latin, Greek, and other tongues. Originally, English was the language of the Germanic tribes known as Angles and Saxons who invaded England in the 5th and 6th centuries C.E. The Norman Conquest of England in the 11th century established French for a time as the language of the English court and the upper classes. Modern English retains the Anglo-Saxon grammatical structure but borrows much vocabulary from French and other languages. English is now the principal language in most parts of the British Isles, having been imposed by conquest or spread by cultural diffusion to areas outside England.

Although the European Union has 23 official languages, English is becoming the de facto lingua franca of this 27-nation organization, with about two-thirds of all EU documents drafted in English. By EU law, and at great expense, all those documents must be translated into all those other languages, which most recently have come to include Ireland's Gaelic. Although some of the key institutions of the EU are in French-speaking cities, French usage across the bloc is declining. One reason is the recent accession of

● **Figure 4.11** The languages of Europe.

EU member countries in eastern Europe, where about 60 percent of the population speaks English as a second language, compared with only 20 percent speaking French as a second language. English is by far the most widely taught second language in non–English-speaking European countries, where it is studied by over 90 percent of secondary school students (compared with 33 percent studying French and 13 percent German).

Slavic languages—also in the Indo-European language group—are dominant in most of eastern Europe. **Russian** and **Ukrainian** are the primary languages of Europe's adjoining region of Russia and other former republics of the Soviet Union along Europe's edge; **Polish, Czech,** and **Slovak** are spoken in central Europe; and **Serbian, Croatian,** and **Bulgarian** prevail in the Balkan Peninsula. They apparently originated in eastern Europe and Russia, spreading and differentiating from each other during the Middle Ages as people traded and migrated. Closely related to the Slavic languages are the **Baltic languages** of **Latvian** and **Lithuanian.**

A few languages in present-day Europe are not related to any of the groups just discussed. Some are ancient languages that have persisted from prehistoric times in isolated,

usually mountainous, locales. Two outstanding examples are **Albanian** (of the Indo-European language family) in the Balkan Peninsula and **Basque** (an indigenous European language without Indo-European roots), spoken in and near the western Pyrenees Mountains of Spain and France. Some languages unrelated to others in Europe have relatives in Russia and reached their present locales through migrations of peoples westward. The prime examples are **Finnish, Saami, Estonian,** and **Hungarian** (also called Magyar). These are all **Uralic languages,** thought to have originated in the Ural Mountain region of Russia and Kazakhstan. **Turkish,** an **Altaic language** that originated in central Asia, is spoken by the ethnic Turkish inhabitants of northern Cyprus. The **Roma (Gypsies)** speak **Romany,** in the **Indic** branch of the Indo-European languages.

There are seven ethnic patterns in Europe, with language serving as important components of ethnicity.[2]

1. Several nations are populated overwhelmingly by a single ethnic group—for example, the Germans of Germany, the **Poles** of Poland, and the Dutch of the Netherlands.

2. There are also substantial populations of ethnic groups that originated in other nations living in "foreign" coun-

140

tries (for example, the Poles of Lithuania), where they are known as **national minorities.**

3. There are groups distinguished more by culture than by ethnicity; they share a common culture but have many different ethnic origins (for example, the French and the Italians).

4. There are a few ethnolinguistic minorities, like the **Saami (Lapps)** of Scandinavia and the **Ladin** of Italy, who have resided in an area for a very long time but have retained cultural distinctiveness.

5. There are also regional ethnic minorities that have a major presence in a particular nation, where they are culturally very different from the country's other ethnic groups; examples are the Basques and Catalans in Spain and the **Bretons** in France. Such groups often call for the devolution of power from the host country to their ethnic subregion (see Insights, page 83).

6. There are recent immigrants from outside Europe, including guest workers mainly from former European colonies in Asia, Africa, and the Americas; examples are the Pakistanis and Indians of Britain and the North Africans of France.

7. Finally, there are ethnic groups with members all across Europe but without a homeland there; the **Jews,** who originated in modern Israel and adjacent areas (see Chapter 6, page 219) and the Roma (Gypsies), who probably originated in India, are the most prominent of these. The characteristics, contributions, and conflicts of Europe's ethnic groups are discussed in more detail in Modules 4.1 and 4.2.

Europeans' Religious Roots

One of the principal culture traits of Europe is the dominance of **Christianity** (see the map of Europe's faiths in ●Figure 4.12 and the discussion of Christianity in Chapter 6, pages 222–223). Initially outlawed by the Romans, Christianity was embraced as the faith of the realm by the Emperor Constantine in the 4th century, and it became a dominant institution in the late stages of the empire. It survived the empire's fall and continued to spread, first within Europe and then to other parts of the world. In total number of adherents, the **Roman Catholic Church,** headed by the pope in Vatican City in Rome, is Europe's largest religious group (with 280 million followers, or 45 percent of Europe's Christians), as it has been since the Christian church was first established. Today, it remains the principal faith of a highly secularized Europe. The main areas that are predominantly Roman Catholic are Italy, Spain, Portugal, France, Belgium, the Republic of Ireland, large parts of the Netherlands, Germany, Switzerland, Austria, Poland, Hungary, the Czech Republic, Slovakia, Croatia, and Slovenia.

During the Middle Ages, a center of Christianity evolved in Constantinople (now Istanbul, Turkey) as a rival to

● **Figure 4.12** The religions of Europe. Sects of Christianity dominate, but Islam has a strong presence in the Balkans.

Rome. From that seat of power, the **Eastern Orthodox Church** spread to Greece, Bulgaria, Romania, Serbia, Montenegro, and Macedonia, as well as in Ukraine and Russia to the east.

In the 16th century, the **Protestant Reformation** took root in various parts of Europe, and a subsequent series of religious wars, persecutions, and counterpersecutions left **Protestantism** dominant in Great Britain, northern Germany, the Netherlands, Denmark, Norway, Sweden, Iceland, and Finland. Except for the Netherlands, where a higher Catholic birth rate has since reversed the balance, these areas are still mainly Protestant, with the **Church of England** and **Calvinism** the strongest sects in Britain and **Lutheran Protestantism** dominant in Germany and Scandinavia. Among the three Baltic countries, Lutheran Protestantism is the dominant faith in Estonia and Latvia, while most Lithuanians are Catholic.

That Christians form a vast majority is a somewhat misleading characterization of religion in Europe. Mainstream Christianity is in decline as Europeans have become increasingly secularized in recent decades. Historic legacy plays a strong role in the trend: over a period of centuries, Europeans paid a costly toll in lives by fighting for causes associated with religion, and they increasingly associate peace with secularism. Recent polls indicate that almost 75 percent of Europeans believe in God, but in most countries, even larger majorities shun traditions like attending church. In predominantly Catholic France, about one person in 20 attends a religious service every week (compared

162

INSIGHTS

Devolution

A notable political process in modern Europe, often related to issues of linguistic and ethnic identity, is **devolution**, the dispersal of political powers to ethnic minorities and other subnational groups. Like that of many other world regions, Europe's historical geography is rife with dominant groups occupying the traditional lands of smaller minority groups. Minorities are sometimes absorbed into the larger culture and political system, but more often than not, they retain some degree of distinctiveness and a desire to control their own affairs.

This text relates numerous examples from around the world of strife and war taking place because the dominant power refuses to yield any measure of authority to minority ethnolinguistic groups.

But in Europe, the process of devolution is occurring peacefully for the most part and more energetically than ever. This may be because the European countries, after centuries of strife, are finally self-assured of their sovereignty and well-being and are so focused on international cooperation that demands for autonomy among their minorities are no longer perceived as threats. However, the European Union, already preoccupied with meeting the diverse needs of 27 member states, sees devolution as a vexing problem. Devolution is an important issue in Wales, Scotland, Northern Ireland, Spain, and elsewhere; ●Figure 4.B is a map of the major locales where devolution has occurred or is being appealed for in modern Europe.

● **Figure 4.B** Devolutionary areas of Europe. Power is being devolved from central governments to regional authorities in many parts of Europe. This map depicts places like Scotland and Wales, where the transfer has taken place, and others like Corsica, where devolution is anticipated or sought.

with one in three in the United States). Various polls suggest that somewhere between 15 and 33 percent of Italians attend church in mainly Catholic Italy. Polls indicate that overall, about 20 percent of all Europeans regard religion as "very important" to them (compared to about 60 percent of Americans). The Christian churches that are growing—even surging—in attendance tend to be independent **Pentecostal churches** attended by immigrants from Nigeria, Sierra Leone, and other African countries.

The faith of **Islam** (practiced by **Muslims**) has the fastest-growing number of adherents of any world religion and is the fastest-growing religion in Europe (for a description of Islam, see Chapter 6, pages 223–226). Muslims are a majority in just two European countries: Albania and Bosnia and Herzegovina on the Balkan Peninsula. Islam was once widespread in the Balkans (as that region is often called), where the Ottoman Turks established it during their period of rule from the 14th to 19th centuries. It was also the religion of the **Moors,** who ruled the Iberian Peninsula from the 8th century until the end of the 15th, when Spain's Catholic monarchy ousted them. Today, Islam is growing throughout the continent, particularly among immigrant communities in western Europe (Table 4.2). Estimates vary, but France has as many as 6 million Muslims, followed by Germany with about 3 million, Britain approaching 2 million, and Spain and the Netherlands with almost 1 million each. Muslim immigrants from the Balkans and from North Africa and the Middle East continue to arrive, and resident Muslims in Europe have higher birth rates than their native counterparts, so the overall numbers of Muslims are poised to rise substantially.

One of humankind's greatest tragedies has been the story of the Jews of Europe (see also Chapter 6, page 222). The Romans forcibly dispersed these followers of **Judaism** to Europe from Palestine in the 1st century C.E., and over time, they became more numerous in Europe than anywhere else. In 1880, before many emigrated to the United States, 90 percent of the world's Jews lived in Europe. In 1939, as World War II erupted, there were 9.7 million Jews, 60 percent of the world total, in Europe. By the end of the war, systematic murder by the Nazis in the **Holocaust** reduced their numbers to 3.9 million. About half of those subsequently emigrated to Israel and elsewhere, and there are about 1 million Jews in Europe today.

European Colonialism and Its Consequences

Far more than any other region, Europe has shaped the human geography of the modern world. Before the late 15th century, Europe played only a minor role in world trade patterns: goods moved from southern France and northern Italy northwest to Britain, for example, and more active trade connected Italy with the margins of the Mediterranean Sea and beyond to southwestern Asia. But the most important global trade routes were much farther east and south. The longest link was the 5,000-mile (8,000-km) **Silk Road** that moved goods overland (and on the Mediterranean Sea) from Chang-an (Xi'an) in China to Venice.

The balance of world affairs started shifting to Europe with the beginning of the Age of Discovery in the 15th century. European sailors, missionaries, traders, soldiers, and colonists burst onto the world scene. Although these agents had different motives, they shared a conviction that European ways were superior to all others, and by various means, they sought to bring as much of the world as possible under their sway. Economic benefits for the European homelands were among their main motivations. By the end of the 19th century, Europeans had created a world in which they were economically dominant and exercised great influence on indigenous cultures.

The process of exploration and discovery by which Europeans filled in the world map began with Portuguese expeditions down the west coast of Africa. In 1488, a Portuguese expedition headed by Bartholomeu Dias rounded the Cape of Good Hope at the southern tip of Africa and opened the way for European voyages eastward into the Indian Ocean. In 1492, North America was brought into contact with Europe when a Spanish expedition commanded by a Genoese (Italian) man named Christopher Columbus (Cristóbal Colón) crossed the Atlantic to the Caribbean Sea. Less than half a century later, these early feats of exploration culminated in the first circumnavigation of the globe by a Spanish expedition commanded initially by a Portuguese man, Ferdinand Magellan. Worldwide exploration continued apace for centuries, with the Dutch, French, and English eventually wresting leadership from the Spanish and Portuguese.

The explorers were the vanguards of a global European invasion that would in turn bring the missionaries, soldiers, traders, settlers, and administrators. They gradually built the frameworks of European colonization of the Americas, Africa, Asia, and the Pacific. The colonial patterns are

58,
73

202,
371

42

TABLE 4.2	European Union Countries with the Largest Muslim Populations	
Country	**Muslim Population**	**% of Population**
France	5–6 million	8–9.6%
Germany	3 million	3.6
United Kingdom	1.6 million	2.8
Spain	1 million	2.3
Netherlands	945,000	5.8
Italy	825,000	1.4
Belgium	400,000	4.0
Austria	339,000	4.1

Source: BBC News, "Muslims in Europe: Country Guide," http://news.bbc.co.uk/2/hi/europe/4385768.stm

described on pages 42–46 in Chapter 3 and will not be repeated in detail here. Briefly, their main manifestations were the transfer of wealth from the colonies to Europe, the movement of great numbers of indigenous laborers to serve European interests in other lands, the settlement of Europeans in agriculturally productive colonies (where their descendants are great majorities or significant minorities today), and the attempted subjugation of indigenous cultures to European ones. These are among the most prominent issues in world geography; every chapter of this text deals with the impacts of European colonization on cultures and landscapes beyond Europe.

Of great importance in reshaping the world's biogeography was the transfer of plants and animals from one place to another following Europe's conquest of the Americas, the so-called **Columbian Exchange.** Major examples include the introduction of hogs and cattle and the reintroduction of horses to the Americas from Europe; of tobacco and corn (maize) from the Americas to Europe and other parts of the world; of rubber from South America to Asia; of the potato from the Americas to Europe, where it became a major food; and of coffee from Africa to Latin America, where it became the principal basis of a number of national economies. More examples are listed in Table 4.3, which shows some of the most significant foodstuffs diffused from the New World to the Old World after 1492, as well as the major New World foodstuffs introduced into world trade after 1492 by the ships and crews that came to the newly found lands of the Americas.

4.4 Economic Geography

Europe had significant material and cultural riches, and the colonial system built on that endowment to make Europe the world's wealthiest region for several centuries. The European capitalist system and huge strides in science and technology helped launch the region into economic supremacy. Parts of Europe had already developed capitalist institutions by the end of the Middle Ages, and the colonial era saw further rapid development of economic enterprise in the region. Profit became an acceptable motive, and the relative freedom of action afforded by the capitalistic system provided opportunities for taking risks and gaining wealth. Energetic entrepreneurs mobilized capital into large companies and used both European and overseas labor.

Europeans reached a level of technology generally superior to that of non-Europeans with whom they came in contact during the early Age of Discovery. Achievements in shipbuilding, navigation, and the manufacture and handling of weapons gave them decided advantages. Nations such as China, once technologically superior, fell quickly behind. Europe gained almost free rein in control and expansion of technological innovation for several centuries. The foundations of modern science were also constructed almost entirely in Europe, and the great names in science

TABLE 4.3	Diffusion of Foodstuffs between the New World and the Old: The Columbian Exchange
Diffused from Old World to New World	**Diffused from New World to Old World**
Wheat	Maize
Grapes	Potatoes
Olives	Cassava (manioc, yuca)
Onions	Tomatoes
Melons	Beans (lima, string, navy, kidney, etc.)
Lettuce	Pumpkins
Rice	Squash
Soybeans	Sweet potatoes
Coffee	Peanuts
Bananas	Cacao
Yams	

between about 1500 and 1900 were overwhelmingly European.

These material, cultural, and intellectual forces combined to launch the Industrial Revolution in Europe, which was the first world region to evolve from an agricultural into an industrial society. In the first half of the 18th century, primarily in Britain, industrial innovations made water power and then coal-fueled steam power increasingly available to turn machines and drive gears in new factories. James Watt's invention of the steam engine in 1769 made coal a major resource and greatly increased the amount of power available. New processes and equipment, including the development of **coke** (coal with most of its volatile constituents burned off), made iron smelting cheaper and more abundant. The invention of new industrial machinery made of iron and driven by steam engines multiplied the output of manufactured products, initially textiles. Then, in the 19th century, British inventors developed processes that allowed steel, a metal superior to iron in strength and versatility, to be made inexpensively and on a large scale for the first time. These developments soon diffused to new hearths of massive, mechanized industrial production and brought unprecedented changes to landscapes and ecosystems around the world (for more on the revolutionary nature of the Industrial Revolution, see Chapter 3, page 42). Among the most immediate and most significant processes accompanying industrialization were rapid population growth and a dramatic shift of people from rural areas into cities.

By 1900, the people and machines of western European cities created about 90 percent of the world's manufacturing output. Europe led the world in the production of textiles, steel, ships, chemicals, and other industries. But in the 20th century (especially after 1960), Europe's preeminence in world trade and industry diminished to about 25 percent of the world's manufacturing output, for several reasons.

One was warfare (•Figure 4.13). Europe suffered enormous casualties and damage in World Wars I and II, which were initiated and fought mainly on that continent. Recovery was eventually achieved, in large part with assistance from the United States, but the region did not regain its former supremacy.

During the 20th century, rising **nationalism**—the quest by colonies and ethnic groups to possess their own homelands—brought the European colonial empires to an end. Taking advantage of a weakened Europe and mounting disapproval of colonialism in the world at large, one European colony after another gained independence quickly in the decades following the end of World War II. Perhaps ironically, opposition to continued European control was often spearheaded by colonial leaders who had been educated in Europe and had acquired nationalistic ideas there.

Europe's predominance was also seriously eroded by the rising economic and political stature of the United States and the Soviet Union. Each far larger than any European nation, these countries outpaced Europe in military power, economic resources, and world influence, particularly in the postwar years 1950–1970.

A major shift took place in global manufacturing patterns as well. Europe once enjoyed a near monopoly in exports of manufactured goods, but manufacturing accelerated in many countries outside Europe in the 20th century. The United States reached industrial maturity by 1950 and became a vigorous competitor of Europe. More recently, the Asian countries of Japan, South Korea, China, and Taiwan have became industrial forces capable of competing with Europe in markets all over the world.

Finally, Europe's leadership was weakened by a new dependence on outside sources of energy. The region's traditional reliance on coal was superseded by technological developments and environmental concerns, and the energy balance shifted to petroleum and natural gas. Despite the development of North Sea oil and gas resources (which benefit primarily the United Kingdom and Norway), Europe became, and remains, highly dependent on imported Middle Eastern oil.

295–299

An Imbalance of Wealth

A striking pattern in Europe's economic geography is that western Europe is wealthier than eastern Europe. This trend dates to at least the 1870s, when per capita incomes in the west were twice those in the east. The gap grew even wider throughout the 20th century. After World War II, eastern European countries were in effect colonized by the Soviet Union, serving as vassal states that gave up human and material resources to service the motherland. Only now, thanks to the dissolution of the Soviet Union in 1991 and the admission of eastern European countries to the European Union, can the eastern countries hope to catch up with their western counterparts. They stand to

139

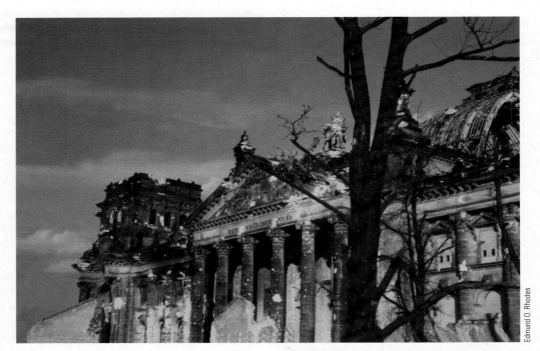

Edmund O. Rhodes

• **Figure 4.13** In the 20th century, two devastating world wars originated in and centered on Europe, causing enormous suffering and eclipsing the region's global economic dominance. These are the ruins of the Reichstag in Berlin in 1945.

benefit enormously from joining the EU, as membership all but guarantees income redistribution from west to east and promises through agricultural subsidies and other instruments to breathe new life into the faltering farms and factories of the east.

Postindustrialization

Europe's economy today, like its population trend, is best characterized as **postindustrial.** In the European context, this means that since about 1960, the region has experienced **deindustrialization,** shifting by choice and by force away from energy-hungry, labor-costly, and polluting industries toward an economy based on production of high-tech goods and on services. The high-technology products include electronic devices, data processing hardware, **robotics,** and telecommunications equipment. Other industries like drug and pesticide manufacturing make use of these technological products to create their own, so there is an important interconnectedness centered on high technology. Europe is also renowned for the production and export of high-quality, expensive luxury goods, including automobiles such as the Ferrari and watches such as the Rolex. In the service sector, growing proportions of Europeans are employed in transportation, energy production, health care, banking, retailing, wholesaling, advertising, legal services, consulting, information processing, research and development, education, and tourism.

Neither high-tech nor service industries employ as many people as the old manufacturing sector did, and like the United States, the European countries are grappling with problems resulting from industrial unemployment. There are concerns over where new jobs will come from and how to make the transition from manufacturing to service roles without too much social dislocation. Many European nations fit the model of the **welfare state,** using the resources collected through high taxation rates to provide generous social services to citizens. That "social model" of the European state is increasingly being subjected to scrutiny and reform as the forces of globalization, stagnant economies, unemployment, and immigration place increasing burdens on national treasuries.

Land and Sea

Despite the dominance of a postindustrial, urban economy, some Europeans continue to live off the land and the sea. Agriculture was the original foundation of Europe's economy, and it is still very important. Food provided by the region's agriculture allowed Europe to become a relatively well-populated area at an early time. After about 1500, a period of steady agricultural improvement began. Introduction of important new crops such as the potato played a part, but so did such practical improvements as new systems of crop rotation and scientific advances that produced better knowledge of the chemistry of fertilizers. The

continuing expansion of industrial cities provided growing markets for European farmers, who received protection through **tariffs** (tax penalties imposed on imports) or direct **subsidies** (payments made to domestic producers) to encourage production and support rural incomes (these are also discussed on page 623 in Chapter 11, in the context of similar protective measures by the United States). As a result of European Union policies, surpluses of many products are common, including grains, butter, cheese, olive oil, and table wines. However, Europe is not agriculturally self-sufficient and depends on a vital trade of foodstuffs (see Insights, page 89).

Europe, like the United States, is finding itself increasingly at odds with the rest of the world as global free-trade agreements insist that its doors be open to more (and generally cheaper) food imports. Europe wants to protect its farmers, particularly those who raise what it calls the "sensitive" products (beef, chicken, and sugar), with high tariffs on imported foods and with subsidies that allow farmers to "dump" their surplus production onto the world market at very low prices. The EU says it is doing enough to open its doors to imports because it imposes no tariffs on imports from the world's 50 poorest countries, but free-trade advocates insist that more trade barriers must fall.

Throughout history, fishing has been an important part of the European food economy, and the coasts still have many busy fishing ports (●Figure 4.14). At times, control of fishing grounds has been a major commercial and political objective of nations, even resulting in warfare. Fisheries are particularly important in shallow seas that are rich in plankton, the principal food of herring, cod, and other fish of commercial value. The Dogger Bank (to the east of Great Britain) in the North Sea and the waters off Norway and Iceland are major historical fishing areas, and Norway, Iceland, and Denmark are Europe's leading nations in total catch. Fishing has long been especially crucial to countries such as Norway and Iceland, which have little agriculturally productive land. However, overfishing of cod, the world's most popular food fish, sometimes known as the "beef of the sea," grew so severe, particularly in the North Sea, that in 2006 the EU legislated cuts of cod fishing by 14 to 20 percent. Environmentalists cried foul; the World Wildlife Fund had called for a complete ban on cod fishing.

4.5 Geopolitical Issues

In the past 100 years, Europe's geopolitical situation changed more profoundly and more violently than that of any other world region. Europe experienced two world wars that wrought unprecedented devastation. World War I (1914–1918), until 1939 known as the Great War, saw Europe fracture on fault lines of military alliances. Germany and the Ottoman Empire (based in what is now Turkey) were solidly defeated by an alliance led by Britain, France, Russia (early on), and the United States (later). The war

Joe Hobbs

● **Figure 4.14** Fishing boats in the harbor of Skagen, Denmark.

accomplished little and destroyed much, and the peace treaty placed a costly burden of reparations on the defeated Germany.

The economic and social humiliation of Germany helped plant the seeds for the rise of Adolf Hitler's fascist Nazi (National Socialist) Party and the outbreak of World War II (1939–1945). Hitler's Germany surged in strength and allied with Italy, the Soviet Union (early), and Japan (later) in a military quest to establish European and Far Eastern empires. Early in the war, Germany advanced swiftly on the European mainland and Scandinavia and set its sights on an occupation of Britain. When Japan bombed a U.S. naval base at Pearl Harbor, in the Hawaiian Islands, in 1941, the United States was drawn into the war firmly on Britain's side. Soon joined by the Soviet Union, these forces, assisted by many smaller ones, began to turn the tide against Germany. In 1945, with German cities in ruin and two Japanese cities nearly obliterated by the first atomic bombs, the war ended. In the European theater, about 50 million people had died. "What is Europe now?" the British prime minister, Winston Churchill, asked. "It is a rubble heap, a charnel house, a breeding ground of pestilence and hate."[3]

Postwar Europe

World War II veterans are now aging and dying, and Europe is establishing, in the form of the European Union, what might be called the "United States of Europe," a federation of nations similar to the United States of America. The European Union is the latest and largest of the various postwar European **supranational organizations,** in which member countries are united beyond the authority of any single national government and are planned and controlled by a group of nations. There are many reasons, largely economic ones, for this push toward greater unity. One of the most important but understated reasons is simply that Europeans want their continent to be warproof, with its member countries so intertwined in economics, foreign policy, and other ways that war among them would be all but impossible.

The Cold War and Its Aftermath

Europe is recovering and reorganizing from a second momentous geopolitical shift, the end of the **Cold War,** which came with the collapse of the Soviet Union in 1991. When World War II ended, Soviet forces and political institutions moved in on most of the countries of eastern Europe, including a portion of defeated Germany. As Churchill described it, an **Iron Curtain** had descended from Poland southward to the Balkan Peninsula. West of it, the capitalist democracies led by Britain, France, West Germany, Italy, and the United States prevailed; to the east, across a 4,000-mile (6,400-km) line of concertina wire, walls, minefields, and guard towers, the Soviet Union controlled authoritarian, state-run regimes among its European "satellites," including East Germany, Poland, Czechoslovakia, Hungary, Romania, and Bulgaria. The United States nurtured western Europe's interests and from 1947 to 1952 pumped billions of dollars into the reconstruction of the devastated European infrastructure in a program known as the **Marshall Plan.**

The United States was also the linchpin of the military alliance known as the **North Atlantic Treaty Organiza-**

INSITES

Read

Genetically Modified Foods and "Food Fights"

The United States is Europe's leading trading partner, but the exchange is not always open or friendly, especially when it comes to food. Countries often pay lip service to the concept of free trade while actually limiting trade in some products to protect their own industries and interests. Nations on both sides of the Atlantic use these trade barriers, sometimes precipitating **"food fights"** and other kinds of **trade wars.** In the late 1990s, for example, the United States wanted to sell bananas that U.S. corporations grew in Latin America to the countries of the European Union. The Europeans, who wanted to protect their investments in banana farming in former European colonies in Africa, the Caribbean, and the Pacific, refused to buy, and the so-called **banana war** ensued.

A current trade dispute involves **genetically modified (GM) foods,** also called **genetically modified organism (GMO) foods,** produced in the United States. These are products such as rice, soybeans, corn, and tomatoes that have been genetically manipulated through biotechnology to be more productive and resilient (•Figure 4.C). European farmers and consumers, with their long-established food cultures, generally disdain GM foods, citing uncertain long-term health and environmental risks (for example, they fear that a GM plant cross-pollinating neighboring plants might have unpredictable ecological consequences). If such foods were to be imported, the Europeans say, they should be labeled as GM foods (which many Europeans derisively refer to as **"Franken-foods"**). The United States replies that labeling would make them almost impossible to sell to GM-conscious Europeans and argues that the ban on GM foods defies the free-trade principles of the World Trade Organization (WTO).

Global trade discussions, particularly in the WTO, have recently forced the European Union to make concessions on GM foods. In 2006, the EU said its member countries should open their doors to GM foods and insisted that conventional and GM crops can "coexist" in the same areas without fear of cross-pollination as long as a safe distance is maintained between them. European farmers and consumers remain skeptical, however, and some member countries (notably Greece) have defied EU leadership and imposed outright bans on GM foods and crops.

Globally, the Europeans are likely to find themselves increasingly isolated on the issue of GM foods and less able to compete in the global agricultural marketplace as China and other large food producers and consumers embrace GM products. Some other

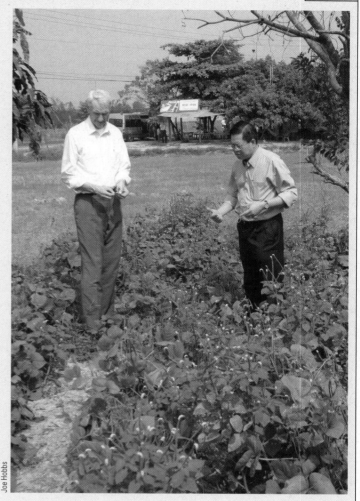

• **Figure 4.C** Biotechnological processes can manipulate the genetic makeup of mung beans (seen here in Vietnam with University of Missouri plant scientists Jerry Nelson, left, and Henry Nguyen) and other food crops to make them more productive, increasing yields—and also, say many Europeans, endangering local plant varieties and evolutionary processes.

countries do share the European view, however: see, for example, how the African nation of Zambia viewed the GM issue in Chapter 9, page 476.

tion (NATO), founded in 1949 between the United States, Canada, most of the European countries west of the Iron Curtain, and Turkey. NATO faced off against the **Warsaw Pact,** an alliance of the Soviet Union and its eastern European satellites. On more than one occasion, notably in the Cuban Missile Crisis of 1962 and in the Arab-Israeli War of 1973, these military alliances came all too close to apocalyptic nuclear war.

In 1989, the Berlin Wall, which separated the German city that epitomized the Cold War, was dismantled in a popular and peaceful uprising. The Soviet grip on East Germany and its other satellites had been slipping for years, and now it crumbled. The Soviet Union itself disbanded in 1991 (see Chapter 5, page 169). An exhilarated public throughout eastern Europe embraced the prospects of democracy, capitalism, and material improvement. The Warsaw Pact was dissolved. The nuclear arsenals of the respective alliances were reduced substantially. Plans were made to turn the path of the Iron Curtain into the **European Greenbelt,** a mosaic of national parks and other protected areas stretching across 18 countries, from the Barents Sea to the Black Sea.

After the end of the Cold War, NATO's attention focused mainly on strife in the Balkans, where it used force to reverse the advance of the Serbs against their neighbors in the former Yugoslavia (see page 147). NATO remains, and in 2004 its 19 members grew to 26 with the addition of seven former Soviet allies (Estonia, Latvia, Lithuania, Slovakia, Slovenia, Romania, and Bulgaria; ●Figure 4.15). Russia very grudgingly accepted this Western expansion into its former sphere of influence, in part because of its own growing alignment with Western defense and security concerns. NATO's future role is uncertain. The Europeans are, however, united in their certainty that they do not want to spend too much money on defense. With the Cold War over, they are also content with far less reliance on the United States for military and other forms of support.

The European Union

The European Union, which maintains its headquarters in the Belgian capital of Brussels, is the most important of Europe's supranational organizations (●Figure 4.16). In 2007,

the organization commemorated 50 years of existence in various forms. The EU began in 1957, when the Treaty of Rome created the **European Economic Community (EEC),** also known as the **Common Market,** made up of France, West Germany, Italy, Belgium, Luxembourg, and the Netherlands. By 1996, three years after it acquired the name European Union, the organization had an additional nine members: the United Kingdom, Denmark, Ireland, Greece, Portugal, Spain, Austria, Finland, and Sweden.

The EEC was initially designed, as the EU is today, to secure the benefits of large-scale production by pooling the resources—natural, human, and financial—and markets of its members. Tariffs were eliminated on goods moving from one member state to another, and restrictions on the movement of labor and capital between member states were eased (in theory, a citizen of any EU country may work in any other EU country, but the process is not simple; a work permit must be obtained from the host country). Monopolies that formerly restricted competition were discouraged.

Meanwhile, a common set of external tariffs was established for the entire area to regulate imports from the outside world, and a common system of price supports for agriculture replaced the individual systems of member states. The founders expected that free trade within such a populous and highly developed bloc of countries would stimulate investment in mass-production enterprises, which could, wherever they were located, sell freely into all member countries. This trade would encourage productive geographic specialization, with each part of the union expanding lines of production for which it was best suited. Each

● **Figure 4.15** European members of the North Atlantic Treaty Organization (NATO).

● **Figure 4.16** Members of the European Union.

country might thereby achieve greater production, larger exports, lower costs to consumers, higher wages, and a higher standard of living than it could achieve on its own.

The European Union, under provisions of the **Maastricht Treaty of European Union** that went into force in 1993, is now pursuing even greater new steps toward unification. These include the removal of nontariff trade barriers, such as "quality standards," that still exist between member countries. Major energy focused on the implementation of a single EU currency, the **euro,** in 1999, as the centerpiece of the **European Economic and Monetary Union (EMU).**

This common currency was promoted on the basis of economic theories suggesting that sharing a currency across borders had many advantages, such as lower transaction costs, more certainty for investors, enhanced competition, and more consistent pricing. A common currency could theoretically also restrain public spending, reduce debt, and tame inflation. In practical terms, consumers spend money in euro-using countries just as Americans do at home; they can compare the price of a car, for example, in Germany and Italy, and decide to buy the cheaper car, just as an American might buy a car in Utah instead of Colorado. Besides sharing a currency, all countries adopting the euro have agreed to allow the strong European Central Bank to make critical decisions on issues such as interest rates, which would be identical for all euro-using nations.

Citizens of various EU countries view the euro—and indeed the European Union itself—differently. The euro is not all that popular; recent polls showed that only 48 percent of its users felt the currency was "advantageous overall." There is a strong aversion to the euro among the Swedes, Danes, and British, and they rejected the currency. Sweden feared a loss of financial and monetary independence. Danes expressed fears that the euro would threaten their generous welfare benefits or even their national unity. Britain's economy has recently been stronger than the economies of Germany and France, suggesting to the British that they might as well shun the euro for now. The new eastern European members of the European Union will have to make wrenching economic reforms to meet the requirements of adopting the euro. Slovenia was the first to do so, in 2007, raising the number of official euro-using nations to 13; these nations comprise the so-called **Eurozone.**

Size and Diversity of the European Union

Ten eastern European nations joined the European Union all at once in what was known as the **"big bang"** of 2004. The newcomers were Poland, the Czech Republic, Hungary, Slovakia, Estonia, Latvia, Lithuania, Slovenia, Malta, and officially all of Cyprus, although de facto EU rule applies only to the Greek portion of the divided island nation. Turkey's application to join was rejected. The big bang created a mega-Europe of 450 million people, stretching from the Atlantic to Russia, with an economy valued at almost

136, 270

$10 trillion, nearly as strong as the U.S. economy. The EU grew enormously, with the addition of 75 million people (about 20 percent, almost half of them in Poland) and a geographic expansion by almost one-third of its prior size. A more modest expansion took place in 2007 when Romania and Bulgaria joined, becoming the EU's poorest member nations. This growth is accompanied by some clear advantages for the organization and its new members but also by some problems. Some thorny issues had to be resolved before the dozen new countries could join, and even thornier problems lie ahead.

Consistent with the historical patterns of west and east, the most outstanding difference between the old and new members is in their economies. The old EU member nations have some of the highest per capita gross domestic product purchasing power parity (GNI PPP) levels in the world—including the very highest, Luxembourg, with $55,970. The new states, by contrast, are much less affluent, and some even approach LDC status (see Figure 3.5 and Table 4.1). Romania has a per capita GNI PPP of just $9,820. The old EU countries have 95 percent of the continent's wealth, and the new ones only 5 percent. When the big bang countries joined in 2004, the EU's average wealth per person fell by 13 percent.

Naturally, questions have arisen about the EU's ability to afford its new members. There is particular concern with the costs of the EU's agricultural policies. The European Union provides generous agriculture subsidies to its member states. Prior to the big bang, $50 billion, fully half of the EU's budget, was spent every year on these subsidies. For example, the EU pays an Austrian farmer $28,000 per year in subsidies for 30 hectares' (c. 75 acres') production of seed potatoes and milk. That farmer, like many in the European Union, would probably not be able to farm profitably without EU subsidies. The new EU members are upset that they began membership receiving only one-fourth of the agricultural subsidies provided to the older members (with an annual 5 percent increase in subsequent years). They fear that their farmers will go out of business, unable to compete with the lower prices from the subsidized, heavily mechanized farms to the west.

For its part, the older member states of the European Union argue that the new countries do not have the high production costs and quality controls of the old countries. Mainly, though, the older European Union members fear that providing full subsidies to the new countries would bankrupt the organization. Germany is especially fearful. It is the wealthiest member of the European Union and carries the largest cost burden. Germany had its own economic pains from absorbing the costs of reunion and integration with the poor former East Germany and is concerned about a far more expensive integration on a much grander scale. Despite hopes among many in the newly joining states that membership will bring them prosperity, there will also be at least short-term pain. Food prices have risen sharply, while salaries have increased slowly. Many

114

western Europeans fear that the lower salaries and taxes of the new eastern member countries will lure industries and jobs away from the west, contributing to the decline of their home economies.

The new countries will for a while be denied another benefit generally enjoyed by the old: freedom of movement across the borders of member countries. That is supposed to come in 2011. For now, western European fears of a tidal wave of cheap eastern labor, terrorists, or human traffickers have produced special rules for the eastern countries, long regarded as abodes for these shady activities. In the meantime, laborers will continue to move illegally—for example, shuttling from Poland to Belgium, where they will work at the low-paying jobs that no one else wants.

In theory, the European Union would like to move toward a situation in which there were no passport, visa, or other control issues at any internal land, sea, and airport frontiers. To date, only a framework outside the European Union has attempted this kind of integration. Known as the **Schengen Agreement,** it allows free circulation of people between the 15 nations that signed the agreement; these are the old EU countries (except for Britain and Ireland), plus non-EU Iceland and Norway: Austria, Belgium, Denmark, Finland, France, Germany, Greece, Iceland, Italy, Luxembourg, the Netherlands, Norway, Portugal, Spain, and Sweden. As of 2007, Switzerland's membership had been approved but not yet enacted. To maintain internal security, within this greater **Schengenland,** the member states are supposed to exercise common visa, asylum, and other policies at their external borders (for example, between Germany and Poland). Some countries, notably the United Kingdom, are skeptical of the effectiveness of control of external borders and have stayed out of Schengen. Truly open borders are probably still far in the future.

The Future

Beyond its unifying economic principles, many questions remain about what the European Union is or should be. Remarkably, most citizens in the EU are not sure what to make of the organization and are not especially excited about its prospects. Many member states seem to be retreating to nationalist interests, and the EU has been experiencing what the Center for European Reform described in 2006 as an "unprecedented malaise."[4]

With the EU's expansion came a struggle to craft a new constitution by which all members would agree on matters such as defense and foreign affairs. The debate was analogous to the effort in the United States in the 1780s to form a confederation of states agreeing to give a national government more power. There were differences about how power should be wielded: some advocated a federal system patterned after that of the United States, but others wanted individual national governments to make most decisions. A basic conflict had to do with distribution of power between big and small states. There were also popular fears that

the constitution might open doors wider to foreign immigrants or make some already bad economies even worse. In national referenda in 2005, such concerns led French and Dutch voters to decisively reject the constitution. Because all 27 EU countries must ratify the constitution for it to take effect, the document instantly became a "dead letter," awaiting revival at some future point when its ramifications are more widely accepted. Predictions were that an acceptable constitution would be ready for ratification in 2008 or 2009.

There were other struggles, too. There was a hard-argued debate on whether or not the preamble to the EU constitution should mention God. The notions was abandoned. There was acrimony over whether France and Germany should be allowed to run budget deficits that exceed statutory limits governing the euro, leading to suspicions that one set of rules existed for big countries and another for small ones. There were concerns with how to articulate a defense policy that did not seem to conflict with that of NATO.

Differences between Europeans and Americans

The rift that developed between the United States and its European peers over the Iraq War raised a lot of questions on both sides of the Atlantic on how much we do or do not have in common. Many of us instinctively equate the United States with Europe when we speak about global patterns; after all, we share many common cultural roots, share a great deal of the world's wealth, and have many of the same measures of quality of life. But there are differences, and these seem to have grown in recent years. "Americans are from Mars and Europeans are from Venus," wrote policy analyst Robert Kagan.[5]

Here are a few exemplary differences. The concept of social justice, in which steps are taken to reduce unemployment and poverty among the least privileged, is a key value in Europe, while it lags in the United States. European governments and people have historically had a stronger sense of the social compact between the state and its citizens. Governments have been expected to provide "free" public services like education and health care, while the people have understood they must foot the bill for these "free" services by paying very high taxes. (Incidentally, undergraduate public education in most European nations is free. However, per capita spending on public education is higher in the United States, and an American college education is widely regarded as superior to that obtainable in Europe.) Europeans accept high taxes on gasoline as an incentive to drive small cars and conserve energy, while for an American politician, advocating higher fuel taxes is suicide; as a result, gas prices in Europe are roughly double those in the United States. Europeans complain that U.S. "cultural industries" such as Hollywood films humiliate other cultures and marginalize languages other than English. Europeans are less inclined than Americans to allow questions of spirituality into political debate; in fact, they are repelled by personal expressions of religiosity in public, which they consider indiscreet. Europeans abhor the death penalty, which is outlawed in all EU countries.

Europeans tend to take apply the **precautionary principle**—the notion that it is appropriate to attempt to reduce or eliminate any risky practice—to things they think may be harmful. Thus there may be risks from GM foods, so many people would like to ban them; global warming is probably occurring, so most people want to take action to reduce greenhouse gases. The United States has adopted more of a "show me" attitude—for example, with respect to global warming; there is not enough scientific proof, so it is better not to impose reductions in carbon dioxide emissions (see page 35).

There are differences on the geopolitical front. Although a stated goal of the European Union is to develop a common voice in foreign affairs, the U.S.-led invasion of Iraq in 2003 underscored how far away that goal was. The governments of Britain, Italy, Spain, and most of the eastern European EU members-to-be expressed support for the United States, while powerful Germany and France raised strong dissenting voices. The United States has leaned toward confrontation with Iran over that country's nuclear program, while the European countries have favored economic and diplomatic approaches to the issue. Europeans want an arms embargo against China lifted, while the United States wants it to remain in place.

For Americans, September 11, 2001, was the watershed time when it reevaluated its place in the world and began to feel less secure. For Europeans, the fall of the Berlin Wall in 1989, leading to the collapse of the Soviet bloc, was the watershed that made them start to feel more secure. It is a sad reflection of relations that following the U.S.-led invasion of Iraq, many Europeans began to feel insecure about the United States. Recent years have seen overtures on both sides to overcome this rift. But a 2007 poll in the five most influential European countries—France, Germany, Italy, Spain, and the United Kingdom—found that most people regarded the United States as the greatest threat to global security—more dangerous than Iran, Iraq, or North Korea.

Modules 4.1 and 4.2 will explore the subregions and individual countries of physically and culturally diverse Europe.

Test will be on main points of chapters and what was discussed in class

SUMMARY

→ Europe is physically part of the great continent of Eurasia, but is generally labeled a continent and is treated as a separate region. Europe's population is about twice that of the United States, and Europe is more densely settled. Europe is demographically postindustrial, with a slowly declining and aging population.

→ Among Europe's most distinctive physical geographic traits are its northerly location, temperate climate, and varied topography.

→ The North European Plain is a major belt of settlement and agricultural productivity. The marine west coast climate, continental climates, and Mediterranean climate are characteristic. The Rhine and the Danube are the two most important rivers.

→ European languages derive primarily from Indo-European roots and include Romance, Germanic, and Slavic languages. English is a Germanic language, and like most European languages, it is enriched by many other tongues. The dominant religion is Christianity, with major followings of Protestant, Roman Catholic, and Eastern Orthodox churches. There are significant populations of Muslims (most of them recent immigrants) and of Jews (whose population is a small remnant of the pre-Holocaust community).

→ Immigration is enriching Europe's ethnic mosaic but also presenting economic and security dilemmas for the European countries.

→ From the beginning of the 16th century until late in the 19th century, Europe was at the center of global patterns of colonization and foreign settlement, long-distance trade, and aricultural and industrial innovation. During this time, Europeans diffused crops and animals between the Old and New Worlds.

→ The Industrial Revolution originated in Europe, with energy derived from coal and factory technology focused on textiles and iron. Industrialization and colonization launched Europe to global economic and political supremacy.

→ Recent decades have seen a global shift in power away from Europe. War dislocation, rising nationalism, the ascendancy of the United States, a shift in world manufacturing patterns, and new energy sources have combined to diminish Europe's global centrality. Europe is nevertheless a very strong force in world economic, political, and social affairs, and its peoples are among the most prosperous in the world.

→ Europe's economy is postindustrial, making the transition from energy-hungry, labor-costly, and polluting industries to leaner high-tech industries and to services. This shift has caused unemployment.

→ European nations often try to protect their domestic industries and have been involved in trade wars with the United States, which likewise wants to protect its industries. Europe is under pressure from international trade agreements to reduce its trade barriers and subsidies that protect its domestic agriculture and other industries.

→ Eastern Europe has long been much poorer than western Europe.

→ In recent decades, Europe has been reorganizing itself to ensure that nothing like the two world wars will happen again, and to strengthen its economies. Its principal military alliance, NATO, is growing in membership and redefining its focus toward peacekeeping.

→ The most important development is the growth of the 27-member European Union (EU), a supranational organization pooling the economic and human resources of its member countries. There have been obstacles to its achievement of common EU policies in money matters, defense, and foreign affairs.

→ The U.S. war in Iraq highlighted major differences among EU members, and overall, there are several marked differences in the ways Europeans and Americans view the world.

KEY TERMS + CONCEPTS

Albanian (p. 81)
Altaic languages (p. 81)
   Turkish (p. 81)
"baby bounty" (p. 72)
banana war (p. 89)
Benelux (p. 68)
Basque (p. 81)
"big bang" (p. 91)
"birth dearth" (p. 71)
Bretons (p. 82)
Christianity (p. 82)
"circular migration" (p. 74)
coke (p. 85)
Cold War (p. 88)
Columbian Exchange (p. 85)
Common Market (p. 90)
culture hearth (p. 67)
deindustrialization (p. 87)
delta (p. 79)
devolution (p. 83)
Eastern Orthodox Church (p. 82)
Euro (p. 91)
European core (p. 68)
European Economic and Monetary Union
   (EMU) (p. 91)
European Economic Community (EEC)
   (p. 90)
European Greenbelt (p. 90)
European Union (EU) (p. 72)
Eurozone (p. 91)
"food fights" (p. 89)
"Frankenfoods" (p. 89)
genetically modified (GM) foods (p. 89)
genetically modified organism (GMO)
   foods (p. 89)
glacial deposition (p. 77)
glacial scouring (p. 77)
glaciation (p. 76)
Gulf Stream (p. 75)

Gypsies (p. 81)
Holocaust (p. 84)
Ice Age (p. 77)
Indo-European languages (p. 80)
   Baltic languages (p. 81)
      Latvian (p. 81)
      Lithuanian (p. 81)
   Celtic (p. 80)
   Germanic languages (p. 80)
      Danish (p. 80)
      Dutch (p. 80)
      English (p. 80)
      Flemish (p. 80)
      German (p. 80)
      Icelandic (p. 80)
      Norwegian (p. 80)
      Swedish (p. 80)
   Greek (p. 80)
   Indic languages (p. 81)
      Romany (p. 81)
   Romance languages (p. 80)
      Catalan (p. 80)
      French (p. 80)
      Italian (p. 80)
      Latin (p. 80)
      Portuguese (p. 80)
      Romanian (p. 80)
      Spanish (p. 80)
      Walloon (p. 80)
   Slavic languages (p. 81)
      Bulgarian (p. 81)
      Croatian (p. 81)
      Czech (p. 81)
      Polish (p. 81)
      Russian (p. 81)
      Serbian (p. 81)
      Slovak (p. 81)
      Ukrainian (p. 81)
Iron Curtain (p. 88)

Islam (p. 84)
Jews (p. 82)
Judaism (p. 84)
Ladin (p. 82)
Lapps (p. 82)
loess (p. 76)
Maastricht Treaty of European Union
   (p. 91)
Marshall Plan (p. 88)
microstates (p. 68)
Moors (p. 84)
Muslims (p. 84)
nationalism (p. 86)
national minorities (p. 82)
North Atlantic Drift (p. 75)
North Atlantic Treaty Organization
   (NATO) (p. 88)
North European Plain (p. 76)
Poles (p. 81)
postindustrial (p. 87)
precautionary principle (p. 93)
Protestant Reformation (p. 82)
Protestantism (p. 82)
   Calvinism (p. 82)
   Church of England (p. 82)
   Lutheran Protestantism (p. 82)
   Pentecostal churches (p. 84)
remittances (p. 74)
robotics (p. 87)
Roma (Gypsies) (p. 81)
Roman Catholic Church (p. 82)
Saami (Lapps) (p. 82)
Schengen Agreement (p. 92)
Schengenland (p. 92)
Silk Road (p. 84)
site (p. 71)
situation (p. 71)
subsidies (p. 87)
supranational organization (p. 88)

tariffs (p. 87)
trade wars (p. 89)
undocumented workers (p. 74)
Uralic languages (p. 81)

Estonian (p. 81)
Finnish (p. 81)
Hungarian (p. 81)
Saami (p. 81)

Warsaw Pact (p. 89)
welfare state (p. 87)
westerly winds (p. 76)

REVIEW QUESTIONS

1. Why is Europe usually treated as a separate region?

2. What terms and trends best describe Europe's population today?

3. Why is immigration so critical to Europe's demographic future, and what problems does immigration pose?

4. What is unusual about Europe's coastline?

5. What are some of Europe's other major physical and environmental characteristics? Why are they consistently described as diverse?

6. What roles have rivers played in Europe's development, and which rivers are the most important ones? Where is industrial and other economic activity concentrated in Europe and why?

7. What are the dominant languages, religions, and other ethnic traits of Europe?

8. What factors led to Europe's global dominance in economic and political affairs? What impacts did that dominance have on other peoples and environments?

9. Explain the origins and significance of the Industrial Revolution, including the role of resources in that revolution. How does Europe's present urban pattern reflect its industrial past?

10. What are Europe's main economic traits? How have these changed in recent decades?

11. What happened to Europe in the 20th century? How are Europe's political and economic institutions of today trying to create a different Europe?

12. What are the main goals and principles of the European Union? What successes and difficulties has the organization had?

13. What are some of the differences between the Europeans and Americans?

NOTES

1. Elisabeth Rosenthal, "European Union's Plunging Birthrates Spread Eastward." *New York Times,* September 4, 2006, p. A3.

2. This characterization is from David Levinson, *Ethnic Groups Worldwide* (Westport, Conn.: Oryx Press, 1998), p. 1.

3. Quoted in A. A. Byatt, "What Is a European?" *New York Times Magazine,* October 13, 2002, p. 58.

4. Richard Bernstein, "In Europe, a Return to Nationalism. *International Herald Tribune,* July 26, 2006.

5. Robert Kagan, *Of Paradise and Power: America and Europe in the New World Order* (New York: Knopf, 2003), p. 1.

Leo Tregenza, a native of Cornwall, in the far southwest of Great Britain.

Joe Hobbs

MODULE 4.1

THE EUROPEAN CORE

Europe has a recognizable **core,** a subregion that has long played a dominant role in the continent's political, economic, and cultural development. Today, the European core is recognizable as the subregion of Europe with the following traits:

* Densest, most urbanized population

* Most prosperous economy

* Lowest unemployment

* Most productive agriculture

* Most conservative politics

* Greatest concentration of highways and railroads

* Highest levels of crowding, congestion, and pollution.[1]

Defined in this way, the core consists primarily of three of Europe's major countries and of smaller nations in the British Isles and the west central portions of the European mainland (•Figure 4.1.1). The United Kingdom (U.K.) and the Republic of Ireland share the British Isles. The countries of west central Europe are Germany and France

■ module
objectives

This module should enable you to

→ Identify the concurrence of economic and other forces that have made this a core region

→ Recognize the spatial relations between heavy industrial resources and cities in Europe and the factors that have caused the deindustrialization of Europe

→ Relate ways in which this portion of Europe represents one of the most heavily modified landscapes in the world—for example, through deforestation and land reclamation

→ Contrast the primacy of Paris with the decentralized urban pattern of Germany and other European countries

→ Become acquainted with the roots and issues of some of Europe's persistent ethnic and political struggles, including the Troubles of Northern Ireland and heightened tensions affecting Muslims

→ Appreciate the significance of a reunified Germany, especially the economic costs of reintegration for the more prosperous west, a process that is significant globally because of its relevance to the two Koreas and other pairs of nations with prospects for reunification

● **Figure 4.1.1** The political geography of Europe's core.

● **Figure 4.1.2** Human activities over long periods of time have transformed the natural landscapes of Europe, especially in the core region. This map depicts areas that once sustained temperate mixed forests and coniferous forests but have lost that tree cover.

and their smaller neighbors the Netherlands, Belgium, Luxembourg, Switzerland, and Austria. Also exhibiting the core traits are roughly the northern half of Italy, Denmark, southern Sweden, and parts of Poland, but other characteristics of these countries place them more appropriately in the European periphery, discussed in Module 4.2. Because they are nestled among the core countries described in this module, the microstates of Liechtenstein, Monaco, and Andorra are included here.

The European core includes some of the world's most geopolitically and economically significant countries (see Table 4.1, pages 70–71). This subregion also has one of the world's most transformed landscapes, shaped by millennia of productive agriculture and hundreds of years of industrial development and affluence (●Figure 4.1.2). The European core is one of just four regions in the world classified as a **major cluster of continuous settlement,** meaning that all habitations lie no more than 3 miles (5 km) from other habitations in at least six different directions. In addition, roads or railroads lie no farther than 10 to 20 miles (16 to 32 km) away in at least three directions.[2]

Europe's cultural landscape is unusually intricate. Different layers of history appear at every turn, and there are sharp differences in human geographies and national perspectives of its peoples. Environments and resources also vary widely. Local resources (especially coal and iron ore) and transportation opportunities were the basis for industrial

• **Figure 4.1.3** Historical industrial concentrations, cities, and seaports of the European core. Older industries such as coal mining, heavy metallurgy, heavy chemicals, and textiles clustered in these congested districts. Local **coal** deposits provided fuel for the Industrial Revolution in most of these areas, which have shifted increasingly to newer forms of industry as coal has lost its economic significance and older industries have declined.

and urban development in past centuries (•Figure 4.1.3), but the modern European core has a distinctly postindustrial geography (•Figure 4.1.4).

Declining and growing industries tend to be occurring in geographically distinct areas. The new, mainly high-tech industries occupy smaller areas than the older industries.

4.1.1 The British Isles

The islands of Great Britain and Ireland, off the northwestern coast of Europe, are known as the British Isles (•Figure 4.1.5). They are home to the Republic of Ireland, with its capital at Dublin, and the much larger United Kingdom of Great Britain and Northern Ireland (U.K., often referred to simply as "Britain"), with its capital at London. The United Kingdom is made up of the entire island of Great Britain, with its political units of England, Scotland, and Wales, plus the northeastern corner of Ireland, known as Northern Ireland, and most of the smaller outlying islands.

The six counties that make up Northern Ireland separated from Ireland in 1922 and maintained their link with the United Kingdom, while the remaining counties achieved full independence as the Republic of Ireland. Altogether, the U.K. encompasses about four-fifths of the area and over 90 percent of the population of the British Isles.

The United Kingdom: Once an Empire

Lying only 21 miles (34 km) from France at the closest point, England is the largest of the United Kingdom's four main subdivisions. England was originally an independent political unit and twice conquered mainly Catholic Ireland, including Northern Ireland.

Scotland was first joined to England when a Scottish king inherited the English throne in 1603, and the two became one kingdom under the Act of Union in 1707. Nearly 300 years later, in 1997, the U.K. government yielded to increasing Scottish demands and allowed for the devolution (dispersal of political power and autonomy) of many administrative powers (including agriculture, education, health, housing, taxation, and lawmaking, but excluding defense and social security) to a new Scottish parliament. In a groundbreaking 2007 election, Scottish nationalists won political dominance in the parliament and promised a referendum on whether Scotland should secede from the U.K. and declare independence. Scots are extremely proud of their distinctive culture, history, and identity and bristle when they are identified as "English," as happens all too often. 83

England conquered Wales in the Middle Ages, but like the Scots, the Welsh managed to preserve their cultural distinctiveness and hold it high—especially in staying true to their Welsh language (a member of the Celtic subfamily). Any thought that the Welsh are English may be set aside by considering this Welsh place name (known as the longest place name in the world): Llanfairpwllgwyngyllgogerychwyrndrobwllllantysiliogogogoch, meaning "The Church of Saint Mary in the hollow of white hazel trees near the rapid whirlpool by Saint Tysilio's of the red cave." A 1997 referendum voted into existence a Welsh assembly dedicated to promoting Welsh interests. While this body has some powers independent of the United Kingdom, growing numbers of Welsh are calling for a stronger devolution of power that would transfer more authority from the national government to a full-fledged parliament, as Scotland has now.

Between the defeat of Napoleonic France in 1815 and the outbreak of World War I in 1914, the United Kingdom was the world's strongest country. Its overseas empire covered a quarter of the earth at its maximum extent; "the sun never sets on the British Empire" was a proud boast. The influence of English law, education, and culture spread still farther. The British take great pride in much of that legacy; the Magna Carta, an English charter written in 1215, granted certain human rights that centuries later would be enshrined in the constitutions of other democracies, including that of the United States. On the other hand, Britain was one of the main forces behind the cruel industry of the enslavement of Africans. Until the late 19th century, the 470 United Kingdom was the world's leading manufacturing and trading nation. Britain's Royal Navy dominated the seas, and the its merchant marine moved half or more of

● **Figure 4.1.4** Deindustrialization and new manufacturing in the European core, 1960–2000.

the world's ocean trade. London became the center of a free-trade and financial system that invested its profits from industry and commerce around the world.

The U.K.'s economic and political decline began late in the 19th century. World War I damaged its free-trade and financial system, and the downward trend accelerated after World War II. All the large colonies of the former British Empire gained independence by the 1960s, and Britain's remaining overseas possessions are scattered and small.

The U.K. is, however, still a very consequential country, in large part because of its lasting impacts from colonial times. It plays a major role in the European Union and is associated with many of its former colonies in the worldwide **Commonwealth of Nations**, a voluntary association of 53 countries that nominally recognizes the British monarch as its head. Its strong allegiance to the United States, even in times of great adversity such as during the war in Iraq, is another of Britain's characteristic traits on the international stage.

Much world culture has British roots; English is by far the leading language of the World Wide Web and serves as a lingua franca in many countries with diverse cultures (including the European Union; see page 80). In the 1960s,

the Beatles, four lads from Liverpool (recently crowned a "European Capital of Culture), revolutionized music in their home country, moved on to conquer a generation in the United States, and left an indelible mark on music everywhere. British pop culture and more traditional fare—Scottish reels, fish and chips, Buckingham Palace Beefeaters—draw fans and visitors from around the world. For many foreigners, the images and artifacts of Britain may become distilled into something homogeneous, but a Briton of Anglo origin is quick to speak about his or her unique piece of the country and what makes it distinct. And then there is multicultural Britain, another legacy of its colonial past, with a bit of Jamaica, India, Kenya, and scores of other lands infused in every corner of the country.

Highland and Lowland Great Britain

The island of Great Britain is about 600 miles (c. 1,000 km) long by 50 to 300 miles (80 to nearly 500 km) wide. For its small size, this island has a wide variety of landscapes and livelihoods. Broadly, the island has two contrasting physical areas: Highland Britain and Lowland Britain.

● **Figure 4.1.5** Principal features of the British Isles. The "Tees-Exe" line connecting the mouths of rivers having those names is the general dividing line between Highland Britain to the northwest and Lowland Britain to the southeast.

The high regions of Highland Britain are the Scottish Highlands, the Southern Uplands of Scotland, the Pennine Chain and the Lake District of northern England, the mountains that occupy most of Wales, and the uplands of Cornwall and Devon in southwestern England (see Figure 4.1.5). These mountains are low by most standards—Ben Nevis in the Scottish Highlands is the tallest at 4,406 feet (1,343 m)—but many appear high because they rise so precipitously from or near sea level. The most important low-lying area in Highland Britain is the Scottish Lowlands, a densely populated industrialized valley separating the rugged Scottish Highlands from the gentler and more fertile Southern Uplands of Scotland. The Scottish Lowlands ac-count for only about one-fifth of Scotland's area but more than four-fifths of its population and its two main cities: Glasgow (population 1.6 million) to the west and Edinburgh, Scotland's political capital (population 700,000), to the east (●Figure 4.1.6).

The highland landscapes have long been popular tourist destinations because of their stark grandeur and quaint settlements, and recent immigration by wealthier English has reversed a long trend of depopulation. The Scots who live there have a history of proud and often defiant distinction from the more powerful English to the south. That independent streak is the stuff of romantic and tragic tales, recounted in history books, novels, and films.

Joe Hobbs

● **Figure 4.1.6** A Scottish flag waves above downtown Edinburgh.

Almost all of Lowland Britain (the English Lowland) lies in England. It has better soils than the highlands and a gentler topography. It is a mosaic of well-kept pastures, meadows, and crop fields fenced by hedgerows (see Insights, page 102) and punctuated by closely spaced villages, market towns, and industrial cities. The largest industrial and urban districts, aside from London, lie in the midlands and in the north, around the margins of the Pennine Chain. Here, clusters of manufacturing cities such as Birmingham, Manchester, Leeds, and Sheffield arose near coalfields, which were a major locational factor in the region's development.

Urbanization and Industrialization in Britain

42 The Industrial Revolution gave a powerful boost to the growth of cities at the same time that Britain's total population was increasing rapidly. In 1851, the United Kingdom became the world's first predominantly urban nation. Except for London, most of the modern large cities developed (and remain) on or near the coalfields that supplied the power for early industrial growth (see Geographic Spotlight, page XX; as on the Continent, geographical situation played a critical role in urban development). 103, 113

Five industries were of major significance in Britain's urban-industrial rise: coal mining, iron and steel manufacture, cotton textiles, woolen textiles, and shipbuilding. Britain's early start in these enterprises made it industrially preeminent through most of the 19th century. Exported surpluses paid for imports of food and raw materials and made Britain a wealthy hub of world trade. Britain has recently had difficulties in these same industries due to growing international competition, changing patterns of demand, and shifting urban populations.

Industrial innovations began in earnest with 17th-century coal mining and associated steam technology. Coal became a major industrial resource between 1700 and 1800, when it fueled blast furnaces and steam engines. With Britain's development of the steam engine, coal replaced human muscle and running water as the principal source of industrial energy. Coal production rose in England and Wales until World War I. Then, between 1914 and the 1990s, output fell by about two-thirds, the number of jobs in coal mining was cut by more than four-fifths, and practically all of the coal export trade disappeared. Major factors contributing to this decline included increasing competition from foreign coal in world markets, depletion of Britain's own coal, the environmental drawbacks of coal (the dirtiest of the fossil fuels), and increasing substitution of other energy sources for coal both overseas and in Britain itself (see Geography of Energy, page 104). By the 1980s, once-dominant coal's principal remaining role in Britain had been reduced to supplying fuel for generating domestic electricity.

Beginning in the late 1960s, large-scale development of newly discovered North Sea oil and gas fields revolutionized Britain's energy situation (see Figure 4.1.4 and Figure 4.2.6, page 126). Those offshore fields continue to be productive, providing welcome revenue to the United Kingdom and especially to the historically marginal Shetland Isles of Scotland. These fields are well into "middle age"; their output is projected to decline by 2010.

Steel was a scarce and expensive metal before the Industrial Revolution. Iron, not too cheap or plentiful itself, was more commonly used. By 1855, Britain was producing about half the world's iron and exporting much of it.

British inventors and entrepreneurs led the world into the age of cheap, mass-produced, and widely used iron and steel, but because of growing resource shortages and rising labor costs, the country lost its leadership in global totals of steel production. The United States and Germany surpassed the British output before 1900. China is now the world's leading steel producer, by a huge margin.

During Britain's period of industrial and commercial ascendancy, cotton textiles were its leading export. The enormous cotton industry became concentrated in the district 45 of Lancashire (especially in Manchester and Liverpool) in

INSIGHTS

Hedgerows

One of the most distinct landscape features of England and Wales is the **hedgerow,** a long, trailing land-use boundary consisting of shrubs and trees or of cleared fieldstones overgrown with heather (●Figure 4.1.A). Farmers and ranchers have used them to mark perimeters, to contain sheep and cattle, and simply to dispose of rocks since very ancient times; the Romans found them when they arrived in Britain. Many hedgerows were created in the **Enclosure Movement** of the 18th and 19th centuries, when large landowners appropriated local "commons," or unfenced land tilled by villagers, and enclosed them with hedges for sheep breeding. This closure of lands that had historically been worked by peasant farmers helped promote the migration of peasant youth to the newly industrializing English cities.

As recently as 1985, there were an estimated 400,000 miles (640,000 km) of hedgerows in England and Wales. But housing and commercial developments, suburban sprawl, and the rearrangement of land for larger farming operations have taken their toll, destroying more than 100,000 miles (160,000 km) of hedge-

rows. Although hedgerow defenders decry the loss of these historical boundaries and wildlife habitats, provincial and national governments have done little to stem the losses.

● **Figure 4.1.A** A hedgerow in Cornwall, southwestern England.

---

northwestern England. Raw cotton imported from Britain's colonies was the basis for this industry. Lancashire inventors first mechanized the spinning and weaving processes. In 1793, an American, Eli Whitney, invented the cotton gin. This machine efficiently separated seeds from raw cotton fibers, making it possible to produce inexpensive cloth. Industry built on this technology acted like a magnet to bring young people from farm to factory.

For more than a century, until World War I, Lancashire dominated world trade in cotton and cotton textiles. But 1913 saw the peak of British production; after that, the decline of the industry was nearly as spectacular as its earlier rise. The root of the trouble was increased foreign competition. Throughout the 20th century, numerous countries surpassed the United Kingdom in cotton textile production and exports, with a resulting slide in employment in Lancashire. Countries such as China and South Korea had lower labor costs and were increasingly able to put their own or imported textile machinery to use.

Its supremacy in cotton eclipsed, Lancashire's economy still languishes and has not yet attracted the high-tech industries that city fathers hope will resurrect it. Generally, England's north is poorer than its south, the reverse of the situation in Italy, as described in Module 4.2.

Britain is well endowed with grazing land, and it exported raw wool and woolen cloth to continental Europe as early as the Middle Ages. Exports of manufactured woolen textiles peaked before World War I. Britain is still a major world exporter, but employment and production in this sector have declined dramatically as China has ramped up production, even of luxury cashmere woolens. The wool manufacturing industry is centered on the east slope of the

Pennines in western Yorkshire (known as the West Riding area), especially in Leeds Bradford.

In the late 19th century, iron and steel ships propelled by steam transformed ocean transportation. By the end of the century, Britain led the way, building four-fifths of the world's seagoing tonnage. Two districts developed as the world's greatest shipbuilding centers: the Northeast England district, with shipyards concentrated along the lower Tyne River between Newcastle and the North Sea, and the western Scottish Lowlands, with shipyards lining the River Clyde for miles downstream from Glasgow. Like the other industries of early importance and current decline in the United Kingdom, shipbuilding lost major significance in the last decades of the 20th century. But shipbuilding rebounded sharply in the early 2000s, especially as a result of new defense contracts.

As a whole, then, resources and industries that put the major British cities (aside from London) on the map are played out or are in serious decline. This process of deindustrialization has hit Britain especially hard, but other regions in the European core, notably the Ruhr in Germany, have also suffered (see Figure 4.1.4). Relatively high unemployment is a major consequence of this trend, but a positive result is that the cities are cleaner and more livable. Many British cities, notably Birmingham—second in size only to London—are replacing old industrial areas with greenbelts and attractive architecture. Ideally, the improved environment will lure in high-tech and other clean industries, services, and tourists that will drive a new economy. The process is under way in selected areas, such as Scotland's Silicon Glen, London's M-4 Corridor, and southeastern England's Silicon Fen, depicted in Figure 4.1.4.

87

GEOGRAPHIC SPOTLIGHT

London Town

Britain's premier city, London (•Figures 4.1.B and 4.1.C), is the country's only significant metropolis that was not located near coalfields. Other important situation factors account for the birth and rise of this great city. It is located in the corner of Great Britain closest to major trading partners on the European continent. It is in a productive agricultural area; even in the preindustrial period, the fertile English Lowland supplied most of Britain's wool for exports and domestic consumption. London's location facilitated transport connections with the surrounding region. The Thames River and its tributaries provided major routes to the interior, where canals connected the Thames drainages with other rivers. Before the 19th-century railroad and the 20th-century truck, water transport had greater advantages over land transport than it has today, and these natural inland waterways were used extensively for transporting goods to and from London. Its river location allowed London to act as an inland seaport, serving more trading territory in the surrounding countryside than a coastal port would.

London's network of highway connections to the rest of the country help facilitate Britain's global trade today. However, getting around the city itself can be difficult. London's traffic jams prompted authorities to begin a program of "congestion charging" in 2003. Except for residents and commuters with special exemptions, motorists using the 8-square-mile (21-sq-km) area of central London must pay a daily fee of £5 (about $10). The system relies on camera surveillance and hefty fines. Another controversial novelty for downtown London is a boom in stylish skyscrapers that promises to transform the city's skyline dramatically by 2010.

Home to 12.6 million people, London is no longer one of the world's major port cities or manufacturing centers. The city's main industries are government, insurance, business administration and services, communications media, tourism, and finance. London is Europe's leading financial center, a status it also held at the height of Britain's colonial empire prior to World War I. London's ethnic geography is a vibrant, living legacy of that colonial past, reflecting the city's function as the empire's core. More than one-third of Londoners belong to an ethnic minority, and collectively, they speak an estimated 300 languages. Most arrived with British decolonization following World War II, drawn by the economic opportunities perceived in their former "mother country." These include more than 500,000 Asians (referring to people of Indian, Pakistani, and Bangladeshi origins), about 300,000 West Indians (mainly Jamaicans), and more than 160,000 West Africans. Tourists in London not only take in the traditional offerings of palaces, museums, and theatrical performances but feel themselves at a cultural crossroads of the world. Entire neighborhoods are ethnic Indian or Arab, for example, and many members of the minorities there wear their homeland's traditional clothing, making the visitor feel "abroad" in London.

• **Figure 4.1.B** The Thames River snakes its way through central London. The Tower Bridge is near the photo's center.

• **Figure 4.1.C** Downtown London and some of its world-famous attractions.

GEOGRAPHY OF ENERGY

Europe's Energy Alternatives

One way in which the European nations are quite distinct from the United States is in their embrace of alternative energy. The EU has taken an official stance that global climate change is a serious problem and must be confronted through dramatic changes in the ways energy is produced and consumed.

In its 2007 recommendations, the European Union Commission called for member countries to cut carbon dioxide emissions to 30 percent below 1990 levels by 2020. However, the EU fears that its members will bear disproportionate costs if other big energy users, notably the United States, do not follow suit; in that case, the EU target would be to cut to 20 percent below 1990 levels by 2020. To achieve its carbon cuts, EU members will promote fuel efficiency in automobiles, encourage the use of public transportation, and expand trading in carbon emissions.

Some EU members are especially aggressive with alternatives. Denmark uses wind power to meet more than 20 percent of its electricity needs—the highest rate in the world from this source. Germany has established itself as the world's leading manufacturer of wind turbines, and more than 12,000 of these mills dot Germany itself, providing about 7 percent of the country's electricity needs. That share is projected to rise to 20 percent by 2020. Seeking to establish itself as the EU leader in cutting carbon dioxide emissions, Great Britain is the most resolute. Going well beyond its pledge in the Kyoto Protocol, Britain vows to derive 10 percent of its electricity from renewable energy by 2010 and 20 percent of its total energy needs from renewable sources by 2020 (as of 2007, less than 5 percent came from renewables) and to cut carbon dioxide emis-

sions to 60 percent below their 1990 levels by 2050. A major component of this effort is a $12 billion project to construct more than 1,000 wind turbines off England's coast, mainly in the Thames estuary, the east coast area, and along the northwest coast. These will supply an estimated 7 percent of the country's energy needs but are not without controversy: even many ecology-minded Britons, recalling William Blake's image of the "dark satanic mills" of early English industry, lament that the "silver satanic blades" will sully the storied Lake District of the northwest.[3]

Nuclear power fell out of favor in Europe after the 1986 Chernobyl disaster in nearby Ukraine, and a decade later, 55 percent of Europeans polled opposed its use. However, the nuclear industry is making an apparent comeback as the region looks for ways to cut carbon dioxide emissions. Sweden, already reliant on renewable energy for about one-third of its power, had resolved to phase out its nuclear energy program but changed course on the issue in 2006, in part because of its ambitious goal of having no dependence on fossil fuels by 2020. Neighboring Finland is building new nuclear reactors, as are the newest EU members, Bulgaria and Romania. Germany had vowed to cease nuclear production completely by 2020, but nuclear proponents are fighting to reverse that decision. The International Thermonuclear Experimental Reactor (ITER), an enormously expensive international effort to produce electricity from nuclear fusion (a technically challenging but environmentally friendly process that would fuse the nuclei of hydrogen atoms rather than split atoms, as nuclear reactors now do) is under construction in Cadarache, France.

Food and Agriculture in Britain

Like other more developed countries, the United Kingdom has few people employed in farming and other rural pursuits. Only about 2 percent of the working population of Great Britain is in agriculture, forestry, and fishing. About one-quarter of the total land area is given over to crops, with nearly half used for pasture. Grass farming for milk and meat production is a major rural land use. Even with its broad **arable** base, more than 60 percent of Britain's food supply is imported. Britain is more self-reliant in fish and seafood, although its fishing industry has been in steady decline since the 1960s. EU policies have led to fishing restrictions and a decline in catch size, but even so, Britain has the third-largest fishing fleet in the EU. Major fishing ports include Hull, Grimsby, Fleetwood, and Plymouth in England and Peterhead in Scotland.

Total food production increased sharply after World War II, mainly through government policies aimed at reducing imports. Like most other EU countries, the U.K. has a policy of subsidized agriculture, with high import tariffs on farm products from non-EU countries. As long as this

policy holds, British farmers will continue to receive incentives in the form of relatively high and protected prices, and the British people, like the other peoples in the European Union, will eat expensive food produced in the home country or elsewhere in the European Union.

Ireland: The Emerald Isle

Ireland (known as Eire in the indigenous Celtic language) is a land of hills and lakes, marshes and peat bogs, cool dampness, and the verdant grassland that earned its epithet the "Emerald Isle"(•Figure 4.1.7). The island consists of a central plain surrounded on the north, south, and west by hills and low rounded mountains. The Republic of Ireland, which occupies a little over four-fifths of the island, is far into its transition from an agricultural to an industrial economy.

Ireland ranked low among western European countries for many decades, but recent economic growth has changed this pattern. High-tech industries such as electronic products and software in particular boosted Ireland's economy. The size of the economy doubled in the 1990s, earning Ire-

GEOGRAPHY OF ENERGY
Europe's Energy Alternatives *continued*

● Figure 4.1.D

● Figure 4.1.E

● Figure 4.1.F

● Figure 4.1.G

● **Figure 4.1.D To G** Europe has a wide variety of energy sources, with Poland almost completely reliant on coal to produce electricity and France nearly so on nuclear power. Traditional and modern sources of energy in Europe include peat from a Scottish bog (●Figure 4.1.D), coal in Poland (●Figure 4.1.E), and windmills, both old (in the Netherlands, ●Figure 4.1.F) and new (in Denmark, ●Figure 4.1.G).

land the nickname **Celtic Tiger** and placing it on par with the United Kingdom in per capita GNI PPP (see Table 4.1, page 70). Unemployment dropped sharply. This development surge was accomplished by a government program luring foreign-owned industries with inexpensive labor, tax concessions, and help in financing plant construction. Hundreds of industrial plants, owned mainly by U.S., British, and German companies, produce computers, electronics, foods, textiles, office machinery, organic chemicals, and clothing. Most of these plants have sprung up around the two main cities, Dublin (population 1 million; ●Figure 4.1.8) and Cork (population 200,000), where they make up an economic region dubbed Silicon Bog (see Figure 4.1.4), and in the vicinity of Shannon International Airport in western Ireland.

Ireland's agriculture, now employing only about 10 percent of the labor force, has had a troubled career. England conquered Ireland in the 17th century and made the island a formal part of the United Kingdom. Ireland was effectively an English colony. Most of the land was expropriated and divided among English landlords into large estates, where Irish peasants worked as tenants. The income from this enterprise generally found its way to Britain, and Ireland became a land of poverty, sullen hostility, and periodic violence.

A crop disease known as blight caused the notorious **potato famine** of 1845–1851. Nearly 10 percent of Ireland's prefamine population of 8 million died of starvation or disease during these years. An even larger number emigrated to England, Wales, Scotland, Australia, and North

● **Figure 4.1.7** A typical Irish landscape, beheld from the Blarney Castle. There is a tradition that kissing the so-called Blarney Stone in one of the castle battlements will make one eloquent in speech.

● **Figure 4.1.8** The Trinity College Library in Dublin is most famous as the repository of the Book of Kells, a beautifully illustrated 9th century version of the Christian gospels.

America. The large Irish American community in the United States traces much of its ancestry to this migration. Today, its members are just a few of the eager consumers abroad of Irish culture, especially dance and music.

Northern Ireland, now a majority Protestant part of the United Kingdom, has had a particularly difficult struggle that has colonial roots. One outcome of the second English conquest of Ireland, in the 17th century, was the settlement of English and Scottish Presbyterians and other Protestants in the north, where they became numerically dominant. They were intended to form a nucleus of loyal population in a hostile, conquered, largely Roman Catholic country. In 1921, when the Irish Free State (later to become the Republic of Ireland) was established in the Catholic part of the island, the predominantly Protestant north, for economic and religious reasons, elected to remain with the United Kingdom. Northern Ireland (also known as Ulster) was given much autonomy, including its own parliament in Belfast. Northern Ireland's Catholic minority felt increasingly marginalized by the pro-British Protestant majority and in the late 1960s began agitating, often violently, for change. The British imposed **direct rule** (in which the U.K.'s central government in London makes all major pol-

icy decisions for Northern Ireland), and Catholic discontent soon focused on the perceived British occupation. The **Irish Republican Army (IRA)** began a campaign of bombings, shootings, and arson, sometimes into the heart of England, ostensibly designed to drive the British army from Northern Ireland. However, **the Troubles,** as the locals refer to them, mainly involved an often violent struggle between **Catholic Republicans** and **Protestant Unionists** in Northern Ireland.

In 1998, by which time more than 3,400 Irish on both sides had died in the conflict, the U.S.-brokered **Good Friday Agreement** was signed. Approved by the IRA (and its political counterpart, **Sinn Fein,** pronounced *shin fayn*) and the Unionists, the agreement called for the devolution of British power. Direct rule by Britain would be replaced by the **Northern Ireland Assembly,** in which Protestants and Catholics would share power. Violent challenges to the peace agreement persisted from both sides, however, leading the British to halt devolution and reimpose direct rule in 2002. In 2005, the IRA announced that it would permanently cease military operations. Despite the refusal of three small splinter groups to lay down arms, Unionists agreed to share power with the IRA, and 2007 saw unprecedented negotiations between the two sides. If achieved, peace could help bring prosperity to Northern Ireland, which has had periods of strong economic growth based on shipbuilding and linen textile industries. The main industrial and residential city is Belfast (population 600,000).

4.1.2 France: *Vive la Différence!*

Thanks to the **Channel Tunnel**, or **Chunnel** (also called the **Eurotunnel**), running 16 miles (26 km) along the floor of the English Channel, it has been possible since 1994 to travel by train from London to Paris in about three hours (•Figure 4.1.9). Britain and France were often enemies in the past, but the Chunnel is an appropriate symbol for how closely linked they are today.

France has long been one of the world's most celebrated countries. Europe's largest country in area, it is also the region's leading agricultural producer. France is a country of great self-esteem, and French culture is prized in many parts of the world. It is one of the world's major industrial and trading nations, is a nuclear power, and is one of the five permanent members of the United Nations Security Council. France has recently distinguished itself in world affairs by breaking some patterns of alignment with the **West**, especially the United States.

Although France has lost its former significance as a major colonial power, it still maintains special relations with a large group of former French colonies. More than a dozen are now **overseas departments** of France and send representatives to the National Assembly in Paris. France has

a tradition—under increasing scrutiny at home, because of its questionable benefits—of posting military forces in unsettled countries, including those that were its former colonies: in 2007, it had about 20,000 soldiers abroad in hot spots including Lebanon, Ivory Coast, the Democratic Republic of Congo, and Chad.

Framed and Fortified

About four-fifths the size of Texas, France is shaped like an irregular hexagon framed on five sides by seas and mountains (•Figure 4.1.10). In the south, the Mediterranean Sea and the Pyrenees Mountains form two sides of the hexagon; in the west and southwest, the Bay of Biscay and the English Channel form two sides; in the southeast and east, the Alps and the Jura Mountains, and farther north, the Vosges Mountains, form a fifth side, with the Rhine River a short distance to the east. Part of the sixth or northeastern side of the hexagon is formed by the low Ardennes Upland, lying mostly in Belgium and Luxembourg, with lowland passageways leading into France from Germany both north and south of the Ardennes.

Aside from the large Massif Central in south central France, all the country's major highlands are on the country's periphery. Even the Massif Central, a hilly upland with low mountains in the east, is skirted by lowland corridors that connect France's extensive plains of the north and west with the Mediterranean coast. These corridors include the Rhône-Saône Valley between the Massif to the west and the Alps and Juras to the east and the Carcassonne Gap between the Massif and the Pyrenees. France is thus a country with no serious internal barriers to movement but with barriers at its borders. These have often failed to prevent the incursions of foreign armies into France; Germany, for example, attacked France three times and occupied the country throughout World War II.

Although they have been breached at times, France has natural fortifications, with two of its three land frontiers lying in rugged mountain areas. The Pyrenees are a formidable barrier, lowest in the west, where the independent-minded Basques, who are neither French nor Spanish linguistically or culturally, occupy an area extending into both France and Spain (Basque issues are discussed in Module 4.2, p. 132). Also straddling the mountainous Pyrenees border is the microstate of Andorra, whose 100,000 inhabitants enjoy a high standard of living from tourist revenues. In southeastern France, the Alps and Jura Mountains follow France's boundaries with Italy and Switzerland. The sparsely populated Alpine frontier between France and Italy is even higher than that of the Pyrenees: Mont Blanc, the highest summit, reaches 15,771 feet (4,807 m). The number of important routes through the mountains is limited, and the Alps have been an important defensive rampart throughout France's history.

The Jura Mountains, on the French-Swiss frontier, are lower but also difficult to cross, as they are arranged in long ridges separated by deep valleys.

• **Figure 4.1.9** Location of the Chunnel and a geological and topographical cross section.

● **Figure 4.1.10** Principal features of France.

Flourishing Agriculture

Although France's population is 77 percent urban, agriculture is a very important part of the French economy, far more than is typical of MDCs. Some 35 percent of the French landscape is agricultural, and another 28 percent is wooded. In 2007, France was the world's largest exporter of wine and barley and a major exporter of wheat, milk, butter, and meat. Since World War II, rural-to-urban migration has accompanied economic change, and increasing mechanization has replaced lost farmworkers.

The French continue, nevertheless, to see theirs as an agricultural nation, and the customs of the family farm still play a strong role in the French identity and sense of well-being. One of the most celebrated efforts to diminish the force of globalization (seen as particularly ruinous to the family farm in France) was the destruction of a McDonald's restaurant under construction outside Paris in 1999. A French sheep farmer named José Bové, long active in efforts to slow down French accommodation and support of American cultural and economic expansion, was jailed for leading a militant farming union in this civil disobedience. Bové remains a national folk hero and a strident opponent of imported and genetically modified foods.

France enjoys some advantages in the agricultural sector. Its membership in the EU gives French farmers free access

to a huge and affluent consuming market and protection by tariffs from non-EU producers. The EU budget has generally provided generous price supports at rates above world prices for many agricultural products, with French farmers being the leading beneficiaries.

France is blessed on its own, independent of the EU. Compared with the rest of continental Europe, the country has a relatively low population density. France's physical assets include superior topographic, climatic, and soil conditions. The main topographic advantage is a large area, especially in the north and west, level enough for cultivation.

Climatically, most of the country has moderate temperatures and year-round moisture provided by a marine west coast climate. The wettest lowland areas are in the northern part of the country, where crop and dairy production is especially intense. Southern France often experiences warmer temperatures, allowing production of corn in the southwestern plains. Productive vineyards are also concentrated in the south and southwest, although **viticulture** is practiced as far north as the Champagne region, east of Paris (●Figure 4.1.11). Finally, northeastern France from the frontier to beyond Paris has fine soils developed from loess. As in much of Europe, these very fertile soils are used mainly to produce wheat and sugar beets. The wheat belt in northeastern France is the country's chief breadbasket.

30,
89

• **Figure 4.1.11** Vineyards in southern France. The French countryside produces about 20 percent of the world's wine.

Beets grown in the same area provide both sugar for humans and, from processing residues, livestock feed for meat and milk production.

French Cities and Industries

From the dawn of the Industrial Revolution until recent decades, France tended to fall behind its European neighbors, except for Spain and Italy, in industrialization and modernization. But the country shifted gears after World War II. Government subsidies supporting children spurred population growth. New energy sources replaced traditionally inadequate coal resources. France's membership in the EU ended the protection of many inefficient producers.

The French government encouraged companies to combine into larger units capable of higher efficiency and international competitiveness. France was well placed geographically and structurally to move into newly developing industries, technologies, and markets. It had a large pool of labor available for transfer from an oversupplied agriculture sector, a relatively modest share of its labor force and capital tied up in older industries, and an excellent educational system.

As the economic pace picked up, the government took steps to allocate investment and development more equitably around the country, deliberately avoiding the traditional tendency to funnel resources into Paris. Planners use restrictions and penalties on new development in the Paris region. They reward firms financially for locating new facilities in regions of greater need and work to locate or relocate government facilities in less populous regional centers.

In industrial output, France ranks only behind the United States, Japan, and Germany. Automobile manufacturing (of Peugeot and Renault cars, for example) is the most important single industry. Based in Toulouse, France, the Airbus Corporation is a major rival to U.S.-based Boeing and is

pinning its market hopes on a huge double-decker aircraft called the A-380 that will carry 555 passengers (•Figure 4.1.12). In addition to automobiles and aircraft, major products include armaments, rubber, chemicals, textiles, and processed foods.

Paris is the greatest urban and industrial center of France, a **primate city** completely overshadowing all other cities in both population and economic activity (see Insights, page 111). With its 11.7 million people, it is by far the largest city on the mainland of Europe. Paris is located at a strategic point on the Seine River relative to natural lines of transportation, but it is more the product of the growth and centralization of the French government and of the transportation system it created. Like London, it has no major natural resources for the industry in its immediate vicinity, yet it is the greatest industrial center of the country. Despite its economic clout, Paris maintains its international reputation as the romantic and monument-studded **"City of Light"** (•Figures 4.1.13 and 4.1.14). It is the world's leading urban tourist destination and the main attraction in what makes France the world's number one tourist destination.

In the Middle Ages, Paris became the capital of a succession of kings who gradually extended their effective control over all of France. As the French monarchy became more absolute and centralized, its seat of power grew in size and came to dominate the cultural and political life of France. As national road and rail systems were built, their trunk lines were laid out to connect Paris with outlying regions. The result was a radial pattern with Paris at the hub, and this is essentially the pattern that exists today (•Figure 4.1.15). As the city grew in population and wealth, it became an increasingly large and rich market for goods.

The local market, plus transportation advantages and proximity to the government, provided the foundations for a huge industrial complex. This development has involved

• **Figure 4.1.12** The A-380 is the largest passenger plane ever placed in regular service. If Airbus can overcome technical problems, delays, and canceled contracts, look for it at major airports worldwide around 2010—and expect long waits at the baggage claim.

• **Figure 4.1.13** Downtown Paris and some of its world-famous attractions.

• **Figure 4.1.15** Paris is the primate city at the hub of France's transportation network.

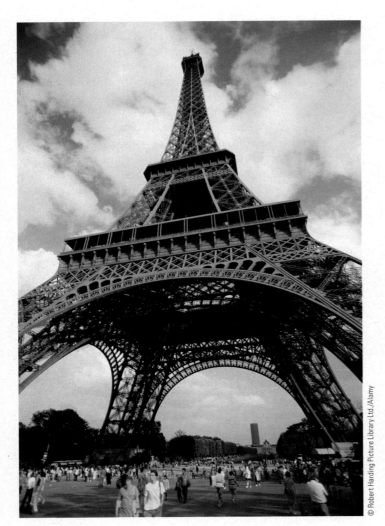

• **Figure 4.1.14** The soaring Eiffel Tower. At 324 meters (1,062 ft), it is the tallest structure in Paris.

two major classes of industries. Paris is the principal producer of the high-quality luxury items (fashions, perfumes, cosmetics, and jewelry, for example) for which that city and France have long been famous. Paris is also the country's leading center of engineering industries, secondary metal manufacturing, and diversified light industries, concentrated in a ring of industrial suburbs that sprang up in the 19th and 20th centuries. Paris inverts the usual pattern of the industrial city by maintaining a core of low-profile historic monuments, while the skyscrapers of the modern economy tower away from the city center.

With the more recent growth of Marseille, Lyon, and Lille, the dominance of Paris has diminished somewhat. Two ports near the mouth of the Seine River handle much of the overseas trade of Paris. Rouen, the medieval capital of Normandy, is located at the **head of navigation** (the farthest inland a ship can go on a river) for smaller ocean vessels using the Seine. With the increasing size of ships in recent centuries, more of Rouen's port functions have been taken over by Le Havre, at the entrance to the wide estuary of the Seine.

France's second-ranking urban and industrial area is the Nord (North), near the Belgian border, stretching along the western edge of the Ruhr coal and industrial region. The largest city, Lille, was established alongside France's once-leading (now closed) coalfield. The Nord was an important contributor to the economy of France, now has major problems. In the 19th and early 20th centuries, it developed a typical concentration of coal-based industries: coal mining, steel production, textile plants, coal-based chemical production, and heavy engineering. As in Britain, these operations have declined in recent decades, and the region suffers from high unemployment.

INSIGHTS

The Primate City

A primate city is a city that is larger than the second- and third-largest cities in that country combined. Paris, for example, has a metropolitan area population of 11.7 million, far exceeding the combined populations of France's next-largest cities, Lille (with 1.8 million people) and Lyon (with 1.7 million people). In practical terms, a primate city dwarfs all others in its demographic, economic, political, and cultural importance. The primate city produces and consumes a disproportionately high share of the country's goods and services, and many national resources—such as bureaucratic offices and educational opportunities—are available only in the primate city. The primate city therefore acts as a great magnet for rural-to-urban migration and tends to absorb more investment than other communities.

Urban primacy is often bad for a country's development because funds that could be spent to improve services and quality of life in smaller cities and rural areas are often diverted instead to the primate city. With greater opportunities available in the primate than in the smaller cities of the hinterland, the most educated and talented people abandon the hinterland for the primate city. This internal brain drain contributes further to underdevelopment outside the primate city. Primate cities are rare in the more developed countries but are quite typical of the LDCs, where regional development is a critical issue. Cairo, Karachi, and Mexico City are exemplary primate cities of Africa, Asia, and the Americas.

France's third-largest urban-industrial cluster centers on the metropolis of Lyon in southeastern France. The valleys of the Rhône and Saône Rivers join here, providing routes through the mountains to the Mediterranean, northern France, and Switzerland. Since prehistoric times, the valley has been a major connection between the North European Plain and the Mediterranean. Today, it forms the leading transportation artery in France, providing links between Paris and the coast of the Mediterranean via superhighways, rails that carry the TGV—one of the fastest trains in the world—and a barge waterway formed by the Rhône, the Saône, and connecting rivers and canals to the north.

France has promoted hydroelectric power and nuclear energy to compensate for inadequate fossil fuel supplies, and the Rhône Valley plays a major role in both. A series of massive dams on the Rhône and its tributaries produce about 10 percent of France's electricity, and more power comes from several large nuclear plants in the Rhône Valley using uranium from the nearby Massif Central and foreign sources. France produces a greater proportion of its electricity from nuclear power—about 80 percent—than any other country in the world. France's most important Mediterranean cities and metropolitan areas—Marseille, Toulon, and Nice (pronounced like *niece*)—are strung along the indented coast east of the Rhône delta. Marseille (population 1.6 million) is France's leading seaport. Recent growth of port traffic and manufacturing in the Marseille area has been associated with increasingly large oil imports and the related petrochemical industry. Toulon, southeast of Marseille, is France's main Mediterranean naval base, and Nice, near the Italian border, is the principal city of the French Riviera, the most famous resort district in Europe.

Along this easternmost section of the French coast, the Alps come down to the sea to provide a spectacular shore-line dotted with the beaches of the Côte d'Azur. Near Nice, a string of resort towns and cities along the coast includes the tiny principality of Monaco, a microstate that is nominally independent but closely related to France economically, culturally, and administratively. About three times the size of the National Mall in Washington, D.C., Monaco makes a living from gambling and other forms of recreation.

The south central wedge of western France and the Massif Central has a rich mixture of manufacturing, agriculture, and tourism. Each of the largest urban centers, Toulouse, Bordeaux, and Nantes, has a population of 1 million or less. Bordeaux is a seaport on the Garonne River, which empties into the Bay of Biscay. The city passed its name to the wines from the surrounding region, exported through Bordeaux for centuries. Nantes is the corresponding seaport where the other main river of western France, the Loire, reaches the Atlantic. Toulouse, a historical city with modern machinery, chemical, and aircraft plants, developed on the Garonne well upstream from Bordeaux at the point where the river comes closest to the Carcassonne Gap. Lying between the Pyrenees and the Massif Central, the Gap provides a lowland route from France's Southwestern Lowland to the Mediterranean.

The Social Landscape

Fine French wines and other exports bestow well-deserved honors on France. However proud and confident the French may seem in the international arena, at home they have their share of troubles. The economy has been sluggish, a problem that recently forced the country to remove its cherished 35-hour cap on the workweek. French regulations make firing an underperforming employee almost impossible, so trimming the fat is not as easy as in most other industrialized countries. The French enjoy their national rights to a six-week paid vacation and comprehensive pensions and health care. Desperate to reduce the economic costs of pensions, the government has been trying to legislate lower retirement payments and more working years for public employees. The workers have responded with national transportation strikes. Caught between their privileges and the social costs these impose, the French are frustrated; a recent poll showed that about two-thirds of the French think their country is in decline.

One of the most serious issues France faces is how to deal with its large immigrant population, especially of Muslim Arabs. France has about 5 to 6 million Muslims, or about 9 percent of the population, which is the highest percentage of Muslims in the European core. Most are either immigrants or their offspring, mainly from the former French colonies of North Africa. France has a vigorously secular constitution, and laws interpreting it have provoked discord with the largely religious Muslim minority. In 2004, France's National Assembly issued a ban against wearing head scarves and other items with religious connotations in public schools, an action affecting Sikhs, Jews, and others but targeted mainly at Muslim women. Except for support by feminist Muslim women, the Islamic community in France responded indignantly, and many Muslim schoolchildren turned to private schools or home schooling. Employment and other indicators of well-being are much lower among the Muslim Arab immigrants than among the majority ethnic French.

Officially, France insists that all of its citizens are simply French and have all the attendant rights and privileges, and the government has taken few steps to address the special problems of its Muslim minority. Muslim immigrant frustration boiled over into ferocious rioting in several French cities in 2006 and 2007. The country is now grappling with its previously colorblind attitude, described by a French sociologist this way: "If you don't attach too much importance to ethnic and racial divisions in society, and don't talk about them, they will shrink and disappear."[4]

4.1.3 Great Germany

The Federal Republic of Germany reappeared on the map of Europe as a unified country in 1990. Between 1949 and 1990, it had been divided into two nations: democratic West Germany (known as the German Federal Republic), formed from the zones occupied by Britain, France, and the United States after Germany's defeat in World War II in 1945, and Communist East Germany (the so-called German Democratic Republic), formed from territory occupied after the war by the Soviet Union. The former capital, Berlin, was similarly divided. West Berlin was a part of West Germany for those decades, even though the city was entirely surrounded by East Germany. The withdrawal of Soviet support for East Germany in 1989 and the destruction of the Berlin Wall led to the rapid collapse of its Communist government in 1990 and to the jubilant reunification of Germany's 16 states (*Länder*).

Reunified Germany is Europe's dominant country. Its population of 82 million is much greater than that of any other nation in the region. Politically, it is seen, along with France, as the cornerstone of the European Union. Economic considerations are even more important; Germany has the world's third-largest economy, after the United States and Japan, and is among the top three, with China and the United States, in exports of goods. The former West Germany alone was Europe's leading industrial and trading country during the latter half of the 20th century. The former East Germany was an advanced country by Soviet bloc standards but an economic disaster by West German standards. Many billions of dollars and much time have been required to rehabilitate Germany's east, and the task is far from done. This area, however, can be expected to contribute increasingly to overall German dominance in the European economy.

A Central Location

Germany arrived at its favorable economic position by skillfully exploiting the advantages of its centrality in Europe together with the advantages of a diverse environment (•Figure 4.1.16). Broadly, the terrain of Germany can be

• **Figure 4.1.16** Principal features of Germany.

divided into a low-lying, undulating plain in the north and higher country to the south. The lowland North German Plain is a part of the much larger North European Plain. The central and southern countryside is much less uniform. To the extreme south are the moderately high German (Bavarian) Alps, fringed by a flat to rolling **piedmont** (foothills) that slopes gradually down to the east-flowing upper Danube River. Between the Danube and the northern plain is a complex series of uplands, highlands, and depressions. The higher lands are mainly composed of rounded, forested hills and low mountains. Interspersed with these are agricultural lowlands draining to the Rhine, Danube, Weser, and Elbe Rivers, which drain into three different seas.

Germany has a maritime climate in the northwest and an increasingly continental climate toward the east and south. The annual precipitation is moderate, with most places in the lowlands receiving an average of 20 to 30 inches (c. 50 to 75 cm) per year. Farmers generally must contend with infertile soils. German agriculture has long been characterized by crops that do well in poor ground, such as rye, cabbages, and potatoes, and by hogs, which can consume farm waste. Two areas are more fertile, supporting the more demanding wheat, sugar beets, and cattle: the southern margin of the North German Plain, forming a belt of fertile loess soils, and the fertile alluvial Upper Rhine Plain, extending from Mainz to the Swiss border and mostly enclosed by highlands on both sides.

German Cities and Industries

Within this environmental framework, Germany's major cities tend to be located in the most economically advantageous areas and those most strategic with respect to transportation. A string of major metropolitan areas is located in the western part of Germany, along and near the Rhine River. This waterway has been a major route since prehistoric times. Four of the eight largest German metropolises—Düsseldorf, Cologne, Wiesbaden-Mainz, and Mannheim-Ludwigshafen-Heidelberg—are on the Rhine itself. So is Bonn, which was made the capital of West Germany in 1949 as a defensive move, just in case the traditional capital city of Berlin were to fall entirely into the Soviet grip.

The other four metropolises—Essen, Wuppertal, Frankfurt, and Stuttgart—are on important Rhine tributaries. This north-south belt of cities also includes several other large metropolitan centers, including Dortmund, Duisburg, and Aachen, and numerous smaller places. The concentration of seven of these cities at or just north of the boundary between the Rhine Uplands and the North German Plain reflects other geographic factors. Historically, some of the seven cities profited from their location on the agriculturally excellent loess soil just north of the Uplands. This concentration of large and medium-sized urban centers is a remarkable contrast with France's urban pattern, dominated by Paris.

Other German cities developed as early industrial centers before the Industrial Revolution, drawing on the Uplands for ores, waterpower, wood for charcoal, and surplus labor. Then, in the 19th and early 20th centuries, all seven cities grew in connection with the development of both the Ruhr coalfield (Europe's largest), located at the south edge of the plain and east of the Rhine, and the smaller Aachen field. Except for Aachen and Cologne, the cities became part of the Ruhr, Europe's greatest concentration of coal mines and heavy industries, linked by a network of active river and canal transport (see Figure 4.1.3). Still other important German cities lie apart from the Rhine and the Ruhr. These include Hanover, Leipzig, and Dresden, spread along the loess belt across Germany; the North Sea ports of Hamburg, located well inland on the Elbe estuary, and Bremen, up the Weser estuary; two cities in the southern uplands—Munich, north of a pass route across the Alps and Nuremberg; and Berlin (population 4 million; ●Figure 4.1.17), whose large size belies its unlikely location in a sandy and infertile area of the northern plain. Berlin is mainly an artificial product of the central governments, first of Prussia and then of Germany, which developed it as their capital. Since reunification, Berlin has once again become Germany's capital.

Much of Germany's present industry is not in the Ruhr or in Saxony, a historically industrial region in the southern part of the former East Germany. Instead, industrial and urban centers are widely scattered. Many cities are large, but none is as nationally dominant as London or Paris. This dispersed pattern came about because Germany was divided for a long time into petty states, many of which eventually became internal states within German federal structures. These often had their own capitals, which would in time become Germany's large cities.

Many of these industrial cities were bombed heavily in World War II, but industrial capacity was not completely devastated. Several factors prompted rapid economic reconstruction of postwar West Germany. The massive U.S.-funded aid package known as the Marshall Plan, supplying $13 billion in capital goods between 1948 and 1952, helped rebuild the country as a strategy to stave off the

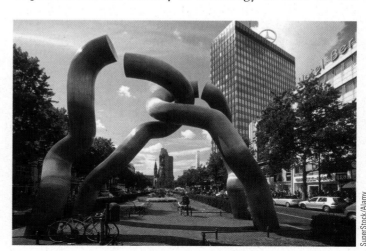

● **Figure 4.1.17** War Memorial; Kaiser Wilhelm Memorial Church; Berlin, Germany.

SuperStock/Alamy

Communist threat from the east. The skills of the population were still there, and labor was inexpensive for a while, as workers were abundant and eager to be working again. Markets were hungry for industrial products. By the 1960s, the pace of growth slowed, but it was still fast enough to provide jobs through subsequent decades for Europe's largest numbers of foreign workers, especially Turks and Kurds from Turkey, who replaced upwardly mobile German workers. The export of German industrial products to world markets became more difficult as Japanese, Chinese, and other foreign competition stiffened.

East Germany was Communist-ruled and Soviet-dominated, but it was still German. Even Soviet exploitation and bureaucratic inefficiency were unable to destroy the traditional German ethics of hard work, efficiency, attention to detail, and high standards. East Germany was the most productive of the Soviet "satellite" countries that lay between Russia and western Europe, but its economy lagged far behind West Germany's. When East Germany collapsed in 1989, a major migration stream began flowing from East to West Germany. By 2007, more than 1 million people had fled westward from what Germans call the "new states" of the former East Germany, and the flow continues, at the rate of about 50,000 a year, leaving what a German professor describes as "a series of ghost towns and enclaves for senior citizens."[5]

The pool of young and educated people who do remain in the east is attracting employers looking for lower-wage workers, and the region's economy may benefit from their presence. There is a growing semiconductor industry in what local promoters like to call Silicon Saxony, especially in Dresden and Leipzig (see Figure 4.1.4). The automakers Volkswagen and BMW have opened plants in Dresden and Leipzig, respectively. Taking advantage of the skilled and inexpensive labor, scores of U.S. chipmaking companies have set up operations in the region.

Billions of dollars of investment are helping to raise economic standards in the east and to clean up the severe environmental damage incurred under Communist rule there. To complete that transition, the former West Germans are bearing much of the tax and social burden of bringing the former East Germans up to the standards long enjoyed in the west. Economically, Germany has been a welfare state that subsidizes its citizens generously. The costs for that support were partly to blame for Germany's economic downturn in the early 2000s, when economic growth was the slowest of all EU countries and unemployment grew to a national average of about 10 percent, with the rate of unemployment in the former East Germany more than twice that in the west. Germany is now making a painful transition from a welfare state to one in which its citizens must pay more out of pocket for health care and other vital services. Other difficult measures, such as cutting jobs, extending work hours without extra pay, and outsourcing work to other countries, helped Germany's economy rebound sharply after the middle of the decade.

4.1.4 Benelux: Tolerance and Trade in the Low Countries

Belgium, the Netherlands, and Luxembourg have been closely associated throughout their long histories. They are known collectively as the Benelux Countries or the Low Countries. The northern two-thirds of Belgium and practically all of the Netherlands consist of a low plain facing the North Sea (•Figure 4.1.18). Luxembourg is also included in the Low Countries because of the long historical connection among the three countries. Large sections near the coast, especially in the Netherlands, are below sea level and have been made into major farming and urban landscapes through the engineering of **polders**—land below sea level that has been reclaimed from the sea and protected from inundation by dikes, canals, and pumps (see Insights, page 115, and Figure 4.1.H). It is not surprising that the Netherlands has played a leading role in negotiations to reduce global emissions of greenhouse gases, which might lead to rising sea levels and increase the threat of catastrophic flooding from the North Sea.

Small Countries, Great Trade

The Benelux countries have long depended on a combination of industry, agriculture, and trade—especially international trade. Successful international trading is reflected by the presence of three of the world's major port cities within

• **Figure 4.1.18** Principal features of the Benelux countries.

INSIGHTS

The Polder

One of the most characteristic Low Country landscapes is land reclaimed from the North Sea. Polder land is low-lying land, often below sea level, that has been surrounded by dikes and artificially pumped dry. The process of turning former swamps, lakes, and shallow seas into agricultural land has been going on for more than seven centuries in the Netherlands. About half of the nation now consists of an intricate patchwork of polders and canals (•Figure 4.1.H), and Belgium has a narrow strip of such lands behind its coastal sand dunes.

In 1953, a monstrous storm from the North Sea wreaked havoc on the Netherlands' defenses, destroying its seawalls and dikes and killing 2,000 people. Since then, the Dutch have built a high-tech system for flood control that sets the standard for what other prone regions, such as New Orleans, might do. They constructed a kind of forward defensive shield that reduced the amount of vulnerable coastline and in the process actually shortened the coast (with all its bends and twists considered) by a total of about 400 miles (640 km). The highest levees are nearly 40 feet (12 m) above water level, about twice the tallest in New Orleans.

Joe Hobbs

• **Figure 4.1.H** This view of the port of Amsterdam reveals the amphibious quality of the Netherlands, which must always be on guard against the sea.

a distance of about 80 miles (c. 130 km): Rotterdam and Amsterdam in the Netherlands and Antwerp in Belgium. Amsterdam is the constitutional capital of the Netherlands. The government is actually located at The Hague, which is important in international circles as the seat of the International Court of Justice, the judiciary branch of the United Nations. Portions of the Netherlands were (and still are) known as Holland, a name that is sometimes used to stand for the whole country, and the people of the Netherlands are known as the Dutch.

Amsterdam was the main port of the Netherlands from the 17th through 19th centuries, when the Dutch colonial empire was expanding. The most important component of this empire was the Netherlands East Indies (now Indonesia). Until Indonesia's independence after World War II, Amsterdam built a growing trade in imported tropical crops and oil from its "Spice Islands." Many of these imports were reexported, often after processing, to other European countries. Profits that accumulated during these centuries supplied funds for the large overseas investments of the Netherlands and made Amsterdam an important financial center, which it continues to be. Amsterdam also remains a major port. Its original approach by sea via the Zuider Zee, a landlocked inlet of the North Sea, proved too shallow for the larger ships of the 19th century. The opening in 1876 of the North Sea Canal solved this problem for a time, but efforts to widen the canal could not give Amsterdam access to the sea comparable to that of Rotterdam.

Continuing ties with its former colony have given Amsterdam a large Indonesian community that imparts a distinctive ethnic flavor to the city. There is also a very large population of Moroccan origin, and today one in three persons in Amsterdam is of non-Western descent. Its recent controversies over immigration notwithstanding, the city has earned an international reputation as a cosmopolitan and tolerant city with a very high quality of life (•Figure 4.1.19). Prostitution is legal and carefully regulated (see Figure 5.1.A, p. 187), and cannabis may be consumed in licensed "cafés." Pedestrian and bicycle traffic replace cars to a large degree in the downtown district, ringed by canals, making for a clean, quiet, and enjoyable urban environment.

Such amenities help compensate for the fact that the Netherlands has a higher population density than any other European country, aside from the small islands and the microstates (see Figure 4.2, page 72). All Dutch cities are trying to limit urban sprawl, provide more outdoor and waterfront recreational areas, and maintain an environment that is pleasant for their residents and attractive to tourists. Planners are especially concerned with maintaining the open space semiencircled by Amsterdam, Utrecht, Rotterdam, The Hague, and other cities. These urban areas are tending to coalesce into what is referred to as Randstad Holland (population 6.6 million), the "Ring City" of Holland.

Rotterdam, the world's third-largest port in tonnage handled, is on one of the navigable distributaries of the Rhine River and is better situated for Rhine shipping than either Amsterdam or Antwerp. Rotterdam controls and profits from the major portion of the river's transit trade, especially from Germany's Ruhr district. A rail freight connection between Rotterdam and the Ruhr, built to maintain Rotter-

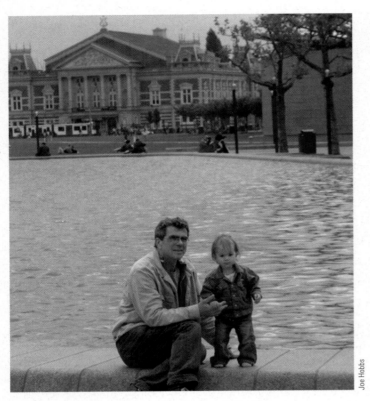

Joe Hobbs

● **Figure 4.1.19** Amsterdam prides itself on its livability.

dam's dominant role in EU trade and to relieve congestion on Europe's roads, opened in 2007. Part of the Rotterdam facilities is the harbor and industrial development of Europoort, designed to handle supertankers and international ferry traffic. Antwerp (population 1.1 million), located about 50 miles (80 km) up the Scheldt River, is mainly a port for Belgium but also has a share of the Rhine transit trade.

Industrial Patterns

The Benelux countries are highly industrialized. Before major discoveries of natural gas were made in the Netherlands in the 1950s, Belgium and Luxembourg had more mineral resources. This helped them initially become more industrialized than the Netherlands. Now, however, the Netherlands, with its advantageous trade position and natural gas, is the leading industrial Benelux country.

The Netherlands has long depended on imported fuels other than natural gas. More oil imports now enter Europe via Rotterdam's Europoort than any other access point, with some arriving as crude oil and some refined further in the Rotterdam area. This activity has generated one of the world's largest concentrations of refineries. Petrochemical plants near the refineries represent the main branch of the Dutch chemical industry. Other leading Dutch industries are food processing, electrical machinery, textiles and apparel, and electrical and electronics industries (of which Philips is the most renowned).

One of Europe's major coalfields crosses Belgium in a narrow east-west belt about 100 miles (160 km) long,

roughly following the valleys of the Sambre and Meuse Rivers and extending into France on the west and Germany on the east. Liège (population 500,000) and smaller industrial cities along this Sambre-Meuse field accounted for most of Belgium's metallurgical, chemical, and other heavy industrial production until those industries followed the familiar decline prompted by depleted reserves and increased international competition. An iron and steel industry developed here even before the Industrial Revolution, using local iron ore and charcoal. Local coal and the early adoption of British techniques soon made Liège the first city in continental Europe to develop modern large-scale iron and steel manufacturing. Newer industrial plants are mainly in northern Belgium, where Belgium's capital and largest city, Brussels (population 2.1 million), is the leading center of a breed of new, less energy-dependent cities. Brussels plays a vital regional role as the location of the European Union's headquarters.

Belgium is internally secure, despite the divide between the majority Flemish population, who speak Dutch, and the large French-speaking Walloon minority. The Flanders region of Belgium, where the Flemish are concentrated, periodically calls for greater devolution of power from Brussels. 83

Luxembourg's most important exports previously came from the iron and steel industry based near the southern border where the Lorraine iron ore deposits of France overlap into Luxembourg. With steelmaking in decline, Luxembourg continues to follow the European pattern in diversifying its manufacturing industries and exports, especially with chemicals and rubber products. It has modified its economy successfully; its annual per capita GNI PPP of almost $56,000 makes it the richest country in the world by that standard.

Benelux Farming

Despite their small size, the Netherlands and Belgium are very productive agriculturally, with high yields per unit of land. Parts of each country have very fertile soils, especially in the polders of the Netherlands and the loess lands of Belgium. Great investments have been made in water control, agricultural machinery, technical expertise, and fertilizers. Both countries have even embraced the concept of coexistence of conventional and genetically modified crops (although some communities are declaring themselves "GMO-free zones"). Agriculture in the Low Countries has profited 89 from European Union tariff protection and subsidies for agricultural producers, stimulating spectacular increases in output to the extent that overproduction of certain products such as milk and butter has become a more worrisome problem than food supply. Belgium and the Netherlands 87 specialize in labor-intensive farm commodities such as commercial flowers (notably the Netherlands' tulips) and dairy products. The dense and affluent populations in and near the Low Countries provide a large market for such specialties. The Netherlands is also the world's largest hub for international trade in fresh ("cut") flowers, a huge industry that links 7,000 nurseries and ships 5 billion flowers each year.

4.1.5 Switzerland and Austria: Prosperous Mountain Fastness

On the high southern margin of Europe's core lie Switzerland and Austria (•Figure 4.1.20). More than half of each country is occupied by the lofty, rugged Alps. North of the Alps, both countries include strips of rolling, once glaciated plains and foothills. The Swiss section of the Alpine foothills, called the Swiss Plateau, extends between Lake Geneva on the French border and Lake Constance (known as the Bodensee in German) on the German border. The Austrian section of the foothills lies between the Alps and the Danube River from Salzburg on the west to Vienna on the east. A third portion of the foothills in southern Germany separates the Swiss and Austrian sections. On their northern edges, Switzerland and Austria have mountains that are much lower and smaller in extent than the Alps. These are the Jura Mountains, on the border between Switzerland and France, and the Bohemian Hills of Austria, on the border with the Czech Republic.

Resources and Development

Economic success is often linked to a generous resource base, but in this case, less well-endowed Switzerland has become somewhat more prosperous than Austria. Based on natural resources alone, Austria should be the more successful country economically. With fewer mountains, it has more arable land than Switzerland and more land per person. It also has more forested land and a better mineral resource base. Both countries depend heavily on abundant hydroelectric potential, but Austria's resource base is greater.

In economic success, however, the advantage lies with Switzerland, whose GNI PPP is about 15 percent higher than that of Austria. The historical developments that favored Switzerland outweighed Austria's natural resource advantage. Switzerland enjoyed peace within stable boundaries, while Austria was rocked and chipped away at by two world wars. Switzerland's peace is the product of its position as a **buffer state** between greater powers, the comparative defensibility of much of the country's terrain, and the relatively small value of Swiss economic production to a would-be aggressor. Switzerland also has historical and political value as an intermediary between belligerents in wartime and as a refuge for people and money. It also has a policy of strict and heavily armed **neutrality**. Surrounded by countries in conflict, Switzerland managed to stay out of both world wars.

Switzerland

With peace prevailing, the Swiss have had the opportunity to develop an economy finely tuned to the country's potentials and opportunities. They have exploited with special skill the fields of banking, tourism, manufacturing, and agriculture. Favored by Swiss law, the country's banks enjoy a worldwide reputation for security, discretion, and service. The result has been a massive inflow of capital—once of dubious origins but less so these days—to Swiss banks. Switzerland's largest city, Zürich (population 1 million), is one of the major centers of international finance, and other Swiss cities are also heavily involved in banking.

Switzerland's scenic resources make it a major tourist destination. The country is a mecca for winter sports, and Alpine resorts such as Zermatt (•Figure 4.1.21), Davos, and Saint Moritz are world famous. Numerous tunnels and costly highway engineering marvels link highland Switzerland with surrounding countries. Because most of its raw materials and fuels must be imported, industrial specialties in Switzerland minimize the importance of bulky materials and derive value from skilled design and workmanship. Swiss watches are the most renowned, but other high-tech instrumentation—for example, in biotechnology and medical devices—is also typical of Switzerland. Swiss-based multinational firms produce foods (Nestlé, the world's largest food company) and pharmaceuticals (Roche). The

• **Figure 4.1.20** Principal features of Switzerland and Austria.

● **Figure 4.1.21** Motorized vehicles are prohibited in the Swiss Alpine village of Zermatt, which is carefully managed as a jewel of international tourism.

country has many small industrial centers, mostly on the Swiss Plateau, and these obtain their electricity from hydropower. Five of the country's six largest cities, including the capital of Bern, are located on the plateau. The remaining city, Basel, lies beyond the Juras at the point where the Rhine River turns north between Germany and France. Basel is the head of navigation for Rhine barges that connect Switzerland with ocean ports and other destinations via river and canal.

Switzerland's internal political organization reflects its linguistic and ethnic diversity. German is the primary language of 63 percent of the population, French of 20 percent, and Italian of 7 percent. A fourth official national language, Romansch, is spoken by only around 1 percent of the population. The city of Geneva in the west has a distinctively French feel, and Zürich in the east is strikingly German. The country was originally a loose alliance of small sovereign units known as *cantons*. With the development of a stronger central authority in the 19th century, the cantons retained many of their powers, except for limited functions assigned to the central government. Central government responsibilities have since grown, but each of the 26 local units has preserved a large measure of authority. Switzerland's government often submits important legislation directly to the people. Guarantees of fundamental rights, local autonomy, and close governmental responsiveness to the will of the people are used to foster national unity, despite the potential difficulties of cultural diversity. Partly because of these strong political traditions, Switzerland has so far resisted joining the European Union. The country also fears it would be a "small fish" in the big EU pond and that its special interests such as banking and farming might lose their privileges within the European Union.

The central government has pushed economic development vigorously. Two outstanding accomplishments are the railroad and hydroelectric power networks. Despite the difficult terrain, the dense railway network carries enough international transit traffic to earn important revenues. The hydroelectric system, taking advantage of the country's

mountainous terrain and abundant precipitation, yields about 60 percent of the country's electric output, making Switzerland one of the highest nations in the world in hydropower produced per capita.

A country with much unrealized hydropower potential is Liechtenstein, a microstate sandwiched between Switzerland and Austria. With just over 30,000 people, Liechtenstein is a kingdom that has prospered. Its decades-long reliance on dubious profits from money laundering and tax evasion has given way to a cleaner financial industry and the unique niche of false-teeth production.

Austria

Austria emerged in its present form as a remnant of the Austro-Hungarian Empire when that empire disintegrated in defeat in 1918, at the end of World War I. Austria's population is essentially German in language and culture—and Austria and Germany have been closely linked historically— but unification with Germany was forbidden by the victors. The economy was racked by the loss of the empire, and the country limped through the interwar period until it was forcefully absorbed by Nazi Germany in 1938. In 1945, after defeat in World War II, Austria was reconstituted as a separate state but was divided, like Germany, into four occupation zones administered, respectively, by the United States, the United Kingdom, France, and the Soviet Union. During 10 years of occupation, the Soviet Union removed a vast amount of industrial equipment as war reparations from its Austrian Soviet Zone in the east (which included Vienna). It was only in 1955 that this occupation was ended by agreement among the four powers, and Austria became an independent, neutral state.

As the heart of the Austro-Hungarian (Hapsburg) Empire in southeastern Europe, Austria developed a diversified industrial economy prior to 1914. Its industries depended on resources and markets provided by the broader empire that extended to much of the North European Plain. Austrian iron ore was smelted mainly with coal drawn from Bohemia and Moravia, now in the Czech Republic. Austria's textile industry specialized in spinning, leaving much of the weaving to be done in Bohemia. Austrian industry had the benefit of a large protected market in which one currency was in use. The producers of foodstuffs and raw materials also had a market within an empire protected by external tariffs.

When the empire disintegrated at the end of World War I, the areas that had formed Austria's protected markets were incorporated into independent countries. Wanting to develop industries of their own, these countries erected tariff barriers, and other industrialized nations competed with Austria in their markets. The resulting decrease in Austria's ability to export made it harder to import the food and raw materials that its unbalanced economy required. The absorption of east central Europe into the Soviet Communist sphere after World War II made the situation worse.

The much needed 1990s reorientation of Austrian industries toward new markets, principally in western Europe, was difficult because Austria did not join the European Union until 1995. Austrian exports proved consistently inadequate to pay for imports. Despite these obstacles, Austria has built a successful economy suited to its status as a small independent state. Major elements in the economy are forest industries, tourism, motor vehicle production, hydroelectric development, manufacturing of traditional clothing and crafts, and larger-scale engineering and textile industries.

Since the end of the empire, Austrian agriculture has become more specialized in dairy and livestock farming. Enlarged acreage for feed crops such as barley and corn has accompanied the increased emphasis on animal products. But Austria is still far behind western Europe's leading countries in agricultural efficiency. Many small and poor farms still exist, and 5 percent of the labor force is still in agriculture—a high proportion by western European standards.

Austria's capital and largest city, Vienna (population 2 million; •Figure 4.1.22) is not as prominent in national and regional affairs as it once was. People and industry have shifted away from the capital to the smaller cities and mountain districts. Vienna's importance, however, has persisted since Roman times and is based on more than purely Austrian circumstances. The city is located at the crossing of two of the European continent's major natural routes: the Danube Valley route through the highlands separating Germany from the Hungarian plains and southeastern Europe, and the route from Silesia and the North European Plain to the head of the Adriatic Sea. The latter follows the lowland of Moravia to the north and makes use of the passes of the eastern Alps, especially the Semmering Pass, to the south. Thus Vienna has long been a major focus of transportation and trade and also a major strategic objective in time of war. In the geography of the early 21st century, Vienna finds itself particularly well positioned to capitalize on the growth potential of the formerly Communist eastern European nations, which are considered in Module 4.2.

• **Figure 4.1.22** The Graben, one of Vienna's main shopping areas.

SUMMARY

→ The nations that make up the European core are the United Kingdom, Ireland, France, Germany, the Low Countries (the Netherlands, Belgium, and Luxembourg), Austria, and Switzerland. The giant nations in this region are Germany, the United Kingdom, and France. They are the primary engines for the economic momentum in this part of the world. Three microstates—Andorra, Monaco, and Liechtenstein—are also included here.

→ The United Kingdom is made up of England, Scotland, Wales, and Northern Ireland. It forms the greater part of two major islands: Great Britain and Ireland, which are called the British Isles. The island of Great Britain has distinctive highlands and lowlands regions.

→ Most of the 18th- and 19th-century industrial and urban development of Britain related to the location of coal and iron ore, the two leading resources of the Industrial Revolution. London arose on the Thames estuary with neither coal nor iron as a local resource and is today a major global financial center.

→ Britain's five leading industries during and after the Industrial Revolution were coal, iron and steel, cotton textiles, woolens, and shipbuilding.

→ While the Republic of Ireland has recently experienced strong economic growth based on high-tech industries, Northern Ireland's growth has been restrained by a long-standing but perhaps nearly resolved conflict between historically indigenous Roman Catholics and more recent residents of Protestant, English background.

→ France is territorially the largest of the European core countries. The nation has an internal coherence and several national borders made up of rivers, mountains, and seacoasts. Agriculture is more important in the French economy and culture than in any other country of the European core. Paris is an unusual example of a primate city in a more developed country. The French are at odds with themselves over quality of life, immigration, and other issues.

→ The EU countries have established an ambitious schedule to replace fossil fuel energy with wind and other renewable energy alternatives. Britain seeks leadership in alternative energy technology and cuts in greenhouse gas emissions.

→ Germany, restored when West Germany reunified with East Germany in 1990, has the largest economy and population of all Europe. German urban centers have a much broader distribution than the industrial cities of France and even Britain. As in Britain, these cities were generally located by the availability of coal and other resources and of rivers for transport, and as in Britain, they have been deindustrializing in recent decades. Germany's economy has had to bear the costs of reunification and the burdens of being a welfare state. The former East Germany remains poorer than the west, but new industrial investment is drawn there by low labor costs.

→ The Netherlands, Belgium, and Luxembourg are known as the Low Countries or the Benelux Countries. They have waged a struggle to hold back the North Sea, and they still derive significant income from agriculture. They have a long history of global trading networks. Three major port cities in the Benelux Countries are Rotterdam (with its Europoort) and Amsterdam in the Netherlands and Antwerp in Belgium.

→ Switzerland and Austria have similar geographic features and economic potential, but Switzerland is slightly more prosper-

ous. Switzerland has gained more stability with its traditional neutrality, whereas Austria has had its fate tied more closely to Germany and its battles on the North European Plain. Both countries are partly in the Alps and are significant tourist destinations. Neither has particularly abundant natural resources. Hydroelectric power is a major source of energy for both countries. Switzerland is world famous for its banking and high-tech instrumentation industries. Austria is a member of the European Union, but Switzerland is not.

→ The microstates of Andorra, Monaco, and Liechtenstein have become prosperous from a combination of financial services and recreation, including tourism and gambling.

KEY TERMS + CONCEPTS

arable (p. 104)
buffer state (p. 117)
Catholic Republicans (p. 106)
Celtic Tiger (p. 105)
Channel Tunnel (p. 107)
Chunnel (p. 107)
"City of Light" (p. 109)
Commonwealth of Nations (p. 99)
core (p. 96)
direct rule (p. 106)

Enclosure Movement (p. 102)
Eurotunnel (p. 107)
Good Friday Agreement (p. 106)
head of navigation (p. 110)
hedgerow (p. 102)
Irish Republican Army (IRA) (p. 106)
major cluster of continuous settlement (p. 97)
neutrality (p. 117)
Northern Ireland Assembly (p. 106)

overseas department (p. 107)
piedmont (p. 113)
polder (p. 114)
potato famine (p. 106)
primate city (p. 109)
Protestant Unionists (p. 106)
Sinn Fein (p. 106)
the Troubles (p. 106)
viticulture (p. 108)
West (p. 107)

REVIEW QUESTIONS

1. What are the countries of the European core, and on what basis have they been designated as belonging to the core?

2. What was the spatial relationship between coalfields and cities in industrial Britain and mainland Europe?

3. What five industries were central to the Industrial Revolution in Britain? To what extent do they exist in modern times?

4. What are the political affiliations of Ireland, Northern Ireland, England, Scotland, and Wales?

5. What are "the Troubles" of Northern Ireland?

6. Why is French agriculture particularly strong when compared with neighboring countries?

7. What is a primate city? What examples exist in Europe and elsewhere?

8. What was the impact of German reunification on the country's economy?

9. What has made Rotterdam the most active and prosperous port in Europe?

10. What role has geography had in the distinct histories of Switzerland and Austria?

11. Why hasn't Switzerland joined the European Union?

12. How do the microstates of Europe make a living?

13. What are the principal physical geographic features of the United Kingdom, France, and Germany, the big three of the European core?

NOTES

1. Terry G. Jordan and Bella Bychova Jordan, *The European Culture Area: A Systematic Geography* (Lanham, Md.: Rowman and Littlefield, 2002), p. 402.

2. Ibid., p. 163.

3. Alan Cowell, "Menacing the Land, but Promising to Rescue the Earth." *New York Times*, July 4, 2005, p. A4.

4. Patrick Simon, quoted in Peter Ford, "Next French Revolution: A Less Colorblind Society." *Christian Science Monitor*, November 14, 2005, p. 4.

5. Quoted in Kevin J. O'Brien, "Last Out, Please Turn Off the Lights: Poor Economy Is Driving East Germans from Home." *New York Times*, May 28, 2004, pp. W1, W7.

Joe Hobbs

A woman of Norway. From the northern tip of Scandinavia to the boot of Italy, Europe boasts an extraordinarily rich range of ethnicities.

MODULE 4.2
THE EUROPEAN PERIPHERY

Just as there is a recognizable European core, where the region's demographic, economic, and political weight is centered, there is also a **European periphery.** This "rimland" consists of countries whose interests are tied closely to those of the core and are strongly influenced by the core countries but have little reciprocal influence. Even integration in the European Union does not make the smaller countries equal partners with the larger ones.

Although they are grouped here under the single heading of "European periphery," these countries are astonishingly diverse. To facilitate our discussion, it is useful to subdivide these countries into three broad categories: those of northern, southern, and eastern Europe (●Figure 4.2.1). Even within those subgroups, the countries have distinct national identities, histories, and experiences. In this module, we explore the geographic personalities and economic and political dynamics of the countries outside the European center. The major traits of each nation are listed in Table 4.1 on pages 70–71.

▪module outline

▪module objectives

This module should enable you to

→ Recognize the geographic, economic, and political factors that have kept these subregions secondary to the European core in power and influence

→ Appreciate the benefits and drawbacks that people realize from different political and economic systems, including the welfare state and collectivization

→ Recognize the impacts of the wars and their aftermath on the political geographies of the region

→ Observe the trend toward greater unity under the European Union and NATO and the concurrent trend of devolution of power from central governments to provinces

→ Relate the often troubled histories of ethnic minorities in Europe and how aims of ethnic peoples have led to war, terrorism, and the redrawing of national boundaries

→ See that some of the environmental obstacles to development (mountains, dry lands, and ice, for example) are also potential tourist assets that may promote development

● **Figure 4.2.1** The subregions and countries of the European periphery.

4.2.1 Northern Europe: Prosperous, Wild, and Wired

Denmark, Norway, Sweden, Finland, and Iceland are the countries of northern Europe (●Figure 4.2.2). The countries are often grouped geographically under the regional term Norden, or "the North," and their traits are qualified by the adjective Nordic. These countries are also known collectively as Scandinavia or the Scandinavian countries. But this regional name is ambiguous. It sometimes refers only to the two countries of the Scandinavian Peninsula, Norway and Sweden. More often, it includes these countries plus Denmark, and sometimes it includes these three plus Iceland. When Finland is included in the group, the term Fennoscandia, or the Fennoscandian countries, is used to acknowledge the cultural and historical differences between the Finns and the other northern nations.

● **Figure 4.2.2** Principal features of northern Europe.

PROBLEM LANDSCAPE

Save the Whales . . . for Dinner

One reason that both Norway and Iceland have refused to join the European Union is that both have economically vital fishing industries, and they fear that common EU policies on fishing might diminish those profits. The two countries, along with Japan, also are distinguished by their unique perspective on another marine resource: the world's whales. People in all three countries maintain that populations of some whale species, especially the minke (pronounced *ming'*-key), have rebounded to levels that should allow regular, limited harvesting for human consumption. Whale meat has long been prized by their palettes; "save the whale for the dinner table" is the Norwegian response to a popular Western bumper sticker (•Figure 4.2.A).

In a landmark agreement honored by most of the world's marine countries, the International Whaling Commission (IWC) banned commercial whaling worldwide in 1986. The Norwegians never accepted the ban and defiantly continue to take about 600 minkes each year. Norwegians eat whale meat without remorse, arguing that the rest of the world is foolish in its insistence that whales are a "sacred cow." Even Norway's famously staunch environmentalists see no problem with the whale harvest.

The other whaling countries are more circumspect. The 1986 ban forbids commercial whaling but allows whales to be killed for scientific purposes and for subsistence (the latter clause allows Native Americans in Alaska, for example, to hunt whales). Iceland and Japan have exploited the loophole of the scientific purpose and in the process put whale meat on the table. They argue that there is a legitimate need for the science: Iceland estimates that its territorial waters are inhabited by more than 40,000 whales that eat perhaps 1 to 2 million tons of fish each year. Only by studying the whales' stomach contents can reliable estimates of the whales' impact on Iceland's vital fishing industry be evaluated. The studies accomplish two other goals: They reduce the pressure on the fish resource, and they allow whale meat to find its way to markets for ordinary consumers.

Killing whales does carry some costs. Iceland resumed commercial whaling in 2006, establishing quotas of 30 minkes and 9 fin whales each year. Most Icelanders support the hunt, but others argue that whale watching—which previously had drawn 90,000 tourists yearly—was the more economically sensible way to manage the resource. Many whale-watching groups have canceled tours to Iceland to protest whaling. And with "save the whales" an international mantra, Iceland, Norway, and Japan are subjected to intense criticism from around the world.

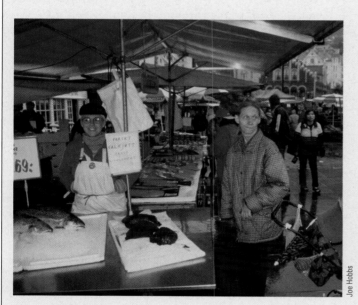

• **Figure 4.2.A** Norwegians are not shy about buying and selling whale meat (read the vendor's sign).

Traits of Northern Europe

No other highly developed countries in the world have their populated areas at such high latitudes (generally those of Alaska; see Figure 4.4, page 76). Westerly winds from the Atlantic, warmed in winter by the North Atlantic Drift, moderate the climatic effects of this northern location. Temperatures average above freezing in winter over most of Denmark and along the coast of Norway. But away from the direct influence of the west winds, winter temperatures average below freezing and are more severe at elevated, interior, and northern locations. In summer, the ocean has a cooling rather than a warming influence, and most of northern Europe's temperatures in July average no higher than the 50s or low 60s Fahrenheit (10°C

to 18°C). The populations of the countries tend to cluster in the southern sections. All except Denmark have large areas of sparsely populated terrain where the rigors of a high-latitude environment pose problems for development.

Historical interconnections and cultural similarities are important factors in the regional unity of northern Europe. Historically, each country has been more closely related to others of the group than to any outside power. In the past, the countries often waged war among themselves. But since the early 19th century, relations among the northern European peoples have been close and peaceful. Recently, they have differed on how to deal with integration into the European Union. Only Norway and Iceland have opted out, due in large part to their interests in marine resources (see the

Problem Landscape—in this case, Seascape—issue on page 123). The other three joined the European Union, with Finland the most enthusiastic to do so.

The languages of Denmark, Norway, and Sweden descend from the same ancient Germanic tongue and are mutually intelligible. Icelandic, although a branch of the same root, has changed less from the original language and has borrowed less from other languages. It derives from Old Norse, the language of the Vikings, and Icelanders have been proud of their linguistic conservatism. Only Finnish, which belongs to the entirely distinct Uralic language family, is completely different from the others. Even in Finland, however, about 6 percent of the population speaks Swedish as a native tongue, and Swedish is a second official language. Remarkably, English is the lingua franca of the Nordic countries, and many businesses insist on conducting all affairs in English.

The five countries have other traits in common. About 90 percent of the population belongs nominally to the Evangelical Lutheran Church, but only a very small percentage attends services; people tend to be quite secular. The Lutheran church is a state church, supported by taxes, and is the most all-embracing organization aside from the state itself. The countries share many similarities in law and political institutions. They all have a long tradition of individual rights, broad political participation, limited governmental powers, and democratic control. They are such peaceful democracies that rare events like the assassination of Sweden's prime minister in 1986 and its foreign minister in 2003 shock Scandinavians to the core.

High standards of health, education, individual security, and creative achievement are common to the Nordic nations. They have increasingly high-tech and service-oriented economies, and per capita, they are the most Internet-connected people in the world. They are the ultimate welfare states, with the governments supplying services such as health care, pensions, and up to four months of sick, disability, or parental leave each year. This service system, known as the **Nordic Model,** is supported by some of the highest tax rates in the world—up to 55 percent in Sweden, for example. Taxes on alcoholic beverages are especially high in an effort to discourage the kind of epidemic alcoholism that plagued Sweden in the 19th century.

Globally, the Nordic countries always rate near the top by quality-of-life measures like the Human Development Index (HDI). For example, Norwegians have the highest overall standard of living, measured by the HDI; Finland has the world's lowest corruption rate, and the World Economic Forum rates its economy the most competitive in the world (just ahead of the United States); the Nordic peoples are happiest with their jobs; and Nordic women have the greatest equality with men. High quality of life may seem improbable given the region's challenging physical environments, but the countries have specialized economies that compensate for their resource limitations. Following is a

summary by country of these specialties and other unique features of the Nordic countries.

Denmark

Denmark is home to the largest city in northern Europe and is the most densely populated of the five countries, yet it is the most dependent on agriculture. The Danish capital of Copenhagen (population 1.6 million; ●Figure 4.2.3), with almost one-third of Denmark's population in its metropolitan area, is clearly recognizable as a primate city. Copenhagen is located on the island of Sjælland (Zealand) at the extreme eastern edge of Denmark (see Figure 4.2.2). Sweden lies only 12 miles (c. 20 km) away across the Sound (Øresund), the main passage between the Baltic Sea and the North Sea. The city grew beside a natural harbor well placed to control traffic through the Sound. Before the 17th century, Denmark controlled adjacent southern Sweden and levied tolls on all shipping passing to and from the Baltic. Copenhagen still does a large transit and **entrepôt** (commercial center) business in North Sea–Baltic Sea trade.

The first land link to connect Denmark and Sweden—and thus the Scandinavian Peninsula with mainland Europe—was opened in 2000. The 10-mile (16-km) Øresund Fixed Link bridge and tunnel for rail and auto traffic connects Malmö, Sweden, with Copenhagen's airport. This link is like the Channel Tunnel in breaching a historical gap between the European mainland and a vital outlying region. Danes cross the bridge to buy Sweden's low-taxed children's shoes and clothing, while Swedes come mainly to buy inexpensive Danish alcoholic beverages. Denmark's food, metal, and chemical exports reach markets across this bridge and around the world from its Copenhagen port and airport.

Membership in the European Union and a location adjacent to the most populous country in western Europe (Germany) have provided a ready market for Danish agricultural

● **Figure 4.2.3** Nyhavn Harbor, Copenhagen

products. Quality farmland is Denmark's greatest natural resource. About three-fifths of Denmark's land is arable, representing the largest proportion of any European country. Although not productive for crops, the Faroe Islands (between Norway and Iceland) and Greenland are considered integral, although self-governing, parts of Denmark. Their peoples have the rights of Danish citizens. The Faroes are a group of treeless islands where some 46,000 people make a living by fishing and grazing sheep. As a geologic extension of North America, Greenland is discussed in the final section of this book.

643

Norway

Norway stretches for well over 1,000 miles (c. 1,600 km) along its west side and around the north end of the Scandinavian Peninsula. The peninsula, as well as Finland, occupies part of the **Fennoscandian Shield,** a block of ancient, stable metamorphic rocks. The western margins of the block in Norway are extremely rugged. Most areas have little or no soil to cover the rock surface, which was scraped bare by the continental glaciation that centered in Scandinavia. Most of Norway is rugged wilderness of coniferous forest and tundra, and only 3 percent of the land is arable.

77

The terrain hinders both agriculture and transportation. Glaciers deepened the valleys of streams flowing from Norway's mountains into the sea, and when the ice melted, the sea invaded the valleys. This process created the famous **fjords**—long, narrow, and deep extensions of the sea into the land, edged by steep cliff faces and valley walls (•Figure 4.2.4). It is difficult to build highways and railroads parallel to a coast like this. The fjords provide some of the finest harbors in the world, but most of them have practically no productive hinterland that would justify port construction. Much of Norway's population is scattered along these waterways in small, isolated clusters, interconnected by boat traffic or long, roundabout roads. One of the largest and most scenic of the settlements is Bergen, established as a trading post for the medieval merchant network known as the **Hanseatic League.**

In Norway's far north, there are about 40,000 Saami, known as Lapps to outsiders. Formerly nomadic reindeer herders, the Saami of Norway, Sweden, Finland, and Denmark—now totaling around 80,000 in the arc of land they call **Sápmi**—have generally settled down to take up manufacturing and service jobs in their home countries. Norway's Saami, recognizing that significant gold and diamond reserves underlie their traditional territories, are calling for self-determination, especially greater rights to minerals and other resources. The Norwegian Saami already have their own parliament, but they want even more devolution of power from Oslo.

83

Norway also controls the Arctic island group of Svalbard, whose largest island is Spitsbergen. Although largely

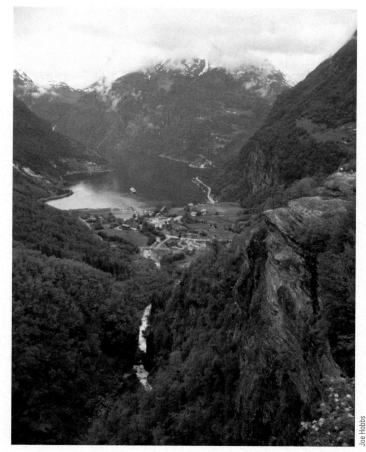

• **Figure 4.2.4** Norway's scenic fjords were carved by glaciers and then filled by rising sea levels. This is Geiranger Fjord.

covered by ice, the main island has the only substantial deposits of high-grade coal in northern Europe. Norwegian and Russian companies now exploit these resources (see Figure 5.5, page 158).

More than half of all Norwegians live in the southeastern core region, centered on the capital and primate city of Oslo (population 800,000). Located at the head of Oslo Fjord, where several valleys converge, Oslo is Norway's principal seaport and industrial, commercial, and cultural center. Valleys are wider and the land is less rugged in this region, where the most extensive agricultural lands and the largest forests in the country are located.

Apart from its sparse agricultural base, Norway has rich natural resources relative to its small population of less than 5 million. Given Norway's beauty, it is difficult to imagine anyone wanting to leave. But the benefits of industrialization came late to this challenging land, and emigration was the ticket to prosperity imagined by many hardscrabble farmers and fishermen. Key resources that keep Norwegians home today include spectacular scenery (a major draw for international tourists), hydropower, forests, and a variety of minerals—notably, oil. Hydroelectricity powers sawmilling, pulp and paper production, metallurgy, electrochemical industries, and industries that produce various goods for Norwegian consumption and some export trade (•Figure 4.2.5). Large oil and natural gas deposits in the Norwegian sector of the North Sea floor prompted an economic boom for Norway, beginning in the 1970s (•Figure 4.2.6). In just a few years, Norway's oil and gas exports grew from nothing to the country's largest by value (representing about 35 percent of exports), and Norway advanced from being reasonably well-off to being one of Europe's richest countries as measured by per capita GNI PPP (see Table 4.1, pages 70–71). Unlike Britain's, Norway's share of the North Sea oilfields are expected to reveal new reserves. Meanwhile, older economic emphases remain

• **Figure 4.2.5** Norway is blessed with mountains and rivers in the right combination for producing hydroelectricity. Hydropower provides almost 100 percent of the country's electricity. The barrier just right of center is part of a small hydropower station.

• **Figure 4.2.6** One of Norway's offshore oil platforms under construction, in the port of Stavanger, for use in the North Sea. Built to withstand the notoriously violent weather of the region, these are also supplied with many amenities to keep the workers comfortable.

important. A seafaring tradition going back to the Vikings is reflected today in Norway's large merchant, recreational cruising, and fishing fleets.

Norway is one of the few European countries that does not belong to the European Union; in 1994, Norwegians voted by a narrow margin not to join. With the economic stability provided by Norway's energy resources, most Norwegians felt they could thrive without EU protection and intervention. Coalitions of farmers, fishermen, environmentalists, and nationalists insisted on the relative independence of Norway from bureaucrats in Brussels, the EU's administrative home. Farmers and fishermen also wanted to retain their Norwegian state subsidies, which are even more generous than those offered by the European Union. Supporters of EU membership warned that Norway's petroleum income would decline in the future and the country would do well to seek the long-term economic security envisioned by EU policy.

Sweden

In the Middle Ages and early modern times, Sweden was a powerful and imperialistic country. By the 17th century, the Baltic Sea was essentially a Swedish lake. During the following century, however, the rising power of Russia and Prussia put an end to Swedish imperialism, aside from a brief campaign in 1814 in which Sweden won control of Norway from Denmark. Although it keeps a strong army as a deterrent, Sweden has not engaged in a war since then, and it has become known as one of Europe's most successful neutral nations. The long period of peace has helped promote economic success and social well-being in what is today the largest and most populous (9.1 million) country of northern Europe. Sweden is also renowned as the most multicultural of the five countries. Whereas Denmark and Norway have imposed strict controls on immigration, Sweden has few restrictions; already 25 percent of the population is of non-Swedish ethnicity, and that number is growing steadily.

In the northwest, Sweden shares the mountains of the Scandinavian Peninsula with Norway. To the north, between the mountains and the shores of the Baltic Sea and the Gulf of Bothnia, there are ice-scoured forested uplands, and a similar small area of this type is around Lake Vättern, south of Stockholm. In its central area of great lakes lies a block of good farmland, and in the south, there are rolling, fertile farmlands like those of Denmark (•Figure 4.2.7). Unlike Denmark, Sweden has not developed a specialized export-oriented agriculture, and unlike Norway, Sweden commands enough good land to supply all but a minor share of its small population's food requirements. The introduction of new laborsaving technologies in the forestry industries has reduced job opportunities in this most traditional occupation in the country's remote lands.

Engineering and metallurgical industries distinguish the Swedish economy. These industries have evolved out of a centuries-old tradition of mining and smelting. The modern Swedish steel industry continues to focus on its famous high-quality steel, especially from large plants on the Baltic and Bothnian coasts. Swedish exports employing this steel include automobiles (Volvo and Saab), industrial machinery (including robots), office machinery, telephone equipment, medical instruments and machinery, agricultural machinery, and electric transmission equipment. The country's reputation for skill in design and product quality also extends to items such as ball bearings, cutlery, tools, glassware, and furniture.

Swedish manufacturing is centered in the Central Lowland, an area favorably situated with respect to minerals, forests, hydropower, labor, food supplies, and trading potential. The Central Lowland is also the historic agricultural and population core of Sweden. The country is now

● **Figure 4.2.7** Long daylight in summer allows crops to do well where good soil is present in Scandinavia. This small farm is near Karlstad in central Sweden.

trying to promote its sparsely settled northlands as centers for growth in the telecommunications industry, beginning with telephone call center services.

The location of Sweden's capital, Stockholm (population 1.7 million; ●Figure 4.2.8), reflects the core role of the Central Lowland and Sweden's early orientation toward the Baltic and trans-Baltic lands. Stockholm is the principal administrative, financial, and cultural center of the country and shares in many of the manufacturing activities typical of the Central Lowland. But in the past century, as Sweden has traded more via the North Sea, Stockholm has been displaced by Gothenburg (Göteborg), located at the western edge of the Central Lowland, as the leading port of the country and in fact of all northern Europe. Gothenburg has a harbor that is ice-free year round, whereas icebreakers are needed to keep Stockholm's harbor open in winter.

Finland

Conquered and Christianized by the Swedes in the 12th and 13th centuries, Finland was ceded by Sweden to Russia in 1809 and remained under Russian control for more than 100 years. Under both the Swedes and the Russians, the country's people developed their own culture and sense of nationality. Finns seized the opportunity for independence provided by the collapse of tsarist Russia in 1917. During World War II, in a bid to regain land lost to the Soviet Union, Finland allied with Germany and fought the Soviets, losing much territory, life, and infrastructure. Finland made a rapid postwar recovery, in part because of pressure applied by the Soviets. Political geographers began to apply the term **Finlandization** to any country that, like Finland, claimed to be neutral but fell under the heavy influence of the Soviet Union. The Soviets required that a large part of the reparations paid to them by Finland after the war were

● **Figure 4.2.8** Stockholm's harbor is the heart of the city.

• **Figure 4.2.9** Finns enjoying nature.

in metal goods. Several small steel plants and other metalworking establishments were built to meet this demand, and they continue to operate today.

One-third of Finland lies within the Arctic Circle. Most of the country is a sparsely populated, glacially scoured, subarctic wilderness of coniferous forest, ancient igneous and metamorphic rocks, lakes, and marshes. About two-thirds of Finland is forested, mainly in pine or spruce, and exports of wood products, including paper, pulp, and timber, are important to the economy. Forest production is concentrated in the south central part of the country, known as the Lake Plateau. The majority of the population lives in the relatively fertile and warmer lowland districts scattered through the southern half of the country. Agriculture focuses on hay, oats, barley, and livestock production, especially dairying. Wherever they live and whatever their occupations, Finns are nature lovers who enjoy retreating to lakeside and seaside saunas, where they take the customary hot steam treatment followed by an invigorating plunge into a pool or pond (•Figure 4.2.9).

Helsinki (population 1.2 million), located on the coast of the Gulf of Finland, is the nation's capital, largest city, main seaport, and principal commercial and cultural center. It is also the country's most important and diversified industrial center. Hydroelectricity originally powered Finland's industries, but this source has now been surpassed by both coal and nuclear energy. Manufacturing is the most important economic sector, with exports including paper and forestry products, metals, electronics, and telecommunications equipment, notably the Nokia products which make up the world's largest market share of cell phones and account for one-quarter of the country's export value.

Iceland

Iceland is a fairly large (39,699 sq mi/102,819 sq km), mountainous island in the Atlantic Ocean just south of the Arctic Circle. Reflecting the island's position astride the Mid-Atlantic Ridge, birthplace of two major plates

of the earth's crust, Iceland's landscapes have been shaped by volcanism (•Figure 4.2.10). A volcanic eruption in 1783 wiped out most of the crops Icelanders were growing at the time, causing a famine that killed 20 percent of the population. An underwater eruption in 1963 gave birth to the brand-new island of Surtsey, off Iceland's southeast coast. The country derives about 10 percent of its electricity from its volcanic resources in the form of geothermal energy; hydropower supplies another 80 percent. One enterprising community in northern Iceland wants to harness its hot waters to develop alligator farming, which the locals say will attract tourists and produce exports of alligator meat and hide.

Because of the proximity of the relatively warm North Atlantic Drift, the coasts of Iceland have winter temperatures that are mild for the latitude. The capital, Reykjavik (64°N), has an average January temperature that is practically the same as New York City's (41°N). Iceland's vegetation is primarily tundra. Summer temperatures average no higher than about 52°F (11°C), limiting agriculture to

• **Figure 4.2.10** Volcanic activity presents a serious natural hazard in Iceland.

potatoes and some other hardy vegetables. There is some cattle and sheep ranching. Mineral resources are almost nonexistent. The economy depends mainly on exports of fish and fish products, which represent about 70 percent of export earnings. Aluminum is the second most valuable export product, but next to fish, tourists bring in the most money. While international visitors soak in Iceland's hot springs and enjoy the nightlife, Icelanders prefer Spain for their holidays.

Despite its environmental limitations, Iceland has been continuously inhabited since at least the 9th century and is now the home of a progressive and democratic republic with a population of about 300,000. Practically all Icelanders live in coastal settlements, with one-third of the population living in and near Reykjavik. Its ice-free ports and strategic location in the North Atlantic have given Iceland geopolitical significance, and it passed the Cold War years as a Western ally. It continues to be a member of the North Atlantic Treaty Organization (NATO).

88

4.2.2 Southern Europe: The Mediterranean World

The four main countries of southern Europe, for the purposes of this text, are Portugal, Spain, Italy, and Greece. They are on peninsulas extending southward from the main body of Europe. Portugal and much larger Spain share the Iberian Peninsula, Italy occupies the Italian Peninsula, and most of Greece is on a subpeninsula, between the Ionian and Aegean Seas, of the larger Balkan Peninsula. Southern Europe also includes the microstates of Vatican City (within the city of Rome), San Marino (entirely surrounded by Italy), and the British colony of Gibraltar. Three of the peninsular countries include islands in the Mediterranean, the largest of which are Italy's Sicily and Sardinia, the Greek island of Crete, and the Spanish Balearic Islands. Some Mediterranean islands lie outside these countries, including Corsica, which is a part of France, and the independent nations of Malta and Cyprus. These are the lands of Mediterranean Europe (see Figure 4.2.1).

The Environment of Mediterranean Agriculture

In terrain, population distribution, climate, and agriculture, the countries of Mediterranean Europe have many similarities. Rugged terrain predominates. Plains tend to be small, facing the sea, and separated from each other by mountainous areas. Population distribution corresponds generally to topography, with the lowland plains being densely populated and the mountainous areas much less so, although some rough areas have surprisingly high densities (see Figure 4.2, page 72). The Alps and the Pyrenees have long served as a permeable barrier between the European core and southern Europe.

A distinctive natural characteristic of southern Europe is its Mediterranean climate, combining mild, rainy winters with hot, dry summers. Characteristics associated with this climate become increasingly pronounced toward the south. The northern extremities of both Italy and Spain have different climatic characteristics. Except for a strip along the Mediterranean Sea, northern Spain has a marine west coast climate like that of northwestern Europe. It is cooler and wetter in summer than the typically Mediterranean climate areas. The basin of the Po River in northern Italy has a relatively wet summer and cold-month temperatures in the lowlands averaging just above freezing. Much of the high interior plateau of Spain, the agriculturally productive Meseta region, is cut off from rain-bearing winds and has less precipitation and colder winters than is typical of the Mediterranean climate. 29

Classic Mediterranean crops are olives, grapes, figs, citrus fruits (especially oranges and lemons), vegetables such as tomatoes and eggplants, and herbs such as thyme and oregano (•Figure 4.2.11). These are typical foods of people all around the Mediterranean basin, and nutritionists regard them as excellent for human health. Olive oil is a major source of fat in the typical **Mediterranean diet,** and virtually all of the world supply is produced in countries along the Mediterranean Sea or in the Mediterranean climate zone (Spain is the largest producer). Grapes go mainly to production of wine, a standard household beverage in southern Europe and a major export product. Winter wheat, which is harvested in May or June, has long been important in the region. Barley replaces it in dry and less fertile soils. Beans and peas are a major source of protein in this climatic zone, where livestock make only a limited contribution to the food supply. Extensive areas too rough for cultivation are used for grazing, mainly by sheep and goats. In many places, grazing depends on **transhumance**—the use of lowland pastures during the wetter winter and mountain pastures during the summer. In some areas, nonfood crops like Greece's tobacco supplement the basic Mediterranean products.

• **Figure 4.2.11** The mild Mediterranean climate offers the opportunity to grow a variety of fresh foods regarded by many as ingredients in a healthy diet.

© SIME s.a.s/eStock Photo

Mediterranean agriculture is most productive where land is irrigated. Since ancient times, people have made extraordinary efforts to deliver water to fields during the dry, hot summer. Aqueducts built at the time of the Roman Empire were constructed particularly well, and many remain today as unique elements of the cultural landscape (•Figure 4.2.12).

An Economic Revival

Though more prosperous than eastern Europe, southern Europe has historically lagged behind northern Europe and the European core in economic terms (see Table 4.1 on pages 70–71). The main reason for this poverty was a lagging development of industry related to unfavorable social and resource conditions during the 19th and early 20th centuries, when many European countries were industrializing rapidly. Lack of industrial employment left too many people in crowded small-farm areas where output per person was low. The absence of large-scale trade (and therefore of capital) to invest in manufacturing fed this underdevelopment. Mainly poor and uneducated populations offered little skilled labor or dynamic market potential. Wealthy landowners often controlled both land and government and had little interest in overhauling the agrarian system that made them rich and powerful. Unstable, undemocratic, and ineffective governments were uninterested in economic development. Resource deficiencies included a shortage of coal and iron ore deposits, scarce wood supplies due to aridity and deforestation, and lack of hydropower in the areas of Mediterranean climate, where streams often dry up in summer. Finally, the rugged topography of much of the region made road and rail construction difficult and costly.

Because of such impediments, only three significant industrial areas developed in southern Europe before World War II. The most important was in the Po Plain of northern Italy (see Figure 4.1.3, page 98). Here imported coal and raw materials together with inexpensive labor boosted production in an area specializing in textiles but also including automobiles, aircraft, and other industries. A smaller industrial area developed in the northeastern corner of Spain in and around Barcelona. A third area developed in the coal- and iron-mining section of northern coastal Spain. All three areas lie near the foothills of mountains that provided the basis for hydroelectricity.

In the second half of the 20th century, the economies of southern Europe finally began to improve. Italy, Spain, Portugal, and Greece have evolved from agricultural to industrial and service-oriented countries. Although the South still lags behind northern and western Europe, recent growth has been shaped by both industrial modernization and an expanding role of the European Union, to which all the countries belong. Industrial and service activities led the way, and there was a sharp decline in the proportion of the population still employed in agriculture.

A number of factors played a part in this revitalization. Improved transportation (especially by sea) made fuel and raw material imports cheaper. Sources of energy changed, with imported oil (and in Italy, geothermal power and domestic natural gas) replacing coal. Market conditions for southern European exports—fruits, vegetables, automobiles, and labor—turned more favorable after World War II, largely as a result of the rising prosperity of northwestern Europe and the expansion of European markets. Streams of temporary migrants from southern Europe found jobs in booming labor-short industries to the north. The absence of the migrants relieved unemployment, and their remittances added purchasing power in their home areas. Foreign money flowed into the southern European countries from other sources as well. Each country received U.S. economic and military aid, and favorable economic circumstances and government policies attracted foreign investment in industrial and other facilities.

The region also saw a tremendous boom in tourism. To increasingly prosperous Europeans and Americans, southern Europe offers a combination of spectacular scenery, historical monuments, pleasant villages, sunny skies, and seacoasts and beaches (•Figure 4.2.13). Following is more insight into each country's attractions, resources, and characteristic features.

Spain

Spanish power and influence reached a zenith in 1492, when a centuries-long struggle, the *Reconquista*, ended with Spain's defeat of the kingdom of Granada, the last part of Iberia held by the Muslim Moors (•Figure 4.2.14). Spanish Catholics expelled both the Moors and a distinct population of **Sephardic Jews** from Iberia and ushered in the **Inquisition**, a repressive program designed to purge Spain of "false Christians." Also in 1492, Christopher Columbus, an Italian navigator from Genoa in the service of Spain, landed in the Americas in search of a sea route to

• **Figure 4.2.12** A Roman aqueduct in Nerja, Spain.

• **Figure 4.2.13** Mediterranean landscapes and settlements are irresistible attractions for tourists. In Greece's Cyclades Islands lies the spectacular archipelago of Santorini (Thera), surrounding a lagoon that fills the caldera (crater) left by a massive volcanic eruption about 3,500 years ago.

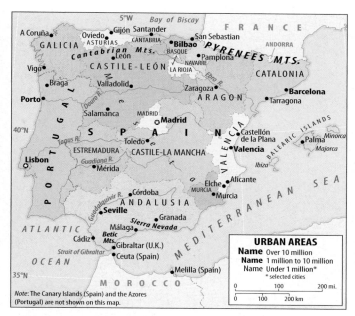

• **Figure 4.2.15** Principal features of Spain and Portugal.

• **Figure 4.2.14** The Alhambra in Granada, dating to the 9th century, is one of the greatest of all Muslim architectural accomplishments.

India and East Asia. For a century thereafter, Spain was the greatest power in Europe and probably in the world. It rapidly built an empire spreading to Italy, the Low Countries, the Americas, Africa, and the Philippines. These possessions were subsequently lost at various times during a long, slow decline of Spanish power over the next three centuries. Of the lands of the empire, only the Canary Islands, off the Atlantic coast of Morocco (see Figure 6.1, page 208), remain an integral part of Spain and attract tourists and retirees from far and wide. The Balearic Islands of Ibiza, Majorca, and Minorca, off Spain's east coast, are possessions of the country and are its main destination for sun-loving tourists.

Most of the population of the Iberian Peninsula lives on discontinuous coastal lowlands (•Figure 4.2.15). Most of the peninsula consists of the less populated Meseta plateau, with a surface lying between 2,000 and 3,000 feet (c. 600–900 m) in elevation. Mountains and deep river valleys interrupt the plateau surface, whose edges are steep and rugged. The Pyrenees and the Cantabrian Mountains border the Meseta on the north, and the Betic Mountains, with their highest elevations in the Sierra Nevada, border it on the south.

The three largest cities of Spain are the capital, Madrid (population 6.1 million), and the Mediterranean ports of Barcelona and Valencia. Madrid was chosen as the capital of Spain in the 16th century because of its location near the geographic center of the Iberian Peninsula, approximately equidistant from the various peripheral areas of dense population. Its position as capital has made Madrid the hub of the Spanish road and rail networks and thus given it business advantages. Its critical function as a railway hub put Madrid in the crosshairs of a ferocious terrorist bombing in 2004. Al-Qa'ida apparently carried out the bombings as revenge for Spain's support of the U.S.-led invasion of Iraq in 2003. In a national election just days later, bereaved and angered Spanish voters ousted the ruling party and ushered in a less pro-American government that quickly removed Spanish troops from Iraq.

Much of Spain's fishing fleet, the largest in the European Union, is based at Europe's largest fishing port, the northwestern city of Vigo. Barcelona, capital of the province of Catalonia, is Spain's major port and the center of a diversified industrial district. So prosperous is this region relative to the rest of Spain that Catalonian politicians have demanded devolution of power from Madrid. Among other things, they want assurances that water from the Ebro River in Catalonia and Aragon will not be diverted south

ETHNIC GEOGRAPHY

The Basque Country

The 2.3 million **Basques** of Spain and 300,000 Basques of France have a unique ethnicity and culture completely unrelated to those of their host country majorities (•Figure 4.2.B). Like other mi-

• **Figure 4.2.B** A Basque shepherd moves her flock of sheep along a road in the Pyrenees Mountains of southern France.

norities, they have often been the targets of discrimination and violence. This is especially true in Spain, where in 1937 a German bombing called for by Spanish fascists led to tragic loss of life among Basque civilians in the city of Guernica. By the 1960s, Basque desire for independence led to the emergence of the militant terrorist group **ETA** (an acronym for **Basque Homeland and Liberty** in Euskara, the Basque language). Spanish authorities gradually allowed limited autonomy for the Basques so that they now control their own affairs in taxation, education, and policing. For more than four decades, these concessions were not enough for ETA, which carried out attacks on Spanish interests and triggered a vigorous crackdown by Spanish security forces. ETA declared a "permanent" cease-fire in 2006, opening the prospect for even more devolution from Spain, but soon rescinded the truce and followed with more bombings.

Basque efforts to retain their unique customs have proved increasingly successful. Basque television now broadcasts on the Spanish side. Across the border, French Basques yearn for at least the kind of autonomy enjoyed by their kinfolk in Spain. For both populations, independence seems a distant and likely inaccessible goal.

to more arid Valencia, Murcia, and Andalusia, as often called for in Spain's national hydrologic planning. They say they want an independent nation within Spain that would have its own voice in the European Union and be able to attract foreign investment independently of Madrid. Spain has recognized Catalonia as an "autonomous community" and in 2006 even as a "nation under Spanish sovereignty." The meaning of *nation* is deliberately vague in this context and does not grant Catalonia independence. Politicians in Galicia, Spain's northwesternmost province, are watching Catalonia to see if they may want to pursue a similar course of devolution. An even more outstanding political challenge for Spain today is the future of its independence-minded Basque population (see Ethnic Geography, above).

After a devastating civil war from 1936 to 1939 and then World War II, Spain gradually developed a stable democratic system. Like the United Kingdom, it is a constitutional monarchy. Relations between Spain and Britain are generally quite good, and they tend to see eye to eye on important international issues. However, there is a thorn in their relations: the tiny enclave of Gibraltar, on the northern side of the Strait of Gibraltar, which Britain has held for three centuries. Britain claims that its sovereignty over Gibraltar helps guarantee the security of one of the most economically critical waterways in the world, while Spain sees Britain's claim as an anachronistic, unjustified vestige of colonialism. To defuse this embarrassing row between

two fellow EU and NATO members, both countries are looking for a sovereignty-sharing formula. This would require approval by Gibraltar's 30,000 inhabitants. Mainly of Genoese, Maltese, and Jewish ancestry, Gibraltarians are overwhelmingly in favor of remaining British colonial subjects. They fear losing the defense, tourism, and tax-free shopping revenues they enjoy under British rule.

Portugal

In the 15th century, Portugal took the lead in seeking a sea route around Africa to southern and eastern Asia. Henry the Navigator of Portugal explored the coast of northwestern Africa between 1418 and 1460. In terms of trade, Vasco da Gama's 1499 voyage to the Indian port of Calicut was even more important than Christopher Columbus's voyage to the Americas seven years earlier. In the century that followed, Portugal dominated European trade with the East, building an empire there and across the Atlantic in Brazil. Then Portugal's fortunes declined rapidly, with Brazil remaining (until 1822) its only major overseas colony. Portugal also ruled large African territories until the 1970s. The Azores, an Atlantic Ocean island group about one-fourth of the distance from the Portuguese coast to the nearest coastline of the United States, still belong to Portugal. Before the days of long-haul aircraft, many transatlantic flights paused in the Azores for refueling.

Portugal's two largest cities are seaports on the lower courses of rivers crossing the Meseta and flowing through Portugal to the Atlantic Ocean. Lisbon (population 2.6 million), on a magnificent natural harbor at the mouth of the Tagus River, is the country's capital and leading seaport. Little architecture survives from before 1755, when an earthquake destroyed the city. The smaller city of Oporto, at the mouth of the Douro River, is the regional capital of northern Portugal and the commercial center for trade in Portugal's famous port wine, which comes from terraced vineyards along the hills overlooking the Douro Valley.

Since joining the European Union in 1986, Portugal has developed a diversified economy based strongly on services, including finance and telecommunications. Key exports are clothing, footwear, machinery, chemicals, and cork products made from its distinct and traditional cork industry.

Italy

80 Italy (●Figure 4.2.16) has bestowed an enormous heritage on Western civilization. Its main period of preeminence spanned five centuries, during which the Roman Empire ruled the entire Mediterranean basin and some lands beyond. Under the empire, Christianity arose and spread, and by the time the empire collapsed in the 5th century C.E., Rome had become the seat of the popes and the Roman Catholic Church. During the Middle Ages, a number of Italian cities, notably Venice, Genoa, Milan, Bologna, and Florence, became powerful independent states. Venice and Genoa controlled maritime empires in the Mediterranean region, and all the cities profited as traders between the eastern Mediterranean and Europe north of the Alps. After unification in 1870, Italy attempted to emulate 75 past glories, but the empire that it acquired in Libya and Ethiopia was lost when Italy was defeated in World War II.

Reflecting its position astride continental plates, Italy has a diverse and often rugged topography. The Alps and the Apennines are the principal mountain ranges. Northern Italy includes most of the southern slopes of the Alps. The Apennines, lower in elevation than the Alps, form the north–south-trending backbone of the peninsula, extending from their junction with the southwestern end of the Alps to the toe of the Italian boot, appearing again across the Strait of Messina in Sicily, where Mount Etna, a volcanic cone, reaches 10,902 feet (3,323 m). Near Naples, volcanic Mount Vesuvius, at 4,190 feet (1,277 m), is renowned for demolishing the thriving Roman cities of Pompeii and Herculaneum, which were frozen in time under a massive ashfall in 79 C.E. (●Figure 4.2.17). Still active, Vesuvius threatens more than 3 million people in the area. West of the Apennines, most of the land between Florence on the north and Naples on the south is occupied by lower hills and mountains, many of volcanic origin. Much of the Mediterranean's monumental architecture is built of Italian marble quarried in this region.

Italy has a population density of about 510 per square mile (c. 197 per sq km), nearly twice that of Portugal, the second most densely populated country of southern Europe. The largest lowland, the Po Plain, has about one-third of the entire Italian population. Some other lowland areas with extremely high densities include the narrow Ligurian Coast centering on Genoa, the plain of the Arno River

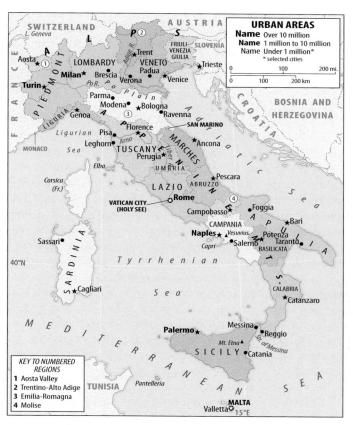

● **Figure 4.2.16** Principal features of Italy.

● **Figure 4.2.17** The eruption of Mount Vesuvius buried the Roman city of Pompeii under a vast deposit of ash, freezing the city in time until modern archaeologists brought it to light.

inland to Pisa and Florence, the lower valley of the Tiber River including Rome, and Campania around Naples.

A country evocative of romance and tradition, Italy is also a major component of the European industrial and postindustrial engine. Italy's per capita GNI PPP of $29,840 is the highest in southern Europe. Its prosperity is, however, threatened by its business culture. Many Italian firms are family-owned, and while their products are of high quality, they are hard pressed to keep up with the output of similar goods made in China. Italy's exports have fallen, and the economy has stagnated to earn the country the unwanted epithet "the sick man of Europe" (a term usually reserved for the early-20th-century Ottoman Empire). A large slice of the economic pie goes into pensions to support Europe's fastest-aging and longest-living population.

The Po Basin is the economic heart of modern Italy and the most highly developed and productive part of southern Europe. With its high-tech and high-fashion output, it has ranked with other western European centers of innovation. Functionally, this region of northern Italy belongs with the rest of the European core described in Chapter 4. Comprising just under a fifth of Italy's area, this zone produces about half of the country's agricultural and two-thirds of its industrial output. A rapid expansion and diversification of Po Basin industry, previously dominated by textiles, occurred after World War II. There was growth in engineering and in the chemical, iron, and steel industries, fueled by domestic natural gas and hydroelectric power and by imported oil.

Many cities in the basin share in the region's industry, but the leading ones are Milan (population 4.3 million), Turin, and Genoa. Milan is Italy's largest industrial city and a dominant center of fashion, finance, business administration, and railway transportation. It is well located between the port of Genoa and important passes, now shortened by railway tunnels through the Alps, to Switzerland and the North Sea countries. Turin is the leading center of Italy's automobile industry (including Fiat) and the heart of its high-tech "Techno City" region (see Figure 4.1.4, page 99).

The port of Genoa, reached from the Milan-Turin area by passes across the narrow but rugged northwestern end of the Apennines, functions as a part of the Po Basin industrial complex. The city dominates the overseas trade of the Po area and is Italy's leading seaport. Genoa is now much larger and economically more important than Venice, its old medieval commercial and political rival at the eastern edge of the Po Plain. Venice is one of the world's most beloved tourist destinations. Its canals, monuments, and people have exerted an irresistible and romantic charm on centuries of visitors. It was built on low-lying and reclaimed land, and today, 40 inches (c. 1 m) lower than it was 250 years ago, it wages a war to hold back the Adriatic as the land continues to sink and the waters rise. In 2003, construction began on what Italy calls the **Moses Project:** 79 giant sea gates that will rise from the ocean floor whenever high tides threaten Venice's lagoon (•Figure 4.2.18).

Jonathan Blair/Corbis

• **Figure 4.2.18** Venice from the air. The amphibious and vulnerable setting of the city is most apparent from this perspective.

The largest cities outside the northern industrial area are the capital, Rome (known as the "Eternal City," population 3.7 million), and Naples (population 3.8 million). Rome has a location that is roughly central within Italy and is a governmental, religious, and tourist center. Entirely surrounded by Rome is the microstate of the Holy See (Vatican City), the independent entity that is the center of the Roman Catholic Church. Yet another microstate, San Marino, lies to the north of Rome. It prospers from tourism and banking. Naples, located south of Rome on the west coast of the peninsula, is the port for the populous and productive, but poor, Campanian agricultural region. It is also the main urban center of one of Italy's major tourist regions, with attractions such as Mount Vesuvius, the ruins of Pompeii, and the island of Capri in the vicinity.

In Italy, a regional distinction has long been made between the north and the south. But recently another regional term has evolved: **Third Italy,** consisting of the relatively prosperous northeast and central portions of the country, with their luxury automobile, jewelry, textile, leather, ceramic, and furniture industries (see Figure 4.1.4, page 99). But the north-south distinction is most deeply rooted and has dimensions beyond the economic. Northerners tend to see themselves as more sophisticated and cosmopolitan than the more earthy and provincial southerners. Southern Italians, whose region is known as the Mezzogiorno, often acknowledge their agrarian roots as the source of their superior kinship values and enjoyment of life. Statistics underscore the economic discrepancies between the two Italies—for example, northern Italy has labor shortages while the south has unemployment; northern labor and industries are more productive, and income levels in the north are much higher.

The feeling among many northerners that they would be better off if they did not have to subsidize the south has even led to a secessionist movement, championed by an organization called the **Northern League,** that calls for the creation of an autonomous or independent state of **Padania** in the north (see Figure 4.B, page 83). Another organization, **Liga Veneta,** wants to create an autonomous region of Veneto, centered around the affluent city of Venice. Regardless of their differences, both northern and southern Italians take pride in the heritage bestowed by civilization from the Romans to the Renaissance and beyond, and both live on some of the most picturesque and storied landscapes on earth.

Greece

Greece, like Italy, has been a vital source area for Western life. Centuries before the Christian era, Greek artists, architects, authors, philosophers, and scholars produced many of the ideas and works that laid the foundation for Western civilization (•Figure 4.2.19). Between about 600 and 300 B.C.E., the sailors, traders, warriors, and colonists of the Greek city-states spread their culture throughout the Mediterranean area. A second period of Greek power and influence occurred in the Middle Ages, when Constantinople (modern Istanbul, Turkey) was the capital of the large Byzantine Empire. The Eastern Orthodox branch of Christianity developed in and diffused from this empire.

Greece's rugged topography and seismic history reveal that it is even more a product of tectonic activity than Italy is (•Figure 4.2.20). The collision of the African and European plates of the earth's crust in this region sometimes produces devastating earthquakes and volcanic eruptions. In about 1650 B.C.E., a massive volcanic explosion tore apart the Mediterranean island of Thera (Santorini; see Figure 4.2.13), producing a catastrophic tsunami ("tidal wave") and ash clouds large enough to ruin crops and change cli-

• **Figure 4.2.20** Principal features of Greece and Cyprus.

mates for decades. There is growing evidence that this event and its aftermath brought an end to the Minoan civilization on the island of Crete. Some popular scientific theories link the catastrophe to the legendary lost civilization of Atlantis.

The islands of the Ionian and Aegean Seas, which are some of the most prized holiday destinations in the world, are seaward extensions of the Pindus Mountains that occupy most of the peninsula north of the Gulf of Corinth. The part of Greece south of the Gulf of Corinth, known as the Peloponnesus, is composed mainly of the Arcadian mountain knot. Along the coasts of Greece, most people live in small lowlands facing the sea between mountain spurs. An especially famous lowland is the Attica Plain, still dominated as in ancient times by Athens and its seaport, Piraeus.

Athens (population 3.8 million), the primate city of Greece, is home to a third of the country's population. It is the capital, main port, and main industrial center. Its development is based on imported fuels, and manufactured products are sold almost entirely in a domestic market whose main center is Athens itself. Tourism is the mainstay of the Greek economy, with nearly $15 billion a year coming from this trade. Greece boasts the beauty of the Aegean Sea, the Parthenon and other monuments of early Greek culture, and vibrant modern arts. Greece prides itself as the birthplace of the Olympics and, in 2004, hosted the summer games.

• **Figure 4.2.19** Ancient Greek architecture is part of the "classic" tradition that is the model for major government and other institutional buildings in the West today. This is the temple of Aphea on the island of Aegina.

The Mediterranean Islands

In addition to Crete, Rhodes, and the many smaller islands of Greece, there are other Mediterranean islands of considerable size and importance. North of Italy's island of Sardinia is the island of Corsica, which was an independent country before France absorbed it in 1768. Corsicans are calling for greater autonomy or even independence. France has answered by agreeing to allow for the devolution of some powers, giving the island's parliament some legislative autonomy, and letting the Corsican language be taught in schools.

Malta is an independent country of less than half a million people, the least populous of the European Union countries. Reflecting their crossroads location between Italy and North Africa, the Maltese are descendents of ancient Carthaginians, Phoenicians, Italians, and other Mediterranean peoples. Most speak both Maltese—an amalgam of languages spoken by its diverse ancestors—and English. Malta was a British colony until 1964. Its economy depends on tourism, financial services, and capitalizing on its strategic mid-Mediterranean location, shipping services.

The large Mediterranean island of Cyprus (see Figure 4.2.20), located near southeastern Turkey, came under British control in 1878 after centuries of Ottoman Turkish rule. In 1960, it gained independence as the Republic of Cyprus. The critical problem on the island is the division between the Greek Cypriots, who are Greek Orthodox Christians and make up about three-fourths of the estimated population of 1 million, and the Turkish Cypriots, who are Muslims and make up about one-fourth of the population. Agitation by the Greek majority for union (*enosis*) with Greece was prominent after World War II and led in the 1950s to widespread terrorism and guerrilla warfare by Greek Cypriots against the occupying British. Violence also erupted between Greek advocates of *enosis* and the Turkish Cypriots, who greatly feared a transfer from British to Greek sovereignty.

A major national crisis erupted in 1974 when a short-lived coup by Greek Cypriots, led mainly by Greek military officers, temporarily overthrew Cyprus's President Makarios, who had embraced a conciliatory policy toward the Turkish minority. Turkey then launched a military invasion that overran the northern part of the island. Cyprus was soon partitioned between the Turkish north and the Greek south. A buffer zone (the **Attila Line** or **Green Line**) sealed off the two sectors from each other, and even the main city of Nicosia was divided. A separate government was established in the north, and in 1983, the Turkish Republic of Northern Cyprus was proclaimed. Only Turkey recognized this state. Meanwhile, the internationally recognized Republic of Cyprus functioned in the Greek Cypriot sector, which comprises about three-fifths of the island. Both republics had their capitals in Nicosia (population 250,000).

The north had dominated the economy prior to the partitioning of Cyprus, but since then, the north has had severe economic difficulties while Greek Cyprus has prospered. Most outside nations have refused to trade directly with Turkish Cyprus since the invasion, and economically weak Turkey has not been able to boost the north's prospects. The depressed north remains tied to Turkey, while the Greek sector makes effective use of economic aid from Greece, Britain, the United States, and the United Nations. Tourism has flourished, and new businesses have thrived.

Events leading to the EU's expansion in the "big bang" of 2004 provided a strong incentive for the two sides to resolve their differences and join the union. The United Nations devised a plan for the two halves to vote separately in a referendum on reunification. Had both sides voted in favor of reunification, a united Cyprus would have taken up membership in the European Union. However, the United Nations and European Union agreed that should either the north or the south, or both, reject the referendum, only the Greek south would actually join the European Union. In the vote, the Turkish north voted overwhelmingly in favor of reunification, while the Greek south overwhelmingly rejected it. The somewhat perplexing result is that in EU terms, "Cyprus" nominally refers to both parts of the island, but only the Greek south is a de facto member of the European Union. This allows the UN, the EU, and other international bodies to continue their policy of not officially recognizing the north as a legitimate, separate political entity. Turkey, long seen as the obstacle to reunification of Cyprus, won rare acclaim in Europe because of the Turkish Cypriots' strong vote in favor of reunification. Turkey's prospect to join the EU seemed to strengthen, but in 2006, it refused to open its ports to Greek Cypriot vessels. The EU halted membership discussions with Turkey over this issue. 91, 268

4.2.3 Eastern Europe: Out from Behind the Curtain

For the purposes of this text, the eastern European countries are those in an arc extending from the Gulf of Finland in the north to the Balkan Peninsula in the south. They are, in the north, Estonia, Latvia, Lithuania, Poland, the Czech Republic, and Slovakia; and in the south, Hungary, Romania, Bulgaria, Slovenia, Croatia, Bosnia and Herzegovina, Serbia, Montenegro, Albania, and Macedonia. Other sources, especially those that follow the strict definition of including western Russia in Europe, refer to these countries as "Central Europe" or "East Central Europe." Based on a definition of Europe that includes a large portion of Russia, the French Geological Institute has calculated that a spot in eastern Lithuania is the geographic center of Europe.

These countries are in the process of reinventing themselves after more than four decades of direct or indirect control by the Soviet Union. Prior to German reunification in 1990, the former East Germany was also part of eastern Europe, and it is brought into the present discussion when appropriate. Three of the countries—Estonia, Latvia, and Lithuania—were incorporated into the Soviet 112

Union during World War II, while many others became **Soviet satellites,** with local Communist governments effectively controlled from Moscow. Exceptions were Albania and the former Yugoslavia (consisting of Slovenia, Croatia, Bosnia and Herzegovina, Serbia, Montenegro, and Macedonia), where national Communist resistance forces took power on their own as German and Italian power collapsed.

Majority Slavic ethnicity, former Communist status, and subjugation to Soviet interests were among the few unifying themes of the region prior to the end of the Cold War. Now the true complexity of the region is more apparent, and this section of the chapter explores various themes of unification and distinction within the eastern European mosaic.

Physical Geography

Eastern Europe is a checkerboard of mountains and plains (see Figure 4.3, page 75). Located mainly in the center and south, the mountains vary widely in elevation and degree of ruggedness. The Carpathian Mountains and lower ranges farther west along the Czech frontiers form a central mountain zone. The Carpathians extend in a giant arc for about 1,000 miles (c. 1,600 km) from Slovakia and southern Poland to south central Romania. Eastern Europe's mountains still contain much forested wilderness and are home to Europe's largest concentrations of carnivores, including wolves, bears, and lynx (●Figure 4.2.21). Under communism, much of this wilderness was untouched state-owned land. There are fears now that the return of land to private hands and the expansion of regional economies will lead to rapid habitat and species loss.

West of the Carpathians, lower mountains enclose the hilly, industrialized Bohemian basin in the Czech Republic. On the north, the Sudeten Mountains and Ore Mountains separate the Czech Republic from Poland and the former East Germany. Eastern Europe's southern mountain zone

occupies the greater part of the Balkan Peninsula. Bulgaria, the Yugoslav successor states, and Albania, which share this zone, are very mountainous but have some productive lowlands. In Bulgaria, the principal mountains are the Balkans, extending east-west across the center of the country, and the Rhodope Mountains in the southwest. These are rugged mountains reaching over 9,000 feet (c. 2,800 m) in a few places.

In Montenegro, southern Serbia, and neighboring areas to the west and south, hills and mountains form a major barrier to travel. Through this difficult region, a historic lowland passage connecting the Danube Valley with the Aegean Sea follows the trough of the Morava and Vardar Rivers. At the Aegean end of the passage is the Greek seaport of Salonika (Thessaloniki). The Morava-Vardar corridor is linked with the Maritsa Valley by an east-west route that leads through the high basin in which Sofia, the capital of Bulgaria, is located. Along the rugged, island-fringed Dalmatian Coast, the Dinaric Alps rise steeply from the Adriatic Sea. The principal ranges run parallel to the coast and are crossed by only a few significant passes.

The plains are concentrated in the north and in the center along the Danube River (see The World's Great Rivers, page 138). Agriculturally, they range from some of Europe's better farming areas to some of the poorest. Between mountain-rimmed Bohemia and the Carpathians of Slovakia, the lowland corridor of Moravia in the Czech Republic provides a busy passageway. Through this corridor run major routes connecting Vienna and the Danube Valley with the plains of Poland. Near the Polish frontier, the corridor narrows at the Moravian Gate between the Sudeten Mountains and the Carpathians. Just beyond this gateway, east central Europe's most important concentration of coal mines and iron and steel plants developed in the Upper Silesian and Moravian coalfield of Poland and the Czech Republic (see Figure 4.1.3, page 98).

Two major lowlands, bordered by mountains and drained by the Danube River and its tributaries, compose the Danubian plains. One of these, the Great Hungarian Plain, occupies two-thirds of Hungary and smaller adjoining portions of Romania, Serbia, and Croatia. The second major lowland is composed of the plains of Walachia and Moldavia in Romania, together with the northern fringe of Bulgaria. The Danubian plains represent the most fertile large agricultural regions in east central Europe.

A handful of Europe's largest rivers carries the region's waters to the Black Sea (via the Danube), the Baltic Sea (via the Vistula), and the North Sea (via the Elbe). Between the Sudeten and Ore Mountains, at the Saxon Gate, the valley of the Elbe River provides a lowland connection and a navigable waterway from Bohemia to the industrial region of Saxony in Germany and, farther north, to the German seaport of Hamburg and the North Sea. To the southwest, the Bohemian forest occupies the border zone between Bohemia and Germany.

In most of eastern Europe, winters are colder and summers are warmer than in western Europe. On the Danubian

● **Figure 4.2.21** Romania's Transylvanian Alps.

Grant Smith/Corbis

151

THE WORLD'S GREAT RIVERS

The Danube

The Danube River, which supplies a vital navigable water connection between the central European lowlands and the outside world, rises in the Black Forest of southwestern Germany and follows a winding course of some 1,750 miles (2,800 km) to the Black Sea. It is the longest river in Europe (second-longest if one considers the Volga River to be European), and passes through the largest number of European countries. It is also one of the most storied, immortalized most by Johann Strauss's waltz *The Blue Danube* (●Figure 4.2.C).

● **Figure 4.2.C** The Danube in Austria. Some of eastern Europe's greatest cities developed on its banks, and its delta is one of the world's premier bird habitats.

Traffic on the Danube is modest compared with the tonnage carried by barges on the Rhine or by road and rail in eastern Europe. The fact that the Danube flows into the Black Sea while the Rhine pours into the North Sea has made a great difference in the economic roles of the two waterways. The industrial world served by the Rhine, and the very active trading world into which it flows, is very distinct from the more agricultural world served by the Danube. The Danube flows mostly through agrarian areas, and heavy industries—the most important generators of barge traffic—are found in only a few places along its banks.

Upstream from Vienna, the river is swift and hard to navigate, although large barges go upstream as far as Regensburg, Germany. Below Vienna, the Danube flows slowly across the Hungarian and Serbian plains past Budapest and Belgrade. In the border zone between Serbia and Romania, the river follows a series of gorges through a belt of mountains about 80 miles (c. 130 km) wide. The easternmost gorge is the Iron Gate gorge (also called Portile de Fier), 2 miles (3.2 km) long with rapids, that serves as the border between Romania and Serbia. Beyond it, the Danube forms the boundary between the plains of southern Romania and northern Bulgaria.

The river then turns northward into Romania and enters the Black Sea through the Danube Delta, Europe's largest wetland. Resided in or transited by more than 300 bird species, this was one of Europe's greatest wilderness areas until in the 1980s, Romania's Communist government began diking and draining the land for new farms and shooting off bird populations as "pests." Efforts are still under way to remove these engineering works and restore life to the marshes.

Jim Zuckerman/Corbis

plains, the humid continental long-summer climate is comparable to that of the corn and soybean region (Corn Belt) in the U.S. Midwest, but the plains of Poland have a less favorable humid continental short-summer climate, comparable to that of the American Great Lakes region. The most exceptional climatic area is the Dalmatian Coast, on the Balkan Peninsula side of the Adriatic Sea. Here temperatures are subtropical, and precipitation, concentrated in the winter, reaches 180 inches (c. 260 cm) per year on some slopes facing the Adriatic—Europe's most plentiful rainfall. The climate of the Dalmatian Coast is classed as Mediterranean (dry-summer subtropical). Together with spectacular scenery, this environment has made the area one of Europe's prime tourist destinations, except during the conflicts of the 1990s.

Political and Ethnic Geography of the "Shatter Belt"

The extension of Soviet power into eastern Europe after World War II was a replay of history. In the Middle Ages, several peoples in the region—the Poles, Czechs, Magyars,

Bulgarians, and Serbs—enjoyed political independence for long periods and at times controlled extensive territories outside their homelands. Their situation deteriorated as stronger powers—Germans, Austrians, Ottoman Turks, and Russians—pushed into east central Europe and carved out empires. These empires frequently collided, and the local peoples were caught in wars that devastated great areas, often resulted in a change of authority, and sometimes brought about large transfers of populations from one area to another. In geopolitical terms, eastern Europe is a classic **shatter belt**—a large, strategically located region composed of conflicting states caught between the conflicting interests of great powers (●Figure 4.2.22).

Large population transfers have occurred in the region since the beginning of World War II. Jews, Germans, Poles, Hungarians, Italians, and others were uprooted by Nazi and Soviet authorities, often without notice, losing all their possessions, and were dumped as refugees in so-called homelands that many had never seen. During the war, Nazi Germany systematically killed approximately 6 million Jews and perhaps as many as 1.5 million Gypsies and other "undesirables" in the Holocaust. The prewar

• **Figure 4.2.22** Positioned between stronger powers to the east and west, eastern Europe is a classic shatter belt with a tumultuous past, reflected in its shifting borders.

populations of Germans in Poland and Czechoslovakia were expelled at the end of World War II and forced into East and West Germany. Many died in this process of forced migration overseen by the Soviet Union. Ethnic minorities now constitute only 2 percent of the population in Poland, which transferred most of its German population to Germany and whose territories containing Lithuanians, Russians, and Ukrainians were absorbed by neighboring countries. A number of countries still have large minorities, including the Roma or Gypsies (see Ethnic Geography, page 140).

These population transfers helped create an ethnic map of eastern Europe that was overwhelmingly Slavic. The Slavs are typically grouped into three large divisions: **East Slavs** of the former Soviet region; **West Slavs**, including **Poles, Czechs,** and **Slovaks;** and **South Slavs**, including **Serbs, Croats, Slovenes, Bulgarians,** and **Macedonians.** Although the various Slavic peoples speak related languages, the languages are often mutually unintelligible; for example, a Czech will not easily understand a Pole, although both are ethnic Slavs. West Serbian and Croatian are essentially the same language in spoken form, but the Serbs, like the Bulgarians and most of the East Slavs, use the Cyrillic alphabet (based on Greek), whereas the Croats use the Latin alphabet, as do the Slovenes and the West Slavs. The language of the **Romanians** is classed with the Romance languages derived from Latin; however, the contemporary Romanian language contains many Slavic words and expressions. There is also a religious division between the

West Slavs, Croats, and Slovenes, most of whom are Roman Catholics, and the East Slavs, Romanians, Bulgarians, Serbs, and Macedonians, who are mainly Eastern Orthodox Christians.

The principal **non-Slavs in eastern Europe** are the Estonians, the Latvians and Lithuanians, the Hungarians, the Albanians, and the Romanians. The **Estonians,** who form an ethnic majority in Estonia, belong to the same Uralic ethnolinguistic group as the Finns and the Hungarians. **Latvians** and **Lithuanians** speak Baltic languages descending from the same ancestry as the Slavic languages but do not consider themselves Slavs. The **Hungarians,** also known as **Magyars,** are the descendants of nomads from the east who settled in Hungary in the ninth century. Roman Catholicism is the dominant religious faith in Hungary, although substantial Calvinist and Lutheran groups have long existed there. The **Albanians** speak an ancient Indo-European language only distantly related to the Slavic languages. Albania and Bosnia and Herzegovina are the only European countries in which Muslims are a majority. Serbia, Montenegro, Macedonia, and Bulgaria have many Albanian and Turkish Muslims, whose presence is a legacy of former Turkish rule.

Wrenching Reform: Communism

In becoming satellites of the Soviet Union, the nations of eastern Europe (and East Germany) were reformed by **communism,** which has these principal traits: one-party

ETHNIC GEOGRAPHY

The Roma

One of Europe's largest ethnic minorities is the Gypsies—properly known as the Roma—who number 8 to 10 million (•Figure 4.2.D). Romania has the highest number—as many as 2.5 million—but estimates of the Roma population vary widely because the Roma live on the road and at the margins of society.

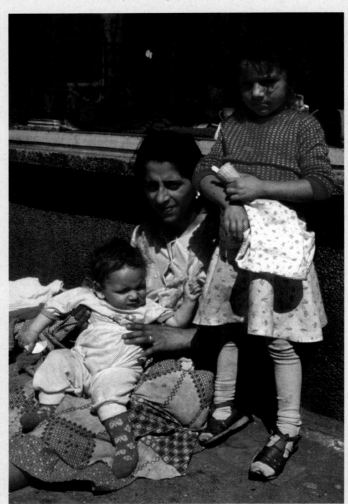

• **Figure 4.2.D** This Roma woman and her children pass their days begging on a street corner in Szczecin, Poland.

Joe Hobbs

The Roma began their great odyssey in what is now India, and their Romany language is very close to languages still spoken on the Indian subcontinent. Having in early times traveled thousands of miles from their homeland, and still often moving in caravans, they have come to be the archetypal people on the move. Throughout the Roma realm, host governments and majority populations have for centuries regarded the Gypsies with disdain. They are typically depicted as a rootless, lawless, and violent people apart, not deserving of the educational and economic opportunities offered to majority populations. Wherever they are, they have more children than the majority populations. Sterilizations of Roma women without their consent have been reported in the Czech Republic, Slovakia, Hungary, Romania, and Bulgaria. The Roma are poorer than the majority populations; in Hungary, for example, the Roma unemployment rate is five times that of non-Roma, and in some Roma communities, the unemployment rate is as high as 90 percent. They were the first to lose their jobs when the eastern European countries gained economic freedom from the Communists.

The Roma typically live in shantytowns without water, sewage, or other services and make their living on scant child benefit payments, minor mechanical work, begging, foraging for food, and petty crime. In 1999, one Czech city built a wall between Gypsy and Czech neighborhoods, insisting it was needed to protect other townspeople from Gypsy criminals, noise, and visual squalor. Most Roma children do not attend schools because their parents do not believe that education will improve their job prospects, given the discrimination against them. Many that want to attend school are funneled into institutions for the mentally disabled. There is also violence; Gypsies are often the targets of attacks by skinheads. These assaults, especially in the Czech Republic and Slovakia, have resulted in recent waves of Roma emigration to Canada, France, Britain, and Finland.

Romany activists have responded to discrimination by demanding more rights to welfare, pensions, and other benefits offered to regular citizens. Already feeling overwhelmed by the economic burden of their welfare services, the wealthier countries of the European core are worried that the new EU membership of eastern European countries will unleash a tide of Roma immigration. The latest 14 EU members have almost 5 million Roma.

dictatorial governments; national economies planned and directed by organs of the state; abolition of private ownership (with some exceptions) in the fields of manufacturing, mining, transportation, commerce, and services; abolition of independent trade unions; and varying degrees of **socialization** (state ownership) of agriculture. Soviet military force crushed attempts by East Germany in 1953, Hungary in 1956, and Czechoslovakia in 1968 to break away from Soviet control. In agriculture, the new Communist governments liquidated the remaining large private holdings and in their place introduced programs of collectivized agriculture on the Soviet model. Some farmland was placed

in large state-owned farms on which the workers were paid wages, but most was organized into collective farms owned and worked jointly by peasant families who shared the proceeds after operating expenses of the collective had been met. This **collectivization**—the bringing together of individual landholdings into a government-organized and government-controlled agricultural unit—met with strong resistance, and it was discontinued in the 1950s in Yugoslavia and Poland. Today, all but a minor share of the cultivated land in Poland is privately owned. The remaining countries are pursuing programs to reprivatize their farmlands. Conversion to private farming after four decades of collectivization presents painful obstacles, and progress has been slow.

Despite recent attempts to diversify and intensify agriculture, farming in eastern Europe remains primarily a crop-growing enterprise based on corn and wheat, the leading crops in the region from Hungary and Romania southward through the Balkan Peninsula. In the Czech and Slovak republics and Poland, with their cooler climates, corn is difficult to grow, but wheat is a major crop as far north as southern Poland's belt of loess soils. Most of Poland lies north of the loess belt and has relatively poor sandy soils; here rye, beets, and potatoes are the main crops. Raising livestock is a prominent secondary part of agriculture throughout the region. The meat supply is much less abundant than in western Europe and comes primarily from pigs. The more southerly areas contain poor uplands that pasture millions of sheep, the main source of meat in Bulgaria, Albania, and parts of the former Yugoslavia.

167 In industry, the ascension of communism in eastern Europe inaugurated a new era, also along Soviet lines. Most industries were taken out of private hands, and national economic plans were developed. Communist planners did not aim at a balanced development of all types of industry and instead stressed those seen as essential to regional industrial development as a whole: mining, iron and steel, machinery, chemicals, construction materials, and electric power. These were favored at the expense of consumer-type industries and agriculture.

The central planning agencies of the Communist governments maintained rigid control over individual industries. Plant managers were directed to produce certain goods in quantities determined by state governmental planners, and the success of a plant was judged by its ability to meet production targets rather than its ability to sell its products competitively and at a profit. Political reliability and conformity to the central plan were qualities much desired in plant managers. This system frequently resulted in shoddy goods that were often in short supply or in mountainous oversupply because production was not being driven or keyed by buyer demand and satisfaction. The planned expansion of mining and industry in eastern Europe under communism produced marked increases in the total output of minerals, manufactured goods, and power. But it also produced inefficient, overmanned industries ineffective in their later competition for the world markets of the post-Communist period—a problem shared by the economies of 168 all the successor states of the Soviet Union.

Under communism, the economies of eastern Europe were closely tied to that of the Soviet Union. Eastern European industries relied on imports of Soviet iron ore, coal, oil, natural gas, and other minerals. In return, the Soviet Union was a large importer of east central Europe's industrial products. During the 1970s and 1980s, however, the region's trade relationship with the Soviet Union weakened. Trade and financial relations with Western countries, companies, and banks expanded rapidly. With relatively little to export, the eastern European countries borrowed massively from Western governments and banks to pay for imports from the West. Poland and Romania pursued this course especially strongly. New industrial plants and equipment acquired in this way were supposed to be paid for by goods that would be produced for export to the West by industries that had learned efficient production and marketing techniques from the West.

But then the Western economies began to slump, lowering the demand for imports. This led to a situation by the 1980s in which large debts to Western governments and banks needed to be repaid if countries were to maintain any credit at all. However, with the help of mismanagement by Communist bureaucrats, exports with which to pay were not being produced or, if produced, could not be sold. The result was a severe impediment to economic expansion and falling standards of living as the governments squeezed out the needed money from their people. In Poland, social action through the formation of an independent trade union called **Solidarity** paralyzed the country and threatened to upset Communist dominance (●Figure 4.2.23). These conditions, along with the Soviet Union's own growing economic and political distress, set the stage for the withdrawal of Soviet control and decommunization in 1989 and 1990.

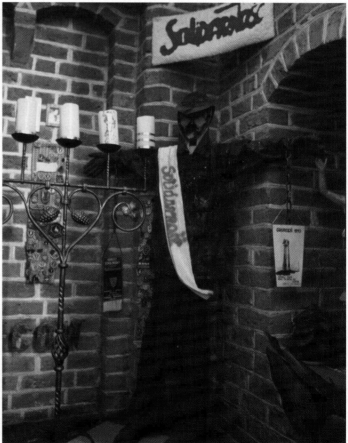

● **Figure 4.2.23** The rise of the Solidarity trade union was the beginning of the end for communism in eastern Europe. This is a shrine, in a Polish Catholic church, dedicated to Solidarity.

More Wrenching Reform: Revolution and Capitalism

As late as the early summer of 1989, East Germany and the eastern European countries seemed firmly in the control of totalitarian Communist governments. Yet suddenly, the Communist order began to crumble. Public demands for freedom, democracy, and a better life gathered momentum in one country after another. Communist dictators who had ruled for many years were forced out, and reformist governments took charge. By mid-1991, democratic multiparty elections had been held in all countries. This liberalizing process continued throughout the 1990s and into the 2000s.

In 1989 and 1990, the Soviet Union withdrew its backing for the region's Communist order. That support was too costly for the USSR, which also recognized the inevitable victory of "people power" in eastern Europe. Several Communist regimes (including East Germany's) quickly collapsed and were replaced by democratically elected non-Communist governments. Governments in the region have been struggling since then to build democracies with capitalist economies out of the economic wreckage of unproductive, unprofitable, uncompetitive, state-owned enterprises. Many countries pursued reforms aggressively to meet qualifications for joining the European Union. Economic restructuring has required many difficult steps, including the **privatization** (shift to nongovernmental ownership) of state-owned enterprises, an increase in the efficiency of state-run enterprises by allowing noncompetitive enterprises to fail and removing support of poor performers, and encouragement of the development of new private enterprises, often with foreign capital and management playing a role. In addition, economic restructuring has seen the termination of price controls so that prices reflect competition in the market, the development of new institutions required by a market-oriented economy (such as banks, insurance companies, stock exchanges, and accounting firms), the fostering of joint enterprises between state-owned firms and foreign firms, and the elimination of bureaucratic restrictions on the private sector. Internationally convertible currencies have also been created to increase trade and thus enhance competition, and access to international communications media of all sorts has been expanded to help local entrepreneurs learn from foreign examples.

One of the most remarkable economic trends has been western European and Chinese **outsourcing** of investments that take advantage of relatively cheap labor in eastern Europe. In the late 1990s, Austrian and German electronics firms established large factories in east central Europe, especially in Hungary. Some of these closed and shifted their assets to China, where labor was even cheaper. But recently, China has opened television and other electronics factories of its own in eastern Europe. The challenge for the eastern European newcomers to the European Union will be to raise their standards of living while still offering competitive terms for international investment.

The New EU Countries and the Balkans

The newest members of the EU are overwhelmingly from eastern Europe. Collectively they a represent dynamic, up-and-coming, and progressive bloc of countries. Each nation is considered here, as are the non-EU Balkan countries that have had a difficult history and whose prospects are less bright.

The Baltics

Estonia, Latvia, and Lithuania, known as the Baltic States, or the Baltics, lie on the eastern shores of the Baltic Sea (•Figure 4.2.24). These countries share many environmental, economic, and historical traits with the Slavic neighbors, including the Russians. However, the indigenous inhabitants of the Baltics are not Slavs, and they have long struggled to safeguard their national identities, especially against Russian encroachment. They were part of the Russian empire before World War I but became independent following the 1917 Russian Revolution. The Soviet Union reabsorbed them in 1940, and in 1991, they regained independence with the collapse of the USSR.

As independent countries again, the Baltics have joined European economic and military organizations. But their

• **Figure 4.2.24** The northern portion of eastern Europe.

geographic past has been hard to shake, and they are still linked to parts of the former Soviet Union in various ways. Russia uses Baltic ports to transport goods abroad. It exports fossil fuels and a variety of other goods and services to the Baltic nations and sometimes cuts oil supplies to the Baltics for political or economic leverage. Estonia, Latvia, and Lithuania want to minimize their ties with their giant neighbor and former occupier to the east. They refused to become members of the Russian-oriented Commonwealth of Independent States (CIS) and instead joined the European Union in 2004. During that same year, they joined the North Atlantic Treaty Organization (NATO), despite Russia's objection that this would pose a serious threat to its security.

Dairy farms alternate with forests in the hilly Baltic landscape. Mineral resources are scarce; the most notable are deposits of oil shale in Estonia, mined for use as fuel in electric power plants. The Baltics have usually had to rely on natural gas and petroleum brought from Russian fields located hundreds and even thousands of miles away. Expansion of industry after World War II helped make the area one of the more prosperous and technically advanced parts of the former USSR. The largest city and manufacturing center is the port of Riga (population 800,000), the capital of Latvia (population 2.3 million). Riga is typical of all of the capitals of eastern Europe in that it serves as the country's primate city.

About 30 percent of Latvia's inhabitants and 85 percent of Latvia's non-Latvian ethnic minorities speak Russian; they are mostly retired Soviet military officers and their offspring. While independent Lithuania and Estonia established procedures early on to grant citizenship to ethnic Russians, Latvia still makes this process very difficult (by insisting for example that they pass an examination on Latvian history and speak Latvian). The Latvian government fears that enfranchised Russians might destabilize the country in future elections. The government is in a bind, however, because after the country joined the EU, more than 100,000 Latvians fled to Ireland and other EU countries in search of well-paying jobs. The resulting labor shortage has led to a flood of illegal workers from Russia, Belarus, and Ukraine, and Latvia must decide whether to welcome them as laborers or oust them as a threat. Ethnic Russians complain of discrimination, and Russia exercises its muscle periodically by cutting off shipments of its oil through Latvia (the transit fees generate important revenue for Latvia).

The presence of Russian troops in the Baltic countries was a serious foreign relations problem after the countries gained independence. With independence, only Lithuania (population 3.4 million; capital: Vilnius, population 500,000) successfully negotiated the immediate withdrawal of Russian troops. A large number of these Russian forces resettled in neighboring Kaliningrad, an anomalous Russian enclave located two countries away from Russia (see Figure 4.2.24).

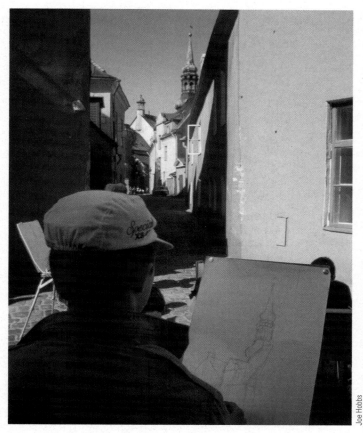

• **Figure 4.2.25** Estonia's capital, Tallinn, is a lively city of the arts.

Since independence, Estonia (population 1.3 million; capital: Tallinn, population 400,000; •Figure 4.2.25) has developed close economic ties with Finland, which quickly replaced Russia as the country's dominant trading partner. Estonia's indigenous people are closely related to the ethnic Finns of Finland, a mere 2½-hour ferry ride away across the Gulf of Finland. Through low taxes and tariff-free trade, mainly with Scandinavia, Estonia has become the most economically prosperous of the three Baltic countries and one of Europe's fastest-growing economies. It wants to be known as "E-Stonia" for its rising high-tech industry; the popular voice-over-Internet protocol system known as Skype was invented here, for example.

Poland

Poland (see Figure 4.2.24) is eastern Europe's largest country in size (about the same as the state of New Mexico) and population (38 million). Most of Poland lies on part of the North European Plain between the Carpathian and Sudeten Mountains on the south and the Baltic Sea on the north. The country's most fertile soils are in the south, where the plain rises to low uplands that touch Europe's loess belt. This is where most of Poland's rural population (about 38 percent of the total) is located (•Figure 4.2.26). The north is wilder, with some wetlands and forests that were protected as hunting reserves by centuries of aristocrats and

● **Figure 4.2.26** Almost 40 percent of Poles are rural. The agrarian nature of eastern Europe contrasts sharply with the more urbanized European core countries to the west.

communists. Some environmentalists fear that commercial logging and development in the new economy will destroy these resources.

Poland's largest rivers, the Vistula (Polish: Wisla) and the Oder (Polish: Odra), wind across the northern plain from Upper Silesia to the Baltic. The Oder is the main internal waterway, carrying some barge traffic between the important Upper Silesian industrial area (to which it is connected by a canal) and the Baltic. Poland's main seaports—Gdansk (formerly Danzig), Gdynia, and Szczecin (formerly Stettin)—are along or near the lower courses of these rivers. Gdansk had a catalyst role in the breakaway of eastern Europe from the Soviet grip, beginning in the early 1980s with Solidarity's first uprising in this port city's shipyards. The **Gdansk Accords** gave Polish workers the right to strike and to form free trade unions.

Gdansk is eclipsed in economic importance by Warsaw (population 2.2 million), Poland's capital. Warsaw was the setting for one of history's worst war crimes when, during World War II, hundreds of thousands of Jews were forced to resettle in the prisonlike **Warsaw Ghetto** and were then systematically shipped out to work and be exterminated in concentration camps like Auschwitz (in southern Poland, near Krakow). Krakow developed as an industrial city on the eastern end of the Upper Silesian–Moravian coalfield, a kind of "Ruhr in miniature" comparable to that area in Germany.

With Germany as its major trading partner, Poland gains about two-thirds of its export revenues from the sale of machinery, transport equipment, and manufactured goods. These exports helped the country's economy stabilize after traditionally large employers like the Szczecin shipyard went bankrupt, but relatively high unemployment remains a problem. Poland became a NATO member in 1999 and was the heavyweight of the countries that joined the European Union in 2004's "big bang."

Czech Republic

The economically diverse Czech Republic (see Figure 4.2.24), where communism had undercut a long tradition

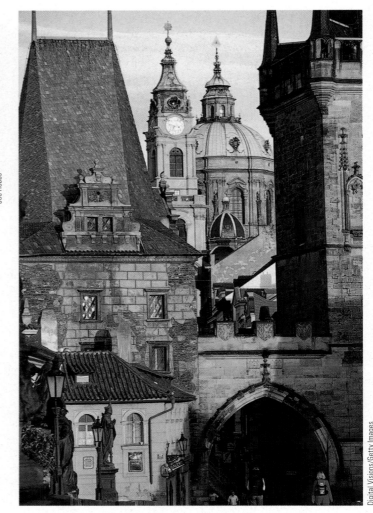

● **Figure 4.2.27** Some of Prague's architectural treasures.

of economic accomplishment, is enjoying an economic revival. The country's transformation began when Czechoslovakia staged a bloodless overthrow of the Communist government during six weeks late in 1989. This so-called **Velvet Revolution** culminated in the presidential election of Vaclav Havel, a gifted playwright. Four years later, Czechoslovakia was peacefully split into the Czech Republic and Slovakia in the **Velvet Divorce.**

Prague is both the capital and primate city of the Czech Republic. This architecturally graceful and lively city of 1.4 million people is located in the country's industrialized Bohemian basin, where most of the country's important output of machinery and transport equipment is based (●Figure 4.2.27). In the extreme east of the Czech Republic is another industrial area, Moravia, that grew up on the Silesian-Moravian coalfield. The eastward migration of manufacturing jobs in Europe has benefited the country. The Czech economy, now more than 90 percent privatized, has enjoyed recent strong growth due to investments in automobile and electronics manufacturing and in information technology.

The borderlands north, west, and south of Prague have long been a sore spot in relations between the Czechs and the far stronger Germany to the west. On the eve of World War II, much to the dismay of Czechoslovakia, Britain and other European powers allowed Germany to annex the Sudetenland and other portions of this borderland, whose inhabitants were primarily ethnic Germans. When Nazi Germany was defeated, vengeful Czechs regained the lands and forced the expulsion of more than 2 million ethnic Germans back into Germany. So far the Czech Republic has refused German demands for apologies or return of properties. A similar stalemate exists between Poland and the ethnic Germans expelled from Poland's borders after the war. These issues became very significant with the 2004 membership of the Czech Republic and Poland in the European Union, as farming interests had to be reassured that prewar claims to the land would not honored.

Slovakia

With the 1993 creation of the Slovak Republic, or Slovakia, the Danube River city of Bratislava (population 400,000) became its national capital (see Figure 4.2.24). The country is one-third arable and two-fifths forested and is more ethnically diverse than either the Czech Republic or Poland. Slovakia has a potential problem of further political separatism within its minority community of about 500,000 ethnic Hungarians. With independence, Slovakia suffered from the fact that many professionals, and much potential foreign investment, stayed with the Czech Republic. Slovakia stagnated until a new government lowered taxes and

brought political stability. These features and the country's low labor costs attracted investors, especially from the automobile industry. Tiny Slovakia became, on a per capita basis, the world's largest auto producer and aims to establish a lasting reputation as the **"Detroit of Europe."** The economy has boomed, earning the country the nickname **"Tatra Tiger"** after its Tatra Mountains. These mountains and the country's many caves and thermal spas also hold much tourism development potential for Slovakia.

666

Hungary

Hungary's move toward free-market capitalism began early and was aided by Western and Japanese investments larger than those received by any other nation in this subregion. After a period of robust growth, overspending in the first years of the 21st century led the government to impose economic austerity measures. Some social unrest ensued. Hungary has an inherent manufacturing edge because of local bauxite and aluminum production capabilities. Industries are concentrated in the beautiful capital city of Budapest (population 2.5 million; ●Figure 4.2.28), dubbed the "Pearl of the Danube" (●Figure 4.2.29). A relatively small but growing class of newly rich private entrepreneurs has emerged, often as participants in joint business ventures with foreign companies. There are still poor people, especially among Hungary's one-third rural population, but officially, they make up less than 10 percent of the population.

● **Figure 4.2.28** The southern portion of eastern Europe.

● **Figure 4.2.29** Budapest's waterfront, on the Danube River.

Romania

Romania is showcasing itself as a strategic, reliable ally of the United States, well located to assist American interests in the Caucasus and the Middle East. The government in Bucharest (population 2.2 million; see Figure 4.2.28) supported the 2003 U.S.-led invasion of Iraq and became a NATO member in 2004. Because of its poor economic status, however, Romania's initial bid to join the European Union was turned down, but it gained membership in 2007. It is well below the average in measures of economic well-being in eastern Europe, however. One-third of its 21 million people live in poverty. The country has to import oil and gas to provide its energy needs. Exports of textiles, minerals, chemicals, and machinery make up approximately 50 percent of Romania's modest export income, with Germany and Italy the primary export destinations. With a long eastern coastline on the Black Sea, hundreds of miles of Danube River landscape, and the storied mountains of Transylvania, Romania has hoped for an influx of foreign exchange from the Western tourist trade. But the tourist traffic has so far been largely from the less affluent east, mainly Russia.

Bulgaria

Between the Rhodope Mountains and the Balkan Mountains is the productive valley of the Maritsa River. Together with the adjoining Sofia Basin, this is the economic core of Bulgaria. Only about a third of the country is arable, and Bulgaria is trying to industrialize. Clothing, footwear, iron and steel, and machinery, mainly produced in the capital Sofia (population 1.2 million; see Figure 4.2.28), are the main exports. Bulgaria has made much progress in restructuring from a state-run to a market economy and economically was an easier candidate than Romania for its recent accession to the European Union. One of the most important economic resources the country possesses is the rail route that links Istanbul, Turkey, with points in central and western Europe.

Albania

After World War II, this small Communist country, populated by non-Slavic ethnic Albanians, linked itself economically with the People's Republic of China. Economic support from its large benefactor failed to move Albania up from its position at the bottom of Europe's economic ladder. Its leader, Enver Hoxha, kept the country isolated from the rest of the world and, to defend it from any threat, had the landscape peppered with almost a million fortified "pillbox" bunkers. The few outsiders who visited Albania described it as a bizarre throwback to a medieval agrarian Europe. Hoxha was finally ousted in 1984, and in the 1990s, following the dissolution of the Communist government and a period of internal political turmoil, Albania began to open up.

More than half the population remains rural, however, and Albania has few resources or other prospects for industrialization. Receipts from remittances—money earned by Albanians working abroad, mainly in Greece and Italy—represent most of the country's wealth. A "pyramid scheme" in the late 1990s swindled much of what little wealth there was from hundreds of thousands of Albanians. Waves of Albanian emigrants have swept into Europe, especially into Italy, where Italians routinely blame the newcomers for theft, robbery, rape, gun-running, prostitution, and a host of other ills.

The Albanian government in the capital, Tirana (population 400,000; see Figure 4.2.28), estimates that since 1991, more than 4,000 Albanian children—mainly Roma—have been sold or traded by their parents, ostensibly for the welfare of the children. However, most are handled by human traffickers who sell them into prostitution rings or to Europeans eager to adopt a child.

Successors of the Fractured Yugoslavia

While economic progress of varying degrees characterized most of eastern Europe in the 1990s, Yugoslavia was plagued by ethnic warfare following the dissolution of the Yugoslav federal state in 1991. Previously, under the Communist regime instituted by Marshal Josip Broz Tito during World War II, Yugoslavia ("Land of the South Slavs") had been organized into six "socialist people's republics" (Serbia, Croatia, Slovenia, Bosnia and Herzegovina, Montenegro, and Macedonia) and two "autonomous provinces" within Serbia (Kosovo and Vojvodina).

As long as Tito and his successors could retain a firm grip, these groups could coexist within the artificial boundaries of a single nation. However, this apparent unity belied many underlying tensions and conflicts based on linguistic, ethnic, and religious distinctions: Orthodox Serb versus Muslim Kosovar, Catholic Croat versus Orthodox Serb, and so on.

As the Iron Curtain dissolved across Europe, Yugoslavia began to fracture along its ancient ethnic fault lines (●Figure 4.2.30). In 1988, Serbia took direct control of Kosovo

• **Figure 4.2.30** Ethnic composition of the Yugoslav successor states.

and Vojvodina as part of its push (under an elected Communist president, Slobodan Milosevic) for greater influence in federal Yugoslavia. The quasi-independent states within Yugoslavia initially demanded a looser federation with more autonomy for each republic, but in 1991 and 1992, Croatia, Slovenia, Macedonia, and Bosnia insisted on independence. When Croatia declared its independence, fighting broke out between ethnic Croats and Serbs in the new country. Croatian Serbs took over one-third of Croatia and embarked on a policy of **ethnic cleansing,** the forced emigration or murder of one ethnic group by another within a certain territory. While Bosnia's Muslims (known as **Bosniaks**) and Croats voted for independence from Yugoslavia, ethnic Serbs in Bosnia, and the Yugoslav government of Slobodan Milosevic, opposed this.

War erupted in 1992 between the Bosnian government and local Serbs. Supported by Serbia and Montenegro, the Bosnian Serbs fought to partition Bosnia along ethnic lines and join Serb-held areas to a "greater Serbia." Bosnian Serbs laid siege to the mainly Bosniak city of Sarajevo. The United Nations withdrew recognition of Yugoslavia because of the government's failure to halt Serbian atrocities against non-Serbs in Croatia and Bosnia. In 1994, the Bosniaks and Croats agreed to create the joint federation of Bosnia and Herzegovina.

In 1995, the presidents of Bosnia and Herzegovina, Croatia, and Serbia signed the **Dayton Accord.** This peace agreement retained Bosnia and Herzegovina's international boundaries and created a joint multiethnic, democratic government charged with conducting foreign and economic policies. The agreement also recognized a second tier of government composed of two similarly sized entities charged with overseeing internal functions: the Bosniak-Croat Federation of Bosnia and Herzegovina and the Bosnian Serb–led Serb Republic (Republika Srpska). To help implement and monitor the agreement, NATO fielded a 60,000-strong force known as IFOR. It was later replaced by a smaller NATO force, in turn replaced by EU peacekeepers numbering just 7,000.

While the Bosnian situation quieted, in 1998 Serbia turned its sights on Kosovo, a majority ethnic Albanian and Muslim region with a minority of ethnic Serbs in the north. Here again, Serbs aspired to cleanse Kosovo of non-Serbs. Serbs asserted that Kosovo was the cradle of their culture and the site of the decisive battle of Kosovo Polje in 1389, which Serbs lost to the Ottoman Turks. NATO responded to the Serb offensive with 11 weeks of bombing. When the bombing stopped, the United Nations took control of Kosovo and oversaw a tense truce between its Serb and Muslim inhabitants. The fate of more than 200,000 ethnic Serbs who fled during the conflict from Kosovo to Serbia and Montenegro has not been resolved. Nominally still a part of Serbia but under United Nations jurisdiction, Kosovo may yet evolve into an independent country.

The ouster of Serbian leader Milosevic in 2000 and his capture and subsequent transfer to The Hague to stand trial for war crimes (he died before the trial was completed), opened the way for a period of stable and peaceful borders in the early 2000s. The international community insisted on no further redrawing of Balkan boundaries. The region had undergone a profound and violent process of **balkanization,** which is political-geographic shorthand for fragmentation into ethnically based, contentious units that took its name from the characteristic disharmony of this region.

358

The Balkan conflicts unfolded on television and other media through the 1990s. They reminded Europeans and Americans of the earlier world wars on the continent and seemed a horrible anomaly in a world region accustomed to peace and prosperity. Western military and political institutions eventually seemed to extinguish the fires of conflict, and the Balkan countries were set on the back burners of Europe.

By all measures of quality of life and economic prosperity, the Balkans are the poor stepchildren of the continent, and most cannot expect to join the European Union for many years. Only Slovenia—the sole Balkan newcomer to the EU in 2004—has done reasonably well, mainly through exports of metallurgical products and electronics. More than 80 percent of the citizens of this outward-looking country are connected to the Internet. Compared with some of the other Balkan countries, Slovenia has enjoyed the advantages of a relatively homogeneous population, proximity to stable Italy and Austria, and access to the sea. Croatia is emerging as a major tourist destination with attractions that including the historical walled city of Dubrovnik, nearby beaches of the Dalmatian Coast, and the inland Plitvice Lakes National Park.

Warfare devastated the still-moribund economy of Bosnia and Herzegovina, which, with Macedonia, is one of the two poorest successor states of Yugoslavia. However, since 1981 and the reported apparition of the Mother of Christ, its site of Medjugorje has grown to be a prospering pilgrimage destination for Catholics from around the world (•Figure 4.2.31). In Macedonia, tensions continue between the majority (70 percent) ethnic Macedonians, who are Slavs and Orthodox Christians, and the mainly Muslim ethnic Albanians who make up 25 percent of the population. The ethnic Albanians, Roma, Jews, and other minorities insist that they are being increasingly marginalized by strength-

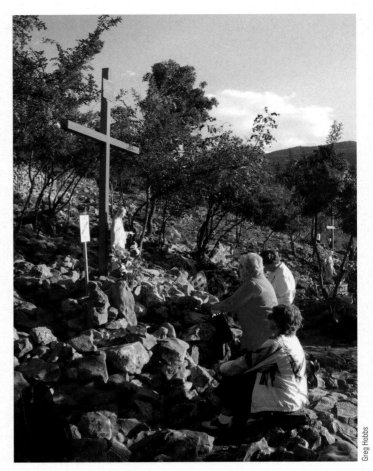

Greg Hobbs

● **Figure 4.2.31** Catholic pilgrims toil up the hill of Medjugorje.

ening Macedonian nationalism. The landlocked country relies on access to the Greek port of Salonika and struggles to maintain good relations with its more powerful neighbor.

Serbia is now also landlocked, thanks to the secession of Montenegro (population 600,000; capital: Podgorica, population 100,000). The post-Yugoslavia nation known as Serbia and Montenegro had been the result of deal making in the West, in which Serbia was called on to surrender more suspects for war crimes trials; in return, the West would not support the independence of Montenegro. The nominal union was also meant to blunt, for the time being, Kosovar desire for independence for Kosovo. But the Montenegrins insisted on attaining the same independence as the other former constituent Yugoslav republics, and the people voted (by a bare majority) in a 2006 referendum to separate from Serbia. Ethnic tensions between the ethnic Slavic Montenegrin majority and the ethnic Albanian minority threaten the young country. Montenegro declares itself Western-oriented; although it lies outside the Eurozone, it has adopted the euro as its currency and wants membership in both the European Union and NATO. Russia has so

90, 91

far made the greatest investments here, especially in resort development on the country's spectacularly beautiful Adriatic coastline.

This concludes the survey of Europe. From here, the text moves eastward to the world's largest country, Russia, and many of the countries within its realm of great influence.

SUMMARY

→ The core region of Europe is ringed by a periphery of three sub-regions: northern, eastern, and southern Europe. These have historically been less integrated into the core, and the eastern and southern countries have been less prosperous.

→ Most of these peripheral countries are members of or are seeking membership in the European Union and NATO. Norway has chosen to remain outside the European Union, and some of the poorer European countries do not qualify for membership.

→ The Northern, or Nordic, countries include the Scandinavian countries of Norway, Sweden, and Denmark, plus Iceland. With Finland, this group is often called Fennoscandia, or the Fennoscandian countries.

→ The northern European countries are the world's most northerly region that is highly settled and economically developed. Human activity is possible at such a high latitude because of the warming waters of the North Atlantic Drift.

→ Except for Finland, the Nordic countries have close connections through language (Finnish has its origins in Asia). Other qualities that link these countries are settlements in the southern reaches of the countries (except Denmark) and sparse populations in the north and in the mountains of all the countries (except flat Denmark). Levels of literacy, education, and Internet connection are high. Of major economic importance are forestry, fishing, skilled manufacturing, and in Denmark and part of Sweden, very productive agriculture. Norway also has significant petroleum reserves in the North Sea. Hydroelectricity is also a major resource in all of the northern countries except Denmark.

→ Southern Europe, or Mediterranean Europe, includes Portugal, Spain, Italy, and Greece, as well as a number of major islands, island countries, and microstates, including San Marino, Vatican City, the British colony of Gibraltar, Malta, Corsica (part of France), and Cyprus. The mainland countries are parts of three peninsulas: the Iberian, Italian, and Balkan.

→ The southern region has lagged in development because of trade deficiency, low educational levels, land tenure patterns, and resource deficiencies. These traditional characteristics have changed somewhat in recent decades due to transportation changes, energy shifts, market expansion, and capital infusion.

→ Southern Europe generally has a Mediterranean climate, with summer drought and winter rain. Irrigation has been an important factor in this region's agricultural history. Important

crops are winter wheat, olives, figs, citrus, grapes and associated winemaking, and vegetables.

→ The Islamic Moors and the Sephardic Jews were important cultures of the Iberian Peninsula until the Spanish ousted them in 1492. Spain and Portugal undertook crucial exploration around the world and were important colonial powers. Italy had some colonial influence in Africa and is currently the most industrial and economically significant nation of the south. The Po Valley is central to industrial northern Italy and houses two-fifths of the nation's population. There is an imbalance between northern and southern Italy, with the north far more prosperous. Some northerners want to develop an autonomous or independent region of Padania.

→ As the result of a 2004 referendum in which both sides of divided Cyprus voted on whether they should be reunited, the Greek south (which rejected reunification) was allowed to join the European Union, while the Turkish north (which favored reunification) was not allowed to join.

→ Eastern Europe is made up of Estonia, Latvia, Lithuania, Poland, the Czech Republic, Slovakia, Hungary, Romania, Bulgaria, Slovenia, Croatia, Bosnia and Herzegovina, Serbia, Montenegro, Albania, and Macedonia. These countries have been shaped powerfully by struggles between stronger countries and make up a geopolitical "shatter belt." Most critical was their role as satellite states of the Communist Soviet Union between World War II and 1990.

→ Slavs are the dominant ethnolinguistic groups in eastern Europe and include Serbs, Croats, Slovenes, Bulgarians, Macedonians, Poles, Czechs, and Slovaks. Important non-Slavic groups include Jews, Roma (Gypsies), Estonians, Latvians, Lithuanians, Hungarians (Magyars), Albanians, and Romanians.

→ The environment of the eastern European countries includes the northern plain, the central mountain zone, the Danubian plains, and the southern mountain zone, with a variety of climates.

→ Collectivization under Communist rule during the Soviet era from the late 1940s until the 1980s was the agricultural model in eastern Europe. Privatization of farmland has taken place since the fall of communism.

→ Mining, iron and steel, machinery production, construction materials, and electrical power were the highlights of the industrial effort during the years of Soviet domination, with the majority of raw materials coming from Russia. The shift to a market economy left many firms uncompetitive, and many were abandoned. Both western European countries and China have taken advantage of inexpensive labor in eastern Europe to outsource some production in the region, and many economies have rebounded.

→ Politically, the disintegration of the former Yugoslavia was the most disabling phenomenon of the post-Soviet era in eastern Europe. Yugoslavia was dismantled and replaced by six countries, some with precarious rivalries among ethnic groups. NATO and the UN have used various military and diplomatic means to prevent further balkanization of the countries. The future of Kosovo, still in the hands of Serbia, remains a concern.

KEY TERMS + CONCEPTS

Attila Line (p. 136)
balkanization (p. 148)
Basques (p. 132)
Bosniaks (p. 147)
collectivization (p. 140)
communism (p. 139)
Dayton Accord (p. 147)
"Detroit of Europe" (p. 145)
East Slavs (p. 139)
enosis (p. 136)
entrepôt (p. 124)
ETA (Basque Homeland and Liberty)
   (p. 132)
ethnic cleansing (p. 147)
European periphery (p. 121)
Fennoscandian Shield (p. 125)
Finlandization (p. 127)
fjord (p. 125)
Gdansk Accords (p. 144)
Green Line (p. 136)
Hanseatic League (p. 125)

Inquisition (p. 130)
Liga Veneta (p. 135)
Mediterranean diet (p. 129)
Moses Project (p. 134)
Non-Slavs in Eastern Europe (p. 139)
   Albanians (p. 139)
   Estonians (p. 139)
   Hungarians (p. 139)
   Latvians (p. 139)
   Lithuanians (p. 139)
   Magyars (p. 139)
   Romanians (p. 139)
Nordic Model (p. 124)
Northern League (p. 135)
outsourcing (p. 142)
Padania (p. 135)
privatization (p. 142)
Reconquista (p. 130)
Sápmi (p. 125)
Sephardic Jews (p. 130)
shatter belt (p. 138)

socialization (p. 140)
Solidarity (p. 141)
South Slavs (p. 139)
   Bulgarians (p. 139)
   Croats (p. 139)
   Macedonians (p. 139)
   Serbs (p. 139)
   Slovenes (p. 139)
Soviet satellites (p. 137)
"Tatra Tiger" (p. 145)
Third Italy (p. 134)
transhumance (p. xxx)
Velvet Divorce (p. 144)
Velvet Revolution (p. 144)
Warsaw ghetto (p. 144)
West Slavs (p. 139)
   Czechs (p. 139)
   Poles (p. 139)
   Slovaks (p. 139)

REVIEW QUESTIONS

1. What traits do the countries of northern Europe share, and which are unique?

2. Describe the climate patterns of the European north, comparing this region with North America at the same latitudes. What geographic feature is most responsible for the regional warmth of the northern region?

3. What is the significance of the Øresund Fixed Link? How do peoples on either end use it?

4. What are the countries, microstates, and islands of southern Europe?

5. What are the typical components of the Mediterranean climate, agriculture, and diet? What technology is critical for successful agriculture in parts of this realm?

6. What geographic, economic, or other factors account for the locations of major cities and belts of industrial activity in southern Europe?

7. Why has southern Europe generally lagged behind the European core in economic development? What countries or parts of countries have run counter to this trend? What threatens Italy's economic future?

8. What are the major countries and ethnolinguistic groups of eastern Europe?

9. Why is eastern Europe considered a "shatter belt" in geopolitical terms?

10. How did communism shape agriculture and industry in eastern Europe? What economic processes have occurred there since 1990?

11. What are the main forces behind the breakup of Yugoslavia and the current borders and ethnic components of its successor countries?

12. What are some of the main urban and natural historical points of interest on the Danube River?

13. Of all the countries in the European periphery, which do not belong to the European Union, and why do they not?

Onion / domes grace Moscow's
Kremlin, the seat of Russia's power.

Joe Hobbs

CHAPTER 5
A GEOGRAPHIC PROFILE OF RUSSIA AND THE NEAR ABROAD

From 1917 to 1991, the huge region known in this text as Russia and the Near Abroad, plus the Baltic countries of Estonia, Latvia, and Lithuania, made up a single Communist-controlled country called (after 1922) the Union of Soviet Socialist Republics (USSR). The Russian-dominated government in Moscow controlled the affairs of many non-Russian peoples in the country and after World War II also effectively controlled its Communist **satellite countries** in eastern Europe. This vast empire—which U.S. President Ronald Reagan dubbed the **"Evil Empire"** in 1983—engaged with the United States in a worldwide contest for political and economic supremacy. For four decades that Cold War between the Soviet bloc of nations and the Western bloc led by the United States dominated world politics. The perspectives and actions of the Soviet Union had major impacts on world events.

With little warning, in 1991, the country split into 15 independent nations (see Table 5.1 and Insights, page 155). Russia—now officially known as the Russian Federation—remained by far the largest in

■ chapter
objectives

This chapter should enable you to

→ Appreciate the environmental obstacles to development in vast areas of the world's largest country and nearby nations

→ Become familiar with the ethnic complexity of a huge region until recently held together—against great odds—as a single country

→ Learn the significant milestones in the historical and geographic development of Russia and the Soviet Union, some of them accompanied by unimaginable loss of life

→ Recognize the differences between command and free-market economies and the post-Soviet difficulties in shifting from one to the other

→ Understand the reasons for the reversal of Russia's progress through the demographic transition

→ Come to know the geopolitical and ethnic forces threatening the unity of Russia and pitting various groups and countries within and outside the region against one another

area, population, and political and economic influence. From the Russian perspective, the other 14 countries, all former Soviet republics, became the **Near Abroad,** in which Russia's special interests and influence should be exerted and preserved. But in 2004, Russian hopes to keep a dominant position in the Baltic region were dashed when Estonia, Latvia, and Lithuania joined the European Union.

What remained was a vast geographic complex consisting of Russia; the similar and mostly or partly Slavic countries of Ukraine, Belarus, and Moldova; the decidedly non-Russian but heavily Russian-influenced "stans" of central Asia—Kazakhstan, Uzbekistan, Turkmenistan, Kyrgyzstan, and Tajikistan; and the fractious countries of Georgia, Armenia, and Azerbaijan in the Caucasus. These 12 countries, designated in this book as Russia and the Near Abroad, are (as of 2007) also the member states (Turkmenistan is an associate member) of the **Commonwealth of Independent States (CIS),** which is essentially an economic rather than a political association.

Geographers are not having an easy time dealing with the successor countries of the Soviet Union. As discussed in Chapter 1, regions are organizing tools, not facts on the ground. There is no acknowledged best way to classify this region. Some geographers have chosen to cleave the central

Asian "stans" from Russia, establishing them as an entirely separate region, or include them as part of a greater Middle Eastern region. Some still consider the Baltics part of the greater Russian realm. The variety of geographers' names for the region suggests how unsettled the terminology is: "Post-Soviet Region," "Russia and Its Neighbors," "Russia and the Newly Independent States," "Russia and Neighboring Countries," "the Commonwealth of Independent States," and this book's "Russia and the Near Abroad" are among them.

The choice to use "Russia and the Near Abroad" in this book was made not to impose a Russian-centered view of the region, as the name might imply, but to reflect the enormous power that Russia wields (or aspires to wield) there. The former Soviet countries have lasting and often uncomfortable strategic and economic associations with Russia. For example, the Soviet-era oil refinery and pipeline system still links Russia with Kazakhstan, Azerbaijan, Georgia, Ukraine, Belarus, and other countries. Since the breakup of the Soviet Union, Russia has periodically shut off the pipelines supplying natural gas (of which it has the world's largest reserves) and oil, or has increased the prices of these commodities dramatically, to obtain political concessions from some of these nations.The needs of many of the Near Abroad countries to buy and sell fossil fuels provide an incentive to keep some kind of political relationship with Russia; otherwise they risk high fuel prices and even in some cases the potential for military conflict with their giant neighbor.

143, 173

Whatever forces favor integration around Russia, political fragmentation and decentralization are the dominant political themes in this region. Without the USSR's Red Army to impose order, long-simmering ethnic conflicts have boiled over. The Russian government in Moscow meanwhile tries vigorously to maintain influence in the former republics. Some analysts fear that Russia may attempt to reexert control over some nations, especially those seen as vital to its economic and political security. The future of this huge territory is uncertain. This chapter is an introduction to a region in which momentous changes are under way.

5.1 Area and Population

With an area (including inland waters) of 8.5 million square miles (22.1 million sq km), the region of Russia and the Near Abroad is the largest world region recognized in this text. A good indication of its staggering size is the fact that this region (even Russia alone) spans 11 time zones; in comparison, the United States from Maine to Hawaii spans seven. Russia is about 1.8 times the size of the United States, including Alaska.

The eventful geopolitical history of Russia and the Near Abroad has given this region land frontiers with 15 countries in Eurasia (●Figure 5.1). Between the Black Sea and the Pacific, the region borders Turkey, Iran, Afghanistan,

TABLE 5.1 Russia and the Near Abroad: Basic Data

Political Unit	Area (thousands; sq mi)	Area (thousands; sq km)	Estimated Population (millions)	Estimated Population Density (sq mi)	Estimated Population Density (sq km)	Annual rate of Natural Increase (%)	Human Development Index	Urban Population (%)	Arable Land (%)	Per Capita GNI PPP ($US)
Slavic States and Moldova	**6,919.1**	**17,920.4**	**201.9**	**29**	**11**	**−0.5**	**0.789**	**73**	**9**	**10,370**
Belarus	80.2	207.7	9.7	121	47	−0.5	0.794	73	29	8,810
Moldova	13.0	33.6	4.0	308	119	−0.2	0.694	45	55	2,880
Russia	6,592.8	17,075.3	141.7	21	8	−0.5	0.797	73	7	11,620
Ukraine	233.1	603.7	46.5	199	77	−0.6	0.774	68	56	7,510
Caucasus Region	**71.8**	**185.9**	**16.1**	**224**	**86**	**0.8**	**0.743**	**53**	**15**	**5,310**
Armenia	11.5	29.8	3.0	261	101	0.6	0.768	64	17	5,880
Azerbaijan	33.4	86.5	8.6	257	99	1.2	0.736	52	19	5,960
Georgia	26.9	69.6	4.5	167	65	0.1	0.743	52	11	3,690
Central Asia	**1,542.3**	**3,994.5**	**59.7**	**38**	**15**	**1.5**	**0.714**	**41**	**7**	**3,450**
Kazakhstan	1,049.2	2,717.4	15.5	15	6	0.9	0.774	57	8	7,780
Kyrgyzstan	76.6	198.4	5.2	68	26	1.6	0.705	35	7	1,980
Tajikistan	55.3	143.2	7.1	128	50	1.9	0.652	26	6	1,410
Turkmenistan	188.5	488.2	5.4	29	11	1.6	0.722	47	3	N/A
Uzbekistan	172.7	447.3	26.5	153	59	1.7	0.696	36	11	2,240
Summary Total	**8,533.2**	**22,100.8**	**277.7**	**33**	**13**	**0.0**	**0.770**	**64**	**8**	**8,590**

Sources: World Population Data Sheet, Population Reference Bureau, 2007; Human Development Report, United Nations, 2006; World Factbook, CIA, 2007.

China, Mongolia, and North Korea. Pakistan and India also lie close by. In the Pacific Ocean, narrow water passages separate the Russian-held islands of Sakhalin and the Kurils from Japan. In the west, the region has frontiers with Romania, Hungary, Slovakia, Poland, Lithuania, Latvia, Estonia, Finland, and Norway.

So much of this vast region is sparsely populated that its estimated population of 278 million in 2007 ranked it seventh among the eight world regions (•Figure 5.2). The average population density of 32 per square mile (12/sq km) is less than half that of the United States. Great stretches of economically unproductive terrain separate many populated areas from one another. The mountainous frontiers in Asia have especially few inhabitants. In contrast, on the frontier between the Black and Baltic Seas, international boundaries pass through populous lowlands that have long been disputed territory between Russia and other countries.

Russia has by far the largest population, about 142 million. The next largest country is Ukraine, with about 46.5 million people. Uzbekistan is the most populous central Asian nation, with about 26.5 million people. All the other countries in the region trail far behind, with populations generally under 15 million apiece. Population growth rates

are highest, around 1.5 percent annually, among the predominantly Muslim populations of the central Asian countries. At the other end of the spectrum, Russia, Ukraine, and Belarus are losing population at a rate of up to 0.5 percent per year. This trend is due to the plummeting quality of life that occurred after the breakup in the Soviet Union in 1991 (see Insights, page 156). In an atmosphere of economic and political uncertainty, couples chose to have fewer children, and there were more broken marriages. More disturbingly, death rates soared due to declining health care; increasing alcoholism, violence, and suicide; and more frequent flawed abortions.

5.2 Physical Geography and Human Adaptations

Stretching nearly halfway around the globe in northern Eurasia, the immense region of Russia and the Near Abroad has problems associated with climate, terrain, and distance. Most of the region is burdened economically by cold temperatures, infertile soils, marshy terrain, aridity, and ruggedness. Natural conditions are more similar to those of

INSIGHTS

Regional Names of Russia and the Near Abroad

Here are some useful geographic terms to know when studying this region. The region of Russia and the Near Abroad consists of what used to be the Union of Soviet Socialist Republics, also known as the Soviet Union or USSR, minus the Baltic countries of Estonia, Latvia, and Lithuania. The USSR came into existence in 1922 following the overthrow of the last Romanov tsar in the Russian Revolution of 1917 and the subsequent civil war. Prerevolutionary Russia is known as Old Russia, tsarist Russia, Imperial Russia, or the Russian Empire.

The name Russia now refers to the independent country of Russia. It is known politically as the Russian Federation, which was the largest of the 15 Soviet Socialist Republics (Union Republics, or SSRs) that made up the Soviet Union. The full name of the Russian Federation during the Communist period was the Russian Soviet Federated Socialist Republic, or RSFSR; the name appears on many older maps.

The loosely aligned Commonwealth of Independent States (CIS) was formed by 12 of the 15 former Union Republics late in 1991. Estonia, Latvia, and Lithuania are not members of this organization. Turkmenistan gave up full membership in 2005 to become an associate member.

The area west of the Ural Mountains and north of the Caucasus Mountains has been known historically as European Russia. The Caucasus and the area east of the Urals have been called Asiatic Russia, Soviet Asia, or the eastern regions. Transcaucasia is the region of the mountains plus the area south of the Caucasus Mountains; Siberia is a general name for the area between the Urals and the Pacific; and central Asia is the arid area occupied by the five countries with large Muslim populations immediately east and north of the Caspian Sea. They were known collectively in pre-Soviet times as "Turkestan" and are known informally today as "the stans" because all the country names end in *-stan*, Persian for "land of."

● **Figure 5.1** Russia and the Near Abroad.

• **Figure 5.2** Population distribution (above) and population cartogram (below) of Russia and the Near Abroad.

Canada than to those of the United States. Interaction with a complex and demanding environment was a major theme in Russian and Soviet expansion and development. Nature continues to provide large assets but also poses great problems for the 12 countries.

The Roles of the Climates and Vegetation

The region of Russia and the Near Abroad has a harsh climatic setting. Severe winter cold, short growing seasons, drought, and hot, crop-shriveling winds are major disadvantages. But there are also advantages in the form of good soils, the world's largest forests, natural pastures for livestock, and diverse wild fauna. However, many of these resources are hard to reach because of the logistical difficulties posed by harsh climates and vast distances.

Most parts of the region have continental climatic influences, with long, cold winters; short, warm summers; and low to moderate precipitation. Severe winters, for which

Siberia (roughly the eastern two-thirds of Russia) is particularly infamous, prevail because the region is generally at a high latitude and has few of the moderating influences of oceans. Four-fifths of the total area is farther north than any point in the conterminous United States (•Figure 5.3). The most extreme continental climate of the world is here; the lowest official temperature ever recorded in the Northern Hemisphere (minus 90°F; minus 68°C) occurred in the Siberian settlement of Verkhoyansk (located at 67°N, 135°E). Westerly winds from the Atlantic moderate the winter temperatures in the west, but their effects become weaker toward the east, where it is much colder.

The average frost-free season of 150 days or less in most areas (except the extreme south and west) is too short for many crops to mature. Most places are relatively warm during the brief summer, and the southern steppes and deserts are hot. Partially offsetting the summer's shortness are the length of the summer days, which encourages plant growth. Summer is also a time of accelerated human ac-

INSIGHTS

The Russian Cross

Russia's economic health plummeted for at least a decade following the breakup of the Soviet Union and has only recently begun to improve. Accompanying the economic downturn were rising unemployment, a decline in health and other services, increasing crime rates, and a growing sense of despair at the individual level. These gave rise to some very unhealthful habits among Russians, which have shown only modest signs of improvement. The rate of alcoholism soared, and one-third of all deaths are alcohol-related (•Figure 5.A). Smoking has long been a national pastime. Organized and petty crime, along with simple drunken brawls, resulted in an alarmingly high incidence of physical violence. Infections of HIV/AIDS have increased more than 10-fold since 2000, first surging through the population of intravenous drug users and then into their sexual partners.

These and other factors led to a drop in Russians' life expectancy, from 68 in 1990 (average for men and women) to 65 in 2007, on a par with the developing nations of Bolivia and Bhutan. Between 1990 and 2007, the death rate rose almost 25 percent, to one of the world's highest levels (15 per 1,000), surpassed only by many countries in Africa, and, outside Africa, only by Afghanistan. The leading causes of death, in descending order, are heart disease, accidents, violence, and cancer. The universal system of health care under the Soviets was not efficient, but a health care system close to collapse replaced it, and hospitals were ill-equipped to help stem the rising tide of deaths.

This clearly is not the benign demographic transition, in which a country's population declines because of falling birth rates that reflect growing affluence. Instead, Russia began moving backward through the transition. As quality of life deteriorated, birth rates plummeted while death rates soared. The birth rate in 2007 was 10 per 1,000, down 16 percent from 12 per 1,000 in 1990. A physician cited "male/female estrangement and a loss of family cohesion" among the reasons for the plummeting birth rate.[1] The trend has been aided by routine abortion: in 2007, there were nearly two abortions for every live birth. Fully 80 percent of all marriages end in divorce.

The disturbing trend lines of increasing death rates and falling birth rates in Russia intersected in the mid-1990s, forming the grim graphic that demographers dubbed the **"Russian cross"** (•Figure 5.B). It is an ominous intersection. Russia's rate of annual population loss of 700,000 (as of 2007) is the highest in the world and has seldom been seen on earth except in times of warfare, famine, and epidemic. One estimate puts Russia's population at less than 100 million in 2050, down nearly 30 percent from 2007. Alarmed, the government in Moscow is studying a package of incentives to increase the birth rate, including cash payouts to mothers, extended maternity leaves, and child care benefits, similar to the incentives offered in Italy and Japan.

• **Figure 5.A** Alcoholism is a major killer in Russia.

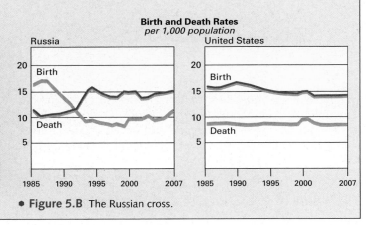

• **Figure 5.B** The Russian cross.

tivity outdoors. Russians are passionate campers and sun worshipers, and they take every opportunity to "get away from it all" to enjoy the fleeting pleasures of summer. Urban residents never lose the traditional Russian passion for nature, and many retreat on weekends and longer holidays to rural cabins called *dachas*.

Aridity and drought also make problems for agriculture. The annual average precipitation is less than 20 inches (50 cm), nearly everywhere except in the extreme west, along the eastern coast of the Black Sea and the Pacific coast north of Vladivostok, and in some of the higher mountains.

There are five main climatic belts in the region of Russia and the Near Abroad: tundra, subarctic, humid continental, steppe, and desert. In this order, these belts, each with its associated vegetation and soils, succeed the other from north to south (•Figure 5.4). There are also smaller scattered areas of subtropical, Mediterranean, and undifferentiated highland climates.

Huge areas in the tundra and subarctic climatic zones are permanently frozen to a depth of a few feet. The frozen ground, called **permafrost,** makes construction difficult. Heat generated by buildings melts the upper layers of permafrost and causes foundations and walls to sink and tilt. Most buildings are elevated on pilings to reduce this risk (•Figure 5.5). Pipelines carrying crude oil (which is hot when it comes from the ground) would likewise melt the permafrost and sink, and so they are heavily insulated or built on elevated supports.

The most productive human activities take place in the subarctic, humid continental, and steppe climatic zones

(•Figure 5.6). The subarctic climate zone corresponds largely with the Russian taiga, or northern coniferous forest, the largest continuous area of forest on earth. The main trees are spruce, fir, larch, and pine, which are useful for pulpwood and firewood but often do not make good lumber. Large reserves of timber suitable for lumber do exist in parts of the taiga, and this is one of the major lumbering areas of the world. Many observers fear that Russia, in its drive to advance the economy, will deplete the taiga (•Figure 5.7). Russia's conifers are believed to act as major carbon dioxide "sinks" that help absorb human-made greenhouse gases, so their removal could contribute to global warming. There is some marginal farming in the taiga, especially toward the south. The dominant soils are the acidic **spodosols** (from the Greek word for "wood ash" and known in Russian as *podzols*), which have a grayish, bleached appearance when plowed, lack well-decomposed organic matter and are low in natural fertility.

A humid continental climate occupies a triangular area south of the subarctic climate, narrowing eastward from the region's western border to the vicinity of Novosibirsk. This area has the short-summer ("cold") subtype of humid continental climate, comparable to that of the Great Lakes region and the northern Great Plains of the United States and adjacent parts of Canada. Here evergreen trees occur in a "mixed forest" with broadleaf deciduous trees such as oak, ash, maple, and elm. Both climate and soil are more favorable for agriculture in the humid continental climate than in the subarctic. Soils developed under broadleaf deciduous or mixed forest are normally more fertile than spodosols, although less so than grassland soils.

The steppe climate characterizes the grassy plains south of the forest in Russia and is the main climate of Ukraine, Moldova, and Kazakhstan. The average annual precipitation is 10 to 20 inches (c. 25 to 50 cm), barely enough for

• **Figure 5.3** Russia and the Near Abroad compared in latitude and area with the continental United States and Canada.

• **Figure 5.4** Climates (left) and biomes (right) of Russia and the Near Abroad.

• **Figure 5.5** In the High Arctic, buildings must be erected on pilings so that they do not melt the permafrost below. This is the Russian coal-mining settlement of Barentsburg in Norway's Svalbard (Spitsbergen) Archipelago.

• **Figure 5.7** The taiga in Russia is the earth's largest continuous forest biome, and Russia's economic growth is based in part on its exploitation. These milled conifers are awaiting shipment from Saint Petersburg's docks.

LAND USE TYPES

- Nomadic herding
- Mixed subsistence
- Mediterranean farming
- Forestry
- Dairy farming
- Grain farming
- Little agricultural activity
- Grazing and stock rearing
- Mixed farming
- Specialized horticulture

• **Figure 5.6** Land use in Russia and the Near Abroad.

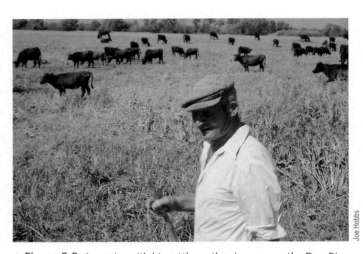

• **Figure 5.8** A grazier with his cattle on the steppe near the Don River in southern Russia.

unirrigated crops. This is the setting of the **black-earth belt,** the most important area of crop and livestock production in the region of Russia and the Near Abroad. The main soils of this belt are known as **chernozem,** which is Russian for "black earth." Among the best soils to be found anywhere, these are thick, productive, and durable **mollisols.** A similar belt of mollisols occupies the eastern portion of the Great Plains of North America. Their great fertility is due to an abundance of **humus** in the topsoil. The major natural threat to productive agriculture here is the occurrence of severe droughts and hot, desiccating winds (*sukhovey*) that damage or destroy crops. The steppe also includes extensive areas of **chestnut** or **alfisol soils** in the zones of lighter rainfall. These are lighter in color than chernozems and lack their superb fertility but are among the world's better soils.

606, 633

The most characteristic natural vegetation of the steppe is short grass. Pastoralists, including the Scythians who in the fourth century B.C.E. produced astonishing artwork of gold in the area that is now Ukraine, grazed their herds from an early time on the treeless steppe grasslands, which stretched over a vast area between the forest and the southern mountains and deserts. Today much of this area is cultivated, with wheat as the main crop. The steppe is also a major producer of sugar beets, sunflowers grown for vegetable oil, various other crops, and all the major types of livestock (•Figure 5.8).

Growing crops is difficult in the desert climate areas east and just north of the Caspian Sea in the central Asian countries. Cotton is the most important crop where irrigation water is available from the Syr Darya, Amu Darya, and other rivers flowing from high mountains to the south and east.

201

The Role of Rivers

In the early history of Russia, rivers formed natural passageways for trade, conquest, and colonization. They were especially crucial in the settlement of Siberia, which is drained by some of the greatest rivers on earth: the Ob, Yenisey, Lena, and Kolyma, all flowing northward to the Arctic Ocean, and the Amur, flowing eastward to the Pacific. By following these rivers and their lateral tributaries, the Russians advanced from the Urals to the Pacific in less than a century.

The Moscow region lies on a low upland from which a number of large rivers radiate like the spokes of a wheel (•Figure 5.9). The longest ones lead southward: the Volga to the landlocked Caspian Sea, the Dnieper (Dnepr) to the Black Sea, and the Don to the Sea of Azov, which connects with the Black Sea through a narrow strait (see The World's Great Rivers, page 160). Shorter rivers lead north and northwest to the Arctic Ocean and Baltic Sea. These river systems are accessible to each other by portages.

A major link in the inland waterway system is the Volga-Don Canal, opened in 1952 to tie together the two rivers where they approach each other in the vicinity of Volgograd (•Figure 5.10). The completion of the canal meant that the

White Sea and Baltic Sea in the north were linked to the Black Sea and Caspian Sea in the south in a single water transport system. The accomplishment was no small task. The Don River, which empties into the Black Sea via the Sea of Azov, is about 150 feet (45 m) higher than the Volga. The ground that separates these two rivers rises to almost 300 feet (90 m). Engineers solved this discrepancy with 13 locks, each with a lift of about 30 feet (9 m), which carry water 145 feet (44 m) above the Don and drop it 290 feet (87 m) to the Volga.

The Role of Topography

Most of the important rivers of Russia and the Near Abroad wind slowly for hundreds or thousands of miles across large plains. Such plains, including low hills, compose nearly all the terrain from the Yenisey River to the western border of the region. The only mountains in this lowland are the Urals, a low and narrow range (average elevation is less than 2,000 ft/600 m) that separates Europe from Asia and European Russia from Siberia. The Urals trend almost due north and south but do not occupy the full width of the lowland. A wide lowland gap between the southern end of

• **Figure 5.9** Physical geography of Russia and the Near Abroad.

THE WORLD'S GREAT RIVERS

The Volga

The Volga is arguably this region's most important river; Russians call it Matushka—"Mother." It rises about 200 miles (320 km) southwest of Saint Petersburg and flows 2,300 miles (3,680 km) to the Caspian Sea. Before railroads supplemented river traffic, the Volga was Russia's premier commercial artery. Traditionally, wheat, coal, and pig iron from Ukraine, fish from the Caspian Sea, salt from the lower Volga, and oil from Baku, on the western shore of the Caspian, traveled upriver toward Moscow and the Urals. Timber and finished products moved downriver to the lower Volga and Ukraine (•Figure 5.C). Boatmen towed barges upstream on a 70-day journey from Astrakhan, near the Volga mouth, to Kazan, on the middle Volga. The steamboat arrived in the late 1800s, bringing an end to the way of life recalled in the famous Russian song "The Volga Boatmen."

The Volga was difficult to navigate until the latter half of the 20th century. There were shoals and shallows, especially in very dry summers, and ice still closes the waterway each winter for 120 to 160 days. During the Soviet era, the **Great Volga Scheme** transformed the river. The goal was to control the flow of the river completely with a stairway of huge reservoirs, each of which reaches upstream to the dam forming the next reservoir, thus assuring complete navigability during the six months when the river is ice-free and supplying hydroelectric power and water for irrigation. Behind the dams that are the backbone of the Great Volga Scheme, the reservoirs are so vast that they are generally called "seas."

With these improvements in navigation, the Volga secured its place as Russia's most important internal waterway. Large numbers of barges and log rafts, as well as a fleet of passenger vessels that carry tourists, use it. Alteration of rivers for power, navigation, and irrigation became an important element of economic development under the Soviets. Large dams were also raised across the main courses or major tributaries of the Dnieper, Don, Kama, Irtysh, Ob, Yenisey, and Angara Rivers.

Joe Hobbs

• **Figure 5.C** Timber moving along the Volga near Saratov. Russian literary depictions of life along the river are sometimes reminiscent of Mark Twain's Mississippi. Down along the Volga wharves at Kazan, Maxim Gorki found "a whirling world where men's instincts were coarse and their greed was naked and unashamed."

the mountains and the Caspian Sea permits uninterrupted east-west movement by land. Cut by river valleys offering easy passageways, the Urals are not a serious barrier to transportation.

Between the Urals and the Yenisey River, the West Siberian Plain is one of the flattest areas on earth. Immense wetlands, through which the Ob River and its tributaries slowly wind their way, cover much of this vast flatland. This waterlogged country, underlain by permafrost that blocks downward seepage of water, is a major barrier to land transport and discourages settlement. Tremendous floods occur in the spring when the breakup of ice in the upper basin of the Ob releases great quantities of water while the river channels farther north are still frozen and

act as natural dams. Russian aircraft sometime bomb ice jams on the rivers to prevent worse flooding.

The area between the Yenisey and Lena Rivers is occupied by the hilly Central Siberian Uplands, which rise 1,000 to 1,500 feet (c. 300 to 450 m) above sea level. Mountains dominate the landscape east of the Lena River and Lake Baikal. Extreme northeastern Siberia is an especially bleak and difficult country for human settlement. High mountains rim the region of Russia and the Near Abroad on the south from the Black Sea to Lake Baikal, and lower mountains rim the region from Lake Baikal to the Pacific. Peaks rise to over 15,000 feet (c. 4,500 m) in the Caucasus Mountains between the Black and Caspian Seas and in the Pamir, Tien Shan, and Altai Mountains east of the Caspian

• **Figure 5.10** A lock in the Volga-Don Canal, a vital link for trade between the Volga watershed and the Black Sea.

Sea. From the feet of the lofty Pamir, Tien Shan, and Altai ranges, and the lower ranges between the Pamirs and the Caspian Sea, arid and semiarid plains and low uplands extend northward and gradually merge with the West Siberian Plain and the broad plains and low hills west of the Urals. "Rugged" and "challenging" thus describe great sections of this huge world region, and the human pattern of the region is neither uniform nor easy.

5.3 Cultural and Historical Geographies

A Babel of Languages

Russia and the Near Abroad form a complex cultural and linguistic mosaic. The region's peoples belong to about 30 major ethnic groups and speak more than 100 languages (•Figure 5.11). In Russia, Belarus, and Ukraine, the majority are Slavs who originated in east central Europe as speakers of an ancestral **Slavic language** (a member of the Indo-European language family) and who now speak **Russian, Belarusian,** and **Ukrainian,** respectively. Along the upper Volga River, in the small regions of Tatarstan, Chuvashiya, and Mariy-El, there are majority speakers of **Finno-Ugric** (a subfamily of the **Uralic** languages) and **Turkic** (in the **Altaic** language family). Another outpost of Turkic speakers is Yakutia (the Sakha Republic) in northeastern Russia. In Moldova, the majority ethnic Moldovans speak **Moldovan,** which is the same language as Romanian, a Romance (**Indo-European**) language.

The Caucasus is extremely diverse in its ethnicity and languages and has long been a magnet for linguistic researchers. The **Armenians** have a unique language within the Indo-European group. Their neighbors in the Caucasus—the Georgians and Azerbaijanis—speak, respectively, a **Kartvelian** (**South Caucasian**) and a Turkic language in the Altaic family. Most smaller ethnic groups in the Caucasus speak languages unrelated to Georgian, belonging to the **Abkhaz-Adyghean** and **Nakh-Dagestanian** families.

The "stan" countries are less complex linguistically. Turkic languages are spoken by most of the ethnic groups in the central Asian countries of Turkmenistan, Kazakhstan, Uzbekistan, and Kyrgyzstan. In Tajikistan, however, the dominant group of **Tajiks** speak an **Iranian** language in the Indo-European family. In far northeastern Russia, there are speakers of the **Chukotko-Kamchatkan** languages, including **Chukchi** and **Koryak,** belonging to the **Proto-Asiatic** language family and related to some of the languages that ancestors of Native Americans carried eastward into the Americas.

This is just a brief overview of the region's tongues; there are many languages spoken by smaller populations. This rich multiethnicity has bestowed great cultural wealth on the region but at the same time has threatened to tear apart the nations within it.

The cultural histories of the minority ethnic peoples have long been influenced and dominated by the Russians, whose origins reach more than 1,000 years into the past. The central figures were Slavic peoples who colonized Russia from the west, interacted with many other peoples, stood off or outlasted invaders, and acquired the giant territory that would become Russia.

Vikings, Byzantines, and Tatars

Slavic peoples have inhabited European Russia since the early centuries of the Christian era. During the Middle

ALTAIC FAMILY
MONGOLIAN
1 Buriat
2 Kalmyk
3 Mongolian
TUNGUSIC
4 Even
5 Evenki
TURKIC
6 Azeri
7 Bashkir

8 Chuvash
9 Karachay
10 Kazakh
11 Kyrghyz
12 Kumyk
13 Tatar
14 Turkmenian
15 Tuvin
16 Uzbek
17 Yakut

CAUCASIAN FAMILY
ABKHAZ-ADYGHEAN
18 Abaza
19 Abkhaz
20 Adyghe
21 Kabardian
KARTVELIAN
22 Georgian
23 Laz
24 Mingrelian
25 Svan

NAKH-DAGESTANIAN
26 Avar
27 Bats
28 Chechen
29 Dido
30 Ingush
31 Lak/Dargwa
32 Lezgian
33 Tabasaran

INDO-EUROPEAN FAMILY
ARMENIAN
34 Armenian
IRANIAN
35 Ossetian
36 Tajiki
37 Talysh
ROMANCE
38 Romanian
SLAVIC
39 Belarusian
40 Russian
41 Ukrainian

PROTO-ASIATIC FAMILY
CHUKOTKO-KAMCHATKAN
42 Chukchi
43 Itelmen
44 Koryak

URALIC FAMILY
FINNO-UGRIC
45 Hungarian
46 Karelian
47 Khanty
48 Komi
49 Mansi

50 Mari
51 Mordvin
52 Saami
53 Votyak
SAMOYEDAN
54 Nenets
55 Nganasan
56 Selkup

OTHERS
57 Ket
58 Nivkh

● **Figure 5.11** Ethnolinguistic distributions in Russia and the Near Abroad, where there is a close correlation between languages and the ethnic groups using them. The Soviet Union had difficulties trying to hold together such a vast collection of culture groups. The Russian Federation faces many of the same challenges.

Ages, Slavic tribes living in the forested regions of western Russia came under the influence of Viking adventurers from Scandinavia known as **Rus** or **Varangians.** The newcomers carved out trade routes, planted settlements, and organized principalities along rivers and portages connecting the Baltic and Black Seas. In the 9th century, the principality of Kiev, ruled by a mixed Scandinavian and Slavic nobility, achieved mastery over the others and became a powerful state. The culture it developed was the foundation on which the Russian, Ukrainian, and Belarusian cultures later arose.

Contacts with Constantinople (modern Istanbul, Turkey) greatly affected Kievan Russia. Located on the straits connecting the Black Sea with the Mediterranean, Constantinople was the capital of the Eastern Roman or Byzantine Empire, which endured for nearly 1,000 years after the collapse of the Western Roman Empire in the 5th century C.E. Constantinople became an important magnet for Russian trade, and the Russians borrowed heavily from its culture. In 988, the ruler of Kiev, Grand Duke Vladimir I, formally accepted the Christian faith from the Byzantines and had his subjects baptized. Following its cleavage from the Roman Catholic Church in 1054, Orthodox Christianity became a permanent feature of Russian life and culture (●Figure 5.12 provides a the map of religions). Moscow eventually came to be known as the **"Third Rome"** (af-

● **Figure 5.12** Religions of Russia and the Near Abroad. The Uniate Church of northwestern Ukraine is an Eastern Catholic Church that accepts the Catholic dogma and the primacy of the pope in Rome but is not subject directly to the pope's control. Sunni and Shiite Islam are described on page 227 in Chapter 6, and the Armenian Church is discussed on page 198.

ter Rome itself and Constantinople) for its importance in Christian affairs.

The Bolshevik Revolution in 1917 began a 75-year period of official repression and neglect of the Orthodox church and other religions. Since the breakup of the Soviet Union, however, there has been a renaissance in religious observance—not just among Christians but also among Muslims, Buddhists, and others—across the vast region of Russia and the Near Abroad (●Figure 5.13). There has even been a small reversal in the outflow of Jews to Israel

● **Figure 5.13** Some of the faithful in a Russian Orthodox church in Vyborg, Russia. Since the breakup of the USSR, there has been a resurgence of organized worship throughout the region.

and the West. Ten million Jews lived in the region before the Nazi Holocaust killed three million of them, and perhaps another million assimilated to escape official Soviet anti-Semitism. In the decade following the breakup of the USSR, half of Russia's Jews fled the country. Some are coming back to today's somewhat more tolerant Russia, which now recognizes four official religions: Orthodox Christianity, Islam, Buddhism, and Judaism.

Yet another cultural influence reached the Russians from the heart of Asia. The steppe grasslands of southern Russia had long been the habitat of nomadic horsemen of Asian origin. During the later days of the Roman Empire and in the Middle Ages, these grassy plains, stretching far into Asia, provided the Huns, Bulgars, and other nomads a passageway into Europe. In the 13th century, the **Tatars** (also known as Tartars) of central Asia took this route. Many steppe and desert peoples of Turkic origins were in their ranks, with the **Mongols** in the lead. In 1237, Batu Khan ("Batu the Splendid"), the grandson of Genghis Khan, launched a devastating invasion that brought all the Russian principalities except the northern one of Novgorod under Tatar rule.

The Tatars collected taxes and tributes from the Russian principalities but generally allowed the rulers of these units to be autonomous. Even the princes of Novgorod, who were not technically under Tatar control, paid tribute to avoid trouble. The Tatars established the khanates (governmental units) of Kazan (on the upper Volga), Astrakhan (on the lower Volga), and Crimea (on the Black Sea). Russians called the Kazan Tatars the **"Golden Horde,"** after the brightly colored tents in which they lived. When Tatar power declined in the 15th century, the rulers of the Moscow principality were able to begin a process of territorial expansion that resulted in the formation of present-day Russia. Today, oil-rich Tatarstan, with its capital at Kazan, is one of Russia's most important "autonomous" political units. Its distinct ethnic Tatar population has proved remarkably resilient to centuries of Russian supremacy.

The Empire of the Russians

The Russian monarchy reached outward from Muscovy, its original domain in the Moscow region. From the 15th century until the 20th, the tsars created an immense Russian empire by building onto this core. This imperialism by land, in the era when the maritime powers of western Europe were expanding by sea, brought a host of alien peoples under tsarist control (see Insights, page 164). Russian motivations for expansion were diverse. Quelling raids by troublesome neighbors, particularly nomadic Muslim peoples in the southern steppes and deserts, was one objective. Many Russian cities, such as Volgograd (originally named Tsaritsyn and then Stalingrad) on the Volga River, were founded as fortified outposts on the steppe frontier. In the wilderness of Siberia, the search for valuable furs and minerals, especially gold, stimulated early expansion, and the missionary impulse of Orthodox priests also played a role. Land hunger and a desire to escape serfdom and taxation led to the flight of many peasants into the fertile black-earth belt of southwestern Siberia, and many landlords also moved there with their serfs.

The initial outward thrust from Muscovy under Ivan the Great (reigned 1462–1505) was mainly northward. The rival principality of Novgorod, near present-day Saint Petersburg, was annexed. This secured a domain that extended northward to the Arctic Ocean and eastward to the Ural Mountains. Later tsars pushed the frontiers of Russia westward toward Poland and southward toward the Black Sea. Peter the Great (reigned 1682–1725) defeated the Swedes under Charles XII to gain a foothold on the Baltic Sea, where he established Saint Petersburg as Russia's capital and its **"Window on the West."** Catherine the Great (reigned 1762–1796) secured a frontage on the Black Sea at the expense of Turkey.

The eastward conquest was even more impressive. Ivan the Terrible (reigned 1533–1584) added large new territories in conquering the Tatar khanates of Kazan and Astrakhan, thus giving Russia control over the entire Volga River. At the end of Ivan the Terrible's reign, traders and pioneers were already penetrating wild and lonely Siberia, where there were only scattered indigenous inhabitants. Among the new settlers were the **Cossacks,** peasant-soldiers in the steppes originally consisting of runaway serfs and others fleeing from tsardom. They eventually gained special privileges as military communities serving the tsars.

A Cossack expedition reached the Pacific in 1639, but Russian expansion toward the east did not stop at the Bering Strait. It continued down the west coast of North America as far as northern California, where the Russian trading post of Fort Ross was active between 1812 and 1841. In 1867, however, Russia sold Alaska to the United States (for 2 cents per acre) and withdrew from North America. The relentless quest of Russian fur traders for sable, sea otter, and other valuable pelts brought great cultural changes to aboriginal peoples in Siberia and Pacific North America. These effects became increasingly pronounced, especially in the last decades of the Russian Empire, when millions of Russians moved into Siberia.

There was one stumbling block in Russia's eastern expansion. In the Amur River region near the Pacific, the Russians were not able to consolidate their hold for nearly two centuries because of opposition by strong Manchu emperors who claimed this territory for China. They were thus barred from the Siberian area best suited to growing food for the fur-trading enterprise and best endowed with good harbors for maritime expansion. This situation changed in 1858–1860 following the victory of European sea powers over China in the Opium Wars.

INSIGHTS

Russia and Other Land Empires

Russia, and later the Soviet Union, developed as a **land empire.** Rather than establishing its colonies overseas, as imperial powers such as Spain and Britain did, Russia established colonies in its own vast continental hinterland (●Figure 5.D). Many of the colonized peoples had little in common with the ethnic Russians ruling from faraway Moscow. Like former colonies of overseas empires such as Britain, non-Russian regions of the periphery of Russia and later the Soviet Union were drawn into a relationship of economic dependence on the imperial Russian core. The central Asian republics of the Soviet Union, for example, followed Moscow's demands to grow cotton, which was shipped to Moscow to be manufactured into value-added clothing that was subsequently sold throughout the empire. This pattern contributed to the growth of the USSR's economy but often inhibited local development. Other entities characterized historically as land empires include the United States, China, Brazil, and the Ottoman, Mogul, Aztec, and Inca Empires.

● **Figure 5.D** The development of Russia's land empire.

Russia was able to add the Amur region to its earlier gains in the Ob, Yenisey, and Lena Basins. Finally, in a series of military actions during the 19th and early 20th centuries, Russian tsars annexed most of the Caucasus region and Turkestan, the name for the central Asia region east of the Caspian Sea.

The Soviet state that succeeded tsarist Russia had to grapple with difficult questions of how to govern and interact with a mosaic of scores of languages and cultures. Soviet authorities permitted the different ethnolinguistic groups to retain their own languages and other elements of their traditional cultures and even created alphabets for those that previously had no written language. Politically, the Soviet Union officially recognized many of the groups as non-Russian nationalities. The USSR established 16 Autonomous Soviet Socialist Republics (ASSRs) as homelands for large ethnic minorities, in theory endowing them with limited autonomous (self-governing) powers. Smaller nationalities were organized into politically subordinate autonomous units, including five autonomous regions (*oblasts*) and ten autonomous areas (*okrugs*). The Soviets' labeling of these units as "autonomous" came back to haunt the Russian nation in the 1990s.

In spite of such apparent concessions to non-Russians, the Russian-dominated regime implemented a deliberate policy of **Russification,** an effort to implant Russian culture in non-Russian regions and to make non-Russians more like Russians. Large numbers of Russians migrated to work in the factories and state farms in non-Russian republics. Russians were prominent in positions of responsibility even within the non-Russian republics; throughout the USSR, they held the majority of top posts in the Communist Party, the government, and the military. But Russian ways had little impact on non-Slavic cultures such as the Georgians and Armenians. Even Slavic kinspeople of the Russians, such as the Ukrainians, insisted on retaining their cultural distinctiveness. In general, the policy of Russification was a failure because of strong nationalist sentiments throughout the Soviet Union.

Russia and the Soviet Union: Tempered by Revolution and War

Russia and the Soviet Union repeatedly triumphed over powerful invaders, notably the Swedish forces led by King Charles XII in 1709, the French and their allied European forces under Napoleon I in 1812, and the German and other European forces sent by Hitler into the Soviet Union in World War II. In each case, the Russians lost early battles and much territory but eventually inflicted a crushing and decisive defeat on the invaders. The success of Old Russia and the Soviet Union in withstanding invasions by such formidable armies was due in part to the environmental rigors (particularly, the brutal Russian winter) that invaders faced, the overwhelming distances of a huge country with

poor roads, and the defenders' love of their homeland. It was also due to talented Russian military leadership and to the willingness of good and poor generals alike to lose great numbers of soldiers in combat. Finally, the successful defense used the **scorched earth** strategy to protect the motherland; rather than leave Russian railways, crops, and other resources to fall into the invaders' hands, the defenders destroyed them.

The **Russian Revolution** of 1917, which set the stage for the formation of the Soviet Union, was really two revolutions that occurred against the backdrop of World War I. In 1914, when a Serb in Sarajevo (in modern Bosnia and Herzegovina) assassinated the heir to the throne of Austria-Hungary, a complicated series of alliances required Tsar Nicholas II to commit Russian troops to fight with Serbia, France, and Britain against Austria-Hungary and Germany in World War I. The first revolution in Russia began early in 1917 as a general protest against the terrible sacrifices of Russian forces on the Eastern Front during this war. That revolt overthrew Nicholas II, the last of the Romanov tsars.

The second was the **Bolshevik Revolution** that came later that year. Led by Vladimir Ilyich Lenin (1870–1924), the Bolshevik faction of the Communist Party seized control of the government. The new regime made a separate peace with Germany and its allies and survived a difficult period of civil war and foreign intervention between 1917 and 1921. Lenin presided over the establishment of Russia's successor state, the Soviet Union, in 1922.

World War II found the Russians again allied with France and Britain in a far more ferocious war against Germany. The Soviet Union's success in withstanding the German onslaught that began as **Operation Barbarossa** in June 1941 surprised many outside observers, who had predicted that the Soviets would prove too weak and disunited to resist for more than a few weeks or months. The crucially important cities of Leningrad (now Saint Petersburg) and Moscow held out through brutal sieges. Late in 1942, Soviet forces halted the German push eastward at the Volga River in the huge Battle of Stalingrad (see Geography of the Sacred, page 166). This engagement was the turning point in the war, when Soviet and Allied forces began to reverse Nazi advances.

The failure of powerful Germany to conquer the Soviet Union was a clear indication that the strength of the Soviet Union had been underrated and that the country's power would henceforth be a major feature of world affairs. But the war's impact on the USSR would linger. The German invasion took an estimated 20 million or more Soviet lives and caused the relocation of millions of people. It did enormous damage to settlements, factories, and livestock. As a strategic precaution during the war, Lenin's successor, Josef Stalin, directed major Soviet industries to relocate eastward away from the front, a move that has had a lasting imprint on the region's economic geography.

GEOGRAPHY OF THE SACRED

Stalingrad

A **sacred space** or **sacred place** may be defined as any locale that people hold in reverence. The most obvious sacred places in the realms of ordinary experience are places of worship such as synagogues, churches, and mosques. Cemeteries, too, evoke a special code of behavior in visitors and are managed as sacred sites. Places on earth where large numbers of people have lost their lives are among the most significant sacred spaces—the Gettysburg Battlefield and Manhattan's Ground Zero are examples in the United States. Russia has the world's most extensive network of sacred places associated with the loss of life.

Perhaps in no other country is the memory of World War II etched so vividly in the national consciousness as it is in Russia, which lost the greatest number of people, perhaps 23 million, including civilians. Even today, millions of pilgrims annually visit a large number of war monuments and cemeteries in an effort to heal deep emotional wounds and to keep alive the memory of Russia's costly wartime resistance. These rites of visitation and commemoration pass to each new generation. Immediately after the wedding ceremony, for example, it is common for a newlywed couple to place flowers at the local tomb of the unknown soldier. Contemporary Russian pride and nationalism have strong roots in wartime sacrifice, and the experiences of those who survived the war continue to influence Russian politics and international relations.

During the war, superiors urged Russian soldiers to fight to the death for the motherland: Russia was sacred ground to be defended at any cost. Today the battlegrounds where those soldiers fell are sacred places. They are kept hallowed by the continuous ritual visitation of veterans, war widows, and three generations of descendants of war survivors.

Stalingrad (now Volgograd), site of the most ferocious battle on World War II's eastern front, is the greatest of all these sites. Germany's Nazi leader Adolf Hitler had a geographic rationale for sending his forces against the city in August 1942. Located on the border of the steppe and semidesert where the Volga and Don Rivers are closest together, it was strategically situated. It was a grain and livestock center, with railway connections to the Don Valley and the Caucasus region. Hitler wanted the vital oilfields of the Caucasus, which would fuel the German war effort, but Rostov-on-Don and Stalingrad stood in his way. Though strategically vital, Stalingrad was to Hitler as much a symbolic as a military prize: Because it bore the name of Russia's leader, its fall would be of great propaganda value to the German war effort. Equally, its salvation from the invader was a goal of nearly religious significance for the Stalin regime.

"Not one step backward," Stalin ordered his troops, and Soviet resistance at Stalingrad is an extraordinary chapter in the history of warfare. The invaders were unprepared for the resolve of the Red Army and for the ferocity of the Russian winter. The five-month battled ended in a devastating defeat for the Germans: Two armies consisting of 24 generals, 2,000 officers, and 90,000 soldiers were taken prisoner. Enough matériel was lost to equip one-fourth of the German army. Two years earlier, the Germans

could not have imagined such a defeat. But victory for the Soviet Union exacted an unimaginable cost. Stalingrad lay in ruins; Soviet authorities dubbed it a "city without an address" and, in honor of its defenders, a "hero town." Fifty years after the battle, Russian military authorities finally released figures on the number of dead. In this sacred ground lay the bodies of 3.5 million soldiers and civilians, of whom 2.7 million were Soviet citizens, mainly Russians.

The visitor to Volgograd cannot help but feel the pain of war that has lingered for more than 60 years. All over the city are monuments to remind the living of the dead. The devastated shell of a mill stands as the only physical artifact of the past, but the monuments built after the war rekindle the emotional losses most strongly. Volgograd's central memorial is the complex on Mamayev Hill, a mecca for Russians. Old soldiers, still wearing their medals, look war-weary even now as they shuffle through. Grandmothers lead small children to place flowers at the feet of statues. Over the mass grave of an estimated 300,000 Soviet and German soldiers, a huge hand raises an eternal torch. The honor guard changes in goosesteps once each hour. This sacred place is dominated by the world's largest statue, a female sword-wielding figure called *The Russian Motherland* (●Figure 5.E).

● **Figure 5.E** The Mamayev Hill monument to the memory of the defenders of Stalingrad. Within the hill lie the bodies of hundreds of thousands of the city's defenders and attackers. This sword-wielding figure, known as *The Russian Motherland,* is the world's largest statue, standing 279 feet (85 m) high and weighing 8,000 tons. *Joe Hobbs*

5.4 Economic Geography

The Soviet Union's collapse dramatically changed the global political landscape and had far-reaching economic consequences. It was in fact the failing economic system that brought the giant country down to begin with. This section discusses the economic events leading up to and in the wake of the Soviet Union's demise and then focuses on the modern economy of the largest successor state, Russia. The economic geographies of 11 successor states of the Soviet Union are taken up in Module 5.1, and those of the Baltic countries are described beginning on page 142.

The Communist Economic System

The Soviet Union's Communist economic system was an attempt to put into practice the economic and social ideas of the 19th-century German philosopher Karl Marx. According to Marx, the central theme of modern history is a struggle between the capitalist class ("the **bourgeoisie**") and the industrial working class ("the **proletariat**"). He forecast that exploitation of workers by greedy capitalists would lead the workers to revolt, overthrow the capitalists, and turn over ownership and management of the means of production to new workers' states. In the classless societies of these states, there would be social harmony and justice, with little need for formal government.

Marx's utopian vision did not materialize anywhere, but it did provide guidelines for antigovernment revolts and Communist political systems in many countries. The ideas of Marx and of Lenin (who died in 1924), as Lenin's successor Josef Stalin interpreted and implemented them, provided the philosophical basis for the Soviet Union's centrally planned **command economy.** Beginning in 1928, a series of five-year economic plans demanded the fulfillment of quotas for the nation: types and quantities of minerals, manufactured goods, and agricultural commodities to be produced; factories, transportation links, and dams to be constructed or improved; and residential areas to be built for industrial workers. The goals were to abolish the old aristocratic and capitalist institutions of tsarist Russia and to develop a strong socialist state equal in stature to the major industrial nations of the West.

In the command economy, an agency in Moscow called **Gosplan (Committee for State Planning)** formulated the national plans, which were then transmitted downward through the bureaucracy until they reached individual factories, farms, and other enterprises. This was an unwieldy and inefficient process in several ways. First, the planners in Moscow were essentially required to act as CEOs of a giant corporation, effectively "USSR, Inc.," that would manage the economy of an area larger than North America—a gargantuan, impossible task. Further, the Soviet planning bureaucracy had no free market to guide it, so the system produced goods that people would not buy or failed to produce goods that people would have liked to buy. Third,

Gosplan stated production targets in quantitative rather than qualitative terms, churning out abundant but substandard products. There was often an obsession with fulfilling huge quotas or implementing grandiose schemes, a Soviet preoccupation sometimes known as **gigantomania.** Finally, fearing reprisals from people higher up the ladder, no one wanted to suggest ways to increase quality and efficiency.

Although cumbersome, this system succeeded in propelling the Soviet Union to superpower status, improved the overall standard of living, prompted rapid urbanization and industrialization, and altered the landscape profoundly. The principal goal of Soviet national planning after 1928 was a large increase in industrial output, with emphasis on heavy machinery and other capital goods, minerals, electric power, better transportation, and military hardware. Masses of peasants were converted into factory workers. New industrial centers were founded, and old ones were enlarged. There were huge investments in defense, and the country grew strong enough to survive Germany's onslaught in World War II. After the war, the Soviets maintained large armed forces and accumulated a massive arsenal of conventional and nuclear weapons in the **arms race** against the world's only other superpower, the United States. By the height of the Cold War in the early 1980s, 15 to 20 percent of the country's GDP was dedicated to the military (in contrast to less than 10 percent in the United States), representing an enormous diversion of investment away from the country's overall economic development.

Farmers and farming had troubled careers under the Soviet system of **collectivized agriculture.** Between 1929 and 1933, about two-thirds of all peasant households in the Soviet Union were collectivized. In this process, their landholdings were confiscated and reorganized into two types of large farm units: the **collective farm (kolkhoz)** and the factory-type **state farm (sovkhoz).** The consolidation of individual farmsteads and villages into fewer but larger communities on the collective farms was supposed to permit the government to administer, monitor, and indoctrinate the rural population and provide services, including education, health care, and electricity, more cheaply and efficiently.

But the rural people fiercely resisted collectivization. In their own version of scorched earth, peasants and nomads slaughtered millions of farm animals and burned crops to avoid turning them over to the "socialized sector." Government reprisals followed, including wholesale imprisonments and executions, together with confiscation of food at gunpoint (often including the peasants' own food reserves and seed). The more prosperous private farmers, known as *kulaks,* were killed, exiled, sent to labor camps, or left to starve. Famine took millions of lives. Soviet leaders disregarded these costs, and collectivization was virtually complete by 1940.

The drive to increase the national supply of farm products also demanded an enlargement of cultivated area. In the 1950s, the Soviet Union began a program to increase the amount of grain (mainly spring wheat and spring barley)

produced by bringing tens of millions of acres of what were called **virgin and idle lands** or **new lands** into production in the steppes of northern Kazakhstan and adjoining sections of western Siberia and the Volga region (•Figure 5.14).

Soviet enterprises in both agriculture and industry harnessed the energies and resources of the whole country to achieve specific objectives. The government called on the people to sacrifice and to make the country strong, especially by working on so-called **hero projects** such as the construction of tractor plants, dams, railways, and land reclamation. The government in turn operated as a collectivized welfare state, providing guaranteed employment, low-cost housing, free education and medical care, and old-age pensions. Social services were often minimal but in some sectors were quite successful; the literacy rate, for ex-

ample, rose from 40 percent in 1926 to 99 percent in 1959. However, military-industrial superpower status was generally achieved at the expense of Soviet consumers, whose needs were slighted in favor of heavy metallurgy and the manufacture of machinery, power-generating and transportation equipment, and industrial chemicals. Consumers lined up in stores to purchase scarce items of clothing and everyday conveniences. The exasperation of shoppers confronted by long lines and empty shelves was an important factor generating dissatisfaction with the economic system and demands that it be reformed.

Economic Roots of the Second Russian Revolution

Internal freedoms and prosperity did not accompany superpower status for Soviet citizens. The flow of goods and services ebbed to a trickle in the 1980s when large demonstrations and strikes underscored public anger at a political and economic system that was sliding rapidly downhill. The Communist system came under open challenge on the grounds that it stifled democracy, failed to provide a good living for most people, and blocked the ambitions of the country's many ethnic groups for a greater voice in running their own affairs. The outpouring of dissent was unprecedented in Soviet history. Worsening economic conditions led to official calls for reform in the mid-1980s.

The revamping of the economic system became an urgent priority during the regime of Mikhail Gorbachev, which began in 1985. Gorbachev proclaimed the new policies of *glasnost* ("openness") and *perestroika* ("restructuring") to allow a more democratic political system, more freedom of expression, and a more productive economy with a market orientation (•Figure 5.15). At the same time,

• **Figure 5.14** Soviet agricultural expansion into the so-called virgin and idle lands, 1954–1957.

• **Figure 5.15** Freedom of expression erupted all over the USSR during the Gorbachev era. This was an artist's kiosk in Saint Petersburg in 1991.

problems of disunity mounted as the various republics and ethnic groups organized by the Soviet system into nominally "autonomous" units took advantage of new freedoms to resurrect old quarrels and demand greater autonomy. Fighting among ethnic groups erupted in several republics. Meanwhile, in all of the republics, declarations of sovereignty and in some cases outright independence challenged the authority of the central government. In 1991, the USSR and most other nations recognized the independence of the three Baltic republics of Estonia, Latvia, and Lithuania. The empire's disintegration had begun.

165, 171

At the center of the crumbling empire, having failed to reverse the downward slide of the economy, Gorbachev faced growing sentiment to scrap the command system and move as rapidly as possible to a market-oriented economy. The reformers' cause was championed by Boris Yeltsin, a leading advocate of rapid movement toward a market-oriented economy and greater control by individual republics over their own resources, taxation, and affairs. Yeltsin's political stature grew when in June 1991 the citizens of Russia chose him president of the giant Russian Soviet Socialist Republic in the Soviet Union's first open democratic election. Gorbachev, himself elected president of the USSR (but by the Congress of People's Deputies rather than by the people as a whole), opposed Yeltsin's proposals for radical reform. He advocated more gradual movement toward a free-market orientation within the state-controlled planned economy and insisted on the need for unity and continued political centralization within the Soviet state.

Matters came to a head in August 1991, when an attempted coup by hardliners failed. Yeltsin gained enhanced stature from his defiant opposition to the coup. Gorbachev resigned his position as head of the Communist Party and began to work with Yeltsin to reconstruct the political and economic order. But his efforts to preserve the Soviet Union and a modified form of the Communist economic system could not stand against the growing tide of change. During the autumn of 1991, the Communist Party was disbanded, and the individual republics seized its property. On December 25, 1991, Gorbachev resigned the presidency, and the national parliament formally voted the Soviet Union out of existence the following day. A powerful empire had quickly and quietly faded away, to be replaced by 15 independent countries, in what was dubbed the **Second Russian Revolution.**

Russia's Period of Misdevelopment

The unprecedented experiment to transform a colossal Communist state into separate bastions of free-market democracy produced strange and often unfortunate results, particularly in the vast Russian Federation. Some observers classified Russia not as a less developed or more developed country but as a **"misdeveloped country."** The disturbing demographic trend of a plummeting population due to rising death rates—the Russian cross described earlier—was

the strongest indicator of Russia's backward momentum in the years following independence. The roots of this trend were mainly economic.

Despite heightened public expectations for a better life in Russia, criticism of the triumphant President Yeltsin and his policy of economic reform toward freer markets increased throughout the course of his two terms in office (1991–1999). Early in his first term, Yeltsin introduced a program of rapid economic reform, known as **economic shock therapy,** designed to replace the Communist system with a free-market economy. It removed price controls and encouraged privatization of businesses. Products and services from the private sector had already been available during the Communist era, but now they were to be the centerpiece of the economy (•Figure 5.16).

The results of shock therapy were generally poor for most people. Former employees of the Communist Party resented the loss of their jobs and privileges. Printing money to meet state obligations caused inflation. Poor people living on fixed incomes struggled to survive high prices for basic necessities. A new consumer-oriented society developed along class lines. Unemployment and homelessness increased, and the gap between rich and poor widened (•Figure 5.17). Access to choice goods and services was no longer forbidden to ordinary citizens, as it had been during the Communist era, but prices were so high that most people could not afford them.

One of the major components that emerged in Russia's new economic geography was the **underground economy,** also known as the *countereconomy* or *second economy* (or

• **Figure 5.16** It was not so long ago that an advertisement for the Rolex watch—one of the world's most lavish luxury items—would have been unthinkable in Russia.

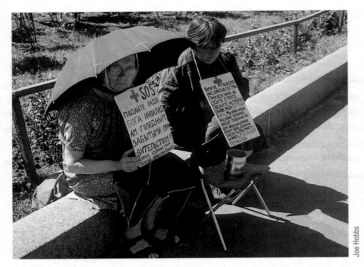

Joe Hobbs

• **Figure 5.17** Having fallen through the cracks in Russia's transition from a command economy to free enterprise, these women are begging for help on a Saint Petersburg sidewalk. Their placards say, "Help us dear brothers and sisters. Please give us money for food, for God's sake. We are invalids with cancer. We have given of ourselves but are now forgotten by the state. We are poor and hungry, with no protection, just left to rot."

economy *na levo,* meaning "on the left"). Rampant black marketeering developed. Although much of this exchange was illegal, there was a general tendency to overlook such transactions because they were essential to the economy. Widespread **barter**—the exchange of goods and services instead of cash—resulted from the declining value of the ruble. Many people resorted to selling personal possessions to buy high-priced food and other necessities. Most Russians grew economically worse off. Many privately owned industries were too unproductive to pay their employees or paid workers under the table to avoid taxation. Unpaid workers could not pay taxes to the government, and unofficially paid workers paid none. Russia's gross domestic product shrank by almost half in the 1990s, the largest fall in production any industrialized country has ever experienced in peacetime. It is little wonder that Russia sought to profit by selling rights to emit carbon dioxide and other greenhouse gases to the richer countries of the West: Its carbon dioxide emissions in 2004 were a third lower than they were in 1990, a sure sign of a deeply troubled industrial sector.

Russia's new private entrepreneurs came to include a large criminal "mafia" that preyed on government, business, and individuals. Organized crime and corruption became pervasive and made "free enterprise" far from free. A company wanting to build a factory, for example, would have to pay off officials at every stage of construction; without the payouts, organized crime would terminate the project. Russia became a **kleptocracy,** an economic and political system based on crime. Corruption is still so rampant that Russians begrudgingly accept it as part of everyday life. Rather than endure hours to pay or contest a traffic violation, for example, most Russians will discreetly hand the equivalent of $15 over to the traffic policeman, and the matter is re-

solved. On its corruption index from 1 to 163, with Haiti the most corrupt at 163, Russia ranked 121 in 2007.

The biggest beneficiaries of the new economic system were the so-called **oligarchs,** Russia's leading businesspeople (*oligarchy* is literally the "rule of the few"). In the 1990s, they acquired massive stakes in former state enterprises through a series of rigged auctions called "loans for shares," finally controlling an estimated 70 percent of Russia's economy. They wielded enormous political power and helped usher Vladimir Putin, a former KGB (state security) officer of the Soviet Union, into Russia's highest office, the presidency, in 2000. Once in office, Putin vowed to crack down on the oligarchs. He purged the largest, especially those in the energy sector, while strengthening the economic and political clout of fellow former servicemen in the Soviet Union's security agency, the KGB. Putin's increasingly authoritarian style proved to be popular with voters, who awarded him a second term in office. He has since grown to have very broad popular support among Russians, with a spectacular approval rating of over 80 percent in 2007. His consolidation of power with little public opposition seemed to reaffirm what Stalin often said: "The Russians need a tsar."

Russia's economic decline following independence raised public fears and prompted some people to call for a return to the relative stability and security of the Soviet state. Others emigrated, including 380,000 from the former Soviet Union to the United States in the 1990s. But most stayed, struggling in creative and often brilliant ways with limited resources to make ends meet and, ideally, help build the middle class and the normal way of life that communism's collapse was supposed to bring.

Recent developments have reversed Russia's decline and have ushered in a new period of self-assurance and economic renewal. Russia's economy and Russians' personal incomes grew steadily in the early 2000s. Between 2000 and 2007, average incomes grew at a rate of about 10 percent each year. Twelve percent of Russia's population still lived in poverty in 2007, but poverty and unemployment had fallen significantly since the start of the decade. Ironically, the improvement resulted partly from the severe economic crisis Russia experienced in 1998. When Russia's currency was devalued that year, sales of imported goods plummeted, and Russian entrepreneurs boosted production of textiles, clothing, and automobiles.

A surge in oil prices after 1999 was the main stimulus to the economy. Russia has enormous energy reserves; its proven oil reserves represent 6 percent of the world total, and its natural gas, 27 percent of proven global reserves, the largest in any single country. After years of belt tightening due to low world market prices, Russia finally began to realize energy revenues. Oil production grew by 50 percent between 1999 and 2007. Energy exports in 2007, when Russia was the world's second-largest oil producer (after Saudi Arabia) and its largest producer of natural gas, accounted for about 64 percent of Russia's export earnings. There was also considerable growth in exports of aluminum and other metals and minerals; by 2006, Russia was tied with

35

Japan as the world's largest steel exporter, for example. The Russian companies selling these resources and products used their growing profits to purchase co-ownership in automobile and other industries and in agricultural companies, helping spark a revival in some sectors of the Russian economy. But as oil revenues continued to rise, investment in other industrial sectors began to lag again.

The biggest danger to Russia's economy lies in excessive dependence on oil and other natural resources, including natural gas, metals, and timber, to drive the economy overall. Should oil prices slump, revenues would plummet and drag the entire economy into a deep recession. And although the oil money is trickling down, there are disparities in where it falls: wealth tends to be concentrated in the larger cities, and nearly a third of all Russians live in smaller communities with little economic activity. For more on Russia's economy, see pages 188–200.

5.5 Geopolitical Issues

Russian President Putin, in a 2005 speech, called the collapse of the Soviet Union "the greatest geopolitical catastrophe of the century."[2] Today there are three concentric spheres of major geopolitical concern in the former Soviet realm: the unity of Russia itself, Russia's relationships with its Near Abroad, and the relations between this region and the rest of the world. There are perhaps more critical geopolitical issues in this world region than in any other, and getting to know them is worth the effort.

Within Russia

Internally, Russia is struggling to maintain a cohesive whole that is fashioned from diverse and sometimes volatile ethnic units. As mentioned earlier, the Soviet Union included a number of "autonomous" units based on ethnicity. When Russia emerged from the Soviet Union, it kept those designations within its borders. The Russian Federation today includes 83 "subjects" divided into six categories: 48 *oblasts* (regions), 7 *krais* (territories), 21 republics (which have varying levels of autonomy), 4 autonomous *okrugs* (ethnic subdivisions of oblasts or krais), 2 federal cities, and 1 autonomous oblast (•Figure 5.18). The 26 **autonomies** (nationality-based republics and lesser units) in total occupy more than 40 percent of Russia's total area and contain about 20 percent of the country's population.

Nearly half of the people in Russia's autonomies are ethnic Russians, who form a majority in many units. The titular nationalities of the autonomies are ethnically diverse, however (see Figure 5.11, page 162). Some, such as the Karelians, Mordvins, and Komi, are Uralic (also known as Finnic, related to the Finns and Magyars); others are Turkic (for example, Tatars, Bashkirs, and Yakuts), Mongolian

• **Figure 5.18** Subject units in the Russian Federation. Russia fears that independence for any of the republics, particularly Chechnya, might fracture the country.

(for example, Buriats near Lake Baikal), and members of many other ethnic groups. Their largest autonomous units form a nearly solid band stretching across northern Russia from Karelia, bordering Finland, to the Bering Strait. The Karelian, Komi, and the largest of all, Sakha (Yakut), Republics are in this group. This band also includes several other large but thinly inhabited units of aboriginal peoples.

What gives some of these units particular geopolitical significance is the presence of major mineral resources: oil and gas in the Volga-Urals fields in Tatarstan and Bashkhortostan, high-grade coal deposits at Vorkuta in the Komi Republic, and a quarter of all diamonds produced in the world in Sakha. After the breakup of the USSR, many of the former Soviet ASSRs, recalling Moscow's long-standing promises of self-rule for them, issued declarations of sovereignty asserting their right to greater self-direction of their internal affairs and greater control over their own resources. Eighteen of Russia's 21 republics signed the 1992 Federation Treaty, which granted them considerable autonomy. The treaty called for the devolution of power centralized in Moscow and for more cooperation between regional and federal governments. Each republic of the Russian Federation was legally entitled to have its own official language, constitution, president, budget, tax laws and other legislation, and foreign and domestic economic partnerships. But many soon complained that Moscow did not honor the treaty, and they began looking for regional solutions to their economic and other problems.

Moscow relented to some of these demands. The Sakha (formerly Yakut) Republic, for example, won the right to keep 45 percent of hard-currency earnings from foreign sales of Sakha diamonds, compared with only a small fraction during the Soviet era. In turn, Sakha must now pay for the government subsidies that Moscow previously gave to local industries. Without credits from Moscow, Sakha no longer pays taxes to Moscow. Salaries and other indexes of living standards in the Sakha Republic have risen since this agreement was implemented. The majority-Muslim, oil-rich Tatarstan, which did not sign the Federation Treaty in 1992, now has "limited sovereignty" with its own constitution, parliament, flag, and official language and has since signed the treaty. The autonomous republic of Chechnya—of particular geopolitical importance because of its crossroads location for oil pipelines—never did sign the treaty, and its future is critical to the Russian Federation (see Problem Landscape, page 173).

As Russia sees it, these regional demands for greater self-rule threaten the unity of the Russian Federation and threaten to deprive Russia of valuable resources. Other autonomous regions are keeping eyes on Tatarstan and Chechnya. Independence for either one might inspire the oil-rich Turkic-speaking Bashkirs and the Chuvash to push for their own independence. This could begin a process that would virtually cut Russia in half. Although some analysts argue that this process of devolution in Russia may actually bring more stability and prosperity to the country,

the government in Moscow worries that what happened to the USSR might happen to Russia itself. To keep the country's vast periphery tied to its core, Moscow created a new layer of bureaucracy with seven administrative regions (Northwestern, Central, Northern Caucasus, Volga, Urals, Siberian, and Far Eastern) appointing a presidential envoy (known as a governor general) to manage national defense, security, and justice matters in each (see Figure 5.1.6, p. 189). In addition, the presidents of Russia's republics, formerly elected, are now appointed by Moscow. The message is that Moscow rules.

Russia and the Near Abroad

A second sphere of geopolitical concerns involves Russia's relations with the other successor states of the USSR. With the collapse of the Soviet Union, the Russian Federation took over the property of the former government within the federation's borders, including Moscow's Kremlin (•Figure 5.19). Russia also took custody of the international functions of the USSR, including its seat at the United Nations. In late 1991, the new countries of Russia, Ukraine, and Belarus (formerly Byelorussia) formed a loose political and economic organization, the Commonwealth of Independent States (CIS), headquartered in Minsk, Belarus. Except for Estonia, Latvia, and Lithuania, the other former Soviet republics eventually joined the CIS, in which Russia took a strong leadership role. To date, the CIS has been concerned mainly with establishing common policies on economic concerns. Russia's President Putin has proposed that its members unite on other issues, particularly security, with the formation of an antiterrorism alliance.

Since the Soviet Union dissolved in 1991, many important links between Russia and the other 14 successor states have persisted. Economic, ethnic, and strategic ties are especially crucial. For example, one major economic link is

• **Figure 5.19** Paul McCartney's 2003 concert in Moscow's Red Square was a fitting symbol of Russia in transition. The show's biggest hit was the Beatles' song "Back in the USSR." On the left is the St. Basil Cathedral and on the right the Kremlin's Spassky Tower.

PROBLEM LANDSCAPE

Chechnya

Nominally part of the Russian Federation, the Caucasus autonomy of Chechnya (population 1 million; see Figure 5.20) has insisted on independence ever since the USSR broke up and is of particular concern to Russia (•Figure 5.F). The Chechens, who are Sunni Muslims, have long resisted Russian rule, and Moscow has fought back. Russian troops attempted but failed to seize control over Chechnya soon after its 1991 declaration of independence. Beginning a second offensive in 1994, and at the cost of an estimated 30,000 to 80,000 lives on both sides, Russian troops succeeded in exerting physical control over most of Chechnya. But in 1996, Chechen forces recaptured the capital city of Groznyy (renaming it Jokhar-Gala). At that time, the Chechen war was deeply unpopular among Russians, and the prospect of further Russian losses led to a peace agreement with the Chechens. The pact required an immediate withdrawal of Russian forces from Chechnya and deferred the question of Chechnya's permanent political status until 2001. Russia allowed Chechens to elect their own president but insisted that Chechnya was, for the time being, still part of the Russian Federation.

The cessation of hostilities was short-lived. Chechen rebels kidnapped and tortured hundreds of Russian citizens in southern Russia and in 1999 carried their campaign to the heart of Russia. They detonated bombs in Moscow and other vulnerable civilian centers, killing more than 300 Russians. The Chechen resistance by this time had become internationalized, with Islamic militant groups from abroad, including from Osama bin Laden's notorious al-Qa'ida organization based in Afghanistan, supplying men and war matériel to their Islamic guerrilla brethren in Chechnya. No longer controlled by Chechen's president, in August 1999 these Chechen Islamists invaded the neighboring province of Dagestan with the hope of creating an independent Islamic state. These insurgents were driven back, but Islamist groups are still active in Dagestan, a potentially volatile area with 37 distinct and fractious ethnic groups.

The 1999 Dagestan invasion and the terrorist bombings in Moscow led to a surge in anti-Chechen sentiment among Russians (already known for strong prejudices against ethnic minorities and immigrants) and to popular calls for resolute action against Chechnya. Russia's newly appointed prime minister, Vladimir Putin, decided to pursue an all-out second Chechen war. In September 1999, Russian forces launched a furious attack that this time put them firmly in control of most of Chechnya. Putin's military success helped secure his victory in the Russian presidential election that followed, but the reaction abroad was mostly negative. Russian troops were documented to have carried out atrocities against Chechen civilians, and Western governments came under pressure to apply sanctions against Russia. None did, however, and were even less inclined to do so after September 11, 2001. After that, President Putin argued that the Chechens were just like the Isla-

mist terrorists who attacked the United States and could only be dealt with by force. His argument was all the more compelling to Washington because many Chechen Islamist militants are among the ranks of al-Qa'ida.

With Chechen guerrilla resistance continuing—still with terrorism directed at Russian civilians, and with some of the more ambitious rebels envisioning a new Islamic caliphate stretching all across the northern Caucasus region—Russia is maintaining a vise grip on Chechnya. In a Russian-sponsored referendum in 2003, Chechens voted that Chechnya should remain part of Russia and in a subsequent election chose a new president. Behind the scenes, Russia ran a rigged election to ensure the victory of their handpicked candidate, reviled by most Chechens; he was assassinated the following year. Prospects for Chechnya's independence from Russia look remote, and Russia is confident enough about keeping the lid on Chechnya that it has begun investing in the reconstruction of its road infrastructure and shattered Groznyy.

© Jenny Matthews/Alamy

• **Figure 5.F** Chechen women rebuilding the train station in war-wracked Groznyy.

the vital flow of Russian oil and gas to other states such as Ukraine (which still gets 80 percent of its energy from Russia), Belarus, the Baltic countries, and Kyrgyzstan, which are highly dependent on this supply of energy. There is some reverse flow of energy, particularly with Turkmenistan's exports of natural gas to Russia.

But the critical factor is the Russian hand on the tap of energy supplies to its neighboring countries. In an energy-hungry world, Russia is using fossil fuels to achieve maximum political clout. Under Vladimir Putin, the Russian government began in 2004 to increase its control of the energy sector, a process of **renationalization** that reversed the trend toward privatization that had been under way. With its state powers in this critical sector, Russia has repeatedly wielded fossil fuel as a political weapon. Loyal former Soviet countries get the "carrot" of large subsidies, while disloyal ones get the "stick"; until 2006, for example, Russia charged Belarus just one-third of what it charged Latvia and Estonia.

It rankles Russia when Western clout grows in Russia's giant backyard, particularly in critical nations like Ukraine, and Russia answers by playing its energy cards. With independence, Ukraine became a nuclear weapon–free nation, and the United States rewarded that move with an increased foreign-aid package. Ukraine's orientation toward the West grew with its membership in NATO's Partnership for Peace program and in its application to become a member of the European Union. Russia's favored candidate then lost Ukraine's 2005 presidential election. Following an eventful campaign in which the opposition attempted to poison him with dioxin, Ukraine's new president, Victor Yushchenko, turned even more solidly westward. In what was widely seen as retaliation for the so-called **Orange Revolution** that brought Yushchenko to power, Russia's mostly state-owned Gazprom company briefly cut natural gas supplies through the so-called Friendship Pipeline to Ukraine in 2006 (and later threatened to do so in Belarus; see page 186).

Ethnic issues are also important in this geopolitical realm. About 25 million ethnic Russians lived in the 14 smaller states at the time of independence. Although some have migrated to Russia since 1991, most stayed, and many have had difficulties finding housing and employment. In certain areas, notably eastern Ukraine and Crimea, many ethnic Russians are causing political instability because they want to secede and form their own state or join Russia. While Russians in the 14 non-Russian nations complain that governments and peoples discriminate against them, the Russian government has said that it has a right and a duty to protect these Russian minorities. Russia also encourages these Russian minorities to maintain and strengthen their ethnic identities, creating prospects for future political secession from host countries. In political geography terms, such a movement by an ethnic group in one country to revive or reinforce kindred ethnicity in another country—often in an effort to promote secession there—is known as **irredentism.**

Ukraine's postindependence struggles with Russia reflect such ethnogeographic problems and other concerns, including control of strategic territory. An early issue of contention was the Crimean Peninsula, a picturesque and verdant resort region. Russia's Catherine the Great annexed the Crimea in 1783, but Soviet leader Nikita Khrushchev returned it to Ukraine in 1954. After the Soviet Union collapsed, Russian irredentists agitated to get back the Crimea, with its 70 percent Russian ethnic population. Ukraine yielded a bit by granting the Crimea special status as an autonomous republic. Crimeans then elected their own president, a pro-Russian politician who advocated the Crimea's reunification with Russia.

For now, Russia seems resigned to the Crimea's belonging to Ukraine but continues to be at odds with Ukraine over a strategic strait separating the Crimea from Russia's Taman Peninsula to the east. This Kerch Strait is the vital sea connection between the Azov Sea and the Black Sea—and therefore between southern Russia and the wider world (see Figure 5.1.3, page 185). Russia wants to share sovereignty of the strait with Ukraine, but Ukraine claims most of the waters as its own and charges Russia several million dollars per year in transit fees. Thanks to a successfully resolved dispute between the two powers, Russia continues to operate its Black Sea naval fleet from the Ukrainian port of Sevastopol, also paying hefty fees for that privilege.

Russia has a number of geopolitical concerns with countries and ethnic groups of the Caucasus south of Chechnya (•Figure 5.20; see also pages 201–202). The Armenians embrace Russia as a strategic ally against their historic enemy, the Turks. The Georgians reject Russia, but several nominally Georgian provinces have populations that are strongly pro-Russian. Russia is keen to nurture the pro-Russian elements and to try to keep the national governments in its orbit.

In 2001 and again in 2006, when Georgia insisted that it was time for Russia to scale back its military presence

• **Figure 5.20** The Caucasus.

in the region, Russia responded by cutting off supplies of natural gas to Georgia. Russia imposed punitive tariffs against imports of Georgian wine and other products and threatened to send home a million Georgian guest workers whose remittances are vital to the Georgian economy. Georgia sees Russia as using such trade, migrant worker, and energy issues to retain leverage in the strategically vital Caucasus region and prevent it from moving further into the Western embrace. Georgia wants to loosen ties with Russia in favor of better relations with the West and its allies (for example, it wants to join the European Union and the North Atlantic Treaty Organization). With the Ukraine, Azerbaijan, and Moldova, Georgia has formed the regional grouping known as **GUAM** (an acronym for its four members) to promote economic integration, democratic reform, and increasing orientation toward Europe—and away from Russia. Georgia was able in 2006 to block Russian accession to the World Trade Organization (WTO), the organization that sets the ground rules for globalization. Most road and rail links between the countries have been severed, and Turkey has already replaced Russia as Georgia's main trading partner. Georgia also wants to reestablish its historically important position in overland trade with the central Asian countries, particularly by serving as an outlet for the export of oil from the Caspian Basin.

For many years, a critical issue between Georgia and Russia was the fate of Batumi, the Black Sea oil-shipping port. It lies in the rebellious province of Ajaria (Adzharia), home to another Russian military base and to a population bitterly opposed to the rule of Georgian President Mikhail Saakashvili, who ousted former President Eduard Shevardnadze in the peaceful **Rose Revolution** of 2003. Saakashvili vowed to reassert firm Georgian control over all of the separatist regions within Georgia, thereby risking a direct military confrontation with Russia. His forces secured Batumi and the rest of Ajaria in 2004, and his attention turned to the even more dangerous South Ossetia and Abkhazia.

Russia supports the aspirations of the primarily Muslim South Ossetians, who would like to free themselves from Georgian control and establish an Ossetian nation (almost 100 percent of the population voted to do so in a 2006 referendum). The mostly Christian North Ossetians, whom they wish to join, live adjacent to them in Russia in a region the North Ossetians call Alania. South Ossetia now exists as a Russian puppet state within Georgia's borders. Since a ceasefire broke down in 2004, the region has been plagued by a violent anti-Georgian insurgency. South Ossetia was also the setting of the horrific civilian deaths at Beslan in 2004 after Islamists from nearby Ingushetia seized hostages at a school.

Also seeking independence from Georgia, with Russian help, is another Muslim people, the Abkhazians, who make up about 2 percent of Georgia's population and are concentrated in the province known as Abkhazia. In 1993, Abkhazian separatists captured the Georgian Black Sea port of Sokhumi in Abkhazia. Russian forces initially aided them, both to regain access to Black Sea resorts and to take

revenge on Georgia's President Shevardnadze, the former Soviet foreign minister whom many Russians held partly responsible for the breakup of the USSR. Russia was also putting pressure on Georgia to rejoin the CIS. Shevardnadze responded by having Georgia rejoin the CIS and agreeing to allow four Russian military bases to remain for a while on Georgian soil and to allow Russian troops to be stationed on Georgia's border with Turkey. Russian forces then put a stop to the Abkhazian offensive but did not drive the rebels from the territory they had captured. Russia began granting Russian citizenship to the majority of Abkhazians—a step that Shevardnadze's successor regarded as Russian annexation of Georgian territory. Fearing that Russia was intent on essentially annexing Abkhazia's 110 miles (180 km) of strategic (and attractive) Black Sea coastline, Georgia invaded Abkhazia's Kodori Gorge in 2006 and began building infrastructure there.

In sum, Russia's perspective on the other former members of the USSR—even the Baltics, which are now firmly in the European orbit but where Russia carried out a cyberattack on Estonia's digital infrastructure in 2007—is that they constitute a special foreign-policy region. Some observers contend that Russia's diplomatic, economic, and limited military involvement in these countries is an effort to establish a buffer zone between Russia and the "Far Abroad." Along the outer frontiers of the cordon of successor states, Russia maintains a chain of lightly-staffed military bases that could be quickly ramped up for combat if conditions warrant. Within the Near Abroad, notably in Moldova, Tajikistan, Georgia, Armenia, and Azerbaijan, Russian troops have become engaged frequently as "peacekeepers" in local ethnic conflicts (•Figure 5.21). In Armenia and Tajikistan, at least, the governments fear what would happen without Russian military support.

Some observers in the West and among Russia's neighbors fear that peacekeeping is merely a euphemism for a plan to restore Russian imperial rule. Some contend that above all, Russia wants access to important resources and

• **Figure 5.21** Russian soldiers outside Russia are variously seen as peacekeepers or conquerors.

GEOGRAPHY OF ENERGY

Oil in the Caspian Basin

Even after recent downward revisions, estimated oil reserves beneath and adjacent to the Caspian Sea make this the world's third most important oil region, behind the Persian-Arabian Gulf and Siberia (there are also large natural gas reserves here). The potential wealth of this resource will be realized only through a difficult and delicate resolution of several geographic and political problems (●Figure 5.G). In the energy-starved world of today, each of these apparently local or regional issues has global implications—even for you, the reader of this book, nowhere near the Caspian.

One problem is how to settle the question of which countries own the fossil fuels under Caspian waters. With smaller reserves of oil lying adjacent to their Caspian shorelines, Iran and Turkmenistan argue that the Caspian should be regarded as a lake, meaning that its lakebed resources should be treated as common property, with shares divided equally among the five surrounding states. Russia, Kazakhstan, and Azerbaijan have signed treaties recognizing the Caspian as a sea consisting of separate sovereign territorial waters, with each country having exclusive access to the reserves in its domain. In this arrangement, Azerbaijan has the largest share.

Azerbaijan's oil situation epitomizes the economic prospects and obstacles facing the southern countries of the region of Russia and the Near Abroad. Planners have studied Azerbaijan's oil with a difficult question in mind: What is the best way to get this oil to market? They considered a variety of export pipeline routes, each of which had a different set of geopolitical, technical, and ecological problems. Since the Caspian is (physically) a lake, Azerbaijan is landlocked, and Azerbaijani oil must cross the territories of other nations to reach market. These territories, however, are embroiled in the wars and political unrest left in the wake of the collapse of the USSR. Russia wants to retain strong influence over Azerbaijan in part by insisting that Azerbaijani oil reach export terminals on the Black Sea by passing through Russian territory, as it now does, through a pipeline to the Russian port of Novorossiysk.

Turkey opposes this strongly on the grounds that oil-laden ships navigating the narrow Bosporus and Dardanelles Straits imperil Turkey's environment (and Turkey would not receive transit fees for those shipments). Environmental concerns have not, however, kept energy-hungry Turkey from importing new supplies of natural gas from Russia via the Blue Stream pipeline (the world's deepest) across the Black Sea floor.

Iran offered a swap with Azerbaijan: Iran would import Azerbaijani oil for Iran's domestic needs and would export equivalent amounts of its own oil from established ports on the Persian Gulf

(Arabian Gulf). Iran also argued that its territory offers the shortest and safest pipeline route for fossil fuel exports from Turkmenistan, Kazakhstan, and Azerbaijan. Iran is also in an excellent geographic position to facilitate transportation of other goods to and from the central Asian nations, using its Gulf of Oman and Persian Gulf (Arabian Gulf) ports. To stimulate this trade, Turkmenistan and Iran have begun linking their rail systems.

Azerbaijan's leaders, however, want more ties with the West (which is generally hostile to Iran and stands firmly against any pipelines routed through Iran) and independence from Moscow. The United States and its Western allies anticipate that Azerbaijan and Kazakhstan will soon be major counterweights to the volatile oil-producing states of the Middle East, and they want to be able to obtain Caspian Sea oil without relying on the goodwill of Russia or Iran. Therefore, in 1995, Azerbaijan signed an agreement with a U.S.-led Western oil consortium to export oil through a combination of new and old pipelines to the Georgian port of Batumi on the Black Sea. To avoid a direct confrontation with Russia over this sensitive issue, some of the initial production passed through the existing Russian pipeline to Novorossiysk.

This compromise applied only to the first stage of production, through 1997. A permanent route still had to be established, and both Turkey and Russia vied for the pipeline to pass through their territories. Russia insisted that the environmental hazard of navigating the Turkish straits could be averted by offloading oil at the Bulgarian Black Sea port of Burgas, shipping it overland through a new pipeline to the Greek Aegean Sea port of Alexandroupolis, and reloading it there. The alternative, a 1,080-mile (1,725-km) Turkish route, would be far from the straits, running across Azerbaijan from Baku, through Georgia (skirting hostile Armenia), and then across Turkey to the Mediterranean port of Dörtyol, near Ceyhan. The Western powers preferred this BTC (Baku-Tbilisi-Ceyhan) pipeline, mainly as a political means to block either Russian or Iranian reassertion of power in the region, and it was built (a natural gas pipeline paralleling its route is under construction). Georgia welcomed the projected $50 million in annual transit fees that the pipeline would generate but also worried about the security costs of protecting its 154 miles (246 km) of the route. Turkey had security in mind when it routed the pipeline to detour around the volatile Kurdish-dominated region of southeastern Turkey; this explains the odd (and costly) bend in the Turkish portion.

Other Caspian Basin countries face similar questions about how best to export their resources. Turkmenistan, on the lake's

---

economic assets in its former republics, such as uranium in Tajikistan, aviation plants in Georgia, military plants in Moldova, and the Black Sea coast and naval fleet in Ukraine's Crimea. Within Russia, there is in fact a body of public opinion favoring reassertion of Russian control over its former empire, but the strength of this feeling is unknown. President Putin spoke of Russia's urgent prior-

ity to draw the countries that were republics of the former Soviet Union closer to Russia. Such reintegration could, of course, take place through peaceful political and economic means. In one case it has, at least nominally; in 2000, Russia and Belarus officially established a political union and elected a joint parliament (but real unity remained elusive; see page 186).

GEOGRAPHY OF ENERGY

Oil in the Caspian Basin *continued*

southeast shore, has large oil deposits and natural gas reserves. At present, these are exported mainly in pipelines passing around the Caspian Sea through Russia, and Turkmenistan is seeking an alternative through Iran to Turkey (it already exports some natural gas to Iran via pipeline), across Afghanistan to Pakistan (a hazardous route, given conditions in Afghanistan), or under the Caspian to link up with Azerbaijani export routes. The country's leadership is in favor of that third alternative, the export route promoted by the United States, with oil and gas traveling westward through pipelines on the bed of the Caspian Sea to Baku and then through Georgian and Turkish pipelines to the Mediterranean Sea.

Of the five central Asian countries, Kazakhstan has the largest fossil fuel endowment. With recent discoveries of huge reserves in the Kashagan oilfield of the Caspian Sea, some have predicted that Kazakhstan will eclipse Russia and Saudi Arabia as the world's leading oil producer by 2015. In the late 1990s, Kazakhstan began exporting its oil from fields around Tengiz to join the Baku-to-Novorossiysk pipeline at Komsomolskaya. Unwilling for the time being to find an alternative to a Russian route, the Kazakhs worked with Omani, Russian, and American firms to build a pipeline bypass from Komsomolskaya to Novorossiysk. Kazakhstan is also studying the possibility of building a 2,000-mile (c. 3,200-km) pipeline to export oil eastward into China, but the construction costs are daunting. Keep your eyes on these pipeline politics; they are among the most critical debates in our world today.

• **Figure 5.G** Numerous physical and political obstacles stand in the way of oil exports from the Caspian region.

The Far Abroad

Finally, there are outstanding geopolitical issues in relations between Russia and the Near Abroad and the rest of the world. In the years following the breakup of the USSR, most of these involved the development of a new model of relations between East and West and a peaceful succession to the Cold War. During those years, important plans were also made about how to ship central Asia's critical oil supplies westward. After 9/11, these issues were complemented by others related to the "war on terrorism," as perceived by both the United States and Russia. After 2004, a popular and confident Vladimir Putin, emboldened especially by Russia's soaring energy revenues, put more distance between Russia and the West and consistently reminded Russians of their greatness.

Russia and the other successors of the Soviet Union have nevertheless generally enjoyed improving relations with the West since the USSR dissolved. Russia has since 1997 been a member of what is known as the **Group of Eight (G-8),** the world's eight most economically powerful and politically influential countries (along with Canada, France, Germany, Italy, Japan, the United Kingdom, and the United States). Russia is a member of the UN Security Council and along with its other permanent members (China, France, the United Kingdom, and the United States) has the enormous veto power that position offers; Russia can therefore have a huge impact on world affairs. Russia has ramped up oil production and advertised itself to the United States and other major industrial powers as a more stable source of energy than the volatile Persian Gulf (Arabian Gulf) countries—only to threaten that status by renationalizing oil and manipulating fuel supplies and prices in its Near Abroad. Countries of the Far Abroad are asking profound questions about how reliable a business partner Russia is and how far it might go in flexing its muscle in the countries of the former Soviet Union. Western Europe frets because of its growing dependence on gas from Russia via a cat's cradle of pipelines; Germany, already the largest foreign consumer of Russian gas, is the most worried because it has been building a new undersea gas pipeline to Russia. On track to start receiving oil from Russia's Sakhalin and Arctic fields in 2010, the United States is also concerned.

Despite geopolitical tensions, the shadow of nuclear doomsday has dissipated. The Warsaw Pact, a military alliance formed around the Soviet Union to counter the West's North Atlantic Treaty Organization, has dissolved, and Russia has very grudgingly accepted the survival and expansion of NATO in its backyard. Both sides have reduced their nuclear arsenals, with Russia taking the lead in calling for even deeper cuts. Despite steps forward in relations, the Western powers increasingly wonder if Russia, apparently in the process of reasserting itself as a world power, could still be a threat to peace. They cite President Putin's threat to retarget nuclear weapons at European countries unless the United States revokes its plans to position a "missile shield" in Poland and the Czech Republic. They also condemn Russia's assistance to the nuclear and would-be nuclear powers North Korea and Iran; its sales of military aircraft to Sudan, Myanmar, and Venezuela; and its support of the Islamist Hamas organization in the Palestinian Territories.

The West is no longer preoccupied by the prospect of a conventional nuclear strike by any of the USSR's successors. However, the United States and its European allies are very fearful that low budgets and lax security at nuclear facilities throughout the region could help divert nuclear weapons or nuclear fuel to terrorists. There have already been hundreds of thefts of radioactive substances at nuclear and industrial institutions in the former USSR. The destination points of these materials remain largely unknown.

The West also fears that underpaid nuclear scientists will sell their know-how to governments or organizations such as al-Qa'ida with nuclear strike intentions. Former Soviet nuclear experts are known to have worked in Iran, Iraq, Algeria, India, Libya, and Brazil, and it is possible that some have cooperated with stateless organizations like al-Qa'ida. Osama bin Laden repeatedly stated his intention to acquire and use nuclear weapons, so the issue of **"loose nukes"** in Russia and the Near Abroad has great importance.

Since 9/11, and particularly since the U.S. invasion of Iraq in 2003, the geopolitical significance of the central Asian "stan" countries has increased enormously. This region, also now empty of nuclear weapons, borders unsettled Afghanistan, the potential flashpoint of Iran, and the great power of China and is itself rich in mineral resources. Diverse kinship, spiritual, economic, and political relations exist between the central Asian Muslims and their neighbors on the other side of the international frontiers. Along with internal ethnic conflicts and dissatisfaction over living conditions, these international affiliations may make the region a major political problem area in the future.

Both Iran and Turkey are vying for increased influence in central Asia. Except in Tajikistan, support for Iran is largely lacking, in part because the majority of central Asians practice Sunni rather than Shi'ite Islam. Ethnicity is also important; the four Turkic states are inclined to orient with Turkey, with which they have already established cultural ties such as educational exchanges and shared media. Turkey also has extended economic assistance and established small business ventures and large construction contracts in central Asia. Turkmenistan has good relations with both Turkey and Iran, and has built a rail link with Iran. Turkmenistan also tries to maintain good ties with Russia.

The region has become a focus for rivalry between the historical enemies Turkey and Russia. Russians are concerned that Turkey is winning too much influence in the young Muslim countries of central Asia. For their part, many Turks believe Russia will try to take over the Transcaucasus region again and then turn to central Asia—a fear shared to some extent by the United States, which is intent on maintaining secure access to the region's oil (see Geography of Energy, pages 176–177). Some Turks uphold a dream of **pan-Turkism,** uniting all the Turkic peoples of Asia from Istanbul to the Sakha Republic in Russia's Siberia region—a prospect that Moscow does not like.

The twin scourges of terrorism and narcotics have put central Asia on the map of geopolitical hot spots. Uzbekistan has erected concrete and barbed-wire barriers and sown land mines on its borders with Tajikistan and Kyrgyzstan. The Uzbek government is trying to prevent the incursion of what it sees as two huge threats: an epidemic in heroin use and a militant Islamic insurgency. Both issues have preoccupied the region's governments. The major routes for exports of heroin from Pakistan and Afghanistan are north through central Asia and then into Russia and western Europe. Along this pathway, large numbers of central Asians have fallen prey to inexpensive, pure heroin. The associated health and social problems, including HIV infection, hepatitis, and prostitution, have also taken root (HIV infection rates doubled here between 2003 and 2005). Kyrgyzstan has an extensive public health education campaign, including a needle exchange program for intravenous drug users; other central Asian countries (except Turkmenistan) are also starting to combat the spread of HIV. Tajikistan's economy is now heavily dependent on drug trafficking, and most Western investors and aid agencies avoid the country.

The potential of an Islamist threat to central Asian stability is uncertain. Russia has expressed great concern about an Islamist wave spreading into central Asia and Russia itself. Tajikistan and Kyrgyzstan have allowed Russia to deploy thousands of Russian troops in their countries, in part to stave off such a threat. Some observers believe Russia is using the Islamist threat in places like Georgia's Pankisi Gorge as a pretense to maintain influence in the region or even to counter the growing U.S. military presence there since 9/11. Others believe the Islamist danger is real. The outlawed Islamist party **Hizb ut-Tahrir** hopes to establish a pan-Islamic caliphate throughout central Asia and is suspected of being responsible for uprisings in Uzbekistan and Kyrgyzstan. Tajikistan is home to a group called the **Islamic Movement of Uzbekistan (IMU)**, which aims to overthrow the Uzbek government of President Islam Karimov and establish an Islamic state in the Fergana Valley. Following a 1999 attempted assassination blamed on the rebels, Karimov cracked down hard on suspected Islamists and even

began to restrict symbols of Islamist sympathies, such as beards on men and headscarves on women. Islamist suicide bombings of government targets carried into the middle of the following decade. Anxious to repress any external support for Islamist insurgency within its borders and to curry political and economic favor with Washington, Uzbekistan initially welcomed the U.S. military to use its land and air space in attacks against al-Qa'ida and the Taliban in 2001 and 2002. Uzbekistan shut down this access in 2005 after the United States and its European partners, citing human rights abuses, slashed economic assistance to the country. The United States, like Russia, continues to use military bases in nearby Kyrgyzstan and Tajikistan.

Most recently, China has sought to build security relationships with the central Asian countries, especially to blunt any aspirations by Muslim insurgents in China's far western region. Not surprisingly, given its geographic proximity and surging wealth, China has growing economic interests in the region, too. Chinese exports to central Asia are surging. To build on this trend, China is upgrading its roads leading into Kyrgyzstan and cooperating with Kazakhstan to build a railway across Kazakhstan that will ultimately link producers and markets in China and Europe. In an apparent effort to counterbalance NATO, China and Russia are urging their central Asian partners (all the central Asian countries except Turkmenistan) in the **Shanghai Cooperation Organization (SCO)** to form a military confederacy.

Module 5.1 provides more detail on the land and life in the subregions and countries that make up Russia and the Near Abroad.

SUMMARY

→ From the Russian perspective, the other 14 countries of the former Soviet Union make up the region of the "Near Abroad," a special realm of policy and interaction. The three Baltic countries have struggled to divorce themselves from this orbit. The remaining countries, along with Russia, are members or associate members of the Commonwealth of Independent States. All have significant ties with Russia, and Russia aspires to increase its influence on them.

→ The area west of the Ural Mountains and north of the Caucasus Mountains is known as European Russia; the Caucasus and the area east of the Urals is Asiatic Russia. Siberia is the area between the Urals and the Pacific. Central Asia is the name for the arid area occupied by the five countries immediately east and north of the Caspian Sea.

→ Russia and the Near Abroad span 11 time zones. Stretching nearly halfway around the globe, the region has formidable problems associated with climate, terrain, and distance.

→ In the years of economic uncertainty after 1991, birth rates fell and death rates rose dramatically in Russia, leading to steep declines in population. Growing revenues from oil have begun to improve standards of living, however.

→ Vast stretches of the region are at very high latitudes, making agriculture difficult. The most productive agricultural area is the steppe region south of the forest in Russia and in Ukraine, Moldova, and Kazakhstan. In early Russia, rivers formed natural passageways for trade, conquest, and colonization. In Siberia, the Russians followed tributaries to advance from the Urals to the Pacific. More recently, alteration of rivers for power, navigation, and irrigation became an important aspect of economic development under the Soviet regime.

→ Large plains dominate the terrain from the Yenisey River to the western border of the country. The area between the Yenisey and Lena Rivers is occupied by the hilly Central Siberian Uplands. In the southern and eastern parts of the region, mountains, including the Caucasus, Pamir, Tien Shan, and Altai ranges, dominate the landscape.

→ Slavic and Scandinavian peoples were prominent in the early cultural development that was the foundation for the Russian,

Ukrainian, and Belarusian cultures. Although Slavs came to dominate economic and political life, the region has large and diverse populations of non-Slavic peoples.

→ Early contacts with the Eastern Roman or Byzantine Empire led to the establishment of Orthodox Christianity in Russian life and culture. Although it was repressed after the Bolshevik Revolution, this and the region's other religions have been experiencing a rebirth since the breakup of the Soviet Union.

→ From the 15th century until the 20th, the tsars built an immense Russian empire around the small nuclear core of Muscovy (Moscow). This imperialism brought huge areas under tsarist control. Russian expansion extended to the Pacific, across the Bering Strait, and down the west coast of North America. Russia has often triumphed over powerful foreign invaders. These achievements have been due in part to the environmental obstacles the invaders faced, the overwhelming distances involved, and the defenders' willingness to accept large numbers of casualties and implement a scorched-earth strategy to protect the motherland.

→ Policies of the Soviet Union permitted most ethnic groups to retain their own languages and other elements of traditional cultures. However, the Soviet regime implemented a deliberate policy of Russification in an effort to implant Russian culture in non-Russian regions. Millions of Russians settled in non-Russian areas, but local cultures were not converted to Russian ways.

→ To meet the needs of its people, the USSR became a collective welfare state. However, military and industrial superpower status was generally achieved at the expense of ordinary consumers.

→ The sudden transition from a command economy to capitalism in Russia and other countries in the region led to a widening gap between the rich and the poor, with growing ranks of poor people. Organized crime and an underground economy emerged. Agricultural and industrial production fell dramatically. However, high oil prices in the early 2000s improved Russia's economic outlook. This overdependence on a single commodity could prove risky for Russia.

→ There are three important realms of geopolitical concern in the region. Within Russia, the aspirations of non-Russian people like the Chechens and Tatars appear to pose a danger to the unity of the Russian Federation. Between Russia and the Near Abroad countries, there are challenging issues including energy shortages and supplies, desires of Russians outside Russia to achieve their own rights and territories, stationing of Russian troops, and Islamist terrorism. Between this region and the rest of the world—the "Far Abroad"—there are concerns about the fate of Soviet-era nuclear materials, whether the oil-rich central Asian countries are more sympathetic to Russia or Turkey, and what should be done to stem the tide of narcotics and terrorism.

→ The Russian government has said that it has the right and duty to protect Russian minorities in the other countries. Some observers contend that Russia is attempting to reestablish the Near Abroad as buffers between Russia and the Far Abroad.

→ The Caspian Sea and adjacent areas make up one of the world's greatest petroleum and natural gas regions. Russia, Iran, and the West, particularly the United States, are vying for influence in this increasingly important area. Of particular concern to the rivals is how the energy supplies should be routed via pipeline to reach ocean terminals.

→ Because of nuclear weapons and other critical issues, relations between Western countries and the countries of the former Soviet Union, particularly Russia, are vital to global security.

KEY TERMS + CONCEPTS

Abkhaz-Adyghean language family
  (p. 161)
alfisol soils (p. 157)
Altaic language family (p. 161)
  Turkic subfamily (p. 161)
arms race (p. 167)
autonomies (p. 171)
barter (p. 169)
black-earth belt (p. 157)
Bolshevik Revolution (p. 165)
bourgeoisie (p. 167)
chernozem soils (p. 157)
chestnut soils (p. 157)
collective farm (*kolkhoz*) (p. 167)
collectivized agriculture (p. 167)
command economy (p. 167)
Commonwealth of Independent States
  (CIS) (p. 152)

Cossacks (p. 164)
economic shock therapy (p. 169)
"Evil Empire" (p. 151)
gigantomania (p. 167)
glasnost (p. 168)
"Golden Horde" (p. 163)
Gosplan (Committee for State Planning)
  (p. 167)
Great Volga Scheme (p. 160)
Group of Eight (G-8) (p. 177)
GUAM (p. 175)
hero projects (p. 168)
Hizb ut-Tahrir (p. 179)
humus (p. 157)
Indo-European language family (p. 161)
  Armenian (p. 161)
  Iranian subfamily (p. 161)
  Moldovan (p. 161)

Tajik (p. 161)
irredentism (p. 174)
Islamic Movement of Uzbekistan (IMU)
  (p. 179)
Kartvelian (South Caucasian) language
  (p. 161)
kleptocracy (p. 170)
land empire (p. 164)
"loose nukes" (p. 178)
"misdeveloped country" (p. 169)
mollisol soils (p. 157)
Mongols (p. 163)
Nakh-Dagestanian language family
  (p. 161)
Near Abroad (p. 152)
new lands (p. 168)
oligarchs (p. 170)
Operation Barbarossa (p. 165)

REVIEW QUESTIONS

1. What are the main climatic belts and corresponding vegetation types of Russia and the Near Abroad? Which are most and least productive for agriculture?

2. What are the region's major river systems?

3. Why has Russia's population been declining?

4. Why is Russia known as a land empire? How did it acquire its empire?

5. What significant conflicts have occurred in this region? Why has Russia so often won them? What was the geographic significance of Stalingrad in World War II?

6. What were the perceived advantages of collectivized agriculture?

7. What major Soviet Communist projects changed the landscape of this region?

8. What were some of the successes and failures of efforts to industrialize the Soviet Union?

9. What were some of the results of Russia's "shock therapy" economic reforms and the subsequent conditions of the country and its people? What accounts for the recent turnaround in Russia's economy?

10. Which of the non-Russian ethnic groups in Russia are particular security concerns to Russia today?

11. How does Caspian Sea oil reach markets abroad? What factors go into decision-making about those exports?

12. Where outside Russia are Russian military troops stationed? What are the reasons for their presence?

NOTES

1. Murray Feshbach, quoted in C. J. Chivers, "Putin Urges Plan to Reverse Slide in the Birth Rate." *New York Times,* May 11, 2006, pp. A1, A6.

2. Quoted in C. J. Chivers, "For Putin and the Kremlin, a Not So Happy New Year." *New York Times,* January 3, 2006, p. A8.

Children at a day care center in Ulyanovsk, Russia.

Joe Hobbs

MODULE 5.1

FRAGMENTATION AND REDEVELOPMENT IN RUSSIA AND THE NEAR ABROAD

Few geographers foresaw the collapse of the Soviet Union. Like most other observers, we assumed that when threatened with disintegration, the core of Russia would assert its military strength to retain the countries of the periphery at any cost. From the geographer's perspective, the survival of the Soviet Union and its status as a global superpower always depended on Russia's ability to control non-Russian resources such as the oil of the Caucasus, the soil of Ukraine, and the cotton of central Asia. Asked prior to 1991 whether Russia would let go of these, most of us answered "no, not without a violent struggle."

But the union of so many disparate peoples was so unlikely that in retrospect, it seems absurd. National unity was, in hindsight, an oxymoron in the Soviet Union. Marxism-Leninism had assumed that the cultural and linguistic differences among more than 200 separate groups would disappear as the ideal, universal proletarian society emerged. In effect, Moscow's rulers hoped they would all become Russian. Russification was supposed to level out the differences, but it could not and did not.

■ module objectives

This module should enable you to

→ Recognize core and peripheral subregions of Russia and the Near Abroad

→ Identify geographic obstacles (for example, vast distances) and opportunities (for example, transpolar routes) for Russia in east-west trade

→ Appreciate the importance of oil in the development of Russia's peripheral regions

→ See that free enterprise in former Communist countries has introduced a new set of environmental problems

→ View cotton as a legacy of Soviet "colonialism" in central Asia

---

Now Russia has lost all but the core: the Russian Federation, itself an amalgam of many non-Russian entities artificially cemented together. The other former republics have gone their own ways as independent countries. How are these entities faring today, and what is their future? What has happened to the Russian and non-Russian resources, agricultures, and industries that once supported a global power? This module will answer these questions and more.

An outstanding feature in the geography of this huge region is that resources are distributed unevenly, favoring development in certain subregions and countries. Agricultural and industrial resources are in fact clustered in a core region encompassing western Russia, northern Kazakhstan, Ukraine, Belarus, and Moldova. Nearly three-fourths of the region's people—the great majority of them Slavic—and an even larger share of the cities, industries, and cultivated land of this immense region are packed into this roughly triangular **Fertile Triangle** comprising about one-fifth of the region's total area. Also known as the **Agricultural Triangle** and the **Slavic Core,** this is the distinctive functional hub of the region (•Figures 5.1.1 and 5.1.2).

A core suggests that there is a **periphery,** and one does exist—the world's largest, in fact. The remaining four-fifths of the region of Russia and the Near Abroad, lying mostly in Asia, consist of land supporting only spotty human settlements in conditions of sometimes-huge environmental adversity. Most of the region's non-Slavic peoples live in these lands, but many immigrant Slavs also reside there, generally in cities. Although these areas are marginal when

• **Figure 5.1.1**   Major regional divisions of Russia and the Near Abroad as discussed in the text.

compared with the well-endowed Fertile Triangle, they do include particular resource and production centers that represent realized or potential wealth. Several countries in the Caspian Sea region are mineral-rich. There are pockets of diverse and productive agriculture in the Caucasus. Russia's Siberia is a storehouse of minerals, timber, and waterpower. The geographic pattern that has emerged is a series of discrete production nodes widely separated by taiga, tundra, mountains, deserts, wetlands, and frozen seas.

5.1.1 Peoples and Nations of the Fertile Triangle

Slavic peoples are the dominant ethnic groups in most of the region of Russia and the Near Abroad in both numbers and political and economic power. The major groups are the Russians (discussed extensively in Chapter 5), Ukrainians, and Belarusians (Byelorussians). Most of these Slavs live in the core region extending from the Black and Baltic Seas to the neighborhood of Novosibirsk in Siberia. Moscow (population 14.6 million) and Saint Petersburg (formerly Leningrad; population 4.8 million) are the largest cities, followed by the Ukrainian capital, Kiev (population 3 million). This is mainly a lowland region; the only mountains are the Urals, a small segment of the Carpathians, and a small range in the Crimean Peninsula. The original vegetation was mixed coniferous and deciduous forest in the northern part and steppe in the south.

Ukraine

Although closely related to the Russians in language and culture, Ukrainians—the second largest ethnic group in the Slavic Core—are a distinct national group. The name

● **Figure 5.1.2** Major economic zones and transportation routes of Russia and the Near Abroad. The map reveals that the Fertile Triangle is also the industrial core of the region. Moscow's function as a transportation hub is evident.

Ukraine translates as "at the border" or "borderland," and this region has long served as a buffer between Russia and neighboring lands (●Figure 5.1.3). Here armies of the Russian tsars fought for centuries against nomadic steppe peoples, Poles, Lithuanians, and Turks, until the 18th century, when the Russian Empire finally absorbed Ukraine. For three centuries, Ukrainians were subordinate to Moscow, first as part of the Russian Empire and then as a Soviet republic. Ukraine's industrial and agricultural assets were always vital to the Soviet Union; the Bolshevik leader, Vladimir Lenin, once declared, "If we lose the Ukraine, we lose our head." Slightly smaller than Texas, Ukraine (population 46.5 million) lies partly in the forest zone and partly in the steppe. On the border between these biomes, and on the banks of the Dnieper River, is the historic city of Kiev, Ukraine's political capital and a major industrial and transportation center (●Figure 5.1.4). Although its agricultural output has plummeted since the USSR's demise, Ukraine was traditionally a great "breadbasket" of wheat, barley, livestock, sunflower oil, beet sugar, and many other products, grown mainly in the black-earth soils of the steppe region south of Kiev.

North of Kiev is **Chernobyl,** where an explosion at a nuclear power station in 1986 rendered parts of northern Ukraine and adjoining Belarus and parts of Russia incapable of safe agricultural production for years thereafter. Encircling the station is still an 18-mile (29-km) "exclusion zone" where farming is prohibited. Recent studies suggest that 100,000 to 200,000 people are still severely affected by the Chernobyl disaster and that 4,000 deaths will ultimately be attributed to it. However, the toll on human health and life is well short of what had been predicted. The psychological trauma to the region's inhabitants has been profound, and the three affected countries have spent large sums to provide benefits to Chernobyl's victims. Having shut down the Chernobyl plant in 2000, Ukraine has promised to close all of its Chernobyl-type nuclear plants within a few years. This will be a costly proposition, considering the country's shortage of affordable alternative fuels.

Ukraine has paid a high economic cost for independence. Previously, the Soviet economy subsidized the provision of Russian oil, gasoline, natural gas, and uranium to Ukraine. But with independence, Ukraine was compelled to

● **Figure 5.1.3** Belarus, Ukraine, and Moldova were among the most productive republics of the Soviet Union and are now struggling to develop their economies independently or in association with Russia.

● **Figure 5.1.4** Kiev, capital of Ukraine.

pay much higher world market prices for these commodities. Ukraine now maintains a costly debt for the natural gas it obtains from Russia and faces the prospect of even higher prices for Russian gas because of its increasing political leaning to the West. Price hikes could prove devastating to Ukraine's natural gas-guzzling chemical plants and steel mills. Ukraine is, however, well endowed with other mineral resources, including large deposits of coal, iron, manganese, and salt.

Since independence, Ukraine has remained far more reliant on heavy industrial production than Russia, so these resources are actively exploited. Coal is concentrated in the Donets Basin (Donbas) coalfield near the industrial center of

Donetsk, where most of Ukraine's iron and steel plants are also located. Iron ore comes by rail from central Ukraine in the vicinity of Krivoy Rog, the most important iron-mining district in the region of Russia and the Near Abroad. Most of the ore is used in Ukrainian iron and steel plants. This area is also a center of chemical manufacturing. Surrounding the inner core of mining and heavy-metallurgical districts in Ukraine is an outer ring of large industrial cities, including Kiev, the machine-building and rail center of Kharkov (east of Kiev), and the seaport and diversified industrial center of Odessa on the Black Sea.

Along with Belarus, Ukraine is the former Soviet republic most like Russia—"Russia writ small," as one author described it.[1] Many Ukrainians, particularly the Russian-speaking inhabitants who are the vast majority in the eastern part of the country (where most of the industries are), would like to cultivate their Slavic distinctiveness and pursue closer ties with their Slavic Russian kin rather than seek a closer orbit around the West. Since achieving independence from Moscow in 1991, Ukraine has restored the Ukrainian language to its educational system. Still, a majority of Ukrainians wants Russian to be a second official language. Ukrainians share many of the Russians' postindependence problems. There are reports of corruption, and much business and accompanying wealth are in the hands of tycoons and organized criminals. There has been a marked brain drain, with the most educated people fleeing west to the promise of higher salaries. They blazed a path for Ukrainian workers in other industries too, and many children have been left behind as their parents have moved west. Ukraine has the highest annual rate of population loss in the world (0.5 percent), with the death rate far exceeding the birth rate.

Belarus

The Kansas-sized country of Belarus (formerly called Byelorussia or White Russia; population 9.7 million) is on Ukraine's northern border. Minsk (population 1.7 million) is the political capital and main industrial city. Belarus developed slowly in the past, mainly because of a lack of fuels other than peat. Recently, however, oil and gas pipelines from Siberia and other areas have provided raw material for oil-refining and petrochemical industries. There is some oil production in Belarus itself. Under Soviet rule, Belarus became an industrial powerhouse, shipping televisions, farming equipment, automobiles, textiles, and machinery to both Moscow and its eastern European satellites.

Belarus's foreign trade is now almost exclusively with Russia, much of it in the form of barter; for example, Belarus has sometimes paid off debts to Russia's natural gas monopoly by supplying it with refrigerators, tractors, and other hard goods. Unlike Russian industries, most manufacturing enterprises in Belarus are still state-run, under what the country's leaders call "market socialism." Industrial output is low, but Belarus did not suffer the extensive economic disruption typical of Russia and most of the other Soviet successor countries.

Belarusian farms raise potatoes (particularly for the production of vodka), small grains, hay, flax, and livestock. Southern Belarus contains the greater part of the Pripyat (Pripet) Marshes, large sections of which were drained for agriculture in Soviet times. Agricultural productivity declined after 1986, when much ground was contaminated by the nuclear accident in nearby Chernobyl. To compensate for these losses, Belarus stepped up production of agricultural machinery for export.

Belarus's strategic geographic location between the Russian core and western Europe brought it enormous suffering in World War II. On its way to Moscow and Leningrad, the Nazi army laid waste to Belarus. One-fourth of the population was killed, and three-quarters of the cities were destroyed. One and a half million Belarusians fled eastward to escape the fighting. The Jews of Belarus were nearly entirely wiped out by Hitler's "final solution."

224

Of all the former Soviet republics, the Slavic nation of Belarus has had the closest ties with Russia. In the mid-1990s, an overwhelming majority of Belarusians voted for economic integration with Russia and for the adoption of Russian as their official language. This led to the official designation of Belarus and Russia as a **union state** with plans to integrate their economies, political systems, defense networks, and cultures. Relations between the countries soured, however, as the proclaimed union failed to take shape. Belarus's increasingly authoritarian president, Alexander Lukashenko (known widely as "Europe's last dictator"), angered and embarrassed Vladimir Putin's Russia, which began cutting its subsidies to Belarus—for example, by doubling the price it charged for its gas exports to Belarus. In 2007, Belarus retaliated by charging higher transit fees for the Russian oil pumped through its territory to Europe. Russia cut off the flow of oil through the pipeline until Belarus dropped the fee and began talking about diverting transit routes away from Belarus. Europe reacted with worry about the reliability of Russian energy supplies.

Moldova

Maryland-sized Moldova (known formerly as Moldavia and, in the 19th and early 20th centuries, as Bessarabia; population 4 million), made up largely of territory that the Soviet Union took from Romania in 1940, adjoins Ukraine at the southwest. The indigenous Moldovan majority (about two-thirds of the country's population) is ethnically and linguistically Romanian, not Slavic. Some Moldovans desire to have their country reunited with Romania, though this sentiment has declined in recent years. There are serious problems of ethnic antagonism, irredentism, and political separatism involving Moldova. In the eastern part of the country, in a sliver of territory along the Dniestr River known as Transdniestria, ethnic Russians and Ukrainians

174

have been agitating, sometimes violently, to form a separate state (see Figure 5.1.3). Their self-proclaimed Trans-Dniester Republic is effectively beyond control of the government and has become a haven for smuggling and money laundering. Despite protests from the European Union and other international organizations, Russia maintains troops there, ostensibly to protect the ethnic Russians. The presence of Russian troops gives Moscow enormous influence in Transdneistria. In a 2006 referendum, a reported 97 percent of this territory's inhabitants voted for independence from Moldova and free association, and possible reunification, with Russia. Although Russia recognized the vote, most other countries (including Moldova itself) have not.

Mainly a fertile, black-earth steppe upland, Moldova is an agricultural country with no major mineral resources. The Moldovan specialty is cultivation of vegetables and fruits, especially grapes for winemaking. Industrial production in Moldova jumped after World War II, especially around the capital of Chisinau (also known as Kishinev; population 0.7 million). When the country gained independence, it embraced free-market enterprise. But Moldova has had little to offer in the world market, and the economy has suffered. Economic dislocation has led to the involvement of many Moldovan women in the sex trade (see Geographic Spotlight, page 187). Voter dissatisfaction with the economy has brought Communists back to power, though Moldova has turned increasingly pro-Western in reaction to Russian influence in Transdniestria.

Kaliningrad

Kaliningrad is the odd piece of Russia stranded between Lithuania and Poland (see Figure 4.2.24, page 142). As the northern half of the former German East Prussia, this territory—then known as Königsberg—was transferred to the USSR at the end of World War II to become the Kaliningrad Oblast. Soviet authorities expelled nearly all of Kaliningrad's Germans to Germany, and Russians now make up more than three-fourths of the population of 1 million. Kaliningrad remains a massive military and naval establishment, with defense workers and their dependents and retired military families making up most of the population. Moscow would like Kaliningrad to continue to be principally a military state, but many in Kaliningrad have other ideas about how to capitalize on their strategic geographic location. Regarding themselves as **Euro-Russians,** they want Kaliningrad to become a more vibrant European economy. Some dream of making Kaliningrad one of the world's great free-trade zones, a sort of **"Hong Kong on the Baltic."**

Kaliningrad does have some geographic and other factors in its favor: Its port seldom freezes, labor is inexpensive, and it is a logical way station for east-west trade. Already much legitimate trade and also smuggling are carried on through Kaliningrad, raising living standards for its people. The fear, however, is now that Lithuania and Poland

GEOGRAPHIC SPOTLIGHT

The Sex Trade

One of the outcomes of the collapse of communism and the dissolution of the Soviet Union has been the unleashing of long-restrained expressions of sexuality. Entrepreneurs in Russia and other former Soviet countries quickly adopted an axiom of Western advertising: sex sells. Russian television has surpassed even the spiciest fare on standard American television.

Many sexual awakenings in the region are applauded or accepted as good, clean fun, but others have a decidedly dark side, and some send ripples far abroad. Women of the former Soviet Union have become commodities (•Figure 5.1.A). Russian, Ukrainian, and Kazakh partners or brides can in effect be purchased by anyone in the world with access to the Internet. Dot-com agencies act as brokers for these New Age "mail order brides." Some of the subsequent unions end happily, of course, but others collapse with physical abuse and heartache. Frequently, the woman ends up abandoned and destitute in a strange land.

The sex trade takes the greatest toll. Primarily in this industry, women become victims of what is known as **human trafficking,** often unwittingly and almost always out of poverty. Criminal gangs use false promises of conventional employment to lure young women into the trade. The typical victim is an 18- to 30-year-old Russian, Ukrainian, Belarusian, or Moldovan—often with children to support—who is approached by a trafficker promising her a job as waitress, barmaid, babysitter, or maid in eastern or western Europe. She pays a fee of $800 to $1,000 for travel and visa costs. Arriving in Bosnia and Herzegovina, the Czech Republic, or Germany, she is met by another trafficker who confiscates her passport and identity papers and then sells her for several thousand dollars to a brothel owner. She is escorted to her guarded apartment. There she is compelled to reside with several other women who have taken a similar journey, and she is informed of her real job: prostitute.

Berlin papers advertise the merchandise: "The Best from Moscow!" "New! Ukrainian Pearls!" The working women meet their clients in clubs or brothels or are driven to their homes. The going rate for their services is $75 per half hour, but less than 10 percent goes to the women. Their pimps and other gang members take the lion's share, and the women use their meager earnings to buy food and pay rent. Most end up in debt. Unable to pay off their debts, lacking identification papers, fearful of going to the police, and terrified the gang might harm family members back home if they try to escape, the women are trapped. Reports of depression, isolation, venereal disease, beatings, and rape abound.

This is a big business with striking demographics. The sex industry in Europe generates revenues of several billion dollars per year. Ukrainian sources estimate that since the country's independence in 1991, more than 400,000 Ukrainian emigrant women have ended up in the European sex trade. Figures vary widely, but tens and perhaps even hundreds of thousands of women of the former Soviet Union immigrate to Europe each year, the great majority of them becoming prostitutes.

There is an overflow into other regions. These so-called Natashas work in Turkey, Israel, the United Arab Emirates, and other countries in the Middle East in similar deplorable circumstances.

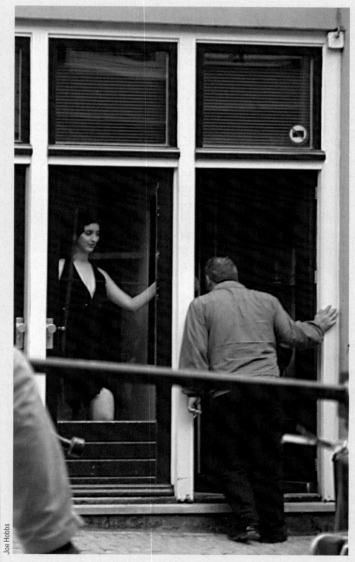

• **Figure 5.1.A** Window shopping in Amsterdam's Red Light District. An estimated 85 percent of the working women here are trafficked, mainly from the former Soviet countries.

Joe Hobbs

have joined the European Union, Kaliningrad will be left to drift in a state of underdevelopment. Its people will also find it harder to cross EU territory en route to the Russian "mainland."

5.1.2 Agriculture and Industry in the Russian Core

Most of the Fertile Triangle is within Russia, which still faces huge problems in transforming state-run into free-market farming. Throughout the second half of the 20th century, the Fertile Triangle was a global-scale producer of farm commodities such as wheat, barley, oats, rye, potatoes, sugar beets, flax, sunflower seeds, cotton, milk, butter, and mutton. The overall high output, however, did not reflect high agricultural productivity per unit of land and labor. There were many inefficiencies in the state-operated and collective farms that dominated the agricultural sector. On the state-operated farms, workers received cash wages in the same manner as industrial workers. Bonuses for extra performance were paid. Workers on collective farms received shares of the income after obligations of the collective had been met. As on the state farms, there were bonuses for superior output.

167

These methods, however, failed to provide enough incentives for highly productive agriculture. Farm machinery stood idle because of improper maintenance; since no individual owned the machine, the incentive to repair it was diminished. There were also shortages of spare parts. Poor storage, transportation, and distribution facilities, plus wholesale pilfering, caused alarming losses after the harvest. Young people were deserting the farms for urban work, often living with the family in the countryside but commuting to their city jobs. With the shortage of younger male workers, the elderly, women, and children did a large share of the farmwork.

In an effort to reverse such disturbing trends, all of the countries of the Fertile Triangle are promoting land reform by privatizing collective and state farms and developing more independent or "peasant" farming. More than 90 percent of Russia's farmland is now privately owned, at least on paper; much of "private" land is still under collective shareholding. Bureaucratic obstacles slow down the purchase and sale of land, and agricultural efficiency is hard to achieve. However, there was unmistakable improvement in the agricultural sector after 2000. Russia made revival of the agricultural sector one of its "national projects," and for the first time since independence, Russia began to record surpluses in grain production. The country continued to be a net importer of meat and dairy products. Even in the dead of Russia's winter, fresh vegetables find their way from greenhouses onto supermarket shelves (•Figure 5.1.5). Most analysts believe that the worst is over for Russia's agriculture.

• **Figure 5.1.5** Fluorescent lighting and artificial climate controls allow year-round harvests in Russian greenhouses.

Russia had the lion's share of Soviet industries and natural resources but lacked certain vital commodities and manufacturing capabilities. Access to those resources was not a problem in Soviet times, when Moscow commanded production of steel in Ukraine or cotton in central Asia for "USSR, Inc." Now, however, Russia must compete in an open market to buy and sell raw materials, components, and manufactured goods. The very abrupt transition from a command to a free-market economy has been a wrenching experience for Russia.

That transition is still very much under way, and the outcome is uncertain. The worst may be over for Russia's industries. If revenues from oil and other natural resource exports continue to be high, the benefits will continue to spread to many areas of manufacturing, and the country will enjoy a more stable, diversified economy. There is no way to know whether that will continue to happen, but in the meantime, it is worthwhile to consider the geography of Russia's industrial assets. Industries and industrial resources are concentrated in four areas: the regions of Moscow and Saint Petersburg, the Urals, the Volga, and the Kuznetsk region of southwestern Siberia (•Figure 5.1.6; see also Figure 5.2).

169

The industrialized area surrounding Moscow is known as the Central Industrial Region, Old Industrial Region, or Moscow-Tula-Nizhniy Novgorod Region. This region has a central location physically within the populous western plains and is clearly the country's economic core, with about 70 percent of Russia's financial wealth located there. It lies at the center of rail and air networks reaching Transcaucasia, central Asia, and the Pacific and is connected by river and canal to the Baltic, White, Azov, Black, and Caspian Seas (Moscow is known as the **Port of Five Seas**). The Moscow region's favorable geographic connections help compensate for its lack of mineral resources. The area's well-developed railway connections, partly a product of political centralization, provide good facilities for an inflow of minerals, materials, and foods and for a return outflow of finished products to all parts of Russia and the

Joe Hobbs

● **Figure 5.1.6** Saint Petersburg was established on the Baltic Sea at the mouth of the Neva River to serve as Russia's "Window on the West." This building served as the former Winter Palace of the Russian tsars and is today the world-renowned Hermitage Museum.

Near Abroad. Moscow, capital of Russia before 1713, of the Soviet Union from 1918 to 1991, and of the Russian Federation today, is the most important manufacturing city, transportation hub, and cultural, educational, and scientific center in the region.

This region ranks with Ukraine as one of the two most important industrial areas in the nations of the former Soviet Union. South of Moscow lies the metallurgical and machine-building center of Tula, and to the east is the diversified industrial center of Nizhniy Novgorod (formerly Gorkiy) on the Volga. Textile milling, mainly using imported U.S. cotton and Russian flax, was the earliest form of large-scale manufacturing developed in the Moscow region. The area still has the main concentration of textile plants (cottons, woolens, linens, and synthetics) in the region of Russia and the Near Abroad, with Moscow and Ivanovo the leading centers.

The Soviet Union's collapse did much damage to this textile industry. The price of central Asian cotton quadrupled with independence, forcing mills to close and consumers to turn to cheap clothing imports—like the rest of the world, mainly from China. The economy of Ivanovo, which had produced 70 percent of the Soviet Union's textiles, went into a deep depression. Now, however, life is stirring again as the mills in Ivanovo and nearby cities seek to exploit global niches in linen products. New technologies allow flax to be worked into linen that is nearly as versatile as cotton. The region's fine linens for kitchens and bedrooms are already selling in Europe and the United States.

The Saint Petersburg area (●Figure 5.1.7) was never as great an industrial center as the Moscow region, Ukraine, or the Urals, but it still developed a significant presence in industrial development and is a critical port. As in the case of Moscow, local minerals did not form the basis of this industrialization. Instead, metals from outside the area provide material for the industries (such as heavy machinery, the manufacture of cars and turbines, and metallurgy) that

are among the leaders in Saint Petersburg's diversified economy. Supported by university and technological-institute research workers, the city's highly skilled labor force played an extremely significant role in early Soviet industrialization. This group pioneered the development of many complex industrial products, such as power-generating equipment and synthetic rubber, and supplied experienced workers and technicians to new industries in other areas. In the first decade of the 2000s, Russia has invested heavily in the development and expansion of port facilities in and around Saint Petersburg. The main goal is to reduce the high transit fees charged by Latvia and Estonia for exports of Russian oil and other commodities through their ports. New port facilities at Primorsk and Ust-Luga, northwest and southwest of Saint Petersburg, and the expansion of the city's own port will allow Russian oil, coal, iron ore, timber, and other cargo to be exported at lower cost. Preservationists worry about Saint Petersburg's graceful architectural legacy, especially if the state-owned energy company Gazprom proceeds with plans to build soaring skyscrapers above the city's modest skyline.

Other industrial cities are strung along the Volga River from Kazan southward, in the region Russians know as the Povolzhye. The largest are Samara (formerly Kuybyshev); Volgograd (formerly Stalingrad); Saratov, heart of the Volga German Autonomous Republic until World War II; and Kazan. These four cities in the middle of the Volga industrial region grew up around diversified machinery, chemical, and food-processing plants. Under the Soviets, Samara was one of the military-industrial "closed cities" (closed to foreigners because of security concerns) whose industries Stalin commanded to be hastily built as a defensive measure against German attack in World War II. Like Nizhniy Novgorod upriver on the Volga, the city is undergoing a process known officially as **conversion**—the retooling of military enterprises for civilian-oriented manufacturing. Even while it converts, Russia hopes to reestablish dominance

KEY TO NUMBERED SUBJECTS
1 Adygeya
2 Karachay-Cherkessia
3 Kabardino-Balkaria
4 North Ossetia
5 Ingushetia
6 Chechnya
7 Mordovia
8 Chuvashia
9 Mari El

URBAN AREAS
Name Over 10 million
Name 1 million to 10 million
Name Under 1 million*
*selected cities

0   150   300 mi
0 150 300 km

• **Figure 5.1.6**   The Russian Federation and its administrative units.

80°N   70°N   60°N

UNITED STATES

Bering Strait

Uelen

St. Lawrence I. (U.S.)

O   C   E   A   N

Wrangel

Chukchi Peninsula

B   E   R   I   N   G

Komsomolets I.

October Revolution I.

Bolshevik I.

Ayon

EAST SIBERIAN

SEA

New Siberia I.

Anadyr Range

C H U K O T K A

Anadyr

Gulf of Anadyr

SEA

L   A   P   T   E   V

Kotelnyy

Lyakhovskiye

Kolyma
Lowland

Omolon R.

KAMCHATKA

Karagin

Komandorskiye

Taymyr Peninsula

Bolshoy Begichev

SEA

Indigirka R.

Kolyma R.

*Shelekhov
Gulf*

anga Mountains

L. Taymyr

Khatanga R.

North Siberian Lowland

Yana R.

Cherskiy   Mts.

M A G A D A N

Kamchatka Peninsula

Putoran

Mts.

Koury R.

Verkhoyansk Range

Magadan

Petropavlovsk

Olenek R.

S   A   K   H   A

C E N T R A L

FAR   EASTERN

Vilyuy R.

Yakutsk

SEA   OF

NOYARSK

S I B E R I A N

Lower Tunguska R.

Lena R.

Aldan R.

OKHOTSK

Paramushir

PLATEAU

Stony Tunguska R.

Lena R.

*Aldan
Plateau*

Shantarskiye

Sahkalin

R   I   A   N

Angara R.

Stanovoy   Mountains

Badzhalsky Mts.

S A K H A L I N

Chuna R.

Ust Ilim

Zeya Res.

Amur R.

Tatar Strait

Bratsk

Bratsk Res.

A M U R

Komsomolsk-na-Amure

Yuzhno-Sakhalinsk

SAYAN MOUNTAINS

IRKUTSK   *L. Baikal*

Vitim R.

Yablonovyy Range

Olekma R.

Zeya R.

Khabarovsk ★

Kunashir

Angarsk   Irkutsk

Ulan Ude   Chita

Shilka R.

Amur R.

Blagoveshchensk

YEVREYSKAYA (JEWISH)

Sikhote-Alin Mts.

PRIMORSKY

J A P A N

A

B U R Y A T I A

ZABAY   KALSKY

L. Khanka

C   H   I   N   A

Ussuriysk

SEA OF JAPAN

Vladivostok   Vostochnyy

(EAST SEA)

Zarubino   Nakhodka

M   O   N   G   O   L   I   A

Ussuri R.

Kuril Islands

180°   170°E   160°E   150°E   140°E   130°E   120°E   110°E   100°E   90°E   80°E

50°N

40°N

Joe Hobbs

• **Figure 5.1.8** Renowned for their drab functionality, most Russian apartment complexes were built in the 1960s and 1970s and are known, not always fondly, as "Khrushchevskies." These apartment blocks for autoworkers in the factories at Togliatti do have their amenities, including fountains enjoyed on hot summer days.

in weapons exporting. Although more than 90 percent of its 1,700 Soviet-era military factories shut down after the USSR broke up, Russia is the world's second-largest arms exporter.

Industries in these cities were powered by the construction of dams and hydroelectric stations along the Volga River and its large tributary, the Kama. The Volga region took early leadership in the automobile industry. Russia's "Detroit" is the Volga city of Togliatti (•Figure 5.1.8), by far the leading center for the production of passenger cars (now including Chevrolets) in the region of Russia and the Near Abroad. The cars come from the Volga Automobile Plant, which was built and equipped for the Soviet government by Italy's Fiat Company; the city itself takes its incongruous name from a former leader of Italy's Communist Party.

Large-scale exploitation of petroleum in the nearby Volga-Urals fields (which stretch from the Volga River to the western foothills of the Ural Mountains) contributed to the industrial rise of the Volga cities. Prior to the opening of fields in western Siberia during the 1970s, these were the Soviet Union's most important areas of oil production. Ufa and Samara are the region's principal oil-refining and petrochemical centers.

The Ural Mountains contain a diverse collection of useful minerals, including iron, copper, nickel, chromium, manganese, bauxite, asbestos, magnesium, potash, and industrial salt. Low-grade bituminous coal, lignite, and anthracite are mined. The former Communist regime fostered the development of the Urals as an industrial region well removed from the exposed western frontier of the Soviet Union. The major industrial activities are a legacy of Soviet investment and include heavy metallurgy, emphasizing iron and steel and the smelting of nonferrous ores; the manufacture of heavy chemicals based on some of the world's largest deposits of potassium and magnesium salt; and the

165

manufacture of machinery and other metal-fabricating activities. Yekaterinburg (formerly Sverdlovsk), located at the eastern edge of the Ural Mountains, is the largest city of the Urals and the region's preeminent economic, cultural, and transportation center. The second most important center is Chelyabinsk, located 120 miles (193 km) to the south. This grimy steel town was infamous in Old Russia as the point of departure for exiles sent to Siberia.

In western Siberia, between the Urals and the large industrial and trading center of Omsk, rail lines from Yekaterinburg and Chelyabinsk join to form the Trans-Siberian Railroad, the main artery linking the Far East with the Russian Core (•Figure 5.1.9; see also Figure 5.1.2). Omsk is a major metropolitan base for Siberian oil and gas development. It is also Siberia's most important center of oil refining and petrochemical manufacturing.

From Omsk, the Trans-Siberian Railroad leads eastward to the Kuznetsk industrial region, the most important concentration of manufacturing east of the Urals. The principal reason for industrial location here is an enormous reserve of coal. The manufacture of iron and steel is also a major industrial activity of the Kuznetsk region, whose main center is Novokuznetsk. The industry draws its iron ore from various sources in Siberia and Russia's central Asian neighbors. The largest urban center of the Kuznetsk region is Novosibirsk, a diversified industrial, trading, and transportation center located on the Ob River at the junction of the Trans-Siberian and Turkestan-Siberian (Turk-Sib) Railroads. Sometimes called the **"Chicago of Siberia,"** Novosibirsk has grown from a town with a few thousand residents at the turn of the 20th century to a city of more than a million inhabitants today. In Soviet times, hundreds of factories produced mining, power-generating, and agricultural machinery, tractors, machine tools, and a wide range of other products.

NASA

• **Figure 5.1.9** View of Omsk and the Trans-Siberian Railway from space. The railway draws a line through the April snow, and Omsk—the dark smudge just below the center of the photo—lies where it crosses the Irtysh River.

The city of Akademgorodok ("Academy Town" or "Science Town"), Siberia's main center of scientific research, is an outlying satellite community of Novosibirsk. The city was established in 1957 as a parklike haven where some of the Soviet Union's greatest scientific minds could research and develop both civilian and military-industrial innovations. Reflecting a general crisis in the post-Soviet military-industrial establishment, Akademgorodok's funding shrank dramatically, forcing many prominent scientists into menial jobs. Then, in the late 1990s, venture capitalists began moving in to take advantage of the city's infrastructure and remaining brain power. Pioneering Russian software companies were soon followed by multinationals, including Intel and Schlumberger. The city revived.

5.1.3 The Russian Far East

The Far East is Russia's mountainous Pacific edge (see Figure 5.9). Most of it is a thinly populated wilderness in which the only settlements are fishing ports, lumber and mining camps, and the villages and camps of aboriginal peoples. Port functions, fisheries, and forest industries provide the main support for most Far Eastern communities, and until now, the output of coal, oil, and a few other minerals has been small. Most of the Russians and Ukrainians who make up the majority of its people live in a narrow strip of lowland behind the coastal mountains in the southern part of the region. This lowland, drained by the Amur River and its tributary, the Ussuri, is the region's main axis of industry, agriculture, transportation, and urban development.

Several small to medium-sized cities form a north-south line along two important arteries of transportation: the Trans-Siberian Railroad and the lower Amur River. At the south on the Sea of Japan is the port of Vladivostok, which is kept open throughout the winter by icebreakers. About 50 miles (80 km) east of the city, the main commercial seaport area of the Far East has developed at Nakhodka and nearby Vostochnyy (East Port). Both ports are nearly ice-free and well positioned for trade in goods manufactured in Korea, China, and Japan. Seaborne shipments of cargo containers from northeast Asia to Scandinavia require 40 days, but the cargo can make the 6,000-mile (9,600-km) journey from Vostochnyy to Finland by rail in only 12 days. Business in Vostochnyy is picking up. A new oil terminal is being built at nearby Krylova Cape to support exports of oil, to be brought in initially on the Trans-Siberian Railroad, to Japan and elsewhere.

North of Vladivostok lies a small district that is the most important center of the Far East's meager agriculture, producing cereals, soybeans, sugar beets, and milk for Far Eastern consumption. The Far East is far from self-sufficient in food and consumer goods; large shipments from the Russian Core and from overseas supplement local production.

The diversified industrial and transportation center of Khabarovsk is located at the confluence of the Amur and Ussuri Rivers, where the main line of the Trans-Siberian Railroad turns south to Vladivostok and Nakhodka and the Amur River turns north toward the Sea of Okhotsk. Before World War II, the Soviet Union and Japan held the northern and southern halves, respectively, of the large island of Sakhalin. The island has much pristine wilderness with some forest, fishing, and coal mining but is particularly important for its offshore petroleum and natural gas. Russian energy officials are describing it as a **"Second Kuwait,"** with about 1 percent of global oil reserves and enough natural gas to supply the U.S. import market for about 25 years. Some of the liquefied natural gas will make its way to the United States from a terminal in northern Mexico, but most of the gas will be exported to China, Japan, and South Korea.

At the end of the war, the USSR annexed southern Sakhalin and the Kuril Islands and repatriated the Japanese population. Russian control of these former Japanese territories continues to be a problem in relations between the two countries and has kept them from reaching a formal post–World War II treaty. In 2007, the countries were trying to break the impasse, negotiating about how to divide the islands the Russians call the southern Kurils and the Japanese know as the "Northern Territories."

Both countries are trying to put these differences aside in favor of economic cooperation, particularly in the energy realm. Japan has negotiated contracts to buy natural gas from Sakhalin and to build an oil pipeline from the eastern Siberian fields near Irkutsk, in which it would also like to invest, to the port of Vostochnyy or of nearby Nakhodka (•Figure 5.1.10). Russia is happy to have the foreign investment, and Japan, the world's second-largest oil consumer, is happy to have the energy from a source outside the volatile Middle East. With its booming economy, China is desperate for more energy, especially from closer and therefore

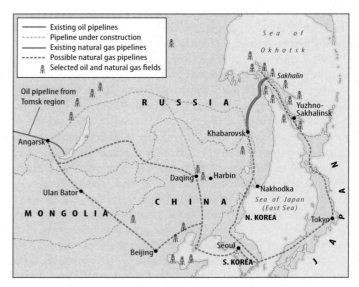

• **Figure 5.1.10** Existing and proposed pipelines for eastern Siberian and Far Eastern oil and gas exports.

● **Figure 5.1.11** The Koryaksky volcano towers over the port of Petropavlovsk on the Kamchatka Peninsula.

cheaper sources, and is also negotiating with Russia to obtain eastern Siberian oil through a proposed pipeline. Russian sources initially estimated that there was not enough oil in the eastern Siberian fields to supply both markets, and the total reserves are still unknown.

The Kurils are small volcanic islands that screen the Sea of Okhotsk from the Pacific. Fishing is the main economic activity. At the north, the Kurils approach the mountainous peninsula of Kamchatka (total population 300,000, with about 60 percent living in Petropavlovsk). Like the islands, Kamchatka is located on the Pacific Ring of Fire—the geologically active perimeter of the Pacific Ocean—and has 23 active volcanoes (●Figure 5.1.11). Soviet authorities protected Kamchatka for decades as a military area, and no significant development activities occurred there. A California-size land of rushing rivers, extensive coniferous forests, and gurgling hot springs, Kamchatka is now one of the last great wilderness areas on earth, resembling the U.S. Pacific Northwest landscape of a century ago. A struggle has been under way between those who want to develop Kamchatka's gold and oil resources and those who wish to set the land aside in national parks and reserves where ecotourism would be the only significant source of revenue. Much to the surprise of environmentalists abroad, Russia's conservationists seem to have prevailed. Fossil fuel development will take place, but so will an ambitious effort to set aside seven wilderness tracts totaling more than 9,400 square miles (c. 24,400 sq km), or about three times the size of Yellowstone National Park in the United States. The main goal will be to have multiple uses from Kamchatka's rich salmon stocks—for food, sport, study, and profit—and also protect this resource from the kind of overfishing that has laid waste to the cod fisheries of the Atlantic and North Seas and to Russia's sector of the nearby Bering Sea.

5.1.4 The Northern Lands of Russia

North and east of the Fertile Triangle and west of (and partially including) the Pacific coast lie enormous stretches of coniferous forest (taiga) and tundra extending from the Finnish and Norwegian borders to the Pacific (see Figure 5.9). These outlying wilderness areas in Russia may for convenience be designated the Northern Lands, although parts of the Siberian taiga extend to Russia's southern border.

These difficult lands form one of the world's most sparsely populated vast regions. The climate largely prevents ordinary types of agriculture, but hardy vegetables, potatoes, hay, and barley are grown in scattered places, and there is some dairy farming. A few primary activities, including logging, mining, reindeer herding, fishing, hunting, and trapping, support most people. Significant towns and cities are limited to a small number of sawmilling, mining, transportation, and industrial centers, generally along the Arctic coast to the west of the Urals, along the major rivers, and along the Trans-Siberian Railroad between the Fertile Triangle and Khabarovsk.

The oldest city on the Arctic coast is Arkhangelsk (Archangel), located on the Northern Dvina River inland from the White Sea. Tsar Ivan the Terrible established the city in 1584 for seaborne trade with England. Today, the city is the most important sawmilling and lumber-shipping center in the region of Russia and the Near Abroad. Despite its restricted navigation season from late spring to late autumn, it is one of Russia's more important seaports.

Another Arctic port, Murmansk (population 300,000, the world's largest city within the Arctic Circle; ●Figure 5.1.12), is located on a fjord along the north shore of the

• **Figure 5.1.12** The far northern city of Murmansk is situated on an ice-free fjord at the northern end of the Kola Peninsula.

Kola Peninsula west of Arkhangelsk. It is the headquarters for important fishing trawler fleets that operate in the Barents Sea and North Atlantic. Murmansk also has a major naval base and cargo port and is home port to the icebreakers that escort cargo vessels and carry Western tourists to the North Pole and other high Arctic destinations (•Figure 5.1.13). It is connected to the Russian Core by rail. The harbor is open to shipping all year thanks to the warming influence of the North Atlantic Drift, an extension of the Gulf Stream. Murmansk and Arkhangelsk played a vital role in World War II, when they continued to receive supplies by sea from the Soviet Union's Western allies after Nazi forces captured and closed the other ports of the western Soviet Union.

Murmansk and Arkhangelsk are western terminals of the **Northern Sea Route,** a waterway the Soviets developed to provide a connection with the Pacific via the Arctic Ocean (see Figure. 5.1.2). Until recently, navigation along the whole length of the route was possible for only up to four months per year, despite the use of powerful icebreakers (including nine nuclear-powered vessels; see Problem Landscape, page 196) to lead convoys of ships. Areas along the route provide cargoes such as the timber of Igarka and the metals and ores of Norilsk. Some supplies are shipped north on Siberia's rivers from cities along the Trans-Siberian Railroad and then loaded onto ships plying the Northern Sea Route for delivery to settlements along the Arctic coast.

The Northern Sea Route is now open to international transit shipping, and if not for the capricious sea ice, it would be an attractive route. In recent winters, typical sea ice coverage in the Arctic Ocean has retreated substantially—apparently one of the most accelerated trends in overall global warming—and the Russian shipping industry is hoping this trend will continue. The distance between Hong Kong and all European ports north of London is shorter via the Northern Sea Route than through the Suez Canal, and Russia could earn valuable transit fees from this route. Russia is also anxious to develop the Northern Sea Route for its own exports because Russian exports by land to western Europe must now pass over the territories of Ukraine, Belarus, and the Baltic countries, which collect transit fees from Russia.

The railway connecting Murmansk with the Russian Core serves important mining districts in the interior of the Kola Peninsula and logging areas in Karelia, which adjoins Finland. The Kola Peninsula is a diversified mining area, holding nickel, copper, iron, aluminum, and other metals, as well as phosphate for fertilizer. The peninsula and Karelia

• **Figure 5.1.13** The Russian nuclear-powered icebreaker *Yamal* at the North Pole. The former Soviet icebreaker fleet still escorts commercial ships on the Northern Sea Route but during the summer carries Western tourists such as these to the High Arctic.

PROBLEM LANDSCAPE

Nuclear Waste in Russia's High Arctic

On the map, the Russian islands of Novaya Zemlya ("New Land"), separating the Barents and Kara Seas in the High Arctic, appear remote, wild, and untouched. These ice- and tundra-covered islands are indeed an isolated wilderness, which is precisely why Soviet authorities perceived them as prized grounds for nuclear weapons testing and nuclear waste dumping.

Soon after the breakup of the Soviet Union in 1991, a former Soviet radiation engineer stepped forward with disturbing information: for decades, Soviet authorities had ordered the dumping of highly radioactive materials off the coast of Novaya Zemlya (●Figure 5.1.B). From the 1950s through 1991, the Soviet navy and icebreaker fleet secretly and illegally used the shallow waters off the eastern shore of Novaya Zemlya as a nuclear dumping ground. At least 12 nuclear reactors removed from submarines and warships, 6 of which still contained their highly radioactive nuclear fuel, were sent to the shallow seabed. Some of these individual reactors contain roughly seven times the nuclear material contained in the Chernobyl reactor that exploded in 1986. Between 1964 and 1990, some 11,000 to 17,000 containers of solid radioactive waste were also dumped here at shallow depths of 200 to 1,000 feet (60 to 300 m). Soviet seamen reportedly cut holes in those "sealed" containers to get them to sink. This dumping occurred even though the Soviet Union joined other nations in a 1972 treaty that allowed only low-level radioactive waste to be dumped at sea and then only at depths greater than 12,000 feet (3,660 m).

The danger posed by the nuclear material is not limited to the islands' shores but extends through the entire ecosystem of Russia's Arctic seas and possibly beyond. Contamination can begin at the base of the food chain (the phytoplankton that "bloom" in the area each spring) and pass upward to higher levels, where fish, seals, walrus, polar bears, and people eat. Alaskan authorities are worried, and Norwegians even more so. Norway's prime minister described the dumping as a "security risk to people and to the natural biology of northern waters." Western European and North American markets for Norwegian and Russian fish, which now represent significant exports for these countries, may refuse purchases if radiation levels rise.

Due to the long half-life of the nuclear materials, the consequences of 30 years of haphazard dumping near Novaya Zemlya may last thousands of years. The threat can be reduced greatly by retrieving the waste and disposing of it in terrestrial sites deep underground, which are considered safer. (The Russian military supports a proposal to convert some of the shafts used in 1950s atomic weapons tests to permanent radioactive waste repositories.) The major obstacle to this step is the extreme costliness of such an operation, estimated to be in the hundreds of billions of dollars. An arms-agreement obligation for Russia to decommission about 80 nuclear submarines based in Murmansk is also costly

and, because of the radioactive materials involved, potentially dangerous. Confessing and confronting the nuclear legacy of the Soviet Union, Russia has asked the United States and other nations to help pay for an environment with fewer nuclear hazards. The United States is in turn asking Russia not to export its nuclear power technologies to Iran and other countries where the West fears they may be used to develop weapons.

The predicament of Novaya Zemlya and the future of the vast wilderness in Russia's Far East reflect important questions about environment and development all across Russia and the Near Abroad. Like the United States and other industrialized countries, the Soviet Union experienced problems of environmental pollution caused by rapid economic growth. Russia and the other countries must now deal with the Soviet legacy of decades of environmental neglect. Environmental cleanup is hindered by the massive cost of repairing past damage and a reluctance to bear the expenses of stringent enforcement of pollution abatement measures. Some Russian sources estimate that 80 percent of all industrial enterprises in Russia would go bankrupt if forced to comply with environmental laws. Following the breakup of the Soviet Union, a desperate search for hard currency and widespread corruption led to other environmental problems, including the selling of natural resources such as timber and the poaching of animals.

● **Figure 5.1.B**   Sites around Novaya Zemlya used as Soviet nuclear waste dumping grounds between 1964 and 1990. A curie is a measure of the intensity of radioactivity, approximately equal to the radiation in 1 gram of radium.

● **Figure 5.1.14** Laying an oil pipeline in Russia's Northern Lands. Immense quantities of steel have been required for thousands of miles of oil and gas pipelines that have been built in Russia and the Near Abroad since World War II.

are physically an extension of Fennoscandia, located on the same ancient, glacially scoured, granitic shield that underlies most of the Scandinavian Peninsula and Finland. Karelia, with its thousands of lakes, short and swift streams, extensive softwood forests, timber industries, and hydroelectric power stations, bears a close resemblance to nearby parts of Finland (which annexed some of this area during World War II).

128

Important energy resources, including good coking coal, petroleum, and natural gas, are present in the basin of the Pechora River, east of Arkhangelsk. The main coal-mining center in Soviet times was Vorkuta, infamous for Stalin's labor camps where great numbers of political prisoners died toiling in the mines. More than half of the mines have shut in recent years, and half the city's population relocated to points further south. Meanwhile, Russian, U.S., and Norwegian oil companies are working together to explore for and develop oil for export from the Pechora region. The proven reserves of this region are significant; the Timan Pechora field (see Figure 5.1.2) contains an estimated 9 billion barrels or more of oil. Natural gas reserves in this area and nearby in the Barents Sea are also huge. One of the big challenges is to find a way to transport these hydrocarbons to market. Plans are under way to construct a pipeline to send oil and gas south to the Gulf of Finland near Saint Petersburg.

East of the Urals, the swampy plain of western Siberia north of the Trans-Siberian Railroad saw a surge of petroleum and natural gas production after the 1960s. The western Siberian fields (see Figure 5.1.2, page 184) are the largest producers in the region of Russia and the Near Abroad.

However, an estimated half of the reserves of these fields have already been pumped, and so Russia looks to the Arctic for its energy future. Extraction and shipment of oil and gas take place in western Siberia under frightful difficulties caused by severe winters, permafrost, and swampy terrain. Huge amounts of steel pipe and pumping equipment have been required to connect the remote wells with markets in the Russian Core and Europe (●Figure 5.1.14).

193, 642

Still farther east, the town of Igarka, located about 425 miles (c. 680 km) inland on the deep Yenisey River, is an important sawmilling center accessible to ocean shipping during the summer and autumn navigation season. Northeast of Igarka lies Norilsk, the most northerly mining center of its size on the globe and the site of a huge metal-refining complex. Rich ores yielding nickel, copper, platinum, and cobalt justified the large investment in this Arctic city, including a railway connecting the mines with the port of Dudinka on the Yenisey downstream from Igarka. During the Soviet era, thousands of political prisoners labored and died in the nickel mines of Norilsk. Russian authorities have resisted appeals from surviving laborers to erect a memorial to those lost in this hub of the **gulag**, Soviet Russia's network of forced-labor camps where as many as 30 million prisoners toiled in the 1930s.

In central and eastern Siberia, the taiga extends southward beyond the Trans-Siberian Railroad. A vast but sparsely settled hinterland has a few cities spaced at wide intervals along the railroad, notably Krasnoyarsk on the Yenisey River and Irkutsk on the Angara tributary of the Yenisey near Lake Baikal. From east of Krasnoyarsk, a branch line of the Trans-Siberian leads eastward to the Lena River.

• **Figure 5.1.15** Fisherman at work on Lake Baikal, Earth's deepest lake.

Years of construction in the 1970s and 1980s under very difficult conditions resulted in continuation of this line to the Pacific under the name Baikal-Amur Mainline, or BAM (see Figure 5.1.2, page 184). The railroad was built to open important mineralized areas east of Lake Baikal and to reduce the strategic vulnerability caused by the close proximity of the Trans-Siberian Railroad to the Chinese frontier. The Pacific terminus is located on the Tatar Strait, which connects the Sea of Okhotsk with the Sea of Japan.

There is growing support in Russia, the United States, and Canada to construct a rail link between Siberia and North America. This would require laying another 2,000 miles (c. 3,200 km) of track from the BAM to the Bering Strait. An underwater tunnel would carry the track across the Bering Strait, where another 750 miles (c. 1,200 km) of track would be laid to Fairbanks, Alaska. New lines would connect Alaska with western Canada and the lower 48 United States. Proponents of this route say that a Russia–North America railway link would reduce trans-Pacific shipping times by weeks and that it could carry 30 billion tons of cargo each year.

The BAM passes through Bratsk, site of a huge dam and hydroelectric power station on the Angara River. Lake Baikal, which the Angara drains, lies in a mountain-rimmed rift valley and is the deepest body of freshwater in the world, more than 1 mile (1.6 km) deep in places (•Figure 5.1.15). Lake Baikal is at the center of a long-running environmental controversy over wastewater that pours into it from two large paper and pulp mills on its shores. Mill managers say they are taking measures to reduce pollution, but environmentalists insist that the waste jeopardizes Baikal's unique natural history, as the lake contains an estimated 1,800 endemic plant and animal species. "Nature as the sole creator of everything still has its favorites, those for which it expends special effort in construction, to which it adds finishing touches with a special zeal,

and which it endows with special power," wrote Valentin Rasputin, a native of Siberia. "Baikal, without a doubt, is one of these. Not for nothing is it called the pearl of Siberia."[2]

Proposed routes for exporting oil eastward from the Angarsk region (see Figure 5.1.10) also threatened this natural gem. Each route would have crossed a number of rivers flowing into Lake Baikal. Responding to fears that an earthquake in this seismically active region could result in calamity, Russia's president in 2006 decreed that the proposed northern route should be moved still farther north, away from Baikal's watershed. As major oil production comes online in the eastern Siberian oilfields around Irkutsk, there will be new environmental concerns.

Siberia's development is based in large part on hydroelectric plants located along the major rivers. The largest hydropower stations are on the Yenisey (at Krasnoyarsk and in the Sayan Mountains) and its tributary the Angara (at Bratsk and Ust Ilim). These power stations and dams are among the largest in the world. They supply inexpensive electricity, consumed mainly in mechanized industries that use power voraciously. These include cellulose plants that process Siberian timber and aluminum plants making use of aluminum-bearing material (alumina) derived from ores in Siberia, the Urals, and other areas and imported from abroad via the Black Sea.

Scattered throughout central and eastern Siberia are important gold-mining centers, mainly in the basins of the Kolyma and Aldan Rivers and in other areas in northeastern Siberia. In Stalin's time, the Kolyma mines, located in one of the world's harshest climates, developed an evil reputation as death camps for great numbers of political prisoners forced to work there. Large coal deposits are mined along the Trans-Siberian Railroad at various points between the Kuznetsk Basin and Irkutsk, Russia's leading center for the production of military aircraft.

● **Figure 5.1.16** Yakut reindeer herders, Sakha.

Aside from these scattered centers of logging, mining, transportation, and industry, where most people are Slavs, the Northern Lands are home mainly to non-Slavic peoples who make a living by herding reindeer, trapping, and fishing and in the more favored areas of the taiga by raising cattle and precarious forms of cultivation. The domesticated reindeer is especially valuable to the tundra peoples, providing meat, milk, hides for clothing and tents, and draft power. The Yakuts, a Turkic people living in the basin of the Lena River, are among the most prominent of the non-Slavic ethnic groups in the region (●Figure 5.1.16). Their political unit, Sakha (formerly Yakutia), is larger in area than any former Soviet Union republic except the Russian Federation, within which it lies. Sakha has only about 900,000 people inhabiting an area of 1.2 million square miles (3.1 million sq km). The capital city of Yakutsk on the Lena River has road connections with the BAM and the Trans-Siberian Railroad and with the Sea of Okhotsk. Riverboats on the broad and deep Lena provide a connection with the Northern Sea Route and the southern rail system

during the warm season. Sakha will figure prominently in future mineral and timber development, with its large reserves of diamonds, natural gas, coking coal, other minerals, and timber.

One of the most interesting trends to watch in coming years will be Moscow's efforts toward **"de-Russification"** of some of the Northern Lands, particularly those that do not have strategic importance (because they are located away from Russia's frontier with China or its oil-producing areas). Soviet development of many parts of this vast wilderness was often based on political or demographic ambitions rather than on anticipated economic returns. Moscow must pay huge subsidies to support the heating, fuel, food, and social program costs for people to live in these cold and remote settlements. To reduce that burden, the Russian government wants the region's population to decline by 600,000 to 800,000 people, or roughly 10 percent. In 2003, the government began offering housing and other subsidies of up to $18,000 to each family willing to relocate southward and westward from cities like Norilsk and Vorkuta. So far, most residents have turned down the offer as insufficient to cover the cost of leaving the Northern Lands.

5.1.5 The Caucasus

The Caucasus is the name of the isthmus between the Black and Caspian Seas (for the Caucasus map, see Figure 5.20, page 174). In the north of the isthmus is the Greater Caucasus Range, forming an almost solid wall from the Black Sea to the Caspian (●Figure 5.1.17). It is similar in age and character to the Alps but much higher, up to 18,510 feet (5,642 m) on Mount Elbrus, on the Russia-Georgia border. The southern Caucasus is known as Transcaucasia. In southern Transcaucasia is the mountainous, volcanic Armenian Plateau. Between the Greater Caucasus Range and the

● **Figure 5.1.17** Mountains and glaciers seen from Mount Elbrus in the Caucasus.

Armenian Plateau are subtropical valleys and coastal plains where the majority of people in Transcaucasia live. A humid subtropical climate prevails in coastal lowlands and valleys of Georgia, south of the high Caucasus Mountains. Mild winters and warm to hot summers combine with the heaviest precipitation found anywhere in the region of Russia and the Near Abroad—50 to 100 inches (c. 125 to 250 cm) a year. The lowland area bordering the Black Sea in Georgia is the most densely populated part of Transcaucasia. Specialty crop products such as tea, tung oil, tobacco, silk, and wine grapes, together with some citrus fruits, grow there.

The Caucasian isthmus has been an important north-south passageway for thousands of years, and the population includes dozens of ethnic groups that have migrated into this region (see Figure 5.11). Most of these are small ethnic populations confined to mountain areas that became their refuges in past times. The tiny republic of Dagestan, within Russia on its southeastern border with Georgia and Azerbaijan, is by itself home to 34 different ethnic groups. Some mountain villages, particularly in southeastern Azerbaijan, have achieved international fame as the abodes of the world's longest-lived people—up to the challenged claim of 168 years!

In addition to Russians and Ukrainians, most of whom live north of the Greater Caucasus Range, the most important nationalities are the Georgians, Azerbaijanis, and Armenians, each represented by an independent country. Throughout history, all have defiantly maintained their cultures in the face of pressure by stronger intruders. These nationalities have ethnic characteristics and cultural traditions that are primarily Asian and Mediterranean in origin. Their religions differ. The Azerbaijanis (also known as Azeri Turks) are Shiite Muslims. Most Georgians belong to one of the Eastern Orthodox churches. The Armenian Apostolic Church is a very ancient, independent Christian body that is an offshoot of Eastern Orthodoxy; in 301 C.E., Armenia became the world's first Christian country when its government pronounced the young religion as its national faith (•Figure 5.1.18).

There has been a history of animosity between the Armenians and Azeri Turks (including a war between them in 1905) growing out of the Turks' persecution of Armenians in the Ottoman Empire prior to and during World War I. Turkey and Azerbaijan still have not accepted responsibility for their roles in the **Armenian genocide**, in which an estimated 1.5 million Armenians died between 1915 and 1918. The descendants of Armenians who fled this holocaust have established themselves in many places around the world (for example, there is a large population in Hollywood's film industry) and today outnumber Armenians living within the country by almost two to one. One indicator of Armenia's declining economic fortunes following independence in 1991 is that an estimated 40 percent of the country's population has moved abroad since then.

There are some outstanding problems among the Caucasus countries, in addition to the international geopolitical issues discussed in Chapter 5 (pages 174–175). Early in the 1990s, historical animosity flared into massive violence between Armenians and Azeris over the question of Nagorno-Karabakh, a predominantly Armenian enclave of 1,700 square miles (4,420 sq km) within Azerbaijan, governed at the time by Azerbaijan but claimed by Armenia (see Figure 5.20). From 1992 until 1994, when Armenian forces finally secured the region (which is still technically part of Azerbaijan), fighting over Nagorno-Karabakh took 30,000 lives and created nearly a million refugees. Large numbers of Armenians fled from Azerbaijan to Armenia as refugees. There was also a flight of Azeris from Armenia to Azerbaijan, where hundreds of thousands continue to live as refugees. Armenia and Azerbaijan have not yet made peace, but for several years, they have had a working framework for an agreement. The Armenians would return to Azerbaijan six of the seven regions they captured. Nagorno-Karabakh and the adjacent Lachin region would be granted self-governing status. Azerbaijan's compensation would be the construction of an internationally protected road linking Azerbaijan with its enclave of Nakhichevan, currently cut off from Azerbaijan by Armenian territory. With no agreement reached, however, in a mainly symbolic gesture, Nagorno-Karabakh declared itself a sovereign nation in 2006. No other country, including Armenia, recognized its sovereignty. Once again there was talk of war between Armenia and Azerbaijan over the issue of Nagorno-Karabakh.

Oil and gas are the most valuable minerals of the Caucasus region. The main oilfields are along and underneath the

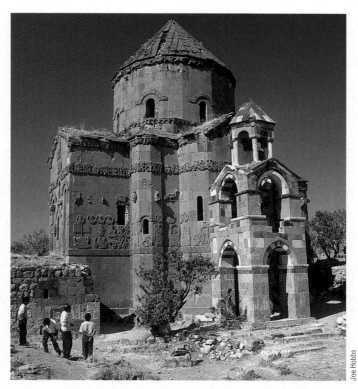

• **Figure 5.1.18** The Armenian Church of the Holy Cross on the island of Akhtamar, now in eastern Turkey. Armenia was the world's first Christian country, and its ancient borders included Mount Ararat and other land later lost to Turkey.

Caspian Sea around Azerbaijan's capital, Baku (population 2.1 million) and north of the mountains near Groznyy in Chechnya. Dry and windswept Baku, the largest city of the Caucasus region, was the leading center of petroleum production in Old Russia and at the dawn of the 20th century was the world's leading center of oil production. Under the Soviets, its leadership was decisively surpassed in the 1950s by the Volga-Urals fields and again in the 1970s by the western Siberian fields. However, petroleum development is now reestablishing the Caucasus as a vital world region (see Geography of Energy in Chapter 5, pages 176–177). Thanks to a gusher of oil wealth, Azerbaijan has one of the hottest economies. It also is ranked as having one of the world's most corrupt governments, so prospects for the overall development of the country are uncertain.

Under the Soviets, diverse manufacturing industries developed in Transcaucasia. Most of these factories, which now produce only a fraction of their Soviet-era output, are located in and near the political capitals: Azerbaijan's Baku, Georgia's Tbilisi (population 1.2 million), and Armenia's Yerevan (population 1.4 million).

Georgia's economy has fared the worst in the post-Soviet years. By the time the Rose Revolution swept the old regime out of power in 2003, more than half of the population lived below the poverty line, and unemployment exceeded 25 percent. Up to 80 percent of the country's wealth was generated by a "shadow economy" fueled by smuggling and trafficking in drugs, guns, oil, and prostitutes. The new government began dealing with the daunting challenge of reversing Georgia's reputation as one of the world's most corrupt countries. Despite its troubles with Russia, and due in part to Western economic assistance, Georgia's economy has improved. Growing revenues are anticipated from the transit fees Georgia will collect from the fossil fuels pipelines crossing its territory.

177

The Caspian Sea, which is the world's largest freshwater lake in area, has resources other than oil. It is an important fishery, especially for its three commercial species of sturgeon, which historically supplied over 70 percent of the world's caviar (sturgeon eggs) (•Figure 5.1.19). When only the USSR and Iran bordered the Caspian, this resource was managed sustainably. Now, however, the smaller, recently independent countries bordering the shallow Caspian are exerting tremendous pressure on this resource in an attempt to boost their market shares in competition with Russia and Iran. The result has been a rapid drop in the lake's sturgeon population (the most prized species, beluga, was down 45 percent in a single year, from 2004 to 2005, for example) and soaring prices for caviar on the world market (a pound from the prized beluga species retails for as much as $6,000, earning it the epithet "black gold"). The Conference on International Trade in Endangered Species (CITES) imposed an outright ban on all caviar exports from the Caspian during 2006 but lifted the ban in 2007 and set an export quota (the United States, however, continues to ban imports of all Caspian caviar). Despite police patrols, sturgeon poaching is big business, netting far more than the licensed harvest of these fish. There are growing fears about what an oil spill would do to the caviar industry.

5.1.6 The Central Asian Countries

Across the Caspian Sea from the Caucasus region lie large deserts and dry grasslands, bounded on the south and east by high mountains. Like Transcaucasia, this central Asian region is inhabited by ethnic groups that outnumber Slavs. The largest populations are four Muslim peoples speaking closely related Turkic languages—the Uzbeks, Kazakhs, Kyrgyz, and Turkmen—and a Muslim people of Iranian origins, the Tajiks (•Figure 5.1.20). Each has its own independent nation (see Table 5.1, page 153). Their total population of 59 million is growing faster than that of any other part of the Near Abroad.

Division and Conquest

The peoples of this region are heirs of ancient oasis civilizations and of the diverse cultures of nomadic peoples. Small-scale social and political units such as clans and tribes were traditionally associated with particular oases. Conquerors

• **Figure 5.1.19** Caviar from the prehistoric-looking beluga sturgeon (*Huso huso*) of the Caspian Sea is one of the world's most prized and expensive delicacies. This is a captive beluga.

• **Figure 5.1.20** The elders of this community in Tajikistan are about to partake in the traditional meal celebrating Novruz, the Muslim New Year.

● **Figure 5.1.21** Principal features of central Asia.

have repeatedly sought to possess this area because of its strategic location on the famed Silk Road, an ancient route across Asia from China to the Mediterranean. Among them were Alexander the Great, the Arabs who brought the Muslim faith in the 8th century, the Mongols, and the Turks. By the time of the Russian conquests in the 19th century, Turkestan (as this region was then known) was contested between feuding Muslim khanates with political ties to China, Persia (Iran), Ottoman Turkey, and British India. Cultural and political fragmentation continued up to the Russian Revolution and into the period of civil war that followed.

Nationalist identities existed historically in central Asia, but nation-states did not. During World War I, central Asian peoples revolted against Russia's tsarist government, which had begun to draft the region's Muslims for menial labor at the front. The Bolshevik government gained control in the early 1920s and organized the area into nationality-based units, effectively setting the boundaries for what are now independent countries.

During the 20th century, Russians and Ukrainians poured into central Asia, mainly into the cities. Meant to Russify this non-Russian area, they included political dissidents banished by the Communists, administrative and managerial personnel, engineers, technicians, factory workers, and in the north of Kazakhstan, farmers in the "new lands" wheat region. Most of the Slavic newcomers lived

apart from the local Muslims, as most of those who remain today still do.

Ethnic Russians have problems in post-Soviet central Asia, especially in Kazakhstan. At the time of independence in 1991, Kazakhstan's population of 17 million was roughly 40 percent Kazakh and 39 percent Russian, with the rest a mix of nationalities. Ethnic Kazakhs, who generally believe they were treated as second-class citizens until independence, have been working to diminish the Russian cultural footprint. They are increasingly predominant in government and business. Kazakh is now the official language, and Russian is the language of "interethnic communication." Even ethnic Kazakhs have a lot of catching up to do; half of them cannot speak Kazakh. With ethnic tensions rising throughout the 1990s, about 1 million Russians and 700,000 ethnic Germans emigrated from Kazakhstan. Meanwhile, the government is calling for an estimated 3 million ethnic Kazakhs who live abroad to come home, though few have done so.

Environment and Agriculture

The five central Asian "stans" are mainly flat, with plains and low uplands, except for Tajikistan and Kyrgyzstan, which are spectacularly mountainous and contain the highest summits in the region of Russia and the Near Abroad in the Pamir and Tien Shan Ranges (● Figure 5.1.21; see also

Figure 5.9). Central Asia is almost entirely a region of interior drainage. Only the waters of the Irtysh, a tributary of the Ob, reach the ocean; all the other streams either drain into enclosed lakes and seas or gradually lose water and disappear in the central Asian deserts.

Many of this region's people were previously pastoral nomads who grazed their herds on the natural forage of the steppes and in mountain pastures in the Altai and Tien Shan Ranges. Over the centuries, they slowly drifted away from nomadism. The Soviet government accelerated this process by forcibly collectivizing the remaining nomads and settling them in permanent villages. Raising livestock became a mainly sedentary form of ranching in the process. Sheep and cattle are the principal livestock animals, followed by goats, horses, donkeys, and camels.

Horses remain central to the national personality, if not the livelihoods, of the indigenous people of Kyrgyzstan. In the post-Soviet period, they have resurrected the oral epics that tell the story of Manas, a great warrior horseman who, 1,000 years ago, fought off the enemies of the Kyrgyz and unified their disparate tribes for the first time. The Kyrgyz are passionate horsemen, and even townspeople keep horses in the countryside for weekend races and other games.

Most people today live in irrigated valleys at the base of the southern mountains. There the wind-blown loess and other soils are fertile, the growing season is long, and rivers flowing from mountains provide irrigation water. The principal rivers in the heart of the region are the Amu Darya (Oxus) and Syr Darya, both of which empty into the enclosed Aral Sea.

Only northern Kazakhstan, which extends into the black-earth belt, receives enough moisture for dry farming (unirrigated agriculture). Irrigation is needed everywhere else, providing water to mulberry trees (grown to feed silkworms), rice, sugar beets, vegetables, vineyards, and fruit orchards.

Commercial orchard farming is focused on the area around Almaty ("City of Apples"; population 1.3 million), the former capital of Kazakhstan. But most of the irrigated land is in Uzbekistan. The most important farming area is the fertile Fergana Valley on the upper Syr Darya, nearly enclosed by high mountains, and the oases around Uzbekistan's cities of Tashkent (the nation's capital and central Asia's largest city; population 3.1 million), Samarkand, and Bukhara. These old cities became powerful political and economic centers on the Silk Road. Samarkand was the capital of Tamerlane's empire, which encompassed much of Asia in the 14th century. Samarkand and nearby Bukhara have superb examples of Islamic architecture that the government of Uzbekistan has recently restored at great cost (•Figure 5.1.22).

Cotton began to be cultivated here during the American Civil War, when U.S. exports abroad virtually ceased. The Soviets later fostered a "plantation economy" of their own in central Asia, exploiting the region's potential to produce large quantities of this valuable crop (known for a time as

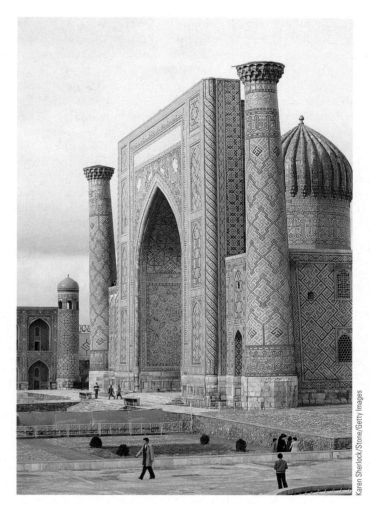

• **Figure 5.1.22** A 16th-century *madrasa* (Islamic school) on Registan Square in the heart of the storied ancient city of Samarkand, Uzbekistan.

"**white gold**") for the central Russian textile mills. Cotton remains the region's major crop. Production flows to textile mills in Tashkent, the Moscow region (at levels far below those of Soviet times), and locations outside the region.

There was a significant environmental price to pay for the development of central Asian agriculture. Driven by the need for foreign exchange from export sales, Soviet planners reshaped the basins of the Syr Darya and Amu Darya with 20,000 miles (c. 32,000 km) of canals, 45 dams, and more than 80 reservoirs. As the area under irrigated cotton grew, the rivers and the Aral Sea became highly polluted with agricultural chemicals draining from the fields. Diversion of water from the rivers caused the Aral Sea to shrink rapidly in volume and area, virtually destroying a vast natural ecosystem and an important regional fishery. Since the 1950s, the Aral Sea has lost 60 percent of its surface area and 80 percent of its volume and has divided into three small lakes (see the former and present shorelines in Figure 5.1.21). With assistance from the World Bank, Kazakhstan recently built a dam to contain the smaller lake, known as the North Aral Sea, completely within its borders. It is engineered to reduce salinity in the lake and divert any excess water to the larger, parched Aral Sea. The North Aral Sea's waters have risen faster than anticipated, and there is much optimism about a revitalization of its fishing industry.

Both Kazakhstan and Uzbekistan use tremendous quantities of water to flood paddies to produce rice—an extraordinary land use decision in such a thirsty region. Rice cultivation is one reason Uzbekistan is desperate for water and accuses Kyrgyzstan of hoarding it. Kyrgyzstan insists it needs to store water for hydroelectric power because the downstream countries will not provide it with the fossil fuels it needs. Meanwhile, Turkmenistan has declared that it will harness Amu Darya waters to create a new "Lake of the Golden Turkmen" to irrigate an additional 1,875 square miles (4,860 sq km) of land. Where all the water will come from is not clear, especially since the five "stans" signed an agreement in the 1990s promising to freeze in place their Soviet-era water distribution allotments.

Another environmental problem—this one with international security dimensions—exists on the former island of Vozrozhdeniye, on Uzbekistan's side of the Aral Sea. In 1988, as the Soviet Union hastened to build new bridges with the West, Soviet biological warfare specialists buried hundreds of tons of living anthrax bacteria, encased in steel canisters, on the island. As the Aral Sea continued to shrink, the island became a peninsula, raising fears that these more accessible biohazards could be exposed and blown into populated areas or be carried off by would-be terrorists. Following a number of incidents of anthrax-tainted letters reaching U.S. Senate offices and other facilities in the wake of the 9/11 attacks, the U.S. government provided the funds needed to clean up "Anthrax Island"; all anthrax burial sites were decontaminated in 2002.

Economic and Political Prospects

In a familiar pattern, the economies of the five central Asian countries worsened after an initial period of exuberance and growth following independence. The long-term prospects for some of the countries are bright, however. Kazakhstan favors close coordination with Russia, and it is a member (along with Russia, Belarus, and all the central Asian nations except Turkmenistan) of an agreement now known as the **Eurasian Economic Community,** which calls for the creation of a common market to ensure the free flow of goods, services, and economic capital among them and to promote coordinated industrial and agricultural policies.

Central Asia has a generous endowment of minerals, particularly natural gas, coal, oil, iron ore, nonferrous metals, and ferroalloys. Kazakhstan (population 15.5 million), the region's richest country, has the greatest natural resource wealth, including perhaps as much as 70 percent of the Caspian Sea's oil reserves. Political stability under the nominally democratic (but quite repressive) government of

Nursultan Nazarbaev since independence has made it the biggest target for Western investment in central Asia. Easterners are coming too: gambling is a booming business, and most casino visitors are Chinese. Large reserves of bituminous coal are at Karaganda in central Kazakhstan. Despite Kazakhstan's coal wealth, however, equipment and management problems have resulted in declining production since the country's independence. In the late 1990s, the country's new capital was established at Astana (formerly Akmola and Tselinograd; population 300,000), northwest of Karaganda, as part of a larger plan to redevelop Kazakhstan's industrial heartland. Prospecting has begun in Kazakhstan in what may be the world's third-largest gold deposit. Kazakhstan also has impressive reserves of iron ore, chrome, and nickel. Industrial growth would help compensate for the plummeting productivity of Kazakhstan's agriculture; without the Soviet-era subsidies they once enjoyed, farmers have not been able to buy pesticides and fertilizers or maintain equipment. Since independence, about half of the country's farmland has been abandoned.

After Kazakhstan, Turkmenistan (population 5.4 million; capital, Ashgabat, population 800,000) has the most promising economic resources. The country borders the Caspian Sea, with its rich natural gas and petroleum fields. Natural gas exports by pipeline to western Europe are the country's main source of hard currency; the rest of the economy has been in shambles since independence. Unlike Kazakhstan, Turkmenistan has had no pretensions of democracy. The country's first post-Soviet president, Saparmurat Niyazov, declared himself leader for life and dubbed himself the Turkmenbashi ("Father of the Turkmen"), renaming cities and even calendar months after himself. Until his death in 2006, his regime maintained an iron grip on the Turkmen majority and the ethnic Russian and Uzbek minorities.

Uzbekistan, central Asia's most populous country (population 26.5 million) and Kazakhstan's main rival for dominance in central Asia, does not benefit from Caspian oilfields. The country is largely agricultural, and cotton exports are the most important source of hard currency. To diversify its important agricultural sector, the country is trying to reverse the ecological damage done to the Aral Sea and its drainage basin and to cut the area planted in cotton in favor of food crops.

Kyrgyzstan (population 5.2 million; capital, Bishkek, population 900,000), the poorest country of the region after Tajikistan, is struggling to modernize. With independence, the country's first president, Askar Akaev, was hailed as a Thomas Jefferson–like figure who would bring freedom and prosperity to his country. Most of the Soviet-era state-owned businesses were quickly privatized, and a stock exchange opened. But Akaev lost the Jeffersonian label when he staged and won a rigged election and turned increasingly corrupt and authoritarian. A bloodless coup known as the **Tulip Revolution** ousted him from power in 2005. Hopes for an economic boom have yet to be realized. Kyrgyzstan has few industrial resources on which to plan its future. Fully

233, 316

40 percent of the country's export revenues come from the Kumtor gold mine. There are hopes for growth in the tourism sector. Central Asia's greatest scenic assets are here. The Tien Shan—the "Celestial Mountains"—dominate the land and hold much potential to draw Western ecotourists interested in rafting and trekking, particularly in the region around Lake Issyk Kul, one of the world's largest bodies of freshwater. At the same time, there is much concern about the environmental hazards posed by uranium waste at several locations in Kyrgyzstan where uranium had been processed for Soviet nuclear weapons.

Tajikistan (population 7.1 million; capital, Dushanbe, population 500,000) has the backing of Washington and Moscow, which prefer its largely authoritarian government to the potential alternative, a radical Islamic regime. After the country gained independence, a coalition of prodemocracy and Islamist rebels, comprised largely of ethnic groups that had little economic and political power in the country, lost a bloody civil war against the Russian-backed government. More than 100,000 people died in the fighting between 1992 and 1997. Thereafter, as part of a United Nations–brokered peace agreement, the former rebel fighters were absorbed into the national army, and their political leaders were brought into the government.

Tajikistan is the poorest of the Soviet Union's successor countries, and there is little to fuel its economy other than foreign aid, smelted aluminum, and cotton. Uranium was once mined in Tajikistan for Soviet nuclear weapons, and cotton was produced there for Moscow's textile mills, but those markets have nearly evaporated. Tajikistan now sells its aluminum and cotton to the Netherlands and other distant markets.

The central Asian countries are like those of the Middle East in many respects, particularly in their overwhelmingly Muslim populations, their traditions of pastoral nomadism and irrigated agriculture, and their petroleum endowments. The text now turns to that critical world region. The discussion of Islam in Chapter 6 provides especially useful background for further appreciation of central Asia and the Caucasus.

SUMMARY

→ Nearly 75 percent of the people, and an even larger share of the cities, industries, and cultivated land of the region of Russia and the Near Abroad, are packed into the Fertile Triangle (Agricultural Triangle or Slavic Core) comprising one-fifth of the total area. The rest (mostly in Asia) consists of land where environmental obstacles limit settlement.

→ Ukraine is one of the most densely populated and most productive areas of the region. Ukraine's industrial and agricultural assets were vital to the Soviet Union. Post-Soviet Ukraine relies more on heavy industry than most of the other successor countries.

→ Under Soviet rule, Belarus became an industrial power. Belarus's trade is now almost exclusively with Russia, much of it in the form of barter. Belarus and Russia technically formed a union state to more closely link the two predominantly Slavic countries, but political problems have prevented its implementation.

→ Moldova is made up of territory that the Soviet Union took from Romania in 1940. Moldova has serious ethnic and political problems, with the ethnic Moldovan people oriented toward their kin in Romania, while the ethnic Russian population has formed a self-proclaimed Trans-Dniester Republic. Agriculture is the economic base.

→ Kaliningrad is an enclave of Russia separated from the rest of the country by the Baltic countries. It is populated mainly by ethnic Russians who would like to become Euro-Russians, developing stronger ties with western Europe while loosening dependence on Russia. Historically a military base, Kaliningrad has a favorable location for future east-west trade.

→ Young women from Moldova, Ukraine, Russia, and other successor states of the Soviet Union are among the main participants in and victims of sex trafficking to Europe.

→ In an effort to reverse the inefficiencies left by the state-run system, all of the countries are promoting land reform by privatizing collective and state farms and developing more independent or peasant farming. After a period of decline following independence in 1991, Russian agricultural production has improved.

→ Russian industrial production plummeted after independence. Profits from the recent boom in oil exports have begun to revitalize industries. The industrialized area around Moscow has a central location within the populous western plains and is functionally Russia's most important industrial region.

→ Petroleum development in the Volga-Urals fields contributed to the industrial rise of the Volga cities.

→ Coal made the Kuznetsk area the most important industrial region east of the Urals. Commerce in this region is facilitated by the Trans-Siberian Railroad and lines linked to it.

→ Most of the Russian Far East is a thinly populated wilderness with few settlements. Several small to medium-sized cities form a north-south line along the Trans-Siberian Railroad and the lower Amur River, the two main transportation arteries. The coastal ports of Vladivostok, Nakhodka, and Vostochnyy are coming to life as shipments of oil destined for Japan and other foreign markets have picked up. Oil from the eastern Siberian fields will flow through new pipelines to China and Vostochnyy. The island of Sakhalin is becoming an important oil and gas producer for the Japanese market.

→ The Northern Lands comprise one of the world's most sparsely populated areas, with coniferous forest (taiga) and tundra extending from the Finnish and Norwegian borders to the Pacific. Murmansk and Arkhangelsk are ports on the Northern Sea Route, a transpolar route for ocean shipping that can be kept open in winter with icebreakers.

→ Soviet-era dumping of nuclear waste off the Arctic island of Novaya Zemlya has created a long-lasting environmental problem that will be very expensive to remedy.

→ Siberia's development, especially of mechanized industries, has been based partly on hydroelectric plants on the major rivers. These power stations and dams are among the largest in the world. The world's deepest lake, Lake Baikal, is a unique natural ecosystem apparently threatened by effluent from paper and pulp mills and by oil production and pipeline exports.

→ Siberia's Northern Lands have rich mineral resources that began to be exploited in Soviet times, often with forced labor. Now it is expensive for Russia to maintain populations in some of the northern settlements, and the government is trying to attract people to more southerly latitudes.

→ The far southern Caucasus region borders Russia between the Black and Caspian Seas. It includes the Caucasus Mountains, a fringe of foothills and level steppes to the north, and the area south known as Transcaucasia. Russians and Ukrainians are the majority north of the Greater Caucasus Range. To the south, the important nationalities are the Georgians, Armenians, and Azerbaijanis, each represented by an independent country and maintaining their unique cultures. Numerous smaller ethnic groups also inhabit the region, which has been a cauldron of conflict since independence from the Soviet Union. One of the outstanding internal problems is the fate of Nagorno-Karabakh, a region claimed by Armenia but controlled by Azerbaijan.

→ Caspian Sea sturgeon, source of the world's most prized caviar, are threatened by the overfishing that is in part a legacy of the USSR's breakup.

→ The central Asian countries, east and north of the Caspian Sea, are home to a variety of non-Slavic and Muslim peoples. The largest populations are the Uzbeks, Kazakhs, Kyrgyz, and Turkmen—all speaking related Turkic languages—and a people of Iranian origins, the Tajiks. Most live in irrigated valleys at the base of the southern mountains that lie along the ancient Silk Road. These countries have different economic fortunes, with the wealthiest, Kazakhstan, enjoying large natural resource wealth while the poorest, Tajikistan, has few resources and is unstable politically.

→ Soviet-era colonization of central Asia for a cotton plantation economy had pronounced and lingering effects on the environment. Most notable was the diversion of water for irrigated agriculture and the subsequent shrinkage of the Aral Sea. The post-Soviet countries of the region are locked in a critical competition over the region's water supplies from the Amu Darya and Syr Darya Rivers.

REVIEW QUESTIONS

1. What resources and boundaries are associated with the Fertile Triangle?

2. To what extent are Ukraine, Belarus, and Moldova dependent on Russia or interested in developing further ties with Russia? Which are most and least like Russia? What problems in relations have occurred between Russia and each of these counties?

3. What is Ukraine's main industrial region, and how does it correspond to the linguistic and political landscapes of the country? What is the main threat to Ukrainian industry? Which Russian industrial regions are within the Fertile Triangle? What industries do these regions and their major cities specialize in?

4. Why is Moscow known as the Port of Five Seas?

5. Which city is dubbed Russia's "Detroit," and which is its "Chicago"? Why?

6. What are the physical aspects of Russia's Far East? What are the main transportation arteries there?

7. What are the resources and economic development prospects of Sakhalin and Kamchatka? Why are the Kurils important in Russian-Japanese relations?

8. What two distinct climate and vegetative zones dominate Russia's Northern Lands? What are the main settlements and resources of the Northern Lands?

9. What are the major countries, cultures, and products of the Caucasus? What major dispute between two groups is focused on Nagorno-Karabakh, and what is the source of that dispute?

10. What are the major physical aspects of the central Asian countries? What are the main economic activities of these "stans"? Which countries are better and worse off economically, and why?

KEY TERMS + CONCEPTS

Agricultural Triangle (p. 183)
Armenian genocide (p. 200)
Chernobyl (p. 184)
"Chicago of Siberia" (p. 192)
conversion (p. 189)
"de-Russification" (p. 199)
Eurasian Economic Community (p. 204)

Euro-Russians (p. 186)
Fertile Triangle (p. 183)
gulag (p. 197)
"Hong Kong on the Baltic" (p. 186)
human trafficking (p. 187)
Northern Sea Route (p. 195)
periphery (p. 183)

Port of Five Seas (p. 188)
"Second Kuwait" (p. 193)
Slavic Core (p. 183)
Tulip Revolution (p. 204)
union state (p. 186)
"white gold" (p. 203)

NOTES

1. "The Ukrainian Question." *Economist,* November 29, 1999, p. 20.

2. Valentin Rasputin, "Baikal." In *Siberia on Fire: Stories and Essays* (De Kalb: Northern Illinois University Press, 1989), pp. 188–189.

Egypt's Saint Katherine Monastery, built in the 6th century, lies at the foot of Jebel Musa— a mountain believed by many Christians and Muslims to be Mount Sinai of the Bible and the Qur'an.

CHAPTER 6
A GEOGRAPHIC PROFILE OF THE MIDDLE EAST AND NORTH AFRICA

The *Middle East* is a fitting designation for the places and the cultures of this vital world region because they are literally in the middle. This is a physical crossroads, where the continents of Africa, Asia, and Europe meet and the waters of the Mediterranean Sea and the Indian Ocean mingle. Its peoples—Arab, Jew, Persian, Turk, Kurd, Berber, and others—express in their cultures and ethnicities the coming together of these diverse influences. Occupying as they do this strategic location, the nations of the Middle East and North Africa have through five millennia been unwilling hosts to occupiers and empires originating far beyond their borders. They have also bestowed on humankind a rich legacy that includes the ancient civilizations of Egypt and Mesopotamia and the world's three great monotheistic faiths of Judaism, Christianity, and Islam.

chapter objectives

This chapter should enable you to

→ Understand and explain the mostly beneficial relationships between villagers, pastoral nomads, and city dwellers in an environmentally challenging region

→ Know the basic beliefs and sacred places of Jews, Christians, and Muslims

→ Recognize the importance of petroleum to this region and the world economy

→ Identify the geographic chokepoints and oil pipelines that are among the world's most strategically important places and routes

→ Appreciate the problems of control over freshwater in this arid region

→ Know what al-Qa'ida and other Islamist terrorist groups are and what they want

People outside the region tend to forget about such contributions because they associate the Middle East and North Africa with war and terrorism. These negative connotations are often accompanied by muddled understanding. This is perhaps the most inaccurately perceived region in the world. Does sand cover most of the area? Are there camels everywhere? Does everyone speak Arabic? Are Turks and Persians Arabs? Are all Arabs Muslims? The answer to each of these questions is no, yet popular Western media suggest otherwise. This chapter and its accompanying module attempt to make the misunderstood, complex, sometimes bloody, but often hopeful events and circumstances in the Middle East and North Africa more intelligible by illuminating the geographic context within which they occur.

6.1 Area and Population

What is the Middle East, and where is it? The term itself is Eurocentric, invented by the British, who placed themselves in the figurative center of the world. They began to use the term prior to the outbreak of World War I, when the *Near East* referred to the territories of the Ottoman Empire in the eastern Mediterranean region, the *East* to India, and the *Far East* to China, Japan, and the western Pacific Rim. With *Middle East,* they designated as a separate region the countries around the Persian Gulf (known to Arabs as the Arabian Gulf and in this text as the Persian-Arabian Gulf or simply the Gulf). Gradually, the perceived boundaries of the region grew.

Sources today vary widely in their interpretation of which countries are in the Middle East. For some, the Middle East includes only the countries clustered around the Arabian Peninsula. For others, it spans a vast 6,000 miles (9,700 km) west to east from Morocco in northwest Africa to Afghanistan in central Asia, and a north-south distance of about 3,000 miles (4,800 km) from Turkey, on Europe's southeastern corner, to Sudan, which adjoins East Africa (•Figure 6.1). The larger area is the region covered in this chapter and its module, where it is referred to as "the Middle East and North Africa." The North African peoples of Morocco, Algeria, and Tunisia generally do not consider themselves Middle Easterners; they are, rather, from what they call the *Maghreb*, meaning "western land." Many geographers would place Sudan in sub-Saharan Africa and Afghanistan in central Asia or South Asia. Both are border or transitional countries in regional terms, and in this text, they are placed in the Middle East and North Africa with consideration given to their characteristics reflective of the other regions.

Thus defined, the Middle East and North Africa include 21 countries, the Palestinian territories of the West Bank and Gaza Strip, and the disputed Western Sahara (Table 6.1), occupying 5.9 million square miles (15.3 million sq km) and inhabited by about just over 500 million people in 2007. This area is about 1.8 times the size of the lower 48 United States and is generally situated at latitudes equivalent to those between Boston, Massachusetts, and Bogotá, Colombia (•Figure 6.2).

These half-billion people are not distributed evenly across the region but are concentrated in major clusters (•Figure 6.3). Three countries contain the lion's share of the region's population: Turkey, Iran, and Egypt, each with more than 70 million people. One look at a map of precipitation or vegetation explains why people are clustered this way (see Figure 2.3, page 23, and Figure 6.5). Where water is abundant in this generally arid region, so are people. Egypt has the Nile River, and parts of Iran and Turkey have bountiful rain and snow. Conversely, where rain seldom falls, as in the Sahara of North Africa and in the Arabian Peninsula, people are few.

The Middle East and North Africa as a whole have a high rate of population growth. The rapid growth is a general indication that this is a developing rather than industrialized region, and also reflects the majority Muslim culture that favors larger families. The average annual rate of population change for the 21 countries, the Palestinian territories, and Western Sahara was 1.9 percent in 2007. The lowest rate of population growth (1.1 percent) is in Tunisia.

50

• **Figure 6.1** The Middle East and North Africa.

The highest is 3.2 percent in Yemen, followed closely by the Palestinian territories (consisting of the West Bank and Gaza Strip) at 2.9 percent. These are some of the highest population growth rates in the world. In the Palestinian territories, such rapid growth may be ascribed in large part to the Palestinians' poverty and perhaps to the wishes of many Palestinians to have more children to counterbalance the demographic weight of their perceived Israeli foe.

Between these extremes are countries with modest rates of population growth of 1.1 to 1.5 percent per year; these include Israel, Morocco, Iran, Turkey, the United Arab Emirates, Lebanon, and Qatar. Generally, their governments have regarded most of these countries as too populous for their resources and economic base and have encouraged family planning. They have been successful in lowering birth rates. On the other hand, oil-rich Saudi Arabia, with its 2.7 percent annual growth rate, is a good example of how rapid population growth is not always a sign of poverty. In this case, an oil-rich nation has encouraged its citizens to give birth to more citizens so that in the future they will not need to import foreign laborers and technicians and will be more self-sufficient in their development. Some oil-rich countries of the Gulf region have more foreigners than citizens living in them; about 75 percent of the working-age population of the United Arab Emirates, for example, is nonnative, mainly from India and other South Asian countries. Many of these workers complain of dreams broken by poor living conditions and low wages.

Many developing countries have economies largely dependent on subsistence agriculture and have low percentages of urban inhabitants. Perhaps surprisingly, however, the Middle East and North Africa have more urbanites than country folk. The average urban population among the 23 countries and territories is 56 percent. The most prosperous countries are also the most urban. Essentially a city-state, Kuwait is 98 percent urban. The other oil-wealthy Gulf countries also have urban populations over 70 percent. Consistent with its profile as a Western-style industrialized country without oil resources, Israel is 92 percent urban. At the other end of the spectrum, desperately poor Afghanistan and Yemen have urban populations of less than 30 percent.

6.2 Physical Geography and Human Adaptations

The margins of the Middle East and North Africa are mainly oceans, seas, high mountains, and deserts (•Figure 6.4). To the west lies the Atlantic Ocean; to the south, the Sahara and the highlands of East Africa; to the north, the Mediterranean, Black, and Caspian Seas, together with mountains and deserts lining the southern land frontiers of Russia and the Near Abroad; and to the east, the Hindu Kush mountains on the Afghanistan-Pakistan frontier and the Baluchistan Desert straddling Iran and Pakistan. The land is composed mainly of arid plains and plateaus, together with large areas of rugged mountains and isolated "seas" of sand. Despite the environmental challenges, this region has given rise to some of the world's oldest and most influential ways of living.

TABLE 6.1 Middle East and North Africa: Basic Data

Political Unit	Area (thousands; sq mi)	Area (thousands; sq km)	Estimated Population (millions)	Estimated Population Density (sq mi)	Estimated Population Density (sq km)	Annual rate of Natural Increase (%)	Human Development Index	Urban Population (%)	Arable Land (%)	Per Capita GNI PPP ($US)
Middle East	**2,630.7**	**6,813.5**	**308.5**	**117**	**45**	**1.9**	**0.738**	**60**	**8**	**9,540**
Afghanistan	251.8	652.2	31.9	127	49	2.6	N/A	20	12	N/A
Bahrain	0.3	0.8	0.8	2,667	1,030	1.8	0.859	100	3	18,770
Iran	630.6	1,633.3	71.2	113	44	1.2	0.746	67	8	8,480
Iraq	169.2	438.2	29.0	171	66	2.5	N/A	67	13	N/A
Israel	8.1	21.0	7.3	901	348	1.5	0.927	92	16	25,470
Jordan	34.4	89.1	5.7	166	64	2.4	0.760	82	2	6,200
Kuwait	6.9	17.9	2.8	406	157	1.9	0.871	98	1	29,200
Lebanon	4.0	10.4	3.9	975	376	1.5	0.774	87	16	5,460
Oman	82.0	212.4	2.7	33	13	2.2	0.810	71	0	14,570
Palestinian Territories	2.4	6.2	4.0	1,667	644	2.9	0.736	72	17	N/A
Qatar	4.2	10.9	0.9	214	83	1.5	0.844	100	1	N/A
Saudi Arabia	830.0	2,149.7	27.6	33	13	2.7	0.777	81	1	16,620
Syria	71.5	185.2	19.9	278	107	2.5	0.716	50	25	3,920
Turkey	299.2	774.9	74.0	247	95	1.2	0.757	66	31	9,060
United Arab Emirates	32.3	83.7	4.4	136	53	1.5	0.839	74	0	23,990
Yemen	203.8	527.8	22.4	110	42	3.2	0.492	26	2	920
North Africa	**3,286.0**	**8,510.7**	**194.7**	**59**	**23**	**1.9**	**0.664**	**50**	**5**	**5,500**
Algeria	919.6	2,381.8	34.1	37	14	1.7	0.728	58	3	6,900
Egypt	386.7	1,001.6	73.4	190	73	2.1	0.702	43	3	6,900
Libya	679.4	1,759.6	6.2	9	4	2.0	0.798	85	1	N/A
Morocco	172.4	446.5	31.7	184	71	1.5	0.640	55	19	5,000
Sudan	967.5	2,505.8	38.6	40	15	2.2	0.516	41	7	2,160
Tunisia	63.2	163.7	10.2	161	62	1.1	0.760	65	18	8,490
Western Sahara	97.2	251.7	0.5	5	2	2.0	N/A	92	0	N/A
Summary Total	**5,916.7**	**15,324.2**	**503.2**	**85**	**33**	**1.9**	**0.709**	**56**	**6**	**7,980**

Sources: World Population Data Sheet, Population Reference Bureau, 2007; Human Development Report, United Nations, 2006; World Factbook, CIA, 2007.

A Region of Stark Geographic Contrasts

Aridity dominates the Middle East and North Africa (• Figure 6.5). At least three-fourths of the region has average yearly precipitation of less than 10 inches (25 cm), an amount too small for most types of dry farming (unirrigated agriculture). Sometimes, however, localized cloudbursts release moisture that allows plants, animals, and small populations of people—the Bedouin, Tuareg, and other pastoral nomads—to live in the desert. Even the vast Sahara, the world's largest desert, supports a surprising diversity and abundance of life. Plants, animals, and even people have developed strategies of **drought avoidance** and **drought endurance**. Migrating to avoid drought is a coping strategy that pastoral nomads and many animals use. Other organisms, such as trees, must endure drought by such adaptations as having very deep roots and small leaves. Populations of people, plants, and animals are all but nonexistent in the region's vast **sand seas,** including the Great Sand Sea of western Egypt (•Figure 6.6) and the Empty Quarter of the Arabian Peninsula.

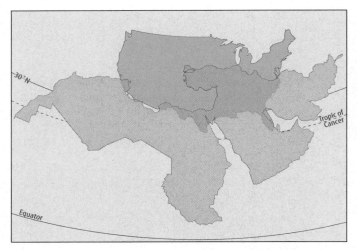

• **Figure 6.2** The Middle East and North Africa compared in latitude and area with the conterminous United States.

The region's climates have the comparatively large daily and seasonal ranges of temperature characteristic of dry lands. Desert nights can be surprisingly cool. Most days and nights are cloudless, so the heat absorbed on the desert surface during the day is lost by radiational cooling to the heights of the atmosphere at night. Summers in the lowlands are very hot almost everywhere. The hottest shade temperature ever recorded on earth, 136 degrees F (58 degrees C), occurred in Libya in September 1922. Many places regularly experience daily maximum temperatures over 100°F (38°C) for weeks at a time.

Human settlements located near the sand seas often experience the unpleasant combination of high temperatures and hot, sand-laden winds, creating the sandstorms known locally by such names as *simuum* ("poison") and *sirocco*. Only in mountainous sections and in some places near the sea do higher elevations or sea breezes temper the intense midsummer heat. The population of Alexandria, on the Mediterranean Sea coast, explodes in summer as Egyptians flee from Cairo and other hot inland locations. In Saudi Arabia, the government relocates from Riyadh to the highland summer capital of Taif to escape the lowland furnace.

Lower winter temperatures bring relief from the summer heat, and the more favored places receive enough precipitation for dry farming of winter wheat, barley, and other cool season crops (•Figure 6.7). In general, winters are cool to mild. But very cold winters and snowfalls occur in the high interior basins and plateaus of Iran, Afghanistan, and Turkey. These locales generally have a steppe climate. Only in the southernmost reaches of the region, notably Sudan, do temperatures remain consistently high throughout the year. A savanna climate and biome prevail there.

Most areas bordering the Mediterranean Sea have 15 to 40 inches (38 to 102 cm) of precipitation a year, falling almost exclusively in winter, while the summer is dry and

warm—a typical Mediterranean climate pattern. Throughout history, people without access to perennial streams have stored this moisture to make it available later for growing crops that require the higher temperatures of the summer months. The Nabateans, for example, who were contemporaries of the Romans in what is now Jordan, had a sophisticated network of limestone cisterns and irrigation channels (•Figure 6.8). Rainfall sufficient for dry farming during the summer is concentrated in areas along the southern and northern margins of the region. The Black Sea side of Turkey's Pontic (also known as Kuzey) Mountains is lush and moist in the summer, and tea grows well there (•Figure 6.9). In the southwestern Arabian Peninsula, a monsoonal climate brings summer rainfall and autumn harvests to Yemen and Oman, probably accounting for the Roman name for the area: *Arabia Felix* ("Happy Arabia"). Mountainous areas in the region, like the river valleys and the margins of the Mediterranean, play a vital role in supporting human populations and national economies. Due to orographic or elevation-induced precipitation, the mountains tend to receive much more rainfall than surrounding lowland areas.

There are three principal mountainous regions of the Middle East and North Africa (see Figure 6.4). In northwestern Africa between the Mediterranean Sea and the Sahara, the Atlas Mountains of Morocco, Algeria, and Tunisia reach over 13,000 feet (3,965 m) in elevation. Mountains also rise on both sides of the Red Sea, with peaks up to 12,336 feet (3,760 m) in Yemen. These are the result of tectonic processes that are pulling the African and Arabian plates apart, creating the northern part of the Great Rift Valley. The hinge of this crustal movement is the Bekaa Valley of Lebanon, where the widening fault line follows the Jordan River Valley southward to the Dead Sea (• Figure 6.10). This valley is the deepest depression on the earth's land surface, lying about 600 feet (183 m) below sea level at Lake Kinneret (also known as the Sea of Galilee or Lake Tiberius) and nearly 1,300 feet (400 m) below sea level at the shore of the Dead Sea, the lowest point of land on the planet. The rift then continues southward to the very deep Gulf of Aqaba and Red Sea before turning inland into Africa at Djibouti and Ethiopia.

In tectonic processes, one consequence of rifting in one place is the subsequent collision of the earth's crustal plates elsewhere. There are several such collision zones in the Middle East and North Africa, particularly in Turkey and Iran, where mountain building has resulted. These are seismically active zones—meaning earthquakes occur there—and rarely does a year go by without a devastating quake rocking Turkey, Iran, or Afghanistan.

A large area of mountains, including the region's highest peaks, stretches across Turkey, Iran, and Afghanistan. These mountains are products of the collision of continental plates. On the eastern border with Pakistan, the Hindu Kush Range has peaks over 25,000 feet (7,600 m) high. The loftiest mountain ranges in Turkey are the Pontic, Taurus

20, 21

459, 466, 482

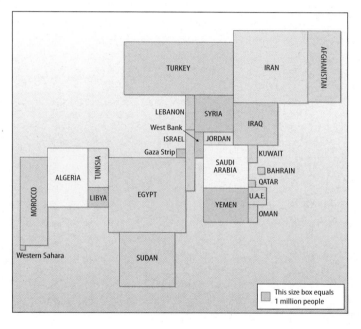

• **Figure 6.3** Population distribution (top) and population cartogram (bottom) of the Middle East and North Africa.

(•Figure 6.11), and Anti-Taurus, and in Iran, the Elburz and Zagros Mountains. These chains radiate outward from the rugged Armenian Knot in the tangled border country where Turkey, Iran, and the countries of the Caucasus meet. Mount Ararat is an extinct, glacier-covered volcano of 16,804 feet (5,122 m) towering over the border region between Turkey and Armenia. Many biblical scholars and explorers think the ark of Noah lies high on the mountain (and some go in search of it), for in the book of Genesis, this boat was said to have come to rest "in the mountains of Ararat."

Extensive forests existed in early historical times in the Middle East and North Africa, particularly in these mountainous areas, but overcutting and overgrazing have almost eliminated them. Since the dawn of civilization in this area,

around 3000 B.C.E., people have cut timber for construction and fuel faster than nature could replace it. Egyptian King Tutankhamen's funerary shrines and Solomon's Temple in Jerusalem were built of cedar from Lebanon (•Figure 6.12). So prized has this wood been through the millennia that only a few isolated groves of cedar remain in Lebanon. Described in ancient times as "an oasis of green with running creeks" and "a vast forest whose branches hide the sky," Lebanon is now largely barren. Lumber is still harvested commercially in a few mountain areas such as the Atlas region of Morocco and Algeria, the Taurus Mountains of Turkey, and the Elburz Mountains of Iran, but supply falls far short of demand.

Villager, Pastoral Nomad, Urbanite

In the 1960s, the American geographer Paul English developed a useful model for understanding relationships between the three ancient ways of life that still prevail in the Middle East and North Africa today: villager, pastoral nomad, and urbanite. Each of these modes of living is rooted in a particular physical environment. Villagers are the subsistence farmers of rural areas where dry farming or irrigation is possible; pastoral nomads are the desert peoples who migrate through arid lands with their livestock, following patterns of rainfall and vegetation; and urbanites are the inhabitants of the large towns and cities, generally located near bountiful water sources but sometimes placed for particular trade, religious, or other reasons. Describing these ways of life as components of the **Middle Eastern ecological trilogy**, English explained how each of them has a characteristic, usually mutually beneficial, pattern of interaction with the other two (•Figure 6.13).

The peasant farmers of Middle Eastern and North African villages (the **villagers**) represent the cornerstone of the trilogy. They grow the staple food crops such as wheat

• **Figure 6.4** Physical geography of the Middle East and North Africa.

• **Figure 6.5** Climates (left) and biomes (right) of the Middle East and North Africa.

and barley that feed both the city dweller and the pastoral nomad of the desert. Neither urbanite nor nomad could live without them. The village also provides the city, often unwillingly, with tax revenue, soldiers, and workers. And before the mid-20th century, villages provided **pastoral nomads** with plunder as the desert dwellers raided their settlements and caravan supply lines. Generally, however, the exchange is beneficial.

The nomads provide villagers with livestock products, including live animals, meat, milk, cheese, hides, and wool, and with desert herbs and medicines. Educated and progres-

sive **urbanites** provide technological innovations, manufactured goods, religious and secular education and training, and cultural amenities (today, including films and music).

There is little direct interaction between urbanites and pastoral nomads, although some manufactured goods such as clothing travel from city to desert and some desert folk medicines pass from desert to city. Historically, the exchange has been violent, as urban-based governments have sought to control the movements and military capabilities of the elusive and sometimes hostile nomads. Pastoral nomads once plundered rich caravans plying the major overland

• **Figure 6.6** Seas of sand cover large areas in Saudi Arabia, Iran, and parts of the Sahara. This is the edge of the Great Sand Sea near Siwa Oasis in Egypt's Western Desert.

• **Figure 6.7** Land use in the Middle East and North Africa.

• **Figure 6.8** The Treasury, a temple carved in red sandstone, probably in the 1st century D/F/, by the Nabateans at their capital of Petra in southern Jordan. The English poet Dean Burgen called Petra "a rose red city, half as old as time." The Nabateans built sophisticated networks for water storage and distribution. For scale, note the man at the lower right.

trade routes of the Middle East and North Africa. Governments did not tolerate such activities and often cracked down hard on the nomads they were able to catch.

Later, in the 1970s, Paul English wrote an article marking the "passing of the ecological trilogy." He noted that cities were encroaching on villages, villagers were migrating into cities and giving some neighborhoods a rural aspect, and pastoral nomads were settling down—thus the trilogy no longer existed. In reality, although the makeup and interactions of its parts have changed somewhat, the trilogy model is still valid and useful as an introduction to the major ways of life in the Middle East and North Africa. It is especially significant that a given man or woman in the region strongly identifies as either a villager, a pastoral nomad, or an urbanite. This perception of self has an important bearing on how these people of very different backgrounds interact, even when they live in close prox-

imity. Urban officials may work in rural village areas, but they remain at heart and in their perspectives city people and usually live apart from farmers. Extended families of pastoral nomads may settle down and become farmers, but they continue to identify themselves by affiliation with the nomadic tribe. Many continue to harvest desert resources on a seasonal basis and retain marriage and other ties with desert-dwelling relatives.

The Village Way of Life

Agricultural villagers historically represented by far the majority populations in the Middle East and North Africa; only in recent decades have urbanites begun to outnumber them. In this generally dry environment, the villages are located near reliable water sources with cultivable lands nearby. They are usually made up of closely related family

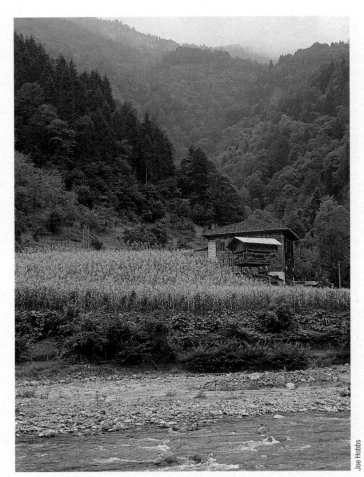

• **Figure 6.9** Rainfall is heavy on the Black Sea side of Turkey's Pontic (Kuzey) Mountains. Note that roofs are pitched to shed precipitation; in the region's drier areas, most roofs are flat. The crop growing here is corn.

• **Figure 6.11** The Taurus Mountains of southeastern Turkey.

• **Figure 6.12** Ancient Egyptians believed that the solar boat of King Cheops of Egypt (c. 2500 c/b/F/), builder of the Great Pyramid, would carry the pharaoh's spirit through the firmament. It had to be made of the best wood—cedar from Lebanon. The solar boat was interred next to the pyramid and now stands in a specially constructed museum.

• **Figure 6.10** The surface of the Dead Sea (upper left) is the lowest point on earth, a consequence of the rifting of continental crust. The escarpment on the right is in the Israeli-occupied West Bank, not far from Jerusalem; to the left, out of view, is a counterpart escarpment in Jordanian territory. This photo reveals the strategic advantage Israel gained by capturing the West Bank in 1967: any force invading Israel from the east would have to travel through a steep, easily-defended pass.

groups, with many fields owned by an absentee landlord. Villagers typically live in closely spaced homes made of mud brick or concrete blocks. Production and consumption focus on a staple grain such as wheat, barley, or rice. As land for growing fodder is often in short supply, villagers keep only a small number of sheep and goats and rely in part on nomads for pastoral produce. Residents of a given village usually share common ties of kinship, religion, ritual, and custom, and the changing demands of agricultural seasons regulate their patterns of activity.

These patterns of village life have been increasingly exposed to outside influences since the mid-18th century. European colonialism brought significant economic changes, including the introduction of cash crops and modern facilities to ship them. Improved and expanded irrigation, financed initially with capital from the West, brought more

Villagers

Live animals, meat, milk, cheese, hides, wool

Food crops, clothing, some manufactured goods

Technical innovations, manufactured goods, education, cultural amenities

Food crops, tax revenue, soldiers, workers

Pastoral nomads

Urbanites

• **Figure 6.13** People, environments, and interactions of the ecological trilogy (here, villagers harvesting sugarcane in upper Egypt, Bedouin at camp in Egypt's Eastern Desert, and shoppers at the main gate to the historic city of San'a, Yemen). This relationship is generally symbiotic, although historically both urbanites and pastoral nomads preyed on the villagers, who are the trilogy's cornerstone.

land under cultivation. Recent agents of change have been the countries' own government-supported doctors, teachers, and land reform officers. Modern technologies such as sewing machines, motor vehicles, gasoline-powered water pumps, radio, television, and most recently, the Internet and cell phones have modified old patterns of living. The educated and ambitious, as well as the unskilled and desperately poor—motivated, respectively, by pull and push factors (see Chapter 3, page 57)—have been drawn to urban areas. Improved roads and communications have in turn carried urban influences to villages, prompting villagers to become more integrated into the national society.

The Pastoral Nomadic Way of Life

Pastoral nomadism emerged as an offshoot of the village agricultural way of life not long after plants and animals were first domesticated in the Middle East (about 7000 B.C.E.). Rainfall and the wild fodder it brings forth, although scat-

tered, are sufficient resources to support small groups of people who migrate with their sheep, goats, and camels (and in some locales, cattle) to take advantage of this changing resource base (see Perspectives from the Field, page 218). In mountainous areas, they follow a pattern of **vertical migration** (sometimes referred to as *transhumance*), moving with their flocks from lowland winter to highland summer pastures. In the flatter expanses that comprise most of the region, the nomads practice a pattern of **horizontal migration** over much larger areas where rainfall is typically far less reliable than in the mountains. In addition to selling or trading livestock to obtain food, tea, sugar, clothing, and other essentials from settled communities, pastoral nomads hunt, gather, work for wages, and where possible, grow crops. Many now work in the tourist industry, as Westerners hungry for insight into traditional cultures seek them out. Their multifaceted livelihood has been described as a strategy of **risk minimization** based on the exploitation of multiple resources so that some will support them if others fail.

PERSPECTIVES FROM THE FIELD
Way-Finding in the Desert

The research for my doctoral dissertation in geography at the University of Texas–Austin was an 18-month journey with the Khushmaan Ma'aza, a clan of Bedouin nomads living in the northern part of Egypt's Eastern Desert. I studied their perceptions, knowledge, and uses of natural resources and tried to understand how their worldviews and kinship patterns apparently helped them devise ways of protecting these scarce resources. I have worked with these people from the early 1980s to the present and have always been astonished by how different they are from me and how much better they are in their ability to "read" the ground and recall landscape details. In the following passage from my book *Bedouin Life in the Egyptian Wilderness*, I marvel at their way-finding abilities.

The process by which the Khushmaan nomads have developed roots in their landscape, fashioning subjective "place" from anonymous "space," and the means by which they orient themselves and use places on a daily basis deserve special attention: these are essential parts of the Bedouins' identity and profoundly influence how they use resources and affect the desert ecosystem.

The Bedouins have little need or knowledge of maps. On the other hand, when shown a map or aerial image of countryside they know, the nomads accurately orient and interpret it, naming the mountains and drainages depicted. Saalih Ali (●Figure 6.A) especially enjoyed my star chart. He would tell me what time the Pleiades would rise, for example, and ask me what time the chart indicated. The two were almost always in agreement.

Many desert travelers have been astonished by the nomads' navigational and tracking ability, calling it a "sixth sense." The Khushmaan are exceptional way-finders and topographical interpreters able, for instance, to tell from tracks whether a camel was carrying baggage or a man; whether gazelle tracks were made by a male or female; which way a car was traveling and what make it was; which man left a set of footprints, even if he wore sandals; and how old the tracks are. Bedouins are proud of their geographical skills, which, they believe, distinguish them from settled people. Saalih told me, "Your people don't need to know the country but we do, to know exactly where things are in order to live." Pointing to his head, he said, "My map is here."

The difference in the way-finding abilities of nomads and settled persons may be due to the greater survival value of these skills for nomads and to the more complex meanings they attribute to locations. Bedouin places are rich in ideological and practical sig-

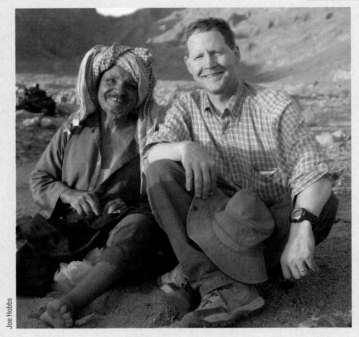

Joe Hobbs

● **Figure 6.A** Saalih Ali, a Bedouin of the Khushmaan Ma'aza, with the author in 2005, 25 years after they began fieldwork together.

nificance. The nomads interpret and interact with these meanings on a regular basis and create new places in their lifetimes. Theirs is an experience of belonging and becoming with the landscape, whereas settled people are more likely to inherit and accommodate themselves to a given set of places.

The nomads' homeland is so vast, and the margin of survivability in it so narrow, that topographic knowledge must be encyclopedic. Places either are the resources that allow human life in the desert or are the signposts that lead to resources. A Khushmaan man pinpointed the role of places in desert survival: "Places have names so that people do not get lost. They can learn where water and other things are by using place names." Tragedy may result if the nomad has insufficient or incorrect information about places: many deaths by thirst are attributed to faulty directions for finding water.*

*From Joseph Hobbs, *Bedouin Life in the Egyptian Wilderness* (Austin: University of Texas Press, 1989), pp. 81–82. Copyright © 1989 University of Texas Press. Reprinted with permission.

Although renowned in Middle Eastern legends, pastoral nomads have been described as "more glamorous than numerous." It is still impossible to obtain adequate census figures on the number living in the deserts of the Middle East and North Africa, though estimates range from 5 to 13 million. Recent decades have witnessed the rapid and progressive settling down, or **sedentarization,** of the nomads—a process attributed to a variety of causes. In some cases, prolonged drought virtually eliminated the resource base on which the nomads depended. Traditionally, they were able to migrate far enough to find new pastures, but modern national boundaries now inhibit such movements. Some have returned with the rains to their desert homelands, but others have chosen to remain as farmers or wage laborers in villages and towns. On the Arabian Peninsula, the prosperity and technological changes prompted by oil revenues made rapid inroads into the material culture—and then the livelihood preferences—of the desert people; many preferred the comforts of settled life. Some governments, notably those of Israel and prerevolutionary Iran, were frustrated by their inability to count, tax, conscript, and control a sizable migrant population and therefore compelled the nomads to settle.

Pastoral nomads of the Middle East and North Africa identify themselves primarily by their tribe, not by their nationality. The major ethnic groups from which these tribes draw are the Arabic-speaking Bedouin of the Arabian Peninsula and adjacent lands, the Berber and Tuareg of North Africa, the Kababish and Bisharin of Sudan, the Yoruk and Kurds of Turkey, the Qashai and Bakhtiari of Iran, and the Pashtun of Afghanistan. Members of a tribe claim common descent from a single male ancestor who lived countless generations ago; their kinship organization is thus a patrilineal descent system. It is also a segmentary kinship system, so called because there are smaller subsections of the tribe, known as clans and lineages, that are functionally important in daily life. Members of the most closely related families comprising the lineage, for example, share livestock, wells, trees, and other resources. Both the larger clans, made up of numerous lineages, and the tribes possess territories. Members of a clan or tribe typically allow members of another clan or tribe to use the resources within its territory on the basis of **usufruct,** or nondestructive mutual use.

Although some detractors have depicted pastoral nomads as the "fathers" rather than "sons" of the desert, blaming them for wanton destruction of game animals and vegetation, there are numerous examples of pastoral nomadic groups who have developed indigenous and very effective systems of resource conservation. Most of these practices depend on the kinship groups of family, lineage, clan, and tribe to assume responsibility for protecting plants and animals.

The Urban Way of Life

The city was the final component to emerge in the ecological trilogy, beginning in about 4000 B.C.E. in Mesopotamia (modern Iraq) and 3000 B.C.E. in Egypt. Though they resembled villages in many ways, the early cities were distinguished by their larger populations (more than 5,000 people), the use of written languages, and the presence of monumental temples and other ceremonial centers. The early Mesopotamian city and, after the 7th century C.E., the classic Islamic city, called the **medina,** had several structural elements in common (•Figure 6.14). The medina had a high surrounding wall built for defensive purposes. The congregational mosque and often an attached administrative and educational complex dominated the city center. Although Islam is often characterized as a faith of the des-

• **Figure 6.14** An idealized model of the classic medina, or Muslim Middle Eastern city.

ert, religious life has always been focused in, and diffused from, the cities. The importance of the city's congregational mosque in religious and everyday life is often emphasized by its large size and outstanding artistic execution.

A large commercial zone, known as a **bazaar** in Persian and a **suq** in Arabic and recognizable as the ancestor of the modern shopping mall, typically adjoined the ceremonial and administrative heart of the city (•Figure 6.15). Merchants and craftspeople selling various commodities occupied separate spaces within this complex, and visitors to an old medina today can still find sections devoted exclusively to the sale of spices, carpets, gold, silver, traditional medicines, or other particular goods. Smaller clusters of shops and workshops were located at the city gates.

Residential areas were differentiated as quarters not by income group but by ethnicity; the medina of Jerusalem, for example, still has distinct Jewish, Arab, Armenian, and non-Armenian Christian quarters (see Figure 6.B, page 223). Homes tended to face inward toward a quiet central courtyard, buffering the occupants from the noise and bustle of the street. The narrow, winding streets of the medina were intended for foot traffic and small animal-drawn carts, not for large motor vehicles, a fact that accounts for the traffic jams in some old Middle Eastern and North African cities today and for the wholesale destruction of the medina in others.

The medinas that survive today are gently decaying vestiges of a forgotten urban pattern. Periods of European colonialism and subsequent nationalism changed the face and orientation of the city. During the colonial age, resident Europeans preferred to live in more spacious settings at the outer edges of the city, and later, the national elite followed this pattern. In recent times, independent governments have adopted Western building styles, with broad traffic arteries cutting through the old quarters and large central squares near government buildings. This opening up of the cityscape has spread commercial activity along the wide avenues, diluting the prime importance of the central bazaar as the focus of trade.

Rural-to-urban migration and the city's own internal growth contribute to a rapid rate of urbanization that puts enormous pressure on services in the region's poorer countries. Governments often build high-rise public housing to accommodate the growing population, contributing to a cycle in which the urban poor move into the new dwellings, only to leave their old quarters as a vacuum to draw in still more rural migrants. In Cairo, millions of former villagers now live in the "City of the Dead," an extraordinary urban landscape composed of multistory dwellings erected above graves—a last resort for the poor who have no other place to go. The overwhelmingly largest city, or primate city so

253

characteristic of Middle Eastern and North African capitals, thus grows at the expense of the smaller original city.

Much of the rural-to-urban migration and subsequent urban gridlock and squalor could probably be avoided if governments invested more in the development of villages and smaller cities. The oil-rich countries with relatively small populations generally enjoy an urban standard of living equaling that of affluent Western countries. Modern industrial cities such as Saudi Arabia's Jubail and others founded on oil wealth were built virtually overnight, providing fascinating contrast to the region's colorful, complex ancient cities. With the recent resurgence of oil revenue into Saudi Arabia and the other Gulf countries, some of the cities boast the most elegant, sophisticated, and expensive amenities and architecture to be found anywhere in the world (•Figure 6.16)

258–261

● **Figure 6.15** A typical Middle Eastern *suq* or *bazaar* (market). This one is in Cairo, Egypt.

● **Figure 6.16** July in Dubai, one of the city states of the United Arab Emirates. Wealth built initially on oil revenues has made it possible for some Gulf countries to build almost anything, including a ski resort located in a shopping mall.

6.3 Cultural and Historical Geographies

Cultures of the Middle East and North Africa have made many fundamental contributions to humanity. Many of the plants and animals on which the world's agriculture is based were first domesticated in the Middle East between 5,000 and 10,000 years ago in the course of the Agricultural Revolution. The list includes wheat, barley, sheep, goats, cattle, and pigs, whose wild ancestors were processed, manipulated, and bred until their physical makeup and behavior changed to suit human needs. The interaction between people and the wild plants and animals they eventually domesticated took place mainly in the well-watered **Fertile Crescent,** the arc of land stretching from Israel to western Iran.

By about 6,000 years ago, people sought higher yields by irrigating crops in the rich but often dry soils of the Tigris, Euphrates, and Nile River Valleys. Their efforts produced the enormous crop surpluses that allowed civilization—a cultural complex based on an urban way of life—to emerge in Mesopotamia (literally, "the land between the rivers," the Tigris and Euphrates) and in Egypt. Accomplishments in science, technology, art, architecture, language, mathematics, and other areas diffused outward from these centers of civilization. Egypt and Mesopotamia are thus among the world's great culture hearths.

The Middle East and North Africa are sometimes mistakenly referred to as the "Arab world"; in fact, the region has huge populations of non-Arabs. It is true that a majority of the region's inhabitants are Arabs. An **Arab** is best defined as a person of Semitic Arab ethnicity whose ancestral language is **Arabic,** a **Semitic language** spoken by about 245 million, or 48 percent, of the region's people. Originally, the Arabs were inhabitants of the Arabian Peninsula, but conquests after their majority conversion to Islam took them, their language, and their Islamic culture as far west as Morocco and Spain.

The region is also the homeland of the **Jews,** whose definition today is complex. Originally, Jews were both a distinct ethnic and linguistic group of the Middle East who practiced the religion of Judaism. It is possible (although difficult) for non-Jews to convert to Judaism (the convert is known as a proselyte, or "immigrant"), so a strictly ethnic definition does not apply today. Many Jews do not practice their religion but still consider themselves ethnically or culturally Jewish. Whatever debate exits about what defines a Jew, Jewish identity is strong and resilient.

It is important to recognize that although political circumstances made them enemies in the 20th century, Jews and Arabs lived in peace for centuries, and they share many cultural traits. Both recognize Abraham as their patriarch. Arabic is a Semitic language in the same **Afro-Asiatic language family** as **Hebrew,** which is spoken by most of the 5.5 million Jewish inhabitants of Israel (see the language map, •Figure 6.17).

There are other very large populations of non-Semitic ethnic groups and languages in the region. The greatest are the 60 million **Turks** of Turkey, who speak **Turkish,** a member of the **Altaic language family,** and the 33 million **Persians** of Iran, who speak **Farsi** (also called Persian), in the **Indo-European language family.** Both Persian and Arabic are written in Arabic script and so appear related, but they are not. Turkish was also written in Arabic script until early in the 20th century, but since then, it has been written in a Latin script. The ethnic **Pashtun** majority of Afghanistan speaks **Pashto,** a language closely related to Persian, and the country's official language is **Dari** (Afghan Persian). About 30 million **Kurds**—a people living in Turkey, Iraq, Iran, and Syria—speak **Kurdish,** which is also an Indo-European language. Many people in North Africa speak **Berber** and **Tuareg** (in the Afro-Asiatic language family). Sudan is ethnically and linguistically a transition zone between the Middle East and North Africa region and sub-Saharan Africa. In Sudan, particularly in the south, there are many speakers of **Chari-Nile languages,** in the **Nilo-Saharan language family.**

The Middle East also gave the world the closely related monotheistic faiths of Judaism, Christianity, and Islam. It is impossible to consider the human and political geographies of this region without attention to these religions, the ways of life associated with them, and their holy places, so here is an introduction (•Figure 6.18).

The Promised Land of the Jews

Judaism, the first significant monotheistic faith as far as we know, is today practiced by about 14 million people worldwide, mostly in Israel, Europe, and North America. September 2007 marked the start of the year 5768 in the Jewish calendar, but unlike its kindred faiths—Christianity and Islam—Judaism does not have an acknowledged starting point in time. Also unlike Christianity, Judaism does not have a fixed creed or doctrine. Jews are encouraged to behave in this life according to God's laws, which, according to the Torah (the Jewish holy scripture, which is known as the Old Testament in the Christian Bible), God gave to Moses on Mount Sinai as a covenant with God's "chosen people." The coming of a savior known as the Messiah ("Anointed One" in Hebrew) is prophesied in the Torah; Christians believe that Jesus was that savior, as described in the New Testament; Jews do not recognize Jesus as the fulfillment of that prophecy and so do not accept the New Testament.

Another distinction that sets Christianity and Islam apart from Judaism is that Judaism is not a proselytizing religion; it does not seek converts. Jewish identity is based strongly on a common historical experience shared over thousands of years. That historical experience has included deep-seated geographic associations with particular sacred places in the Middle East—particularly with places in Jerusalem, capital of ancient Judah (Judea), the province from

• **Figure 6.17** Languages of the Middle East and North Africa.

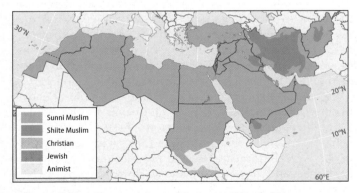

• **Figure 6.18** Religions of the Middle East and North Africa

which Jews take their name. Tragically, the Jewish history also has included unparalleled persecution.

Depending on one's perspective, the Jewish connection with the geographic region known as Palestine, essentially the area now composed of Israel, the West Bank, and the Gaza Strip, is most significant on a time scale of either about 4,000 years or about 100 years. According to the Bible, around 2000 B.C.E., God commanded Abraham and his kinspeople, known as **Hebrews** (later as Jews), to leave their home in what is now southern Iraq and settle in Canaan. God told Abraham that this land of Canaan—geographic Palestine—would belong to the Hebrews after a long period of persecution. The Bible says that the Hebrews did settle in Canaan until famine struck that land. At the command of Abraham's grandson Jacob, the Hebrews—known then as **Israelites**—relocated to Egypt, where grain was plentiful. That began the long sojourn of the Israelites in Egypt, which, according to the Bible, ended in about 1200 B.C.E. when Moses led them out in the journey known as the **Exodus.**

According to Jewish history, the prophecy of Abraham was first fulfilled when the Israelites settled once again in Canaan, their **Promised Land.** The Jewish King Saul unified the 12 tribes that descended from Jacob into the first united Kingdom of Israel in about 1020 B.C.E. In about 950 B.C.E. in Jerusalem—the capital of a kingdom enlarged by Saul's successor, David—King Solomon built Judaism's **First**

GEOGRAPHY OF THE SACRED

Jerusalem

The old city of Jerusalem is filled with sacred places and is one of the world's premier pilgrimage destinations. Over thousands of years, Jerusalem has been coveted and conquered by people of many different cultures and faiths. In the process, a place held sacred by one group has often come to be held sacred by a second and even a third.

The most important thing to notice on this map is that the location of the First and Second Jewish Temples is identical to that of the Muslims' Dome of the Rock. Judaism, Islam, and Christianity are very close in their origins and many of their basic precepts, but their faiths are not conjoined or syncretized in practice, as, for example, the Maya and Catholic faiths have been in Mesoamerica (see page XXX) or Buddhism and other traditions have been in East Asia (see page 294). In each faith, there has been a long tradition of quarreling over ownership of sacred space. Nowhere in the world does contested sacred space have more potential to ignite violence and outright war than in Jerusalem (which means, in Hebrew, "City of Peace"). Module 6.1 offers more insight into role of Jerusalem's holy places in the region's turbulent history. •Figure 6.B depicts the major places sacred to Jews, Christians, and Muslims and also the city's ethnic quarters; the walled old city is a classic medina.

• **Figure 6.B** Sacred sites and ethnic quarters of the old city of Jerusalem.

Temple. He located it atop a great rock known to the Jews as *Even HaShetiyah,* the **Foundation Stone,** plucked from beneath the throne of God to become the center of the world and the core from which the entire world was created (see Geography of the Sacred, above). The Ark of the Covenant, containing the commandments that God gave to Moses atop Mount Sinai, was placed in the Temple's Holy of Holies.

The united Kingdom of Israel lasted only about 200 years before splitting into the states of Israel and Judah. Empires based in Mesopotamia destroyed these states: the Assyrians attacked Israel in 721 B.C.E., and the Babylonians sacked Judah in 586 B.C.E. The Babylonians destroyed the First Temple (at which point the Ark of the Covenant disappeared) and exiled the Jewish people to Mesopotamia, where they remained until conquering Persians allowed them to return to their homeland. In about 520 B.C.E., the Jews who returned to Judah rebuilt the temple (the **Second Temple**) on its original site. A succession of foreign empires came to rule the Jews and Arabs of Palestine: Persian, Macedonian, Ptolemaic, Seleucid, and around the time of Jesus, Roman. Herod, the Jewish king who ruled under

Roman authority and was a contemporary of Jesus, greatly enlarged the temple complex.

The Jews of Palestine revolted against Roman rule three times between 64 and 135. The first revolt broke out as the profoundly monotheistic Jews refused to acknowledge the Roman emperor as a god. The Romans quashed these rebellions in a series of famous sieges, including those of Masada and Jerusalem. The Romans destroyed the Second Temple, and a third has never been built. All that remains of the Second Temple complex is a portion of the surrounding wall built by Herod. Today, this **Western Wall,** known to non-Jews as the Wailing Wall, is the most sacred site in the world accessible to Jews (•Figure 6.19). Some religious traditions prohibit Jews from ascending the **Temple Mount** above, the area where the temple actually stood, because it is too sacred. After the temple's destruction, that site was occupied by a Roman temple and then in 691 by the Muslim shrine called the **Dome of the Rock,** which still stands today (also in Figure 6.19). The mostly Muslim Arabs know the Temple Mount as *al-Haraam ash-Shariif,* the **Noble Sanctuary.** Supercharged with meaning, this place

• **Figure 6.19** Jerusalem's Western Wall, Temple Mount, and Noble Sanctuary. In this view are some of Judaism's and Islam's holiest places. At left, below the golden dome, is the Western Wall, which is all that remains of the structure that surrounded the Jews' Second Temple. The dome is the Muslims' Dome of the Rock. At far right, with the black dome, is the al-Aqsa Mosque, another very holy place in Islam.

has in modern times often been the spark of conflagration between Jews and Palestinian Arabs.

The victorious Romans scattered the defeated Jews to the far corners of the Roman world. Thus began the Jewish exile, or **Diaspora.** In their exile, the Jews never forgot their attachment to the Promised Land. One Jewish prayer, prayed during the weeklong Passover holiday, ends with the words "Next year in Jerusalem!" In Europe, where their numbers grew to be greatest, Jews were subjected to systematic discrimination and persecution and were forbidden to own land or engage in a number of professions. Known as **anti-Semitism,** hatred of Jews developed deep roots in Europe. This sentiment in part grew out of the long-simmering perception that Jews were responsible for the murder of Jesus Christ and in part out of the fact that Christian Europeans were generally prohibited from practicing moneylending. Jews were permitted to loan money to the Christians, who accepted the loans but then resented having to pay the Jews interest.

In the 1930s, anti-Semitism became state policy in Germany under the Nazis, led by Adolf Hitler. Many German Jews fled to the United States, and others emigrated to Palestine in support of the **Zionist movement,** which had since the late 1890s aimed at establishing a Jewish homeland in Palestine with **Zion** (Jerusalem) as its capital. Most Jews were not as fortunate as the Zionist emigrants. Within the boundaries of the Nazi empire that conquered most of continental Europe during World War II, Hitler's regime implemented its "final solution" to the "Jewish problem" by shipping to prison camps and eventually murdering all the Jews they could round up. The Nazi Germans and their al-

lies killed an estimated 6 million Jews, along with millions of other minorities they deemed "inferior," including Roma (Gypsies) and homosexuals. It was this Holocaust that prompted the victorious allies of World War II, from their powerful position in the newly formed United Nations, to create a permanent homeland for the Jewish people in Palestine. A seemingly endless cycle of violence, discussed beginning on page 241, ensued.

Christianity: Death and Resurrection in Jerusalem

Nearly 1,000 years after Solomon established the Jews' First Temple, a new but closely related monotheistic faith emerged in Palestine. This was **Christianity,** named for Jesus Christ (*Christ* is Greek for "Anointed One," the equivalent of the Hebrew word for Messiah). Jesus, a Jew, was born near Jerusalem in Bethlehem, probably around 4 B.C.E. Tradition relates that when he was about 30 years old, Jesus began spreading the word that he was the Messiah, the deliverer of humankind long prophesied in Jewish doctrine. A small group of disciples accepted that he was the Messiah and followed him for several years as he preached his message. He taught that love, sacrifice, and faith were the keys to salvation. He had come to redeem humanity's sins through his own death. To his followers, he was the Son of God, a living manifestation of God Himself, and the only path to eternal life was by accepting his divinity.

Jesus Christ's teachings denied the validity of many Jewish doctrines, and around 29 C.E., a growing chorus of Jewish protesters called for his death. Palestine was then under

Roman rule, and Roman administrators in Jerusalem placated the mob by ordering that he be put on trial. He was found guilty of being a claimant to Jewish kingship, and Roman soldiers put him to death by the particularly degrading and painful method of crucifixion. The cornerstone of Christian faith is that Jesus Christ was resurrected from the dead two days later and ascended into heaven. Christians believe that he continues to intercede with his Father on their behalf and that he will come again on Judgment Day, at the end of time.

After a period of relative tolerance, the Romans began actively persecuting Christians. Nevertheless, Christian ranks and influence grew. The turning point in Christianity's career came after 324 C.E., when the Roman Emperor Constantine embraced Christianity and established it as the official religion of the empire. The Christian Byzantine civilization that developed in the **"New Rome"** that Constantine established—Constantinople, now Istanbul, Turkey—created fine monuments at places associated with the life and death of Jesus Christ. These include the place long acknowledged as the center of the Christian world, Jerusalem's **Church of the Holy Sepulcher** (●Figure 6.20). This extraordinary, sprawling building, now administered by numerous separate Christian sects, contains the locations where tradition says Jesus Christ was crucified and buried. Incidentally, several of these Christian denominations have a long history of squabbling over turf within the church, and the key to this place is in the hands of a trusted, neutral party: a Muslim family.

Christianity has seldom been the majority religion in the land where it was born; only until Islam arrived in Palestine in 638 C.E. was the region primarily Christian. From then until the 20th century, most of Palestine's inhabitants were Muslim, and since the mid-20th century, Muslims and Jews have been the major groups. Between the 11th and 14th centuries, European Christians dispatched military expeditions to recapture Jerusalem and the rest of the Holy Land from the Muslims. These bloody campaigns, known as the **Crusades,** resulted in a series of short-lived Christian administrations in the region.

There are significant minority populations of Christians throughout the Middle East, including members of distinct sects such as the **Copts** of Egypt and the **Maronites** of Lebanon. Their share of the population has generally been declining, both because of emigration and lower birth rates than those of the majority Muslims. Nevertheless, the Middle East remains the cradle of their faith for Christians the world over, and Jerusalem and nearby Bethlehem are the world's premier Christian pilgrimage sites.

The Message of Islam

Islam is by far the dominant religion in the Middle East and North Africa; only Israel within its pre-1967 borders has a non-Muslim majority. Because of Islam's powerful influence not merely as a set of religious practices but as

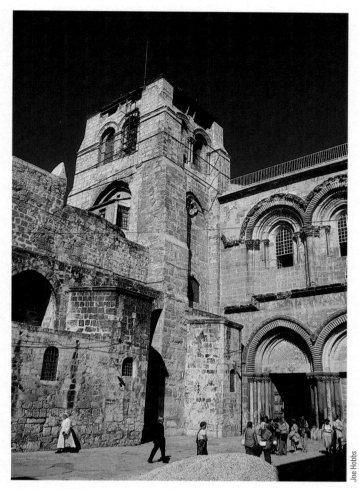

● **Figure 6.20** Jerusalem's Church of the Holy Sepulcher, containing the locations where many Christians believe Jesus Christ was crucified and buried.

a way of life, an understanding of the religious tenets, culture, and diffusion of Islam is vital for appreciating the region's cultural geography.

Islam is a monotheistic faith built on the foundations of the region's earliest monotheistic faith, Judaism, and its offspring, Christianity. Indeed, **Muslims** (people who practice Islam) call Jews and Christians **People of the Book,** and their faith obliges them to be tolerant of these special peoples. Muslims believe that their prophet, Muhammad, was the very last in a series of prophets who brought the Word of God to humankind. Thus they perceive the Bible as incomplete but not entirely wrong—Jews and Christians merely missed receiving the entire message. Muslims do not accept the Christian concept of the divine Trinity (God manifested in the form of the Father, his son Jesus, and the Holy Spirit) and regard Jesus as a prophet rather than as a manifestation of God.

Muhammad was born in 570 C.E. to a poor family in the western Arabian (now Saudi Arabian) city of Mecca. Located on an important north-south caravan route linking the frankincense-producing area of southern Arabia (now Yemen and Oman) with markets in Palestine (now Israel) and Syria, Mecca was a prosperous city at the time. It was also a pilgrimage destination because more than 300 deities were venerated in a shrine there called the **Ka'aba** (the Cube; •Figure 6.21). Muhammad married into a wealthy family and worked in the caravan trade. Muslim tradition holds that when he was about 40 years old, Muhammad was meditating in a cave outside Mecca when the Angel Gabriel appeared to him and ordered him to repeat the words of God that the angel would recite to him. Over the next 22 years, the prophet related these words of God (whom Muslims call **Allah**) to scribes who wrote them down as the **Qur'an** (or **Koran**), the holy book of Islam.

During this time, Muhammad began preaching the new message, "There is no god but Allah," which the polytheistic people of Mecca viewed as heresy. As much of their income depended on pilgrimage traffic to the Ka'aba, they also viewed Muhammad and his small band of followers as an economic threat. They forced the Muslims to flee from Mecca and take refuge in Yathrib (modern Medina), where a largely Jewish population had invited them to settle. There were subsequent skirmishes between the Meccans and Muslims, but in 630, the Muslims prevailed and peacefully occupied Mecca. The Muslims destroyed the idols enshrined in the Ka'aba, which became a pilgrimage center for their one God.

The Ka'aba is Islam's holiest place, and Mecca and Medina are its holiest cities. Jerusalem is also sacred to Muslims. Muslim tradition relates that on his **Night Journey,** the Prophet Muhammad ascended briefly into heaven from the great rock now beneath the Dome of the Rock (the same rock Jews regard as the Foundation Stone). Nearby on the Temple Mount (*al-Haraam ash-Shariif*) is **al-Aqsa Mosque,** a sacred congregational site. The proximity, even the duplication, of holy places between Islam and Judaism came to be the most difficult issue in peace negotiations between Palestinians and Israelis, as explained on page 245 in Module 6.1. 223, 224

After Muhammad's death in 632, Arabian armies carried the new faith far and quickly. The two decaying empires that then prevailed in the Middle East and North Africa—the Byzantine or Eastern Roman Empire, based in Constantinople (Istanbul), and the Sassanian Empire, based in Persia (Iran) and adjacent Mesopotamia (Iraq)—put up only limited military resistance to the Muslim armies before capitulating. Local inhabitants generally welcomed the new faith, in part because administrators of the previous empires had not treated them well, whereas the Muslims promised tolerance. Soon the Syrian city of Damascus became the center of a Muslim empire. Baghdad assumed this role in 750 C.E.

Arab science and civilization flourished in the Baghdad immortalized in the legends of *The Thousand and One Nights*. Important accomplishments and discoveries were made in mathematics, astronomy, and geography. Scholars translated the Greek and Roman classics, and if not for their efforts, many of these works would never have survived to become part of the modern European legacy. It was an age of exploration, when Arab merchants and voyagers visited China and the remote lands of southern Africa. Many important discoveries by the Arab geographers were recorded in Arabic, a language unfamiliar to contemporary

• **Figure 6.21** The black-shrouded cubical shrine known as the Ka'aba (just right of center) in Mecca's Great Mosque is the object toward which all Muslims face when they pray and is the centerpiece of the pilgrimage to Mecca required of all able Muslims. Note the carpet of humanity spread across this image.

Nabeel Turner/Stone/Getty Images

INSIGHTS

Sunni and Shiite Muslims

A schism occurred very early in the development of Islam, and it persists today. The split developed because the Prophet Muhammad had named no successor to take his place as the leader (caliph) of all Muslims. Some of his followers argued that the person with the strongest leadership skills and greatest piety was best qualified to assume this role. These followers became known as **Sunni,** or orthodox, Muslims. Others argued that only direct descendants of Muhammad, specifically through descent from his cousin and son-in-law Ali, could qualify as successors. They became known as **Shia,** or **Shiite,** Muslims.

The military forces of the two camps engaged in battle south of Baghdad at Karbala in 680 D/F/, and in the encounter, Sunni troops caught and brutally murdered Hussein, a son of Ali (Karbala is thus sacred to Shiites, as is nearby al-Najaf, where Ali is buried). The rift thereafter was deep and permanent. The martyrdom of Hussein became an important symbol for Shiites, many of whom still today regard themselves as oppressed peoples struggling against cruel tyrants, including some Sunni Muslims.

Today, only two Muslim countries, Iran and Iraq, have Shiite majority populations. Significant minority populations of Shiites are in Syria, Lebanon, Yemen, and the Arab states of the Persian-Arabian Gulf (see Figure 6.18 on page 222).

264, 268

Europeans, and had to be rediscovered centuries later by the Portuguese and Spaniards. Arab merchants carried their faith on the spice routes to the East. One result, surprising to many today, is that the world's most populous Muslim country is not in the Middle East and North Africa, and its people are not Arabs; it is Indonesia, 5,000 thousand miles (8,000 km) east of Arabia.

Whether in Arabia or Indonesia, whether they be Sunni Muslims or Shiite Muslims (see Insights, page 227), all believers are united in support of the five fundamental precepts, or **Pillars of Islam.** The first of these is the **profession of faith:** "There is no god but Allah, and Muhammad is His Messenger." This expression is often on the lips of the devout Muslim, both in prayer and as a prelude to everyday activities.

The second pillar is **prayer,** required five times daily at prescribed intervals. Two of these prayers mark dawn and sunset. Business comes to a halt as the faithful prostrate themselves before God. Muslims may pray anywhere, but wherever they are, they must turn toward Mecca. There also is a congregational prayer at noon on Friday, the Muslim Sabbath.

The third pillar is **almsgiving.** In earlier times, Muslims were required to give a fixed proportion of their income as charity, similar to the concept of the tithe in the Christian church. Today, the donations are voluntary. Even Muslims of very modest means give what they can to those in need.

The fourth pillar is **fasting** during Ramadan, the ninth month of the Muslim lunar calendar. Muslims are required to abstain from food, liquids, smoking, and sexual activity from dawn to sunset throughout Ramadan. The lunar month of Ramadan falls earlier each year in the solar calendar and thus periodically occurs in summer. In the torrid Middle East and North Africa, that timing imposes special hardships on the faithful, who, even if they are performing manual labor, must resist the urge to drink water during the long, hot days.

The final pillar is the *hajj,* or **pilgrimage to Mecca,** Islam's holiest city. Every Muslim who is physically and financially able is required to make the journey once in his or her lifetime. A lesser pilgrimage may be performed at any time, but the prescribed season is in the 12th month of the Muslim calendar. Those days witness one of earth's greatest annual human migrations as about 3 million Muslims from all over the world converge on Mecca. Hosting these throngs is an obligation the government of Saudi Arabia fulfills proudly and at considerable expense. However, there has been some trepidation in recent years because of the security threat foreign visitors may pose to the host country and because accidents such as stampedes and tent city fires have cost numerous lives. Many pilgrims also visit the nearby city of Medina, where Muhammad is buried. Most Muslims regard the *hajj* as one of the most significant events of their lifetimes. All are required to wear simple seamless garments, and for a few days, the barriers separating groups by income, ethnicity, and nationality are broken. Pilgrims return home with the new stature and title of *hajji* but also with humility and renewed devotion.

All Muslims share the Five Pillars and other tenets, but they vary widely in other cultural practices related to their faith, depending on the country they live in; whether they are from the desert, village, or city; and their education and income. The governments and associated clerical authorities in Saudi Arabia and Iran insist on strict application of **Islamic law** (known as *sharia*) to civil life; in effect, there is no separation between church and state. The Qur'an does not state that women are required to wear veils, but it does urge them to be modest, and it portrays their roles as different from those of men. Clerics in Saudi Arabia insist that women wear floor-length, long-sleeved black robes and black veils in public, that they travel accompanied by a male member of their families, and that they not drive cars. In Egypt, by contrast, Muslim women are free to appear in public unveiled if they choose. However, in most Muslim countries, conservative ideas about the role of women are still very strong: they should be modest, retiring, good mothers, and keepers of the home. The Qur'an portrays women as equal to men in the sight of God, and in principle, Islamic teachings guarantee the right of women to hold and inherit property.

Most Muslim women argue that what others often see as "backward" cultural practices are in fact progressive. For example, they say, their modest dress compels men to evaluate them on the basis of their character and performance, not their attractiveness. They argue that segregation of the sexes in the classroom makes it easier for both women and men to develop their confidence and skills. Sexual assault is rare. A married woman retains her maiden name. These apparent advantages can be weighed against the drawbacks that in many cases women are generally subordinate to men in public affairs and have fewer opportunities for education and for work outside the home.

6.4 Economic Geography

Overall, the Middle East and North Africa is a poor region; per capita GNI PPP for the 23 countries and territories averages only $7,980 (see Table 6.1). This may seem surprising in view of the "rich Arab" stereotype. Only the oil-endowed states around the Persian-Arabian Gulf deserve the reputation for wealth, and by most measures other than per capita wealth, only non-Arab Israel is truly a more developed country (MDC). Israel's prosperity comes from its innovation in computer and other high-technology industries, the processing and sale of diamonds, large amounts of foreign (mostly U.S.) aid, and investment and assistance by Jews and Jewish organizations around the world.

Vital to the industrialized countries as a source of fuels, lubricants, and chemical raw materials, petroleum is one of the world's most important natural resources. A crucial feature of world geography is the concentration of approximately two-thirds of the world's proven petroleum reserves in a few countries that ring the Persian Gulf (Arabian Gulf).

By coincidence, the countries rich in oil tend to have relatively small populations, whereas the most populous nations have few oil reserves; Iran is an exception. All but about 1 percent of the Persian Gulf (Arabian Gulf) oil region's proven reserves of crude oil are located in Saudi Arabia, Iraq, Kuwait, Iran, and the United Arab Emirates (UAE), with smaller reserves in Oman and Qatar. Saudi Arabia, by far the world leader in reserves, has about 20 percent of the proven crude oil reserves on the globe. Canada is in second place with almost 14 percent of the world's proven oil; the next four are all along the Persian Gulf (Arabian Gulf): Iran (10 percent), Iraq (9 percent), Kuwait (8 percent), and the UAE (8 percent). Venezuela has 6 percent of world reserves, Russia another 5 percent, and the world's largest oil consumer, the United States, only 2 percent.

In the Gulf region, the great thickness of the region's oil-bearing strata and high reservoir pressures have made it possible to secure an immense amount of oil from a small number of wells. The productivity of each well makes each barrel inexpensive to extract and makes it simple to increase

or reduce production quickly in response to world market conditions. Gulf oil is thus "cheap" to produce unless expenditures to maintain huge military forces to defend it are factored in. In 2007, the United States maintained about 190,000 military personnel in the region (including Iraq) at a cost of $100 billion per year. Some analysts—including a former U.S. Navy secretary—like to calculate the "real price" of oil with military expenditures factored in. If one accepts the premise that maintaining a costly military presence in the region is part of the United States' long-term strategy to maintain the flow of Gulf oil, that oil is no longer cheap by any measure. With these external costs factored in, Gulf oil would be priced at over $250 per barrel in September 2007, not the roughly $80 per barrel that was the world market price at that time. 63

Production, export, and profits of Middle Eastern and North African oil were once firmly in the hands of foreign companies. That situation changed after 1960 when most of the Gulf countries and other exporting nations formed the **Organization of Petroleum Exporting Countries (OPEC)** with the aim of taking joint action to demand higher profits from oil. It changed again after 1972 when the oil-producing countries began to nationalize the foreign oil companies. OPEC was relatively obscure until the Arab-Israeli war of 1973, after which the organization began a series of dramatic price increases. In 1980, the organization's price reached $37 (U.S.) per barrel, compared with $2 a barrel in early 1973. Adjusted for inflation, that $37 would have been the equivalent of $100 per barrel in 2007, so the price hikes were truly titanic.

These events had enormous repercussions for the world economy. Immense wealth was transferred from the more developed countries to the OPEC countries to pay for indispensable oil supplies. The skyrocketing cost of gasoline and other oil products helped cause serious inflation in the United States and many other countries and contributed to the 1973 **energy crisis** in the United States. Desperately poor, less developed countries (LDCs) found that high oil prices not only hindered the development of their industries and transportation but also reduced food production because of high prices for fertilizer made from oil and natural gas. In the Gulf countries, the oil bonanza produced a wave of spending on military hardware, showy buildings, luxuries for the elite, and ambitious development projects of many kinds. Per capita benefits to the general populace were greatest in the Arabian Peninsula, where small populations and immense inflows of oil money made possible the abolition of taxes, the establishment of comprehensive social programs, and heavily subsidized amenities such as low-cost housing and utilities, including freshwater distilled from the salty Gulf in desalination plants.

Then, as the 1980s began, the era of continually expanding OPEC oil production, sales, and profits seemed to come to an end. After 1973, the high price of oil stimulated oil development in countries outside OPEC. Oil conservation measures such as a shift to more fuel-efficient vehicles and furnaces were instituted. Substitution of cheaper fuels for oil increased. Coal replaced oil in many electricity-generating stations. Oil refineries were converted to make gasoline from cheaper "heavy" oils rather than the more expensive "light" oils previously used. Meanwhile, the world entered a period of economic recession due in part to high oil prices. Decreased business activity reduced the demand for oil. Profits of the world oil industry (and taxes paid to governments) were severely cut, large numbers of refineries had to close, and much of the world tanker fleet was idled.

Oil prices rose temporarily in 1990–1991 when the flow of Iraqi and Kuwaiti oil was cut off following Iraq's military takeover of Kuwait, but the prices soon fell again. In the 1990s, Saudi Arabia and other oil-rich Gulf states adopted economic austerity measures for the first time. However, the immense oil and gas reserves still in the ground guaranteed that the Gulf region would continue to have a major long-term impact on the world and would remain relatively prosperous as long as these finite resources are in demand in the MDCs. The lasting economic clout of the region was apparent early in the 21st century as the price of oil once again climbed to record levels, due especially to OPEC decisions to reduce production, and to explosive economic growth and energy demands in China and India.

There are of course other resources and industries in the region's economic geography—remittances (earned income) sent home by guest workers in the oil-rich countries, revenues from ship traffic through the Suez Canal, and exports of cotton, rice, and other commercial crops, for example—but oil dominates the region's economy and is central to the global economy. The economies of the respective countries are described in more detail in Module 6.1. Here the focus remains on Middle Eastern oil and its crucial role in geopolitical affairs.

6.5 Geopolitical Issues

This has long been a vital region in world affairs and a target of outside interests. Its strategic crossroads location often has made it a cauldron of conflict. From very early times, overland caravan routes, including the famous Silk Road, crossed the Middle East and North Africa with highly prized commodities traded between Europe and Asia. The security of these routes was vital, and countries at either end could not tolerate any threat to them. In more recent times, geopolitical concerns have focused on narrow waterways, access to oil, access to freshwater, and terrorism.

Chokepoints

One of the striking characteristics of the geography of the Middle East and North Africa is how many seas border and penetrate the region. In many cases, these seas are connected to one another though narrow straits and other passageways. In geopolitical terms, such constrictions are known as **chokepoints**—strategic narrow passageways on land or sea that may be easily closed off by force or even the threat of force (•Figure 6.22). Chokepoints must be unimpeded if world commerce is to carry on normally. Keeping them open is therefore usually one of the top priorities of regional and external governments. Similarly, closing them is a priority to a combatant nation or a terrorist entity seeking to gain a strategic advantage. Many notable events in military history and the formation of foreign policy in the Middle East focus on these strategic places.

One of the world's most important chokepoints is the Suez Canal, which opened in 1869. Slicing 107 miles (172 km) through the narrow Isthmus of Suez, the British- and French-owned Suez Canal linked the Mediterranean Sea with the Indian Ocean, saving cargo, military, and passenger ships a journey of many thousands of miles around the southern tip of Africa (•Figure 6.23). Keeping the Suez Canal in friendly hands was one of Britain's major military concerns during World War II. That objective led to a hard-fought and eventually successful British and Allied campaign against Nazi Germany in North Africa. Egyptian President Gamal Abdel Nasser's nationalization of the canal in 1956 led immediately to a British, French, and Israeli invasion of Egypt and a conflict known as the **Suez Crisis**

• **Figure 6.22** Chokepoints in the Middle East and North Africa.

• **Figure 6.23** The strategically vital Suez Canal zone saw bitter fighting in the Middle East wars of 1956, 1967, and 1973.

GEOGRAPHY OF ENERGY

Middle Eastern Oil Pipelines

The oil-exporting countries of the Middle East, often with the financial support of the United States and other leading oil-consuming countries, have invested enormous resources to ensure the safe passage of oil to world markets. The Middle East is not landlocked like the oil-rich "stans" of central Asia are, but some of the same kinds of geopolitical issues are involved in oil export planning. These concerns often point toward pipelines as the best means of routing oil shipments (●Figure 6.C). Pipelines are attractive because they shorten the time and expense involved in seaborne transport, and they bypass chokepoints. However, they are vulnerable to disruption. Weapons as small as grenades can disrupt supplies. A major challenge is how to route a pipeline so that it will not cross through a potential enemy's territory, and another is how to maintain friendships with the countries the oil crosses.

One of the first major pipelines in the region was the 1,100-mile (1,760-km) Trans-Arabian Pipeline (Tapline), leading from the oil-producing eastern province of Saudi Arabia to a terminal on the Mediterranean coast of Lebanon. It has the advantage of bypass-

ing three chokepoints (Strait of Hormuz, Bab el-Mandeb, and Suez Canal). However, at the time it was completed (1950), it could not have been anticipated that it would cross two of the worst conflict zones in the Middle East: the Golan Heights of Syria (which fell to Israel in the 1967 war) and southern Lebanon (a major theater of the Lebanese civil war of the 1970s and the Israeli–Lebanese Shiite struggles of the 1980s and 1990s). Tapline was knocked out early by its unfortunate geography.

Another Middle East conflict, the 1967 Six-Day War between Israel and its neighbors, led to the closure of the Suez Canal (for about a decade) and thus to the construction of two new pipelines to bypass Suez: one across southern Israel from the Gulf of Aqaba to the Mediterranean Sea and another across Egypt from the Gulf of Suez to the Mediterranean (thus its acronym, SUMED).

No Middle Eastern country has been more dependent on pipelines than Iraq, and none has so systematically experienced the liabilities of pipelines. The Iran-Iraq war of 1980–1988 had major impacts on pipeline geography. Saudi Arabia supported Iraq in

● **Figure 6.C** Principal oilfields and pipelines in the heart of the Middle East. Vulnerable chokepoints and volatile political relations have led to the construction and often indirect routing of many pipelines.

GEOGRAPHY OF ENERGY

Middle Eastern Oil Pipelines *continued*

the war and so tried to ensure it could get oil to market without being threatened by Iran; this meant bypassing the Strait of Hormuz. That led to the construction of Petroline (opened in 1981), the east-west pipeline across the Arabian Peninsula. Iraq built a pipeline to ship some of its oil south to Petroline, thereby bypassing the Strait of Hormuz as well. Beginning in 1961, Iraq was also able to ship oil through pipelines almost due west though Syria to the Mediterranean Sea. But when it attacked Iran in 1980, Iraq lost that route because Syria was Iran's ally.

Iraq quickly responded to that setback by building a new pipeline leading almost due north to bypass Syrian territory and then making a sharp 90-degree turn westward through southern Turkey to the Mediterranean Sea. But in 1990, Iraq attacked Kuwait, threatened Saudi Arabia, and prompted a U.S.-led counterattack. Saudi Arabia responded by closing the southern link with Petroline. Turkey, a NATO ally of the United States, reacted by closing the northern link. The United Nations responded by issuing strict controls on Iraqi oil exports. In sum, by its military actions, Iraq did almost everything imaginable to deprive itself of oil export capabilities.

---

and the **Arab-Israeli War of 1956.** Egypt effectively won that war when international pressure caused the invading forces to withdraw, leaving the canal in Egypt's hands.

242 Nearby, another chokepoint played a critical role in the most important Arab-Israeli war, the Six-Day War of 1967. One of the events that precipitated the war was President Nasser's closure of the Strait of Tiran (at the southern end of the Gulf of Aqaba) to Israeli shipping. Israel had won the right of navigation through the strait after its war of independence and would not accept its closure; therefore, it attacked Egypt.

On either side of the Arabian Peninsula are two more critical chokepoints. One is extraordinarily vital to the world's economy: the Strait of Hormuz, connecting the Persian-Arabian Gulf with the Gulf of Oman and the Arabian Sea. Much of the world's oil supply passes through here in the holds of giant supertanker ships. Closure of the Strait of Hormuz would have devastating effects on the world's industrial and financial systems. Iran's plans to station Chinese-made missiles on the Strait of Hormuz and thus threaten international oil shipments led to a new level 264 of U.S. involvement late in the Iran-Iraq War of 1980–1988. That war also prompted a flurry of new oil pipeline construction designed to bypass the Strait of Hormuz chokepoint (see Geography of Energy, page 230). One of these pipelines, the Petroline route running east-west across the Arabian Peninsula, was built to bypass another important chokepoint, the Bab el-Mandeb, which connects the Red Sea with the Gulf of Aden and the Indian Ocean.

Turkey controls two more chokepoints that together are known as the Turkish Straits. The northernmost strait is the Bosporus, which cleaves the city of Istanbul into western (European) and eastern (Asian) sides, and the southern strait is called the Dardanelles. Their security has long been critical to the successful passage of goods between Europe and Asia and even more so to the successful passage of vessels between Russia (and the Soviet Union) and the rest of the world. Throughout the 20th century, one of the Soviet Union's constant strategic priorities was the right of navigation through the Turkish Straits.

Finally, it should be noted that the Strait of Gibraltar, connecting the Mediterranean Sea with the Atlantic Ocean, is also a chokepoint. Here, too, maintaining and monitoring the flow of maritime traffic have long been important concerns. Britain's insistence on maintaining control over its enclave of Gibraltar, decades after the decolonization of 254–most of the world, is an excellent indicator of how critical 255 this chokepoint is.

Access to Oil

The Suez Canal and other chokepoints, the cotton of the Egypt's Nile Delta, and the strategic location of the region were important during colonial times and have remained so ever since. But oil has been and will remain (as long as fossil fuels drive the world's economies) what keeps the rest of the world interested in the Middle East and North Africa. The region's oil flows to many countries, but most of it is marketed in western Europe and Japan. The United States also imports large amounts of Gulf oil, but has a smaller relative dependence on this source than Japan and Europe do; 22 percent of its imported oil came from the Gulf in 2007. However, the Gulf region is very important to the United States because of the heavy dependence of close American allies on Gulf oil and because of the importance of the oil as a future reserve. American companies are also heavily involved in oil operations and oil-financed development in the Gulf countries. Gulf "petrodollars" are spent, banked, and invested in the United States, contributing significantly to the U.S. economy. Maintaining a secure supply of Gulf oil has therefore been one of the long-standing pillars of U.S. policy in the Middle East.

The United States has long had a precarious relationship with the key players in the Middle Eastern arena. The United States has pledged unwavering support for Israel while at the same time courting Israel's traditional enemies such as oil-rich Saudi Arabia. That Arab kingdom and its neighbors around the Persian-Arabian Gulf possess more than 60 percent of the world's proven oil reserves. Thus they are vital to the long-term economic security of the Western industrial powers and Japan. The United States and its Western allies made it clear they would not tolerate any disruption of access to this supply when Iraqi troops, directed by Iraqi President Saddam Hussein, occupied Kuwait on August 2, 1990. U.S. President George Bush drew a "line in the sand," proclaiming, "We cannot permit a resource so vital to be dominated by one so ruthless—and we won't." In what came to be known as the **Gulf War,** the United States and a coalition of Western and Arab allies mounted a massive array of military might that ousted the Iraqi invaders within months and secured the vital oil supplies for Western markets. When the United States invaded and occupied Iraq in 2003, ostensibly to extinguish Iraq's ability to develop and deploy **weapons of mass destruction (WMD),** including chemical, biological and nuclear weapons, many critics insisted that this was just another example of America's determination to control Middle Eastern oil.

The Gulf War was not the first time the United States expressed its willingness to use force to maintain access to Middle Eastern oil. In the wake of the revolution in Iran in 1979, the Soviet Union invaded neighboring Afghanistan. U.S. military analysts feared that the Soviets might use Afghanistan as a launch pad to invade oil-rich Iran. The United States deemed this prospect unacceptable, and President Jimmy Carter issued the policy statement that came to be known as the **Carter Doctrine:** the United States would use any means necessary to defend its vital interests in the region. "Vital interests" meant oil, and "any means necessary" meant that the United States was willing to go to war with the Soviet Union, presumably nuclear war, to defend those interests.

Middle Eastern wars had already become proxy wars for the superpowers, with oil always looming as the prize. In the 1967 and 1973 Arab-Israeli wars, for example, Soviet-backed Syrian forces fought U.S.-backed Israeli troops. American support of Israel in this war prompted Arab members of OPEC to impose an **oil embargo,** refusing to sell their oil to the United States, thus precipitating the nation's first energy crisis. During the 1973 war, the United States put its forces on an advanced state of readiness to take on the Soviets in a nuclear exchange if necessary. All of these events illustrate that the Middle East and North Africa form, as eastern Europe did, a shatter belt—a large, strategically located region composed of conflicting states caught between the conflicting interests of great powers.

Access to Freshwater

Some of the most serious geopolitical issues in the Middle East and North Africa relate to **hydropolitics,** or political leverage and control over water. In this arid region, where most water is available either from rivers or from under-

ground aquifers that cross national boundaries, control over water is an especially difficult and potentially explosive issue. An estimated 90 percent of the usable freshwater in the Middle East crosses one or more international borders.

Water is one of the most problematic issues in the Palestinian-Israeli conflict. Freshwater aquifers underneath the West Bank supply about 40 percent of Israel's water. Palestinians point to Israel's control over West Bank water as one of the most troubling elements of its occupation. The average Jewish settler on the West Bank uses 74 gallons (278 l) per day, whereas the average West Bank Palestinian uses 19 (72 l). (The World Health Organization calculates that 13 gallons or 50 liters per person per day is needed for minimal health and sanitation standards.) Israeli policies prohibit Palestinians from increasing their water usage. Many Israeli policymakers insist that water resources in the West Bank must remain under strict Israeli control and on these grounds oppose the creation of a Palestinian state in the West Bank. Critically, it is estimated that the West Bank aquifers will not contain enough water to support the region's population at current levels of consumption for more than a few more years (even taking into account anticipated replenishment from rainfall).

More promisingly, Jordan and Israel are working on agreements to share waters from the Jordan River (which forms a portion of their common border) and its tributary, the Yarmuk River. They are discussing a joint venture to build the **Red-Dead Peace Conduit,** which would send seawater from the Gulf of Aqaba to the Dead Sea via a network of canals and pipelines (see Figure 6.1.4, page 244). This would replenish the Dead Sea, which has retreated by about 3 feet (c. 1 m) per year for the last 25 years. The gravity flow of seawater to the Dead Sea would run generators to produce electricity, some of which would be used to desalinate the water. The two nations, and potentially the Palestinian Authority, could share this water and power.

Water is a critical issue blocking a peace treaty between Israel and Syria. If Syria were to recover all of the Golan Heights (which it lost to Israel in the 1967 war), it would have shorefront on Lake Kinneret and therefore, presumably, rights to use its water. That prospect is unacceptable to Israel. This body of water, also known as Lake Tiberias and the Sea of Galilee, is Israel's principal supply of freshwater, feeding the National Water Carrier system that transports water south to the Negev Desert. In occupying the Golan Heights, Israel also controls some of the northern bank of the Yarmuk River on the border with Jordan. For many years, Israel has stated it would never allow Syria and Jordan to construct their proposed Unity Dam, intended to store waters of the Yarmuk River to be shared between those countries. Israel thus implied it would bomb the dam rather than allow it to deprive Israel of Jordan River water.

A useful way to think about the geography of hydropolitics is in terms of **upstream** and **downstream countries.** Simply because water flows downhill, an upstream country is usually able to maximize its water use at the expense of a downstream country (a situation described as a *zero-sum game,* where any gain by one party represents an equivalent loss to the other). However, the situation between Israel and the countries upstream on the Yarmuk shows that this is not always true. Although downstream, Israel is far more powerful militarily and can use the threat of force to wrest more water out of the system.

Historically, the same has been true of Egypt. It is the ultimate downstream country, at the mouth of a great river that runs through five countries and sustains about 160 million people (a population that is expected to double in about 20 years; •Figure 6.24). However, it has long been the strongest country in the Nile Basin and has threatened to use its greater force if it does not get the water it wants. In 1926, when the British ruled Egypt and many other colonies in Africa, 10 countries located on the Nile or its tributaries upstream of Egypt were compelled to sign the **Nile Water Agreement.** This guaranteed Egyptian access to 56 billion cubic meters of the Nile's water, or fully two-thirds of its 84 billion cubic meters—even though barely a drop of the Nile's waters actually originates in Egypt. The treaty forbids any projects that might threaten the volume of water reaching Egypt, prohibits use of Lake Victoria's water without Egypt's permission, and gives Egypt the right to inspect the entire length of the Nile to ensure compliance. In recent years, however, one country after another has defied the treaty, calling it an outmoded legacy of colonialism. Kenya and Tanzania have plans to build pipelines to carry Lake Victoria waters to thirsty towns and villages inland. Uganda is building its controversial Bujagali Dam on the Nile, mainly for hydroelectricity production. With Chinese assistance, Ethiopia is building the huge Tekaze Dam, for hydropower and irrigation, on a tributary of the Blue Nile. Sudan is building the Merowe (Hamdab) and Kajbar Dams on its northern stretch of the Nile. Predictably, Egypt has had a bellicose response to these developments. For exam-ple, Egypt called Kenya's stated intention to withdraw from the Nile Water Agreement an "act of war."

Meanwhile, Egypt's demands on Nile waters are increasing. Egypt recently built the multibillion-dollar Toshka Canal, which transports water from Lake Nasser over a distance of 100 miles (160 km) to the Kharga Oasis of the Western Desert (see Figure 6.24). Proponents of the canal insist that it will result in the cultivation of nearly 2,350 square miles (c. 6,100 sq km) of "new" land and provide a living for hundreds of thousands of people. Critics argue that it is a waste of money and that salinization and evaporation will take a huge toll on the cultivated land and the country's water supply.

Turkey—the source of four-fifths of Syria's water and two-thirds of Iraq's—exercises its upstream advantage on the Tigris and Euphrates Rivers (•Figure 6.25). Turkey's position has long been that water in Turkey belongs to Turkey, just as Saudi Arabian oil belongs to Saudi Arabia. Not surprisingly, downstream Syria and Iraq reject this position and are distraught by the diminished flow and quality of water resulting from Turkey's comprehensive Southeast Anatolia Project. When completed, the project is expected to reduce Syria's share of the Euphrates waters by 40 percent and Iraq's by 60 percent. Also increasing the likelihood of serious future tension is a history of strained relations among Turkey, Syria, and Iraq, accompanied by the fact that no commonly accepted body-of-water law governs the allocation of water in such international situations.

Turkish leaders have said they will never use water as a political weapon, but Turkey has already wielded water to its advantage. In 1987, for example, Turkey increased the Euphrates flow into Syria in exchange for a Syrian pledge to stop support of Kurdish rebels inside Turkey. Turkey now says it wants to use water to promote peace in the Middle East by shipping it in converted supertankers for sale to such thirsty (and therefore potentially combative) countries as Israel, Jordan, Saudi Arabia, Libya, and Algeria.

270

316, 353

• **Figure 6.24** Recent and proposed water developments in the Nile Basin. Note that this map is oriented such that north is at the left rather than at the top.

• **Figure 6.25** The Tigris and Euphrates Rivers rise in Turkey, giving this non-Arab country control over a resource vital to the lives of millions of Arabs in downstream Syria and Iraq. This waterfall is on a tributary of the Tigris in far eastern Turkey.

Terrorism

Viewed from the perspective of the U.S. administrations of President George W. Bush, almost all of the geopolitical issues related to this region—oil, economic development, trade, aid, the Arab-Israeli conflict, and more—were subsumed beneath the broader rubric of the president's declared **war on terror.** Even the 2003 invasion of Iraq was explained in part as essential to that new kind of war (see that discussion in Module 6.1, pages 265–266). How the United States pursues that war in the coming years will have enormous effects on societies and economies in the Middle East and North Africa, South Asia, and perhaps in the United States as well.

The terrorists pursued by the United States are almost without exception Islamist militants—best known as **Islamists**—and so it is useful to understand the nature of Islamic "fundamentalism" and radicalism. Not all Islamists are militant or terrorist, but they all reject what they view as the materialism and moral corruption of Western countries and the political and military support these countries lend to Israel. Both Sunni and Shiite Muslims have advanced a wide range of Islamic movements, notably in Iran, Lebanon, Egypt, Afghanistan, Sudan, and Algeria.

Although nominally religious, the more radical of these movements have political and cultural aims, particularly the destabilization or removal of U.S. and Israeli interests in the region and abroad. In 1993, followers of the radical Egyptian cleric Sheikh Umar Abdel-Rahman bombed New York City's World Trade Center as a protest against American support of Israel and Egypt's pro-Western government. In an attempt to destabilize and replace Egypt's government, which it viewed as an illegitimate regime too supportive of the United States, another Egyptian Islamist group attacked and killed foreign tourists in Egypt in the 1990s. In the 1980s, members of the pro-Iranian **Hizbullah,** or Party of God, in Lebanon kidnapped foreign civilians to use as bargaining chips for the release of comrades jailed in other Middle Eastern countries.

Within Israel and the autonomous Palestinian territories of the West Bank and Gaza Strip, Palestinian members of **Hamas** (an Arabic acronym for the Islamic Resistance Movement) and another organization called the al-Aqsa Martyrs Brigade carried out terrorist attacks on Israeli civilians and soldiers in an effort (apparently successful) to derail implementation of the peace agreements reached in the 1990s between the Israeli government and the Palestine Liberation Organization (PLO).

In Algeria, years of bloodshed followed the government's annulment of 1991 election results that would have given the **Islamic Salvation Front (FIS)** majority control in the parliament. Muslim sympathizers carried the battle to France, bombing civilian targets in protest against the French government's support for the Algerian regime.

In Western capitals, concern about "state-sponsored terrorism" has long focused on Iran, one of the three countries (along with Iraq and North Korea) that President Bush in 2002 dubbed the **"axis of evil."** Iran has extended direct or clandestine assistance to a variety of Islamist terrorist groups, including Hizbullah and Hamas. There is great concern about Iran's nuclear weapons potential because such weapons might find their way to terrorist groups or be delivered by Iran itself on its own missiles against Israel or another target. Iran denies that it is developing the weapons, but U.S. and Israeli intelligence agencies believe Iran is on course to create them some time before 2016. Israeli officials have said publicly that Israel (which has its own nuclear weapons) regards Iran as an "existential threat" and will prevent that development, presumably by an air strike like the one Israel carried out against Iraq's nuclear facility in 1981. The United States, Israel, and the United Nations' **International Atomic Energy Agency (IAEA)** are particularly watchful of Iran's acquisition of centrifuges, which are used to enrich uranium. The enrichment process is necessary for the production of fuel for civilian use in nuclear power production, which Iran is permitted to do, but can also be expanded for the production of nuclear weapons.

Prior to a late 2007 release of a U.S. intelligence finding that Iran suspended its nuclear weapons development program in 2003, the United States was weighing two choices. It could engage in diplomatic dialogue with Iran, as it had done with North Korea, offering a package of incentives and security guarantees in exchange for Iran's pledge to halt weapons development. This is the option that Russia, China, and European allies of the United States favored. The United States and Iran do share some common interests, notably stability in Iraq; Iran does not want to deal with strong new Sunni or Kurdish states there or with the tide of refugees that more Iraqi conflict would send its way. Alternatively, the United States could launch military strikes on Iran's suspected nuclear weapons development sites, beginning at Natanz. Most analysts believed that the military option would have far-ranging and dangerous consequences. Iran's nuclear program would be set back a few years, but the country would portray itself as a victim of U.S. aggression, galvanizing support across the Muslim world (even among Iran's traditional enemies, the Sunni Arabs), perhaps inciting its Hizbullah allies in Lebanon and Hamas allies in the Palestinian territories to attack Israel, encouraging al-Qa'ida-style terrorism against the West, and discouraging Iranians (particularly the younger ones) who favor peaceful relations with the United States. A military strike would probably also lead to a sharp reduction in oil exports from the Gulf, perhaps triggering a global economic slowdown. Russia and China would meanwhile have an opportunity to enhance their geopolitical interests by offering at least diplomatic support to Iran throughout such a crisis. Both are already gaining more leverage in oil-rich Iran.

In 1998, the world began to hear about Osama bin Laden, a former Saudi businessman living in exile in Afghanistan, whose **al-Qa'ida** organization bombed U.S. embassies in Kenya and Tanzania as part of a worldwide armed struggle, or *jihad,* against American imperialism and immorality (see Geography of Terrorism, pages 236–237). Bin Laden's organization was also responsible for the 2000 bombing of the American naval destroyer U.S.S. *Cole* in Yemen's harbor of Aden. However shocking those assaults were, they pale in comparison to al-Qa'ida's attacks against targets in the United States on September 11, 2001. In the

most ferocious terrorist actions ever undertaken to that date, members of al-Qa'ida "cells" (small groups of agents) in the United States hijacked four civilian jetliners and succeeded in piloting two of them into the twin towers of New York City's World Trade Center and one into the Pentagon, home of the U.S. Department of Defense, outside Washington, D.C. More than 3,000 people, mainly civilians, perished. Bin Laden and his followers cheered the carnage as justifiable combat against an infidel nation whose military troops occupied the holy land of Arabia, where Mecca and Medina are located.

As the U.S. counterattacked, ousting al-Qa'ida from its bases in Afghanistan and cracking down hard on its leadership around the world, al-Qa'ida evolved into a far more geographically diffuse organization. It carried out or supported attacks in Indonesia, Morocco, Algeria, Tunisia, Turkey, Spain, Uzbekistan, Egypt, Saudi Arabia, Jordan, and the United Kingdom (•Figure 6.26). A succession of al-Qa'ida videotapes and audiotapes promised an unrelenting and costly continuation of *jihad* against the United States and its allies. It will be many years before Americans in particular might be able to emerge from the shadow of this threat. Al-Qa'ida clearly will not hesitate to use the most devastating weapons, even against large numbers of civilians; this com-

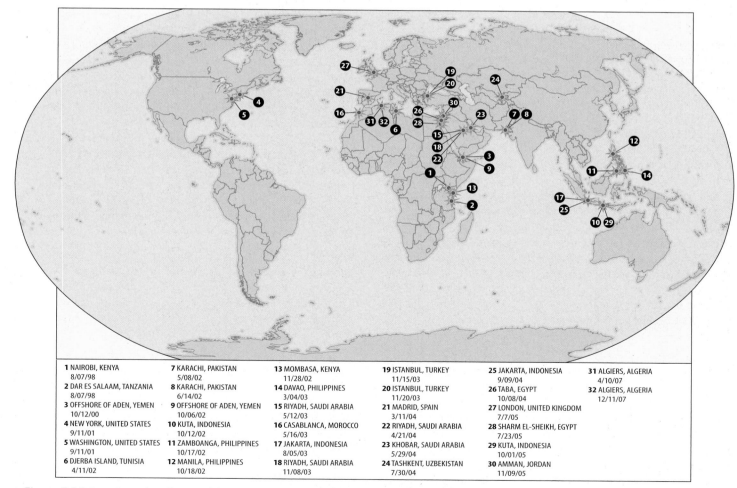

1 NAIROBI, KENYA 8/07/98	**7** KARACHI, PAKISTAN 5/08/02	**13** MOMBASA, KENYA 11/28/02	**19** ISTANBUL, TURKEY 11/15/03	**25** JAKARTA, INDONESIA 9/09/04	**31** ALGIERS, ALGERIA 4/10/07
2 DAR ES SALAAM, TANZANIA 8/07/98	**8** KARACHI, PAKISTAN 6/14/02	**14** DAVAO, PHILIPPINES 3/04/03	**20** ISTANBUL, TURKEY 11/20/03	**26** TABA, EGYPT 10/08/04	**32** ALGIERS, ALGERIA 12/11/07
3 OFFSHORE OF ADEN, YEMEN 10/12/00	**9** OFFSHORE OF ADEN, YEMEN 10/06/02	**15** RIYADH, SAUDI ARABIA 5/12/03	**21** MADRID, SPAIN 3/11/04	**27** LONDON, UNITED KINGDOM 7/7/05	
4 NEW YORK, UNITED STATES 9/11/01	**10** KUTA, INDONESIA 10/12/02	**16** CASABLANCA, MOROCCO 5/16/03	**22** RIYADH, SAUDI ARABIA 4/21/04	**28** SHARM EL-SHEIKH, EGYPT 7/23/05	
5 WASHINGTON, UNITED STATES 9/11/01	**11** ZAMBOANGA, PHILIPPINES 10/17/02	**17** JAKARTA, INDONESIA 8/05/03	**23** KHOBAR, SAUDI ARABIA 5/29/04	**29** KUTA, INDONESIA 10/01/05	
6 DJERBA ISLAND, TUNISIA 4/11/02	**12** MANILA, PHILIPPINES 10/18/02	**18** RIYADH, SAUDI ARABIA 11/08/03	**24** TASHKENT, UZBEKISTAN 7/30/04	**30** AMMAN, JORDAN 11/09/05	

• **Figure 6.26** Locations of terrorist attacks attributed to al-Qa'ida and affiliated organizations, 1998–2007.

GEOGRAPHY OF TERRORISM

What Does al-Qa'ida Want?

This is an important question. Perhaps if an answer could be found, there would be a way either to defeat this organization or to address the root causes of its existence in such a way that it would no longer have a reason to exist. Here we will look at the question mainly from the inside, examining what al-Qa'ida says it wants. The answer may be fundamentally a geographic one.

To begin, it is useful to know what al-Qa'ida is. Al-Qa'ida ("the Base" in Arabic) is a transnational organization that seeks to unite Islamist militant groups worldwide in a common effort to achieve its goals. An Islamist or *jihadist* organization is one that employs Islamic faith, culture, and history to legitimize its philosophy and actions.* The principal sources Islamists cite are the Qur'an, the Hadith (sayings of the Prophet Muhammad), and the writings of earlier Islamic militants such as Ibn Taymiyya and Sayyid Qutb. According to Rohan Gunaratna, the leading academic authority on the organization, al-Qa'ida is "above all else a secret, almost virtual, organization, one that denies its existence in order to remain in the shadows."† This desire for secrecy explains why al-Qa'ida seldom takes direct responsibility for terrorist acts. Instead, these are attributed to the other names and identities employed by al-Qa'ida, particularly the "World Islamic Front for the Jihad against the Jews and the Crusaders," a coalition of seven Islamist militant groups (three Egyptian, two Pakistani, one Bangladeshi, and one Afghan). Osama bin Laden, formerly a Saudi national, emerged as the figurehead of this organization.

Al-Qa'ida certainly is a terrorist organization if one employs the U.S. Department of State's definition of **terrorism**: "premeditated, politically motivated violence perpetrated against noncombatant targets by subnational groups or clandestine agents, usually intended to influence an audience."‡ Ayman al-Zawahiri, al-Qa'ida's second in command, called for an escalation of attacks with "the need to inflict the maximum casualties against the opponent, for this is the language understood by the West, no matter how much time and effort such operations take."§ Osama bin Laden, as evident in his own words, has called for the killing of American and other Western civilians. Gunaratna regards al-Qa'ida as an extremely unusual terrorist group in the category he calls "apocalyptic," one that believes "it has been divinely ordained to commit violent acts, and likely to engage in mass casualty, catastrophic terrorism."** He warns that al-Qa'ida "will have no compunction about employing chemical, biological, radiological and nuclear weapons against population centers."††

But why? Why do they want to kill Western, and particularly American, civilians? Al-Qa'ida leaders have been very explicit in their rationale. Osama bin Laden explained that the main reason for the terrorist attacks is to convince the United States that it should withdraw its military forces and other interests from the Islamic Holy Land. In the following statement, the "Land of the Two Holy Places" means Saudi Arabia. The two holy places are the Saudi Arabian cities of Mecca (site of the Ka'aba, and Islam's most sacred city) and Medina (Islam's second holiest place, where the Prophet Muhammad is buried). The Dome of the Rock is the shrine in Jerusalem, described earlier, containing the sacred rock from which the Prophet Muhammad was said to have ascended into heaven on the Night Journey.

> The Arabian Peninsula has never—since God made it flat, created its desert, and encircled it with seas—been stormed by any forces like the crusader armies spreading in it like locusts, eating its riches and wiping out its plantations. . . . The latest and greatest of these aggressions incurred by the Muslims since the death of the Prophet . . . is the occupation of the Land of the Two Holy Places—the foundation of the house of Islam, the place of the revelation, the source of the message, and the place of the noble Ka'aba, the *qibla* [direction faced during prayer] of all Muslims—by the armies of the American crusaders and their allies. We bemoan this and can only say "No power and power acquiring except through Allah." . . . To push the enemy—the greatest *kufr* [infidel]—out of the country is a prime duty. No other duty after Belief is more important than this duty. Utmost effort should be made to prepare and instigate the *umma* [Islamic community] against the enemy, the American-Israeli alliance—occupying the country of the two Holy Places . . . to the al-Aqsa mosque in Jerusalem. . . . The crusaders and the Jews have joined together to invade the heart of Dar al-Islam—the Abode of Islam: our most sacred places in Saudi Arabia, Mecca and Medina, including the prophet's mosque and Dome of the Rock in Jerusalem, *al-Quds*.‡‡

In what was arguably his most important policy statement, announcing the formation of the Islamic World Front for the Jihad against the Jews and the Crusaders, bin Laden used the occupation of Islamic sacred space as the principal justification for war against the United States:

246

bination is in fact its tactical priority. There is also growing consensus that al-Qa'ida is no longer a solitary severe threat to U.S. interests; instead, the *jihadist* cause has spread to numerous smaller or less renowned groups that are also are keen to take on the West.

In 9/11 and other atrocities, a tiny minority of Muslims carried out terrorist actions that the great majority of Muslims condemned. Islamic scholars and clerics pointed out in each case that the murder of civilians is prohibited in

Islamic law and that the attacks had no legitimate religious grounds. Mainstream Islamic movements are not military or terrorist organizations but have distinguished themselves through public service to the needy and through encouragement of strong moral and family values. For most Muslims, the growing Islamist trend means a reembrace of traditional values like piety, generosity, care for others, and Islamic legal systems, which have proved effective for centuries. For them, these values pose a reasonable alternative

GEOGRAPHY OF TERRORISM

What Does al-Qa'ida Want? *continued*

In compliance with God's order, we issue the following *fatwa* [religious injunction] to all Muslims: The ruling to kill the Americans and their allies—civilians and military—is an individual duty for every Muslim who can do it in any country in which it is possible to do it, in order to liberate the al-Aqsa mosque and the holy mosque [in Mecca] from their grip, and in order for their armies to move out of all the lands of Islam, defeated and unable to threaten any Muslim.[§§]

In an earlier *fatwa* (1996), bin Laden laid out these goals for al-Qa'ida: to drive U.S. forces out of the Arabian Peninsula, overthrow the Saudi government, and liberate the holy places of Mecca and Medina. Overthrowing the Saudi government and the other autocratic dynasties of the Persian Gulf (Arabian Gulf) region, along with other secular and pro-Western regimes of the Middle East—notably Mubarak's Egypt—is a theme that is very often articulated in al-Qa'ida ideology.

What is al-Qa'ida's ultimate goal? Is the organization satisfied now that U.S. troops actually have been withdrawn from Saudi Arabia? What would be achieved if all Western interests were driven from the Islamic Holy Land and if all of the governments sympathetic to the West were overthrown? According to Gunaratna, the ultimate aim is to reestablish the **caliphate**—the empire of Islam's early golden age—and thereby empower a formidable array of truly Islamic states to wage war on the United States and its allies.[***]

With these goals in mind, what should the United States do? Al-Qa'ida's strategists clearly hope that terrorism will inflict unacceptable losses of American lives, forcing the United States to withdraw its troops from Iraq and neighboring countries. That conviction may stem from the withdrawal of U.S. troops from Somalia after the loss of 18 American soldiers in Mogadishu in 1993 (the "Blackhawk Down" episode; see page 506) and the withdrawal of U.S. forces from Lebanon following the suicide bombing of the Marine barracks in Beirut in 1983 (see page 250). Al-Qa'ida thinks the United States has no stomach for sustained sacrifice.

Al-Qa'ida poses a dilemma for the United States. If the United States were to withdraw its troops from the region, it might only reaffirm al-Qa'ida's belief that its adversary is weak and vulnerable, thus encouraging more attacks. And if al-Qa'ida's ultimate goal is to take on the United States once it has established an Islamic empire, there is no reason to believe that a unilateral withdrawal would bring a cessation of hostilities.

On the other hand, if the United States continues to conduct a war on al-Qa'ida and a broader war on terror in a host of Muslim countries, al-Qa'ida can use the American presence, and especially unintended civilian losses, to incite widespread hatred and violence directed against the United States. Al-Qa'ida used the U.S. occupation of Iraq to argue its case that "the crusaders" *are* waging a war against Muslims and the Islamic world. President Bush argued that Iraq is "the central front on the war on terror,"[†††] but many analysts argue that the U.S. invasion of Iraq served as a catalyst to recruit uncounted numbers to the *jihadist* cause, turning Iraq into a "superbowl of terror" where new militant groups could cut their teeth on American soldiers and civilian contractors and their Iraqi allies.

Using geographic information systems and a variety of other tools, many geographers are interpreting the geographic dimensions of terrorism. One of the most profound questions for geographers is where future attacks will take place. Gunaratna estimates that al-Qa'ida maintains a reserve of at least 100 targets worldwide.[‡‡‡] Where are most of these? Suleiman Abu Ghaith, one of bin Laden's top aides, offered a clue, along with a geographic answer to the question of what al-Qa'ida wants: "Let the United States know that with God's permission, the battle will continue to be waged on its territory until it leaves our lands."[§§§]

* John L. Esposito, *Unholy War: Terror in the Name of Islam* (Oxford: Oxford University Press, 2002), p. 28.

† Rohan Gunaratna, *Inside al Qaeda: Global Network of Terror* (New York: Cambridge University Press, 2002), p. 3.

‡ Quoted in Rex A. Hudson, *Who Becomes a Terrorist and Why: The 1999 Government Report on Profiling Terrorists* (Guilford, Conn.: Lyons Press, 1999), p. 18.

§ Ayman al-Zawahiri, "Why Attack America? (January 2002)." In *Anti-American Terrorism and the Middle East,* Barry Rubin and Judith Colp Rubin, eds. (Oxford University Press, 2002), p. 133.

** Gunaratna, *Inside al Qaeda,* p. 93.

†† Ibid., p. 11.

‡‡ Osama bin Ladin in three selections from *Anti-American Terrorism and the Middle East:* "Declaration of War (August 1996)," pp. 137, 139; "Statement: Jihad against Jews and Crusaders (February 23, 1998)," p. 149; and "Al-Qa'ida Recruitment Video (2000)," p. 174.

§§ Bin Laden, "Statement," note vii, p. 150.

*** Gunaratna, *Inside al Qaeda,* note ii, p. 55; p. 89.

††† George W. Bush, "President Discusses War on Terror and Operation Iraqi Freedom," March 20, 2006, http://www.whitehouse.gov/news/releases/2006/03/20060320-7.html.

‡‡‡ Gunaratna, *Inside al Qaeda,* p. 188.

§§§ Suleiman Abu Ghaith, "Al-Qa'ida Statement (October 10, 2001)." In *Anti-American Terrorism and the Middle East,* p. 251.

to Western cultural influences and often repressive political and administrative systems. For many people outside the region, however, "Muslim" and "terrorist" have become synonymous—an erroneous association that can be overcome in part by careful study of the complex Middle East and North Africa.

This concludes our overview of land and life in the Middle East and North Africa. Module 6.1 offers more insight into the peoples and nations of this vital region. It begins with a continuation of the geopolitical theme, examining one of the world's most problematic, persistent, and influential conflicts, the one between Israel and its Arab neighbors.

SUMMARY

→ Misleading stereotypes about the environment and people of the Middle East and North Africa are common, as people outside the region often associate the region solely with military conflict and terrorism.

→ The region has bestowed on humanity a rich legacy of ancient civilizations and the three great monotheistic faiths of Judaism, Christianity, and Islam.

→ Middle Easterners include Jews, Arabs, Turks, Persians, Pashtuns, Berbers, people of sub-Saharan African origin, and other ethnic groups who practice a wide variety of ancient and modern livelihoods.

→ Arabs are the largest ethnic group in the Middle East and North Africa, and there are also large populations of ethnic Turks, Persians (Iranians), and Kurds. Islam is by far the largest religion. Jews live almost exclusively in Israel, and there are minority Christian populations in several countries.

→ Population growth rates in the region are moderate to high. Oil wealth is concentrated in a handful of countries, and as a whole, this is a developing region.

→ The Middle East has served as a pivotal global crossroads, linking Asia, Europe, Africa, and the Mediterranean Sea with the Indian Ocean. These countries have historically been unwilling hosts to occupiers and empires originating far beyond their borders.

→ The margins of this region are occupied by oceans, high mountains, and deserts. The land is composed mainly of arid and semiarid plains and plateaus, together with considerable areas of rugged mountains and isolated "seas" of sand.

→ Aridity dominates the environment, with at least three-fourths of the region receiving less than 10 inches (25 cm) of yearly precipitation. Great river systems and freshwater aquifers have sustained large human populations.

→ Many of the plants and animals on which the world's agriculture depends were first domesticated in the Middle East.

→ The Middle Eastern "ecological trilogy" consists of peasant villagers, pastoral nomads, and city dwellers. The relationships among them have been mainly symbiotic and peaceful, but city dwellers have often dominated the relationship, and both pastoral nomads and urbanites have sometimes preyed on the villagers, who are the trilogy's cornerstone.

→ About two-thirds of the world's oil is here, making this one of the world's most vital economic and strategic regions.

→ Since World War II, several international crises and wars have been precipitated by events in the Middle East. Strong outside powers depend heavily on this region for their current and future industrial needs. Unimpeded access to Persian-Arabian Gulf oil is one of the pillars of U.S. foreign policy.

→ The region of the Middle East and North Africa is characterized by a high number of chokepoints, strategic waterways that may be shut off by force, triggering conflict and economic disruption.

→ Oil pipelines in the Middle East are routed both to shorten sea tanker voyages and to reduce the threat to sea tanker traffic through chokepoints, but are themselves vulnerable to disruption.

→ Access to freshwater is a major problem in relations between Turkey and its downstream neighbors, Egypt and its upstream neighbors, and Israel and its Palestinian, Jordanian, and Syrian neighbors.

→ Al-Qa'ida and affiliated Islamist terrorist groups aim to drive the United States and its allied governments from the region and to replace them with an Islamic caliphate. Al-Qa'ida is an apocalyptic group that seeks to inflict mass casualties on its enemies, particularly on Americans in their home country.

KEY TERMS + CONCEPTS

Afro-Asiatic language family (p. 221)
  Berber subfamily (p. 221
  Semitic subfamily (p. 221)
    Arabic (p. 221)
    Hebrew (p. 221)
  Tuareg (p. 221)
al-Aqsa Mosque (p. 226)
Allah (p. 226)
al-Qa'ida (p. 235)
Altaic language family (p. 221)
  Turkish (p. 221)
anti-Semitism (p. 224)
Arab (p. 221)

Arab-Israeli War of 1956 (p. 231)
"axis of evil" (p. 234)
bazaar (p. 219)
caliphate (p. 237)
Carter Doctrine (p. 232)
chokepoint (p. 229)
Christianity (p. 224)
Church of the Holy Sepulcher (p. 225)
Copts (p. 225)
Crusades (p. 225)
Diaspora (p. 224)
Dome of the Rock (p. 223)
downstream countries (p. 232)

drought avoidance (p. 211)
drought endurance (p. 211)
energy crisis (p. 228)
Exodus (p. 222)
Fertile Crescent (p. 221)
First Temple (p. 222)
Foundation Stone (p. 223)
Gulf War (p. 232)
Hajj (p. 227)
Hamas (p. 234)
Hebrews (p. 222)
Hizbullah (p. 234)
horizontal migration (p. 217)

hydropolitics (p. 232)
Indo-European language family (p. 221)
    Dari (Afghan Persian) (p. 221)
    Farsi (Persian) (p. 221)
    Kurdish (p. 221)
    Pashto (p. 221)
International Atomic Energy Agency
    (IAEA) (p. 234)
Islam (p. 235)
Islamic law (*sharia*) (p. 227)
Islamic Salvation Front (FIS) (p. 234)
Islamists (p. 234)
Israelites (p. 222)
Jews (p. 221)
jihad (p. 235)
Ka'aba (p. 226)
Kurds (p. 221)
Maronites (p. 225)
medina (p. 219)
Middle Eastern ecological trilogy (p. 213)
    pastoral nomads (p. 214)
    urbanites (p. 214)
    villagers (p. 213)

Muslims (p. 225)
"New Rome" (p. 225)
Night Journey (p. 226)
Nile Water Agreement (p. 233)
Nilo-Saharan language family (p. 221)
    Chari-Nile languages (p. 221)
Noble Sanctuary (*al-Haraam ash-Shariif*)
    (p. 223)
oil embargo (p. 232)
Organization of Petroleum Exporting
    Countries (OPEC) (p. 228)
Pashtun (p. 221)
People of the Book (p. 225)
Persian (p. 221)
Pillars of Islam (p. 228)
    almsgiving (p. 227)
    fasting (p. 227)
    pilgrimage to Mecca (*hajj*) (p. 227)
    prayer (p. 227)
    profession of faith (p. 228)
Promised Land (p. 222)
Qur'an (Koran) (p. 226)
Red-Dead Peace Conduit (p. 232)

risk minimization (p. 217)
sand sea (p. 211)
Second Temple (p. 223)
sedentarization (p. 218)
Shia (p. 227)
Shiite (p. 227)
Suez Crisis (p. 229)
Sunni (p. 227)
suq (p. 219)
Temple Mount (p. 223)
terrorism (p. 236)
Turks (p. 221)
upstream countries (p. 232)
usufruct (p. 219)
vertical migration (p. 217)
war on terror (p. 234)
weapons of mass destruction (WMD)
    (p. 232)
Western Wall (p. 223)
Zion (p. 224)
Zionist movement (p. 224)

REVIEW QUESTIONS

1. What countries constitute the Middle East and North Africa? Which are the three most populous? Which encourage and discourage population growth?

2. What are the major climatic patterns of the Middle East? Where are the principal mountains, deserts, rivers, and areas of high rainfall?

3. Why can this region be described as a culture hearth? What major ideas, commodities, and cultures originated there?

4. What are the major ethnic groups, and in which countries are they found? What is an Arab? A Jew? A Turk? A Kurd? A Persian? A Muslim?

5. What are the principal beliefs and historical geographic milestones of Jews, Christians, and Muslims?

6. What is the difference between Shiite and Sunni Islam?

7. Where is oil concentrated in this region?

8. What is the Carter Doctrine?

9. What options does the United States weigh in dealing with Iran's suspected nuclear weapons development program?

10. What are upstream and downstream countries, and which are usually the more powerful? What are the exceptions to this rule?

11. What are Hizbullah, Hamas, and al-Qa'ida?

A sheikh (headman) of the Bisharin, camel-raising pastoral nomads of Egypt's southern Eastern Desert. Bisharin men carry their swords for ornamentation, and the photographer asked to see it.

MODULE 6.1
THE MIDDLE EAST AND NORTH AFRICA: MODERN STRUGGLES IN AN ANCIENT LAND

Conflict between Israel and its neighbors has been one of the most persistent and most dangerous problems in global affairs since the end of World War II. This module begins with an introduction to the key players and issues in the conflict. This depiction reveals why the Palestinian-Israeli dispute is so difficult to solve and how beneficial its resolution would be. The module continues with a survey of the major subregions, nations, and geographic problems of the diverse, fascinating Middle East.

■ module outline

■ module objectives

This module should enable you to

→ Understand the major issues of the Arab-Israeli conflict and the obstacles to their resolution

→ Appreciate how Lebanon's political system failed to represent the country's ethnic diversity, setting the stage for civil war

→ Balance the pros and cons of Egypt's Aswan High Dam, Libya's Great Manmade River, and other large water engineering projects

→ View the Sudanese civil war as an effort by one ethnic group to dominate others

→ Recognize how the U.S. war on terrorism has transformed political systems and international relations

→ Trace the trajectory of Persian Gulf (Arabian Gulf) countries from underpopulated, wealthy welfare states to populous, vulnerable states overly dependent on a single commodity

→ See the U.S. wars on Iraq as the result of tragic miscalculations by Saddam Hussein, combined with strategic American interests in the region

→ Recognize Turkey as an "in-between" country, a less developed Asian nation aspiring to be a European power

→ Consider how both superpower preoccupation and profound neglect sowed the seeds of conflict in Afghanistan

6.1.1 The Arab-Israeli Conflict and Its Setting

The Arab-Israeli conflict persists as one of the world's most intractable disputes. It has not been resolved in part because the central issues are closely tied to such life-giving resources as land and water and to deeply held religious beliefs.

The Arab-Israeli conflict is above all a conflict over who owns the land—sometimes very small pieces of land—and is therefore of extreme interest in the study of geography. It is also a conflict that has repercussions far beyond the boundaries of the small countries and territories involved. As long as it simmers or boils, there are other countries and entities that will use the unresolved Arab-Israeli conflict to advance their interests at the expense of others; both al-

Qa'ida and Iran, for example, derive much beneficial propaganda value from it. A United Nations–sponsored group called the Alliance of Civilizations concluded in 2006 that the Palestinian-Israeli conflict is the largest force behind global tensions.

A geographic understanding of the Middle East today requires familiarity with the events leading up to the creation of the state of Israel and with the wars that have followed, particularly as they have rearranged the boundaries of nations and territories.

The modern state of Israel was carved from lands whose fate had been undetermined since the end of World War I. The Ottoman Empire, based in what is now Turkey, had ruled Palestine (roughly the area now made up of Israel and the Palestinian territories) and surrounding lands in the eastern Mediterranean since the 16th century. After the British and French defeated the Ottoman Turks in World War I and destroyed their empire, they divided the region between themselves (●Figure 6.1.1). The British received the "mandate" (authority to establish a government) for Palestine, Transjordan (modern Jordan), and Mesopotamia (Iraq), while the French received the mandate for Syria (now Syria and Lebanon).

During World War I, British administrators of Palestine had made conflicting promises to Jews and Arabs. They implied that they would create an independent Arab state in Palestine and yet at the same time vowed to promote Jewish immigration to Palestine with an eye to the eventual establishment of a Jewish state there. The **Palestinians**—Arabs who historically formed the vast majority of the region's inhabitants—did not welcome the ensuing Jewish immigration and rioted against both the migrants and the British administration. Militant Jews attacked British interests in Palestine, hoping to precipitate a British withdrawal.

Placing themselves in a no-win position with these conflicting promises and under increasing pressure from both Jews and Arabs, in 1947 the British decided to withdraw from Palestine and leave the young United Nations with the task of determining the region's future. The United Nations responded in 1947 with the **two-state solution** to the problem of Palestine. It established an Arab state (which would have been called Palestine) and a Jewish state (Israel). The plan was deeply flawed. The states' territories were long, narrow, and fragmented, giving each side a sense of vulnerability and insecurity (●Figure 6.1.2). When Israel declared itself into existence in May 1948, the armies of the neighboring Arab countries of Transjordan, Egypt, Iraq, Syria, and Lebanon mobilized. In what Israelis call the War of Independence and Palestinians call the Catastrophe (*al-Nakba*), the smaller but better-organized and more highly motivated Israeli army defeated the Arab armies, and Israel acquired what have come to be known as its **pre-1967 borders** (the 1949 Armistice Agreement borders shown in Figure 6.1.2).

Prior to and during the fighting of the **1948–1949 Arab-Israeli war,** approximately 800,000 Palestinian Arabs chose or were forced to flee from the new officially Jewish state

● **Figure 6.1.1** The Middle East and North Africa in 1920. The victorious allies of World War I carved up the Middle East among themselves. Growing difficulties of administration would drive them from the region within a few decades.

of Israel to neighboring Arab countries (●Figure 6.1.3). The United Nations established refugee camps for these displaced persons in Jordan, Egypt, Lebanon, and Syria. Little was done to resettle them in permanent homes, and both the local Arab governments and the refugees themselves continued to insist on the right of the refugees to return to Israel and the restoration of their property there. This Palestinian **right of return** is one of the central problems of the ongoing peace process.

Other countries assumed control of those parts of the proposed Palestine that Israel did not absorb. Egypt occupied the Gaza Strip, a piece of land on the Mediterranean shore adjacent to Egypt's Sinai Peninsula that was inhabited mostly by Palestinian Arabs. Transjordan, renaming itself simply Jordan, occupied the predominantly Arab hilly region of central Palestine on the west side of the Jordan River, known as the West Bank, and the entire old city of Jerusalem, including the Western Wall and Temple Mount, Judaism's holiest sites. The Palestinian Arab state envisioned in the UN partition plan was thus stillborn.

The **Six-Day War** of 1967 fundamentally rearranged the region's political landscape in Israel's favor, setting the stage for subsequent struggles and the peace process. This conflict was precipitated in part when Egypt, by positioning arms at the Strait of Tiran chokepoint, closed the Gulf of Aqaba to Israeli shipping. Egypt's President Nasser and his Arab allies took several other belligerent but nonviolent steps toward a war they were ill prepared to fight. Israel elected to make a

preemptive strike on its Arab neighbors, virtually destroying the Egyptian and Syrian air forces on the ground. Israel gave Jordan's King Hussein an opportunity to stay out of the conflict. However, Jordan went to war and quickly lost the entire West Bank and the historic and sacred Old City of Jerusalem. The entire nation of Israel was transfixed by the news that Jewish soldiers were praying at the Western Wall. The Israeli army (Israeli Defense Forces, or IDF) also seized the Gaza Strip, Egypt's Sinai Peninsula (shutting down the Suez Canal), and the strategic Golan Heights section of Syria overlooking Israel's Galilee region. These three pieces of land would henceforth be known to the world as the **Occupied Territories.** Israel had tripled its territory in six days of fighting (●Figure 6.1.4).

The Palestinian refugee situation became more complex as a result of the war. Israel took over the areas where most of the camps were located. Many persons in the camps again took flight, and Palestinians fleeing villages and towns in the newly Occupied Territories joined them as refugees (see Figure 6.1.3). Some later returned to their homes, but an estimated 116,000 either did not attempt to return or were denied permission by Israel authorities to do so.

The **1973 Arab-Israeli War** was a multifront Arab attempt to reverse the humiliating losses of 1967. On October 6, 1973, the Jewish holy day of Yom Kippur, Egypt and Syria launched a surprise attack on Israel with the hope of rearranging the stalemated political map of the Middle East. Egypt's army initially showed surprising strength, penetrat-

• **Figure 6.1.3** Palestinian refugee movements in 1948 and 1967. Many who fled in the first conflict relocated in the second.

• **Figure 6.1.2** The 1947 UN partition plan for Palestine and Israel's original ("pre-1967") borders. The 1948–1949 war, which began as soon as Britain withdrew from Palestine and Israel proclaimed its existence, aborted the UN plan and created a tense new political dynamic in the region.

ing deep into the Sinai Peninsula and overturning Israel's image of invulnerability. Israeli troops soon reversed the tide and surrounded Egypt's army, but in the ensuing disengagement talks, Egypt won back the eastern side of the Suez Canal and by 1975 was able to reopen it to commercial traffic. In 1979, Israel agreed to return Sinai to Egypt under the U.S.-sponsored **Camp David Accords**, and Egypt recognized Israel's rights as a sovereign state. The Gaza Strip, West Bank, and Golan Heights continued to be held by Israel. Different Israeli administrations regarded them as lands that rightfully belong to Jews or as cards to be played in the regional game of peacemaking.

Arabs and Jews: The Demographic Dimension

In addition to issues of land, water, politics, and ideology, the Palestinian-Israeli conflict is about sheer numbers of people. Each side has wanted to maximize its numbers to the disadvantage of the other. To realize the Zionist dream of establishing a Jewish state, Jews began immigrating to Palestine around the start of the 20th century. Jews made

up 11 percent of Palestine's population in 1922, 16 percent in 1931, and 31 percent in 1946, on the eve of Israel's creation. In keeping with national legislation known as the **Law of Return,** the state of Israel has always granted citizenship to any Jew who wishes to live there. Following the 1948–1949 war, waves of new immigrants from Europe (**Ashkenazi Jews**) joined the Middle Eastern **Sephardic Jews** who had inhabited Palestine and other parts of the Middle East (and until 1492, Spain) since early times.

In the 1990s, more waves of Jewish immigration followed the dissolution of the Soviet Union and a change of government in Ethiopia, where an ancient Jewish group called **Falashas** had lived in isolation for thousands of years. In 2007, Jews made up 76 percent of the population of 7.3 million people living within Israel's pre-1967 borders. About 1.6 million Palestinian and non-Palestinian Arabs, or 23 percent of the country's population, are Israeli citizens and residents; they are known as "Israeli Arabs." Palestinian Arabs without Israeli citizenship, along with a variety of other ethnic groups, comprise the balance of the population within Israel's pre-1967 borders.

Another 4 million people live in the West Bank, Gaza Strip, and Golan Heights—areas occupied by Israel in the wars of 1967 and 1973. About 187,000 of those are Jewish settlers, and most of the rest are Palestinian Arabs (mainly in the West Bank and Gaza Strip), **Druze** (members of a small offshoot of Shiite Islam, living mainly in the Golan

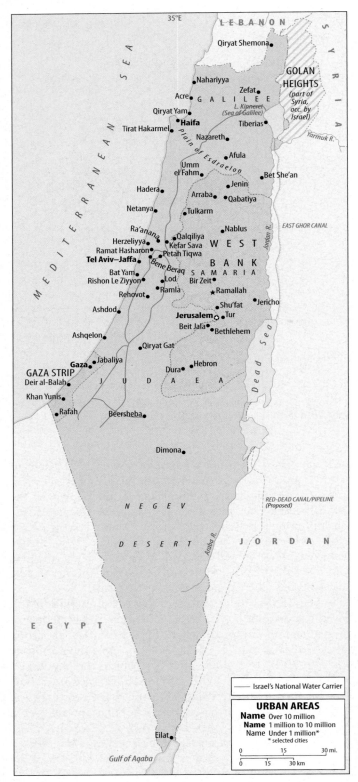

• **Figure 6.1.4** Israel and Arab territories as of 2007. Israel occupied the Sinai Peninsula in the 1967 war but returned it to Egypt following the Camp David Accords of 1979. Israel withdrew settlers and soldiers from the Gaza Strip in 2005 but kept them in the West Bank and the Golan Heights.

Heights), and other Arabs. Another 177,000 Jewish settlers live in East Jerusalem, which is also part of the territories Israel conquered in 1967. There are approximately 9 million more Palestinians living outside Israel and the Occupied Territories. Jordan has the largest number (3 million), followed by Syria and Lebanon (with 400,000 each). The remaining 1.5 million are in other Arab countries and in numerous nations around the world, comprising a sizable diaspora of their own that Palestinians call the *ghurba*. In this respect, the Palestinians are in much the same situation as the Jews prior to Israel's creation.

After 1967, and especially after the election of its conservative **Likud Party** to power in 1977, Israel moved to strengthen its grip on the Occupied Territories through security measures and the government-sponsored establishment of new **Jewish settlements**. Most of these settlements were, and still are, spread through the West Bank (see Figure 6.1.B, page 246). Much smaller numbers of Jews settled in the troubled Golan Heights, where about 15,000 remained in 2007. These are generally not frontier farming settlements but communities inhabited by relatively prosperous middle-class Jews who commute to jobs in Israel. They were not built as temporary encampments but as permanent fixtures meant to create what have been described as **"facts on the ground"**—an Israeli presence so entrenched that its withdrawal would be almost inconceivable (•Figure 6.1.5). Some of the motivation for this settlement has been ideological, as the West Bank is composed of the biblical lands of **Judaea and Samaria,** which many devout Jews regard as part of Israel's historical homeland.

The proliferation of Jewish settlements worsened relations between Israel and its Arab neighbors and even drove a wedge into the Jewish population of Israel itself. Many Israelis strongly oppose further settlement in the Occupied Territories and annexation of these territories to the Jewish state because if the territories were annexed, Israel

• **Figure 6.1.5** The Jewish settlement of Maaleh Adumim has a commanding position overlooking the strategic West Bank road linking Jerusalem with the Dead Sea rift valley.

PROBLEM LANDSCAPE

Land for Peace

They loom large in the headlines and in world affairs, but a closer look reveals that Israel and the Occupied Territories are almost unfathomably small. Within its legally recognized, pre-1967 borders, Israel's area is only 8,019 square miles (20,770 sq km), about the size of New Jersey or Slovenia (•Figure 6.1.A). Israelis often cite the small size of their country, nestled within the vastness of the Arab Middle East and North Africa, to highlight their vulnerability on the world stage. When and if Palestinians acquire their country, Palestine will be even smaller. The question "Whose lands are these?" has always been at the heart of the conflict between Israel and its neighbors.

President Anwar Sadat of Egypt set the precedent in 1979: Arabs could make peace with Israel on the formula of "land for peace," with Israel swapping Arab lands it occupied in 1967 in exchange for peace with its neighbors. The United States, Russia, and Norway initiated negotiations in the 1990s that made such a prospect appear possible. The process began in earnest in 1993, when Israel recognized the legitimacy of and began to negotiate with the **Palestine Liberation Organization (PLO),** long recognized by the Palestinians as their legitimate governing body. In return, the PLO, under the leadership of Yasser Arafat, recognized Israel's right to exist and renounced its long-standing use of military and terrorist force against Israel. U.S. President Bill Clinton orchestrated a historic handshake between Arafat and the Israeli prime minister, Yitzhak Rabin, and the leaders signed an agreement known as the **Gaza-Jericho Accord** (in its implementation, it came to be known as the **Oslo I Accord,** after the Norwegian capital where it was negotiated). It was designed to pave a pathway to peace by

Israel's granting of limited autonomy, or self-rule, in the Gaza Strip and in the West Bank town of Jericho. Autonomy meant that the Palestinians were responsible for their own affairs in matters of education, culture, health, taxation, and tourism. Israel also pulled its troops out of these areas. Palestinians were allowed to form their own government, known as the **Palestinian Authority (PA),** and they elected Yasser Arafat as its first president. The accord established a five-year timetable for the resolution of much more difficult matters, the so-called **final status issues.** These included the political status of Jewish and Muslim holy places in Jerusalem and of the city itself, the possible return of Palestinian refugees, and Palestinian statehood.

In a deal known as the Wye Agreement or **Oslo II Accord,** three successive Israeli prime ministers (Yitzhak Rabin, Benjamin Netanyahu, and Ehud Barak) promised to transfer more West Bank lands from Israeli to Palestinian control, and Arafat promised increased Palestinian efforts to crack down on Palestinian terrorists and so guarantee the security of Israelis in the Palestinian territories and in Israel. If finally implemented (it had not been as of 2007), this arrangement would have brought the total of West Bank lands under complete Palestinian control to 17 percent, leaving 57 percent completely in Israeli hands and 26 percent under joint control (•Figure 6.1.B).

During his final year in office in 2000, U.S. President Clinton sought to solidify his legacy as peacemaker by brokering a historic final settlement between Israelis and Palestinians. PLO Chairman Arafat and Israeli Prime Minister Barak huddled with Clinton and his advisers in the presidential retreat at Camp David, a site chosen because of its historical significance in Middle East peacemaking. Over weeks of tough negotiations, the two sides came close to a deal. It would have included the creation of an independent Palestinian country in the Gaza Strip and approximately 95 percent of the West Bank, with the other 5 percent representing clusters of Jewish settlements that would remain under Israeli control while Palestine received 5 percent of Israel's land in exchange. However, the talks broke down over two thorny issues. One was the status of Palestinian refugees abroad; Israel was unwilling to permit the right of return of the large numbers demanded by the Palestinians. But the deal breaker was Jerusalem, which both Israelis and Palestinians proclaim as their capital. Israel agreed to have the Old City of Jerusalem divided, with Israel assuming sovereignty over the Jewish and Armenian quarters and Palestinians assuming sovereignty over the Arab and Christian quarters (see Figure 6.B, page 223). These issues were not as problematic as those involving the places held holy by each side. Israel agreed to allow the Palestinians "soft sovereignty" over the Temple Mount (*al-Haraam ash-Shariif*), meaning that Palestinians could administer the area and fly a flag over it, but Israel would technically "own" it. Israel saw this point as essential because the First and Second Temples had stood on this spot. The Palestinians, however, insisted on full sovereignty over the Temple Mount and walked out on the negotiations.

• **Figure 6.1.A** Israel and the Occupied Territories compared in size with New England and the Benelux countries.

PROBLEM LANDSCAPE

Land for Peace *continued*

ZONES OF CONTROL

▨ Zone A: Full Palestinian control	— "Green Line"
▨ Zone B: Palestinian civil control, Israeli security control	— Israeli security fence (existing and proposed)
▨ Zone C: Full Israeli control	········ Jerusalem city limits
△ Jewish settlement	

● **Figure 6.1.B** The West Bank. Israeli and Palestinian areas of control were delimited in the peace process of the 1990s, but because of recurrent violence, almost all areas are effectively controlled by Israel. Also depicted are Jewish settlements, the completed and planned portions of the Israeli-built security fence, and the Green Line delimiting the internationally recognized border between sovereign Israel and the occupied West Bank.

Tragically, within weeks, this historic opportunity for peace evaporated and was replaced by a state of war. On September 28, 2000, the right-wing Israeli opposition leader (and subsequently prime minister) Ariel Sharon walked up to the Temple Mount under heavily armed escort. He meant to make the point that this

sacred precinct belonged to Israel. Many observers claim he also meant to disrupt the peace process. Whether or not it was his intention, a disruption in the peace process resulted. Beginning that day and for the next few years, Palestinians took to the streets in the **al-Aqsa *Intifada*** to challenge Israeli forces with rocks, bottles, and guns. Hundreds of Palestinians and Israelis died.

Following the collapse of the Camp David talks and an escalation of suicide bombings of Israeli civilians, Israel chose to effectively suspend the peace process. Prime Minister Sharon branded Arafat an untrustworthy partner for peace negotiations, isolating him physically in the West Bank town of Ramallah and systematically dismantling the Palestinian Authority he headed. Israel also persuaded the United States to stop negotiating with Arafat, who died a forlorn figure in 2004. Israel and the United States insisted on the creation of the new post of Palestinian prime minister, hoping it would be filled by someone they could negotiate with. The United States led an effort joined by the other members of the so-called Quartet—the European Union, the United Nations, and Russia—to establish what they called a **"road map for peace"** between the Israelis and the Palestinian Authority. It was based largely on the 2000 Camp David formula and, if successful, would presumably lead to an independent Palestinian state consisting of the West Bank and the Gaza Strip.

Prime Minister Sharon, while resuming dialogue with the Palestinian Authority, decided unilaterally to withdraw Israeli settlers and troops from the Gaza Strip in August 2005 and to build an immense **security fence** separating the West Bank from Israel. Israel's stated goal in constructing the new barrier was to prevent Palestinian suicide bombers from reaching Israel. Noting that the security fence did not follow the internationally recognized **Green Line** of Israel's pre-1967 border but instead penetrated significant portions of the West Bank to pick up Jewish settlements there, Palestinians condemned it as a "land grab." The International Court of Justice concluded that the wall violated customary international law and several conventions on human rights to which Israel is a signatory.

Work on the barrier continued after Sharon lapsed into a coma and was succeeded as prime minister by a Likud Party colleague, Ehud Olmert, in 2006. Olmert vowed to continue Sharon's consolidation of Israeli territory, holding on to three major blocs of Jewish settlements in the West Bank and to "united Jerusalem." On the Palestinian side, a political earthquake occurred that year with the surprising majority vote for parliamentary candidates from the Islamist group Hamas, giving Hamas control of the Palestinian Authority. Palestinians had grown disillusioned with the showy wealth of many elected Palestinian Authority officials and by the PA's failure to deliver an independent country or any other tangible benefits. Hamas, meanwhile, won hearts and minds through its network of social services, often providing what the Palestinian Authority failed to, including education, medicine, and food rations.

Hamas's stunning victory in the parliamentary elections added another layer of uncertainty and discord to the region's political

234

landscape. In the subsequent power-sharing agreement, Palestinian president Mahmoud Abbas of the PLO retained his post, but a Hamas leader, Ismail Haniyeh, took office as the Palestinian prime minister presiding over the Hamas-dominated parliament. Because the Hamas charter does not recognize Israel and in fact calls for its destruction, Israel and the Quartet suspended financial aid and discontinued dialogue with the Palestinian Authority. These sanctions would be lifted on three conditions: the Hamas-led PA must recognize Israel's right to exist, renounce violence, and abide by previous agreements reached between Israel and the Palestinians.

As the Palestinian economy deteriorated even further, clashes erupted between Hamas and PLO factions led by Abbas's Fatah Party. Hamas emerged victorious in fighting to control the Gaza Strip in 2007. Expelled from the territory, Abbas withdrew his recognition of Haniyeh as prime minister, and the Palestinian Authority government was dissolved. Abbas's Fatah Party consolidated its hold in the West Bank while the international community including numerous Arab countries, convened in Annapolis, Maryland, in an effort to resurrect the peace process.

---

would become a country whose population was only about 60 percent Jewish. With an Arab minority that has a far higher birth rate than that of the Jews, Jews could eventually become the minority population in their own country. In 2006, the Israeli advocacy group Peace Now leaked official Israeli government documents showing that 39 percent of the land on which Jewish settlements in the West Bank were built is privately owned by Palestinians. There is also international pressure against the settlements because they violate Geneva Conventions on activities allowed in territories captured in war. The International Court of Justice in the Hague has declared them illegal. **United Nations Resolutions 242 and 338** (passed after the wars of 1967 and 1973, respectively) called on Israel to withdraw completely from the Occupied Territories.

The settlements and other aspects of Israeli occupation, including the periodic closing off of the territories that prevents many Palestinians from reaching their jobs inside Israel proper, have fostered widespread and long-lasting Palestinian resistance against Israel. In 1987, Palestinians instigated a popular uprising in the Occupied Territories known as the *Intifada* (Arabic for "shaking"). Initially, they relied on rocks and bottles to engage Israeli troops, who answered with rubber and metal bullets. International media coverage of a Palestinian "David" fighting an Israeli "Goliath" did much to damage Israel's image abroad, and many Israelis came to question the justice and importance to security of Israel's continued occupation. Popular opinion and, by 1992, a new Israeli government led by the liberal **Labor Party** began to consider the previously unthinkable: giving the Palestinians at least some control over land and internal affairs in the Occupied Territories (see Problem Landscape, pages 245–247).

A Geographic Sketch of Israel

Israel is much more than a country in a perpetual state of conflict. It is a physically and culturally diverse and vibrant nation with a promising future, especially if relations with its Arab neighbors improve.

One of Israel's greatest achievements since independence has been the expansion and intensification of agriculture within its small territory; it is almost self-sufficient in agriculture, excluding grains. Production concentrates on citrus fruits (including the famous Jaffa oranges), which provide export revenue, and on dairy, beef, and poultry products, as well as flowers, vegetables, and animal feeds. Israel's intensive, mechanized agriculture resembles the agriculture of densely populated areas in western Europe. Collectivized settlements called **kibbutzim** (sing., *kibbutz*) are a distinctively Israeli feature on the agricultural landscape. Many of these lie near the frontiers and have defensive as well as agricultural and industrial functions. They have also been important in the development of Israeli identity, instilling a sense of work for the common good and reminding Israelis that their ancestors were often deprived of the right to own and till the land. Far more numerous and important in Israel's agricultural economy are other types of villages, including the small farmers' cooperatives called **moshavim** (sing., *moshav*) and villages of private farmers.

222

Agricultural development is concentrated in the northern half of the country, with its heavier rainfall (concentrated during the winter in this Mediterranean climate; see Figure 6.5, page 214) and more ample supplies of surface and underground water for irrigation. Annual precipitation averages 20 inches (50 cm) or more in most sections north of the city of Tel Aviv and in the interior hills where Jerusalem lies. Due to the rain shadow effect, very little rain falls in the deep rift valley of the Jordan River. From Tel Aviv southward to Beersheba, annual rainfall decreases to about 8 inches (20 cm), and in the central and southern Negev Desert, it drops to 4 inches (10 cm) or less. To expand agriculture along the coastal plain and in the northern Negev, Israel transfers large quantities of water from the north by the aqueduct and pipeline network of the National Water Carrier (see Figure 6.14, page 244). The largest source for this network is Lake Kinneret (Lake Tiberias, or Sea of Galilee), which is fed by the upper Jordan River. Since occupying the West Bank in 1967, Israel has intensively developed irrigated agriculture along the lower Jordan River and in scattered areas of the drier West Bank hills. Israel would lose most of this production to any future Palestinian state.

Jerusalem (population 700,000, excluding East Jerusalem, which international law regards as occupied territory) is the country's capital, but few countries recognize it as such because of the political sensibilities involved. Tel Aviv–Jaffa (population 3.2 million), on the Mediterranean coast, is Israel's greatest industrial center. The leading seaport is Haifa in northern Israel. The ancient town of Beersheba is the administrative and industrial center of the Negev Desert region. To the south lies Israel's small Red Sea port and resort town of Eilat. Thanks to the successful peace accords with Egypt and Jordan, open doors now link Eilat with the neighboring resort towns of Taba in the Egyptian Sinai and Aqaba in Jordan. Whenever conflict subsides, tourism booms, and developers are building and promoting a three-nation "Red Sea Riviera" to satisfy mainly European sun worshipers. Repeated episodes of conflict since 2000 have taken a significant toll on Israel's economy, which relies heavily on the exports of high-technology equipment, cut diamonds, and agricultural products and on assistance from the United States.

Jordan: Between Iraq and Israel

The economic and political situation of Jordan (•Figure 6.1.6) has always been precarious. Throughout his long reign, ending with his death in 1999, the country's King Hussein withstood various crises. Many of these related to the large numbers of Palestinian refugees in Jordan. Pales-

tinians still make up about 52 percent of Jordan's population of 5.7 million, but most are refugees without citizenship. After the 1967 war, Jordan was the chief base for military operations of the PLO against Israel, but in 1970 and 1971, the king expelled the armed Palestinian forces, which relocated to Lebanon. Jordan's monarchy has always walked a tightrope between its pro-Western stance and the need to satisfy its anti-Israeli and anti-American inhabitants and neighbors. Its support of the U.S. war in Iraq put Jordan on al-Qa'ida's target list; in 2005, the organization carried out merciless bombings (including one on a wedding party) at three Western-owned hotels in Amman.

Jordan is a land of desert and semidesert. The main agricultural areas and settlement centers, including the largest city and capital, Amman (population 1.3 million), are in the northwest. Prior to the 1967 war with Israel, northwestern Jordan included two upland areas—the West Bank on the west (captured during the war) and the East Bank, or Transjordanian Plateau, on the east—separated by the deep rift valley occupied by the Jordan River, Lake Kinneret, and the Dead Sea (see Figure 6.1.4). Most of the valley bottom lies in deep rain shadow and receives less than 4 inches (10 cm) of precipitation a year. The bordering uplands, rising to more than 2,000 feet (610 m) above sea level and lying in the path of moisture-bearing winds from the Mediterranean, receive precipitation averaging 20 inches (50 cm) or more annually. But overall, rainfall is so scarce that only small areas can be cultivated without irrigation, and the water available for irrigation is limited. Only about 2 percent of Jordan is cultivated, but larger areas are devoted to the grazing of sheep and goats. Irrigated vegetables, olives, citrus fruits, and bananas provide substantial exports, but some foods must be imported.

Jordan's natural resources are limited, and the country has little manufacturing, though it does export textiles. Rock phosphate and potash (both used for fertilizers) extracted from the Dead Sea are shipped out of Jordan's small port of Aqaba on the Gulf of Aqaba (known to Israelis as the Gulf of Eilat). Tourism is a growing industry, with the Nabatean rose-red city of Petra the chief attraction (see Figure 6.8, page 215). Al-Qa'ida probably sought to destabilize Jordan's regime by attacking hotels frequented by foreign tourists.

Lebanon and Syria

Israel's immediate neighbors on the north are Lebanon and Syria (see Figure 6.1.6). Their physical and climatic features are similar to those of Israel and Jordan. As in Israel, narrow coastal plains backed by highlands front the sea. The highlands of Lebanon and Syria are loftier than those of Israel and Jordan. Mount Hermon and the Lebanon and Anti-Lebanon Ranges rise above 10,000 feet (c. 3,000 m). Cyclonic precipitation brought by air masses off the Mediterranean is supplemented by orographic precipitation as rain and snow fall along the mountain chains paralleling the coast.

• **Figure 6.1.6** Principal features of Jordan, Syria, and Lebanon.

Lebanon and Syria have agricultural patterns suited to the Mediterranean climatic pattern of winter rain and summer drought. The coastal plains and seaward-facing mountain slopes are settled by villagers growing Mediterranean crops such as winter grains, vegetables, and drought-resistant trees and vines. Lebanon's Bekaa Valley, drained in the south by the Litani River, lies between the Lebanon and Anti-Lebanon Ranges and is a northward continuation of the Jordan rift valley and the African Great Rift Valley. Baalbek, its chief city and the site of an important ancient Roman temple, became a Syrian military stronghold in 1976 and a center of pro-Iranian Hizbullah activities against Israel after 1982. The political and military volatility of the Bekaa Valley was conducive to the emergence of a flourishing drug industry, producing both hashish (from marijuana plants) and opium (from poppies), from the mid-1970s until the 1990s. Marijuana cultivation made a comeback in the early 2000s as alternative crops proved to bring farmers far less income.

Syria is more industrialized than Lebanon, but neither country compares with Israel industrially. Textile milling, agricultural processing, and other light industries predominate. Both countries are deficient in minerals, although Syria has an oilfield in the east. Its reserves are expected to be exhausted by 2010.

There are three large, historically important cities in Syria and Lebanon. Damascus (population 2.6 million), Syria's capital, is in a large irrigated district at the foot of the Anti-Lebanon Range. Its Umayyad Mosque, originally a Byzantine cathedral, is one of the great monuments of the Middle East. The main city of northern Syria, Aleppo, is an ancient caravan center located in the Syrian Saddle between the Euphrates Valley and the Mediterranean. The old Phoenician city of Beirut (population 2 million), on the Mediterranean, is Lebanon's capital, largest city, and main seaport. Equipped with modern harbor facilities and an important international airport, Beirut before 1975 was one of the busiest centers of transportation, commerce, finance, and tourism in the Middle East.

For its prosperity and its natural beauty, Lebanon was once known as the **"Switzerland of the Middle East."** Then a precarious political balance broke down, and this Connecticut-size country was plunged into civil war in 1975 and 1976. The origins of that war and the country's suffering ever since are rooted in Lebanon's mixture of strongly localized religious communities, an ethnogeographic recipe for tragedy.

What the French carved out of greater Syria and put together as Lebanon in 1920 was a collection of small geographic areas, each dominated by an ethnic or religious group. Seventeen separate groups, known as *confessions,* are recognized officially (•Figure 6.1.7). Major confessions include the Maronite Christians, who predominate in the Lebanon Range north of Beirut; Sunni Muslims close to the northern border with Syria; Shiite Muslims in the Bekaa Valley and along Lebanon's southern border with Israel;

• **Figure 6.1.7** Generalized map of the distribution of religious groups in Lebanon.

and Druze Muslims in the Shouf Mountains, part of the Lebanon Range south of Beirut. Eastern Orthodox, Greek Catholic, and nearly a dozen other Christian sects are also present in Lebanon. Beirut itself is divided between the predominantly Muslim west and Christian east.

A 1932 census revealed a majority of Christians in Lebanon, and the country's 1946 constitution distributed power according to those census figures. Maronite Christians held a majority of seats in parliament, followed by Sunni Muslims and Shiite Muslims. Top leadership positions were allocated by a formula still followed today, with the president a Maronite Christian, the prime minister a Sunni Muslim, and the speaker of parliament a Shiite Muslim.

The central government did not try to unify the disparate geographic and ethnic enclaves of this mountainous land. Muslim populations grew and eventually outnumbered the Christians, who were unwilling to relinquish their political and economic privileges. Palestinians became part of the volatile mix. Pushed from Jordan in 1971, PLO guerrillas established a virtual "state within a state" in the southern slums of Beirut and in Shiite-controlled areas of southern Lebanon, from which they launched attacks on Israel.

The **Lebanese Civil War** started in 1975 as the have-not Sunnis, soon joined by Shiites and Palestinians, took on the more prosperous Maronite Christians. Syria and Israel supported opposing sides. By the end of the war the following year, as many as 100,000 people, most of them civilians, had been killed (•Figure 6.1.8). The war accomplished little. Lebanon effectively split into a number of geographically distinct microstates or fiefdoms—a politically unstable situation that set the stage for the next war in 1982.

The **1982 Lebanese war** was one facet of a larger war in Lebanon involving both civil strife and international intervention. Lebanon had become the main stronghold of the Palestine Liberation Organization, which used the country as a base for guerrilla operations against Israel. In June 1982, Israel invaded Lebanon with the avowed intention of smashing the PLO and establishing security for the northern Israeli frontier. Syria resisted the Israeli advance and gave support to the PLO. Israeli troops routed and inflicted heavy casualties on both the Syrians and the PLO. The Israeli push continued northward to the suburbs and vicin-

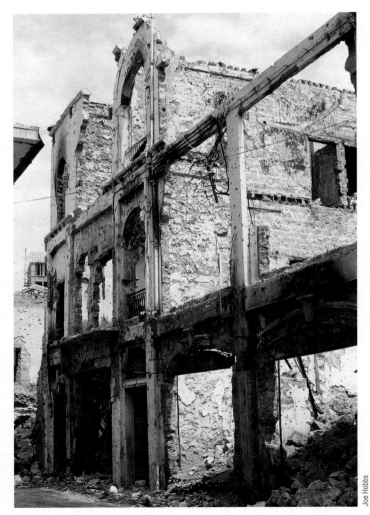

Joe Hobbs

• **Figure 6.1.8** Much of the fighting in the Lebanese civil war of 1975–1976 took place on streets like this, along the line separating Muslim west from Christian east Beirut. This portion of what had been a thriving, prosperous city became a desolate moonscape. A gunfight was under way here as the photographer took this picture in 1976; a taxi driver dumped him here and sped off.

ity of Beirut, culminating in a ferocious aerial and artillery bombardment of the city—the **Battle of Beirut**—that ended only when U.S. President Ronald Reagan exerted pressure on Israel to declare a cease-fire. Under supervision by U.S. and other international peacekeeping troops, the PLO withdrew from Beirut to Tunisia. After the multinational troops withdrew, Muslim assailants assassinated Lebanon's Christian president, Bashir Gemayal. Angry Christian forces then massacred hundreds of Palestinians and other civilians in the Sabra and Shatila refugee camps south of Beirut. U.S. and other multinational forces were again deployed to Beirut, and American ships used firepower to support the new Christian president in solidifying his authority.

Lebanese Shiite Muslim guerrillas, seeking to avenge U.S. attacks on Lebanon and its support of Israel, soon bombed American and French targets in Beirut, killing over 200 U.S. Marines in a barracks near Beirut's airport in 1983. The multinational forces withdrew, but Syrian troops remained in Lebanon and allowed Iranian forces and Iran's Shiite Muslim allies in Lebanon to establish a stronghold in the Bekaa Valley town of Baalbek. Subjected to persistent guerrilla attacks, Israeli troops withdrew from all of Lebanon except a self-declared "security zone" in the south, inhabited by both hostile Shiite Muslims and Israel's de facto Christian Lebanese allies. Israel exerted control over this zone and engaged in regular skirmishes with Hizbullah, a   234 Shiite Muslim organization sympathetic to Iran and hostile to the peace process between Israel and its Arab neighbors. With growing numbers of casualties among Israeli forces in the security zone, the occupation became increasing unpopular in Israel, and troops were finally withdrawn in 2000.

Buoyed by a peace pact signed between the warring factions in 1989, Lebanon embarked on a steady path of reconciliation and reconstruction. Beirut rose from the ashes in an ambitious, expensive reconstruction program that was the brainchild of its wealthy Sunni prime minister, Rafiq al-Hariri. Political power was redistributed to reflect the new demographic composition in which Muslims outnumber Christians, although a new, official census was such a sensitive issue that none was conducted. Once dominated by Christians, the Lebanese army enrolled soldiers from all of Lebanon's ethnic confessions. Social and education programs got under way to mix children from different ethnic groups, with the hope they would learn tolerance. Remarkably, however, Lebanon's textbooks still avoid any mention of the ethnic divisions and civil strife of the country, as if not discussing these rifts might help them go away.

Its latest war over, Lebanon was still not in control of its own destiny, however. The Syrians had never accepted the 1920 French decision to cleave the separate state of Lebanon from greater Syria, and since the civil war of the 1970s had stationed tens of thousands of soldiers in Lebanon. In 2005, a powerful suicide bomb killed former Lebanese Prime Minister al-Hariri, who had been outspokenly critical of Syria's presence in and influence over Lebanon. Many Lebanese concluded that Syria was behind the assassination, and hundreds of thousands of them took to the streets in the so-called **Cedar Revolution.** Their loud calls for Syria's withdrawal were answered within weeks. But the call for Syria's withdrawal was not unanimous. Supporters of Lebanon's Shiite Hizbullah—traditionally aligned with both Syria and Iran—massed in the streets to protest Syria's pullout. With Syria gone, Iran's direct influence on Hizbullah grew.

The stage was set for another conflagration in Lebanon. In July 2006, Hizbullah forces kidnapped two Israeli soldiers. Israel's usual response to this situation would have been to arrange a prisoner swap, but this time, Israeli Prime Minister Ehud Olmert decided to crush Hizbullah in Lebanon. The Israeli Defense Force unleashed a ferocious 34-day assault on Hizbullah and other Shiite targets, attacking more than 400 cities, towns, and villages across Lebanon. Israeli warplanes concentrated on targets in the southern part of the country but also hit the mainly Shiite southern suburbs of Beirut hard. Extensive damage was done to the country's bridges and other infrastructure that had been rebuilt following years of warfare. Israel's superior military forces proved incapable of halting a steady barrage of Hizbullah rocket fire into northern Israel, where Israeli civilian casualties mounted. A rising drumbeat of Israeli public opinion against the fighting led to a cease-fire between Israel and Hizbullah.

Israel had failed to defeat or even substantially weaken Hizbullah, which declared itself the victor of the **2006 Israeli-Hizbullah war.** Hizbullah sought to build on its momentum, moving quickly through Lebanon's devastated Shiite communities to help rebuild services. Rallying behind Hizbullah, Lebanon's long-marginalized Shiites sensed an opportunity to increase their political presence, insisting that they now outnumbered both Sunnis and Christians and beginning to challenge the prominence of the country's Sunni Muslim and Christian establishments. A divide reminiscent of Lebanon's civil war days emerged, with the Shiite Hizbullah, backed by its proxies Iran and Syria (and joined by one Christian faction in Lebanon) squaring off against Sunni, Druze, and other Christian groups. Supporters of each side gathered in huge demonstrations in Beirut. There were fears that an all-out civil war might reignite.

There were some silver linings in all this regional tension, including the prospect of a long-delayed peace agreement between Syria and Israel. The countries revealed that they had been conducting "back door" negotiations since 2004. Their formula for peace would likely include a Syrian cessation of support for Hizbullah in Lebanon in exchange for an Israeli withdrawal from the Golan Heights, occupied since 1967. A buffer zone designated as a peace park to be shared by Israelis and Syrians would run down the eastern shore of Lake Kinneret.

Syria, like Lebanon, is a country beset by religious factionalism. In Syria, Muslims outnumber Christians nine to one, and the major rifts are between Sunni and Shiite Muslims. In 1970, a coup overthrew the ruling Sunni Mus-

lim establishment and brought to power Hafez al-Assad, a member of the **Alawite** sect, an offshoot of Shiite Islam. Under his authoritarian rule as president, dominance of the government quickly passed to the Shiite Alawites, who make up only 12 to 15 percent of the country's population. The police and army crushed opposition in other communities, notably Sunni Islamists, who reject Alawites as heretics.

For three decades, until his death in 2000, President Assad was a skilled politician who played Middle Eastern affairs cautiously. His regime survived the economic and political setbacks accompanying the downfall of its benefactor, the Soviet Union. During Iran-Iraq War of the 1980s, Alawite-controlled Syria supported its kindred Shiite Iran, and Syria promoted the rise of the Shiite Hizbullah organization in Lebanon. These alignments earned Syria pariah status in neighboring Arab countries and especially in the United States, which views Syria as a proxy for Iran's interests in the region. Unlike the PLO leaders, President Assad did not pursue formal reconciliation with Israel. His son and successor, Bashar al-Assad, has continued the country's generally unaccommodating approach to Israel and the West. There is always a danger that Israeli or U.S. confrontations with Shiite interests in Middle East will spark a wider conflict, as the governments in Syria and Iran and Shiite factions in Lebanon could all be drawn in.

264

6.1.2 Egypt: The Gift of the Nile

Southwest of Israel lies the Arab Republic of Egypt, the most populous Arab country (73 million) (•Figure 6.1.9). This ancient land, strategically situated at Africa's northeastern

• **Figure 6.1.9** Principal features of Egypt.

corner between the Mediterranean and Red Seas, is utterly dependent on a single river. The Nile has created the flood-plain and delta on which more than 95 percent of all Egyptians live (see Figure 6.24, page 233). The Greek historian and geographer Herodotus aptly called Egypt "**the gift of the Nile**." The river's bountiful waters and fertile silt helped Egypt become a culture hearth that produced remarkable achievements in technology and cultural expression over a period of almost 3,000 years until about 500 B.C.E., when a succession of foreign empires came to rule the land.

The Nile Valley may appropriately be described as a "river oasis," for stark, almost waterless desert borders this lush ribbon. Only 3 percent of Egypt is cultivated, and nearly all this land lies along and is watered by the great river. The conversion of the original papyrus marshes and other wetlands along the Nile to the thickly settled, irrigated landscape of today is a process that has been unfolding for more than 50 centuries. It is difficult to imagine that in the time of the pharaohs, crocodiles and hippopotamuses swam in the Nile and game animals typical of East Africa roamed the nearby plateaus.

Ancient Egyptian civilization was based on a system of **basin irrigation**, in which the people captured and stored in built-up embankments the floodwaters that spread over the Nile floodplain each September. Egyptian farmers allowed the water to stand for several weeks in the basins, where it supersaturated the soil and left a beneficial deposit of fertile volcanic silt washed down by the Blue Nile from Ethiopia. They drained the excess water back into the Nile and planted their winter crops in the muddy fields. They were able to harvest one crop a year with this method. In limited areas, they could use **perennial irrigation** devices to obtain a second or even a third crop.

Achieving perennial irrigation on a vast scale—and thus boosting crop production enormously—required the construction of barrages and dams, which were innovations of the 19th and 20th centuries in this part of the world. Late in the 19th century, French and British colonial occupiers of Egypt, anxious to raise Egypt's exports of cotton (a summer crop demanding perennial irrigation), began Egypt's conversion from basin to perennial irrigation by constructing a number of **barrages,** or low barriers designed to raise the level of the river high enough that the water flows by gravity into irrigation canals. Barrages are not designed to store large amounts of water, a function now performed by two dams in Upper Egypt (the old Aswan Dam, completed in 1902, and the huge Aswan High Dam, completed in 1970; see Figure 6.24) and several others along the Nile and its tributaries in Sudan and Uganda.

The enormous Aswan High Dam (●Figure 6.1.10) is located on a stretch of the Nile known as the First Cataract (the Nile's many **cataracts** are areas where the valley narrows and rapids form in the river). Its reservoir, Lake Nasser, stretches for more than 300 miles (480 km) and reaches into northern Sudan. Like all giant hydrological projects, the Aswan High Dam has generated benefits and liabilities, and its construction was controversial. On the plus side, by storing water in years of high flood for use in years of low volume, the High Dam provides a perennial water supply to the Nile Valley's 11,000 square miles (28,500 sq km) of irrigated fields, thus extending the cultivation period. The dam also made possible the "reclamation" of

● **Figure 6.1.10** Egypt's Aswan High Dam, on the First Cataract of the Nile River.

Compliments of NASA/Earth Sciences and Image Analysis Laboratory at Johnson Space Center.

previously uncultivated land on the desert fringe adjoining the western Nile Delta. Navigation of the Nile downstream from (north of) Aswan is much easier, and the country's industries rely heavily on the hydroelectricity produced at the dam. Egypt weathered a two-year drought in the late 1980s thanks to the excess waters stored in Lake Nasser. The dam also eliminates the prospect of catastrophic flooding, which sometimes occurred before it was built.

The Aswan High Dam has also had drawbacks. The dam has caused the water table to rise, making it harder for irrigated soil to drain properly. When farmers use too much water, standing water evaporates and leaves a deposit of mineral salts—a problem known as **salinization**—that causes the once-fertile soil to lose its productivity. In addition, perennially available canal waters are ideal breeding grounds for the snail that hosts the parasite that causes schistosomiasis, or bilharziasis, a debilitating disease affecting a large proportion of Egypt's rural population. Mediterranean sardine populations, now deprived of the rich silt that nurtured their feeding ground, have plummeted off the Nile Delta, and the sardine industry has faltered. Without the free fertilizer the silt offered, Egypt's bill for artificial fertilizers has increased. Generally, however, Egyptians are very proud of the High Dam. It has increased domestic food and crop export production dramatically.

Egypt's most serious problem is that its population has grown even more dramatically. The country was once a great exporter of crops such as corn, wheat, barley, rice, millet, fruit, vegetables, and sugarcane, but its population has grown so explosively that food now comprises over 30 percent of Egypt's import expenditures. Cotton has long been the main commercial export crop for the hard currency Egypt desperately needs. Rural population densities are already among the highest in the world, and the crowding is growing worse despite a steady migration to Egypt's primate city, Cairo (population 15.9 million; •Figure 6.1.11)—the largest metropolis in Africa and the Middle East—and other cities. So far, the government's efforts to lure urban dwellers and factories out of the Nile Valley to desert "satellite

cities" have borne little fruit. Egypt is much more successful in exporting its brains and brawn abroad: Egyptians are prominent among the teachers, doctors, lawyers, engineers, and laborers working in the wealthy Persian-Arabian Gulf countries, and the remittances they send home pump much-needed cash into Egypt's economy.

Egypt's industries are concentrated in Cairo and in the Mediterranean port of Alexandria. They include cotton textiles, food processing, clothing manufacture, chemical fertilizers, and cement. Automobiles and other consumer durables are assembled primarily from imported components. The country is self-sufficient in some vital minerals, including oil, and exports some oil. But Egypt lacks the mineral wealth to support extensive industrialization, and for its development, the country depends heavily on foreign loans and grants. Income from tourism, particularly to the outstanding temples and tombs of ancient Egypt and to the world-famous scuba-diving sites on the Red Sea and Gulf of Aqaba, helps redress the unfavorable trade balance, as do transit fees from ships using the Suez Canal. However, tourism is a very vulnerable resource, as the number of visitors plummets each time violence—including al-Qa'ida attacks on resorts in the Egypt's Sinai Peninsula—rocks the Middle East.

Egypt is the giant of the Arab world, its largest country in terms of population, and its most influential in regional and international politics. The Western powers recognize its key role in mediation between Palestinians and Israelis. The West is anxious to see that Egypt does not become poorer than it already is and that the already huge gap between the rich and poor does not widen, perhaps throwing its population into the embrace of militant Islam. And Egypt's strategic location astride the Suez Canal is also a major concern for the West.

The United States is Egypt's chief non-Arab ally, lavishing huge sums of civilian and military aid in support of the autocratic regime led by President Hosni Mubarak. To quell real and perceived threats by Islamists, Mubarak has maintained a state of emergency for more than two decades and has effectively stifled most opposition. The ranks of outlawed and grassroots Islamist movements have grown at the same time. Egyptian society has generally reembraced traditional Islamic values in dress, education, relations between the sexes, and other social and cultural practices. A radical fringe has periodically attacked the interests of Egypt's 6 million Coptic Christian minority and foreign targets such as the tourism industry.

• **Figure 6.1.11** Cairo, on the banks of the Nile River.

6.1.3 Sudan: Bridge between the Middle East and Africa

Egypt is bordered on the south by Africa's largest country, the vast, tropical, and sparsely populated republic of Sudan (population 38.6 million) (•Figure 6.1.12). Formerly controlled by Britain and Egypt, Sudan gained its independence

• **Figure 6.1.12** Principal features of Sudan.

in 1956. This southern neighbor is vitally important to Egypt, for it is from Sudan that the Nile River brings the water that sustains the Egyptian people. The Nile's major tributaries join in Sudan, and storage reservoirs exist there on the Blue Nile, the White Nile, and the Atbara River, all of which benefit Egypt as well as Sudan itself (see Figure 6.24, page 233).

The Nile's main branches—the White Nile and Blue Nile—originate at the outlet of Lake Victoria in Uganda and of Lake Tana in Ethiopia, respectively, and flow into Sudan. In southern Sudan, the river passes through a vast wetland called the Sudd, where much water is lost to evaporation. The Jonglei Canal, an artificial channel to straighten the river's course and increase the volume and velocity of its flow through the Sudd, was under construction for a long time. Civil war in Sudan halted its progress in 1983. Predicting the death of the Sudd wetlands if the canal is completed, many environmentalists would be pleased if work did not resume.

Sudan is overwhelmingly rural. The largest urban district, formed by the capital, Khartoum (population 8.3 million, including the adjoining cities of Omdurman and Khartoum North), is located strategically at the junction of the Blue and White Niles. Transportation in Sudan as a whole is poorly developed. Several cataracts prohibit navigation on the Nile north of Khartoum, but the White Nile is continuously navigable at all seasons from Khartoum to

Juba in the deep south. Most of the country lacks railways or good highways. The main route for rail freight connects Khartoum with Port Sudan, a modern, well-equipped port that handles most of Sudan's seaborne trade.

About 7 percent of Sudan is cultivated, but agricultural productivity is low. The largest block of irrigated land is found in the Gezira region between the Blue and White Niles. The country's main cash crop is irrigated cotton. Many Sudanese support themselves by raising cattle, camels, sheep, and goats. The country's industries are meager. The most important resource is oil, which Sudan began exporting in modest quantities in 1999. Sudan's reserves are not large, but a sudden inflow of oil money ever since then has made Sudan one of the fastest-growing economies in the world. The government has used substantial resources to improve the country's infrastructure, but deep poverty—and a growing gap between the newly wealthy minority and the vast poor majority—remains widespread.

Most of Sudan is an immense plain, broken frequently by hills and mountains. Climatically and culturally, the area is transitional between the Middle East and sub-Saharan Africa. A contrast exists between Sudan's arid Saharan north, peopled mainly by Arabic-speaking Muslims, and the seasonally rainy equatorial south, with its grassy savannas, papyrus wetlands, and non–Arabic-speaking peoples of many tribal and ethnic affiliations who practice both Christian and animistic faiths. About three-fourths of Sudan's total population is Muslim.

Much political friction exists between north and south. A civil war raged between the two sections from 1983 to 2003, with southern factions of the **Sudanese People's Liberation Army (SPLA)** engaging government army troops from the north. As many as 2 million people, mostly civilians, died, and another 4.5 million were driven from their homes. Among the reasons for this conflict were historical antagonisms and economic disparities between the more developed north and the less developed and poorer south, along with efforts by the Arab-dominated government to impose Islamic law (*sharia*) and the Arabic language on the south and to exploit the south economically. The conflict helped keep Sudan mired in poverty. SPLA rebels repeatedly attacked the country's single export pipeline leading from the oil-producing area around Bentui to the coast at Port Sudan (see Figure 6.C, page 230). Bentui is close to the southern SPLA stronghold, and the county's largest oil reserves are within SPLA territory.

Negotiations between the government and SPLA rebels led in 2005 to a peace agreement and the tentative end to Sudan's long civil war. The government in Khartoum now recognizes the autonomy (but not independence) of southern Sudan, also known as New Sudan, and allows oil revenues to benefit both southerners and northerners. They parties agreed to hold a referendum in 2010 in which southerners would decide if they want independence separate from the north. Meanwhile, some of the 3 to 5 million refugees displaced from southern Sudan by two decades of civil conflict began going home. Most are internally displaced persons (IDPs, meaning they sought refuge within their own borders), of which Sudan has more than any other country in the world. They are rebuilding a homeland in which virtually all infrastructure was destroyed.

Drought has regularly visited southern Sudan and adjacent Ethiopia during recent years, and Sudan has periodically hovered near catastrophic famine. Huge tides of refugees have flowed back and forth between Sudan and Ethiopia. International relief agencies have been allowed to provide food and medical aid in only a small portion of the stricken zone. Sudan's military regime has meanwhile been accused of gross human rights violations, including ethnic cleansing of the Nuba, Nuer, Dinka, and other tribespeople in the south. Human rights groups have accused the government of promoting the abduction and enslavement of non-Muslim (mostly Dinka) southerners, who usually end up serving Arab masters in the north.

A new humanitarian crisis emerged in 2004 when government-backed militias called **Janjawiid** and Sudanese army regulars began to carry out ethic cleansing in Sudan's westernmost province, Darfur. The victims (as many as 400,000 dead by 2007) were blacks of the Zaghawa, Massaliet, and Fur tribes. They were targeted because of competition with Arabs over access to land and water and initially because of government fears that these groups might link up with antigovernment rebels in the south. Early on, the crisis in Darfur went relatively unnoticed by the international community, but it gradually acquired visibility as the situation worsened. The United States and other countries accused Sudan's government of genocide. African Union (AU) forces were called in as peacekeepers to quell the violence, but their effectiveness proved limited.

Sudan's foreign relations have changed dramatically twice since 1989, when an Islamic military regime overthrew an elected government. The new rulers allied themselves with militant Iran and also briefly provided refuge for Osama bin Laden. In 1993, after obtaining information linking Sudanese officials with the first bombing of New York City's World Trade Center, the United States branded Sudan a terrorist nation and halted economic aid. Following the analysis of intelligence information about bombings of U.S. embassies in Kenya and Tanzania, in August 1998 American cruise missiles obliterated a pharmaceutical factory, alleged to be a chemical weapons factory, in Sudan's capital. After that, and especially after 9/11, Sudanese relations with the West improved considerably, and the United States came to regard Sudan as an ally in its war on terrorism.

6.1.4 Libya: Deserts, Oil, and a Reformed Survivor

Egypt's neighbor to the west, Libya—formerly an Italian colony—became an independent kingdom in 1951. In 1969, Libya became a republic after Colonel Muammar al-Qaddafi and other army officers led a coup against the monarchy. Some 97 percent of the country's people are Muslim Arabs and Berbers.

Libya has three major geographic divisions: Tripolitania in the northwest, Cyrenaica in the east, and the Fezzan in the southwest (•Figure 6.1.13). The principal cities are the capital, Tripoli (population 1.9 million), in Tripolitania and Benghazi in Cyrenaica.

Most of Libya lies in the Sahara and is too dry to support cultivation except in scattered oases watered by wells and springs. The country is sparsely inhabited; its 6.2 million people are heavily concentrated in coastal lowlands and low highlands along the Mediterranean, where limited winter rains and irrigation with groundwater allow small-scale cultivation of typical Mediterranean crops. Thanks to **center-pivot irrigation**—a method in which the irrigation equipment rotates around a pivot, creating a circular crop pattern—wheat is grown in southern Libya. Deep wells there tap nonreplenished **fossil waters** that accumulated in deep limestone aquifers in ancient times when rains fell abundantly in what is now the Sahara. In an ongoing project called the **Great Manmade River** (•Figure 6.1.13), these same fossil waters are diverted from wells 1,155 feet (347 m) deep around Tazirbu in east central Libya and Hasquina in west central Libya by pipelines 13 feet (4 m) in diameter stretching hundreds of miles to the Mediterranean (see Figure 6.14). The goal of this effort, which Colonel Qaddafi calls "the eighth wonder of the world" and critics abroad decry as "Libya's pipe dream," is agricultural self-sufficiency through the irrigation of fruits, vegetables, and wheat in northern Libya.

Crude oil, petroleum products, and natural gas represent nearly all of Libya's exports (primarily to western Europe) and have brought prosperity to the country since the 1960s. The government spends much of the oil revenue on public works, education, housing, aid to agriculture, and other projects for social and economic development. The country has also spent large sums on weapons.

Libya's revolutionary government was long involved in an exchange of hostile rhetoric and sometimes hostile action with the United States, its Western allies, and Israel. In 1986, U.S. warplanes bombed Tripoli and Benghazi in reprisal for alleged terrorist acts against American citizens instigated by Libya. After Western intelligence authorities linked two Libyans to the 1988 bombing of Pan American Airlines' Flight 103 over Lockerbie, Scotland, Libya was subjected to an international air embargo.

Apparently conceding that his country's isolation was devastating the economy, Colonel Qaddafi decided to make amends with the West. Libya admitted to its role in the Flight 103 bombing and agreed to pay billions of dollars in restitution to the victims' families. Libya also announced that it was giving up development of weapons of mass destruction, including nuclear weapons. That proclamation led to much praise from Western governments and to the progressive renewal of diplomatic and economic relations.

● **Figure 6.1.13** Principal features of Libya, the Maghreb countries, and Western Sahara.

Akwa Betote/H2O/Corbis

● **Figure 6.1.14** Laying pipe for Libya's Great Manmade River.

The U.S. removed Libya from its list of "terrorist states" in 2006, and American oil companies are getting back to work in this oil-blessed country. Growing numbers of international tourists are visiting Libya's outstanding Roman sites, pristine beaches, and desert wilderness.

6.1.5 Northwestern Africa: The Maghreb

Arabs know the northwestern fringes of Africa as the Maghreb (meaning "western land"). Most of this area's people are Muslim Arabs. The ethnically distinct Berbers, who are most numerous in Morocco and Algeria, converted to Islam after the Arabs brought the new faith into the region in the 7th century. Many Berbers now speak Arabic, and most have adopted Arab customs. Majority Arabs discriminate against them in some countries. Berbers in Algeria, who make up 10 percent of the population, have demanded that the government recognize the Berber language and help the economically disadvantaged Berbers achieve better working and social conditions.

The Maghreb includes four main political units: Morocco, Algeria, Tunisia, and the disputed Western Sahara (see Figure 6.1.13. Morocco is geographically and culturally the closest to Europe; only the Strait of Gibraltar, a mere 11 miles (18 km) wide, separates it from Spain (●Figure 6.1.15). Islamic influences crossed the strait beginning early in the 8th century and prevailed in southern Spain until 1492, when Spain's monarchs ordered both Jews and the Muslim Moors out of the country. France became the domi-

● **Figure 6.1.15** The Strait of Gibraltar (center) from space, with the southern portion of the Iberian Peninsula at the top of the image and extreme northwestern Africa at the bottom.

● **Figure 6.1.16** The western Maghreb has breathtaking physical landscapes, from the Mediterranean regions in the north to the Atlas Mountains and Sahara in the south. Traditional village and urban architecture lends a unique character to the human environment. This is Vallée du Dades in Morocco, with the Atlas Mountains in the background.

nant power in the Maghreb during the 19th and early 20th centuries. French investment developed mines, industries, irrigation works, power stations, railroads, highways, and port facilities, but French rule was very unpopular. In 1956, after long agitation by local nationalists, Tunisia and Morocco secured independence. In Algeria, independence in 1962 came only after a bitter civil war lasting eight years.

Most people in the Maghreb live in the coastal belt of Mediterranean climate (called "the Tell" in Algeria), with its cool-season rains and hot, dry summers. Here, in a landscape of hills, low mountains, small plains, and scrubby Mediterranean vegetation, people grow the familiar products associated with the Mediterranean climate (●Figure 6.1.16). The region's major cities are on the coasts: on the Mediterranean, Tunisia's capital, Tunis (population 2 million), and Algeria's capital, Algiers (population 5.9 million), and on the Atlantic, Morocco's fabled Casablanca (population 3.2 million), and the capital, Rabat (population 1.7 million).

Inland, the Atlas Mountains extend in almost continuous chains from southern Morocco to northwestern Tunisia, with the highest peaks in the High Atlas Mountains of Morocco. They block the path of moisture-bearing winds from the Atlantic and the Mediterranean, and some mountain areas receive 40 to 50 inches (c. 100 to 130 cm) of precipitation annually, sometimes more. The precipitation nourishes forests of cork oak or cedar in some places. In Algeria, the mountains form two east-west chains: the Tell Atlas nearer the coast and the Saharan Atlas farther south.

All the countries reach the Sahara to the south. Some pastoral nomadic tribespeople, including the legendary Tuareg, still migrate through the Sahara with their camels, sheep, and goats. Clusters of oases exist in a few places, such as in the dramatic sandstone Ahaggar Mountains of southern Algeria, which rise high enough to catch moisture from the passing winds. A line of oases fed by springs, wells, mountain streams, and *foggaras* (tunnels from mountain water sources, known as *qanats* in Iran) lies along or near the southern base of the Atlas Mountains. Several large areas of sandy desert (*ergs*) and a barren gravel plain, the Tanezrouft, occupy portions of the Algerian Sahara.

The Maghreb economies benefit from a valuable mineral endowment. Algeria's oil and natural gas, located in the Sahara, are the region's greatest assets (see Figure 6.1.13). The oil flows through pipelines to shipping points on the Mediterranean in Algeria and Tunisia, and two pipelines carry Algerian natural gas to Europe. Petroleum products, crude oil, and natural gas make up almost all of Algeria's exports by value. Morocco's exports also include some minerals, notably phosphate, plus agricultural exports and clothing. Consumer industries, including plants that assemble foreign-made components, are the second-leading industries in the Maghreb countries.

Algeria's oil boom went bust in the 1980s, but population growth continued, fueling an economic crisis and shortages of food and consumer goods that contributed to widespread unrest in the 1990s. Algeria built up a staggering foreign debt that exacerbated the instability. Discord in Algeria reverberated through the entire Maghreb and across the Mediterranean to the region's former colonial power, France. During almost eight years after 1992, when Algeria's military government nullified the results of parliamentary elections that would have given a majority of

234 the seats to the Islamic Salvation Front (FIS), civil strife claimed an estimated 150,000 Algerian lives. Violent sympathizers of the FIS also carried their insurgency to France in the 1990s, bombing government and civilian targets to protest France's support of the military government in Algiers. The 2000s have seen Algeria's 34 million people reconciling with the West and enjoying the strong economic growth prompted by high energy prices.

Morocco and Tunisia viewed events in neighboring Algeria with anxiety. Citing the violence in Algeria, Tunisia banned Islamist political parties in 1994. The government of Tunisia (population 10 million) has promoted family planning and women's rights and is often cited as an "oasis" of openness and progressive change in the region—but it also closely monitors its citizens' e-mails and keeps a tight lid on Internet access. Tunisia also has the Maghreb's strongest economy, with active agricultural, mining, energy, tourism, and manufacturing sectors.

In Morocco, King Muhammad VI rules over a constitutional monarchy that promises political accommodation of Islamic groups, in contrast with Algeria's experience, but human rights monitors are increasingly alarmed by the young king's muzzling of opposition voices. Morocco's close alliance with the United States in the war on terrorism made it a target of an al-Qa'ida attack in 2003. Morocco is the poorest of the Maghreb countries. But there has been some progress in this nation of 31 million in recent years. The country has shown exceptional success in bringing down birth rates; whereas the average Moroccan woman had seven children in 1980, she now has two or three.

Some significant international problems exist in this region. Morocco would like Spain to return to Moroccan control the two Spanish enclaves of Ceuta and Melilla on Morocco's Mediterranean coast. Many African refugees find their way into these enclaves and eventually into Spain itself and onward into other destinations in Europe. About 30,000 Moroccans emigrate to Spain each year. The two countries have contesting claims to a number of small islands near the Moroccan coast.

Both Morocco and Algeria have long been at odds over the future of Spain's former dependency known as Western Sahara. This sparsely populated coastal desert (estimated population 500,000), which contains rich phosphate deposits, adjoins Morocco on the south and is claimed by that country. When Spain relinquished control in 1975, the territory was partitioned between Morocco and Mauritania, but in 1979, Mauritania withdrew its presence. Algeria objected to Moroccan control and, together with other outside parties, gave assistance to an armed resistance movement in Western Sahara against Morocco called the **Polisario Front.** In 1989, Morocco and the Polisario Front agreed to a referendum on independence, and a peace treaty was signed in 1991. To influence the outcome of the referendum in its favor, Morocco relocated 40,000 Moroccans into the region in 1991; they remain there today under Moroccan guard. The Moroccan government has subsequently used tax breaks and other subsidies to attract new Moroccan settlers there.

Such questionable actions led the United Nations to repeatedly suspend the referendum and finally call it off in 2000. Morocco continued to assert its claim to Western Sahara. It sold Western Sahara's offshore fishing rights to European countries. To stave off Algerian interests, Moroccan technicians also erected a 1,500-mile (2,400-km) sand barrier parallel to Western Sahara's border with Algeria, reinforcing it with land mines and 150,000 troops (see Figure 6.1.13). West of this berm, in Morocco-held territory, is the world's largest reserve of phosphate, used in fertilizers, toothpastes, and detergents. The United Nations wants to help to establish limited Western Saharan autonomy under Moroccan rule, but 75 countries recognize the Polisario-proclaimed Saharawi Arab Democratic Republic as a country with a government in exile. While the status of Western Sahara remains unresolved, more than 165,000 **Saharawis** (as the indigenous people of Western Sahara are known) live in refugee camps in the Algerian desert, as they have for decades. Despite their bleak conditions, the Saharawis have achieved a 95 percent literacy rate and have a representative government with a constitution that insists on religious tolerance and gender equality.

6.1.6 The Gulf Oil Region

As discussed in Chapter 6, the Persian Gulf (Arabian Gulf) region has by far the world's largest oil reserves. Foreign companies, mainly British and American, originally developed the oil industry in this region. These firms made huge profits from their oil concessions, but with a few exceptions, after 1972 the companies saw their properties nationalized by the local governments. However, many outside companies still work closely with national oil companies in the Gulf countries, providing them with technical and managerial expertise, skilled workers, equipment, marketing, and varied services. 231

Gulf oil is shipped out by a combination of pipelines and ships. Most oil produced in the Gulf region is transported to market by oceangoing supertankers. These ships load oil at terminals spaced along the Gulf. Some crude oil moves by pipelines from oilfields in the Gulf countries to tanker terminals on the Mediterranean and Red Sea coasts (see Figure 6.C, page 230). A growing proportion of extracted crude oil is refined in the Gulf countries for local use and export. Petrochemical industries exist in most of the producing countries, with the largest development in Saudi Arabia. The industries use local natural gas as their primary source of energy. Huge gas reserves are associated with oil deposits, and other major reserves exist apart from the oilfields, in "nonassociated" deposits. Growing amounts of gas are exported in liquefied form; Iran and Qatar have the largest reserves of natural gas in the Middle East.

Saudi Arabia: The Oil Colossus

With an area of roughly 830,000 square miles (2.1 million sq km), Saudi Arabia occupies the greater part of the Arabian Peninsula, the homeland of Arab civilization and the faith of Islam (•Figure 6.1.17). The interior of the country is an arid plateau fronted on the west by mountains that rise steeply from a narrow coastal plain bordering the Red Sea. Most of the plateau lies at an elevation of 2,000 to 4,000 feet (c. 600 to 1,200 m), with the lowest elevations near the Gulf and in the eastern part of the gigantic sand sea called the Empty Quarter (the Rub' al-Khali desert). Mountains (Arabic: *jebels*) rise locally from the plateau, and the plateau surface is cut by many valleys (Arabic: *wadis*) that carry water whenever it happens to rain; Saudi Arabia has no permanent rivers. Most interior sections average less than 4 inches (10 cm) of precipitation annually, but years may pass at a given locale with no precipitation at all. Rainfall is greatest along the mountainous western margin of the peninsula, especially in the southwestern Asir region, near the border with Yemen, which benefits from summer monsoon rains. About 95 percent of Saudi Arabia's crop production depends on irrigation. Cultivated land makes up only 1 percent of the country's area, yet surprisingly, Saudi Arabia is self-sufficient in many agricultural staples. It is an exporter of wheat, irrigated mostly in the central part of the country by gigantic center-pivot sprinkler systems tapping fossil waters.

Saudi Arabia's oil reserves make up more than one-fourth of the world's total. Most of the country's oilfields, including the world's largest (the Ghawar field, with enough reserves to supply U.S. oil needs for 9 years at current consumption levels), are located in the eastern margins of the country, including its adjacent seabed. Some Saudi Arabian oil is processed in refineries within the country and in nearby Bahrain, but most of the production leaves Saudi Arabia in crude form on tankers, using loading terminals that jut into both the Gulf and the Red Sea. To meet anticipated future demand, huge refining and petrochemical complexes have been developed at Jubail on the Gulf and Yanbu on the Red Sea. These cities sprang up almost overnight in the 1980s. Saudi Arabia's best-known city is ancient Mecca, Islam's holiest city. The interior city of Riyadh (population 4.4 million) is Saudi Arabia's capital. The Red Sea port of Jidda, some 40 miles (64 km) from Mecca, is the main point of entry for foreigners who make the pilgrimage to Mecca.

Saudi Arabia's population of 27 million is young and growing; half of all Saudis are younger than 21, and the annual growth rate is 2.7 percent. About one-fifth of the population consists of foreign workers. The country was long concerned that its native population was too small to run the country's economy and stave off cultural domination by foreign workers. The government promoted population

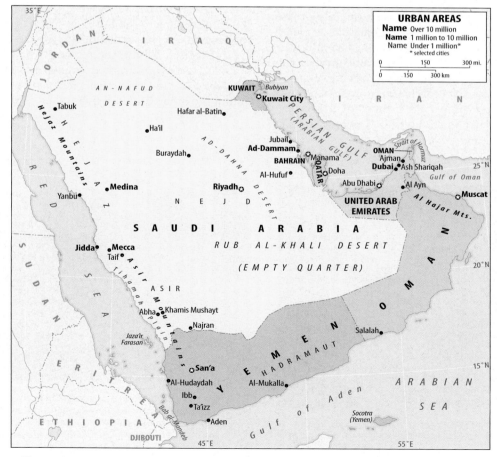

• **Figure 6.1.17** Principal features of the Arabian Peninsula.

growth and passed **Saudization** laws to put Saudis in jobs previously held by foreigners. Now the fear is that the indigenous Saudi population is growing much faster than job opportunities will, perhaps creating a young, unemployed, and disaffected population that might agitate for change. Since the height of the oil boom in 1981, the population has more than doubled, and per capita income, despite the recent surge in oil revenues, has fallen.

There are striking income discrepancies in this desert kingdom; although the sharing of oil wealth brought once undreamed-of prosperity to most of the population, the Shiite Muslim minority in the kingdom's Eastern Province remains poor (•Figure 6.1.18). The country's transition to a standard of living comparable to the more developed countries is not complete; the infant mortality rate, for example, remains high. The oil will of course run out one day, so the biggest challenge to Saudi Arabia's development is to diversify away from its near complete dependence on that commodity. The government is working to strengthen Saudi Arabia's global roles in finance, business services, trade, and manufacturing.

Despite the Western-style luxuries that oil wealth has brought to Saudi Arabia, the conservative Saudi family that rules this absolutist monarchy insists on maintaining strict local Muslim traditions. The legal and religious systems are based on interpretations of Islam by the Wahhabist sect, followers of the 18th-century Islamic reformer Muhammad bin Abd al-Wahhab, who joined forces with the founder of the Saudi dynasty. **Wahhabism** emphasizes fastidious piety, rejection of non-Muslim beliefs and cultures, corporal and capital punishment, and complete segregation of men and women in public life. Women are subject to many restrictions; for example, they are prohibited from driving automobiles, and although almost 60 percent of university graduates are women, they make up only 5 percent of the workforce (•Figure 6.1.19).

The vast Saudi royal family (which has more than 7,000 princes) keeps a wary eye on many internal and external threats. Al-Qa'ida has bombed Saudi civilian compounds and U.S. military facilities in an effort to purge the region of secular and "imperialist" Western interests and to destabilize the monarchy in order to replace it with an Islamic government. U.S. forces did withdraw from Saudi Arabia after the 2003 Iraq War began. Countercurrents are now buffeting the desert kingdom. While struggling to prevent outside influences from changing Saudi ways, the government is finally allowing tourists to visit, encouraging foreign investment, and signing international trade and human rights treaties.

Bahrain, Qatar, and the United Arab Emirates

The sheikdom of Bahrain (ruled by a monarch called a *sheikh* or *emir*; population 800,000) occupies a small archipelago in the Gulf between the Qatar Peninsula and Saudi Arabia (see Figure 6.1.17), to which its main island is connected by a modern causeway. In addition to being an oil producer, Bahrain is a center of oil refining, seaborne trade, ship repair, air transportation, business administration, international banking, and aluminum manufacturing using imported alumina and local natural gas. Like Saudi Arabia, Bahrain is an oil-financed welfare state. However, in contrast to its giant neighbor, Bahrain's petroleum reserves are expected to be exhausted within two decades. Bahrain is home to the U.S. Navy's Fifth Fleet, making this tiny country a vital strategic ally of the United States in a critical region. Within the region, largely Shiite Bahrain is notable for having granted women the rights to vote and run for office; in Qatar, women can vote but not be elected, and in the other Gulf countries, they can do neither.

• **Figure 6.1.18** Saudi Arabia's Shiite Muslim minority is significantly poorer than the Sunni majority. These are Shiites in the Eastern Province town of Qatif.

• **Figure 6.1.19** Education of women in the Gulf countries has increased dramatically in recent decades, but strong traditions keep most women out of the workforce. This is a geographic information systems (GIS) class in the Department of Geography, United Arab Emirates University.

East of Bahrain along the Gulf lie the small emirates of Qatar (population 900,000) and the United Arab Emirates (UAE) (population 4 million). Qatar is atop the world's third-largest reserve of natural gas. Only about one-quarter of the country's inhabitants are indigenous Qataris; the rest are guest workers, mainly from South and Southeast Asia, brought in to turn that mineral wealth into tangible assets. Qatar established itself as a vital U.S. ally in the 2003 war against Saddam Hussein's Iraq, permitting American warplanes and warships to be based there. Qatar is also the home of al-Jazeera, the satellite television network that upsets many Gulf autocracies and the United States with its frank criticisms and that serves as the media outlet for al-Qa'ida videotapes and audiotapes. Qatar is also the only Gulf country to maintain an open dialogue with Israel.

The United Arab Emirates is a loose confederation of seven sheikhdoms once known as the Trucial States that until 1971 were British protectorates: Abu Dhabi, Ajman, Dubai, Fujeira, Sharjah, Ras al Khaima, and Umm al Quwain. Only three of these produce oil, and only Abu Dhabi pumps it in major quantities. Qatar and the UAE are extremely arid and prior to the oil era supported only a meager subsistence economy based on fishing, pearl diving, seafaring, herding, and a little agriculture, augmented by smuggling and piracy. Today, the UAE is capitalizing on its central location between Asia, Africa, and Europe with the construction of sophisticated airport and seaport facilities. The residential and commercial developments of Dubai and Abu Dhabi have dramatically changed the landscape (•Figure 6.1.20). With the aid of an overwhelmingly foreign workforce, the UAE is successfully diversifying its economies beyond the oil industry. Dubai is a well-established international trade and banking center, featuring the world's largest shopping mall, the world's tallest building, and a massive port that U.S. naval ships are permitted to use. Abu Dhabi, the largest and richest unit of the seven emirates, is hoping to establish regional supremacy as a financial center. It also aspires to become one of the great cultural centers of the Middle East, with a host of new museums, galleries, and performing arts centers. Unlike its more conservative neighbors, the UAE has promoted itself as a destination for sun-worshiping international tourists and shoppers. Dubai is the most liberal of the emirates and, in marked contrast to its teetotaling neighbors, tolerates bars and brothels.

The Southern Margins of Arabia

Known to the ancient Romans as *Arabia Felix*, or "Happy Arabia," because of their verdant landscapes, Yemen (population 22 million) and Oman (population 2.7 million) occupy, respectively, the mountainous southwestern and southeastern corners of the Arabian Peninsula (see Figure 6.1.17). The present Republic of Yemen was created in 1990 by the union of the former North Yemen and South Yemen. They remain united despite a brief civil war in 1994. Its political capital and largest city is San'a (population 2 million), famed for its distinctive indigenous architecture (•Figure 6.1.21). Until the 1960s, North Yemen was an autocratic monarchy, extremely isolated and little known to the outside world. In the mid-1960s, a republican government supported by Egypt and opposed by Saudi Arabia took control of large sections at a cost of much bloodshed.

The former South Yemen, known until 1967 as Aden, a British colony and protectorate, lies along the Indian Ocean between the former North Yemen and Oman. The main city, Aden, is located on a fine harbor near the straits. Aden is perfectly located for refueling vessels that transit the Suez Canal and is close to the strategic chokepoint of Bab el-Mandeb, connecting the Red Sea with the Indian Ocean. Possession or use of this port has therefore long been prized by outside powers. After British colonial times, South Yemen became a Marxist state supported by the Soviet Union and other Communist bloc nations. Since the two Yemens were unified, Aden has been an important refueling station for the United States and its Western allies. Security at the port of Aden has been stepped up since the bombing there of the U.S.S. *Cole* in 2000, and Western governments have pressured Yemen's government to crack down on Islamist groups that advocate struggle against Western interests in the region. Yemen gave the United States permission to use a remote-controlled predator drone to kill an important al-Qa'ida leader on a remote Yemeni road in 2002. Some of Yemen's tribal peoples live in areas well beyond the effective control of the government in San'a and are known to be strongly sympathetic to al-Qa'ida. The mountainous north-central region of Yemen, known as the Hadramaut, is the ancestral homeland of Osama bin Laden's family.

210

Compliments of NASA/Visible Earth.

• **Figure 6.1.20** Perhaps nowhere on earth is humanity's transformation of nature so visible from space. The United Arab Emirates' city state of Dubai boasts the world's tallest building, the artificial deepwater part of Jebel Ali, and an artificial island complex shaped like a palm tree. The Palm Island development offers very high priced homes and apartments.

Yemen receives the heaviest rainfall in the Arabian Peninsula, enjoying a summer drenching derived from the same monsoonal wind system that brings summer rain to the Horn of Africa and the Indian subcontinent. Its main subsistence crops are millet and sorghum. Most of the country's best agricultural land consists of terraces and is dedicated to the cultivation of the valuable cash crop qat (Cathya edulis), a stimulant shrub chewed by virtually all adults in the country on a daily basis. Most of the country's people live in the rainier highlands, and the coastal plain is extremely arid and sparsely populated.

Yemen's meager agricultural output and lack of industrial production have been compensated for during recent decades by remittances, the earnings sent home by Yemeni workers in the oilfields and services of the wealthy Gulf countries. The support of Yemen's government for Saddam Hussein's regime in Iraq during the 1991 Gulf War prompted Saudi Arabia to expel millions of these expatriate Yemeni workers, leading to a precipitous decline in Yemen's economy.

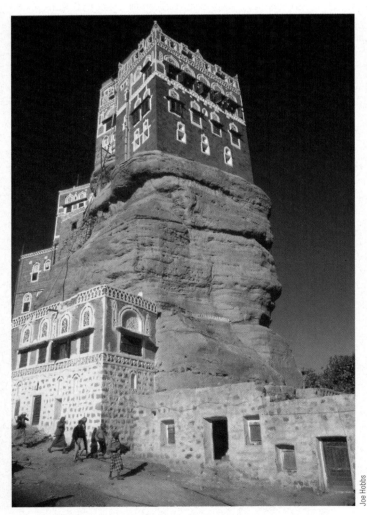

• **Figure 6.1.21** The Palace of the Imam Yahya in Wadi Dhar, near San'a, is typical of the architectural gems that have made old San'a a World Heritage Site.

Oil production began in Yemen itself in the 1980s. The fields are located in a contested border area between Yemen and Saudi Arabia and subjected periodically to kidnappings and killings of foreign oil workers and tourists by groups opposed to Yemen's government, so production has been slow to increase.

Most of the Sultanate of Oman lies along the Indian Ocean and the southern shore of the Arabian Sea, with a small, detached part bordering the Strait of Hormuz and the Gulf. As in Yemen, the coastal region is arid, and most people live in interior highlands where mountains induce a fair amount of rain. For about 2,000 years, beginning around 1000 B.C.E., Oman was the source of most of the world's frankincense, a prized aromatic commodity that made the region a hub of important oceanic and continental trade routes. The capital, Muscat (population 900,000), has long been a vital port in trade between the Middle East and South Asia. Crude oil now furnishes almost all of Oman's exports, but with the easiest-to-reach oil already tapped, production has been declining since 1997. The country's strategic position on the Strait of Hormuz, through which much of the oil produced by the Gulf countries is transported, has made it an object of concern in industrial nations, particularly Britain, with which it has historic ties, and the United States. These Western powers have rights of access to base facilities in Oman in the event of any military emergency that threatens the continued flow of oil from the Gulf region. Allied forces used the bases after Oman joined the United Nations coalition against Iraq in the Gulf War of 1991 and again in the U.S.-led invasion of Iraq in 2003.

Kuwait and the Gulf War of 1991

With 8 percent of the world's proven oil reserves within its borders, tiny Kuwait (6,900 sq mi/17,900 sq km—a little smaller than the state of New Jersey; see Figure 6.1.17) has the world's fifth-largest store of proven petroleum resources (after Saudi Arabia, Canada, Iran, and Iraq). Associated with the oil is much natural gas, which Kuwait both uses and exports. Oil revenues have changed Kuwait (population 2.8 million) from an impoverished country of boat builders, sailors, small traders, and pearl divers to an extraordinarily prosperous, completely urbanized welfare state. Generous benefits have flowed to Kuwaiti citizens, but not to the Asian and other non-Kuwaiti expatriates drawn by employment opportunities to this emirate. About half of the people residing in Kuwait are foreigners. To reduce possible security threats and become more self-reliant, the government is trying to reduce dependence on foreign workers. A huge challenge will be to provide jobs for its young population (half of its people are under 25) in non-oil-related industries and services. As Kuwait is almost completely dependent on fossil fuel commodities, it is vulnerable to developments such as alternative energy and, of course, to the eventual depletion of its own reserves.

Kuwait took center stage in world affairs in August 1990 when Iraqi military forces invaded the country. Claiming that it was historically part of Iraq and calling it Iraq's "19th province," Saddam Hussein's regime attempted to annex Kuwait. Saudi Arabia and other Gulf oil states seemed in danger of invasion. Worldwide opposition to Iraq's aggression resulted in a sweeping embargo on Iraqi foreign trade imposed by the Security Council of the United Nations. Led by the United States, a coalition of 28 UN members (including several Arab states) staged a rapid buildup of armed forces in the Gulf region.

Beginning in January 1991, the coalition's crushing air war drove the Iraqi air force from the skies, massively damaged Iraq's infrastructure, and hammered the Iraqi forces in Kuwait and adjacent southern Iraq with intensive bombardment. In the ensuing ground war, the Iraqis were driven from Kuwait within 100 hours. Overall, an estimated 110,000 Iraqi soldiers and tens of thousands of Iraqi civilians were killed, while the technologically superior coalition forces suffered 340 combat deaths, of which 148 were Americans. During the conflict, which came to be known as the Gulf War, the Iraqis also made war on the environment, setting fire to hundreds of oil wells in Kuwait and allowing damaged wells to discharge huge amounts of crude oil into the Gulf.

Iraq: Vortex of Violence

Iraq (population 29 million, about 67 percent urban) occupies a broad, irrigated plain drained by the Tigris and Euphrates Rivers, together with fringing highlands in the north and deserts in the west (•Figure 6.1.22). This riverine land once known as Mesopotamia has a complex geopolitical history. It has been a seat of empires, a target of conquerors, and in the 20th and early 21st centuries, a focus of oil development and political and military contention involving many other nations as well as groups within Iraq itself.

A critical (and now unfortunate) feature of Iraq's geography is its ethnic and religious complexity, with three major groups present: the Shiite Arabs, mainly in the south,

• **Figure 6.1.22** Principal features of Iraq.

who make up about 60 percent of the country's population; the Sunni Arabs, about 35 percent of the total and living mainly in the center (especially in the so-called Sunni Triangle); and the Kurds, mainly in the north, most of whom are Sunni Muslims and who represent 15 to 20 percent of the total (•Figure 6.1.23; see also Ethnic Geography, page 265). There are numerous smaller minorities, including Yazidi and Sunni Turkoman.

Internal sectarianism, foreign intervention, and costly conflict are not new to Iraq, and the modern problems there may be better appreciated through some understanding of the past. Iraq was a province of the Ottoman Empire (based in what is now Turkey) for almost 400 years. British forces defeated the Turks in Iraq in World War I, and Britain had a League of Nations mandate over the country until it gained its independence in 1932. Iraq's government was a pro-Western monarchy from 1923 to 1958, when an army coup made the country a republic. By the mid-1970s, Iraq had become an authoritarian one-party state ruled by President Saddam Hussein in the name of the Arab Baath Socialist Party. After 1973, increasing oil revenues from the country's huge reserves (about 9 percent of the world's total) underwrote new social programs, large construction projects, and the creation of new businesses. New high-

rise buildings, busy expressways, and an active commercial and social life gave metropolitan luster to ancient Baghdad. Subsequent wars caused massive destruction in the city and in Iraq's economy. This country should have become rich from its oil, but it squandered that wealth on hapless military adventures.

Saddam Hussein's misadventures were based to some extent on the country's geography. Baghdad (population 8 million), Iraq's political capital, largest city, and main industrial center, is located strategically on the Tigris River near its closest approach to the Euphrates River. It is not a port city; Basra is Iraq's main seaport, although it is located upriver on the Shatt al-Arab, the waterway formed by the confluence of the Tigris and Euphrates. Iraq suffers from the fundamental geographic disadvantage of not having natural deep-water ports in its tiny 35-mile (55-km) window on the Gulf. One of the main reasons Iraq took on its immediate neighbors in two separate wars was to widen this window. In 1980, perceiving internal weakness in the neighboring fledgling Islamic Republic of Iran and a means of securing land and mineral resources on the Iranian side of the Shatt al-Arab waterway, Saddam Hussein launched an invasion of his more populous neighbor. Thus began an eight-year war that proved enormously costly to both sides, with an estimated 500,000 people killed or wounded. Iran mounted a far better defense than Iraq anticipated, expending its greater military personnel in bloody "human wave" assaults to counter Iraq's greater firepower from tanks, artillery, machine guns, and warplanes. A complex web of geopolitical relationships made the **Iran-Iraq War** a storm center in the political world, particularly as the superpowers and other countries sold sophisticated arms to both sides. The war ended in stalemate in 1988, with neither side gaining significant territory or other assets.

Just two years later, Saddam Hussein's military invaded Kuwait. A far more internationalized conflict, that war proved disastrous for Iraq. A formal cease-fire to the so-called **Gulf War** came in 1991, with harsh conditions imposed on Iraq by the UN Security Council. UN-sponsored economic sanctions created shortages of food and medical supplies, bringing great hardship to the country's population. Iraq agreed to pay reparations for the damage its forces had caused, and it renounced its claims to Kuwait. Saddam Hussein's surviving forces then mercilessly put down rebellions by Shiite Arabs in southern Iraq and by Kurds in the north. To deter Iraq's military and provide a safe haven for the Kurds in the north and the Shiites in the south, the United Nations established "no-fly zones," where Iraqi aircraft were prohibited from operating. But there was widespread condemnation of the Western coalition in the Arab world for seemingly encouraging the Kurds and Shiites to rebel and then failing to help them.

Iraq's Shiites fared especially badly. One of the great environmental and cultural tragedies of the late 20th century was Iraq's systematic destruction after 1991 of the marshes adjoining the southern floodplains of the Tigris and Eu-

263

• **Figure 6.1.23** Ethnicity in Iraq and neighboring regions.

ETHNIC GEOGRAPHY

The Kurds

A mostly Sunni Muslim people of Indo-European origin, the Kurds are by far the largest of several non-Arab minorities in Iraq. They have revolted against Iraqi central authority on numerous occasions, including during the Gulf War of 1991. The Kurdish question has international implications because the area occupied by an estimated 30 million Kurds also extends into Iran, Turkey, Syria, and the southern Caucasus region (see Figure 6.1.23); an estimated 2 million more live outside the Middle East. European powers promised an independent state of Kurdistan at the end of World War I, but the governments concerned never took steps to create it. Kurds today remain the world's largest ethnic group without a country.

A book about the Kurds is appropriately titled *No Friends but the Mountains.** Turkish officials have long downplayed the identity of Kurds (who make up about 20 percent of Turkey's population), often referring to them as "Mountain Turks," and even banned or restricted Kurdish-language media in Turkey until 2002.

A Kurdish rebellion against Turkish rule, rising in part from the great poverty of the Kurds relative to the rest of Turkey's population, has smoldered and sometimes flamed since 1984. The main Kurdish resistance comes from the **Kurdistan Workers Party,** known by its Kurdish acronym **PKK.** Under Saddam Hussein, Iraq tried to "Arabize" its Kurds by forcing them to renounce their ethnicity and sign forms saying they were Arabs. However, after a no-fly zone and "safe haven" established after the Gulf War put them beyond Saddam Hussein's reach, Iraq's Kurds came to enjoy relative autonomy and prosperity. Even in the tumultuous years since the U.S.-led invasion in 2003, the Kurdish region of the north has escaped most of the bloodshed and economic chaos characteristic of the south, although recent resurgence of PKK activities have raised the prospect of an incursion by Turkey.

*John Bullock, *No Friends but the Mountains* (London: Oxford University Press, 1993).

---

phrates Rivers. To quell any opposition by the ancient Shiite culture of the **Ma'adan,** or **"Marsh Arabs,"** the Iraqi military built massive embankments and canals that drained and destroyed the marshes, forcing the Ma'adan to abandon their historical homeland. After 2003, the embankments were demolished, and restoration of the marshlands began. Under Saddam Hussein, Iraq's government had expanded irrigation, flood control, and hydropower through the construction of dams on the Tigris and Euphrates and their tributaries. Agricultural output in Iraq remains low, however, due mainly to the problem of overwatering, which causes the salinization of once-fertile soils.

Throughout the 1990s, the regime of Saddam Hussein played a cat-and-mouse game with UN inspectors whose duty was to eliminate Iraq's ability to produce weapons of mass destruction. United States, British, and other forces repeatedly struck Iraq as the country refused to allow full access by the UN inspectors and, in 1998, attacked scores of military targets when access was denied. The Baghdad regime flouted the no-fly zones and restated its wartime position that Kuwait was an illegitimate country. International support for the sanctions against Iraq weakened greatly, particularly in the Arab world. President Saddam Hussein endured, much to the chagrin of his foes, and might have been able to hang on to power indefinitely if not for the 9/11 al-Qa'ida attacks on the United States.

Immediately after the Gulf War and during the Clinton administration of 1993–2001, the reasoning in Washington was that a unified Iraq under Saddam Hussein was better than the alternative: its possible fragmentation into the three entities of a Shiite south, a Sunni center, and a Kurdish north, some or all of which might be hostile to U.S. interests in the region. Although unsavory, Saddam

Hussein could be contained. Under the new Bush administration, and especially after 9/11, the downfall of Saddam Hussein became an urgent priority, ranking alongside the war on terrorism. The Bush administration attempted to link Saddam Hussein with al-Qa'ida, particularly to garner public American support for a new war on Iraq, but such connections have never been decisively established.

Publicly, Washington built the case that Iraq was harboring and continuing to develop weapons of mass destruction that might be used against the United States and its allies. The United States was able to prompt further United Nations pressure on Iraq, which finally allowed UN weapons inspectors to resume their work. When these inspections failed to turn up weapons, the United States insisted that Iraq was hiding them and that only the use of force could remove their threat. It was necessary to effect "regime change" in Baghdad. To take on Saddam Hussein, President Bush assembled what he called a "coalition of the willing" consisting mainly of Britain, Australia, Denmark, and Poland. It fell far short of the coalition assembled during the Gulf War. France, Germany, and other traditional allies of the United States opposed the impending war against Iraq, and even within the United States there was considerable dissent against what was perceived as a war of choice rather than one of necessity.

The U.S.-led ground and air assault on Iraq (dubbed **Operation Iraqi Freedom**) that began in March 2003 led quickly and with relatively few casualties to the downfall of the Iraqi regime. With about 130,000 troops, concentrated mainly in the Sunni-dominated center of the country around Baghdad, the United States settled in for a long and troubled occupation of the country. The United States handpicked what it called the Governing Council of Iraqis, broadly representative of the country's ethnic composition, and ran many of Iraq's security and other affairs with its own Coalition Provisional Authority (CPA) until June 2004, when it turned over sovereignty to an interim Iraqi government.

The United Nations, with strong U.S. support, organized a January 2005 election for a 275-member transitional national assembly. In this, Iraq's first-ever free election, Iraq's largest ethnic population unsurprisingly won the lion's share of the seats: the Shiites, who turned out in huge numbers to vote, won 48 percent for their main political party. Anticipating the Shiite success and condemning the election as an illegitimate U.S. action, most Sunnis boycotted the vote. That made it possible for the main political party of the Kurds, a smaller ethnic group, to capture another large bloc of the assembly votes—26 percent. Iraq's Sunnis ended up with just 2 percent of the assembly seats. Having been the dominant force in Iraq for decades under Saddam Hussein (who was convicted of war crimes by an Iraqi court and hanged in 2006), the Sunnis now found themselves deprived of power and increasingly angry about the American occupation. The pro-U.S. government meanwhile faced a perplexing array of challenges in cementing a constitution for the fragile country: how much autonomy the Kurds and Shiites might receive within the country, how oil revenues would be split among the geographic and ethnic regions, and what the role of Islam should be in national affairs.

Sunni resistance against the United States grew (●Figure 6.1.24). Using roadside bombs, car bombs, small arms, and rocket-propelled grenades, Sunni militants were able to inflict American casualties on an almost daily basis. Predictably, al-Qa'ida used the U.S. occupation as a rallying cry for intensified resistance against the United States. A well-organized insurgency under the leadership of Jordanian-born Abu Mus'ab al-Zarqawi announced that it was officially linked with al-Qa'ida. His **al-Qa'ida in Iraq** organization kept up a steady stream of bloody attacks on U.S. targets even after al-Zarqawi was killed in 2006. Also in the crosshairs were the Iraqis perceived to be collaborating with the Americans, particularly those who joined the new Iraqi army. The U.S. presence became increasingly confined to its fortified "Green Zone" in Baghdad. While the American death toll mounted—to more than 3,700 by late 2007, so did the Iraqi toll, with at least 75,000 dead to that date. Sunni insurgents relentlessly targeted Shiites in particular, even twice bombing one of the their holiest sites, the golden-domed Askariya Mosque in Samarra.

These Sunni assaults on Shiite sacred space helped push Iraq into **civil war,** defined as an armed conflict "between the government of a sovereign state and domestic political groups mounting effective resistance in relatively continuous fighting that causes high numbers of deaths."[1] Increasingly, this internecine violence took the form of ethnic cleansing, in which one group systematically murdered members of the other. Moderate Shiite leaders such as Ayatollah Ali al-Sistani were no longer able to restrain a Shiite backlash. The more militant anti-Sunni and anti-U.S. Shiite

factions, rallying chiefly behind a cleric named Moqtada al-Sadr and his Mahdi Army (Mahdi Militia), ascended. Shiites took retaliation against Sunnis, and a ceaseless cycle of revenge killings ensued.

At home, the Bush administration found itself in an increasingly difficult position. Costly in lives and dollars—with thousands of soldiers and supporting civilians killed and a price tag of over $100 billion per year—the war grew increasingly unpopular among Americans, and the president's approval rating slid. Many opposition politicians, academics, and others spoke increasingly of Iraq as "Vietnam," referring to a previous war that proved untenable and unwinnable for the United States. The Bush administration faced a dilemma: "stay the course," as President Bush insisted with nearly 170,000 U.S. soldiers still on the ground in Iraq, or pull out (as the United States did in Vietnam) and risk Iraq's plunge into further chaos. Fearing anarchy, the Bush administration in 2007 opted for a "surge" in U.S. troops to maintain order, even as its main ally, Britain, drew down its military presence (primarily in Shiite southern Iraq).

That anarchy could take several forms. A U.S. withdrawal in the absence of strong Iraqi security might allow

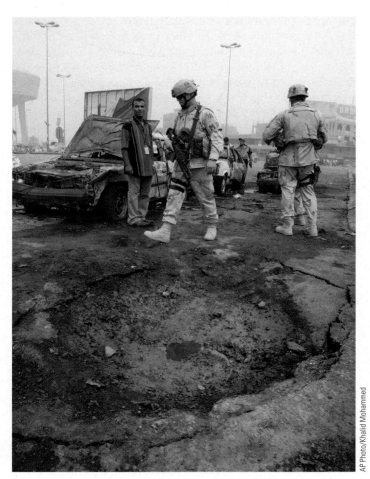

● **Figure 6.1.24** U.S soldiers inspect the scene of a massive suicide bomb targeting a U.S. military convoy in April 2005.

AP Photo/Khalid Mohammed

al-Qa'ida and other militant Islamists to operate with impunity as they did in Afghanistan prior to 9/11, perhaps even using the country as a training ground for more attacks on the West. Al-Qa'ida and its allies would thus accomplish what they promised to do in Iraq—a very unsavory outcome for the West. The country could fragment along ethnoreligious lines into sectarian ministates, potentially including a militant Sunni state and an Iranian-leaning Shiite state. The country's fragmentation could be carefully calculated to reduce the potential fallout. The broad geographic distributions of Sunnis, Shiites, and Kurds seen in Figure 6.1.23 suggest what the boundaries might look like. The biggest problems would come in the big cities like Baghdad and Mosul, where neighborhoods of contentious groups live cheek by jowl.

Oil and Upheaval in Iran

Iran, formerly known as Persia (•Figure 6.1.25), was the earliest Middle Eastern country to produce oil in large quantities and has the world's third-largest reserves. The main oilfields form a line along the foothills of the Zagros Mountains in southwestern Iran. British-built pipelines linked the large refinery at Abadan (on the Shatt-al Arab waterway) with the inland oilfields. Now nearly all exports of crude oil from Iran pass through a newer terminal at Kharg Island (see Figure 6.C, page 230), which allows supertankers to load in deep water. Iran is also endowed with vast natural gas reserves, second only to those of Russia.

Oil is the leading source of revenue for Iran, a country where a dry and rugged habitat makes it hard to provide an agriculture-based livelihood for its population of 71 million. Arid plateaus and basins, bordered by high, rugged mountains, are Iran's characteristic landforms (•Figure 6.1.26). Encircling mountain ranges prevent moisture-bearing air from reaching the interior, creating a classic rain-shadow desert. Only about 10 percent of Iran is farmed. Enough rain falls in the northwest during the winter to permit unirrigated cropping (dry farming), but rains sufficient to support intensive agriculture fall mainly on a densely populated lowland strip between the high, volcanic Elburz Mountains and the Caspian Sea. The western half of this lowland receives more than 40 inches (c. 100 cm) annually and produces a variety of subtropical crops such as wet rice, cotton, oranges, tobacco, silk, and tea. Agriculture elsewhere in Iran depends mainly on irrigation.

Many irrigated lands are on gently sloping alluvial fans (triangular deposits of stream-laid sediments) at the foot of the Elburz and Zagros Mountains; the capital, Tehran (population 12.5 million), is in one of these areas south of the Elburz. *Qanats* (underground sloping tunnels) supply water to a large share of the country's irrigated acreage. However, they are expensive to build and maintain, and an increasing share of water comes from stream diversions and from deep wells. Wheat is the most important irrigated crop.

The leading revenue earner, and Iran's third most important export, after oil and its famous carpets, is the pistachio nut. Cattle are raised in areas where there is enough moisture to grow forage for them, but drier areas support sheep and goats. Many livestock belong to seminomadic mountain peoples—the Qashqai, Baktiari, Lurs, Kurds, and others—whose independent ways have long been a source of friction between these tribespeople and the central government. Before the Islamic revolution in Iran in 1979, it was government policy to sedentarize Iran's nomads so that they did not threaten the state.

Before World War I, Iran was a poor and undeveloped country, in marked contrast to its imperial grandeur when Persian kings ruled a great empire from Persepolis and to its architectural florescence under earlier Muslim leaders. In the 1920s, a military officer of peasant origin, Reza Khan, seized control of the government and began a program to modernize Iran and free it from foreign domination. He had himself crowned as Reza Shah Pahlavi, the founder of the new Pahlavi dynasty. Influenced by the modernizing efforts of Mustafa Kemal Atatürk in neighboring Turkey, the new shah (king) introduced social and economic reforms. In 1941, his young son took the throne as Mohammed Reza Shah Pahlavi and set out to expand the program of modernization and Westernization begun by his father. In the 1970s, Iran played a leading role in raising world oil prices. Mounting oil revenues underwrote explosive industrial, urban, and social development.

In the countryside, the shah's government tried to upgrade agricultural productivity and rural standards of living through land reform and better technologies. But agricultural reforms generally failed, and poverty-stricken

• **Figure 6.1.25** Principal features of Iran.

• **Figure 6.1.26** Aerial view of farms and settlements on an alluvial fan pouring from the western Zagros Mountains, Iran.

families from the countryside poured into Tehran and other cities. From this devoutly Muslim group of new urbanites came much of the support for the revolution that ousted the shah in 1978–1979. Dissent also grew because the regime spent vast sums on its military, mainly for U.S.-made weapons, thus undercutting benefits to the people.

The revolution that overthrew the shah took Iran abruptly out of the American orbit. The central figure among the revolutionaries was an **ayatollah** ("sign of God") named Ruhollah Khomeini, an elderly critic of the regime whom the shah had forced into exile in 1964. From Paris, Khomeini sent repeated messages to Iran's Shiites that helped spark the revolution. A furious tide of Shiite religious sentiment rose against corruption, police heavy-handedness, modernization, Westernization, Western imperialism, the exploitation of underprivileged people, and the monarchy, which Shiite beliefs regard as illegitimate. Growing numbers of people staged strikes and demonstrations in 1978, forcing the shah to flee the country and abdicate his throne. After hospitalization in the United States and life in exile in Panama and Egypt, the shah died in Cairo in 1981 and is buried there. The Ayatollah Khomeini returned in triumph to Iran in 1979. His followers seized and held 52 U.S. diplomatic personnel as hostages in Tehran for 444 days. U.S. President Jimmy Carter proved unable to resolve this crisis, which contributed to his defeat in his 1980 bid for reelection.

Under Khomeini (who died in 1989), the country became an Islamic republic governed by Shiite clerics, who included a handful of revered and powerful ayatollahs at the head of the religious establishment, and an estimated 180,000 priests called **mullahs.** Khomeini served as the Guardian Theologian, an infallible supreme leader enshrined by the principle of **divine rule by clerics** (*velayet-e-faqih*). These religious leaders continue to supervise all aspects of Iranian life and perform many functions allotted to civil servants in most countries. They base their authority on Shiite interpretations of Islam (about 89 percent of Iran's population is Shiite).

In contrast to their status during the shah's reign, when they were encouraged to take on larger public roles, women in post-revolutionary Iran have lost many freedoms. Under the new regime, "polluting" influences of Western thought and media were purged. However, recent years have seen a general softening of restrictions on personal freedom of expression, and the Internet and other media have introduced broader worldviews to Iran's youth, a very important cohort, as about 60 percent of the population is under the age of 30 (•Figure 6.1.27). Young people led an unprecedented series of prodemocracy protests in 2002. Five years later, Iranians of different generations took to the streets to protest government hikes on gasoline prices; ironically, this oil-rich country has a low oil refining capacity and needs to import one-third of its gasoline. Iran imports more gasoline than any country in the world except for the United States.

Prior to 2005, Iran watchers saw a tug-of-war between the moderate and reform-minded president Muhammad Khatami and the less tolerant supreme religious leader, Ayatollah Ali Khamanei (Khomeini's successor), who has lifetime control over the military and the judicial branch of government. There was speculation that the country's younger and more Western-oriented sectors might win the day and that relations with the wider world—and the United States in particular—would improve. But the winner of Iran's 2005 presidential election (for a five-year term

© Abbas/Magnum Photos

• **Figure 6.1.27** Iranian women demanding the release by the Iranian government of Jamil Bassam and Abrahim Khodabandeh

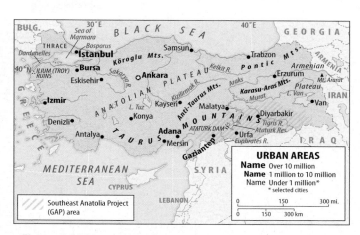

• **Figure 6.1.28** Principal features of Turkey.

of office) was Mahmoud Ahmadinejad, who outspokenly opposed American policies in the region, goaded Israel by insisting that the Holocaust never happened and calling for the country's destruction, and accelerated Iran's nuclear power program. The level of rhetoric and tension between Iran and the United States escalated. American officials accused Iran of supplying weapons used to kill U.S. soldiers in Iraq, and Iran began gearing up for the possible imposition of economic sanctions by the United States and its European allies.

6.1.7 Turkey: Where East Meets West

Like Iran, Turkey is a demographic heavyweight (population 74 million people) and has long had an enormously important presence in the region (•Figure 6.1.28). The Turks who organized the Ottoman Empire beginning in the 14th century originated as pastoral nomads from central Asia, where Turkic cultures still prevail. From the 16th century to the 19th, their empire, based in what is now Turkey, was an important power. Modern Turkey was created from the wreckage of the old empire after World War I. Its founder, Mustafa Kemal Atatürk, was determined to Westernize the country, raise its standard of living, and make it a strong and respected national state. He inaugurated social and political reforms designed to break the hold of traditional Islam and open the way for modernization and Turkish nationalism.

Islam had been the state religion under the Ottoman Empire, but Atatürk divorced church from state, and to this day, Turkey is the only Muslim Middle Eastern country to officially separate them. The wearing of the red cap called the *fez,* an important symbolic act under the Ottoman caliphs, was prohibited, and state-supported secular schools replaced the all-pervasive religious schools. To facilitate

public education and remove further traces of Muslim culture, Latin characters replaced the Arabic script of the Qur'an. Slavery and polygamy were outlawed, and women were given full citizenship. Legal codes based on those of Western nations replaced Islamic law, and forms of democratic representative government were instituted, although Atatürk ruled in a dictatorial fashion.

Turkey has continued to have trouble establishing a fully democratic system, and there have been frequent periods of military rule (•Figure 6.1.29). Turkey's constitution invests the military with the responsibility of protecting the country's secular government, and the army has used its powers to overthrow four elected governments since 1960. In 2007, Turkish voters elected a president, Abdullah Gul, with a strong background in Islamic politics. Questions immediately arose over how long the army might tolerate his rule. There was also great controversy about the fact that President Gul's wife wore an Islamic headscarf in public: the wearing of the headscarf in public offices and schools is officially banned, and Turkey's secularists see the scarf as an affront to the country's identity.

Turkey is composed physically of two units: the Anatolian Plateau and associated mountains occupying the interior of the country and the coastal regions of hills, mountains, and small plains bordering the Black, Aegean, and Mediterranean Seas. Turkey in general has a physical presence quite unlike the popular image of the Middle East and North Africa as a desert region; it is in fact the only country in the region that has no desert, and cultivated land occupies about a third of its land area. The Anatolian Plateau, an expanse of wheat and barley fields and grazing lands, is bordered on the north by the lofty Pontic (Kuzey) Mountains and on the south by the Taurus Mountains. The plateau ranges from 2,000 to 6,000 feet (c. 600 to 1,800 m) in elevation and is highest in the east where it adjoins the mountains of the Armenian Knot. Its surface is rolling and windswept, hot and dry in summer and cold and snowy in winter, with a natural vegetation of short steppe grasses and shrubs. Production of cereals (especially barley, wheat,

• **Figure 6.1.29** The military has a strong presence is secular Turkey. These soldiers are guarding an archaeological site in eastern Turkey, a region of Kurdish rebellion against government.

• **Figure 6.1.30** Where there had been a bridge across the Euphrates River, there is now a ferry across the Ataturk Reservoir, upstream from the huge Ataturk Dam.

and corn) and livestock (especially sheep, goats, and cattle), employing both traditional and mechanized means, prevails in Anatolia. Coastal plains and valleys along and near the Aegean Sea, Sea of Marmara, and Black Sea are generally Turkey's most densely populated and productive areas. Here are grown cash crops such as hazelnuts, tobacco, grapes for sultana raisins, and figs. The slopes of the Pontic Mountains facing the Black Sea have copious rains in both summer and winter and support prolific tea harvests (see Figure 6.9, page 216).

Relative poverty compared with the nearby European countries is a striking characteristic of present-day Turkey. A strong rural and agricultural component in Turkish life persists, and about one-third of the population is still classified as rural. However, the country has embarked on a course of change that promises to modernize its agriculture, expand industry, and raise general standards of living. The value of output from manufacturing (mainly textiles, agricultural processing, cement manufacturing, simple metal industries, and assembly of vehicles from imported components) is already greater than that from agriculture. Some of Europe's most recognizable brand names, like Bosch, Renault, and Fiat, have factories in Turkey. Turkey does produce some oil.

Early in the 1990s, Turkey began an impressive $32 billion agricultural effort called the **Southeast Anatolia Project** (known by its Turkish acronym **GAP**). Its aim is to convert the semiarid southeastern quarter of the country into the "breadbasket of the Middle East." Surpluses of cotton, wheat, and vegetables are exported, along with value-added manufactures such as cotton garments and textiles. The centerpiece of this massive irrigation project is the Atatürk Dam on the Euphrates River. About 20 other dams on the Turkish Euphrates and Tigris are also part of the ongoing project to provide hydroelectricity as well as water (•Fig-

ure 6.1.30). The GAP is controversial. It is inundating historic settlements and priceless archaeological treasures and is compelling people to move. It is being developed in the part of the country where a restive Kurdish population is seeking recognition, autonomy, and among some factions, independence from Turkey. In addition, Turkey's Tigris and Euphrates waters flow downstream into neighboring states, raising serious questions about downstream water allocation and quality.

Turkey is an "in-between" country. Economically, it is well below the level of the MDCs but above most of the world's LDCs. Culturally, it is between traditional Islamic and secular European ways of living. Part of Turkey—the section called Thrace—is actually in Europe, and Turkey as a whole aspires to become more European. It has been a member of the North Atlantic Treaty Organization (NATO) since 1952. It is an associate member of the European Union and is a candidate for full membership. The European Union, however, would first like to see Turkey become more European politically, for example, by diminishing the army's influence in government and by improving its human rights record. Turkey insists it has made huge strides toward satisfying the minimal democratic and human rights norms for EU member countries. It has abolished the death penalty and eased restrictions it imposed on the cultural expressions of its minority Kurdish population.

At expansion talks in 2002 and 2006, EU leaders nevertheless rejected Turkey's bid and would not promise a date for membership negotiations. They argued that Turkey has still not gone far enough with democratic and human rights reforms. They probably feared the costs of absorbing Turkey's mainly low-income population into the union. Many Turks feel that their membership in the European Union was denied because they are a Muslim people. It is possible that the main reason Turkey was rejected was because EU leaders feared that Islamist terrorists would make their way from an EU Turkey into the heart of continental Europe.

The EU's rejection of Turkey may fan the growing flames of sentiment in the Islamic world that the "crusader West" is waging a systematic campaign against Muslims. Turkish public opinion has become decidedly more anti-European; a 2006 poll showed only 35 percent support for the EU, half that of 2004. A Turkey shunned by Europe may seek alliances eastward in Russia, India, and China.

Istanbul (population 12.2 million), founded as Constantinople, is Turkey's main metropolis, industrial center, and port (•Figure 6.1.31). One of the world's most historic and cosmopolitan cities, Istanbul was for many centuries the capital of the Eastern Roman (Byzantine) Empire. It became the capital of the Ottoman Empire when it fell to the Turks in 1453. However, the capital of the Turkish Republic was established in 1923 at the more centrally located and more purely Turkish city of Ankara (population 3.6 million) on the Anatolian Plateau. Turkey's third-largest city is the seaport of Izmir on the Aegean Sea.

82, 162

Istanbul is located at the southern entrance to the Bosporus Strait, the northernmost of the Turkish Straits that connect the Mediterranean and Black Seas. The straits have long been a focus of contention between Turkey and Russia, who fought several wars in the past; their relations are much better now. Turkey's military posture has also been shaped by fragile relations with its NATO neighbor, Greece. The main questions recently at issue have been the status of Cyprus and conflicting claims to hundreds of islands in the Aegean Sea and that sea's oil and gas reserves. Already an EU member, Greece pressured Turkey to revise its policy toward Cyprus and to settle the Aegean Sea dispute as preconditions for Turkey's joining the organization.

229

136

6.1.8 Rugged, Strategic Afghanistan

High and rugged mountains dominate Afghanistan (population 32 million). This is the only landlocked nation of the Middle East and North Africa, and like Sudan and Turkey, it is a country that is transitional between regions; some geographers place it in central Asia or South Asia (•Figure 6.1.32). Historically, it has occupied an important strategic location between India and the Middle East. Major caravan routes crossed it, and a string of empire builders sought control of its passes. Today, it is a land of limited resources, poor internal transportation, little foreign trade, and much conflict; Afghanistan is one of the poorest of the world's poor countries.

Most Afghans live in irrigated valleys around the fringes of the high mountains that occupy a large part of the country. The country's second most populous area is the northern side of the central mountains, where people live in oases in foothills and steppes. Northern Afghanistan borders three of the five central Asian countries, and millions of people on the Afghan side are related to peoples of those countries. The most populous area is the northeast, particularly the fertile valley of the Kabul River, where the capital and largest city, Kabul (population 3.2 million; elevation 6,200 ft/1,890 m), is located. Most of the inhabitants of the southeast are Pashtuns (also known as Pushtuns or Pathans). Their language, Pashto, is related to Persian. The Pashtuns are the largest and most influential of the numerous ethnic groups that make up the Afghan state; the next largest are the Tajik and Hazara, respectively.

221, 222

The independent-minded tribal Pashtun people have always been loath to recognize the authority of central governments. In the days of Britain's Indian Empire, the area saw warfare among tribes, tribal raids on British-controlled areas, and British punitive expeditions against the tribes. In those days, Peshawar—on the Indian (now Pakistani) side of the Khyber Pass into Afghanistan—became the hotbed for British, Russian, and other agents playing what came to be known as the **Great Game** of vying (and spying) for strategic influence in this part of the world.

327

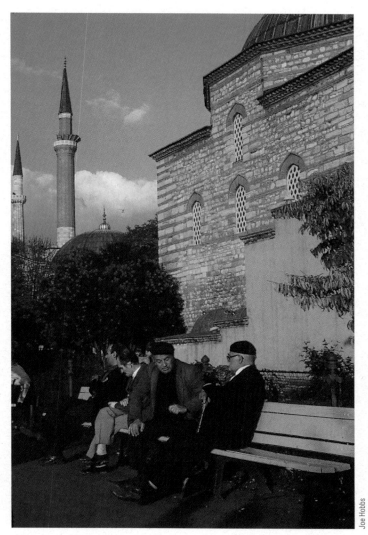

Joe Hobbs

• **Figure 6.1.31** The people of Istanbul live amidst architectural treasures with roots in many civilizations.

• **Figure 6.1.32** Principal features and ethnic groups of Afghanistan.

Afghanistan is overwhelmingly a rural agricultural and pastoral country, and its agriculture bears many of the customary Middle Eastern earmarks: traditional methods, simple tools, limited fertilizer, and low yields. It is so mountainous and arid that only about 12 percent is cultivated. Enough rain falls in the main populated areas during the winter for dry farming of grains (•Figure 6.1.33). A variety of cultivated crops, fruits, and nuts are important locally. Many people raise livestock on a seminomadic or nomadic basis.

The most successful crop has been the opium poppy, which routinely accounts for between one-third and two-thirds of the country's gross domestic product. A succession of rebel armies and then Afghanistan's government under the Taliban (see next page) used revenues from opium and heroin to obtain their arms. Afghanistan was by far the world's largest opium producer when in 2001, the Taliban reversed course and imposed a ban on cultivation. There was virtually no opium harvest that year. After the United States ousted the Taliban from power, Afghanistan regained its former status, producing fully 92 percent of the world's opium in 2007. Most of the production takes place in the southern province of Helmand, where the Taliban have retrenched themselves most successfully. Once again, the Taliban are gaining funds and strength from the drug economy. Partly in order to restrain the Taliban, Britain and the United States have stepped up efforts to eradicate opium. A policy split has developed among foreign forces: the NATO mission in Afghanistan worries that in the absence of crop alternatives, elimination of the poppy crop could turn destitute farmers into vengeful fighters. One option rarely called for but perhaps most appropriate in Afghanistan's future, would be to allow the country to produce opium legally for

pharmaceutical needs, as a handful of countries, including Turkey and India, are allowed to do.

Through most of the 20th century, this highland country was remote from the main currents of world affairs. After the Islamic revolution in Iran in 1979, however, Afghanistan's location next to that oil-rich country made it once again the target of foreign interests. The Soviet military intervention of 1979 and the ensuing devastation catapulted the country into world prominence. The USSR's motives for its invasion of Afghanistan may have included a desire to prevent by force the spread of Iranian-style Islamic fundamentalism into Afghanistan (which is 80 percent Sunni and 19 percent Shiite) or into the central Asian Soviet republics deemed vital to the superpower's security. For its part, the United States warned the Soviet Union that it would not tolerate further Soviet expansionism.

The ensuing war entailed the widespread killing and maiming of civilians, destruction of villages, burning of crops, killing of livestock, destruction of irrigation systems, and sowing of land mines over vast areas (see Insights, page 274). Soviet ground and air forces caused several million Afghan refugees to flee into neighboring Pakistan and Iran. Arms from various foreign sources filtered into the hands of the *mujahidiin,* the anti-Soviet rebel bands that kept up resistance in the face of heavy odds. The United States was one of the powers supporting the rebels and in that sense waged a proxy war against the Soviet Union in Afghanistan. In its waning years, the USSR recognized that it could not win its "Vietnam War," and its troops withdrew from Afghanistan by 1989. It is estimated that of the 15.5 million people who lived in Afghanistan when it was invaded in 1979, at least 1 million died, 2 million were displaced from their homes to other places within the country, and 6 million fled as refu-

THOMAS J. ABERCROMBIE/National Geographic Image Collection.

● **Figure 6.1.33** The Bamiyan Valley and the Koh i Baba Range of mountains Afghanistan.

gees into Pakistan and Iran. By 2007, about two-thirds of those refugees had returned.

Support that the United States and moderate Arab states such as Egypt and Saudi Arabia lent to the *mujahidiin* and sympathetic Arab fighters soon came back to haunt those countries in a phenomenon that came to be known as **blowback.** Emboldened by their victory against the Soviets, the most militant Islamists among the fighters turned their attention to the United States and its Middle Eastern allies. Chief among them was Osama bin Laden, who developed "the base" (*al-Qa'ida* in Arabic) of thousands of Afghans, Arabs, and other anti-Soviet war veterans he could call on to wage a wider *jihad*. Bin Laden's organization trained an estimated 10,000 fighters in al-Qa'ida camps in Afghanistan and from this unlikely, remote setting devised spectacular acts of terrorism: an assassination attempt on Egyptian President Hosni Mubarak, the bombing of U.S. military barracks in Saudi Arabia, the bombing of U.S. embassies in Kenya and Tanzania, the attack on the U.S.S. *Cole*, and the 9/11 attacks. The East African bombings in 1998 led within days to a cruise missile strike by the United States on six al-Qa'ida camps near Khost, Afghanistan. Many al-Qa'ida personnel were killed, but not its top leaders.

The *mujahidiin* succeeded in overthrowing the Communist government of Afghanistan in 1992, but after that, rival factions among the formerly united rebels engaged in civil warfare. During this period, most other countries utterly neglected Afghanistan, inadvertently promoting the growth of militant movements (which tend to thrive in remote, underdeveloped, and ungoverned regions). By 1996,

one of the rebel factions, the **Taliban** (backed by Saudi Arabia and Pakistan), gained control of most of the country, including the capital of Kabul. Proclaiming itself the sole legitimate government of Afghanistan, the Taliban imposed a strict code of Islamic law in the regions under its control and gained international notoriety for its austere administration. The Taliban removed almost all women from the country's workforce, forbade public education of girls, and outlawed "un-Islamic" practices such as dancing, flying kites, watching television, keeping birds, and trimming beards.

The Taliban continued to make advances against its opponents inside Afghanistan (particularly the **Northern Alliance**, whose leader, Ahmed Shah Massoud, was assassinated just days before and in apparent preparation for the 9/11 attacks) and, by 2001, controlled 95 percent of the country's territory. The neighboring central Asian countries, Russia, and even Iran grew increasingly fearful of the spread of the Taliban's extreme interpretation of Islam into their nations. Russia ended up supporting some of the rebels it had fought because they were now fighting the Taliban. Russia's once unlikely ally in supporting those rebels was the United States, which successfully lobbied the United Nations to levy economic sanctions against Afghanistan in 1999. The United States hoped that economic losses would pressure the Taliban into turning over bin Laden for prosecution and also announced a $5 million reward for information that would lead to bin Laden's capture or death.

The bin Laden bounty rose into the tens of millions of dollars after September 11, 2001. Named as the mastermind of the attacks against New York and Washington,

INSIGHTS

Land Mines

The scars of war will last a long time in Afghanistan. One of the country's urgent reconstruction tasks is to rid the landscape of millions of land mines sown during decades of war. These weapons have already killed or maimed more than 200,000 Afghans since 1979, and they continue to kill or maim about 60 Afghans, the great majority of them civilians, every month.

Land mines are antitank and antipersonnel explosives, usually hidden in the ground, designed to kill or severely wound the enemy. There are many kinds: Soviet troops planted at least nine varieties in Afghanistan, including seismic mines, which are triggered by vibrations created by passing horses or people; mines with devices that pop up and explode ("Bouncing Betties") when a person approaches; "butterfly" mines meant to maim rather than kill; and explosives disguised as toys, cigarette packs, and pens (●Figure 6.1.C).

There may be as many as half a million land mines in Afghanistan, or 2 per square mile (5/sq km). Reconstruction of this agricultural country cannot begin in earnest until the mines are cleared. The obstacles are enormous in rugged Afghanistan. Mines typically remain active for decades after they are planted and tend to last longer in arid places. The mountainous terrain prevents systematic clearance of the explosives. Many mines were scattered

Joe Hobbs

● **Figure 6.1.C** Inexpensive to manufacture and deploy, land mines are a scourge that outlast conflict and bring suffering to civilians.

indiscriminately from the air and will be difficult to locate. Heavy rains move the mines downhill and away from known locations. And it is expensive to clear mines; Kuwait spent $800 million to rid its landscape of them after the Gulf War. So far, Western countries have donated more than $100 million toward the expense of clearing Afghanistan's land mines, and over 200,000 been cleared; Russia has contributed nothing.

Land mines are a global problem. Up to 70 million mines contaminate the land in over 80 countries, and they kill or maim more than 15,000 civilians each year. The numbers and concentrations in some countries are staggering. The leader in numbers is Egypt, with 22 million mines; in density of mines, Cambodia leads with about 100 per square mile (40/sq km). The United Nations estimates that with current technology, it would take over $30 billion and 1,100 years to clear the world's land mines.

Several developments may portend an eventual end to the land mine scourge. Paul McCartney and other celebrities have led high-profile campaigns against these weapons (see http//:landmines .org). The world's most visible critic of land mines, Princess Diana of Wales, was killed in an automobile accident in Paris in 1997. In her wake, there were renewed calls for a moratorium on the manufacture and use of antipersonnel land mines. Two months after Diana's death, anti–land mine activist Jody Williams won the Nobel Peace Prize. Soon after, representatives from 125 nations met in Ottawa, Canada, to sign a treaty strictly banning the use, production, storage, and transfer of antipersonnel land mines. Many of the largest producers, users, and exporters of these explosives, including the United States, Russia, China, India, Pakistan, Iran, Iraq, and Israel, refused to sign. The United States argued that land mines were necessary to protect South Korea against an invasion from North Korea and failed to win an exemption on this count that would have allowed it to sign the treaty.

Too many countries regard these as legitimate weapons to make the prospect of a mine-free world likely. Costing as little as $2 each, they are very affordable weapons for cash-strapped warring LDCs (and so have been dubbed the "Saturday night specials of warfare"). For now, these hidden horrors continue to be planted and to be cleared, as one demining specialist put it, "one arm and one leg at a time."

bin Laden became the "world's most wanted" as U.S. President George W. Bush evoked a Wild West vow to have him captured "dead or alive." In crafting its pronounced war against terrorism (which it promised to carry anywhere in the world necessary), the U.S. administration identified the Taliban as al-Qa'ida's mentor and targeted both organizations for elimination.

Within a month of the attacks on the United States, American warplanes and special forces struck in **Operation**

Enduring Freedom against Taliban and al-Qa'ida facilities and personnel around Afghanistan. The British military assisted in the air campaign, and most of the ground fighting was carried on by the Northern Alliance, the Taliban's rival and nemesis. The United States successfully persuaded Pakistan to drop its support of the Taliban and join the effort, and much of the military campaign was based on U.S.-Pakistani cooperation. One after another of the Taliban's urban strongholds fell, including Kabul and the Taliban spiritual capital, Kandahar. Withering assaults from U.S. warplanes crushed the tunnels and other hideouts used by al-Qa'ida guerrillas, and many al-Qa'ida prisoners were taken to the U.S. naval base in Guantánamo Bay, Cuba. The elusive bin Laden, however, along with the Taliban leader Mullah Omar, slipped away, perhaps seeking safe harbor in the ethnic Pashtun region of western Pakistan.

The reconstruction of Afghanistan began almost as soon as the Taliban were driven from power. From the billions of dollars in aid pledged, a new highway has been built connecting Kabul with Kandahar. Hospitals are being built in a country that has some of the highest infant and maternal mortality rates in the world. The numbers of Afghans enrolled in primary schools rose from 25 percent in 1999 to over 90 percent in 2007, and girls and women are again being educated. There are plans to replant landscapes that have been almost completely deforested. Afghanistan has resources of its own. Further prospecting may reveal that it

has the world's largest copper deposits, and it is known to have large stores of high-grade iron ore, along with some gas, oil, coal, and precious stones.

Progress is slow because the country is so devastated and because of the tenuous security situation. American forces and the Afghan government headed by Hamid Karzai effectively took control of only a small area around Kabul, and NATO troops were stationed in the troubled south of the country. Much of the country remains in the hands of warlords whose allegiance is up for grabs. Al-Qa'ida and Taliban fighters operate in the rugged countryside, sometimes carrying fighting into Kabul and other cities. There are fears of a resurgent Taliban engaging Western forces in a conflict that would parallel the long war against the Soviets in Afghanistan. But if peace could prevail, it would be possible to rebuild this shattered land and bring hope to a country that is all too symptomatic of the troubled Middle East and North Africa.

SUMMARY

→ The United Nations Partition Plan of 1947 attempted a two-state solution to the dilemma Britain had created by promising land to both Arabs and Jews in Palestine. It envisioned geographically fragmented states, making each side feel vulnerable to the other. War prevented the plan's implementation. The Arab-Israeli conflict has continued from that time, with Israel gaining more territory and the indigenous Palestinians failing to acquire a country of their own. Principal obstacles to peace between Israelis and Palestinians are the status of Jerusalem, the potential return of Palestinian refugees, the future of Jewish settlements in the Occupied Territories, and Israeli construction of a security barrier that penetrates into portions of the Palestinian West Bank.

→ Israel's eastern neighbor, Jordan, has a majority Palestinian population.

→ In the mid-1970s, Israel's northern neighbor, Lebanon, was plunged into a devastating civil war that sparked when minority Christians failed to yield power to majority Muslims and was protracted by an Israeli invasion to crush the Palestine Liberation Organization. Lebanon's "Cedar Revolution" of 2005 led to Syrian troop withdrawal from the country, but the following year saw Israel attempt to crush the Shiite organization Hizbullah. That war destroyed much of Lebanon's rebuilt infrastructure and apparently strengthened Hizbullah politically.

→ Egypt is a populous country with a limited inhabitable area along the Nile. It has transformed the Nile Valley with thousands of years of habitation and recent large engineering works, but it remains a poor country dependent on outside assistance.

→ Sudan is a transition zone between the Middle East and sub-Saharan Africa. Its Muslim Arab government has tried to impose its will forcefully on the non-Arab south, promoting civil war. The conflict may have ended, but Sudanese Arabs have carried out ethnic cleansing against non-Arabs in the Darfur region of western Sudan.

→ Most of Libya lies in the Sahara, and its primary natural resource is oil. A massive irrigation project is tapping ancient underground waters. Colonel Muammar Qaddafi has given up support for international terrorism and weapons of mass destruction programs, leading to much improved relations with the West.

→ The Maghreb ("western land") is composed of Morocco, Algeria, Tunisia, and the disputed Western Sahara. Morocco is the region's poorest country, Algeria has the greatest mineral wealth, and Tunisia is the most progressive. Morocco controls Western Sahara, whose Saharawi people want independence.

→ Two-thirds of the world's proven petroleum reserves are concentrated in a few countries that ring the Persian-Arabian Gulf. Saudi Arabia controls more than one-quarter of the world's oil. Growing populations and the prospect of the oil running out present some critical challenges. Some of the countries, notably Saudi Arabia, are very traditional, whereas the United Arab Emirates and Bahrain have tolerated social change. On the southern end of the Arabian Peninsula, Oman has enjoyed oil wealth, but Yemen remains poor.

→ Under Saddam Hussein, Iraq squandered its oil wealth on military misadventures, including invasions of Iran and Kuwait. The Kuwait invasion led to an enormous U.S.-led counterattack that devastated the country's infrastructure and subjected it to years of economic sanctions. The U.S.-led invasion of Iraq in 2003 resulted in Saddam Hussein's downfall and death, and an uncertain future for this ethnically diverse country (with large Shiite Arab, Sunni Arab, and Kurdish populations) engaged in civil war.

→ Iran's oil revenues were used to modernize and Westernize the nation during the Pahlavi dynasty. Rapid social change and uneven economic benefits precipitated a revolutionary Islamic movement, which forced the shah to abdicate and flee in 1979. This theocratic state has had difficult relations with the United States, most recently over its alleged nuclear weapons program. Iran's young population is agitating for change, but the ruling clerics are resisting.

→ Turkey was formerly the seat of power for the Islamic Ottoman Empire. It is a secular nation with membership in NATO, and hopes to join the European Union. It is blessed with freshwater resources that it has mustered with the hope of creating the "breadbasket of the Middle East."

→ Afghanistan is one of the world's poorest nations. It is land-locked and has limited resources, poor internal transportation, and little foreign trade. Opium is its main export. Its location along the ancient caravan routes and adjacent to large oil reserves have made it the target of stronger powers. Backed by the United States, rebels succeeded in driving out a Soviet occupation force in the 1980s. That costly conflict was followed by a period of civil war during which outside countries neglected Afghanistan. The militant Islamist Taliban came to power and protected the presence and training of al-Qa'ida militants. Osama bin Laden and others planned anti-Western attacks, including 9/11, from Afghanistan. Al-Qa'ida was driven into hiding from there in U.S.-led attacks following 9/11, and the Taliban were deposed from power. Reconstruction of the country is progressing, but a Taliban insurgency continues.

KEY TERMS + CONCEPTS

1948–1949 Arab-Israeli war (p. 241)
1973 Arab-Israeli war (p. 242)
1982 Lebanese war (p. 250)
2006 Israeli-Hizbullah war (p. 251)
al-Aqsa *Intifada* (p. 246)
Alawite (p. 251)
al-Qa'ida in Iraq (p. 266)
Ashkenazi Jews (p. 243)
ayatollah (p. 268)
barrage (p. 252)
basin irrigation (p. 252)
Battle of Beirut (p. 250)
blowback (p. 273)
Camp David Accords (p. 243)
cataract (p. 252)
Cedar Revolution (p. 251)
center-pivot irrigation (p. 255)
civil war (p. 266)
divine rule by clerics (p. 268)
Druze (p. 243)
erg (p. 257)
"facts on the ground" (p. 244)
Falashas (p. 243)
final status issues (p. 246)
foggara (p. 257)
fossil waters (p. 255)
Gaza-Jericho Accord (p. 245)
ghurba (p. 244)

"gift of the Nile" (p. 252)
Great Game (p. 271)
Great Manmade River (p. 255)
Green Line (p. 246)
Gulf War (p. 264)
Intifada (p. 247)
Iran-Iraq War (p. 264)
Janjawiid (p. 255)
Jewish settlements (p. 244)
Judaea and Samaria (p. 244)
kibbutzim (p. 247)
Kurdistan Workers Party (PKK) (p. 265)
Labor Party (p. 247)
land mines (p. 274)
Law of Return (p. 243)
Lebanese civil war (p. 249)
Likud Party (p. 244)
Ma'adan ("Marsh Arabs") (p. 265)
moshavim (p. 247)
mujahidiin (p. 272)
mullah (p. 268)
Northern Alliance (p. 273)
Occupied Territories (p. 242)
Operation Enduring Freedom (p. 274)
Operation Iraqi Freedom (p. 265)
Oslo I Accord (p. 245)
Oslo II Accord (p. 245)

Palestine Liberation Organization (PLO) (p. 245)
Palestinian Authority (PA) (p. 245)
Palestinians (p. 241)
perennial irrigation (p. 252)
Polisario Front (p. 258)
pre-1967 borders (p. 241)
qanat (p. 257)
right of return (p. xxx)
"road map for peace" (p. 246)
Saharawis (p. 258)
salinization (p. 253)
Saudization (p. 260)
security fence (p. 246)
Sephardic Jews (p. 243)
Six-Day War (p. 242)
Southeast Anatolia Project (GAP) (p. 270)
Sudanese People's Liberation Army (SPLA) (p. 254)
"Switzerland of the Middle East" (p. 249)
Taliban (p. 273)
two-state solution (p. 241)
United Nations Resolutions 242 and 338 (p. 247)
Wahhabism (p. 260)

REVIEW QUESTIONS

1. What precipitated the various conflicts between Israel and its Arab neighbors? How did each of these conflicts rearrange the political map?

2. What lands did the peace process of the Oslo Accords yield to Palestinian control? What is the current situation in those areas? Who controls the Gaza Strip and the West Bank?

3. What are the major ethnic groups and territories of Lebanon?

4. How did Egypt transform the Nile Valley from a system of basin irrigation to one of perennial irrigation?

5. In what ways is Sudan's north different from its south? What significant conflicts have affected the country?

6. What is the role of the Great Manmade River?

7. What are the main resources of the Maghreb countries? What issues do they have with countries outside the region?

8. Which Persian-Arabian Gulf countries have the most oil and natural gas? Have Oman and Yemen benefited from the oil boom?

9. What happened when Saddam Hussein's troops invaded Kuwait? What happened when they invaded Iran? What critical decisions must the United States make in Iraq?

10. What is the ethnic composition of Iraq? How did that consideration cause the United States to avoid invading the country in 1991?

11. Who are the Kurds, and where do they live?

12. What were some of the grievances against the Pahlavi dynasty that led to revolution in Iran? What changes has the Islamic regime brought to the country?

13. In what ways is Turkey an "in-between" country?

14. What were Afghanistan's shifting fortunes between 1979 and the present?

NOTES

1. Quoted in Nicholas Sambanis, "It's Official: There Is Now a Civil War in Iraq." *New York Times,* July 23, 2006, p. WK-13.

View across the Swat Valley in northern Pakistan, with the Hindu Kush Mountains rising along the Afghan border in the distance.

Joe Hobbs

CHAPTER 7
A GEOGRAPHIC PROFILE OF MONSOON ASIA

Monsoon Asia is a great triangle that runs from Pakistan in the west to Japan in the northeast and the island of New Guinea in the southeast (•Figure 7.1). In this land area of less than 8 million square miles (20,761,000 sq km)—about twice the size of the United States (•Figure 7.2), or less than one quarter of the earth's landmass—lives more than half of the world's population (•Figure 7.3). The cultural makeup of these 3.6 billion people is as diverse and expressive as the physical environments they inhabit. The world's highest mountain peaks, some of the longest rivers, and the earth's most highly transformed and densely settled river plains are the settings for some of the oldest civilizations and the most modern economies. Just as this region has played a major role in human development over many millennia, it is a major force in world affairs in the 21st century. Some observers say this will be known as the **Asian Century** for the ways in which the region will reshape the globe.

chapter objectives

This chapter should enable you to

→ Recognize the role of seasonal monsoon winds and rains in people's livelihoods and perceptions

→ Appreciate China and India as the demographic giants and surging economies of early-21st-century Asia

→ Learn how unique spatial and spiritual considerations have influenced the traditional layout of Asian settlements

→ Know the basic beliefs of Hinduism, Buddhism, Confucianism, and Daoism

→ Become familiar with the pros and cons of the Green Revolution

→ Understand the geopolitical dimensions of the tensions between India and Pakistan, North Korea and the West, and Islamists and governments in Pakistan and Indonesia

The countries of eastern Eurasia were at one time referred to as "the Orient," from a Latin word meaning "facing the east" or "facing the sunrise." (The West was then known as "the Occident," derived from a word meaning "facing the sunset.") But this terminology reflected the view from Europe and became associated with Western stereotypes and misconceptions about Asia. Consequently, those terms have fallen out of favor. This text refers to the region shown in Figure 7.1 as Monsoon Asia because it connotes a major physical attribute that affects most of this region and also in acknowledgment of the achievements of the geographer George Cressey, who coined the name and wrote so well of the region.

7.1 Area and Population

Monsoon Asia encompasses the following subregions, which in some geography books are classified as separate regions:

✴ East Asia, which includes Japan, North and South Korea, China, Mongolia, Taiwan, and many near-shore islands

✴ South Asia, which includes Pakistan, India, Sri Lanka, Bangladesh, the mountain nations of Bhutan and Nepal, and the island country of the Maldives

✴ Southeast Asia, which includes both the peninsula jutting out from the southeast corner of the Asian continent, on

which are located Myanmar (Burma), Thailand, Laos, Cambodia, Vietnam, Malaysia, and Singapore, and the island world that rings this peninsula, which includes the countries of Indonesia, the Philippines, Brunei, and Timor-Leste (Table 7.1; see also Figure 7.1).

Demographic Heavyweights of Monsoon Asia

Perhaps the most critical fact of Monsoon Asia's geography is that this region is home to about 54 percent of the world's population (•Figure 7.3; see also Figures 3.14 and 3.15, pages 57 and 58). Two countries alone, India and China, together have 2.4 billion people, or 37 percent of the world total. Their extraordinary demographic weight reflects thousands of years of human occupation of productive agricultural landscapes, combined with 20th-century advances in health technology that reduced death rates. The highest population densities are found in the areas of most abundant precipitation or surface water and with the most productive soils; rivers and coastal plains are especially densely settled. Rugged mountains, almost waterless deserts, and until recently, tropical rain forest vegetation have limited populations elsewhere. 50

Some of the world's highest urban densities are found in the cities of Hong Kong, Macao, and Singapore, but in Mongolia and Nepal, population densities are extremely low. Singapore is one of the few nations to claim a 100 percent urban population; in contrast, only 8 percent of Timor-Leste's population is urban. A mere 3 percent of Bhutan is arable, whereas farmers in India and Bangladesh cultivate more than 50 percent of their lands.

Population Growth Patterns

It is not possible to generalize about population growth in Monsoon Asia. Growth rates range from 0 percent per year in Japan to a very high 3.3 percent per year in Timor-Leste. The region does reflect the general worldwide trends of poorer countries having higher growth rates and wealthier countries having lower growth rates. The region is demographically and otherwise recognizable as one of mainly less developed countries (LDCs), with some very noticeable exceptions. Japan may be seen as a classic postindustrial country and, like Italy, must worry about a declining, aging population in which young people will have to shoulder an increasing economic burden to support their elders. As discussed later in this chapter, there are other strong economies in the region. 72, 297

One of the outstanding anomalies in Asia's population equation is China. With a per capita GNI PPP of just $7,730, China might be expected to have a rather high population growth rate like India, which has a similar economic ranking. But in the 20th century, China's population was so alarmingly large and rapidly growing that its Communist leaders decided to restrain its growth drastically. China adopted a coercive "one-child policy" that has brought its growth rate to just 0.5 percent per year, representing

● **Figure 7.1** Monsoon Asia.

a doubling time of 140 years (see Geographic Spotlight, pages 286–287). That is a remarkable turnabout from the situation in 1964, when China had a growth rate of 3.2 percent and a doubling time of 22 years.

To reduce its poverty, India has made great strides in reducing its population growth, but not in China's coercive fashion. India is projected to overtake China as the world's most populous country in 2040, when it will have an estimated 1.5 billion people. A major wild card in India's population deck is HIV/AIDS. The scale of the problem there is not certain; 2007 estimates put the number of Indians infected with HIV at between 2 and 5 million, with the U.S. Central Intelligence Agency projecting up to 25 million cases by 2010. The potential for a pandemic in the world's second-largest country is strong, and much will depend on how India's government chooses to fight the disease. About 80 percent of India's HIV infections are spread

by heterosexual contact, and public health authorities have so far been shy to mount a strong campaign about sexual ethics and practices. China is thus far not in the peril that India is; 2007 estimates put the number of people infected at under 1 million, and plans were announced to provide free HIV tests for all Chinese citizens.

7.2 Physical Geography and Human Adaptations

Monsoon Asia's physical geography may be described broadly as consisting of three concentric arcs, or crescents, of land: an inner western arc of high mountains, plateaus, and basins; a middle arc of lower mountains, hill lands, river plains, and basins; and an outer eastern arc of islands and seas (●Figure 7.4, page 284).

TABLE 7.1   Monsoon Asia: Basic Data

Political Unit	Area (thousands; sq mi)	Area (thousands; sq km)	Estimated Population (millions)	Estimated Population Density (sq mi)	Estimated Population Density (sq km)	Annual rate of Natural Increase (%)	Human Development Index	Urban Population (%)	Arable Land (%)	Per Capita GNI PPP ($US)
South Asia	**1,732.6**	**4,487.4**	**1,499.3**	**865**	**334**	**1.7**	**0.595**	**28**	**47**	**3,480**
Bangladesh	55.6	144.0	149.0	2,680	1,035	1.9	0.530	23	62	2,340
Bhutan	18.1	46.9	0.9	50	19	1.3	0.538	21	3	N/A
India	1,269.3	3,287.5	1,131.9	892	344	1.6	0.611	28	54	3,800
Maldives	0.1	0.3	0.3	3,000	1,158	1.6	0.739	27	13	N/A
Nepal	56.8	147.1	27.8	489	189	1.9	0.527	14	21	1,630
Pakistan	307.4	796.2	169.3	551	213	2.3	0.539	34	28	2,500
Sri Lanka	25.3	65.5	20.1	794	307	1.2	0.755	30	14	5,000
Southeast Asia	**1,735.3**	**4,494.4**	**574.4**	**331**	**128**	**1.4**	**0.716**	**38**	**14**	**5,420**
Brunei	2.2	5.7	0.4	182	70	1.6	0.871	74	1	N/A
Cambodia	69.9	181.0	14.4	206	80	1.7	0.583	16	21	2,920
Indonesia	735.4	1,904.7	231.6	315	122	1.4	0.711	42	11	3,950
Laos	91.4	236.7	5.9	65	25	2.4	0.553	19	3	2,050
Malaysia	127.3	329.7	27.2	214	82	1.8	0.805	62	5	11,300
Myanmar	261.2	676.5	49.8	191	74	0.9	0.581	28	15	N/A
Philippines	115.8	299.9	88.7	766	296	2.1	0.763	48	19	5,980
Singapore	0.2	0.5	4.6	23,000	8,880	0.6	0.916	100	1	31,700
Thailand	198.1	513.1	65.7	332	128	0.7	0.784	31	29	9,140
Timor-Leste	5.7	14.8	1.0	175	68	3.3	0.512	8	4	N/A
Vietnam	128.1	331.8	85.1	664	256	1.3	0.709	25	20	3,300
East Asia	**4,545.6**	**11,773.1**	**1,543.0**	**339**	**131**	**0.5**	**0.770**	**48**	**13**	**10,320**
China	3,696.1	9,572.9	1,318.0	357	138	0.5	0.768	44	15	7,730
Japan	145.9	377.9	127.7	875	338	0.0	0.949	79	12	33,730
Korea, North	46.5	120.4	23.3	501	193	0.9	N/A	60	20	N/A
Korea, South	38.3	99.2	48.5	1,266	489	0.4	0.912	82	17	23,800
Mongolia	604.8	1,566.4	2.6	4	2	1.2	0.691	59	1	2,280
Taiwan	14.0	36.3	22.9	1,636	632	0.3	N/A	78	24	N/A
Summary Total	**8,013.5**	**20,754.9**	**3,616.7**	**451**	**174**	**1.1**	**0.688**	**38**	**20**	**6,710**

Sources: World Population Data Sheet, Population Reference Bureau, 2007; Human Development Report, United Nations, 2006; World Factbook, CIA, 2007.

The inner arc includes the world's highest mountain ranges, interspersed with plateaus and basins. In the south, the great wall of the Himalaya, Karakoram, and Hindu Kush Mountains overlooks the north of the Indian subcontinent. In the north, the Altai, Tien Shan, Pamir, and other towering ranges separate this region of Asia from the countries of central Asia. Between these mountain walls lie the sparsely inhabited Tibetan Plateau, at over 15,000 feet

(c. 4,500 m) in average elevation, and the dry, thinly populated basins and plateaus of Xinjiang (Sinkiang) and Mongolia.

The middle arc, between the western inner highland and the sea, is occupied by river floodplains and deltas bordered and separated by hills and low mountains. The major features of this area are the immense alluvial plain of northern India, built up through ages of meandering and deposition by the Indus, Ganges, and Brahmaputra Rivers; the

• **Figure 7.2** Monsoon Asia compared in latitude and area with the conterminous United States.

hilly uplands of peninsular India, geologically an ancient plateau; the plains of the Irrawaddy, Chao Praya (Menam), Mekong, and Red Rivers in the Indochinese Peninsula of Southeast Asia, together with bordering hills and mountains; the uplands and densely settled small alluvial plains of southern China; the broad alluvial plains along the middle and lower Chang Jiang (Yangtze River) in central China and the mountain-girded Red Basin on the upper Chang Jiang; the large delta plain of the Huang He (Yellow River) and its tributaries in northern China, backed by loess-covered hilly uplands; and the broad central plain of northeastern China (Manchuria), almost enclosed by mountains.

The outer arc is an offshore fringe of thousands of islands, mostly grouped in great **archipelagoes** (clusters of islands) bordering the mainland. On these islands, high interior mountains (including volcanoes) are flanked by coastal plains where most of the people live. Most of the islands are in three major archipelagoes: the East Indies, the Philippines, and Japan. Sri Lanka, Taiwan, and Hainan are large, densely populated islands outside these archipelagoes. Between the archipelagoes and the mainland lie the East and South China Seas and, to the north, the Sea of Japan (known in Korea as the East Sea). At the southwest, the Indian peninsula projects southward between two immense arms of the Indian Ocean: the Bay of Bengal and the Arabian Sea.

Climate and Vegetation

Monsoon Asia is characterized generally by a warm, well-watered climate, but there are nine distinct types of climate in the region: tropical rain forest, tropical savanna, humid subtropical, warm humid continental, cold humid continental, desert, steppe, subarctic, and undifferentiated highland (•Figure 7.5, left). These are generally associated with predictable patterns of vegetation, which have been heavily modified by millennia of human use (•Figure 7.5, right).

The tropical rain forest climate zone and vegetation types are typically found within 5 to 10 degrees of the equator and are characteristic of most parts of the East Indies, the Philippines, and the Malay Peninsula. Precipitation is spread throughout the year so that each month has considerable rain. The amount of precipitation is at least 30 to 40 inches (c. 75 to 100 cm) and often 100 inches (c. 250 cm). Average temperatures vary only slightly from month to month; Singapore, for example, has a difference of only 3°F between the warmest and coolest months. High temperatures prevail year-round, although some relief is afforded by a drop of 10° to 25°F (6° to 14°C) at night, and sea breezes refresh coastal areas. Winter in climates such as this, it can be said, "comes only at night."

The tropical savanna climate, like the tropical rain forest climate, is characterized by high temperatures year-round but is generally found in areas farther from the equator, and the average temperatures vary more from month to month. The savanna has a better-defined dry season, lasting in some areas as long as six or eight months each year, creating a problem for agriculture. Tropical savanna climates occur in southern and central India, in most of the Indochinese Peninsula, and on eastern Java and the smaller islands to the east.

In Asia, the characteristic natural vegetation associated with this climate is a deciduous forest of trees smaller than those of the tropical rain forest. Tall, coarse grasses such as those found on African and Latin American savannas grow only in limited areas.

The humid subtropical climate zone prevails in southern China, the southern half of Japan, much of northern India, and a number of other countries. It is characterized by warm to hot summers, mild or cool winters with some frost, and a frost-free season lasting 200 days or longer. The annual precipitation of 30 to 50 inches (c. 75 to 125 cm) or more is well distributed throughout the year, although monsoonal trends produce a dry season in some areas. The natural vegetation is a mixture of evergreen hardwoods, deciduous hardwoods, and conifers.

The humid continental climates characterize the northern part of eastern China, most of Korea, and northern Japan. They are marked by warm to hot summers, cold winters with snow, a frost-free period of 100 to 200 days, and less precipitation than the humid subtropical climate. Most areas experience a dry season in winter. The main natural vegetation is a mixture of broadleaf deciduous trees and conifers, although prairie grasses probably formed the original cover in parts of northern China. Extreme northeastern China and northern Mongolia have a subarctic climate associated with coniferous trees.

Steppe and desert climates are found in Xinjiang and Mongolia and in parts of western India and Pakistan. Due

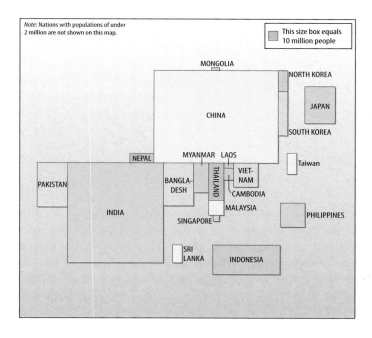

• **Figure 7.3** Population distribution (top) and population cartogram (bottom) of Monsoon Asia.

to limitations on agriculture, along with very high mountains, they are the region's least inhabited climate zones. The undifferentiated highland climate is most extensive in the Tibetan Highlands and adjoining mountain areas. It is made up of a broad range of montane microclimates that limit plant growth and vary from place to place according to elevation, aspect, and latitude.

The Monsoons

The **monsoons** are the prevailing sea-to-land and land-to sea winds that are the dominant climatic concern for people living in this world region (•Figure 7.6). They play a significant role in both wet and dry environments and are especially influential in the coastal plains and lowlands of South Asia, the peninsula and islands of Southeast Asia, and the eastern third of China. The ways in which the sea and land absorb **insolation** (heat from the sun) are different, causing the instability in air masses that creates monsoonal winds. If coastal waters and the adjacent coastline receive about equal amounts of warmth from the sun, the land becomes warm or hot much more quickly than the seawater. As a result, air over the land begins to become unstable and to rise. This ascending air creates a low-pressure attraction for the air masses over the water, and marine winds begin to blow toward the land. In Asia, this is the **summer monsoon,** characterized by high humidity, moist air, and generally predictable rains. Because of the moisture carried by such air masses, these wind shifts are the sources of major rainfall in the late spring, summer, and early fall seasons. Agriculture and patterns of human activity are keyed to these incoming rains.

If there is a little bit of elevation in the landscape—hills or mountain flanks—the moist air is driven higher, where it cools and releases even more rain in the pattern of orographic precipitation.

In the **winter monsoon,** the land loses its relative warmth while the sea and coastal waters stay warm longer. As a

● Figure 7.4 Physical geography of Monsoon Asia.

result, the wind shifts and air masses begin to flow from the inland areas of Asia toward the sea. However, because there is little moisture in the source areas over land for the winter monsoon, little moisture is picked up, and much less rain results. A monsoon climate, overall, is characterized by spring and summer precipitation and a long dry season in the low-sun (winter) cycle. Japan, however, experiences an unusual pattern of winter precipitation related to the monsoon wind shifts. The winter winds blow out of China and East Asia, sweep across the Sea of Japan (East Sea), and pick up moisture, dropping heavy, wet snow on the west coast of the Japanese islands.

The wet summer monsoon is of critical importance to the livelihoods of a large portion of humanity, particularly the people of India. By early May in India, the land has typically been parched by months of drought, and the rising temperatures become almost insufferable. No thought can

be given to farming until the rains come. Anticipation and discomfort rise steadily until the monsoon finally "breaks," bringing great relief in the form of much needed rain and lower temperatures (●Figure 7.7). Sometimes the onset of the rain comes with particular ferocity, as related by this Indian writer:

> That year the monsoon broke early and with an evil intensity such as none could remember before. It rained so hard, so long and so incessantly that the thought of a period of no rain provoked a mild wonder. It was as if nothing had ever been but rain, and the water pitilessly found every hole in the thatched roof to come in, dripping onto the already damp floor. If we had not built on high ground the very walls would have melted in that moisture.[1]

India and Bangladesh have experienced particularly heavy early monsoonal rains of this kind for the past three

CLIMATE TYPES

Subarctic	Humid continental (cold)	Humid continental (warm)
Desert	Semiarid / steppe	Humid subtropical
Tropical savanna	Tropical rain forest	Undifferentiated highland

BIOME TYPES

Coniferous forests	Prairie and steppe	Temperate mixed forests
Desert and desert shrub	Tropical deciduous forest	Savanna
Tropical rain forest	Undifferentiated highland	

• **Figure 7.5** Climates (left) and biomes (right) of Monsoon Asia.

→	Summer (southwestern) monsoon flows
→	Winter (northeastern) monsoon flows

• **Figure 7.6** The monsoons at work.

• **Figure 7.7** Torrential monsoon rains are regular and usually welcome features of land and life in much of the region. This is a flooded road in south central Sri Lanka.

decades, with significant loss of life. Inevitably, some scientific fingers point to global climate change as the culprit.

Agricultural Adaptations

The wet and warm climates associated with the monsoons might be expected to offer outstanding opportunities for agriculture (•Figure 7.8). However, the high temperatures and heavy rains promote rapid leaching of mineral nutrients

and decomposition of organic matter, so many soils are infertile. The lush vegetation cover in these low-latitude areas is deceiving because most of the system's nutrient matter is in the vegetation itself rather than in the underlying soil. When Asian farmers clear the tree cover, they often find that the local soils will not support more than one or two poor harvests (see Insights, page 289).

The Importance of Rice

Agriculture in Monsoon Asia—with the principal exception of Japan—is characterized by the steady input of arduous manual labor. Known as **intensive subsistence agriculture**, it is built around the growing of cereals, especially

34

GEOGRAPHIC SPOTLIGHT

China's One-Child Policy

China's population of 1.3 billion represents nearly one-fifth of the world's people and is increasing by 6.6 million each year. Population is a very serious matter for a less developed country whose area of about 3.7 million square miles (9.4 million sq km) is only slightly smaller than the United States but whose inhabitants outnumber the U.S. population by about five to one. But a glimpse at China's age structure diagram, with its wide middle ages but tapering youth, reveals that China's is one of the world's great success stories in terms of population management (●Figure 7.A).

The drive for smaller families was motivated by the fact that China's population was surging while per capita food output was increasing at only a very modest rate. In 1976, China's Communist regime instituted one of the most stringent and most controversial programs of birth control ever attempted. It culminated in the **one-child campaign,** which aims to limit the number of children per married couple to one (●Figure 7.B). The government tracks individual family birthing patterns—especially in the cities—and maintains surveillance through local authorities. It dispenses free birth control devices, free sterilization operations, free hospital care in delivery, free medical care for the child, free education for the child, an extra month's salary each year for parents who comply, and other favors and preferences to induce compliance with the one-child-per-family norm.

Those who violate the norm are subjected to constant social and political pressure, denial of the privileges accorded one-child families, pay cuts, and fines. A woman who becomes pregnant after having had one child is pressured to have a free state-supplied abortion, with a paid vacation provided. China's government hopes to continue to decrease childbearing so that population will stabilize

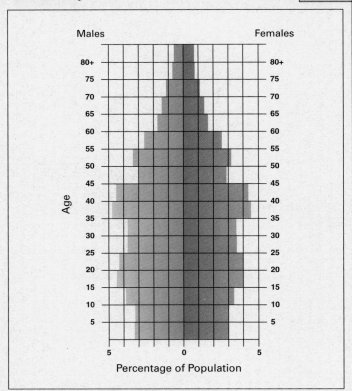

● **Figure 7.A** The age structure diagram of China (2008) reveals the country's success in slowing population growth.

● **Figure 7.8** Land use in Monsoon Asia.

rice, which is the premier staple of Monsoon Asia and the crop of choice in areas with adequate rainfall or where irrigation waters are available (see the areas of rice subsistence in Figure 7.8). Farmers use organic fertilizers from both their animals and their own latrines and irrigate to the extent that they can or need to. Where natural conditions are not suitable for irrigated rice, grains such as wheat, barley, soybeans, millet, sorghum, or corn (maize) are raised. Where possible, farmers grow cash crops for local markets. Large-scale, commercial plantation agriculture is described in Module 7.2 on Southeast Asia, the subregion in which cash cropping has had the strongest role.

Shifting cultivation represents a precarious adaptation to local soil and climate conditions, capable of sustaining only small populations for brief periods in particular locales. At the other end of the spectrum of human adaptation to Asian lands is **wet rice cultivation,** capable of producing two or three crops per year and sustaining large populations over long periods of time. Wet rice agriculture takes place both in lowland floodplains and on carefully constructed upland **terraces** requiring enormous inputs of human labor (the term **paddy,** also spelled **padi,** may refer to the rice itself or the

GEOGRAPHIC SPOTLIGHT

China's One-Child Policy *continued*

Joe Hobbs

● **Figure 7.B** Strong disincentives mean one child is the norm for Chinese couples.

before reaching projections of 1.4 billion by 2010 or 1.5 billion by 2025. The higher number may be more likely because government officials report widespread disregard of the "one couple, one child" policy. Reasons for noncompliance vary. In some cases, parents feel wealthy enough to pay the fines of 50,000 yuan ($6,200) or more per extra child. Many rural dwellers think their extra children will escape detection by authorities, who are most effective in the cities. Anticipating their years as elders, some couples want the "safety net" of additional children—especially boys. If a young urban couple has just one child and it is a girl, it is likely that she will marry and become, by Chinese tradition, such a part of the husband's family that she may not be available to care for her parents as they get older. China has traditionally used the son's family as the accommodation of choice for senior citizens.

Some parents want to produce a son if their first child was a girl, and many girls are given to orphanages (China is the world's leading source country of adopted children, about 90 percent of them girls). Female feticide is increasingly associated with the desire to have a son. Couples often insist on an ultrasound test to determine the baby's gender, and about 90 percent of girls detected in this way are aborted. In some provinces where the practice is widespread, three boys are born for every two girls. As in India, this practice is beginning to have serious repercussions, reflected in a shortage of eligible brides. Rural women have been abducted and trafficked as virtual slaves to other parts of the country where they become brides. On the positive side, the relative shortage of women in China has helped enhance their status, particularly in urban areas, where they are also becoming more influential in the workplace.

321

field in which it is grown). Terraces have been built almost everywhere in Asia where there is a combination of sloping land, available irrigation water, and productive soil (●Figure 7.9). The dramatic stairstep terraces of the island of Luzon in the Philippines are the most famous, but equally dramatic cultural landscapes can be found in Pakistan's Punjab, in South China, and in many parts of Southeast Asia.

Terraces have the important function of preventing or dramatically slowing soil erosion, as the terrace walls allow heavy rainfall to run off without carrying topsoil away. Doing what the Chinese call "teaching water," people use canals and pipes to direct irrigation water from a source to the terrace, which acts like a kind of dam to hold the water in place. Wet rice is actually immersed in standing water for extended periods, creating an environment unfavorable to many crop pests. The productivity of wet rice agriculture is sometimes exceptionally high, supporting as many as 500 people per square mile (200/sq km) on the volcanic soils of Java and Bali in Indonesia. This extraordinary output is a contributing factor to the high densities and overall populations of some of the Asian countries and regions, including Java. One anthropologist has observed that most

of Monsoon Asia's rice is eaten within walking distance of where it is grown.[2]

Is There a Correlation between Agriculture and Culture?

Anthropologists have observed some unique cultural adaptations to opportunities for terraced wet rice. Group welfare takes precedence over individual interests, for example. Individual farmers cannot cultivate at will but must work with the community to ensure that crops are staggered over time. This allows water to be shared fairly and efficiently and gives some fields the chance to dry out while others are irrigated. This agricultural system also prevents the use of tractors and other mechanical technology. Over much of Asia today, just as in ancient times, wet rice cultivation is the productive of intensive work by many human hands.

In some cases, population growth has outpaced the ability of even the most productive wet rice farming to feed the people, creating the classic scenario of "people overpopulation." In Nepal, for example, high rates of population growth on limited areas of moderately sloped, terraceable   61

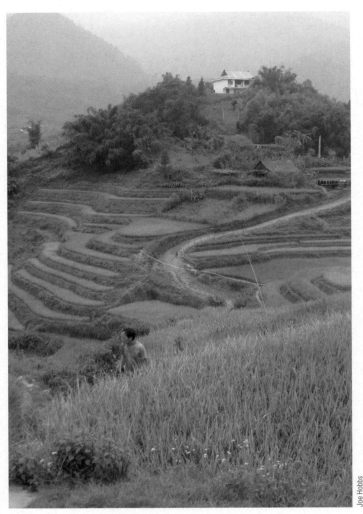

• **Figure 7.9** These meticulously crafted and maintained terraces produce a high yield of rice for the Hmong people of northern Vietnam.

57 land have been a push factor behind migration. After the 1950s, when malaria began to be eradicated in the lowland Terai region, millions of Nepalis migrated there, cleared the forests, and planted crops. For decades, the Terai served as a virtual "safety valve" for overpopulation in Nepal. Now that the Terai itself is mostly settled, there is nowhere for landless Nepalis to go but up. They clear and cultivate increasingly steep lands on which it is impossible to build terraces. Without terraces in place to hold the topsoil, heavy monsoon rains cause relentless erosion of their farmlands. Locally, the consequences are that the eroded plots cannot be cultivated again, and landslides occur downslope, often causing loss of life.

Much farther away, in the lowlands of India and Bangladesh, the increased sediment load contributed by erosion in Nepal causes rivers to swell out of their banks and bring extensive flooding—again, with losses of crops and lives. This sequence of events, beginning with overpopulation in Nepal and ending with flooding in Bangladesh, has

been described as the **theory of Himalayan environmental degradation.**[3]

Where Asians Live

Monsoon Asia has some of the world's largest cities, including the greatest of all, Tokyo, with a metropolitan population of 37 million. About 1.3 billion people, or 38 percent of the region's total, live in urban settlements. But 2.3 billion, or 62 percent, do not. The main unit of Asian settlement is the village. Most of Asia's roughly 2 million villages are essentially groups of farmers' homes, although some villages are home to other occupational groups, such as miners or fisherfolk. Clusters of houses bunched tightly together are typical, and inexpensive and simple structures—often made of local clays and other building materials—are characteristic. Indoor plumbing continues to be the exception in village homes, but electricity lines have been extending steadily throughout Asia. So have cell phones and satellite dishes; it is not unusual today to see what appear to be very "traditional" Asians enjoying some of the latest innovations in global communications (•Figure 7.10). In Monsoon Asia, as elsewhere in the developing world, many people are making the jump directly from 19th- to 21st-century technology. Wireless technology is especially 47 appropriate where landlines and other infrastructure are difficult and expensive to install and where bureaucracies often slow down applications for access to utilities.

The original site selection for Asian villages was usually closely related to natural conditions and to perceived spiritual circumstances as well (see Geography of the Sacred, pages 290–291). With consideration always given to the possibility of flooding monsoon rains, lowland villages tend to be situated on natural **levees** (raised riverbanks built up by deposition of sediments during floods), dikes, or raised mounds. Early villages in Indonesia were often built in defensible mountain sites. European colonists sometimes

• **Figure 7.10** Across Monsoon Asia, satellite dishes sprout incongruously from traditional dwellings.

INSIGHTS

Shifting Cultivation

Many Asian farmers, particularly in tropical rain forest regions, practice **shifting cultivation** (also known as **swidden cultivation** or **slash-and-burn cultivation**) or land rotation between crop and fallow years (•Figure 7.C). The procedure begins with people clearing forest cover and then burning the dried vegetation during the dry season. Ash from the burn provides short-lived fertility to the soil in which farmers plant their subsistence crops. Most tropical soils lose their natural fertility quickly when they are cropped and after two or three harvests must be rested for several years (often 10 to 15 years) before they will again produce a crop. During this **fallow** period, the land reverts to wild vegetation.

The system of shifting agriculture is widespread in the tropics. Although it is not very productive, it does provide a bare living for people too poor to afford fertilizer or farm machinery. Traditionally, this system has minimized soil erosion because most of the land is not in cultivation at any given time. Unfortunately, there has been a recent widespread trend for farmers to reduce the length of the fallow period in the cycle of shifting cultivation or to abandon it altogether. This is a direct result of a surging growth in population: with more mouths to feed, in many places there is simply not enough land to provide the "luxury" of letting it lie fallow.

The tragic consequence is that short-term overuse of the land is eroding the soil. On the **lateritic soils** characteristic of many tropical regions, excessive exposure of the land to the sun's ultraviolet radiation and to repeated wetting and drying turns the ground to a bricklike substance called **laterite** or **plinthite,** which is essentially impossible to cultivate. When the best available lands are exhausted, farmers are often "marginalized" to places unsuitable for farming, such as semiarid lands or slopes that are too steep to till without causing disastrous erosion. Such dilemmas of land use are typical of the world's tropics from Indonesia to West Africa and Amazonia.

46, 48

Joe Hobbs

• **Figure 7.C** Shifting cultivation (swidden or slash-and-burn cultivation) is widespread in the world's tropical regions and can be sustained as long as the land is given sufficient time to recover.

57, 357

rejected traditional settlement patterns to serve their own interests. Dutch colonial administrators in Indonesia, for example, required that villages be built along main roads and trails in the lowlands, making it easier to exercise control, collect taxes, and draft soldiers or laborers for roadwork and other projects.

The majority of people in Monsoon Asia are rural, but the region's future, in terms of demographic weight, economic power, and cultural change, is unquestionably focused on the cities. Some of the countries are already highly urbanized, notably South Korea (82 percent), Japan (79 percent urban), and Taiwan (78 percent) and entities like Singapore and Brunei that are effectively city-states. The layouts and architectures of the cities span a continuum from the ancient, walled fortress city to the modern Western-style metropolis, and many, like Beijing in China and Lahore in Pakistan, accommodate both. A striking pattern across the region is the steady stream of rural-to-urban migration, consisting of people motivated by both push and pull factors to leave the relative poverty of the countryside

57

in search of employment and even prosperity in the city. Many Asian cities are hard-pressed to provide accommodation and services to their swelling populations, and where the political system allows (notably in China), restrictions on internal migrations are imposed.

7.3 Cultural and Historical Geographies

Monsoon Asia has been home to some of the most important cultural developments of humankind in landscape transformation, settlement patterns, religion, art, and political systems. From Korea (not Gutenberg's Germany, as is widely thought) came the first movable printing type. From China came gunpowder, paper, silk, and china (porcelain). From India came the great faiths of Hinduism and Buddhism. Handmade textiles, artwares, bone and leather and paper products, and precious metals and gems have flowed from Asia into global trade for millennia. Spices, foodstuffs, and exotic plants and products from the East Indies have been major commodities in Asian-European trade and commerce for more than five centuries. Some of the world's most important domesticated plants and animals originated in Monsoon Asia, including rice, cabbage, chickens, water buffalo, zebu cattle, and pigs.

Ethnic and Linguistic Patterns

As would be expected of such a vast region, in many places fragmented by islands and mountain barriers, the ethnic and linguistic composition is rich and complex (•Figure 7.11).

GEOGRAPHY OF THE SACRED

The Korean Village

In the West, people are accustomed to communities laid out on a perfect grid of streets intersecting at right angles, with their homes and businesses symmetrically fronting those streets. But in large parts of Asia, settlement and residence considerations are far more varied, especially because they are tied to spirituality, aesthetics, and topography. Villages, and the homes in particular, tend to be arranged for maximum harmony with the natural and spiritual environments—a practice known as **geomancy**—and constitute a special realm that may be considered sacred space. The traditional Korean village illustrates these unique Asian qualities.*

From ancient times, Koreans developed ways of determining the most auspicious village sites. The principles of *feng shui* and of Confucianism played strong roles in how they adapted settlements and homes to geographic conditions. **Feng shui** (pronounced "fung-*shway'*") is a traditional theory for selecting favorable sites for buildings, homes, and cities and for guiding many other urban and architectural planning decisions.† It is essentially an ordering device for environmental planning and for daily living. *Feng shui* (the words mean "wind" and "water") attempts to balance **yin** and **yang** to achieve harmony. *Yin* (conceived of as negative, dark, and feminine) and *yang* (representing the positive, bright, and masculine) stand for all the dualistic, opposing forces that may be reconciled harmoniously, including female and male, night and day, moon and sun, cold and hot, and right and left. *Feng shui* principles have begun to appear in American suburban home architecture, as in houses built to conceal overhead beams and sharp protrusions (which are disharmonious) and provided with windows opening to the east (a harmonious situation).

The theory of *feng shui* and the Confucian view of nature have some common characteristics that affect site, settlement, and resi-

dence designs. Confucianism reflects the sociopolitical philosophy of the Chinese philosopher known in the West as Confucius (see page 296). One of the main Confucian principles is that architecture and the lived environment should be in harmony with nature. This can be seen in the Confucian institutes for higher learning, called *seowon,* which in Korea peaked in the Joseon Dynasty, around 1400 C.E. Each *seowon* had a lecture hall where Confucian scholars gathered to learn philosophy. The buildings were laid out in such a way that from the inside of the complex, the scholars could see distant landscapes. The scholars' minds and spirits were enhanced as they gazed at the mountains and contemplated nature's grandeur. They believed that the earth's power was an invisible force that would help them become men of dignity and eminence.

The principles of *feng shui* and Confucianism are also apparent in the design of Korean villages. Each village is laid out to achieve harmony with the mountains—an important consideration in a land that is 70 percent mountainous. Rising behind many villages is a mountain, called *jinsan,* which is a spiritual focus for the villagers. *Jinsan* serves as the village symbol and guardian. Typically, a small stream flows in front of the village, and beyond it, there is a flat expanse of farmland. Beyond those fields is another mountain, called *ansan,* facing the village (●Figure 7.D).

These geographic conditions are economically advantageous to the community, providing it with productive fields and a good supply of water. But just as important, this situation is psychologically, spiritually, and symbolically beneficial to the villagers. The natural features are meant to embrace the village, as Koreans say a mother would hold a child. Korean historical records describe the traditional village as a place of comfort and serenity that safeguards against negative natural forces.

Central and South Asia gave birth to the **Indo-European language family,** which contains tongues as diverse as English and Hindi. **Hindi** and **Urdu,** which are major languages in India and Pakistan, respectively, are in the **Indo-Iranian subfamily** that includes such other major South Asia languages as **Pashto** (spoken in western Pakistan and Afghanistan), **Sanskrit** (which is extinct as a vernacular language but is still used in Hindu ritual), **Sinhalese** (the majority language of Sri Lanka), **Punjabi** (spoken on both sides of the northern frontier between Pakistan and India), Nepal's national language of **Nepali, Bengali** (the language of Bangladesh), and Romany, the tongue of the Roma (Gypsies) of Europe.

When the first of the Indo-European languages was carried from central Asia into India—apparently by the ethnic group known as the **Aryans,** between 2000 and 1000 B.C.E., languages in the ancient and indigenous **Dravidian language family** were being spoken there by peoples who

had much darker skin than the newcomers. The Dravidian languages, today found almost exclusively in southern India and northern Sri Lanka (with the exception of a small pocket in Pakistan), include **Tamil, Malayalam,** and **Telegu.** Along the eastern side of the Himalayan mountain wall separating South Asia from China, people speak languages in the **Tibeto-Burman subfamily** of the **Sino-Tibetan language family. Tibetan,** spoken farther north in Tibet, and **Burmese,** spoken to the east in Myanmar (Burma), are also in the Tibeto-Burman subfamily.

The greatest numbers of people in China speak languages in the **Sinitic subfamily** of the Sino-Tibetan languages, overwhelmingly **Chinese** and its variants, including **Mandarin, Xiang, Min,** and **Yue.** The Chinese government has adopted Mandarin (Putonghua) as the country's official dialect, but only about half the population can speak it. Almost all of these more than 1 billion speakers of about 1,500 Chinese dialects belong to the **Han Chinese** ethnic

81, 140

GEOGRAPHY OF THE SACRED

The Korean Village *continued*

Symbolic guardians are associated with the topography. Ideally, there is the Green Dragon to the east, the White Tiger to the west, the Black Tortoise to the south, and the Red Bird to the north. These guardians, named for figures in Chinese astronomy, could have varied manifestations. The Green Dragon, for example, could be a hill or a river to the east of the village, while the White Tiger could be a mountain or a road on the west side. One enters the village through a symbolic portal. For many villages, a pair of "spirit posts" stands at the foot of a nearby mountain. The posts mark the preliminary boundary to the outside, and they ward off evil spirits.

Not only is the village site selection process different from that in the West, but the settlement pattern is uniquely Asian too. Unlike in Western settlements, no more than two houses should be built side by side, they should not stand parallel to each other, and their gates should never face each other. The side street should curve naturally between the home lots. Overall, then, Koreans choose to emphasize the surrounding topography rather than geometric order in the construction of their communities and homes.

The traditional Korean home is adapted to the natural environment, embracing its surroundings while offering protection, comfort, and order. As in the *seowon,* the layout brings outdoor scenery into the house itself. Homes are also adapted to the environment in their construction materials. A thatched roof and earthen walls effectively block heat in the summer and insulate in winter, a perfect combination for Korea's sweltering summers and raw winters. The mulberry paper covering lattice windows diffuses sunlight, providing a pleasant natural light while absorbing sound and allowing some ventilation. The home's eaves are built so that sunlight penetrates deep into the house during the cold weather but never during summer—a rather recent innovation in the West, where it is known as "passive solar" design.

● **Figure 7.D** The Korean village of Yangdong is nestled between mountains that are separated by a river and lush rice fields.

*This discussion is based largely on two unpublished manuscripts, "Living Spaces of Korean Architecture" and "In Tune with Nature: Rural Villages and Houses in Korea," by the Korean geographer Sang-Hae Lee, of Sungkyunkwan University, and personal talks with him. I am grateful for his contributions.

†A useful introduction to *feng shui* may be found at FengShuiHelp .com, http://www.fengshuihelp.com/history.htm.

---

group, which is culturally, politically, and economically the dominant ethnic group of China. Northern and northwestern China, along with Mongolia, are home to speakers of languages in the **Altaic language family,** including **Uighur, Kazakh, Turkic, Mongolian,** and **Tungus.** Cut off from their Altaic-speaking kin are the **Korean** speakers of the Korean Peninsula and the **Japanese** speakers of Japan. Japan's indigenous **Ainu** language has no known linguistic relatives and is nearly extinct.

The people of Taiwan include Chinese-speaking migrants from China and indigenous speakers of an unrelated language. Most linguists assign this tongue to the **Austronesian subfamily** of the more widespread **Austric language family.** Extreme southern China and adjacent portions of Vietnam and Laos are linguistically complex, reflecting the presence of many minority ethnic groups. Several are kin to the peoples of peninsular Southeast Asia, including the **Hmong** of Laos and Vietnam, who speak a language in the **Hmong-**

Mien subfamily of the Austric language family. Most people in Laos and neighboring Thailand speak **Lao** and **Thai** in the **Tai-Kadai subfamily** of the Austric family. The majority languages of Cambodia and Vietnam, respectively, are **Khmer** and **Vietnamese,** in the **Mon-Khmer** branch of the **Austro-Asiatic subfamily** of the Austric languages.

Southward and eastward into peninsular Malaysia, Indonesia, and the Philippines, the linguistic map is exceptionally rich and fragmented. The eight languages of the Philippines belong to the Austronesian group. Also in the Austronesian branch is a multitude of languages spoken in Indonesia, including **Malay, Dayak, Kayan-Kenyah, Barito, Javanese, Bugis, Tomini,** and **Timorese.** In many cases, a single language is spoken almost exclusively by an ethnic group inhabiting a single island. In other cases, as will be seen in Module 7.2, different ethnolinguistic groups sometimes live in conflicted proximity to one another on a single island. The large island of New Guinea and some of

• Figure 7.11 Languages of Monsoon Asia.

ALTAIC FAMILY	15 Santali	32 Ida'an	50 Seram	DRAVIDIAN FAMILY	81 Rajasthani	93 Xiang
1 Japanese	16 Vietnamese	33 Ilocano	51 Sundanese	66 Brahui	82 Sindhi	94 Yue
2 Kazakh	17 Wa	34 Javanese	52 Tagalog	67 Kannada	83 Sinhalese	TIBETO-BURMAN
3 Korean	AUSTRONESIAN	35 Kaili-Pamona	53 Timorese	68 Malayalam	84 Urdu	95 Burmese
4 Kyrgyz	18 Acehnese	36 Kapampangan	54 Tomini	69 Tamil	IRANIAN	96 Dzonkha
5 Mongolian	19 Apo-Duat	37 Kayan-Kenyah	HMONG-MIEN	70 Telugu	85 Baluchi	97 Ladakhi
6 Torgut	20 Aru	38 Lampung	55 Hmong	INDO-EUROPEAN FAMILY	86 Pashto	98 Loloish
7 Tungus	21 Bali-Sasak	39 Madurese	56 Miao	INDIC	87 Tajik	99 Manipuri
8 Uighur	22 Barito	40 Malay	57 Yao	71 Assamese		100 Mizo
	23 Batak	41 Mentawai	TAI-KADAI	72 Bengali	PAPUAN LANGUAGES	101 Tibetan
	24 Biak	42 Mongondow	58 Dai	73 Bihari	88 *Papuan languages*	
AUSTRIC FAMILY	25 Bikol	43 Murutic-Dusunic	59 Kam-Tai	74 Gujarati		
AUSTRO-ASIATIC	26 Bima-Sumba	44 Nias-Simeulue	60 Lao	75 Hindi	SINO-TIBETAN FAMILY	
9 Aslian	27 Bugis	45 Pangasinan	61 Li	76 Kashmiri	SINITIC	
10 Khasi	28 Buru	46 Rejang	62 Pak Tay	77 Marathi	89 Gan	
11 Khmer	29 Cebuano	47 Saluan	63 Shan	78 Nepali	90 Mandarin	
12 Khmu	30 Dayak	48 Samar-Leyte	64 Thai	79 Oriya	91 Min	
13 Mon	31 Hiligaynon	49 Sangir-Minahasan	65 Zhuang	80 Punjabi	92 Wu	
14 Nicobarese						

its smaller island neighbors are home to the 750 different **Papuan languages,** which are not related to Austronesian tongues.

Colonialism and regional economic enterprise led to the infusion of distant tongues into many parts of Monsoon Asia. British colonists brought English to India, where it still serves as a lingua franca in this linguistically complex land. The British also introduced Tamil-speaking Dravidians from south India into what is now Malaysia, where they worked the rubber plantations and tin mines and

354

354

where their descendents remain as an important minority today. Another small but economically prosperous minority in several countries (and a majority in Singapore) is ethnic Han Chinese. Some people in Indonesia speak Dutch, reflecting that country's former colonial status, and the Philippines' former status as a U.S. commonwealth is reflected in the fact that English remains one of that country's two official languages.

Religions and Philosophical Movements

Monsoon Asia is, along with the Middle East, one of the world's two great hearths of religion. Hinduism, Buddhism, and related traditions of Confucianism and Daoism—collectively practiced or observed by a whopping 1.7 billion, or 25 percent of the world's population—originated in this region. The basics of these **belief systems** are introduced here. The precepts of Islam and Christianity, also practiced in the region, are described in Chapter 6. The faiths with small numbers of adherents, like Zoroastrianism and Sikhism, are described in the contexts of particular countries in the following modules. Religion is sometimes, unfortunately, at the root of domestic or international conflict, and these cases are also discussed in the appropriate places.

The geographic distribution of the region's faiths is depicted in •Figure 7.12. As this map reveals, Hinduism is dominant in India, although many other religions are practiced there as well. Islam prevails in Pakistan, Bangladesh, most parts of the East Indies, parts of the southern Philippines, parts of western China, and among the Malays of western Malaysia. Various forms of Buddhism are domi-

nant in Myanmar, Thailand, Cambodia, Laos, Tibet, and Mongolia. Most people in Sri Lanka and Nepal are Buddhists or Hindus. The religious patterns of the Chinese, Vietnamese, and Koreans are more difficult to describe. Among them, Buddhism, Confucianism, and Daoism have been influential, often in the same household. Growing prosperity among the Chinese has been accompanied by a renewed interest in spiritual matters. China officially recognizes and even encourages five faiths (Buddhism, Islam, Taoism, Catholicism, and Protestantism) but discourages or forbids some sects and belief systems, including Tibetan Buddhism; the quasi-Buddhist movement known as Falun Gong (Falun Dafa, a practice of calisthenics and meditation that the government calls an "evil cult"), and "house church" Protestantism. In Japan, religious affiliations often overlap, with an estimated 84 percent of the Japanese population sharing Buddhism with the strongly nationalistic religion of **Shintoism,** itself a blend of nature reverence, Japanese folklore, and Chinese ritual. The Philippines and Timor-Leste, with large Roman Catholic majorities, are the only predominantly Christian nations in Asia, although Christian groups are found in all other countries in this region. About one-third of South Koreans are Christian, and Korean missionaries work in many countries of the world.

Other customs include **ancestor veneration,** which is especially prominent among Chinese, Japanese, Vietnamese, and Koreans. Most homes in these countries contain small shrines with photographs and other memorabilia of deceased ancestors at which family members place fresh fruits and other offerings on a daily basis. It is crucial in these cultures that the ancestors never be forgotten (•Figure 7.13).

354

354

• **Figure 7.12** Religions of Monsoon Asia.

Legend:

- Daoist/Confucianist
- Shinto & Buddhist
- Sikh
- Mahayana Buddhist
- Hindu
- Protestant
- Animist
- Therevada Buddhist
- Sunni Muslim
- Roman Catholic

Joe Hobbs

• **Figure 7.13** Ancestor veneration is not a religion per se but an important spiritual tradition that requires the living to remember and pay tribute to their forebears. This is a shrine to the deceased parents of an elderly woman in a southern Thai village.

GEOGRAPHY OF THE SACRED

The Sacred Cow

Many attitudes, beliefs, and practices associated with cattle make up the world-famous but often poorly understood "sacred cow" concept of India. There are nearly 200 million cattle in India, representing about 15 percent of the world total and the largest concentration of domesticated animals anywhere on earth. India's dominant religion of Hinduism forbids the slaughter of cows but allows male cattle and both male and female water buffalo to be killed. Reverence for the cow is well founded. Indians favor cow's milk, ghee (clarified butter), and yogurt over dairy products from water buffalo. They value cows as producers of male offspring, which serve as India's principal draft animals. Both cows and bullocks provide dung, an almost universal fuel and fertilizer in rural India.

Reverence for the cow in particular and cattle in general pervades Hindu religion and mythology. The bull Nandi is associated with the Hindu god Shiva, the cow is linked to the goddess Lakshmi, and Krishna (an incarnation of the god Vishnu) was a cowherd. People allow cattle to roam the streets of Indian cities (●Figure 7.E). As a symbol of fertility, cows are associated with several deities. The mother of all cows, Surabhi, was one of the treasures churned from the cosmic ocean. Hindus honor cows at several special festivals. They use cow's milk in temple rituals. They believe that the "five products of the cow"—milk, curds, ghee, urine, and dung—have unique magical and medicinal properties, particularly when combined. In India, there are about 4,000 *gaushalas*, or "old folks' homes," for aged and infirm cattle.

Remarkably, however, the subcontinent countries of Pakistan, Bangladesh, and India are major exporters of leather and leather goods. In India, the leather industry is mainly in the hands of Muslims, who do not share Hindus' restrictions on killing or eating cows. Muslims are forbidden to eat pork; therefore, pigs, a major food resource in many developing countries, are of little importance here. They are eaten mainly by Christians, very low-caste Hindus (some of whom are pig breeders), and tribal peoples.

Joe Hobbs

● **Figure 7.E** People and cows share street space throughout India.

Many members of hill tribes in Southeast Asia believe that natural processes and objects possess souls, a doctrine known as **animism.** Such indigenous animistic beliefs have been incorporated into a number of mainstream religions, creating vibrant and highly varied belief systems by Western standards. This blending, known as **syncretism,** has parallels elsewhere in the world, for example, among the Maya of Central America.

Hinduism is a regionally varied faith and a belief system so complex that it is extremely difficult to summarize. However, it has the following characteristic elements. It lacks a definite creed or theology. It is very absorptive, encompassing an unlimited pantheon of deities (all of which, most Hindus say, are simply different aspects of the one God, Brahman) and an infinite range of types of permissible worship (●Figure 7.14). Most Hindus recognize the social hierarchy of the caste system, deferring authority to the highest (Brahmin) caste. They practice rituals to honor one or another of the principal deities (Brahman the creator, Vishnu the preserver, and Shiva the destroyer) and thousands of their manifestations (*avatars*) and lesser deities, attending temples to

worship them in the form of sanctified icons in which the gods' presence resides. They believe in reincarnation and the transmigration of souls and revere many living things (see Geography of the Sacred, above). They are supposed to be tolerant of other religions and ideas. They participate in folk festivals to commemorate legendary heroes and gods. To earn religious merit and to struggle toward liberation from the bondage of repeated death and rebirth, they make pilgrimages to sacred mountains and rivers.

The Ganges is a particularly sacred river to Hindus, who believe it springs from the matted hair of the god Shiva. The pilgrimage to celebrate the festival of Ardh Kumbh Mela at Allahabad, at the confluence of the Ganges and Yamuna Rivers, drew 75 million pilgrims in January 2007. It was the largest pilgrimage in human history. The most enduring Hindu pilgrimage destination is the Ganges city of Varanasi (Benares) in the state of Uttar Pradesh. Many elderly people go to die in this city and be cremated where their ashes may be strewn in the holy waters. Indian authorities have passed laws, so far ineffective, to clean up the Ganges, polluted in part by incompletely cremated corpses; many of the faithful

• **Figure 7.14** Hinduism has a complex and very colorful pantheon. This is Vishnu, preserver god, and his consort, Lakshmi, goddess of wealth and prosperity.

poor cannot afford to buy the fuel needed for thorough immolation. Scavenging water turtles released into the river at Varanasi help dispose of the cadavers.

Buddhism is based on the life and teachings of Siddhartha Gautama, known as the Buddha or the Enlightened

One (•Figure 7.15). The Buddha was born a prince in about 563 B.C.E. in northern India.[4] Although presumably born a Hindu, he came to reject most of the major precepts of Hinduism, including caste restrictions. When he was about 29 years old, he renounced his earthly possessions and became an ascetic in search of peace and enlightenment. The Buddha eventually settled for a "middle path" between self-denial and indulgence, and he meditated through a series of higher states of consciousness until he attained enlightenment (*bodhi*). From that point, he traveled, preached, and organized his disciples into monastic communities. In his sermons, he described the "Four Noble Truths" revealed in his enlightenment: life is suffering; all suffering is caused by ignorance of the nature of reality; suffering can be ended by overcoming ignorance and attachment; and the path to the suppression of suffering is the Noble Eightfold Path, made up of right views, right intention, right speech, right action, right livelihood, right effort, right mindedness, and right contemplation.

An important Buddhist concept is the doctrine of **karma,** a person's acts and their consequences. A person's actions lead to rebirth, in which good deeds of the previous life are rewarded (for example, by which one could be reborn a human) and bad deeds punished (for example, by being reborn as a resident of hell). The goal of the practicing Buddhist

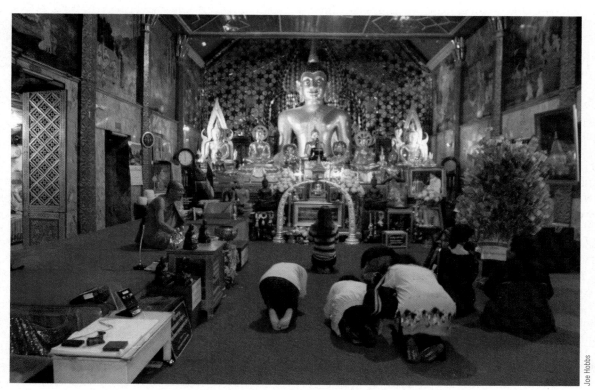

• **Figure 7.15** The Buddha has many manifestations. This temple of the Theravada faithful is in Chiang Mai, Thailand.

is to attain **nirvana,** a transcendent state in which one is able to escape the cycle of birth and rebirth and all the suffering it brings.

Buddhism evolved into two separate branches: **Theravada** ("Way of the Elders") and **Mahayana** ("Great Vehicle"). (Mahayanists sometimes apply the term Hinayana, "Lesser Vehicle," in a derogatory fashion to the Theravada.) Theravada Buddhism is strongest in Sri Lanka, Myanmar, Laos, Cambodia, and Thailand. Theravada Buddhists claim to follow the true teachings and practices of the Buddha. Mahayana Buddhism originated in India and then diffused along the Silk Road to central Asia, Tibet, and China and eventually into Vietnam, Korea, Japan, and Taiwan. Mahayana Buddhists accept a wider variety of practices than those espoused by the Theravada, have a more mythological view of the Buddha, and are interested in broader philosophical issues.

Confucianism is not as much a belief system as it is a sociopolitical philosophy that serves on its own or is blended with other religions, particularly among ethnic Han Chinese. It is based on the writings of Confucius (Kung Futzu, or "Master Kung," who lived from 551 to 479 B.C.E.), collected primarily in his *Analects.* Mencius (c. 371–288 B.C.E.) later became a major force in the widespread diffusion of this belief system throughout China. Following a period in which Communist authorities disavowed them, Confucian ideals are making a major comeback in China today.

The major tenets of Confucianism are embodied in a system of ethical precepts for the proper management of society, emphasizing honor of elders and other authorities, hierarchy, and education. Confucius never proclaimed his beliefs to constitute a religion; he was simply attempting to create a social contract between different classes central to Chinese government and society. He viewed his philosophy as secular, and its diffusion to Korea, Japan, and Vietnam has been part of the cultural baggage taken abroad by a steady stream of Chinese emigrants to those places.

Another important Chinese school of thought, second only to Confucianism and also widely blended with Buddhism among ethnic Han Chinese, is **Daoism** (sometimes spelled **Taoism;** •Figure 7.16). Its philosophy comes from a body of work known as the *Daodejing (Tao-te Ching),* or in English, *Classic of the Way and Its Power,* ascribed to Chinese Lao-tzu (Laozi), who lived in the sixth century B.C.E. Daoism encourages the individual to reject Confucian-style social conformity and seek to conform only to the underlying pattern of the universe, the "Way" (Dao, or Tao). To follow that way, the individual should "do nothing," meaning nothing unnatural or artificial. One can become rid of all doctrines and knowledge to achieve unity with the Dao and thereby gain a mystical power. With that power, the individual can transcend everything ordinary, even life and death. Daoists hold the simple earthly life of the farmer in high esteem.

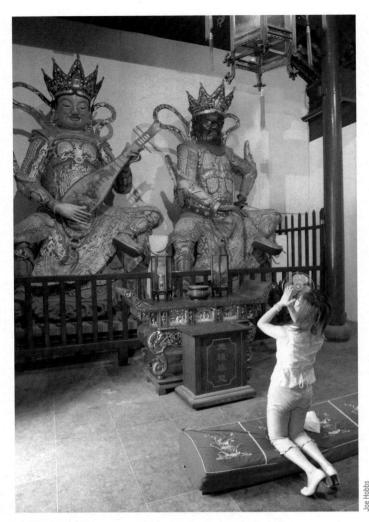

• **Figure 7.16** Daoism is imbued with images of heroic figures associated with such attributes as good health, long life, prosperity and bountiful harvests. This is a temple in Shanghai, China.

Effects of European Colonization

Like much of the developing world, Monsoon Asia is a region where Western colonialism reshaped many traditional geographic patterns (•Figure 7.17). By the end of the 15th century, Portugal and Spain had begun to extend economic and political control over some islands and mainland coastal areas of South and Southeast Asia. In the 18th and 19th centuries, the pace of colonization and economic control quickened, and large areas came under European domination. By the end of the 19th century, Great Britain was supreme in India, Burma (modern Myanmar), Ceylon (today's Sri Lanka), and Malaya and northern Borneo (now parts of Malaysia); the Netherlands possessed most of the East Indies (Indonesia); France had acquired Indochina (the region containing Vietnam, Laos, and Cambodia) and small holdings on the coasts of India; and Portugal had Goa and

Legend:
- British colonies
- French colonies
- Dutch colonies
- Portuguese colonies
- American colonies
- Independent

• **Figure 7.17** Colonial realms and independent countries in Monsoon Asia at the beginning of the 20th century.

ties. Some of the region's most important cities, including Shanghai in China, Kolkata (Calcutta) in India, and Singapore, were developed mainly by Western capital as seaports serving Western colonial enterprises.

Western domination of Asia ended in the 20th century for a variety of reasons. The two world wars weakened the West's ability to conduct its colonial affairs in the region, Japan rose to great power status and challenged the West early in World War II, and effective anticolonial movements arose in nearly all areas subject to European control. After World War II, all colonial possessions in Asia gained independence. The last to revert were Hong Kong (returned by Britain to China in 1997) and Macau (returned by Portugal to China in 1999). Until this era of independence, however, 20th-century Asia was marked by revolution, war, and considerable turmoil, as described in the following modules.

7.4 Economic Geography

Despite any stereotypes that might linger in the West, Monsoon Asia is not a quaint, old-timey backwater—especially in economic terms. It is a dynamic region with some of the world's strongest and fastest-growing economies. The world's second- and fourth-largest economies are there: Japan and China, respectively. The economies of China and India are surging, with both goods and services inundating the global economic system. Their clout, their accomplishments, and their ambitions are enormous. Some economists and equities marketers have taken to conflating the two countries and calling them "Chindia," a double-barreled engine of economic growth that cannot be stopped and that should attract all serious investors. "Chindia is at the intersection of emerging markets and natural resources," one mutual fund solicitation writes, "Chindia is already consuming a quarter of the world's copper, a third of the world's steel production and more than half of the world's cement. That consumption is projected to rise dramatically by 2020."[5]

It must be noted, however, that this growth is occurring against the backdrop of extraordinary poverty, and many hundreds of millions of people remain desperately poor. While Monsoon Asia as a whole has taken great strides at reducing poverty rates, the gap between rich and poor is growing in many countries. This is particularly true in India, where the overall economy has grown about 8 percent annually since 2000; yet in that same period, there has been almost no reduction in childhood malnutrition, which afflicts almost half of India's young. Poverty tends to be deepest in the rural areas, fueling the classic push factor of rural-to-urban migration.

Despite being generally a region of less developed countries, Monsoon Asia is home to the strong, industrialized, export-oriented economies of South Korea, Taiwan, Hong Kong and Singapore, which late in the 20th century came to

Diu in India, Macau in China, and Timor (today's Timor-Leste) in the East Indies. Although retaining a semblance of territorial integrity, China was forced to yield possession of strategic Hong Kong to Britain in the 1842 Treaty of Nanking and to grant special trading concessions and extraterritorial rights to various European nations and the United States through the latter half of the 19th century. In 1898, the Philippines, held by Spain since the 1500s, came under the control of the United States.

A few Asian countries escaped domination by the Western powers during the colonial age. Thailand (historically known as Siam) formed a buffer between British and French colonial spheres in peninsular Southeast Asia. Japan withdrew into almost complete seclusion in the mid-17th century but emerged in the late 19th century as the first modern, industrialized Asian nation and soon acquired a colonial empire of its own. Korea also followed a policy of isolation from foreign influences until 1876, when Japan began to colonize it. It is noteworthy that Japan, South Korea, and Thailand would emerge as very or relatively prosperous countries, whereas most of the former colonies lagged in economic development. Dependency theorists point to these discrepancies to affirm their argument that colonization hampered development.

Monsoon Asia was an extraordinarily profitable region for the colonizers. Western nations extracted vast quantities of tropical agricultural commodities and in turn found large markets for their manufactured goods. Westerners also invested heavily in plantations, factories, mines, transportation, communication, and electric power facilibe known as the **Asian Tigers** (or "Little Dragons"). Some economists consider the rapidly industrializing Southeast Asian countries of Thailand, Malaysia, Indonesia, the Philippines, and Vietnam the **New Asian Tigers** or **Tiger Cubs,** following the path of rapid economic growth blazed by their neighbors.

Japan was the first Asian country to develop modern cities and modern types of manufacturing on a large scale. It has been a major industrial power for more than a century. China and India are much better supplied with mineral resources than Japan. With the world's largest populations, they also have cheaper labor. But Japan's labor force is, at least for the time being, generally more skilled, and it leads Asia in high **value-added manufacturing**—the process of refining and fabricating more valuable goods from raw or semiprocessed materials. Now, however, there are signs that Japan's preeminence is waning as China's economy surges. The biggest economic story in Monsoon Asia is the rise of China. The highlights of that story are told here and the rest in Module 7.3, while India's economic geography is discussed in Module 7.1.

China's Surging Economy

In conventional terms of gross domestic product (total output of goods and services), China has the world's fourth-largest economy (after the United States, Japan, and Germany). The more intriguing measure of purchasing power parity—which takes account of price differences between industrialized countries like the United States and emerging-market countries like China—places China as the second-largest economy, after the United States (India is fourth by this measure). China, incidentally, is second only to the United States in energy consumption; Russia is third, Japan is fourth, and India is fifth.

However it is weighed, China's economic growth is soaring, with an average annual rate of 10 percent between 1990 and 2007. China's exports (along with imports) increased over 900 percent during those years, growing an average of 50 percent a year. China's share of the world's goods and services also increased 250 percent between 1990 and 2007 to 7 percent (compared with 11 percent for both the European Union and the United States). Just a few years ago, China was notable for making low-value-added products—cheap toys, for example—that the United States and other richer countries really had no interest in making. There were shirts and shoes, too, most of them manufactured in more than 20,000 factories in the special economic zone (SEZ) of Shenzhen, adjoining Hong Kong.

Now China is making at least a little bit of almost everything, including about three-quarters of the toys sold in the United States (•Figure 7.18). The country is making especially rapid progress in manufactures and exports of information technology (IT) hardware. Already China is the world's third-largest manufacturer of personal computers, sold under such brand names as Dell and IBM. China is poised to displace Taiwan as the regional leader of PC manufacturing and is projected to be the world's largest producer of IT hardware by 2012. The roots of China's boom in IT may be traced to the dot-com bubble that burst in the United States in 2000. Many of the talented Chinese who worked in California's Silicon Valley at that time went back to China, taking their skills with them in a kind of "reverse brain drain."

Recent changes in China's global status and in government economic policy have fueled the boom. China joined the World Trade Organization in 2001, prompting a surge in foreign investment. China sweetens the deal for investors with incentives like subsidized loans, tax exemptions, and 50 percent discounts on land prices. The addition of Chinese brawn and brains creates a nearly perfect investment climate. Chinese labor is much less expensive than is typical of most more developed countries, and China has a rapidly growing pool of university-trained engineers and other innovators.

• **Figure 7.18** Mechanized silk production in China.

China's Economic Impact

China's economic boom is having enormous consequences for other countries, particularly its Asian neighbors. Japanese electronics companies are cutting costs by moving increasing amounts of production to China. Surging investment in China is linked to disinvestment elsewhere, especially in Southeast Asia. That region's Tigers and Tiger Cubs—notably Malaysia, Singapore, and Thailand—dominated the 1990s with manufactures and exports of medium- and high-tech industrial products. With its growing consumer appetite, China was buying such goods from these countries. But now the manufacturing is shifting to China from those countries—some of them still smarting from an economic downturn in the late 1990s. South Korea's Samsung will soon make China the main base for production of its flat screens and computers. Japan's Toshiba makes televisions only in China. Dell recently moved some of its computer-making facilities from Malaysia to China. Between 1999 and 2004, Singapore lost more than 35,000 jobs, most of them to China. Motivated by lower costs in China, Japan's largest electronics manufacturer, Matsushita, has eliminated about 40 percent of its production and sales contractors in Southeast Asia and shifted new assets to China.

With China's increasing gravity, Southeast Asian economies may at least temporarily be relegated to supplying low-end products, especially food and raw materials, to China. In turn, they may end up buying cheap Chinese manufactured goods, harming their own fledgling industries. They may try to counter this trend by developing higher-value niche products to meet demands in China. Thailand, for example, wants to boost manufactures of health care products, while Singapore focuses on biomedical products and financial services. Poorer Indonesia will probably be able to count on helping satisfy China's growing appetite for natural gas to fuel its southern industrial cities. Meanwhile, the critical economies of Japan, Taiwan, and South Korea are increasingly vulnerable to China's ascendancy as investors are lured from those countries by China's lower labor costs and its potential 1.3 billion consumers.

Cheap labor is unlikely to make China a permanent magnet, however. As affluence rises there, so will wages, and China's less prosperous neighbors will again attract investment. China's one-child policy also means it will have fewer industrial workers in the future. Mindful of this demographic problem, farsighted multinational companies are boosting manufacturing investment in India, where the pool of young workers is likely to remain large and growing for decades to come.

China's rise is chipping away at U.S. economic dominance in the region. In one country after another (including the most critical, Japan and South Korea), China has eclipsed the United States as Asia's most essential trading partner. In the meantime, the U.S.-China relationship is growing, mainly in China's favor. Of China's exports, about 20 percent go to the United States (its largest single trading partner), but its imports from the United States lag far behind. This U.S. trade deficit with China threatens to become a sore point in relations between the countries. Textile and other industries in the United States are pressuring the U.S. government to pass punitive tariffs on Chinese imports, a move that could precipitate a costly trade war. Problems with tainted food and toy imports have already led to bans of some Chinese products in the United States.

The United States is also furious at China's reluctance to stop the counterfeiting or "piracy" of American products, particularly computer software and DVDs of Hollywood films. China is the epicenter of prolific Asian trade in pirated products (•Figure 7.19). Authorities in the region periodically crack down on sales of such goods, but the vendors are usually on a hair trigger, ready to disappear on a moment's notice with their goods in tow, only to reappear moments later when police leave. There is also growing concern in the United States over the perceived loss of jobs through outsourcing to India (see Insights, page 300).

The Green Revolution

Most Asian countries do not want to improve their economies only by enhancing their outputs of services and manufactured goods. They also want to boost agricultural self-sufficiency and crop exports. Revolutionary changes in crops are at the heart of efforts to reshape Asia's agricultural economies. One of the most significant agricultural innovations in Asia during the past century is the development of new seed types and innovations in planting, cultivating, harvesting, and marketing of crops, efforts known collectively as the Green Revolution. These developments are not exclusive to Monsoon Asia, but this is where most of the growth and investment have taken place.

Ever since 1962, when the International Rice Research Institute (IRRI) was founded in the Philippines, there have

• **Figure 7.19** Counterfeited goods made in China and sold throughout Asia bear the brand names of U.S. and other manufacturers but are sold to happy consumers at a fraction of the cost of the genuine product. This is a CD and DVD vendor in Laos.

INSIGHTS

Outsourcing

Outsourcing (also known as **offshore outsourcing** or **offshoring**) is the flight of technology and other jobs from countries with high manufacturing and service costs to countries with low manufacturing and service costs. India is the world's largest recipient of these jobs, and the United States is the leading outsourcer.

In the United States, there is growing concern and even alarm over the trend of outsourcing jobs to India. Many of these jobs are in computer programming and a wide variety of technical support, from computer problem phone calls to X-ray diagnosis. Many are mundane, like bill processing and order taking. Increasingly, however, high-paying white-collar employment is being outsourced to India, in such fields as investment banking, aircraft engineering, and pharmaceuticals research. Top Western corporations like the United States' Cisco Systems, the world's leading maker of communications equipment, are beginning to relocate substantial numbers of their senior executives to India, where much of the firm's work is actually done.

India is an especially attractive venue for outsourcing for several reasons. It has a large population of well-educated people, most of whom speak excellent English. Those who work the telephone call centers typically receive training in vernacular American English and strive to perfect their American accents while shedding their Indian and British pronunciations. The bottom line is another major advantage: Indian skilled labor is cheap. The average computer programmer in the United States earns $75,000 per year, while his or her counterpart in India earns $9,000 per year.

Finally, the Internet and superb telephone communications, ironically combined with India's physical distance from the United States, provide some remarkable opportunities. While people in the United States sleep, the day shift is on in India. An American doctor can X-ray a patient in the afternoon and by early the next morning have the analysis—performed overnight in India. On the down side, when an American consumer places a technical service call for a computer problem, he or she may be speaking with a very tired Indian employee working the graveyard shift. There are many accounts of fatigue and burnout among Indian call center employees.

What about the loss of American jobs due to outsourcing? By conservative estimates, 3.3 million U.S. service jobs will move offshore by 2015. Although this sounds alarming, it amounts to just over 200,000 jobs per year, which is a small fraction of the approximately 3.5 million jobs that are created in a typical year in the United States. But some projections are more alarming. In 2007, a former official of the U.S. Federal Reserve described outsourcing as a "third Industrial Revolution," predicting it would take jobs from as many as 42 million American workers in the coming years.* Outsourcing companies generally argue that reducing the costs of their services reduces inflationary costs for their consumers and improves efficiency and productivity. This rationale is, of course, cold comfort to the American worker whose job has moved offshore.

*Alan S. Blinder, quoted in Anand Giridharadas, "India's Edge Goes beyond Outsourcing." *New York Times*, April 4, 2007, p. C4.

---

been efforts to use science to increase food yields (particularly of rice) to stave off hunger and generate export income. Notable success has been achieved in breeding the new high-yield varieties of seed stock, and there has been a large upsurge of production in certain areas where the new strains have been widely introduced. Many of the new varieties are bioengineered or genetically modified (GM) crops. This means that through precise manipulation of the genetic components of a crop, it can not only produce a higher yield but also be more resistant to drought, flood, or pests or generate a higher amount of a desired nutritional component such as vitamin A. China, second only to the United States in the global biotechnology industry, has produced genetically engineered rice, corn, cotton, tobacco, sweet peppers, petunias, and poplar trees.

Because the underlying premise of the Green Revolution is economic—more crops will be produced at lower cost for higher revenue—most countries in the region are racing to increase their biotechnological output, especially in an effort to catch up with China and ensure they do not lose positions in the global marketplace. Malaysia hopes to build Biovalley, a biotechnology research park near Kuala Lumpur, and Indonesia is planning a similar Bio-

island. One of Biovalley's goals is to create palm oil trees genetically modified to produce the raw material for specialized plastics used in medical devices (•Figure 7.20). South Korea, Japan, and India have large and well-funded

• **Figure 7.20** Palm oil trees in Malaysia are being genetically modified to produce specialized plastics. The palm oil comes from the cluster of nuts at the base of the fronds.

biotechnology research programs. Japan has so far resisted production of GM food crops out of fears of possible health hazards, a concern that most Europeans and their governments share.

For Asian farmers to capitalize fully on the Green Revolution, they must overcome many obstacles. Success requires levels of capital that are often beyond the means of peasant farmers, landlords, and governments. Such expenditures are needed for water supply facilities (for example, the tubewells that now number hundreds of thousands in the Indo-Gangetic Plain of the northern Indian subcontinent), chemical fertilizers, and chemicals to control weeds, pests, and diseases. And as agriculture becomes more mechanized, considerable increases in the costs of machinery, fertilizers, and fuel have to be borne by farmers who have, in many cases, had little experience with the cash economy. Governments must improve transportation so that the large quantities of fertilizer required by the new seed varieties can be delivered in a timely fashion. Not only must there be these associated infrastructural changes to support this "revolution," but the crop calendar of the farmer becomes much less forgiving because many of the newest seed grains demand more precise water, fertilizer, and cultivation requirements than traditional grains. Grain storage facilities, now subject to plundering by rats, must be improved.

Overcoming the financial obstacles is rendered more difficult by the widespread system of **share tenancy.** Farmers who are share tenants generally have no security of tenure on the land and thus cannot be sure that money they invest in the Green Revolution will benefit them in the future. They may not wish to assume any additional risk, even if credit on reasonable terms is available. If they remain on their holdings, landlords may take up to half of the increased crop but bear little or none of the additional expense. Landlords may in turn be content to collect their customary rents without expending the additional capital necessary in this new mode of farming, or they may try to turn tenants off the land to create larger spatial units that they themselves can farm more profitably with machinery and hired labor. In fact, landowners with large holdings often become the chief beneficiaries of the new technology, with many smaller farmers becoming a class of landless workers hired for low wages on a seasonal basis or migrants in the rural-to-urban migration stream.

There are other problems associated with the Green Revolution. Economic dislocations result when rice-importing countries become more self-sufficient, causing hardships for rice exporters. Large infusions of the agricultural chemicals associated with high-yield varieties have negative repercussions on natural ecosystems. Of much wider scope is concern that the development of a limited number of high-yield crop varieties grown in vast monocultures will dramatically reduce the genetic variability of crops, in essence interfering with nature's ability to adjust to environmental changes.

7.5 Geopolitical Issues

In the Middle East and North Africa and in Russia and the Near Abroad, principal geopolitical concerns focus on the production and distribution of energy resources. In Monsoon Asia, by contrast, some of the most serious geopolitical issues are prospects for what may be done with weapons created from a particular energy source: nuclear energy. There are also concerns about traditional fossil fuels (see Geography of Energy, page 302), Islamist terrorism, and the security of shipping lanes.

The preceding discussion of the mushrooming economic clout of China and India also raises another critical geopolitical reality: Asia is emerging as center of gravity that will seriously challenge the century-long primacy of the United States in world affairs. Japan long stood alone as a pillar of strength in Asia, but with China and even India surging ahead, regional might has been reshuffled. Eyes in the region and around the world will be focused especially on China, which some analysts fear may use military power to enhance its economic clout.

Nationalism and Nuclear Weapons

As discussed in Module 7.1, after its independence from Britain in 1947, India emerged as an avowedly secular democratic state. But after the 1998 victory of the Hindu nationalist Bharatiya Janata Party (BJP), hopes for new privileges emerged among India's Hindu majority. Among BJP supporters, there was hope for a "Hindu bomb," a counterweight to a long-feared "Islamic bomb" in neighboring Pakistan.

On May 11, 1998, much to the surprise of U.S. and other Western intelligence agencies, India conducted three underground nuclear tests in the Thar Desert. With the blasts, India's government seemed to be trying to stake India's claim as a great world power, exhibit its military muscle to Pakistan and China, and garner enough political support for the BJP to form an outright parliamentary majority in the future. Initial reactions among India's vast populace were highly favorable; 91 percent of residents polled within three days of the event supported the tests. Outside India, there was alarm. India had defied an informal worldwide moratorium on nuclear testing that went into effect in 1996, when 149 nations (not including India and Pakistan) signed the **Comprehensive Test Ban Treaty** (also known as the **Nuclear Nonproliferation Treaty,** or **NPT**), which prohibits all nuclear tests.

The world's eyes quickly turned to India's neighbor to the west. Governments pleaded with Pakistan to refrain from answering India with nuclear tests of its own, arguing that Pakistan would have a public relations triumph it if exercised restraint: It would appear to be a mature and responsible power, whereas India revealed itself as a dangerous rogue state. But Pakistan's leaders felt obliged by their population's demands for a tit-for-tat response to India's

GEOGRAPHY OF ENERGY

The Spratly Islands

About 60 islands make up the Spratly Island chain, which lies in the South China Sea between Vietnam and the Philippines (see Figure 7.2.1, page 336). They are an idyllic tourist destination where divers can hire luxury boats to explore the coral reefs and palm-lined beaches of remote atolls. However, the islands are much more significant for their strategic location between the Pacific and Indian Oceans. During World War II, Japan used the islands as a base for attacking the Philippines and Southeast Asia. Still more significant, as much as $1 trillion in oil and gas may lie beneath the seabed around the Spratlys.

Not surprisingly, many nations covet control of the Spratlys. Six countries claim some or all of the islands, and international law is not equipped to sort out these conflicting claims (see the U.N. Convention on the Law of the Sea, page 407). China, Vietnam, and Taiwan claim sovereignty over all of them, while Malaysia, Brunei, and the Philippines claim some of them. During the Cold War, the competing nations felt it was too hazardous to push their claims on the islands. As the Cold War drew to a close, however, the situation became more volatile. All the contenders except Brunei placed soldiers, airstrips, and ships on the islands. In 1988, the Chinese navy invaded seven of the islands occupied by Vietnam, killing about 70 Vietnamese soldiers. In 1992, China again landed troops in the islands and began exploring for oil in a section of the seabed claimed by Vietnam. In 1995, China moved to expand its territorial designs on islands already claimed by the Philippines and has since built what it calls "shelters," which look like fortifications to the Filipinos. Late in the 1990s, Malaysia occupied two disputed Spratly Islands and began building on one of them. In 2004, Vietnam began renovations of an airport on one of the islands, saying it was necessary to boost tourism there. China condemned the construction as a violation of its territorial sovereignty, an accusation it repeated in 2007 when Vietnam negotiated natural gas development with foreign companies in the Spratlys.

Indonesia, which has no claim on the Spratlys, sponsored unofficial workshops on joint efforts in oil exploration among the six contenders in an effort to ward off a potential crisis. But so far, China, which has proclaimed not only the Spratlys but also the entire South China Sea as its own, has refused to discuss the issue at the official level. There are fears that as the countries' petroleum needs grow, each will seek more aggressively to gain control of the Spratlys (●Figure 7.F). The smaller powers fear that China could turn the islands into a kind of permanent aircraft carrier that could be used to dominate them militarily. An incident in these remote islands could trigger a much wider and more serious conflict in Asia.

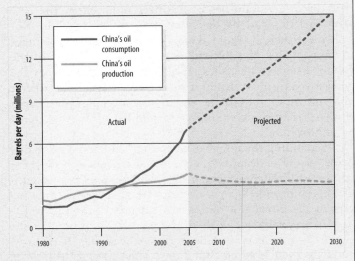

● **Figure 7.F** With China's oil appetite expected to grow enormously, every potential source of oil—like the Spratly Islands—takes on unprecedented importance.

blasts and soon followed with six nuclear tests. India and Pakistan thus joined only five other nations—the United States, Russia, China, the United Kingdom, and France—in acknowledging that they possess nuclear weapons.

There is disagreement about what this recent regional and global shift in the balance of power means. Some analysts fear an escalating nuclear arms race that, perhaps ignited by a border skirmish in Kashmir, could lead to the **mutually assured destruction (MAD)** of Pakistan and India. There are concurrent fears of nuclear war between China and India. Others, particularly in South Asia, argue that the weapons represent the best deterrent against conflict, as they did for decades between the countries of NATO and the Warsaw Pact. However, few people anywhere argue with the contention that the nuclear rivalry between India

and Pakistan has taken an enormous economic toll in two nations that need to wage war on poverty.

To the United States, both India and Pakistan are **pivotal countries,** defined by several influential historians as ones whose collapse would cause international migration, war, pollution, disease epidemics, or other international security problems.[6] The disposition of their nuclear arsenals is one of the main reasons for their being considered pivotal. If the government of Pakistan were to fall, Pakistan's nuclear weapons could come into the possession of a rogue element or a new government hostile to the West. India also has much to fear. Either a right-wing or an Islamist government in Pakistan would be far more likely than the current regime to take India on in the disputed Kashmir region. Such a confrontation could set the stage for a nuclear exchange

314

between Pakistan and India, not only decimating those countries but also sending economic shock waves around the world. India's recently surging economy has been a major force in India's recent diplomatic overtures to Pakistan; nothing discourages investment like war or the fear of it.

U.S.-Pakistan Relations since 9/11

Pakistan became even more pivotal with the events of 9/11. Up to that point, the United States had slapped tough economic sanctions against Pakistan (and less severe ones against India) for the nuclear weapons tests. The United States pursued a diplomatic courtship with India, both as a counterweight against China (India's longtime foe in the region) and as a means of expressing displeasure with Pakistan (China's ally) for helping Afghanistan's Taliban harbor Osama bin Laden. But the United States and Pakistan did an abrupt diplomatic about-face and embraced one another after 9/11. Pakistani President Pervez Musharraf instantly dropped support for the Taliban and quietly allowed the United States to use its territory to prepare for the assault on the Taliban and al-Qa'ida in Afghanistan. In return, the United States forgave much of Pakistan's debt to the United States and lifted its sanctions against Pakistan. To avoid isolating India, the United States also lifted its post–nuclear test sanctions against that country.

The United States has dramatically strengthened military ties with both countries, recognizing Pakistan in 2004 as a major non-NATO ally—one of just 11 countries to have that designation—and acknowledging India the same year as a "strategic partner." These designations opened the doors of both countries to major new shipments of American war matériel. But India would be favored: in 2007, Washington singled out India as its most critical strategic ally in the region by offering nuclear fuel and technology to New Delhi.

Under U.S. and allied attack in Afghanistan after 9/11, Taliban and al-Qa'ida fighters retreated to and regrouped in western Pakistan, particularly in the semiautonomous federally administered tribal areas (FATA; see Figure 7.1.17, page 326). Here the populace is mainly Pashtun and is sympathetic to the causes of their Taliban ethnic kin and their al-Qa'ida spiritual kin (●Figure 7.21). This is an area where Pakistani government authority has long been kept at bay and where it was considered extremely difficult to insert Pakistani forces on a sustained, aggressive basis. Bowing to American pressure, Pakistan's government periodically undertakes military operations in the FATA but considers sustained effort there potentially detrimental to its survival. Even without the direct insertion of U.S. troops, assertive Pakistani intervention in the FATA could lead to a popular uprising, a military coup, or both. Not just the Pashtuns but the vast majority of Pakistan's ethnic groups are strongly against U.S. interventions in the region, and two factions in the army (one right-wing, another Islamist) stand ready to challenge the government's authority. If the

221

Joe Hobbs

● **Figure 7.21** The Pashtun of western Pakistan are a proudly independent people who are opposed to American interests in the region.

government were to fall, there would be grave concerns about Pakistan's nuclear weapons.

What Does North Korea Want?

There are also major concerns about nuclear weapons in Northeast Asia. Ever since suffering Hiroshima and Nagasaki, Japan has had an official policy never to develop or use atomic weapons. But Japan worries about the potential nuclear threat from three adversaries, all of which possess nuclear weapons: Russia, China, and North Korea. The Japanese feel that the West dismissed potential threats from Russia too readily when the Soviet Union dissolved. Japan still has territorial disputes with Russia, particularly involving the four Kuril Islands of Kunashiri, Etorofu, Shikotan, and Habomai, which the government of Josef Stalin seized at the end of World War II and lie just off the northeastern coast of Hokkaido (see Figure 7.4.1, page 393). Japan and China also have a territorial dispute concerning the East China Sea, and if major oil reserves are discovered there, as anticipated, relations between the two countries could deteriorate. China continued to test nuclear weapons through 1996, adding to tensions in Japan. Finally, Japan fears reunification in Korea, which, as a former Japanese colony from 1910 to 1945, has a historical dislike of Japan. There is speculation that Japan may do the unthinkable and the officially disavowed—develop nuclear weapons—to counter the perceived threat of North Korea's nuclear weapons program. 406

The world looks nervously at the troubled relations between North and South Korea and between North Korea and the West. A crisis flared in 1994 when North Korea refused to permit full inspection of its nuclear facilities by the **International Atomic Energy Agency (IAEA).** The country was suspected of separating plutonium that could be used in making nuclear bombs. Over time, the United States, South Korea, and Japan worked out an agreement in which North Korea would agree to freeze development of nuclear weapons in exchange for the others' providing fuel oil and assistance in building nuclear power plants. Those nuclear reactors would be of the "light water" variety, much less likely than North Korea's plutonium-based reactors to be "dual use" for both military and civilian purposes. The Clinton administration trumpeted the agreement as a success, but it had several flaws. One was that North Korea did not have to dispose of its existing nuclear fuel; it simply had to stow it safely away—meaning that it could quickly be reactivated.

North Korea pressed ahead with a program to build missiles capable of carrying nuclear warheads and in 1998 launched such a missile (minus the warhead) over Japan. This incident prompted the United States to renew its resolve to establish an antimissile defensive "shield" over the United States—the so-called **Strategic Defense Initiative (SDI),** which detractors soon dubbed the "Star Wars Initiative," dating to the Reagan administration of the 1980s. Some analysts believe that the United States is reluctant to see the two Koreas reunite because it would remove much of the justification on which the missile shield program is based. It might also put pressure on the United States to reduce its military presence in the western Pacific. They add that Japan, too, would not want to see Korean reunification because it would remove Japan's justification for building up its defenses, which are designed less for confrontation with North Korea than with China.

President George W. Bush came into office with a hard line against North Korea, effectively suspending dialogue and technical assistance for the still uncompleted nuclear power plants. In 2002, he proclaimed that North Korea was one of three countries making up an "axis of evil," along with Iraq and Iran, signaling that the United States was more interested in confronting than accommodating North Korea. Late that year, North Korea dropped a virtual bombshell on the United States by admitting—after being confronted with evidence collected by U.S. intelligence agencies—that it did indeed have an active nuclear weapons program (developed, as it turns out, with Pakistani technical assistance). This admission followed earlier strident denials of such a program by North Korean officials. The American side, focused on developing the military campaign in Iraq, had been content to let the problem of North Korea's weapons program simmer in the background. By most measures, North Korea should also have been content to keep quiet; the United States, South Korea, and Japan had for almost a decade been providing food and fuel to the economically beleaguered country, and any belligerence by North Korea could end that aid.

What, then, was North Korea hoping to accomplish by revealing its nuclear weapons program? There are several possibilities. The least likely is that North Korea was instigating a military confrontation with the United States and its allies South Korea and Japan. In any war scenario, those allies would obliterate North Korea. The United States is obliged to defend South Korea in the event of a war with the North and would effectively wield its huge military advantage (•Figure 7.22). However, that victory would come at an enormous price for South Korea and possibly Japan. North Korea has a standing army of nearly 1 million soldiers, an impressive military arsenal including short- and medium-range missiles, a stockpile of chemical and biological weapons that it might not be shy to use, and now, presumably, at least a few nuclear weapons that it could deploy. The casualties in South Korea would be enormous, whether from a conventional or an unconventional assault from the North.

It is much more likely that by disclosing its nuclear weapons program, North Korea sought guarantees that it could avoid war and also gain even more assistance from the West. Since the early 1990s, its nuclear weapons program has been the only leverage North Korea has had to coax desperately needed supplies from abroad. North Korea had only to mention the prospect of reactivating its nuclear weapons program to get more concessions from the United States and its allies—especially food aid during a succession of droughts and floods in the 1990s. In effect, North Korea was extorting money from the United States and its allies, which have been happy to ante up as a means of containing the rogue nation. By admitting to the nuclear weapons program, North Korea may have felt it could get even more aid in technology, fuel, food, or other forms. It may have been right.

A series of on-again, off-again negotiations known as the "Six Party Talks" (held between North and South Korea, the United States, China, Japan, and Russia) culminated in 2007 with an agreement: the United States would help unfreeze North Korean funds in a Macau bank, take steps to remove North Korea from its list of countries supporting terrorism, lift trade sanctions in place since the Korean War ended in the 1950s, and with along with its negotiating partners restart the flow of fuel oil to North Korea; in return, North Korea would close its nuclear plants and allow them to be inspected by international monitors. In the deal, North Korea did not promise to dismantle its nuclear facilities at Yongbyon but only "abandon" or "disable" them, which it did do by the end of that year. North Korea is not required to destroy any existing stocks of nuclear weapons, and much uncertainty remains. The United States apparently hopes that the involvement of China and the other countries will make North Korea less likely to restart its nuclear program this time.

383,
400

234,
265

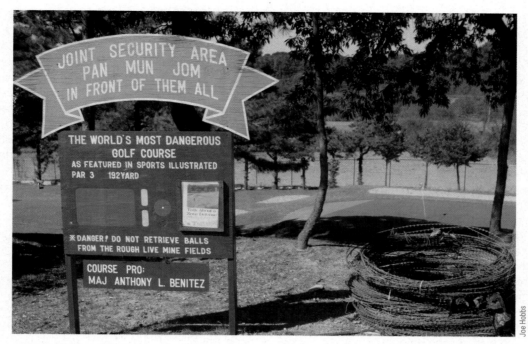

Joe Hobbs

• **Figure 7.22** A little humor eases the tension for U.S. troops serving along what may be the world's tensest border, the "no-man's-land" separating North and South Korea. Even a minor incident here could touch off a conflagration that might take a huge toll in human lives.

Islands, Sea Lanes, and Islamists

As related in Module 7.2, Indonesia—another pivotal country from the U.S. perspective—is a very ethnically complex nation whose integrity is threatened from a variety of secessionist movements in locales as far apart as Aceh in the west and Papua (formerly Irian Jaya) in the east (see Figure 7.2.18, page 357). A possible **domino effect** scenario for Indonesia concerns foreign powers: If one province falls or proclaims its independence, many others will, and the largest country in the region will break apart. Countries like the United States fear what would happen to vital international shipping lanes should several new and perhaps militant states emerge from Indonesia's fragmentation. There have already been scores of pirate attacks on vessels plying the strategic Strait of Malacca, a critical chokepoint through which a quarter of the world's trade passes, including two-thirds of the world's shipment of liquefied natural gas and half of all sea shipments of oil (it ranks second only to the Strait of Hormuz as an oil shipping lane). Security in the Strait of Malacca is a source of great concern to Japan, which imports 80 percent of its oil on ships that pass through the strait. Aceh's vast natural gas deposits and Papua's copper and gold are also seen as critical in the global economy, so like Indonesia, many world powers are anxious to see that they stay in the "right" hands. The United States gives generous amounts of foreign aid to Indonesia, but the American presence is increasingly unwelcome because so many Indonesians, who constitute the world's largest Muslim population, perceive the post-9/11 United States as anti-Muslim.

U.S. intelligence agencies fear that Southeast Asia, and Indonesia in particular, will emerge as an Islamist terrorist hearth to replace Afghanistan. Al-Qa'ida and its affiliate organizations are known to be active in Southeast Asia. Two of the 19 men who carried out the 9/11 attacks planned part of the operation in Malaysia's capital, Kuala Lumpur. Malaysia is an unlikely refuge for groups like al-Qa'ida, mainly because it sees itself as a progressive engine of high-tech economic growth and would be loath to suffer the sanctions or other fallout associated with harboring terrorists. However, Malaysia is a convenient transit and staging point for terrorist interests, and some members of al-Qa'ida and affiliated groups are Malaysian nationals. Citizens of any predominantly Muslim country may enter Malaysia without obtaining a visa (and thus escape most opportunities for a background check).

Mainly Chinese, prosperous, aggressively policed Singapore is the region's most unlikely terrorist refuge of all. But with its resident population of 17,000 Americans and its numerous Western commercial and military activities (including regular visits by U.S. naval vessels), Singapore is an attractive target. Evidence uncovered in an al-Qa'ida "safe house" in Afghanistan that was examined by U.S. intelligence experts revealed a plot that would have carried out ammonium nitrate truck bombings against U.S. and British embassies and other facilities in Singapore in 2002. An al-Qa'ida "sleeper cell" based in Singapore had been activated from Malaysia by Riduan Isamuddin (also known as Hambali), an Indonesian citizen who hosted the two 9/11 hijackers on their visits to Kuala Lumpur, was a suspect in the U.S.S. *Cole* bombing, and was arrested by intelligence authorities in 2003. These sleepers, eight of whom had been trained by al-Qa'ida in Afghanistan, were Indonesians.

Indonesia is a secular rather than a religious state, and a moderate form of Islam prevails in politics and in everyday life. What concerns Western intelligence authorities and Indonesia's moderate leadership are the more militant Islamist organizations that have emerged in the past decade and have apparently gathered strength since 9/11. One of these is **Laskar Jihad,** created in 2000 to mobilize Indonesian Muslims to fight Christians in Indonesia's Sulawesi and Muluku Islands and Americans throughout the country. Another organization with suspected terrorist links is **Jemaah Islamiah (JI),** also known as the Islamic Group, led by Abu Bakar Baasyir (Abu Bakar Bashir), whose nominal role was preacher in an Islamic school in Solo on the island of Java. (Western authorities are much concerned about such religious schools, known as *pesantrens* in Indonesia. Like the *madrasas* of Pakistan and the Arab heartland of the Middle East, these are venues for both religious and secular learning, and some of the schools' curricula are imbued with virulent anti-Western content.) Baasyir admitted to having tutored 13 of the men arrested in connection with the plot to blow up Western targets in Singapore. These included Hambali, al-Qa'ida's point man in Southeast Asia, whose regional affiliation is with JI. Intelligence authorities believe that Baasyir was also the ringleader for a series of attacks that would have been carried out against American embassies and naval vessels in several Southeast Asian countries on the first anniversary of the 9/11 attacks. Finally, after the ferocious bombings of nightclubs on the Indonesian island of Bali in 2002 in which both al-Qa'ida and Baasyir were implicated, Indonesian authorities imprisoned

Baasyir. Amid furious international criticism, they released him after two years.

Indonesia has much going for it as a potential new hearth for al-Qa'ida affiliated activities. Much like Afghanistan was prior to 9/11, it has a predominantly Muslim population, including some extremist factions that would be willing to provide safe haven; it has a largely poor and otherwise disaffected population (in theory, providing recruits for a militant cause); it includes remote locales that make suitable weapons and tactics training grounds; and it has regions that the government is unwilling or unable to control. In 2003, the director of Indonesia's national intelligence agency acknowledged that al-Qa'ida had set up training camps in the country. U.S. authorities believe that Southeast Asia, with Indonesia as its centerpiece, has the world's highest concentration of al-Qa'ida operatives outside Pakistan and Iraq.

Why, then, does the United States not have troops on the ground in Indonesia or advisers there (as in the Philippines) to train Indonesian forces in antiterrorism tactics? The problem is that Indonesia is even more difficult than Pakistan for U.S. antiterrorism forces to operate in. It is more geographically complex and even less receptive to a U.S. presence. Perceived intervention by the United States could be enough to spark a now largely neutral populace into violent anti-Americanism and perhaps bring down Indonesia's government. Washington fears that many states, some of them anti-Western, could emerge in Indonesia's place and threaten American interests in the country—notably its oil, natural gas, and copper resources and its location astride vital shipping lanes.

This concludes an introduction that has set the stage for further exploration of land and life in the subregions and countries of Monsoon Asia.

SUMMARY

→ Monsoon Asia includes the countries of Japan, North and South Korea, China, Taiwan, Pakistan, India, Sri Lanka, Bangladesh, Bhutan, Nepal, Maldives, Myanmar (Burma), Laos, Thailand, Cambodia, Vietnam, Malaysia, Singapore, Indonesia, the Philippines, Brunei, Timor-Leste, and numerous islands scattered along the edges of this major continental bloc. The term *monsoon* is used because of the central role this wind and precipitation pattern plays throughout this region.

→ This is the most populous world region, with 54 percent of the world's people. It includes the world's most populous countries, China and India. Population growth rates in the region vary widely, from zero growth in Japan to over 3 percent per year in Timor-Leste.

→ Three concentric arcs make up the broad physiography of the region: an inner arc of the high mountain ranges, the Himalaya, Karakoram, and Hindu Kush; a middle arc of major floodplains, deltas, and low mountains; and an outer arc of thousands of islands, including the archipelagoes of the East Indies, Philippines, and Japan.

→ Major rivers include the Indus, Ganges, and Brahmaputra in South Asia; the Irrawaddy, Chao Praya (Menam), Mekong, and Red in Southeast Asia; and the Chiang Jiang (Yangtze) and Huang He (Yellow) in East Asia.

→ The region's climate types and biomes include tropical rain forest, savanna, humid subtropical, humid continental, steppe, desert, and undifferentiated highland. Shifting cultivation and wet rice cultivation are important forms of agriculture. Wet rice cultivation produces very high yields and is associated with dense human populations.

→ Although Monsoon Asia has some of the world's largest cities, about 62 percent of the region's population is rural. Many uniquely Asian traditions helped shape village settlement planning and home design.

→ Monsoon Asia's ethnic and linguistic compositions are diverse. Major language families are Indo-European, Dravidian, Sino-Tibetan, Altaic, Austric, and Papuan.

→ Major religions and sociopolitical philosophies of Monsoon Asia are Hinduism, Islam, Buddhism, Confucianism, Daoism, and Christianity.

→ Great Britain, the Netherlands, France, and Portugal were the most important colonial powers in this region. Most of these domains were relinquished by the middle of the 20th century, and the later British return of Hong Kong and the Portuguese return of Macao (both to China) closed the colonial period.

→ Japan has the region's strongest economy, second only to that of the United States. Recent decades have seen strong economic growth among the Tigers and Tiger Cubs of the region, including South Korea, Taiwan, Singapore, Thailand, and Malaysia. The most important emerging economic power is China, whose large and inexpensive labor force has recently been attracting investment away from other parts of the region. India's well-educated and inexpensive labor force is contributing to its strong economic growth. There are fears in the United States about the outsourcing of American jobs to India.

→ The Green Revolution is a broad effort to increase agricultural productivity in dominant crops. Biotechnology has produced crops that are more drought- and pest-resistant and capable of creating much higher yields, but genetic engineering of crops has perceived risks. Profits and other benefits from the Green Revolution are not uniformly spread.

→ There are several major geopolitical issues in Monsoon Asia. The traditional enemies Pakistan and India (both pivotal countries from the U.S. point of view) now possess nuclear weapons. There are fears that destabilization in Pakistan might allow the weapons to fall into the wrong hands. Pakistan's support of the United States in its war on terrorism is very risky because a popular backlash could bring down the government. North Korea has nuclear weapons and is apparently trying to use them as bargaining chips to get more food aid and other assistance from the West. Antigovernment and anti-Western Islamists are active in Indonesia, another country viewed as pivotal by the United States. Strife and fragmentation in Indonesia could threaten vital oceanic shipping lanes.

KEY TERMS + CONCEPTS

Ainu (p. 293)
Altaic language family (p. 293)
   Japanese (p. 293)
   Kazakh (p. 293)
   Korean (p. 293)
   Mongolian (p. 293)
   Tungus (p. 293)
   Turkic (p. 293)
   Uighur (p. 293)
ancestor veneration (p. 294)
animism (p. 294)
archipelago (p. 282)
Aryans (p. 291)
Asian Century (p. 278)
Asian Tigers (p. 298)
Austric language family (p. 293)
   Austro-Asiatic subfamily (p. 293)
      Mon-Khmer (p. 293)
         Khmer (p. 293)
         Vietnamese (p. 293)
   Austronesian subfamily (p. 293)
      Barito (p. 293)
      Bugis (p. 293)
      Dayak (p. 293)
      Javanese (p. 293)
      Kayan-Kenyah (p. 293)
      Malay (p. 293)

      Timorese (p. 293)
      Tomini (p. 293)
   Hmong-Mien subfamily (p. 293)
      Hmong (p. 293)
   Tai-Kadai subfamily (p. 293)
      Lao (p. 293)
      Thai (p. 293)
belief systems (p. 293)
Buddhism (p. 295)
Comprehensive Test Ban Treaty (p. 301)
Confucianism (p. 296)
Daoism (Taoism) (p. 296)
domino effect (p. 305)
Dravidian language family (p. 291)
   Malayalam (p. 292)
   Tamil (p. 292)
   Telegu (p. 292)
fallow (p. 288)
feng shui (p. 290)
geomancy (p. 290)
Han Chinese (p. 293)
Hinduism (p. 294)
Indo-European language family (p. 291)
   Indo-Iranian subfamily (p. 291)
      Bengali (p. 291)
      Hindi (p. 291)
      Nepali (p. 291)

      Pashto (p. 291)
      Punjabi (p. 291)
      Sanskrit (p. 291)
      Sinhalese (p. 291)
      Urdu (p. 291)
insolation (p. 283)
intensive subsistence agriculture (p. 286)
International Atomic Energy Agency (IAEA) (p. 304)
Jemaah Islamiah (JI, or the Islamic Group) (p. 306)
karma (p. 295)
Laskar Jihad (p. 306)
laterite (p. 288)
lateritic soils (p. 288)
levee (p. 290)
Mahayana Buddhism (p. 296)
monsoon (p. 283)
mutually assured destruction (MAD) (p. 302)
New Asian Tigers (p. 298)
nirvana (p. 296)
Nuclear Nonproliferation Treaty (NPT) (p. 301)
offshore outsourcing (offshoring) (p. 300)
one-child campaign (p. 284)
outsourcing (p. 300)

paddy (padi) (p. 287)
Papuan languages (p. 293)
pivotal countries (p. 302)
plinthite (p. 288)
shifting cultivation (p. 288)
share tenancy (p. 301)
Shintoism (p. 294)
Sino-Tibetan language family (p. 292)
   Sinitic subfamily (p. 292)
      Chinese (p. 292)
         Mandarin (p. 293)

Min (p. 293)
Xiang (p. 293)
Yue (p. 293)
Tibeto-Burman subfamily (p. 292)
   Burmese (p. 292)
   Tibetan (p. 292)
slash-and-burn cultivation (p. 288)
Strategic Defense Initiative (SDI) (p. 304)
summer monsoon (p. 284)
swidden cultivation (p. 288)
syncretism (p. 294)

terraces (p. 287)
theory of Himalayan environmental
   degradation (p. 289)
Theravada Buddhism (p. 296)
Tiger Cubs (p. 298)
value-added manufacturing (p. 298)
wet rice cultivation (p. 286)
winter monsoon (p. 284)
yang (p. 290)
yin (p. 290)

REVIEW QUESTIONS

1. What countries make up Monsoon Asia?

2. Use the concept of the three arcs to trace the outlines of Monsoon Asia's geography. What are the largest plains and river valleys of the region? What are the most important mountain ranges? What are the major island groups?

3. What are the region's most populous countries? Where are the highest and lowest population densities? Which countries have the highest and lowest population growth rates?

4. What are the monsoons? What roles do they play in climates and human activities?

5. What types of environments are shifting cultivation and wet rice cultivation practiced in? What are some of the basic methods and land use considerations of both types of agriculture?

6. What roles have *feng shui* and Confucianism played in the site selection and design of homes, schools, and villages in parts of Asia?

7. What are some of the innovations and products that originated in Monsoon Asia and diffused from there?

8. What are the major linguistic and ethnic groups of Monsoon Asia? What are the principal religions and sociopolitical philosophies?

9. Who were the major colonial powers in Monsoon Asia? What regions did they colonize?

10. What are the region's most significant economic trends?

11. Why are India, Pakistan, and Indonesia considered pivotal countries for U.S. interests? Has the United States been forced to take sides in the traditional enmity between India and Pakistan?

NOTES

1. Kamala Markandaya, *Nectar in a Sieve* (New York: Day, 1954), p. 57.

2. John Reader, British anthropologist and traveler, cited in Fred Pearce, "Terraces: The Other Wonders of the World," *Eurozine,* March 2001, http://www.eurozine.com/articles/2001-03-01-pearce-en.html. Accessed July 11, 2007.

3. Jack D. Ives and Bruno Messerli, *The Himalayan Dilemma: Reconciling Development and Conservation* (New York: Routledge, 1989).

4. Descriptions of Buddhism and Daoism are based on the *Microsoft Encarta Reference Library,* 2004.

5. U.S. Global Investors, "What You Need to Know about Chindia." Postcard from U.S. Global Investors, Inc., 2006. See also Frank E. Holmes, *Chindia and Its Global Impact* (San Antonio, Tex.: U.S. Global Investors, 2007), http://www.usfunds.com/docs/presentations/chindia/viewhtml.asp. Accessed October 12, 2007.

6. Robert Chase, Emily Hill, and Paul Kennedy, "Pivotal States and U.S. Strategy," *Foreign Affairs,* January-February 1996, pp. 33–51.

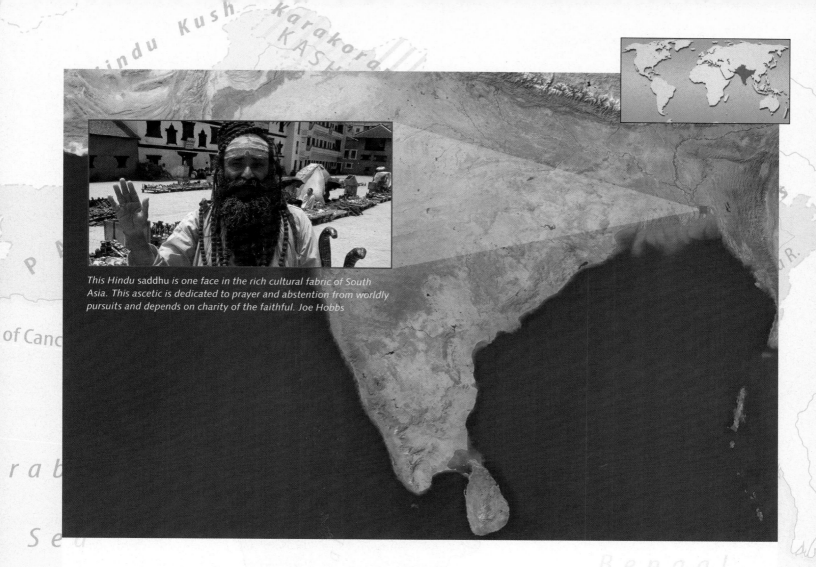

This *Hindu* saddhu is one face in the rich cultural fabric of South Asia. This ascetic is dedicated to prayer and abstention from worldly pursuits and depends on charity of the faithful. Joe Hobbs

MODULE 7.1
COMPLEX AND POPULOUS SOUTH ASIA

A triangular peninsula thrusts southward 1,000 miles (c. 1,600 km) from the Asian landmass, splitting the northern Indian Ocean into the Bay of Bengal and the Arabian Sea. The peninsula is bordered on the north by the alluvial plain of the Indus and Ganges (Ganga) Rivers, and north of these rise the highest mountains on earth. The entire unit—peninsula, plain, and fringing mountains—is known as the Indian subcontinent. It contains the five countries of India, Bangladesh, Pakistan, Nepal, and Bhutan in an area about half the size of the 48 conterminous United States (●Figure 7.1.1). Off the southern tip of India, across the narrow Palk Strait, lies the island nation of Sri Lanka, which shares many physical and cultural traits with the subcontinent. Farther to the southwest is the Maldives, a country of some 1,100 small, low-lying islands. These seven countries make up the subregion of South Asia.

■ Module outline

■ Module objectives

This module should enable you to

→ Recognize the tensions between Muslims and Hindus in South Asia that often give rise to violent conflict

→ Appreciate how the partition of India created lasting problems in political relations, resource use and allocation, and industrial development

→ See that food production has kept pace with enormous population growth in South Asia

→ Consider the role of high technology in reshaping the economies and outlooks of South Asia's peoples

→ Recognize the serious consequences of political insurgencies in Kashmir, India, Pakistan, Nepal, and Sri Lanka

● **Figure 7.1.1** Political units of South Asia. The status of Kashmir, disputed between India and Pakistan, with some parts occupied by China, remains unresolved.

A world-class giant, India outranks the other South Asian nations in both population (1.1 billion) and area (1.2 million square miles, or 3 million square kilometers, about the combined size of the states of Alaska, Texas, California, and Colorado). Bangladesh (149 million) and Pakistan (169 million) are much smaller than India in area and population, but both are among the world's most populous countries. Nepal and Bhutan, on the southern flank of the Himalaya Mountains, are small, rugged, and remote, with fewer people (27 and 1 million, respectively). They serve as buffer states between India and China, which have engaged in sometimes violent border disputes since the early 1950s. Strife has not spared Sri Lanka, whose 20 million people have experienced a civil war for more than two decades. The 300,000 people of the Maldives enjoy relative peace and tranquillity on their isolated islands. Altogether, about 1.5 billion people, more than one of every five on earth, lived in South Asia in 2007.

South Asia is a paradox. The true South Asia is found in the villages, where 7 of every 10 persons live and where many peasant farmers live and work much as their ancestors have for millennia. But the true South Asia is also found in its great cities (which in India the government is progressively renaming): Kolkata (formerly Calcutta),

Mumbai (formerly Bombay), Bengaluru (formerly Bangalore), Karachi, and Dhaka are central to the identities and, in some cases, the cutting edge of technology economies of this region's people. South Asia is thus both ancient and modern, rural and urban.

7.1.1 The Cultural Foundation

An enormous range of ethnic groups, social hierarchies, languages, and religions can be found among regions of South Asia and even within a single settlement. The subcontinent is the most culturally complex area of its size on earth, and its civilizations have roots deep in antiquity. The Indian subcontinent is one of the world's culture hearths. 41 Among the animals first domesticated in the region were zebu cattle (Indian humped cattle), which today are important livestock in many world regions. After 3000 B.C.E., some of the world's earliest cities (notably Mohenjo-Daro) developed on the banks of the Indus River in what is now Pakistan (●Figure 7.1.2). The peoples of the **Harappan civilization** were probably the ancestors of the Dravidians, represented today mainly by the Tamils of southern India.

Between 2000 and 1000 B.C.E. came the Aryans (or **Indo-Aryans**), herders of cattle and horses from central Asia who

● **Figure 7.1.2** Mohenjo-Daro, on the banks of the Indus River in what is now central Pakistan, was one of the world's earliest cities, with an estimated 50,000 people in 2500 C/B/F/ The written language of the Harappan civilization has not yet been deciphered.

were the source of many ethnic and cultural attributes that are dominant in the subcontinent today. These include the lighter complexions than are typical of the Dravidians, the numerically dominant Indo-European languages such as Hindi, and the faith of Hinduism, which developed mainly from the beliefs and customs of the Aryans. In the eighth century, Arabs introduced the religion of Islam to what is now Pakistan. Waves of Muslim conquerors who came through mountain passes from Afghanistan and central Asia spread Islam farther afield in later centuries. The last great Islamic invasion, by a Mongol-Turkish dynasty called the **Moguls (Mughals)**, began in the early 16th century, soon after the Europeans first came to India by sea. Before it succumbed to rebellious local Hindus and Sikhs and to the British Empire, the Mogul empire ruled nearly the entire subcontinent from its capital at Delhi. Some of the region's most spectacular artistic and scientific achievements date to the peak of Mogul power in the 16th and 17th centuries, including the Red Fort in Delhi and the tomb of Emperor Shah Jahan and Empress Mumtaz Mahal—the astonishing Taj Mahal in Agra (●Figure 7.1.3).

There are major social and political divisions within the subcontinent that relate to religion. As will be seen later in this module, religious differences between the two largest religious groups—Hindus (whose religion is described in Chapter 7, pages 294–295) and Muslims (whose religion is described in Chapter 6, pages 225–227)—are often troublesome. Hinduism was the dominant religion at the time Islam made its appearance in this region. It continues to have the largest number of adherents, including an estimated 13 per cent of the population of India and 4 percent of Nepal. Muslims are majorities in Pakistan (97 percent) and Bangladesh (83 percent). Muslims make up a seemingly small percentage of India's population, but they number nearly 150 mil-

lion people, making India home to the world's third largest Muslim population, behind Indonesia and Pakistan.

Seldom in history have two large groups with such differing beliefs lived in such close association with each other as Hindus and Muslims in the subcontinent. Islam holds to an uncompromising monotheism and prescribes uniformity in religious beliefs and practices. Hinduism is monotheistic for some believers but polytheistic for others and asserts that a variety of religious observances is consistent with the differing natures and social roles of humans. The exuberant and loud celebrations of the Hindu faith are a striking contrast to the austere ceremonies of Islam. Islam has a mission to convert others to the true religion, whereas most Hindus regard proselytizing as essentially useless and wrong; one is born or reborn as a Hindu, so conversion is not an issue. Islam's concept of the essential equality of all believers is in total contrast to the inequalities of the caste system endorsed in Hinduism (see page 325). Islam's use of the cow for food is anathema to Hinduism. And then there are religion-based politics, which often turn differences into discord and conflict, as will be seen later.

In addition to Muslims, there are other significant religious minorities in India. About 20 million **Sikhs** are concentrated mainly in the prosperous state of Punjab. **Sikhism** is India's fourth-largest religion, after Hinduism, Islam, and Christianity. The faith emerged in the 15th century as a religious reform movement intent on narrowing the differences between Punjab's Hindus and Muslims. Sikhs regard men and women as equals and disavow the caste system recognized by Hindus. Sikh men are recognizable by their turbans, which contain hair that should never be shorn (although a growing number of Sikhs insist on doing so), by a bracelet called a "bangle" worn on the right arm, and by their ubiquitous surname Singh, meaning "lion."

There are about 25 million Christians in India, most of whom live in the south of the peninsula; Buddhists, Jains, Parsis, and members of a variety of tribal religions make up the numerous smaller remaining religious minorities. Most of the estimated 7 million Buddhists are recent converts from among India's lowest castes. Numbering perhaps 75,000 and concentrated mainly in Mumbai, the famously entrepreneurial **Parsis** have attained wealth and economic power far out of proportion to their number. Their religion is the pre-Islamic Persian faith of **Zoroastrianism,** a monotheistic faith known mainly for its reverence of fire. Zoroastrians follow the teachings of the prophet Zarathustra, who was born around 600 B.C.E. in Persia (modern Iran). He preached that people should have good thoughts, speak good words, do good deeds, and believe in a single god, Ahura Mazda ("Wise Lord"). Persia's Zarathustras, as the Zoroastrians are also known, fled Persia in two waves, first during the conquest by Alexander the Great around 300 B.C.E. and then after Islam swept into Persia in the seventh century. They settled in China, Russia, and Europe, where they became assimilated with other cultures. In India, they

● **Figure 7.1.3** For Indians and foreigners alike, the Taj Mahal in Agra symbolizes the cultural achievements of the Indian subcontinent.

maintained a separate existence until quite recently; now about one in three takes a spouse of another religion and abandons Zoroastrianism. Many scholars consider theirs an endangered religion, with only 124,000 to 190,000 followers worldwide.

India's 6 million **Jains** also have influence beyond what their numbers suggest, as they control a significant share of India's business. They live mainly in the western state of Gujarat. **Jainism,** founded on the teachings of Vardhamana Mahavira (a contemporary of the Buddha, c. 540–468 B.C.E.), is renowned for its respect for geographic features and animal life. Jains believe that souls are in people, plants, animals, and nonliving natural entities such as rocks and rivers. Jainism has taken the principle of nonviolence (*ahimsa*, also present in Hinduism) to mean they should not even harm microbes. Therefore, Jain worshipers often wear masks to prevent inhalation of microscopic organisms, and Jains cannot be farmers because they would have to destroy plant life and living organisms in the soil (●Figure 7.1.4).

Sri Lanka has two major ethnic groups, distinguished from each other by religion and language: the predominantly Buddhist **Sinhalese,** making up about 74 percent of the population, and the Hindu **Tamils,** about 9 percent (●Figure 7.1.5). The light-skinned Sinhalese are an Indo-Aryan people who settled in Sri Lanka about 2,000 years ago. The dark-skinned Tamils, whose main area of settlement is the Jaffna Peninsula and adjoining areas in the north, are descendants of early invaders and more recent imported tea plantation workers from southern India. The migrations of Tamil tea workers from India to Sri Lanka, and their subsequent travails there, are part of the legacy of British colonial rule.

328–329

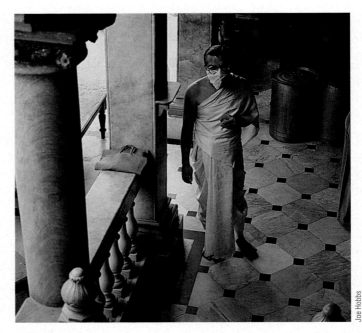

● **Figure 7.1.4** A priest in a Jain temple in Mumbai wears a mask so as not to breathe in and thereby harm microbes.

7.1.2 Geographic Consequences of Colonialism and Partition

On a physical map, the Indian subcontinent looks like a discrete unit, marked off from the rest of the world by its mountain borders and seacoasts. But the social complexity of this area is so great that it has never been unified politi-

Parvinder Sethi

● **Figure 7.1.6** The central rail station in New Delhi, India, is one of the hubs in a vast network dating to the British colonial period.

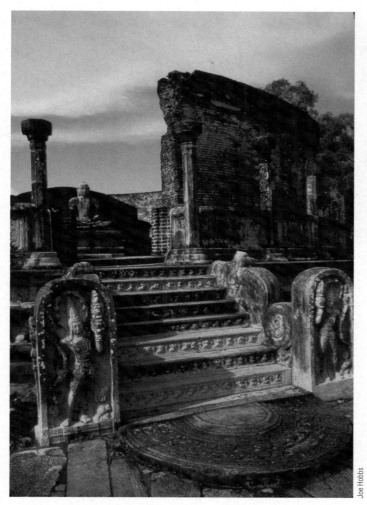

Joe Hobbs

● **Figure 7.1.5** Buddhism is the majority faith in Sri Lanka. This is the Vatadage Temple at Polonnaruwa, where the relic of the tooth of the Buddha was said to have been kept in the twelfth century.

cally, except when Britain ruled. For more than a century, until 1947, India was the most important possession of the British Empire—the **"jewel in the crown."** India then consisted of the entire area of the present India, Pakistan, and Bangladesh. It had a classic colonial relationship with Britain, supplying the motherland with an almost endless pool of low-cost labor and valuable raw commodities like cotton and opium while providing Britain with a huge market for manufactured goods.

British colonialism left indelible marks on the subcontinent's cultural and economic landscapes, some favorable and others decidedly not. To serve the needs of their empire, the British developed the great port cities of Karachi, Bombay (now known as Mumbai), Madras (now Chennai), and Calcutta (now Kolkata) and India's current inland capital of New Delhi. The British created a lasting positive legacy in developing India's infrastructure, particularly the superior rail transportation networks that still assist industrial development in India, Pakistan, and Bangladesh.

The vast rail network created by the British has been expanded and improved by the independent nations. In addition to transporting goods to internal markets and to ports for export, this system carries millions of passengers daily (●Figure 7.1.6).

Around the world, the colonizing power frequently favored one ethnic or socioeconomic group over another, sowing the seeds for eventual discord. This was the case in British India, where the formerly subordinate Hindus came to dominate the civil service and most businesses. Many Muslims feared the results of being incorporated into a state with a Hindu majority, and their demands for political separation led in 1947 to the creation of two independent countries: the secular but predominantly Hindu nation of India and the Muslim nation of Pakistan. The respective secular and religious identities of both countries remain critical issues today. Pakistan had two parts, West Pakistan and East Pakistan, separated by 1,000 miles (c. 1,600 km) of Indian territory. In 1971, an Indian-supported revolt in East Pakistan led to the birth there of the new independent country of Bangladesh.

The partition of colonial India precipitated all kinds of problems for the new countries, some of them still prominent today. The largest problem is the disputed Kashmir region (see Problem Landscape, pages 314–315). Immediately preceding and following partition in 1947, violence broke out between Hindus and Muslims, and hundreds of thousands of lives were lost in wholesale massacres. More than 15 million people migrated between the two countries. Pakistan today has a large population known as **Mohajirs**, or "migrants," the Muslim immigrants and their descendants who poured into the region from India when partition occurred. Particularly in Karachi and elsewhere in the southern province of Sind, violent conflict frequently occurs between rival factions among the Mohajirs and between the Mohajirs and other ethnic groups, including Pathans (Pushtuns) and Biharis.

PROBLEM LANDSCAPE

Kashmir

Since 1947, India and Pakistan have been in conflict over the status of Kashmir (called Jammu and Kashmir in India), a disputed province straddling their northern border (●Figure 7.1.A and 7.1.1, page 310). Before independence, Kashmir was a princely state administered by a Hindu maharaja. About three-fourths of Kashmir's estimated population of 15 million is Muslim, which is the basis of Pakistan's claim to the territory. But under the partition arrangements, the ruler of each princely state was to have the right to join either India or Pakistan, as he chose. Kashmir's Hindu ruler chose India, which is the legal basis of India's claim.

After partition, fighting between India and Pakistan led to the demarcation of a cease-fire line—the "line of control" that still separates the forces—leaving eastern Kashmir, with most of the state's population, in India and the more rugged western Kashmir in Pakistan. Beginning in the mid-1950s, China pressed its own claims to remote northern mountain areas of Kashmir and occupied some of the border territories. Pakistan ceded part of the occupied territory to China and established friendly relations with its giant northern neighbor, but India rejected China's claims (relations between these countries remain chilly). India's subsequent small-scale military actions failed to dislodge Chinese forces. India now holds about 55 percent of the old state of Kashmir, Pakistan 30 percent, and China 15 percent. India has another dispute with China over China's control of 34,750 square miles (90,000 sq km) in India's northeastern state of Arunachal Pradesh.

Three wars between India and Pakistan have effected little change in Kashmir. In 1965, conflict began in Kashmir, spread to the Punjab, and escalated to a brief but indecisive full-scale war involving tanks, airborne forces, and widespread air raids. Renewed hostilities in Kashmir in 1971, as part of the war in which India supported the revolt of Bangladesh against Pakistan, again did not

alter the political landscape of Kashmir. During that conflict, the Siachen Glacier in the Karakoram Mountains, at about 20,000 feet (2,800 m), earned the title of "world's highest battlefield."

Kashmir's Muslim majority escalated the campaign for secession from India in 1989, and the strife has continued ever since. India and Pakistan came perilously close to the brink of war over Kashmir in 2002. Late in 2001, armed insurgents attacked India's parliament in New Delhi in a failed attempt to assassinate Indian politicians. India's government identified the assailants as members of two Muslim rebel movements (Lashkar-e-Taiba and Jaish-e-Muhammad) based in Kashmir. India further accused Pakistan's

● **Figure 7.1.A** Political boundaries of Kashmir. See Figure 7.1.1 on page 310 for its context.

As a consequence of the partition, India received almost all of the subcontinent's modern industries and most of its mineral wealth and energy supplies. Pakistan, including the present Bangladesh, was left with a small industrial infrastructure and fewer natural resources. Pakistan and Bangladesh are therefore still very far behind India in both high-tech and traditional industrial output. The burlap industry is a good example of how partition frustrated development in the region. The province of Bengal, in colonial India's eastern Ganges-Brahmaputra Delta region, is the world's greatest production area for jute, the raw material for burlap. With independence, most of the jute-growing areas were isolated on the Pakistani (now Bangladeshi) side, while the jute mills, which had developed in the Calcutta area, are in India. After independence, India struggled to promote jute cultivation and Bangladesh to promote jute processing. There were also problems in how to divide up

fresh water among the newly independent countries (see The World's Great Rivers, page 316).

7.1.3 Natural Regions and Resources

Physical conditions are very important in the lives of the rural majorities of South Asia. People's well-being depends largely on local agricultural resources, and industrialization is based mainly on other locally available natural resources. Fortunately, South Asia has some generous endowments of natural resources and a huge and talented labor force to work with them. It is also prone to colossal natural hazards, including cyclones, floods, droughts, landslides, and earthquakes. The subcontinent may be roughly divided into three natural areas: the outer mountain wall, the northern plain, and peninsular India (see Figure 7.4, page 286).

PROBLEM LANDSCAPE

Kashmir *continued*

government of supporting these rebel groups and therefore being ultimately responsible for the attack on India's parliament. Indian military forces began amassing on the line of control in Kashmir, and Pakistani troops responded in kind. Tremendous international diplomatic pressure was applied to both sides in an effort to prevent all-out war.

The pressure was particularly intense for Pakistan. Pakistani President Pervez Musharraf responded by outlawing and beginning to crack down on five militant groups, including the two blamed for the December attack. It was a risky move, especially because many people in Pakistan believed their leader was forsak-

ing a legitimate struggle in Kashmir and "selling out" to the United States and other Western powers in the wake of the 9/11 attacks. While Musharraf's actions did at least temporarily defuse the explosive situation in Kashmir, many observers remained concerned that Pakistan would prove unable to rein in Muslim attacks against Indian targets in Kashmir. There were and remain many questions about how much control Pakistani leaders have over their own intelligence service, the ISI (Inter-Service Intelligence Agency), which may contain rogue elements intent on continued support of the militants fighting India.

The Kashmir conflict has been very costly to both sides, with more than 65,000 people killed since 1965, about half of those since 1989. India's international reputation suffered because of the country's unyielding position on the region. Pakistan has steadily spent about a third of its budget on defense, mostly focused on a potential engagement with India over Kashmir. There would be economic benefits for both India and Pakistan if lasting peace were achieved. "Kashmir is a Paradise on Earth," wrote Samsar Kour.* The stunning snow-crowned peaks and flower-laden valleys of this western Himalayan region once lured millions of Indian tourists and ranks of Western trekkers each year, and the tourism development potential is enormous (•Figure 7.1.B). In 2003, Indian and Pakistani leaders finally met to speak about resolving the Kashmir problem and declared a formal cease-fire in the region for the first time in 14 years. Both countries, and the wider world, await more moves toward a permanent peace agreement.

• **Figure 7.1.B** Kashmir, "Paradise on Earth" but for the bloodshed there. This is Mt. Rakapashi, 22,500 ft. (7,788 m) high.

*Samsar Chand Kour, *Beautiful Valleys of Kashmir and Ladakh* (Mysore, India: Wesley Press, 1942), p. 2.

The Outer Mountain Wall

Between 40 and 50 million years ago, a triangular piece of the earth's crust completed its tectonic journey across the Indian Ocean and collided with southern Asia (•Figure 7.1.7). The collision resulted in the thrusting up of a massive and almost solid wall of mountains. That wall, made up mainly of the Himalaya Mountains and the Karakoram Range, forms an inverted U to define the northern boundary of the subcontinent. From each end of this massive wall, lower mountain ranges trend southward to the sea.

Pakistan's borders with Iran and Afghanistan traverse the rugged western section of the mountain wall. Desert and steppe climates reach upward to high elevations. These mountains provide only a few crossing places, including the storied Khyber Pass through the Hindu Kush Mountains on the border between Pakistan and Afghanistan (see Figure

7.1.17). These passes have long been gateways into India for conquerors from the outer world, including the Aryans and the Arabs.

The towering northern segment of the mountain wall has been much less passable, but a trickle of trade and cultural exchange has crossed these mountains for millennia. The Himalayas extend about 1,500 miles (c. 2,400 km) from Kashmir in the west through the countries of Nepal and Bhutan to the northeastern corner of India (•Figure 7.1.8). Paralleling the Himalayas on the northwest and separated from them by the deep gorge of the upper Indus River in Kashmir lies another lofty range called the Karakoram. The two ranges contain 92 of the approximately 100 world peaks rising higher than 24,000 feet (7,315 m), including the two tallest: Mount Everest (29,035 ft, or 8,850 m), in the Himalayas on the border between Nepal and China, and K2, or

THE WORLD'S GREAT RIVERS

Indus, Ganges, Brahmaputra

South Asia's Indus, Ganges, and Brahmaputra Rivers are the waterways on which some of the earth's most significant civilizations developed, and they are still vital to the livelihoods and spirituality of a large portion of humanity. The allocation of these precious and meaningful waters since India's independence from Britain has been problematic.

Since 1947, the region of the subcontinent called the Punjab, meaning "Five Waters" (referring to the five Indus River tributaries), has been split between two countries, with its rivers flowing through Indian territory before entering Pakistan (•Figure 7.1.C; see also Figures 7.1.13 and 7.1.17). In geographic and geopolitical terms, India is an upstream country and Pakistan is a downstream country where Punjab waters are concerned. There are no internationally accepted laws regarding water sharing between upstream and downstream states, and the upstream country often exercises its geographic advantage to siphon off what the downstream country views as too much water. **Riparian** (river-owning or river-bordering) countries often come close to, or actually go to, war over the problem. In many cases, formal treaties have averted what could have become major international conflicts.

After 1947, relations between India and newly independent Pakistan deteriorated in part because of the disputed waters of the Indus and the five Punjab rivers. For many years, the parties could not agree on how much of the water India could divert and how much should remain for Pakistan. The countries finally reached a satisfactory resolution in the 1960 **Indus Waters Treaty.** The agreement allocates the water of the three eastern rivers of the Punjab to India, which in return allows unrestricted and undiminished flow of the two others, and the Indus River itself, into Pakistan. The two countries have abided by the agreement ever since, despite their periodic wars and constant state of tension. Today, the biggest problem over Indus waters is within Pakistan,

with downstream Sind province complaining that upstream Punjab province is committing "water robbery."

Another landmark water-sharing agreement was signed between India and Bangladesh in 1996. For two decades, these nations had a bitter disagreement about the allocation of water from the Ganges River. In the 1996 accord, which is meant to be in effect for 30 years, upstream India guarantees a larger share of water to downstream Bangladesh. This achievement may open the way to the integrated development of what is known as the Ganges-Brahmaputra-Barak (GBB) river basin, a watershed that is home to more than 500 million people in Bangladesh, Bhutan, China, India, and Nepal. Cooperation to settle the vital issue of water allocation among these countries could promote peace and reduce poverty in the region.

Joe Hobbs

• **Figure 7.1.C** The confluence of the Gilgit (left) and Indus (right) Rivers, in Pakistan's section of the disputed Kashmir region.

233, 353

34

Mount Godwin Austen (28,268 ft, or 8,616 m), in the Karakoram within the Pakistani-controlled part of Kashmir.

Geographic surveys recently quieted a brewing debate over whether K2 might actually be higher than Everest: The Himalayan peak is the highest. Both great ranges, however, are young, dynamic, and still rising, so K2 could conceivably outgrow Everest. While the peaks are climbing, their glaciers are retreating. There is more ice here than anywhere else on earth outside Greenland and the polar regions, but it is melting quickly. Some fingers point to human-made greenhouse gases as the culprit, and by one estimate, at current rates of melting, these glaciers would be gone by 2035.

The mountains along and near the subcontinent's border with Myanmar (Burma) are much lower than the Himalayan ranges, but cloaked with dense vegetation and soaked with rain, they are also nearly impenetrable. Cherrapunji

in the Khasi Hills of India has an average annual rainfall of 432 inches (11 m), more than nine-tenths of it falling in the summer. It ranks with a station on Kauai in the Hawaiian Islands as one of the two spots with the greatest annual rainfall ever recorded. Cherrapunji reportedly has the highest alcoholism rate in South Asia because so many people stay indoors and drink on rainy days!

The Plains of India, Pakistan, and Bangladesh

Just inside the outer mountain wall lies the subcontinent's northern plain. This large alluvial expanse contains the core areas of the three major countries. In the west, the plain is split between Pakistan and India. The Indus River and its tributaries and **distributaries** cross the portion lying in Pakistan and water this country's two most populous provinces, Punjab and Sind, and its major cities of Kara-

322, 324

• **Figure 7.1.7** Most of India was once an island. Tectonic movement transported it across the Indian Ocean. When it collided with the Eurasian landmass, the world's highest mountains were formed.

• **Figure 7.1.8** The roof of the world, near Mount Everest in eastern Nepal. Generations of Sherpa farmers and herders have cleared all the forest that once cloaked the hillsides in the foreground.

chi and Lahore. The climate of the western plains is desert and steppe; the dry region straddling the two countries is known as the Thar Desert, or the Great Indian Desert. Irrigation water from the rivers sustains highly productive agriculture and millions of people here.

East of the Punjab lies the part of the northern plain traversed by the Ganges (Ganga) River and its tributaries, especially the Jumna (Yamuna). The western portion of this region is known as the Doab, "the land between the two

rivers." The northern plain is India's core region, containing more than two-fifths of the country's population as well as its capital at New Delhi, many other major cities, and sacred places in the religions of Hinduism, Buddhism, and Jainism. This region of India is relatively well watered, especially toward the east, where agriculture based on irrigated rice supports high population densities. The southeastern part of the northern plain lies in the region of Bengal, which is essentially the huge delta formed by the combined flow of the Ganges and Brahmaputra Rivers. Bengal is divided between India and Bangladesh and has an extremely dense population, with the major cities of Kolkata on the Indian side and Dhaka on the Bangladeshi side.

Peninsular India

The southern peninsular portion of the subcontinent, which is entirely within India, consists mainly of a large volcanic plateau called the Deccan. It is generally low, less than 2,000 feet (c. 600 m) above sea level. Its rivers run in valleys cut well below the plateau surface. This topographic problem, together with the rivers' seasonal flow pattern, makes it difficult to use their waters for irrigation of the upland surface without large inputs of capital and technology. Wells and rain catchment ponds called tanks are used, but water is in short supply and population density is low compared with the better-watered parts of the northern plain (•Figure 7.1.9).

Hills and mountains outline the edges of the Deccan. In the north, belts of hills (the Central Indian Hills) separate the plateau from the northern plain. On both east and west, the plateau is edged by ranges of low mountains called *ghats* ("steps") overlooking alluvial plains along the coasts of the

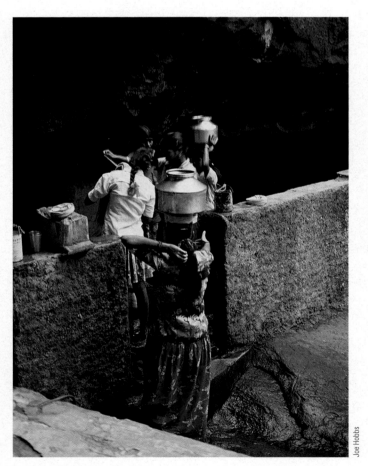

Joe Hobbs

• **Figure 7.1.9** It is usually the task of women to draw water from communal supplies such as this one in a small settlement near Mumbai.

Bay of Bengal and the Arabian Sea. The low and discontinuous eastern mountains are known as the Eastern Ghats. The higher and almost continuous western mountains are the Western Ghats. The two ranges merge near the southern tip of the peninsula in clusters known locally as the Nilgiri, Palni, and Cardamom Hills. They are really rather high mountains, with some peaks rising above 8,000 feet (c. 2,400 m). The colonizing British founded several "hill stations" in them for recreation and administration during summer months, when the climate on the subcontinent's lowlands is stifling. They are now popular tourist destinations for Indians during that season. The coastal plains between the ghats and the sea support very dense populations, especially in the far south along the Malabar Coast of western India and on the corresponding Coromandel Coast in the east. India's largest city, Mumbai, is on the peninsula's northwest coast, and the large city of Chennai is on its southeast coast. Inland are the major cities of Hyderabad and Bengaluru.

Climate and Agricultural Conditions

Climatic conditions in South Asia range from Himalayan ice to the tropical heat of peninsular India and from some of the world's driest climates to some of the wettest. Climatic and biotic types in the subcontinent include undifferentiated highland climates in the northern mountains; desert and steppe in Pakistan and adjacent India; humid subtropical climate in the northern plains; tropical savanna in the peninsula; and rain forest along the seaward slopes of the Western Ghats and in parts of the Ganges-Brahmaputra Delta and the eastern mountains near Myanmar (see Figure 7.5, page 287). Most of the region's original vegetation cover is gone, but there are numerous large national parks and protected areas, especially in India, designed to ensure that important remnants will remain. The region's wildlife still includes tigers, elephants, rhinoceroses, and in western India, the sole surviving population of the once widespread Asiatic lion (see Figure 2.12g, p. 29).

Heat is nearly constant except in the mountains. The tropical peninsula is hot all year. The subtropical north has stifling heat before the arrival of the wet monsoon season, and there are warm conditions even in the winter. Some of the warmest temperatures on the globe occur in the plains of Pakistan and northwestern India. Delhi averages 94°F (34°C) in both May and June, the months when the highest temperatures occur over most of the subcontinent. High humidities accompanying the impending monsoon rains combine with the heat to create almost intolerable conditions. The monsoon rains bring cooling relief to the subcontinent—but sometimes also catastrophic flooding.

India's monsoons include both the **wet southwest monsoon,** which brings the short, wet summer between June and September, and the **dry northeast monsoon,** which creates mostly arid conditions the rest of the year; these are the typical monsoon patterns of Asia described on pages 283–285. There are two main arms of the wet, or summer,

• **Figure 7.1.10** Aridity and deforestation are among the urgent environmental problems faced by Pakistan. Foraging by goats and cutting trees for fuel in an already dry environment have created a lunar landscape in the country's Karakoram Mountain region.

monsoon. One, approaching from the west off the Arabian Sea, strikes the Western Ghats and produces heavy rainfall on these mountains and the coastal plain. The amount of rain diminishes in the interior Deccan rain shadow region to the east of the mountains, where barely enough rain falls to support dry farming, and drought sometimes brings crop failures. The second arm of the summer monsoon approaches from the Bay of Bengal, bringing moderate rainfall to the eastern coastal areas of the peninsula and heavy rains to the Ganges-Brahmaputra Delta region and northeastern India. This is the monsoon that feeds cyclones (hurricanes). Wet monsoon winds pass up the Ganges Valley to drop moisture that diminishes in quantity from east to west.

Both arms of the monsoon bring some rain to Pakistan, but it is so little that semiarid or desert conditions prevail in most areas (•Figure 7.1.10). Blowing mainly from land

to sea, the dry monsoon brings dry and cooler weather to most parts of the subcontinent. An exception is in the far south of the peninsula, where the heaviest rainfall of the year falls along the eastern coast and in adjacent uplands from October through January.

With precipitation concentrated in a short part of the year, people must extend the crop season by irrigating. They go to great lengths to get additional water onto the land, and groundwater is in increasingly short supply. Spurred on by cheap electricity and the absence of any legal constraints, farmers pump out as much water as they can afford the technology to reach. India is considering an overhaul of its agricultural policies, especially by revising water laws and switching to more drought-tolerant crops. Drought conditions in a country often visited by flooding monsoon rains may seem ironic, but the problem is that India's water storage capacity remains low. The conventional solution—building more and larger dams—is problematic because of environmental and social concerns.

Large dams and associated canal systems have been built in the steppe and desert areas of the Punjab and the lower Indus Valley (Sind province) in Pakistan, and more are planned. A massive irrigation and hydroelectric power project is under way in the north central part of peninsular India, where the Sardar Sarovar Dam was recently constructed on the Narmada River (•Figure 7.1.11). Largely because the dam project would displace 320,000 tribal and rural people (known as "oustees" in the project language) and inundate about 44 square miles (114 sq km) of cropland and 50 square miles (130 sq km) of forest, environmentalists and human rights advocates both in India and abroad fiercely opposed the dam. The World Bank withdrew its financial support in response to their protests, but asserting that its benefits will greatly outweigh its costs, India pressed ahead with the dam's construction. It is the centerpiece of the massive Narmada Valley Development Project,

• **Figure 7.1.11** The controversial Sardar Sarovar Dam, on India's Narmada River.

where 30 dams will displace about 1 million people by the projected completion date of 2025.

Correlations between people and food supplies, both spatially and in raw numbers, are always critical in South Asia (see Regional Perspective, page 321). The spatial distributions of food crops and population densities correlate 282 with the amount of water available in the subcontinent. Rice is the basis of life in the wetter areas and, with its high caloric yields per acre, is associated with the highest population densities. Rice dominates in the delta area of Bengal (in both India and Bangladesh), in the adjacent lower Ganges Valley, and in the coastal plains of the peninsula. India's Punjab (which produces just over half of India's food crops) and Pakistan's Sind and Punjab provinces produce surplus rice, including the basmati variety, which is an important export. Irrigated wheat is the staple crop and food in the drier upper Ganges Valley of India and in the dry Punjab of India and Pakistan. In these areas, crop yields and population densities are intermediate. Unirrigated plantings of sorghum and millet prevail over most of the Deccan plateau and in other areas where it is difficult to supplement the low rainfall through irrigation. Yields from these crops are low, and so are population densities.

Maize (corn) is an important grain in the lower regions of the Himalayas. But rice is the preferred food almost everywhere, and where enough water is available, there are patches of irrigated rice. Rice, cotton, jute, and tea are the main crops grown in the subcontinent for industrial use and for export. Of the four, only tea is exported on a sizable scale, and only tea is a plantation crop. The plantations were developed, and are still largely owned, by British interests. Production is greatest in the northeastern state of Assam, with another center in the mountainous far south of peninsular India (●Figure 7.1.12). Cotton is grown mainly in the interior Deccan. India's large textile industry absorbs most of the production. Coffee, rubber, and coconuts are minor plantation crops, and India exports products made from coir (coconut fiber).

● **Figure 7.1.12** A tea plantation in the Western Ghats of eastern Kerala state. These mountains have been widely cleared of native forest by logging and agricultural expansion.

The remainder of this module deals with economic and other important geographic characteristics and problems of each of the South Asian countries.

7.1.4 India: Power, Courage, and Confidence

Despite its overwhelmingly rural population (72 percent, living in about 600,000 villages), India boasts some of the world's **megacities** (extremely large cities), notably Mumbai and Kolkata (●Figure 7.1.13). These and many smaller cities are at the heart of India's current economic boom, which is based more on information technology than on old-fashioned heavy industry.

It is difficult to make generalizations about India in economic or most other matters because there is such a vast continuum of Indian experience. Visitors to India may be astounded to see both the latest computer technologies and the labors of human hands performing, on a large scale, the tasks routinely done by machines in more developed countries—for example, people making gravel by hammering boulders into pebbles. And always looming over India are questions of poverty, caste, and religious strife. Poverty is arguably the most prominent issue: nearly one in three Indians is among the official ranks of the poor, and the country is ranked 126th of the 177 countries on the Human Development Index. 44

Urban and Economic Geography

India's recent high rate of economic growth (an average of 8 percent per year between 2001 and 2007) owes largely to important reforms and gives new meaning to its ancient "three lions" symbol, standing for power, courage, and confidence.

Until the 1990s, India was intent on keeping out most foreign manufacturing in favor of industrial self-sufficiency. Most of the economy was socialized, with many state-owned industries. With these policies, India's industries expanded and diversified, but profitable exports were limited. After that, India opened up to more imports, privatized many state-run industries, and boosted exports (but in notable contrast with China, India still imports more goods than it exports). India's vast pool of well-educated people responded in creative ways to new opportunities. The result was remarkably strong economic growth—so strong, in fact, that there is nervousness about a "bubble" that could burst; some observers fear that skyrocketing real estate prices, stock indexes, inflation, and borrowing all point in that direction. 295

India's diverse and rapidly expanding manufactures include old-economy products like cars, trucks, motorcycles, bicycles, and farm machinery, along with high-tech data processing equipment, silicon chips for electronics, and computer software. India's leading factory industries are cotton textiles, jute products, food processing, and iron and steel. Unlike China, India has abundant reserves of iron ore

REGIONAL PERSPECTIVE

Keeping Malthus at Bay

Poverty and human health are serious problems in South Asia, and the prospect of continued high population growth raises the question of whether growth in food supplies can avert the proverbial Malthusian crisis (see Chapter 3, page 60).

India's population has surged since independence, from 352 million in 1947 to 1.1 billion in 2007. In recent decades, the overall trend has between toward lower population growth rates. Fifty years ago, the average Indian woman had six children; now she has three. That rate is still well above the natural replacement rate, however, that would keep the population steady. The population base is already so vast that even modest growth will add huge numbers. India is predicted to overtake China as the world's most populous country by 2040.

For many people, the name "India" invokes an image of grinding poverty, and with good reason: it is estimated that fully one-third of the population is "abjectly poor," defined as living on less than $1 per day, and almost half of India's children are malnourished. But perhaps in defiance of Malthus, an increased population has so far succeeded in producing more economic resources overall. A few people in India are very wealthy, and there is an emerging middle class of as many as 300 million people—the world's largest middle class (●Figure 7.1.D). India's economy has grown impressively in recent years. An apparent problem is that the benefits of the growing economy are unevenly distributed. Most of the economic growth benefits those who are already better off, widening the gap between rich and poor. About 800 million Indians are either unemployed or underemployed. In its bid for reelec-

tion in 2004, India's ruling party, the BJP, promised to make the country an economic superpower by 2010. In a stunning setback viewed as a referendum by the poor, who felt that India's growth would continue to pass them by, the BJP lost the election to the opposition Congress Party. Population growth in the long run may further distort India's economic imbalances.

The issue of population growth is also critical for Pakistan, which has never mounted an effective family planning program, mainly because of opposition from Muslim religious authorities, who regard birth control as an intervention against God's will (Pakistan's birth rate in 2007 was 31 per 1,000, compared with 24 per 1,000 in India). Muslim Bangladesh, along with Nepal and Bhutan, also have high birth rates, but they are much lower in the most critical nation of Bangladesh than they once were. In the 1970s, the average woman had seven children, while now she has three, and contraception use nationwide has increased from 4 percent then to more than 50 percent now. South Asia's success story in bringing population growth under control is Sri Lanka, with a birth rate of just 18 per 1,000 and annual population growth of 1.2 percent.

Agricultural output in South Asia has increased since independence. Most notably, despite its huge and growing population, India has achieved self-sufficiency in grain production, with a fourfold increase in output since 1950. Almost half a million "ration shops" sell subsidized food staples to the country's poorest people. The successes of agriculture in the subcontinent have been due mainly to the increased use of artificial fertilizers, the introduction of new high-yield varieties of wheat and rice associated with the Green Revolution, more labor provided by the growing rural population, the spread of education, the development of government extension institutions to aid farmers, and better irrigation.

One of the challenges confronting India is to improve the status of women. Despite a 1961 ban on dowries (money and gifts given by a bride's parents to the groom), the practice is continuing. So is the rate of killing brides who do not provide enough dowry. The burden placed on the bride's family is one reason that many prospective parents choose to abort female fetuses, which can be detected with ultrasound technology. India banned this use of ultrasound in 1996, but the practice continues. Female feticide—particularly prevalent among India's more educated and affluent, who can afford the ultrasound screening—and infanticide have taken tens of millions of young and unborn girls. In some of India's states, especially Haryana, Punjab, Delhi and Gujarat, one result is a serious shortage of local brides; for every 1,000 men, there are 900 women. The gap is being filled by marriages arranged with women from distant Indian states and Bangladesh.

Joe Hobbs

● **Figure 7.1.D** This Indian family visiting the Red Fort in Agra is among the growing ranks of the subcontinent's middle class.

to draw on for its steel production, which has surged in recent years. India is both a major user and major exporter of cotton textiles and clothing. Favoring cotton manufacturing are a huge home market, inexpensive labor, and prolific domestic production of cotton.

India makes a wide range of chemical products, from vitamins to fertilizers. The pharmaceutical industry is flourishing. Much to the chagrin of Western drug companies, their Indian counterparts routinely replicated foreign drug formulas and produced copycat medications for sale at a

● Figure 7.1.13 Principal features of India.

fraction of their cost in the West. India finally stopped arguing that it was a poor country needing low-cost medicines and in 2006 bowed to the pressure to join global patent standards. India's success in providing lifesaving medications to its population cannot be disputed; life expectancy has risen by 20 years since the country became independent. One of the big questions is if such improvements in the quality of life can be sustained for India's growing population.

While manufacturing is growing quickly, the economic sector with the greatest current growth is information technology, which employed 1.3 million of India's 400 million workers in 2007. India's software exports, two-thirds of which go to the United States, have been growing in recent years at rates of up to 50 percent annually, and technology and service jobs have been increasingly outsourced to India from the United States.

Information technology is helping bridge the digital divide separating the rich and poor in India. Internet cafés are turning up even in remote villages, where their operators use diesel generators to keep computer terminals from failing during frequent power outages. The Internet is debuting in many rural areas with Wi-Fi access, thanks to high-range wireless transmitters. Increasing Internet usage (now only 16 per 1,000 in India, versus 57 per 1,000 in all of Asia and 96 per 1,000 worldwide) may bring significant social and economic improvements to the vast numbers of India's rural poor. For example, villagers might be able to use this technology to gain unbridled access to land records, rather than having to bribe officials to get them. Those seeking work abroad could download passport applications instantly rather than spend time and money working through intermediaries. Farmers could get information on crop prices in different towns to improve their bargaining positions during harvests. These kinds of advances are already happening in parts of rural India, where a private company called ITC has set up Internet access so that farmers can check the price of soybean futures on the Chicago Board of Trade. Armed with this knowledge, the farmer can decide whether to sell his soybeans immediately or wait a while longer until prices rise. In this arrangement, known as the **e-Choupal initiative** (a *choupal* is the leading farmer in a district), ITC is encouraging farmers to sell their better soybeans directly to the company rather than to the village traders or government markets from which ITC previously had to buy the crops. Mobile phones are also aiding entrepreneurship in all sectors of India's population, including farmers. At less than 1 U.S. cent per minute, India's cell phone rates are the lowest in the world, and half of the country's population is expected to own cell phones by 2010.

Some of India's intellectual productivity is lost to other countries, perhaps slowing down potential economic growth. India suffers from a brain drain, the emigration of highly trained and highly educated individuals abroad. Chances are that the American reader of this text has been or will be treated in a hospital or clinic by a physician who emigrated from India, attracted by higher U.S. wages. The United States' "brain gain," in the form of highly qualified medical personnel, is inevitably India's loss because better development requires better human resources. Indian computer scientists, engineers, nurses, and other professionals have in recent years followed the trend in "exports" of doctors to the United States and other countries. In part reflecting the fact that Indian universities are not yet up to world-class standards, American universities currently train more students from India (about 80,000) than from any other foreign country. Most return to India, and others are eventually wooed into returning. Altogether, 20 to 30 million Indian citizens live abroad, with about 2 million in the United States and about 3 million in the oil-exporting countries of the Middle East. Another 2 million Bangladeshis and 1 million Pakistanis work in the Middle East, many in semiskilled or unskilled jobs. The **remittances** they send home are vital to South Asia's economies.

India faces several challenges to growth. One is upgrading energy sources and distribution. India has some hydropower, coal deposits, and a modest oil output from fields in Assam, Gujarat, and offshore near Mumbai. Nuclear power produces about 3 percent of its electricity. Like the other South Asian countries, however, India is heavily dependent on imported energy. Infrastructure needs improvement. Long neglected, the country's decrepit road system is getting a major makeover, comparable to the interstate highway system built in the United States in the mid-20th century. At the heart of this effort is the **Golden Quadrilateral,** a four- and six-lane highway system running 3,625 miles (5,833 km) through 13 Indian states and connecting 19 of the 35 Indian cities with more than a million inhabitants (see Figure 7.1.13). The highway's purpose is to speed up the pace of development and integration in this vast and complex country. It will certainly facilitate social and economic exchange, and health experts warn that the road will also be a major transmission route for HIV/AIDS. An estimated 11 percent of India's truckers are HIV-positive. India's labyrinthine bureaucracy, by far the world's largest and a legacy of the country's past socialistic policies, also challenges the country's development. In an effort to lure foreign investors who might otherwise be scared off by India's red tape, the country has set up numerous Special Economic Zones (SEZs) patterned after those of China. Located near the larger cities, these industrial sites promise tax breaks, speedy paperwork, and reliable electricity supplies to companies that agree to locate here.

India's industries are concentrated in its cities. On the peninsula's northwest coast is the largest, Mumbai (population 20.4 million). The country's cotton mills are concentrated mainly in the metropolitan area of Mumbai and neighboring areas. Mumbai is known as **India's Wall Street** and is also its Hollywood (it calls itself **Bollywood**), the center of a prolific and accomplished film industry (the world's largest) that produces movies of international stature as well as the regular fare that a huge domestic market consumes passionately (•Figure 7.1.14). In the south central peninsula is Bengaluru, known as **India's Silicon Valley,** where bustling operations of Microsoft, Motorola, Intel, Texas Instruments, and a host of other high-tech firms present a startling contrast to the largely rural and unchanging image that many outsiders have of India. The largest metropolis of the far south, Chennai, is a seaport on the Coromandel Coast.

At the western end of the Ganges-Jumna section of the subcontinent's northern plain, the old fortified capital of Delhi arose as a bastion against invaders from beyond the western mountains (•Figure 7.1.15). The present adjoining capital of New Delhi was founded after 1912 as the capital of British-controlled India. Today, the population of Delhi and New Delhi is an estimated 18 million, making it India's second-largest metropolis. Its main industries are

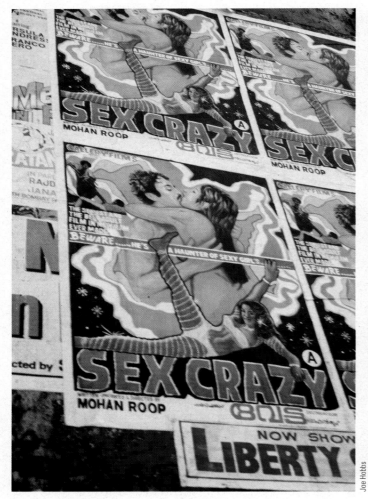

• **Figure 7.1.14** Mumbai's Bollywood produces the romance and action films that are the staples of Indian cinema. The advertising is misleading; Indian films tend to show little flesh.

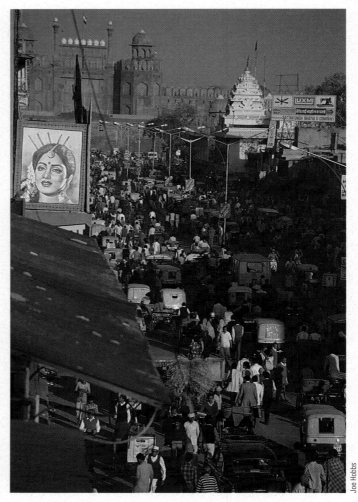

• **Figure 7.1.15** The Mogul Emperor Shah Jahan established Delhi as India's capital early in the 17th century. His monumental Red Fort (top left) still dominates this old city. Adjacent New Delhi is India's capital today.

manufacturing, textiles, electronics, and chemicals. Far to the southeast is the English-founded port of Kolkata in the state of West Bengal. A city of 14.9 million, Kolkata has a worldwide reputation for teeming slums and dire poverty—Rudyard Kipling called it "The City of Dreadful Night"—and it was this Indian city to which the extraordinary nun Mother Teresa dedicated her life's work of aiding the poor. Within India, Kolkata is recognized as a great center for culture and the arts. The surrounding Bengal region is an important industrial area where the country's jute, iron, and steel industries are concentrated. The most valuable mineral resources in India, including iron ore, coal, manganese, chromium, and tungsten, are found in this region. These resources have led to the rise of small to medium-sized industrial and mining cities (notably Jamshedpur) in an area that otherwise is one of the least developed parts of the country.

Social and Political Challenges

Partition following independence did not solve India's problems. Serious issues have persisted among the country's ethnic and religious groups. Some of the most problematic involve language, nationalism, and caste (see Insights, page 325). Whatever its travails, India is admired for its tireless and record-setting democracy. Every time a national election is held there, it is the largest in human history.

India has 17 major languages and more than 22,000 dialects. Many people speak two or even three languages, but overall literacy rates are low (about 61 percent for the country as a whole), and communication is often a problem. Since British colonial times, English has been the lingua franca and the de facto official language of the subcontinent. However, only a small percentage of the population

INSIGHTS

The Caste System

One ancient Hindu belief is that every individual is born into a particular **caste**, a social subgroup that determines the individual's rank and role in society. Castes form a hierarchy, with the Brahmin caste at the top (comprising about 5 percent of all Hindus) and three others (Kshatriya, Vaisya, and Sudra) below. Caste membership is inherited and cannot be changed. Particular castes are associated with certain religious obligations, and their members are expected to follow traditional caste occupations. Marriage outside the caste is generally forbidden, and meals are usually taken only with fellow caste members.

At the bottom of the social ladder are the **Dalits,** or *scheduled castes,* once known as **untouchables,** accounting for about 20 percent of all Hindus. They are not part of the caste system but are "outcastes," and according to Hindu belief, they are not reincarnated. The untouchables were so called because they traditionally performed the worst jobs, such as the handling of corpses and garbage; therefore, their touch would defile caste Hindus.

This ancient, rigid social system is changing. India's constitution outlawed the caste system in 1950, but a single decree could not undo thousands of years of tradition. Brahmin privileges are now being increasingly challenged and restricted by more laws. Indian law explicitly forbids recognition of Dalits as a separate social group. Confronting the fact that upper castes account for less than one-fifth of India's population but command more than half of the best government jobs, the Indian government has instituted an affirmative action program to allocate more jobs to members of lower castes. Indian leaders of high caste have championed the Dalits' cause for both moral and practical reasons. Politicians use caste divisions to their advantage to bring out voters, promising if elected to act in the interests of their respective large social blocs. In 1997, a Dalit was elected president of India for the first time in the country's history. Since then, Dalits have won more regional and local offices.

The Dalits are voting in greater numbers than ever before with the conviction they can use India's democratic system to overcome ancient discrimination against them. They are making progress. Disintegration of the caste system has been especially rapid in the cities, where people of different castes must mingle in factories, in public eating places, and on public transportation. In India's vast rural areas, the caste system is more entrenched.

---

is literate in English. This fact, along with nationalist sentiments, has prompted calls for indigenous languages to be adopted as official languages. After independence and partition, disputes arose in each country over which language should be chosen. Pakistan's leadership insisted that Urdu be the sole official language of the country, even in East Pakistan, where Bengali prevailed and Urdu was virtually unknown. The linguistic divide would become one element 291 in the eventual strife between the two Pakistans. India decided on Hindi, an Indo-European language spoken by the largest linguistic group, and made it the country's national language in 1965. This government decision resulted in strong protests, especially in the Dravidian south. The result has been a proliferation of official languages to satisfy huge populations; India now has 15, all of which appear on the country's paper currency. Rioting between speakers of different languages has occurred repeatedly, and demands by language groups have reshaped the boundaries of several Indian states.

Hindu nationalism is a serious issue in a country with such large religious minorities. The country is supposed to be secular, but in recent years, Hindu nationalists have heavily influenced its politics. The explicitly sectarian Bharatiya Janata Party (BJP) held a majority of parliamentary seats, and the position of prime minister, between 1998 and 2004. This Hindu nationalist party would like India's Hindu majority to have the strongest say in the country's affairs. Until recently, the BJP platform included the adoption of a code of civil law that would strip Muslims of separate laws in matters of marriage, divorce, and property rights.

Sectarian violence (known in the region as **communal violence**) between Hindus and Muslims is a constant threat to India's social and political fabric. Contention over places sacred to both faiths sometimes ignites widespread violence. The 16th century Babri Mosque in the city of Ayodhya in Uttar Pradesh was a place of prayer for Muslims on a spot revered by the Hindus as the birthplace of the god-king Ram, an incarnation of the god Vishnu. Backed by Hindu nationalists in the provincial government, a mob of about 250,000 Hindu fundamentalists demolished the mosque in December 1992 with the intention of building a Hindu temple on the site. The ensuing communal violence between Hindus and Muslims throughout India, even among the cosmopolitan and usually tolerant urbanites of Mumbai, left thousands dead.

Ever since this incident, as a means of garnering votes from Hindu Indians, Hindu politicians have made veiled promises to proceed with construction of the temple. In 1998, the BJP platform included a pledge to build a Hindu temple atop the ruins of the mosque at Ayodhya. His apparent support for the temple helped sweep the BJP leader, Atal Vajpayee, into power as prime minister that year. The conflict over this multireligious sacred site in India has parallels with the Temple Mount dispute between Israel and the Palestinians. Politicians in India today are increasingly 245 aware of the threat that Hindu-Muslim discord may pose to the country's development. Acknowledging the widespread

• **Figure 7.1.16** The Golden Temple (established in 1577 and rebuilt in 1764), in Amritsar, is the holiest shrine of the Sikh religion.

• **Figure 7.1.17** Principal features of Pakistan.

discrimination against Muslims in Indian society, they are considering affirmative action programs to improve the stature of the country's Muslims.

In India's Punjab, where Sikhs make up 60 percent of the population and Hindus 36 percent, the 1980s and early 1990s were violent years during which about 20,000 people died in armed clashes. Sikh factions were intent on establishing their own homeland of **Khalistan**, prompted to do so in part by New Delhi's plans to divert water from the Punjab. They challenged Indian authority and turned Amritsar's Golden Temple, the holiest site in the Sikh faith, into their military stronghold (•Figure 7.1.16). In a controversial move to quell the revolt, Prime Minister Indira Gandhi ordered troops to storm the Golden Temple in June 1984. The resulting deaths and desecration led directly to Gandhi's assassination by her own Sikh bodyguards later that year. Indian authorities accused Pakistan of arming the Sikh militants.

Other trouble spots for India are the northeastern states of Assam, Manipur, Nagaland, and Arunachal Pradesh, an extraordinarily diverse region with more than 200 ethnic groups and followers of many different religions. Several rebel groups have been active in this region for years, agitating for autonomy and independence from India. The government has begun investing more resources in the area, especially infrastructure to boost trade ties with Myanmar. There are also troubles in half of India's states from Maoist insurgents, known as **Naxalites**, who claim to champion the causes of the rural poor and promise to carry violent rebellion into India's cities and ultimately expand their "compact revolutionary zone" to include the entire country.

7.1.5 Pakistan: Faith, Unity, and Discipline

Pakistan (•Figure 7.1.17) is not the demographic giant that India is, but its 169 million people face many challenges, especially owing to a small resource base relative to the pop-

ulation, a general shortage of industrial output, and local and geopolitical tensions that are destabilizing forces. Unlike India, Pakistan is not a secular country but an Islamic one whose motto is "faith, unity, and discipline."

Urban and Economic Geography

As mentioned earlier, the 1947 partition left Pakistan with far fewer industries and less mineral wealth than India received. The country has yet to make up for that setback. Many of the country's bills are paid by foreign aid donors, with a large share of those contributions tied directly to Pakistan's perceived importance in the U.S.-led war on terrorism.

Pakistan does have some assets to build on. Its cotton-textile industries are well established, and irrigated plantations of cotton in the Punjab and lower Indus Valley make Pakistan the world's fourth-largest cotton producer (after China, the United States, and India). Textiles, yarn, ready-made garments, and raw cotton constitute about two-thirds of Pakistan's exports. Rice is another important export. Substantial deposits of natural gas may further enhance the nation's development.

Opium was once a major illegal export crop of Pakistan, but stepped-up law enforcement has dramatically cut production. However, processing of the raw opium into heroin is still a significant industry in the rugged mountain region of the Northwest Frontier Province and nearby Federally Administered Tribal Areas, situated along the frontier with Afghanistan and inhabited mainly by Pashtun peoples. Domestic heroin addiction is a growing problem, with an estimated 3 million Pakistanis dependent on the drug.

Pakistan's major cities, which are also the main industrial centers, are on plains near the Indus River: the seaport of Karachi (population 12.3 million), located at the western edge of the Indus River Delta in Sind province, and the cultural and textile manufacturing center of Lahore (population 6.6 million; ●Figure 7.1.18) in the Punjab. Farther north, at the foot of the Himalayas, is the capital city of Islamabad ("City of Islam"), a much smaller, very modern administrative metropolis.

Social and Political Challenges

Pakistan has internal problems centering on religious divisions, mainly between the majority Sunni Muslims (77 percent of the country's population) and Shiite Muslims (20 percent). Shiite Muslims opposed the government's efforts in the 1980s to establish an avowedly Sunni Islamic state, and there was violent conflict between members of the two sects. Pakistani factions who preferred a Western-style secular democracy also rioted in protest against this **Islamization** of the government. Democratic elections in the 1980s and 1990s twice installed Benazir Bhutto, a liberal, Western-educated woman, as the country's prime minister. Her regimes were so riddled with corruption that she was twice voted out of office to be replaced by Nawaz Sharif, who promised to cleanse the country of corruption and embrace the traditional Islamic system of law known as **sharia**. Sharif was in turn ousted in a military coup in 1999.

Sharif's successor, General Pervez Musharraf, took the helm of a country torn between the secular wishes of its Western-oriented middle class and popular calls for greater Islamic influences and fewer ties with the West. The growing Islamist sentiment is focused both on ousting India from disputed Kashmir and on resisting the U.S. and other Western intervention in the region in the wake of the 9/11 terrorist attacks. Much of the dissent is based in the Federally Administered Tribal Areas (FATAs) and in the Northwest Frontier Province, especially in and around its principal city, Peshawar (●Figure 7.1.19). At the time of the attacks on the United States, there were already 3 million mainly ethnic Pashtun refugees from Afghanistan living in camps outside the city. Hundreds of thousands more poured in from Afghanistan in the weeks following 9/11, but a reverse flow has taken place since then. Within these impoverished, desolate encampments, much anti-U.S. and generally anti-Western sentiment still simmers.

An important breeding ground for anti-Western (and anti-Indian) thought and action in Pakistan is the institution known as the **madrasa**, an Islamic school or seminary (●Figure 7.1.20). There are as many as 8,000 *madrasas* in Pakistan, where approximately 700,000 children (mainly boys) are educated. Their curriculum is based largely on study of the Qur'an (Koran) and other Islamic texts, and some instructors promote an interpretation of the Islamic concept of *jihad* ("struggle"), in which Muslims are urged to fight India, the United States, and other entities branded as "the infidel." Many of the *jihadis* ("holy warriors") who join groups like al-Qa'ida, Lashkar-e-Taiba, and Jaish-e-Muhammad have been educated in these schools. But *madrasas* also offer much legitimate instruction and help compensate for Pakistan's poor public educational system. Pakistan's leaders are in a difficult position, at once called on by the outside world to cleanse the *madrasas* of militant content and by their own people to uphold the long-cherished

(margin numbers: 272, 306, 235)

● **Figure 7.1.19** Pakistan's Northwest Frontier Province city of Peshawar has long been a haven for spies, smugglers, and insurgents.

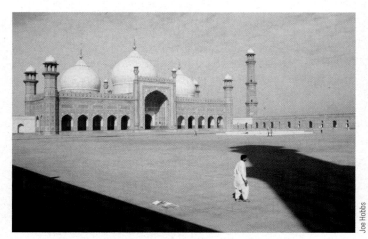

● **Figure 7.1.18** Lahore in the Punjab region of Pakistan, is the country's cultural center. It boasts many fine Islamic monuments, including this seventeenth-century Badshahi Mosque, built by the Mogul emperor Aurangzeb.

• **Figure 7.1.20** Islamic education in Pakistan's *madrasas* begins at an early age. These boys are memorizing the Qur'an.

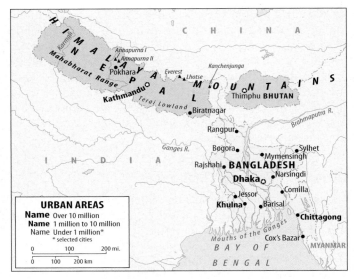

• **Figure 7.1.21** Principal features of Bangladesh, Nepal, and Bhutan.

institution of the religious school. Public opinion against President Musharraf surged after July 2007, when he ordered a bloody armed invasion of a *madrasa* held by militants in the nation's capital. His position became even more untenable following the assasination of Benazir Bhutto in December 2007, during her final bid to be elected the country's prime minister. A Pakistani group affiliated with al-Qa'ida claimed responsibility for her killing, but many Bhutto supporters held Musharraf responsible.

7.1.6 Vulnerable Bangladesh

Formerly East Pakistan, Bangladesh is about the size of the state of Iowa or the nation of Greece—not an important fact until one considers that it is home to 149 million people (•Figure 7.1.21). With its limited resources, people-overpopulated Bangladesh is South Asia's third-poorest country on a per capita GNI PPP basis, after Bhutan and Nepal. Poverty is so deep that many Bangladeshi see neighboring India as a land of opportunity. About 20 million Bangladeshis are illegal immigrants living mainly in India's West Bengal province, where authorities fear they will form permanent Muslim-majority enclaves and perhaps threaten national security. Legal migrant labor is one of Bangladesh's major "exports"; more than 6 million Bangladeshis work abroad, sending home an estimated $6 billion in remittances each year. But India has had enough; in 2007, it completed construction of an iron fence 8 feet (2.5 m) high along its 2,500-mile (4,100-km) border with Bangladesh.

The country's most notable success has been the development of cotton textile industries. If you are American, you almost certainly own clothing that was made in Bangladesh. Textile, jute, leather, and seafood export industries are concentrated in Dhaka (formerly Dacca; population 12.3 million), the capital and largest city, and Chittagong, the main seaport.

The Bangladesh portion of the Ganges-Brahmaputra Delta ranks, along with the Indonesian island of Java, as one of the two most crowded agricultural areas on earth. This area is subject to catastrophic flooding (•Figure 7.1.22). Massive tropical cyclones (hurricanes) ravage Bangladesh regularly in September and October. Deforestation upstream on the steep slopes of the Himalayas has increased water runoff and sediment load in the Ganges, also contributing to Bangladesh's flood problems.

Bangladesh and adjacent portions of India must also keep a wary eye on the potential for the sea level to rise if the world's temperatures increase, as predicted by many climate change models. The country's environment minister has insisted that 20 percent of Bangladesh will be under water by 2015 if nothing is done to control global warming. The Intergovernmental Panel on Climate Change recognizes this as one the world's most endangered areas, predicting a rise in sea level there of 23 inches (c. 58 cm) by 2100. On the Indian side of the delta, scientists are documenting the shrinkage of the river delta islands known as the Sundarbans. Renowned as habitat for the Bengal tiger, the Sundarbans are succumbing to a combination of sea level rise and deforestation that makes them very prone to flooding and erosion.

7.1.7 Nepal and Bhutan: Mountain Kingdoms

The kingdoms of Nepal (population 27 million), with its capital at Kathmandu (•Figure 7.1.23), and Bhutan (population 1 million), with its capital at Thimphu, are perched on South Asia's mountain wall (see Figure 7.1.21). Their landlocked locations and difficult topographies have contributed

● **Figure 7.1.22** Bangladesh is both a low-lying coastal land and a downstream country of several large watersheds. As a result of storms blowing in from the Bay of Bengal and of accelerated runoff from rainfall and snowmelt on deforested mountains upstream, it is subjected to devastating floods.

● **Figure 7.1.23** Nepal's capital, Kathmandu, sprawls across a plain in the country's "middle" mountain range. In the foreground is the Royal Palace.

to their slow economic development. Both countries have productive agricultural regions in the "middle ranges," the bands of foothills between the vast lowland of the subcontinent and the towering Himalayan peaks. In Nepal, when the capacity of this productive area was saturated, many poor peasants moved to the lowland region known as the Terai. The shortage of quality farmland for Nepal's mainly rural (86 percent) population of 24 million is a critical problem. Nepal's mostly deforested landscape bears witness to relentless pressure to provide new land for agriculture and wood for fuel and construction (●Figure 7.1.24).

The conservative Buddhist kingdom of Bhutan has been more careful in opening its doors to tourism, placing restrictions on the numbers permitted in and on the routes they may follow. Tourism and hydroelectric power (mainly for Indian needs) are the most promising areas for future development of this poor country. This is the only country in the world whose politicians claim to disdain conventional economic growth; their mantra is **"gross national happiness"** for the people of Bhutan. They measure happiness using criteria such as equality, good government, and harmony with nature.

But the Bhutanese "Shangri-la" is not without its religious and political problems. Bhutan is ruled from the capital by a Buddhist king of the Drukpa tribe, to which most of the government ministers also belong. The king wants the monarch's power to become more symbolic, with real authority turned over to a multiparty parliament of elected officials. Critics complain that the king will not tolerate any real opposition, however, particularly from the ethnic Nepalis, who are Hindus and who for decades have been migrating out of their own overcrowded Himalayan kingdom to work the productive soils of Bhutan. Ostensibly as a means of promoting national unity, Bhutan's king in the 1990s outlawed the use of the Nepali language in Bhutan's

● **Figure 7.1.24** People are constantly collecting, hauling, storing, and using wood in Nepal, creating tremendous pressure on the country's remaining forests.

schools and insisted that all inhabitants of Bhutan wear the national dress and hairstyle of the Drukpa tribe. Violent clashes ensued between ethnic Nepalis and Drukpas, and more than 100,000 ethnic Nepalis fled back to Nepal, where they languish in refugee camps.

Nepal itself has suffered violent consequences of one-party rule and of palace intrigue. The country was an absolute monarchy until 1991, when prodemocracy riots and demonstrations forced King Birendra to accept a new role as constitutional monarch and permit elections for a national parliament. King Birendra and many other royal family members were shot to death at a dinner gathering in 2001, killed by Nepal's crown prince in a drunken rage over his mother's refusal to allow him to take the bride of his choice. Birendra's successor, King Gyanendra, soon declared a state of emergency and dismissed the democratically elected parliament. He claimed that these steps would quell a Maoist insurgency that began in 1996 with promises to alleviate rural poverty and rid the country of its caste system, which funnels wealth, power, and land ownership into relatively few hands. By 2004, conflict between government troops and the Maoist rebels of the **Communist Party of Nepal** had taken more than 9,000 lives, and the rebels claimed to control 80 percent of the country. The violence snuffed out the country's vital tourism industry.

As his rule stood at the brink of collapse in 2006, Nepal's king made concessions. His government signed a peace agreement with the Maoist rebels, and a temporary constitution and parliament (which included Maoist members) were instituted to address demands for more democracy. Despite the Maoist concessions, other violent opposition has emerged, particularly among groups wanting to end the long-standing political and economic supremacy of the top two Hindu castes in Nepal. The Hindu majority is angry that the government's deal with the Maoists included a declaration that Nepal was henceforth a secular rather than a Hindu nation. Betting on peace, however, tour agencies are clamoring to get their clients back on Nepal's mountain trails, especially in the Everest region of the east and the Annapurna area of the west.

7.1.8 Sri Lanka: Resplendent and Troubled

Sri Lanka (population 20 million), formerly known as Ceylon, is a tropical island country with affinities to the Indian subcontinent. Its name conveys its physical beauty: Sri Lanka means "resplendent isle" in the native Sinhalese language, and medieval Arabs knew it as Serendip, source of the concept of serendipity (•Figure 7.1.25). The island consists of a coastal plain surrounding a knot of mountains and hill lands (•Figure 7.1.26). Most people live in the wetter southwestern portion of the plain, in the south central hilly areas, and in the drier Jaffna Peninsula of the north. Coconuts and rice are the major crops of the low southwestern coast and Jaffna Peninsula, while tea and rubber plantations dominate the economy of the uplands. Colombo (population 2.5 million), in the southwest, is the capital, chief port, and only large city.

Sri Lanka's economy is highly commercialized. Three cash crops—rubber, coconuts, and the world-famous Ceylon tea—supply about 11 percent of export earnings. Sri Lanka is the world's second-largest exporter of tea. Textiles and clothing manufactured in the Colombo area make up about two-thirds of the country's exports by value. Sri Lanka is nearly self-sufficient in rice production, and with declining rates of population growth, its development prospects are good, particularly if its civil conflict can be ended.

External influences have diversified the island's population and sown seeds of unrest. Centuries of recurrent invasion from India were followed by domination by the Portuguese in the early 16th century, the Dutch in the 17th, and the British from 1795 until the country was granted independence in 1948 as a member of the British Commonwealth. This eventful history, plus the long-standing commercial importance of the sea route around southern Asia, has given Sri Lanka a polyglot population that includes Burghers (descendants of Portuguese and Dutch settlers) and Arabs. 296

Sri Lanka's civil conflict reflects antagonisms between ethnic groups and discontent with economic and political conditions, especially among the minority Tamils (9 percent of the population). Since 1983, more than 70,000 people have been killed and many more made homeless as the Tamils of the country's north have fought for autonomy or independence from Sri Lanka's majority Sinhalese gov-

• **Figure 7.1.25** Landscapes made verdant by abundant rains are typical of all but northernmost Sri Lanka. This is Sigiriya, in the central part of the island.

• **Figure 7.1.26** Principal features of Sri Lanka.

ernment. Tamils complain that the Sinhalese treat them as second-class citizens and deprive them of many basic rights. Tamil fighters want to establish their own homeland, **Tamil Eelam,** along the east, north, and west of the country. In 1987, guerrilla attacks by the organization of Tamil separatists called the **Tamil Tigers,** officially the **Liberation Tigers of Tamil Eelam (LTTE),** led to military action by Indian troops invited to help restore order. India withdrew in 1990, but the troubles continued, and India worried that Tamil war refugees from Sri Lanka could flood southern India or that Sri Lankan infiltrators could inspire Tamil separatism within India itself.

The 1991 assassination in southern India of Indian Prime Minister Rajiv Gandhi, son of the slain Indira Gandhi, was linked to Indian sympathizers with the Tamil separatist movement in Sri Lanka. In 1995, Sri Lankan government troops succeeded in retaking the Tamils' geographic stronghold on the Jaffna Peninsula. The Tamil guerrillas retreated to the wild forests just to the south and continued their insurgency with the sinking of Sri Lankan naval ships, a devastating car bombing of downtown Colombo, an attack on the Sinhalese Buddhists' holiest site (the Temple of the Tooth in Kandy), an assassination attempt on the Sri Lankan prime minister, and a spectacular assault on Colombo's airport that destroyed most of the national airline's planes. Many of these were suicide bombings, a favored tactic of the Tigers.

The country's leadership is being variously advised to crush the revolt militarily or to pursue a course of devolution that would turn over federal powers to outlying states—a possible track for granting autonomy in the disputed territories to the Tamils. A cease-fire was declared by both sides in 2002, but the Tamil Tigers promised more violence if their basic demands for self-rule were not met. They have delivered, including another bombing of Colombo's airfield in 2007, and intensified fighting on both sides makes it clear that the cease-fire has expired. Until a formal agreement is reached, the rebels will govern their de facto state of Tamil Eelam; they have their own tax, legal, health, and educational systems and operate banks, newspapers, and radio stations. Unrest in rebel-held areas has made it particularly difficult to rebuild northern sections of the area hit hard by the December 2004 tsunami (see pages 340–341). Early hopes that this catastrophe would have led the warring parties to resolve their differences have been dashed.

7.1.9 The Laid-Back, Low-Lying Maldives

The roughly 1,100 islands that make up the Maldives (population 300,000; capital, Male, population 90,000), an independent country since 1965, appear to be the very essence of tropical paradise (•Figure 7.1.27; for location,

• **Figure 7.1.27** This aerial view of Male (pronounced *mah*-lay), the Maldives' capital, reveals how vulnerable the site is to storms and sea level rise. This is the world's most densely populated city.

see Figure 7.1.1). So prized are its palm-blessed and coral-fringed beaches that more than half a million tourists, mainly Europeans, visit the country each year. More than 60 percent of the country's foreign-currency earnings come from tourism, with fishing and clothing providing most of the rest. Some sources list the Maldives as South Asia's wealthiest country on a per capita basis. The greatest wealth is in the tourist sector, where it is concentrated in so few hands that a potentially dangerous rich-poor divide has developed.

There may be other troubles in paradise. More than 80 percent of this country's very limited land area consists of limestone atolls less than 3 feet (90 cm) above the sea. If sea levels were to rise, as predicted by many common climate change scenarios, the entire country could rapidly be submerged. This prospect led to the Maldives' president's famous cry at a global warming conference: "We are an endangered nation!"

This concludes a detailed look at the geographies of South Asia. Despite linguistic, social, religious, and political difficulties, South Asia's accomplishments in recent decades have been remarkable. India has lowered its birth rate, and its economy has blossomed. Despite its internal divisions and frictions, India has retained a representative democracy. But sectarian violence between Hindus and Muslims, a growing Hindu nationalist movement that threatens India's remarkable diversity, Sikhs' desire for autonomy, and the problem of Kashmir all pose long-term difficulties for this South Asian giant. Pakistan must walk the tightrope between support for its own majority Muslim interests and support for the secular West in its campaign against terrorism while also avoiding war with India. Both Pakistan and Nepal must confront the problems of rapid population growth, deforestation, and land degradation where the balance between people and resources is already precarious. Bangladesh faces the huge challenge of dealing with the environmental consequences of upstream deforestation and repeated flooding assaults from the Bay of Bengal. The Maldivians hope and pray that sea levels remain constant but also lobby industrial powers to cut greenhouse gas emissions. Sri Lanka must accommodate its Tamil minority. All of these nations struggle with the chronic problem of poverty, which is ultimately the source of many of the region's most severe environmental and political dilemmas.

SUMMARY

→ Most of South Asia is a triangular peninsula splitting the northern Indian Ocean into the Bay of Bengal and the Arabian Sea. The entire unit is often called the Indian subcontinent.

→ The Indian subcontinent is one of the world's culture hearths. Some of the world's earliest cities developed on the banks of the Indus. The Aryans arrived between 2000 and 3000 B.C.E. and were the source of many of the region's dominant cultural attributes. The Tamils and other Dravidians predated the arrival of the Aryans.

→ South Asia encompasses the countries of India, Bangladesh, Pakistan, Nepal, Bhutan, Sri Lanka, and the Maldives in an area about half the size of the conterminous United States.

→ The subcontinent is home to many different faiths, including Hinduism, Sikhism, Christianity, Islam, Buddhism, Zoroastrianism, and Jainism. Religious, ethnic, and other differences underlie serious and often violent conflicts. The main ones have been Hindus versus Muslims in India, Sikhs versus the government of India, and Tamils versus Sinhalese in Sri Lanka.

→ Until 1947, what are now Pakistan, Bangladesh, and India formed the single country of India. For over a century, it was the most important unit in the British colonial empire—the "jewel in the crown."

→ With independence, India was partitioned between avowedly Muslim Pakistan (which later divided into two countries, Pakistan and Bangladesh) and secular, primarily Hindu India. The partition had many consequences, including violence between Hindus and Muslims and their large-scale migrations, disputes over the sharing of Indus River waters, and resources and industries stranded on either side of the new partition lines.

→ The most severe product of the partition has been lasting conflict in Kashmir. Several wars have been sparked or fought in the still-volatile frontier province, which, despite its Muslim majority, was joined within India in 1947.

→ The Himalaya and Karakoram ranges, the world's highest mountains, enclose the subcontinent on its northern borders. They were produced by the tectonic collision of the former island of India with Asia.

→ The subcontinent's northern plain contains the core areas of the three major countries of India, Bangladesh, and Pakistan. Three great rivers water this plain: the Indus River, the Ganges River, and the Brahmaputra River. Although mostly desert and steppe in Pakistan, the plain is relatively well watered in India, where it supports unusually high population densities.

→ The southern portion of the subcontinent is mainly a large volcanic plateau called the Deccan. Generally low in elevation, it is rimmed by highlands: the Western Ghats, the northern Central Indian Hills, and the Eastern Ghats.

→ Climates vary, with undifferentiated highland in the northern mountains to desert and steppe in Pakistan and western India. A humid subtropical climate occurs in the northern plain, tropical savanna is in the peninsula with the exception of the rain shadow in the lee of the Western Ghats, and there are rain forests in the Western Ghats' seaward slopes, the coastal plain to the south, and part of Bangladesh.

→ The wet southwest monsoon is at its height from June to September, blowing in from the Arabian Sea and Bay of Bengal. The dry northeast monsoon occurs in the winter, with dry, cooler winds from the northeast coming from high-pressure systems over the continent.

→ Although it is an overwhelmingly poor and rural country, India has many modern industries that, along with privatization of former state-run firms, have contributed to its substantial rate of economic growth in recent years. These include the software and computer industries of Bengaluru and the film industry of Mumbai. Pakistan and Bangladesh are much less industrialized and rely heavily on textile exports. India and Bangladesh have been successful in lowering birth rates, but population growth is a serious issue for both countries. India has managed to feed its huge population.

→ South Asia provides large numbers of workers to other countries, ranging from professional to semiskilled and unskilled. Remittances sent home by these workers are important to the South Asian economies. Many of the professionals do not return, contributing to the region's brain drain.

→ Some of India's large minorities have had serious difficulties with the majority Hindu population. Hindu nationalism is a problem in India, fanning the flames of communal violence between Hindus and Muslims. In the 1980s, some of the Punjab's Sikhs agitated for an independent country.

→ Pakistan's government faces challenges from its Pashtun population in the country's west and northwest and from strong Islamist sentiment throughout the country.

→ Bangladesh and the Maldives are threatened by serious natural hazards from the sea. Strong typhoons (hurricanes) regularly devastate large parts of low-lying Bangladesh, and most of the Maldives would be inundated if sea levels were to rise by 3 feet (90 cm).

→ Nepal and Bhutan are picturesque but poor mountain kingdoms. Both have strong tourism potential, but Bhutan restricts tourists, and for a time a violent Maoist insurgency in Nepal scared many tourists away.

→ Sri Lanka, the world's largest tea exporter, has a Buddhist Sinhalese majority and a Hindu Tamil minority. For more than 20 years, Tamil factions have led a violent struggle to establish an autonomous homeland on the island's east, north, and west periphery. The violence has been punctuated by peace talks that have not yet arrived at a permanent solution.

KEY TERMS + CONCEPTS

Bollywood (p. 323)
caste (p. 325)
communal violence (p. 325)
Communist Party of Nepal (p. 330)
Dalits (p. 325)
distributaries (p. 316)
dry northeast monsoon (p. 318
e-Choupal initiative (p. 323)
Golden Quadrilateral (p. 323)
"gross national happiness" (p. 329)
Harappan civilization (p. 310)
India's Silicon Valley (p. 323)
India's Wall Street (p. 323)
Indo-Aryans (p. 310)

Indus Waters Treaty (p. 316)
Islamization (p. 327)
Jainism (p. 312)
Jains (p. 312)
"jewel in the crown" (p. 313)
Khalistan (p. 326)
Liberation Tigers of Tamil Eelam (LTTE)
  (p. 331)
madrasa (p. 327)
megacity (p. 320)
Moguls (p. 311)
Mohajirs (p. 313)
Mughals (p. 311)
Naxalites (p. 326)

Parsis (p. 311)
remittances (p. 323)
riparian (p. 316)
sectarian violence (p. 325)
sharia (p. 327)
Sikhism (p. 311)
Sikhs (p. 311)
Sinhalese (p. 312)
Tamils (p. 312)
Tamil Eelam (p. 331)
Tamil Tigers (p. 331)
untouchables (p. 325)
wet southwest monsoon (p. 318)
Zoroastrianism (p. 311)

REVIEW QUESTIONS

1. What groups of peoples have been most significant in forming the cultural foundation of South Asia?

2. Using maps and the text, locate the outer mountain wall, the northern plain, peninsular India, and Sri Lanka. What are the major climates, resources, and other natural attributes of these areas? What and where are the three major rivers of the northern plain?

3. How were questions of water allocation between the region's countries resolved after independence?

4. Where are the main production areas for the export crops rice, cotton, jute, and tea?

5. What are the influences of the wet and dry monsoons in South Asia? What particular problems do Bangladesh and the Maldives have in terms of climate and climate change?

6. Why are there far fewer women than men in some Indian provinces? What problems are associated with this phenomenon?

7. What are the major nonagricultural industries of South Asia? Where is India's Silicon Valley? Its Hollywood? Its Wall Street?

Where are the other major industries located? Which country is the most industrialized? Which lag behind, and why?

8. What is the caste system? How does it affect employment and social relations in India? What evidence of change in this ancient system is there?

9. Where is Ayodhya? What site there is critical in relations between Hindus and Muslims in India?

10. What are Pakistan's *madrasas*? Why is their curriculum of so much concern to the government of Pakistan and to the West?

11. What ethnic and political differences have given rise to strife in Sri Lanka?

12. What issues are particularly important in the mountain countries of Nepal and Bhutan and in the lowland country of Bangladesh?

13. Why is climate change such a serious issue for the Maldives?

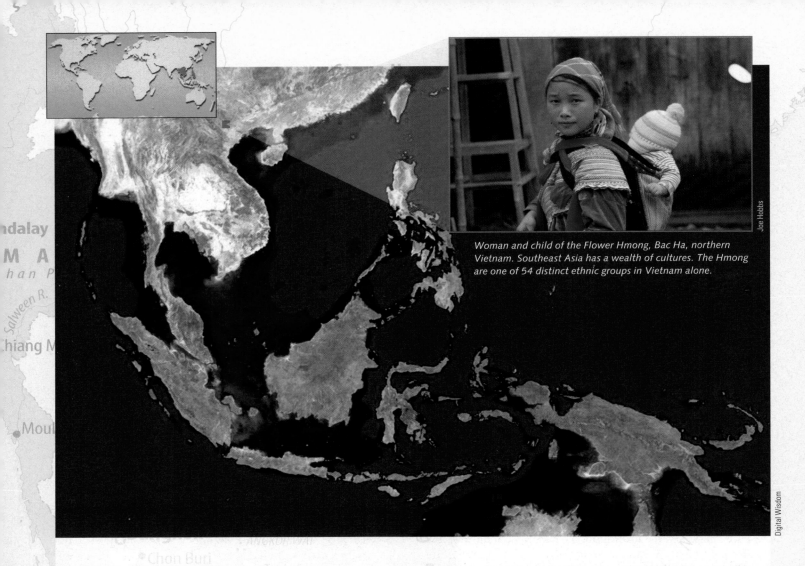

Woman and child of the Flower Hmong, Bac Ha, northern Vietnam. Southeast Asia has a wealth of cultures. The Hmong are one of 54 distinct ethnic groups in Vietnam alone.

Joe Hobbs

Digital Wisdom

MODULE 7.2
SOUTHEAST ASIA: FROM SUBSISTENCE FARMING TO SEMICONDUCTORS

Southeast Asia, from Myanmar (known until 1989 as Burma) in the west to the Philippines in the east, is a fragmented region of peninsulas, islands, and seas. East of India and south of China, between the Bay of Bengal and the South China Sea, the large Indochinese Peninsula projects southward from the continental mass of Asia (●Figure 7.2.1). From it, the long, narrow Malay Peninsula extends another 900 miles (c. 1,450 km) toward the equator. To the south and east are thousands of islands, the largest of which are Sumatra, Java, Borneo, Sulawesi (Celebes), Mindanao, and Luzon. Another large island, New Guinea, is culturally a part of the Melanesian archipelagoes of the Pacific World, but its western half, held by Indonesia, is politically part of Southeast Asia.

Southeast Asia contributed domesticated plants (including rice), animals (including chickens), and the achievements of civilizations to a wider world. Innovations and accomplishments diffused from several centers, which themselves were influenced by foreign cultures. Indian and Chinese traits have been especially strong in shaping the

■ module objectives

This module should enable you to

→ Recognize the correlation between environmental opportunity, environmental constraint, and population densities

→ Appreciate the catastrophic reach of the December 2004 tsunami and some of the region's other natural hazards

→ See how the interconnectedness of the global economic system led to "economic meltdown" in Southeast Asia in 1997

→ Appreciate the economic strength of minority Chinese populations in many of the countries and how that prosperity provokes resentment in times of difficulty

→ Understand some of the major political obstacles and environmental consequences associated with large dams, in this case on the Mekong River

→ Consider how China's economic ascendancy has forced some of the countries to retool their industries

→ Understand how demands for self-rule threaten the cohesion of Indonesia and raise alarm there and abroad

---

294

region's cultural geography. Hinduism came from India, and Theravada Buddhism came from India through Sri Lanka (Ceylon). The Chinese cultural influences of Confucianism, Daoism, and Mahayana Buddhism were particularly strong in Vietnam. China periodically demanded and received tribute from states in Burma, Sumatra, and Java. The Islamic way of life came with Arab traders from the west from the 13th through 15th centuries and took permanent root. Southeast Asia is the eastern margin of the world of Islam, and Indonesia has more Muslims than any other country in the world today.

The historical and contemporary monumental architectures of Southeast Asia reflect these diverse origins. In the 9th century, the advanced Sailendra culture of the Indonesian island of Java built Borobudur, the world's largest Buddhist *stupa* (shrine) and still the largest monument in the Southern Hemisphere. In the 12th century, a Hindu king built the extraordinary complex of Angkor Wat as the capital of the Khmer empire based in Cambodia (•Figure 7.2.2). The first independent Thai kingdom emerged in the

13th century, and in the 14th century, Thai kings ruling from the great city of Ayuthaya (near modern Bangkok) extinguished the Khmer empire and extended their own control into the Malay Peninsula and Burma. The world's largest mosque is in modern Brunei, a tiny oil-rich country near the northern tip of the island of Borneo, and for many years Malaysia boasted the world's tallest buildings, the Petronas Towers in Kuala Lumpur (•Figure 7.2.3).

Southeast Asia today is composed politically of 11 countries: Myanmar (formerly Burma), Thailand, Laos, Cambodia, Vietnam, Malaysia, Singapore, Indonesia, Timor-Leste (formerly East Timor), Brunei, and the Philippines. With the exception of Thailand (which was never colonized) and Timor-Leste (which gained independence first from Portugal and then from Indonesia), all of these states became independent from colonial powers after 1946. These 11 nations are fragmented both politically and topographically. Many of the countries are also culturally fragmented and have experienced strife between different ethnic groups. Outside intervention has sometimes complicated and worsened local discord, producing enormous suffering in the region and lasting trauma among many of the foreign soldiers who fought the determined inhabitants of Southeast Asia.

7.2.1 Area, Population, and Environment

It is about 4,000 miles (c. 6,400 km) from western Myanmar to central New Guinea and 2,500 miles (c. 4,000 km) from northern Myanmar to southern Indonesia. Despite these great distances, the total land area of Southeast Asia is only about 1.7 million square miles (4.4 million sq km), or about half the size of the United States (including Alaska) or China (see Figure 7.2, page 282). The estimated population of the region was 574 million in 2007, giving it an overall density of 331 people per square mile (128/sq km) (see Figure 7.3, page 282). To put this in perspective, Southeast Asia has nearly four times the overall population density of the United States but well under half that of the Indian subcontinent or the densely populated humid eastern part of China, south of the Great Wall.

Southeast Asia has been less populous than India and China throughout history. Only about 10 million people lived in the region in the year 1800, and there were no widespread dense populations. Historically, geographic and environmental obstacles probably helped keep population growth low in Southeast Asia. By land, the region is rather isolated, cut off to some extent by the rugged mountains of the northern peninsula of Indochina. Over many centuries, the ancestors of most of the present inhabitants entered the area from the north as recurrent thin trickles of population, crossing this mountain barrier as refugees. Southeast Asia is tropical, with continuous heat in the lowlands, torrential rains, prolific vegetation that is difficult to clear and

● **Figure 7.2.1** Principal features of Myanmar, Thailand, Cambodia, Laos, Vietnam, Malaysia, Brunei, Singapore, and the Philippines. (See Figure 7.2.18, page 357 for Indonesia and Timor-Leste.)

keep cleared, soils that are generally leached and poor. Historically, people in the region suffered a high incidence of disease. A four- to six-month dry season on the Indochinese Peninsula and on some of the islands presents further challenges.

Violent storms known locally as **typhoons** (elsewhere called cyclones or hurricanes) strike often, particularly in Vietnam and the Philippines (●Figure 7.2.4). More spectacular environmental difficulties result from the location of part of the region on the Pacific Ring of Fire. subjecting Indonesia and the Philippines especially to earthquakes, the tsunamis that earthquakes unleash, and volcanic eruptions (see Natural Hazards, pages 340–341). Two of earth history's most cataclysmic volcanic eruptions took place here: the 1815 eruption of Mount Tambora on the Indonesian island of Sumbawa killed 117,000 people, and its ash lingering in the atmosphere caused the worldwide "year without

a summer" in 1816. The 1883 eruption and explosion of nearby Krakatoa (which scientists believe caused the loudest sound ever heard) also killed tens of thousands and affected the global climate for half a decade.

Typical of less developed countries (LDCs), the introduction of modern medical, food-producing, and other technologies into the region softened the blows of environmental adversity and lowered death rates, resulting in a tremendous population increase in the last century and a half. This increase, which accelerated after World War II, is now slowing but still rapid. Between 1973 and 2007, population grew from an estimated 316 million to 574 million, a 34-year increase of 258 million people, or 81 percent. As will be seen later in this module, the combination of Dutch colonial practices and localized, very productive soils gave rise to an extraordinarily dense population on the island of Java. This Indonesian island's 125 million people represent over one-fifth of the entire population of Southeast Asia. In recent times, the race between population growth and food supply has seen food supply winning in some countries but falling behind in others. Since the 1970s, increases in food production outpaced population growth in Malaysia and Indonesia. At the other end of the scale were Cambodia and Laos, in which war, repression, and inefficient communalization of agriculture slowed progress. These discrepancies are typical of Southeast Asia, home to a remarkable range of economies, from quite poor to very prosperous.

7.2.2 Livelihood Patterns

Southeast Asia as a whole is among the world's poorer regions, yet some of its countries are rather prosperous. The countries' mean per capita GNI PPP of $5,420 in 2007 was about average for the world's less developed countries. In contrast, Singapore, Malaysia, Brunei, and Thailand were well above average for LDCs. Because of their high rates of industrial productivity and economic growth, Singapore is generally included in the ranks of the economically dynamic "Asian Tigers," while Malaysia, Thailand, and sometimes Indonesia, the Philippines, and Vietnam are described as either "New Asian Tigers" or "Tiger Cubs." The commercial-industrial city state of Singapore has about the same per capita GNI PPP as Germany. High-tech industries brought spectacular economic growth to these countries through much of the 1990s. In 1997, a series of events pre-

298

cipitated an economic crisis that roared through the region and temporarily declawed these emerging economies, but conditions have since improved (see page 344).

This region's economic activities span the entire continuum from Stone Age hunting and gathering (in Papua on the island of New Guinea) to slash-and-burn farming—one of humankind's earliest agricultural techniques—to the commercial agriculture typical of colonial times and from the export of raw materials like rubber to manufactured goods like clothing, shoes, and computer products. As Southeast Asians see it, their economic future will depend not so much on rubber, coffee, or even oil, of which some countries have abundant supplies, but on high-tech exports like semiconductors and software. But despite pockets of technological innovation, rapid urbanization, and industrialization, the majority of the people in Southeast Asia are still farmers, with 62 percent of the people classified as rural.

Jerry Alexander/Stone/Getty Images

• **Figure 7.2.2** Angkor Wat, the 12th-century capital of the Khmer empire, is a symbol of the rich cultural history of Southeast Asia.

Joe Hobbs

• **Figure 7.2.3** Modernity and prosperity characterize some nations in Southeast Asia today. At 1,462 feet (446 m), the twin Petronas Towers in Kuala Lumpur were until 2005 the tallest buildings in the world.

Shifting and Sedentary Agriculture

Permanent agriculture has been difficult to establish in many parts of this largely mountainous, tropical realm. Much land is used only for shifting cultivation, in which a migratory farmer clears and uses fields for a few years and then allows them to revert to secondary growth vegetation while he moves on to clear a new patch. Since an abandoned field needs about 15 years to restore itself to forest that can again be cleared for farming, only a small portion of the land can be cultivated at any one time. This extensive form of land use can support only sparse populations. Unirrigated rice is the principal crop of shifting agriculture, with corn, beans, and root crops such as yams and cassava (manioc) also grown. In some forested areas, people supplement their farming by hunting and gathering. These migratory subsistence farmers are often of different ethnic groups from adjacent settled populations, having historically sought refuge in the backcountry from stronger invaders (•Figure 7.2.5).

Most of Southeast Asia's people live in widely scattered, dense clusters of permanent agricultural settlements (see Figure 7.3 top, page 282). These core regions stand in striking contrast to the relatively empty spaces of adjoining districts. Superior soil fertility is the main locational factor in these areas. However, there are a few districts of dense populations without high-quality soils to support them. Most of these rely on plantation agriculture of rubber and other less demanding cash crops.

The typical inhabitant of the areas of permanent sedentary agriculture is a subsistence farmer whose main crop is wet rice, grown with the aid of natural flooding or irrigation. In some drier areas, unirrigated millet or corn replaces rice. Secondary crops that can be grown on land unsuitable for rice and other grains supplement the major crop. These secondary food crops include coconuts, yams, cassava, beans, and garden vegetables. Some farmers grow a secondary cash crop such as tobacco, coffee, or rubber. Fishing is important both on the coasts and along inland streams and lakes. Because of their importance in the food supply, many fish are harvested from artificial enclosures and flooded rice fields, in which they are raised as a supplementary "crop" (•Figure 7.2.6).

Subsistence agriculture in Southeast Asia exists alongside tropical cash crops grown for export. The region is one of the world's major supply areas for such crops, which include palm oil, rubber, and cloves. This highly commercial type of production is a legacy of Western colonialism, which began in the 16th century. Until the 19th century, Westerners were interested mainly in the region's location on the route to China. They were content to control patches of land along the coasts and to trade for some goods produced by local people. During the 19th century, however, the Industrial Revolution in the West greatly expanded the demand for tropical products. Europeans consequently

• **Figure 7.2.4** Typhoon Xangsane slammed ashore in central Vietnam on October 1, 2006, causing extensive wind and flood damage. This is the city of Hoi An on the following morning. *Joe Hobbs*

extended their control over almost all of Southeast Asia. European colonists used their capital and knowledge, along with indigenous and imported labor, to increase export production rapidly. They focused on exporting some local produce such as copra (dried coconut meat, from which oil is pressed) and spices. They also introduced some new commercial crops.

The usual method of introducing commercial production was to establish large estates, or **plantations,** managed by Europeans but worked by indigenous labor or labor imported from other parts of Monsoon Asia. There were often problems finding available local hands, so Europeans imported contract labor from India and China. Large migrations from these countries complicated an already complex ethnic mixture, setting the stage for today's unusual distribution of ethnic groups and their sometimes discordant relations.

● **Figure 7.2.5** This couple of the ethnic group known as the Black Tai are farmers near the village of Mai Chau in northern Vietnam. They have just served tea and mangoes to guests in their single room stilt home. They are not wearing their traditional clothing, which they don only for ceremonial occasions.

Many export cash crops, some raised by small landowners, are still produced in Southeast Asia. The major ones are rubber, palm products, and tea. Nearly three-fourths of the world's natural rubber comes from the region, mainly Malaysia, Indonesia, and Thailand; Thailand is in fact the world's largest producer and exporter (see Insights, page 342).

Palm products include palm oil, coconut oil, and copra. Malaysia and Indonesia are the world leaders in producing palm oil, which now accounts for about 20 percent of edible oils worldwide and is used in a variety of detergents, soaps and cosmetics. The Netherlands and other European countries had promoted the use of palm oil for biofuels to replace fossil fuels. It has recently become apparent, however, that to meet this European demand, Malaysia and Indonesia are accelerating the clearing of natural vegetation to replace them with palm plantations. In Indonesia, peat-covered lands are being replaced with palm oil farms. The draining and burning of peat there releases large quantities of carbon dioxide into the atmosphere. By this unusual means, Indonesia became the world's third-largest emitter of carbon dioxide (after China and the United States) in 2007. Alarmed energy planners in the Netherlands and elsewhere have revised their views of palm oil as a "green" source of energy.

103, 589

Indonesia dominates the world output of coconut palm products. It is also the region's leading producer of tea and is among the world's top five producers. Other export crops grown in Southeast Asia include coffee from Indonesia, cane sugar from the Philippines and Indonesia, and pineapples (of which the Philippines and Thailand have become the world's leading producers and exporters). However, the countries of Southeast Asia are generally coming to depend less on agricultural exports and more on mining and manufacturing.

One lasting impact of Western colonialism on the economies of Southeast Asia was the stimulation of commercial rice farming in formerly unproductive areas. Three areas became prominent in commercial rice growing: the deltas of the Irrawaddy River in Myanmar, the Chao Praya in Thailand, and the Mekong in Vietnam and Cambodia. Almost impenetrable wetlands and uncontrolled floods had

● **Figure 7.2.6** A marine fish farm off the southern coast of Cambodia. This type of aquaculture is an increasingly important in subsistence and commerce in Southeast Asia and elsewhere.

NATURAL HAZARDS

The Great Tsunami of 2004

On December 26, 2004, the most cataclysmic natural event of modern times afflicted a vast portion of the earth. Beneath the Indian Ocean, just off the northwestern coast of the Indonesian island of Sumatra, a gargantuan "megathrust" earthquake measuring magnitude 9.3—with the force of more than 32 billion tons of TNT, or the equivalent of as many as a million nuclear explosions—struck at 6:58 B/N / local time. In the Java (Sunda) Trench, as much as 750 miles (1,200 km) of fault line slipped 60 feet (over 18 m) along the subduction zone where the Indian Plate is sliding beneath the Burma Plate (part of the greater Eurasian Plate; see Figure 2.1, page 21).

As the seabed of the Burma Plate instantly rose several meters vertically above the Indian Plate, the massive displacement

of seawater created a series of tsunamis (*tsunami* is Japanese for "harbor wave"; the phenomenon is sometimes popularly but incorrectly referred to as a "tidal wave") that pulsed across the Andaman Sea and the Indian Ocean at up to 500 miles (800 km) per hour. On the open ocean, such waves are barely perceptible, but they break on coastlines with the force of giant storm surges. The tsunami roared ashore, at heights up to 45 feet (14 m), in 14 countries: Indonesia, Malaysia, Thailand, Myanmar, Bangladesh, India, Sri Lanka, the Maldives, Seychelles, Somalia, Kenya, Tanzania, Yemen, and South Africa (●Figure 7.2.A).

These were the deadliest tsunamis in history. The total number of dead exceeded 200,000. By far the greatest number of deaths (over 130,000) were in Indonesia, where Aceh's principal city of

● **Figure 7.2.A**   The epicenter of the earthquake, which registered magnitude 9.3, was located off the coast of Sumatra, Indonesia. It created a tsunami of tremendous force and scope.

kept these deltas thinly settled, but incentives and methods for settlement have turned them into densely populated areas within the past century. Thailand, Vietnam, and Myanmar are the region's leading rice exporters. Other Southeast Asian nations import much of this produce.

Deforestation

Near areas of dense settlement in Southeast Asia, there are still some large areas covered in primary forest, but they are shrinking as population expands and development progresses. Many environmentalists view the destruction of the region's tropical rain forest as an international environmental problem (●Figure 7.2.7). In 2007, only 28 percent of Thailand's original forest cover remained. Poorer neighbor Myanmar had a 49 percent forest cover. Vietnam lost more than one-third of its forest between 1985 and 2007, by which time only 37 percent of the original forest cover

remained. Deforestation is advancing most rapidly in Malaysia, where 63 percent of the land area is still forested (official statistics often include palm oil plantations in forest estimates, so these figures should be evaluated carefully), and in Indonesia, where 49 percent is forested. Malaysia is now the world's largest exporter of tropical hardwoods. Virtually all of the primary forests in Peninsular Malaysia have already been cleared, and it is projected that commercial logging will decimate those remaining outside protected areas in Malaysian Borneo soon after 2010 (●Figure 7.2.8).

Although Southeast Asia's tropical forests are smaller in total area than those of central Africa and the Amazon Basin, they are being destroyed at a much faster rate, most significantly by commercial logging for Japanese and Chinese markets rather than by subsistence farmers. Environmentalists fear the irretrievable loss of plant and animal species and the potential contribution to global warming that deforestation in this region may cause. Malaysia and

NATURAL HAZARDS

The Great Tsunami of 2004 *continued*

Banda Aceh was obliterated (●Figure 7.2.B). Sri Lanka lost 35,000; India, 12,000; and Thailand, over 5,000. Almost all the deaths occurred within a mile of the shoreline and were caused by blunt force from debris and by drowning. An estimated one-third of the fatalities were children. There were reports of children running out to investigate fish stranded by retreating seas—a characteristic precursor to a tsunami—only to be struck seconds later by a wall of water.

As many as 2 million persons were made homeless by the disaster. An unprecedented international relief effort, designed to feed and house these refugees and to contain the spread of epidemic diseases, began within hours.

Tsunami experts concur that countless thousands of lives would have been saved had the affected Indian Ocean region been equipped with the same kind of tsunami early warning system now in place in the Pacific Ocean.[1] That investment had not been made because historically about 90 percent of the world's tsunamis have occurred in the Pacific and because many of the nations afflicted lacked the financial resources for the wave sensors and other infrastructure required by the system. Now, however, installing an early warning system in the Indian Ocean basin is an urgent priority. In this event, Indonesia would have had just a few minutes' warning, but Thailand would have had two hours and Somalia seven hours to prepare for the disaster.

● **Figure 7.2.B** Banda Aceh before and after the tsunami.

● **Figure 7.2.7** In Southeast Asia, natural forest cover has been rapidly reduced by many human uses, especially commercial logging and farming.

Indonesia have passed laws to slow the destruction, with Indonesia taking the aggressive step of banning the use of fire to clear forests. However, enforcing such legislation has so far proved impossible. Hundreds of Indonesian and Malaysian companies, most of them large agricultural concerns (for example, palm oil plantations and pulp and paper companies) with close ties to the government, continue to use fire as a cheap and illegal method of clearing forests. The World Wildlife Fund estimates that 70 percent of Indonesia's exported trees are illegal, many of them removed by "timber barons" in the country's national parks while bribed park and police officials look the other way. The urgent need for construction materials after the devastating tsunami of 2004 further increased the pressure on the country's forests.

There is a growing movement in the consuming countries to boycott the tainted timber by selling only certified "good wood." The U.S. company Home Depot, for example, sells only imported woods bearing the Forest Stewardship Council (FSC) logo.

INSIGHTS

Rubber

The earliest major cash crop introduced into Southeast Asia was rubber, first cultivated there in the 1870s. An Englishman, Sir Joseph Priestley, gave the name "rubber" to the latex of the *Hevea brasiliensis* tree in 1770, when he discovered he could use it to rub errors off the written page. Extensive exploitation of this tree in its native Brazil gave rise to the "rubber boom" of the late 19th century, when Brazilian rubber traders were so wealthy that they used banknotes to light cigars and sent their shirts to be laundered in Europe.

In 1876, the Englishman Henry Wickham smuggled 70,000 rubber seeds out of Brazil. Delighted botanists at London's Kew Gardens cultivated them and shipped them to Ceylon (Sri Lanka) and Southeast Asia for experimental commercial planting. Rubber trees proved exceptionally well adapted to the climate and soils of South Asia and then Southeast Asia (•Figure 7.2.D). Back in Brazil, however, a fungus known as leaf blight wiped out the rubber plantations, and by 1910, bats and lizards inhabited the mansions of Brazil's rubber barons. Fortunately, the Kew Gardens stock was free of the fungus; by 1940, fully 90 percent of the world's natural rubber came from Asian plantations. The desire to secure this vital resource was Japan's principal reason for occupying much of Southeast Asia at the outset of World War II.

Perhaps surprisingly, real rubber is more important today than ever. Synthetic rubber, most of which is made from petroleum, now meets the greater part of the world's demand for products made originally of natural rubber. However, natural rubber has superior elasticity, and aircraft and truck tires are made almost exclusively from it. China's boom is helping ensure a strong future for future rubber producers, and in an unusual turnabout, many Southeast Asian workers are leaving the factory for the rubber farm.

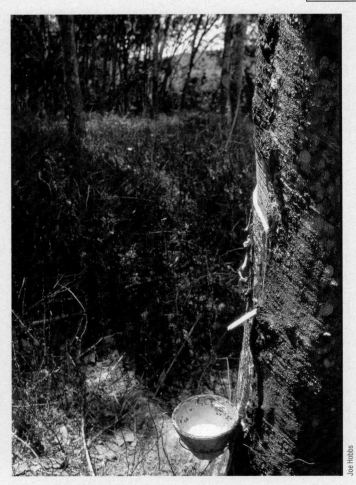

Joe Hobbs

• **Figure 7.2.D** Rubber latex bleeding down a fresh incision and into the collection cup on a Malaysian plantation.

Deforestation in Southeast Asia has many actual and potential transboundary consequences. In 1997, a widespread drought attributed to the warming of Pacific Ocean waters by the phenomenon known as El Niño turned the annual July-to-October burning season into a holocaust of human origin. Deliberately set fires raced out of control over large areas of Sumatra and Borneo, resulting in a choking haze that shut down airports, closed schools, deterred tourists, and caused respiratory distress to millions of people in Indonesia, Malaysia, Singapore, the Philippines, Brunei, and Thailand. Less widespread but still severe smog crises struck Malaysia and Indonesia, particularly the island of Sumatra, in subsequent years, and Indonesian authorities have fought back by revoking the licenses of plantation companies.

Many of the plants and animals in these forests are **endemic species** (found nowhere else on earth). Indonesia, which contains 10 percent of the world's tropical rain forests, is known, after Brazil, as the world's second most important **megadiversity country,** with about 11 percent of all the world's plant species, 12 percent of all mammal species, and 17 percent of all bird species within its borders. The Southeast Asian region is also particularly significant in biogeographic terms because of the so-called **Wallace's Line**—named for its discoverer, the English naturalist Alfred Russel Wallace—that divides it (see Figure 7.2.18). Nowhere else on earth is there such a striking local change in the composition of plant and animal species in such a small area on either side of this divide. East of the line (separating the Indonesian islands of Bali and Lombok, for

• **Figure 7.2.9** Existing and proposed natural gas pipelines for Southeast Asia.

• **Figure 7.2.8** Commercial logging has removed all but small portions of the tropical rain forest (as in the lower left) in East Malaysia's states of Sarawak, seen here, and Sabah. The land is then used for oil palm plantations. The same process is underway in neighboring Kalimantan on Indonesia's side of the island of Borneo.

example), marsupials are the predominant mammals, and placental mammals prevail west of the line. Similarly, bird populations are remarkably different on either side of the line.

Mineral Production and Reserves

Petroleum is the most important mineral resource in Southeast Asia. Seven of the region's countries—Indonesia, Malaysia, Brunei, Myanmar, Thailand, Vietnam, and the Philippines—produce oil. Collectively, their proven oil reserves amount to about 1.1 percent of the world total. Their combined output is about 3 percent of world oil production, with Indonesia the leading producer. Indonesia's main oilfields are in Sumatra and Indonesian Borneo (known as Kalimantan).

Although it does not loom large on the world scene, Southeast Asian oil is very important to the countries that own and produce it. Oil and associated natural gas provide nearly all the exports of Brunei and make that tiny Muslim country one of the world's wealthiest nations. That wealth is precarious, as Brunei has little to offer world markets except its finite fuels. Oil, gas, and refined products supply almost one-fourth of Indonesia's exports and about one-tenth of Malaysia's. Singapore profits as a processor of oil. The main customer for Southeast Asian oil and natural gas is Japan, the leading trade partner of several Southeast Asian countries. Natural gas is so important domestically for electricity production in several Southeast Asian countries that efforts are under way to link them in a network of pipelines, tapping mainly Indonesian and Malaysian gas reserves (•Figure 7.2.9). About 70 percent of Thailand's power production comes from natural gas from Myanmar and its own offshore fields. This relatively clean source of energy has helped clear Thailand's air of its long-standing smog problem.

There is other mineral wealth in Southeast Asia. Tin is the most important ore mined, with about one-fifth of world output from Indonesia. The Philippines produces a variety of metals—copper, chromite, nickel, silver, and gold—in significant amounts. The most serious mineral deficiency in the region is the near absence of high-grade coal. However, more energy could come from the region's additional potential for hydroelectric power. Gemstones are important exports from Myanmar and Thailand (•Figure 7.2.10).

Joe Hobbs

● **Figure 7.2.10** Star sapphires from Myanmar are among Southeast Asia's most prized gemstones.

Economic Meltdown and Its Aftermath

Following years of spectacular growth, in which economies across the region grew at rates up to 8 percent annually, 1997 saw a stunning reversal of fortunes in Southeast Asia. Until then, the region's economic tigers and cubs were praised for high rates of savings, balanced budgets, and low inflation. Export-driven growth rates were climbing, particularly with the increased manufacturing of computers and other electronics. Gradually, however, the appetite of the main consumer of Southeast Asia's products, the United States, began to wane, and trade deficits developed in the producing countries. At the same time, the region's emerging economies pursued expensive public works projects, such as Malaysia's new administrative capital of Putrajaya (see Figure 7.2.15, page 355).

In some countries, corruption added to the potential for economic crisis as insolvent companies were propped up with infusions of cash from sympathetic politicians (the phenomenon is known as **crony capitalism**). Banks extended risky loans for both private and public businesses and construction projects. The loans were generally made in U.S. dollars. When local currencies rapidly lost value against the dollar, the borrowers were unable to repay these loans. Nervous foreign investors in the region's stock markets withdrew their investments; theirs was the "hot money" that can send stock markets soaring but can also be withdrawn quickly, with devastating results.

Economies melted down. Stock markets, which had seen explosive growth, imploded; Malaysia's market, for exam-

ple, lost 80 percent of its value between September 1996 and September 1998. Among the region's stronger economies (excluding Vietnam, Cambodia, Laos, and Myanmar, which had far less to lose), Malaysia, Thailand, and Indonesia fell into **economic recession**, characterized by a relatively short fall in economic output and a moderate rise in unemployment. As the crisis dragged on, it threatened to tip into a full-scale **economic depression**, a devastating breakdown of the economy. Only the Philippines and Singapore escaped the bloodletting.

Economies gradually began to improve, but the economic meltdown had lasting impacts. Largely unchallenged in prosperous times, Southeast Asian ways of running governments and businesses have been under strong scrutiny ever since. Indications of this shift run the gamut from large-scale revolts against or impeachments of national leaders to parents fighting favoritism in school admissions, farmers protesting dam-building projects, peasants protesting industrial pollution, and patients suing doctors for malpractice. Examples of these events are described in the discussions of the individual countries. Despite their internal problems and differences, the countries hoped that their common membership in the Association of Southeast Asian Nations (ASEAN) might unite them against future economic calamities (see the Regional Perspective, page 345).

Although the countries of Southeast Asia have many similarities, each has its own distinctive qualities, with unique environmental features, indigenous and immigrant peoples, culture traits, and economic activities. Many of the characteristics and problems of these countries are the outcome of European imperialism, but in no two countries have the results of colonialism been the same. Here are portraits of the countries.

7.2.3 Myanmar (Burma)

Myanmar (population 50 million; see Figure 7.2.1, page 336) is centered in the basin of the Irrawaddy River and includes surrounding uplands and mountains. Within the basin are two areas of dense population: the so-called Dry Zone, around and south of Mandalay (population 1.2 million), and the Irrawaddy Delta, which includes the former capital and major seaport of Yangon (Rangoon; population 4.6 million). The Dry Zone is the historical nucleus of the country. The annual rainfall of this area—34 inches (87 cm) at Mandalay—is very low for Southeast Asia. The people are supported by mixed subsistence and commercial farming, with millet, rice, and cotton the major crops. During the past century, the Dry Zone was surpassed in population by the delta, where a commercial rice-farming economy now provides one of the country's two leading exports (forest products is the other). The Irrawaddy forms a major artery of transportation uniting the two core areas.

The indigenous Burman people, most of whom live in the Irrawaddy Basin, make up an estimated 68 percent of

REGIONAL PERSPECTIVE

The Association of Southeast Asian Nations (ASEAN)

The countries of Southeast Asia have sought to strengthen their economic and security interests by forming a regional organization. Known as the **Association of Southeast Asian Nations (ASEAN),** the coalition was formed in 1967 to promote regional trade and minimize the need for alignment with either of the two world superpowers, the United States and the Soviet Union. Singapore, Malaysia, Thailand, Indonesia, Brunei, and the Philippines were founding members. Vietnam joined in 1995 with the promise of strengthening regional ties and adopting the more open economic system that the Vietnamese call *doi moi,* or renovation. Myanmar and Laos joined in 1997 and Cambodia in 1999.

After U.S. troops withdrew from bases in the Philippines in 1991, ASEAN members grew concerned that they lacked a security "umbrella" and that China or possibly even Japan might become more aggressive in the region. Several ASEAN members, notably Indonesia, Malaysia, and Singapore, therefore strengthened ties with the U.S. military to deter other powers from attempts to dominate the region. The organization thus aspired to become both a large trading bloc and a substantial security counterweight to China.

By the end of the century, however, ASEAN members moved closer to China, increasingly seen as a benign and economically advantageous ally, and created an informal economic cooperation alliance with China, Japan, and South Korea known as **ASEAN+3.** China even proposed that a China-ASEAN Free Trade Area should be established by 2015. ASEAN's main goal is to adopt policies that would avert the type of economic meltdown that shook all the group's members in 1997 and 1998. ASEAN nations aspire to have, by 2015, a European-style common market that would eliminate member country barriers to trade and investment. The countries are also trying to reduce dependence on the International Monetary Fund and other Western monetary agencies and countries in the event of future economic crises.

Solidarity among the member countries will demand that they first resolve quarrels among themselves. Many tensions focus on the most prosperous member, Singapore, which other members view as an arrogant economic giant that grows at the expense of the other members. Indonesia notably has cut back exports of some commodities to Singapore to register its discontent.

the country's population. About 90 percent of Myanmar's citizens are Theravada Buddhists. Distinct ethnic groups live in the Arakan Mountains of the west and the northern highlands. Ethnic Shans live in the northern part of the Shan Plateau of eastern Myanmar, and ethnic Karens live in the southern Shan Plateau and in the delta of the Salween River. These ethnic groups account, respectively, for about 9 and 7 percent of the country's population. The Shan Plateau in northeastern Myanmar is part of the **Golden Triangle**—the frontier region of Myanmar, Thailand, and Laos—which was the world's leading source area for opium until Afghanistan pushed ahead of it in the 1990s (see the Regional Perspective, pages 346–347).

Britain conquered Burma in three wars between 1824 and 1885. It became an independent republic outside the British Commonwealth in 1948. Civil war has been a constant since independence. Communism, targeted persecution, and ethnic separatism have motivated rebellions by ethnic Karens and Shans, a number of less numerous hill peoples, and Muslims of the Rohingya ethnic group who fear the establishment of a pure Buddhist state. Beginning around 1999, however, Myanmar's ruling military government reached cease-fire agreements with most of these ethnic groups (except the Karens) and gave these groups what it called **contingent sovereignty,** offering them more civil rights and economic opportunities than previously. The government also encouraged profits made by some of these groups in the drug business to be "invested" in national development.

The economy has been badly damaged by fighting and for years was mismanaged under a form of rigid state control promoted as the **"Burmese Way to Socialism."** Myanmar became the poorest non-Communist country in Southeast Asia, a sad contrast with its reputation in early independence years for having the best health care, highest literacy rate, and most efficient civil service in Southeast Asia. It used to be the world's largest rice exporter; it now exports less than 1 percent of the world total. Myanmar's government spends comparatively less on health care for its citizens than all but three countries in the world.

In the 1990s, Myanmar began to abandon socialism in favor of a free-market economy and enjoyed considerable economic growth, with China its main trading partner and military ally. Most of the fruits of this economic growth, however, lined the pockets of the ruling military. The **State Law and Order Restoration Council (SLORC),** the military government that seized power in Burma in 1988 (and changed the country's name to Myanmar), has yielded little to popular pressure for democratization. Myanmar remains one of the world's most repressive places to live. Access to the Internet was prohibited until 1999 and is still strictly regulated. Foreign journalists are banned, and citizens may not allow foreigners into their homes. It is illegal to gather outside in groups of more than five. Giant green billboards across the country read: "Crush all internal and external destructive elements as the common enemy."

A much publicized contest, followed around the world, has been raging between the government and one of its citizens, the Nobel Peace Prize laureate Aung San Suu Kyi, who had been placed under house arrest after leading her antimilitary political party to an overwhelming victory in

REGIONAL PERSPECTIVE

Sex, Drugs, and Health in Southeast Asia

Southeast Asia has long been one of the world's main source areas for opium and its highly addictive semisynthetic derivative, heroin. For decades, global concern focused on combating the drug trade originating in this region and on dealing with the problem of heroin addiction in the consuming nations. Now, owing to the emergence of HIV/AIDS, the other traditional Southeast Asian industries of prostitution and tourism have woven drugs into a leviathan problem of global concern.

The region's drug production is centered in the Golden Triangle, where Laos, Thailand, and Myanmar historically exercised little control over their territories. Thailand has recently increased its authority substantially in the area. As in Afghanistan and Pakistan, the absence of a strong central government presence and ideal growing conditions facilitated explosive growth in drug production (•Figure 7.2.E). But as Thailand and Laos exercised increasing authority in the area, drug production plummeted. Today, Myanmar is the only one of the three countries with a strong opium poppy economy, but even there, production is beginning to show signs of decline. The reason is that neighboring China, with an eye to more trade with Myanmar, is pressuring Myanmar authorities to crack down near their common border.

In recent years, with new heroin-producing areas cropping up in Mexico and elsewhere, the price of heroin for the consumer has dropped and its quality has risen. In the United States, these trends are noticeable in the growing appeal of heroin among middle and upper income groups, by smoking and other previously rare forms of heroin consumption, and by the increasing numbers of deaths due to heroin overdose.

In this region, the U.S. Drug Enforcement Administration and other international antidrug authorities focus their concerns on Myanmar, where refining of heroin from opium is done in remote laboratories. In northeastern Myanmar, armed ethnic minority groups have produced both heroin and a potent methamphetamine known as *ya baa*. They smuggle raw opium into both China

Joe Hobbs

• **Figure 7.2.E** Raw opium oozes from slices in a poppy pod. Heroin, which is derived from opium, has a hearth in Southeast Asia's Golden Triangle region.

and India and amphetamines into Thailand, providing investment capital to contribute to Myanmar's "development." Laundered drug money has paid for the country's recent construction boom, and the government encourages such investment from the drug sector to make up for shortfalls in legitimate export revenues.

Not all of the associated problems are exported. A flood of cheap and potent heroin hit the market in Myanmar after the current regime came to power in 1988. Heroin addiction has soared ever since. Primary delivery is by needles shared in tea stalls. The numbers of users continue to grow; perhaps 1 percent of the adult population is addicted. Accompanying the heroin use is an AIDS

the 1990 elections (which the military government nullified). In 1995, the government released "The Lady," as Suu Kyi is popularly known. She then called for dialogue with the military junta and appealed for the parliament that was elected in 1990 to be convened. The government responded by again placing her alternatively under house arrest and in prison, triggering strong economic sanctions by the United States.

Aung San Suu Kyi has called for would-be tourists to avoid Myanmar, which has abundant sites of interest but few visitors, arguing that tourism revenue would help strengthen the repressive government. She has also been successful in gaining support abroad for boycotts of U.S. and other Western companies doing business with Myanmar. American corporations including PepsiCo and Conoco-

Phillips have withdrawn from Myanmar, and the U.S. government has banned new investments. The United States regards the country's government as illegitimate and therefore rejects the change of the country's name to Myanmar. There is pressure on the American firm Unocal to suspend its contract to join a French firm in developing Myanmar's considerable offshore natural gas reserves in the Yadana Field of the Andaman Sea. The two companies have already built a pipeline from the field to a port west of Bangkok, Thailand, where the gas is used to generate electricity (see Figure 7.2.9). Human rights groups complain that slave labor was used to build the pipeline and that profits from gas sales help prop up Myanmar's repressive government. The pipeline has to be well defended in view of the government's many domestic opponents who might try to destroy

REGIONAL PERSPECTIVE

Sex, Drugs, and Health in Southeast Asia *continued*

epidemic, with again about 1 percent of the adult population infected. The best available figures indicate that the HIV infection rate among heroin users in Myanmar is around 37 percent.

The epidemic that began among heroin users spread quickly to Myanmar's sex industry. Recent estimates put the HIV infection rate among prostitutes at 26 percent, three times the rate in neighboring Thailand. Many young men with AIDS are shunned by their families and go to monasteries to die. Tragically, their virus has spread among the populations of Buddhist monks (there are now 400,000 in Myanmar), who share razors to shave their heads. AIDS prevention is virtually unheard of. The government had outlawed condoms until 1993. They are now legal, and over 30 million have been distributed by the government since 2001, but this is still short of what is needed. Practically no anti-HIV drugs find their way to Myanmar.

Myanmar's prosperous, cosmopolitan neighbor Thailand also has the scourge, with a 1.4 percent HIV infection rate among adults in 2007. This is a substantial reduction from the peak infection rate the country recorded in 1994. Thailand's liberal social climate has given rise to the unique phenomenon of sex tourism, with package tours catering to an international clientele (•Figure 7.2.F). Thailand's overall annual income from prostitution is estimated to be between $2 and $4 billion, and prostitutes send as much as $300 million each year to relatives in the villages from which they migrated. However, the spread of HIV infection among intravenous drug users and among prostitutes in the red-light districts of Bangkok and the popular beach resorts of Pattaya and Phuket threatens both the country's public health and its lucrative tourist trade. The Thai government mounted an all-out war on the drug trade in 2002 that resulted in the deaths of more than 2,000 people, most of them suspected drug dealers.

There are striking differences in how this region's countries are dealing with HIV/AIDS. Thailand, regarded globally as a leader in efforts to combat the problem, has run an aggressive and in-

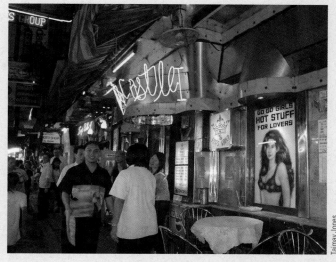

• **Figure 7.2.F** Bangkok's red-light district was a breeding ground for HIV/AIDS until the government and nongovernmental agencies aggressively promoted a safe-sex campaign.

creasingly successful anti-AIDS public awareness campaign, resulting in a dramatic decrease in the number of new HIV infections. Thailand distributes antiretroviral drugs to all AIDS patients at extremely low cost. Other Southeast Asian countries lag far behind. The government of the Philippines has been unsuccessful in pursuing an anti-AIDS campaign because of opposition from the country's powerful and anticontraception Catholic clergy. Church officials have denounced the government's anti-AIDS program as "intrinsically evil" and have set boxes of condoms on fire at antigovernment demonstrations. Similarly, in the conservative and largely Islamic nations of Indonesia and Malaysia, Muslim clerics denounce their governments' anti-AIDS campaigns as efforts to encourage promiscuity.

464

it as a means of weakening the regime. In an apparent move to gain strategic depth from its own population and from any outside threats, the government in 2005 established a new capital at Naypyidaw, in the country's remote central region. The move did nothing to quell popular discontent with the regime. Widespread antigovernment protests, with broad participation of Buddhist monks, broke out all over the country late in 2007. Authorities responded with deadly force and the insurgency faded—at least temporarily.

7.2.4 Thailand

Governed today as a democracy in which military influence is strong and in which a traditional monarchy has mainly symbolic functions, Thailand (historically known as Siam;

population 65 million) was the only Southeast Asian country to preserve its independence throughout the colonial period. This unique status was due to Siam's position as a buffer between the British and French colonial spheres.

Thailand is located in the delta of the Chao Praya River, known to Thais as the Menam ("River"). Annual floods of the river irrigate the rice that is the most important crop in this Southeast Asian land of relative abundance and progress. Thailand is the world's largest exporter of rice and dominates the niche market in the prized, aromatic and expensive jasmine variety.

On the lower Chao Praya is Bangkok (population 10.2 million), the capital, main port, and primate city. It is notoriously congested and was until recently a smoggy city. Environmental blight prompted the Thai government to crack down on carbon emissions and also to envision construction of

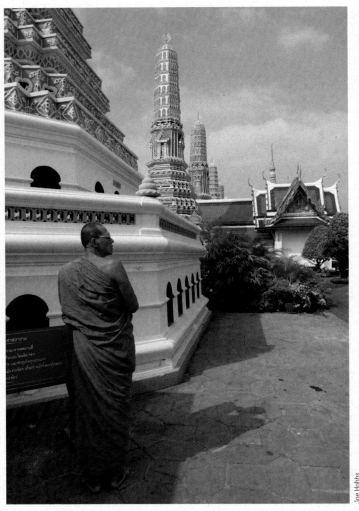

● **Figure 7.2.11** The Buddhist monastery and temple of Wat Phra Kaew ("Temple of the Emerald Buddha") is one of the architectural gems of Bangkok, which is also a city of modern skyscrapers.

"Heaven City," a satellite of Bangkok designed to have little pollution or traffic gridlock.

Bangkok is a city of stunning architectural marvels, particularly Buddhist temples (●Figure 7.2.11). Bangkok's networks of canals (*klongs*), thronged with both commercial and residential activities, give it a distinctively amphibious character; it has been dubbed the "**Venice of the East**." Its "floating market" is especially renowned. Many of Bangkok's canals have been filled in with paved roads and buildings, thereby making the city vulnerable to flooding. Due to the pumping out of groundwater beneath the city, Bangkok's soils are settling downward, and the entire capital is sinking at a rate of about 2 inches (5 cm) a year. Thai officials forecast that Bangkok will be below sea level by 2050. If nearby sea levels rise, as predicted by many climate change models, the city would be imperiled.

Areas outside the river's delta are more sparsely populated and include the mountainous territories in the west and north, inhabited mainly by Karens and other ethnic groups, and the dry Korat Plateau to the east, populated by Thais, the related Laotians, and Cambodians. To the south in the Kra Isthmus, a part of the Malay Peninsula, live more than 2 million ethnic Malays. Most Malays are Muslims, but an estimated 95 percent of the people of Thailand are Buddhists whose faith includes elements borrowed from Hinduism and local spiritualism. Spectacular Buddhist temples are a prime attraction for one of Southeast Asia's largest tourist industries, as are ceremonies celebrating Thailand's monarchy and other historical traditions. The Thai people are extremely reverent of the country's king, and criticism of the monarchy is illegal. In 2007, a drunken Swiss tourist who spray-painted on images of King Bhumibol Adulyade was sentenced to ten years in jail. The king pardoned him.

The main ethnic minority in Thailand is the Chinese, an estimated 9 million, or 14 percent of the population. They control much of the country's business, a situation that many Thais resent and that makes the loyalty of the Chinese an important issue.

Although Thailand, dubbed the "Land of Smiles," has enjoyed more internal tranquillity than most of Southeast Asia, it has had its troubles, including a 2006 coup that ousted an unpopular prime minister but also diminished foreign confidence in the Thai economy. Since the late 1960s, guerrilla insurrections motivated by communism and ethnic separatism have persisted in the northeast and south. In the late 1970s, the country dealt with a mass migration of refugees from adjoining Cambodia and still hosts about 2 million refugees. Recently, Muslim ethnic Malay guerrillas seeking independence for Thailand's southernmost region have burned schools, killed Buddhist monks, and ambushed government troops in that area. Thailand annexed this region, which was formerly a Muslim sultanate, in 1902. Thai authorities today are divided on how to deal with the unrest. Some want to build a security barrier along the Malaysian border and crack down on the insurgency north of that line, but others—including the country's prime minister—would like to accommodate some of the ethnic Malay demands, including the application of Islamic law (*sharia*) in the region.

In spite of conflict and refugee problems, Thailand has had one of the strongest economies in Southeast Asia. From 1993 to 1996, the economy grew a robust 8 percent annually. The country shifted away from its traditional exports of garments, textiles, cut gems, and seafood in favor of medium-tech industries such as computer assembly. Tourism was the leading source of foreign exchange. Then came 1997, when Thailand was the first "domino" to fall in the Asian economic crisis (see page 344). Economic growth had attracted tens of thousands of the rural poor to the factories of Bangkok, where they made garments for such major U.S. retailers as Gap, Nike, and London Fog. Now they were laid off and joined the swelling ranks of Thailand's poor. Even before the slump, Thailand had a huge gap between the wealthy and the poor, with more than 50 percent of the country's wealth in the hands of the richest 10 percent of its population. By the beginning of the new century,

Thailand's economy was back on its feet. In recent years, it has recorded strong growth based largely on surging demand in China for electronic parts, petrochemicals, and rubber.

7.2.5 Vietnam, Cambodia, and Laos

Vietnam, Cambodia, and Laos (see Figure 7.2.1) are the three countries that emerged from the former French colony of Indochina. Vietnam is the largest and most geographically complex, with three main regions. Tonkin, the northern region, consists of the densely populated Red River Delta and surrounding, sparsely inhabited mountains; Annam, the central region, includes the wild Annamite Cordillera and small, densely populated pockets of lowland along its seaward edge; and Cochin China, the southern region, lies mainly in the delta of the Mekong River and has a high population density.

Almost 90 percent of Vietnam's people are ethnic Vietnamese (also known as Kinh), closely related to the Chinese in many cultural respects, including language and the shared religious elements of Confucianism, Buddhism, and ancestor veneration. A Catholic minority reflects the French occupation, and there are small indigenous religions. Fifty-four minorities enrich the country's ethnic makeup. Various ethnic groups known collectively to outsiders as the **hill tribes** or **Montagnards** live in the central highlands of Vietnam, where some have recently agitated for independence and religious freedom. A large Chinese minority lives in the major cities of all three countries.

Cambodia and Laos occupy interior areas except for Cambodia's short coast on the Gulf of Thailand. They are very different countries physically, and a particular culture dominates each. Cambodia is mostly plains along and to the west of the lower Mekong River, with mountain fringes to the northeast and southwest. Except for half a million ethnic Vietnamese, almost all the 14 million people of Cambodia are ethnic Cambodians, also known as Khmer. Their language is Khmer, and their predominant religion is Buddhism.

Laos is mountainous, sparsely populated with just 6 million people, and landlocked. It has 68 ethnic groups, the largest of which is the ethnic Lao, who are linguistically and culturally related to the Thais and whose dominant religion is Buddhism (•Figure 7.2.12). Their largest numbers are in the lowlands, especially in and near the capital, Vientiane. Most of the ethnic minorities of Laos live in the mountain regions, including the country's second-largest ethnic group, the Hmong.

The Vietnam War

The Vietnam War, which is such an important chapter in the American experience, had roots that preceded U.S. interests in the region. This conflict profoundly affected Southeast Asia and has a lingering legacy there today.

• **Figure 7.2.12** The stupa of That Louang in Vientiane is Laos' most important religious building, and its national symbol.

France conquered Indochina between 1858 and 1907. The French first extinguished the Vietnamese empire, which covered approximately the territory of today's Vietnam, and defeated the Chinese, whom the Vietnamese called on for help. Later, the French took Laos (in 1893) and western Cambodia (in 1907) from Thailand. France's administration of Indochina, centered in Hanoi and Saigon, left a big cultural imprint. Its major economic success lay in opening the lower Mekong River area to commercial rice production, converting both Vietnam and Cambodia into important exporters of rice.

Japanese forces overran French Indochina in 1941, initiating five decades of warfare in the area. During World War II, a Communist resistance movement led by a Vietnamese nationalist named Ho Chi Minh carried on guerrilla warfare against the Japanese. When French forces attempted to reoccupy the area after the end of war, these forces fought to expel them. In 1954, Ho's forces destroyed an entrenched French garrison at Dien Bien Phu, in the northwestern mountains. With this defeat, France withdrew from Indochina, ending a century of colonial occupation.

Four countries came into existence with France's departure. Laos and Cambodia became independent non-Communist states. North Vietnam, where resistance to France had centered, emerged as an independent Communist state. South Vietnam gained independence as a non-Communist state that received many anti-Communist refugees from the north. Vietnamese Catholics were prominent among those who fled southward.

The partition of Vietnam into two countries was largely the work of U.S. power and diplomacy. South Vietnam was from the outset a client state of the United States, whereas North Vietnam became allied first with Communist China and then with the Soviet Union. Warfare in Vietnam gained momentum. Between 1954 and 1965, the **Viet Cong,** an insurrectionist Communist force supported by North Vietnam, achieved increasing successes in the south in its bid to reunify the country. Increasing intervention by the United States in South Vietnam escalated in 1965 and eventually involved the commitment of half a million U.S. military personnel. North Vietnam responded by fielding regular army forces against American and South Vietnamese troops. The United States never invaded North Vietnam, although American air strikes there were devastating, with more ordnance dropped on Vietnam than on all of Europe during all of World War II. American bombs, napalm (an explosive), and defoliants such as Agent Orange (dioxin) caused enormous damage to the natural and agricultural systems of Vietnam, destroying over 8,400 square miles (nearly 22,000 sq km) of forest and farmland in an effort described in military parlance as **"denying the countryside to the enemy."** Agent Orange also left a legacy of birth deformities among Vietnam's postwar generation.

The United States avoided invading North Vietnam partly because of the risks posed by Chinese and Soviet support of the North. In limiting the theater of ground warfare, however, the United States found itself unable to expel from the South both the Viet Cong and the North Vietnamese army. Those combined forces were determined, skillfully commanded, increasingly better equipped by its allies, accomplished in guerrilla tactics, and willing to bear heavy losses. More than 3 million Vietnamese soldiers and civilians died in the conflict, and about 58,000 U.S. soldiers and support staff perished. In 1973, almost all American forces were withdrawn from the costly war, which was extremely divisive at home as growing ranks of Americans questioned its rationale and protested its conduct. There are still 1,948 Americans listed as missing in action in Indochina.

North Vietnam completed its conquest of South Vietnam in 1975. Saigon was renamed Ho Chi Minh City (although most of its people still call it Saigon), after the country's leader (who died in 1969). Vietnam entered a period of relative internal peace, but there were ongoing conflicts, including Vietnam's 1979 conquest of Cambodia to oust the brutal Khmer Rouge regime (see page 352), continued warfare against Cambodian and Laotian resistance groups, and a brief border war with China. Within its borders, the Communist government of the reunited Vietnam unleashed a repression against those who had assisted the Americans, including Montagnards and members of the Cao Dai religion (•Figure 7.2.13) This backlash caused a massive outpouring of refugees, including the "boat people" (see Ethnic Geography, page 351). Pressure on perceived opponents intensified, especially for the country's ethnic Chinese after China's brief invasion of Vietnam in 1979. The two countries gradually came to terms, however, finally normalizing relations in 1991 and solving long-standing disputes such as the demarcation of their territorial waters and corre-

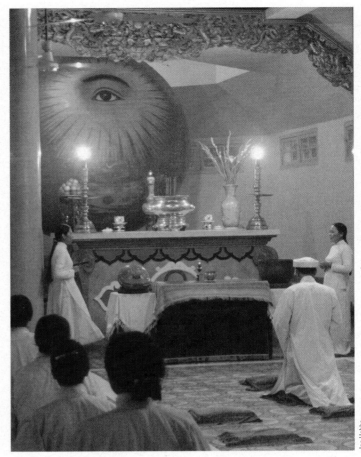

Joe Hobbs

• **Figure 7.2.13** The "Divine Eye" is the symbol of God for members of the Cao Dai faith. Founded in Vietnam in 1926, the Cao Dai religion synthesizes many religions, and its pantheon includes Jesus Christ, Louis Pasteur and Victor Hugo. Cao Dai fighters resisted the French occupation, and also supported South Vietnam in its war against the North. This is the Cao Dai Temple in Danang, Vietnam.

sponding exclusive economic zones in the Gulf of Tonkin. They have, however, been unable to resolve disputed claims to two island groups in the South China Sea: the Paracel Islands (with rich fishing grounds and potential for oil and gas), which China took from Vietnam by force in the early 1970s, and the Spratly Islands (see Chapter 7, page 302).

Vietnam Today

Vietnam (population 85 million), with its capital at Hanoi (population 2.6 million; •Figure 7.2.14) is restoring its war-torn landscape through large-scale reforestation, agricultural reclamation, and nature conservation programs, aided by international organizations such as the World Wildlife Fund. Animal species new to science have been discovered in the rugged Annamite Cordillera (Truong Son Mountains). Such natural treasures are attracting international ecotourists who provide much-needed revenue, and tourism in general is booming.

ETHNIC GEOGRAPHY

The Boat People

The desperation of many people in Vietnam and its neighboring war-torn countries created the crisis of the **boat people**—refugees who fled onto the open ocean on large rafts, with chances of survival variously estimated at 40 to 70 percent. After 1974, almost 2 million Vietnamese fled their homeland, and the United States provided permanent refuge for more than half of those who resettled. Canada, France, and Australia also took in significant numbers of immigrants (•Figure 7.2.G). Large numbers of Vietnamese and Laotian refugees occupied camps in China near its borders with those countries. Boat people also filled refugee camps in Hong Kong, Malaysia, Indonesia, Thailand, and the Philippines.

The United States and other Western governments long ago reduced the size of the welcome mat for Southeast Asian refugees, but Vietnamese continue to immigrate to the United States at a rate of about 25,000 a year; they are now the fifth-largest immigrant group in the United States. Most of these immigrants are typical of an **expanding pyramid,** in which family members in the United States bring extended family members over from Vietnam. About 2,000 are classified as political refugees who survived Vietnam's postwar "education camps." Many of these are **Amerasians,** the offspring of American soldiers who served in the Vietnam War. Vietnam is also among the leading foreign sources of children adopted by American couples. Altogether, the United States is now home to about 1.2 million ethnic Vietnamese. Approximately 100,000 Vietnamese Americans visit Vietnam each year, bringing much welcome hard currency with them into Vietnam.

In 1996, Vietnamese refugee camps around Asia closed, and about 100,000 Vietnamese were compelled to return to their homeland. Just as those countries were closing their doors to Vietnam, Vietnam was undertaking political and economic reforms to provide incentives for the expatriates to return home and for resident Vietnamese to stay home. Since 1990, about 1 million overseas Vietnamese ("Viet Kieu"), mostly from the United States and France, have returned to Vietnam to stay. Their repatriation has also pumped economic capital into Vietnam's fledgling free-market economy. Many Vietnamese who once aspired to leave their home country have chosen to remain. Growing numbers leave only to earn doctorates and other professional degrees to carry home and invest in the country's growth.

• **Figure 7.2.G** Many ethnic Vietnamese refugees settled in Canada, introducing their cuisine and culture to the new homeland.

Since its embrace in 1986 of free-market reforms in the policy known as *doi moi*, Vietnam's economy has improved markedly. The poverty rate fell by half in the 1990s, and in recent years, Vietnam has had Asia's second-fastest-growing economy, after China's. The Communist government halted collectivized farming in 1988 and turned control of land over to small farmers under 20-year lease agreements. The results have been impressive; Vietnam since 1988 has become a major exporter of rice, second in the world after Thailand. This is a vulnerable commodity, however, especially because production is concentrated along the Mekong River, which is prone to flooding in the May-to-October rainy season (see The World's Great Rivers, page 353). Vietnam is also pinning hopes on exports of its proven reserves of 2.5 billion barrels of oil, especially to energy-hungry China.

In addition to joining ASEAN, Vietnam in 1995 restored full diplomatic relations with the United States, and U.S. businesses began scrambling for consumers in this large, promising market. After the two countries signed a trade agreement in 2000, Vietnam began exporting shoes, finished clothing, and toys to the United States. Both its imports and its exports boomed when the country joined the World Trade Organization (WTO) in 2007. The dual personalities of North and South are being perpetuated in the new economic climate, with most investment and infrastructural improvement focused in the South. The North is a more austere economic landscape, and northerners dominate the civil service. Along with a growing gap between the rich and the poor, the differences between North and South could slow the country's overall development.

• **Figure 7.2.14** The heart of Hanoi's Old Quarter.

In an effort to better integrate the nation, Vietnam recently completed a road running the entire 1,000-mile (1,600-km) length of the country, along the historic Ho Chi Minh Trail. Environmentalists fear the highway's impacts on wildlife and point out that annual heavy rains regularly bring landslides and make this mainly mountainous route unusable. Anthropologists are worried about its effects on the traditional peoples inhabiting remote areas along the road. And many in Vietnam's growing private business sector lament the road as a costly white elephant; the country already has a more practical lowland north-south highway and railway, they complain, and resources would be better spent on "new economy" infrastructure such as the Internet.

Like that of many LDCs, Vietnam's progress has been threatened by the specter of uncontrolled population growth. Since the war with the United States ended in 1975, Vietnam's population has increased by 70 percent. In view of the country's limited size and resource base, the government has stepped in with strong economic incentives for couples to have no more than two children. Birth rates have begun to fall sharply. Nevertheless, 29 percent of the population is under the age of 15, and foreign investors are pouring into Vietnam to seek this young brainpower and muscle power.

Cambodia: Away from the Killing Fields

For a time, Cambodia escaped the ills known to Vietnam, but then it suffered even worse catastrophes. After five years of fighting, a Communist insurgency overcame Cambodia's American-backed military government in 1975. Known as the **Khmer Rouge** and led by a man with the *nom de guerre* of Pol Pot, this Communist organization ruled the country for four years with exceptional savagery. In 1973, the population had been estimated at 7.5 million, but by 1979, it was estimated at only about 5 million; a third of the population had been murdered. Some intended victims managed to escape as refugees from what became known as Cambodia's **killing fields** to adjoining Thailand. The Cambodian Communists' stated policy was to build a "new kind of socialism" by eradicating the educated and the rich, emptying the cities, and breaking up families. The population of Phnom Penh, the capital and largest city, was expelled on a death march to the countryside. The city's population dropped from more than 1 million in 1975 to an estimated 40,000 to 100,000 in 1976 (its population in 2007 was 1.7 million).

The Communist-Nationalist ambitions in Vietnam came to dominate Cambodia's political landscape in 1979. Border conflicts between the two countries had been escalating, and in 1979, Vietnamese troops swiftly conquered most of Cambodia, ousting the Khmer Rouge and installing a puppet Communist government. The Khmer Rouge, supported by China (Vietnam had by then become a close ally of China's rival, the Soviet Union), began guerrilla resistance against the Vietnamese. Two non-Communist Cambodian guerrilla forces also took the field. These three resistance forces often quarreled among themselves, but all maintained the conflict with the Cambodian-Vietnamese army of the Phnom Penh government, which periodically pursued them into Thailand and engaged Thai border military units.

THE WORLD'S GREAT RIVERS

The Mekong

The Mekong River, known as "Water of Stone" in Tibet, "Great Water" in Cambodia, "Nine Dragons" in Vietnam, and "Mother of Waters" (Mae Nam Khong, contracted to Mekong) in Laos and Thailand, is Southeast Asia's great river (•Figure 7.2.H). From its source on the Tibetan Plateau in the Chinese province of Qinghai, the Mekong snakes along a 2,600-mile (4,200-km) path through China's Sichuan and Yunnan provinces, forms the boundary between Laos and its neighbors Myanmar and Thailand, winds across Cambodia, and fans out across Vietnam's densely populated Mekong Delta region before empting into the South China Sea.

Unlike the Chao Praya in Thailand, the Red River in Vietnam, and the Irrawaddy in Myanmar, the Mekong is not a great magnet for cities and civilizations. The biggest city on its banks is Cambodia's Phnom Penh, with just over a million people. The main factor limiting more settlement along the Mekong has been the difficulty in navigating the river above Phnom Penh. Historically, very little trade has been carried along much of its length; the oceans and overland routes have provided superior opportunities.

The Mekong is not economically unimportant, however. It is one of the world's most prolific fisheries, yielding about 2 million tons of fish each year, or twice the annual North Sea catch. There are more species of fish in the river (about 1,200) than in any other rivers but the Amazon and the Congo. Unique creatures inhabit its waters, including about 70 Irrawaddy dolphins and the giant catfish that attain 10 feet (c. 3 m) in length and 660 pounds (300 kg) in weight. Overfishing and the construction of dams have all but eliminated this superlative catfish throughout the Mekong Basin.

A big push is being made to capitalize on the Mekong's potential to produce hydroelectric power, with plans to build about 100 dams on the river and its tributaries. The first dam on the river itself (the Man Wan in China) was completed in 1993. China has since built another and has plans for four more. Vietnam is building five dams on the Mekong tributary called the Hoyt, which forms part of its border with Cambodia. Thailand has dammed its main tributaries to the Mekong. Laos plans to build the most dams of all—20—and thereby become mainland Southeast Asia's biggest supplier of electricity. With funding from the World Bank—despite the criticism this organization has received for supporting large dams—Laos is proceeding with construction of the gi-

ant Nam Theun 2 dam. The World Bank insists it has gone to extraordinary lengths to minimize environmental impacts from this project.

These dams will nevertheless have enormous impacts, some of them unforeseeable. It may be reliably predicted that the fish catch will fall dramatically because the main reason for the Mekong's productive fishery is its regular flooding, which will be dramatically reduced. Fishermen are already complaining of smaller catches since the dam building began. There are concerns that dams will ruin the ecosystem of Cambodia's Tonle Sap Lake, which is drained by a river that actually changes direction when downstream Mekong floodwaters inundate its basin. More than a million people make their living by fishing this lake.

As always with shared river systems, there are serious questions about how the Mekong waters should be divided. Vietnam, Cambodia, Laos, and Thailand have joined to form the **Mekong River Commission,** whose biggest responsibilities are to fix the minimum amount of water each country must discharge downstream on the Mekong and its tributaries and to agree on rules to ensure water quality. Both quality and quantity of water are threatened by upstream Myanmar and China, which have refused to join the Mekong River Commission.

233, 316

• **Figure 7.2.H** Ferries on the Mekong near Can Tho, Vietnam.

Joe Hobbs

At first, the Vietnamese rejected international food aid and systematically starved Cambodia in an effort to weaken resistance. Later, they promoted the country's recovery.

Peace negotiations among Cambodia's contending factions led in 1991 to an agreement resulting in the restoration of democratic government under United Nations supervision. A new constitution with new leaders took effect in 1993, and Cambodia's former King Norodom Sihanouk returned to the throne. In 1997, Hun Sen, a former Khmer Rouge leader who had defected and then opposed the

group, ousted the popularly elected royalist leader Prince Norodom Ranariddh in a violent coup. Under much international pressure, Hun Sen permitted elections in 1998. Finally succeeding in forming a government legitimately and peacefully, he secured Cambodia's membership in ASEAN and regained Cambodia's seat at the UN.

Cambodia is a very poor country, but more stable conditions have helped the economy improve recently. Oil and gas reserves have recently been discovered offshore, giving cause for optimism. Garment manufacturing is the largest industry, accounting for 90 percent of export earnings. Fears that Cambodia would prove unable to compete with growing exports of cheaper clothing from China have so far not been realized, thanks to large foreign contracts from companies like Gap and Abercrombie & Fitch that certify Cambodia's textiles as "sweatshop-free." Tourism is the fastest-growing industry. Few visitors miss Angkor Wat, still coveted by some in Thailand because it once stood within that country's borders. Cambodians resent Thai claims to the site and also Thai ownership of many prominent tourism, telecommunications, and forestry businesses in Cambodia. That resentment led to riots in 2003—sparked by a Thai actress's claim that Angkor Wat belongs to Thailand—that did millions of dollars in damage to Thai businesses and the Thai embassy in Cambodia.

386,
540

Even in this relatively peaceful time, the people of Cambodia must deal with a persistent scourge of war: land mines. Sown by hostile forces during years of warfare, the antipersonnel explosives remain active indefinitely. Mines have maimed more than 35,000 Cambodians, and about 50 more casualties occur each month. People bury new mines, imported mainly from China and Singapore, to protect their property in Cambodia.

274

The Lao Way

Laotians also suffer from ordnance dating to the Vietnam War. Between 1964 and 1973, U.S. warplanes dropped more than 2 million tons of bombs—more than the United States dropped on Germany in World War II—on the Laotian frontier with Vietnam in an effort to disrupt Communist supply lines on the Ho Chi Minh Trail (a network of supply lines snaking through the mountains of Vietnam and Laos) and to prevent Communist troops from entering Laotian cities. This Plain of Jars region retains the distinction of being the most heavily bombed place on earth. An estimated 30 percent of the bombs failed to detonate, and today, millions of unexploded American cluster bombs remain. Smaller than a tennis ball, each cluster bomb contains more than 100 steel ball bearings. Many curious Laotian children who play with them are killed or dismembered. About 200 people are killed or maimed yearly by these explosives. Of Laos's 17 provinces, 15 contain **unexploded ordinance (UXO)**, according to UN studies, creating a serious deterrent to farming in a country with inadequate food supplies. Efforts are under way to clear the explosives, with only minimal economic assistance from the United States. The agency in charge of the cleanup hopes to clean up enough land to produce 10,000 tons of rice yearly, which would feed 50,000 people.

The administrative capital of Laos is Vientiane (with only 200,000 people), located on the Mekong River where it borders Thailand. This sleepy city poses a remarkable contrast to the nearby boomtowns of Vietnam (•Figure 7.2.15). Vientiane replaced the royal capital of Luang Prabang ("Royal Holy Image"), located upriver on the Mekong when Laos came under Vietnamese occupation in 1975. That year, the Pathet Lao organization overthrew the Laotian monarchy and established a new "people's democratic republic." The Communist dictatorship now rules Laos in a benign fashion, encouraging a market economy and foreign investment. The government calls its conservative approach to economic reform and social change the **"Lao Way."** To slow the potential flood of foreign ideas and materials into the country, Laos has permitted the building of only two bridges to Thailand across the Mekong River. However, there is an already important and growing casino gambling industry in Laos, patronized mainly by visitors from Thailand.

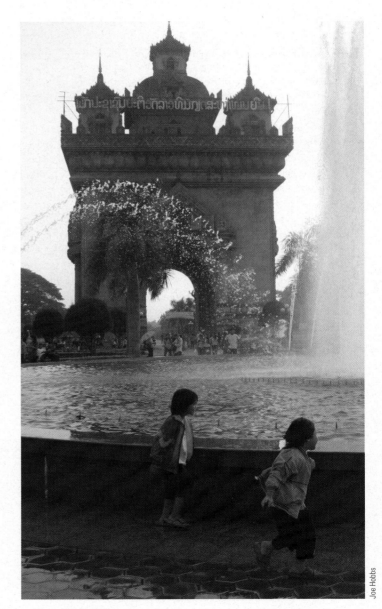

• **Figure 7.2.15** Vientiane's Pataouxi, or Victory Arch, built in the 1950s, is often likened to Paris' Arc de Triomphe.

Despite gambling revenues, landlocked Laos is now one of Asia's least developed countries, ranking 133rd of the 177 countries on the United Nations Human Development Index. About 80 percent of its people are peasant farmers. Exports are limited to timber, hydroelectric power (to Thailand), and garments and motorcycles assembled by inexpensive Laotian labor using imported materials. Gold and copper mining, which began in the early 2000s, may bring sorely needed revenue to Laos.

As in Vietnam and Cambodia, the Laotian economy has suffered from a brain drain prompted by years of warfare, political turmoil, and underdevelopment—conditions that caused the best-educated and most talented of the country's inhabitants to flee. After the 1975 takeover, 343,000 Laotians, mostly members of the Hmong ethnic minority who had U.S. support in fighting Laotian government troops during the war, fled to neighboring Thailand. Many Hmong refugees subsequently emigrated to the United States, where 250,000 ethnic Hmongs now live, generally as an underprivileged and misunderstood underclass in cities of the Midwest and West. Back in Laos, a low-level Hmong insurgency continues against the Vientiane government.

Given the conditions of warfare, political turmoil, oppression, and inefficient economic systems, it is not surprising that the three Indochinese countries are, with Myanmar, the poorest in Southeast Asia. However, their outlook seems to be improving as tensions with the outer world relax, privatization of Communist economies increases, the supply and trade of goods and services improves, and the number of returning refugees exceeds emigrants.

7.2.6 Malaysia and Singapore

The small island of Singapore (•Figure 7.2.16; see also Figure 7.2.1) sits just off the southern tip of the mountainous Malay Peninsula. It lies at the eastern end of the Strait of Malacca, the major passageway for sea traffic between the Indian Ocean and the South China Sea. For centuries, European sea powers competed to control this passageway. Britain gained control in 1824 and used the newly established settlement of Singapore to monitor the strait. With its central position among the islands and peninsulas of Southeast Asia, British Singapore developed into a large naval base and the region's major entrepôt. Singapore is the world's second-largest port (after Shanghai) in tonnage of goods shipped.

From Singapore, the British extended their political hold over the adjacent southern end of the Malay Peninsula. This expansion gave Britain control over the part of the peninsula that is now included, along with northwestern Borneo (except Brunei), in the independent country of Malaysia. This southern end of the peninsula is known as Malaya, Peninsular Malaysia, or West Malaysia. The Borneo section includes the states of Sarawak and Sabah, together known as East Malaysia.

During the British period, Malaya developed a commercialized and prosperous economy. Tin and rubber were the major products, with oil palms and coconuts of secondary importance. Chinese and British companies pioneered the tin-mining industry in the late 19th century. The building of railroads gave access to a line of tin mines and rubber plantations in the foothills between Malacca and the hinterland of Penang (George Town). Along this line, the inland city of Kuala Lumpur (population 6.9 million) became Malaya's leading commercial center and then Malaysia's capital (see Figure 7.2.3, page 338).

Although Kuala Lumpur will remain Malaysia's official capital, all government administration has been transferred to Putrajaya, 25 miles (40 km) to the south (•Figure 7.2.17).

• **Figure 7.2.16** Singapore, located at a crossroads where the Pacific and Indian Oceans meet, is a very modern, prosperous city.

• **Figure 7.2.17** The Prime Minister's department complex and city mosque in Putrajaya, Malaysia's new and controversial administrative capital.

It is a high-tech capital stitched together by fiber-optic cable. Malaysian officials boast that the Internet and other forms of digital communication will replace paper in the country's new administrative center.

Nearby is Cyberjaya, a new city that, as its name suggests, hopes to lure technology companies and become Asia's Silicon Valley. These cities are symbols of how Malaysia sees itself, a country leaving behind its colonial legacy as a "rubber republic" to become an engine of technical innovation and export. Putrajaya's $5 billion construction cost drew much criticism in the wake of Malaysia's financial crisis, however.

The development of a colonial economy gave Malaya an ethnically mixed population. The native Muslim Malays played only a minor role in this development. Chinese and Indian immigrants and their descendants became the principal farmers, wage laborers, and businesspeople of the tin and rubber belt. So heavy was the immigration that 23 percent of Malaysia's population of 27 million is Chinese and 7 percent is Indian. Today, the Chinese minority is quite prosperous in business, and Indians dominate many professional ranks. The growth of Singapore brought even heavier immigration so that now over three-fourths of Singapore's population of 4.6 million is Chinese.

Ethnic antagonisms between Malays and Chinese shaped the political geography of Malaysia and Singapore. Malaya accepted independence in 1957 only on the condition that Singapore not be included. This was to ensure that ethnic Malays would be a majority in the new state. In 1963, however, the Federation of Malaysia was formed, composed of Malaya, Singapore, and the former British possessions of Sarawak and Sabah in sparsely populated northern Borneo. The non-Chinese majorities in Borneo were counted on to counterbalance the admission of Singapore's Chinese. This experiment in union lasted only until 1965, when Singapore was expelled from the federation and left to go its own way as an independent state.

Within Malaysia, the government passed legislation to ensure that at least 60 percent of the university population was ethnic Malay (who with small indigenous groups are known as **Bumiputra**), with measures to ensure that 30 percent of business ownership went to Malays. Although ethnic Chinese and Indians still complain that such policies continue to shut doors of opportunity to them, the economic gap between these minorities and the majority Malays has narrowed considerably since the 1970s. But tensions between Muslims and non-Muslims in this country bear monitoring. Among Malaysia's majority Muslims, there are growing complaints that the government is too secular, although it is nominally Islamic (•Figure 7.2.18). The leading champion of the adoption of Islamic law (*sharia*) as the law of the land is the **Parti Islam se-Malaysia (PAS)**, which previously administered state governments of Kelantan and Terengganu in Peninsular Malaysia.

Malaysia and Singapore are among the more economically successful of the world's formerly colonial states that

• **Figure 7.2.18** Most ethnic Malays, like this woman, are Muslim.

became independent after World War II. Singapore is an outstanding center of manufacturing, finance, and trade along the Pacific Rim. Its industries include electronics, chemicals, oil refining, machine building, and tourism. With its electronics sector now suffering intense competition from China, Singapore is hoping to develop a strong new niche in biomedical research and other biotechnology industries; a huge biotechnology research complex called **Biopolis** has been built there. Due to restrictions on federal spending on stem cell research in the United States, many top American scientists have relocated there. Singapore suffered less than Malaysia, Thailand, and Indonesia during the region's economic crisis of the late 1990s. This was due in part to a geographic advantage: its ability to conduct trade with a vast regional hinterland in other countries without having to attend to the struggles of a large and poor rural population of its own. Perceived as an island of stability in an unsettled region, the country has also long been a magnet for investment. Wealthy Indonesians buy up real estate there, and ethnic Chinese capital and talent from all over Southeast Asia flow into Singapore. The country's prosperity is accompanied by strict attitudes toward propriety and secularism. Chewing gum is banned, graffiti artists are flogged, headscarves and other apparent religious symbols are banned in schools, and crows—famously loud, messy, and unruly—are shot.

Malaysia still exports rubber and tin but has a diversified export portfolio that includes crude oil (from northern Borneo), timber, palm oil, automobiles, and electronic components. Malaysia is the world's largest exporter of semiconductors and air conditioners. Its annual per capita GNI PPP is only about one-third that of Singapore's but is about double the average figure for the world's LDCs (see Table 7.1).

Malaysia has a national mantra, "Wawasan 2020" (Vision 2020), referring to the year by which national economic planners intend Malaysia to be a fully developed country. The vision became clouded after 1997, however, when following a decade in which the economy grew at a robust 8 percent annually, Malaysia fell into recession. The country quickly froze its ambitious and expensive development projects and adopted economic austerity measures. Like the other emerging markets, Malaysia had been benefiting from huge and often speculative foreign capital investments. And like the others, Malaysia suffered in 1997 and 1998 when worried foreign investors rapidly withdrew their capital.

Malaysia effectively pulled out of the orbit of globalization in 1998 when it took drastic actions to protect itself from the whims of foreign investment. Malaysian politicians blamed the International Monetary Fund and other institutions in the West for Asia's financial troubles, and these institutions in turn condemned Malaysia for retreating from the principles of the global free market. Despite the ominous downturn, by the end of the decade, Malaysia was once again registering economic growth—particularly as exports of computer chips and other electronics resumed—and was pulled out of the abyss in which neighboring Indonesia was still mired.

47, 344

7.2.7 Indonesia and Timor-Leste

Indonesia (•Figure 7.2.19) is by far the largest and most populous country of the region (population 231 million) and is the fourth most populous country on earth (after China, India, and the United States). Stretching across 3,000 miles (4,800 km), its 13,600 islands comprise a land area about three times the size of Texas. Indonesia's land represents over 40 percent of Southeast Asia's land and is home to the same percentage of its population, due primarily to the dense concentration of people on the island of Java, whose 125 million people represent about 54 percent of Indonesia's population. The island's population density is about 2,500 per square mile (965/sq km). In contrast, the remainder of Indonesia, which is about 13 times larger than Java in land area, has an average of only about 150 people per square mile (58/sq km).

This extraordinary concentration of population on this island is in part a reflection of the superior fertility of Java's volcanic soils (•Figure 7.2.20). Other Indonesian islands (with the notable exception of Borneo) have areas of fertile volcanic soil but never developed the extremely high population densities found on Java. Unfriendly coastlines, coral reefs, cliffs, and extensive wetlands have hampered the economic development of some islands.

Java's extreme population density also has cultural and historical roots in the concentration of Dutch colonial activities there. The Dutch East India Company, after lengthy hostilities with Portuguese and English rivals and with indigenous states, secured control of most of Java in the 18th century. Large sections of the other islands, however, were not brought under colonial control until the 19th and early 20th centuries. The Netherlands exploited the natural

JULIO ETCHART/Peter Arnold

• **Figure 7.2.19** Volcanic landscape in the region of Mt. Bromo, eastern Java, Indonesia.

● **Figure 7.2.20** Principal features of Indonesia and Timor-Leste.

wealth of Java with the introduction of the culture system in 1830 (see Chapter 3, page 57). Under this system, Dutch colonizers forced Javanese farmers to contribute land and labor for the production of export crops. From the Dutch point of view, this harsh system was successful. The culture system was abolished in 1870, but commercial agriculture continued to expand, and its success supported increasing numbers of people. The introduction of the culture system thus coincided with an enormous increase in Java's population, which is more than 20 times larger than it was a century and a half ago.

Java's historical leadership and the island of Sumatra's current supremacy in export production are seen in the distribution of Indonesia's larger cities. Of the 15 cities having estimated metropolitan populations of more than 1 million, 9 are on Java. These include Jakarta, the country's capital, its main seaport, and the largest metropolis in Southeast Asia (population 18.2 million), and Indonesia's second-largest city, Bandung. Of the remaining Indonesian cities with more than a million people, four are on Sumatra.

The Indonesian state has had a turbulent career. Increasing nationalism during the Japanese occupation of World War II led to a bitter struggle for independence from the Netherlands following the war. Indonesia finally won independence in 1949. Another confrontation with the Netherlands in 1962 took western New Guinea (West Irian) out of Dutch control. A United Nations plebiscite in 1965 ratified Indonesian control, and relatively friendly relations were established between Indonesia and the Netherlands.

In 1965, an attempted Communist coup against the long-standing regime of Indonesian President Sukarno was unsuccessful. The ensuing retaliation resulted in the massacre of about 300,000 Communists and their supporters by the Indonesian army and by Islamic and nationalist groups.

From 1968 to 1998, Indonesia's president was an army general named Suharto, whose regime was regularly criticized for human rights abuses. Suharto ruled Indonesia undemocratically and often repressively but argued that he was a benevolent figure acting on behalf of his people. Indonesians generally tolerated his authority because the country's economy was advancing rapidly. The economic boom of the 1980s and early 1990s centered on the oil industry (in the hands of the state firm Pertamina) and on food production. Oilfields located mainly on Sumatra and Kalimantan now produce about 1.3 percent of the world's oil output. Crude oil and liquefied natural gas account for about 25 percent of Indonesia's exports and have tied its economy very closely to that of Japan, which is the main market for these products. Indonesia is a member of OPEC, the Organization of Petroleum Exporting Countries. In addition to oil and gas, important contributions to the export trade include clothing, wood products, various tropical agricultural products, and increasingly, electronics.

Tourism has also brought welcome foreign exchange earnings to Indonesia. The country's principal attraction by far is the legendary island of Bali, home of Indonesia's only remnant of the India-born Hindu culture that permeated the islands from the 4th through 16th centuries. Bali

228

relentless pressure and averting a potential bloodbath, Suharto resigned from office in May 1998. He was succeeded by his vice president, who in turn stepped down in 1999 following widespread protests.

Indonesia's first truly democratic elections in 40 years subsequently brought to power President Abdurrahman Wahid, a widely respected Islamic scholar. His honeymoon with the public was short-lived, however, and by 2000, an official investigation revealed that he had embezzled millions of dollars of government funds. Wahid denied the allegations and insisted he would finish his term of office in 2004. He did not. Against a backdrop of massive protests across Indonesia, the president was impeached by Indonesia's supreme court, and he fled the country in 2001. He was replaced in office by Megawati Sukarnoputri, daughter of former President Sukarno, who was in turn ousted by the election of the secularist general Susilo Yudhoyono in 2004.

All of these regime changes reveal that Indonesia faces some huge problems. There is a constant threat that this vast country might splinter along ethnic lines (see Problem Landscape, pages 360–361). Many observers believe that if Indonesia's secular leadership does not deliver economic improvement to the tens of millions of disadvantaged citizens, the country will turn to Islamic rule. Rapid population growth is a serious challenge to the prospects for development. Indonesia's population grew by 90 million, or more than 63 percent, in the three decades between 1977 and 2007.

Lacking an aggressive family planning program, the government instead compelled some of its "surplus" population to move from Java, Bali, and other densely populated islands to some of the sparsely inhabited outer islands of the vast Indonesian archipelago. Begun under Dutch colonial rule, the **Transmigration Program** was one of the word's largest, most costly, and most controversial resettlement efforts. The Indonesian government picked up the tab for moving more than 6 million people, most of them landless agricultural workers and their families, to new islands

is world-famous for its unique culture and forms of art and dance, which have endured and become even richer despite a massive influx of international tourism to this legendary island (●Figure 7.2.21). Bali's popularity as a destination for Western tourists put it in the crosshairs of the al-Qa'ida-affiliated Jemaah Islamiah, which in 2002 carried out twin bombings of the nightclub district of Kuta. More than 200 people were killed, most of them young Australians; the Bali bombings have been called **"Australia's September 11."**

Indonesia's economy collapsed under the force of the spreading Asian economic crisis of the late 1990s. The rupiah, Indonesia's currency, lost 80 percent of its value in less than a year. The Suharto government removed expensive subsidies on gasoline and kerosene. Resulting protests quickly grew into vast riots. The economic troubles were compounded by a prolonged drought attributed to El Niño's warming of the Pacific's waters, which caused significant declines in the country's rice harvest. Skyrocketing prices of rice and widespread shortages of the grain helped fuel widening popular discontent, which focused on two targets. First were the ethnic Chinese, who represent 3 percent of Indonesia's population but control half the economy, and second was President Suharto. Rioters looted and destroyed hundreds of Chinese-owned shops in Indonesia's major cities, forcing thousands of ethnic Chinese to flee the country. Since the Chinese had dominated the food production business in Indonesia, food shortages intensified in the wake of the disturbances. Protesters and rioters called ever more loudly for the resignation of President Suharto, on whom they also blamed their economic troubles. Bowing to the

● **Figure 7.2.21** The cultural arts of Bali's Hindu people are world-renowned, making this Indonesian island a popular tourist destination.

PROBLEM LANDSCAPE

The Balkanization of Indonesia?

The national credo of Indonesia, seen on banners and placards throughout the country, is "One country. One people. One language." The government's constitution, in an effort to promote national unity, officially recognizes four faiths: Islam, Christianity, Hinduism, and Buddhism. But slogans and laws belie a grim reality: Indonesia may be in a process of balkanization, fragmenting into ethnically based, contentious units as happened in the Balkans of southeastern Europe. The presence of some 300 different ethnic groups, combined with religious frictions, physical fragmentation, and economic problems, has made it difficult to attain peace, order, and unity in Indonesia. There is an official national language (Malay, known locally as Bahasa Indonesian), but more than 200 languages and dialects are in use. The largest ethnic group is the Javanese, who make up about 41 percent of Indonesia's population. About 86 percent of all Indonesians are Muslim, about 9 percent are Christian, and 2 percent are Hindu, living mainly in Bali.

Various groups in the outer islands have resented the dominance of the Javanese, who have traditionally asserted their power in colonialist fashion over the larger, more resource-wealthy outer islands. Such animosities have escalated at times to armed insurrections. The Suharto regime countered these militant expressions with a state ideology called **Pancasila.** Aimed mainly at suppressing militant Muslim aspirations, *Pancasila* was a pan-Indonesian nationalist ideology designed to neutralize all ethnic identities.

For a time, Indonesia's government vowed to break with the repressive ways of Suharto and appeased would-be separatists with promises of **liberal autonomy,** meaning that the provinces would have greater control over local administration and take more profits from the sales of local natural resources. With growing economic problems, pressure from nationalists, and accusations of corruption leveled against him, however, the government shifted to using an iron fist against every Indonesian province that might aspire to follow the path the former enclave of East Timor took.

Now the independent country of Timor-Leste (population 800,000; capital, Dili), East Timor was a Portuguese possession that Indonesia occupied after the collapse of the Portuguese colonial empire in the 1970s. The government carried on a long and bitter struggle against an independence movement led by East Timor's Catholics. In 1998, Indonesia and Portugal reached an autonomy agreement that would give the Timorese the right to local self-government.

Indonesia's incoming Wahid administration was bolder, however, and in 1999 allowed the people of East Timor to vote on whether or not they wanted outright independence. The result was overwhelmingly in favor of independence.

Despite official Indonesian blessing of the result, progovernment militias reacted to the vote with a violent rampage that left 1,000 Timorese dead and tens of thousands homeless. The United Nations dispatched a peacekeeping force to oversee the complete transition to East Timor's full independence in 2002. The country's first president was José Alexandre Gusmao, a Timorese rebel leader whom Indonesia had imprisoned in 1992 and released in 1999.

Since its birth, this tiny young country has had many woes: a breakdown of law and order, interethnic violence, and growing unemployment. The economic future of this poor nation could be brightened by resolution of an international legal wrangle over resources. Between Timor-Leste and Australia, in the portion of the Timor Sea known as the Timor Gap, lies a huge field of petroleum and natural gas. Because the distance between Timor-Leste and Australia is less than 400 miles (640 km)—making it impossible to apply the international standard of a 200-mile (320-km) offshore territorial limit—the two countries must come to an agreement on how to divide these undersea riches. Any deal is likely to yield significant wealth to Timor-Leste.

Of outstanding concern for Indonesia's unity is the province of Aceh (population 4.3 million), at the northernmost tip of Sumatra. In 1976, the inhabitants of Aceh, who are a predominantly Muslim people of Malayan ethnicity, began seeking independence from Indonesia. The central government had made them a promise of autonomy with Indonesia's 1949 independence from the Netherlands but failed to keep it. The Acehnese insisted that the Jakarta government is too secular for their Muslim tastes and that too little of the revenue from sales of Aceh's abundant natural gas reserves remained in the province (which also has a wealth of oil, gold, rubber, and timber).

where they received houses, farmland, and subsistence payments. The results generally fell short of expectations, however. Java's population and poverty levels remained high, and the settlers' crop yields were low, reflecting lack of experience in the new lands, shortages of family labor, and difficulties in marketing crops. The new farming was also a major contributor to Indonesia's deforestation. The World Bank supported the effort but withdrew in the 1990s when human rights groups criticized the program's compulsory nature, violence flared between the resettled Javans and the indigenous peoples of the outer islands, and Jakarta beefed up its control over the outer islands, reducing the influence of indigenous peoples. Indonesia abandoned the Transmi-

gration Program in 2000, and a much smaller—and this time, voluntary—resettlement program has taken its place.

7.2.8 The Philippines

The Philippine archipelago includes over 7,000 generally mountainous islands. The two largest islands, Luzon and Mindanao, make up two-thirds of the total area.

The Philippines were a Spanish colonial possession governed from Mexico from the late 16th century until 1898, when the United States achieved victory in the Spanish-American War and gained control of the Philippines. With

PROBLEM LANDSCAPE

The Balkanization of Indonesia? *continued*

The secessionists' main voice was the **Free Aceh Movement** (known by the acronym **GAM**), which wanted to install a member of its own indigenous royal family as president of a free Aceh. Seeing Aceh as vital to the nation's unity and economic viability and fearful that Aceh might inspire separatism elsewhere in the country, the government refused to discuss independence for the province. Violent clashes persisted, and the death toll grew—to 12,000 between 1976 and 2004. But then an unlikely arbiter appeared. The tsunami of December 2004 concentrated its unspeakable sorrow on Aceh. The international community stepped in with a massive relief effort, and Indonesian authorities listened with unprecedented interest to Acehnese concerns. GAM dropped its demands for independence, and Indonesia agreed to allow GAM to participate in local elections. Nominally secular Indonesia allowed Aceh to become the only one of its 33 provinces to adopt Islamic *sharia* law. Redefining itself as the Aceh Transitional Committee, GAM laid down its weapons, and Indonesia scaled back its military presence in Aceh. The government gave amnesty to the rebels and, critically, pledged that 70 percent of the revenues from Aceh's natural resources would return to the provincial government.

Progress in Aceh raised hopes for a similar future for Papua, at the extreme eastern end of the Indonesian archipelago. Known formerly as Irian Jaya and West Irian, Papua is home to 2.7 million people of 200 different tribes and speaking 100 different languages. Most are of Melanesian origin, and most are at least nominally Christian. Collectively known as **Papuans,** they have little in common with the Javanese Muslims who control them from 2,500 miles (4,000 km) away in Jakarta. Their homeland on the western half of the island of New Guinea is of great importance to the economies of Indonesia and the United States because it contains the world's largest copper and gold mines. The American company Freeport-McMoRan, which provides more taxes to Indonesia's treasury than any other foreign source, operates these mines. There are also large assets of oil, natural gas, and timber. Indonesia has never recognized the province's 1961 unilateral declaration of independence. It is little wonder that Jakarta refuses to let its poorest but most resource-rich province go; the Dutch,

coveting this wealth, originally refused to relinquish the area when it granted independence to Indonesia. Pressure from the United States led to a UN-sponsored plebiscite on independence in 1969. The vote in favor of the region's becoming part of Indonesia was widely regarded as rigged. Recently, Indonesia has promised some autonomy and a greater share of resource revenues to Papua, but Papuans complain that these promises are unfulfilled.

West of Papua lie North Maluku and Maluku, once known as the fabled Spice Islands of the Moluccas. Islamists there, who are bent on establishing *sharia* in as many locales as possible in Indonesia, have targeted the islands' majority Christian inhabitants. Over a period of three years, until a fragile peace was achieved in 2002, these Muslim fighters—often aided by regular Indonesian government troops—ethnically cleansed Christians from the North Maluku capital. Thousands of people became refugees. Humanitarian crises have become a familiar product of Indonesia's unrest. Scores of overcrowded refugee camps have sprung up in more than half of the country's 33 provinces.

Events in the Malukus reflect a broader wave of Islamist activism across Indonesia. Advocates of *sharia* have succeeded in closing bars and discos across Java and have terrorized people they consider hostile to Islam. They have scapegoated Christians, who are generally wealthier than Indonesia's Muslims, for many of the country's economic and political problems. Jakarta keeps a nervous eye on the central part of the island of Sulawesi, where tensions between Christians and Muslims are high and where the militant Jemaah Islamiah is known to be active.

Will Timor-Leste's independence be followed by independence for Aceh, heralding the balkanization of Indonesia? Some argue that Timor-Leste is unique; it was the only province of Indonesia colonized by the Portuguese rather than the Dutch and is the only one forcibly brought into becoming part of Indonesia. Others point out that Indonesia itself is an artificial union inherited by Dutch colonialism and is bound for devolution, the dispersal of political power and autonomy to smaller spatial units. The geopolitical concerns of larger powers may also influence the fate of some Indonesian provinces (see that discussion on page 305 in Chapter 7).

---

the exception of the Japanese occupation of 1942 to 1944, the Philippines were governed by the United States from 1898 until 1946, when the country became independent.

Spanish missionary activity in the Philippines made it the only Christian nation in Asia. The country's Spanish legacy is still important. Ever since its founding in 1571, the Spanish-oriented capital of Manila (population 18.5 million) has been the major metropolis of the islands (•Figure 7.2.22). The society created by Spain was composed of a small upper class of Hispanicized Filipino landowners and a great mass of landless peasants. Problems created by this uneven distribution of agricultural land have persisted in the Philippines.

American economic, military, and other ties with the Philippines continued after independence. Preferential treatment in the U.S. market stimulated the growth of major Philippine export industries: coconut products, sugarcane, and abaca (a fiber used in making rope and fabrics). Education in English grew so that even today, English is spoken by about 50 percent of the people and serves as a bridge between many of the diverse linguistic groups in the nation. After independence, Tagalog, one of the most widely used indigenous Austronesian languages, became the principal base for the official national language, Filipino, spoken by 40 percent of the populace. English also remains an official language. Despite the colonial history, very few people

● **Figure 7.2.22** Downtown Manila.

speak Spanish. This owes largely to the fact that Spanish missionary work in the Philippines was carried on in the local languages rather than in Spanish.

The Philippine economy has expanded considerably since independence. Until the 1970s, the most striking growth was in agricultural exports such as coconut products, sugar, bananas, and pineapples and in copper and other metal exports. More recently, there has been growth in manufacturing and food production. Labor-intensive manufacturing, such as that of electronic devices and clothing, now accounts for about 80 percent of all exports. The English-language skills of Filipinos have made the country an attractive outsourcing center for business processing and Web site building and editing. Japanese and American capital, together with American agricultural science and technology, has been very important in the Philippines' economic development. The United States and Japan buy more than half of all Philippine exports. Inexpensive, skilled labor is another of the country's principal exports; about 7 million Filipino workers—one of every seven—is an **overseas Filipino worker**, or **"OFW."** About one-fourth of the world's seafarers are from the Philippines. In modern Greek, the word for "maid" is *Filipenza*. Filipinos swell the ranks of workers in Taiwan, Japan, Hong Kong, the United Arab Emirates, and Saudi Arabia. From these and other countries, where they often complain of discrimination and abuse, they pump $15 billion of remittances back into their country's economy each year. This represents about 14 percent of the country's gross domestic product.

Despite encouraging progress in recent decades, the Philippine republic remains poor and potentially explosive politically. It has not joined the ranks of the Asian Tigers, although some economists consider it a Tiger Cub. The country has taken a conservative approach to its development and has borrowed money carefully to invest in modest projects, avoiding the aircraft industry, the building of a new national capital, and other such showy and expensive enterprises undertaken in other Southeast Asian nations in the 1990s. Since the mid-1970s, the Philippines has also adhered closely to the terms of loans and financial reforms imposed by the International Monetary Fund. The country did not boom, but neither did it bust in the economic crisis that rolled across Asia in 1997 and 1998. During the regional economic recovery that followed, however, the Philippines' growth continued to lag. An estimated 30 percent of the population lived below the poverty line, and unemployment stood at 9 percent in 2007.

Economic expansion in the Philippines has been hard pressed to stay ahead of the more than 70 percent growth in population since 1975. The predominant Roman Catholic culture (Catholics make up 81 percent of the population, with 10 percent adhering to other Christian denominations) encourages large families. The country's longtime leading cleric, Cardinal Sin, responded this way to a question about the population explosion: "The more the merrier." Church officials have labeled the government's current promotion of family planning as "demographic imperialism" masterminded by the United States. Unsubsidized condoms and other forms of birth control are too expensive for many to afford in a country where 17 percent of the population makes less than $1 per day. The collective result is that the Philippines has a rather high annual population growth rate of 2.1 percent. This is a marked contrast with Thailand, with its growth rate of 0.7 percent. Both countries had about the same population size in 1975, but the Philippines now has about 23 million more people than Thailand.

Compounding the problem of population growth, land and wealth are very unevenly distributed as Philippine society continues to be dominated by a few hundred wealthy families. The gap between haves and have-nots has inspired rebellion for decades. Discontented peasants supported a Communist-led revolt after World War II. It was eventually suppressed, in large part through granting land (generally on sparsely populated Mindanao) to surrendered rebels. Revolts flared up again in the late 1960s, however, and have continued ever since. A Communist group known as the **New People's Army** leads an active insurgency in many parts of the country, frequently targeting military patrols and civilian infrastructure.

On the island of Mindanao, rebels of the **Moro Islamic Liberation Front (MILF)** and a splinter organization, **Abu Sayyaf,** have long fought the government in an effort to establish a separate Islamic state. In 2001, the MILF and the government reached a cease-fire, but Abu Sayyaf accelerated its campaign, employing terrorist tactics such as kidnapping and beheading civilians. In its declared worldwide war on terrorism, the United States in 2002 began providing military advisers and weaponry to Philippine troops fighting Abu Sayyaf.

This unrest has political and religious as well as economic roots. Despite a wealth of oil, natural gas, fisheries, and forests, Mindanao is the Philippines' poorest region, but the predominantly Christian central government invests little development aid in this mainly Muslim area. Hundreds of thousands of Mindanao's Muslims have fled as illegal immigrants to more prosperous Sabah, in the Malaysian portion of the island of Borneo.

The Philippines has seen some momentous political changes in recent years. In 1986, the country's dictator, Ferdinand Marcos, was ousted in a popular uprising and replaced by a democratically elected president, Corazón Aquino. Marcos and his wife, Imelda, had looted the country of billions of dollars over a period of 20 years and headed a corrupt and abusive regime virtually ignorant of the needs of the ordinary Filipino. The Marcos path seemed to be followed by Philippine President Joseph Estrada, who allegedly enriched himself in the 1990s through bribes, kickbacks, and embezzlement of public funds. Impeachment proceedings against Estrada crawled along, but he resisted their inevitable conclusion. Only massive street protests succeeded in forcing his resignation in 2001, when Gloria Macapagal-Arroyo replaced him as president.

The United States has maintained a concerned watch on the country's struggles. During the 1980s, many Filipino legislators resisted the prospect of renewing the leases under which the United States held the two military bases of Subic Bay and Clark Field on Luzon. Then, in 1991, the eruption of volcanic Mount Pinatubo put an end to the debate over the fate of these two American outposts; volcanic ash forced their closure. The Philippines has 22 active volcanoes. With volcanic eruptions, plus an average of 20 typhoon strikes yearly and regular monsoonal flooding, the Philippines earned the distinction of having more natural disasters in the 20th century than any other country. Mount Pinatubo also ejected enough ash high into the earth's atmosphere to cool temperatures globally over the following several years, at least temporarily confounding efforts to track the trend in global warming. Meanwhile, the U.S. base at Subic Bay was converted into a thriving free-trade port that the Filipinos hope will become Asia's "New Hong Kong."

The Philippines is Southeast Asia in a nutshell: There are humanized landscapes never free of nature's whims; struggles in transforming a mercantile colonial economy to one based on high-tech, global trade; seemingly intractable odds to overcome in unifying diverse peoples under one flag; and wars on poverty that have yet to be fought.

SUMMARY

→ Southeast Asia is composed politically of 11 countries: Myanmar, Thailand, Laos, Cambodia, Vietnam, Malaysia, Singapore, Indonesia, Timor-Leste, Brunei, and the Philippines. With the exception of Thailand, all of these were formerly colonies of foreign powers.

→ Southeast Asia contributed several domesticated plants and animals (including rice and chickens) and the achievements of civilization to a wider world.

→ Relative to other parts of Asia, especially China and India, Southeast Asia has historically not had large populations or population densities. The most notable exception is on the Indonesian island of Java, where good soils and Dutch colonial polices produced a very dense population.

→ This region poses numerous environmental challenges for its inhabitants, including tropical climates, poor soils, disease, long dry seasons, deforestation, heavy rains and erosion, and tectonic processes. A megathrust earthquake off the northeastern coast of Sumatra caused the great tsunami of 2004, which killed more than 200,000 people in 14 countries.

→ There are more Southeast Asian villagers than urbanites. Farmers use shifting cultivation in the more marginal areas and sedentary cultivation in more productive regions.

→ Plantation cash crops such as oil palm and rubber are grown for export and are the legacy of Western colonialism.

→ Southeast Asia's tropical forests are being destroyed at a faster rate than others of the world, mostly for commercial logging for Japanese markets. In biodiversity terms, Indonesia is a mega-diversity country whose forests are under intense pressure.

→ Southeast Asia as a whole is among the world's poorer regions. The poorest countries—Myanmar, Cambodia, Laos, and Vietnam—have suffered from warfare and, in the case of Myanmar, from unwise political management. Peace has finally come to Cambodia, Laos, and Vietnam. Singapore, Malaysia, and Thailand, because of their high rates of industrial productivity and economic growth, have often been described as Asian Tigers or Tiger Cubs.

→ The region's mineral wealth includes oil, gas, and a wide variety of metal-bearing ores. Indonesia has the most oil.

→ Economic problems in Thailand in 1997 led to an economic meltdown in the region due to the interconnectedness of the regional and global economies. Most of the countries have since recovered from that setback, and several have experienced strong economic growth.

→ Drug production, drug abuse, and HIV/AIDS are serious problems in Myanmar. Neighboring Thailand has some of these problems but has dealt more effectively with them.

→ Prodemocracy efforts in Myanmar have been thwarted by official repression.

→ Thailand was never colonized and is more prosperous than its neighbors. French-colonized Cambodia, Laos, and Vietnam are much poorer. U.S. forces succeeded the French withdrawal from Vietnam in the 1950s. American efforts to win a war against communism in Vietnam failed. Normal relations now exist between the United States and Vietnam, and although Communist, Vietnam encourages private enterprise and has seen recent strong economic growth.

→ Numerous dams are being constructed on the Mekong River. These will produce hydroelectricity but will damage the productive fishery of this basin.

→ Cambodia has recovered politically from the devastating days of the killing fields and has joined or rejoined significant international organizations, including the United Nations. Landlocked Laos is very poor but hopes to benefit from hydropower and mineral exports.

→ Malaysia and Singapore are former British colonies with rather strong economies based on high-tech industries. Singapore has a large and prosperous Chinese population and is developing a new niche in biotechnology.

→ The region's demographic giant is Indonesia, which is very complex culturally and has had a turbulent political past.

→ The Philippines is a mainly Catholic country and a former colony of Spain. Population growth there is relatively high, and a gap between rich and poor threatens overall development. Large numbers of Filipinos work abroad.

→ Much unrest in the region is related to tensions between ethnic groups. Following the regional economic crisis of 1997–1998, frustrations in Indonesia were directed against the prosperous Chinese minority. There are insurgents among the Muslim, ethnic Malay minority of southern Thailand. Indonesia is in danger of fragmentation. Catholic Timor-Leste has gained impendence from Indonesia, and the provinces of Aceh and West Irian are seeking self-rule or independence. Muslim insurgents in the southern Philippines and Communist rebels throughout the country have stepped up their campaigns against the government.

KEY TERMS + CONCEPTS

Abu Sayyaf (p. 362)
Amerasians (p. 351)
ASEAN+3 (p. 345)
Association of Southeast Asian Nations (ASEAN) (p. 345)
 "Australia's September 11" (p. 359)
Biopolis (p. 356)
boat people (p. 351)
Bumiputra (p. 356)
"Burmese Way to Socialism" (p. 345)
contingent sovereignty (p. 345)
crony capitalism (p. 344)
"denying the countryside to the enemy" (p. 350)
doi moi (p. 345)
economic depression (p. 344)

economic recession (p. 344)
endemic species (p. 342)
expanding pyramid (p. 351)
Free Aceh Movement (GAM) (p. 361)
Golden Triangle (p. 345)
hill tribes (p. 349)
Khmer Rouge (p. 352)
killing fields (p. 352)
"Lao Way" (p. 354)
liberal autonomy (p. 360)
megadiversity country (p. 342)
Mekong River Commission (p. 353)
Montagnards (p. 349)
Moro Islamic Liberation Front (MILF) (p. 362)
New People's Army (p. 362)

(Overseas Filipino Worker (OFW) (p. 362)
Pancasila p. 360)
Papuans (p. 361)
Parti Islam se-Malaysia (PAS) (p. 356)
plantation (p. 339)
State Law and Order Restoration Council (SLORC) (p. 345)
Transmigration Program (p. 360)
typhoon (p. 336)
unexploded ordnance (UXO) (p. 354)
"Venice of the East" (p. 348)
Viet Cong (p. 350)
Wallace's Line (p. 342)

REVIEW QUESTIONS

1. What are the countries of Southeast Asia? What are their major physiographic features?

2. What caused the great tsunami of 2004, and what were some of its effects?

3. What generalizations can be made about population numbers and densities in this region? What factors account for these characteristics?

4. What agricultural practices are typical of Southeast Asia? What are the main subsistence and commercial crops?

5. What countries are the largest producers and consumers of Southeast Asian tropical hardwoods?

6. What is the region's most important mineral resource? Which countries have most of it?

7. What are the major goals of ASEAN and ASEAN+3?

8. What major changes are coming to the Mekong River basin?

9. What political events have hampered development in Vietnam, Cambodia, and Laos?

10. Who are the boat people? Why did they leave their homelands? Where are they now?

11. How did Malaysia transform itself from a tin and rubber exporter to a high-tech economy?

12. What are the major political and resource-related issues faced by Indonesia and the Philippines?

NOTE

1. See UNESCO, *Assessment of Capacity Building Requirements for an Effective and Durable Tsunami Warning and Mitigation System in the Indian Ocean: Consolidated Re-* *port for the Countries Affected by the 26 December 2004 Tsunami* (Paris: UNESCO, 2005), http://unesdoc.unesco.org/images/0014/001445/144508e.pdf. Accessed July 15, 2007.

Schoolchildren in Guilin, China.

MODULE 7.3
CHINA:
THE GIANT

Historically, China identified itself as the **Middle Kingdom,** positioned centrally in the known world. Chinese culture is very rich, very old, and very influential. China has been home of the dominant culture of East Asia since ancient times. Nearby countries and regions carry the profound influences of Chinese culture in their languages and writing, religion, agriculture, arts, crafts, institutions, ways of thinking, and histories. The culture hearth of China has contributed much to humanity at large. Pioneering Chinese inventions and discoveries included gunpowder, paper, the wheelbarrow, and the magnetic compass. Over thousands of years, Chinese civilizations have also accomplished great engineering feats, from the Great Wall (which is, incidentally, *not*, as one urban myth would have it, the only human-made object visible from space) to the Three Gorges Dam.

■ module
objectives

This module should enable you to

→ Recognize China as a land empire in which a single ethnic group added vast peripheral areas to its eastern core

→ Appreciate the constraints imposed by China's Communist system on personal, political, and economic freedoms and the recent substantial easing of those restrictions

→ Balance the pros and cons of China's hugely ambitious plans to transform its natural environment through dams, canals, forest restoration, and other projects

→ Consider the economic discrepancies between China's rural and urban populations and the forces behind large-scale rural-to-urban migration

→ Understand the regional and geopolitical risks of Taiwan's aspirations for an identity independent of China

From the earliest dynasties in the second millennium B.C.E. to the present, the story of China has been one of manipulation of the landscape to make it more productive. There are ever-present challenges from nature, including floods and droughts. Today, while trying to feed its enormous population and raise fiber for its huge textile sector, China is pulling away from its past as a monumental agricultural nation and is aggressively pursuing industrial development to reach its goal of becoming a "well-off society" by 2040. With China's export-driven economy booming, its 2.5 million–soldier army backed by nuclear weapons and by military spending second only to the United States, and its political influence represented with a permanent place on the UN Security Council, even the world's sole superpower, the United States, must weigh most of its geopolitical decisions mindful of this world giant. Above all, China—officially the People's Republic of China (PRC)—must be recognized as the world's most populous country, home to 1.3 billion people, or one in every five on earth. China is arguably the most important country on the planet today.

After a brief sketch of China's history, this module examines, by subregion, the physical resources for its people and some of their culture and livelihood traits. It discusses Chinese agriculture, industries, and cities. The module ends with depictions of Taiwan and Mongolia, countries whose histories and modern interests are closely linked to China.

7.3.1 Accomplishment, Subjugation, and Revolution

China (•Figure 7.3.1) was already a major political entity by the time of the Shang Dynasty (c. 1523–1027 B.C.E.). The Chinese had by then developed an agricultural system able to support large numbers of people on small areas of intensely worked land. During the Shang Dynasty, Chinese peasants occupied the great alluvial plain of the Huang He (Yellow River) in North China. There they had a garden type of agriculture based on the hoe, irrigated crops, fertilizers, intensive use of flat alluvial lands, and the terracing of fertile hillsides. From its origins in the Huang He Basin, this system diffused by migration and conquest throughout the eastern part of the country and farther afield over a period of many centuries. The Huang He region is therefore recognized as core of the Chinese culture hearth.

In the 3rd century B.C.E., the Qin (Ch'in)* Dynasty accomplished the first unification of the Chinese people into one empire (the country's Western appellation, China, is thought to be derived from the name of this dynasty, as is the combining form *Sino-*; the Chinese call their country Zhongguo). Thus began China's development as a land empire that colonized its own vast hinterland. The conquerors were, as the dominant culture in China today is, ethnic Han Chinese. Under the Han Dynasty (206 B.C.E.–220 C.E.), China's influence spread as its leaders extended their authority through direct conquest and by accepting the fealty and tribute of less significant powers around them. Chinese armies pushed far westward across central Asia and nearly made contact with the Romans in the vicinity of the Caspian Sea. Chinese ideas and materials traveled west along caravan routes to the Middle East and Europe.

During times of internal disorder and weakness, China was conquered by nomadic invaders, notably the Mongols from the northwest and Manchus from the northeast. But China always rebounded. Under the Yuan (Mongol) Dynasty (1280–1367), the Chinese empire controlled Burma (modern Myanmar), Indochina, Korea, Manchuria, and Tibet. The leaders of China's last dynasty, the Manchu (Qing) of 1644–1911, were of the nomadic Jurchen tribes of northeastern China. They held large areas that eventually became part of the Russian empire.

164

293

*From the late-1800s until the 1970s, Western spellings and pronunciations of Chinese words were governed by a system known as the Wade-Giles romanization system. In the 1970s the Communist Chinese government deemed this system somewhat arbitrary and embraced the more accurate Pinyin system for representing Chinese names; since then Pinyin has been standard, but many reference materials, such as older books and maps, use the Wade-Giles spellings, which is why both are given in this text: Pinyin first, with Wade-Giles in parentheses.

China under the Manchus was increasingly subjected to unwelcome pressures from the West. European ships became frequent visitors to China in the 16th century, and Europeans began to press for more trade. The Chinese, who had high-quality cottons, silks, teas, and other products, were not much interested in what the West offered, and they felt culturally superior to these intrusive "barbarians." The Chinese government strictly controlled trade, limiting it to the single port of Guangzhou (Canton) in southern China.

As their science, technology, and industrialization advanced at home, Westerners became more powerful and assertive in Asia. A British expedition in the **Opium War** of 1839–1842 defeated the faltering Manchu government, forcing China to open its doors to opium (from British-controlled India) and to other British goods. Fully aware of the drug's destructive effects, Britain had carefully encouraged and fed the appetite for highly addictive opium in China, creating an enormous and very profitable market for itself.

Other European countries, Russia, and the United States joined the search for riches. China became a kind of joint colony of many competitive foreign exploiters, who got their way by occasional military actions and frequent threats. China was thus opened to Western products and then to Western-owned factories in its ports. Under the impact of Western machine-produced goods, China's handicrafts industries began to disintegrate. Foreigners in China were exempted from the control of Chinese law in an arrangement known as **extraterritoriality,** and the Western powers gradually gained control of China's taxation system.

The foreigners also extorted territorial concessions but imposed explicit colonial status on only a small part of China (see the map of colonial Asia on page 297). Britain secured the vital coastal port of Hong Kong. Portugal had earlier obtained nearby Macau. France and Germany won naval bases. Japan took Taiwan and the Ryukyu Islands, and Russia conquered parts of central Asia, Mongolia, and Siberia. **Treaty ports** were established, including Shanghai and Tianjin (Tientsin). In these ports, the foreign powers were granted special settlement areas under their own control. Chinese people could not normally reside in these areas and in some cases could not even enter them. European and American naval vessels patrolled China's coasts and great rivers to enforce foreign privileges.

The Chinese did not accept this situation passively. Disorder mounted within the country, but Western military power kept the submissive Manchu sovereigns on the throne until they were finally overthrown by the **Chinese Revolution of 1911,** led by Sun Yat-sen (sometimes called the "George Washington of China"). However, the new Republic of China had only nominal control over much of the country. Local warlords controlled many areas until in the late 1920s, the army of the **Nationalist Party** (**Kuomintang,** or **KMT**) under Chiang Kai-shek fought northward from Guangzhou and managed to impose control on most of the country.

The Nationalist regime, which established itself at Nanjing (Nanking), abolished some foreign privileges, but its strength was sapped by almost continuing civil war against the **Chinese Communist Party (CCP)** led by Mao Zedong (Mao Tse-tung) and then by war against Japan. Chiang finally drove the Communists from their bases in the hills of southeastern China in 1934. Thus began the two-year **Long March**—a more than 6,000-mile (9,660-km) flight on foot of nearly 100,000 Communists, pursued by Nationalist soldiers, from south central China north to Yenan, in Shaanxi province, which became their wartime home base.

The Sino-Japanese War of 1937–1945 compelled Mao's Communists and the Nationalists to fight in common cause against Japan. But with the defeat of Japan in World War II, the CCP and the Nationalists resumed their civil war, in which Mao's forces triumphed. In 1949, 2 million Chinese Nationalists under Chiang Kai-shek fled to Taiwan, and Mao and the CCP claimed victory on October 1, 1949. Subsequent historical events in China are discussed throughout this module, and Taiwan has its own section beginning on page 386.

7.3.2 The Setting

China's area of about 3.7 million square miles (9.4 million sq km) is only slightly larger than that of the lower 48 United States (see Figure 7.2, page 282), but its inhabitants outnumber the U.S. population by more than 4 to 1. Two main environmental regions make up this populous country: an eastern core region, known as **Humid China** or China Proper, in which the country's population, developed resources, and productive capacity are heavily concentrated, and a western periphery, or **Arid China** (see Figure 7.3.1). A rough boundary between the two extends from China's southern border with Myanmar to its northeastern border with Russia. This line corresponds generally to the stretch of land with average annual rainfall of about 20 inches (c. 50 cm), with Humid China to the east and Arid China to the west (see Figure 2.3, page 23, and Figure 7.5, page 287). Land uses of the regions are depicted in Figure 7.8 on page 288.

Arid China

Much of China's unproductive and thinly settled land is in the western half of the country. Here are high mountains and plateaus, together with arid and semiarid plains, where rainfall is generally insufficient for agriculture. This arid region, sometimes called Frontier China, contrasts markedly with the better-watered, more densely settled eastern half of the country and is home to only about 6 percent of China's population (see Figure 7.3a, page 282).

The principal regions of Arid China are the Tibetan Highlands, Xinjiang (Sinkiang), and Inner Mongolia (Nei Mongol). All three are identified historically with ethnic groups that are not Han Chinese (see Figure 7.11 and the ethnic geography discussion on pages 291–293). Now, after decades of politically motivated domestic migration from the east, Han Chinese comprise the great majority in Inner Mongolia and a slight majority in Xinjiang.

The thinly inhabited Tibetan Highlands, including the province of Tibet (Xizang) itself and fringes of adjoining provinces, occupy about one-fourth of China's area. Most of this vast area is a very high, barren, and mountainous plateau averaging nearly 3 miles (c. 5 km) in elevation. To the northwest, there are many high basins with internal drainages, some containing large salt lakes. To the southeast, the plateau is cut into ridge and canyon country by the upper courses of great rivers such as the Tsangpo (which becomes the Brahmaputra of India), the Mekong, and China's own Chang Jiang (Yangtze Kiang, or Yangtze River) and Huang He (Hwang Ho, or Yellow River).

Lower elevations, warmer temperatures, and greater precipitation in parts of the southeastern plateau support some extensive grasslands and stands of conifers. Around the edges of the Tibetan Highlands are huge mountain ranges: the Himalayas on the south, the Karakoram and Kunlun Shan ranges to the north and northwest, and the Qinling Shan and others to the east (•Figure 7.3.2).

Most of the 2.9 million people of the Tibetan Highlands are sedentary farmers and animal herders. A few hardy crops, especially barley and root crops, do well in this challenging environment. In grasslands at higher elevations, some pastoral nomads graze their flocks of yaks, sheep, and goats. Yaks are particularly important not only as durable beasts of burden in these high elevations but also as providers of meat, milk and butter, leather, and hair traditionally woven into cloth.

Xinjiang (meaning "New Dominion") adjoins the Tibetan Highlands on the north. It has an area of roughly 635,000 square miles (c. 1.6 million sq km) and a population of about 20 million. It consists of two great basins: the Tarim Basin to the south and the Dzungarian Basin to the north, separated by the lofty Tien Shan range. The Tarim Basin is rimmed to the south by the mountains bordering Tibet, and the Dzungarian Basin is enclosed on the north by the Altai Shan and other ranges along the southern borders of Russia and Mongolia. Both basins are arid and semiarid. The Tarim Basin is particularly dry because it is almost completely enclosed by high mountains that block rain-bearing winds. The smaller adjoining Turpan Depression drops to 928 feet (283 m) below sea level and is the second-lowest land spot on earth, after the Dead Sea. These basins include the greater part of the desert of Taklamakan, whose name means "Place of No Return," often rendered as "Once you get in, you'll never get out."

From the Taklamakan and Mongolia's Gobi Desert, and from adjacent areas despoiled by ill-conceived agricultural reclamation projects of the past, vast quantities of sand are airborne in spring sandstorms that darken skies and choke lungs downwind in Humid China and even Korea and Japan. In an effort to stem the desert's spread, the Chinese have since 1982 planted more than 40 billion jujube and

● **Figure 7.3.1** Principal features of China and Mongolia. On the inset map, the line represents the division between Humid China, or China Proper, to the east and Arid China, or Frontier China, to the west. Humid China has the great majority of the Chinese population and economic activity, and Arid China is home to many minority peoples.

● **Figure 7.3.2** At 24,757 feet (7546 m), Muztagh Ata is one of the highest mountains of the Kunlun Shan, at the northern edge of the Tibetan Plateau.

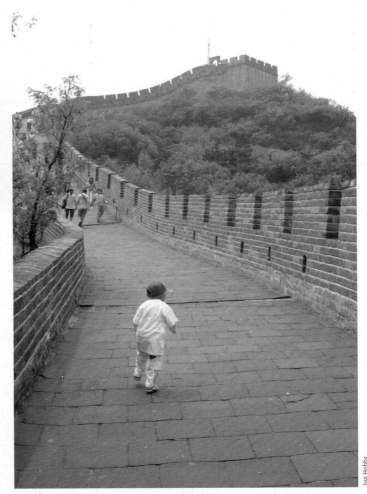

● **Figure 7.3.3** The Great Wall of China at Badaling, near Beijing. Most of the wall standing today dates to the Ming Dynasty of the 15th century. The wall proved ineffective in keeping out the Ming enemies to the north, the Qing.

other trees as "green walls" in Arid China. The government's goal for the decade 2010–2020 is to plant enough trees in this region to cover 170,000 square miles (440,000 sq km), an area about the size of California or Sweden. There are also plans use licorice plants to hold back the desert. Long valued in traditional Chinese medicines and therefore of potential economic value, licorice is drought-tolerant and helps bind sandy soils.

The great majority of Xinjiang's population is concentrated near oases located mainly around the edges of the basins where streams from the mountains enter the basin floors. For many centuries, these oases were stations on caravan routes crossing central Asia from Humid China toward the Middle East and Europe on the historic Silk Road (see Regional Perspective, page 371). The ancient routes have been superseded by modern transportation. A railroad connects Lanzhou (Lanchow; population 1.8 million), a major industrial center and supply base in northwestern China Proper, with Urumqi (Urumchi; population 1.5 million), the capital of Xinjiang. Major roads and airline routes link the oasis cities (see Figure 7.3.14).

These transport links are key elements in the latest drive to grow the economic importance of this remote part of China and bring it under firmer political control. Early in the Communist era of the mid-20th century, demobilized soldiers and urban youth from eastern China were organized to expand irrigated land on state-operated farms in this arid landscape, promoting cotton as a major crop.

Xinjiang is China's most mineral-rich region, with deposits of coal, petroleum, iron ore, and other minerals. Mining and manufacturing have expanded there. It continues to be a major focus in Chinese efforts to relocate significant numbers of ethnic Han Chinese away from the coasts (see Ethnic Geography, pages 372–373).

Northwest of the Great Wall, rolling uplands, barren mountains, and parched basins stretch into the arid interior of Asia. Here lie slightly more than 1 million square miles

(2.6 million sq km) of dry, thinly grassed terrain divided almost equally between Nei Mongol (Inner Mongolia), the area nearest the Great Wall, and the country of Mongolia—known as Outer Mongolia in older literature. Most of the 24 million people of Nei Mongol, which is an autonomous region of China, are ethnic Han Chinese concentrated in irrigated areas along and near the great bend of the Huang He. Much of the region is still home to a sparse population of ethnic Mongol herders. From these austere, arid landscapes, the Mongols emerged to control China and, by the 13th century, a huge swath from the Korean Peninsula in the east to the margins of Poland in the west. The desert landscape bears many traces of its rich past: caravan routes, temples hidden in desert caves along those trails, and remnants of the Great Wall (●Figure 7.3.3). The earliest version of the Great Wall was built in the 3rd century B.C.E., but the Ming Dynasty constructed it in earnest in the 15th century C.E. in an unsuccessful bid to stave off incursions from the north.

REGIONAL PERSPECTIVE
The Silk Road

In the early Han Dynasty (c. 206 B.C.E.–8 C.E.), the Chinese began a steady exchange of goods for items from the Mediterranean trading ports. For centuries to follow, China exported silk, china, paper, and spices, and the West exported wool, gold, silver, glass, and metalware. Not produced in the West until the 6th century, silk is made from the cocoon of a caterpillar known as a silkworm and is still prized for its breathability, texture, and beauty (•Figure 7.3.A). China is still the world's leading silk producer.

The overland route for this exchange of goods—called the Silk Road or Silk Route—was approximately 4,000 miles (c. 6,400 km) long, connecting the early Chinese capital of Chang'an (now Xi'an) to ports of the eastern Mediterranean (•Figure 7.3.B). The road went alongside the Great Wall and the Taklamakan Desert, over the Pamir Mountains, and through Baghdad to the Mediterranean Sea. From there, goods were shipped to various European ports. This route also played a role in the diffusion of Buddhism into China in the 1st century C.E. and subsequently in the spread of languages, customs, and technologies. Until the Venetian explorer and trader Marco Polo and his entourage did it in the 1270s, apparently no one had ever followed the whole route. Travel was dangerous, and every caravan load had to pass numerous checkpoints and pay heavy tolls.

• **Figure 7.3.A** Silk production is a laborious process, in which some steps must be done by human hands. Here, a Chinese worker removes each pupa (far left) from the silk cocoon (center), then stretches the silk over a frame to prepare it for drying and processing.

• **Figure 7.3.B** The Silk Road.

The traditional picture of nomadic Mongol tribespeople with their sheep, goats, camels, and horses is fast disappearing. The older economic core of Nei Mongol is made up of agricultural areas near the Huang He. The irrigated areas produce mainly oats and spring wheat, while unirrigated fields are used primarily for drought-resistant millet and kaoliang, a sorghum grain used for animal feed and for liquor. The main city is Baotou (Paotow; population 1.5 million), an expanding industrial center on the Huang. One of China's major iron and steel works was opened at Baotou with Soviet help in the 1950s to take advantage of nearby reserves of coal and iron ore. Nei Mongol's economy today is growing faster than that of any other Chinese province because of a boom in the production of coal, natural gas, copper, gold, and other natural resources.

Humid China

Humid China, sometimes referred to as Eastern China, Monsoon China, or China Proper—the core region—includes the most densely settled parts of the country (see Figure 7.3a, page 282). Humid China is commonly recognized in geographic literature as having two major subregions, North China and South China, which differ from each other in several physical, economic, and cultural respects. The Northeast, or Manchuria, is another subregion that straddles the line separating Humid China from Arid China, but it is important economically to the county and is therefore treated within the discussion of Humid China as the country's core region.

The Northeast

China's extreme Northeast region, better known to the West as Manchuria, lies north and east of the Great Wall. It is made up of three provinces: Liaoning, Jilin, and Heilongjiang. It has an area of about 310,000 square miles (c. 800,000 sq km) and a population of approximately 107 million. To the west are the Mongolian steppes and deserts; to the east, the area is separated from the Sea of Japan (East Sea) by the Korean Peninsula and by Russian terri-

ETHNIC GEOGRAPHY

Han Colonization of China's "Wild West"

China's growth as a land empire has involved both subjugation of people who are not ethnic Han and the colonization of those ethnic areas by ethnic Han. There are at least 56 non-Han ethnic groups in China, in total about 100 million people living mainly in the west and southwest of the country. As the Soviets did in the Soviet Union, China's Communist government granted at least token recognition of the distinctive identity and rights of five large minorities by creating **autonomous regions.** These are Guangxi, bordering Vietnam; Nei Mongol (Inner Mongolia) and adjacent, small Ningxia in the north; Xizang (Tibet); and Xinjiang in the far west (see Figure 7.3.1).

The Hui of Ningxia are Muslims of many different ethnicities who absorbed Chinese language and many Han customs. In contrast, the Uighurs of Xinjiang and the Tibetans have defiantly resisted becoming Chinese—and so Han colonization, or **Hanification,** focuses especially on their territories. The Chinese government aims either to assimilate these minority peoples into a broader, Han-based Chinese culture or to establish a strong enough Han demographic presence among them to dispel any hopes of autonomy or independence.

Chinese authorities have also initiated what they call the **Western Big Development Project** with the stated objective of improving locals' livelihoods enough to diminish their desire for ethnic and political separatism. Focused on the autonomous regions and four other provinces in which there are large minority populations, this project seeks to improve infrastructure like airports and roads and to boost output from factories and farms. The Uighurs, Tibetans, and other minorities generally fear that this "development" is just a cover for demographic, political, and economic domination by the Han.

Buddhism reached Tibet in the middle of the 8th century, and its arrival was followed by decades of conflict between Chinese and Indian Buddhists. The Chinese were defeated in this struggle and expelled from Tibet at the end of the 8th century. China sometimes reexerted loose control over Tibet, but Chinese author-

ity vanished again with the overthrow of the Qing (Manchu) Dynasty in 1911. From 1912 until the Chinese Communists' violent conquest of 1951, Tibet existed as an independent state, with its capital and main religious center at Lhasa (•Figure 7.3.C).

After the Chinese completed roads to Tibet in late 1954, an increase in restrictive measures by the Communist government contributed to the rise of Tibetan guerrilla warfare. This culminated in a large-scale Tibetan revolt in 1959 and the flight of the **Dalai Lama** (the spiritual and political leader of Lamaism, or Tibetan Buddhism) and many other refugees to India. Subsequently, the Chinese drove most of the monks from their monasteries, expropriated the large monastic landholdings, implemented socialist programs, and prohibited organized religion in the rest of China. Large-scale immigration by ethnic Han picked up; about half of Lhasa's population is now Chinese.

Beijing insists it is helping Tibet with major development investment, including the $4 billion Qinghai-Tibet railway recently constructed between Beijing and Lhasa. But Tibetan resistance to Chinese rule continues. From his place of exile in Dharamsala, India, and on frequent international road trips, the Dalai Lama continues to appeal for Tibetan freedom. He asks China to grant what he calls **genuine autonomy** to Tibet, meaning that Tibetans would have authority over most matters except foreign affairs and defense, which the Chinese government would control. His cause has many prominent backers, including the American actor Richard Gere. However, China's government has so far refused to negotiate with the Dalai Lama. For many years, India had promoted the cause of Tibetan autonomy, but as its trade ties with China have grown more vital, its advocacy for Tibet has quieted.

China's government regards Xinjiang's Muslims, mainly from the 8-million-strong Uighur ethnic group, who speak a Turkic language and are kin to the Kazakhs to the west, as the most problematic minority groups. Many Uighurs would like to pull at least a part of Xinjiang out of China's orbit to create an independent Chinese Muslim state. Two short-lived independent Uighur republics,

tory. The Northeast consists of a broad, rolling central plain surrounded by a frame of mountains rising to about 6,000 feet (c. 1,800 m). The mountains on the east and north have much valuable hardwood and softwood timber. The central plain, oriented northeast to southwest, is about 600 miles (966 km) long by 200 to 400 miles (322 to 644 km) wide. The soils are very fertile, and there is usually enough rainfall for summer crops. Conditions are not favorable for irrigated rice, and the winters are too severe for winter grains. But soybeans, kaoliang millet, corn, and spring wheat do well during the short frost-free season of 150 to 180 days.

This was once a sparsely populated frontier region, but political instability in the early 20th century prompted the

ruling Manchu Dynasty to encourage migration to the relatively open lands of the Northeast. Millions of Chinese farmers moved to these provinces, pioneered its agricultural frontier, and gave it an ethnic Han majority by the start of the Sino-Japanese War in 1937. As described on page 383, Japanese intervention initiated Manchuria's development as an important industrial region.

North China

Affected mightily by the monsoon wind patterns that characterize so much of Asia, North China has hot, humid, and rainy summers and cold, dry winters—traits typical of a continental climate. When the winter monsoon blows out of

ETHNIC GEOGRAPHY

Han Colonization of China's "Wild West" *continued*

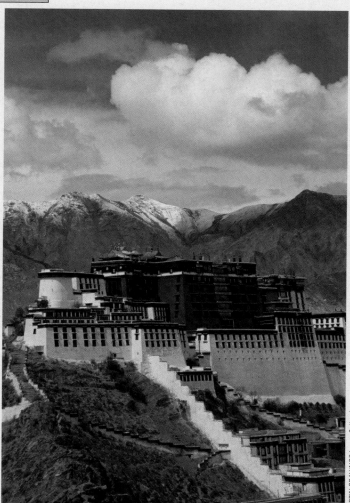

both known as the Eastern Turkestan Islamic Republic, existed in 1933 and again from 1944 to 1949, and some Uighur separatists are fighting for one again. Beyond execution of those separatists deemed to be terrorists, China's response has been to press ahead with development projects and Han immigration.

At the outset of the Communist takeover of China in 1949, the Han population in Xinjiang amounted to 6 percent of the province's population. Today, it accounts for over 40 percent, and some 250,000 more Han Chinese immigrate to the cities (especially Urumqi and Yining) annually, enticed in part by relatively high salaries and other incentives. Many of these Han migrants are in Xinjiang only a short time; they stay long enough to make some money and then return to eastern China.

Human rights monitors report systematic Han abuses of the Uighurs. Scores of mosques have been torn down, Uighur literature has been burned, Muslim clerics have been forced into "reeducation" training, Islamic schools have been closed, public prayer is now forbidden, head scarves have been banned, and prodemocracy demonstrations are brutally quashed. China's struggle to maintain a grip on its ethnically distinct regions is reminiscent of Soviet efforts in non-Russian areas, and it remains to be seen whether the land empire of China might someday crumble as the Soviet empire did.

165

MARTIN GRAY/NGS Image Collection

● **Figure 7.3.C** The Potala Palace in Lhasa, Tibet, is the ceremonial home of the 14th Dalai Lama, now in exile in India.

interior Asia, long, dark, cold winter days dominate North China. Many people cope with the cold with no more than the heat from the cooking stove or from the fire-heated brick bed called the *kang*.

North China receives 15 to 21 inches (c. 38–53 cm) of precipitation annually, but this is highly variable from year to year. Principal crops adapted to this climate include winter wheat, millet, and kaoliang, with some summer rice crops, corn, and a variety of vegetables.

There is a terrible history of droughts and floods. Floods in North China relate mainly to the problem of managing the Huang He (●Figure 7.3.4). The Huang drops down from its source on the Tibetan Plateau and courses for more than

1,000 miles (1,600 km) through the loess soils of North China on its 2,500-mile (4,000-km) journey to the sea. The yellow loess soil blown from the Gobi Desert collapses easily into the river system, which carries it as silt and deposits it along the final 500 to 600 miles (800–960 km) of the river's flow to its delta.

This physical process is a mixed blessing. On the plus side, it has built up a broad and fertile arable area at the mouth of the river. On the negative side, this sedimentation has elevated the actual stream channel of the river above the surrounding landscape, creating a perfect setup for flooding. Before the recent construction of a series of dams, a critical system of dikes followed the river all the way from

• **Figure 7.3.4** The Huang He (Yellow River) passes through areas of intensively cultivated loess soils, and in this stretch has carved a deep gorge.

the Tai Hang Mountains to the delta. But when flooding breached these dikes, the river poured out of the channel and down to the lower, densely settled farmlands and cities of the North China Plain. Even after the ruptured dike was repaired and the river was contained once again, the great flow that issued forth from the elevated stream bed had nowhere to go but down into ponds, fields, and communities.

This historical pattern on the North China Plain earned the Huang He its nickname of **"China's Sorrow."** Despite these natural hazards, the Huang He and its Wei He tributary (east of Xi'an) were the historical focus for Chinese civilization. Today, many cities along the Huang He are jostling for position as newly important urban-industrial centers. The resulting building boom is putting further pressure on the water supply, now dwindling both because of increased irrigation for agriculture and because of reduced glacial cover and melting permafrost at the river's headwaters; as the ground thaws, more water saturates the soil, and less runs into the Huang He.

Much trade between North and South China has been carried along the Grand Canal, a monumental civil engineering project of the Sui Dynasty (581–618 C.E.). This 1,050-mile (1,700-km) inland waterway—the longest ever built—links Hangzhou in the Chang Jiang Delta region to the area just southwest of Beijing. It was built principally to ensure a steady and safe supply of rice from the productive, subtropical southern lands to the north and to supply the south with wheat and coal. The canal was built at a great cost in tax and human labor but this north-south artery accomplished the major feat of linking China's great west-east rivers. The canal's northern third eventually fell into disuse, but the remaining portion is heavily used and is being rehabilitated and in places even gentrified. More than 100,000 vessels use it every year, carrying hundreds of millions of tons of goods, including coal and construction materials. The Grand Canal will be even more important when the eastern portion of the Jiang Chang Water Transfer Project is complete (see Figure 7.3.E, page 377).

South China

The Qinling Mountains, reaching elevations of about 10,000 feet (c. 3,000 m), mark the traditionally recognized dividing line between North and South China. These mountains run east to west and divide stream drainages between the Huang He to the north and the Chang Jiang (Yangtze River) to the south. South China's agricultural dominance owes mainly to the 3,400-mile (c. 5,440-km) Chang Jiang, flowing from its headwaters in the Tibetan Plateau to the coastal delta just north of Shanghai. This river has played a role similar to that of the historic Missouri River in the United States, serving as a major east-west transit corridor. 614

There are three large basins on the Chang Jiang. The two eastern ones—the Lower Chang Basin at the river's mouth and the Middle Chang Basin centered around Wuhan—are broad, densely settled areas with agricultural villages, small towns, and some expanding industrial cities (Wuhan being the most significant). Above the Middle Chang are the deep gorges recently transformed by the Three Gorges Dam (see Problem Landscape, pages 376–377). Above the gorges, and also partly inundated by the Three Gorges Dam reservoir, is the Red Basin, home to more than 100 million people. The Red Basin served as the center of the Chinese government during World War II because of its isolation and distance from Japanese bombers.

Nature has provided little level land in the basin, but rice and other crops are grown on enormous numbers of small fields in narrow ribbons of valley land and on laboriously terraced hillsides. The most prolific of China's tea-growing areas lies south of the middle and lower stretches of the Chang Jiang. China is the world's largest tea producer, and tea has long had an important place in Chinese culture (•Figure 7.3.5). A Chinese tradition holds that the seven basic necessities of daily life are fuel, rice, oil, salt, soy sauce, vinegar, and tea.[1]

• **Figure 7.3.5** China grows some of the world's best tea and has built a rich culture around it. Chinese tea is prepared with great care and served ceremoniously.

● **Figure 7.3.7** Rural China.

7.3.3 Issues in Chinese Agriculture

China has had productive agriculture for thousands of years, helping both to promote and to sustain the country's vast population. Today nearly 870 million Chinese (about two-thirds of China's population) live in rural settlements and are supported by agriculture (●Figure 7.3.7). These peasants must wrest from the landscape food for themselves and for the ever-expanding urban population, a burden made daunting by the fact that China has 20 percent of the world's population but less than 1 percent of its arable land.

Fully 85 percent of China is not in cultivation because of steep slopes, dryness, or short growing seasons. The country's arable area, concentrated in the plains and river valleys of the east, is about 70 percent of the arable area of the United States, but through the practice of **multiple cropping** (growing successive crops on a given field throughout the year), the sown acreage may actually be larger than the area farmed in the United States. China's territory provides, on average, about two-thirds of an acre (2,700 sq m) of arable land per person to the rural population and only one-fourth of an acre (1,000 sq m) per person to the total population. For comparative purposes, arable land in the United States provides 7 acres (over 28,000 sq m) of arable land per person to the rural population and 1.5 acres (over 6,000 sq m) per person to the total population. However, the process of desertification, caused mainly by human misuse of the landscape, is threatening China's arable land, which declined by 6 percent between 1996 and 2006.

China's agricultural history has been punctuated by extraordinary efforts to farm new lands and to raise yields. In the 1960s, China's Communist leader, Mao Zedong, proclaimed that "there are no unproductive regions; there are only unproductive people." During that decade, Mao launched a national campaign promoting self-reliance, especially through more productive agriculture. His policy of self-sufficiency in grain remains a central tenet of China's

488

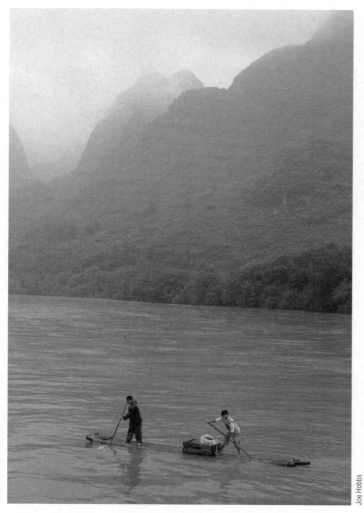

● **Figure 7.3.6** Tower karst along the Li River near Guilin.

South of the Chang Jiang basins is the real cultural heartland of South China: the delta region at the mouth of the Pearl River in the Guangdong lowlands. Its central features are the city of Guangzhou (Canton; population 5.7 million) and the numerous small, densely farmed, and settled lowlands on the Xi Jiang and its tributaries and distributaries. West of the Guangzhou lowlands, mostly in Yunnan and Guizhou provinces, is a deeply dissected limestone plateau 6,000 feet (1,800 m) high that features some of the world's most dramatic landscapes of **karst,** characterized by limestone pinnacles, caves, sinkholes, and "losing streams" that disappear from the surface into cavities below (●Figure 7.3.6). The area is home to many of China's small ethnic minority populations, who support themselves by hillside farming. The isolation of the region and its distance from the easy reach of national authority made it a perfect locale for opium poppy cultivation until the Communist government cracked down in the 1950s. Today the government has a development strategy to make this the world's largest source area for cut flowers, especially roses, to be exported to markets abroad.

PROBLEM LANDSCAPE
The Three Gorges Dam

The Chang Jiang ("Long River") has been central to China's identity and welfare for thousands of years. The river delivers precious water and fertile soils, creating environments that have been intensively settled and farmed, especially for rice and wheat. But this great river has also delivered tragedy in the form of devastating floods. Flooding in August 1998 affected 300 million people along the river and did an estimated $24 billion in damage, for example. Inevitably, questions arose about the advantages to be gained—especially in flood control, drought relief, and hydroelectricity production—if the river were to be dammed. The concept of a giant dam on the river dates to 1919, when it was proposed by Sun Yat-sen. His vision has been realized on a scale far grander than he could have imagined in the massive Three Gorges (Sanxia) Dam, begun in 1994 and scheduled for completion in 2009 (●Figure 7.3.D).*

As with any large dam project, the Three Gorges may best be evaluated by considering its pros and cons. On the plus side, this largest dam ever built—1.3 miles (2.1 km) and 610 feet (186 m) high—is creating a reservoir 385 miles (620 km) long that will dramatically reduce the threat of downstream flooding. Stored waters will also help alleviate drought. As part of the larger **Chang Jiang Water Transfer Project,** the dam will also assist in the provision of water to more arid North China through three canals (●Figure 7.3.E). China has one of the world's lowest per capita water supplies and most uneven distributions of water. More than 40 percent of China's people live in the north, but less than 15 percent of the country's water is there. Chinese engineers see water transfer on a massive scale as the way to redress this imbalance.

The Three Gorges Dam will improve navigation and therefore enhance trade, especially by connecting the burgeoning inland city of Chongqing to the world abroad. Historically, the Chang Jiang was able to accommodate freighters of 10-foot (3-m) draft all the way from the East China Sea to Wuhan. Above Wuhan, the shallower river was traditionally plied by smaller freighters to Yichang, situated just below the river's spectacular gorges. Yichang was a classic **break-of-bulk point,** with freight and passengers offloaded

*Much of the information in this feature is from the film *Three Gorges: The Biggest Dam in the World,* first broadcast on the Discovery Channel on December 2, 2000.

NASA/Visible Earth

1 km   July 17, 2000   May 14, 2006

● **Figure 7.3.D** The Three Gorges Dam, viewed from space. Construction is underway in the image dating to 2000 (left), and approaching completion in 2006 (right).

and put on flat-bottomed junks and sampans drawn upstream through the gorges to Wanxian by human trackers who pulled and rowed the vessels. With the completion of the Three Gorges Dam project, vessels with much deeper draft will be able to sail 1,500 miles (2,400 km) upstream from Shanghai all the way to Chongqing in the Sichuan Basin. Small vessels will be able to pass over the dam in a ship elevator, a vast steel box that will lift the ship the height of the dam in 42 minutes. Most ship traffic will use a system of locks with five levels, taking three hours in transit. It is projected that shipping to Chongqing will increase fivefold. Some observers say the increased commerce will transform Chongqing into a new Hong Kong. Chinese authorities view Chongqing as a new model for urban growth, as discussed on page 386.

Also on the plus side, the world's largest hydropower plant that is built into the dam will provide about 85 billion kilowatt-hours of energy annually, enough to satisfy 15 percent of China's electricity needs. Some of the power may be sold to private users, offsetting some of the dam's huge construction costs.

Most of the negatives are related to the vast reservoir forming behind the dam. When filled in 2009, the new lake, 500 feet

agricultural economy today, often compelling peasants to grow grain when they could make more money by producing and selling fruits and vegetables.

Since Mao's early appeals, many efforts have been made to increase yields on already farmed land. Irrigation, water conservation, and flood control measures have been part of this effort. The Chinese have struggled both to prevent water from running off in floods and to conserve it in reservoirs for when it is needed. To achieve these aims,

local communities have built large numbers of small dams, ponds, canals, and dikes. These are increasingly overshadowed by large dams and associated waterworks, such as the Three Gorges project.

A major objective of water conservancy measures has been to increase the amount of land under irrigation and thereby increase output. China has more than doubled the amount of arable land under irrigation since 1949, when Mao came to power. In arid areas, salinization is a chal-

PROBLEM LANDSCAPE

The Three Gorges Dam *continued*

• **Figure 7.3.E** The Three Gorges Dam and the Chang Jiang Water Transfer Project.

(150 m) deep, will inundate 4,000 villages, 140 towns, 13 cities, numerous archeological sites, and nearly 160 square miles (over 400 sq km) of farmland. Because their homes have been or will be drowned, as many as 2 million people are being resettled. Many of the settlements to be vacated are quite ancient, and their inhabitants have deep cultural roots in them. The old villages are typically situated close to fertile farmlands that are also being inundated. In the new, higher communities into which people are being located, soil conditions are generally inferior.

The reservoir will have aesthetic consequences too, replacing the world-class wonder of a wild river passing through scenic canyons. The natural marvels of the three gorges drew 17 million Chinese tourists and hundreds of thousands of foreign visitors every year. There are also practical concerns about the silt that will build up behind the dam. It is possible that this increased soil load will have the ironic effect of causing flooding upstream from the dam, even as the dam prevents flooding below. Finally, there are fears of what is almost unthinkable: that the dam could burst, causing unimaginable devastation and suffering downstream.

Although recently admitting some environmental consequences, the government of China has pursued the mighty Three Gorges Dam project without remorse and with no tolerance for dissent within China. The enormous cost of the project, nearly $30 billion, is born solely by China. Like megadams elsewhere, the Three Gorges Dam is a statement that its builder is a great power to be reckoned with on the world stage.

252, 319

lenge; too much irrigation deposits mineral salts from the water and reduces the land's productivity. Ongoing efforts at reforestation, like those described on page 370, also relate closely to water control. Only 18 percent of China is forested. Under steady human pressure over thousands of years, most primary forests in China have been cut down. Large areas of the country are without significant vegetative cover, contributing to loss of water through rapid runoff. Planting trees is not the answer in every situation, how-

ever: in China's arid lands, too many trees would deplete the available groundwater.

Another approach to improving yields from existing lands lies in the application of scientific knowledge and technology to develop better seed, equipment, and farming techniques, together with the development of administrative means of promoting their adoption—China's version of the Green Revolution, described on pages 299–300. With more trained agricultural scientists and technicians, insecticides and herbicides are coming into wider use. These follow more rudimentary systems of pest control practiced during the 1950s and 1960s, during Mao's **Four No's Campaign** to eradicate rats, sparrows, flies, and mosquitoes. Observers were amazed at the sight of Chinese peasants fanatically hunting down the pests, motivated by government radio broadcasts over speakers placed in rural areas. A major objective was to reduce the annual losses of many millions of tons of grain. It was an unintended consequence that so many birds were killed that their natural prey, caterpillars, infested crops. Campaigns against flies, mosquitoes, and rats had additional public health benefits. They were also politically important as they demonstrated the government's capacity to marshal enormous numbers of people in a wide variety of national campaigns, many of which were focused on modifying the environment.

Increased fertilization has boosted farm productivity. The use of chemical fertilizers is growing alongside more traditional means. Chinese farmers have long fertilized their fields with pond mud, plant residues and ashes, oilseed cakes, nitrogen-rich crops raised to be plowed under for soil enrichment, and animal and human manures ("night soils"). Such traditional fertilizers, and the close contact between people and animals in rural China, play a role in the country's status as an "incubator" for disease (see Medical Geography, pages 380–381).

Beginning in the 1950s, the Communists looked for ways to increase land cultivation. Small gains in the amount of cultivated land were made through reclamation of poorly drained land and extension of irrigation into deserts. As discussed on page 375, these measures did not always succeed in the long run. Mechanization has developed slowly. The use of tractors is growing, especially in the northeastern and western regions, where more extensive farm fields are located, but human and animal power and simple hand tools are only slowly fading from the rural scene.

After 1949, when the Communist Party gained control of China, agriculture was not merely a matter for scientific and technological attention but also an engine for profound ideological change. China's leaders mandated a series of reorganizations of Chinese agriculture in successive attempts to solve the country's food problems using Communist ideology and control. Collectivization of agriculture began in 1955, with private farms reverting to state control in the form of agricultural cooperatives. The cooperatives were merged into even larger units in the agricultural **commune system** that began in 1958 as the cornerstone of an economic program called the **Great Leap Forward.** In the commune, 20 to 30 cooperatives consisting of over 20,000 people and 40 to 100 villages were merged into a single unit. Residents lived communally, sharing bathing, dining, and residence facilities. Any land and equipment of the former cooperative and any money or property still owned by peasants became the commune's property. An economic and administrative body controlled the labor and production and supervised all aspects of education and defense.

Small-scale industries, most notably millions of "backyard steel furnaces," were also set up in the rural communes.

The legacy of this effort to bring industries to the countryside is still apparent in the Chinese economy today. There are about 25 million rural industrial enterprises with more than 125 million employees, representing about 65 percent of China's industrial labor force. Their rural-based, small-scale industries generate about half of China's annual industrial output, one-quarter of its total economic value, and more than one-third of China's exports. These successes are largely the result not of the commune movement but of more recent economic reforms.

Collectivization and communes had disappointing results, related mainly to inefficiencies inherent in trying to manage such large units and to the lack of incentives for individual farmers. An estimated 30 million Chinese died from famine attributable to the failures of the Great Leap Forward. Reforms began in the 1960s, with communes decentralized and some even divided into private farms. After the death of Mao and the 1978 ascension of Deng Xiaoping to the major leadership role in the Communist Party, free-market reforms began to take place in China. Individual households acquired long-term (30-year) "leases" on the farms, which continued to belong to the state but farmers now viewed as effectively their own. They paid fixed amounts of their production to the state and were allowed to consume or sell the rest. Chinese farmers were finally allowed to sublet land, hire labor, own machinery, and make their own land use decisions. They still lacked outright ownership of property, however, and in recent years, the government has appropriated farmland for housing and factories, creating unrest and uncertainty in the countryside.

The overall results of this experiment with a freer semi-capitalist system in agriculture have been profound. Chinese agricultural output expanded by about 50 percent during the 1980s and 1990s, and rapidly enough—when coupled with the steady decrease in birth rate—to generate an increase of 20 percent in the per capita food supply. It is remarkable that the world's largest country has generally been able to keep agricultural growth apace with population growth and provide more than 90 percent of its own supplies of rice, corn, and wheat. The challenge will be for China to continue this remarkable trend, especially as the Chinese develop new appetites for beef, beer, and other luxuries that require more inputs of energy and agriculture. Chinese agricultural planners are also alarmed by the accelerating loss of farmland to urban and other uses and by farmers' decisions to forsake grains in favor of more lucrative cash crops.

Chinese scientists are among the world's most fearful in forecasting the consequences of global warming. They have documented a rapid (7 percent per year) melting of glaciers on the Tibetan Plateau, a trend that threatens future water supplies to large parts of China and adjacent regions of southern Asia. According to a 2007 International Energy Agency (IEA) report, climate change could cut China's harvests of wheat, rice, and corn by 10 percent by 2030 and by 40 percent after 2050. China, which produced about one-fifth of the world's carbon dioxide emissions in 2007—slightly more than the United States—will be the industrial engine behind the largest share of global greenhouse gases for decades to come.

If nothing were done to reduce China's CO_2 footprint, the IEA warns, China's emissions would be double those of all industrialized nations by 2032. Citing its own need to develop as developed countries have, and thus refusing to come on board with the existing Kyoto Protocol, China has set its own target of producing 16 percent of its energy needs from renewable resources (mainly hydropower) by 2020. The country has already taken aggressive steps to reduce energy use in factories and automobiles. The latter is a huge concern in a country that had 13 million private cars in 2007, with a projected 10-fold increase by 2020—by which time China will also be well established as a major automobile exporter. The country will also increase its output of nuclear power substantially.

7.3.4 China's Industrial Geography

China today is one of the world's great industrial powers, ranking fourth in overall economic output (•Figure 7.3.8). Remarkably, production of industrial goods did not start in China until the beginning of the 20th century. Prior to that time, imperial China's handicraft industries produced high-

• **Figure 7.3.8** Major industrial resources, industries, and industrial cities of China.

quality goods that attracted Western traders in spite of China's official disinterest in such trade. To this day, handicraft and shop-scale industries continue to supply many of China's simpler needs.

China's early factory industries were mainly foreign-owned and were attracted to China by the huge market for cheap goods, by inexpensive labor, and sometimes by certain natural resources. As in so many countries, the first factory industry to develop on a large scale was the manufacture of cotton textiles. Japanese and other foreign firms, together with some native Chinese companies, developed a major cotton industry by the 1930s. This industry, and others, developed in coastal cities such as Shanghai and Tianjin. Meanwhile, the Japanese used the minerals of the Northeast (in their puppet state of Manchukuo) to make that area the first center of heavy industry in China in the 1930s.

Communist China made major efforts to industrialize after the 1949 victory over the Chinese Nationalists. First, the industrial infrastructure, which had been badly hurt by invasion and civil war from 1937 to 1949, was brought back to prewar production levels under the new system of state ownership. China also expended considerable capital and human effort in its involvement in the Korean War of 1950–1953. Aided by the Soviet Union, throughout the 1950s, the Chinese emphasized coal, oil, iron and steel, electricity, cement, certain machine-building industries, and armaments. China's resources for these industries were considerable, and the country's centrally planned command economy—a system in which major economic activity is initiated by government planners—achieved major successes in developing them, as was the case in many Communist countries. There was also a strategic spatial component to this drive, with the Soviets advising China to reduce its concentration of industrial plants along the coast. This strategy, meant to reduce the vulnerability of Chinese industry in the event of military conflict and to promote development outside the historic core, led to expansion into new industrial landscapes at Wuhan on the Chang Jiang, at Baotou in Nei Mongol, and into the far west near the Soviet border in Xinjiang.

In the 1960s, after China broke ranks with the Soviet Union, the pace of technical development, quality of output, and modernization slowed. This was a consequence both of China's near-isolation from more advanced industrial countries and of the Cultural Revolution from 1966 into the early 1970s (see Insights, page 383). This slowdown also mirrored the difficulties that non-market-based, centrally planned economies on the Soviet model typically encountered as they moved toward more consumer-oriented industries and advanced technologies.

The latest phase of China's industrialization began after Mao's death in 1976, when another veteran of the Long March, Deng Xiaoping, became the country's leader. China opened its doors to the West and devoted more attention to personal incentives for agricultural and industrial success.

MEDICAL GEOGRAPHY

Deadly Viruses: Origins and Diffusion

Seemingly out of nowhere, **severe acute respiratory syndrome (SARS)** burst onto the world scene in spring 2003. But this extraordinarily virulent virus did come from somewhere: southeastern China.

As any recipient of a flu shot knows, the vaccine is typically targeted against a host of flu strains bearing Asian, usually Chinese, names such as "Shanghai" and "Beijing." SARS, like these influenza varieties, originated in a region where there is frequent close contact between people and animals. Farmers in China often cohabit with ducks, pigs, geese, and other domesticated animals and make extensive use of animal excrement as fuel and fertilizers (●Figure 7.3.F). In these close quarters, the influenza strain often starts in a duck or other animal and jumps the species barrier to affect humans. Once the sickness manifests itself in a human host,

● **Figure 7.3.F** Situations like this raise fears that the bird flu will jump the barrier between poultry and humans.

it can spread rapidly to major cities in China and thence by airliner to distant parts of the globe.

The SARS virus had similar origins, although it probably arose from close contact between animals and animal merchants in southeastern China. In Guangdong province, there are numerous large animal markets where poorly paid migrant workers carry out the sordid job of killing, skinning, plucking, gutting, and packing a wide variety of wild and domesticated creatures, including snakes, frogs, turtles, chickens, badgers, cats, rats, and civets, all bound for Chinese dinner tables.

Between November 2002 and January 2003, several of these animal merchants, along with food handlers and chefs further down the supply chain, began developing flu-like and pneumonia-like symptoms, including high fever. They had contracted SARS, an illness new to the human world and to the coronavirus family, which also includes the common cold. Some sufferers, perhaps 5 percent, died of the illness. Some sought medical attention in Guangdong hospitals. There they infected doctors, nurses, and other staff (30 percent of China's early SARS victims were hospital workers).

On February 21, 2003, a doctor from one of the hospitals managed, despite his high fever, to attend his nephew's wedding in Hong Kong. He stayed at the Metropole Hotel, where he passed the virus (probably by sneezing or coughing) on to two Canadians, an American businessman on his way to Vietnam, a man from Hong Kong, and three women from Singapore. In the days that followed, most of these sick people flew to other countries. One of the Canadian women carried the illness home to Toronto, where she died after infecting her son and several health workers. At that point, SARS was poised to become a global epidemic. By June, nearly 7,000 probable SARS cases had been reported in 29 countries, and 495 victims had died. In the United States, about 200 suspected SARS cases were reported in 38 states, with no fatalities.

The liberalization of agriculture has already been described, and related measures were introduced into trade and industry, where emphasis shifted toward light industry, consumer products, and profit-oriented enterprises. Much more private enterprise was allowed, especially after the mid-1990s. State factories were required to operate at a profit, part of which they were allowed to retain and invest in expansion. The factories introduced pay incentives for superior performance and were no longer obligated to provide lifetime jobs regardless of output. The Chinese welcomed foreign investment capital and technical expertise from Japan and the West. Great numbers of Chinese began to study abroad in graduate and professional programs; they account for the second-largest percentage of foreign students in the United States and the largest in Australia. Turning its back on the isolationist days of Mao, China sought to become a global

trading power. It has succeeded handsomely in this goal (see Chapter 7, pages 298–299).

So successful has China's industrial drive been that the country is now a gigantic, insatiable consumer of domestic and imported energy and other commodities. For example, China is today the world's largest producer of steel, aluminum, magnesium, and tungsten. It is the world's leading consumer of copper, iron ore, soybeans, cotton, and wood. Heavy industry has been built on a rich domestic resource base, but it is not yet big enough to meet all the country's needs.

Because of local shortfalls, China imports vast amounts of iron ore, lead, alumina, nickel, and copper to feed its booming industries. For a time, oilfields in the Northeast, Gansu, Sichuan, and Xinjiang met China's needs, even providing some surpluses for exports. But as the economy

MEDICAL GEOGRAPHY

Deadly Viruses: Origins and Diffusion *continued*

In less than six months, SARS had become an enormous problem that posed several threats. Of most immediate concern was its impact on human health. There was no vaccine for the disease. It is especially difficult to develop a vaccine for an illness that affects both people and animals. There was no effective treatment; SARS, like all viruses, does not respond to antibiotics. In the absence of a preventive or cure, the disease follows its course: a high fever of 100°F or more is followed by chills, muscle aches and headaches, a dry cough, difficulty in breathing, and occasionally, severe pneumonia and even death.

Despite taking relatively few lives, SARS had a strong impact on human psyches and societies. The disease could have appeared anywhere and infected or killed anyone; consequently, many people—especially near the Asian epicenter—assumed the worst during the outbreak. As reports of the disease spread in spring 2003, untold numbers of people in China began wearing masks to reduce their chances of contracting the disease. They no longer shook hands. Schools, theaters, dance halls, and Internet cafés closed in many communities. Shopping malls had few shoppers. Several Asian countries stopped granting visas to Chinese visitors.

The economic costs for such a limited outbreak were also extraordinary. In the globalized world of business, many firms halted interpersonal dealings with China. Toronto lost $30 million a day in canceled conventions and other events during the brief period when the World Health Organization (WHO) posted a warning that travel there should be avoided. In spring 2003, China and South Korea lost $2 billion in the tourism, retail sales, and manufacturing sectors, and Hong Kong, Taiwan, and Singapore lost perhaps $1 billion each. Sixty percent of U.S. corporations banned travel to Asia altogether. North American airline bookings to Hong Kong fell 85 percent, and international arrivals in Singapore were down by 60 percent. Anyone diagnosed as having a fever in Singapore airport's arrival hall was unapologetically led off to two weeks of quarantine.

In summer 2003, the disease peaked and began to decline in Canada, Singapore, and Hong Kong and then subsided in its birthplace, China. SARS receded from the public imagination. So why is it the focus of this essay? Because the SARS experience may provide a preview of something of far greater consequence, also originating from the same part of the world.

Fresh on the heals of SARS came the specter of this new menace, **bird flu** (the H5N1 subtype of avian influenza A). Like SARS, this respiratory virus originated in animal species—both wild and domesticated birds—and has the potential to jump the species barrier and afflict large numbers of people. Some people working with ducks, geese, turkeys, and other domesticated fowl in East and Southeast Asia did come down with the disease, and health officials in several countries ordered hundreds of millions of birds to be killed to staunch the virus's spread.

The threat persisted. By late 2007, about 200 people, mainly in Vietnam, Indonesia, Thailand, and China, had died from bird flu. That number is small, but what most concerns health officials is that it represented 70 percent of the people who had the virus. With such a high fatality rate, health experts fear that bird flu could be the "big one": a virus like the Spanish flu that in 1918 and 1919 killed 40 to 50 million people worldwide, including 850,000 in the United States (where it apparently first manifested itself, in a soldier at Camp Funston, Kansas, on March 11, 1918). According to the WHO, even in a best-case scenario, a bird flu **pandemic** (an epidemic over a wide geographic area, affecting a large percentage of the population) would kill up to 7 million people worldwide and leave tens of millions in need of medical attention. Indonesia, China, and Vietnam have developed a flu vaccine for birds and are inoculating vast numbers, but other countries argue that the vaccine is not completely effective. Work to develop an avian flu vaccine to protect humans is ongoing.

---

zoomed into overdrive, China became a major oil importer, with imports increasing an average of 30 percent per year—and by 2007 was second only to the United States.

China's imported commodities come particularly from Africa and Latin America. While securing vital commodities from developing countries in these and other regions, China is also pouring economic aid into them, building roads, bridges, dams, and other infrastructure that serves the interests of those countries and, of course, China itself. China tends to attached fewer strings to its aid than the World Bank and other traditional development agencies.

478, 480, 481, 538

Coal is king in China, and this is one commodity China does not have to seek abroad. The country has about 12 percent of the world's coal reserves (behind the United States and Russia) and is the world's largest producer and consumer of coal. The reserves are located mainly under the

North China Plain and its bordering hills and in Liaoning province in the Northeast. This resource concentration has made the southern area of the Northeast the country's leading heavy-industrial region. China uses coal to meet about 80 percent of its overall energy needs and to provide 82 percent of its electricity. This is the dirtiest of all the fossil fuels, producing far more particulate matter and carbon dioxide than oil or natural gas.

What's Next for Industrial China?

There are risks for China's future industrial landscape, and it cannot be taken for granted that the country will persist as the world's economic dragon. China's extraordinary recent successes are tempered by a few concerns. One is freedom of expression. In 1989, the Chinese government cracked down brutally on prodemocracy demonstrations centered on Tiananmen Square in Beijing. Years later, there continue to be strictures on academic, media, artistic, and other freedoms that some observers believe are conducive to long-term economic well-being. China's government is one of few in the world (including those of Myanmar, Singapore, Turkmenistan, Uzbekistan, Saudi Arabia, Syria, the United Arab Emirates, Tunisia, and Iran) to censor the Internet, which is arguably one of the most important tools of economic development today. The government also monitors mobile phones, the use of which is growing 30 percent annually in China, to ensure that they too do not become instruments of protest.

Another potential problem is that China's leaders are breathlessly trying to feed what they call the "socialist market economy" with gigantic projects, thereby hoping to create jobs and stimulate economic growth. They reckon that the country needs to register at least 7 percent economic growth annually just to prevent massive unemployment and social unrest. And so one of the world's most impressive building booms ever is now under way: new subway systems, railroads, highways, bridges, sewage systems, dams, water diversion projects, a natural gas pipeline, a national electricity project, urban improvements, airports, industrial parks, and more. These efforts cost a lot of money, and the government is drawing down its national treasury to fund them. State banks are pouring billions of dollars into federal projects whose returns are uncertain.

A further hazard is that with its galloping growth, China is developing a classic "economic bubble" that will eventually burst, probably with costly results. Historically, few economies have been able to maintain torrid growth for long, especially when that growth is fueled by speculative

cash flows and questionable bank loans, as has been the case recently in China.

There are also environmental costs for such spectacular growth. Due to generally lax environmental quality standards, China's skies and waters are often badly polluted (•Figure 7.3.9). Twenty of the world's 30 most polluted cities are in China. The country's State Environmental Protection Administration (SEPA) reported in 2007 that 26 percent of the water in China's largest river systems was so polluted that it was dangerous for people and other living things to come into contact with or had "lost the capacity for basic ecological function." Polluted water was killing tens of thousands of people every year and reducing crop yields. SEPA warned that China's approach of "growth through industrialization" was pushing its environment "close to the breaking point."[2]

Finally, there is concern that the benefits of China's economic growth have not been distributed evenly throughout the population. Prosperity and consumption have increased dramatically, but in a remarkably unequal manner. Less than 10 percent of China's people controlled about 45 percent of the country's private wealth in 2007. Meanwhile, 17 percent of all Chinese lived on less than $1 a day—the World Bank's measure of abject poverty. These inequalities are comparable to those that existed when the revolution brought Mao's Communists to power on the promise of redistributing wealth. In 1980, China had one of the world's most even distributions of wealth, but that equation has changed dramatically. In sum, despite its phenomenal growth, China is still relatively poor (with a per capita GNI PPP of $7,730, comparable to that of Kazakhstan and Panama) and largely agricultural (see Table 7.1, page 281). The agricultural face of China is changing rapidly, though, as the cities boom.

• **Figure 7.3.9** Blessed with coal reserves, China is also cursed by them. Its coal-fired power plants are responsible for calamitous domestic air pollution and contribute significantly to global atmospheric warming.

INSIGHTS

Mao's Cultural Revolution

Between 1949 and his death in 1976, Mao Zedong (●Figure 7.3.G), chairman of the Chinese Communist Party, led the People's Republic of China through a demanding and sometimes counterproductive series of experiments in social reorganization. In 1957, he initiated the Great Leap Forward and the commune movement discussed earlier. The **Cultural Revolution** (1966–1972) was Mao's attempt to maintain an ongoing "permanent revolution" in China. Politically, the movement was meant to throw a newly established urban and intellectual elite off balance by promoting purges of virtually anyone who had offended Communist Party functionaries.

Local youth who donned the red armbands of the Red Guards became the agents of change in this campaign. The Red Guards were set loose on the populace to harass those who were accused of putting learning, skill, expertise, or personal matters above revolutionary enthusiasm or Marxist and Maoist goals and principles. Government oppression led to perhaps a million assassinations, executions, or suicides and to lifetime jobs for those deemed most loyal to the Maoist revolution, regardless of their performance. Reforms initiated in the late 1970s finally began removing unproductive people from their positions, but people whose lives were turned upside down by the Cultural Revolution continue to speak of themselves as a lost generation.

Despite some monumental setbacks and miscalculations, China under Mao regained a global prominence that had been denied for a century and a half under Western colonialism and interference in the nation's domestic affairs. Even though much of Mao's legacy has been dismantled, he continues to be a major icon of China's presence on the world stage.

● **Figure 7.3.G** Chairman Mao greeting a crowd in 1971.

7.3.5 China's Urban and Transportation Geography

As recently as 1978, about 80 percent of China's population lived in the countryside. Although there were massive cities such as Shanghai, Beijing, and Guangzhou, most Chinese were peasants living in small villages. As a result of the greatest migration in human history, China's population is now on track to be 50 percent urban by 2010 (for comparative purposes, India is 28 percent urban, and the United States 79 percent).

As might be expected, despite the rural industries described earlier, China's industrial production is centered mainly in its cities. The country's 12 largest metropolitan cities are located in geographic clusters. This is an important spatial characteristic of China's economy. Four major clusters of cities are especially notable, in the Northeast, North China, the Chang Jiang region, and the area south of the Chang Jiang (see Figure 7.3.8, page 379).

Strong neighbors have long coveted the cities and industries of China's Northeast. After a treaty signed in 1858, tsarist Russia landlocked Manchuria by extending Russia's coastal territory all the way to northern Korea. Even today, China is eager to open a northeast access to the sea, even if only by leasing the Russian port of Zarubino. During the 20th century, the Northeast became a region of even more

contention among China, Russia, and Japan. Russian interest focused on the Chinese Eastern Railway—a shortcut across the northern two provinces between Vladivostok and the Trans-Siberian Railroad east of Lake Baikal—and on the naval and port facilities of Dalian (formerly known as Darien–Port Arthur and as Lüda), which the Russians had used at various times.

Japan, even today, is widely resented in eastern Asia for coveting and conquering the resources of China and its neighbors. Japan was long interested especially in developing Northeast China's impressive industrial resources and its potential to produce surplus food. After the Russo-Japanese War of 1904–1905, fought mainly in Liaoning, Japan became increasingly active in the area. Finally, in 1931, Japan took over with the establishment of the puppet state of Manchukuo and in the course of about 15 years created much industrial and transportation infrastructure. In 1945, during the final days of World War II, Soviet forces wrested the area from the Japanese, stripping many of Manchuria's factories and shipping them home. Eventually, the Chinese Communists incorporated the Northeast into their own People's Republic, making this region central to their industrial development.

The cities of the Northeast include another four of China's 20 largest cities: the port of Dalian; Shenyang (Mukden), the main metropolis of a group of heavy-industry centers;

184

Changchun, a center of automobile and other transportation manufacturing; and Harbin, a railway center with connections to Russia and Korea. After the Communists took over in 1949 and expanded China's industrial base, these cities of the Northeast lost some of their industrial importance, but they continue to be centers of manufacturing, chemical production, and agricultural marketing and processing. The industries are centered around Shenyang. The largest center of iron and steel production is Anshan, south of Shenyang, where industry is based on deposits of iron ore and some coal in a belt crossing the southern part of the region. The Daqing oilfields, China's largest, are north of Shenyang near Harbin.

The cities of North China include the historic national capital of Beijing (Peking; population 11.8 million; ●Figure 7.3.10); the port city of Tianjin (Tientsin); and far to the southwest, the very ancient city of Xi'an (Sian). Each is located at an entryway to the North China Plain: Tianjin, by the sea; Beijing, near the end of a pass leading through the uplands to the north toward Mongolia and the Great Wall; and Xi'an, in the Wei River Valley, which leads upstream to interior China and downstream to the Huang Valley and the North China Plain (●Figure 7.3.11).

The Chang Jiang Valley city group is a string of cities along and near the river. This group includes five of China's 20 largest cities, reflecting the importance of the Chang Jiang basins as a major axis of the national economy. The Chang Valley group includes a metropolis of 4 to 7 million inhabitants in each productive basin, plus a much larger overseas global connecting point as well as a transfer and connecting point for interior China. At the western end of the axis in the Red Basin is Chengdu, a provincial capital and node for transport lines connecting the axis with Arid China. It lies away from the river. On the river is the Red Basin's booming metropolis of Chongqing.

Downstream lie Wuhan in the Middle Basin, an important manufacturing, transportation, and agricultural marketing center, and Nanjing in the Lower Basin. Nanjing

386

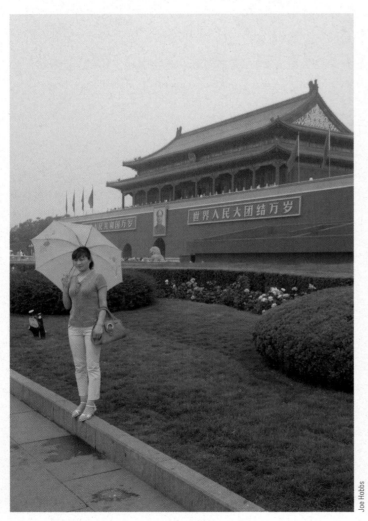

● **Figure 7.3.10** The entrance to Beijing's Forbidden City, from Tiananmen Square.

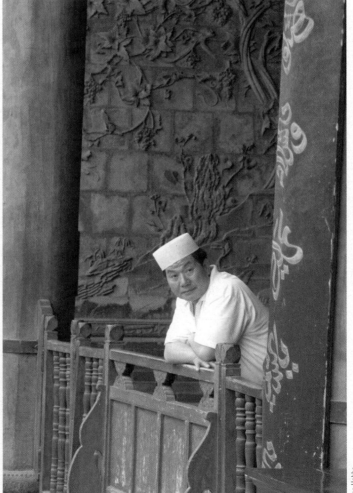

● **Figure 7.3.11** The 60,000 Muslims who worship in this Great Mosque of Xi'an (established 742 C.E.) are living evidence of the city's historic position at the eastern end of the Silk Road.

400 (a temporary capital for China during the Sino-Japanese War in the mid-20th century) is a major agricultural marketing center.

On a navigable tributary (the Huang Po Creek) near the mouth of the Chang in the Lower Basin is the port of Shanghai, the traditional center of Chinese business life and a contender with Hong Kong for the position of China's financial capital (population 15.1 million). With its location at the seaward end of this urban axis, Shanghai serves the whole Chang Valley and other areas. Built from a fishing village by foreign enterprise in the 19th century, Shanghai is the largest city in China and has the largest port in the world by tonnage. The piercing skyscrapers of its futuristic section of Pudong are just opposite the colonial waterfront known as the Bund (●Figure 7.3.12). Many Chinese and outsiders lament the loss of Shanghai's traditional and historical architecture in this city of nonstop construction.

The region south of the Chang Jiang has two of the country's 12 largest cities, including the port of Guangzhou (population 5.7 million) in the Xi (Pearl) River Delta. Guangzhou and its vicinity now form a rising center of new industries, particularly around Shenzhen, a young city that in the 1980s

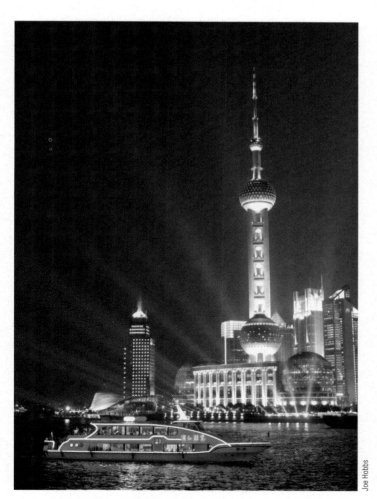

● **Figure 7.3.12** The futuristic Pudong section of Shanghai is China's modern face to the world.

Joe Hobbs

was designated as one of five **Special Economic Zones (SEZs)** designated to attract investment and boost production through tax breaks and other incentives (see Figure 7.3.8). 323 The other SEZs, all with access to the sea, are nearby Zhuhai; the entire island of Hainan; farther north, Shantou and Xiamen on the Taiwan Strait; and Pudong, linked to Shanghai. Many of China's coastal cities specialize in the factory production of a single category of goods; for example, moving from Shangdong province in the north to Guangdong province in the south, one would pass through cities known for men's wear, casual wear, down-filled products, ties, socks, underwear, wedding dresses, evening gowns, and jeans.

Close to the Xi River cluster of SEZs are Hong Kong (Xianggang; population 9 million), which reverted from British control in 1997, and Macau (population 1 million), returned from Portuguese control in 1999. Each is now a **Special Administrative Region (SAR)** governed by China under a unique **"one country, two systems"** formula. In this arrangement, China's central government rules Hong Kong and Macau but grants each considerable autonomy and political liberalization. Each has the right to continue its capitalist economic development, while China proper maintains its socialist economic principles.

This formula has generally been successful, although both SARs would like to have more democratic freedoms. More than a decade after returning to Chinese control, Hong Kong continues to hold its position one of the world's busiest and most prosperous trading and financial centers. Macau's economy is based on trade with inland China, fishing, tourism, and especially gambling, which is forbidden elsewhere in China. Macau's gaming industry is more valuable than that of Las Vegas. It has helped increase Macau's overall wealth enormously, but it has also led to high prices for real estate and consumer staples. Many citizens are not receiving benefits from the new flush of cash, and thousands of them took to the streets in 2007 to protest the gaming industry.

One of the results of the clustering of cities, and the economic power they represented, was a serious problem of regional imbalance. Most of the country's industrial development and urban growth occurred along its east coast, near the traditional port and river cities that were strongly influenced by Western economic interests from the mid-19th century on. Even the rural areas of eastern China experienced much more growth relative to those in the west, especially because they were well linked to expanding markets and prospering cities. Meanwhile, people in western regions of China, mountain villages all through Humid China, and landscapes lacking in surface transportation in the southwest and parts of upland China were frustrated. They saw relatively little evidence of China's much touted economic prosperity.

Consequently and somewhat predictably, large numbers of people migrated from the less prosperous regions of the country to the promising SEZs and other urban centers of the east and south. Millions of migrant workers left their villages—now populated primarily by children and the

● **Figure 7.3.13** China's urban workforce has a large contingent of migrants from rural areas, performing both skilled and unskilled labors. This man is making cloisonné vessels in Beijing.

elderly—to work in construction or menial labors in the coastal cities, sending much of their meager monthly earnings of $50 to $100 home as remittances (●Figure 7.3.13). People in the cities often discriminated against these migrants in education, housing, and jobs, treating them like illegal aliens. Many technically were illegal because they failed to sign work contracts that would both register them and require factories to pay fees to local cities. Police would routinely "shake down" workers without papers, seeking bribes from the frightened migrants. Some of the fac-

tories were world-class, but others were **sweatshops** where working conditions bordered on the nightmarish. Management typically suppressed strikes, banned independent trade unions, and failed to enforce minimum-wage laws for workers who toiled through 18-hour workdays and caught brief rest in prisonlike dormitories. It is little wonder that many of the eastern boomtowns saw as much as 10 percent annual turnover in the workforce and soaring crime rates.

Some of these conditions may persist for some time, but there are also signs of rapid change, reflecting the pace of China's growth. Chinese authorities are convinced that urbanization is the ticket to greater prosperity for China's huge, poor, rural population. But they want the larger cities of the interior, such as Chongqing, to absorb much of the migrant traffic that has been flowing farther east and south. Already home to 8 million people, Chongqing is poised to grow much more through its "one-hour economy circle plan." In this plan, by 2012, 2 million rural residents will be moved into newly urbanized urban areas within an hour's drive of the city center. Two million more are to be accommodated by 2017. Millions of new jobs are thus being created in the long-neglected urban centers of China's interior. At the same time, wages are rising sharply all over the country. Once pushed and pulled toward the coast, workers are now finding competitive jobs closer to home. The result is a sudden and surprising shortage of wage laborers in the coastal factories. Once-confident sweatshop managers now worry that the global marketplace will shift jobs to lower-wage countries like nearby Vietnam.

● **Figure 7.3.14** One of China's most ambitious efforts in expanding its infrastructural base for continuing economic development has been its transportation system.

• **Figure 7.3.15** Principal features of Taiwan.

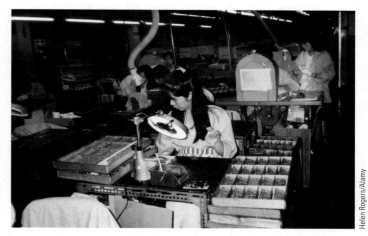

• **Figure 7.3.16** High-tech manufacturing is central to Taiwan's economy. This worker is assembling cameras.

In another effort to redress the regional imbalances that drew so many people to the east and southeast and to integrate China's widespread resources and widely scattered producing and consuming centers into a functioning whole, the government has placed great importance on improving the transportation system (•Figure 7.3.14). The country is investing $250 billion in a national highway system to be completed by 2037 and billions more in the construction of dozens of airports. Much of the air travel will be abroad, especially as the Chinese expand their segment of the international tourism market. Estimates are that at least 50 million Chinese will travel overseas each year by 2010 and 100 million by 2020.[3]

The government has also made major efforts to improve and extend railways. Before 1949, the Northeast had an extensive rail system developed by tsarist Russia and expanded by Japan, but most of China south of the Chang Jiang floodplain and all of the western provinces were without rail lines. The postwar Communist government made South China accessible by building the first three bridges ever to cross the Chang Jiang. They pushed rail lines into South China and the Red Basin and also into Xinjiang and across Mongolia to connect with the Trans-Siberian Railroad. Since 1949, the length of China's rail network has more than tripled, and the expansion continues. However, railroad development lags behind that of roads. Except for transport over short distances, railroads are much more efficient for hauling goods than trucks are. With China's economy booming, the huge bottlenecks at ports and factories suggest that more railways need to be built.

7.3.6 Taiwan

Known by Westerners for centuries as Formosa (Portuguese for "beautiful"), Taiwan is separated from South China by the Taiwan, or Formosa, Strait, 100 miles (160 km) wide (•Figure 7.3.15). The island is nearly 14,000 square miles (c. 36,200 sq km) in area and is home to 23 million people, nearly all of Chinese origin. High mountains rise steeply from the sea on the eastern side of the island but slope on the west to a broad coastal plain. Taiwan lies on the northern margin of the tropics, and its temperatures are moderated by the warm, northward-flowing waters of the Kuroshio, or Japanese, Current. Irrigated rice, the main food crop, is favored by the hot, humid summers in the lowlands and by abundant irrigation water from mountain streams. Sugar and pineapples are also important. The Japanese governed the island from 1895 to 1945, developing it mainly as a sugar-producing colony.

Driven from the mainland in 1949, the Chinese Nationalist government fled to Taiwan with remnants of its armed forces and many civilian followers. Nearly 2 million Chinese were in this migration. Here, protected by American sea power, the government reestablished itself as the Republic of China with its capital at Taipei (population 8.2 million). The Nationalists, originally an authoritarian regime but with increasing elements of political democracy, were very successful in fostering industrial development. They united inexpensive Taiwanese labor with foreign capital to build one of Asia's first urban-industrial countries and one of its Asian Tigers (•Figure 7.3.16). Its major exports are machinery, electronics (notably computers, cell phones, modems, routers, and global positioning systems), metals, textiles, plastics, and chemicals, especially to mainland China and the United States. The population is 78 percent urban, its infant mortality rate is very low, and its birth rate is one of the lowest in Asia (see Table 7.1). The average Taiwanese citizen

298

is four times wealthier than the average mainland Chinese citizen. In almost all respects, Taiwan is clearly recognizable as a more developed country. A major hurdle against even stronger growth has been a lack of energy resources. Some coal and natural gas exist, and some hydropower has been harnessed, but the island depends heavily on imported oil.

Taiwan's Republic of China continues its claim as the legitimate government of China. Meanwhile, the People's Republic of China claims Taiwan as part of its own territory. The United States backed the Nationalist claim for a time, but during the 1970s, the United States and the mainland People's Republic developed closer relations. After years of opposition, in 1971, the United States supported the revocation of Taiwan's seat in the United Nations and its replacement by the People's Republic. In 1979, bowing to the wishes of the People's Republic, the United States withdrew its official recognition of Taiwan, instead recognizing China's claim of sovereignty over Taiwan. This **One China Policy** prevents the United States from having formal diplomatic relations with Taiwan. However, the United States opposes the annexation of Taiwan to China by force, vowing to defend Taiwan from attack, just as it opposed Taiwan's call for a Nationalist effort to retake the Chinese mainland during periods of internal weakness there. In the meantime, the United States is a strong backer of Taiwan, supplying it with generous packages of weapons and economic aid.

Taiwan continues to resist mainland China's overtures for reunification, which include promises of broad autonomy, and is moving increasingly away from any type of reunion. That trend reflects Taiwan's ethnic makeup. Less than 15 percent of the island's people descend from the Nationalist refugees, who with their recent ties to the mainland are known as "mainlanders." Native non-Chinese people make up about 2 percent of the population. The vast majority of the island's inhabitants are descendents of Han peoples who emigrated from China as many as four centuries ago, notably the Fujianese, who came from the southeastern coastal province of Fujian, and the Hakka, from Guangdong province. This majority is much less mainland-oriented in political outlook. Despite discrimination by the Nationalists, these "indigenous" Chinese of Taiwan have enjoyed increasing political power. In 2000 elections, their Democratic Progressive Party defeated the Nationalists for the first time in the country's history. Though a majority of Taiwanese support continuing the status quo indefinitely, President Chen Shui-bian of the Democratic Progressive Party came under pressure in 2004 by advocates of Taiwanese independence to fulfill his campaign vows to achieve formal independence for Taiwan within four years.

With Chen's election, the relationship between Taiwan and China became very tense. The two countries must confront serious choices about economic and political relations. Taiwan's economy has become very dependent on mainland China, where Taiwanese companies take advantage of low-cost mainland labor (about one-fifth the cost in Taiwan for the manufacture and assembly of products) and where they have a huge market for their products. But there is strong pressure on President Chen to sever most ties with China and assert Taiwan's independence more forcefully. These steps could have large economic consequences and invoke military intervention from Beijing—and an American military response.

7.3.7 Mongolia

Mongolia (until recently known as the Mongolian People's Republic and popularly referred to as Outer Mongolia) was once part of the Chinese empire, but it became a separate Communist country in 1924. Relations with the former Soviet Union were close, both politically and economically. Mongolia had the distinction of being the first Asian country to abandon communism (in 1990). It has an area of about 600,000 square miles (960,000 sq km) and a population of about 2.6 million. Roughly 85 percent of the population is ethnic Mongol, and the largest minority (7 percent) is Turkic. Almost all residents of the country follow Tibetan (Lamaist) Buddhism. About 59 percent of Mongolia's population is urban. Ulan Bator (population 900,000) is the capital and only major urban center.

Mongolia was the home of Genghis Khan and his grandson, Kublai Khan, who in the second half of the 13th century assembled the largest land kingdom history has ever known. From the Korean Peninsula in the east, all across the top of Asia to near the eastern border of the North European Plain, the Mongols—without benefit of a written language, cities, or most of the trappings associated with "civilization"—controlled an empire of major magnitude. And although they could not maintain that expansive empire, the Mongols were able to take over China in 1279 and control it until 1368, when the Ming Dynasty ascended.

The world's largest landlocked country, Mongolia contains large desert plains in the south and east, locally termed *gobis* (hence the Gobi Desert), mountain ranges in the west, and wooded hills and mountains to the north. Most of the grassy valleys that are the archetypal steppes of Asia are also in the north (●Figure 7.3.17). These steppes are favored lands for the vast herds of livestock, mainly sheep and goats, that Mongolians graze. Much of the population still derives its livelihood from animal husbandry. The Communist government made major efforts to promote farming (especially of wheat, barley, and potatoes), but even today the total amount of cultivated land amounts to no more than 1 percent of Mongolia's area. Mongolia's export income comes mainly from mineral products (especially copper), livestock and animal products, and textiles, including cashmere wool.

Political and economic changes are progressing rapidly. In 1990, the first multiparty elections ever held resulted in a coalition government committed to changing the Soviet-style command economy to a market economy. New private businesses are being established, and many state-owned enterprises and properties are being privatized. Commercial

163

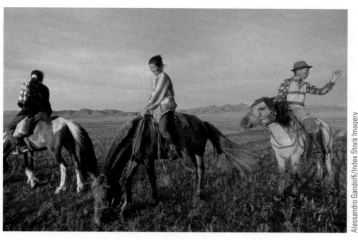

● **Figure 7.3.17** Many lives and much lore of the Mongolian steppe are focused on horses.

banks, business schools, and a stock market have opened, and most Russian technical and military personnel have left.

Much economic hardship has accompanied the changeover. There is tension between the Mongolian majority and the Chinese minority that so far has prospered from capitalist reforms. In 2003, the government began the privatization of what was once the exclusively government-owned land, with every Mongolian now entitled to a free plot of land. There are fears that the purchase and parceling out of the grasslands will bring an end to Mongolia's ancient nomadic way of life, which is an extensive form of land use requiring unbridled access to pasture.

Module 7.4 turns to two lands inhabited mainly by distant kin of the Mongols: Japan and Korea. Both have had complex and often tense relations with China, which cast a large shadow on the western Pacific.

SUMMARY

→ China is home to approximately 20 percent of the world's population on about 7 percent of the world's land. With 3.7 million square miles (9.6 million sq km) in area, it is the world's third-largest nation and the largest in East Asia.

→ China's language, literature, government, agriculture, religion, and other major cultural traits diffused from China to other parts of East Asia during the last several millennia.

→ China was subjected to colonial interests in much of the 19th century, including Britain's successful bid to create large-scale and profitable opium addiction. Several European countries and Japan won territorial and trade concessions from China that did not end until after World War II with the Communist defeat of the Nationalists.

→ China may be physically divided into the arid west and the humid east. An arc drawn from China's border with Myanmar northeast to Harbin is the rough dividing line between the arid west and the humid east. The majority ethnic Han are associated with the more densely populated humid east, and non-Han realms are primarily in the arid west. The Huang He (Yellow River) and the Chang Jiang (Yangtze River) descend from the Tibetan highlands.

→ Humid China, also known as China Proper, is densely settled, focused on irrigated rice production, and highly dependent on the Chang Jiang and other river systems. This region is further subdivided into North China and South China. In subtropical South China, the agriculture is centered on rice. North China, which is drier than South China and has a more continental climate, is more dependent on wheat, corn, soybeans, and sorghum. The Huang He deposits beneficial silt on the North China Plain but because of regular flooding is also called "China's Sorrow."

→ The Northeast, historically called Manchuria by the West, was the early industrial heartland of China. It has coal, iron ore, petroleum, and well-developed transportation systems.

→ The Three Gorges Dam, under construction on the Chang Jiang, is designed to control floods, provide hydropower, and allow transfer of water to the more arid north. The project has been both celebrated and lamented because of the scale of environmental change and social dislocation associated with it.

→ Revolutionary campaigns in the first three decades of government after the 1949 defeat of the Chinese Nationalists led to experiments in communal agriculture, migration control, and backyard industrialization, all of which were less significant than China's opening to the Western free-market economy after the death of Mao Zedong in 1976. Since then, China has modeled much of its most dynamic economic growth after capitalist rather than Communist economic and political models.

→ Chinese agriculture in the past four decades has involved expansion of farmland, increase in irrigated land, opening new land in areas not previously farmed, and the use of more chemical fertilizers, new plant stock, and slowly increasing agricultural mechanization. Much capital has also been spent on expanding the railroad network and highway systems and upgrading the canal and river systems.

→ China's urban centers have continued to grow, with China now about 44 percent urban. Inexpensive labor helped fuel the booming industries of the coastal cities. There is still poverty in China, with the cities generally wealthier than rural areas. The most prosperous areas are in the southeast, along the east coast (including the Special Administrative Regions of Hong Kong and Macau, and several Special Economic Zones), and in farmlands near large cities, while regions in the west, southwest, uplands, and Arid China have had less economic development. Jobs and wages are now growing in more areas, and labor shortages have developed in the coastal boomtowns.

→ Taiwan (the Republic of China) continues to claim to represent the true government of China and has a tense relationship with the mainland People's Republic of China. Taiwan's current president leans toward less political accommodation with

the mainland, but his country is heavily invested in mainland China's development. The United States does not have diplomatic relations with Taiwan but supports it militarily and economically.

→ One of the first countries to abandon communism in the final decade of the 20th century, Mongolia has privatized its industries and lands, perhaps threatening the pastoral nomadic livelihood for which the country is famous.

KEY TERMS + CONCEPTS

Arid China (p. 367)
autonomous regions (p. 372)
bird flu (p. 381)
break-of-bulk point (p. 376)
Chang Jiang Water Transfer Project
   (p. 376)
"China's Sorrow" (p. 374)
Chinese Communist Party (CCP) (p. 367)
Chinese Revolution of 1911 (p. 367)
commune system (p. 378)
Cultural Revolution (p. 383)
Dalai Lama (p. 372)

extraterritoriality (p. 367)
Four No's Campaign (p. 378)
genuine autonomy (p. 372)
Great Leap Forward (p. 378)
Humid China (p. 367)
Hanification (p. 372)
karst (p. 375)
Long March (p. 367)
Middle Kingdom (p. 365)
multiple cropping (p. 375)
Nationalist Party (Kuomintang, or KMT)
   (p. 367)

One China Policy (p. 388)
"one country, two systems" (p. 385)
Opium War (p. 367)
pandemic (p. 381)
severe acute respiratory syndrome (SARS)
   (p. 380)
Special Administrative Region (SAR)
   (p. 385)
Special Economic Zones (SEZs) (p. 385)
sweatshops (p. 386)
treaty ports (p. 367)
Western Big Development Project (p. 372)

REVIEW QUESTIONS

1. What are some of the innovations that began in China and spread through eastern Asia and even farther afield?

2. What were some of China's experiences with foreign powers prior to 1949?

3. What were some of the major events and other milestones of China in the 20th century? Who were the major political figures associated with these events?

4. What are some of the major differences in physical and cultural geographies of Arid and Humid China? Of North and South China? What are the distinguishing characteristics of China's Northeast?

5. For what reasons were the Great Wall and the Grand Canal built?

6. What were the Great Leap Forward, the commune movement, and the Cultural Revolution?

7. What factors may threaten the continuation of China's economic boom?

8. How can China's economy be characterized? What is its "average" standard of living based on per capita GNI PPP? How effectively is the wealth distributed? Where is it concentrated geographically, and how is the government hoping to change that pattern?

9. Where are the main clusters of China's cities? What geographic and historical factors help explain these concentrations?

10. Where are China's Special Economic Zones? What does their distribution suggest about China's economic geography? What is the history and status of each of China's Special Administrative Regions?

11. What are the major environmental, economic, and political attributes of Taiwan? Of Mongolia?

NOTES

1. Senicyang, "Chinese Tea," February 18, 2006, http://my.opera.com/chinaculture/blog/show.dml/148037. Accessed October 12, 2007.

2. Jamil Anderlini and Mure Dickie, "Taking the Waters: China's Gung-Ho Economy Exacts a Heavy Price." *Financial Times,* July 24, 2007, p. 11.

3. Howard W. French, "Next Wave of Camera-Wielding Tourists Is from China." *New York Times,* May 17, 2006, http://travel.nytimes.com/2006/05/17/world/asia/17travel.html. Accessed October 13, 2007.

Friends take a self-portrait after making a temple offering, Tokyo.

Joe Hobbs

MODULE 7.4
JAPAN AND THE KOREAS: ADVERSITY AND PROSPERITY IN THE WESTERN PACIFIC

On a cluster of islands off the eastern coast of Asia, the 127 million people of Japan operate an economy smaller only than that of the United States. This powerful economic engine is on mountainous, resource-poor, crowded lands about the size of California that are home to almost half as many people as the entire United States. Japan's dramatic rise from the ashes of World War II has drastically changed the world's economic geography. This module explains that phenomenal achievement and some of the challenges associated with it.

Although the Korean Peninsula has a seemingly marginal position on the Pacific Rim and the Eurasian landmass, it, too, has a central place in geopolitical affairs, mostly because of its strategic location between Japan, China, and Russia; the prosperity of South Korea; and the sometimes dangerous backwardness of North Korea. This module examines the problems of divided Korea and its prospects for reunification.

module outline

module objectives

This module should enable you to

→ Recognize the pattern of natural hazards associated with locations where plates of the earth's crust collide

→ Appreciate how countries burdened by a lack of natural resources can become prosperous by marshaling their human resources

→ Understand the unique challenges associated with a postindustrial society that has a prosperous and aging population

→ See how a country's geographic location can make it a target of division and conquest

→ Appreciate how different political and economic systems can produce dramatically different results for almost identical peoples

7.4.1 The Japanese Homeland

There are four main islands in the Japanese archipelago (●Figure 7.4.1). On the west, the Sea of Japan—which Koreans know as the East Sea—and the Korea Strait separate the main islands from Russia and Korea. Japan faces the open Pacific on the east. Hokkaido, the northernmost island, is close to Russian-controlled islands to its north. The scenic and busy Inland Sea separates Honshu, the largest island, from Shikoku and Kyushu. South of Kyushu, the smaller Ryukyu Islands run southwestward almost to Taiwan.

All of the main islands are mountainous, with the higher peaks mostly between 5,000 and 9,000 feet (c. 1,500–2,750 m) above sea level. The islands are volcanic (●Figure 7.4.2) and are subject to frequent and sometimes severe earthquakes (see Natural Hazards, pages 396–397). Located west of Tokyo, volcanic Mount Fuji is Japan's highest mountain at 12,388 feet (3,776 m). An **active volcano** (one that has had an eruption in recorded history), in the early 2000s Mount Fuji began experiencing "swarms" of earthquakes that many geologists interpreted as signs of an approaching eruption; it would be the first since 1707. Mount Fuji was sacred to the aboriginal Ainu people of Japan as long as 2,000 years ago. It is still a holy mountain in the national Shinto faith, and about 300,000 pilgrims and tourists visit its lofty summit each July and August, when it is free of snow. The journey to the peak of this most prominent symbol of Japan is a difficult one, as suggested in the Japanese proverb "It is as foolish not to climb Fuji-san as to climb it twice in a lifetime."

Climate and Resources

Japan's location east of the Eurasian landmass at latitudes comparable to those from South Carolina to Maine strongly influences its climates (see Figure 7.2, page 282). A humid subtropical climate extends from southernmost Kyushu to north of Tokyo, a humid continental long-summer climate characterizes northern Honshu, and Hokkaido has a humid continental short-summer climate. In lowland locations, where most of the people live, the average temperatures for the coldest month range from about 20° F (–7°C) in northern Hokkaido to about 40°F (4°C) in the vicinity of Tokyo and about 44°F (7°C) in southern Kyushu. Averages for the hottest month are only about 64°F (18°C) in northern Hokkaido but rise southward to about 80°F (27°C) around Tokyo and even higher farther south. Conditions are cooler in the mountains, with snowfalls common in winter in Honshu and Hokkaido. Hokkaido has twice hosted the Winter Olympics.

Japan is a humid country. Annual precipitation of more than 60 inches (c. 150 cm) is common from central Honshu south, and 40 inches (c. 100 cm) or more fall in northern Honshu and Hokkaido. Their location off the east coast of Eurasia puts the islands in the path of the inward-blowing summer monsoon from the Pacific, and summer is the main rainy season in eastern Japan. But the islands are also exposed to the seaward-blowing winter monsoon after it has crossed the Sea of Japan (East Sea), so much precipitation also falls in winter in western Japan. Japan is subjected in summer and fall to powerful typhoons, the great low-pressure storm systems known as hurricanes in the western Atlantic region. These add to the woes of this hazard-prone country. 283 338, 531, 602

Japan's natural resource base is not large enough to meet the country's growing needs. Until the middle of the 19th century, the country was nearly self-sufficient economically, although the standard of living was low. Japan was already crowded with about 30 million people, but disease, starvation, abortion, and infanticide slowed the rate of natural increase. Industrialization stimulated rapid population growth in the second half of the 19th century and later. Population growth has recently plateaued, however, as delayed marriages and the rising costs of child rearing have lowered birth rates. Meanwhile, affluence has increased dramatically. The material demands of so many people enjoying a high standard of living in a country with limited natural resources require Japan to import an astonishing 97 percent of its raw materials and energy supplies and about 60 percent of its food.

Forests, which now cover about 70 percent of the land surface, illustrate the limitations of Japan's natural resources. When it was a preindustrial and early industrial society with a much smaller population, the nation's heavily forested

● **Figure 7.4.1** Principal features of Japan.

mountainsides supplied the wood it needed, but if today's demand for forest products were met with domestic supplies, the country would be rapidly deforested. A wide range of valuable trees grow in Japan, mainly broadleaf deciduous toward the south and needleleaf coniferous toward the north. Many mountainsides are quilted with rectangular plots of trees planted in rows. Most first-growth timber is long gone, but some primary forests are still protected in preserves. Forests are important in Japanese culture, and this affluent country does not wish to lose them. With its enormous demand for wood products, Japan is therefore the world's second-largest importer of wood (after China), mostly hardwoods from the tropical rain forests of Southeast Asia.

Michael S. Yamashita/Corbis

● **Figure 7.4.2** Japan has numerous volcanoes—some active and others extinct. In 1991, Mount Unzen, near Nagasaki on the island of Kyushu, erupted for the first time since 1792. Thirty-five people were killed as a pyroclastic flow inundated the village of Shimabara below.

Elizabeth Whiting & Associates/Alamy

● **Figure 7.4.3** The wood and paper characteristic of Japanese interior design contrast sharply with the drywall and metals of the West. This is a traditional Japanese guest room, or *ryokan*. The futons used for sleeping, stuffed with fragrant dried vegetation, are rolled up and put away in the daytime.

Kerosene stoves have largely replaced the charcoal formerly used for household cooking and heating, but the Japanese still use wood for many purposes. Japan's large publishing industry and other industries and handicrafts require a great deal of paper. Single-family dwellings continue to be built largely of wood, which provides more safety from earthquake shock waves than more rigid construction (●Figure 7.4.3). However, wood makes the dwellings more vulnerable to the fires that typically sweep through Japanese cities after earthquakes rupture gas lines.

Small deposits of minerals, including coal, iron ore, sulfur, silver, zinc, copper, tungsten, and manganese, supported early industrialization in Japan. These minerals are in short supply today, and like most other raw materials, they must be imported. Japan's mountain streams provided much of the power for early industrialization, but they are so swift, shallow, and rocky that they are of little use for navigation. Most of the water used for irrigation in Japan comes from these streams.

Only about one-eighth of this small country's area is arable; Japan is not one of the world's major crop-growing nations. Most cultivable land is in mountain basins and in small plains along the coast. Terraced fields on mountainsides supplement this level land. Irrigated rice, the country's basic food and most important crop, was once nearly everywhere on arable land (●Figure 7.4.4). Production of this grain is heavily subsidized by the Japanese government and promoted by a powerful farm lobby, which aims at rice self-sufficiency. The country's farmers usually produce a net export of rice. However, in the 1990s, a scarcity of rice throughout East Asia and pressure on Japan to reduce its massive trade surplus with the United States prompted Japan to open its rice market to foreign imports for the first time. Now with silos overflowing with rice and with the

Japanese diet more dependent on wheat (in the form of bread and noodles) than on rice, the government is paying Japanese farmers not to grow rice. Japan's lavish subsidies of its declining, aging population of farmers are the subject of much condemnation by free-trade advocates.

Mild winters and ample moisture permit **double cropping**—the growing of two crops a year on the same field—on most irrigated land from central Honshu south. Rice is grown in summer, and wheat, barley, or another winter crop is planted after the rice harvest. Overall, about one-third of Japan's irrigated rice fields are sown to a second crop, and more than half of the unirrigated fields are double-cropped. **Intertillage**—the growing of two or more crops simultaneously in alternate rows—is also common. In addition, dry farming produces a variety of crops: sugar beets in Hokkaido; all sorts of temperate-zone fruits and vegetables; potatoes, peas, and beans, including soybeans; and in the south, tea, citrus fruits, sugarcane, tobacco, peanuts, and mulberry trees to feed silkworms, which eat their leaves.

Changes in Japan's agricultural sector have accompanied a surge in urban-industrial employment and general affluence in recent decades. Farmers have been moving to towns, and the number of agricultural workers has dropped dramatically (from about 25 percent of the labor force in 1962 to 4 percent in 2007). A huge increase in farm mechanization has made it possible to till Japan's tiny farms with far less human labor. With this technology and the heavy

● **Figure 7.4.5** Fish and seafood provide the major source of protein for the Japanese. *Joe Hobbs*

● **Figure 7.4.4** Rice fields near Tokyo. Note that people maximize the amount of land dedicated to rice by confining their settlements to the hills and that the forests on those hills have been preserved.

applications of fertilizers, plus improved crop varieties, Japanese agriculture has achieved the world's highest yields per unit of land.

The island people of Japan view the ocean as an important resource and a promising frontier. The country has an ambitious "inner space" program of deep-sea exploration to mine the sea's living and mineral riches. Many food fish species are in the waters around Japan, and Japanese fishermen range widely through the world's major ocean fishing grounds. Japan takes more fish from the sea than any other nation—about 15 percent of the world's catch and the Japanese have the world's highest per capita consumption of fish and other marine foods, including *sashimi,* or raw seafood (●Figure 7.4.5). Some 40 to 50 percent of the protein consumed by the Japanese is in the form of fish (the figure is closer to 5 percent in the United States).

Historically, the Japanese have also enjoyed whale meat, but pressure from international environmentalists compelled Japan in 1985 to support a worldwide moratorium on commercial whaling. However, Japan has resumed small-scale whaling of minke, Bryde's, and sperm whales for what are called "scientific research" purposes. This loophole allows whale meat to be widely sold. Subjected to strong international condemnation for putting whale on the dinner table, Japanese whaling interests have accused outsiders of practicing "culinary imperialism."

In recent years, the Japanese have begun to eat less fish and rice in favor of a more Western diet. Although ocean fish continue to be the main sources of animal protein in the Japanese diet, increased consumption of meat, milk, and eggs is reflected in the growing number of cattle, pigs, and poultry on Japanese farms. The animals require heavy imports of feed, and the increasing variety in people's tastes means many more food imports. Despite the country's agricultural efficiency and in part because of its diversifying tastes, Japan is the world's largest agricultural importer country, relying on imports for 60 percent of its food needs. The United States is Japan's largest supplier of imported feeds and food, and the United States exports more food to Japan than to any other country.

Japan's Core

Most of Japan's economic activities and people are packed into a corridor about 700 miles (c. 1,100 km) long. This **megalopolis** (several large cities that make up an urban complex) extends from Tokyo, Yokohama, and the surrounding Kanto Plain on the island of Honshu in the east through northern Shikoku to northern Kyushu in the west (see Figure 7.4.1). This highly urbanized core contains more people than live in the Boston-to-Washington, D.C., megalopolis on the U.S. East Coast. All of Japan's greatest cities and many less important ones have developed here, on and near harbors along the Pacific and the Inland Sea. They are sited typically on small, agriculturally productive alluvial plains between the mountains and the sea.

NATURAL HAZARDS

Living on the Ring of Fire

194, 336

The people of Japan must cope with one of the most hazard-prone regions on earth. By far the greatest threats are related to Japan's location on the Pacific **Ring of Fire**, the name given to the boundaries of the Pacific plate of the earth's crust where earthquakes and volcanic eruptions are commonplace and often catastrophic (•Figure 7.4.A). In the scheme of plate tectonics, Japan is in an area of subduction where one plate (the Pacific) is being pushed, as if by a conveyor belt, under another plate (the Eurasian), creating the country's characteristic volcanoes, earthquakes, and related features (see Figures 2.1 and 2.2, page 21).

Japan is most threatened by earthquakes on and near its islands but must also be on guard for the effects of distant earthquakes, even as far away as the shore of Nort h America. A strong quake there can produce a tsunami that may travel thousands of miles to strike and flood coastal Japan (see pages 340–341). Japan has an extensive (and expensive) system of seawalls to protect coastal communities from tsunamis. Critics argue that the walls are not high enough and promote a false sense of confidence that people can control nature.

Japan has a long history of seismic assaults. One of the most significant recent events took place on January 17, 1995, when a devastating earthquake released the tension that had built up on a transverse fault near historic Kobe, a port city of 1.5 million people. The Japanese soon labeled the 6.9-magnitude episode the Great Hanshin Earthquake, establishing its place in the annals of Japanese disasters alongside such events as the Great Kanto Earthquake (magnitude 7.9 on the Richter scale, which killed 143,000 people in and around Tokyo in 1923).

The Kobe quake was the worst disaster to hit Japan since World War II, killing 6,400 people and injuring thousands more, damaging 190,000 buildings, displacing 300,000 people, and causing at least $100 billion worth of damage. Kobe's remarkable lack of disaster preparedness and relief focused attention on a common Japanese complaint: that the government invests too much in commerce-oriented development and not enough in the welfare of its people.

Japanese authorities did not predict the Great Hanshin Earthquake. After the fact, there were reports that large schools of fish swam uncharacteristically close to the surface of the sea off Kobe in the days preceding the quake. Crows and pigeons were reportedly noisier than usual and flew erratically in the hours before the ground moved. In the weeks preceding the quake, levels of radon gas rose in a water aquifer below the city and then declined sharply after January 9. Animal behavior, water level and composition, and many other variables are often debated but still unproved barometers of an impending earthquake.

Since 1965, Japan has spent more than $1 billion in an effort to develop means of predicting earthquakes. During the 1990s, more than $100 million yearly went to this effort. But by 1998, no earthquake had been successfully predicted, and Japan suspended this expensive research program. Its critics argue that earthquakes are chaotic events and will always be impossible to predict. They also point out the hazards of issuing an earthquake warning that is not followed by a real quake. A false alarm in the Tokyo region would cost an estimated $7 billion per day in lost business and other economic disruptions. The bigger question, also applicable to such cities as San Francisco and Mexico City, is how such a great metropolis could be evacuated safely, efficiently, and with minimal panic.

With the prediction program scrubbed, Japan has redoubled its efforts to prepare for future "big ones." New structures must meet a tough safety code that aims at "earthquake-proof" construction. Much of the focus is on the Japanese capital, Tokyo, which is underlain by a complex system of active fault lines and typically suffers a big quake every 70 years (in 2007, it had been more than 80 years since the last one). Current estimates are that a big one nearing 8.0 on the Richter scale would kill 6,717 people and destroy 300,000 structures (mostly by fire) in Tokyo. These numbers do not include the effects of the tsunami an estimated 27 feet (8 m) high that would strike low-lying Tokyo within minutes of an offshore quake. Much of the devastation in Kobe resulted from the city's location on soft coastal soils and reclaimed land, which also underlie large parts of Tokyo.

With such threats in mind, the government began serious discussions about moving most national administrative functions inland to a new capital, where tsunamis are not an issue and where mountain bedrock would reduce the devastating impacts of seismic shock waves. Two sites on the island of Honshu have the most

The largest urban complex in the world—Tokyo, Yokohama, and the surrounding suburbs (population 37 million by one estimate; figures differ according to how the metropolitan area is defined)—has developed on the Kanto Plain, Japan's most extensive lowland. Tokyo grew as Japan's political capital and Yokohama as the area's main seaport. This urban complex functions today as Japan's national capital, one of its two main seaport areas (the other is Osaka-Kobe), its leading industrial center, and the commercial center for northern Japan (•Figure 7.4.6).

Beyond Tokyo, in the entire northern half of Japan, the only major city is Sapporo (population 2.5 million), on Hokkaido. About 200 miles (c. 320 km) west of Tokyo lies Japan's second-largest urban cluster, in the Kinki District at the head of the Inland Sea. Here three cities—Osaka, Kobe, and Kyoto—form a metropolis, together with their suburbs, of about 17.5 million people.

The inland city of Kyoto was the country's capital from the 8th century C.E. until 1869 (when it was moved to Tokyo) and is now preserved as a shrine city where modern

NATURAL HAZARDS

Living on the Ring of Fire *continued*

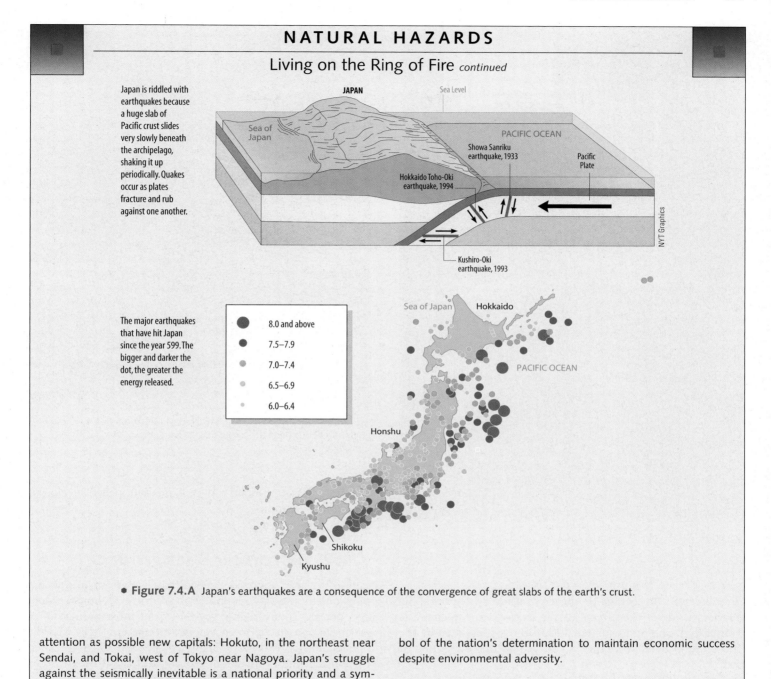

Japan is riddled with earthquakes because a huge slab of Pacific crust slides very slowly beneath the archipelago, shaking it up periodically. Quakes occur as plates fracture and rub against one another.

JAPAN
Sea Level
Sea of Japan
PACIFIC OCEAN
Showa Sanriku earthquake, 1933
Pacific Plate
Hokkaido Toho-Oki earthquake, 1994
Kushiro-Oki earthquake, 1993
NYT Graphics

The major earthquakes that have hit Japan since the year 599. The bigger and darker the dot, the greater the energy released.

8.0 and above
7.5–7.9
7.0–7.4
6.5–6.9
6.0–6.4

Sea of Japan
Hokkaido
PACIFIC OCEAN
Honshu
Shikoku
Kyushu

• **Figure 7.4.A** Japan's earthquakes are a consequence of the convergence of great slabs of the earth's crust.

attention as possible new capitals: Hokuto, in the northeast near Sendai, and Tokai, west of Tokyo near Nagoya. Japan's struggle against the seismically inevitable is a national priority and a symbol of the nation's determination to maintain economic success despite environmental adversity.

industrial disfigurement is prohibited (•Figure 7.4.7). Some U.S. military planners wanted to target Kyoto with a nuclear weapon in August 1945, but the decision that it should be avoided because of its historical importance spared the city. Kyoto continues to be the destination of millions of pilgrims and tourists annually.

Osaka developed as an industrial center and Kobe as the district's deepwater port. Both cities now combine port and industrial functions. Another large metropolis, Nagoya (population 8.8 million), is situated directly between Tokyo and Osaka on the Nobi Plain at the head of a bay. It is both a major industrial center and a seaport.

West of the Kinki District, smaller metropolitan cities spot the core. The largest are metropolises of more than 1.5 million people: the resurrected Hiroshima on the Honshu side of the Inland Sea, Kitakyushu (a collective name for several cities) on the Strait of Shimonoseki between Honshu and Kyushu, and Fukuoka on Kyushu. Hiroshima was destroyed on August 6, 1945, when the United States dropped the first atomic bomb ever used in warfare, detonating it

Joe Hobbs

● **Figure 7.4.6** Tokyo's Ginza District.

Bruce Burkhardt/Westlight

● **Figure 7.4.7** Kyoto has a unique status in Japanese history and religious life. This is the 14th-century Kinkakuji Temple, also known as the Golden Pavilion.

bombing, has an official policy that it will never develop or use atomic weapons.

303

directly over the center of the city. With no topographic barriers to deter the effects of the explosion, the city was obliterated, and an estimated 140,000 residents died by the end of 1945, when radiation sickness had taken its greatest toll.

Among the several smaller cities of northern Kyushu is Nagasaki, which, on August 9, 1945, was hit by a second U.S. atomic bomb, the last used in warfare. Nagasaki was chosen during the flight mission after clouds obscured the primary target, Kokura. The bomb detonated over the suburbs rather than the city center, and hills helped diminish the explosion's impact, but even so, an estimated 70,000 people died by year's end. The bombing of Nagasaki brought Japan's surrender and the end of World War II. American decision makers insisted that the use of these weapons spared many thousands of lives, both American and Japanese, that would have been lost if the United States had instead undertaken an invasion of the Japanese homeland. Japan, which has always utterly condemned the

7.4.2 Historical Background

Japan has seen successive periods of isolationism and expansionism and of economic and military accomplishment and defeat. Today, it is near the top of most indicators of prosperity in the more developed countries (MDCs). The remarkable success story of Japan can best be appreciated by considering its turbulent history and the difficult home environment in which the Japanese have lived.

The Japanese, Japan's dominant ethnic group, are descended from peoples who reached Japan from other parts of eastern Asia in the distant past, especially after 7000 B.C.E. Most of these ancestors of the modern Japanese may have arrived at about the same times as the **Ainu.** Often described as the original indigenous group of Japan, the Ainu (also called **Utari**), numbering perhaps 20,000 today, are unrelated to the ethnic Japanese. Scholars dispute their ethnic and geographic origins. The majority Japanese gradually drove the Ainu into outlying areas, mainly in Hokkaido.

The oldest surviving Japanese written records date from the 8th century C.E. These depict a society strongly influenced by Chinese cultural and ethnic traits, often reaching

293 Japan by way of Korea. A distinct Japanese culture gradually evolved, and today, it is linguistically and in other ways different from its Chinese and Korean antecedents. Traditions about Japanese origins extend back to the reign of Jimmu, the first emperor, whose accession is ascribed to the year 660 B.C.E. but who may be a mythical rather than a historical figure. Traditions of kingship are important in the Japanese worldview, which holds that the islands and their emperor are of divine origin. Jimmu was said to have descended from the Sun Goddess, and the Japanese have known their homeland as the "Land of the Gods." The Japanese concept of Japan as unique and invincible developed early, shaped by legends such as that of the *kamikaze,* the "divine wind" that repelled a Mongol attack on Japan in the 13th century.

The early emperors gradually extended control over their island realm. A society organized into warring clans emerged. By the 12th century, powerful military leaders called *shoguns,* who actually controlled the country, diminished the emperor's role. Meanwhile, nobles, or *daimyo,* whose power rested on the military prowess of their retainers, the *samurai,* ruled the provinces. This structure resembled the European feudal system of medieval times.

Adventurous Europeans reached Japan early in the mid-16th century, beginning with the Portuguese in the 1540s. Most of these early arrivals were merchants and Roman Catholic missionaries. Japanese administrators allowed the merchants to trade and the missionaries to preach freely. There were an estimated 300,000 Japanese Christians by the year 1600.

Contacts with the wider world were curtailed as Japan entered a period when military leaders imposed central authority on the disorderly feudal structure. After winning the battle of Seikigahara in 1603 with arms purchased from the Portuguese, the warlord Ieyasu Tokugawa proclaimed himself Nihon Koku Taikun ("Tycoon of All Japan") and became the first *shogun* to rule the entire country. During the era of the **Tokugawa Shogunate,** from 1600 to 1868, the Tokugawa family acquired absolute power and shaped Japan according to its will. To maintain power and stability, the early Tokugawa *shoguns* wanted to eliminate all disturbing social influences, including foreign traders and missionaries. They feared that the missionaries were the forerunners of attempted conquest by Europeans, especially the Spanish, who held the Philippines. The *shoguns* drove the traders out and nearly eliminated Christianity in persecutions during the early 17th century. After 1641, a few Dutch traders were the only Westerners allowed in Japan. Their Japanese hosts segregated them on a small island in the harbor of Nagasaki, even supplying them with a brothel so that they would not be tempted to venture out and pollute Japan's ethnic integrity. Under Tokugawa rule, Japan settled into two centuries of isolation, peace, and stagnation.

Foreigners succeeded in reopening the country to trade two centuries later, when Japan was ill prepared to resist Western firepower. Visits in 1853 and 1854 by American naval squadrons of black ships under Commodore Matthew Perry (who wanted to establish refueling stations in Japan for American whaling ships) resulted in treaties opening Japan to trade with the United States; this became known as **gunboat diplomacy.** The major European powers were soon able to obtain similar privileges. Some feudal authorities in southwestern Japan opposed accommodation with the West. United States, British, French, and Dutch ships responded in 1863 and 1864 by bombarding coastal areas under the control of these authorities. These events so weakened the Tokugawa Shogunate that a rebellion overthrew the ruling shogun in 1868. The revolutionary leaders restored the sovereignty of the emperor, who took the name Meiji, or "Enlightened Rule." The revolution of 1868 is known as the **Meiji Restoration.**

The men who came to power in 1868 aimed at a complete transformation of Japan's society and economy. They felt that if Japan were to avoid falling under the control of Western nations, its military impotence would have to be remedied. This would require a reconstruction of the Japanese economy and social order. They approached these tasks energetically and abolished feudalism, but only after a bloody revolt in 1877. The Meiji leaders forged the model of a strong central government that would modernize the country while resisting foreign encroachment. Under this system, which lasted until 1945, democracy was strictly limited, with small groups of powerful men manipulating the government's machinery and the emperor's prestige. Military leaders were prominent in the power structure.

Following their fact-finding tours of Europe and the United States, Meiji leaders pressed their people to learn and apply the knowledge and techniques that Western countries had accumulated during the centuries of Japan's isolation. Foreign scholars were brought to Japan, and Japanese students were sent abroad in large numbers. Japan adopted a constitution modeled after that of imperial Germany. The legal system was reformed to be more in line with Western systems. The government used its financial power, which it obtained from high land taxes, to foster industry. New developments included railroads, telegraph lines, a merchant marine, light and heavy industries, and banks. Wherever private interests lacked capital for economic development, the government provided subsidies to companies or built and operated plants until private concerns could acquire them. Rapid urbanization accompanied a process of rapid industrialization.

So spectacular were the results of the Meiji Restoration that within 40 years, Japan had become the first Asian nation in modern times to emerge as a world power. Outsiders often spoke of the Japanese in derogatory terms as mere imitators. The Japanese, however, knew which elements of Western technology they wanted, and they adapted them successfully to Japanese needs.

The shift from Tokugawa isolationism to Meiji openness put Japan on a pathway to empire and eventually to war.

Japan was bent on expansion between the early 1870s and World War II, and by 1941, it controlled one of the world's most imposing empires (•Figure 7.4.8). Between 1875 and 1879, Japan absorbed the strategic outlying archipelagoes of the Kuril, Bonin, and Ryukyu Islands. Japan won its first acquisitions on the Asian mainland by defeating China in the Sino-Japanese War of 1894–1895, which broke out over disputes in Korea. In the humiliating 1895 Treaty of Shimonoseki, China ceded the island of Taiwan (Formosa) and Manchuria's Liaotung Peninsula to Japan and recognized the independence of Korea.

Feeling threatened by Japan's emerging strength so close to its vital port of Vladivostok, Russia demanded and won a Japanese withdrawal from the Liaotung Peninsula and leased the strategic peninsula from China. Japan went to war with Russia in 1904 after attacking the peninsula's main settlement of Port Arthur. Russia lost the ensuing Battle of Mukden, which involved more than a half-million soldiers, making it history's largest battle to that date. Russia's loss of the Russo-Japanese War of 1904 represented the first defeat of a European power by a non-European power. Victorious Japan regained the Liaotung Peninsula, secured southern Sakhalin Island, and established Korea as a Japanese protectorate; Korea was formally annexed to Japan in 1910.

At the end of World War I, the Caroline, Mariana, and Marshall Islands, in the western Pacific Ocean, were transferred from defeated Germany to Japan as mandated territory. Japanese encroachments on China's region of Manchuria followed in the 1930s. In 1937, Japan began an all-out attack that overran the most populous parts of China. Japanese troops committed many atrocities in the assault on China, even employing biological warfare by airdropping plague-infested fleas on the city of Ningbo. The most notorious was the **Rape of Nanking** (now known as Nanjing) in 1937, when the invaders took control of China's temporary capital, killing an estimated 300,000 Chinese civilians and soldiers and raping about 20,000 Chinese women. All told, an estimated 2.7 million Chinese civilians were killed in Japan's effort to "pacify" China.

The encroachments continued. Japan seized French Indochina in 1940. After Japan's carrier-based air attack devastated the United States' Pacific Fleet at Pearl Harbor, Hawaii, on December 7, 1941 ("a day which will live in infamy," declared President Franklin Roosevelt), Japanese forces rapidly overran Southeast Asia as far west as Burma (Myanmar), as well as much of New Guinea and many smaller Pacific islands.

The motives for Japanese expansionism were mixed. They included a perception of national superiority and "manifest destiny," perceived security needs, a desire for recognition by the great powers, the wishes of military leaders to inflate their importance and gain control of the Japanese government, and the ambitions of industrialists to gain sources of raw materials (Manchurian and Korean coal, for example) and markets for Japanese industries.

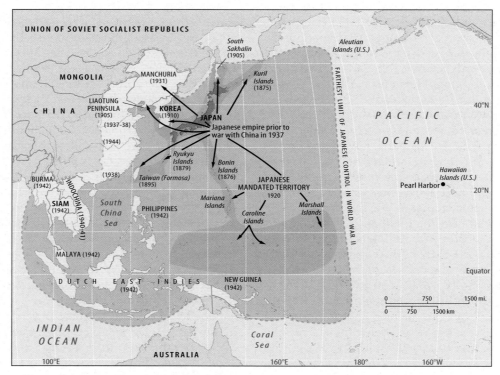

• **Figure 7.4.8** Overseas areas held by Japan prior to 1937 and the line of maximum Japanese advance in World War II.

The desire for materials and markets was based solidly on need. Expansion of industry and population on an inadequate base of domestic natural resources had made Japan dependent on sales of industrial products outside the homeland. Such sales provided, as they do now, the principal funds for buying necessary imported foods, fuels, and materials. As worldwide depression hurt the trade-dependent prewar Japanese economy, Japan chose to create a Japanese-controlled Asian realm that would insulate the Japanese economy from the volatile world economy and make the country a leading world power. In 1938, Japan therefore proclaimed its "New Order in East Asia," and in 1940 and 1941, this widened into the **Greater East Asia Co-Prosperity Sphere,** a euphemism for Japanese political and economic control over China and Southeast Asia.

Japan's militaristic and colonial enterprises proved disastrous at home and abroad. Japanese soldiers inflicted great suffering on civilians throughout the western Pacific. Even today, nations are demanding formal apologies from Japan. In August 1995, on the 50th anniversary of Japan's defeat in World War II, Japan's prime minister did issue a formal general apology for his nation's role in the war. Critics say Japan has not fully atoned for its war crimes and are trying, among other efforts, to block Japan's quest to gain a permanent seat on the United Nations Security Council. Annual visits by the Japanese prime minister to the Yasukuni War Memorial are accompanied by annual protests by the Chinese and Koreans, angry that 14 of the 2.5 million people enshrined there are convicted war criminals. There continue to be official protests by the governments of China, Korea, Taiwan, the Philippines, and other countries about Japan's reluctance to admit that it enslaved women as prostitutes for Japanese soldiers during World War II. An estimated 200,000 of these "comfort women" were put to work in brothels throughout the Japanese war empire.

August 1945 found Japan completely defeated. Its overseas territories, acquired during nearly 70 years of successful imperialism, were lost. Its great cities, including Tokyo—where 100,000 people died in napalm bombings— were in ruin. In addition to 1.8 million deaths among its armed services, Japan had suffered some 8 million civilian casualties from the American bombing of the home islands; most of its major cities were over half destroyed. Tokyo's population had fallen from nearly 7 million people to about 3 million, most of them living in shacks. In 1950, five years after the beginning of the U.S. military occupation of Japan, national production stood at only about one-third of its 1931 level, and annual per capita income was $32.

7.4.3 Japan's Postwar Miracle

Stripped of its colonies and its empire, Japan has astonishingly become an economic superpower since World War II. The nation's explosive economic growth after its defeat was one of the most remarkable developments of the late 20th century and is known widely as the "Japanese miracle."

Japan watchers cite different reasons for the country's economic success. Proponents of dependency theory argue that Japan, never having been colonized, escaped many of the debilitating relationships with Western powers that hampered many potentially wealthy countries. Some analysts believe that the country's postwar economic accomplishments grew from an intense spirit of achievement and enterprise among the Japanese. Notably, many Japanese attribute this industrious spirit to Japan's geography as a resource-poor island nation. To overcome the constraints nature has placed on them, the Japanese people feel they must work harder.

Still others attribute Japan's postwar achievements to its association with the United States following World War II. Some of the postwar U.S.-imposed reforms worked well economically, and relations between the two countries since the war have also generally stimulated Japan's industrial productivity. The U.S. military occupation of Japan from 1945 to 1952 brought about major changes in Japanese social and economic life. The divine status of the emperor was officially abolished. When Emperor Hirohito announced Japan's surrender on the radio, it was the first time the stunned Japanese public had ever heard the voice of this mythic figure. A new U.S.-written constitution (which has remained unchanged ever since) made Japan a constitutional monarchy with an elected parliamentary government. With the rejection of divine monarchy, the strongly nationalistic Shinto faith lost its status as Japan's official religion, although worship at Shinto shrines was allowed to continue; most Japanese observe a faith that is a syncretism of Shinto and Buddhist beliefs and rituals (•Figure 7.4.9). There was land reform in the countryside to do away with a feudal-style landlord system. Some major Japanese companies were broken up to reduce their monopolistic hold on the economy. Women were enfranchised. A democratic trade union movement was established.

• **Figure 7.4.9** Prayers at the Meiji Jinju Shrine, Tokyo. This Shinto shine is dedicated to the deified spirits of the Emperor Meiji and his spouse, Empress Shoken, who died early in the 20th century.

Perhaps most significant for the economy, the United States forbade Japan to rearm, except for its small **Self-Defense Forces (SDFs)**. When the occupation ended in 1952, leaving U.S. bases in Japan but returning control of Japanese affairs to the Japanese, Japan was placed under American military protection. Without large military expenditures, much of Japan's capital was freed to invest in economic development. Only recently has Japan sought a wider role for its military—for example, by posting some forces to serve in Iraq. As early as 2010, the Japanese may hold a referendum to change its pacifist constitution. The United States, which wrote the constitution, is in favor of this change because it would allow the two countries to form a stronger military counterweight against prospective foes in the region.

The U.S. military umbrella over Japan remains controversial today. Many Japanese regard it as an outmoded vestige of colonialism and the war. There have been several incidents of Japanese women being raped by U.S. soldiers stationed on the island of Okinawa, where three-quarters of the U.S. bases and more than half of the 50,000 U.S. troops in Japan are stationed. The resulting protests and diplomatic appeals led the United States to agree to return control of a major air base in Okinawa to Japan, to shift base locations on the island, and to relocate some troops from Okinawa to mainland Japan. In a subsequent referendum, the people of Okinawa voted overwhelmingly in favor of further U.S. military withdrawal from the island.

Japan's special relationship with the United States boosted the economy long after the American occupation ended. The postwar economic recovery accelerated when the United States called on Japanese production of steel, textiles, and clothing to support American forces in the Korean War of 1950–1953. American economic aid continued to flow to Japan. The United States permitted many Japanese products to have free access to the American market while allowing Japan to protect its economy from imports. Japanese firms also had relatively free access to American technologies, which they often improved on and marketed even before American firms did.

By the 1970s, Japan had become an industrial giant with a gross domestic product far exceeding that of any country except the United States and the Soviet Union. The United States and European nations found themselves at a disadvantage to Japanese competitors in many industries. They busily sought ways to match this competition or to protect themselves from it without doing too much damage to their exporting firms, consumers, and trade relations. By the 1980s, the United States and Japan were at odds over Japan's massive trade surplus with the United States.

Observers often point to several unique features of Japanese management and employment to explain the country's meteoric postwar gains. One has been recruitment through an extremely challenging (some say brutal) educational system that emphasizes technical training. While recent years have seen a move to a more lenient and more comprehen-

sive education, a rigorous and stressful testing system still determines access to higher education. Japanese management strategies emphasize benevolence toward employees, encouragement of employee loyalty, and participation of workers in decision making. About 20 percent of Japanese workers enjoy guarantees of lifetime employment in their firms, and many large Japanese companies help provide housing and recreational facilities for their employees.

One essential factor in Japan's postwar economic growth was a high level of investment in new and efficient industrial plants. Investment capital was freed by government policies that cut expenditures on amenities and services such as roads, antipollution measures, parks, housing, and even higher education. Japanese industrial cities grew explosively and became highly polluted areas of dense and inadequate housing, with few public amenities and snarled transportation. The money "saved" by not being spent on amenities and services was made available for investment in industrial growth, but the costs to Japanese society were high.

Some analysts cite elements of Japan's political culture to explain the country's economic successes. One political party, the **Liberal Democratic Party (LDP)**, has with few exceptions been elected to power since Japan regained its sovereignty in 1952. It is a conservative and strongly business-oriented and business-connected organization. The party has promoted Japanese exports with policies to keep the yen (the Japanese unit of currency)—and thus Japanese goods—inexpensive. The party has also cooperated closely with Japanese business in the development of new products and new industries. However, that very coziness between government and business—the "crony capitalism" seen in several other East and Southeast Asian countries—was partly to blame for the economic malaise that spread over Japan in the 1990s, and support for the LDP began to erode.

Despite all of these factors in Japan's economic favor, the Japanese miracle did not last. The peak of Japan's postwar success came in the 1980s. A powerful economic boom led to speculative rises in stock prices and land prices. At the end of the 1980s, the total value of all land in Japan was four times greater than that in the United States. The Tokyo Stock Exchange was the world's largest, based on the market value of Japanese shares. Then the **bubble economy** burst, and real estate and stock prices fell by more than 50 percent. Japan had near-zero economic growth through most of the 1990s and into the early 2000s. Japanese industries whose growth had seemed unstoppable experienced an increasing loss of market share to U.S. and European producers. Part of the reason the economy could not regain its footing was official Japanese commitment to prop up faltering industries and farms with huge subsidies. Because of the disproportionate political influence of small towns and rural areas, the Japanese tried to modernize remote villages and islands with expensive and underutilized public works projects. Due to such projects, Japan is now the world's most indebted country. Overall, the system succeeds in subsidizing its past at the expense of its future.

7.4.4 Japanese Industry

The gradual evolution of industry over more than three centuries that characterized Europe and the United States has been compressed into little more than a century in Japan. Early iron and steel development was concentrated in northeastern Kyushu, on and near the country's main coal reserves. The products went mainly into a new railway system and into shipping. Japanese coal, supplemented increasingly by imports, powered the country's factories, railways, and ships. (•Figure 7.4.10). Hydroelectricity was developed on mountain streams and became important in the energy economy.

Japan's petrochemical and other oil-based industries reflect the country's great dependence on imports. Almost devoid of petroleum reserves, the country imports nearly all of its oil and gas. With about 75 percent of its oil coming from the Persian-Arabian Gulf region—and no sign that that level of dependence will decline—Japan downplays its alignment with the West in most Middle East disputes. Japan briefly committed ground troops after the 2003 Iraq War began and maintained a naval presence to support U.S.- and NATO-led operations in Iraq and Afghanistan for what it insisted were humanitarian reasons.

Despite the unique stigma Japan associates with the power of the atom, the country has come to rely on nuclear power, with 55 plants supplying about 30 percent of the country's electricity needs. Hydropower supplies 8 percent, but with fossil fuels almost absent, the critical lack of energy has compelled the nation to develop the world's most energy-efficient economy. The government imposes high taxes on fossil fuels and uses much of that revenue to invest in and subsidize energy alternatives such as solar power and voltaic cells. Japan has also persisted with the nuclear technology to which many of its people are opposed. Small fires, explosions, and other accidents in nuclear power and reprocessing plants in recent years—including a nuclear

• **Figure 7.4.10** The Port of Tokyo. The island nation of Japan thrives on seaborne trade.

plant fire triggered by a 2007 earthquake—have prompted new domestic concerns about the safety of nuclear energy, forcing a slowdown in Japan's nuclear industry.

There are also international fears. Japan imports large quantities of plutonium, primarily from Europe, for use in its nuclear power industry, and the long-distance ocean shipment of this very hazardous material has contributed to Japan's poor reputation in environmental matters. International protests against the plutonium shipments have caused Japan to postpone the construction of a series of nuclear breeder reactors, which use and create recyclable plutonium.

The phenomenal advance of industry since the early 1950s has been marked by a series of booms and declines in various sectors. In the 1960s, electronics, cameras and optical equipment, petrochemicals, synthetic fibers, and automobiles (notably those produced by Mazda, Honda, Toyota, Mitsubishi, and Nissan) became boom industries. In the 1970s and 1980s, Japan stepped up the manufacture of computers and robots. By the early 1980s, Japan led the world in the robotization of industry. For the first time in the postwar era, Japanese industry in the 2000s—when Japan began offshoring a growing share of its manufacturing to China—was not characterized by world supremacy in any single manufacturing sector. Japanese industry must now look for new directions in which to establish leadership. As always, to overcome the severe limitations imposed by small, resource-poor islands, the Japanese must live by their wits.

7.4.5 The Industrious People behind Japanese Industry

Japan has one of the world's most homogeneous populations; it is 99.5 percent ethnic Japanese. The largest non-Japanese group is the 600,000 strong community of ethnic Koreans, the people and their descendants whom Japan imported during World War II to do mainly hard manual labor. These ethnic Koreans have long suffered from discrimination in Japan. So have the ethnic Japanese social outcasts known as the **Burakim,** descendants of people who practiced "impure" trades like butchering and street sweeping as early as the 8th century. They still tend to be marginalized in low-paying jobs and forced to live in ghettos. Other than Koreans, few non-Japanese have settled in the country. Only with the country's booming prosperity and growing labor shortages of the 1980s did Japan open its doors to a few unskilled and low-skilled immigrants from countries such as Pakistan, Bangladesh, Thailand, Peru, and Brazil.

This lack of racial and ethnic diversity has had mixed results for Japan. Many observers believe that it has helped the country achieve a sense of unity of purpose, allowing the Japanese to persist through periods of adversity, especially the postwar years of reconstruction. However, the Japanese have also earned a reputation for intolerance of ethnic minorities. This attitude may be costly for Japan's future economic development. One way to increase productivity in the face of Japan's declining population would be to import skilled workers from abroad, but the country remains averse to immigration. Inward-looking tendencies have also left Japan short of people capable of and interested in learning English, a language routinely sought by aspiring businesspeople in nearby South Korea and China.

Capitalist Japan has had a remarkably egalitarian society, with about 80 percent of the population solidly middle-class. The richest third of the population has a total income just three times as great as that of the poorest third (compared with a fivefold differential in the United States). The Japanese have historically embraced the concept of *wa,* or harmony, based in part on the principle of economic equality. However, the recessions of the 1990s and early 2000s, followed by an economic rebound, resulted in growing joblessness and homelessness in Japan and a widening gap between haves and have-nots (•Figure 7.4.11). The country's official unemployment rate stood at 3.8 percent in 2007, compared with 2.1 percent in 1990. There are growing ranks of "permanent temporary workers," mainly young people who live with their parents to keep their expenses low.

The government has balked at many suggestions for economic reform because of the threat they pose to a socially harmonious Japan. There have been loud international appeals for Japan to reform its economic system by allowing inefficient businesses to collapse (rather than prop them up with expensive subsidies) and to resist the temptation to satisfy marginal rural populations with costly education and public works projects.

The legendary Japanese work ethic has had its advantages and drawbacks. It has helped the Japanese create a prosperous country, but it has also created a nation of workaholics beset with the same problems—stress, suicide, depression, and alcoholism—experienced by workers in the world's other MDCs. There is a growing incidence of what Japanese call *karoshi,* or death by overwork. The educational system encourages children to be highly successful but also to conform; critics say this stifles creativity and innovation. College graduates who are not hired during the annual recruiting season face difficult obstacles to entering the job market. And for those it affects, the practice of lifetime employment makes it difficult to change jobs. The country's welfare system, pensions, and social security are inadequate, compelling workers to work harder for savings. In their long hours at work, men are accustomed to spending little time with their spouses and children.

Women have not achieved parity with men in the workplace, and they complain increasingly of discrimination and sexism. In struggling to increase their footing in Japanese society, growing numbers of Japanese women are both working longer hours and marrying later, contributing to Japan's remarkably low birth rate (9 per 1,000 annually, one of the lowest in the world). Ironically, Japan's falling birth rates will only increase the burdens of its workers (•Figure 7.4.12). The population is both shrinking (it is expected to go from 127 million in 2007 to 100 million in 2050 and 67 million in 2100) and aging (already one in five people were over age 65 in 2007, and 1 million people a year join those ranks). The Japanese of working age will face an increasing load of taxes and family obligations to meet the needs of older citizens.

Crowding and overdevelopment, accompanied by a surprising lack of amenities, are inevitable facts of life in Japan. Officially designated parks comprise 14 percent of Japan's land area, but many of these are developed with roads, houses, golf courses, and resorts. There has been enormous growth in the popularity of winter sports destinations in northern Honshu and Hokkaido and of hot springs and mineral bath resorts throughout the country. Reflecting dissatisfaction with too much growth and development, the conservation ethic is strong and growing in Japan. However, it is difficult for the Japanese to truly "get away from it all" at home.

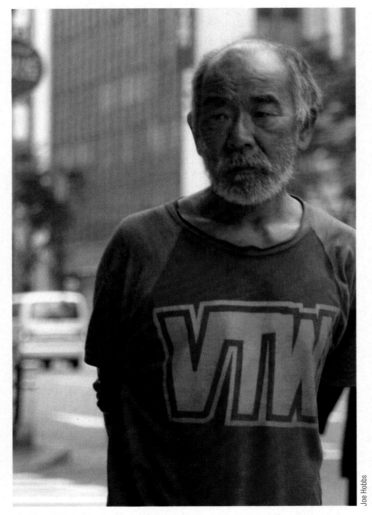

Joe Hobbs

• **Figure 7.4.11** In a society that has traditionally prided itself for its harmony and egalitarianism, some have fallen through the cracks in recent years.

Daily life for the Japanese is devoid of many of the benefits associated with the MDCs. Apartments and homes are generally very small. Only about half of the country's homes are connected to modern sewage systems. Smog from automobile exhaust and industrial smokestacks is a persistent health hazard. The ever-present traffic jams are known locally as "traffic wars." Japan's much-vaunted rail system, including its high-speed *Shinkansen,* or bullet trains, that carry traffic between northern Honshu and Kyushu at speeds averaging 106 miles (170 km) per hour, is chronically overburdened with passengers (•Figure 7.4.13). Summarizing these living conditions, the Japanese politician Ichiro Ozawa described Japan as having "an ostensibly high-income society with a meager lifestyle." Japan has long emphasized production of manufactured exports to compete with the United States and other large economies rather than sought to raise the material standard of living of its own citizens.

Japan had a troubled period of almost continuous economic recession from the early 1990s to the early 2000s. By the middle of the latter decade, the economy's worst troubles seemed to be over, thanks largely to painful cuts in payrolls, in inefficient businesses, and in subsidies to rural communities. Japan appeared to turn recent adversity into opportunity, as it has in the past. Japan's highly educated and skilled workforce is a major resource. Although natural assets are lacking, the country's favorable geographic location should also help continue its recovery. As the western Pacific region experiences the dynamic economic growth that is propelled by China (Japan's largest trading partner), Japan is well positioned to benefit from trade.

Japan will certainly continue to be an important member of the international community. Even during recession, it was the world's largest donor of foreign aid, eclipsed only in 2001 by the United States. Its long-term economic well-being is seen as crucial to the overall stability of the global economy. The Japanese consume more goods than the roughly 2 billion people of the rest of East Asia combined. Any slowdown in Japanese spending harms production in the factories of Indonesia, Malaysia, Thailand, and China that make goods for Japanese consumption and reexport; conversely, an economically stronger Japan raises all these economic boats. The United States supplies Japan with a wide range of goods, including semiconductors, software, heavy machinery, movies, apparel, and agricultural goods, and it buys

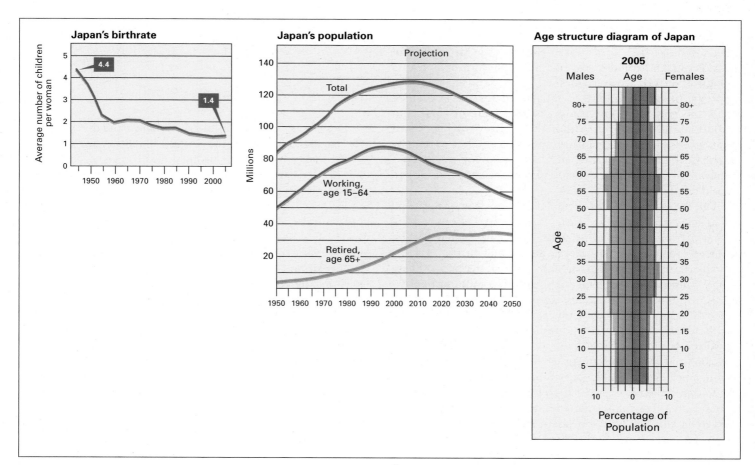

• **Figure 7.4.12** Many countries face problems of overpopulation, but Japan is concerned about not having enough people to keep its economic engine running in the future.

● **Figure 7.4.13** The Tokyo subway.

even more goods from Japan, so these two economies—the world's largest and second largest—are also extremely interdependent on one another's health.

7.4.6 Unfortunately Located Korea

Just 110 miles (177 km) from westernmost Japan is the Korean Peninsula, roughly the size of Minnesota or Portugal (●Figure 7.4.14). Korea's political history during the 20th century, first as a colony of Japan and then as a divided land, has tended to obscure its distinctive culture and contributions to the world. Although the Koreans have been influenced by Chinese culture and to a smaller extent by Japanese culture, they are ethnically and linguistically a separate people that have made important contributions to world civilization.

The two Koreas have occupied an unfortunate location in historic geopolitical terms. North Korea adjoins China along a frontier that follows the Yalu and Tumen Rivers; it faces Japan across the Korea Strait; and in the extreme northeast, it borders Russia for a short distance. These small countries are thus located near larger and more powerful neighbors—China, Russia, and Japan—that have frequently been at odds with one another and with the Koreans. China and Russia traditionally feared Japanese rule in Korea because it might have served as a springboard for invasion of their home countries. Meanwhile, to Japan, Korea has been seen as a dagger aiming at its heart to be used by China or Russia. Japan therefore wanted to occupy or block Korea. For many centuries, the Korean Peninsula has served as a bridge between Japan and the Asian mainland. From an early time, both China and Japan have been interested in controlling this bridge, and Korea was often a subject or vassal state of one or the other.

● **Figure 7.4.14** Principal features of the Korean Peninsula.

Despite external ambitions of conquest and division, from the late 7th century to the mid-20th century, Korea was a unified state, sometimes invaded and forced to pay tribute but never destroyed as a political entity. The decline of Chinese power in the 19th century was accompanied by the rise of modern Japan, whose influence grew in Korea, and from 1905 until 1945, Korea was firmly under Japanese control. In 1910, it was formally annexed to the Japanese empire. A legacy of hostility and occupation continues to cast a shadow over relations between the Koreas and Japan, and disputes periodically emerge between them over issues such as control of islands and fishing rights (see Insights, page 407).

Japan lost Korea at the end of World War II, setting the stage for the division between North and South Korea. To understand Korea's split, it is important to consider the Korean War, sometimes in the United States called "America's Forgotten War." The conflict cost the lives of an estimated 2 million North Koreans, perhaps 300,000 South Koreans, and 33,629 Americans.

In the closing days of World War II, the Soviet Union entered the Pacific war as an ally of the United States against Japan. The two sides drew up plans to accept Japan's surrender on the Korean Peninsula, arbitrarily drawing a line

303

INSIGHTS

The Law of the Sea

Some 90 miles (145 km) between the shores of Japan and South Korea in the Sea of Japan (East Sea) lie 34 small, inhospitable islands known to outsiders as the Liancourt Rocks, Takeshima to the Japanese, and Dokdo (or Tokdo) to the Koreans (see Figure 7.4.14). South Korea has controlled the islands since 1956, but with only a handful of Korean civilians and some Korean coast guard personnel living on the islands, Japan has periodically asserted its right to them. Since the 1990s, these remote rocks have been the focus of a dispute between Japan and Korea, not because of any riches they contain but because of a 1970s United Nations treaty known as the **Convention on the Law of the Sea**. This treaty would permit their sovereign power to have greater access to surrounding marine resources.

The Law of the Sea began as an effort to apportion ocean resources as equitably as possible and to avoid precisely the kind of dispute that developed between Japan and South Korea. The treaty gives a coastal nation mineral rights to its own continental shelf, a territorial water limit of 12 miles (19.3 km) offshore, and the right to establish an **exclusive economic zone (EEZ)** of up to 200 miles (320 km) offshore (in which, for example, only fishing boats of that country may fish). The power that controls offshore islands such as Takeshima/Dokdo can extend its EEZ even further.

By early 1996, some 85 nations had ratified the treaty. South Korea ratified it late in 1995. Japan was preparing to ratify it in 1996 when news reached Tokyo that South Korea had plans to build a wharf on the islands. To avoid provoking Japan, North Korea, and China, South Korea avoided declaring an exclusive economic zone off its waters that would include the islands. Japan ratified the treaty. But fears that South Korea may build facilities and station more people on the islands as a step toward establishing an EEZ and thereby excluding Japanese fishermen from the area caused Japan to restate its claim to the islands in 1996. South Korean officials answered with military exercises near the islands, and South Korean civilians staged loud demonstrations outside Japan's embassy in Seoul. South Korea declared the islands a national monument in 2002. Long simmering, the issue boiled up again in 2006 when Japan announced it would send ships to survey the area and South Korea answered that it would form a naval blockade against the Japanese vessels. The problem may have to be resolved in the international legal arena.

Even more important than Japan's competition with South Korea over territorial waters and islands is its contest with China. Both lay claims to the islands in the East China Sea that the Japanese know as the Senkakus and the Chinese as the Diaoyu Islands (see Figure 7.4.1). Again, both are interested in staking out an EEZ in which oil reserves might be found. China does not accept the UN definition of the EEZ beginning from a country's coastline; it insists on a wider zone that starts at the continental shelf.

The Senkaku or Diaoyu Islands are uninhabited, but their control is championed by nationalists on both sides and is therefore a potentially explosive issue. In 2004, Chinese activists landed on one of the islands to stake China's claim there, but Japanese authorities promptly arrested them, and Japan took formal possessions of the islands the following year. Since then, there have been several encounters between Japanese survey ships and Chinese coast guard vessels near the islands. Japan's formal claim to the islands dates to 1895, when Japan declared that the islands had previously been *terra nullius,* or no one's property. But China insists that there is Chinese documentation about control of the islands dating to the Ming Dynasty in the 15th century. The islands are close to Taiwan, which also lays claim to the islands, so any move by mainland China to secure access to them could also precipitate conflict between the two Chinas.

302

308

at the 38th parallel. The Soviet Union would accept Japan's surrender north of that line and the United States south of that line. This was not meant to be a permanent boundary, but it became one. On either side of the line, the Soviet Union and the United States moved to set up governments that would be friendly to them. By 1948, the Soviet-style, Soviet-backed, Soviet-armed Democratic People's Republic of Korea was established in the north, while in the south, the Western-oriented Republic of South Korea was created under the auspices of the United Nations.

War came soon. On June 25, 1950, North Korean troops invaded the south and quickly took Seoul. In those Cold War days, the United States feared Soviet expansionism into and beyond South Korea. President Truman sent American forces in, and they soon fought under the United Nations flag with support from 16 other countries. Truman selected General Douglas MacArthur to serve as supreme commander of the effort. By September, MacArthur's forces had pushed the North Korean advance all the way back to the Yalu River on the Chinese border. The general proposed using nuclear weapons to press forward from there, but the president refused, and his dissent over the issue cost MacArthur his job.

China, seeing enemy forces on its doorstep, reacted by sending huge numbers of forces across the border, driving the allies south across the 38th parallel and taking Seoul on January 4, 1951. Seoul was recaptured on March 15, and the battlefront stabilized generally along the 38th parallel. An armistice was signed at the border site of Panmunjom on July 27, 1953, by the Chinese, the North Koreans, and the United Nations command. It was only a cease-fire, and to this day, there has not been a full treaty or reconciliation between the Koreas.

The border between the two Koreas, often cited as the tensest boundary on earth, still follows this armistice line.

• **Figure 7.4.15** The DMZ between North and South Korea is an unlikely nature refuge, but wherever a half-century passes without significant human manipulation of the landscape, plants and animals thrive.

• **Figure 7.4.16** The two Koreas are like night and day. (a) A satellite image taken at night shows the profound economic distinctions, with prosperous South Korea awash with urban and industrial lights while North Korea is nearly dark.

The **demilitarized zone (DMZ)** between them, 150 miles (240 km) long and 2.5 miles (4 km) wide, is a virtual no-man's-land of mines, barbed wire, tank traps, and underground tunnels. The world's largest concentration of hostile troops faces off on either side of it. Remarkably, with the near absence of human activity, rare birds like the Manchurian crane and other endangered animal species have taken refuge in this narrow strip (•Figure 7.4.15). The DMZ thus has the potential to become an international "peace park" like that taking shape along the old Iron Curtain in Europe.

7.4.7 Contrasts between the Two Koreas

The Korean armistice line divides one people, united by their ethnicity and language, into two very different countries (•Figure 7.4.16; also see Table 7.1, page 281). North Korea has an approximate area of 46,500 square miles (120,400 sq km) and is inhabited by about 23 million people, with a density of 501 per square mile (193/sq km). South Korea's corresponding figures are 48 million people in only 38,000 square miles (98,390 sq km), averaging 1,266 per square mile (489/sq km).

South Korea is a republic that has fluctuated between attempts at democracy and a repressive military dictatorship. It has a capitalist economy heavily dependent on relationships with the United States and Japan. North Korea is a rigid and very tightly controlled Communist state that promotes what it calls *juche* (pronounced "*djoo*'-cheh"), or self-reliance. North Korea's leadership has used this political philosophy's three principles of political independence, economic self-sustenance, and national self-defense to limit

• **Figure 7.4.16** (b) Affluence, open trade, and freedom of expression are apparent in South Korea

• **Figure 7.4.16** (c) while military security preoccupies austere Stalinist North Korea.

economic and other exchanges with the outside world. North Korea's principal Cold War ties shifted between 88 China and the Soviet Union With the collapse of the USSR, China became North Korea's main ally and benefactor, and the Russian Federation is emerging as one of North Korea's few trading partners.

From World War II until his death in 1994, the dictator Kim Il Sung, whom his citizens called "The Great Leader," governed North Korea. His son and successor, Kim Jong Il (dubbed "The Dear Leader"), has perpetuated the country's militaristic character. North Korea spends a staggering 23 percent of its gross domestic product on its military, compared with South Korea's expenditure of only 2.7 percent, China's 4.3 percent, and Japan's 0.8 percent. North Korea's lavish expenditures on defense have been one cause of its severe economic decline. North Korea's economy is only 7 percent the size of South Korea's. No GNI PPP data are available for North Korea, but South Korea's is $23,800, similar to that of Greece.

There are marked physical contrasts between the two countries. Although both are mountainous or hilly, North Korea is more rugged. North Korea has a humid continental long-summer climate, with hot summers and cold winters. South Korea is mostly a humid subtropical area with shorter and milder winters. Both are monsoonal, with precipitation concentrated in the summer.

Rice is the staple food in both countries. Rice-growing conditions are best in South Korea and diminish northward. Unlike the North, South Korea is able to double-crop irrigated rice fields by growing a dry-field winter crop such as barley after the rice has been harvested. The main supplementary crop in the North is corn, which must be grown in the summer and so cannot occupy the same fields as rice.

Beginning in 1995, successive waves of flood and drought brought famine to North Korea. The natural calamities only contributed to the already plummeting economic health of the country that came in the wake of the collapse of the Soviet Union, formerly North Korea's main benefactor and trading partner. North Korea's almost impenetrable veil of secrecy made it difficult to calculate the losses, but estimates of the number of people who died in the famine range from 900,000 to 2.4 million, or up to 10 percent of the country's prefamine population. As one crop after another failed, North Korea gradually and reluctantly sought food aid from abroad. It coaxed emergency supplies of wheat from the United States in part by threatening to withdraw from its 1994 agreement with the United States to halt its nuclear weapons program (see the discussion on pages 303–304 in Chapter 7). Although food aid from the United States, China, and South Korea helped alleviate the immediate crisis, the poor conditions of North Korea's health system, drinking water supplies, and electrical generation have perpetuated malnutrition and disease.

Another physical contrast between the two Koreas is that most of the nonagricultural natural resources are in North Korea. All of the peninsula's major resources—coal, iron ore, some less important metallic ores, hydropower potential, and forests—are more abundant in the North than in the South. North Korea was thus originally the more industrialized state, featuring a typical Communist emphasis on mining and heavy industry, together with hydroelectric production and timber products. Viewed strictly in terms of resource potential for development, North Korea should have become the more prosperous power. Its decision to close its doors to the outside world and insist on strict state control of industry and agriculture provides a fascinating 46 opportunity to appreciate how different political systems produce different economic results.

While the North focused on defense and strict socialism, the South took over industrial leadership in the 1970s and 1980s by developing a dynamic and diversified capitalist industrial economy.[1] South Korea's economy first began to boom between 1961 and 1980 under a state capitalism approach rather than free-market and free-trade capitalism. This was a time during which the government favored and strongly funded a few small companies that it nurtured into giant conglomerates called **chaebols.** Hyundai is a good example. Originally a small rice milk company, it diversified into trucks and buses. The government believed in its potential and supported Hyundai financially, supervising its growth into a world-class corporation renowned for the manufacture of automobiles and ships. Samsung, now one of the best electronics companies in the world, followed a similar path (•Figure 7.4.17). Altogether about 15 of these gigantic, interlocking, family-controlled conglomerates came to dominate the economy. Large investments from Japan and the United States, as well as access to markets in those countries, aided in the explosive development of *chaebol* industries. Also important was the availability of inexpensive and increasingly skilled Korean labor. Finally, the factor South Koreans most frequently point to in their success has been an emphasis on education. They sell valued possessions if need be to afford their children's educations at home or abroad, and their culture prizes teachers and professors. South Korea has one of the world's highest numbers of Ph.D.'s per capita.

Democratic and free-market challenges came to South Korea's economic, political, and social systems in the 1980s, with differences emerging between labor and management, military and civilian rule, and even men and women. In the late 1990s, an economic crisis occurred in part because of these challenges and also because South Korea's was hit by the shock waves of Asia's spreading economic crisis. The 344 pain was almost universal because so much of the country's wealth was in the hands of the *chaebols,* which employed the majority of South Korea's working population and owned most of the banks. During the country's explosive period of growth, the banks had lent money back to their parent companies for risky investments all over the world, particularly throughout Asia. As the Asian economic crisis spread, those projects were halted, and the borrowers defaulted on their loans. Foreign investors rushed to pull their capital out of South Korea, whose national currency and foreign reserves deflated.

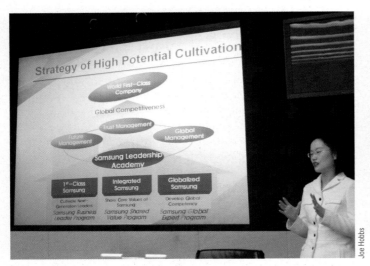

• **Figure 7.4.17** Directing a training session, a supervisor discusses Samsung's corporate work ethics and global ambitions.

• **Figure 7.4.18** Seoul, capital of South Korea.

With financial help from the International Monetary Fund (IMF), South Korea responded quickly to the crisis, especially by getting rid of bad loans and by restructuring the companies that borrowed too much. In the early 2000s, the country was once again experiencing economic growth, bringing sighs of relief around the world. With the world's 11th-largest economy, South Korea is now among the top five countries worldwide in the production of automobiles, ships, steel, computers, and electronics. The development of its high-tech industries has concerned competitors in Japan and the United States; for example, Samsung has established an increasing market share in an industry dominated by Japanese firms. South Korea is the seventh-largest trading partner of the United States, which has long subscribed to a "domino theory" that if South Korea's economy fell, so would Japan's and perhaps even its own. In 2007, the United States and South Korea signed a free-trade pact that will even more strongly enmesh their economies.

South Korea's cities are visible evidence of the country's industrialization. The capital city of Seoul has a population of 22.6 million, ranking it among the five most populous metropolitan areas in the world (•Figure 7.4.18). Forty miles (64 km) east of Seoul, on an artificial island near the port of Inchon, is the free-trade zone of New Songdo, being built to capitalize on trade with China. Its designers call New Songdo a **ubiquitous city,** or **U-city,** where all major residential, medical, business, government, and other information systems are integrated. Other cities with more than 2 million population are the port of Pusan, which rivals Shanghai as the largest container port in northeastern Asia, and the inland industrial center of Taegu.

By contrast, the only large city in North Korea is the capital, Pyongyang, with a population of 3 million. In an effort to begin to close the gap, North Korea has built free-trade zones intended to attract foreign investment and in-

dustries to favorable locations near the borders with China, Russia, and South Korea. As unlikely as it may seem, the North Koreans have a vision for one of these, the city of Sinuiju (in the extreme northwest, just across the Yalu River from China), as a "new Hong Kong." Just north of the DMZ, North Korea has built a trade zone in the city of Kaesong. Many South Korean companies have taken advantage of North Korea's rock-bottom wages (5 percent of those of South Korea) and have set up factories there, producing kitchenware, semiconductor parts, and a range of other products.

In sum, South Korea is one of Asia's success stories, an Asian Tiger that made an extraordinary postwar recovery, especially by using brainpower to overcome its resource limitations, and has enjoyed periods of rocketing economic growth. South Korea is, however, still something of an "in-between" country, not entirely an MDC or an LDC, not all urban but not all rural either, and both industrial and agricultural. The government still lavishes huge subsidies to protect its rice growers, for example, while also promoting the virtues of free trade in industry. The transition from LDC to MDC has brought some growing pains, seen especially in the social changes of more women in the workforce, higher divorce rates, and falling birth rates. Like Japan, South Korea wants to maintain its cohesive society and does not want to invite foreigners to fill its impending shortage of productive young people. The government once encouraged family planning but is now asking its people to have more babies.

South Koreans tend to debate contemporary social and political issues more openly than many of their Asian neighbors, and raucous demonstrations for and against a wide range of causes are commonplace. South Korean pop culture and Christian mission movements are making inroads across China and Southeast Asia. The former, known as the "**Korean Wave**" and taking the form of music, television, films, and fashion, has generally been embraced but has produced

402

a backlash in Japan, where some people steadfastly refuse to accept South Korea's rising clout.

Meanwhile, despite its experimental free-trade zones, North Korea is a poor country almost frozen in time, intent on pursuing the command economy model that has failed in every other nation that employed it. Despite its motto of self-reliance, it depends on huge imports of food and energy, especially from Russia. Its principal exports are apparently missiles and other military hardware; heroin, amphetamines, and other synthetic and semisynthetic drugs; and counterfeit money. Its people, whose sentiments and hopes are unknown because they are not allowed contact with the outside world, periodically hover close to famine. Only a select few are allowed access to the Internet, although the infrastructure is there: much of the country is wired with a fiber-optic intranet system. The North Korean leadership has a dilemma: If it gives up on *juche* and opens its doors, the economy might improve, but the people of this country often called the **Hermit Kingdom** will discover the bounties just across the border and might revolt.

7.4.8 Sunshine for Korea?

States of cold war and periodic border incidents have persisted between North and South Korea. Relations between the two countries have periodically thawed, only to ice over again. A warming trend came to be known as the **Sunshine Policy** in 2000, when the leaders of North and South Korea held a historic, first-ever summit meeting between the countries. Their substantive focus was on reconciliation rather than reunification. The South proclaimed that it wanted to help protect North Korea's national security and assist in its economic development. In exchange for this help, South Korea expected North Korea to cease threatening the South and to provide reassurances of abandoning its nuclear weapons and long-range missile programs. The two countries agreed to restore a long-defunct rail link built during Japan's occupation of Korea. South Korea is hoping its resurrection will provide a large economic boost: it could then ship its products overland to China and across the Trans-China Railway or Trans-Siberian Railway to Europe (Koreans are thus wistfully calling the link the **Iron Silk Road**).

186

Koreans on both sides of the DMZ were delighted by the family reunions that occurred in the summit's wake. Seven million South Koreans who fled south before and during the Korean War have relatives in the North. Since 2000, small numbers of North Koreans have been able to make short visits with relatives in South Korea, and vice versa, enjoying emotional reunions that have given hope on both sides of better times to come. And as discussed earlier, South Korea has begun to capitalize on North Korean labor in the free-trade zone of Kaeson, an arrangement benefiting both countries.

The prospect for Korean reunification appears to be very remote, even as relations between the two countries im-

prove. North Korea continues to argue that the two countries should unite peacefully and without foreign influence. Outside observers generally believe, however, that North Korea is not really interested in reunification because it would probably doom the country's regime. For its part, South Korea is fearful of the enormous economic cost it would have to pay for absorbing its much poorer neighbor, just as West Germany did in absorbing East Germany. Reunification would probably begin with a huge stream of poor northerners moving south. The nearby great powers are also quietly pleased with Korea's lingering division. The two Koreas provide dual buffer zones between the historical adversaries China and Japan. And if the two Koreas united, finally taking advantage of North Korea's strong natural resource base, China and Japan could suddenly have a rival great power to contend with.

112

For now, there is little thought of Korean unification. Even with the progress of the Six Party Talks and North Korea's pledge to disable its nuclear weapons facilities (see page 304), there is much more concern with simply avoiding war. Any escalation of the tension between the United States and North Korea, or even an incident in the DMZ, could be a tripwire for a great conflict (●Figure 7.4.19). U.S. troops could not avoid being involved, and Chinese troops might be brought in. Some analysts fear that Korea may yet become a nuclear battlefield. Japan worries that it would also be drawn into any conflict between the Koreas. The economic and political stability of these three small countries on the western Pacific Rim is critical to the well-being of the global system.

● **Figure 7.4.19** Two South Korean soldiers keep a wary eye on their North Korean counterpart, just steps away in the Joint Security Area of the Panmunjom "truce village" in the DMZ. The low concrete slab running left-to-right on the left of the photo marks the border between the two countries. From time to time, a North Korean defector runs across the border, sometimes pursued by North Korean soldiers. Firefights involving both U.S. and South Korean troops against North Korean soldiers have accompanied some of these incidents.

SUMMARY

→ Japan's economy is the world's second largest, after that of the United States.

→ There are four main islands in the Japanese archipelago: Hokkaido, Honshu, Shikoku, and Kyushu. The smaller Ryukyu Islands extend almost to Taiwan.

→ All of the main islands of Japan consist of volcanic mountains and are subject to frequent and sometimes severe earthquakes. Other natural hazards also plague Japan.

→ Japan went through periods of relative isolation. Then it opened up to the outside world and is today one of the most influential players in world affairs.

→ Japan experienced phenomenal economic growth after its crushing defeat in World War II, but it has recently suffered economic recession.

→ Due to its relative poverty in natural resources, Japan must import most of its raw materials, most energy supplies, and a large share of its food.

→ The people of Japan view the ocean as an important resource and a promising frontier. The Japanese have the world's highest per capita consumption of fish and other marine foods.

→ The quest for possession of offshore resources near islands is a source of tension between Japan and Korea and between Japan and China.

→ Most of Japan's economic activities and a majority of its people are packed into a corridor about 700 miles (c. 1,100 km) long. This megalopolis extends from Tokyo, Yokohama, and the surrounding Kanto Plain on the island of Honshu through northern Shikoku to northern Kyushu.

→ The hardworking Japanese are experiencing many symptoms of workaholism, and many complain about their low level of amenities relative to that of other prosperous countries.

→ The Korean Peninsula has an unfortunate location in geopolitical terms, sandwiched between the greater powers of Russia, China, and Japan. Korea has had a turbulent recent political history, first as a colony of Japan and then as a land divided between the Communist North and the more dynamic and democratic South.

→ Although the Koreans have been influenced by Chinese culture and to a lesser extent by Japanese culture, they are ethnically and linguistically a separate people.

→ There are marked physical and economic contrasts between North and South Korea, with North Korea being much more rugged and having more natural resources for industry. North Korea has a humid continental long-summer climate, with hot summers and cold winters, whereas South Korea is mostly a humid subtropical area, with much shorter and milder winters. Capitalist South Korea is far more prosperous than Communist North Korea.

→ North Korea has suffered famine recently but has been reluctant to open up to trade and assistance because North Koreans might revolt.

→ There are prospects for reconciliation and, more remotely, reunification but also for war between North and South Korea. Reunification is not favored by the South because it would be too costly. China and Japan would apparently be happier to have a weaker divided Korea than a stronger unified Korea.

KEY TERMS + CONCEPTS

active volcano (p. 392)
Ainu (p. 398)
bubble economy (p. 402)
Burakim (p. 403)
chaebol (p. 409)
Convention on the Law of the Sea
    (p. 407)
daimyo (p. 399)
demilitarized zone (DMZ) (p. 408)
double cropping (p. 394)
exclusive economic zone (EEZ) (p. 407)

Great East Asia Co-Prosperity Sphere
    (p. 401)
Hermit Kingdom (p. 411)
gunboat diplomacy (p. 399)
intertillage (p. 394)
Iron Silk Road (p. 411)
juche (p. 408)
kamikaze (p. 399)
karoshi (p. 404)
Liberal Democratic Party (LDP) (p. 402)
megalopolis (p. 395)

Meiji Restoration (p. 399)
Rape of Nanking (p. 400)
Ring of Fire (p. 396)
samurai (p. 399)
Self-Defense Forces (SDFs) (p. 402)
shogun (p. 399)
Sunshine Policy (p. 411)
Tokugawa Shogunate (p. 399)
ubiquitous city (U-city) (p. 410)
Utari (p. 398)
wa (p. 404)

REVIEW QUESTIONS

1. What are Japan's four main islands? What are the outlying islands? Which islands are contested with other countries?

2. What elements of Japan's physical geography are related to the islands' location on the Pacific Ring of Fire?

3. What are Japan's principal climate types? In which parts of the country do they occur?

4. What are the principal features of Japan's agriculture? What position does rice have in the agricultural and economic systems?

5. To what extent does Japan draw on the resources of the sea?

6. What area is defined as Japan's core? What urban, industrial, and population characteristics definite it as a core region?

7. What is Japan's ethnic makeup? What relations have the Japanese majority had with internal minorities and foreign groups at various times?

8. How has Japan been able to overcome its poverty in natural resource assets by commercial, military, or other means?

9. What are some of the features of Japanese society today, especially those related to the country's economy?

10. What precipitated the division of the Korean Peninsula into North and South Korea?

11. Why are relations tense between Japan and South Korea?

12. What are the major differences between North and South Korea in population numbers, natural resources, political systems, and economic development?

13. What are the prospects for reunification of the Koreas? How do some outside powers and the Koreans themselves view reunification?

NOTE

1. I thank the Korea Society for sponsoring my field study of Korea, on which much of this discussion is based. Professor Byong Man Ahn of the Hankuk University of Foreign Studies and Professor Taeho Bark of the Graduate School of International Studies at Seoul National University graciously provided much of the information on Korean economies and politics.

Australia's Aborigines say that spirits "dreamed" the world and all its features into existence. This is a rock painting of one of those creator beings, the wandjina, a deity of cyclones (hurricanes) and monsoons.

Joe Hobbs

Tom Van Sant/Geosphere Project/SPL/Photo Researchers, Inc.

CHAPTER 8
A GEOGRAPHIC
PROFILE OF
OCEANIA

Covering fully one-third of the earth's surface, **Oceania** (pronounced "oh-shee-*an*-ee-uh") is mostly water. The world's largest ocean, the Pacific, defines and dominates this region. It is bigger than all the earth's continents and islands combined. Before World War II, the Western world spawned legends about this ocean and its islands as a kind of utopia. There has long been trouble in this paradise, however. On many islands, foreign traders, whaling crews, labor recruiters, and other opportunists exploited the indigenous peoples, reducing their numbers and disrupting their cultures. The military battles of World War II further shattered the idyllic qualities of many islands. Yet today, the Pacific mystique, which has been perpetuated in books and films, forms part of the allure for booming tourism development across the region. This chapter presents the cultural and natural histories of Oceania, summarizes the impacts of colonial enterprises in the region, and discusses some of the obstacles to development confronting the countries of this watery realm.

chapter objectives

This chapter should enable you to

→ Appreciate the economic prominence of larger, Europeanized Australia and New Zealand in a vast sea of small, mainly indigenous political units

→ Recognize the associations between the physical geographies of islands, their typical social and political organizations, and their economic characteristics

→ Consider the impacts of human agency, especially deforestation and the introduction of exotic species, on island ecosystems

→ Understand how minority control of most of the wealth generated by mineral and other resources has led to discontent and rebellion among majorities

→ Hear the concern expressed by low-lying island countries about the production of greenhouse gases in faraway industrialized nations

→ Recognize the peculiar dependence of some Pacific populations on the military interests of distant nations

8.1 Area and Population

Oceania, also known as the Pacific World, includes Australia, New Zealand, and the islands of the mid-Pacific lying mostly between the tropics (Table 8.1). The Pacific islands nearer the mainlands of East Asia, Russia, and the Americas are excluded here on the basis of their close ties with the nearby continents. Hawaii belongs to this region but is dealt with mainly in Module 11.2's discussion of the United States. Large areas of the eastern and northern Pacific that contain few islands are also not considered here. Australia and New Zealand have strong political and economic interests in the smaller tropical islands, have a similar insular character in many ways, and share ethnic affiliations with the original inhabitants of the smaller islands. But they are also uniquely European in character and dwarf the other countries in economic and political clout. Therefore, these two countries are examined in more detail in Module 8.1.

Major Divisions of the Region

Geographers have long divided the Pacific islands into three principal regions: Melanesia, Micronesia, and Polynesia (•Figure 8.1). The islands of **Melanesia** (Greek for "black islands"), lying northeast of Australia, are relatively large. New Guinea, the largest, is about 1,500 miles (c. 2,400 km) long and 400 miles (c. 650 km) across at the broadest point. It is divided between two countries. The western section, called Papua (formerly Irian Jaya), is part of Indonesia and is discussed in Module 7.2. The eastern section is an independent country, Papua New Guinea. The other Melanesian countries are the Solomon Islands, Vanuatu, and Fiji. New Caledonia, also in Melanesia, is controlled by France.

Micronesia (Greek for "tiny islands") includes thousands of scattered small islands in the central and western Pacific, mostly north of the equator. The Micronesian countries are Palau, the Federated States of Micronesia, the Republic of the Marshall Islands, Nauru, and the western part of the nation of Kiribati (pronounced "keer-uh-*bahss*"). Micronesia also includes two possessions of the United States: Guam and the Mariana Islands.

Polynesia (Greek for "many islands") occupies a greater expanse of ocean than Melanesia or Micronesia. It is shaped like a rough triangle with corners at New Zealand, the Hawaiian Islands, and remote Easter Island. Excluding Hawaii and New Zealand, Polynesia's independent countries are the eastern part of Kiribati, Tuvalu, Samoa, and Tonga. Possessions of other countries in Polynesia include French Polynesia and France's Wallis Islands, American Samoa, New Zealand's Tokelau and Cook Islands, Britain's Pitcairn Island, and Chile's Easter Island.

The typical Pacific island country (excluding Australia and New Zealand) has about 100,000 to 150,000 people in an area of 250 to 1,000 square miles (c. 650 to 2,600 sq km); consists of a number of islands; is poor economically; is an ex-colony of Britain, New Zealand, or Australia; and depends heavily on foreign economic aid. The total land area, including Australia and New Zealand, is 3.3 million square miles (8.5 million sq km), or about 90 percent of the size of the entire United States (•Figure 8.2).

The People and Where They Live

The region's total population is 34.4 million (•Figure 8.3). Aside from Australia's approximately 21 million people, populations range from 6.3 million in Papua New Guinea, which is exceptionally large in population and area for this region but has a low population density, to 12,000 on tiny Tuvalu. The highest population densities are in the smallest island groups, notably the Tuamotu Islands in French Polynesia and the Ellice Islands of Tuvalu. Population growth rates vary widely, from the predictably low 0.6 percent and 0.7 percent in prosperous Australia and New Zealand, respectively, to a very high 3.3 percent in the much poorer Marshall Islands.

There are some apparent problems of people overpopulation, especially in Polynesia. Considerable emigration occurs, especially from the Cook Islands, Samoa, Tonga, and Fiji to New Zealand and North America. Many educated

TABLE 8.1 Oceania: Basic Data

Political Unit	Area (thousands; sq mi)	Area (thousands; sq km)	Estimated Population (millions)	Estimated Population Density (sq mi)	Estimated Population Density (sq km)	Annual rate of Natural Increase (%)	Human Development Index	Urban Population (%)	Arable Land (%)	Per Capita GNI PPP ($US)
Australia and New Zealand	**3,093.4**	**8,011.9**	**25.2**	**8**	**3**	**0.6**	**0.953**	**89**	**6**	**30,810**
Australia	2,988.9	7,741.3	21.0	7	3	0.6	0.957	91	6	31,680
New Zealand	104.5	270.7	4.2	40	16	0.7	0.936	86	5	26,470
Others	**212.3**	**549.8**	**9.2**	**43**	**16**	**2.1**	**0.657**	**24**	**1**	**3,050**
American Samoa (U.S.)	0.2	0.5	0.05	250	97	0.0	N/A	90	10	N/A
Cook Is. (N.Z.)	0.09	0.2	0.02	222	86	N/A	N/A	70	17	N/A
Fiji	7.1	18.3	0.9	127	49	1.4	0.758	39	11	6,200
French Polynesia (Fr.)	1.5	3.8	0.3	200	77	1.3	N/A	53	1	N/A
Guam (U.S.)	0.2	0.5	0.2	1000	386	1.6	N/A	93	9	N/A
Kiribati	0.3	0.7	0.1	333	129	2.3	N/A	43	2	8,970
Marshall Islands	0.1	0.2	0.1	1000	386	3.3	N/A	68	16	N/A
Micronesia, Federated States of	0.3	0.7	0.1	333	129	2.0	N/A	22	5	7,830
Nauru	0.009	0.02	0.01	1111	429	1.9	N/A	100	0	N/A
New Caledonia (Fr.)	7.2	18.6	0.2	28	11	1.3	N/A	71	0	N/A
Northern Mariana Islands (U.S.)	0.2	0.5	0.07	350	135	2.7	N/A	94	13	N/A
Palau	0.2	0.5	0.02	100	39	0.7	N/A	70	8	N/A
Papua New Guinea	178.7	462.8	6.3	35	14	2.2	0.523	15	0	2,410
Solomon Islands	11.2	29.0	0.5	45	17	2.6	0.592	16	1	2,170
Tonga	0.3	0.7	0.1	333	129	2.1	0.815	32	23	8,580
Tuvalu	0.01	0.02	0.01	1000	386	1.7	N/A	47	0	N/A
Vanuatu	4.7	12.1	0.2	43	16	2.5	0.670	21	2	3,280
Summary Total	**3,305.7**	**8,561.7**	**34.4**	**10**	**4**	**1.0**	**0.881**	**71**	**5**	**23,990**

Sources: World Population Data Sheet, Population Reference Bureau, 2007; Human Development Report, United Nations, 2006; World Factbook, CIA, 2007.

young Australians are also lured abroad, especially to the United States. Because birth rates are low, people in both Australia and New Zealand share the typical postindustrial fear of not having enough people to support the countries' economies and their aging populations. Both countries, however, are averse to liberal immigration policies that would boost their populations.

8.2 Physical Geography and Human Adaptations

The climates, vegetation, landscapes, and cultural geographies of the islands scattered across the vast Pacific vary, but a few notable patterns emerge (•Figures 8.4 and 8.5; note that only the islands of the southwestern Pacific are large enough to have meaningful data depicted at this map scale).

Climates and Biomes

Climatically, most of the region is tropical; all of Micronesia and Melanesia and most of Polynesia (except for northern Hawaii and New Zealand and its outlying islands) lie in the tropics. Most of New Guinea, which lies entirely in the tropics, has tropical rain forest climate and biome types, but its high mountains have undifferentiated climates and vegetation types, varying with elevation all the way to tundra above 11,000 feet (3,300 m). The great majority of the smaller Pacific islands lying in the tropics are influenced by the easterly trade winds that bring abundant precipitation for tropical rain forest climate and biome types.

• **Figure 8.1** Principal features of Oceania. Note the peculiar jog in the International Date Line near Kiribati (along the equator), discussed on page 429.

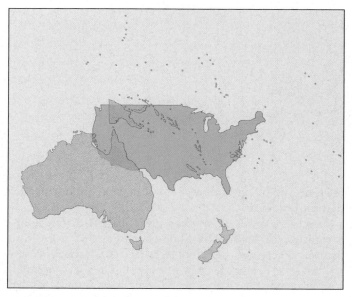

• **Figure 8.2** Oceania compared in area (but not latitude) with the conterminous United States.

The Tropic of Capricorn bisects Australia. Cooling mid-latitude westerly winds bring much of New Zealand and parts of coastal southern Australia a marine west coast climate, with associated temperate mixed forests. Some coastal areas of southern Australia also have Mediterranean and semiarid steppe climates, with associated Mediterranean scrub vegetation and prairie and steppe grasses. While much of the interior of Australia has a desert climate and vegetation, coastal regions of northern Australia generally have climates and vegetation of tropical savanna, with small patches of tropical rain forest climate and vegetation. More details on the climates and biomes of Australia and New Zealand can be found in Module 8.1.

Island Types

Topographically, Oceania features three general types of islands. **Continental islands** are continents or were attached to continents before sea level changes and tectonic activities isolated them. These include New Guinea, New Britain,

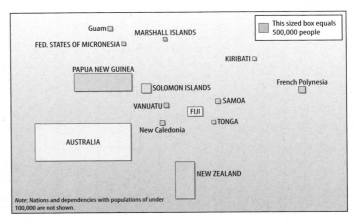

• **Figure 8.3** Population distribution (left) and population cartogram (right) of Oceania.

• **Figure 8.4** Physical geography of Oceania.

and New Ireland in the Bismarcks; New Caledonia, Bougainville, and smaller islands in the Solomons; the two main islands of Fiji; Australia; and the North and South Islands of New Zealand.

Some of these have lofty peaks, including several over 16,000 feet (c. 4,900 m) in New Guinea, while Australia is overall rather low. The region's other islands are categorized as either **high islands** or **low islands** (•Figure 8.6). Most high islands are the result of volcanic eruptions. The low islands are made of coral, a material composed of the

skeletons and living bodies of small marine organisms that inhabit tropical seas.

These very different settings offer sharply different opportunities for human livelihoods (•Figure 8.7). Overall, however, except in urban, industrialized Australia and New Zealand, the Pacific peoples are rural. They live in small farming or fishing villages where they have long depended on root crops (such as taro and yams), tree crops (such as breadfruit and coconuts), ocean fish, and pigs. Outside Australia and New Zealand, there are only four cities with

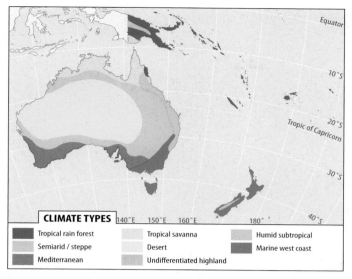

• **Figure 8.5** Climates (left) and biomes (right) of Oceania.

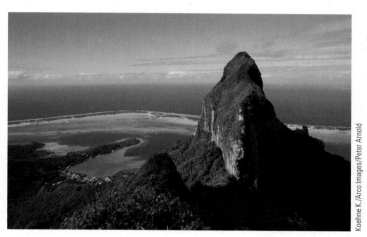

• **Figure 8.6** (a) Bora Bora, in Tahiti, is a classic and much romanticized high island.

Koehne K./Arco Images/Peter Arnold

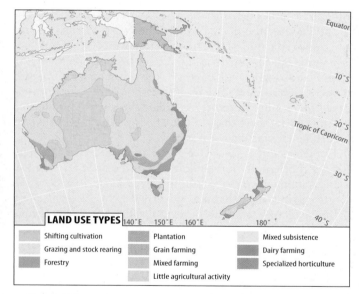

• **Figure 8.7** Land use in Oceania.

• **Figure 8.6** (b) Low islands, such as the one seen here, are most associated with the "coconut civilizations" of the Pacific. Low islands are extremely vulnerable to the dangers that would be posed by global warming.

Karl & Jill Wallin

a population greater than 100,000: Port Moresby in Papua New Guinea, Suva in Fiji, Nouméa in New Caledonia, and Papeete in Tahiti.

Most of the region's high islands are not continental but volcanic in origin and have a familiar pattern. There is a steep central peak with ridges and valleys radiating outward to the coastline. Permanent streams run through the valleys. A coral reef surrounds the island, and between the shore and the reef is a shallow lagoon. Many of the high volcanic islands are spectacularly scenic (see Figure 8.6a). Classic high islands in the region include the Polynesian groups of Hawaii, Samoa, and the Society Islands.

Some of the volcanic high islands of the Pacific comprise island chains. These are formed when the oceanic crust slides over a stationary **geologic hot spot** in the earth's mantle where molten magma is relatively close to the crust

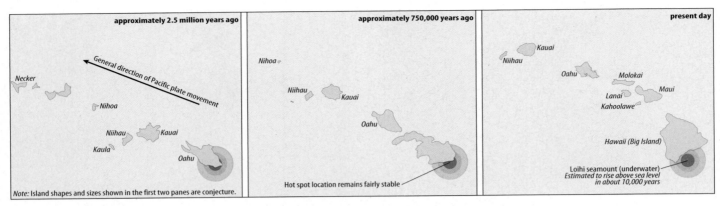

approximately 2.5 million years ago

General direction of Pacific plate movement

Necker

Nihoa

Niihau

Kaula

Kauai

Oahu

Note: Island shapes and sizes shown in the first two panes are conjecture.

approximately 750,000 years ago

Nihoa

Niihau

Kauai

Oahu

Hot spot location remains fairly stable

present day

Kauai

Niihau

Oahu

Molokai

Lanai

Maui

Kahoolawe

Hawaii (Big Island)

Loihi seamount (underwater)
Estimated to rise above sea level
in about 10,000 years

● **Figure 8.8** The Hawaiian Islands were formed as a piece of the earth's crust slid over a geologic hot spot.

683 (●Figure 8.8). As the crust slides over the geologic hot spot, magma rises through the crust to form new volcanic islands. The Hawaiian Island chain is an excellent example. The big island of Hawaii is now situated over the hot spot and is still active; it is one of the chain's youngest members. The Pacific Plate of the earth's crust is moving northwestward here; older volcanic islands that were born over the hot spot have moved to the northwest and have become inactive. Still older islands farther northwest in the chain have submerged to become **seamounts,** or underwater volcanic mountains. Other seamounts have yet to break the surface, including one in the southeast that will emerge in about 10,000 years to become Hawaii's newest island.

The geographer Tom McKnight explained that the high island's topography has historically influenced its cultural and economic patterns.[1] Each drainage basin formed a relatively distinct unit that was governed by a different chief. Within each unit, there were several subchiefs, each of them controlling a bit of each type of available habitat: coastal land, stream land, forested land, and gardening land. Since available resources were distributed relatively equally among and between chiefdoms, there were few power struggles or monopolies over resources. It is important to note that these are general rather than universal patterns of association between physical geography and human adaptations in the Pacific. Geographers reject the outdated notion of **environmental determinism,** which insisted that certain environmental conditions always correlate with specific culture traits.

Today, the main cities and seaports of Oceania, and the largest populations, are on the volcanic high islands and the continental islands. Valuable minerals are scarce, but the rich soils (typically of volcanic origin) support a diversity of tropical crops. Generally, these islands are more prosperous than the low islands.

The low islands are formed of coral. Most take the shape of an irregular ring surrounding a lagoon; such an island is called an **atoll.** Generally, the coral ring is broken into many pieces, separated by channels leading into the lagoon, but the whole circular group is commonly considered one island. Charles Darwin devised a still widely accepted explanation

(a) Volcanic island with fringing reef

(b) Slight subsidence, barrier reef

(c) Atoll

● **Figure 8.9** Charles Darwin's explanation of the development of an atoll. (a) First, a fringing reef of coral is attached to the volcanic island's shore. (b) As the island subsides, a barrier reef forms. (c) With continued subsidence, the coral builds upward, and the volcanic center of the island finally becomes completely submerged, forming an atoll.

for the three-stage formation of atolls (●Figure 8.9). First, coral builds what is known as a fringing reef around the edge of a volcanic island. Then as the island slowly sinks, the coral reef builds upward and forms a barrier reef separated from the shore by a lagoon. Finally, the volcanic island sinks out of sight, and a lagoon occupies the former land area, whose outline is reflected in the roughly oval form of the atoll.

The low islands are generally smaller than the volcanic high islands and lack the resources to support dense populations. Their shorelines are typically dotted with the waving coconut palms that are a mainstay of life and the trademark of the South Sea isles. The Gilbert Islands of Kiribati in Micronesia are typical of the picturesque low island atolls idealized by Hollywood and travel brochures (see Figure 8.6b). Other major atoll groups in Micronesia are the Caroline and Marshall Islands. The Tuamotu Archipelago is a Polynesian atoll group, and atolls are also scattered across Melanesia.

Despite their idyllic appearance, these islands pose many natural hazards to human habitation. The lime-rich soils are often so dry and infertile that trees will not grow, and shortages of drinking water limit permanent settlement. Their low elevation above sea level provides little defense against storm waves and tsunamis, the waves generated by undersea earthquakes. Coastal regions of higher islands can also suffer from these events. In 1998, an underwater landslide triggered by an earthquake created a 30-foot (9-m) tsunami that swept over a 15-mile (24-km) stretch of northern Papua New Guinea, killing 2,100 people. A tsunami of equal height struck the Solomon Islands in 2007, but casualties in the lightly populated affected region were far fewer.

Before the intrusion of outsiders, the low-island economies were based heavily on subsistence agriculture (with strong reliance on coconuts), gathering, and fishing. Tom McKnight explained that sparse resources were distributed with relative uniformity among the low islands' populations.[2] The political and social units were smaller and less structured than those of the high islands. These resource-poor low islands have a long history of population limiting factors such as war, infanticide, and abortion.

Many low-island countries now rely heavily on modern commercialized versions of the coconut and fishing economy. The coconut is so important in their subsistence that many of these islands have developed what has been characterized as a **"coconut civilization"** (•Figure 8.10). Coconuts provide food and drink for the islanders, and the dried meat, known as copra, is the only significant export from many islands. The husks and shells of the nuts have many uses, as do the trunks and leaves of the coconut palms. People make baskets and thatching from the leaves, for example, and use timber from the trunk for construction and furniture.

Why Are Oceania's Ecosystems So Vulnerable?

Island ecosystems in the Pacific region, like those across the globe, are typically inhabited by endemic species of plants and animals—those found nowhere else in the world. They result from a process in which ancestral species colonize the islands from distant continents and, over a long period of isolation and successful adaptation to the new environ-

• **Figure 8.10** Most coconut harvesting is still done by hand, and can be a hazardous occupation. Falling coconuts are also dangerous, and locals know not to sleep or linger under these trees.

ments, evolve to become new species. Gigantic and flightless animals, like the moa bird of New Zealand and the giant tortoises of the Galápagos (in the Pacific) and the Seychelles (in the Indian Ocean), are typical of endemic island species. More than 70 percent of the native plant species of New Caledonia and Hawaii are endemic.

Because they have developed in the absence of natural predators and inhabit relatively small areas, island species are especially vulnerable to the activities of humankind. People have intentionally or accidentally introduced to the island alien plant and animal species, known as **exotic species** and including rats, goats, sheep, pigs, cattle, and fast-growing plants, that often end up preying on or overtaking the endemic species. Habitat destruction and deliberate hunting also lead to extinction. For example, New Zealand's moas and the dodo birds of the Indian Ocean island of Mauritius are now, well, "dead as dodos." Indigenous inhabitants of New Zealand set fires to hunt the giant ostrichlike moa and had already killed off most of the birds by the time the Maori people colonized New Zealand around 1350 C.E. By 1800, the moa was extinct.

From one end of Oceania to the other, concern about such human-induced environmental changes has been mounting. The Hawaiian Islands have become known to ecologists as the **"extinction capital of the world"** because of the irreversible impacts that exotic species, population growth, and development have had on indigenous wildlife there. Environmentalists are also concerned about the impact of commercial logging on the island of New Guinea (•Figure 8.11). About 22,000 plant species grow there, fully 90 percent of them endemic. And ecologists consider nearby New Caledonia one of the world's biodiversity hot spots because of the ongoing human impact on its unique flora and fauna.

Human-induced extinctions are just one of a long list of natural hazards to island species. For example, island animals are particularly prone to volcanic eruptions, typhoons (hurricanes), and rises in sea level (see page 432). Most animals find it difficult or impossible to evacuate in the event of a natural catastrophe, and the island's distance from other lands may hinder or prevent recolonization. So although island habitats have a higher percentage of endemic species than would be found on mainlands, environmental difficulties sometimes cause the number of different kinds of species (that is, species diversity) to be lower. Far more than natural hazards, however, it is the human agency that altered the Pacific island ecosystems, in some cases completely transforming the natural landscape (see Insights, page 423).

• **Figure 8.11** The luxuriant forests of New Guinea contain an exceptional wealth of plant and animal species.

STEPHEN ALVAREZ/National Geographic Image Collection

8.3 Cultural and Historical Geographies

Their indigenous cultures having been greatly diminished in numbers and influence, Australia and New Zealand today are mainly European in culture and ethnicity. The region's other cultures, however, are overwhelmingly indigenous and quite diverse. Aside from Australia and New Zealand, with their majority European populations, and Fiji, New Caledonia, and Guam, each of which has populations about half indigenous and half foreign, approximately 80 percent of the Pacific's people are indigenous. The balance is 13 percent Asian and 7 percent European. Of the indigenous populations, Melanesians make up about 80 percent, with 14 percent Polynesian and 6 percent Micronesian.

The Indigenous Peoples of Oceania

Settlement in the Pacific region began about 60,000 years ago. Although the archaeology is far from definitive on this point, it is generally agreed that this is when the first people settled in Australia. They were the ancestors of today's **Aborigines.** Some ethnographers refer to the Aborigines as a distinct Australoid race. However, they share racial characteristics with other indigenous groups in Asia, including the Mundas of central India, the Veddas of Sri Lanka, and even the Ainu of Japan. All of these groups probably descended from a common ancestral race in Asia, and the Aborigines developed their distinctive features over tens of thousands of years of living in Australia.

Recent DNA analysis suggests that they arrived in Australia in a single major migration about 50,000 years ago. During this migration, sea levels were much lower and straits were narrower and easier to navigate. The settlers probably came to Australia across the land bridge that linked New Guinea and Australia until about 10,000 B.C.E. (the historical continent made up of these two landmasses is known as Sahul). Once established in Australia, they developed a unique and enduring culture. Their **Aboriginal languages,** classified into two major groups—**Pama-Nyungan** in the southern 90 percent of Australia and **non-Pama-Nyungan** in the north—are not clearly related to any languages outside the continent.

New waves of migrants known as **Austronesians** began a long process of diffusion across the western Pacific and eastern Asia after 5000 B.C.E. Their ancestral stock probably originated in Taiwan and southern China. Their descendants migrated to the mainland and islands (initially the Philippines and Indonesia) of Southeast Asia. Their livelihoods were based on fishing and simple farming, especially of taro, yams, sugarcane, breadfruit, coconuts, and perhaps rice. Their domesticated animals included pigs and probably dogs and chickens, but they had no herd animals like cattle, sheep, or goats. From their Southeast Asian bases, over a period of several thousand years,

INSIGHTS

Deforestation and the Decline of Easter Island

"Easter Island is Earth writ small," wrote the naturalist and geographer Jared Diamond. Recent archaeological and paleobotanical studies suggest that the civilization that built the island's famous monolithic stone statues destroyed itself through overpopulation and abuse of natural resources (●Figure 8.A). Deforestation has serious repercussions wherever it occurs, but perhaps nowhere are these impacts so apparent as on Easter Island.

When the first colonists from eastern Polynesia reached remote Easter Island (known as Rapa Nui to locals) in about 400 D/F/, they found the island cloaked in subtropical forest. Plant foods, especially from the Easter Island palm, and animal foods, notably porpoises and seabirds, were abundant. The human population grew rapidly in this prolific habitat. The complex, stratified society that emerged on the island grew to an estimated 7,000 to 20,000 people between 1200 and 1500, when most of the famous statues were built. Apparently in association with their religious beliefs, Easter Islanders erected more than 200 statues, some weighing up to 82 tons and reaching 33 feet (10 m) in height, on gigantic stone platforms. At least 700 more statues were abandoned in their quarry sites and along roads leading to their would-be destinations, "as if the carvers and moving crews had thrown down their tools and walked off the job," Diamond observed.*

"Its wasted appearance could give no other impression than of a singular poverty and barrenness," the Dutch explorer Jacob Roggeveen wrote of the island on the day he discovered it—Easter Sunday 1722.† Not a single tree stood on the island. The depauperate landscape bears testimony to the fate of the energetic culture that built the great statues. By 800 D/F/, people were already exerting considerable pressure on the island's forests for fuel, construction, and ceremonial needs. By 1400, people and the rats they introduced to the island caused the local extinction of the valuable Easter Island palm. Continued deforestation to make room for garden plots and to supply wood to build canoes and to transport and erect the giant statues probably eliminated all of the island's forests by the 15th century. By then, people had hunted to extinction many terrestrial animal species and could no longer hunt porpoises because they lacked the wood needed to build seagoing canoes. Crop yields declined because deforestation led to widespread soil erosion. There is evidence that in the ensuing shortages, people turned on each other as a source of food. By about 1700, the population began a precipitous decline to only 10 to 25 percent of the number who once lived on this isolated Eden. Today about 4000 live there, welcoming more than 45,000 tourists every year.

THOMAS J. ABERCROMBIE/National Geographic Image Collection

● **Figure 8.A** The Polynesian people who erected at least 900 *moai* statues between 1200 and 1500 D/F/ deforested most of Easter Island—in part to supply levers and rollers for the statues—and thus precipitated the demise of their civilization.

* Jared Diamond, "Easter's End." *Discover, 16*(8), 1995, p. 64.
†Quoted in ibid.

---

the Austronesians embarked on voyages to Madagascar, New Zealand, Easter Island, Hawaii, and perhaps Chile (around 1400 C.E.), "island-hopping" all the way and accomplishing extraordinary feats of navigation in their outrigger canoes. Their settlement of Polynesia came rather late in these adventures, within the last 1,500 years. Today's Micronesians and Polynesians are mainly their descendants, but considerable mixing has occurred over the centuries.

These seafaring peoples spread their characteristic agriculture and a language called **Proto-Austronesian**—the ancestor of all modern Austronesian languages—across the Pacific. The Pacific region's linguistic diversity today is extraordinary (●Figure 8.12). Most of the indigenous peoples of the region speak languages in the **Austronesian language family,** which is divided between the **Formosan** and the **Malayo-Polynesian language subfamilies.** The Melanesian peoples, however, speak **Papuan languages,** which are not a distinctive linguistic family but an amalgam of often unrelated and mutually unintelligible tongues. There are about 700 Papuan languages.

Papua New Guinea is home to 860 languages (mainly Papuan but also Austronesian), or 20 percent of the world's total number of languages, making it by far the world's most linguistically diverse country, with an average of roughly one language per 5,000 people. Vanuatu has fewer than 200,000 people and 105 identified languages, earning that country's population the distinction as the world's most linguistically diverse on a per capita basis, with roughly one language per 2,000 people. There are fewer of the mainly Austronesian languages in Micronesia and Polynesia, and more of them

● **Figure 8.12** Languages and settlement routes of Oceania.

are mutually intelligible. Reflecting the colonial past, English and French are official languages in some of the islands and are spoken widely across the region. Another lingua franca is **pidgin,** which consists of English and other foreign words mixed with indigenous vocabulary and grammar. Pidgin is the official language of Papua New Guinea, where it is known as Tok Pisin. Its kinship to English may be appreciated in these phrases: *kanu i kapsait* means "the canoe capsized," and *mi lukim dok* means "I saw the dog."[3]

Europeans in Oceania

Europeans began to visit the Pacific islands during the 16th century, and their impact was especially profound after 1800. Spanish and Portuguese voyagers were followed by Dutch, English, French, American, and German explorers. Many famous names are connected with Pacific exploration, including Ferdinand Magellan, Abel Tasman, Louis-Antoine de Bougainville, Jean-François de La Pérouse, and the most famous of all, British Captain James Cook, who undertook three great voyages in the 1760s and 1770s, only to be killed by Hawaiian Islanders in an unexpected dispute

in 1779. Europeans were interested mainly in the high islands because of their relative resource wealth. They sought sandalwood, oyster shells, whales, and local people to indenture as whaling crews and as manual laborers. On island after island, European penetration decimated the islanders and disrupted their cultures. The intruders introduced venereal and other infectious diseases, alcohol, opium, forced labor, and firearms, which worsened the bloodshed in local wars.

They also came as Christian missionaries (see Perspectives from the Field, pages 426–427). Some ancient indigenous belief systems thrive in parts of the region, often nestled within Christianity and other recently introduced faiths (Table 8.2). There were even new faiths that arose based on the premise that foreigners were preventing locals from acquiring their just rewards. These so-called **cargo cults** emerged in New Guinea and elsewhere in Melanesia, especially during World War II, as "Stone Age" people were exposed for the first time to chocolates, cigarettes, soft drinks, bottles, metals, manufactured cloth, and other material trappings brought by American soldiers and other Westerners. There were many variations among the cargo cults, but most were

TABLE 8.2   Religions of Oceania

Country or Territory	Major Religions
American Samoa (U.S.)	Christian Congregationalist, Roman Catholic, Protestant
Australia	Anglican, Roman Catholic, other Christian
Cook Is.(N.Z.)	Cook Islands Christian Church
Fiji	Hindu, Methodist, Roman Catholic, Muslim
French Polynesia (Fr.)	Protestant, Roman Catholic
Guam (U.S.)	Roman Catholic
Kiribati	Roman Catholic, Protestant
Marshall Islands	Protestant
Micronesia, Federated States of	Roman Catholic, Protestant
Nauru	Protestant, Roman Catholic
New Caledonia (Fr.)	Roman Catholic, Protestant
New Zealand	Anglican, Presbyterian, Roman Catholic, Methodist, other Christian
Northern Mariana Islands (U.S.)	Roman Catholic
Palau	Roman Catholic, other Christian, Modekngei religion
Papua New Guinea	Roman Catholic, Lutheran, Presbyterian, other Christian, indigenous beliefs
Solomon Islands	Anglican, Roman Catholic, United, Baptist, other Christian
Tonga	Christian
Tuvalu	Church of Tuvalu
Vanuatu	Presbyterian, Anglican, Roman Catholic, indigenous beliefs

Source: *World Factbook*, CIA, 2004.

ropeans, became more powerful and wealthier than other island leaders. Eventually, he became high chief and often expanded his control to other islands; thus the first Polynesian kings emerged.

Europeans, North Americans, and East Asians changed the natural history of the islands by introducing exotic crops and animals, including arrowroot and cassava (manioc); bananas; tropical fruits such as mangoes, pineapples, papayas, and citrus; coffee and cacao; sugarcane; and cattle, goats, and poultry. They established extensive sugar plantations on some islands, notably in the Hawaiian Islands, Fiji Islands (where sugar is still the major export), and Saipan in the Marianas. The valuable Honduran mahogany trees planted by British colonists in Fiji early in the 20th century are only now beginning to mature, and competition for export profits from this resource has become intense.

It is remarkable that as late as 1840, almost all of the Pacific islands had yet to be formally claimed by outside powers; only Spain had taken the Marianas. Then, rapidly, Britain, France, Germany, and the United States vied for power. By 1900, the entire Pacific Basin was in European and American hands. Germany held Western Samoa (today's independent Samoa), part of New Guinea, the Bismarck Archipelago, Nauru, and Micronesia's most important islands. Britain controlled part of New Guinea, the Solomon Islands, Fiji, and smaller island groups and held the kingdom of Tonga as a protectorate. France controlled New Caledonia, French Polynesia, and with Britain, the condominium (jointly ruled territory) of the New Hebrides. The United States held Hawaii, Guam, and eastern Samoa.

When Germany lost its Pacific territories at the end of World War I, Japan took control of German Micronesia, New Zealand took Western Samoa, and Britain and Australia took the German section of New Guinea and the Bismarcks. Nauru came to be administered jointly by Britain, New Zealand, and Australia.

With the outbreak of World War II, Japan quickly overran much of Melanesia. The ensuing ferocious battles between Japan and the Allies for control of New Guinea, the Solomon Islands, and much of Micronesia brought enormous changes to the region. Many of the cultures were in the Stone Age as the war began and by its end had seen the Atomic Age; Micronesia had become a nuclear proving ground. Everywhere, traditional peoples were drawn into cash economies and affected by other global forces originating far beyond their watery horizons.

Most of the colonists are gone, but some of the seeds of strife they planted are still germinating. Many conflicts in the region today reflect the demographic changes Europeans wrought to meet their economic interests. In many cases, the Europeans introduced laborers from outside the region because local peoples were too few in number or refused to work in the cane fields. Successive infusions of Chinese, Portuguese, Japanese, and Filipinos brought a polyglot character to Hawaii, with little apparent ethnic discord. But ethnic strife is ongoing in Fiji and the Solomon

millenarian movements that maintained that a cataclysmic set of events would trigger a new and more prosperous age for their followers. Intervention by white foreigners would end, and the locals would be mystically delivered the massive "cargo" of material possessions long denied them. Some of the movements built jetties and storehouses in anticipation of those events. Most withered with time.[4] Overwhelmingly, the peoples of the Pacific adopted and continue to practice Christianity, with Methodists, Mormons, and Catholics the largest denominations. Where Asians have settled, there are pockets of Hinduism, Buddhism, and Islam; Fiji, for example, is a distant outpost of Hindu India.

Europeans created new settlement patterns, disrupted old political systems, and rearranged the demographic and natural landscapes. In Polynesia, for example, arriving Europeans typically sought shelter for their vessels on the lee side of an island. That safe place became the island's European port. Whatever chief happened to be in control of that spot typically, because of his association with the Eu-

PERSPECTIVES FROM THE FIELD

Aborigines and Missionaries in Northwestern Australia

In the austral winter of 1987, I served as leader and lecturer aboard a small passenger vessel that took Western tourists to the remote coastline of the Kimberley region in northwestern Australia. Except for the towns of Wyndham and Broome on the northeastern and southwestern edges of this region, this is a wilderness, virtually devoid of settlements, roads, and other infrastructure. The major population, small in number, is Aboriginal, but there are a few European missionaries and cattle ranchers. It is a land of sandstone, spinifex grass, eucalyptus trees, and abundant wildlife, including the fearsome saltwater crocodile.

My journal entries reflect on past and present cultural encounters in this tropical wilderness:

Last night we sailed from the Bonaparte Archipelago and made our way to the northern tip of the Kimberley. This part of the mainland is known as the Mitchell Plateau. It has long been an important Aboriginal region, and today most of the lands in this coastal area are actually Aboriginal Reserve Lands. Traditionally very resistant to outsiders, these indigenous people succeeded in driving off Malays, Indonesians, and Europeans over a period of several hundred years. Within the past century, however, various Christian groups have successfully established several missions in the area.

Peter [Sartori, the vessel's captain] shifted the ship around Mission Bay, and we went ashore. We were received by a small but very enthusiastic party of Aboriginal children, who enjoyed themselves immensely clambering in and out of the inflatable boats. Their parents were our hosts for a very interesting couple of hours.

These people are members of the Wunambal tribe. They had come down from the Kalumburu Mission, about 18 miles inland on the King Edward River. Its predecessor, the Pago Mission, was founded by Benedictine (Spanish Catholic) monks in 1908. The missionaries' relationship with the Aborigines was not always peaceful; there were a few hostile encounters in which the missionaries came out the worse. However, the parties came to trust one another and enjoyed good relations right up until the Pago Mission had to be abandoned in the 1930s, when its water supply ran out. The Kalumburu Mission was established to take its place.

Our hosts were very congenial and went out of their way to make us feel comfortable. We spoke freely with them and found their English quite good; they had been educated at the Mission. The men, who had been wearing Western clothes, disappeared for a while and then returned decked out with loincloths and their finest white ocher dance paint. Just before dusk, a signal was given, and a traditional corroboree, or ceremonial dance, began (●Figure 8.B). The women danced too but wore Western clothing and had a minor role in the corroboree.

The dance leader, Basil, was a real ham, boasting about his talents at every opportunity. But mainly he explained how each dance or skit told a story about some aspect of daily or ritual life of the Wunambal. It was quite clear that although they had been in a Christian mission environment for decades, these people held their ancestral beliefs about their origins and the world around them. Of the Kalinda dance, Basil said, "It has to do with Cyclone Tracy [the same hurricane that devastated Darwin, Australia, on Christmas

● **Figure 8.B** The Wunambal (with Basil at far left) performing a corroboree by the shore of Mission Bay, Kimberley, Australia.

Islands, which became independent from Britain in 1970 and 1978, respectively.

The Solomon Islands' troubles date to 1942, when U.S. Marines fought to push Japanese troops off the island of Guadalcanal. The American forces enlisted the support of thousands of people from the neighboring island of Malaita. After the successful offensive, many of these Malaitans elected to stay on Guadalcanal, where under British colonial rule they came to dominate economic and political life. The indigenous people of Guadalcanal resented their domi-

nance and in particular what they saw as the loss of their lands to the Malaitans (who purchased the lands). Conflict between the groups finally erupted in 1999, with the majority Guadalcanalans murdering scores of Malaitans and driving tens of thousands off the island; most returned to Malaita. An Australian-led peacekeeping mission arrived in 2003 to quell the unrest and put civil servants to work rebuilding the country's shattered economy.

The violence in the Solomon Islands was inspired by ethnic unrest in Fiji. In the early 1900s, colonial British planta-

PERSPECTIVES FROM THE FIELD

Aborigines and Missionaries in Northwestern Australia *continued*

Eve, 1974]. Kalinda is the song spirit. When we see people coming in a line to the Pruru, or Dancing Place, with the totems, we know it is the Song Spirits coming. When we know there are strong winds or cyclones coming, we take the totems to the hills for safety. When we are going to the hill where the Rolling Stone is, and we are halfway up the hill, the big stone rolls down and all the people stop until the stone falls to the ground. One man ventures out to see if it is safe; then all people come out to dance." Then there was the Palgo; Basil explained: "The spirits catch fish at night time. They light a bundle of sticks and take them to the water hole. The fish are attracted by the light, and so we catch them." The last dance that told a story was the Djarlarimirri ("Witch Doctor"): "A man is killed by a spear, and the Witch Doctor is called. He performs his magic, and the man is restored to life." Finally, there was the Djalurru, a dance without spiritual significance, performed just for fun.

Later I wrote of the Kalumburu Mission:

I visited the Kalumburu Mission at the invitation of one of the nuns working there, the redoubtable Sister Scholastica, known affectionately as "Sister Scholly" (●Figure 8.C), who sported about on an all-terrain vehicle. Instead of a cross as her motor vehicle's masthead, she had a bark painting of a wandjina, the aboriginal ancestral spirit of cyclones and monsoon rains. The mission was celebrating the 55th anniversary of its founding. The entire community proceeded in single file under a column of white smoke that symbolized purification. Everyone wore Western clothes—one young girl even had a Mickey Mouse T-shirt. And each wore distinctive mission and Aboriginal marks on this feast day: a blue headband symbolizing the mission and traditional Wunambal white face paint. They all took their seats on a grassy area, and Father Saunders' outdoor Mass got under way. The ceremony was quite interesting. It opened with a reading by Mary Pundilow, a 75-year-old woman who was the last child born at the old Pago Mission. She talked about how things use to be before the "white men in black cloth" appeared and about the gradual acceptance of the missionaries and their peaceful efforts.

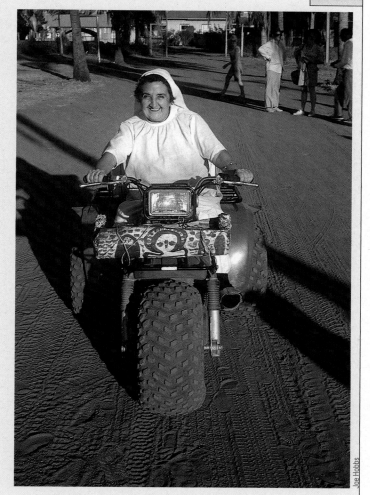

Joe Hobbs

● **Figure 8.C** "Sister Scholly," Kalumburu Mission.

I walked around during part of the service, admiring the thriving gardens and grounds where peanuts, corn, tomatoes, mangoes, pineapples, papayas, and cashews grew. There were pens for pigs, geese, guinea fowl, and Brahman cattle. When the Mass ended, the mission's population socialized on the tidy grounds around the clinic, church, and school.

tion owners imported mainly Hindu Indians as indentured laborers to work Fiji's sugarcane plantations. Today, indigenous Fijians now only slightly outnumber ethnic Indians (**Indo-Fijians**). Here, too, the indigenous inhabitants resent the dominance of the foreign ethnic group; ethnic Indians control most of Fiji's economic life. A coup led by ethnic Fijians in 2000 succeeded in ousting Fiji's ethnic Indian prime minister and throwing out a 1987 constitution that granted wide-ranging rights to ethnic Indians. Accused of lawlessness, the coup leaders were threatened with economic sanctions by the European Union, which pays higher than market prices for most of Fiji's sugar crop, and by the United States, Australia, and New Zealand. The prospect of economic crisis led the army to take control of Fiji's affairs, imprison the coup leader, and promise a general election and return to civilian government in 2002. The unrest caused a crash in tourism, Fiji's most valuable industry. Prior to the coup, about 400,000 visitors came each year. Tourist numbers rebounded after a peaceful election that brought the indigenous Fijian political party back to power, but a military-led coup in 2006 temporarily diminished the vital industry again.

8.4 Economic Geography

A lack of industrial development characterizes the Pacific island economies (other than those of Australia and New Zealand), and the poverty typical of less developed countries (LDCs) prevails in the region. Contributing to this underdevelopment are what could be described as tyrannies of size and distance. Many of the countries consist of little more than small islands of limestone sprinkled with palm trees and situated hundreds and often thousands of miles from potential markets and trading partners. Most of the countries suffer large trade deficits and must import far more than they can export.

Making a Living in Oceania

There are six major economic enterprises in the Pacific region, only one of them involving value-added processing of raw materials: exports of plantation crops (especially palm products), exports of fish, exports of minerals, services for Western military interests, textile production, and tourism. Significant manufacturing outside Australia and New Zealand is limited mainly to the Cook Islands and Fiji. Until worldwide quotas on textile trading were lifted in 2005, giving an immediate advantage to China, Saipan in the Northern Marianas had a flourishing garment industry owned by South Korean companies and operated by Chinese laborers. Polynesia's Cook Islands, a self-governing territory in free association with New Zealand, has an industrial economy based on exports of labor-intensive clothing manufacture. Fiji also began exporting manufactured clothes after the mid-1990s, when Australia granted preferential tariff treatment to Fijian-made garments.

A few Pacific countries are beginning to participate, in rather unlikely and sometimes unscrupulous ways, in the information technology revolution that has generated so much global wealth since about 1990. One example is tiny Tuvalu in Polynesia. In 2000, an American entrepreneur agreed to pay the Tuvalu government $50 million to acquire Tuvalu's two-letter Internet suffix, tv. While the new owner of the .tv domain has been busily profiting by selling Internet addresses, Tuvalu is delighted with its windfall. It used part of the money to finance the process of joining the United Nations, build new schools, add electricity and other infrastructure to outlying islands, and expand its airport runway. Vanuatu enjoys significant hard-currency revenue by permitting its country's phone code to be used for 1-900 sex calls. And Nauru and other island nations use telephone and digital technology to practice money laundering (more on this shortly).

Mining takes place on only a few islands. The major reserves are on the continental islands, notably petroleum, gold, and copper on the main island of Papua New Guinea; copper on Bougainville (an island belonging to Papua New Guinea); and nickel on the French-held island of New Caledonia. French Polynesia exports some of the world's finest and largest sea pearls, which are classified as gems. The 13,000 people of the tiny oceanic island of Nauru, which gained independence from Australia in 1968, once enjoyed a high average income from the mining of phosphate, the product of thousands of years of accumulation of seabird excrement, or **guano,** which is used as fertilizer. Until recently, Nauru's economy depended totally on annual exports of 2 million tons of phosphate and on revenues earned from overseas investments of profits from phosphate sales. This resource is approaching total depletion, however, and because the mining of phosphate has stripped the island of 80 percent of its soil and vegetation, turning it into a virtual lunar landscape, many people in Nauru are now looking for a new island home (•Figure 8.13).

Others are content with what they see as Nauru's most recent and, they hope, future source of revenue: **money laundering.** With the phosphate stocks dwindling, Nauruan entrepreneurs have undertaken offshore banking, in which they can operate loosely controlled banks anywhere but in Nauru. In Nauru's case, these are **shell banks,** existing only on paper and specializing in turning money earned in illegal ways into money that looks legal. Nauruan law does not require any official oversight of offshore banking transactions, so no official records exist; therefore, any dirty funds moved through a bank registered in Nauru cannot be traced. Russian clients have registered about half of Nauru's hundreds of shell banks, and the assumption is that much of the Russian "Mafia" money is laundered through them. Nauru has now started cracking down on this practice, but money laundering and tax-sheltering schemes have also been reported in Vanuatu, Palau, the Cook Islands, the Marshall Islands, Niue, Tonga, and Samoa.

After 2001, hard-up Nauru began earning revenue by serving yet another unlikely function: detention center for refugees from the war in Afghanistan. By 2004, more than 1,000 Afghan refugees who were apprehended at sea in transit to Australia, where they hoped to gain asylum, had been diverted to Nauru. Australia paid the government of Nauru about $15 million yearly to hold the detainees while their cases were reviewed in the Australian courts.

The most dynamic element of the Pacific economies is the rapid growth of tourism, with most visitors coming from the United States, Australia, and Japan and a rapidly growing number from China. Tourism is the largest industry in the Pacific region, providing substantial income and employment and stimulating overall economic growth by encouraging foreign investment and facilitating trade. The Pacific islands continue to exert an irresistible appeal to people awash in the conveniences and congestion of the industrialized world. One online entrepreneur, hoping to capitalize on disenchantment with the modern world, is building an e-community of people who will contribute funds to, and sometimes visit, his idyllic Fijian island (see http://tribewanted.com). The most popular destinations are

APF/Getty Images

● **Figure 8.13** The 10 square miles (26 sq km) of land that is Nauru have been devastated by the phosphate mining that once made Nauruans among the wealthiest people per capita on earth.

Hawaii, Australia, New Zealand, and French Polynesia, but there has been strong growth in ecotourism, archaeological tourism, and other tourism niches in destinations including Palau, Vanuatu, and Easter Island. Some of the countries have developed unique appeals for international tourists. In a bid to attract visitors by being the first country in the world to greet the new millennium, Kiribati decided unilaterally in 1997 to shift the International Date Line eastward by more than 2,000 miles (3,200 km). The impact of this decision stands out on any map depicting the date line (see Figure 8.1).

Mining Brings Strife

Issues of control over the region's considerable mineral wealth have brought unrest to New Caledonia and Bougainville. Separatists demanding independence for nickel-rich New Caledonia clashed with French police in the 1980s. An accord signed after the confrontation guaranteed a 1998 referendum for New Caledonia's people (known as **Kanaks**), who would then choose to become independent or remain under French rule. In 1998, however, apparently to maintain its interests in the colony's nickel, France postponed the vote until 2014. In the meantime, France is allowing New Caledonia to have considerable autonomy; since 1999, a legitimate New Caledonian government has been permitted to look after civil affairs.

In the late 1980s, a crisis shook Bougainville, a former Australian colony that in 1975 became, at Australia's insistence, part of the newly independent Papua New Guinea. Most Bougainvilleans were unhappy about their association with Papua New Guinea and wanted to be united again

with the Solomon Islands, to whose people they are culturally and ethnically related. Angry landowners calling themselves the Black Rambos destroyed mining equipment and power lines to force the closure of the world's third-largest copper mine, the Australian-owned Panguna mine. These generally younger landholders were receiving none of the royalties and compensation payments made by New Guinean and Australian mining interests to an older generation of the island's inhabitants. Residents of Bougainville who were concerned about the ruinous environmental effects of copper mining, disturbed that most mine laborers were imported from mainland Papua New Guinea, and angry that little mining revenue remained on the island, joined the opposition. The growing crisis drove world copper prices to record-high levels.

Opposition to the Australian-backed mining interests of Papua New Guinea in Bougainville grew after 1988 into a full-fledged independence movement led by the **Bougainville Revolutionary Army (BRA)**. War broke out in 1989 when the BRA leader declared independence from Papua New Guinea. Papua New Guinea responded with an economic blockade around Bougainville to prevent supplies from reaching the rebels, who salvaged World War II–era weapons and ammunition to fight superior troops. In a decade of fighting, Bougainville's 200,000 people suffered huge losses, with nearly 20,000 killed and 40,000 left homeless. A 1998 cease-fire finally brought to an end the longest conflict in the Pacific since World War II. Bougainville remains a part of Papua New Guinea but has been granted greater autonomy.

Not only mineral but military interests of outside powers, particularly the United States and France, generate much revenue for some of the Pacific economies, notably

REGIONAL PERSPECTIVE

Foreign Militaries in the Pacific: A Mixed Blessing

The remote locales and nonexistent or small human populations of some Pacific islands have made them irresistible sites for weapons tests by the great powers controlling them (•Figure 8.D). In the 1940s and 1950s, U.S. authorities displaced hundreds of islanders from Kwajalein atoll in the Marshall Islands to make way for intercontinental ballistic missile (ICBM) tests. During those years, the United States also used nearby Bikini atoll as one of its chief nuclear proving grounds (the island's fame at the time led French designers to name a new two-piece bathing suit after it). In 1946, U.S. government authorities relocated the indigenous inhabitants of Bikini to another island, promising they could return when the tests were complete.

Over the next 12 years, there were 23 nuclear blasts on Bikini, out of a total of 67 conducted as **Operation Crossroads** in the

• **Figure 8.D** The nuclear explosion, code-named *Seminole*, at Eniwetok atoll in the Pacific Ocean on June 6, 1956. This atomic bomb was detonated at ground level and had the same explosive force as 13,700 tons (13.7 kilotons) of TNT. It was detonated as part of Operation Redwing, an American program that tested systems for atomic bombs.

Marshall Islands. A 15-megaton hydrogen bomb detonation on Bikini in 1954 had the power of 1,000 Hiroshima bombs, blowing a mile-wide crater in the island's reef and producing radioactive dust that fell downwind on the Rongelap atoll, where children played in the dust "as though it were snow," one observer wrote.* For decades, the U.S. Atomic Energy Commission claimed that a sudden shift in winds took the fallout to inhabited Rongelap, but recently revealed records proved that authorities knew of the wind shift three days prior to the test. The Marshallese today suffer a legacy of cancers and deformities linked to the weapons tests on Bikini and Eniwetok atolls. So far, the United States has provided more than $60 million to compensate Marshall Islanders made ill by the blasts.

The Bikinians have never accepted their exile. But returning to Bikini has been a hazardous prospect, given its legacy of radioactive soil. A repatriation of the Bikinians in the 1960s was found to be premature, as radiation persisted, and the nuclear nomads moved again. More recent tests have determined that Bikini has little ambient radioactivity and is essentially habitable, except for the cesium-137 that permeates the soil and becomes concentrated in fruits and coconuts. Bikinians do not want to go home to stay until they have assurances from the United States that the problem of poisoned produce can be resolved and that the island is completely safe. Along with the people of Eniwetok, they are seeking massive funding from the United States for complete environmental decontamination of their homeland and for further compensation for their travails.

Now that Bikini is at least safe to visit, some Bikinians are capitalizing on its formerly off-limits status. The island's lack of human inhabitants for more than 50 years has had the remarkable effect of rendering Bikini a true marine wilderness. In the late 1990s, it opened as a tourist destination, with naval ships sunk by the atomic blasts a highlight for many divers. The dive masters and tour guides are Bikinians.

*Nicholas D. Kristof, "An Atomic Age Eden (but Don't Eat the Coconuts)." *New York Times*, March 5, 1997, p. A4.

U.S. Navy/SPL/Photo Researchers, Inc.

Guam, American Samoa, and French Polynesia. This military presence is a mixed blessing, however, because of its environmental impacts and because it perpetuates dependence on foreign powers. These are among the geopolitical concerns discussed next.

8.5 Geopolitical Issues

Once entirely colonial, Oceania today is a mix of units still affiliated politically with outside countries and others that have become fully independent. Since the end of World War

II, the United States, Britain, Australia, and New Zealand have abandoned most of their colonies in the region. Only France has insisted on holding on to all of its colonies.

Why Are Foreign Powers Interested in the Pacific?

In some cases, colonial powers have delayed independence to would-be island nations. They explain that such territories are too small and isolated or are not economically viable enough for independence. Some islands remain dependent because they confer unique military or economic

REGIONAL PERSPECTIVE

Foreign Militaries in the Pacific: A Mixed Blessing *continued*

Like the United States, France has had a nuclear stake in the Pacific. In 1995, after a three-year hiatus, France resumed underground testing of nuclear weapons on the Mururoa ("Place of Deep Secrets") atoll in French Polynesia. There was an immediate backlash from some national governments and from environmental and other organizations. Crew members of the *Rainbow Warrior 2,* the flagship of the environmental organization Greenpeace, confronted French Navy ships near Mururoa in July 1995. French commandos used tear gas to overwhelm the crew and seize the ship. Greenpeace managed to televise the event, inflicting enormous public relations damage on France. Strongly antinuclear New Zealand protested loudly against the nuclear tests. In Australia, customers boycotted imported French goods and local French restaurants, and postal employees refused to deliver mail to the French embassy and consulates. Antinuclear protesters firebombed the French consulate in Perth. Australia joined New Zealand in filing a case at the World Court in The Hague aimed at halting French nuclear tests. Worldwide, informal boycotts diminished sales of French wines. In French Polynesia itself, activists for independence from France used the issue to highlight international attention to their demands. Over several days in September 1995, hundreds of anti-French demonstrators in Tahiti rampaged through French Polynesia's capital city of Papeete, looting and setting parts of the city on fire and causing millions of dollars of damage to the island's international airport, a vital tourist hub. Although French authorities continued to insist that the tests and the atoll itself were safe, France soon bowed to international pressure, ceasing its nuclear tests in the Pacific and signing the Comprehensive Test Ban Treaty in 1996.

Ironically, the greatest fear in the region now is that France will reduce its military—and hence economic—presence even more. Despite the nuclear tests, most inhabitants of French Polynesia have long favored continued French rule. Tourism and other local businesses contribute only about 25 percent of the territory's revenue; the remainder comes from French economic assistance, largely in the military sector. Many locals, however, whose parents and grandparents gave up subsistence fishing and farming for jobs in the military and its service establishments, are worried about their future.

Similar fears stalk Palau, the Marshall Islands, and the Federated States of Micronesia (FSM), three independent nations carved from the former U.S. Trust Territory of the Pacific. The United Nations awarded that territory, which includes the Bikini and Eniwetok atolls, to the United States in 1946 as the world's first and only political-military zone. In 1986, the United States granted independence to the Marshall Islands and the FSM in an agreement known as the **Compact of Free Association.** In this agreement, the countries agreed to rent their military sovereignty to the United States in exchange for tens of millions of dollars in annual payments and access to many federal programs. The United States uses its leases to conduct military tests—for example, on Strategic Defense Initiative (SDI, also known as "Star Wars") technology. The United States also maintains a military presence in Palau and provides large cash grants and development aid to the nation, which became independent from the United States in 1994 but remains in free association with it.

In these cases, the American government handed over the monies with no strings attached. The funds have been embezzled and misspent, and there is little development to show for them. The United States has begun to reassess the payments with an eye to their eventual curtailment. Many islanders fear that their economies will subsequently plummet. The people of the Marshall Islands and the Federated States of Micronesia are especially worried because more than half of the gross domestic product of each country is comprised of economic aid from the United States. The Marshall Islands' leaders, however, are taking a more defiant stand, saying that the United States has no right to question what becomes of the aid money and pointing out that the United States gets much in return in the form of extensive military rights. The United States is hard-pressed here because the Kwajalein Missile Range in the Marshalls is the only reasonable place in the world where, Pentagon officials say, they can test many of the SDI program components.

---

advantages on the governing power. For example, French Polynesia was until recently the venue for French atomic testing (see Regional Perspective, above), and Guam and American Samoa are still useful to the United States for military purposes. Guam, which has been a U.S. possession since 1898 and a U.S. territory since 1950 (its residents are American citizens), is especially vital in U.S. strategic thinking. Pentagon planners classify it as a **power projection hub** that can be used as a forward base five full days' sailing time closer to Asia than Hawaii is. It has been used as a jumping-off point for American B-52 jets on bombing missions to Iraq and would be used in any U.S. military ac-

tion in Korea or elsewhere in East Asia. In the event of such hostilities, the United States would need Japan's permission to operate from bases in Japan, but no such approval would be necessary to act from Guam. A generally unspoken but widely acknowledged U.S. military objective in building up readiness in Guam is that the island would be critical to any confrontation with China, should that giant country ever become a belligerent power.

Australia and New Zealand have Pacific-oriented defense and security agreements, somewhat precariously balanced with U.S. interests in the region. In 1951, they joined the United States to form the **Australia New Zealand**

United States (ANZUS) security alliance. Australian troops supported American forces in the Vietnam War. Since 1987, however, when its liberal government declared that New Zealand would henceforth be a nuclear-free country, New Zealand's relations with the United States have been strained. Legislation banned ships carrying nuclear weapons or powered by nuclear energy from New Zealand's ports. The United States refused to confirm or deny whether its ships violated either restriction, so New Zealand denied port access to them. This action removed New Zealand from ANZUS. New Zealand paid some heavy political and economic costs for its nonnuclear position; for example, it was pointedly left out of negotiations that led to the Australia–United States Free Trade Agreement (AUSFTA) that came into force in 2005.

Despite its small army (only 50,000 strong), Australia has begun to provide a selective security blanket over its interests in the region and elsewhere. In 1999, Australian troops were deployed to head up the United Nations peacekeeping force to oversee the referendum that gave independence to East Timor and its subsequent transi-

360 tion to full independence as Timor-Leste. Australia (and New Zealand) also lent troops and other support in the U.S. buildup to and execution of the war against Iraq in 2003. Al-Qa'ida vowed revenge and apparently delivered it

358 with a bombing in Bali, Indonesia, that killed many young Australian tourists in 2002. Australia's apparent concern about a long-term threat from North Korea also prompted it to join the United States in the development of the Strategic Defense Initiative missile shield program. Australia is sometimes snubbed by its Southeast Asian neighbors in political and economic affairs because it is seen as something of a regional "police officer" or as a United States "deputy."

Oceania's Environmental Future

Another set of geopolitical concerns in the Pacific region is environmental. The remote islands of Oceania, so far from the industrial cores of North America, Europe, China, and Japan, are actually the places most likely to suffer first and most from the most feared impacts of industrial carbon dioxide emissions. The islands would be the among the earliest and hardest-hit victims of any rise in sea level due to global warming. Sea levels have risen in recent years at a rate of 34 0.1 inch (3 mm) annually, and there are reports of unprecedented tidal surges on Pacific island shores. If the trend continues, the first Pacific islands to be totally submerged would be Kiribati, the Marshall Islands, and Tuvalu, while Tonga, Palau, Nauru, Niue, and the Federated States of Micronesia would lose much of their territories to the sea.

These island nations are among the 39 countries comprising the **Alliance of Small Island States,** which argued forcefully but unsuccessfully at the 1997 Kyoto Conference on Climate Change that by 2005, global greenhouse gas emissions should be reduced to 20 percent below their 1990 36 levels. What they did get was a pledge by most industrialized nations (not including the United States), in the form of the Kyoto Protocol, to cut these emissions to at least 5 percent below their 1990 levels by 2012. Already claiming to be a victim of lost coastline, higher storm surges, and more storms as a result of global warming, Tuvalu in the early 2000s lobbied other countries to join it in a lawsuit against the United States and Australia. Tuvalu wanted to make the case that these large countries' failure to ratify the Kyoto Protocol is a principal cause of global warming. As so often in the past, the peoples of the Pacific today feel vulnerable to the actions and policies of more powerful nations far from their tranquil shores.

SUMMARY

→ Oceania encompasses Australia, New Zealand, and the islands of the mid-Pacific lying mostly between the tropics. Tropical rain forest climates and biomes are most common, but Australia and New Zealand have several temperate climate and biome types.

→ The Pacific islands are commonly divided into three principal regions: Melanesia, Micronesia, and Polynesia.

→ Oceania is ethnically complex, having been settled by peoples of various Asian origins. Polynesia was the last to be populated. Papua New Guinea is the world's most linguistically diverse country. Christianity is the majority faith in this region.

→ Although countless islands are scattered across the Pacific Ocean, there are three generally recognized types: continental

islands, high islands, and low islands. Continental islands are either continents themselves (such as Australia) or were connected to continents when sea levels were lower (such as New Guinea). Most high islands are volcanic. Low islands are typically made of coral, a material composed of the skeletons and living bodies of small marine organisms that inhabit tropical seas.

→ The island ecosystems of the Pacific region are typically inhabited by endemic plant and animal species—species found nowhere else in the world. Island species are especially vulnerable to the activities of humans, such as habitat destruction, deliberate hunting, or the introduction of exotic plant and animal species.

→ Europeans began to visit and colonize the Pacific islands early in the European Age of Exploration and brought mainly negative impacts to island societies. However, a steady process of decolonization has accompanied a recent surge of Western interest and investment in the region.

→ There are ethnic conflicts, related mainly to maldistribution of income, between Malaitans and indigenous Guadalcanalans on the Melanesian island of Guadalcanal and between indigenous Fijians and Indo-Fijians in Fiji. Interest in securing more income from minerals has pitted the people of New Caledonia against the ruling power, France, and the people of Bougainville against Australian corporate interests in Papua New Guinea.

→ Aside from a few notable exceptions, the poverty typical of less developed countries prevails throughout most of the Pacific region.

→ In general, the Pacific islands' economic picture is one of non-industrial economies. Typical economic activities include tourism, plantation agriculture, mining, and income derived from activities connected with the military needs of occupying powers. Several countries are profiting from offshore banking and telemarketing.

→ During the 1940s and 1950s, the United States used the Bikini atoll in the Marshall Islands as one of its chief testing grounds for nuclear weapons. Strong negative reaction arose throughout the region in the 1990s as the French resumed underground testing of nuclear weapons on the Mururoa atoll in French Polynesia. That testing has since ceased. The United States relies on the region for testing of its Strategic Defense Initiative missile technology. Many inhabitants of French and American military zones are fearful of the economic repercussions of the withdrawal of foreign military presence.

→ Some of the low-island countries are fearful that global warming might raise sea levels and inundate them, and Tuvalu considered legal recourse against the United States and Australia for failing to ratify the Kyoto Protocol.

KEY TERMS + CONCEPTS

Aboriginal languages (p. 422)
   non-Pama-Nyungan (p. 422)
   Pama-Nyungan (p. 422)
Aborigines (p. 422)
Alliance of Small Island States (p. 422)
atoll (p. 420)
Australia New Zealand United States
   (ANZUS) security alliance (p. 431)
Austronesians (p. 422)
Austronesian language family (p. 423)
   Formosan language subfamily (p. 423)
   Malayo-Polynesian subfamily (p. 423)
   Proto-Austronesian (p. 423)
Bougainville Revolutionary Army (BRA)
   (p. 429)

cargo cults (p. 424)
"coconut civilization" (p. 421)
Compact of Free Association (p. 431)
continental islands (p. 417)
environmental determinism (p. 420)
exotic species (p. 421)
"extinction capital of the world" (p. 422)
geologic hot spot (p. 419)
guano (p. 428)
high islands (p. 418)
Indo-Fijians (p. 427)
Kanaks (p. 429)
low islands (p. 418)
Melanesia (p. 415)
Micronesia (p. 415)

millenarian movements (p. 425)
money laundering (p. 428)
Oceania (p. 414)
Operation Crossroads (p. 430)
Papuan languages (p. 423)
pidgin (p. 424)
Polynesia (p. 415)
power projection hub (p. 431)
seamount (p. 420)
shell banks (p. 428)

REVIEW QUESTIONS

1. What three principal subregions comprise Oceania? What are some of the countries and colonial possessions of each?

2. Describe the major differences between high and low islands. According to the geographer Tom McKnight, how did the physical characteristics of the islands relate to social and political developments? What are the typical demographic and economic qualities of each?

3. What is a geologic hot spot, and how does it relate to Pacific islands?

4. What unique ecological attributes do islands generally have? What are the major threats to them?

5. What is an atoll, and how does it form?

6. What are the major ethnic groups, languages, and religions of Oceania?

7. What are two notable ethnic conflicts on mineral-rich islands in Oceania?

8. How was Nauru able for a time to avoid the poverty typical of most Pacific islands? How is Nauru's economy diversifying now that its principal asset is dwindling?

9. What are the major economic activities in the Pacific islands, excluding Australia and New Zealand? Why and where have military interests been prominent?

10. What explains the peculiar jog in the International Date Line?

11. What fears do some Pacific countries have about potential global warming?

NOTES

1. Tom L. McKnight, *Oceania: The Geography of Australia, New Zealand and the Pacific Islands* (Englewood Cliffs, N.J.: Prentice Hall, 1995), p. 181.

2. Ibid.

3. Jeff Siegel, "Tok Pisin," http://www.une.edu.au/langnet/definitions/tokpisin.html.

4. The notable exception is John Frum, the last surviving cargo cult. See Paul Raffaele, "In John They Trust," *Smithsonian*, February 2006, http://www.smithsonianmagazine.com/issues/2006/february/john.php.

Australia is known as the Red Continent. This is the Indian Ocean coast near the northwestern town of Broome

Joe Hobbs

MODULE 8.1
AUSTRALIA AND NEW ZEALAND: PROSPEROUS AND NO LONGER SO REMOTE

Australia and New Zealand are unique. Far removed from Europe, their dominant cultures are nevertheless European. In the less developed world region of Oceania, they are prosperous. Distinctive plants and animals inhabit the often-odd landscapes of these islands, stirring a sense of wonder among indigenous peoples, European settlers, and foreign tourists alike. The writer Alan Moorehead described the reactions of early European visitors: "Everything was the wrong way about. Midwinter fell in July, and in January, summer was at its height. In the bush there were giant birds that never flew and queer, antediluvian animals that hopped instead of walked or sat munching mutely in the trees. Even the constellations in the sky were upside down."[1] This module explores these singular countries. There is also a brief discussion of the continent of Antarctica, the ultimate "land down under."

■ module
outline

■ module
objectives

This chapter should enable you to

→ Recognize the parallels between the settlement of Australia and the settlement of the United States

→ Understand the obstacles faced by indigenous people in winning legal recognition of ancestral claims

→ Evaluate the impacts of exotic species on island ecosystems

→ Appreciate the process by which Australia and New Zealand are loosening ties with their ancestral European homeland and strengthening their regional orientation

→ Consider the consequence of restrictions on immigration in these affluent countries with aging populations

→ Learn about issues on the continent of Antarctica

8.1.1 Peoples and Populations

These two unusual countries are akin in population, cultural heritage, political problems and orientation, type of economy, and location. Australia and New Zealand are among the world's minority of prosperous countries. Australia's per capita GNI PPP of $31,680 is comparable to that of France and Germany, although well below that of the United States and the most prosperous European countries. Australia has been dubbed the "Lucky Country," and a recent World Bank study ranked Australia as the world's wealthiest country based on its natural resource assets divided by the total population (●Figure 8.1.1). New Zealand is less affluent but is still prosperous compared to most nations of the world. In both countries, there are relatively few people among whom to spread the wealth; Australia's 21 million people and New Zealand's 4 million together amount to only about 70 percent of the people living in California.

Both countries owe their prosperity to the wholesale transplantation of culture and technology from the industrializing United Kingdom to the remote Pacific beginning in the late 18th century. Australia (established originally as a penal colony for British convicts) and New Zealand are products of British colonization and strongly reflect the British heritage in the ethnic composition and culture of their majority populations. They speak English, live under British-style parliamentary forms of government, acknowledge the British sovereign as their own, and attend schools patterned after those of Britain. Despite their independence (for Australia in 1901 and New Zealand in 1907), both

● **Figure 8.1.1** For many people at home and abroad, Australia epitomizes the "good life." This is Bondi Beach near Sydney.

Matt Jones

countries maintain loyalty to Britain. They still belong to the British Commonwealth of Nations. In both world wars, the two countries immediately came to the support of Britain and lost large numbers of troops on battlefields far from home. In World War I, Australia lost 60,000 of the 330,000 soldiers it sent to fight in Europe, by far the highest proportion of dead among any of the Allied countries. In World War II, U.S. forces helped prevent a threatened Japanese assault on Australia. Since that time, Australia and New Zealand have sought closer relations with the United States, and British influence has waned (see Regional Perspective, page 448).

Both Australia and New Zealand have minorities of indigenous inhabitants. The native Australian people are known as Aborigines. They have an extraordinarily detailed knowledge of the Australian landscape and its natural history and a complex belief system about how the world came about. According to the Aborigines, the earth and all things on it were created by the "dreams" of humanity's ancestors during the period of the **Dreamtime.** These ancestral beings sang out the names of things, literally "singing the world into existence." The Aborigines call the paths that the beings followed the **Footprints of the Ancestors** (or the **Way of the Law;** whites know them as **Songlines**). In a ritual journey called **Walkabout,** the Aboriginal boy on the verge of adulthood follows the pathway of his creator-ancestors. "The Aborigine clings to his native soil with every fiber of his being," wrote the ethnographer Carl Strehlow. "Mountains and creeks and springs and water holes are to the Aborigine not merely interesting or beautiful scenic features. They are the handiwork of ancestors from whom he himself has descended. The whole country is his living, age-old family tree."[2]

An estimated 300,000 to 1 million Aborigines inhabited Australia when the first Europeans arrived in the 17th century. Colonizing whites slaughtered many of the natives and drove the majority into marginal areas of the continent (see

422

• **Figure 8.1.2** Principal features of Australia.

Ethnic Geography, pages 438–439). The Aborigines today number about 510,000, living mainly in the tropical north of the country. Some carry on a traditional way of life in the wilderness, but most live in generally squalid conditions on the fringes of majority-white settlements.

When Aboriginal populations began plunging precipitously, many white Australians assumed that they would die out entirely. When they did not, white Australia tried a program of **assimilation** of the Aborigines by settling them down, particularly in mission stations, and educating them in the service of whites. From 1910 to 1970, Australian authorities abducted about 100,000 Aboriginal children (between a tenth and a third of all Aboriginal children) from their parents and, hoping to assimilate them fully, put them in the care of white foster parents. Today, members of the so-called **stolen generations** are seeking to identify their real kin and obtain apologies and compensation from the government. Thus far, the government has declined to make an official apology because that acceptance of responsibility might mean it would have to pay billions of dollars in reparation to the stolen children.

In New Zealand, the dominant indigenous culture is the **Maori** (pronounced "*mow*-ree," to rhyme with *dowry*), a Polynesian people whose native language is in the Malayo-Polynesian subfamily and who live mainly on North Island. Whites (known as *Pakeha* to the Maori) broke the Maori

hold on the land in a bloody war between 1860 and 1870. The Maori, who in 1900 seemed destined for extinction, have rebounded to make up almost 15 percent of New Zealand's population. Although their socioeconomic standing is poor, it is much better than that of the Australian Aborigines.

The Maori are being increasingly integrated, both socially and racially, with New Zealand's white population. A Maori cultural and political movement is resisting the loss of traditional culture that this integration may cause. Like the Aborigines, the Maori are increasingly filing claims against the government to recover what they regard as traditional tribal lands. They assert that New Zealand has violated the spirit and letter of the **Treaty of Waitangi**, signed in 1840 by Maori tribes and the British government. In the agreement, the Maori gave up political autonomy in exchange for "full and undisturbed" rights to lands, forests, fisheries, and the "national treasures" of New Zealand. In one small but symbolically important concession to such rights, the government has made it legal for Maori to fish without a license.

8.1.2 The Australian Environment

Australia is truly a world apart. Located in the Southern Hemisphere, where landmasses are few and far between, it is both a large island and a small continent (•Figure 8.1.2).

ETHNIC GEOGRAPHY

Australia's Indigenous People Reclaim Rights to the Land

As in the United States, newcomers to Australia forged a nation by dispossessing the ancient inhabitants of the land. And as in the United States, Australia has a long history of white racism, discrimination, and abuses against people of color, who in Australia's case were the original inhabitants. Aborigines were not mentioned in the 1901 constitution, not allowed to vote until 1962, and not counted in the national census until 1967. The most disadvantaged group in Australian society, Aborigines suffer from high infant mortality (four times that of the country's whites), low life expectancy (15 to 20 years less than that of white Australians), and high unemployment (officially 20 percent, although Aboriginal sources claim 50 percent, versus 4 percent for whites). Like Native Americans, a disproportionately large share of Aborigines fall prey to the economic and social costs of alcoholism. Statistics indicate they are 13 times more likely than non-Aborigines to end up in jail. Once in jail, they are more likely than non-Aborigines to commit suicide or be killed by guards. The Australian government now has a program to improve the justice system for native Australians.

One of the largest issues of contention between Australia's indigenous people and the white majority (known to the Aborigines as "whitefellas") is land rights. As in the United States, European newcomers pushed the native people off potentially productive ranching, farming, and mining lands into special reservations on inferior lands. The Europeans used a legal doctrine called *terra nullius,* meaning "no one's land," to lay claim to the continent.

Under the 1976 Land Rights Law, Aborigines were permitted to seek title to vacant state land ("Crown land") to which they could prove a historical relationship, but few succeeded. Following new legislation in 1992, the government began to return titles on a parcel-by-parcel basis—generally, in very marginal lands—to Aborigines who had argued their rights successfully. However, most Aborigines were unhappy with the terms of the returned titles, which allowed the native title to coexist with, but not supersede, the established Crown title. The government retained mineral rights to the land and could therefore lease native land to mining companies.

In 1993, after the longest senate debate in the country's history, Australia's foreign minister pushed the **Native Title Bill** through Australia's parliament. The new legislation addressed the Aborigines' major objections and provided the following concessions:

1. Aborigines have the right to claim land leased to mining concerns once the lease expires.

2. Aborigines have the right to negotiate with mineral leaseholders over development of their land.

3. Where their native title has been extinguished, Aborigines are entitled to compensation paid by the government.

4. With the help of a land acquisitions fund, impoverished Aborigines can buy land to which they have proved native title.

The Native Title Bill is of great consequence in the states of South Australia and Western Australia, where there is much vacant Crown land and many Aborigines are able to file claims on it. Those states are also heavily dependent on mining. Members of Australia's Mining Industry Council are unhappy with this legislation and worry that their mine leases will run out before the minerals do, which will require them to negotiate with the Aboriginal titleholders.

Potential Aboriginal claims to land expanded vastly in 1996 with another Australian High Court ruling on what is known as the **Wik case** (named after the indigenous people of northern Queensland who initiated it). The court concluded that Aborigines could claim title to public lands held by farmers and ranchers under long-term "pastoral leases" granted by state governments. These lands comprise about 42 percent of Australia's territory. The number of relevant Aboriginal land claims awaiting judgment in the courts is 700 and growing.

What Aboriginal claims to these lands would mean is bitterly debated. Technically, in light of this court decision, Aborigines have a "right to negotiate," meaning they have a decision-making voice in how non-Aboriginal leaseholders use the land. They also have a right to a share of the profits those leaseholders earn from their land uses. White farmers and ranchers protest that they would be

---

Its natural wonders and curiosities have been sources of endless inspiration, and sometimes consternation, for its inhabitants and for travelers and visiting writers. On arriving in Botany Bay, near Sydney Harbor, in 1788, the British major Robert Ross declared, "I do not scruple to pronounce that in the whole world there is not a worse country than what we have yet seen of this. All that is contiguous to us is so very barren and forbidding that it may with truth be said here that nature is reversed; and if not so, she is nearly worn out."[3] Mark Twain marveled at the Land Down Under in his book *Following the Equator:* "To my mind the exterior aspects and character of Australia are fascinating things to look at and think about. They are so strange, so

weird, so new, so uncommonplace, such a startling and interesting contrast to the other sections of the planet."[4]

Including the offshore island of Tasmania, Australia has an area of nearly 3 million square miles (7.7 million sq km), about equal to the area of the contiguous United States (see Figure 8.2). On the basis of climate and relief, Australia has four major natural regions: the humid eastern highlands, the tropical savannas of northern Australia, the Mediterranean climate lands of southwestern and southern Australia, and the dry interior (see Figure 8.5). Mainly because of extreme aridity, most of the continent is sparsely populated, and its people are concentrated where there is enough freshwater and arable land. Most people live in the

ETHNIC GEOGRAPHY

Australia's Indigenous People Reclaim Rights to the Land *continued*

ruined economically if they had to share profits. Generally, however, Aborigines have insisted that their title rights would not mean running whites out of business but would simply confer nominal recognition of their title and access to the land, especially for visitation to sacred sites. Under pressure from white farmers, ranchers, and miners, Australia's conservative government under Prime Minister John Howard reacted to the Wik case with a 10-point plan. The plan would make it much more difficult for Aborigines to make native title claims and to challenge farming, ranching, and mining activities.

In the Northern Territory, which has its own land rights legislation, vast tracts of land have already been given back to the Aborigines. Aborigines there make up one-quarter of the population and now control more than a third of the land. Aborigines also hold title to Australia's two greatest national parks, Uluru (Ayers Rock) and Kakadu, both in the Northern Territory. In a **comanagement** arrangement that has become a model worldwide for management of national parks where indigenous people reside, the Aboriginal owners of the reserves have leased them back to the Australian government park system, which manages them jointly with the Aborigines.

Tourism programs in Uluru and Kakadu now highlight the cultural resources of the parks in addition to their natural wonders. At both parks, visitors can enjoy interpretive natural history walks that also emphasize Aboriginal culture and that are conducted by the land's oldest and most knowledgeable inhabitants (●Figure 8.1.A). At Uluru, the Anangu Aborigines who own the rock hope that cultural awareness will cut down the number of tourists who climb the rock—estimated at 70 percent of the 400,000 annual total—because it is sacred to the Anangu.

The Aborigines' struggle for legal recognition of traditional land tenure is representative of a growing trend among indigenous peoples in many parts of the world. These cultures are increasingly enlisting the aid of geographers, anthropologists, and other so-

cial scientists to document, measure, and analyze traditional land claims. The objective is to produce harder evidence of traditional land ownership that can be successfully presented in national and international court cases to win land rights. Such strategies are discussed at international meetings like the **International Forum on Indigenous Mapping.** Among the contributors at these meetings are geographers fresh from the field with global positioning system (GPS) data and maps testifying to traditional land claims. Many belong to the active Indigenous Peoples Specialty Group within the Association of American Geographers.

● **Figure 8.1.A** This Aboriginal woman of the Anangu clan is a guide at Uluru (Ayers Rock) National Park.

---

humid eastern highlands and coastal plains (especially in the cooler south), in the areas of Mediterranean climate, and in the more humid grasslands adjoining the southern part of the eastern highlands and the Mediterranean areas. In only about three centuries of habitation, the mainly European peoples of Australia have transformed the island's ecosystems in profound ways, sometimes intentionally to suit their agricultural needs and sometimes accidentally—and disastrously (see Biogeography, page 441).

The Humid Eastern Highlands and Tasmania

Australia is the world's flattest continent. The only major highlands run down the east coast from just north of the Tropic of Capricorn to southern Tasmania in a belt 100

to 250 miles (160 to 410 km) wide. These highlands are generally lower than 3,000 feet (914 m), but their highest summit and the highest point in Australia is Mount Kosciusko, reaching 7,310 feet (2,228 m). The highlands and the coastal plains at their base are the country's only normally drought-free areas. South of Sydney and at higher elevations to the north, the climate is classified as marine west coast (despite the east coast location). North of Sydney, higher summer temperatures change the classification to humid subtropical, while still farther north, beyond approximately the parallel of 20 degrees south latitude, hotter temperatures and greater seasonality of rain create tropical savanna conditions.

Tasmania is a rugged, beautiful island off Australia's southeast coast (●Figure 8.1.3). About 30 percent of its land

● **Figure 8.1.3** Large parts of Tasmania are rugged and wild. This is Wineglass Bay in Freycinet National Park.

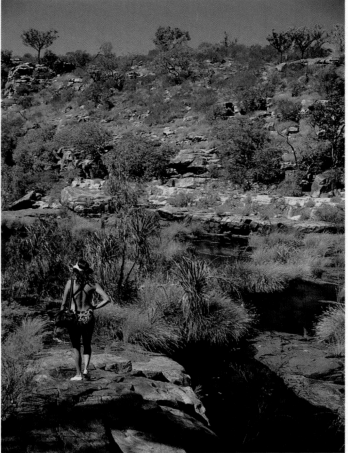

● **Figure 8.1.4** Tropical savanna prevails in northern Australia. This is typical vegetation and terrain of the Kimberley region in the continent's extreme northwest.

is dedicated to national parks and other protected areas. The state's western region is heavily forested and dotted with lakes. Numerous rivers cut rapid, short courses to the sea, representing potential sources of hydroelectric power. At this latitude—roughly that of the U.S. state of Oregon— snow sometimes falls on the high peaks even in the height of summer (November–January). Tasmania has the unfortunate distinction of having lost its entire Aboriginal population to the afflictions of white colonialism.

The island's economic prospects changed dramatically after 2001. Until then, it was a backwater losing population and employment. The September 11 terrorist attacks in the United States, however, had the unlikely impact of making Australians feel less secure even in their vast continent. Many found what they perceived as safe harbor in Tasmania and snapped up homes and other real estate there.

The Tropical Savannas of Northern Australia

Northern Australia, from near Broome on the Indian Ocean to the coast of the Coral Sea, receives heavy rainfall during the season locals call **"the wet"**—the Southern Hemisphere summer. The coastal zone is subjected in this season to hurricanes, known as cyclones in Australia. As in the United States, there are fears that because of warmer ocean waters, future cyclones will have greater strength than ever. On Christmas Eve in 1974, powerful Cyclone Tracy flattened the city of Darwin, which was subsequently rebuilt. After such deluges, northern Australia experiences almost complete drought during the winter. This highly seasonal distribution of rainfall is the result of monsoonal winds that blow onshore during the summer and offshore during the winter. The seasonality of the rainfall, combined with the

tropical heat of the area, has produced a savanna vegetation of coarse grasses with scattered trees and patches of woodland (●Figure 8.1.4). The long season of drought, the poverty of the soils, and the lack of highlands that would nourish perennial streams all combine to limit agricultural development. The alluvial and volcanic soils that support large populations in some tropical areas are almost absent in northern Australia.

The "Mediterranean" Lands of the South and Southwest

The southwestern corner of Australia and the lands around Spencer Gulf have a Mediterranean or dry-summer subtropical type of climate, with subtropical temperatures, winter rain, and summer drought. In winter, the Southern Hemisphere belt of the westerly winds shifts far enough north to bring precipitation to these districts, while in summer, this belt lies offshore to the south and the land is dry. Crops introduced from the Mediterranean climate lands of Europe generally do well in these parts of Australia, but the shortage of highlands to catch moisture and supply irrigation

• **Figure 8.1.5** Australian wine production is on the increase. This vineyard is in the Hunter Valley north of Sydney.

• **Figure 8.1.6** Looking rather like a town in Nevada, this is Australia's frontier town of Alice Springs.

water to the lowlands limits agricultural development. Australian wines produced in this region, particularly around Adelaide, are becoming increasingly popular abroad. Approximately 1,300 wineries in Australia now produce about 2 percent of the world's total wine output (•Figure 8.1.5).

The Dry Interior

The huge interior of Australia is desert surrounded by a broad fringe of semiarid grassland (steppe) that is transitional to the more humid areas around the edges of the continent. This is the **outback** that has lent so much to Australian life and lore. Locals know it as the **bush,** the **back of beyond,** and the **never-never.** Altogether, the interior desert and steppe cover more than half of the continent and extend to the coast in the northwest and along the Great Australian Bight in the south. This huge area of arid and semiarid land is too far south to get much rain from the summer monsoon, too far north to benefit from rainfall brought by the westerlies in winter, and deprived of Pacific winds by the eastern highlands. The region is sparsely populated. The chief city is Alice Springs (population 25,000; •Figure 8.1.6), a famous frontier settlement that is the jumping-off point for visitors to Uluru (Ayers Rock).

The western half of Australia is occupied by a vast plateau of ancient igneous, metamorphic, and hardened sedimentary rocks. Its general elevation is only 1,000 to 1,600 feet (c. 300 to 500 m), and its few isolated mountain ranges are too low to influence the climate or to supply perennial streams for irrigation. To the east, between the plateau and the eastern highlands, the land is even lower. The Lake Eyre Basin, the lowest part of this lowland—and of Australia—is the driest part of the continent. However, another part of the lowland, the Murray-Darling Basin, contains Australia's only major river system. In normal years, it produces half

of Australia's agricultural wealth and fully three-quarters of its irrigated crops. Some of its water comes from a massive engineering project, undertaken in the 1950s, to divert 99 percent of the waters of the Snowy River, which flowed into the Pacific, via underground tunnels to the basin. One unforeseen consequence was a dramatic rise in the water table, which forced salts to the surface. This salinization proved to be enormously destructive to the basin's cotton and rice. Consideration is now being given to growing more drought-tolerant olives and grapes and to returning more of the Snowy's water to its original drainage.

Drought is a regular visitor to Australia, which in 2007 was weathering its worst "big dry" on record, with seven years of searing temperatures and unprecedented low rainfall. Harvests of cotton, oranges, tangerines, and other crops plummeted as the Murray-Darling waters all but dried up. Australians took up an intense debate on global warming. Many criticized their country for being, along with the United States, one of the few MDC holdouts against the Kyoto Protocol on Climate Change (a position that was reversed with the election of Kevin Rudd as Prime Minister in 2007). Some agricultural planners proposed moving farming operations from their southern strongholds to the country's north, where abundant seasonal rains still fell. Critics, however, pointed out that poor soils, the superabundance of precipitation in a short time, and the potential for conflict with Aboriginal land claims in the north canceled out any benefits that might be anticipated from such an ambitious move.

Australia's wheat, rice, dairy, and other agricultural exports consume enormous quantities of water. The country is the world's largest exporter of "virtual water"—water consumed in producing these exports. Inevitably, the long drought in this already dry environment raised difficult questions about whether water use should be reallocated from the agricultural to the urban sectors.

BIOGEOGRAPHY

Exotic Species on the Island Continent

Exotic species are nonnative plant and animal species introduced through natural or human-induced means into an ecosystem. Their impact tends to be pronounced and often catastrophic to native species, particularly on island ecosystems where resources and territories are limited. In Ecuador's Galápagos Island National Park, for example, cats, rats, pigs, goats, wasps, and other animals introduced by people have disrupted nests and food supplies and thereby reduced populations of endemic tortoises and other species.

The severe impacts of exotic species demonstrate that Australia truly is an island. Both inadvertently and purposely, people have introduced foreign plants and animals that have multiplied and affected the island, sometimes in biblical plague proportions. English settlers imported the rabbit for sport hunting, confident in the belief that as in England, there would be natural checks on the animal's population. They were mistaken (●Figure 8.1.B). The rabbits bred like rabbits, eating everything green they could find, with ruinous effect on the vegetation. Observers described the unbelievable concentrations of the lagomorphs as "seething carpets of

brown fur." In the mid-20th century, the government mounted a huge eradication effort, employing fences, snares, dogs, guns, fires, and numerous other means and exhorting combatants with the slogan "The Rabbit War Must Be Won!" An introduced virus called "white blindness" eventually helped reduce the vast numbers, but the animal is still prolific throughout large parts of the country.

Other remedies have been applied, with varying success, to correct numerous other problem species, including foxes, mice, water buffalo, cane toads, and the prickly pear cactus. Many of the problem populations, notably cats and water buffalo, are **feral animals**—domesticated animals that have abandoned their dependence on people to resume life in the wild. An estimated 12 million feral cats infest the country and are held responsible for causing the extinction or endangerment of 39 native animal species. One Australian politician promoted legislation that would kill all cats on the continent by 2020.

Exotic livestock such as sheep and cattle are also a problem in the Australian environment. These hard-hoofed imports cut into the soil, promoting erosion and desertification. Some range management scientists and conservationists believe that the future of Australian ranching lies with the country's 25 million kangaroos. These soft-hoofed marsupials are adapted to the Australian landscape and do not have such a damaging impact on the soil. Farmers and ranchers have traditionally eradicated about 2 million of them each year as pests. However, if more consumers at home and abroad could learn to appreciate kangaroo meat, there would be a strong incentive for ranchers to reduce their cattle and sheep herds and allow kangaroos to proliferate and be harvested. South Africa already imports about 20 tons of kangaroo meat yearly from Australia, and nearly 100 more tons go each year to countries including Russia, France, the Netherlands, and Germany. Kangaroo skin is also used to make athletic shoes.

Australia is an unlikely exporter of another quadruped, the dromedary camel. The more than 200,000 camels in Australia are feral descendants of those brought from South Asia and Iran early in the 20th century to provide transport across the arid continent. Australian camels are exported even to Saudi Arabia, where their wild temperaments suit them to racing.

● **Figure 8.1.B** Rabbits were one of Australia's many human-caused scourges and are still a problem, but not nearly as much as in the 1920s, when people went on "rabbit drives" like these to chase, corral, and kill the declared vermin.

Hulton-Deutsch Collection/Corbis

8.1.3 Australia's Natural Resource–Based Economy

Due to widespread aridity and the presence of highlands in its only humid area, Australia has little good agricultural land (see Figure 8.7). Although as much as 7 percent of the total land area is potentially cultivable, only about one-half of 1 percent of Australia is actually cultivated today. Fourteen percent of Australia's area is classified as forest and woodland, mostly of little economic value. More than 50 per

cent is classified as natural grazing land and about 20 percent as unproductive land. Despite its small proportion of arable area, however, Australia has one of the most favorable ratios of crop-producing land to population in the world.

The sectors of the Australian economy that regularly produce the country's leading exports are mining, ranching, and even with its small proportion of cultivable land, agriculture (which accounts for 3 percent of the country's gross domestic product and 15 percent of its total exports). Several factors support the country's emphasis on surplus

Bill Bachman/Photo Researchers, Inc.

● **Figure 8.1.7** Sheep are big business in both Australia and New Zealand. These motorized shepherds are mustering Merino sheep at the Narolah Station in southwestern Western Australia, on land that is productive for both sheep and wheat.

production and export of primary commodities. There are large and dependable markets for Australian primary products in Japan, South Korea, China, the United States, Europe, and elsewhere. Growth has been particularly strong in recent years because of China's booming economy and the accompanying high demand for commodities and raw materials like copper, gold, and iron ore. What used to be Australia's liability—its "tyranny of distance" from its major trading partners in the United States and Europe—has now become one of its greatest assets. Its proximity to Asia has made it a crossroads of globalization.

Australia's exports and its service-oriented economy have given the country consistent growth, sometimes bordering on the spectacular. Australia has not experienced a recession since 1990, a remarkable achievement considering the country's close economic ties with Asia and the United States, which did experience turbulent economic times.

Sheep and Cattle Ranching

In founding the first settlement around the excellent natural harbor of Sydney in 1788, the British government intended merely to establish a penal colony. There was to be enough agriculture on this "Fatal Shore" to make the colony self-sufficient in food and perhaps provide a small surplus of some products for export. Although self-sufficiency was elusive in the early days due to the poverty of the leached soils around Sydney, the colonists soon discovered that sheep did well, particularly the Merino breed first imported

from Spain in 1796. The market for wool expanded rapidly with the mechanization of Britain's woolen textile industry. Australia thus found its major export staple in the early years of the 19th century, and until recently, sheep ranching was a growing rural industry (●Figure 8.1.7). Sheep grazers rapidly penetrated the interior of the continent, and by 1850, Australia was the largest supplier of wool on the world market, a position it has never lost.

People have tried to raise sheep everywhere in Australia that seemed to offer any hope of success. Much of the interior has proved too dry, parts of the eastern mountain belt are too rugged and wet, and most of the north is too hot and wet in summer. The sheep ranching industry has thus become localized, concentrated mainly in a crescent-shaped belt of territory following the gentle western slope of the eastern highlands, from the northern border of New South Wales to the western border of Victoria. Beyond this concentration, the sheep industry spreads northward on poorer pastures into Queensland and westward into South Australia. Still another area of production rims the west coast, especially near Perth (see Figure 8.7). The sheep ranches, or "stations," in Australia vary in size from a few hundred to many thousands of acres.

The main product of the sheep stations is wool, which has been of great importance in Australia's development. Through most of the country's history, wool was the principal export, amounting to as much as a quarter of all exports as late as the 1960s. Recently, however, it has declined rapidly in importance, partly because of the rise of other commodities in the Australian economy and partly because of wool's slow losing battle with synthetic fibers. Today, Australian sheep are raised increasingly for meat. Each year, Australia exports large numbers of live animals to Saudi Arabia and other Muslim countries of the Middle East, where the devout slaughter them on the Feast of the Sacrifice commemorating the end of the pilgrimage to Mecca.

Australian lands not suited for sheep or intensive agriculture are devoted mainly to cattle ranching. Cattle stations are spread across the northern lands, with their hot, humid summers and coarse forage. The main belt of cattle ranching extends east-west across the northern part of the country, with the greatest concentration in Queensland (see Figure 8.7). On many of the remote ranches, especially in the Northern Territory, Aborigines are the principal ranch hands, known as "drovers" and "jackaroos."

Beef is Australia's leading agricultural export, and Australia is the world's leading beef-exporting nation, a status achieved by the growing demand for meat that has accompanied increasing affluence in East Asia, Europe, and the United States. A single case of mad cow disease in the United States in 2003 also boosted Australian exports when several Asian countries banned American beef imports. Japan imports more Australian beef than any other nation, with the United States in second place. Australian cattle are pasture-fed rather than grain-fed and tend to be leaner than their U.S. counterparts. Americans prefer fattier meats, so much of the Australian beef is ground into hot dogs or mixed with American beef in hamburger.

Wheat Farming

Australia is also one of the world's leading wheat exporters (•Figure 8.1.8). Wheat was the key crop in early attempts to make the settlement at Sydney self-sufficient. In the early years, the colony almost starved as a result of wheat failures, and the problem of grain supply was not solved until the development of the colony (now the state) of South Australia. Settlers there found fertile land and a favorable climate near the port of Adelaide, where wheat could be grown and shipped by sea to other coastal points.

Australian wheat production today is an extensive and highly mechanized form of agriculture, characterized by a small labor force, large acreages, and low yields per acre but very high yields per worker. The main belt of wheat production spreads from the eastern coastal districts of South Australia into Victoria and New South Wales, generally following the semiarid and subhumid lands that lie inland from the crests of the eastern highlands (see Figure 8.7). The main wheat belt has thus come to occupy nearly the same position as the main belt of sheep production. The two types of production are in fact often combined in this area.

• **Figure 8.1.8** Despite its limited arable land, Australia is a prodigious producer of wheat. This is an aerial view of center pivot irrigation for wheat near Wyndham in northwestern Australia. Note the irrigation canal as well.

Dairy and Sugarcane Farming

Although less important than grazing and wheat farming for export revenues, dairy farming is Australia's leading type of agriculture in numbers of people employed. It has developed mainly in the humid coastal plains of Queensland, New South Wales, and Victoria (see Figure 8.7). Dairy farming has grown largely in response to the needs of the country's rapidly growing urban populations.

Australia is a sugar exporter with a unique history. The Queensland sugar-growing area is unusual in that it was a tropical area settled by Europeans doing hard manual work without the benefit of indigenous labor. Sugar production began on the coastal plain of Queensland in the middle of the 19th century to supply the Australian market. Laborers were initially imported from Melanesia to work as virtual slaves in the cane fields. However, when the Commonwealth of Australia was formed in 1901 by the union of the seven states (New South Wales, Victoria, Queensland, Northern Territory, South Australia, Western Australia, and Tasmania), Queensland was required to expel these imported "Kanaka" laborers in the interest of preserving a "white Australia." Anticipating that production costs would soar with the employment of exclusively white labor, the commonwealth gave Queensland a high protective tariff on sugar. Operating behind this protective wall, the sugar industry has prospered there, though most of the original large estates have been divided into family farms worked by individual farmers.

Australia's Mineral Wealth

Unification of the separate Australian colonies into the independent federal Commonwealth of Australia encouraged the rise of industry, creating a unified internal market and tariff protection for manufacturing as well as for agricultural enterprises. Australia is rich in mineral resources (see Figure 8.1.10), and that wealth has facilitated some industrial development. In addition to supplying most of its own needs for minerals, Australia has long been an important supplier to the rest of the world. In recent years, large new discoveries of iron ore, diamonds, and gold have been expanding this longtime role in international trade.

Mining has long had an impact on Australian life and landscapes. Gold played a large role in Australia's history, as it did in America's; some of the California Forty-Niners even continued their prospecting careers in Australia. Gold strikes in Victoria in the 1850s set off one of the world's major gold rushes and produced a threefold increase in the continent's population in just 10 years.

Ample supplies of coal also have aided Australian industrialization. Every Australian state has coal, but the major reserves are near the coast in New South Wales and Queensland. Most of Australia's steel-making capacity is located near the coal mines north of Sydney at Newcastle and at Wollongong to the south. Iron ore is brought from

distant deposits along and near other sections of the Australian coast. Major new iron ore discoveries and developments in Western Australia have made Australia the world's leading exporter (•Figure 8.1.9).

Australia is by far the world's leading producer of bauxite, the ore from which aluminum, the world's second-most-used metal (after iron), is derived. Major reserves of bauxite are near the north Queensland coast, in the Darling Range inland from Perth, and on the coast of Arnhem Land. Aluminum plants supply the Australian market, and large quantities of the partially processed material called alumina are exported for final processing into aluminum metal at overseas plants.

Among Australia's other important mineral products are copper, lead, titanium, zinc, tin, silver, manganese, nickel, tungsten, uranium, and diamonds. Australia now produces 18 percent of the world's diamonds, most of them industrial cutting diamonds rather than gemstones. A diamond rush is under way, with prospecting and new finds in the country's northwest, both onshore and at sea. The country has over one-third of the world's proven uranium reserves—the largest of any country—making it of critical importance in global security. The United States constantly

• **Figure 8.1.9** Australia is an important exporter of iron ore. This open-pit mine is in Western Australia.

encourages Australia to see that this potentially dangerous element is not exported to the wrong hands. But uranium-mining companies, eager to meet surging demand in numerous Asian countries, are itching to repeal their government's long-standing restriction that Australia should have only three operating uranium mines.

Australia's most famous mining centers are three old settlements in the dry interior: Broken Hill in New South Wales, Kalgoorlie in Western Australia, and Mount Isa in Queensland, all of which are still active. The Broken Hill Proprietary Company, which grew with the mines of that town, later branched into iron and steel production and other businesses to become Australia's largest corporation.

A shortage of petroleum blemishes Australia's general picture of mineral abundance. There are only small oil-fields. Most of these are on the mainland, but the main producing field is offshore Victoria. Although local supplies have aided the country's economy, Australia continues to import oil and to look for more domestic sources. Australia is blessed with a large endowment of natural gas, of which it is the world's seventh-largest exporter, with China the main importer. A particularly large natural gas field called the Gorgon, off the northwest coast, is now being developed.

Urbanization and Industrialization Down Under

Although mining, grazing, and agriculture produce most of Australia's exports, most Australians—about 91 percent—are city dwellers. This high degree of urbanization is one of the country's most notable and perhaps most surprising characteristics.

About 13 million people, or 61 percent of the population, live in the five largest cities: Sydney (the largest; population 4.5 million; •Figure 8.1.11), Melbourne, Brisbane, Perth, and Adelaide. All of these are seaports, and each is the capital of one of the five mainland states. The country's capital, Canberra (meaning "meeting place" in the Aboriginal language), is one of its smaller cities (population 300,000). Located inland about 190 miles (310 km) southwest of Sydney, Canberra is widely regarded as one of the world's most beautiful modern planned cities.

The Australian government has encouraged industrial development to support the country's growing population, to provide more adequate armaments for defense, and to increase stability through a more diversified and self-contained economy. A wealthy, urbanized, and resource-rich country such as Australia should, like most MDCs, be a major exporter of finished industrial products. However, approximately two-thirds of the country's exports come directly, or with only early-stage processing, from mining, agriculture, and grazing.

Far from being an international industrial power of consequence, Australia does not meet its own demands for manufactured goods and relies heavily on overseas produc-

• **Figure 8.1.10** Australia's mineral resources.

tion. Australia is a big importer of manufactured goods from the United States, Japan, and China. The Labor Party government was elected in the 1980s and early 1990s in part on its promise to correct this problem and to transform Australia from a protected agriculture- and mining-based economy to a lean, high-technology export economy. Its failure to achieve this goal contributed to the party's defeat in the country's four elections prior to that of 2007.

Representing about 5 percent of the country's gross domestic product, or about the same as mining, tourism has become an important industry in Australia, particularly since the worldwide acclaim of the 1986 film *Crocodile Dundee*. (Crocodiles themselves occupy an important place in Australia's tourist business. In Kakadu National Park and other sections of northern and western Australia, many tours promise visitors a close-up look at "salties," the great saltwater crocodiles that are ancient and often deadly neighbors of people in the region.) Australia's attractions are sufficiently diverse to draw many different styles of international tourists, who number about 5 million yearly, to various destinations on the continent. The 1,200-mile (1,920-km) Great Barrier Reef, located off the northeast coast, is one of the world's most prized scuba-diving areas. Hotels and other services on nearby Queensland beaches of the Gold Coast cater to those looking for an idyllic, tropical, restful holiday. The geologic features of Uluru (Ayers Rock) and the nearby Olgas in the country's center are on the itinerary of most international tourists. For many visitors, Sydney is as lovely a port city as San Francisco or Venice. Adventurous tourists raft the whitewaters and hike the mountains of Tasmania.

● **Figure 8.1.11** One great attraction to international tourists is the beautiful port city of Sydney. This is the famous Sydney Opera House, designed by a Danish architect and completed in the 1970s.

8.1.4 New Zealand: Pastoral and Urban

Located more than 1,000 miles (c. 1,600 km) southeast of Australia, New Zealand consists of two large islands, North Island and South Island (separated by Cook Strait), and a number of smaller islands (●Figure 8.1.12). North Island is smaller than South Island but has the majority (about 77 percent) of the country's population of 4.2 million. With a total area of 104,000 square miles (269,000 sq km), the population density averages only 40 per square mile (16/sq km). New Zealand is thus a sparsely populated country, although not as sparsely populated as Australia.

Environment and Rural Livelihood

Much of New Zealand is rugged. The Southern Alps, often described as one of the world's most spectacular mountain ranges, dominate South Island (●Figure 8.1.13). Filmmakers prize them; they featured prominently in the *Lord of the Rings* trilogy, for example. These mountains rise above 5,000 feet (c. 1,500 m), with many glaciated summits over 10,000 feet, and in the southwest, they are cut deeply by coastal fjords. The mountains of North Island are less imposing and extensive, but many peaks exceed 5,000 feet.

New Zealand's highlands lie in the Southern Hemisphere belt of westerly winds and have abundant precipitation. While lowlands generally receive more than 30 inches (76 cm), highlands often have more than 130 inches

● **Figure 8.1.12** Principal features of New Zealand.

REGIONAL PERSPECTIVE

Australia and New Zealand: Less British, More Asian-Pacific

Australia and New Zealand have been remote satellites of Britain throughout most of their short histories. Even with independence, they pledged loyalty to the British Commonwealth of Nations. The United Kingdom was far more important than any other country in their trade relations, and as recently as the early 1970s, it accounted for about 30 percent of New Zealand's trade and 15 percent of Australia's.

Their orientation is changing, however, as Australia and New Zealand seek stronger roles in the economic growth projected for the Pacific Basin. Britain is now a rather insignificant trading partner of both Pacific nations. Seven of Australia's 10 largest markets are in the Asia-Pacific region, with China, Japan, and the United States its leading trading partners (in that order). Australia, the United States, and Japan have become New Zealand's leading trade partners.

Perhaps the best symbol of Australia's new Pacific perspective is the debate over whether or not Australians should convert the country into a republic, ending more than 200 years of formal ties with Britain. Even now, according to Australia's constitution (which was written in 1901 by the British), Australia's head of state cannot be an Australian; it has to be England's king or queen (this continues to be the case in 16 of the 53 countries that make up the Commonwealth). Republic status would change that and allow Australians to have their own elected or appointed president. The monarch's portrait would be removed from Australian currency. The British Union Jack would be removed from its position in the upper left corner of the Australian flag.

In 1998, an Australian constitutional convention voted to adopt this republican model, and Australia's avowedly monarchist prime minister, John Howard, reluctantly agreed to put the issue to a referendum in 1999. In essence, the referendum asked Australians where they preferred their economic and political future to rest—either with the other nations of Asia and the Pacific or with Britain and the Commonwealth. A narrow majority (55 percent) voted to keep Australia's lot with Britain and the Commonwealth.

The debate has not gone away, however. Polls showed that at the time of the referendum, 60 percent of Australians favored a republic. Not all of these voted for republic status, however, because the referendum stated that the republic's president would be chosen by the parliament rather than elected by popular vote. Since the 1999 referendum, polls have shown the percentage of Australians favoring abandoning the Commonwealth declining, down to 45 percent in 2007, but promoters of both sides of the issue remain active.

The monarchy-versus-republic debate has led to something of an identity crisis for Australia and has focused more scrutiny on the sensitive issue of nonwhite immigration. Between 1945 and 1972, some 2 million people, mainly British and continental Europeans, emigrated to Australia. They were allowed in by a **"white Austra-**

lia" policy that excluded Asian and black immigrants because of fears of invasion from Asia and a desire to increase the country's population with white, skilled, English-speaking immigrants. That policy was eventually dropped, and doors opened considerably to skilled nonwhites. About half of the annual quota of 90,000 legal immigrants are Asian, who now make up about 7 percent of Australia's population (about 92 percent are Caucasian). But Asians will form up to one-fourth of Australia's population by 2025, according to some projections. Australia has recently passed legislation requiring new immigrants to settle in cities other than Sydney, whose services threaten to be overwhelmed by growth.

Recent years have seen a marked surge in racism, targeted mainly at Asians but also at Middle Easterners; a wave of white violence targeted at ethnic Lebanese swept over the country in 2005, for example. There has been a growing tide of illegal immigration from new sources, mainly Afghanistan and Iraq. These boat people are generally trafficked by Indonesian racketeers, who charge a refugee as much as $7,000 for the service. Prime Minister John Howard's vow to tighten restrictions against refugees and asylum seekers helped win him a rare third term in office in 2001. Howard won a fourth term in 2004 but went down to defeat in 2007.

Ethnic tensions aside, Australia and New Zealand continue to forge important ties with their neighbors around the vast Pacific and with each other. Both countries adopted a free-trade pact in 1990 called the **Closer Economic Relations Agreement,** which eliminated almost all barriers between them to trade in farm and industrial goods and services. Both countries belong to the 18-member **Asia-Pacific Economic Cooperation (APEC)** group, which has agreed to establish "free and open" trade and investment between member states by the year 2020. Australia and the United States are the most aggressive member states in arguing for the abolition of trade quotas and the reduction of tariffs imposed on imported goods.

Australians refer to their Asian neighbors not as the Far East but as the **Near North.** Australia's Northern Territory city of Darwin is well positioned geographically to take advantage of new trading relations with the Near North. Darwin is closer to Jakarta, Indonesia's capital, than it is to Sydney, and Darwin is promoting itself as Australia's "gateway to Asia." The city has established a trade development zone to provide manufacturing facilities and incentives for overseas companies and for Australian companies wanting to do business overseas. A deep-water port was built in the 1990s, and a recently completed transcontinental Darwin-Adelaide railway called the Ghan is expected to help fuel trade between Australia and its neighbors to the northeast (see Figure 8.1.10, page 446). Australia's Asia-oriented perspective has already changed the nature of one of the country's export staples—beef—as ranchers have been shifting to breeds of cattle better suited for live shipment by sea to Indonesia, the Philippines, and Thailand.

Matt Jones

• **Figure 8.1.13** The Seaward Kaikoura Range of the Southern Alps.

Taimay Jones

• **Figure 8.1.14** A pastoral scene near Wellington.

(330 cm). Precipitation drops to less than 20 inches (50 cm) in small areas of rain shadow east of South Island's Southern Alps. Temperatures typical of the middle latitudes are moderated by a pervasive maritime influence. The result is a marine west coast climate, with warm-month temperatures generally averaging 60°F to 70°F (16°C to 21°C) and cool-month temperatures of 40°F to 50°F (4°C to 10°C). Highland temperatures are more extreme, and a few glaciers flow on both islands.

Rugged terrain and heavy precipitation in the Southern Alps and in the mountainous core of North Island account for the almost total absence of people from one-third of New Zealand. About one-fifth of the country is completely unused, except for tourists drawn to the wilderness. The attractions in New Zealand's growing tourist industry include golfing, fishing, hunting, hiking, and camping in spectacular mountainous national parks and watching geysers around the Rotorua thermal area of North Island.

Despite all of this wilderness, New Zealand has suffered the impacts that typically occur as a result of human activities in island ecosystems. Nearly 85 percent of the country's lowland forests and wetlands have been destroyed since the Maori arrived about 800 years ago. Of New Zealand's presettlement 93 bird species, 43 have become extinct and 37 are endangered (including the kiwi, the national bird and the informal namesake of all New Zealanders). As in Australia, exotic plant and animal species are a major threat to the indigenous varieties.

Populous areas are restricted to lowlands around the periphery of North Island and along the drier east and south coasts of South Island. The climate of New Zealand's lowlands is ideal for growing grass and raising livestock (•Figure 8.1.14). More than half of the country's total area is in pastures and meadows supporting major sheep and cattle industries. Pastoral industries contribute much to the country's export trade. Meat, wool, dairy products, and hides together account for about half of what New Zealand sells abroad. The country has developed a niche market in exports in high-quality wines, allowing New Zealand to be competitive in an industry with a glut of cheaper vintages.

The earning power of its pastoral exports has given New Zealand a moderately high standard of living (per capita GNI PPP of $26,470, comparable to that of Israel). The United Kingdom was once the country's main market. When the United Kingdom joined the European Common Market in the 1970s, however, a new situation emerged because the Common Market (now the European Union) included countries with surpluses competitive with New Zealand's. In response, New Zealand tried to foster industrial growth and to expand trade with other partners, particularly in Asia. These efforts were not entirely successful. The economy suffered during the 1970s and 1980s, and the country slipped downward in the ranks of the world's affluent countries. A sizable emigration took place, mainly to Australia. These emigrants complained of not only the economic troubles but also boredom in an isolated, placid society in which tax policies made it difficult for almost anyone to earn a high income or accumulate wealth.

Today the flow of migrants has reversed, spurred particularly by Asians moving to New Zealand's cities. Like Australia, New Zealand has begun to tighten up on immigration. Also like Australia, New Zealand is seeking to loosen or break its political and economic ties with the British Commonwealth of Nations and to reorient its economic orbit more toward Asia. New Zealand's economy felt the negative impact of the late-1990s Asian economic crisis as its exports to the region declined, but it has rebounded with renewed growth in Asia.

Industrial and Urban Development

In New Zealand, as in Australia, a high level of urbanization and attempts to develop manufacturing accompany a basic dependence on pastoral industries (•Figure 8.1.15).

Matt Jones

● **Figure 8.1.15** Queenstown, situated on the South Island, displays both the urbanity and the wilderness that are New Zealand.

populations of more than 100,000: Auckland (the largest; population 1.3 million; ●Figure 8.1.16), on North Island; the capital, Wellington (population 400,000), at the opposite end of North Island; Napier-Hastings on the southeastern coast of North Island; Tauranga, on the Bay of Plenty; Hamilton, not far south of Auckland; and Christchurch and Dunedin on the drier east coast of South Island. Christchurch is the main urban center of the Canterbury Plains, which have the largest concentration of farmlands in New Zealand. About half of the country's population lives in these seven modest-sized cities (in comparison, Australia's Sydney metropolitan area has about the same population as all of New Zealand). Note that most of these cities have English names; as evidence of New Zealand's struggle to reidentify itself, there is a strong movement afoot to replace many of the country's Anglicized place names with their original Maori designations (for example, the Maori name for the country, Aotearoa (pronounced "*ow*-tay-uh-*roh*-uh")), is gaining popularity and can often be seen in the names of organizations along with the name New Zealand).

The two countries are similar in their conditions and objectives of urban and industrial development. New Zealand's raw materials do not equal those of Australia, but the nation has modest reserves of coal, iron, and other minerals. The potential for hydropower development is high. New Zealand has magnificent natural forests, and production and export of forest products have become increasingly important as the country tries to reduce its dependence on sheep and dairy exports.

New Zealand is not as urbanized as Australia. The country has only seven urban areas with estimated metropolitan

8.1.5 Antarctica: The White Continent

Antarctica is the world's fifth-largest continent, with an area of 5.5 million square miles (14,245,000 sq km) lying south of the tip of South America and virtually filling the Antarctic Circle (●Figure 8.1.17). It is the setting of enormous human dramas in exploration, bravery, and foolhardiness as people have crossed the continent with dogsleds, on skis, on foot, and in airplanes and helicopters. Some, including

Taimay Jones

● **Figure 8.1.16** Auckland, on North Island, is New Zealand's largest city.

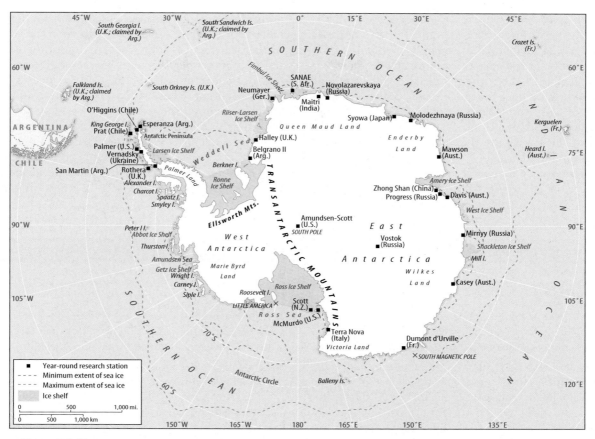

● **Figure 8.1.17** Principal features of Antarctica.

all members of the Scott Expedition of 1910–1912—which had hoped to beat the Norwegians to the South Pole but arrived a few days too late—did not return alive.

Antarctica's distinctiveness comes in part from its winter of darkness, alternating with summer "whiteouts" caused by light refraction on snow and ice surface covering about 95 percent of the continent. Average temperatures during the summer months barely reach 0°F (−18°C). Winter averages are the coldest in the world, with winter mean temperatures averaging −70°F (−57° C). It is also the world's windiest and driest continent and is very sunny during its half year of light; the South Pole on average receives more sunlight each year than the world's equatorial regions.

Antarctica has a central position in global concerns about climate change. The continent is capped with glacial ice up to 10,000 feet (c. 3,000 m) thick, yet only 2 to 10 inches (5 to 25 cm) of precipitation fall each year. Records show that the Antarctic Peninsula has warmed by 4.5°F (2.5°C) since 1944, causing both the peninsula's fringing ice sheet of sea ice and its inland glacial ice to melt. Further melting of the land ice would lead inevitably to rising sea levels worldwide. The "ozone hole" in the earth's stratosphere is concentrated seasonally over Antarctica, and scientists keep constant watch, hoping that it does not enlarge, bringing more ultraviolet radiation to inhabited areas of nearby continents.

There is virtually no human settlement beyond the research teams that have built facilities on small areas of exposed land that lie near the outer edges, on the island of Little America in the Ross Sea, and at the South Pole itself. About 1,000 researchers are on the continent in the winter and about 4,000 in the summer. The U.S. has blazed a 1,000-mile (1,600-km) "ice highway" from its McMurdo Research Station to the South Pole, angering Antarctic environmentalists.

Interest in this distant world is increasing. Expensive voyages by sea bring 30,000 visitors each year (as of 2007, up from 5,000 in 1990) to mingle with penguins and enjoy some of the wildest scenery on earth. Seven countries have staked claims (some of which overlap) to portions of the continent as national territories: nearby Chile and Argentina, the Southern Hemisphere powers of Australia and New Zealand, the United Kingdom, France, and Norway. No other countries recognize these claims. Both the United States and Russia have avoided laying claim to parts of Antarctica but maintain that they have the right to do so. Such

36

claims, which have been technically "frozen" since 1961, may be moot. The Antarctic Treaty, signed by 45 countries including the seven laying claims, forbids any exploitation of the continent's natural resources until 2048. There has long been speculation about the potential for considerable resource wealth lying beneath the ice, but to date, no mineral deposits of significant economic value have been found.

523   Antarctica is ringed by a number of small, mainly volcanic islands. The largest are the Falklands, owned by the United Kingdom but also claimed by Argentina (which calls them the Malvinas). These two countries fought a brief war 592   over control of the islands in 1982. It ended with a humili-

ating defeat of Argentina, which, with an eye on the potential for offshore oil and gas, has recently raised the level of rhetoric about Britain's occupation of the islands. Fewer than 3,000 people live here, the majority employed in a lucrative fishing industry. Other British possessions include South Georgia Island and the South Sandwich Islands, which were important whaling and sealing ports until 1965 but are now uninhabited. France controls Kerguelen and a host of smaller islands in the southern Indian Ocean, all uninhabited except for small teams of scientific researchers. These islands do contribute to France's enormous exclusive economic zone (EEZ) here, second largest in the world.

SUMMARY

→ Australia and New Zealand are products of British colonization. The British heritage is strongly reflected in the ethnic composition and culture of the two nations' majority populations.

→ Both Australia and New Zealand have minorities and indigenous inhabitants. Native Australians are known as Aborigines, and the dominant indigenous group in New Zealand is the Maori.

→ In Australia, one of the largest issues of contention between the Aborigines and the white majority is land rights. In 1993, the Native Title Bill addressed the Aborigines' major concerns and made several land rights concessions. New Zealand's Maori have won some concessions on resource use.

→ Australia has four major natural regions: the humid eastern highlands, the tropical savannas of northern Australia, the "Mediterranean" lands of southwestern and southern Australia, and the dry interior.

→ Exotic species and human pressures have had catastrophic effects on native species and natural environments in Australia and New Zealand.

→ The sectors of the Australian economy that regularly produce the country's leading exports are mining, ranching, and agriculture. Most exports go to Japan, the United States, and China.

→ Most Australians—more than 90 percent—live in cities. About 60 percent of the population lives in the five largest cities: Sydney, Melbourne, Brisbane, Perth, and Adelaide.

→ New Zealand is located more than 1,000 miles (1,600 km) southeast of Australia. The country consists of two large islands, North Island and South Island, and a number of smaller islands. South Island is very mountainous, and most people live on North Island.

→ Pastoral industries, such as meat, wool, dairy products, and hides, contribute greatly to New Zealand's export trade.

→ Australia and New Zealand have been reorienting their focus toward the Pacific Rim and away from the United Kingdom. An important expression of this change in focus is the ongoing debate about whether or not Australia should become a republic, thus ending more than 200 years of formal ties with Britain.

→ Antarctica is the world's coldest, driest, and least populated continent. There is no permanent population, only scientific researchers. The region's melting sea and land ice may reflect global warming, and the "hole" in the earth's stratospheric ozone layer concentrates over Antarctica. Seven countries stake claims to the continent, but no other countries recognize these claims, and so far they have not had any economic significance.

KEY TERMS + CONCEPTS

Asia-Pacific Economic Cooperation
(APEC) group (p. 448)
assimilation (p. 437)
back of beyond (p. 440)
bush (p. 440)
Closer Economic Relations Agreement
(p. 448)
comanagement (p. 439)
Dreamtime (p. 436)

feral animals (p. 441)
Footprints of the Ancestors (p. 436)
International Forum on Indigenous Mapping (p. 439)
Maori (p. 437)
Native Title Bill (p. 438)
Near North (p. 448)
never-never (p. 440)
outback (p. 440)

Songlines (p. 436)
stolen generations (p. 437)
terra nullius (p. 438)
"the wet" (p. 438)
Treaty of Waitangi (p. 437)
Walkabout (p. 436)
Way of the Law (p. 436)
"white Australia" policy (p. 448)
Wik case (p. 438)

REVIEW QUESTIONS

1. What features do Australia and New Zealand have in common?

2. Who were the first nonindigenous settlers of Australia? What brought them there?

3. What is the significance of the Dreamtime to Australia's Aborigines?

4. What are the four natural regions of Australia?

5. What advantages might kangaroo husbandry have over that of conventional livestock?

6. What are the major agricultural regions of Australia? What products come from them?

7. What factors support Australia's production and export of commodities and raw materials? What countries import these products?

8. What are Australia's largest cities? What characteristics do they share?

9. What are the notable landform features and natural resource assets of New Zealand?

10. What are the major physical characteristics of Antarctica? What is the continent's political status?

NOTES

1. Alan Moorehead, *Cooper's Creek* (New York: St. Martin's Press, 1963), p. 7.

2. Quoted in Geoffrey Blainey, *Triumph of the Nomads* (Melbourne: Sun Books, 1987), p. 181.

3. Howard Ensign Evans and Mary Alice Evans, *Australia: A Natural History* (Washington, D.C.: Smithsonian Institution Press, 1983), p. 12.

4. Mark Twain, *Following the Equator: A Journey around the World* (Hartford, Conn.: American Publishing, 1897), p. 146.

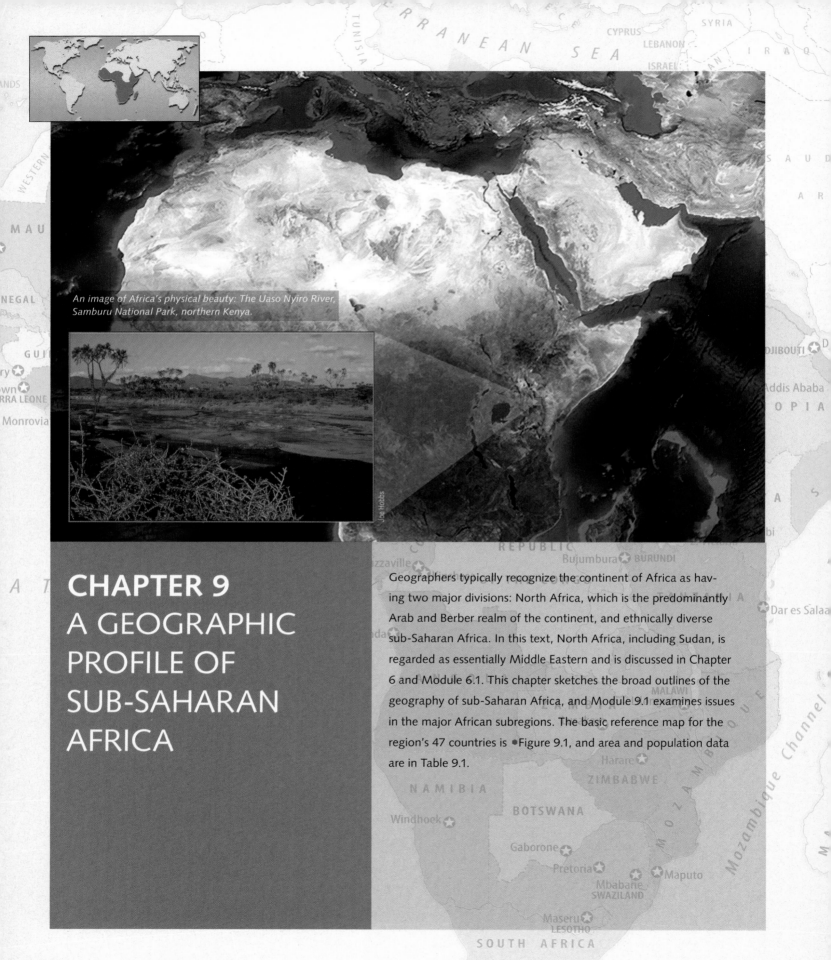

An image of Africa's physical beauty: The Uaso Nyiro River, Samburu National Park, northern Kenya.

Joe Hobbs

CHAPTER 9
A GEOGRAPHIC PROFILE OF SUB-SAHARAN AFRICA

Geographers typically recognize the continent of Africa as having two major divisions: North Africa, which is the predominantly Arab and Berber realm of the continent, and ethnically diverse sub-Saharan Africa. In this text, North Africa, including Sudan, is regarded as essentially Middle Eastern and is discussed in Chapter 6 and Module 6.1. This chapter sketches the broad outlines of the geography of sub-Saharan Africa, and Module 9.1 examines issues in the major African subregions. The basic reference map for the region's 47 countries is •Figure 9.1, and area and population data are in Table 9.1.

■ chapter
objectives

This chapter should enable you to

→ Understand what caused sub-Saharan Africa to become and remain the world's poorest region

→ Know how the region came to have the highest HIV/AIDS infection rates in the world and how the epidemic could be reversed there

→ Appreciate the pressures on African wildlife and the unique approaches taken to protect the animals

→ Know what is uniquely African about African cultures

→ Recognize why, after more than a decade at the sidelines, sub-Saharan Africa is considered important again in geopolitical affairs

---

The region of sub-Saharan Africa is culturally complex, physically beautiful, and problem-ridden. This introduction to Africa surveys the region's diverse environments, peoples and ways of living, population distributions, European colonial legacies, and major current problems. Sub-Saharan Africa ranks at the very bottom of every statistical indicator of global quality of life. About half of this region's people subsist on less than $1 per day (the World Bank's benchmark for extreme poverty). This is the only world region that has grown poorer in the last 25 years.

It is also a region that has been plagued by strife, but there is good news here: while there were 16 conflicts in the region in 2002, there were just 5 in 2007. Africa's plight may seem ironic because by some measures, the region has the world's greatest variety of natural resources. But as this chapter will show, some of Africa's recent and current conflicts, in many cases fought by child soldiers, have not been about ideology or ethnicity but simply about control over resources such as diamonds, oil, and other minerals. A recent study of the world's civil wars since 1960 determined that there were three important risk factors for such conflicts: poverty, low economic growth, and high dependence on natural resources. Sub-Saharan Africa has been richly endowed with these ingredients of war.

For many people, sub-Saharan Africa is the most poorly understood or misunderstood of the world's regions. It seems to have so many countries, so much violence and disease, and so much poverty that it is simply too difficult

to think about. Hollywood still perpetuates stereotyped images of an Africa rich in wildlife but with dangerous or backward peoples. This chapter will show that getting to know the real Africa is manageable, rewarding, and interesting. This region's diverse cultures and natural environments are sources of endless wonder, and conditions for many Africans are improving in a number of ways. There are many reasons for hope and much to celebrate in the study of this fascinating region. Just ask the people: a 2006 Gallup survey of 50,000 citizens around the world found that Africans are the most optimistic.[1]

9.1 Area and Population

Sub-Saharan Africa (including Madagascar and other nearby Indian Ocean islands) has the second-largest land area of all the major world regions described in this book. It covers 17.4 million square miles (21.8 million sq km) and so is more than twice the size of the United States (•Figure 9.2).

People overpopulation is apparent in some areas, and yet much of the region is sparsely populated. With a population of 749 million as of 2007, the region's average population density is slightly more than that of the United States. Even with the loss of population due to AIDS, the rate of natural population increase in sub-Saharan Africa is 2.5 percent per year, or about four times that of the United States. As in most LDCs, African parents generally want large families for several reasons: to have extra hands to perform work; to be looked after when they are old or sick; and in the case of girls, to receive the "bride wealth" a groom pays in a marriage settlement. Large families also convey status. However, there are signs of significant change in this pattern: birth rates have been dropping in every country in this region over the past two decades.

Most of this region's people live in a few small, densely populated areas (•Figure 9.3). The main areas are the coastal belt bordering the Gulf of Guinea in West Africa from the southern part of Africa's most populous country, Nigeria, westward to southern Ghana; the savanna lands of northern Nigeria; the highlands of Ethiopia; the highland region surrounding Lake Victoria in Kenya, Tanzania, Uganda, Rwanda, and Burundi; and the eastern coast and parts of the high interior plateau of South Africa.

Sub-Saharan Africa is the world's most rural region (69 percent), with rural populations of most countries between 65 and 85 percent. The most rural are two East African nations with very fertile soils: Burundi (8 percent urban) and Uganda (12 percent urban). The most urbanized are Djibouti (82 percent), where there is no arable land and a small number of people are clustered in a port city, and Gabon (73 percent), where people are flocking to partake in an oil boom that is benefiting urbanites. The region's major cities, discussed in Module 9.1, are magnets that attract many poor, rural people. But life in villages is the rule,

61

51

● **Figure 9.1** Sub-Saharan Africa.

where a typical rural home is a small hut made of sticks and mud, with a dirt floor, thatched roof, and no electricity or plumbing (●Figure 9.4).

Africa's Population Prospects

This is the world's youngest population, where 43 percent of the region's people are under 15 years old (compared with about a third of the populations of Latin America and Asia). This is another indicator of how rapidly the region's population should grow, barring the vagaries of disease or famine. But the Malthusian scenario does seem to loom over Africa. Analysts fear the consequences of what they call the **"1 percent gap"**: Since the 1960s, the population of sub-Saharan Africa has grown at a rate of about 3 percent annually, while food production in the region has grown at only about 2 percent annually. This is the only world region where per capita food production is declining. The

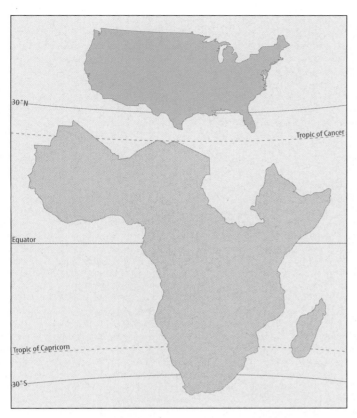

• **Figure 9.2** Sub-Saharan Africa compared in area and latitude with the conterminous United States.

with many areas at 5,000 feet (1,520 m) and higher (see Regional Perspective, page 466). The highest peaks and largest lakes of the continent are located in this belt. The loftiest summits lie within 250 miles (c. 400 km) of Lake Victoria. They include Kilimanjaro (19,340 ft/5,895 m) and Kirinyaga (Mount Kenya, 17,058 ft/5,200 m; •Figure 9.6), which are volcanic cones, and the Ruwenzori range (up to 16,763 ft/5,109 m), a nonvolcanic massif produced by faulting. Lake Victoria, the largest lake in Africa, is surpassed in area among inland waters of the world only by the Caspian Sea and Lake Superior. It is relatively shallow, however, and the large numbers of people living on its shores are taxing its resources. Other very large lakes in East Africa include Lake Tanganyika and Lake Malawi.

The physical structure of Africa has influenced the character of African rivers. The main rivers, including the Nile, Niger, Congo, Zambezi, and Orange, rise in interior uplands and descend by stages to the sea. At some points, they descend abruptly, particularly at plateau escarpments, with rapids and waterfalls interrupting their courses. These often block navigation a short distance inland. Helping offset this problem, Africa's discontinuous inland waterways are interconnected by railroads and highways more than on any other continent. The Congo is used more for transportation than any other river in the region.

The many waterfalls and rapids do have a positive side: They represent a great potential source of hydroelectric energy. There are major power stations on the Zambezi River at the Cabora Bassa Dam in Mozambique and at the Kariba Dam (•Figure 9.7), which Zimbabwe and Zambia share; at the Inga Dam on the Congo River, just upstream from Matadi; at the Kainji Dam on the Niger River in Nigeria; and at the Akosombo Dam on the Volta River in Ghana. But only about 5 percent of Africa's hydropower potential has been realized (compared to about 60 percent in North America). Many of the best sites are remote from large markets for power. In some cases, geopolitical considerations pose obstacles to dam construction. Downstream Egypt, for example, has expressed concern and even hostile rhetoric about dams and water diversions of the Nile and its tributaries by upstream Sudan, Ethiopia, Uganda, Kenya, and Tanzania (see page 231 in Chapter 6).

wild card in Africa's population deck is the **human immunodeficiency virus (HIV)**, and the disease it causes—**AIDS (acquired immunodeficiency syndrome)**—has inevitably been identified by some as the Malthusian "check" to the region's population growth (see Medical Geography, pages 464–465).

9.2 Physical Geography and Human Adaptations

Sub-Saharan Africa is both rich in natural resources and beset with environmental challenges to economic development. It is home to some of the world's greatest concentrations of wildlife and to some of the most degraded habitats.

The Landscapes of Africa

Most of Africa is a vast plateau, actually a series of plateaus, with a typical elevation of more than 1,000 feet (c. 300 m) (•Figure 9.5). Near the Great Rift Valley in the Horn of Africa and in southern and eastern Africa, the general elevation rises 2,000 to 3,000 feet (c. 600 to 900 m),

Africa's Biomes and Climates

The equator bisects Africa, so about two-thirds of the region lies in the low latitudes, having tropical climates and biomes; Africa is the most tropical of the world's continents (•Figure 9.8). Areas of tropical rain forest climate center on the great rain forest of the Congo Basin in central and western Africa. The forest merges gradually into a tropical savanna climate on the north, south, and east. This is the climatic and biotic zone supporting the famous large mammals of Africa. The savanna areas in turn trend into steppe and desert on the north and southwest. A broad belt

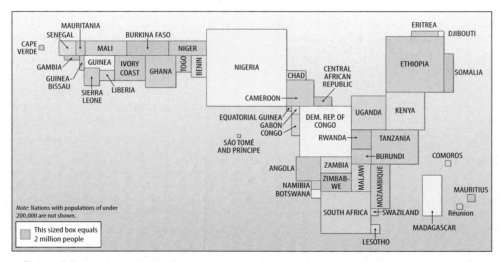

● **Figure 9.3** Population distribution (top) and population cartogram (bottom) of sub-Saharan Africa.

of drought-prone tropical steppe and savanna bordering the Sahara on the south is known as the Sahel. There is desert on the coasts of Eritrea, Djibouti, and Somalia in the Horn of Africa. In South Africa and Namibia, a coastal desert, the Namib, borders the Atlantic. The Kalahari Desert, which lies inland from the Namib, is better described as steppe or semidesert than as true desert. Along the northwestern and southwestern fringes of the continent are small but productive areas of Mediterranean climate, while eastern coastal sections and adjoining interior areas of South Africa have a humid subtropical climate. Bordering the subtropical climate region is an area of marine west coast climate.

Joe Hobbs

● **Figure 9.4** Homes in African villages are usually elevated above the ground to minimize the risk of flooding during the wet season. Their roofs are typically made of thatch or sheet metal. Most village roads are unpaved.

Total precipitation in the region is high but unevenly distributed; some areas are typically saturated, while others are perennially bone dry. Even in many of the rainier parts of the continent, there is a long dry season, and wide fluctuations occur from year to year in the total amount of precipitation. One of the major needs in Africa is better control over water. In the typical village household, women carry water from a stream or lake or a shallow (and often polluted) well. Use of more small dams would help provide water storage throughout the year.

Drought is a persistent problem in most of the countries. Although all droughts create problems, some last for years with devastating effects in this heavily agricultural region. Droughts have been particularly severe in recent decades in the Sahel and in the Horn of Africa. The number of food

488, 499

● **Figure 9.5** Major topographic features, rivers, lakes, and seas of sub-Saharan Africa.

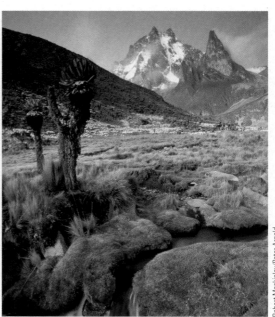

• **Figure 9.6** Africa's highest mountains, both volcanic, are Tanzania's Kilimanjaro (above) and Kenya's Kirinyaga (right), also known as Mt. Kenya.

emergencies, most of them tied to drought, has tripled since the mid-1980s in sub-Saharan Africa. The consensus among scientists is growing that this region may pay the highest price for global climate change, even though it contributes the fewest greenhouses gases to the atmosphere. United Nations studies predict even more severe droughts and crop losses, along with the destruction of coastal infrastructure that will accompany rising sea levels. Such dangerous environmental changes may be contributing to conflict in the region as well. Pastoralists and farmers already compete for scarce land and water resources in many places, and droughts intensify those struggles.

34

Living off the Land

The patterns of Africa's land use (•Figure 9.9) reveal that the most productive lands are on river plains, in volcanic regions (especially the East African and Ethiopian highlands), and in some grassland areas of tropical steppes (notably the High Veld in South Africa). Soils of the deserts and regions of Mediterranean climate are often poor. In the tropical rain forests and savannas, there are reddish, lateritic tropical soils that are infertile once the natural vegetation is removed and can support only shifting cultivation.

288

To support growing populations, farmers across sub-Saharan Africa have shortened fallow periods and pressed their lands to yield more crops. The result has been an unprecedented degradation of the resource. In 2006, fully 75 percent of the region's farmland was severely low in the nutrients needed to grow crops, up from 40 percent a decade earlier. Studies suggest that at current rates, crop yields will fall as much as 30 percent by 2022. Most African farmers cannot afford fertilizers and are not familiar with the soil conservation techniques that would help reverse this ominous trend.

Africa's soils favor subsistence agriculture (people farming their own food but producing little surplus for sale) and pastoralism, and over half of the region's people practice these livelihoods. Women do a large share of the farm-

• **Figure 9.7** The Kariba Dam straddles the Zambezi River between Zimbabwe and Zambia and provides hydropower to both countries.

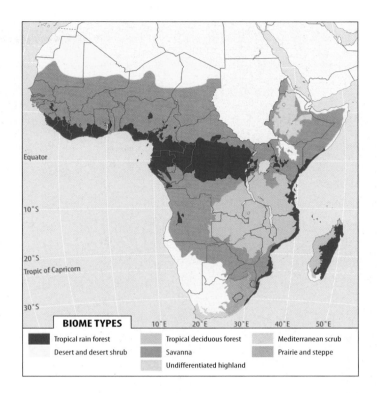

• **Figure 9.8** Climates (left) and biomes (right) of sub-Saharan Africa.

• **Figure 9.9** Land use in sub-Saharan Africa.

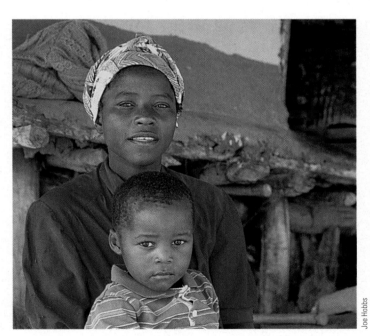

• **Figure 9.10** Mother and child in Zimbabwe. Women plant and harvest most of Africa's food and care for most of its children.

work—they produce 80 to 90 percent of Africa's food—in addition to household chores and the bearing and nurturing of children (•Figure 9.10). Mechanization is rare, fertilizers are expensive, and so crop yields are low. In the steppe of the northern Sahel, both rainfall and cultivation are scarce.

The more dependable rainfall of the southern Sahel creates a major area of rain-fed cropping, with unirrigated millet, sorghum, corn (maize), and peanuts the major subsistence crops. In the tropical savannas south of the equator, corn is a major subsistence crop in most areas, with manioc (cassava) and millet also widely grown. Corn, manioc, bananas, and yams are the major food crops of the rain forest areas.

TABLE 9.1   Sub-Saharan Africa: Basic Data

Political Unit	Area (thousands; sq mi)	Area (thousands; sq km)	Estimated Population (millions)	Estimated Population Density (sq mi)	Estimated Population Density (sq km)	Annual rate of Natural Increase (%)	Human Development Index	Urban Population (%)	Arable Land (%)	Per Capita GNI PPP ($US)
The Sahel	**2,047.6**	**5,303.3**	**69.6**	**31**	**12**	**3.1**	**0.372**	**27**	**3**	**1,370**
Burkina Faso	105.8	274.0	14.8	140	54	3.0	0.342	15	14	1,330
Cape Verde	1.6	4.1	0.5	313	121	2.5	0.722	53	9	5,980
Chad	495.8	1,284.1	10.8	22	8	3.1	0.368	24	3	1,230
Gambia	4.4	11.4	1.5	341	132	2.7	0.479	26	25	1,970
Mali	478.8	1,240.1	12.3	26	10	3.3	0.338	30	4	1,130
Mauritania	396.0	1,025.6	3.1	8	3	2.7	0.486	40	0	2,600
Niger	489.2	1,267.0	14.2	29	11	3.4	0.311	21	3	830
Senegal	76.0	196.8	12.4	163	63	3.0	0.460	43	12	1,830
West Africa	**812.4**	**2,104.1**	**224.1**	**255**	**98**	**2.5**	**0.444**	**37**	**20**	**1,340**
Benin	43.5	112.7	9.0	207	80	3.0	0.428	40	18	1,160
Ghana	92.1	238.5	23.0	250	96	2.3	0.532	44	16	2,640
Guinea	94.9	245.8	10.1	106	41	2.9	0.445	33	3	2,410
Guinea-Bissau	13.9	36.0	1.7	122	47	3.1	0.349	32	10	830
Ivory Coast	124.5	322.5	20.2	162	63	2.4	0.421	46	9	1,550
Liberia	37.2	96.3	3.8	102	39	3.1	N/A	45	4	N/A
Nigeria	356.7	923.9	144.4	405	156	2.5	0.448	36	31	1,050
Sierra Leone	27.7	71.7	5.3	191	74	2.8	0.335	37	7	850
Togo	21.9	56.7	6.6	301	116	2.8	0.495	33	46	1,490
West Central Africa	**1,576.0**	**4,081.8**	**90.8**	**53**	**20**	**2.8**	**0.423**	**35**	**4**	**1,200**
Cameroon	183.6	475.5	18.1	99	38	2.3	0.506	48	13	2,370
Central African Republic	240.5	622.9	4.3	18	7	1.9	0.353	39	3	1,280
Congo Republic	132.0	341.9	3.8	29	11	2.7	0.520	52	1	940
Congo, Democratic Republic of	905.4	2,345.0	62.6	69	27	3.0	0.391	30	3	720
Equatorial Guinea	10.8	28.0	0.5	46	18	2.3	0.653	45	4	10,150
Gabon	103.3	267.5	1.3	13	5	1.6	0.633	73	1	5,310
São Tomé and Príncipe	0.4	1.0	0.2	500	193	2.7	0.607	38	6	N/A
East Africa	**703.0**	**1,820.8**	**121.9**	**155**	**60**	**2.8**	**0.464**	**22**	**9**	**1,120**
Burundi	10.7	27.7	8.5	794	307	3.0	0.384	8	35	710
Kenya	224.1	580.4	36.9	165	64	2.8	0.491	36	8	1,300
Rwanda	10.2	26.4	9.3	912	352	2.7	0.450	17	40	1,270
Tanzania	364.9	945.1	38.7	106	41	2.6	0.430	22	4	740
Uganda	93.1	241.1	28.5	306	118	3.1	0.502	12	26	1,490

continued

TABLE 9.1   Sub-Saharan Africa: Basic Data *continued*

Political Unit	Area (thousands; sq mi)	Area (thousands; sq km)	Estimated Population (millions)	Estimated Population Density (sq mi)	Estimated Population Density (sq km)	Annual rate of Natural Increase (%)	Human Development Index	Urban Population (%)	Arable Land (%)	Per Capita GNI PPP ($US)
Horn of Africa	**727.0**	**1,882.9**	**91.9**	**152**	**59**	**2.6**	**0.340**	**17**	**6**	**1,200**
Djibouti	9.0	23.3	0.8	89	34	1.8	0.494	82	0	2,540
Eritrea	45.4	117.6	4.9	108	42	3.0	0.454	19	5	1,090
Ethiopia	426.4	1,104.4	77.1	181	70	2.5	0.371	15	10	1,190
Somalia	246.2	637.7	9.1	37	14	2.9	N/A	33	1	N/A
Southern Africa	**2,310.8**	**5,985.0**	**129.3**	**52**	**20**	**1.6**	**0.515**	**38**	**5**	**5,600**
Angola	481.4	1,246.8	16.3	34	13	2.7	0.439	33	2	2,360
Botswana	224.6	581.7	1.8	8	3	-0.1	0.570	54	1	12,240
Lesotho	11.7	30.3	1.8	154	59	0.3	0.494	17	11	4,340
Malawi	45.7	118.4	13.1	287	111	2.8	0.400	14	23	720
Mozambique	309.5	801.6	20.4	66	25	2.1	0.390	29	5	1,220
Namibia	318.3	824.4	2.1	7	3	1.5	0.626	33	1	8,110
South Africa	471.4	1,220.9	47.9	102	39	0.8	0.653	53	12	11,710
Swaziland	6.7	17.4	1.1	164	63	-0.1	0.500	25	10	5,170
Zambia	290.6	752.7	11.5	40	15	1.9	0.407	35	7	1,000
Zimbabwe	150.9	390.8	13.3	88	34	1.1	0.491	32	8	1,940
Indian Ocean Islands	**229.7**	**595.0**	**21.4**	**93**	**36**	**2.5**	**0.506**	**29**	**5**	**1,790**
British Indian Ocean Territory (U.K.)	0.02	0.05	0.003	150	58	N/A	N/A	100	0	N/A
Comoros	0.9	2.3	0.7	778	300	2.9	0.566	33	36	2,010
Madagascar	226.7	587.2	18.3	81	31	2.7	0.509	26	5	960
Mauritius	0.8	2.1	1.3	1,625	627	0.7	0.800	42	49	13,500
Mayotte (Fr.)	0.1	0.3	0.2	2,000	772	3.6	N/A	28	0	N/A
Réunion (Fr.)	1.0	2.6	0.8	800	309	1.3	N/A	89	13	N/A
Seychelles	0.2	0.5	0.1	500	193	1.0	0.842	50	2	16,590
Summary Total	**8,406.5**	**21,772.9**	**749.0**	**89**	**34**	**2.5**	**0.439**	**31**	**6**	**2,020**

Sources: World Population Data Sheet, Population Reference Bureau, 2007; Human Development Report, United Nations, 2006; World Factbook, CIA, 2007.

Many peoples, particularly in the vast tropical grasslands both north and south of the equator, are pastoral. Herding of sheep and hardy breeds of cattle is especially important in the Sahel. An increasing problem is that farmers often drive pastoralists from traditional grazing lands. Confined to smaller areas in which to browse and graze, the nomads' cattle, sheep, and goats often overgraze vegetation and compact the soil.

Although cattle raising is widespread throughout the savannas, cattle are largely excluded from extensive sections both north and south of the equator by the disease called **nagana,** which is carried by the tsetse flies that also transmit sleeping sickness to humans (see Insights, page 468). In tropical rain forests, tsetse flies are even more prevalent and few cattle are raised, but goats and poultry are common (as they are in tsetse-frequented savanna areas).

MEDICAL GEOGRAPHY

HIV and AIDS in Africa

AIDS is a global problem. It had killed over 25 million people worldwide by 2007, and there are an estimated 4 million new infections around the world each year. Sub-Saharan Africa is where the virus originated (apparently among chimpanzees; the first human case was reported in the Belgian Congo—today's Democratic Republic of Congo—in 1959) and is today the epicenter of this health crisis. The statistics are startling. In 2005 (the latest year for which data are available), 62 percent of the world's estimated 39 million persons infected with HIV lived in sub-Saharan Africa. In this region, 5.9 percent of adults were HIV-positive. In 2006, 2 million died in the region. Some public health authorities estimate that 25 percent of the region's more than 700 million people will have died of AIDS by 2010, leaving behind 20 million orphans and slashing the regional economy by 25 percent.

The epidemic is most severe in southern Africa. In some countries, including Botswana and Zimbabwe, as much as one-quarter of the adult population is HIV-positive. South Africa has more HIV cases (about 5 million) than any other country in the world. Eighteen percent of adult South Africans were HIV-positive in 2005. United Nations officials predict that if the epidemic continues apace, as many as half of South Africa's 15-year-olds will die of AIDS in the coming years. Similar impacts are predicted for South Africa's neighbors. Demographically, the effect would change the pyramid-shaped age-structure diagram typical of South Africa,

Botswana, and Zimbabwe to a more tapered structure that demographers call "chimney-shaped" (●Figure 9.A).

The earth has not seen a **pandemic** (very widespread epidemic) like this since the bubonic plague devastated 14th-century Europe and smallpox struck the Aztecs of 16th-century Mexico. This scourge is causing sharp reductions in life expectancy in Africa and, unless contained, will dramatically alter projections of the region's population growth. Life expectancy in Botswana was 61 years at birth in 1993, but because of the virus, it was 34 in 2007 and is projected to be 29 in 2010. Life expectancies in Zimbabwe (37 years) and Namibia (52 years) are projected to be 33 and 42 years, respectively, in 2010; and in South Africa (51 years), it is projected to be 44. In Zimbabwe, the population growth rate in 1998 was 1.5 percent rather than the projected 2.4 percent because of AIDS-related deaths. By 2007, Zimbabwe's growth rate was down to 1.1 percent. If the disease stays on course, the populations of Zimbabwe and South Africa will soon begin declining; Botswana and Swaziland are already experiencing population declines.

HIV/AIDS is darkening the prospects for development in this least developed of the world's regions. Although it may seem logical that a lower population would mean greater prosperity, the incidence of HIV infection is particularly high among the region's most educated, skilled, and ambitious young urban professionals, including teachers, white-collar workers, and government employ-

● **Figure 9.A** HIV/AIDS has had a devastating impact in sub-Saharan Africa and threatens to redraw the demographic profiles of some countries in unprecedented ways.

MEDICAL GEOGRAPHY

HIV and AIDS in Africa *continued*

ees. These are the ones, in addition to truckers and merchants, who travel the most, have higher incomes, and are therefore the most likely to have many sexual liaisons. The effects on human capital are incalculable but probably huge. They affect not only education but corporate profits, research, health care, tax revenues, and other indicators of progress.

AIDS's effects also contribute to underdevelopment. Sick people, notably women (who have the higher infection rates and who do most of Africa's farming) may be too weak to grow food. Sick workers are less productive. Family members must use precious funds to buy life-prolonging drugs rather than goods that would boost the economy. People anticipating short lives are less inclined to save and invest money and less inclined to send their children to school. They may join gangs or engage in other reckless behavior.

Why is the HIV infection rate so high in this region? A number of customs and cultural factors come into play, although the examples given here do not necessarily apply to every African culture and country. First, in some African societies, little stigma is attached to adultery or multiple sexual partners. Some countries are trying to crack down on the "sugar daddies" who seduce young women with money and promises, only to exploit and inadvertently infect them. Some male teachers, often city-educated and perceived as powerful, similarly transmit the disease to girls in the villages where they are sent to teach. Second, partners are often reluctant to use condoms. A tradition of subservience to male partners means that women often feel they have no choice when a man insists on unprotected sex. Both these factors help reveal why women are three times as likely as men to contract HIV in this region. Third, the stigma associated with having AIDS is pervasive. A person who is known to have the disease is often shunned, even by family members, and often suffers and dies in neglect and isolation. The family or community will insist that the person died of tuberculosis or some other ailment, so the real problem is never confronted and goes untreated. Fourth, despite the intensity of the problem, public awareness of the disease and ways to combat it still fall dramatically short. Where leadership on this issue is needed most, it is sometimes absent. South Africa's President Thabo Mbeki, for example, came to office professing that HIV does not cause AIDS, instead blaming poverty and malnutrition. His health minister discouraged the use of **antiretroviral (ARV) drugs** in favor of garlic to treat the infected. Finally, the health measures that have been taken to fight the epidemic have been inadequate. Simple tests to determine the presence of HIV are seldom available, so people who have the virus pass it on unknowingly (the United Nations estimates that 90 percent of HIV-positive Africans do not know they have the virus). And mainly because of their high cost, anti-AIDS drugs have been scarce.

What can be done about these difficult problems? Two countries that have had remarkable success in stemming the AIDS tide provide some answers. Uganda and Senegal have used relatively inexpensive public health tools, especially education, in the fight against the disease. To reduce the stigma associated with AIDS and to encourage counseling, testing, and condom use, the countries' political and religious leaders have been outspoken about AIDS. The results have been impressive. Uganda's adult infection rates dropped from 14 percent in the early 1990s to 8 percent by 2000 and 6.7 percent in 2005. Senegal's president spoke out forcefully about AIDS in the late 1980s, and infection rates have stayed below 2 percent ever since.

The wealthier countries outside Africa could invest more in combating the disease. The United Nations in 2001 established the Global Fund to Fight AIDS, Malaria and Tuberculosis, but struggles to raise enough resources to fight the epidemic. Canada and the European Union have been criticized for not donating more. The United States has its own effort to combat AIDS in 12 countries in Africa and 2 (Haiti and Guyana) in the Americas. The Bush administration pledged over $45 billion to the effort—an enormous amount, but one that is dwarfed by the scale of the problem: it is enough to treat only about one in four persons infected with HIV. Drug companies have a unique opportunity to address Africa's AIDS problems. Some of them manufacture life-prolonging ARV drugs containing protease inhibitors that suppress replication of HIV but do not kill the virus. These drugs reduce the likelihood of transmission of the virus to another person. The companies could help make such drugs more accessible; in 2005, only 18 percent of the 24 million HIV-positive Africans were receiving these medications. That would mean making the drugs inexpensive, which drug companies in the West have until recently been reluctant to do, fearing that there would be public pressure to reduce prices on all drugs.

However, other countries, notably Brazil, India, and Thailand, have taken advantage of a loophole in international patent rules to manufacture their own cheaper copies of the U.S.-made drugs. Called **compulsory licensing,** the loophole allows the countries to breach patents during national emergencies to manufacture generic versions of patented drugs. Citing AIDS as an emergency, these countries began producing ARV drugs at a small fraction of their cost in the West. Pressure grew on U.S. companies to lower their prices on anti-AIDS drugs sold in Africa and elsewhere in the developing world. Soon some U.S. drug firms relented. Merck, for example, offered to make one of its key drugs available in Africa at cost, making no profit. The U.S. administration promised to keep the door open for even cheaper anti-AIDS drugs by declaring it would not seek sanctions against poor countries stricken with AIDS, even if American patent laws were being broken. The U.S. Food and Drug Administration also approved some generic AIDS drugs to be produced inexpensively in the United States for use abroad. Such measures are helping to get ARVs to increasing numbers of the afflicted, but there are still big challenges to help the majority gain access to these drugs.

REGIONAL PERSPECTIVE

The Great Rift Valley

One of the most spectacular features of Africa's physical geography is the **Great Rift Valley,** a broad, steep-walled trough extending from the Zambezi Valley (on the border between Zimbabwe and Zambia) northward to the Red Sea and the valley of the Jordan River in southwestern Asia (see Figure 9.5 and also Figure 2.1, page 21). Its relationship to the tectonic movement of crustal plates is still poorly understood. However, most earth scientists believe it marks the boundary of two crustal plates that are rifting, or tearing apart, causing a central block between two parallel fault lines to be displaced downward, creating a linear valley. This movement will eventually cut much of southern and eastern Africa away from the rest of the continent and allow seawater to fill the valley.

The Great Rift Valley has several branches. Lakes, rivers, seas, and gulfs already occupy much of it. It contains most of the larger lakes of Africa, although Lake Victoria, situated in a depression between two of its principal arms, is an exception. Some, like Lake Tanganyika—at 4,823 feet (1,470 m), the world's second-deepest lake, after Russia's Lake Baikal—are extremely deep. Most have no surface outlet. Volcanic activity associated with the Great Rift Valley has created Kilimanjaro, Kirinyaga, and some of the other great African peaks, along with lava flows, hot springs, and other thermal features. Faulting along the Great Rift Valley in Ethiopia, Kenya, and Tanzania has also exposed remains of the earliest known ancestors of *Homo sapiens*.

Most Africans who live by tilling the soil also keep some animals, even if only goats and poultry. Among African peoples such as the Maasai of Kenya and Tanzania and the Tutsi (Watusi) of Rwanda and Burundi, livestock not only contribute to the daily diet but are also an indispensable part of customary social, cultural, and economic arrangements. Cattle are particularly important, with sheep and goats playing a smaller role. Traditional Maasai pastoralists are probably the most famous African example of close dependence on cattle. They milk and carefully bleed the animals for each day's food and tend them with great care. The Maasai give a name to each animal, and herds play a central role in the main Maasai social and economic events through the year.

Madagascar is a good example of an African country in which cattle represent status, wealth, and cultural identity. Malagasy livestock owners tend to want higher numbers of the animals, rather than better-quality stock, to enhance their standing (●Figure 9.11). A Malagasy family practicing the ritual commemoration of deceased ancestors sacrifices a large number of zebu cows for fellow villagers, and similar feasts accompany other important ritual dates. Due to such demands, the population of zebu cattle on the island is about 11 million. Their forage needs have grave consequences for Madagascar's rain forests and other wild habitats. People clear the forests and repeatedly set fire to the cleared lands to provide a flush of green pasture for their livestock, causing a rapid retreat of the island's natural vegetation.

● **Figure 9.11** Zebu cattle are extremely important in the cultures and household economies of rural Madagascar. There are about 11 million cattle in this country of 18 million people.

Africa's Wildlife

Africa has the planet's most spectacular and numerous populations of large mammals. The tropical grasslands and open forests of Africa are the habitats of large herbivorous animals, including the elephant, buffalo, zebra, giraffe, and many species of antelope, as well as carnivorous and scavenging animals, such as the lion, leopard, and hyena. The tropical rain forests have fewer of these "game" animals (as Africans call them); the most abundant species here are insects, birds, and monkeys, with the hippopotamus, the crocodile, and a great variety of fish in the streams and rivers draining the forests and wetter savannas.

• **Figure 9.12** Daggers are a nearly universal dress accessory for men in the Arabian Peninsula nation of Yemen. The most prized dagger handles are of rhino horn, a custom that has had a devastating impact on African rhinos thousands of miles away.

Although documentary films promote a perception outside Africa that the continent is a vast animal Eden, the reality is less positive. Human population growth, urbanization, and agricultural expansion are taking place in Africa, as elsewhere in the world, at the expense of wildlife. Hunting and competition with domesticated livestock also take their toll. Many species are safeguarded by law, especially in the protected areas that make up about 15 percent of the region. Such laws are difficult to enforce, however, and poaching has devastated some species. Still, Africa remains home to some of the world's most extraordinary and successfully managed national parks, including South Africa's Kruger National Park, Tanzania's Ngorongoro Crater National Park, and Kenya's Amboseli National Park (the location of Kilimanjaro). International tourism to these parks is a major source of revenue for some countries, particularly South Africa, Tanzania, Kenya, Namibia, and Botswana.

Elephants and rhinoceroses are Africa's largest herbivores and the most vulnerable to global market demand for wildlife products. In 1970, there were about 2.5 million African elephants living on the continent. Poaching and habitat destruction had reduced their numbers to an estimated 680,000 in 2007. Poaching for ivory was depleting the herds of African elephants at a rate of 10 percent annually when delegates of the 112 signatory nations of the Convention on International Trade in Endangered Species (CITES) met in 1989. The organization succeeded in passing a worldwide ban on the ivory trade, and since then, the precipitous decline has halted.

CITES member states won the ivory ban over the strong objections of several southern African states, led by Zimbabwe, that were experiencing what they saw as an elephant overpopulation problem. At the 1989 CITES meeting, these countries appealed for an exemption from the ivory ban so that they could earn foreign export revenue from a sustainable yield of their elephant populations. The majority of CITES members rejected this appeal, arguing that any loophole in a complete ban would subject elephants everywhere to illegal poaching.

CITES later reversed its policy, allowing Zimbabwe, Namibia, Botswana, and South Africa to sell ivory under rigorous monitoring as a reward for their positive wildlife policies. All but South Africa continue to cull (kill) their "excess" elephants, with the meat going to needy villagers and crocodile farms.

If elephant ivory is like gold, rhinoceros horns are like diamonds. Men in the Arabian Peninsula nation of Yemen prize daggers with rhino horn handles (•Figure 9.12). Although Western scientists deny the medicinal efficacy of powdered rhino horn, traditional medicine in East Asia makes wide use of it, including as an aphrodisiac. These demands, and the current black-market value of tens of thousands of dollars per horn, have led to a precipitous decline in population of black rhinoceroses in Africa. There were an estimated 65,000 black rhinos in Africa in 1982; in 2007, there were an estimated 3,700, but strident antipoaching programs are under way in the most critical countries.

9.3 Cultural and Historical Geographies

Many non-Africans are unaware of the achievements and contributions of the cultures of sub-Saharan Africa. The African continent was the original home of humankind.

INSIGHTS

Africa's Greatest Conservationist

Not much larger than the common housefly, the tsetse fly of sub-Saharan Africa packs a wallop. This insect carries two diseases, both known as **trypanosomiasis,** which are extremely debilitating to people and their domesticated animals. People contract **sleeping sickness** from the fly's bite, and cattle contract nagana. Since the 1950s, widespread efforts have been made to eradicate tsetse flies so that people can grow crops and herd animals in fly-infested wilderness areas (●Figure 9.B). Where the efforts have been successful, people have cleared, cultivated, and put livestock on the land. The results are mixed. While people have been able to feed growing populations in the process of opening up these lands, they have also eliminated important wild resources and in many cases caused erosion, desertification, and salinization of the land. Where the tsetse fly has been eliminated, so has the wilderness. The diminutive tsetse fly thus may be characterized as a **keystone species**—one that affects many other organisms in an ecosystem. The loss of a keystone species—in this case, a fly that keeps out humans and cattle—can have a series of destructive impacts throughout the ecosystem. For its role in maintaining wilderness in sub-Saharan Africa, some wildlife experts call the tsetse fly "Africa's greatest conservationist."

● **Figure 9.B** Strenuous and successful efforts to eradicate tsetse flies have opened vast new areas of Africa to human use. This is a tsetse fly trap with two components: a liquid chemical called "simulated cow's breath" that attracts the fly and a pesticide-soaked tarp that kills it.

Recent DNA studies suggest that the first modern people (*Homo sapiens*) to inhabit Asia, Europe, and the Americas were descendants of a small group that left Africa via the Isthmus of Suez about 100,000 years ago. After about 5000 B.C.E., indigenous people were responsible for agricultural innovations in four culture hearths: the Ethiopian Plateau, the West African savanna, the West African forest, and the forest-savanna boundary of West Central Africa. Africans in these areas domesticated important crops such as millet, sorghum, yams, cowpeas, okra, watermelons, coffee, and cotton. From Africa, these diffused to populations in other world regions.

Civilizations and empires emerged in Ethiopia, West Africa, West Central Africa, and South Africa. In the first century C.E., a Christian empire based in the Ethiopian city of Axum controlled the ivory trade from Africa to Arabia. Ethiopian tradition holds that a shrine in Axum still contains the biblical Ark of the Covenant and the tablets of the Ten Commandments, which disappeared from the Temple in Jerusalem in 586 B.C.E. Several Islamic empires, including the Ghana, Mali, and Hausa states, emerged in West Africa between the 9th and 19th centuries. All of these agriculturally based civilizations controlled major trade routes across the Sahara. They profited from the exchange of slaves, gold, and ostrich feathers for weapons, coins, and cloth from North Africa. Three kingdoms arose between the 14th and 18th centuries in what are now the southern Democratic Republic of Congo and northern Angola. These included the Kongo kingdom, which had productive agriculture and was the hub of an interregional trade network for food, metals, and salt. In what is now Zimbabwe, the Karanga kingdom of the 13th to 15th centuries built its capital city at the site known as Great Zimbabwe. Its skilled metalworkers mined and crafted gold, copper, and iron, and merchants traded these metals with faraway India and China.

The Languages of Africa

Sub-Saharan Africa exhibits great linguistic diversity (●Figure 9.13). The peoples of this region speak more than 1,000 languages, which generally belong to one of four broad language groups.

The **Niger-Congo language family** (sometimes considered a subfamily of the Niger-Khordofanian family) is the largest. It includes the many West African languages and the roughly 400 Bantu subfamily languages that fall into seven branches: **Benue-Congo, Kwa, Atlantic, Mandé, Gur, Adamawan,** and **Khordofanian.** Most of these are spoken south of the equator. The **Bantu** language of the Benue-Congo branch is the most widespread.

The **Afro-Asiatic language family** includes **Semitic** languages (such as the **Amharic** language of Ethiopia and **Arabic**) and tongues of the **Cushitic** (**Oromo** and **Somali** of the Horn of Africa, for example) and **Chadic** (especially the **Hausa** of

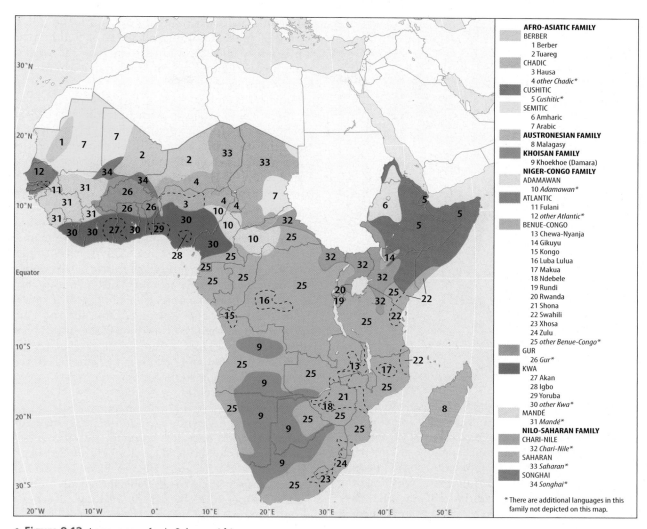

AFRO-ASIATIC FAMILY
BERBER
  1 Berber
  2 Tuareg
CHADIC
  3 Hausa
  4 *other Chadic**
CUSHITIC
  5 *Cushitic**
SEMITIC
  6 Amharic
  7 Arabic
AUSTRONESIAN FAMILY
  8 Malagasy
KHOISAN FAMILY
  9 Khoekhoe (Damara)
NIGER-CONGO FAMILY
ADAMAWAN
  10 *Adamawan**
ATLANTIC
  11 Fulani
  12 *other Atlantic**
BENUE-CONGO
  13 Chewa-Nyanja
  14 Gikuyu
  15 Kongo
  16 Luba Lulua
  17 Makua
  18 Ndebele
  19 Rundi
  20 Rwanda
  21 Shona
  22 Swahili
  23 Xhosa
  24 Zulu
  25 *other Benue-Congo**
GUR
  26 *Gur**
KWA
  27 Akan
  28 Igbo
  29 Yoruba
  30 *other Kwa**
MANDÉ
  31 *Mandé**
NILO-SAHARAN FAMILY
CHARI-NILE
  32 *Chari-Nile**
SAHARAN
  33 *Saharan**
SONGHAI
  34 *Songhai**

* There are additional languages in this family not depicted on this map.

● **Figure 9.13** Languages of sub-Saharan Africa.

northern Nigeria) branches. People living in the area adjoining the Sahara, from West Africa to the Horn of Africa, speak these languages. Even some of the Niger-Congo languages originating south of the Sahara, such as the **Swahili (Kiswahili)** tongue spoken widely in East Africa, have borrowed much from Arabic and other languages with roots elsewhere. The prominence of Arabic words in Swahili reflects a long history of Arab seafaring along the Indian Ocean coast of Africa; in fact, Swahili means "coastal" in Arabic.

The **Nilo-Saharan language family** of the central Sahel region, the northern region of West Central Africa, and parts of East Africa includes about 100 languages in three branches: **Songhai** (a single language, spoken mainly in Mali) and the **Saharan** and **Chari-Nile** language groups.

The **Khoisan languages** of the San and related peoples in the western portion of southern Africa are nearly extinct.

The largest, **Khoekhoe (Damara),** is spoken by fewer than 200,000 people in Namibia.

Quite distinct from those four groups, the people of Madagascar speak a language without African roots. This is **Malagasy,** an **Austronesian** tongue that originated in Southeast Asia.

Africa's list of lingua francas—the languages most likely to be recognized on the continent and those that African can use to speak to the wider world—is provided by the six official languages of the African Union, the continent's supranational organization. They are English, French, Portuguese, Spanish, Swahili, and Arabic. Use of English is growing most rapidly, even in the formerly French-dominated "Francophone" countries. Countries that formerly were British colonies have a distinct advantage in the world of outsourced services: South Africa, Kenya, and Ghana, for example, have already opened call centers that they hope

292

493

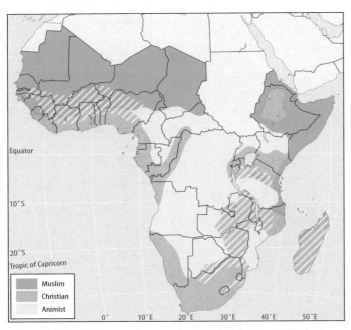

• **Figure 9.14** Religions of sub-Saharan Africa.

will rival those of India. Some Francophone countries also have call centers catering to French industries and clients.

The dominance of foreign languages in digital technology could endanger African languages, many of which are already in peril because they are spoken by so few people. However, many linguists in sub-Saharan Africa see the computer as a tool for saving traditional languages. They argue that the computer will make it much easier for people to learn and preserve their traditional tongues.

Africa's Belief Systems

The religious landscape of Africa is complex and fluid. Spiritualism is extremely strong, but spiritual affiliations and practices are more interwoven and flexible than in most other world regions. It is not uncommon for family members to follow different faiths or for an individual to change his or her religious beliefs and practices in the course of a lifetime.

Broadly, however, some dominant patterns of religious geography can be discerned (•Figure 9.14). Islam is the dominant religion in North Africa and the countries of the Sahel on the southern fringe of the Sahara. The **Ethiopian Orthodox Church,** closely related to the Coptic Christian faith of Egypt, makes Ethiopia an exception to the otherwise Islamic Horn of Africa region. Islam is also the prevailing religion of the East African coast, where Arab traders introduced the faith. Muslims are a majority or strong minority in rural northern Nigeria and Tanzania, and there are minority Muslim populations in cities and towns across

the continent. Christians are a majority in southern Nigeria, Uganda, Lesotho, and parts of South Africa. Both Christianity and Islam are strongest in the cities, whereas traditional religions prevail or overlap with these monotheistic faiths in rural areas.

Christians and Muslims have mounted efforts to dispose of "heretical" notions in African traditional belief systems, but these concepts are very resilient. Even where the large monotheistic faiths nominally prevail and their followers are devout, a strong substrate of indigenous beliefs has persisted and usually comfortably merged with the "official" faith—a syncretism also present in other cultures of the world.

Indigenous African religions have many animistic elements, emphasizing that spiritual forces are manifested everywhere in the environment. Many spirits are tied to particular places in the landscape. Such beliefs contrast with the introduced Christian and Muslim faiths, which tend to see nature as separate from God and people as apart from and superior to nature.[2] Across the spectrum of African cultures, the use of mediums to contact spirits, either of deceased ancestors, creator beings, or earth genies that can intercede on one's behalf, is widespread. Reverence for ancestors is strong, reflected also in great respect for the elderly who are living.

Outsiders have tried, seldom successfully, to give accurate names to indigenous African belief systems, practices, and personnel: ancestor worship, animism, force vitale ("life force"), living dead, and witch doctor among them. While it would be gratifying to have precise names and generalizations for African beliefs, they are unique and can best be understood by careful reading, observation, and discussion on a case-by-case basis.

The Origins and Impacts of Slavery

Until about 1,000 years ago, the cultures of Africa south of the Saharan desert barrier remained largely unknown to the peoples north of the desert. Egyptians, Romans, and Arabs had contact with the northern fringes of this region, and some trade filtered across the Sahara, but to most outsiders, Africa was the "Dark Continent," a self-contained, tribalized land of mystery. Even at the opening of the 20th century, vast areas of interior tropical Africa were still little known to Westerners.

The tragic impetus for growing contact between Africa and the wider world was slavery (•Figure 9.15). Over a period of 12 centuries, as many as 25 million people from sub-Saharan Africa were forced to become slaves, exported as merchandise from their homelands. The trade began in the 7th century, with Arab merchants using trans-Saharan camel caravan routes to exchange guns, books, textiles, and beads from North Africa for slaves, gold, and ivory from sub-Saharan Africa. As many as two-thirds of the estimated 9.5 million slaves exported between the

● **Figure 9.15** Slave export trade routes from sub-Saharan Africa.

years 650 and 1900 along this route were young women who became concubines and household servants in North Africa and Turkey. Male slaves usually became soldiers or court attendants (some of whom eventually assumed important political offices). From the 8th to 19th centuries, about 5 million more slaves were exported from East Africa to Arabia, Oman, Persia (modern Iran), India, and China. Again, most were women who became concubines and servants.

The notorious and lucrative traffic in slaves provided the initial motivation for European commerce along the African coasts, inaugurating the long era of European exploitation of Africa for profit and political advantage. The European-controlled slave trade was the largest by far, particularly with the development of plantations and mines in the New World. Between the 16th and 19th centuries, the capture, transport, and sale of slaves was the exclusive preoccupation of trade between the European world and West Africa. Portuguese and Spaniards began the trade in the 15th century, and within 100 years, English, Danish, Dutch, Swedish, and French slavers were participating. The peak of the transatlantic slave trade was between 1700 and 1870, when about 80 percent of an estimated total of 10 million slaves made the crossing. In escape attempts, in transit, and in the famines and epidemics that followed slave raids in Africa, probably more than 10 million others died.

Slaves were a prized commodity in the **triangular trade** linking West Africa with Europe and the Americas (see Figure 3.4, page 43). European ships carried guns, ammunition, rum, and manufactured goods to West Africa and exchanged them there for slaves. They then transported the slaves to the Americas, exchanging them for gold, silver, tobacco, sugar, cotton, rum, and tropical hardwoods to be carried back to Europe. As "raw material" and as the labor working the mines and plantations of Latin America, the West Indies, and Anglo America, slaves generated much of the wealth that made Europe prosperous and helped spark the Industrial Revolution.

While Europeans carried out the triangular trade, their physical presence was limited to coastal shipping points. Africans were the intermediaries who actually raided inland communities to capture the slaves and assemble them at the coast for transit shipment. West African kingdoms initially acquired their own slaves in the course of waging local wars. As the demand for slaves grew, these kingdoms increasingly went to war for the sole purpose of capturing people for the trade. As the exports grew, so did the practice of Africans keeping African slaves. Even after Britain (in 1807) and the other European countries abolished slavery (1870), it continued to flourish in Africa. By the end of the century, slaves made up half the populations of many African states.

Slavery has not yet died out in the region (and indeed, a modern-day form of slavery exists even in the United States and other MDCs; see pages 187 and 601). In Mauritania, some Moors still enslave blacks, although the national government has outlawed this practice several times. Enslavement of children persists in West Africa. The typical pattern is that impoverished parents in one of the region's poorer

528, 536, 537, 615

countries, such as Benin, are approached by an intermediary who promises to take a child from their care and see that the boy or girl is properly educated and employed. This involves a fee (as little as $14) that, unbeknown to the parents, is a sale into slavery. The intermediary sells the child (for $250 to $400, on average) to a trafficker who sees that the child is transported to one of the region's richer countries, such as Gabon. There the child ends up working without wages and under the threat of violence as a domestic servant, plantation worker (especially on cacao and cotton plantations), or prostitute. Benin is the leading slave supplier and trafficking center; Gabon, Ivory Coast, Cameroon, and Nigeria are the main buyers of slaves. Slavery also exists in Sudan. In the 1990s, well-intentioned church groups and other organizations in the West began paying hard cash to buy freedom for slaves in Sudan. While many real slaves were freed, Sudanese profiteers also moved in with "counterfeit slaves," ordinary people hired to act as slaves until their "rescue" had been paid for.

The Impact of Colonialism

In the 16th century, European colonialism began to overshadow and inhibit the growth of indigenous African civilizations. This is a diverse region, but one thing its peoples have in common is a sense that in the past they were humiliated and oppressed by outsiders.[3]

Portugal was the earliest colonial power to build an African empire. The epic voyage of Vasco da Gama to India in 1497–1499 via the Cape of Good Hope was the culmination of several decades of Portuguese exploration along Africa's western coasts. During the 16th century, Portugal controlled an extensive series of strong points and trading stations along both the Atlantic and Indian Ocean coasts of the continent. European penetration of the African interior began in 1850 with a series of journeys of exploration. Missionaries like David Livingstone, as well as traders, government officials, and adventurers and scientific explorers such as James Bruce, Richard Burton, and John Speke, undertook these expeditions. By 1881, when Africans still ruled about 90 percent of the region, foreign exploits had revealed the main outlines of inner African geography, and the European powers began to scramble for colonial territory in the interior. Much of the carving up of Africa took place at the Conference of Berlin in 1884 and 1885, when the French, British, Germans, Belgians, Portuguese, Italians, and Spanish established their respective spheres of influence in the region. By 1900, only Ethiopia and Liberia had not been colonized. For more than half a century, sub-Saharan Africa was a patchwork of European colonies (•Figure 9.16), and Europeans in these possessions were a privileged social and economic class.

At the outbreak of World War II in 1939, only three countries—South Africa, Egypt, and Liberia—were independent. The United Kingdom, France, Belgium, Italy, Portugal, and Spain controlled the rest. But after the war, mainly in the 1960s and 1970s, a sustained drive for independence was mounted. Ceasing to be a colonial region, Africa emerged with more than a quarter of the world's independent countries. This was a peaceful process in most instances, but bloodshed accompanied or followed independence in several countries.

Dependency theorists often point to Africa as a prime example of how colonialism created lasting disadvantages for the colonized. European colonization produced or perpetuated many negative attributes of underdevelopment. These included the marginalization of subsistence farmers, notably those who colonial authorities, intent on cash crop production, displaced from quality soils to inferior land. In addition, European use of indigenous labor to build railways and roads often took a high toll in human lives and disrupted countless families. Furthermore, the colonizers often corrupted traditional systems of political organization to suit their needs, sowing seeds of dissent and interethnic conflict.

The European colonial enterprise did have some positive impacts. The colonies, and the independent nations that succeeded them, were the beneficiaries of new cities and the transport links built with forced or cheap African labor; new medical and educational facilities (often developed through Christian missions); new crops and better agricultural techniques; employment and income provided by new mines and modern industries; new governmental institutions; and government-made maps useful for administration and planning. Such innovations were very helpful, but they were distributed unequally from one colony to another and were inadequate for the needs of modern societies when independence came.

One of the persistent problems in Africa's political geography is that many modern national boundaries do not correspond to indigenous political or ethnic boundaries. In most cases, this is another legacy of colonialism: British, French, and other occupying powers created arbitrary administrative units that were transformed into countries as the colonial powers withdrew. Nigeria is a good example of an ethnically complex, unnaturally assembled nation. On the political map, it appears as an integral unit, but its boundaries have no logical basis in physical or cultural geography. Some Nigerians still refer to the **"mistake of 1914,"** when British colonial cartographers created the country despite its ethnic rifts (see Figure 9.1B on page 491). The greatest divide is between Muslim north and Christian south, but there are at least 250 ethnic groups within the country. The British colonizers invested more in education and economic development in the south and built army ranks among northerners, whom they thought made better fighters; so the subsequent pattern is that the northern military leaders have tried to blunt the economic clout of southerners. Colonial administration and boundary drawing thus sowed seeds of modern conflict in Nigeria, a problem described in more detail in Module 9.1.

• **Figure 9.16** Colonial rule in 1914. Germany lost its colonies to the British, French, and Belgians after World War I.

Although formal political colonialism has vanished, most countries still have important links with the colonial powers that formerly controlled them, and many foreign corporations that operated in colonial days still maintain a significant presence. France has a long history of postindependence intervention in the political and military affairs of its former African colonies. France is the only ex-colonial power to keep troops in Africa (with the highest numbers in Djibouti, Ivory Coast, the Central African Republic, and Chad). France took steps to ensure that most of its former colonies trade almost exclusively with France and in turn supported national currencies with the French treasury. But France found this paternalistic approach to be extremely expensive and has been reducing its military presence and other costly assistance to its African clients.

9.4 Economic Geography

Great poverty is characteristic of sub-Saharan Africa; 25 of the world's 30 poorest countries are there (see Table 9.1). All of the economies except South Africa's are underindus-trialized. Africa's place in the commercial world is mainly that of a producer of primary products, especially cash crops and raw materials (particularly minerals), for sale outside the region. In most nations, one or two products supply more than 40 percent of all exports—for example, oil in Angola and coffee and tea in Kenya (•Figure 9.17). Such a country is vulnerable to international oversupply of an export on which it is vitally dependent. The value of imports far exceeds that of exports in sub-Saharan Africa, with imports consisting mainly of manufactured goods, oil products, and food.

Social and structural problems contribute to the region's underdevelopment. Most African societies lack a substantial middle class and the prospect of upward economic mobility. Instead, most are hierarchical, and any significant income tends to flow into the hands of a small elite controlling the lion's share of the nation's wealth. There are not enough schools to promote the economic welfare that can accompany literacy, and attendance in many is poor. Even with recent growth in attendance, only 70 percent of all children attend primary school, the lowest percentage of any world region. Bureaucratic obstacles and corruption

• **Figure 9.17** Kenya's economy is highly dependent on the export of coffee. Cash crops and other raw materials are typical exports of Africa south of the Sahara.

can make starting up a new business a long, painful, and costly ordeal, discouraging investment both by Africans and foreigners.

Despite the overall grim picture of African economies, recent decades have seen some changes for the better, including improved health and literacy, growth in transportation infrastructure, and increased manufacturing. Such changes have affected some peoples and areas more than others, and their total impacts have only begun to lift the region from underdevelopment. Many problems remain unresolved, and much of the potential for development in agriculture, mining, and infrastructure for communication and transportation goes unexploited.

Agriculture

Per capita food output in most of sub-Saharan Africa has declined or remained flat since independence. The average African consumes 10 percent fewer calories than two decades ago, and malnutrition afflicts almost half the region's children. Rapid population growth and drought are partly responsible for this trend. Many regimes have also invested more in their militaries than in getting food to their citizens. Food shortages also relate to government preference for cash crops over subsistence food crops. Coffee, cotton, and cloves, for example, provide a means of gaining foreign exchange with which to buy foreign technology, industrial

equipment, arms, and consumption items for the elite. The proportion of crops grown for export to overseas destinations and for sale in African urban centers has therefore risen significantly in recent times.

Most export crops are grown on small farms rather than on plantations and estates. Large plantations have never become as established as they are in Latin America and Southeast Asia. Political and economic pressures forced many of them out of business during the period of transition from European colonialism to independence. Governments of the newly independent countries nationalized and subdivided white-owned plantations, and many whites were obliged to sell or give their land to Africans. Now operated by Africans, many of these farms provide vital tax revenue and foreign exchange. One exception is South Africa, where the most important producers and exporters are the ethnic Europeans, who raise livestock, grains, fruit, and sugarcane. But throughout tropical Africa as a whole, the trend is toward export production from small, black-owned farms. The most valuable export crops are coffee, cacao, cotton, peanuts, and oil palm products. Secondary cash crops include sisal (grown for its fibers; see Figure 10.21), pyrethrum (used in insecticides), tea, tobacco, rubber, pineapples, bananas, cloves, vanilla, cane sugar, and cashew nuts.

Although cash crops are important sources of revenue in these poor countries, excessive dependence on them can

● **Figure 9.18** Minerals, oil pipelines, and transportation links in sub-Saharan Africa.

MINERALS

Ab	Asbestos
Ag	Silver
Au	Gold
Bx	Bauxite
C	Coal
Cb	Columbium
Co	Cobalt
Cr	Chromite
Cu	Copper
Dm	Diamonds
Fe	Iron ore
Ls	Limestone
M	Mica
Mn	Manganese
NG	Natural gas
Ni	Nickel
Oil	Crude petroleum
P	Phosphate
Pb	Lead
PG	Platinum group metals
RE	Rare earth metals
Sb	Antimony
Se	Selenium
Sep	Sepiolite
Sn	Tin
Ta	Tantalum
U	Uranium
V	Vanadium
W	Tungsten
Zn	Zinc

Ngalaba oil pipeline
Major railroad
Potential West Africa-to-Europe roadway (when completed)

be harmful to a country's economy. The income they bring often goes almost exclusively to already prosperous farmers and corporations. The prices they fetch are vulnerable to sudden losses amid changing world market conditions. The plants are susceptible to drought and disease (see Biogeography, page 476). Finally, they are often grown instead of food crops, and cash crops do not feed hungry people.

Mineral Resources

Mineral exports have had a strong impact on the physical and social geographies of sub-Saharan Africa. The three primary mineral source areas are South Africa and Namibia; the Democratic Republic of Congo–Zambia–Zimbabwe region; and West Africa, especially the areas near the Atlantic Ocean (●Figure 9.18). Notable mineral exports from these regions include precious metals and precious stones (particularly diamonds; the region has 60 percent of the world's supply—see Problem Landscape, page 478), iron alloys, copper, phosphate, uranium, petroleum, and high-grade iron ore, all destined principally for Europe, the United States, and China.

Large multinational corporations, financed initially by investors in Europe or the United States and most recently by China, do most of the mining in Africa. Mining has attracted far more investment capital to Africa than any other economic activity. Money is invested directly in the mines, and many of the transportation lines, port facilities, power stations, housing and commercial areas, manufacturing plants, and other elements in the continent's infrastructure have been developed primarily to serve the needs of the mining industry.

Great numbers of workers in the mines are temporary migrants from rural areas, often hundreds of miles away. The recruitment of migrant workers has had important cultural and public health effects. Millions of Africans who work for mining companies have come into contact with Western ideas as well as those of other ethnic groups and have carried these back to their villages. Unfortunately, the phenomenon of migrant labor has also helped spread HIV, the virus that causes AIDS. Most miners are single or married men who spend extended periods away from their wives or girlfriends, and it is common that they contract the virus from prostitutes and then return home to their

BIOGEOGRAPHY

Crop and Livestock Introductions: "The Curse of Africa"

Many important food and export crops of Africa are not native to the continent. For example, corn (maize), manioc (cassava), peanuts, cacao, tobacco, and sweet potatoes were introduced from the New World. Cattle, sheep, chickens, and other domesticated animals now characteristic of sub-Saharan Africa also originated outside the region.

Some nonnative species, such as manioc and chickens, have adapted well to the environments and human needs in sub-Saharan Africa. Others are more problematic. Known as "mealie-meal," milled corn has become the major staple food for southern Africans. However, the people of southern Africa came to regard corn as more of a scourge than a blessing during the severe regional drought of 1992. Introduced from relatively well-watered Central America, corn has flourished in African regions as long as rain or irrigation water has been sufficient. But the crop failed completely during the 1992 drought. The situation was particularly critical in Zimbabwe, where the government had just sold its entire stock of corn reserves in an effort to repay international debts. Zimbabwe suddenly became a large importer of corn. Many in the country proclaimed corn to be the **"curse of Africa,"** lamenting that with its high water requirements, it should never have been allowed to become the principal crop of this drought-prone region. Agricultural specialists argued that native sorghum and cowpeas are much better adapted to African conditions and should be cultivated as the major foods of the future.

Corn was a major issue again in a severe drought in 2002. This time, the United States and other nations reacted early with major shipments of corn to especially hard-hit Zambia. Thousands of tons of American corn were refused by Zambian authorities or locked up in silos while Zambians went hungry. The reason: Zambia's concern that the U.S. corn was genetically modified (GM). Zambian authorities feared that if unmilled corn were to mix with Zambian seed stock, the anti-GM European Union might refuse future imports of corn from Zambia. They also worried about the potential health impacts of the milled GM corn.

Concern is also growing about whether the cow is an appropriate livestock species for Africa. Cattle frequently overgraze and erode their own rangelands. The traditional approach to improving cattle pastoralism in sub-Saharan Africa, and to boosting livestock exports, has focused on the need to develop new strains of grasses that can feed cattle and other non-indigenous strains of livestock because the native grasses of African savannas suit the wild herbivores but generally not the domesticated livestock.

A fresh approach turns this problem around by asking whether it might be more appropriate to commercially develop the indigenous animals already suited to the native fodder and soil conditions. Kenya, Zimbabwe, South Africa, and other countries are now involved in **game ranching** in which antelopes—such as eland, impala, sable, waterbuck, and wildebeest—along with zebras, warthogs, ostriches, crocodiles, and even pythons are bred in semicaptive and captive conditions to be slaughtered for their meat, hides, and other useful products (●Figure 9.C). Safari hunting and tourism provide supplemental revenue in some of these operations.

So far, these enterprises have demonstrated that large native herbivores produce as much or more protein per area than domestic livestock do under the same conditions. In commercial terms, the systems that have incorporated both wild and domestic stock have proved more profitable than systems employing only domestic animals. The World Wildlife Fund and other proponents of this **multispecies ranch system** argue that it is an economically and ecologically sustainable option for land use in Africa. They emphasize that ranchers who keep both wild and domesticated animals will be better able financially to weather periods of drought and will require a much lower capital input than they would expend on cattle or other domesticated livestock alone.

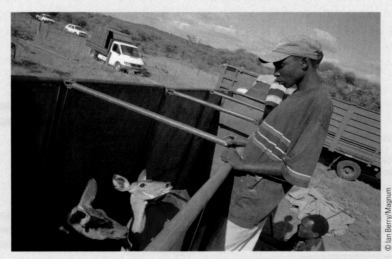

● **Figure 9.C** Game ranching in Africa. In addition to raising game as an alternative to domestic livestock, there are a number of ranches where tourists come to shoot animals for trophies with guns or bows and arrows.

Joe Hobbs

● **Figure 9.19** Many crucial African river crossings have no bridges, where people must rely on ferries instead. Here vehicles and people board a ferry to cross an estuary north of Mombasa on the Kenyan coast. Backup of traffic at these bottlenecks may take hours or even days to clear.

villages, where they spread it still further. The HIV infection rate among migrant miners working in South Africa is more than 40 percent.

Africa's Fragile Infrastructure

Poor transportation hinders development in sub-Saharan Africa. Few countries can afford to build extensive new road or railroad networks, and much of the colonial infrastructure has deteriorated (●Figure 9.19). The region critically needs a good international transportation network, together with a lowering of trade barriers, to enlarge market opportunities. Poor transportation is also often a contributing factor to famine. In Ethiopia, for example, the road network is so poor that it is extremely difficult to get food from the western part of the country, which often has crop surpluses, to the eastern part, which has chronic food shortages. On more than one occasion, it has proved cheaper to ship food from the United States to eastern Ethiopia than to truck it across the country! Similarly, a recent World Bank study showed that it cost $50 to ship a metric ton of corn from Iowa 8,500 miles (13,600 km) to the Kenyan port of Mombasa but $100 to move it from Mombasa 550 miles (880 km) inland to Kampala, Uganda. Transportation problems contribute to the high costs of agricultural inputs like fertilizers, which in this region cost two to three times what they do in Asia; indeed, fertilizer consumption in Asia is 10 times what it is in sub-Saharan Africa.

Poor communication, especially in digital technology applications, also hinders development. The expanding ability to receive radio and television broadcasts across the continent, even in remote villages, is providing Africans with

an unprecedented wealth of information about the wider world. But more high-tech, interactive forms of communication are much in need. Telephone, fax, e-mail, and other technologies are critical to successful participation in the globalizing economy. Africa has not been able to take part because it has been critically short of most of these assets. For example, there are more telephone land lines in the city of Tokyo than in all of Africa.

Internet cafés are springing up, and the Internet may prove to be Africa's answer to many of its communications problems. There are hopes that voice-over-Internet technology will make telephone calling more affordable for hundreds of millions of Africans. In 2007, the entire continent of Africa had only 44 million Internet users (of 1.2 billion worldwide), but the number is growing quickly. Private companies and individuals cooperated to build a 20,000-mile (32,000-km) undersea fiber-optic cable system, completed in 2002, that provided a ring of connections around Africa.

Mobile phones may also help bridge the digital divide. So far, the main obstacle has been that many of the telecommunications companies are government-owned and therefore profitable for governments, which are reluctant to allow competition by private companies with cellular technologies. Some governments also like to monitor phone calls and e-mail, and more communication makes that a more challenging task. In some countries, the barriers are falling, however, and mobile phones are beginning to flourish in some of the world's most isolated places. Mobile phone network coverage was available to only 10 percent of sub-Saharan Africa's people in 1999, but that grew to 60 percent by 2007 and is expected to reach 85 percent in 2010, by which time there should be more than 100 million users. The simple but effective technology of text messaging is already improving

PROBLEM LANDSCAPE

Cleaning Up the Dirty Diamonds

As recently as 2004, as much as 15 percent of the world's annual production of rough diamonds was made up of **dirty diamonds** (also called **conflict diamonds** or **blood diamonds**), defined by the United Nations as "rough diamonds used by rebel movements or their allies to finance conflict aimed at undermining legitimate governments." Gems of such questionable origin financed at least three African wars. But diplomatic, nongovernmental, and business efforts have largely stemmed the tide of Africa's dirty diamonds.

In 2002, following four years of negotiations, 45 countries endorsed a UN-backed certification plan, called the **Kimberley Process,** designed to ensure that only legally mined rough diamonds, untainted by violence, reach the market. Rough diamonds must be sent in tamper-proof containers with a certificate guaranteeing their origin and contents. The importing country (the biggest of which are China and India, which have replaced Belgium and Israel as the world's leading diamond cutters and polishers) must certify that

• **Figure 9.D** Many consumers are unaware that this symbol of love and beauty may have had a sinister origin in Africa.

the shipments have arrived unopened and reject any shipments that do not meet the requirements. Only countries that subscribe to the Kimberley Process are allowed to trade in rough diamonds.

Corporate interests and consumer ethics have cleaned up the diamond trade. Diamond certification has been spearheaded by the world's leading diamond company (with about two-thirds of the market), the South African–based multinational named De Beers. De Beers was embarrassed by a report that it had bought $14 million worth of diamonds from Angolan rebels in a single year. Perceiving that a public relations debacle could lead to a business disaster, as happened with an organized boycott against fur products in the 1980s, De Beers seized the initiative. In the name of Africa's welfare, but certainly also a means of increasing demand and profit, De Beers introduced **branded diamonds,** certified as coming from nonconflict areas. Other diamond companies followed suit, especially in an effort to please Americans—who buy half the world's diamonds (•Figure 9.D).

---

farmers' access to markets through an integrated 12-country network in West Africa. Web-enabled cell phones could make it much easier for millions of Africans to jump to the Internet without having to use computers.

Africa's Place in World Markets and Economies

Africa's status as a supplier of raw materials to world markets has channeled significant new wealth into countries that have oil and other resources for which demand is high in China and other major consuming countries. In the decade 1996–2006, the commodities boom brought annual economic growth rates of about 5 percent to 16 sub-Saharan African countries. This windfall does not guarantee economic development, however, and there are several critical areas where reforms are needed.

One of the major obstacles to African economic development is that many countries outside the region have effectively closed their doors to African imports. These restrictions typically take the form of subsidies, high tariffs, or low quotas imposed on agricultural products (like cotton)

or manufactured goods (like textiles). Potential importing countries, such as the United States, impose these restrictions to protect their own industries. For instance, the United States and Africa are the world's leading exporters of raw cotton. The U.S. government spends $2 billion each year to subsidize its 25,000 cotton farmers. This means higher U.S. production and exports and consequently lower cotton prices and incomes for African farmers.

An example of what reduced trade barriers might do for Africa is provided by the **African Growth and Opportunity Act (AGOA),** passed by the U.S. Congress in 2000 and due to expire in 2017. AGOA reduced or ended tariffs and quotas on more than 1,800 manufactured, mineral, and food items that could be imported from Africa. African exports to the United States have surged, especially in the clothing industry. Apparel made in Lesotho, Kenya, Mauritius, South Africa, Madagascar, and elsewhere in Africa is now sold in Target and Wal-Mart stores. The United States is considering extending AGOA's benefits with the group of countries comprising the **Southern African Customs Union (SACU):** South Africa, Lesotho, Swaziland, Botswana, and Namibia.

INSIGHTS

Microcredit

A promising alternative to the problem of national debt is **microcredit,** the lending of small sums to poor people to set up or expand small businesses. Typically, poor people have no collateral—assets to pledge against loans they seek—and therefore cannot borrow from commercial banks. This drives them into the arms of local loan sharks, who may charge huge interest rates of 10 to 20 percent per day. But in the interest of fostering development, microlending organizations have begun to encourage poor borrowers to form a small group that will cross-guarantee all members' loans. One member of the group may take out a loan of $25, for example, to help start up a small business such as a restaurant. Only when that individual has paid it back may the next person in the group receive a loan. Peer pressure minimizes the default rate, and the accumulated history of timely paybacks increases the entire group's creditworthiness.

Microlenders generally prefer to lend to women because women are more likely to use the money in ways that will feed, clothe, and otherwise benefit their children, whereas men may be more likely to spend the money on alcohol or other frivolous pursuits. One fear in Africa is that microloans will not be repaid if the borrower or someone in the borrower's family contracts HIV or some other debilitating illness. Therefore, the loans are often extended only after the borrower receives some amount of public health training, for example, in the use of condoms and in general food preparation hygiene.

Jobs and incomes for Africans have grown because of these new opportunities. Much of the profit, however, went to the Asian firms that set up clothing factories in Africa because of limitations on clothing exports from their home countries. Those export quotas were set by the World Trade Organization (WTO). After they expired in 2005, any country was able to export as much as it could from the home country. African countries now have to compete with Cambodia and other poor countries as manufacturers look for the cheapest labor they can find. When lower costs have been found elsewhere, the up-and-coming African factories have been threatened. The garment industry in Lesotho, for example, immediately plummeted in 2005, only to be resurrected by the AGOA and by cultivating a niche market in "sweatshop-free" garments.

For decades, almost all of the countries of sub-Saharan Africa were heavily in debt to foreign lenders. Young nations undertook costly development projects with borrowed money, particularly after the mid-1970s. Western financiers, planners, and contractors gave optimistic assessments of the benefits to be expected from such projects, and African leaders were ready to accept loans as a way to reap quick benefits from newly won independence. Countries soon found they had great difficulty in meeting even the interest payments on their debts, and their efforts to do so often resulted in further economic woes and destructive environmental effects. The International Monetary Fund (IMF), World Bank, and other lenders generally resisted requests of African debtor nations for rescheduled interest payments and new loans.

As a condition of further support, foreign lenders often demanded that African debtors put their finances in better order by such measures as revaluing national currencies downward to make exports more attractive, increasing the prices paid to farmers for their products, and reducing corruption, especially among government officials and urban elites. Although many African debtor nations attempted to meet these demands, internal political factors slowed progress. Loss of economic privileges angered urban elites on whom governments depended for support. Relaxation of price controls on food to stimulate production was often met by rioting among city dwellers with tight budgets already allocated for other costs of living.

In the West, increased foreign aid in the form of gifts or grants to Africa has been controversial. Critics argue that foreign aid only increases dependence and deflects national attention away from problems underlying famine. International aid monies often free up funds from national treasuries to be spent on unfortunate uses. In the Horn of Africa, for example, when foreign money became available for famine relief, it allowed Ethiopia and Eritrea to spend more on weapons used in their war. Aid monies are often simply stolen, and monitoring by donors of how they are spent has been inadequate.

Since about 1990, donors have proceeded more cautiously, more often funneling money through private and religious groups working in health care, education, and small-business development using microcredit practices (see Insights, above). Foreign aid has been very effective in advancing primary and secondary education in the region. However, many African governments view universities as potential hotbeds of dissent and discourage investment in higher education. Africa's universities are crumbling, and the classic brain drain of African talent to Western institutions means there is that much less intellectual capital left at home to invest in the region's development.

Africa's untenable debt situation, and the slow flow of aid to the region, began to change around 2005. Two major

352,
428

46

59

factors were at work. First, in the West, awareness of the humanitarian costs of not directing more resources to Africa was growing. Media, political and corporate stars like the singer Bono (of U2), former U.S. President Bill Clinton, and Microsoft founder Bill Gates kept up a steady appeal for governments and citizens of the MDCs to be more generous to impoverished Africa. On the eve of the 2005 meeting of the G-8, the seven biggest Western economic powers plus Russia, a multicontinental live music event called "Live 8" rocked public opinion in favor of giving more. Under such scrutiny, the G-8 leaders decided to open their wallets, doubling their aid to the region by 2010. They agreed to forgive more than $40 billion of debt owed by 14 African countries recognized by the **Heavily Indebted Poor Countries (HIPC)** initiative. Those 14 qualified for debt relief because they met such standards as fighting corruption and having more transparent economic systems. Critics pointed out that this excluded critical, heavily indebted countries like Nigeria. But by 2007, it was clear that the West was falling behind on its promised course to double aid, and an Eastern giant stepped into the gap.

Thus the second and more powerful shift in Africa's debt, trade, and aid picture came after 2006 with the meteoric rise of China's engagement with the region. China's trade with sub-Saharan Africa increased 10-fold between 1999 and 2006, when China displaced Britain as the region's third most important trading partner (after the United States and France). In its 2007 dialogue with the **African Development Bank (AfDB)**, China pledged $20 billion in infrastructure and trade financing in the region. Those funds, to be expended over a three-year period, represent a doubling of China's previous inputs and rival the value of those pledged by Africa's traditional financiers in the West. There is an important difference, however. As will be seen, the Western institutions tend to have strings attached to loans and aid, insisting that African countries adopt democratic and other reforms to be eligible for the monies. China, on the other hand, has a "no strings attached" policy. It is not insisting on political reforms, is not expressing concern about environmental or social responsibilities, and much to the chagrin of the West is even engaging with repressive governments that the West is trying to isolate. Western governments and financial institutions are increasingly alarmed that China is making rapid inroads into African markets, possibly at the expense of Western economies and geopolitical interests.

Even Africans look with some trepidation on China's advance. China is unashamed about its interests in sub-Saharan Africa: it needs and wants the region's raw materials—especially oil, iron ore, copper, and cotton—to feed its insatiable industrial appetite. To many Africans, China is building the same kind of exploitive mercantile relationship that characterized the colonial West: it takes Africa's raw materials, transforms them into value-added products, and sells them mainly to its own advantage.

South African President Thabo Mbeki warned in 2007 that Africa risked becoming an economic colony of China. But desperate for assistance, even the wary are inclined to accept China's terms.

A Legacy of Failed States

Africa scholars speak of the **failed-state syndrome,** a pernicious process of economic and political decay that is eating away at countries including the Democratic Republic of Congo, Ivory Coast, Somalia, and Zimbabwe. Some African countries are little more than **shell states.** A shell state appears to have all the institutions of a country: a constitution, a parliament, ministries, and so on. But most of the positions are occupied by friends and family of the country's ruler, who are happy to line their pockets rather than pursue development for their fellow citizens. Nine of the world's 15 most corrupt countries are in this region, according to Transparency International, a leading watchdog group on corruption. How bad can corruption be for a country's development? In Liberia, former President Charles Taylor passed a law allowing him to personally sell any of the country's "strategic commodities," including mineral resources, forest products, art, archaeological artifacts, fish, and agricultural products. The wealth taken by such countries' leaders generally does not find its way back into the economy but is either spent on grandiose personal accommodations or parked in foreign bank accounts.

One of the assumptions of international lenders and aid agencies is that democracy helps weed out such economic bloodletting. True democracy, they observe, is still in short supply in the region. Foreign lenders and donors want to see more, telling African leaders that if they do not institute democratic reforms, hold fair elections, or improve bad human rights records, there will no longer be development aid or lending. Often the leaders make just enough concessions to win the aid without instituting real reform—a phenomenon known as **donor democracy.** Some Africa observers say that positive development assistance must go where it is needed most, channeled directly into rebuilding destroyed institutions such as the civil service, courts, police, and the military.

Foreign powers might not be able to promote such reforms without becoming guilty of or being accused of neo-colonialism, but new African institutions are emerging to confront the problems themselves. In 2002, the 53-member **African Union (AU)** was formed to replace the old and often inconsequential Organization of African Unity (OAU). One of the African Union's main articulated goals is to implement self-inspection or "peer review" to help promote democracy, good governance, human rights, gender equity, and development. It has formed an agency responsible for this monitoring, known as the **New Partnership for Africa's Development (NEPAD).** NEPAD's premise is that improvements in such human affairs will improve the climate for international investment and assistance to Africa.

The African Union has also created a 265-member **Pan-African Parliament** that will begin making regional laws by 2009. To help ensure the stability needed for economic and political progress, the African Union is also establishing a continental military force, the **African Standby Force,** that takes orders from the organization's Peace and Security Council. The force will, in theory, field peacekeepers or peacemakers to ensure that the map of Africa, so filled with deadly conflicts in the past, has far fewer trouble spots in the future. The African Union has plans to create an African Court on Human and People's Rights and, like the European Union after which it is modeled, may also eventually seek a common currency.

9.5 Geopolitical Issues

Sub-Saharan Africa has waxed and waned as a theater of geopolitical interest since the end of World War II. Initially, the great powers were interested in the region. To boost their competing aims during the Cold War years, the Soviet Union and the United States played African countries against one another, arming them with weapons with which to wage **proxy wars.** The superpowers also extended aid generously to many African nations.

Great power concerns about Africa changed with the end of the Cold War around 1990, and the nature of conflict in Africa changed. The great powers withdrew support, but their weapons remained to prolong smoldering conflicts. The United States helped build the arsenals of eight of the nine countries involved in the recent Democratic Republic of Congo conflict, for example. Cold War–era weapons like AK-47 assault rifles were "dumped"—sold at low prices—in Africa, where they were not considered obsolete and where they fanned the flames of conflict. Increasingly, fighting began to spread across international borders. Previously, the United States and the Soviet Union had maintained a kind of security balance that kept warfare from becoming internationalized, but that restraint no longer existed. The scenario of regional or even Africa-wide wars was realized with Africa's first "world war," centered on the Democratic Republic of Congo. In the absence of regional or international powers to keep the peace, some countries (for example, Sierra Leone) simply imploded, fragmenting into fiefdoms run by factions that claimed to be revolutionaries but were essentially profiteers. Other national governments (as in Ethiopia) spent vast sums on expensive military aircraft while their people suffered from malnutrition.

The end of the Cold War constricted (and in the case of Russia, ruptured) aid pipelines. The United States cut its economic assistance to sub-Saharan Africa by 30 percent between 1985 and 1992 and reduced it another 10 percent between 1992 and 2001. Why the waning concern? U.S.

President George W. Bush was quoted as saying, "While Africa may be important, it doesn't fit into the national strategic interests, as far as I can see them."[4]

Then came 9/11. Al-Qa'ida's attack on the United States and the U.S. counterstrike against al-Qa'ida in Afghanistan raised fears that the Islamist organization might find new recruits and training grounds in Africa. Al-Qa'ida had already bombed U.S. embassies in Kenya and Tanzania in 1998 and in 2002 struck again in the Kenyan city of Mombasa. The United States maintains a focus on **terrorism hot spots** in the region, including Kenya, Somalia, Djibouti, Niger, Chad, and Mali, all of which are regarded as real or potential training grounds for al-Qa'ida–affiliated groups. With a relatively low profile, U.S. forces have made the tiny Horn of African country of Djibouti a major military center for potential action against nearby Yemen or other countries where Islamist militants may be active. So far, U.S. interest in the hot spots has translated into military expenditures and partnerships with African governments but not wider development assistance.

In addition to the terrorism hot spots, the United States has strong geopolitical interests in the region's oil-producing countries. These, too, have become more critical since 9/11 and the U.S. war on Iraq, which only heightened the sense that Middle Eastern oil supplies are vulnerable. The United States is seeking more stable oil supplies and has pinned many hopes on Africa. Petroleum-rich Nigeria and Angola together supply more than 15 percent of America's oil, a figure projected by the U.S. Central Intelligence Agency to rise to 25 percent by 2015. (Nigeria is also the United States' largest trading partner in Africa.)

Increasingly, the United States will be competing with China for these dwindling finite reserves, and given current trends, the geopolitical landscape of sub-Saharan Africa will see Asia on the rise and American clout in decline. Countries that are not pleased with Washington's wishes or demands will find partners elsewhere, particularly in Asia. Some military analysts describe the emerging U.S.-China rivalry in Africa as a modern Great Game, comparable to the British-Russian contest over central Asia in the 19th century.

For many years following the deaths of U.S. service personnel in Somalia in 1993, there was great reluctance to project U.S. military power in African trouble spots. In 2007, the Pentagon reshuffled its bureaucracy and created a command center dedicated exclusively to the region. There are plans to establish U.S. military bases in several key countries, beginning with Cameroon, Equatorial Guinea, and Gabon, and possibly to secure air base rights in Benin, Ivory Coast, and Nigeria. The U.S. Navy would like to establish bases in São Tomé and Príncipe and in Equatorial Guinea. Some Africa observers say the Iraq War will dampen U.S. ambitions in the region; not only are military resources stretched thin, but anti-American sentiment related to the war is entrenched, particularly in the Muslim countries.

The United States views South Africa as the region's economic powerhouse and as the key to regional stability. South Africa's military strength, which included nuclear weapons until they were dismantled in 1990, contributes to its global significance. It is strategically important for its frontages on both the Atlantic and Indian Oceans, its possession of Africa's finest transport network (vitally important to many African countries that trade with and through South Africa), and its diversified mineral wealth.

Finally, HIV/AIDS in the region is also a geopolitical concern. Despite Africa's apparent remoteness from the United States, the two are linked by hundreds of air traffic routes traveled by thousands of people each day. Hundreds of HIV-positive Africans, perhaps most of them unaware of their virus, pass through U.S. airport gateways daily. Some African strains (subtypes) of the virus are more easily transmitted by heterosexual contact than the strains prevalent in the United States, so the risk to the population at large is greater. There are other dimensions of the potential burden on the United States. As long as the epidemic rages in Africa, there may be large flights of "medical refugees" seeking asylum in the United States. There could also be AIDS-related political instability or civil wars that invite U.S. military intervention in the region. How the world should respond to poverty and recurrent crises is one of the key questions facing the huge, resource-rich, problem-ridden region of sub-Saharan Africa. The following module examines these challenges in greater detail.

SUMMARY

→ Africa is the cradle of humankind where hominids originated and from where they diffused. The region has seen many indigenous civilizations and empires and is ethnically and linguistically diverse.

→ The majority of the region's people are rural. There are several clusters of dense population.

→ HIV/AIDS is taking a huge toll, dramatically lowering life expectancy and projections for population growth, especially in southern Africa.

→ Sub-Saharan Africa has a relatively low population density overall, but a majority of this region's people live in a small number of densely populated areas.

→ Most of Africa consists of a series of plateau surfaces dissected by prominent river systems.

→ One of the most spectacular features of Africa's physical geography is the Great Rift Valley, a broad, steep-walled trough. The feature marks the boundary of two crustal plates that are rifting, or tearing apart.

→ Although about two-thirds of the region lies within the low latitudes and has tropical climates and vegetation, sub-Saharan Africa contains a great diversity of climate patterns and biomes, some resulting from elevation rather than latitudinal position.

→ Africa's diverse wildlife is often threatened by human population growth, urbanization, and agricultural expansion. Management of elephants and rhinoceroses involves international market demand and legal restrictions.

→ In the four culture hearths of sub-Saharan Africa, early indigenous people were responsible for several agricultural innovations, including the domestication of millet, sorghum, yams, cowpeas, okra, watermelons, coffee, and cotton.

→ The four major language families of the region are Niger-Congo, Afro-Asiatic, Nilo-Saharan, and Khoisan. Malagasy, spoken only on Madagascar, is an Austronesian language. Islam and Christianity are major faiths, but indigenous belief systems are often mixed with them or exist on their own in some locales.

→ The tragic impetus for growing contact between Africa and the wider world was slavery. Pockets of slavery still exist in the region.

→ Most of sub-Saharan Africa fell under European colonialism after the Conference of Berlin in 1884 and 1885.

→ The majority of the people of sub-Saharan Africa are poor, live in rural areas, and practice subsistence agriculture. In most countries, per capita food output has declined or has not increased since independence. Export crops include coffee, cacao, cotton, peanuts, and oil palm products.

→ Frequent droughts, lack of education, poor transportation, and serious public health issues have hindered development in sub-Saharan Africa. Most countries are underindustrialized and overly dependent on the export of a few primary products.

→ The export of minerals has had a particularly strong impact on the physical and social geography of sub-Saharan Africa. The notorious trade in dirty diamonds has been largely cleansed.

→ Exports of clothing from Africa to the United States boomed after the U.S. Congress lowered trade barriers on numerous products, but the suspension of WTO-imposed export quotas in 2005 has forced the region to compete with other low-cost producers.

→ Most countries are heavily in debt to foreign lenders. Economic and humanitarian assistance to the region slowed considerably after the end of the Cold War, but it has picked up again since

9/11. China is the latest major economic power to engage in Africa and has a "no strings attached" policy when it comes to aid and trade with the region.

→ Although many countries have been under authoritarian governments since independence, there has been some progress toward democracy. Serious political instability is characteristic of many African countries. Important links with the colonial powers that formerly controlled them remain strong in many countries of the region.

→ The recently established African Union is a supranational organization dedicated to solving African problems without the intervention of outside powers.

→ So often judged to be marginal in world affairs, sub-Saharan Africa deserves and is receiving increased international attention because of its humanitarian problems, the global implications of its public health and environmental situations, problems in the management of its natural resource wealth, its oil reserves, and concerns about terrorism.

KEY TERMS + CONCEPTS

African Development Bank (AfDB) (p. 479)
African Growth and Opportunity Act (AGOA) (p. 477)
African Standby Force (p. 481)
African Union (AU) (p. 480)
Afro-Asiatic language family (p. 468)
 Semitic subfamily (p. 468)
  Amharic (p. 468)
  Arabic (p. 468)
 Chadic subfamily (p. 468)
  Hausa (p. 468)
 Cushitic subfamily (p. 468)
  Oromo (p. 468)
  Somali (p. 468)
AIDS (acquired immunodeficiency syndrome) (p. 458)
antiretroviral (ARV) drugs (p. 463)
Austronesian language family (p. 468)
 Malagasy (p. 468)
blood diamonds (p. 478)
branded diamonds (p. 478)
compulsory licensing (p. 463)
conflict diamonds (p. 478)

"curse of Africa" (p. 476)
dirty diamonds (p. 478)
donor democracy (p. 480)
Ethiopian Orthodox Church (p. 469)
failed-state syndrome (p. 480)
game ranching (p. 476)
Great Rift Valley (p. 466)
Heavily Indebted Poor Countries (HIPC) (p. 479)
human immunodeficiency virus (HIV) (p. 458)
keystone species (p. 468)
Khoisan language family (p. 468)
 Khoekhoe (Damara) (p. 468)
Kimberley Process (p. 478)
microcredit (p. 480)
"mistake of 1914" (p. 473)
multispecies ranch system (p. 476)
nagana (p. 461)
New Partnership for Africa's Development (NEPAD) (p. 480)
Niger-Congo language family (p. 468)
 Adamawan subfamily (p. 468)
 Atlantic subfamily (p. 468)

 Benue-Congo subfamily (p. 468)
  Bantu (p. 468)
  Swahili (Kiswahili) (p. 468)
 Gur subfamily (p. 468)
 Khordofanian subfamily (p. 468)
 Kwa subfamily (p. 468)
 Mandé subfamily (p. 468)
Nilo-Saharan language family (p. 468)
 Chari-Nile subfamily (p. 468)
 Saharan subfamily (p. 468)
 Songhai (p. 468)
"1 percent gap" (p. 456)
Pan-African Parliament (p. 481)
pandemic (p. 462)
proxy war (p. 481)
shell state (p. 480)
sleeping sickness (p. 468)
Southern African Customs Union (SACU) (p. 478)
terrorism hot spot (p. 481)
triangular trade (p. 470)
trypanosomiasis (p. 468)

REVIEW QUESTIONS

1. Where are the five principal areas of population concentration in sub-Saharan Africa?

2. What factors account for the high HIV infection rates in sub-Saharan Africa? What can be done to fight the epidemic? What is HIV/AIDS doing to life expectancy and the age structure profiles in countries where it is epidemic?

3. What are the principal climatic zones and biomes of sub-Saharan Africa?

4. Where were the major slave trade routes in and from sub-Saharan Africa?

5. Where were the major colonial possessions of various European powers in sub-Saharan Africa?

6. What is the significance of cattle in many cultures of sub-Saharan Africa?

7. What are the region's major export crops? Why does dependence on them make a country vulnerable?

8. What is a keystone species? Why is the tsetse fly considered a keystone species in sub-Saharan Africa?

9. What is the "1 percent gap"? What are its implications?

10. Why have clothing exports from Africa to the United States boomed in recent years?

11. What do the terms *shell state, failed-state syndrome, donor democracy,* and *donor fatigue* mean, and what do they reveal about Africa's political geography?

12. What impacts did the end of the Cold War have on many African countries?

13. How is China's ascendancy affecting Africa's economies and geopolitical status?

NOTES

1. Lydia Polgreen, "Misery Loves Optimism in Africa." *New York Times,* March 5, 2006, p. WK-1.

2. Robert Stock, *Africa South of the Sahara,* 2nd ed. (New York: Guilford Press, 2004), p. 41.

3. Ibid.

4. Quoted in Ian Fisher, "Africans Ask if Washington's Sun Will Shine on Them." *New York Times,* February 8, 2001, p. A3.

Sarah Tsialiva, from the village of Mahamasina in northern Madagascar. Almost half of the people in sub-Saharan Africa are of her generation, under the age of 15.

Joe Hobbs

MODULE 9.1
THE ASSETS AND AFFLICTIONS OF THE SUB-SAHARAN COUNTRIES

In this chapter, sketches of the major subregions and their countries illustrate some of the characteristics and unique challenges of sub-Saharan Africa. The major subregions are the Sahel, West Africa, West Central Africa, East Africa, the Horn of Africa, southern Africa, and the Indian Ocean islands (●Figure 9.1.1). The relevant area and population data are in Table 9.1 on pages 457–458.

■ module outline

■ module objectives

This module should enable you to

→ Recognize how European colonial favoritism of some ethnic groups over others sowed seeds of modern strife and warfare

→ Appreciate what corrupt leadership has done to impoverish people in countries with enormous oil and other natural resource wealth

→ Recognize the political, economic, and demographic clout of a select number of countries—notably South Africa and Nigeria—in African affairs

→ Understand how drought regularly triggers environmental degradation and famine and becomes a political as well as humanitarian issue in national and international affairs

→ Consider regional and international efforts to prevent recurrence of the genocide and "ethnic cleansing" that marred Africa in the 1990s

→ Evaluate the efforts to redistribute farmland from minority white to majority black control

9.1.1 The Sahel: On the Shore of a Great Desert

The Sahel region extends eastward from the Cape Verde Islands to the Atlantic shore nations of Mauritania, Senegal, and the Gambia and inland to Mali, Burkina Faso ("Land of the Upright Men," formerly Upper Volta), Niger, and Chad (●Figure 9.1.2). In colonial times, the Gambia was a British dependency; Mauritania, Senegal, Mali, Burkina Faso, Niger, and Chad were French; and Cape Verde was Portuguese. Cape Verde's dry volcanic islands lie well out in the Atlantic off Senegal, but in population, culture, economy, and historical relationships, the islands are so akin to the adjacent mainland that they fit the Sahelian context.

Environments, Resources, and Settlements

Climatically and floristically, the Sahel region ranges from desert (parts of the Sahara) in the north through belts of tropical steppe and dry savanna in the south (see Figure 9.8). One country is exceptional: nearly surrounded by Senegal, the Gambia consists of a narrow strip of savanna

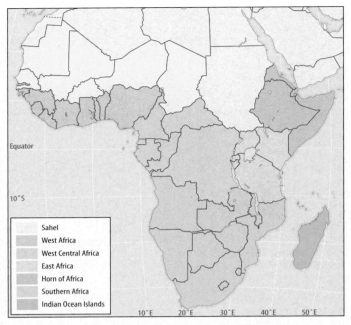

● **Figure 9.1.1** Regional groupings of the countries of sub-Saharan Africa.

grassland and woodland only 20 miles (32 km) wide, extending inland 300 miles (c. 500 km) along the banks of the its namesake river. The name Sahel in Arabic means "coast" or "shore," referring to the region as a front on the great desert "sea" of the Sahara. Since the late 1960s, the area has been subjected to severe droughts, which, in combination with increased human pressure on resources, have prompted a process of desertification (see Regional Perspective, pages 488–489).

This area has seen many dramatic changes in climate and vegetation. There is abundant evidence, particularly in the form of prehistoric rock drawings, that much of the now extremely arid northern Sahel and Sahara region was a grassy, well-watered savanna some 6,000 to 10,000 years ago. Many of the large mammals now associated with East Africa frequented the region. Now, even though under much less favorable conditions, the raising of sheep, goats, camels, and cattle, combined where possible with subsistence farming, is the major livelihood for peoples of the Sahel. There are new hopes in the region that farmers will be able to expand cultivation of a drought-tolerant plant called jatropha in order to feed the world's growing demand for biofuels. Jatropha beans, from which the fuel is made, can be produced with far less water, fertilizers, and pesticides than the corn, soybeans, and sugarcane conventionally used to make plant-based fuels. European aid agencies and investors are already working with Sahelian farmers to boost their jatropha output.

One of the most prominent features of the Sahel's landscape is Lake Chad, located north of landlocked Chad's capital, N'Djamena (population 1.4 million). This large,

589, 619

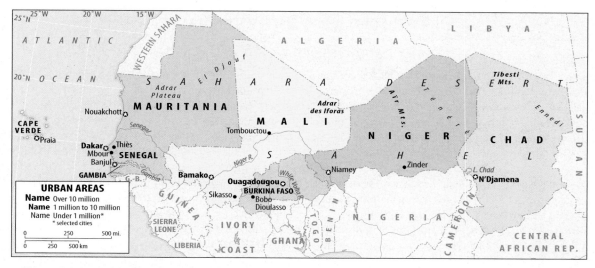

• **Figure 9.1.2** Principal features of the Sahel.

shallow lake, only 7 to 23 feet (2 to 7 m) deep, is situated in a vast open plain. It has no outlets and yet is not very saline because groundwater carries most of the salts away. Its waters are important economically for fish and irrigated agriculture. Characteristic of shallow lakes in arid lands, its surface area historically increased dramatically in rainy periods and shrank in times of drought. However, the expansion of irrigation along the northward-flowing rivers that feed the lake has caused a huge and enduring shrinkage. Lake Chad's surface area in 2007 was 500 square miles (1,300 sq km), compared to 9,700 square miles (25,125 sq km) in 1963; that is a 95 percent loss. Four countries share the lake and its drainage area, so decisions to reverse the lake's shrinkage are difficult to reach, and the likelihood of future conflict brought about by water shortages is high. The situation is reminiscent of that of the Aral Sea in central Asia, which is slowly disappearing.

Mining, manufacturing, and urbanization are modest in this subsistence-oriented region. Mauritania exports iron ore, and Senegal exports phosphates. Mali's cotton production has faded, but the country is now Africa's third-largest producer of gold and is hopeful of producing oil soon. France controls virtually all of Niger's uranium, which is used for France's nuclear power program. Salt is mined on Niger's sterile landscape, but its importance today pales in comparison with centuries ago, when a bar of Niger salt could be traded for a bar of gold. Niger ranks last on the UN's Human Development Index.

The most remarkable asset in the region's economy is the oil of southern Chad (see Figure 9.18, page 475). These reserves, with an estimated life span of 25 years, began to be exported in 2003 through a controversial World Bank–financed pipeline to Cameroon's Atlantic coast. Chad's government promised the World Bank that almost all the oil revenues would be used to assist the country's poor or be set aside in a "future generations fund." Critics warned the

World Bank that Chad's government was among the most corrupt on the planet and should not be trusted. In a classic example of the **"resource curse,"** Chad's government proved them correct by nullifying the laws it had passed to satisfy the World Bank's demands, redirecting the oil wealth away from the poor. The World Bank responded by freezing all future loans.

The Sahel region was historically a crossroads for caravan traffic (especially northbound slaves, food, and gold and southbound salt, weapons, and cloth) between West Africa and North Africa. Ancient Mali was an especially important trading intermediary. Its legendary port of Timbuktu on the Niger River was an intellectual as well as a commercial center. In 1989, it was recognized by the United Nations Educational, Scientific and Cultural Organization (UNESCO) as a **world patrimony site.** Today, much of the traffic through this region is a northward migration of poor people from Nigeria, Niger, Guinea, Ghana, Cameroon, and the Democratic Republic of Congo seeking work in oil-rich Libya. Many subsequently use Libya as a jumping-off point for settlement in Europe. Some Europeans are fearful of a potential new stream of immigrants who might use a highway that, when complete, will for the first time allow vehicles to travel from Nigeria, Mali, and Mauritania all the way to the Strait of Gibraltar (see Figure 9.18).

The largest metropolises are Senegal's main port and capital of Dakar (population 2.4 million) and Bamako (population 1.7 million), the capital and main city of Mali. Dakar was once the capital of the immense group of eight colonial territories known as French West Africa. It is the commercial outlet for much of the peanut-producing area, known as the peanut belt, in the Sahelian and West African savanna and is a considerable industrial center by African standards. Dakar is also a tourist and resort destination with fine beaches and an exotic blend of French and Senegalese cultures.

202

REGIONAL PERSPECTIVE

Drought and Desertification in the Sahel

The Sahel region, like much of sub-Saharan Africa, experiences periodic drought. This is a naturally occurring climatic event in which rain fails to fall over an area for an extended period, often years. When rain does return, the arid and semiarid ecosystems of the Sahel come to life with a profusion of flowering plants, insects, and herbivores like gazelles, whose populations climb when foods are abundant.

These ecosystems have high **resilience,** meaning they are able to recover from the stress of drought and have mechanisms to cope with a natural cycle that includes periods of dryness and rain.

Desertification is the destruction of that resilience and the biological potential of arid and semiarid ecosystems. It is an unnatural, human-induced condition that has afflicted the Sahel periodi-

cally and severely since the late 1960s, and thus it is different from the natural phenomenon of drought (●Figure 9.1.A). However, the recent history of the Sahel suggests that drought can be the catalyst that initiates the process of desertification. During decades of good rains prior to the late 1960s, the Fulani (Fulbe), Tuareg, and other Sahelian pastoralists allowed their herds of cattle and goats to grow more numerous. The animals represent wealth, and people naturally wanted their numbers to increase beyond the immediate subsistence needs of their families. However, due to declining death rates, the numbers and sizes of families keeping livestock were also very large. Thus the unprecedented numbers of livestock that built up during the rainy years made the Sahelian ecosystem vulnerable to the impact of drought on an unprecedented scale.

Steve McCurry/Magnum

● **Figure 9.1.A** Desertification in progress is shown graphically in this photo from Mali in the Sahel. As growing human populations intensify the impact of naturally occurring drought, the Sahara crept southward in the 1970s and early 1980s.

Ethnic Tensions

The Sahel is the border region between mainly Arab and Berber North Africa and mainly black sub-Saharan Africa. Relations between the ethnic groups are good in most countries, but there are problems in some. In Mauritania, for example, the government—led by the country's majority Arabic-speaking **Moors**—expelled black Mauritanians across the border into Senegal in 1995 and has not allowed them to come back. Analysts continue to fear that the Mauritanian government aims eventually to "cleanse" the country of its 40 percent black population. Laws passed to outlaw the enslavement of blacks have yet to be enforced.

Chad has north-south tensions that are reminiscent of those of Sudan and Nigeria. The northerners, who now

dominate Chad's political life, are mostly Arabic-speaking Muslims whose historical mainstay has been cattle pastoralism. The southerners are non-Arab Christians (mainly of the **Sara** ethnic group) whose livelihood is farming. The French favored the Christian southerners and empowered them more than the northerners, and now that the tables are turned, the northern elite is discriminating against southerners. Historically, northerners have enslaved southerners, and there are reports that they continue to do so. There are frequent land disputes as northern pastoralists overrun southern farms. The government's recent backpedaling on its promise to share oil wealth with the poor also fuels fears that as in Sudan and Nigeria, the dominant northern power

REGIONAL PERSPECTIVE

Drought and Desertification in the Sahel *continued*

Drought struck the region in 1968, persisted through 1973, and returned over many years through 1985. With the drought, annual plants failed to grow, so the large herds of cattle and goats turned to acacia trees and other perennial sources of fodder. They ate all of the palatable vegetation. This destruction was a critical problem because plants play an important role in maintaining soil integrity by helping intercept moisture and funnel it downward through the root system. Plant litter and decomposer organisms working around the plant contribute to soil fertility and help stabilize the soil. With hungry livestock in the region eating the plants, the landscape changed. No longer anchored and replenished, good soils eroded. "Junk plants" like Sodom apple (*Calotropis*) replaced palatable plants, and an almost impermeable surface formed on the land. This degraded ecosystem lost its resilience so that even when rains returned, vegetation did not recover.

Meteorologists studying the Sahel drought discovered another important link between precipitation and the removal of vegetation. They noticed that when people and livestock reduced the vegetative cover, they increased the **albedo,** the amount of the sun's energy reflected by the ground. Fewer plants meant more solar energy was deflected back into the atmosphere, lowering humidity and reducing local precipitation. This connection is known as the **Charney effect,** named for Jules Charney, the scientist who documented it in the early 1970s. It was a tragic sequence: People responding to drought actually perpetuated further drought conditions.

The 1968–1973 drought devastated the great herds of the Sahel. An estimated 3.5 million cattle died. Two million pastoral nomads lost at least 50 percent of their herds, and many lost as much as 90 percent. The Niger and Senegal Rivers dried up completely. The cost in human lives was estimated at 100,000 to 250,000. Environmental refugees poured into towns unprepared to deal with such a large and sudden influx. Many pastoral nomads gave up herding and settled down to become wage earners and, where possible, farmers.

This crisis resulted in an international effort to combat desertification using recommendations developed by the United Nations in 1977. The UN report, however, soon became a case study in how *not* to deal with an environmental crisis in the developing world. Most of the recommendations were universal, high-technology solutions that proved impossible to implement in the villages and degraded pastures of the Sahel. For example, many international agencies attempted to relieve the suffering of Sahelian pastoralists using techniques such as digging deep wells to water livestock. But these agencies failed to anticipate that these wells would act like magnets for great numbers of people and animals; although there was plenty to drink, many animals starved to death when they decimated what vegetation remained around the new water supplies.

Since the disappointment of the 1977 UN plan, and with the lessons relief organizations learned as a result, there has been greater focus on sustainable development, with its local rather than universal solutions. In Niger, for example, hit by a devastating drought as recently as 2005, farmers have ceased the traditional practice of clearing trees before planting crops and are now protecting trees and growing crops under and between them. They are also harvesting and selling many useful tree products. A key factor for the farmers' new concern for trees is that Niger's government released its long-held claim that all trees belonged to the state; now farmers recognize the trees as private property. New attitudes and practices like these, and a sustained period of good rains (which unfortunately have prompted periodic locust swarms), have combined to produce a remarkable regreening of much of the Sahel since 1985. Satellite imagery shows significant regrowth of vegetation in southern Mauritania, northern Burkina Faso, northwestern Niger, central Chad, and southern Sudan. Annual millet and sorghum harvests have risen in some cases as much as 70 percent above their pre-1985 levels. This change in the Sahel's fortunes is an important lesson that desertification is not inevitable or irreversible.

63

will skim off the benefits from the newly developing oil industry and leave the oil-bearing region in poverty.

Chad is deeply engaged in the struggles of neighboring Sudan. Chad's president is from the same Zaghawa ethnic group as some of the rebels fighting Sudanese government forces and militia in Sudan's Darfur province. He has allowed refugees from Darfur to settle in camps inside Chad. Sudanese military and paramilitary forces have made periodic raids on those camps and are also accused of supporting an internal rebellion against the Chadian government. In explaining why he is diverting oil wealth away from Chad's poor, the president said the funds are needed to fight this rebellion.

9.1.2 West Africa: Populous and Struggling to Leave Strife Behind

West Africa extends from Guinea-Bissau eastward to Nigeria (●Figure 9.1.3). Its nine political units make up about 800,000 square miles (c. 2 million sq km), or nearly one-fourth of the area of the United States. These nine countries are Guinea-Bissau, Guinea, Sierra Leone, Liberia, Ivory Coast, Ghana, Togo, Benin, and Nigeria. The spatial, demographic, political, and economic giant of the region is Nigeria.

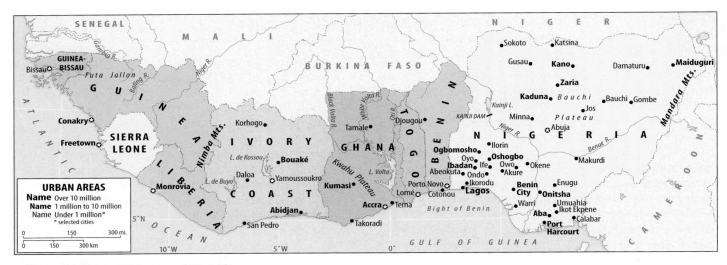

● **Figure 9.1.3** Principal features of West Africa.

Environments and Resources

Climatic and biotic contrasts within this region are extreme, ranging from tropical steppe, dry savanna, and wetter savanna to areas of tropical rain forest along the southern and southeastern coasts (see Figure 9.8). An estimated 90 percent of the region's tropical forest has been lost to commercial and subsistence deforestation. Most of the countries' more populous areas are in the zone of tropical savanna climate.

As is typical of much of Africa, the West African countries depend heavily on subsistence farming and livestock grazing, along with selected agricultural or mineral exports. In the wetter south, subsistence agriculture relies mainly on root crops such as manioc and yams and on maize, the oil palm, and in some areas, irrigated rice. In the drier, seasonally rainy grasslands of the north (in northern Nigeria and northern Ghana, for example), nomadic and seminomadic herding of cattle and goats coexists with subsistence farming of millet and sorghum.

The major export specialties of West African agriculture are cacao (from which cocoa and chocolate are made), coffee, and oil palm products in the wetter, forested south and peanuts and cotton in the drier north. Ivory Coast leads the world in cacao exports, producing almost half the global total in 2007. Nigeria was once the world leader in exports of palm oil and peanuts, but these exports almost vanished as Nigerian agriculture was neglected during the oil boom of recent decades.

The other main export industry of West Africa is mining (see Figure 9.18). The most important resource is petroleum, with OPEC member Nigeria leading the region in production and export. In 2007, when Nigeria stood as the world's sixth-largest oil exporter (and the source of 9 percent of the United States' imports), petroleum accounted for xx percent of the country's foreign export earnings. Oil drilling takes place mainly in the mangrove-forested Niger River Delta and offshore in the adjacent Gulf of Guinea (see Problem Landscape, page 491).

Other prominent mineral products of West Africa include iron ore, bauxite, diamonds, phosphate, and uranium. Iron ore exports are very important to Liberia. The main bauxite resources lie near the southwestern bulge of the West African coast; Guinea is the main producer and exporter. Togo exports phosphate. Gold, bauxite, and manganese are important to Ghana, as are bauxite, titanium, and diamonds to Sierra Leone.

Population and Settlements

West Africa has an impressive total population of 224 million, or about three-fourths that of the United States. The most striking single aspect of West Africa's population distribution is that about two-thirds of the region's people reside in one country, Nigeria (whose 144 million also represent 19 percent of the population of the entire region of sub-Saharan Africa; see Figure 9.3). This unusual concentration of people, which existed even before European contact, is related in part to the region's agricultural productivity and resource wealth.

Only about 37 percent of West Africa's people are city dwellers, but the urban population is increasing rapidly. The world's large cities are associated mostly with trade, centralized administration, and large-scale manufacturing, but West Africa's big cities are different. They include national capitals and seaports, but as the region has only minor manufacturing, they have developed mainly in areas of productive commercial agriculture. In nearly every country in the region, the capital is a primate city, far larger than the nation's second-largest city.

Nigeria has the most impressive urban development: Of the 11 largest metropolitan cities in West Africa, three are

460

111

PROBLEM LANDSCAPE

The Poor, Oil-Rich Delta of Nigeria

Oil is the economic lifeblood of populous Nigeria. Most of the production is concentrated in the Niger River Delta, home to about 7 million mostly Christian people (•Figure 9.1.B). They belong to several different ethnic groups who have one thing in common: they say they have derived few benefits and have suffered greatly from oil development in their homeland. They complain that for more than 30 years, oil spills have tainted their croplands and water, destroying their crops and fisheries, while the flaring off of natural gas has polluted their air and caused acid rain. They note that despite the enormous revenue generated from oil drilled on their land, little money has returned to the area; most goes to foreign oil companies like Shell and Chevron and to the Muslim-dominated government in the north. Meanwhile, most of them live in palm-roofed mud huts; half of the delta region lacks adequate roads, water supplies, and electricity. Schools have few books, and clinics have few medical supplies.

Since 1990, several ethnically based organizations have been agitating for change. **Ogoni** activists founded the **Movement for the Survival of the Ogoni People (MOSOP)**, issuing a bill of rights in which they declared the right to a safe environment and more federal support of their people. One of the movement's leaders, a popular author, playwright, and television producer named Ken Saro-Wiwa, also called for self-determination for the Ogoni. De-

spite international condemnation, Nigeria's government executed him after a questionable murder trial.

Despite the return to civilian rule in Nigeria that came after Saro-Wiwa's death, dissent in the delta continued to grow. Ogoni defiance and contempt of government and the oil companies spread to the region's other major ethnic groups: the **Ibo (Igbo), Yoruba,** and **Ijaw.** The Ijaw (who make up about 5 percent of Nigeria's population) have been especially strident in demanding that some of the oil money be used to fund roads, electricity, running water, and medical clinics. Ijaw activists have spoken of secession and self-determination if their demands are not met. They have periodically registered dissatisfaction by seizing onshore stations and offshore rigs belonging to foreign oil companies, temporarily disrupting Nigeria's oil exports on both occasions. In 2006, a mainly Ijaw group calling itself the **Movement for the Emancipation of the Niger Delta (MEND)** escalated these tactics into frequent kidnappings of foreign oil workers. Opportunistic profiteers without well-established political views also joined in the action, and kidnapping became big business. An oil worker is typically freed unharmed upon payment of a ransom of several hundred thousand dollars.

This militancy in the delta has sent shockwaves through the world economy. Nigeria had touted itself as the safe alternative to Middle Eastern oil but has had to cut production by as much as 25 percent when tensions have been particularly high. The resulting shortages on world markets have helped send oil and gasoline prices to record levels.

Questions of equity in a potentially rich country thus pose a threat to national, regional, and international stability. The peoples of the delta are waiting for implementations of former President Obasanjo's promise to allot the oil-producing states 13 percent of onshore oil revenues (compared with 5 percent previously), along with some portion of offshore receipts, and to shut down the region's foul gas flares. The government insists the money is coming in, but locals say it is not reaching them, and watchdog groups say corrupt officials are siphoning it off.

One of Nigeria's ironies is that in this oil-rich country, there are continuous shortages of gasoline and electricity. Neglect, mismanagement, corruption, and theft are all responsible. One of the cruelest and most repetitive Nigerian news stories is the immolation of scores and even hundreds of poor delta villagers who, in an activity known as "scooping" or "bunkering," illegally puncture gasoline or oil pipelines in an effort to sell the hydrocarbons, only to inadvertently ignite the fires that consume them.

Daniel Laine/Corbis

• **Figure 9.1.B** Nigeria's ethnic landscape.

clustered in the Yoruba-dominated southwestern part of Nigeria, in and near the densely populated belt of commercial cacao production. The largest by far is Lagos (population 11.6 million; •Figure 9.1.4). It had long been Nigeria's political capital, but in the 1990s, most governmental func-

tions were moved to a new capital at Abuja in the interior. The growth of Lagos surged in the early 20th century when the British colonial administration chose the city as the ocean terminus for a trans-Nigerian railway linking the Gulf of Guinea with the north. During the oil boom of recent

• **Figure 9.1.4** Lagos is Nigeria's vibrant and chaotic capital.

decades, migrants from all parts of Nigeria (and many from outside the country) flooded into the city. With its severe traffic congestion, rampant crime, inflated prices, and inadequate public facilities, Lagos gives the impression of urbanism out of control. It is also indicative of a disorderly national economy.

Close to the border with Niger is Kano, the most important commercial and administrative center of Nigeria's north and a major West African focus of Islamic culture. Immediately west of Nigeria on the Guinea Coast lies Cotonou, the main city and port of Benin. Farther west, in southern Ghana—another agricultural area with a high degree of commercialization—is Accra (population 3.5 million), Ghana's national capital. Until recently, it was the main seaport, despite the lack of deep-water harbor facilities. A modern deep-water port was built east of Accra at Tema, and that city is now Ghana's main port and industrial center. Well inland from Accra is the country's second-largest city, Kumasi, in the heart of Ghana's cacao belt, which was once the world's greatest cacao-producing area.

Today, the world's leading cacao producer and exporter is Ivory Coast (the country's official name is its French name, Côte d'Ivoire), with another of West Africa's largest cities, Abidjan (population 5.3 million; •Figure 9.1.5). Although the inland city of Yamoussoukro is the official capital, Abidjan (dubbed the "Paris of Africa" for its architectural and cultural graces) is the de facto capital of a country that in the 1990s—prior to its descent into chaos—vowed to achieve economic prosperity comparable to that

• **Figure 9.1.5** Only about one-third of the people in sub-Saharan Africa live in cities, but with considerable rural-to-urban migration under way, that number is growing. This is very modern Abidjan, the leading city of Ivory Coast, with Saint Paul's Cathedral in the foreground.

of the Asian Tiger countries. Also on the Guinea Coast are three more seaport capitals: Monrovia in Liberia; Freetown in Sierra Leone, on West Africa's finest natural harbor; and Conakry in Guinea.

Many of these port cities have poor connections with their countries' hinterlands. Rapids and lower water during

298

the dry season limit the utility of rivers for transport. Some rivers do carry traffic, notably the Niger and its major tributary, the Benue. Railways are more important than rivers in the region's transport structure, but they are few and widely spaced. Instead of forming a network, the rail pattern is a series of individual fingers that extend inland from different ports without connecting with other rail lines (see Figure 9.18).

Ethnic and Political Complexity and Conflicts

West Africa was an advanced part of sub-Saharan Africa when the European Age of Discovery began. A series of strong pre-European kingdoms and empires developed there, both in the forest belt of the south and in the grasslands of the north. In colonial times, it became a French and British realm, except for a small Portuguese dependency and independent Liberia. Of the present countries, Sierra Leone, Ghana (formerly Gold Coast), and Nigeria were British colonies; Guinea, Ivory Coast, Togo, and Benin (formerly Dahomey) were French; and Guinea-Bissau was Portuguese. Liberia is unique, having become Africa's first republic in 1822, when 5,000 freed American slaves sailed there and settled with support from the U.S. Treasury. The country's capital, Monrovia, was named for U.S. President James Monroe. Its flag is a variant of the U.S. stars and stripes, and its currency is the Liberian dollar. Many Liberians speak fondly of the United States as their "big brother."

In West Africa today, the English, French, and Portuguese languages continue in widespread use, are taught in the schools, and are official languages in the respective countries. European languages are a useful means of communication in this region, where hundreds of indigenous languages and dialects, often unrelated, are spoken. Economic, political, and cultural relationships between West African countries and the respective European powers that formerly controlled them continue to be close. Britain maintains ties with **Anglophone** (English-speaking) **Africa** through meetings of the Commonwealth of Nations, to which all its former colonies in sub-Saharan Africa belong. France supplies economic and military aid to its former colonies in this and other parts of **Francophone Africa.** France perpetuates relations with its former dependencies by inviting their leaders to summit conferences to discuss matters of common interest. France also has some colonial-style economic relations with some of its former colonies.

As is often the case in Africa, the governments of the region must deal with serious problems related to ethnicity. European governments contributed initially to some present troubles by drawing arbitrary boundary lines around colonial units without proper concern for ethnicity. As a result, the typical West African country today, like most countries elsewhere in sub-Saharan Africa, contains a collection of ethnic groups that have little sense of identifica-

tion with the national unit. Nigeria, as described on page 473 in Chapter 9, is a good example.

The largest of Nigeria's 250 or more ethnic groups are the mostly Muslim Hausa and Fulani (together, 29 percent of the population) mainly in the north, the mostly Christian Yoruba (21 percent) in the southwest, the Ibo (Igbo) (18 percent) mainly in the southeast, and the Ijaw (10 percent) of the southern delta region. Tensions and violence have erupted between some of these groups, especially between the mainly Muslim Hausa and mainly Christian Yoruba, and each has formed militias. Muslim northerners have traditionally dominated Nigerian politics, and the long rule of Christian Yoruba President Obasanjo was a concession by the northerners to prevent the country from fragmenting. For the 2007 elections, the northern Muslim power base insisted that one of theirs should be president, and Obasanjo handpicked a successor from their ranks. The subsequent election of his man, Umaru Yar'Adua, was widely viewed inside and outside Nigeria as fraudulent. Polls show that Nigerians have gone from overwhelmingly optimistic to extremely pessimistic about their future, and the rigged election raises serious concerns about the country's future stability.

Mainly as a means of blunting former President Obasanjo's perceived Christian leanings, most of the Muslim northern states in 1999 adopted sharia (Islamic) law. The Christian minorities of these regions fear persecution and intolerance (see Figure 9.1.B, page 491). For their part, Christians have attacked and killed Muslims elsewhere in the country, and violence between the groups has taken thousands of lives. Some of the unrest here is related to livelihood as well as religion: longtime residents are mainly Christian farmers, and they resent a recent influx of Muslim cattle herders. Cattle theft and crop burning are typically part of their skirmishes. The prospect of violent devolution hangs over Nigeria, reviving memories of deaths of a million people in 1967–1970 when the country's ethnic Ibo population struggled to establish the separate nation of Biafra.

A diamond purge and civil war plagued Sierra Leone between 1991 and 2002, taking more than 50,000 lives. Almost two-thirds of the country, including all of the diamond-producing area, came under the control of the rebel **Revolutionary United Front (RUF)**. The RUF projected its power with a brutal combination of mutilation, rape, and murder. The country's prospects improved when its first democratically elected president, Ahmed Tejan Kabbah, took office in 1996. However, a military junta assisted by the RUF seized power a year later, forcing Kabbah into exile. Asserting its self-appointed identity as West Africa's regional superpower, and perhaps with an eye on Sierra Leone's mineral wealth, Nigeria responded in 1998 with a military attack on Sierra Leone that in turn forced the junta to flee and restored President Kabbah to office. RUF resistance continued. The 1999 **Lomé Agreement** gave the RUF some seats in the national cabinet and established RUF leader Foday

Sankoh as the country's vice president. But the RUF broke the peace agreement and took hundreds of United Nations peacekeepers hostage, reigniting the war. As the fighting continued, British troops intervened, capturing Sankoh (who died while awaiting a war crimes trial) and helping Kabbah reassert government authority over this devastated country. In 2002, Kabbah was reelected to office, and the war was declared over; all UN peacekeepers had left the country by 2005.

In neighboring Liberia, a regional kingpin, Charles Taylor, long provided arms to Sierra Leone's RUF in exchange for diamonds. In 1989, the U.S.-educated Taylor ignited a civil war that persisted for seven years. In 1997, Liberians elected him president because, analysts say, they feared that Taylor would restart the war if not elected. After that, he led the country on a path to ruin. Public services such as electricity were all but nonexistent. Schools and hospitals were downsized or closed. Taylor's family enriched itself by selling the county's assets, including its tropical hardwoods, and in exchange for Guinean diamonds, Taylor supplied guns to rebels in neighboring Guinea. By 2003, anti-Taylor rebels calling themselves the **Liberians United for Reconciliation and Democracy (LURD)** had mounted an effective insurgency, and Liberia was once again plunged into war. This time, the United States opposed its onetime ally Taylor, who fled the country, and briefly fielded a peacekeeping force to Liberia. An interim administration, in which the government and the rebels shared power, was established to control the country. An election in 2005 brought to power the world's first elected black female head of state, President Ellen Johnson-Sirleaf. Her principal task is to rebuild the economy of a country with 25 percent unemployment. There is much discussion of reviving the timber industry, but under strict supervision to prevent the rapid depletion of the rain forest that covers nearly one third of Liberia.

There was violence in Ivory Coast too. The conflict there was particularly unfortunate because it took place in a country long heralded as a beacon of hope and relative prosperity, with the region's second-strongest economy (behind Nigeria's). Ironically, Ivory Coast's economic promise helped sow the seeds of conflict and economic decline. The country grew prosperous from cacao and coffee exports in the 1990s. There were new roads, electricity and phone services, and schools. The middle class grew. Ivory Coast's fortunes made it a magnet for settlement by people from nearby, poorer nations. The government actively encouraged immigration because it needed extra hands and skills to assist in the country's development. Large numbers of mainly Muslim immigrants poured in from Burkina Faso and Mali; most of them settled in the predominantly Muslim northern part of Ivory Coast. By the end of the decade, these newcomers made up an astonishing one-third to one-half of the country's population of 16 million.

Cacao prices took a nosedive after 2000, and so did Ivory Coast's economic good times. The once-welcomed Muslim immigrants in the north began to be resented and even de-

spised by the mainly Christian southerners. The government, essentially a mirror of that southern population, began an active program of discrimination against the north. Southerners were portrayed as the only "true" Ivoirians—a concept they call *Ivoirité*—and laws were passed to exclude northerners from participation in many aspects of the country's economic and political life.

By 2002, three rebel groups in the north were sufficiently organized to pose a challenge to the government. They were poised to take control of the rest of the county and might well have succeeded. But France, the country's former colonial power, intervened, sending 2,500 Legionnaires to block the rebel advance. The legendary fighting force quickly separated the warring sides and was joined by peacekeepers from the **Economic Community of West African States (ECOWAS)**. The French government managed to persuade leaders from both sides to attend peace talks in Paris as a means of averting all-out war and perhaps the country's partition.

In a peace treaty signed in Paris in 2003, the southern-based government under President Laurent Gbagbo made large concessions to the rebels, especially in agreeing to share power with them. The new "national reconciliation" government was supposed to allow the rebels to achieve their main aim in a nonmilitary fashion: an end to the official discrimination against the mainly Muslim northerners. In a subsequent pact signed in 2007, the main rebel leader became the country's prime minister. A degree of normalcy has returned to Ivory Coast; road networks have reopened, and cotton and sugar produced in the north is again being trucked to Abidjan. The European Union is promising generous economic aid if the peace can hold.

9.1.3 West Central Africa: Colonial "Heart of Darkness"

The subregion of West Central Africa is flanked by Cameroon and the Central African Republic to the north and the Democratic Republic of Congo to the south (•Figure 9.1.6). It contains seven countries: Cameroon, Gabon, Central African Republic, Congo Republic (often known by the shorthand Congo-Brazzaville), Democratic Republic of Congo (known by the shorthand Congo-Kinshasa or DRC), São Tomé and Príncipe, and Equatorial Guinea.

Environments and Resources

Parts of West Central Africa are mountainous, with the most prominent mountain ranges lying in the eastern part of the Democratic Republic of Congo near the Great Rift Valley and in the west of Cameroon along and near the border with Nigeria. Volcanism has been important in mountain building in both instances (see Natural Hazards, page 496).

● **Figure 9.1.6** Principal features of West Central Africa.

As in West Africa, the coast is fairly straight and has few natural harbors. Most of the region is composed of rolling or hilly plateaus. An exception to the generally uneven terrain is a broad area of flat land in the inner Congo Basin, once the bed of an immense lake. The greater part of West Central Africa lies within the drainage basin of the Congo River.

The heart of the region has tropical rain forest climate and vegetation grading into tropical savanna to the north and south. The biological resources are immense. The tropical forests that cover much of the two Congos, Gabon, Cameroon, Equatorial Guinea, and a small corner of the Central African Republic make up one-fifth of the world's remaining tropical rain forests (●Figure 9.1.7).

One of the region's most problematic environmental and economic issues is the fate of these forests. They are being cut, mainly for commercial logging, at a rate of about 3,000 square miles (8,000 sq km) per year. At that pace, they will be gone by 2020. In an effort to slow the destruction, the World Wildlife Fund has reached an agreement with the World Bank whereby the bank will not extend loans in this region unless the borrowing governments can demonstrate they are using, or planning to use, their tropical forest resources in a sustainable manner.

These forested countries do not like being, as they see it, held hostage by Western environmentalists, but they are making concessions. A good example is Gabon, which has an extensive area of almost untouched tropical rain forest. As its oil reserves dwindle, Gabon sees its future in selling

tropical hardwoods. But in a recent pact, it has promised that part of its pristine rain forest, in the Lopé Reserve of central Gabon, will not be harvested by the Malaysian and French logging companies that have extensive concessions in the country.

Subsistence agriculture supports people in most areas of West Central Africa. There is small-scale production of export crops, mainly in rain forest areas, and some cotton is exported from savanna areas. Southern Cameroon is an important cash crop region, with coffee and cacao the main export crops, supplemented by bananas, oil palm products, natural rubber, and tea. The farms occupy an area of volcanic soil at the foot of Mount Cameroon, a volcano standing 13,451 feet (4,100 m) tall that is still intermittently active. The mountain has extraordinarily heavy rainfall, averaging around 400 inches (1,000 cm) a year.

The island of Bioko has long been notable for its cacao exports and more recently for its oil boom. Bioko (earlier known as Fernando Pó) and the adjacent mainland territory of Río Muni, both former Spanish colonies, received independence in 1968 as the Republic of Equatorial Guinea. Cacao production in Bioko developed on European-owned plantations, but the newly independent country's leaders expelled the plantation owners.

Exports of minerals, principally oil, aid the financing of some countries, notably Gabon, Congo-Brazzaville, Cameroon, and most recently, Equatorial Guinea, where oil revenues since production began in 1996 have spurred enormous economic growth rates. Oil exploration began in the waters off São Tomé and Príncipe in 2004, and the 150,000 people of this cocoa-producing country are hoping for a bonanza.

Cameroon won a protracted oil-related dispute with Nigeria over possession of the Bakassi Peninsula, which juts into the Gulf of Guinea near the border between the two countries. The two nations nearly went to war in 1981 over control of Bakassi; the International Court of Justice ruled that it was part of Cameroon in 2002. Control of the peninsula gives Cameroon ownership of promising offshore oil reserves of the Gulf of Guinea. Cameroon is also receiving royalties and lost-land compensation for the shipment of Chadian petroleum in a pipeline to the Atlantic through Cameroon's territory. Copper from the mineral-rich Katanga province in the southeast part of the Democratic Republic of Congo was long the country's main export, but civil war disrupted the industry. There are also rich cobalt deposits there, and diamonds are in the adjoining Kasai region (see Figure 9.18). Another important resource in the DRC is coltan, a rare mineral used in the manufacture of cell phones and fighter jets.

Population, Settlements, and Infrastructure

In West Central Africa as a whole, population densities are low and populated areas are few and far between. Rough terrain, thick forests, wetlands, and areas with tsetse flies

NATURAL HAZARDS

Deadly Lakes

The Cameroonian volcanoes region is also its "deadly lakes" region. In two volcanic-cone lakes, Nyos and Monoun, carbon dioxide gas generated by volcanic activity deep in the earth builds up in the watery depths below a layer of gas-free freshwater (●Figure 9.1.C). The boundary between these layers is called a **chemocline.** Sometimes an event such as a windstorm or landslide disturbs the top layer deeply enough to puncture the chemocline and release the carbon dioxide from its solution. It erupts explosively, rocketing gas and water upward and outward. On August 21, 1986, such an explosion in Lake Nyos killed 1,700 people who were suffocated by a roving cloud of carbon dioxide.

In recent years, U.S. and Cameroonian scientists and engineers have cooperated to install pipes that dissipate the carbon dioxide from these deadly lakes before it builds up to potentially catastrophic levels. The authorities are pleased with the progress but uncertain that enough venting pipes have been installed. Similar steps have yet to be taken in East Africa's Lake Kivu, where the same threat exists.

Volcanism also poses more conventional hazards in this region. In 2002, the eruption of Mount Nyiragongo, near the northern shore of Lake Kivu, sent rivers of molten lava into the Congolese

city of Goma, incinerating the city and causing almost its entire population of 400,000 to flee as refugees.

● **Figure 9.1.C** Cameroon's Lake Nyos.

● **Figure 9.1.7** Tropical rain forest covers a large portion of West Central Africa. This bridge provides rare access to the forest canopy in Gabon.

tend to isolate clusters of people from each other. The savanna lands and highlands are more densely populated than the rain forests.

Most of the countries have an overall population density well below that of sub-Saharan Africa as a whole. The Democratic Republic of Congo is sub-Saharan Africa's largest country; it is about the size of western Europe, or Alaska, Texas, and Colorado combined, but its estimated population in 2007 was only 62 million, or well under one-half that of the continent's most populous country, Nigeria. Within the subregion, however, it is by far the most populous country. Nearly half of the Democratic Republic of

Congo is an area of tropical rain forest in which the population is extremely sparse.

Cities are few in these countries. Most of the larger ones are political capitals or seaports, including Cameroon's capital, Yaoundé (population 1.9 million), and main seaport, Douala (1.8 million), and the Congo Republic's capital city of Brazzaville (population 1.3 million). Brazzaville, located directly across the Congo River from the much larger capital city of Kinshasa in the Democratic Republic of Congo, was once the administrative center of French Equatorial Africa. Kinshasa (population 8.4 million), once known as Leopoldville, is the DRC's capital, largest city, and principal manufacturing center. Another important city is Lubumbashi (formerly Elisabethville), in the copper-mining region of the southeast.

Transportation infrastructure is generally poor due to political mismanagement, warfare, neglect, and poverty. Aside from a handful of rail fingers reaching inland from seaports, West Central Africa is largely devoid of railways, and roads only poorly serve most areas. Isolation and poor transport are major factors limiting economic development. There is not yet a paved road between Brazzaville and Pointe Noire, the Congo Republic's main seaport, but the 320-mile (515-km) Congo-Océan Railway does connect the cities. There are only 300 miles (c. 500 km) of paved roads in the massive Democratic Republic of Congo. The country's major transportation artery is a combination of railroad and river connecting Kinshasa with Lubumbashi.

Colonialism and Modern Struggles

In colonial days, France held three of the present units—Central African Republic, Congo-Brazzaville (its colonial name was Middle Congo), and Gabon—in the federation of colonies called French Equatorial Africa. Cameroon consisted of two trust territories administered by France and Britain. Belgium held what is now the Democratic Republic of Congo, Spain held Equatorial Guinea, and Portugal held the islands of São Tomé and Príncipe. During the last quarter of the 19th century, the Congo Basin was virtually a personal possession of Belgium's King Léopold II, whose agents ransacked it ruthlessly for wild rubber, ivory, and other tropical products gathered by Africans. This lawless era inspired Joseph Conrad's famous novel *Heart of Darkness* (1902), in which the trader Kurtz, at the point of death, evokes the ravaged Congo region with the cry "The horror! The horror!" In 1908, the Belgian government formally annexed the greater part of the Congo Basin, creating the Belgian Congo colony.

Pressure for independence built rapidly in 1959, and Belgium yielded to the demands of Congolese leaders. In 1960, three months before adjacent French Equatorial Africa became the independent People's Republic of the Congo, Belgium's Congo colony became the independent Republic of the Congo. In 1971, it took the name Zaire, meaning "river." Following the overthrow of Zaire's government in 1997, the country was renamed Democratic Republic of Congo.

Hundreds of ethnic groups speaking many languages and dialects live in West Central Africa. The majority speak Bantu languages of the Niger-Congo family, while Nilo-Saharan language speakers predominate in the drier areas of grassland north of the rain forest. Small bands of Pygmies (Twa) live in the rain forest of the Congo Basin. The population of the Democratic Republic of Congo is especially complex, with some 250 ethnolinguistic groups speaking more than 75 languages.

In recent decades, the region's troubles have focused on its major power, the DRC. During his long tenure in office (1965–1997), Zaire's autocratic ruler, Mobutu Sese Seko, had neglected the country's development but acquired a massive fortune. Political opposition and public discontent with unemployment and inflation led to unrest in the early 1990s. All-out war began after 1997, when Zairean army troops attempted to expel a force of indigenous Tutsis living in the country's far eastern region (for discussion of the region's Tutsis and Hutus, see pages 501–502). The assault backfired when these Tutsis, supported by neighboring Tutsi-dominated governments in Rwanda and Burundi and by other sympathetic countries in the region, took up yet more arms and initiated fierce attacks on government forces.

The Tutsi counterattack quickly widened into an antigovernment revolt comprised of many ethnic factions and led by a non-Tutsi guerrilla named Laurent Kabila. Kabila's forces pushed westward virtually unopposed through the huge country and in May 1997 occupied Kinshasa with little loss of life. President Mobutu fled on the eve of Kabila's triumph. Kabila then declared himself president of the country he now called the Democratic Republic of Congo and promised that political reforms would live up to the country's new name. Foreign companies rushed to sign mining contracts with the fledgling government, and throughout the country, there was hope that decades of neglect and poverty could be reversed in this resource-rich giant of West Central Africa.

But it was not to be. President Kabila's new regime proved to be a bitter disappointment to most of the DRC's citizens and neighbors and to the international community. Like his predecessor Mobutu, he funneled much of the country's wealth and power into the hands of family and friends. He also obstructed international human rights investigations into the deaths and disappearances of thousands of Hutu refugees during the heady days of rebellion against Mobutu. And he failed to bring stability to the far east of his huge country, where an environment of chaos had given birth to his own bid for power.

In 1998, Hutu forces based in the eastern DRC launched successive devastating attacks on Tutsi interests in neighboring Rwanda. Although the Tutsis of Rwanda had helped bring Kabila to power, they now turned against him, forming the backbone of a rebel alliance bent on bringing down Kabila's young regime. The unrest led to something rather rare in the region, a broader fight dubbed **"Africa's first world war"** that involved numerous countries, some of which were not even the DRC's neighbors. Each player became involved to protect its own national, separatist, or economic interests, largely at the expense of anything that would resemble national cohesion and stability in the Democratic Republic of Congo.

On one side was the so-called Rebel Coalition, which sent forces to oppose the government in Kinshasa: Rwanda, Uganda, and Angola's rebel UNITA forces. On the other side, the Congo Coalition sent forces to support the government in Kinshasa: Zimbabwe, the Angolan government, Zambia, Namibia, Sudan, and Chad. Although the fighting was sometime ferocious, it was not warfare that caused the loss of more than 3 million lives—the most in any war since World War II—in this struggle between 1998 and 2001. Most victims were civilians who died from starvation and disease or in widespread massacres directed against ethnic groups, usually Hutus killing Tutsis but sometimes the reverse.

The country appeared to be on the verge of dissolution when Laurent Kabila was assassinated by a bodyguard in 2001. He was succeeded by his son Joseph, a Congolese leader seemingly, finally, committed to halting the country's decline. A formal peace treaty was reached in April 2003 when Joseph Kabila agreed to have several formal rebel groups and political parties join him in a government of national unity, meant to pave the way for national elections in 2006. Kabila kept his post as president in what was widely seen as a fair election.

Conflict in the Democratic Republic of Congo was enormously profitable for the warring factions, foreign governments, and some foreign companies. United Nations investigators singled out Uganda and Rwanda for looting the eastern part of the country of its gems, minerals, timber, agricultural produce, and wildlife, including elephant ivory from some of the country's national parks. African, European, and U.S. companies were able to exploit the country's mineral wealth without paying corporate taxes. Despite fighting by Rwandan and Ugandan rebels operating inside the country and by forces loyal to the presidential candidate who lost the 2006 election, hopes rose that the long nightmare of Congo's plunder would end. A United Nations peacekeeping mission known by the French acronym MUNOC stayed on the ground to help ensure the country's stability, but rebellion simmered in the east.

9.1.4 East Africa: Mauled but Healing

Five countries make up East Africa: Kenya and Tanzania, which front the Indian Ocean, and landlocked Uganda, Rwanda, and Burundi (•Figure 9.1.8). Their total area is 703,000 square miles (1.8 million sq km), roughly the size of the U.S. state of Texas, with a total population in 2007 of 122 million (more than one-third the population of the United States).

• **Figure 9.1.8** Principal features of East Africa.

Environments and Resources

East Africa's terrain is mainly plateaus and mountains, with the plateaus generally at elevations of 3,000 to 6,000 feet (c. 900 to 1,800 m). All five countries include portions of the Great Rift Valley. International frontiers follow the floor of the Western Rift Valley for long distances and divide Lakes Tanganyika, Malawi, and Albert and smaller lakes in different countries. The more discontinuous Eastern Rift Valley crosses the heart of Tanzania and Kenya. Lake Victoria, lying in a shallow downwarp between the two major rifts, is divided between Uganda, Kenya, and Tanzania.

The two highest mountains in Africa, Kilimanjaro (19,340 ft/895 m) and Kirinyaga (Mount Kenya, 17,058 ft/ 5,200 m; see Figure 9.6), are extinct volcanoes. Both are majestic peaks crowned by snow and ice and visible for great distances across the surrounding plains; recent scientific observations, however, suggest that Kilimanjaro's legendary snows may be melted completely by 2015. Most of East Africa's productive agricultural districts are in the fertile soils of these volcanic areas (•Figure 9.1.9).

Volcanism was also important in building the fertile highlands of Rwanda and Burundi. The Virunga Mountains along the Rwanda-Uganda border include many active volcanoes. One of the region's most active is Mount Elgon (14,176 ft/ 4,321 m), on the Uganda-Kenya border. On the Uganda-DRC border is Margherita Peak (16,762 ft/5,109 m), part of

465

a nonvolcanic range known as the Ruwenzori. This range is often identified as the "Mountains of the Moon" depicted by early Greek geographers on their maps of interior Africa, but it should be noted that these geographers had little real knowledge of the region. Tanzania's nonvolcanic Eastern Arc mountains, although cleared of more than 70 percent of their forests, remains one of the world's greatest concentrations of endemic plant and animal species and is a biodiversity hot spot (see page 31).

The climate and biome types of most of East Africa are classified broadly as tropical savanna (see Figure 9.8). In Uganda, Kenya, and Tanzania, park savanna—composed of grasses and scattered flat-topped acacia trees—stretches over broad areas and is the habitat for some of Africa's largest populations of large mammals. Great herds of herbivores and their predators and scavengers migrate seasonally across these countries to follow the changing availabil-

ity of food and water. The countries manage many parks and wildlife reserves, and in Kenya and Tanzania, these are responsible for a huge share of the nations' hard-currency revenues. Among the most famous are the Serengeti Plains and Ngorongoro Crater of Tanzania and the Maasai Mara and Amboseli National Parks of Kenya (•Figure 9.1.10).

Over much of East Africa, moisture is too scarce for nonirrigated agriculture. There are great variations in the quantity, effectiveness, and dependability of precipitation and in the length and time of occurrence of the dry season. Long and sometimes catastrophic droughts occur. From 1998 to 2001 and again in 2006, Kenya suffered historic droughts. Millions of people were in need of food aid as crops shriveled and livestock died. As the country's hydropower reserves were depleted, Kenyan cities suffered long periods without electricity. Tanzanian, Kenyan, and Ugandan efforts to stave off future emergencies in part by damming and redistributing waters of the Nile Basin for irrigation, drinking water, and hydropower have aroused the ire of downstream Egypt (see discussion and map on page 233).

The main subsistence crops are maize (corn), millet, sorghum, sweet potatoes, plantains, beans, and manioc (cassava). Elevation has such a strong effect on temperatures that in some areas, midlatitude crops like wheat, apples, and strawberries do well. Kenyan tourist brochures boast that Nairobi (elevation 5,500 ft/1,676 m) has a springlike climate year round. Some cultivators depend exclusively on crops for subsistence, and others also keep livestock. Cattle are the most important livestock animals.

The most valuable export crop for Kenya, the region's leading exporter, is tea, followed closely by cut flowers and fresh vegetables air-freighted to Europe. Coffee exports are also very important to Kenya. A niche market in high-quality gourmet coffee, produced in ideal conditions, has

• **Figure 9.1.9** East Africa is blessed with extensive areas of volcanic soil where commercial and subsistence crops thrive. This is a wheat field in the central highlands of Kenya.

• **Figure 9.1.10** Savanna grasslands in eastern and southern Africa host the world's greatest concentrations of large mammals. This is Kenya's Maasai Mara National Park in part of the greater Serengeti ecosystem that also extends into northern Tanzania.

helped Rwanda's economy experience robust economic growth since 1995. Many specialty coffee stores abroad sell the coffees with pitches of benefits for people and wildlife in Rwanda. Rwandans themselves do not drink coffee; they see it as a bitter reminder of the colonial Belgians who forced them to grow it. However, they are thankful that the Europeans did not uproot some of the world's best arabica coffee trees, which were planted in that time. Kenya and Tanzania produce sisal for its valuable fiber. The islands of Zanzibar and Pemba support the millions of clove trees that provide much of world's supply for spices and for Indonesia's aromatic kretek cigarettes.

Tanzania, Kenya, and Uganda in 2001 revived the **East African Community,** vowing to reduce trade barriers among its member states, establish a stock exchange, adopt a common currency for a unified market (by 2009), and even elect a common president (by 2010). Many tariffs between the countries have been eliminated, and products, services, and people can move more freely between them. First established in 1967, the community fell apart 10 years later because of regional strife. The bloc showed confidence that Rwanda's and Burundi's years of conflict were over by admitting these countries in 2007.

The East African nations generally lack mineral reserves and production. A project to mine titanium, very controversial because it will require the relocation of thousands of people, is under way near the Kenyan coast. Manufacturing is limited mainly to agricultural processing, some textile milling, and the making of simple consumer items. There is almost no petroleum. There is a large cotton textile mill at Jinja, Uganda, that uses hydroelectricity from a station at the nearby Owen Stanley Dam on the Nile. The Tana River project in Kenya produces hydropower and supplies irrigation waters.

Populations, Settlements, and Infrastructure

The highest population densities are in a belt along the northern, southern, and eastern shores of Lake Victoria; in south central Kenya around and north of Nairobi; and in Rwanda and Burundi, which have among the world's highest population densities, aside from those of city-states and some islands (see Figure 9.3a). As recently as the early 1990s, population growth rates in this region were so high that East Africa was a common academic example of the Malthusian scenario. But after sharp reductions as a result of falling birth rates and increasing AIDS-related deaths, the population growth rates of the five East African countries moderated, averaging 2.4 percent as of 2007. There were some exceptions. Rwandan women bear an average of six children, and the government was delicately trying to bring the birth rate down in a country that lost nearly a million people in the 1994 genocide (to be discussed shortly). The region's populations are overwhelmingly rural. Only about 8 and 12 percent of the people of Burundi and Uganda, respectively, live in cities. Kenya and Tanzania are the most urbanized at 36 and 22 percent, respectively.

The most agriculturally productive parts of East Africa are bound together by railways that form a connected system leading inland from the seaports of Mombasa in Kenya and Dar es Salaam and Tanga in Tanzania. Mombasa is the most important seaport in East Africa. From Mombasa, the main line of the Kenya-Uganda Railway leads inland to Kenya's capital, Nairobi (population 3.9 million), the largest city and most important industrial center in East Africa. It is also the busiest crossroads of international air traffic and the main outfitting and departure point for safaris into East Africa's world-renowned national parks. It is a gritty, crime-ridden city that many locals call "Nairobbery." Nairobi was founded early in the 20th century as a construction camp on the railway, which continues westward to Kampala (population 1.7 million), the capital and largest city of Uganda. Dar es Salaam ("House of Peace" in Arabic; population 2.9 million) is the main city, main port, and main industrial center of Tanzania, with rail connections to Lake Tanganyika, Lake Victoria, and Zambia. Dar es Salaam had been the capital of Tanzania since colonial times, but the government is in the process of moving to the small inland city of Dodoma. Zanzibar wants to become a free economic zone, modeled after Singapore and Hong Kong, that would boast the region's most modern city. Rwanda's capital is in the highlands at Kigali (population 800,000), and Burundi's capital is on the northernmost point of Lake Tanganyika at Bujumbura (population 400,000). All of the East African capitals are primate cities.

Political and Social Issues

The five East African countries face many social and political dilemmas. None has had much internal peace, order, or prosperity, and none has had easy relations with its neighbors. Ethnic rivalries and conflicts have beset all of them, but prospects are also now better for most of them than at any time in recent decades.

Aside from small minorities of Asians, Europeans, and Arabs, the population of East Africa is composed of a large number of African ethnic groups; for example, Tanzania alone has about 120. Among the large groups are the Kikuyu and Luo of Kenya, the Baganda of Uganda, the Sukuma of Tanzania, the pastoral Maasai of Kenya and Tanzania, and the Tutsi and Hutu of Rwanda and Burundi. Most East African peoples speak Bantu tongues of the Niger-Congo language family, but some in northern Uganda, southern and western Kenya, and northern Tanzania speak Nilo-Saharan languages. Swahili, a Bantu language drawing heavily on Arabic for vocabulary, is a widespread lingua franca (and is the national language of Tanzania), as is English. Christian, Muslim, and animist faiths thrive in all the countries. The Muslim religion reflects a long history of Arab commercial enterprise in the region, including slave trading.

All of the countries are former European colonies. Rwanda, Burundi, and mainland Tanzania (known as Tanganyika) were part of German East Africa until the end of World War I. At the end of the war, Tanganyika joined Kenya and Uganda as British dependencies, and Rwanda and Burundi, at the time the single Belgian colony known as Ruanda-Urundi, became independent countries in 1962. The islands of Zanzibar and Pemba, which lie north of the city of Dar es Salaam and some 25 to 40 miles (40 to 64 km) off the coast, were once a British-protected Arab sultanate. The city of Zanzibar was the major political and slave-trading center and entrepôt of an Arab empire in East Africa. While Arabs formerly controlled political and economic life, the majority of the population was, and is, African. Africans have been in control since 1964, when a revolution overthrew the ruling sultan. That same year, the political union of Tanganyika and Zanzibar was formed, taking the name United Republic of Tanzania. However, Zanzibar and Pemba have a parliament and president separate from Tanzania's.

Prior to independence, East Africa was home to about 100,000 Europeans, 60,000 Arabs, and 300,000 Asians. Arabs were primarily a mercantile class. Europeans were a professional, managerial, and administrative class. In the so-called White Highlands of southwestern Kenya, about 4,000 European families owned and operated estates that were worked by African laborers to produce export crops and cattle. With independence in 1963, Kenya made a point of inviting the British and other Europeans to maintain their presence in the commercial crop sector and to train the black majority in the managerial and technical skills that had served white interests well. Many Europeans left, and Africans acquired their farms, but those who stayed formed the core of the still-influential white Kenyans of today. Some still have vast tracts of land coveted by the Maasai and black Kenyan communities. With Zimbabwe's economically disastrous policies of confiscation of white lands in mind, however, Kenya's government refuses to help push 515 white owners off their lands. It is urging that land leases to whites—some written to last as long as 1,000 years—be revised dramatically downward to less than 100.

Collectively known as the **Wahindi** in Swahili, the Asians included Indians, Pakistanis, and Goans (from the former Portuguese colony of Goa on the west coast of India). They dominated retail trade, owned factories, and produced export crops. Some were wealthy, and practically all were more prosperous than most Africans. Since independence, the Asian minority has lost much of its commercially predominant position. The Ugandan leader Idi Amin abruptly expelled most of the 80,000 ethnic Indians and Pakistanis from Uganda in 1972. Because they were so vital to the country's economy, it nearly collapsed. Many thousands have returned in recent years, have successfully reclaimed lost land and other assets, and once again own and operate successful service and manufacturing businesses. The more recent economic decline of Kenya led to a slow but steady exodus of Asians, mainly to Australia, Britain, and the United States.

Episodes of bloodshed have punctuated East Africa's history. From 1905 to 1907, when Tanganyika was under German control, German authorities put down a great revolt against the colonial power, costing 75,000 African lives. Immediately prior to its independence, Kenya experienced the bloody guerrilla **Mau Mau Rebellion**, led by a secret society within the large Kikuyu ethnic group. Antagonism directed against British colonialism in Kenya cost the lives of about 11,000 Africans and 95 white settlers over the period of rebellion from 1951 to 1956.

From the 1960s through the 1980s, Kenya was regarded as a relative economic success story in sub-Saharan Africa. However, corruption, ethnic strife, crime, terrorism (including the 1998 bombing of the U.S. embassy in Nairobi, apparently by associates of al-Qa'ida, and another al-Qa'ida-linked bombing in Mombasa in 2002), drought, and floods then combined to reduce Kenya's prospects for economic development. After a sharp drop, tourism rebounded after 2005, with over 1.5 million people visiting 235 the country in 2007, bringing in some much-needed cash and helping the Kenyan economy grow by 6 percent before turmoil resumed.

Through much of the latter half of former President Daniel arap Moi's 24 years in office, which ended in 2002, Kenya suffered from corruption and human rights abuses. Britain, the United States, and other international lenders cut off economic aid because Moi refused to implement democratic reforms. Mwai Kibaki, the candidate opposing Moi's political party, won the 2002 presidential election with a vow of "zero tolerance" of corruption. Hopes evaporated, however, when his administration proved to be as corrupt as his predecessor's. Wary international aid agencies focussed on strictly monitored projects in Kibaki's Kenya. Kenya plunged into anarchy following a 2008 election, widely denounced as fraudulent, in which Kibaki was declared victorious over opposition leader Raila Odinga.

In postcolonial Uganda, a savage orgy of murder, rape, torture, and looting took place under the notorious regimes of Idi Amin and Milton Obote. The ravaged country had a respite after 1986 when Yoweri Museveni, the leader of a southern guerrilla force that had developed in opposition to Obote, took over the government. Despite some simmering conflicts among Uganda's 40 ethnic groups and despite the country's controversial interventions in regional conflicts, Uganda under Museveni has generally been regarded as a relatively progressive model for development in sub-Saharan Africa. International scrutiny focuses on Uganda's government to ensure that it will not promote ethnic cleansing of the Acholi people. Some Acholi supported the rebel **Lord's Resistance Army (LRA),** which fought government interests in the country until a 2006 cease-fire.

Rwanda and Burundi have had tragic disputes between their majority and minority populations. About 85 percent of the population in Burundi and 90 percent in Rwanda are composed of the **Hutu (Bahutu),** and most of the remainder of the two populations is **Tutsi (Watusi).** These were not originally separate tribal or ethnic groups; the peoples speak the same language, share a common culture, and often intermarry. The distinction between Hutu and Tutsi was instead based on socioeconomic classification. Historically, the Tutsi were a ruling class who dominated the Hutu majority. Tutsi power and influence were measured especially by the vast numbers of cattle they owned. Tutsi who lost cattle and became poor came to be identified as Hutu, and Hutu who acquired cattle and wealth often became Tutsi.

Colonial rule in East Africa attempted to polarize these groups, increasing antagonism between them. German and Belgian administrators differentiated between them exclusively on the basis of cattle ownership: Anyone with fewer than 10 animals was Hutu, and anyone with more was Tutsi. Europeans mythologized the wealthier Tutsis as "black Caucasian" conquerors from Ethiopia who were a naturally superior, aristocratic race whose role was to rule the peasant Hutus. The colonists replaced all Hutu chiefs with Tutsi chiefs. These leaders carried out colonial policies that often imposed forced labor and heavy taxation on the Hutus. Education and other privileges were reserved almost exclusively for the Tutsis.

Ferocious violence between these two peoples marred Rwanda and Burundi intermittently after the states became independent in the early 1960s. After independence, the majority Hutus came to dominate the governments of both nations. In 1994, the death of Rwanda's Hutu president in a plane crash (in which Burundi's president also died) sparked civil war in Rwanda between the Hutu-dominated government and Tutsi rebels of the **Rwandan Patriotic Front (RPF),** based in Uganda. Hutu government paramilitary troops, still vengeful about decades of domination by Tutsis, systematically massacred Tutsi civilians with automatic weapons, grenades, machetes, and nail-studded clubs. Women, children, orphans, and hospital patients were not spared. An estimated 800,000 Tutsis and "moderate" Hutus (those who did not support the genocide) died. International media tracked these horrors, but outside nations did nothing to stop the fighting.

Well-organized and motivated Tutsi rebels seized control of most of the country and took power in the capital, Kigali, in 1994. That precipitated a flood of 2 million Hutu refugees mainly into neighboring Zaire, now the Democratic Republic of Congo. Although the Tutsi-dominated government promised to work with the Hutus to build a new multiethnic democracy, Hutu insurgents, based mainly in eastern Zaire, continued to carry out attacks against Tutsi targets in Rwanda. That violence and Rwandan interests in the Democratic Republic of Congo's civil war kept Rwandan forces active outside its borders until 2002.

Rwanda now has a national unity policy aimed at reconciling Hutus and Tutsis; both occupy important government posts, and although the Tutsis still hold the political upper hand, they have encouraged the resettlement of Hutu refugees who fled the country during its troubles (•Figure

• **Figure 9.1.11** Burundian Hutu refugees at Likole camp, Ngar.

9.1.11). In 2004, marking the 10-year anniversary of the genocide with a new policy meant to prevent its recurrence, Rwanda's government outlawed ethnicity. Now any Rwandan who speaks or writes of Hutus and Tutsis may be fined or imprisoned for practicing "divisionism." The country's past nevertheless continues to shape its future; for example, there is an ongoing defection of the faithful from Rwanda's majority Catholic church to other religions, especially Islam but also smaller Christian denominations. Most of the country's worst massacres took place in Catholic churches, with the apparent knowledge and even participation of church leaders.

In Burundi, the Tutsi minority traditionally dominated politics, the army, and business. In 1993, there was a brief shift of power to the Hutu majority when Burundians elected the country's first Hutu president. Tutsis murdered him in October 1993, and a large-scale Hutu retaliatory slaughter of Tutsis ensued. An estimated 300,000 died in a decade of violence. Burundi's Tutsis and Hutus now have a power-sharing agreement in which the presidency rotates between them. The peace has been maintained by the first-ever deployment of the African Union's African Standby Force.

Collective shame and embarrassment in the international community about the failure to stop the killing in Rwanda and Burundi led in subsequent years to numerous new antigenocide measures. These include a United Nations "early warning" system to prevent genocide and a new forum for crimes against humanity in the **International Criminal Court,** established in 2002 in The Hague, Netherlands.

9.1.5 The Horn of Africa: Refuge for Judaism, Christianity, Islamist Militancy

In the extreme northeastern section of sub-Saharan Africa, a great volcanic plateau rises steeply from the desert. This highland and adjacent areas occupy the greater part of the

URBAN AREAS
Name Over 10 million
Name 1 million to 10 million
Name Under 1 million*
* selected cities

• **Figure 9.1.12** Principal features of the Horn of Africa.

• **Figure 9.1.13** The rugged Ethiopian highlands have many good soils, but frequent drought and poor transportation infrastructure set back Ethiopia's agricultural productivity.

Horn of Africa, named for its projection from the continent into the Indian Ocean. This subregion includes the countries of Ethiopia, Eritrea, Somalia, and Djibouti (•Figure 9.1.12).

Environments and Resources

Much of the Horn of Africa lies at elevations above 10,000 feet (c. 3,000 m), and one peak in Ethiopia reaches 15,158 feet (4,620 m). The highlands where the Blue Nile, Atbara, and other Nile tributaries rise receive their rainfall during the summer. Temperatures change from tropical to temperate as elevation increases. Bananas, coffee, dates, oranges, figs, temperate fruits, and cereals can be produced without irrigation. Highland pastures support sheep, cattle, and other livestock. The region's people are concentrated in the relatively well-endowed highlands; most of Ethiopia's 77 million people live there (•Figure 9.1.13).

East of the mountain mass, in Eritrea, Djibouti, and Somalia, lower plateaus and coastal plains descend to the Red Sea, the Gulf of Aden, and the Indian Ocean. Extreme heat and aridity prevail at these lower levels, and nomadic and seminomadic peoples make a living by herding camels, goats, and sheep. Dwellers of scattered oases carry on a precarious agriculture. The arid lowland sections have many characteristics more typical of the Middle East than other areas of sub-Saharan Africa. The populations of these countries are small compared to Ethiopia: Somalia with 9 million, Eritrea with 5 million, and Djibouti with less than a million.

Ethiopia has substantial resources, perhaps including significant mineral wealth. However, its resource base re-mains largely undeveloped, and the country is poor. Coffee and animal hides are the main exports. With foreign assistance, a few textile and food-processing factories have been built, and with Chinese help, the massive Tekaze Dam is being built on a tributary of the Blue Nile to harness part of the country's large hydroelectric potential.

A poor transportation system—whose best feature is the 3,000 miles (c. 4,800 km) of good roads built by the Italians during their occupation in the 1930s—is a major factor keeping the country isolated and underdeveloped. The French-built railroad from Djibouti to Ethiopia's capital, Addis Ababa (population 2.8 million), lacks feeder lines and has never been as successful as hoped. Many roads into the Ethiopian heartland are rough tracks, although better connections have gradually been developing. Air services help overcome the deficiencies in ground transportation. Many of Ethiopia's people live in villages on uplands cut off from each other and from the outside world by precipitous chasms. These formidable valleys were cut by streams carrying huge amounts of fertile volcanic silt to the Nile. Before the Aswan High Dam was built, they provided material for the nourishing floods in Egypt.

Eritrea has gold, copper, and marble reserves but lacks oil, and little of its mineral potential has been developed. Its Red Sea coastline has superb coral reefs, and the country is hoping to attract international tourists to dive and snorkel among them. There is some cotton and oilseed agriculture, and fish are a mainstay in the diet.

Djibouti supports pastoralists but has almost no arable land. The country has little to offer economically except its strategic position near the mouth of the Red Sea, and this has been a windfall for Djibouti since the terrorist attacks on the United States in 2001. Its fine port, also named Djibouti, is a major staging ground in the U.S. military forces in the region. France has long maintained a significant military presence and has pumped much-needed economic aid into the country.

Somalia is extremely arid, and only 2 percent of the country is cultivated. Nomadic and seminomadic pastoralism is the main economic activity. Until an outbreak of livestock disease in the 1990s, Somalia's main exports were live animals (goats, sheep, camels, cattle), shipped mainly to Saudi Arabia, Yemen, and the United Arab Emirates. Fish products and bananas are still shipped, but despite a ban in effect since 2002, charcoal is the main export to the Arabian Peninsula. The resulting environmental impacts on Somalia are severe because trees never were abundant, and there are fights between the tree cutters and the pastoralists, whose camels feed on tree fodder. An estimated 11 percent of Somalia is forested.

Droughts have stricken this region, particularly Ethiopia, almost incessantly since the late 1960s. Ethiopian villagers are especially vulnerable to drought because of their heavy dependence on rain-fed agriculture. The volcanic soils are fertile, but the streams are entrenched so far below the upland fields that irrigation is virtually impossible. In 1984 and 1985, drought in Ethiopia reached the point of widespread catastrophe. An estimated 1 million people perished. Only a massive and sustained international relief effort, some of it promoted by British and American rock musicians, helped reduce the suffering. Successive droughts have revisited the country, and chronic hunger reportedly affects between 4 and 5 million Ethiopians each year. Ethiopia's multimillion-dollar effort to resettle 2.2 million farmers from the apparently overcrowded highlands to the less populous lowlands has run into major problems. The lowlands were less crowded to begin with because of malaria and sleeping sickness, which have taken a big toll on settlers.

Local Cultures, European Imperialism, and Regional Struggles

Ethiopia's people are ethnically and culturally diverse. About 45 percent, including the politically dominant **Amhara** peoples, practice **Ethiopian Orthodox Christianity,** an ancient branch of Coptic Christianity that came to Ethiopia in the 4th century from Egypt. The entire area has had important cultural and historical links with Egypt, the Fertile Crescent, and Arabia. The Ethiopian monarchy based its origins and legitimacy on the union of the biblical King Solomon and the Queen of Sheba, who, tradition holds, gave birth to the first Ethiopian emperor, Menelik. Until a Marxist coup brought an end to the emperorship in the 1970s, Ethiopia's rulers were always Christian. Ethiopia has many outstanding Christian artistic and architectural treasures, including the 11 churches of Lalibela, carved from solid rock in the 12th and 13th centuries (•Figure 9.1.14). Most of the rest of Ethiopia's people are either Muslims (who make up about 40 percent of the population) or members of Protestant, Evangelical, and Roman Catholic churches. There are still small numbers of **Falashas,** or **Ethiopian Jews,** in Ethiopia, a remnant of a very ancient and isolated Jewish population. The majority, about 100,000, have fled to Israel since the mid-1980s. Because this mountainous country has long served an isolated refuge for such unique groups, it has been nicknamed the **"Galápagos Islands of Religion."** 578

Eritrea's population, like Ethiopia's, is about half Muslim and half Christian. Djibouti is about 95 percent Muslim and 5 percent Christian. In Somalia, about 99.8 percent of

• **Figure 9.1.14** The churches of Lalibela, carved from volcanic rock in the 12th and 13th centuries, are among Ethiopia's many Christian cultural treasures. Note the man at the left center for scale.

the people are Sunni Muslims and 85 percent are ethnic Somalis. They share the same language and the same nomadic culture. This is one of the most homogeneous populations in the world, a fact that would suggest peace and stability, but rivalries between clans tore the country apart in the 1990s, as will be related shortly.

Colonialism had a big impact on the region's peoples. European powers seized coastal strips of the Horn of Africa in the late 19th century (see Figure 9.16). Britain was first with British Somaliland in 1882, and then France annexed French Somaliland in 1884. Finally, Italy took control of Italian Somaliland and Eritrea in 1889. These areas along the Suez–Red Sea route have never had much economic importance, but they have had great strategic significance.

Italy tried to extend its domain from Eritrea over more attractive and potentially valuable Ethiopia in 1896, but Ethiopians annihilated the Italian forces at Aduwa. Forty years later, in 1936, the Italians succeeded in taking Ethiopia. In World War II, however, British Commonwealth forces defeated Italian troops throughout the region. Ethiopia regained independence, and Eritrea was federated with it as an autonomous unit in 1952. Ethiopia incorporated Eritrea as a province in 1962.

After World War II, Italian Somaliland was returned to Italian control to be administered as the trust territory of Somalia under the United Nations, pending independence in 1960 as a republic. The present country of Somalia officially includes both the former Italian territory and the former British Somaliland, which chose to unite with Somalia. The people of much smaller French Somaliland remained separate and eventually became independent as the Republic of Djibouti, with its capital at the seaport of the same name.

In 1974, Ethiopia's aging Emperor Haile Selassie (whom 537 **Rastafarians,** who live mainly in Jamaica, regard as God) was deposed by a civilian and military revolt against the country's feudal order. A Marxist dictatorship, strongly oriented to the former Soviet Union, emerged. American influence, which had been strong under Haile Selassie, was eliminated. Munitions and advisers from the Soviet bloc poured in to equip and train one of Africa's largest armies, and the Soviets gained access to base facilities in Ethiopian (now Eritrean) ports. The United States, which had previously supported Ethiopia in its struggles against neighboring Somalia, now switched sides to back Somalia. Such foreign interventions and proxy conflicts contributed to the strife that has hindered development in the region.

Ethiopia had many internal troubles following independence. In what was then the province of Eritrea, both Muslims and non-Amhara Christians resented their political subjugation to Ethiopia's Christian Amhara majority. Terrorism and guerrilla actions escalated into civil war in the early 1960s. The Ethiopian army tried but failed to stamp out the revolts of two Eritrean "liberation fronts." At the same time, there was also a rebellion in Tigre province adjoining Eritrea. In 1991, these combined rebel forces captured Addis Ababa and took over the Ethiopian government. The rebels in Eritrea established an autonomous regime, and Eritrea became independent in 1993. Eritrea's main city and manufacturing center, Asmara (population 1 million), is located in highlands at an elevation of more than 7,000 feet (2,130 m) and has a rail link with the Red Sea port of Massawa.

Many geographers subsequently pointed to Eritrea as a model for sustainable development in sub-Saharan Africa. After independence, the country reduced its foreign debt, cut its population growth rate, and took other promising steps toward eliminating the burdens of underdevelopment. However, in 1998, Eritrea dimmed its bright prospects by launching a war against Ethiopia. The conflict emerged from economic rivalries and unresolved territorial issues between the nations. With Eritrea's independence, Ethiopia had agreed to become landlocked, as long as it could freely use Eritrea's Red Sea ports of Massawa and Assab. The concession was in large part Ethiopia's reward to Eritreans for helping with the overthrow of Ethiopia's Marxist regime. The countries' leaders in effect agreed on an economic union in which Eritrea would be the industrial power and Ethiopia the agricultural one. They promised to allow the free movement of people and goods between the two states and to share a single currency, the Ethiopian birr.

Trouble began when Eritrea introduced its own currency and announced steep duties and fees on goods shipped to and from Ethiopia through Eritrean ports. Ethiopia then turned to Djibouti and Kenya to find new outlets to the sea. Deprived of an anticipated major source of revenue, Eritrea attacked Ethiopia. The war finally ended in 2000 when Ethiopia gained the upper hand by securing a border region disputed between the two countries. A demilitarized zone monitored by United Nations troops was established along the border. An estimated 100,000 Eritreans and Ethiopians had perished in the conflict. Eritrea's economy was hit particularly hard. With the Ethiopian market gone, Eritrea's exports declined 80 percent from their 1997 high. The port of Assab shut down for some time. As the years passed, the two countries failed to agree on the demarcation of their common border. Anti-Ethiopian forces became active inside Eritrea, and war threatened to engulf these neighbors again.

Somalia has been in ruin since a rebellion overthrew the government in 1991 and ignited a civil war between regional Somali clans. The country has six major clans and numerous subclans. These are family-based units based on a patrilineal (male-line) descent system, essentially an extended 219 family that defines one's identity, loyalties, and obligations. There is little sense of nationhood in Somalia because a person's primary affiliations are with clan rather than any larger political entity. Most people are educated, socialized, and married within their clan. There is a history of antagonism between the clans, worsened by easy access to light arms and more deadly weapons.

In the 1991 conflict, contending clan militias inflicted great damage on the capital, Mogadishu (population 2.8 million), from which large numbers of people fled to the country-side as refugees. Massive famine threatened the devastated country in mid-1992. Law and order disintegrated, and humanitarian assistance became difficult. United States military forces intervened in **Operation Restore Hope,** a mission first designed to help protect relief workers and aid shipments. However, policymakers of the Clinton administration decided to use military force to neutralize the power of one of the country's leading clan leaders, or "warlords," Mohamed Farrah Aidid. In a gruesome sequence of events remembered today as the **Blackhawk Down incident,** 18 U.S. servicemen stranded in Aidid's territory in Mogadishu were killed in a single day in 1993.

Osama bin Laden's al-Qa'ida organization had supplied the weapons used to assault the American troops. American public outcry at the tragedy brought an end to this U.S. military mission in 1994. Blackhawk Down had two lasting repercussions. It made the United States more reluctant to assist in humanitarian crises. It also emboldened al-Qa'ida, which became convinced that the United States could not stomach the loss of American lives. Osama bin Laden frequently cited the American withdrawal from Somalia as evidence of his group's successful *jihad* against the United States.

For its part, the United States kept a vigilant watch on Somalia as a potential base for al-Qa'ida and on its waters as a dangerous, pirate-ridden threat to international trade. In 2006, a loose coalition of Islamist groups, including a small number of al-Qa'ida personnel, succeeded in gaining control of Mogadishu and some other cities in Somalia. With their characteristic motto of "Islam is the solution," they argued that clan differences should be subservient to a broader unity of faith. This philosophy attracted a growing number of war-weary Somalis. The United States was alarmed that Somalia could fall to radical Islam. Neighboring Ethiopia worried that Somali Islamists, with support from Ethiopia's enemy Eritrea, would encourage Muslim rebellion within its borders. In late 2006, from bases in Ethiopia, Djibouti, and Kenya, the United States mounted air assaults on the Somalia Islamists, while Ethiopian troops moved in on the ground. As the Islamist forces were killed or melted back into the countryside, and hundreds of thousands of civilians fled by sea to Yemen, Ethiopian troops took control of Mogadishu. Somali resentment of the occupation by the county's historical Ethiopian enemy led to a rapid withdrawal of Ethiopian troops and the deployment of an African Union peacekeeping force made up mainly of Ugandans. A provisional government was established in Mogadishu with the aim of restoring order to the country. But with the country's clan-based factionalism again at the fore, the prospect of peace seemed very distant.

Somalia since 1991 has fractured into three de facto states (see Figure 9.1.12). The southernmost, centered on Mogadishu (still much in ruin and without services), is Somalia, the only one with international recognition as a sovereign country. To the north is Puntland, which includes the tip of the Horn of Africa and considers itself autonomous (with a president of its own) but not independent of Somalia. To the northwest, bordering Djibouti and Ethiopia, is Somaliland (the former British Somaliland). It proclaimed itself independent of Somalia in 1991 and has a president and broadly representative government based in the city of Hargeysa. Fearing fragmentation of their own countries, even African nations do not recognize its independence. Its port city of Berbera has served as Ethiopia's main trade outlet to the sea since Ethiopia's conflict with Eritrea began. Somaliland's economic mainstay was exports of cattle to the Muslim nations of the Arabian Peninsula until an outbreak of Rift Valley fever led to a complete cessation of the trade in 2001. Now Somaliland is pinning its hopes on exports of gemstones, frankincense, vegetable dyes made from henna, and products made from qasil plants, such as hair conditioners and body cleansers.

9.1.6 Southern Africa: Resource-Rich, Finally Free

In the southern part of sub-Saharan Africa, five countries—Angola, Mozambique, Zimbabwe, Zambia, and Malawi—share the basin of the Zambezi River and have long had important relations with each other. The first two were former Portuguese colonies, and the latter three were formerly British. Still farther south are South Africa and four countries whose geographic problems are closely linked to South Africa: the three former British dependencies of Botswana, Swaziland, and Lesotho, and the former German colony of Namibia.

These 10 states are grouped here as the countries of southern Africa (●Figure 9.1.15). Other than their geographic location, little economic or political integration binds these diverse states together. This may change as a result of goals established by the **Southern African Development Community (SADC),** a regional bloc composed of these 10 countries plus Tanzania, Democratic Republic of Congo, and Mauritius. The SADC's member nations have established an economic free-trade zone in which goods can cross borders without tariffs and taxes. The dominant power in that group, in the region, and on the continent is South Africa, and so it is treated in greatest detail here (with a dedicated map, ●Figure 9.1.16).

Environment and Resources: Water, Natural Regions, and Agriculture

Water is all-important in this drought-prone region of a dry continent. South Africa is fairly well endowed with water, as two river systems—the Orange (including its tributary, the Vaal) and the Limpopo—drain much of the country's high (3,000 to 6,000 ft/c. 900 to 1,800 m) interior plateaus. Many dams have been constructed on these and

• **Figure 9.1.15** Principal features of southern Africa.

• **Figure 9.1.16** Principal features of South Africa.

other South African rivers to provide irrigation water and hydroelectricity.

The Zambezi is another of the region's great rivers. Rising in Zambia, it flows eastward to empty into the Indian Ocean on the Mozambican coast. On the middle course of the Zambezi River, separating Zambia from Zimbabwe, is the spectacular Victoria Falls, regarded traditionally as one of the seven natural wonders of the world. It is a significant international tourist destination today. Downstream from the falls, hydropower from the Kariba Dam (completed in 1962; see Figure 9.7, page 466) flows to both countries.

The driest countries are Namibia and Botswana. They can support pastoralism and, in some areas, limited subsistence agriculture but not commercial agriculture. The Namib Desert extends the full length of Namibia in the coastal areas (•Figure 9.1.17). Inland is a broad belt of semiarid plateau country, merging at the east with the Kalahari semidesert of Botswana. The dry zone extends southward into South Africa's Cape provinces and northward into western Angola. Chronic shortages of freshwater in Namibia have led the government to consider desalinating Atlantic Ocean

● **Figure 9.1.17** A surprising abundance and diversity of wild animals and plants are adapted to the harsh conditions of the Namib Desert in Namibia. This is an oryx antelope known as the gemsbok.

waters (this option was rejected as too expensive); building the Epupa Dam on the Cunene River, which divides Namibia from Angola (this project was shelved after being opposed by environmentalists and by human rights groups, who argued that the homeland of the Himba ethnic group would be inundated); and diverting water by pipeline southward from the Namibian portion of the Okavango River to Windhoek (widely regarded as Africa's cleanest and safest capital; population 300,000). Arguing that the World Heritage Site of its Okavango Delta would be damaged or destroyed, Botswana is strongly opposed to this last Namibian plan. With a wealth of wildlife attractions in the Okavango Delta and adjacent national parks, Botswana is a growing destination for international nature-loving tourists. Namibia has its own wetlands resource of international importance: large wildlife populations responding to dramatic fluctuations in wet and dry conditions make Etosha National Park an attraction for the international tourists, who represent an important source of hard currency for Namibia.

Not being landlocked, Namibia enjoys an additional resource that Botswana lacks. The arid western coast of southern Africa, swept by the cold, northward-moving Benguela Current, supports Africa's most important fisheries. **Upwelling** cool water brings up nutrients from the ocean floor and provides a fertile habitat for plankton, the basic food of fish. Namibia has established an exclusive economic zone (EEZ; see page 407) in the Atlantic stretching 200 miles (320 km) offshore, in which fishing and fish processing based at Walvis Bay have become important industries.

More than 90 percent of the region's annual rainfall comes in the cooler six months from November to April. The long dry period creates ideal conditions for the tropical savanna prevailing over most of southern Africa, including much of Zimbabwe, Zambia, Malawi, Mozambique, eastern Angola, and northern South Africa (see Figure 9.8, page 461). Malawi had extensive tropical forests, but at

least half have fallen to the axes of woodcutters and charcoal burners in recent years. This extremely poor country, where only 2 percent of the population has electricity, has the unfortunate distinction of having the region's highest deforestation rate.

There is tallgrass savanna on the High Veld of Zimbabwe. The High Veld lies generally at elevations of 4,000 feet (c. 1,200 m) and higher and forms a band about 50 miles (80 km) broad by 400 miles (c. 650 km) long, trending southwest-to-northeast across the center of Zimbabwe. The High Veld was historically the main area of European settlement, and most of Zimbabwe's postcolonial population of 50,000 whites lived on it. Until the year 2000, these included about 4,500 white farmers whose large holdings, worked by black laborers, produced most of the country's agricultural surplus to feed urban populations and to export. Tobacco from these farms was the country's most important export, followed by cotton, cane sugar, corn (maize), and roses. As discussed on page 515, that export sector collapsed with the subsequent eviction of the white farmers.

South Africa's physical geography has played a major role in its unusual course of development. Lying almost entirely in the middle latitudes, it is a cooler land than most of sub-Saharan Africa, with annual temperatures generally between 60°F and 70°F (16°C–21°C). In combination with the country's natural resource wealth, this climate made the area very attractive for European settlement; it was one of the "neo-Europes" of the colonial era and was historically Africa's main center of European settlement and advanced economic development.

Europeans favored the extreme east of the country's interior plateau, a region also known as the High Veld. The original vegetation here was a grassy steppe, similar to the midwestern prairies of the United States. This is South Africa's leading area of crop and livestock production. The principal farm animals are beef cattle, dairy cattle, and

• **Figure 9.1.18** The Mediterranean climate of Western Cape province is ideal for viticulture. This is a vineyard near Stellenbosch.

• **Figure 9.1.19** Gold miners at work near Johannesburg, South Africa.

sheep, and the main crops are maize and wheat. Maize is the staple diet for South Africa's blacks. Surpluses are exported and used as livestock feed. In KwaZulu-Natal province, there is hilly terrain between the Indian Ocean and the Drakensberg, a mountainous escarpment at the edge of the interior plateau. The leading commercial crop of that area is sugarcane.

South Africa's Western Cape province has a Mediterranean climate. It is the only African area south of the Sahara with this climate type (•Figure 9.1.18). Average temperatures are similar to those of the same climate area in southern California. The early settlers established vineyards in this region, and today, the area's grapes and wines sell both in South Africa and abroad, where they are acquiring a growing share of the world market. The province also produces oranges and other citrus fruits, deciduous fruits, and pineapples. Sheep ranching is a major occupation in the semiarid interior.

Mining and Manufacturing

Minerals, including petroleum, are by far the most valuable material assets of this region. South Africa has the greatest mineral endowment, and the country's relative prosperity is owed in large part to that wealth. Immense British and other foreign investments pour into South Africa's mining industry. South Africa's diverse and abundant mineral wealth has fueled its continental leadership in manufacturing. The country has about a third of Africa's manufacturing and a third of its electric generation. Its diversified industries include iron and steel, machinery, metal manufacturing, weapons, automobiles, tractors, textiles, chemicals, and processed foods. Many thousands of jobs in Western countries are directly linked to mining and manufacturing in South Africa. The mines employ mainly black African workers recruited domestically and from other African countries, particularly Lesotho, Swaziland, Mozambique, Botswana, and Malawi. There is a large illegal immigration to South Africa, too; as many as 10 million undocumented migrant workers may be in this relatively prosperous country.

South Africa is the world's largest producer of gold (•Figure 9.1.19), and gold has often been its most important export (platinum and coal exports have recently brought in more revenues). The gold-mining industry is concentrated in the High Veld, especially the Witwatersrand (formerly Rand), an area in the province of Gauteng. Johannesburg, South Africa's largest city (population 7.3 million), the adjacent impoverished and overcrowded black township of Soweto (population 1.8 million), and nearby Pretoria, the national capital (population 2.1 million), are all located in Gauteng. Limpopo and Mpumalanga provinces and adjacent areas make up one of the world's most vital mineral-exporting regions; exports include copper, asbestos, chromium, platinum, vanadium, antimony, and phosphate. Recovery of uranium from mine tailings and as a by-product of current gold mining has made South Africa the largest uranium producer on the continent. And since 1867, the country has been an important source of diamonds from alluvial deposits along the Vaal and Orange Rivers and from kimberlite "pipes" (circular rock formations of volcanic origin). Kimberley in Northern Cape province is the administrative center for the diamond industry.

South Africa's mining industries require large amounts of electricity. Most of this is supplied by generating plants powered by bituminous coal, which is present in vast quantities near the surface on the interior plateau. These coal deposits, the largest in Africa, are critical to South Africa, as the country has only small petroleum and natural gas reserves and modest hydroelectric potential. Coal is a major export, and it has facilitated development of Africa's leading iron and steel industry, concentrated at Pretoria and around Johannesburg. Iron ore comes by rail from large deposits in the northeast and central parts of the country.

Some of South Africa's neighbors also have considerable mineral wealth (see Figure 9.18). Partly for this reason, South Africa is much engaged in their affairs, especially in neighboring Swaziland and in Lesotho, which South Africa entirely surrounds. Botswana and Lesotho gained independence from Britain in 1966, and Swaziland, the last absolute monarchy in southern Africa, became independent in 1968. South Africa has occasionally exerted power on behalf of its interests in these weaker, mineral-rich countries, particularly Lesotho. In 1998, for example, the prime minister of Lesotho perceived that a coup might be imminent in his nation and called on South Africa for assistance. South African troops immediately invaded Lesotho, but rather than quelling the unrest, this intervention led to widespread condemnation of South Africa for meddling in the internal affairs of a sovereign nation. Mining of coal, gold, and tin is an important source of revenue in Swaziland, as is tourism—which features gambling casinos—and sugar production. Lesotho's economy depends largely on exports of shoes and textiles, as well as remittances from miners working in South Africa.

Diamonds make up about four-fifths of the total value of exports from Botswana, which after South Africa has the region's strongest economy. South African, British, and U.S. mining companies export Namibian diamonds, uranium, copper, and various other metals. These exports contribute to the high (by African standards) GNI PPP levels of Botswana and Namibia ($12,240 and $8,110, respectively).

Zimbabwe has an extraordinary geological feature called the Great Dike, an intrusion of younger, mineral-rich material into older rock. Ranging in width from 2 to 7 miles (3 to 11 km), the Great Dike crosses the entire nation and is readily visible from space as a giant stripe on the landscape. Along and near the Great Dike are the principal areas where commercial farms thrived and mines producing gold, nickel, asbestos, copper, chromium, titanium, and other minerals provided valuable exports until Zimbabwe plunged into political and economic crisis (described on page 515). Along the Great Dike, the historical axis of economic development in Zimbabwe, lies a railway connecting the largest city and capital, Harare (formerly Salisbury; population 2.9 million), in the northeast with the second-largest city, Bulawayo, in the southwest. Iron ore mined at Kwekwe, about midway between Harare and Bulawayo, was used there to produce steel. The Wankie coalfield in western Zimbabwe supplied coal to smelt the ore. Until 1998, Zimbabwe's manufacturing industries were more important, both in value of output and in diversity of production, than those of any other African nation south of the Sahara except Nigeria and South Africa. Manufacturing since then has been reduced to a trickle because of the country's crisis, which has made it very difficult for factories to purchase fuel and spare parts.

Zambia's economy depends heavily on copper exported from several large mines developed with British and American capital. An urban area has developed at each mine, and the mining area, known as the Copperbelt, consists of a series of separate population nodes strung close to Zambia's frontier with the adjacent copper-mining Shaba region of southeastern Democratic Republic of Congo. The Cop-

perbelt, which contains the majority of Zambia's manufacturing plants and copper mines, is a primarily urban region. Boosted by a global commodities boom and Chinese and other foreign investment, copper (now produced exclusively by private companies) made up 70 percent of Zambia's exports in 2007. The government wants to diversify the country's economy so that it does not remain overly reliant on this single product. Other mineral production includes cobalt, a by-product of copper mining; zinc and lead mined along the railway between the Copperbelt and Zambia's capital and largest city, Lusaka (population 2.3 million); and coal production from extreme southern Zambia.

The region's leader in petroleum production is Angola, which is increasingly important on the global energy stage and is OPEC's newest member. One of its vital oil-producing areas is the coastal exclave of Cabinda, separated from the rest of the country by the mouth of the Congo River (which is controlled by the Democratic Republic of Congo). Other fields are scattered along the Angolan coast between the Congo mouth and the city of Lobito. Major new reserves have been discovered recently offshore Angola; in fact, more new petroleum reserves have been detected there in the past decade than in any other place in the world. In addition to oil and its products, Angola exports small amounts of diamonds, coffee, and other items. Diamonds are concentrated in the northeast, where UNITA rebels mined them for years to finance their antigovernment operations (see below).

Ethnicity, Colonialism, and Strife
South Africa: A Special Case

Visitors from western Europe and North America will find in South Africa most of the institutions and facilities to which they are accustomed. However, the European-ized cultural landscape does not reflect the majority culture of this unusual country. Whites represent only about 10 percent of the total population, and blacks make up 79 percent. There are two other large racial groups: the so-called **colored,** of mixed origin (9 percent), and the Asians (2 percent), most of whom are Indians. Many peoples make up the black African majority. The **Zulu** and **Xhosa** (pronounced "*khoh*-suh") are both concentrated in hilly sections of eastern South Africa near the Indian Ocean and are the most populous of the nine officially recognized tribal groups.

A huge economic gulf separates South Africa's impoverished black Africans from about 5 million whites of European descent who dominate the economy and enjoy a lifestyle comparable to that of people in western Europe and North America. South Africa's overall per capita GNI PPP of $11,710 is comparable to that of Russia. For the unskilled and semiskilled labor needed to operate their mines, factories, farms, and services, the whites of South Africa have always relied mainly on low-paid black workers in most of the country and colored workers in Western Cape province. The white-dominated economy could not oper-

228

ate without them, and the Africans are in turn extremely dependent on white payrolls. Although the races are interlocked economically, relations between them have historically been very poor and have been structured to benefit the whites. A country whose people were legally segregated by race, South Africa was one of the world's most controversial nations until its new beginning in 1994. Following is a summary of how the country's racial structures—a critical issue in its modern geography—evolved.

South Africa's white population is split between the British South Africans and the **Afrikaners,** or **Boers** (Dutch for "farmers"; pronounced *"boorz'"*). The Afrikaners speak Afrikaans, a derivative of Dutch, as their preferred language. They outnumber the British by approximately three to two. The Afrikaners are the descendants of Dutch, French Huguenot, and German settlers who began coming to South Africa more than three centuries ago. Their impact on indigenous Africans was catastrophic. The white settlers killed, drove out, and through the unintentional introduction of smallpox decimated most of the original **Khoi** and **San** inhabitants (derisively called **"Hottentots"** and **"Bushmen,"** respectively, by the settlers). They put the survivors to work as servants and slaves. When these proved to be insufficient in number, the colonists brought slaves from West Africa, Madagascar, East Africa, the Malay Peninsula, India, and Ceylon (Sri Lanka).

Most of the workers in the cane fields of the Natal colony were indentured laborers brought from India beginning in the 1860s. The majority chose to remain in South Africa when their terms of indenture ended, and some Indians came as free immigrants. About 85 percent of South Africa's Indian population is still in KwaZulu-Natal province, mainly in and near Durban (population 3.3 million), the province's largest city and main industrial center and port. Most Indians today are employed in commercial, industrial, and service occupations.

Like KwaZulu-Natal, the Cape provinces have a distinctive racial group: South Africa's coloreds (widely known as the "Cape coloreds"). Nearly nine-tenths live in the three provinces, primarily in and near Cape Town (population 4.7 million), the country's second-largest city and one of its four main ports. This group originated in the early days of white settlement as a product of relationships between Europeans (Dutch East India Company employees, settlers, and sailors) and non-Europeans, including slaves and Khoi, Malagasy, West African, and various South Asian peoples. The resulting mixed-race people vary in appearance from those with pronounced African features to others who are physically indistinguishable from Europeans. About nine-tenths speak Afrikaans as a customary language, and most of the others speak English.

Culturally, the coloreds are much closer to Europeans than to Africans, and many with light skins have "passed" into the European community. Coloreds have always had a higher social standing and greater political and economic rights than black South Africans, although they have ranked below Europeans in these respects. Most coloreds work as domestic servants, factory workers, farm laborers, and fishers and perform other types of unskilled and semi-skilled labor. There is a small but growing professional and white-collar class of coloreds. White owners of grape vineyards are donating portions of their fields to an increasing number of the coloreds who once worked as indentured servants on the white lands, leading to a sharp increase in the region's output and exports of wines. Such progress is a sign of how truly different the new South Africa is from the old.

Racial issues in South Africa have a long history of producing strife, even among the ruling white powers. Britain occupied South Africa—which it called the Cape Colony—in 1806, during the Napoleonic Wars. Friction developed between many Boers and the British authorities, who imposed tighter administrative and legal controls than the Boers were accustomed to. The British abolished slavery throughout their empire in 1833, contributing to the anti-British sentiments of the slaveholding Boers. Boer discontent resulted in the **Great Trek,** a series of northward migrations by which groups of Boers, primarily from the eastern part of the Cape Colony, sought to find new interior grazing lands and establish new political units beyond British reach. After some earlier exploratory expeditions, the main trek by horse and ox-wagon began in 1836. It resulted in the founding of the Orange Free State, the Transvaal, and Natal as Boer republics. Britain annexed Natal in 1845 but recognized Boer sovereignty in the Transvaal and the Orange Free State in 1852 and 1854, respectively.

While Boer disaffection with Britain was building, British settlers were coming to South Africa in increasing numbers. The area's natural resource wealth helped shape subsequent events. The British colonies and Boer republics might have developed peaceably side by side if diamonds had not been discovered in the Orange Free State in 1867 and gold in the Transvaal in 1886. The discoveries set off a rush to these republics of prospectors and other fortune hunters and entrepreneurs from outside the region, particularly from Britain. Ill feeling led to the Anglo-Boer War in 1899. British troops defeated the Boer forces decisively, ending the war in 1902 but leaving behind a reservoir of animosity that exists to this day.

The Union of South Africa was organized in 1910 from four British-controlled units as a self-governing constitutional monarchy under the British crown. Dutch became an official language on par with English (this was later altered to specify Afrikaans rather than Dutch). Pretoria in the Transvaal became the administrative capital as a concession to Boer sentiment (however, the national parliament now meets at Cape Town and the supreme court sits at Bloemfontein in the Free State). In 1961, following a close majority vote by the ruling white population, South Africa became a republic outside the British Commonwealth. After 1910, and particularly after the Nationalist Party took power in 1948, Afrikaners dominated the political life of South Africa.

Racial segregation characterized South African life from 1652 onward, but the Afrikaners after 1948 systematized it with laws supporting the official policy of "separate development of the races," subsequently called "multinational development" and commonly known as **apartheid** (pronounced "uh-*pahr*-tayt"). The new laws imposed racially based restrictions and prohibitions on the entire population, but they weighed most heavily on black Africans and denied them political power. The laws fragmented and displaced families and ethnic groups, rearranging the country's cultural landscape, and culminated in the establishment of makeshift tribal states, or **homelands** (also known as **native reserves, Bantustans,** and **national states**), which were nominally intended to achieve "independence" as black nations. Whites ejected several million Africans not wanted in "white" South Africa from their homes and transferred them to the homelands, which were located on the most undesirable and least productive lands. Thus great numbers of blacks were removed from cities and white farms and dumped in poverty-stricken, crowded, resource-poor areas, often far from where they had ever lived before.

The apartheid system drew furious criticism from many other nations. Most of the world community ostracized South Africa, and many countries, including the United States, imposed economic sanctions against it. Black unrest directed against apartheid and the general underdevelopment of the African majority became so widespread and violent between 1984 and 1986 that the government declared a state of emergency. But even after the state of emergency was lifted in 1990, violence continued. Much fighting took place between two large, tribally based rival factions: the Xhosa-dominated **African National Congress (ANC),** led by Nelson Mandela, and the Zulu-dominated **Inkatha Freedom Party (IFP),** led by Chief Mangosuthu Buthelezi.

With the country on the brink of anarchy, a complete official turnabout on the issue of apartheid resolved South Africa's ongoing racial crisis. It began in 1989 after Frederik W. de Klerk, an insider in the Afrikaner political establishment, became president. Urged on by white South African business leaders, antiapartheid activists, the powerful force of South African and international political opinion, the impact of international economic boycotts against South Africa, and his own stated convictions concerning the injustices and unworkability of apartheid, de Klerk launched a broad program to repeal the apartheid laws and put South Africa on the road to revolutionary governmental changes under a new constitution. Years of negotiations led to an all-race election in 1994, in which voters turned out in massive numbers to elect a new national parliament. The parliament then chose Nelson Mandela as president. (The white government had freed Mandela only four years earlier after a 27-year period of political imprisonment.) This historic election ended white European political control in the last bastion of European colonialism on the African continent. The apartheid laws became null and void, and the homelands were abolished.

Mandela, "South Africa's George Washington," retired in 1999 and was succeeded by a second black president, Thabo Mbeki. The country's political landscape is remarkably stable, but challenges remain. The races are continuing on their long path of reconciliation. Some of the healing process has been formalized in the national **Truth and Reconciliation Commission,** which allows those who were party to racial violence during the apartheid era to confess their misdeeds and, in a sense, be absolved of them.

There is still a huge economic gulf between the haves and have-nots, but there are now more blacks—over 10 million—in the ranks of the middle and upper classes. Growing black wealth has boosted racial integration in neighborhoods. The black underclass continues to grow, however, and an overall unemployment rate of 25 percent fuels an ongoing epidemic of petty and violent crime, especially in the cities. Crime, the world's worst HIV/AIDS epidemic, a low level of education among workers, and relatively high labor costs have dampened foreign investment and economic growth in South Africa. No longer restrained by apartheid-era economic sanctions, South Africa has at the same time become a major investor in other African economies, buying banks, railways, cell phone networks, power plants, and breweries across the region. 462

There is a real need for land reform in South Africa because the white minority owns about 70 percent of the productive land. A plan is in place to help reduce the imbalance by redistributing 30 percent of the country's farmland from white to black hands by 2014 on a willing-seller, willing-buyer basis. Progress is slow, however, and some landless blacks vow a takeover of white farms as occurred in Zimbabwe (see page 515). A program is also under way to compensate the estimated 3.5 million blacks forcibly displaced by the government to the homelands between 1960 and 1982. More broadly, South Africa has an aggressive affirmative action program to help redress the economic imbalance between the races. The **Employment Equity Act** does not impose quotas but requires employers to move toward "demographic proportionality" based on the national proportions of race and gender. In another effort to secure more income for the black majority, the South African government assumed control of all the country's mineral resources in 2002. Mining companies now can exploit these resources only under state license and, presumably, with more supervision to ensure that profits are not drained excessively abroad or to whites.

Troubles for Portugal's Ex-Colonies and Independent Namibia

Although Portugal was the first colonial power to establish interests in sub-Saharan Africa, stronger European powers came to prevail in the region. However, Portugal managed to retain footholds along both the Atlantic and Indian Ocean coasts. By 1964, Portugal's principal African possessions, Portuguese West Africa (now Angola) and Portuguese East Africa (now Mozambique), were the most populous

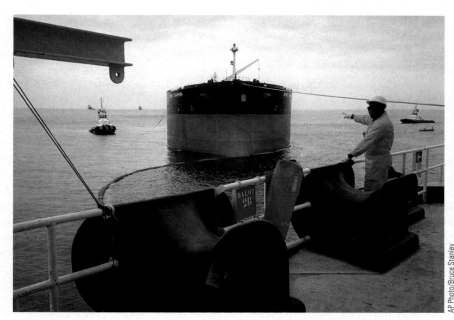

• **Figure 9.1.20** Andre Lina, a training manager for the oil company TotalFinaElf, points out a Spain-bound tanker taking a cargo of crude oil from the production platform at Girasso.

European colonies still remaining in the world. Portugal fielded a large military force to combat guerrilla opposition in both areas. In 1974, Portugal's resistance broke, and an army coup in Portugal installed a regime in Lisbon that granted the two colonies independence.

When the Portuguese withdrew from Angola in 1975, a Marxist-Leninist liberation movement supported by the Soviet Union and Cuba took over the central government in the seaport capital and largest city, Luanda (population 4 million). But this movement's authority was not recognized by other liberation movements, one of which, the **National Union for the Total Independence of Angola (UNITA)**—rooted in the country's largest ethnic group, the Ovimbundu—came to control large parts of the country. South Africa and the United States supplied aid to this anti-Marxist movement. Large quantities of Soviet arms, thousands of Cuban troops, and many Soviet and East German military advisers supported the government in Luanda.

88 This Cold War proxy conflict raged for years. Finally, in 1988, representatives of Angola, Cuba, and South Africa signed a U.S.-mediated agreement under which Cuba and South Africa would withdraw militarily from Angola, and South Africa would grant independence to Namibia. Fighting continued, but in a 2002 peace agreement, UNITA agreed to demobilize and disarm, bringing an end to 27 years of civil war. Efforts are ongoing to clear the country 274 of millions of land mines.

Like Nigeria, Angola is a resource-rich country that so far has little to show for its oil wealth (•Figure 9.1.20). Oil revenues have made the country's wealthy elite even wealthier but have done little for most Angolans, 65 percent of whom live in poverty. Multinational oil companies, including Chinese and American companies, are at work developing this resource; Angola is China's leading oil supplier. Crude oil and refined products made up 90 percent of Angola's exports in 2007, and in that year, Angolan oil held the fastest-growing share of all U.S. oil imports as well. Desperate for low-interest loans and more investment, Angola has promised the international community to triple its expenditures on health, education, and other social services. With its "no strings attached" approach, China is already heavily invested there, helping with the construc- 480 tion of both a new railway and a new capital city south of Luanda.

Namibia became independent from South Africa only recently. South African forces overran this former German colony during World War I, and after the war, the League of Nations mandated it to South Africa under the name of South West Africa. Only the war in Angola released South Africa's grip. Backing UNITA, South African military forces had actively intervened in the Angolan struggle in the mid-1970s. Eventually, South African forces withdrew but continued to attack Angolan bases of the **South West Africa People's Organization (SWAPO)**, a guerrilla group fighting for Namibia's independence. In 1988, South Africa agreed to grant independence to Namibia, but only if the would-be Namibian government under SWAPO leadership could convince its main source of support, the government of Angola, to expel Cuban troops from Angola. The South African government granted independence to Namibia in 1990.

SWAPO is now the major political party, but its Marxist leanings have not prevented the country from promoting capitalist investment. White Namibians look anxiously at the confiscation of white farmlands in nearby Zimbabwe;

just 6 percent of Namibia's whites own nearly half of Namibia's arable land. Like South Africa, Namibia has a willing-seller, willing-buyer policy designed to encourage the sale of white farms to blacks, but not much land has changed hands to date. Black farmers warn that Zimbabwe-style land invasions could ensue (see page 515).

The anomalous sliver of territory in Namibia's extreme northeast known as the Caprivi Strip, which extends 280 miles (450 km) eastward to the waters of the Zambezi River, was important to the German colonists who ruled Namibia early in the 20th century. They insisted on this access to the Zambezi in the mistaken belief that it would allow navigation all the way to the Indian Ocean (alas, Victoria Falls stands in the way). Independent Namibia was permitted to retain the strip, despite protests from neighboring Botswana. The Caprivi is home to the **Barotse (Lozi)** ethnic group, whose tribal lands also extend into Botswana, Angola, and Zambia. Barotse efforts both in the Caprivi Strip and in Zambia to establish an autonomous or independent homeland are strongly opposed by both Namibia and Zambia. With its scattered livestock forage, the Kalahari section of Namibia is peopled mainly by Bantu-speaking herders and by remnants of the Khoi and San populations. Groups of San are among the world's last remaining hunters and gatherers, but the majority have given up the nomadic way of life in recent decades and have settled as farmers and wage laborers.

Portugal's other ex-colony, Mozambique, is one of the world's poorest countries, but it has rather bright prospects. It suffered from years of guerrilla warfare, which first involved African insurgencies against the Portuguese and then warfare by African dissidents against the Marxist government that took over from the Portuguese in Maputo, the country's capital (population 1.9 million). The highly destructive civil war between the Maputo government and the South African– and Zimbabwean-backed rebel organization called **RENAMO (Mozambique National Resistance)** dragged on, at the cost of 1 million lives, until 1992, when a peace accord was signed and the United Nations sent a military peacekeeping force to Mozambique. Elections took place peacefully in 1994, and a large share of the 5 million people displaced by the war began returning to their villages. Only periodic disputes between the government and RENAMO have occurred since their reconciliation.

Mozambique is being rebuilt, although recurrent devastating floods bring setbacks. Transportation facilities in the Beira Corridor to Zimbabwe are being upgraded, and a regional economic link with South Africa has been established. Known as the **Maputo Development Corridor,** this toll road joins Johannesburg's industrial and mining center with the deep-water port at Maputo. Low-tax zones were created to attract new businesses to the corridor. The country's estimated 1 million land mines are being cleared, and war refugees are returning to farm this fertile land. Remarkably, Mozambique has become self-sufficient in food production. Major exports are prawns, cashew nuts, cotton, tea, sugar, and timber. Multinational efforts are under way to process

titanium for export and to develop the country's pristine beaches into world-class tourist destinations. Impressed with Mozambique's progress, foreign lenders have agreed to cancel almost all of the country's debt. Mozambique has acted aggressively on this windfall, making dramatic investments in health, education, and other social services.

Zimbabwe's Struggles

If Mozambique symbolizes the new, hopeful Africa, Zimbabwe represents a slide backward into the abyss. It is a cruel reversal of fortune for a country that was one of the region's beacons of postcolonial progress.

Prior to the post–World War II movement for African independence, the countries of Zimbabwe, Zambia, and Malawi (then known as Southern Rhodesia, Northern Rhodesia, and Nyasaland, respectively) were British colonies. In 1953, the three became linked politically in the Federation of Rhodesia and Nyasaland, also known as the Central African Federation. It had its own parliament and prime minister but did not achieve full independence. The racial policies and attitudes of Southern Rhodesia's white-controlled government caused serious strains within the federation, which led to its dissolution in 1963. Northern Rhodesia then achieved independence as the Republic of Zambia, and Nyasaland became the independent Republic of Malawi.

The government of Southern Rhodesia, the main area of European settlement in the federation, was unable to come to an agreement with Britain concerning British demand for political participation of the colony's African majority. In 1965, Southern Rhodesia issued a "unilateral declaration of independence," and in 1970, it took a further step in separation from Britain by recasting itself as Rhodesia, a republic that would no longer recognize the symbolic sovereignty of the British crown.

Britain and the Commonwealth responded with a trade embargo. Negotiators for Britain and the Rhodesian government unsuccessfully attempted to arrive at a formula to allow political rights and participation for country's 95 percent black population. While economic sanctions were ineffective, guerrilla fighting against the white government increased in the mid-1970s. Independence as a parliamentary state within the Commonwealth of Nations finally came in 1980, when Rhodesia became Zimbabwe.

From the start, Zimbabwe has faced several ethnic dilemmas. First, the course of relations between the country's two largest ethnic groups—the majority **Shona** (82 percent of the total population) and the **Ndebele** (14 percent)—remains uncertain. Many Ndebele people want autonomy for the Ndebele stronghold in the southwest known as Matabeleland. In addition, although most whites left Zimbabwe when it became independent, about 100,000 remained. The black government allowed whites to participate in Zimbabwe's political life, setting an early example to South Africa of an effective transition to majority-black rule. But the apparent harmony has since shattered, particularly over the issue of land ownership. By the mid-

1990s, millions of rural blacks were crowded onto agriculturally inferior **communal lands,** while whites owned about 70 percent of Zimbabwe's arable land. A poorly handled effort to address this imbalance has threatened to tear the country apart.

The problem of disproportionate white land ownership became explosive in 1997, when Zimbabwe's president, Robert Mugabe (of the Shona ethnic group), used the issue to boost his party's prospects in impending parliamentary elections. He announced that the government would seize half of Zimbabwe's white-owned farms and, without compensating the farmers, simply distribute the land to black peasants living in the overcrowded communal areas. Mugabe argued that the former colonial power, Britain, illegally occupied black lands in the first place and should now pay compensation to whites losing those lands. Insisting that the seizures violated its agreements with the government and angry about Zimbabwe's support of the Kinshasa government in the Democratic Republic of Congo's civil war, the International Monetary Fund (IMF) and most other foreign lenders cut off aid to Zimbabwe in 1999.

An economic crisis, so far unabated, followed; the value of the Zimbabwean dollar crashed, inflation soared (at the blistering rate of 7,600 percent in 2007), gold mines shut down, and tourism all but disappeared. Unemployment reached 80 percent, and a tide of illegal Zimbabwean immigrants flowed into South Africa and other countries. Mugabe won a fraudulent election, postponed another, and cracked down hard on political opponents.

Meanwhile blacks, mostly either Mugabe loyalists or war veterans having little or no experience with farming, moved onto the white-owned farms. Virtually all of the country's 4,500 white farmers had been uprooted by 2007, and the 350,000 black workers on the former white farms had lost their jobs. Zimbabwe's supreme court ruled that the farm seizures were illegal, but Mugabe encouraged the squatters to stay. Few had the requisite farming abilities or materials, and the country's agricultural output plummeted. Some white farmers fled to Zambia, where the government welcomed the new agricultural investors with open arms. Zambia hopes that their expertise with tobacco production in particular will boost exports and revenues. Zimbabwe's loss in tourism numbers has also been Zambia's gain, as visitors can view Victoria Falls from either country, and few now think of going to Zimbabwe.

9.1.7 The Indian Ocean Islands: Former Edens

The islands and island groups off the Indian Ocean coast of Africa are unique in their cultures and natural histories. Madagascar, Comoros, Réunion, Mauritius, and the Seychelles have had African, Asian, Arab, European, and even Polynesian ethnic and cultural influences (•Figures 9.1.21 and 9.1.22). As island ecosystems, they are home to many endemic plant and animal species.

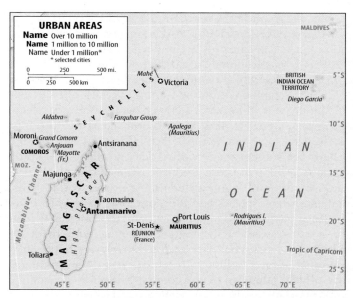

• **Figure 9.1.21** Principal features of the Indian Ocean islands.

• **Figure 9.1.22** The people of Madagascar have a unique culture, with roots in Southeast Asia, the Pacific, and Africa.

Madagascar and the tiny Comoro Islands are former colonies of France, and Réunion remains an overseas possession of France. Mauritius became independent of Britain in 1968 as a constitutional monarchy under the British crown. The Seychelles, which gained independence from Britain in 1976, is a republic within the Commonwealth of Nations.

Madagascar (known in French as La Grande Île) is the fourth-largest island in the world, nearly 1,000 miles (1,600 km) long and about 350 miles (c. 560 km) wide. It lies off the southeast coast of Africa and has geologic formations similar to those of the African mainland. Its distinctive flora and fauna include most of the world's lemur and chameleon species (see Biogeography, page 517). Some of Madagascar's early human inhabitants migrated from the southwestern Pacific region, bringing with them the cultivation of irrigated rice and other culture traits. There was also a later influx of Africans from the mainland. The official language of the country's 18 million people is Malagasy, an Austronesian tongue that has acquired many French, Arabic, and Swahili words.

The east coast of Madagascar rises steeply from the Indian Ocean to heights over 6,000 feet (1,830 m). Because the island lies in the path of trade winds blowing across the Indian Ocean, the east side receives the heaviest rain and has a natural vegetation of tropical rain forest. The remainder of the island has vegetation consisting primarily of savanna grasses with scattered woody growth. An eastern coastal strip is backed by hills and then by escarpments of the central highlands. The most densely populated part of the island is the central High Plateau, where the economy is a combination of growing rice in valleys and raising cattle on higher lands. Antananarivo (formerly Tananarive; population 1.7 million), the capital and largest city, is located in the central highlands. It is connected by rail with Toamasina (formerly Tamatave), the main seaport, located on the east coast.

Madagascar is not as economically developed as other large tropical islands such as Java, Taiwan, and Sri Lanka. Deep economic troubles and political dissent have marked its recent history under a socialist government. Unwise decision making made Madagascar fertile ground in the 1990s for international drug money launderers, loan sharks, and con artists who purloined much of this poor nation's financial assets. Paddy rice is the principal food, and coffee, vanilla (Madagascar is the source of 70 percent of the world supply), cloves, and sugar represent most of the country's limited exports. There is a growing clothing export industry, thanks in large part to the American AGOA initiative (see pages 477–478). Given Madagascar's extraordinary wildlife and scenic beauty, the ecotourism sector could grow considerably, but so far, the government has taken little interest in developing an infrastructure to encourage and sustain tourism.

The Comoro Islands, located about midway between northern Madagascar and northern Mozambique, are of volcanic origin. They have a complex mixture of African, Arab, Malaysian, and European influences. Aside from the island of Mayotte, which has a Christian majority and which, by its own choice, remains a dependency of France, the Comoro group forms the Union of Comoros (usually called just the Comoros), in which Islam is the official religion and both Arabic and French are official languages. The capital is Moroni (population 40,000) on the island of Grand Comoro. France has a defense treaty with the Comoros, which it invoked in 1995 to put down a coup led by a French mercenary. In 1997, two more Comoros islands (Mohéli and Anjouan) declared secession from the national government in Moroni and pleaded to become French dependencies in the hope that they could enjoy the somewhat higher standard of living prevailing on nearby Mayotte. The country implemented a new constitution in 2002, giving each island its own president and self-governance on most issues. The federal presidency is rotated among the three island presidents.

Fishing, subsistence agriculture, and exports of vanilla provide livelihoods to the 700,000 inhabitants of these islands. The coelacanth (pronounced "*see*-luh-kanth"), a primitive fish thought to have become extinct soon after the end of the Mesozoic era (the age of the dinosaurs), has been observed alive in Comoro Islands waters since 1938 (and more recently in Indonesian waters).

Réunion (capital, Saint-Denis; population 130,000), east of Madagascar in the Mascarene island group, is the tropical island home to 800,000 people mainly of French and African origin. The island was uninhabited until the French arrived in the 17th century. Elevations reach 10,000 feet (3,050 m). The volcanic soils are fertile, and tropical crops of great variety are grown. Cane sugar from plantations supplies 75 percent of all exports by value, with rum and molasses (both derived from cane) supplying additional export revenues. Réunion also has a significant international tourist industry.

Sugar is the mainstay of the larger island of nearby Mauritius (capital, Port Louis; population 150,000), also in the Mascarenes. Grown mainly on plantations, sugarcane occupies most of the cultivated fields on this volcanic landscape. Mauritius has a larger proportion of lowland area than Réunion. The island has developed a clothing industry that supplies more export value than sugar. International tourism also produces valuable revenues (about $250 million annually) for its 1.3 million people. The people of Mauritius are mainly descendants of Indians brought in by the British to work the cane fields, along with descendants of 18th-century French planters as well as some blacks, Chinese, and racial mixtures. Mauritius is also home to about 5,000 Ilois, or "islanders," relocated from their homes in the Chagos Islands 1,250 miles (2,000 km) to the northwest. The Chagos make up a British protectorate known as the British Indian Ocean Territory (BIOT). The British administered the islands from Mauritius until 1965, when at the height of the Cold War, it leased the largest island, Diego Garcia, to the U.S. military and required all the Ilois to move out. Since then, U.S. forces have used Diego Garcia

292

BIOGEOGRAPHY

Madagascar and the Theory of Island Biogeography

The entomologist Edward Wilson and the biologist Paul Ehrlich have forecast that if tropical rain forests continue to be cut down at the present rate, a quarter of all of the plant and animal species on earth will become extinct by 2040.* They base their estimate on a model that correlates habitat area with the number of species living in the habitat. This **theory of island biogeography** emerged from observations of island ecosystems in the West Indies. It states that the number of species found on an individual island correlates with the island's area, with a 10-fold increase in area normally resulting in a doubling of the number of species. If island A, for example, is 10 square miles in area and has 50 species, 100-square-mile island B may be expected to support 100 species.

What makes the theory useful in projecting species losses is the inverse of this equation: a 10-fold reduction in area will result in a halving of the number of species; therefore, 1-square-mile island C can be expected to hold only 25 species. In applying the model, ecologists treat habitat areas as if they were islands. Thus if people cut down 90 percent of the tropical rain forest of the Amazon Basin, for example, the theory of island biogeography suggests that they would eliminate half of the species of that ecosystem. Scientists caution that the theory is only a tool meant to help in making rough estimates; the actual number of species lost with habitat removal may be higher or lower.

As a rough guideline, the theory of island biogeography is useful in projecting and attempting to slow the rate of extinction in the world's biodiversity hot spots, such as Madagascar. More than 90 percent of Madagascar's plant and animal species are endemic, occurring nowhere else on earth. Extinction of species was well under way soon after people arrived on the island; the giant flightless elephant bird (*Aepyornis*) was among the early casualties. But human activities, particularly the clearing of forests to grow rice and provide pasture for zebu cattle, are eliminating habitat areas on the island at a faster rate than ever before. Meanwhile, scientists are anxious to learn whether some of Madagascar's remaining plants might be useful in fighting diseases such as AIDS and cancer. Already Madagascar's rosy periwinkle has yielded compounds effective against Hodgkin's disease and lymphocytic leukemia. Other species could become extinct before their useful properties ever become known.

How urgent is the task to study and attempt to protect plant and animal species in Madagascar? Scientists turn to the theory of island biogeography for an answer. Although people have lived on Madagascar for less than 2,000 years, they have succeeded in removing 90 percent of the island's forest, setting the stage for some of the most ruinous erosion seen anywhere on earth (•Figure 9.1.D). The theory of island biogeography suggests that in the process, they have caused the extinction of roughly half of the island's species. With Madagascar's human population on track to double in 23 years, and with pressure on the island's remaining wild habitats expected to increase accordingly, the task of conservation is extremely urgent.

• **Figure 9.1.D** The subsistence needs of a growing human population have had ruinous effects on Madagascar's landscapes and wildlife. Before people came some 2,000 years ago, most of the island was forested, as in the Montagne d'Ambre National Park (top). Today, less than 10 percent of the island is forest, and a characteristic landscape feature of its High Plateau is the erosional feature known locally as *lavaka* (bottom).

*Edward O. Wilson, "Threats to Biodiversity." *Scientific American*, 261(3), 1989, pp. 60–66.

on many military forays, particularly against Iraq during the Gulf War and the later Iraq War and against Afghanistan in the wake of the 9/11 attacks. Now many of the Ilois want to return to their home islands, a move opposed by the United States.

The unusual core of the Seychelles island group is composed of granites, which typically are found only on continental mainlands. Many foods have to be imported to the islands, which are generally steep and lack cultivable land. The country's largest exports consist of petroleum products, although the islands have no crude petroleum and no refining industry; the products are simply imported and then reexported. Fish products are the only other important exports. The Aldabra atoll, far to the south of the main group of islands, is home to a population of giant tortoises and other endemic animals (•Figure 9.1.23).

A tourist boom began in 1971 when an airport was built in the Seychelles (capital, Victoria; population 20,000). The exceptional physical beauty of the islands and the renowned friendliness of its 100,000 people of French, Mauritian, and African descent have made the Seychelles a prime destination among tourists wealthy enough to afford travel to this remote area. Outsiders are drawn to the Seychelles, as

• **Figure 9.1.23** A giant Aldabran tortoise on the beach of Aldabra atoll in the Seychelles. As in the Galápagos Islands, unique life forms have evolved through long periods of isolation from other animal populations.

they have long been to much of sub-Saharan Africa, as an imagined Eden. The realities of land and life in the region actually exceed the imagination and are indeed worthy of the world's attention.

SUMMARY

→ Sub-Saharan Africa can be divided into seven subregions: the Sahel, West Africa, West Central Africa, East Africa, the Horn of Africa, southern Africa, and the Indian Ocean islands.

→ The Sahel region extends eastward from the Cape Verde Islands to the Atlantic shore nations of Mauritania, Senegal, and the Gambia and inland to Mali, Burkina Faso, Niger, and Chad. The name Sahel in Arabic means "coast" or "shore," referring to the region as a front on the great desert "sea" of the Sahara.

→ Between the late 1960s and 1985, the Sahel was subjected to severe droughts, which, in combination with increased human pressure on resources, prompted a process of desertification. Human and natural factors have helped restore the region recently.

→ Oil production in Chad is the most significant recent development in the Sahel countries' economies.

→ West Africa is the region extending from Guinea-Bissau eastward to Nigeria. Most West African countries maintain strong economic, political, and cultural relationships with the European powers that formerly controlled them.

→ The West African countries are heavily dependent on subsistence crop farming, livestock grazing, and a few agricultural and export specialties.

→ The most significant mining development in West Africa has been Nigeria's emergence as a producer and major exporter of oil. There are serious conflicts among ethnic, religious, and political groups in Nigeria, some resulting from the maldistribution of income from the country's oil wealth.

→ West Africa's largest cities are also national capitals and seaports and have developed primarily in its areas of major commercial agriculture.

→ Civil wars have recently taken place in Liberia, Sierra Leone, and Ivory Coast, but all have ended or subsided.

→ The subregion of West Central Africa is flanked by Cameroon and the Central African Republic to the north and the Democratic Republic of Congo to the south.

→ Subsistence agriculture supports people in most areas of West Central Africa. Exports of cash crops and minerals aid the financing of some countries.

→ Originally a colony of Belgium, the Democratic Republic of Congo has faced a number of challenges since its independence in 1960. Due to political mismanagement and war, the resource-rich country is in a state of extreme poverty. It was the epicenter of "Africa's first world war," which concluded with a hopeful peace.

→ East Africa includes the countries of Kenya, Tanzania, Uganda, Rwanda, and Burundi.

→ Subsistence farming is a mainstay of East Africa. Export crops, such as coffee, tea, and cut flowers, provide important revenues. East African nations generally lack mineral reserves.

→ Ethnic rivalries and conflicts have beset all of the East African countries. Warfare between the Hutus and Tutsis of Rwanda and Burundi in the 1990s was the most serious.

→ Kenya suffers from corruption that has prevented the long-suspended resumption of significant foreign aid.

→ Comprised of a great volcanic plateau and adjacent areas, the Horn of Africa includes the countries of Ethiopia, Eritrea, Somalia, and Djibouti.

→ After nearly 30 years of terrorism, guerrilla actions, and civil war, the former Ethiopian province of Eritrea became independent in 1993.

→ In addition to the devastation wrought by war with Eritrea, droughts have stricken Ethiopia periodically since the late 1960s.

→ The United States and Ethiopia recently intervened militarily to suppress Eritrean-backed Islamist forces in Somalia, a country divided along clan lines.

→ Southern Africa includes the countries of Angola, Mozambique, Zimbabwe, Zambia, Malawi, South Africa, Botswana, Swaziland, Lesotho, and Namibia.

→ The dominant power in southern Africa is South Africa. Its economic power, based on its industries and mineral wealth, is comparable to that of the rest of Africa combined.

→ Racial segregation characterized South African life from 1652 onward, but it was systematized after 1948 under a body of laws known as apartheid.

→ In 1994, Nelson Mandela was elected president of South Africa after an all-race election was held in the country. After years of conflict, this historic election ended apartheid and white European political control. Reconciliation between the races is continuing. The ranks of middle- and upper-class blacks are growing, but the gap between the haves and have-nots in South Africa is huge.

→ After a long civil war, peace has returned to Angola. This country is becoming an important supplier of oil to the West, but oil exports have benefited few Angolans so far.

→ Zimbabwe has experienced an economic crisis of its own making, triggered when President Mugabe redistributed commercial farms owned by the white minority to blacks who were not farmers.

→ The Indian Ocean islands of Madagascar, the Comoros, Réunion, Mauritius, and the Seychelles exhibit African, Asian, Arab, European, and even Polynesian ethnic and cultural influences. Madagascar has a wealth of endemic species under tremendous pressure by humans, who have cleared 90 percent of the original vegetation cover in the past 2,000 years.

KEY TERMS + CONCEPTS

African National Congress (ANC) (p. 513)

"Africa's first world war" (p. 497)

Afrikaners (p. 512)

albedo (p. 489)

Amhara (p. 504)

Anglophone Africa (p. 493)

apartheid (p. 513)

Bantustans (p. 513)

Barotse (p. 511)

Blackhawk Down incident (p. 506)

Boers (p. 512)

"Bushmen" (p. 512)

Charney effect (p. 489)

chemocline (p. 496)

coloreds (p. 512)

communal lands (p. 515)

desertification (p. 488)

East African Community (p. 500)

Economic Community of West African States (ECOWAS) (p. 494)

Employment Equity Act (p. 514)

Ethiopian Orthodox Christianity (p. 504)

Falashas (Ethiopian Jews) (p. 504)

Francophone Africa (p. 493)

"Galápagos Islands of Religion" (p. 504)

Great Trek (p. 512)

homelands (p. 513)

"Hottentots" (p. 512)

Hutu (Bahutu) (p. 501)

Ibo (Igbo) (p. 491)

Ijaw (p. 491)

Inkatha Freedom Party (IFP) (p. 513)

International Criminal Court (p. 502)

Ivoirité (p. 494)

Khoi (p. 512)

Liberians United for Reconciliation and Democracy (LURD) (p. 494)

Lomé Agreement (p. 493)

Lord's Resistance Army (LRA) (p. 501)

Lozi (p. 511)

Maputo Development Corridor (p. 511)

Mau Mau Rebellion (p. 501)

Moors (p. 488)

Movement for the Emancipation of the Niger Delta (MEND) (p. 491)

Movement for the Survival of the Ogoni People (MOSOP) (p. 491)

National Union for the Total Independence of Angola (UNITA) (p. 510)

national states (p. 513)

native reserves (p. 513)

Ndebele (p. 515)

Ogoni (p. 491)

Operation Restore Hope (p. 506)

Rastafarians (p. 505)

RENAMO (Mozambique National Resistance) (p. 511)

resilience (p. 488)

"resource curse" (p. 487)

Revolutionary United Front (RUF) (p. 493)

Rwandan Patriotic Front (RPF) (p. 502)

San (p. 512)

Sara (p. 488)

Shona (p. 515)

Southern African Development Community (SADC) (p. 506)
South West Africa People's Organization (SWAPO) (p. 511)
theory of island biogeography (p. 517)

Truth and Reconciliation Commission (p. 513)
Tutsi (Watusi) (p. 501)
upwelling (p. 508)
Wahindi (p. 501)

world patrimony site (p. 487)
Xhosa (p. 512)
Yoruba (p. 491)
Zulu (p. 512)

REVIEW QUESTIONS

1. What are the seven subregions of sub-Saharan Africa? What are some of the most significant countries in each of these subregions?

2. In what ways is the Sahel both a physical and cultural boundary zone in Africa?

3. What country has the greatest demographic, political, and economic clout in West Africa? What unique difficulties does this country face?

4. What are the notable natural resource assets of the Democratic Republic of Congo? Why isn't this country richer than it is? What kinds of relationships has it had with its neighbors?

5. How did Idi Amin bring economic ruin to Uganda? What did Daniel arap Moi do to reverse the bright prospects of Kenya? Has his successor lived up to his promises?

6. What was the original distinction between the Hutus and the Tutsis? What issues brought them into conflict in the 1990s?

7. Why is Ethiopia known as the "Galápagos Islands of Religion"?

8. How has Somalia's instability involved foreign powers?

9. What are the main ethnic groups in South Africa? What were their traditional social and economic roles? How have these changed since the end of apartheid?

10. What were the main causes of the Anglo-Boer War of 1899?

11. What tensions emerged between blacks and whites over the issue of land tenure in Zimbabwe? What effects has government policy had on land reform?

12. Who owns and who uses the island of Diego Garcia? What is its importance in military affairs? What are the principal characteristics and livelihoods of the other Indian Ocean islands?

A "typical" Latin American cultural landscape: A Roman Catholic Cathedral dominates a settlement in a rugged environment (here, the southern Sierra Madre range of Mexico).

CHAPTER 10
A GEOGRAPHIC PROFILE OF LATIN AMERICA

The land portion of the Western Hemisphere south and southeast of the United States is commonly known as Latin America. This name reflects the importance of cultural traits inherited mainly from the European colonizing nations of Spain and Portugal, whose languages evolved from Latin. Native American cultures were flourishing when Columbus arrived in the Americas in 1492, soon to be followed by an onslaught of other Europeans. The native cultures still persist to varying degrees and are especially strong in Mexico and Guatemala and in several South American countries. But "Latin" influences predominate today in the region as a whole.

This chapter acquaints readers with the varied indigenous and European cultures that have shaped the region. The exceptional topographic and environmental variations in Latin America are introduced here too. The economic prospects for this developing region are good and are tied particularly closely to relationships with the United States. These issues are described here, along with broader

chapter
objectives

This chapter should enable you to

→ Appreciate how topographic variety creates a predictable range of environmental conditions and livelihood opportunities

→ Know how diverse and accomplished the indigenous cultures of the region were, and how European conquest and colonization decimated and changed these cultures

→ Recognize the predominant ethnic patterns of the region and how ethnicity correlates with livelihood, wealth, and political power

→ Evaluate the region's efforts to shift from raw-material exports to manufacturing through participation in free-trade agreements

→ Understand how U.S. interests have shaped the region's political and economic systems

involvement of the United States in affairs "south of the border." Exploration of the subregions and countries of Latin America continues in Modules 10.1 and 10.2.

10.1 Area and Population

With its 38 countries, Latin America consists of a mainland region that extends from Mexico south to Argentina and Chile, together with the islands of the Caribbean Sea (•Figure 10.1; Table 10.1). Its maximum latitudinal extent of more than 85 degrees, or nearly 5,900 miles (c. 9,500 km), is greater than that of any other major world region. Its maximum east-west measurement, amounting to more than 82 degrees of longitude, is also impressive.

Latin America's two main subregions are neatly offset from one another geographically. The northern part, known as Middle America, includes Mexico; the countries of Central America (Guatemala, Belize, El Salvador, Honduras, Nicaragua, Costa Rica, and Panama); Haiti and the Dominican Republic on the island of Hispaniola; Cuba; and the smaller islands of the Caribbean. It trends sharply northwest to southeast before intersecting with the north-south orientation of the continent of South America. South America protrudes much farther into the Atlantic Ocean than the Caribbean realm or Latin America's northern neighbor, North America. In fact, the meridian of 80 degrees west

longitude (80°W), which touches the west coast of South America in Ecuador and Peru, passes through Pittsburgh, Pennsylvania. Brazil also lies less than 2,000 miles (3,200 km) west of Africa, to which it was joined in the supercontinent of Pangaea until around 110 million years ago.

Latin America has a land area of slightly more than 7.9 million square miles (20.5 million sq km), with South America accounting for 6,893,000 square miles (17,869,000 sq km) and Middle America 1,047,000 square miles (2,713,000 sq km). The total is about 2.5 times the size of the conterminous United States (•Figure 10.2).

Latin America's 2007 population of 569 million represented 8.5 percent of the world total. The population has uneven distributions and densities (•Figure 10.3). Most of Latin America's people are packed into two major geographic alignments. The larger of these two areas, known as "the rim" or "the Rimland," is a discontinuous ring around the margins of South America. The second, generally a highland, extends along a volcanic belt from central Mexico southward into Central America.

About two-thirds of Latin America's people live along the South American Rimland. There are two major segments of the rim. The much larger segment, in terms of both population and area, extends along the eastern margin of the continent from the mouth of the Amazon River in Brazil southward to the humid **pampa** (subtropical grassland) around Buenos Aires, Argentina. The second segment is located partly on the coast and partly in the high valleys and plateaus of the adjacent Andes Mountains. It stretches around the north end and down the west side of South America. This crescent begins in the vicinity of Caracas, Venezuela, on the Atlantic coast and arches all the way around to the vicinity of Santiago, Chile, on the Pacific side of the continent.

This second segment is more fragmented, broken in many places by steep Andean slopes and coastal desert. A strip of hot, rainy, and thinly populated coast lies between Caracas and the mouth of the Amazon River. In the far south on the Atlantic side of the continent, the population rim is again broken in the rugged, relatively inaccessible, rain-swept southern Andes and the dry lands of Argentina's Patagonia in the Andean rain shadow. These territories have only a sparse human population, far exceeded by the millions of sheep that graze in Patagonia and adjacent Tierra del Fuego.

The second major alignment of populated areas in Latin America lies on the mainland of Middle America. It runs along an axis of volcanic land dominating central Mexico and from there reaches southeastward through southern Mexico and along the Pacific side of Central America to Costa Rica. This belt is characterized by good soils, enough rainfall for crops, and elevations high enough in most places to moderate the tropical heat. In Pacific Central America, more people live in the highlands than the lowlands.

Average population densities disguise the fact that nearly every Latin American mainland country reflects a basic spatial configuration: a well-defined population core (or multiple cores) with one or more outlying, sparsely popu-

● **Figure 10.1** Principal features of Latin America.

lated **hinterlands** (peripheries). Densities within the cores are often greater than anything found within areas of comparable size in the United States. On the other hand, most of the mainland countries have a larger proportion of sparsely populated terrain than is true of the United States. Figures on population density for the greater part of Latin America are extraordinarily low, averaging fewer than two persons per square mile over approximately half of the entire region.

TABLE 10.1 Latin America: Basic Data

Political Unit	Area (thousands; sq mi)	Area (thousands; sq km)	Estimated Population (millions)	Estimated Population Density (sq mi)	Estimated Population Density (sq km)	Annual rate of Natural Increase (%)	Human Development Index	Urban Population (%)	Arable Land (%)	Per Capita GNI PPP ($US)
Middle America	**957.5**	**2,479.9**	**147.6**	**154**	**60**	**1.8**	**0.792**	**68**	**12**	**9,700**
Belize	8.9	23.1	0.3	34	13	2.3	0.751	49	3	6,640
Costa Rica	19.7	51.0	4.5	228	88	1.2	0.841	59	4	10,770
El Salvador	8.1	21.0	6.9	852	329	1.9	0.729	58	32	5,340
Guatemala	42.0	108.8	13.4	319	123	2.8	0.673	39	12	4,800
Honduras	43.3	112.1	7.1	164	63	2.1	0.683	47	9	3,540
Mexico	756.1	1,958.3	106.5	141	54	1.7	0.821	75	13	11,330
Nicaragua	50.2	130.0	5.6	112	43	2.3	0.698	58	16	4,010
Panama	29.2	75.6	3.3	113	44	1.5	0.809	62	7	7,680
Caribbean	**90.3**	**233.9**	**40.2**	**445**	**172**	**1.1**	**0.619**	**61**	**25**	**5,630**
Anguilla (U.K.)	0.0	0.1	0.01	250	97	N/A	N/A	100	0	N/A
Antigua and Barbuda	0.2	0.5	0.1	500	193	1.5	0.808	37	18	13,490
Aruba (Neth.)	0.1	0.2	0.07	1,000	386	N/A	N/A	51	10	N/A
Bahamas	5.4	14.0	0.3	56	21	1.0	0.825	89	1	N/A
Barbados	0.2	0.5	0.3	1,500	579	0.6	0.879	50	37	N/A
Cayman Islands (U.K.)	0.1	0.3	0.04	400	154	N/A	N/A	0	4	N/A
Cuba	42.8	110.9	11.2	262	101	0.3	0.826	75	33	N/A
Dominica	0.3	0.8	0.1	333	129	1.7	0.793	71	6	6,490
Dominican Republic	18.8	48.7	9.4	500	193	1.8	0.751	64	22	8,290
Grenada	0.1	0.3	0.1	1,000	386	1.2	0.762	39	6	7,810
Guadeloupe (Fr.)	0.7	1.8	0.5	714	276	1.0	N/A	100	11	N/A
Haiti	10.7	27.7	9.0	841	325	1.8	0.482	36	28	1,490
Jamaica	4.2	10.9	2.7	643	248	1.1	0.724	52	16	4,030
Martinique (Fr.)	0.4	1.0	0.4	1,000	386	0.7	N/A	95	10	N/A
Netherlands Antilles (Neth.)	0.1	0.3	0.2	2,000	772	0.5	N/A	69	10	N/A
Puerto Rico (U.S.)	3.5	9.1	3.9	1,114	430	0.5	N/A	71	4	N/A
St. Kitts and Nevis	0.1	0.3	0.05	500	193	1.0	0.825	33	19	12,680
St. Lucia	0.2	0.5	0.2	1,000	386	0.8	0.790	30	6	6,960
St. Vincent and the Grenadines	0.2	0.5	0.1	500	193	1.1	0.759	44	18	7,000
Trinidad and Tobago	2.0	5.2	1.4	700	270	0.6	0.809	74	14	16,250
Turks and Caicos Islands (U.K.)	0.1	0.3	0.02	200	77	N/A	N/A	46	2	N/A
Virgin Islands (U.S.)	0.1	0.3	0.1	1,000	386	N/A	N/A	47	12	N/A
South America	**6,898.4**	**17,866.9**	**381.1**	**37**	**14**	**1.5**	**0.795**	**79**	**6**	**8,790**
Argentina	1,073.5	2,780.4	39.4	37	14	1.2	0.863	89	12	15,390
Bolivia	424.2	1,098.7	9.8	23	9	2.1	0.692	63	2	2,890
Brazil	3,300.2	8,547.5	189.3	57	22	1.4	0.792	81	7	8,800

(continued next page)

TABLE 10.1 Latin America: Basic Data *(continued)*

Political Unit	Area (thousands; sq mi)	Area (thousands; sq km)	Estimated Population (millions)	Estimated Population Density (sq mi)	Estimated Population Density (sq km)	Annual rate of Natural Increase (%)	Human Development Index	Urban Population (%)	Arable Land (%)	Per Capita GNI PPP ($US)
South America *(continued)*										
Chile	292.1	756.5	16.6	57	22	1.0	0.859	87	2	11,260
Colombia	439.7	1,138.8	46.2	105	41	1.5	0.790	71	2	7,620
Ecuador	109.5	283.6	13.5	123	48	2.0	0.765	61	6	4,400
Falkland Islands (U.K.)	4.7	12.2	0.003	1	0	N/A	N/A	85	0	N/A
French Guiana (Fr.)	34.7	89.9	0.2	6	2	2.7	N/A	75	0	N/A
Guyana	83.0	215.0	0.8	10	4	1.2	0.725	36	2	4,680
Paraguay	157.0	406.6	6.1	39	15	2.1	0.757	54	7	5,070
Peru	496.2	1,285.2	27.9	56	22	1.5	0.767	72	3	6,070
Suriname	63.0	163.2	0.5	8	3	1.4	0.759	69	0	8,120
Uruguay	68.5	177.4	3.3	48	19	0.6	0.851	93	7	11,150
Venezuela	352.1	911.9	27.5	78	30	1.7	0.784	87	3	7,440
Summary Total	**7,946.2**	**20,580.7**	**568.9**	**72**	**28**	**1.5**	**0.781**	**74**	**7**	**8,800**

Sources: World Population Data Sheet, Population Reference Bureau, 2007; Human Development Report, United Nations, 2006; World Factbook, CIA, 2007.

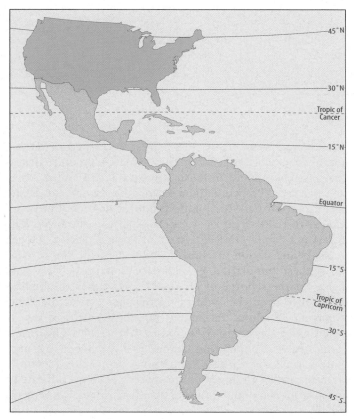

● **Figure 10.2** Latin America's size and latitude relative to the conterminous United States.

The pattern of core and hinterland is especially noticeable in Brazil and Argentina (see Figure 10.3). Most Brazilians live along or near the eastern seaboard south of Belém, a port city in northeastern Brazil, whereas large areas in the interior are still thinly populated. In Argentina, about three-fourths of the population is clustered in Buenos Aires and the adjacent humid pampas. Mexico and the Andean countries also have this kind of spatial imbalance in their populations.

Not every country, however, displays this pattern of core and hinterland. El Salvador and Costa Rica (both with densely settled volcanic lands) and most Caribbean islands are different. The combined populations of the Caribbean is much less than that of central Mexico. But these islands have heavy population densities, grown over a long period of expanding population on restricted territories. The crowding is often aggravated by high rates of natural population increase or steep slopes limiting opportunities for farming and settlement.

Very high rates of urban growth have characterized Latin America in recent times. Overall, the region is 76 percent urban, compared to the world average of 49 percent. A major element in the rapid growth of Latin American cities is rural-to-urban migration caused by both push and pull factors. 57

Most Latin American countries are in the second stage of the demographic transition, clearly recognizable as less developed countries (LDCs) having relatively high (although declining) birth rates and low death rates due to advances in the spread of medical technologies. The highest population 54

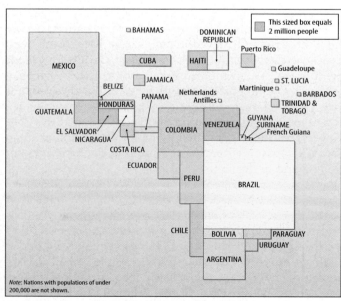

• **Figure 10.3** Population distribution (top) and population cartogram (bottom) of Latin America.

growth rates tend to be in the least developed countries, notably those of Central America; Guatemala and Nicaragua have the highest in Latin America at 2.8 and 2.3 percent per year (French Guiana, owned by France, has a rate of 2.7 percent per year), and Honduras has the lowest per capita GNI PPP of all the Central American countries. The lowest per capita GNI PPP in all of Latin America belongs to the island nation of Haiti. It might be expected to have the highest population growth rate and probably would if not for the scourge of HIV/AIDS. Haiti has the highest infection rate in the Western Hemisphere, affecting some 4 percent of adults. Its annual population growth rate is held down by losses to AIDS to just 1.8 percent. The lowest annual rate of population change is 0.3 percent in Cuba, thanks more to steadfast Communist promotion of family planning rather than general prosperity. The overall rate of increase for the population of Latin America is 1.5 percent, which is the average for the world's less developed countries. This regional rate of increase is down considerably from the high of 3 percent in 1960, thanks in large part to the fact that the region's economies are generally improving, albeit slowly.

10.2 Physical Geography and Human Adaptations

Dramatic differences in elevation, topography, biomes, and climates characterize Latin America (•Figure 10.4). Low-lying plains drained by the Orinoco, Amazon, and Paraná-Paraguay River systems dominate the north and central part of South America and separate geologically older, lower highlands in the east from the rugged Andes of the west. In Mexico, a high interior plateau broken into many basins lies between north-to-south-trending arms of the Sierra Madre. High mountains, largely within the Sierra Madre and the Andes, form a nearly continuous landscape feature from northern Mexico to Tierra del Fuego at the southern tip of South America. Most of the smaller islands of the Caribbean Sea's West Indies are volcanic mountains, although some islands of limestone or coral are lower and flatter. The largest islands of the Caribbean have a more diverse topography, including low mountains.

Climates and Vegetation

The climatic and biotic diversity of Latin America is extraordinary (•Figure 10.5). Even within some single countries (Ecuador for example), conditions range over a very short horizontal distance from sea-level tropical rain forests to alpine tundra. The tropical rain forest climate and biome, with heavy year-round rainfall, continuous heat and humidity, and superabundant vegetation dominated by large broadleaf trees, is generally a lowland type lying mainly along and near the equator, with segments extending to the tropical margins of the Northern and Southern Hemispheres. The largest segment, representing the world's largest continuous expanse of tropical rain forest, lies in the basin of the Amazon River system. Additional areas are in southeastern Brazil, eastern Panama, the western coastal plain of Colombia, on the Caribbean side of Central America and southern Mexico, and along the eastern (windward) shores of some Caribbean islands, particularly Hispaniola and Puerto Rico. Much of the original tropical rain forest vegetation of the islands has been lost to human activity, and even the vast Amazon forest is under considerable pressure (this problem, which is largely a reflection of Brazil's development, is described on pages 585–587).

On either side of the principal region of tropical rain forest climate, the tropical savanna climate extends to the vicinity of the Tropic of Capricorn in the Southern Hemisphere and, with more fragmentation, to the Tropic of Cancer in the Northern Hemisphere. Still farther poleward, in the eastern portion of South America, lies a large area of humid subtropical climate, with cool winters unknown in the tropical zones. Its Northern Hemisphere counterpart is north of the Mexican border in the southeastern United States. In South America, this climate is associated mainly with prairie grasses in the humid pampas of Argentina, Uruguay, and extreme southern Brazil. On the Pacific side of South America, a small strip of Mediterranean or dry-summer subtropical climate in central Chile is similar to that in southern California. This is an ideal climate for wine production, and Chilean wines are now well established on world markets.

The humid climates of Latin America have a fairly predictable spatial arrangement. But the region's dry climates and biomes—desert and steppe—are the product of local circumstances, such as rain shadows or nearby cool ocean water. In northern Mexico, aridity is due partly to the rain shadow effect of high mountain ranges on either side of the Mexican plateau. These dry climates are also partly associated with the global pattern of semipermanent zones of high pressure that create arid conditions in many areas of the world along the Tropics of Cancer and Capricorn.

In Argentina, the extremely high and continuous Andean mountain wall accounts for the aridity of large areas, particularly the southern region called Patagonia. Here, the Andes block the path of the prevailing westerly winds, creating heavy orographic precipitation on the Chilean side of the border but leaving Patagonia in rain shadow. However, in the west coast tropics and subtropics of South America, the Atacama and Peruvian deserts cannot be explained so simply. Here, shifting winds that parallel the coast, cold offshore currents, and other complexities, as well as the Andes Mountains, combine to create the world's driest area. The mountains restrict this area of desert to the coastal strip. Arid and semiarid conditions also prevail along the northernmost coastal regions of Colombia and Venezuela and in the region called the Sertão in northeastern Brazil.

Elevation and Land Use

One of the most significant features of Latin America's physical geography with respect to human adaptations is a series of highland climates arranged into zones by elevation. This zonation results from the fact that air temperature decreases with elevation at a normal rate of approximately 3.6°F (1.7°C) per 1,000 feet (c. 300 m). At least four major zones are commonly recognized in Latin America (•Figures 10.6 and 10.7): the *tierra caliente* (hot country), the *tierra templada* (cool country), the *tierra fría* (cold country), and the *tierra helada* (frost country). At the foot of the highlands, the *tierra caliente* is a zone embracing the tropical rain forest and tropical savanna climates. The zone reaches upward to approximately 3,000 feet (c. 900 m) above sea level at or near the equator and to slightly lower elevations in parts of Mexico and other areas near the margins of the tropics. In this hot, wet environment, the favored crops are rice, sugarcane, bananas, and cacao. Latin America's blacks are concentrated in many of the *tierra caliente* zones, a legacy of

• **Figure 10.4** Reference map of the topographic features, rivers, and seas of Latin America.

the slave trade when they were forced to work the region's plantations.

The *tierra caliente* merges almost imperceptibly into the *tierra templada*, which flanks the rugged western mountain ranges and is the uppermost climate in the lower uplands and highlands to the east. Sugarcane, cacao, bananas, oranges, and other lowland products reach their uppermost limits in the *tierra templada*, but this zone is most fa-

• **Figure 10.5** Climates (left) and biomes (right) of Latin America.

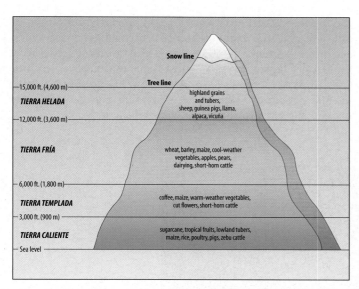

• **Figure 10.6** Altitudinal zonation in Latin America. Specific patterns of land use have evolved to take advantage of the region's diverse altitudinal zones. This graph shows elevations and general land use patterns, which may also be compared with land uses in Figure 10.7. Elevations shown here are for the equatorial region and are somewhat lower toward the poles for the respective land uses.

• **Figure 10.7** Land use in Latin America reflects a remarkable range of opportunities posed by varying latitudes and elevations.

mous as coffee habitat (•Figure 10.8). The upper limits of this zone—approximately 6,000 feet (1,800 m) above sea level—tend to be the upper limit of European-introduced plantation agriculture and modern commercial crops in Latin America. Densely inhabited sections occupy large areas in southeastern Brazil, Colombia, Central America, and

Mexico. Although broadleaf evergreen trees characterize the moister, hotter parts of this zone, coniferous evergreens replace them in parts of the zone's poleward margins. In places such as the highlands of Brazil and Venezuela,

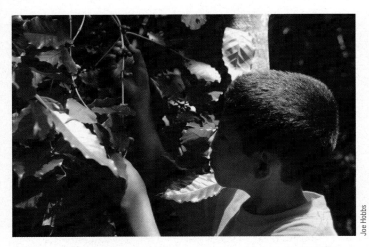

• **Figure 10.8** Tending the coffee crop on a *finca* (plantation) in the *tierra templada* of southern Mexico.

where there is less moisture, scrub forest or savanna grasses prevail.

Seven metropolises exceeding 2 million in population—São Paulo, Belo Horizonte, Brasília, Caracas, Medellín, Cali, and Guadalajara—are in the *tierra templada,* and two others—Mexico City and Bogotá—lie slightly above it. Four smaller cities that are national capitals and the largest cities in their respective countries are located in the *tierra templada:* Guatemala City, San Salvador, Tegucigalpa, and San José. Others, like Rio de Janeiro, which are situated at lower elevations, have close ties with predominantly residential or resort towns in these cooler temperature areas.

The *tierra fría* is comprised of high plateaus, basins, valleys, and mountain slopes within the great mountain chain that extends from northern Mexico to Cape Horn. By far the largest areas are in the Andes, with significant areas also in Mexico (•Figure 10.9). Around the equator, the *tierra fría* begins at about 6,000 feet (c. 1,800 m) and extends upward to the **upper limit of agriculture** (represented by such hardy crops as potatoes and barley) and the **tree line** (the upper limit of natural tree growth) at about 12,000 feet (c. 3,600 m).

The *tierra fría* experiences frost and is often the habitat of a Native American economy with a strong subsistence component. In a classic process of marginalization, European colonization of Latin America drove some Native American settlements upslope and into the *tierra fría* zone, although some major populations (notably the Inca of Peru) had already selected upland locations for their settlement before the arrival of Columbus. These upland settlements are most extensive in Ecuador, Peru, and Bolivia and are also very evident in Colombia, Guatemala, and southern Mexico.

Another zone lies above the other three and consists of the alpine meadows, known as **páramos,** along with still higher barren rocks and permanent fields of snow and ice (•Figure 10.10). This is the *tierra helada.* It generally lies between 12,000 feet (c. 3,600 m) and the lower edge of the snow line. It supports some grains and livestock (llama, alpaca, sheep) but is largely above the mountain flanks that are central to upland native settlement and agriculture. In the highlands of Latin America, the *tierra fría* tends to be a last retreat and a major home of the indigenous peoples, except in Guatemala and Mexico. Most of their settlements are rural villages based on subsistence agriculture. In some cases, notably in Bolivia and Peru, valuable minerals like tin and copper are located here, attracting large-scale mining enterprises into the *tierra fría* and the *tierra helada.*

• **Figure 10.9** Characteristic *tierra fría* landscape and agriculture in Andean South America.

● **Figure 10.10** The *páramo* at about 12,000 feet (c. 3,600 m) in the Andes of Colombia.

Natural Hazards in Latin America

21 Adjoining a large section of the Pacific Ring of Fire and fronting two seasonal **hurricane** regions, Latin America is beset by natural hazards. A belt of land stretching from northern Mexico all the way to Tierra del Fuego, the "Land of Fire" at the southern tip of South America, has a violent history of earthquakes and volcanic eruptions. Hurricanes originating in the Pacific but most especially in the Atlantic have had sometimes devastating impacts. A single storm can wreck the economy of an island nation, as Hurricane Ivan did by wiping out Grenada's nutmeg crop in 2004. The monstrous Hurricane Mitch (category 5, highest on the Beaufort scale) was an unusually late storm that made landfall in Honduras on October 28, 1998, and roamed for five days over that country and Nicaragua, El Salvador, Guatemala, and Mexico. Its ferocious winds and rainfalls of up 2 feet (60 cm) brought floods, landslides, and storm damage—intensified by the region's widespread deforestation—that killed an estimated 15,000 people; inflicted enormous damage on property, crops, roads, power, and other infrastructure; and left disease epidemics in its wake. As if Mitch were not enough, the region had also suffered from drought and fires visited on it in 1997 and 1998 by El Niño (see Natural Hazards, page 532).

603, 662

10.3 Cultural and Historical Geographies

It is perhaps unfortunate that this region came to be known as Latin America because there were no "Latins" among its inhabitants before the end of the 15th century. In 1492, when the first Europeans arrived (perhaps after seafaring Polynesians reached the west coast of South America; see page 423), the region was home to an estimated 50 to 100 million **Native Americans**, also known as **Amerindians** or, as Columbus (who thought he had landed in India) misidentified them, **Indians.** After their early migrations starting around 10,000 B.C.E. from the Asian continent, apparently across a land bridge now drowned by the Bering Strait, they developed many distinctive livelihoods and cultures. Some Native Americans practiced hunting and gathering, while others developed agriculture and took the same kind of pathway to urban life and civilization followed by several Old World cultures. Culture hearths associated with civilization emerged in the Andes region of South America and in southern Mexico and adjacent Central America (●Figure 10.11).

41

Civilizations Predating the Europeans' Arrival

In 1492, as they do today, the **Maya** inhabited southern Mexico, Belize, and Guatemala. They practiced agriculture based on maize (corn), squash, beans, and chili peppers and had by 200 C.E. developed a highly complex civilization in both lowland tropical rain forests and highland volcanic regions. They created monumental religious and residential structures, including pyramids, temples, and astronomical observatories (●Figure 10.12). They built stone roadways through dense forest areas but notably had no wheeled vehicles, nor did they have the sailing vessels, the plow, beasts of burden, and some other tools and trappings associated with civilizations elsewhere. The Maya had highly developed systems of mathematics, astronomy, and engineering and an extremely precise calendar (which ends, some say ominously, on December 21, 2012).

The Maya also had a hieroglyphic writing system that survives on stone monuments and in a handful of books. Regarding them as heretical, Spanish priests put almost all of the written volumes to the torch. On July 12, 1562, Bishop Diego de Landa destroyed at least 30 Mayan books in a bonfire outside a church in Mexico's Yucatán Peninsula. "They contained nothing in which there were not to be seen superstitions and falsehoods of the devil," de Landa later wrote. "We burned them all." Many Maya watched this event, which, de Landa recalled, "they regretted to an amazing degree and which caused them much affliction."[1]

The Maya civilization peaked around 900 C.E. and then began a slow decline. It is still unknown why this advanced civilization apparently abandoned its massive religious and urban centers and melted back into the forests of the Yucatán Peninsula and points south. Theories include natural disasters, agricultural failures, and revolts against ruling elites. When the Spaniards arrived, the great Maya cities were overgrown. Many can be seen in their restored glory today. Chichén Itzá in Mexico's Yucatán Peninsula and Tikal (see Figure 10.12) in northern Guatemala are very significant sources of international tourism revenue for the respective countries.

NATURAL HAZARDS

El Niño

Every few years, things seem to go wrong with the winds and waters of the Pacific Ocean (●Figure 10.A). Normally, the winds blow east to west, helping to promote the upwelling of cold, nutrient-rich water from the depths to the surface of the tropical eastern Pacific. But sometimes, usually beginning in December, the winds reverse direction, blowing from west to east, suppressing the upwelling water and thus raising the surface temperature of the water by as much as 20°F (11°C). Because of its common occurrence in December, that event has been dubbed **El Niño**, meaning "the Baby" in reference to the Christ child, whose birth is celebrated in December. Less prosaically, meteorologists know it as **El Niño Southern Oscillation (ENSO).**

El Niño conditions sometimes last a year or more, causing global climatic disruptions. These vary between events, but the typical El Niño pattern is illustrated in ●Figure 10.B. Closest to the source of the condition, Peru experiences torrential rains. Unusually high rainfall also occurs in southern Brazil, the southeastern United States, western Polynesia, and East Africa. Drought or unusually high temperatures settle in over northern Brazil, southeastern Canada and the northeastern United States, southern Alaska and western Canada, the Korean Peninsula and northeastern China, Indonesia and northern Australia, and southeastern Africa.

When warmed, Pacific waters are far less rich in nutrients, so typically, fish catches plummet and food chains are disrupted, leading to massive wildlife losses. The most severe recent El Niño events occurred in 1982–1983 and in 1997–1998, but it is not unusual for three or four moderate events to occur every decade.

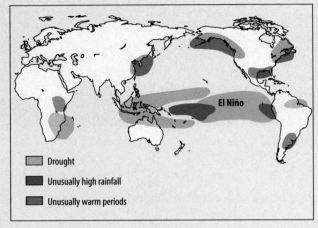

● **Figure 10.B** Climatic impacts of El Niño.

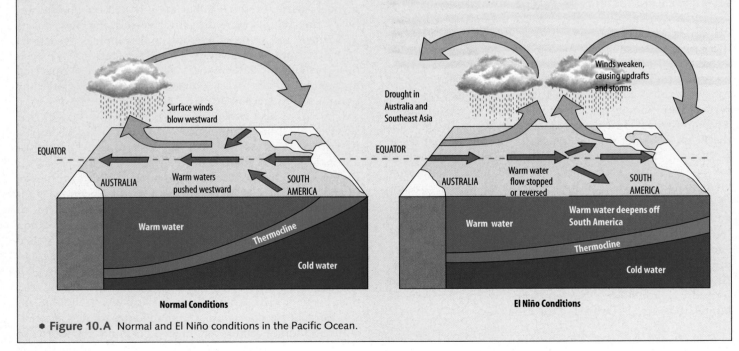

● **Figure 10.A** Normal and El Niño conditions in the Pacific Ocean.

North of the Maya realm, in the Valley of Mexico about 30 miles (40 km) from present-day Mexico City, arose Teotihuacán, the first true urban center in the Western Hemisphere. Its construction began about 2,000 years ago, and by 500 C.E., it covered about 8 square miles (21 sq km) and had 200,000 residents, comparable to London a mil-

lennium later. The city was laid out on a grid aligned with stars and constellations that were central to the ceremonial timekeeping of the **Teotihuacános.** There were some 2,000 apartment compounds, a 20-story Pyramid of the Sun, a Pyramid of the Moon, and a 39-acre (16,000-sq m) civic and religious complex called the Great Compound or Cita-

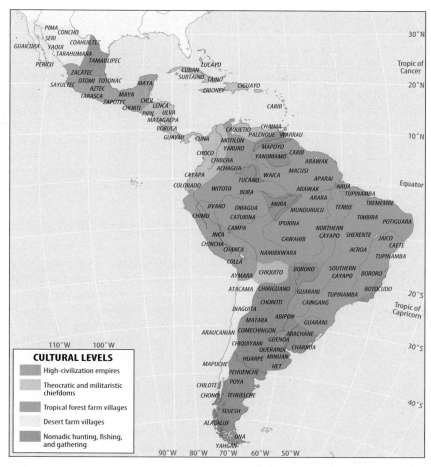

● **Figure 10.11** Major Native American groups and civilizations in Latin America on the eve of the Spanish conquest.

● **Figure 10.12** In the late classical period of the Maya civilization, about 650 D/F/, an estimated 70,000 people lived in Tikal.

● **Figure 10.13** Teotihuacán's Pyramid of the Sun.

del (●Figure 10.13). The origins, language, and culture of Teotihuacános are still not fully understood, and it is hoped that future archaeology will shed more light on these extraordinarily creative people.

The **Aztecs,** who called themselves the **Mexica** (pronounced "meh-*shee*-kuh") and came later than the Teotihuacános, honored the earlier city as sacred space and a religious center when they took over the Valley of Mexico in the 1400s. They founded their capital, Tenochtitlán (now overlain by Mexico City), in 1325. With a population of about 250,000, Tenochtitlán had a sacred center surrounded by active marketplaces, all linked by stone-surfaced roads to other parts of the urban settlement. There were some outlying districts that provided goods for trade, and craftspeople and merchants came to the city to work. In the middle of the 15th century, the Aztecs joined with rulers of two neighboring cities to form the **Triple Alliance.** From this base, the Aztecs and their allies developed an empire stretching across central Mexico. They tried but failed to conquer the kingdom of an adjacent civilization, the **Tarascan.** The Aztecs left a legacy of much merit in metallurgy (which diffused to Mexico from the Andean region), woodworking, weaving, pottery, and especially urban design.

In South America, the **Inca** built an empire about 2,000 miles (3,200 km) long from northern Ecuador to central Chile. They controlled it for only about a century before 1532, when the Spanish conquistadors, under Francisco Pizarro, began their subjugation. The Inca's origins, dating to about 1200 C.E., are not fully known. From their governing center in Cuzco, Peru, the Inca engineered a system of roads, suspension bridges, and settlements that connected and supported their empire. The road system ran more than 2,500 miles (c. 4,100 km) and was busy with movement of trade and tribute. The Inca also achieved great skill in stoneworking, terrace construction, and irrigation networks (•Figure 10.14). Remarkably, they had neither paper nor a writing system but kept mathematical records by knotted ropes called *quipus* (pronounced "*kee*-pooz").

Still another group, the little-known **Nazca** culture, left a unique legacy on the landscape of South America about 2,000 years ago. Their "Nazca lines" in the desert of southern Peru is a complex of nearly 200 square miles (c. 520 sq km) of massive carvings in the sandy surface, depicting birds, insects, and fish, the smallest of which is an animal figure some 80 feet (c. 23 m) in length. These designs can only be appreciated from the air, giving rise to

the improbable theory that they were built as landmarks for "ancient astronauts." No written records of the Nazca culture survive, so the purpose of their lines may never become known. Another of the less celebrated civilizations of South America is the **Chibcha** of what is now Colombia. The Chibcha lived in small agricultural villages rather than large cities, and their exquisite gold work is a major legacy of their culture. Speakers of Chibchan languages now live in the Andes of Colombia and Ecuador and as far north as Costa Rica. Beneath the dominant imported languages of Spanish and Portuguese, the linguistic substrate of Latin America is still rich.

Languages in Latin America

With the exception of the Nazcans (whose record ends about 600 C.E.), the groups just discussed were only the largest major civilizations at the time of the Spanish conquest. There were smaller civilizations and others that predated the conquest. Far more numerous were the groups that practiced hunting and gathering or agriculture but did not build cities and civilizations. Their legacy survives today in the cultures and languages of a rich variety of Native American ethnic groups. Linguistic geographers disagree sharply over Native American language distributions, so only very general patterns are depicted in •Figure 10.15.

Six Native American language families are represented in Mexico and Central America. The **Hokan-Siouan language family** has speakers in Mexico's Baha Peninsula and

• **Figure 10.14** Machu Picchu, the legendary "lost city of the Inca," has a dramatic site above Peru's Urubamba Valley.

IRA BLOCK/National Geographic Image Collection

INDO-EUROPEAN FAMILY

DUTCH

ENGLISH

FRENCH

PORTUGUESE

SPANISH

OTHERS

/// GUARANÍ

/// QUECHUA

/// Additional areas of significant native language speakers

• **Figure 10.15** Indigenous and European languages of Latin America.

in El Salvador and adjacent Honduras. The **Aztec-Tanoan language family** includes the Nahuatl that the Aztecs spoke and that is still spoken in northern Mexico today. The **Oto-Manguean language family** includes seven languages spoken mainly in northern Mexico. There is a small **Totonac language family** region on the Gulf of Mexico. Southern Mexico is home to the **Penutian** and **Mayan language families.** The indigenous cultures these languages represent form a significant minority overall in Mexico at 30 percent, but they comprise over 50 percent in some southern states of the country. The Mayan language family has five major subfamilies, representing more than 60 million speakers living all across the realm of the ancient Maya, especially in Guatemala. Some Maya, particularly in highland Guatemala, never did mix with outsiders and are ethnically little changed from their forebears. The Maya of highland Guatemala tend to speak Mayan as their first language, but many are bilingual with Spanish. With its 23 Maya groups, Guatemala's population is more than 40 percent Native American overall.

South America—which had a Native American population as early as 12,500 years ago—has an even richer language palette, with about 600 languages belonging to 16 families. They include the **Quechu-Aymaran language family,** to which the language of the ancient Inca belonged. More than 7 million people today, mainly in Peru but also in Colombia, Ecuador, Bolivia, and Argentina, speak one of the five languages in this family. Most still speak **Quechua** as a first language, but many are bilingual with Spanish.

Indigenous languages are used far more widely in South America than elsewhere in Latin America, and there are more speakers there that do not speak a second language. Very high proportions of the populations in highland South America are Native Americans. The largest populations of Native Americans are found in Bolivia (55 percent indigenous) and Peru (45 percent), with smaller but still significant numbers in the basin of the Amazon River and in Panama, where scattered groups of lowland peoples maintain their traditional cultures.

The wider world is still harming the indigenous peoples of South America, despite some efforts on the part of governments to protect their lives and ways of life. New constitutions expanded indigenous rights in Colombia in 1991, Peru in 1993, Bolivia in 1994, and Ecuador in 1997. In these countries, Native Americans now have more rights over land, and the governments recognize communal rights over some resources. Brazilian law forbids mineral extraction by outsiders on aboriginal lands, but illegal diamond and gold mining has led to many bloody skirmishes between Native Americans and prospectors in several Indian reserves in the country. On the negative side of this protection, as the Native Americans see it, many resources of Brazilian and other national Indian reserves are placed under government control, and often indigenous peoples are forbidden from setting up their own businesses.

European languages joined the linguistic mix in Latin America at the end of the 15th century (see Figure 10.15). At first, they were spoken only by the colonizers, but they became more widespread as colonial administration progressed and expanded. The language distributions today reflect the pattern of colonial rule. Spanish, now one of the world's **megalanguages** (along with Chinese, English, Hindi, and Arabic), is the most widespread and is the prevalent European language throughout the region, with these exceptions: Portuguese in Brazil; French in French Guiana, Haiti, Guadeloupe, and Martinique; Dutch in Suriname; and English in Guyana and Belize. These languages are also spoken in the dependencies and former colonies of several other small Caribbean islands. The European languages generally became the sole official languages of both the colonial administrations and the independent countries that succeeded them. They serve as welcome lingua francas in countries like Mexico and Peru that have a bewildering diversity of indigenous languages. Bolivia, Peru, Paraguay, and Guatemala have official languages that are both European and indigenous.

Another linguistic product of European colonialism in Latin America was the evolution of **creole languages.** These are tongues that developed among the black slaves and indentured servants brought to work the plantations of the Caribbean islands and the Atlantic and Caribbean coasts of South and Central America. Their vocabularies come mainly from the respective colonizers' languages of Spanish, Portuguese, French, English, and Dutch. Sometimes a speaker of the parent language can recognize the creole; for example, in Costa Rican Creole English, "*Mi did have a kozin im was a boxer, kom from Panama*" is easily enough understood as "I had a cousin from Panama who was a boxer."[2] But it is not always that simple. Creoles are distinct languages and have their own grammars.

None of the indigenous African languages that the slaves brought to the New World survived there to the present day. Some vocabulary and other elements of the ancestral African languages do survive in some of the creoles, such as in the Garifuna spoken on the coast of Belize.

The European Conquest

The arrival of Christopher Columbus and his three Spanish ships in 1492 brought more than a new language to the New World. It marked the beginning of profound changes in almost every aspect of life in what would become Latin America.

First came death, both deliberate and unintended. Series of expeditions to the New World convinced the Spaniards that there were riches worth pursuing in that far-off land. The Spanish conquistador Hernando (Hernán) Cortés landed on Mexico's coast near present-day Veracruz in 1519 and led his cohort of invaders inland, building alliances with ethnic groups opposed to the Aztecs and killing those who would not join him. Unfortunately for the Aztecs, Cortés's arrival seemed to fulfill the prophecy that Quetzalcoatl, a fair-skinned, bearded Aztec god would return one day, approaching from the east. In Tenochtitlán, the Aztec leader, Moctezuma, received Cortés and his army as emissaries of Quetzalcoatl. Cortés and just 30 men took Moctezuma prisoner and were unopposed, apparently because of the widespread belief in Quetzalcoatl's return. Using a divide-and-conquer strategy that enlisted local forces, Cortés achieved a military victory over the Aztecs in 1521, destroying Tenochtitlán and laying the foundation for Mexico City in its ashes.

Spain soon dispatched new expeditions to conquer other peoples of Mexico and Central America. When Francisco Pizarro landed in Peru in 1531 with only 168 soldiers and 102 horses, he was warmly greeted by an Incan leader, Atahualpa. The Spanish conquistador imprisoned Atahualpa, freed him for a huge ransom, imprisoned him again, and then had him hanged (he was to have been burned at the stake, but Pizarro modified the execution because Atahualpa agreed to be baptized). Two years later, Pizarro's forces held the Incan capital Cuzco, and from that point, the colonization of all of South America by the Spanish— and in Brazil, the Portuguese—proceeded swiftly.

Disease brought death to the indigenous people on a vast scale. The Native Americans had never been exposed to, and hence had no natural immunities against, the host of diseases that the Spanish and the Portuguese introduced: bubonic plague, measles, influenza, diphtheria, whooping cough, typhus, chickenpox, tuberculosis, and smallpox, joined later by the malaria and yellow fever that came with African slaves. The worst was smallpox, which killed between one-third and one-half of all Native Americans in affected areas. The diseases, famine, and enslavement that came with Spanish and Portuguese colonization brought what was probably humankind's largest population decline ever. The numbers are elusive but were certainly huge. The Spanish chronicler Bartolomé de Las Casas wrote that in Spanish-controlled Latin America, 40 million Native Americans had died by 1560. The geographer David Clawson estimates that by 1650, 90 to 95 percent of the Native Americans who had come into contact with Europeans had died.

European settlement patterns emerged as the conquest proceeded. The development of ports as bases for subsequent penetration of the interior was the first major settlement task. Large coastal cities had not existed prior to European colonization. Many early ports eventually grew into large cities, and a few became major metropolitan centers. In some cases—notably Lima, Caracas, and Santiago—the main city developed a bit inland but retained a close connection with a smaller coastal city that was the actual port.

Around the ports, agricultural districts developed, spread, and shipped an increasing volume of trade products overseas. Plants and animals domesticated by Native Americans would revolutionize diets and habits in Europe and elsewhere in the Old World: tobacco, potatoes, corn, cacao, and turkeys were among them (●Figure 10.16; Table 4.3, page 85). In time, the European-introduced horses, cattle, sheep, donkeys, wheat, sugarcane, coffee, and bananas would revolutionize New World agriculture. In the second

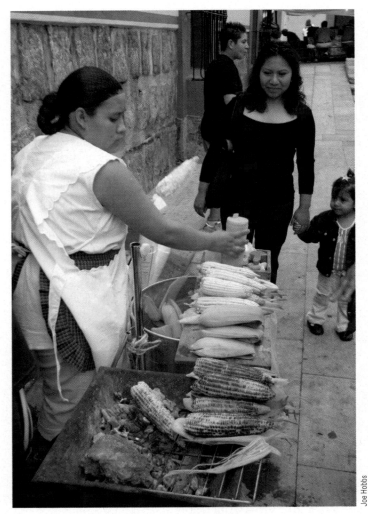

Joe Hobbs

● **Figure 10.16** Corn (maize), one of the world's great staple crops, was domesticated by Native Americans. It continues to be the most important food in much of Latin America. This is a street scene in Oaxaca, Mexico.

half of the 1500s, slave ships began to bring the Africans who provided the principal labor on the plantations created and owned by Europeans, primarily in the tropical lowlands with the generally fertile alluvial soils of the coastal plains.

45, 469

The powerful lure of gold and silver stimulated deeper European penetration of the Andes and the Brazilian Highlands in South America. After capturing the stores of precious metals that had been accumulated by the indigenous peoples for ceremonial use, tribute, and trade, the foreigners opened or reopened mines and established market centers to service them. Some highland settlements became centers of new ranching and plantation enterprises. A few highland cities, of which the largest is Bogotá in Colombia, grew into large metropolises. Separated from the seaports by difficult terrain, these cities were eventually connected to the coasts by feats of great engineering skill. Some of the main seaports and highland cities became important centers of colonial government as well as economic nodes,

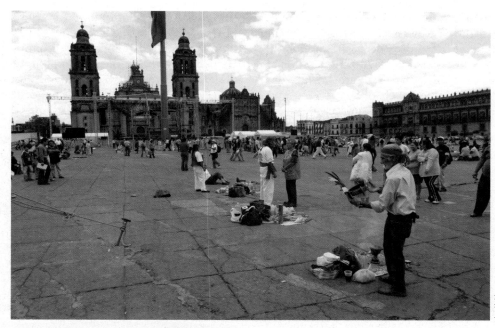

• **Figure 10.17** Plaza and cathedral are at the center of most Latin American cities. This is the *Zocalo* of Mexico City. From lower right to center are three male traditional healers or "shamans" practicing their trade.

and several are national capitals today, including La Paz in Bolivia and Quito in Ecuador.

At the heart of each Spanish and Portuguese city established in Latin America, the newcomers placed their familiar combination of a plaza dominated by a massive Roman Catholic cathedral (•Figure 10.17). Roman Catholicism was introduced—actually, imposed—as the one and only acceptable faith in the New World, and 80 percent of the region's people are Catholic. The British and Dutch brought their Protestant faiths to their possessions. Today, Protestant offshoots, notably Evangelical and Pentecostal religions—characterized by high-energy, participatory forms of worship and their appeal to disrupted families and the poor—are making inroads in traditionally Catholic communities throughout the region Jamaica has embraced Rastafari, a faith championing black empowerment and holistic living dating to the early 20th century, with unlikely links to Ethiopia. Many indigenous beliefs have become syncretized with Catholicism, but some of them dating to **pre-Columbian times** (before 1492) remain alive among Native Americans in Latin America. The Catholic Maya of Guatemala, for example, still make offerings to the "earth lords" who dwell in caves (•Figure 10.18).

82

505,
569

Ethnicity in Latin America

Despite the dominance in Latin America of culture traits derived from Europe, a large majority of the original European settlers or their descendants intermarried with Native Americans or blacks. Only Argentina, Uruguay, and Costa Rica have significant white European ethnic groups today

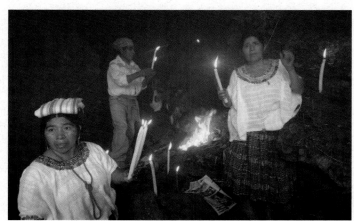

• **Figure 10.18** Cave offering of the Q'eqchi' Maya, Alta Verapaz, Guatemala.

(•Figure 10.19a). These are mainly families of immigrants from Europe who arrived relatively recently, during the late 19th and early 20th centuries. Scattered districts in other countries are also predominantly European.

Black Latin Americans of relatively unmixed African descent live in the greatest numbers on the Caribbean islands and along the Atlantic coastal lowlands in Middle and South America (•Figure 10.19b). These are the areas to which African slaves were brought during the colonial period, mainly to work on sugar plantations. An estimated 3 to 4 million were sold in Brazil alone. Slavery was gradually abolished during the 19th century, although not until the 1880s in Brazil and Cuba. By that time, slavery had generated large fortunes for many owners of plantations and slave ships. The African peoples it introduced to the region have made cultural contributions that today provide a varied, vibrant, and important part of Latin American civilization.

Latin America has escaped many of the racial tensions and violence that have battered much of the world. This may be because a majority of Latin Americans are of mixed heritage, the most common being a mixture of Spanish and Native American that has resulted in a heterogeneous group known as **mestizos** (•Figure 10.19c). A person of mixed African (black) and European (white) ancestry, sometimes with Native American blood as well, is known as **mulatto** (•Figure 10.19d); in some locales, the term used is **Creole**. It must be noted, however, that in most places, socioeconomic discrepancies are strongly associated with ethnicities in the region. A thin stratum of people of mainly European origin often dominate government and business sectors, with mestizos and mulattoes making up the bulk of the middle and lower classes. Native Americans almost always have the lowest standing, and in some countries such as Guatemala, systematic exclusion and violence has been directed against them from time to time.

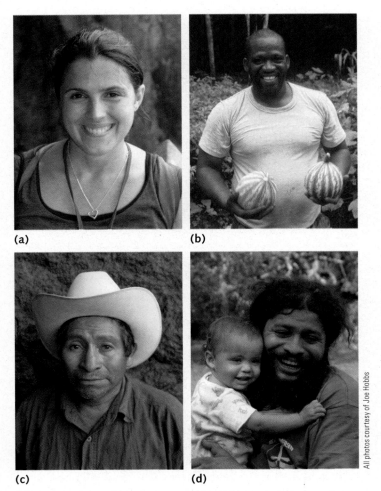

(a)   (b)

(c)   (d)

All photos courtesy of Joe Hobbs

• **Figure 10.19** In addition to the Native Americans of Latin America, the leading racial types are (a) Europeans, (b) blacks, (c) mestizos, and (d) mulattoes.

Mestizos have not lost all traits of their Amerindian forebears. While it can be hazardous to generalize about ethnicities, the mestizos' inheritance from their Native American roots includes the collective exploitation of land, the accumulation of lifelong personal bonds, the devotion to a whole series of supernatural beings (who have Christian names but non-Western origins), the ritualized remembrance of deceased relatives, and reverence for political leaders.[3] As will be seen later in the chapter, reverence for political leaders cannot be taken for granted in this region today.

10.4 Economic Geography

Latin America is generally a region of less developed countries (LDCs) where people do not enjoy a high standard of living. It is not the worst off of the major world regions; in fact, its overall per capita GNI PPP is more than three times that of sub-Saharan Africa (see Table 1.1, page 5). But compared to North America or Europe, the Latin American region as a whole is quite poor, with an estimated one-third of its residents living in poverty. Both poverty and unemployment have diminished in recent years for the region as a whole; thanks to its relative abundance of raw materials, Latin America has benefited from the global boom in commodities related in large measure to China's dynamic growth. As in Africa and elsewhere, Chinese interests are scouting out every possible prospect in Latin America that could help feed that nation's voracious appetite for minerals, energy, and food. In some Latin American countries, this raises the same fears as in Africa: that China is seeking to develop a classic mercantile relationship with Latin America, taking raw materials and profitably selling its own manufactured goods. But for now, the influx of Chinese capital is welcomed. Economic growth for the region averaged 5 percent per year from 2004 to 2007. The wealthiest independent countries on a per capita GNI PPP basis are Trinidad and Tobago and Argentina, while the poorest are Haiti and Bolivia.

Although Latin America's overall situation is improving, the region suffers from notoriously large discrepancies between the haves and the have-nots. The glitter of the great metropolises like Mexico City and Rio de Janeiro, with their forests of new skyscrapers, may distract attention from poverty, but portions of these cities are massive, squalid slums, known as *favelas* in Brazil and *barrios* in Spanish-speaking countries (•Figure 10.20). Such **shantytowns,** which are practically universal in the world's LDCs, are full of unemployed, underemployed, and ill-fed people. Many prefer their plight in them to the conditions of the depressed rural landscapes from which they came. They aspire to the upward mobility that is at least within view in the urban areas.

Trying to overcome their economic problems and constrain popular discontent, many Latin American governments have borrowed heavily from the international banking community. In the early 2000s, unpaid loans had reached staggering proportions, and the Latin American

• **Figure 10.20** Many of Latin America's poorest people live in ramshackle housing. This is the shantytown of Belen in Iquitos, Peru.

debtor nations were having great difficulty in mustering even the annual interest payments, let alone generating surplus capital to pay toward the principal of these troubling loans—a dilemma all too characteristic of the world's LDCs. Like other LDCs, Latin America also has historically been overreliant on non-value-added goods like cash crops and minerals (Table 10.2). Free-trade agreements form the foundations of a recent push to move away from raw materials and toward manufactured exports.

Commercial Agriculture

Although the total number of Latin Americans employed in farming has not declined much in recent decades, the value of agriculture in national economies has experienced a marked percentage decline as economies have become more diversified. But in many countries, more than half of all export revenue is still derived from farm products, and some countries rely mainly on one or two agricultural commodities (see Table 10.2). As in other LDCs, overreliance on a narrow range of exports makes the Latin American countries economically vulnerable to changes in market conditions, competition from other sources, and changing consumer appetites. Reliance on coffee and bananas has whipsawed the economic fortunes of many countries, especially those in Central America that came to be known disparagingly as "banana republics" (see Insights, page 540).

Types of Farms

Farms in Latin America can be divided into two major classes by size and system of production. Large estates with a strong commercial orientation are known as **latifundia**

TABLE 10.2	Primary Exports of Selected Latin American Countries
Country	**Selected major exports**
Argentina	Edible oils, fuels and energy
Bahamas	Fish, rum, salt
Brazil	Transport equipment, iron ore, soybeans, coffee
Chile	Copper, fish, fruits, paper and pulp
Colombia	Petroleum, coffee, coal
Costa Rica	Coffee, bananas, sugar
Cuba	Sugar, nickel, tobacco
Dominican Republic	Ferronickel, sugar
Guatemala	Coffee, sugar, bananas
Mexico	Manufactured goods, petroleum, silver
Nicaragua	Coffee, shrimp and lobster, cotton
Panama	Bananas, shrimp, sugar
Peru	Fish, gold, copper, zinc
Trinidad & Tobago	Petroleum, chemicals, steel products
Venezuela	Petroleum, bauxite, aluminum, steel

Source: Central Intelligence Agency, *World Factbook* (Washington, D.C.: Central Intelligence Agency, 2007).

(sing., *latifundio*). These estates, whether called **haciendas,** plantations, or some other name, are owned by families or corporations (•Figure 10.21). Some have been in the hands of the same family for centuries. The desire to own land as a form of wealth and a symbol of prestige and power

INSIGHTS

Fair Trade Fruits

Bananas are the most popular fruit in the United States, with more than 8 billion sold annually—or about 26 bananas per person per year. Now it is possible to be a discriminating banana buyer. At many North American and European grocery stores, one may find bananas, pineapples, other fruits, chocolates, and coffees bearing "Fair Trade Certified" stickers (•Figure 10.C). These are higher in price than the standard products, but consumers who buy them are generally motivated by a sense of social responsibility. In standard trade transactions, only a tiny fraction—2 cents per pound of bananas, for example—of the final price paid in the supermarket actually goes to the grower. Intermediaries, including importers, distributors, and "ripeners" (who spray ethylene gas to turn the bananas yellow), all get a cut. Fair trade buyers deal directly with the farmer cooperatives they helped organize, avoiding brokers and intermediaries and guaranteeing higher prices—typically, 18 cents per pound for bananas—for the farmers. The fair trade organizations also help the farmers establish schools and health clinics.

Before socially conscious consumerism became popular in the United States, the **fair trade movement** developed in Europe as people began taking notice of the impacts of collapsing coffee prices. Whereas farmers could earn $3.18 per pound for their coffee in 1999, a glut in the world coffee supply, caused largely by surging production in Vietnam, sent prices down to 47 cents per pound by 2001—the lowest in a century, adjusted for inflation. Europeans learned that African and Latin American coffee farmers, never wealthy to begin with, were now destitute, so poor that their young children were compelled to work in the fields to help make ends meet. Green Party politicians and others stepped forward with the fair trade solution, seeking to guarantee what they considered a living wage for the producers of these commodities.

Coffee is the world's second-largest traded commodity (after oil), so fair trade coffee could affect many people. Now more than a million coffee growers in Tanzania, Rwanda, Ecuador, Nicaragua, and other countries belong to the cooperatives that sell their products through fair trade channels instead of to commercial producers. Fair trade products are becoming more mainstream in the United States; all Dunkin' Donuts and some Starbucks and even Wal-Mart coffees, for example, bear the label.

• **Figure 10.C** Fair trade certification of coffee sold in a U.S. health food store reflects changes in consumers' awareness and an improvement in the producers' livelihoods.

479, 500

• **Figure 10.21** These henequen (sisal) plants on a plantation in Mexico's Yucatán region brought great wealth to the ancestors of the hacienda's current owners. Note the elevated irrigation channel at right.

has always been a strong characteristic of Latin American societies. Spanish and Portuguese sovereigns granted huge tracts to members of the military nobles who led the way in exploration and conquest. Some of this land has been reallocated to small farmers by government action from time to time. But in many countries, a very large share of the land is still in the hands of a small, wealthy, landowning class.

Minifundia (sing., *minifundio*) are smaller holdings with a strong subsistence component. The people who farm them generally lack the money to purchase large and fertile properties; they farm marginal plots, often on a sharecropping basis. Individual landowners are often burdened by indebtedness, the threat of foreclosure of farm loans, and the fragmented nature of farms, which become smaller as they are subdivided through inheritance and the reversal of promised governmental programs of land distribution. Such farms produce food for family use and also for the local market (•Figure 10.22). The crops most commonly raised, especially in Middle America, are maize (corn),

● **Figure 10.22** Market scene in Mexico.

beans, and squash, although many other crops are locally important, especially in the different climatic zones of the highland regions.

Although small farmers make up the bulk of Latin America's agricultural workforce, food has to be imported into many areas, and many of the people are poorly nourished. Productivity is low because of the marginal quality of the land and the simple agricultural techniques used in peasant farming. Little capital is available to buy machinery, fertilizers, and improved strains of seeds. Soil erosion and soil depletion take a toll in many areas. As elsewhere in the developing world, increasing numbers of young people are responding to difficult rural conditions and the perceived opportunities of the big city by abandoning the countryside and settling in urban shantytowns. The tide of rural-to-urban migration might be slowed or even reversed if there were effective **land reform**, but this has long been one the most contentious issues of political and social change in Latin America (see Problem Landscape, page 542).

Minerals and Mining

Latin America is a large-scale producer of a small number of key minerals, notably petroleum, iron ore, bauxite, copper, tin, silver, lead, zinc, and sulfur. Only a handful of Latin American nations gain large revenues from exporting these minerals. Even in the countries that do have large mineral output—notably Mexico, Venezuela, Chile, Ecuador, and Brazil—much of the profit appears to be dissipated in the form of showy buildings, corruption, ill-advised development schemes, and enrichment of the upper classes and foreign investors. Latin America has thus suffered from the same **"resource curse"** that has stricken other parts of the developing world. Even though benefits to the broader population from such mineral production is often minimal, that production has funded the development of significant infrastructure, including new highways, power stations, water systems, schools, and hospitals, and hence created jobs.

Most of the extracted ores from Latin America are shipped to overseas consumers in raw or concentrated form. The region has scattered iron and steel plants, as well as smelters of nonferrous ores, which process metals for use in Latin American industries or for export. Deposits of high-grade iron ore in Venezuela and central Brazil are the largest known in the Western Hemisphere and are among the largest in the world; the two countries are Latin America's main producers and exporters of ore. In Latin America, Brazil is the main producer of iron and steel by far, with Mexico second. Most of the region's production of bauxite—the major source for aluminum—comes from Jamaica, Brazil, Suriname, and Guyana. The deposits in these countries are generally located near the sea and are of critical importance to the industrial economies of the United States, Canada, and China. Venezuela has huge bauxite deposits and the sources of energy needed to process the ore; the country is a major aluminum exporter.

Chile is the largest copper producer in Latin America and also ranks number one in the world. Most of Latin America's known reserves of tin are in Peru, Bolivia, and Brazil, and these countries produce most of the region's output. The silver of Mexico, Peru, and Bolivia—sought from the very beginning of Spanish colonization—is found mainly in mountains or rough plateau country. Mexico and Peru

480, 491, 511

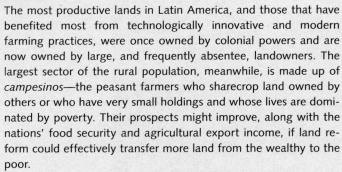

PROBLEM LANDSCAPE

The Troubled Fields of Land Reform

The most productive lands in Latin America, and those that have benefited most from technologically innovative and modern farming practices, were once owned by colonial powers and are now owned by large, and frequently absentee, landowners. The largest sector of the rural population, meanwhile, is made up of *campesinos*—the peasant farmers who sharecrop land owned by others or who have very small holdings and whose lives are dominated by poverty. Their prospects might improve, along with the nations' food security and agricultural export income, if land reform could effectively transfer more land from the wealthy to the poor.

Land reform attempts and outcomes have varied sharply from one Latin American country to another. Most efforts have focused on breaking up existing properties or bringing vacant land (whether owned by the public, by private individuals, or by the Catholic Church) into cultivation, usually by small farmers. Some new farms have been structured as communal holdings, reflecting indigenous traditions or 20th-century revolutionary plans of agrarian reform. In Cuba, for example, the Communist government that took control in 1959 placed the land from expropriated estates in large farms owned and operated by the state. Workers on these farms are paid wages. Also motivated by Communist principles, Nicaraguan Sandinistas implemented land reform as part of their revolutionary efforts in the 1980s. The left-leaning government in Venezuela is undertaking what it calls the "Return to the Countryside" program, designed in principle to purchase land from *latifundia* holders at market value and reallocate it to the more needy. Landowners say the program instead involves army units forcing them off their farms and ranches and rules that make compensation difficult.

In Mexico, there was great fanfare at the introduction of the *ejido* (communally farmed or grazed land) program in the 1930s. The *ejido* program broke up large estates and other holdings and redistributed the land as government property to be worked collectively. The idea was to prevent the emergence of a permanent poor, landless class of peasants. However, the system was inefficient, the *ejidos* proved unproductive, and only large government subsidies kept the program going. By the 1980s, with the population growing so rapidly, the Mexican government essentially ran out of land to redistribute while millions of peasants were still waiting for their plots. Most existing *ejidos* were so small, they could barely sustain the families of the farmers living on them. Rural poverty increased dramatically. Cash-strapped farmers could not make capital improvements (such as buying tractors), and since *ejido* land was government-owned, it could not be sold to private interests. That law was changed in 1992, and since then, *ejido* land that has been sold is now largely in the hands of real estate developers or large agribusinesses.

Virtually every nation in Latin America can point to events, martyrs, and programs birthed in protest or blood in efforts to achieve more equitable land ownership. Nevertheless, the great majority of the arable land in Latin America continues to be owned by wealthy farming families, agribusiness operations (both domestic and foreign), and the church. Even where successful, land reform can help relieve poverty in the countryside but not in the cities, where the majority of Latin America's poor live.

are the largest Latin American producers of silver, lead, and zinc. All but a small proportion of Latin America's petroleum is extracted in the Caribbean Sea and Gulf of Mexico area, particularly the central and southern Gulf coast of Mexico and northern Venezuela. Other oilfields in Latin America are widely scattered, with the main ones located along and near the Atlantic coast in Brazil, Argentina (in Patagonia), and Colombia, and in lowlands along the eastern flanks of the Andes in every country from Trinidad and Tobago to Chile. Natural gas is extracted in many areas that produce oil, but Latin American production is not yet of major world consequence. Mexico, Venezuela, Argentina, and Trinidad and Tobago are the largest gas producers in Latin America. Venezuela is also the only Latin American nation in the Organization of Petroleum Exporting Countries (OPEC) and was one of its founding members in 1961.

Latin America's heavy reliance on unprocessed minerals has periodically dealt crushing economic blows, especially when global recession reduced demand for such commodities and forced prices downward. But strong growth charac-

terized Latin American economies after 2002, particularly because of China's superheated economic growth. Brazilian iron ore, Chilean and Peruvian copper, and Venezuelan oil are in ever-increasing demand to feed China's appetite for commodities. The flip side is that if China's boom were to go bust, Latin American economies would fall too. 298, 479

Free Trade Agreements

Many Latin American countries are trying to reduce their dependence on raw materials and boost their exports of value-added manufactured products, thus improving their financial situations. To do so, they have formed or joined free-trade agreements (FTAs), particularly with their giant neighbor to the north (●Figure 10.23) (see Regional Perspective, pages 544–545).

Five nations in South America—Venezuela, Brazil, Argentina, Uruguay, and Paraguay—belong to the **Southern Cone Common Market,** known as **Mercosur,** which was established in 1991. Mercosur reduced tariff and other long-standing

• **Figure 10.23** Latin American countries belong to a variety of economic associations.

barriers that had existed between the countries, promoting a huge increase in trade. Its members have worked to coordinate economic policies and a better negotiating position with the United States and other powers. The more than 250 million people of this union can live and work in any of the member countries and have the same rights as citizens of those countries. This stipulation brought immediate relief to an estimated 3 million illegal workers. In 2004, Mercosur and the Andean Community (comprised of Bolivia, Colombia, Ecuador, and Peru) agreed to join to form the **Union of South American Nations**, modeled after the European Union.

Mercosur is not a lightweight: In value of products traded, this is the third-largest trade group in the world, after NAFTA and the EU. Mercosur and the European Union negotiated an important trade pact in 2004 in which Europe offered the South American countries more generous import quotas for beef, dairy products, sugar, and coffee. In turn, Mercosur offered the European Union privileged access to investments in telecommunications, banking, and other services. The South American countries embraced the agreement in part at the urging of the regional heavyweight, Brazil, which did not want to be seen as bowing first and foremost to U.S. interests in another pact, the FTAA (to be discussed shortly).

Patterned after NAFTA, another U.S.-brokered trade organization was ratified by the U.S. Congress in 2005. This was the **Dominican Republic–Central America Free Trade Agreement (DR-CAFTA)**, composed of the United States, Costa Rica, El Salvador, Guatemala, Honduras, Nicaragua, and the Dominican Republic. U.S. advocates of DR-CAFTA promised that by eliminating duties on 80 percent of the U.S. goods that could be sold in the region, it would create the second-largest market in Latin America for U.S. exports, after Mexico. Politicians from textile-producing states like South Carolina argued that DR-CAFTA would destroy American jobs by opening the U.S. market to cheap imports, and they promised but failed to defeat ratification of DR-CAFTA in the U.S. Congress. Latin American critics of DR-CAFTA argue that the trade agreement demands too many concessions by the poor Central American countries in protecting U.S. **intellectual property rights (IPR)**; under the agreement, the countries would not be allowed to replicate American pharmaceutical drugs and sell them at lower prices. They also point out that DR-CAFTA will still protect the U.S. sugar industry, hurting that important business in Central America, while flooding Central America with cheap U.S. corn and other grains that will drive Central American farmers out of business. Proponents of DR-CAFTA prevailed, however; as of 2007, the agreement was in force in all signatory countries.

Using NAFTA as its foundation and DR-CAFTA as a building block, the United States is taking the lead in establishing a hemisphere-wide trade organization called the **Free Trade Area of the Americas (FTAA)**. Like the political organization known as the **Organization of American States (OAS)**, it would include all of the countries of North, Middle, and South America, with the exception of Cuba. Originally envisioned for establishment in 2005, this agreement is proving very difficult to achieve because some of its loudest critics in Latin America—Brazil in particular—insist that the United States is including too many restrictions in its vision of "free" trade. Mercosur members including Brazil want to see the United States drop the huge subsidies of many agricultural products and steel that keep American farmers and factory workers in business but that shut out Brazilian products. In return, Mercosur would endorse patent, copyright, and other IPR protection of U.S. industries. With the FTAA stalled, the United States has chosen to negotiate one-on-one trade agreements with members of the proposed FTAA, beginning with Chile in 2004 and Peru in 2007.

There are other regional trade associations, including the Central American Common Market, the Andean Community, and the Caribbean Community (CARICOM) (see Figure 10.23). The large *maquiladoras* of Mexico, churning out automobiles, textiles, electronics, and computers, are the symbols of the new Latin American free-trade-oriented economies. These have become an exception to the previous spatial pattern of Latin American manufacturing enterprises. Larger operations were almost always located in the larger cities, including the greatest producing metropolises of Mexico City, São Paulo, and Buenos Aires, representing the region's three greatest manufacturing countries, Mexico, Brazil, and Argentina, respectively. But *maquiladoras*

REGIONAL PERSPECTIVE

The North American Free Trade Agreement (NAFTA)

The most economically significant FTA is the **North American Free Trade Agreement (NAFTA),** of which Canada, the United States, and Mexico are members (●Figure 10.D). In terms of both economic value and geographic area, it is the largest trading bloc in the world. Citizens of these countries make up about half of the 904 million people of the hemisphere. The essential goal in establishing this free-trade agreement, like most, was to reduce duties, tariffs, and other barriers to trade between the member countries, theoretically strengthening all the economies. The United States had a strong if perhaps understated NAFTA goal of reducing Mexican immigration into the United States by boosting Mexico's prosperity.

NAFTA was already controversial before it went into effect in 1994, and it remains so today. Among U.S. citizens, there were

fears that factories would close down and jobs would flow south to Mexico. American environmentalists feared that the U.S. environment would suffer because Mexican factories and other polluters, unable to afford the cleanup of their industries, would set new and lower standards that U.S. companies could then match. Mexican officials feared that a surge in their consumption of American-made products would so increase their trade deficit with the United States that Mexico's monetary reserves would be exhausted.

More than a decade on, even as the agreement's provisions are still coming into force (they are being phased in gradually through 2009), some of NAFTA's long-term impacts are clear. One of the most profound early consequences was a virtual collapse of clothing industries in Jamaica and other Caribbean islands. Once duties were no longer imposed on Mexican clothing exports to the United States, Mexico immediately had a huge advantage over Jamaican and other Caribbean producers. Job growth in Jamaica stopped, and unemployment surged.

Some U.S. factories did close, in part because production was cheaper south of the border—notably in the *maquiladora* factories (discussed on page 557)—but NAFTA was not the exclusive cause of the shutdowns. Lower-cost manufacturing for American companies shifted to a host of countries outside NAFTA and the Western Hemisphere. Trying to maximize benefits for its three members in the global marketplace, NAFTA does require all its member countries to abide by **rules of origin** specifying that half or more of the components of any manufactured good must originate in Canada, the United States, or Mexico. This is designed to

299, 300

● **Figure 10.D** The trinational NAFTA flag symbolizes the integrated economic ambitions of its members.

---

are being situated where location serves them best: close to the border of the United States—the principal destination for most of their output—or deep in areas like the Yucatán where labor costs are lowest. It should be noted that the most numerous manufacturing establishments in Latin America are not like *maquiladoras* but are household enterprises and small factories that employ fewer than a dozen workers and sell their products—such as textiles, ceramics, and wood products—mainly in home markets (●Figure 10.24).

Sending Money Home

More and more Latin American families, particularly from Mexico, Central America, and the Caribbean, have come to rely on having at least one member work abroad to help out the family economy. Among the largest sources of income for Latin American economies are the remittances, 74, 358 or earned savings, sent home by people working abroad, especially in the United States. In 2006, Latin Americans working typically two jobs in the United States and earning less than $20,000 per year sent home more than $45 billion (50 percent more than in 2004), typically in periodic elec-

● **Figure 10.24** "Panama hats" actually come from Ecuador. This small, family-owned operation produces a modest but high-quality output of these famed *sombreros* for domestic and foreign consumption.

tronic transfers of $200 to $300 each. Mexico is the world's third-largest remittance market (after China and India), with $23 billion infused into the country from Mexican laborers in the United States.

619

89

477, 623

615

REGIONAL PERSPECTIVE

The North American Free Trade Agreement (NAFTA) *continued*

prevent a fourth country from using Mexican labor in the assembly of its own component parts. Critics of NAFTA say the agreement should have demanded that an even larger proportion of the component parts originate in the member countries to boost their economies.

NAFTA did create more manufacturing jobs in Mexico but in the process contributed to a growing disparity between the wealthier north and the poorer south. The higher-paying jobs became concentrated in a *maquiladora* belt in the north. In the more agricultural south of the country, where few of these factories were established, Mexican farmers lost incomes or livelihoods because of falling prices for corn, rice, beans, and pork. Ironically, this pushed a new steam of poor immigrants into the United States. NAFTA gradually eliminated most Mexican tariffs on agricultural imports from the United States. The heavily subsidized American farmer, receiving about $16,000 in government subsidies each year, was able to raise grain or livestock at prices below production cost. A pig, for example, can be raised and sold in Iowa for one-fifth of what it would cost to raise and sell it in Mexico. Cheap American agricultural products flooded the Mexican market, making it impossible for Mexican farmers to compete and driving many out of business; one-third of Mexico's 18,000 swine producers went out of business soon after NAFTA went into effect. On the positive side, Mexican companies have been able to turn the cheaper U.S. pork into sandwich meats sold back at a profit to American consumers.

Corn occupies a cherished place in Mexico's culture and nutrition, and increasing integration of the Mexican and U.S. economies led to a corn crisis there in 2006. The United States began diverting more of its corn crop into the production of ethanol, a biofuel. Subsequent shortages of corn on the food market drove prices higher in the United States. Those increases were passed on in corn exported to Mexico and matched by domestic corn producers in Mexico. The young government of President Calderon found itself besieged by protesters demanding that Mexico's corn producers be shielded from U.S. imports (then accounting for one-fourth of Mexico's corn) and that the government intervene to keep this most basic of all foodstuffs affordable to the poor. At the same time, genetically modified corns began appearing in southern Mexican fields, and environmentalists and other anti-GM activists chimed in their demands for "food sovereignty" with the protection of Mexico's corn heritage. These events swirling around corn came to be known as Mexico's "tortilla wars."

In sum, in the United States and Canada, NAFTA has been beneficial to farmers but not to factory workers, while the reverse is true in Mexico. Both the United States and Canada have trade deficits with Mexico, importing more from that country than exporting to it. The hoped-for reduction of emigration to the United States has not materialized. Despite post-9/11 security concerns along the U.S.–Mexican border, about 1 million illegal immigrants (nearly half of them Mexican, the others from Central and South America) pass into the United States from Mexico each year.

The greatest numbers of these estimated 35 million immigrant workers, of whom perhaps a third are illegal or undocumented, are in the U.S. states of California, New York, Texas, Florida, and Illinois, but all 50 states have at least some. An estimated 75 percent of them send monies home, in many cases to grandparents caring for the working parents' children; families often remain divided for many years while the father or both parents work abroad. Collectively, these remittances are so valuable—far surpassing the amount supplied to Latin America in foreign aid and foreign investment—that home countries are searching for ways to tax the income or otherwise find ways to funnel more of it into national development. Some critics say that Mexico and other Latin American countries have become so dependent on remittances that they neglect to invest their own resources in national development.

One recent and important twist in the traditional one-way flow of remittances is that some experienced workers are returning to their home countries to establish companies that invest in the United States. These so called **multi-Latina companies** are aiding the revival of certain downtrodden sectors of the U.S. economy, including cement and steel mills.

Tourism in Latin America

Tourism has become a major regional economic asset to Latin America, generating critical foreign exchange; only oil exports are more valuable. During what many North Americans see as their too-long winters, they are bombarded by television and print advertisements of vacation in paradise: Jamaica and hosts of other destinations of the Caribbean basin and western Mexico. A wide variety of vacation experiences are available, from the hedonistic resort complexes that tourists seldom leave (one resort on Jamaica actually is called Hedonism) to the insular sea voyage that offers day trips ashore in exotic destinations to whitewater and rain forest adventures of the ecotourist variety (●Figure 10.25).

The region benefits from the proximity of its year-round warmth to the wealthy and mobile and seasonally chilled U.S. and Canadian populations. Tourism revenues reflect that **distance-decay relationship:** the highest tourism receipts in Latin America flow to Mexico, the nearest neighbor to the wealthy countries, but tend to fall off for more far-flung destinations. Countries in which tourist dollars amount to more than 50 percent of the nation's foreign exchange include Antigua and Barbuda, the Bahamas, and Barbados.

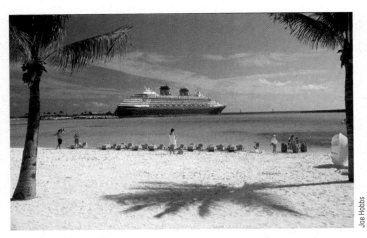

• **Figure 10.25** Disney cruise ships call at the Bahamas' Castaway Cay, an island owned by this American company.

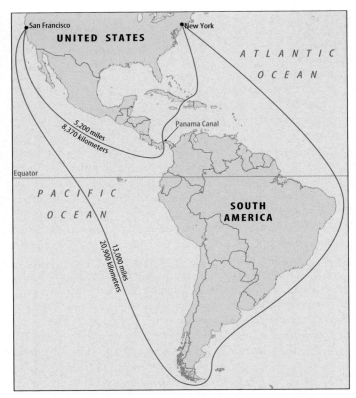

• **Figure 10.26** Distance, time, and money are saved by use of the Panama Canal.

10.5 Geopolitical Issues

The central geopolitical reality of Latin America is that it is in America's backyard. For better or worse, the United States has staked its geostrategic claim to the region. Much of the United States itself is comprised of land wrested from Mexico in conflicts of the early 19th century. In 1823, U.S. intentions for Latin America were formulated in a policy known as the **Monroe Doctrine,** after U.S. President James Monroe announced that the United States would prevent European countries from undertaking any new colonizing activities in the hemisphere. Many subsequent interventions, wars, investments, and other U.S. activities may be seen in light of this seminal policy. Even more consequential was the 1904 **Roosevelt Corollary** to the Monroe Doctrine, whereby the United States declared that it had the right to supervise the internal affairs of Latin American countries to ensure U.S. national security.

Ever since the Monroe Doctrine was proclaimed, the United States has had a particularly firm hand in Central America. Washington has dealt unapologetically and sometimes harshly with perceived threats to the United States from this region. Examples include U.S. actions to support governments against nationalist insurgencies or insurgents against leftist governments: El Salvador and Nicaragua in the 1980s, Cuba in the 1960s, and Nicaragua in the 1950s were the most significant instances.

The Panama Canal is a legacy of American hegemony in Central America (•Figure 10.26). This is one of the world's most critical chokepoints: 1,000 ships a year, traveling 80 shipping routes and representing about 4 percent of the world's cargo volume (and 16 percent of the United States'), use the 50-mile (80-km) shortcut from the Atlantic Ocean to the Pacific. A 14-day sea voyage between New York and San Francisco via the canal cuts 7,800 miles (12,530 km) and 20 days from the trip (Figure 10.26; see also Figure 10.1.A, p. 564). That translates into enormous cost and time savings.

Constructed between 1904 and 1914 by American contractors, the Panama Canal is a product of U.S. commercial and strategic interests. For decades, it was sovereign U.S. territory, carefully monitored by an American military presence, and the United States kept more than half of the transit fees charged to ships using it. In the late 1970s, however, the administration of U.S. President Jimmy Carter accepted the Panamanians' complaint that U.S. control of the canal was an outmoded vestige of colonialism. Carter negotiated the transfer of authority over the canal to Panama in a treaty that went into effect in 1999. Now the revenues from ships transiting the canal are exclusively Panama's (see page 564).

From a strategic and economic defense standpoint, however, the United States is still deeply involved with the canal. The U.S. Navy is concerned that its two access points (Balboa in the Pacific and Cristóbal in the Atlantic) are now controlled by a company with strong backing from China, the canal's second-largest user. The United States has already effectively invoked the Roosevelt Corollary since returning control of the canal; in 1989, U.S. troops invaded Panama and ousted its ruler, Manuel Noriega, whom the Americans accused of facilitating the illegal trade of drugs into the United States. The U.S. **war on drugs** has been another central geopolitical issue of the region since the early 1970s (see Geography of Drug Trafficking, pages 548–549).

The most consequential test of U.S. commitment to its interests in Latin America came with the **Cuban Missile Crisis** of 1962. Cuba is as close as 90 miles (144 km) from the United States. The frosty relations between the United States and Cuba's Marxist regime, led after 1959 by a firebrand revolutionary named Fidel Castro, were emblematic of the Cold War. Those tensions came dangerously close to hot war and even nuclear holocaust in 1962, when the Soviet Union began to build facilities in Cuba that would have accommodated nuclear missiles capable of striking Washington, D.C., New York, and other cities in the eastern United States. The presence of the Soviet missile silos and launchers was discovered through the use of aerial photo interpretation, a geographic research tool that at that time was growing in military and commercial utility.

Although the missiles were ultimately withdrawn under intense pressure from the U.S. government, led at the time by President John F. Kennedy, Cuba has remained a storm center of political affairs in the Western Hemisphere. After several attempts to assassinate Castro—even with the unlikely ploy of an exploding cigar—and a disastrous attempt by Kennedy's administration to overthrow him by military force (in an incident known as the **Bay of Pigs invasion**), the U.S. administration settled on a course of economic and political isolation of Cuba.

Despite the resumption of diplomatic and economic relations with most of the remaining Communist countries in the world—China most notable among them—the United States has remained steadfast in imposing sanctions on Cuba. This anomaly to some extent reflects domestic political realities in the United States, where a large population of **Cuban Americans** (including 800,000 in Florida alone) who fled Castro's Cuba tend to vote for the politicians who keep up the most pressure against Castro. The Cuban Americans are against any relaxation of the virtual ban on American tourism to Cuba, which is in place because the **Trading with the Enemies Act** forbids Americans from spending more than $300 in Cuba. The failing health of Fidel Castro and the discovery of significant oil reserves off Cuba's shores led to unprecedented discussions among U.S. lawmakers about reopening American doors to Cuba. One fear is that U.S. businesses will lose important opportunities if the doors remain closed. Recent legislation allows U.S. states to sell food and medicine to Cuba on a cash-only basis, and about 40 states do so.

The government of Cuba would like to see the country opened to American tourists. It would also like to see the United States withdraw from Guantánamo Bay, an enclave the U.S. secured in the 1903 treaty that addressed Cuba's fate following the Spanish-American War. The United States dutifully goes through the motions of paying Cuba the $4,000 annual rent for the lease on Guantánamo, which it has used as a military base, and Cuba dutifully refuses to accept it. After 9/11, the United States established a prisoner camp for suspected al Qa'ida members and other terrorists at the base. The fact that it lies outside the borders of the United States apparently allows the U.S. government to suspend the usual legal and constitutional rights awarded to criminal suspects on U.S. soil, but this contention is highly controversial.

Several U.S. interests in Latin America converge on the nation of Colombia, now the fifth-largest recipient of U.S. aid, after Israel, Iraq, Egypt, and Pakistan. While the war on drugs was long the dominant concern, U.S. priorities focus increasingly on securing access to Colombia's considerable petroleum reserves. The United States already obtains more of its oil from Latin America than from the Middle East, and it views Colombia, Venezuela, and Mexico in particular as long-term counterweights to the volatile Middle East oil region.

Washington's thinking has been that the best way both to obtain Colombia's oil and to stem the drug tide is to take on Colombia's main rebel army, the Revolutionary Armed Forces of Colombia (FARC; see page 577). FARC became the quintessential **narcoterrorist organization,** deriving most of its funding from the profitable cocaine industry based mainly in the areas it controlled. FARC and another rebel group, the National Liberation Army (ELN), took control of areas containing much of Colombia's roughly 40 billion barrels of petroleum. FARC attacked the petroleum infrastructure, especially the 500-mile (800-km) pipeline between Canon Limon and Covenas. That pipeline is partially owned by Occidental Petroleum of Los Angeles. Occidental and other U.S. companies spent tens of millions of dollars lobbying the U.S. government to help protect Colombia's oil from the rebels and therefore allow American firms to operate more freely and safely in the country.

Their lobbying paid off. With U.S. funding and training as part of the Plan Colombia program (see page 549), which began in 2000, Colombian forces undertook conventional drug war tactics, such as stepped-up herbicide spraying of coca crops, confiscation of traffickers' assets, and shooting down suspected drug-carrying planes. In a related operation dubbed **Plan Patriot,** U.S. military personnel civilian contractors targeted FARC to both reduce drugs and secure oil. To date, these efforts have succeeded in wresting much control from FARC. The government-backed "paramilitary" forces that both fought these rebels and had their own stake in the drugs trade have also been weakened, and the security situation in Colombia is much improved.

Between about 1980 and 2000, U.S. administrations exercised a hands-on policy in Latin America. Through a combination of blunt U.S. military interventions and the countries' own efforts, Latin America witnessed an overall transformation of political systems from authoritarian to multiparty democratic. In the thinking of successive U.S. administrations, those greater political freedoms should have ushered in more prosperity. This was one component of the so-called **Washington Consensus.** In this political-economic philosophy, the United States would press the democratic governments of Latin America to open markets to international trade, reduce tariffs, expand quotas, sell

GEOGRAPHY OF DRUG TRAFFICKING

The War on Drugs

From the Nixon administration of the early 1970s to the present, the United States has led a determined, expensive effort known as the war on drugs. The Nixon administration was the only one to focus significant efforts on combating demand for heroin, cocaine, and other drugs within the United States. All successive administrations have concentrated their efforts on the supply side, hoping to eradicate drugs at their sources—particularly in South America. Although it is far from over, the war on drugs has produced some definitive and sometimes predictable patterns (●Figure 10.E).

One of the most consistent geographic patterns is the so-called **balloon effect,** referring to the fact that when a balloon is pressed in one place, the air shifts elsewhere. In the case of drugs, concentrated eradication efforts in one place almost always shift production elsewhere, sometimes even from one continent to another. During the Nixon years, for example, U.S.-funded efforts succeeded in all but eliminating production of opium poppies (the plant from which heroin is derived) in Turkey. With demand still high in the United States, Europe, and elsewhere, however, poppy

● **Figure 10.E** Production and trafficking of drugs in the Americas.

off state companies to private investors, allow more foreign investment, cut government spending, and reduce bloated bureaucracies. These reforms would bring the prosperity that would solve the region's problems. If serious problems emerged, the United States could help; it did act, for example, to prevent a financial crisis that shook Mexico in 1994 from spreading.

After 2000 and especially after 9/11, aside from its boots on the ground in Colombia, the U.S. government appeared

to adopt a more hands-off policy toward Latin America, perhaps distracted by regions that seemed more threatening in the war on terrorism. Washington watched from the sidelines while the Argentine economy collapsed in 2001 and saw a succession of elections and polls that seemed to show a drift away from American wishes for the region.

Widespread disillusionment with economic conditions led to successive elections of reformist, left-leaning and generally anti–Washington Consensus leaders. First came

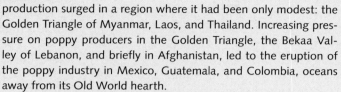

GEOGRAPHY OF DRUG TRAFFICKING

The War on Drugs *continued*

346

production surged in a region where it had been only modest: the Golden Triangle of Myanmar, Laos, and Thailand. Increasing pressure on poppy producers in the Golden Triangle, the Bekaa Valley of Lebanon, and briefly in Afghanistan, led to the eruption of the poppy industry in Mexico, Guatemala, and Colombia, oceans away from its Old World hearth.

In South America, increasingly successful eradication of coca (*Erythroxylon coca,* the shrub from which cocaine is processed) in the 1990s and early 2000s helped diminish cocaine production in Bolivia and Colombia, but production grew in Peru. The total area cultivated in coca in the Andes was greater in 2007 than it was when aerial spraying of defoliants began. In Colombia, eradication efforts led to some remarkable new developments in the industry. Cultivation dispersed into smaller and more widely scattered plots, often in new areas (it is grown in 23 provinces, compared to 3 when the U.S.-funded eradication under a program known as

547

Plan Colombia began, including the lowland Amazon region and in national parks and other protected areas). In addition, growers practiced selective breeding to produce plants that contain more active alkaloids, are more resistant to defoliants, and can grow with more shade cover, thereby escaping detection from the air. The growers who lose their crops to aerial spraying are resilient; an estimated 70 percent of the affected fields are replanted in coca. Drug war officers tout their successes, but despite the billions of dollars spent, Colombia was still producing 90 percent of the cocaine consumed in the United States in 2007.

Recent developments in the geography of methamphetamine also illustrate the balloon effect. U.S. authorities have cracked down on "meth" by removing many of its chemical ingredients from store shelves and by raiding the rural labs that produce the drug. Enterprising Mexicans have stepped into the breach. Most meth consumed in the U.S. now comes from south of the border, along the same trafficking routes used for heroin and cocaine.

Much of the force behind the balloon effect is economic: when eradication is successful in one area, supply falls and even more value is added to the already high value of cocaine and heroin. Growers and traffickers are easily tempted by the money to be made under such market conditions. Prices of heroin and cocaine are low, and the purities of the drugs are high, testifying to a glut of both on world markets and suggesting that the war on drugs is nowhere near being won. Much more money is made in the distri-

bution than in the production of drugs, and here Mexico has taken the lead. About 90 percent of all illegal drugs entering the United States, including those from South America, come from or through Mexico, where organized networks (often aided by corrupt police) arrange deliveries to Mexican distributors inside the United States.

Alternative proposals to the long-fought, conventional war on drugs typically have two elements. First, advocates of these other approaches argue that much more attention needs to be paid to the demand side, using education, treatment, and other means to wean addicts from drugs and discourage others from using them. The United States now spends about two-thirds of its drug war budget on foreign supply sources and one-third on domestic prevention and treatment. The second element focuses on the supply side. Peasant farmers who grow drugs are generally not trying to get rich (they do not, although many others up the supply chain do) but to secure a livelihood where jobs are scarce. Few crops offer the reliable market and certain returns that coca or poppies do, and **crop substitution** schemes to get farmers to switch to flowers or coffee almost always fail. Successful substitutions have occurred only in the context of so-called **alternative development** schemes, in which new crops are just one part of an overall effort to enhance the education, health, and livelihoods of rural communities. Alternative development is not a component of Plan Colombia, which may account for the high replanting rate for coca.

A successful war on drugs, won though efforts on the supply side, the demand side, or both, would have enormous environmental benefits, among others. Clearing of forest for new cultivation of coca has done enormous damage to South American forests, particularly on the steep eastern slopes of the Andes, where probably no other cultivation would ever be attempted. The processing of coca leaves to coca paste and finally powdered cocaine involves the use of highly toxic chemicals that make their way into water supplies, harming ecosystems and human health. A successful war on drugs could also help restructure U.S. interactions with the countries of Latin America. At present, countries deemed to be uncooperative in the war on drugs are subject to **decertification,** meaning that U.S. aid and investment in those countries are curtailed. This punishment is sometimes arbitrary; a country like Mexico often fails all the tests for certification but is too important to U.S. interests to be decertified.

Brazilians' choice in the 2002 and 2006 presidential elections: leftist labor union leader Luiz Inacio da Silva, known as "Lula," who vowed to fight the tide of globalization. Similar changes in administration took place in Argentina, Venezuela, Bolivia, Ecuador, and Nicaragua. In Venezuela, President Hugo Chávez was elected in 1999 and again in 2006, promising to form a regional trade association that would act against U.S. interests. He vowed to use the country's oil wealth to fund social programs. Hoping to do the

same with his country's natural gas and declaring that "the worst enemy of humanity is U.S. capitalism," a former coca grower, Evo Morales, became president of Bolivia in 2006. That same year, Ecuadorian President Rafael Correa was elected on promises to close a base used by U.S. aircraft to bomb drug fields, reject free-trade deals, and spend more oil revenue on the poor. All of these leaders, Chávez most vocal among them, have in common a stated intention to reverse the U.S.-backed push toward free trade and free enterprise, returning to state-owned companies and protected markets that would assist the poorest segments of their populations.

These leaders are supported by the indigenous Americans, who feel that the descendants of European colonizers of the New World, along with U.S. and other multinational firms, are robbing them of both their natural resources and their political clout. The Native American Aymara majority of Bolivia actually succeeded in driving that country's president from office in 2003, setting the stage for the election of Morales, the first indigenous head of state in Bolivia since the European conquest.

Bolivia had been a strong advocate of a proposed development to export Bolivia's natural gas, mainly to the United States, through a pipeline that would cross Chile to a Pacific port. Native American Bolivian peasants, backed by labor unions, student groups, and opposition politicians, complained that royalties for Bolivia from the gas sales would be so low that the common people would never realize any benefit from them. They cited Bolivia's history as a colony stripped of its silver and tin to enrich the Spaniards and local aristocrats and insisted that again only foreigners and Bolivia's elite would gain from the gas project. These opponents wanted the gas to remain in Bolivia, where it could form the basis for new local industries. (They also resented the pipeline's passing through Chile, a focus of Bolivian anger ever since Chile, following an 18th-century war, seized lands that cut Bolivia off from the sea.) Their votes were instrumental in bringing Evo Morales, himself an Aymara, to office. Morales championed their cause by announcing that his government would nationalize (impose government control over) all of Bolivia's critical energy and other resources and use them for the benefit of the people. The successful "ideology of fury" expressed by the Aymara in Bolivia is now rattling nerves among the elites in other countries of Latin America with large indigenous populations, including Ecuador, Peru, Paraguay, Guatemala, and even Mexico.

The following modules explore more issues among the peoples and lands of the region, first in Middle America and then in South America.

SUMMARY

→ Latin America lies south of the United States and includes Mexico, the Central American countries, the island nations of the Caribbean Sea, and all the countries of South America. It is called Latin America because of the post-15th-century dominance of Latin-based languages and the Roman Catholic Church imposed by European colonizers. The term Middle America refers to Mexico, Central America, and the Caribbean.

→ Latin American landscapes and livelihoods are correlated in altitudinal zones, which include *tierra caliente, tierra templada, tierra fría,* and *tierra helada.*

→ Large areas of humid climates and biome types are in the lowlands of Latin America, with tropical rain forests occupying major river basins, especially in Brazil and Venezuela. Grasslands are found poleward of the tropics, and there is a zone of Mediterranean climate on the western coast of Chile.

→ Dry climates and biomes, including desert and steppe, exist because of rain shadow and other factors. There are major desert regions on the west coast of South America and northern Mexico.

→ Latin America's population is about 8.5 percent of the world's total, and its growth rate is above the world average. A majority of the people living on the rim of the landmasses, with major population bands in the mountain valleys and, in some cases, on high plateaus or on the flanks of the Andes Mountains in western South America.

→ Before the arrival of the Europeans in the late 1400s, Latin America had been home to highly advanced indigenous civilizations, notably the Aztecs, Maya, and Inca. There were many other Native American groups that did not develop urban livelihoods. All these cultures declined precipitously after Europeans arrived. Although some of the losses are associated with war and with mining and other land uses, the biggest killer was European diseases, especially smallpox.

→ The region's major ethnic groups today are Native Americans, mestizos, mulattoes, blacks, and Europeans. Political and economic systems still tend to favor people of European extraction, but non-Europeans have found increasing power through democratic political systems. Many reject the free-trade and capitalist systems that they believe favor the ethnic Europeans.

→ Latin America has been urbanizing rapidly, with now more than two-thirds of its population living in cities. There continue to be steady rural-to-urban migration and a steady stream of legal and illegal immigrants northward into the United States. The remittances of those workers are an important resource for Latin American economies.

→ *Latifundia* and *minifundia* are two major agricultural systems. There is a strong plantation economy, especially in Central America, the Caribbean, and coastal Brazil, with sugar and fruits dominant. The fair trade movement raises prices for consumers of coffee, bananas, and other produce but helps ensure better wages for Latin American farmers. Ownership of farmland is concentrated in the hands of relatively few people, creating imbalances that land reform efforts have so far not redressed.

→ Minerals have played a major role in the history, economic development, and trade networks of Latin America. Petroleum is of growing importance, especially in Mexico and Venezuela. The global commodities boom related especially to China's growth has helped Latin American economies in recent years.

→ The Latin American countries have formed or joined several free-trade agreements. The largest in volume and value is the 1994 North American Free Trade Agreement (NAFTA). One of its goals is to promote economic growth in Latin America so that migration to the United States becomes less attractive to unemployed Latin Americans.

→ Tropical Latin America's proximity to temperate and affluent North America has helped make tourism one of the region's most important industries.

→ U.S. interests in Latin America dominate the region's geopolitical themes and issues. The Monroe Doctrine and its Roosevelt Corollary were designed to justify U.S. activities and interventions in the region. Historical efforts have been directed at suppressing Communist or leftist interests in the region and at safeguarding passage through the Panama Canal. Modern interests focus on promoting trade, fighting drug trafficking, and guaranteeing secure access to oil in Venezuela and Colombia.

KEY TERMS + CONCEPTS

alternative development (p. 549)
Amerindians (p. 531)
Aztec-Tanoan language family (p. 535)
Aztecs (p. 533)
balloon effect (p. 548)
barrio (p. 538)
Bay of Pigs invasion (p. 547)
Chibcha (p. 534)
Creole (p. xxx)
creole languages (p. 535)
crop substitution (p. 549)
Cuban Americans (p. 547)
Cuban Missile Crisis (p. 547)
decertification (p. 549)
distance-decay relationship (p. 545)
Dominican Republic–Central America
    Free Trade Agreement
    (DR-CAFTA) (p. 543)
ejido (p. 542)
El Niño Southern Oscillation (ENSO;
    El Niño) (p. 532)
fair trade movement (p. 540)
favela (p. 538)
Free Trade Area of the Americas (FTAA)
    (p. 543)
hacienda (p. 539)
hinterland (p. 522)
Hokan-Siouan language family (p. 534)

hurricane (p. 531)
Indians (p. 531)
intellectual property rights (IPR) (p. 543)
Inca (p. 534)
land reform (p. 541)
latifundia (p. 539)
maquiladora (p. 544)
Maya (p. 531)
Mayan language family (p. 535)
megalanguage (p. 535)
Mercosur (p. 542)
mestizo (p. 538)
Mexica (p. 533)
minifundia (p. 540)
Monroe Doctrine (p. 546)
mulatto (p. 538)
multi-Latina companies (p. 545)
narcoterrorist organization (p. 547)
Native Americans (p. 531)
Nazca (p. 534)
North American Free Trade Agreement
    (NAFTA) (p. 544)
Organization of American States (OAS)
    (p. 543)
Oto-Manguean language family (p. 535)
pampa (p. 522)
páramos (p. 530)
Penutian language family (p. 535)

Plan Colombia (p. 549)
Plan Patriot (p. 547)
pre-Columbian times (p. 537)
Quechu-Aymaran language family
    (p. 535)
    Quechua (p. 535)
resource curse (p. 541)
Roosevelt Corollary (p. 546)
rules of origin (p. 544)
shantytown (p. 538)
Southern Cone Common Market
    (Mercosur) (p. 542)
Tarascan (p. 534)
Teotihuacános (p. 532)
tierra caliente (hot country) (p. 527)
tierra fría (cold country) (p. 527)
tierra helada (frost country) (p. 527)
tierra templada (cool country) (p. 527)
Totonac language family (p. 535)
Trading with the Enemies Act (p. 547)
tree line (p. xxx)
Triple Alliance (p. 534)
Union of South American Nations (p. 543)
upper limit of agriculture (p. 530)
war on drugs (p. 546)
Washington Consensus (p. 547)

REVIEW QUESTIONS

1. Where are the main areas of population concentration in Latin America?

2. What and where are the principal climate and biome types of Latin America?

3. What are the four elevation zones? What human uses are generally associated with them?

4. What causes El Niño? What effects are usually associated with it in particular regions?

5. What and where were the major Native American civilizations prior to the arrival of the Europeans?

6. What are the major pre-Columbian and modern languages of Latin America, and where are they spoken?

7. What are the principal faiths in Latin America?

8. What are the principal cash crops and minerals exported from Latin America?

9. What are some of the main types of landholdings? What crops are associated with them? Why has land reform been an important issue in Latin America?

10. What are the region's most important existing and planned free-trade agreements?

11. What are remittances? What role do they play in Latin American economies?

12. What country is most significant in the geopolitics of Latin America, and how has it exercised its influence in the region?

NOTES

1. Quoted in Alfred M. Tozzer, "Landa's *Relación de las Cosas de Yucatán:* A Translation." In *Papers of the Peabody Museum of American Archaeology and Ethnology,* Vol. 18 (Cambridge, Mass.: Harvard University Press, 1941), p. 78.

2. Bernard Comrie, Stephen Matthews, and Maria Polinsky, *The Atlas of Languages* (New York: Facts on File, 2003), p. 151.

3. Simon Collier, Harold Blakemore, and Thomas E. Skidmore, eds., *The Cambridge Encyclopedia of Latin America and the Caribbean* (Cambridge: Cambridge University Press, 1985), p. 159.

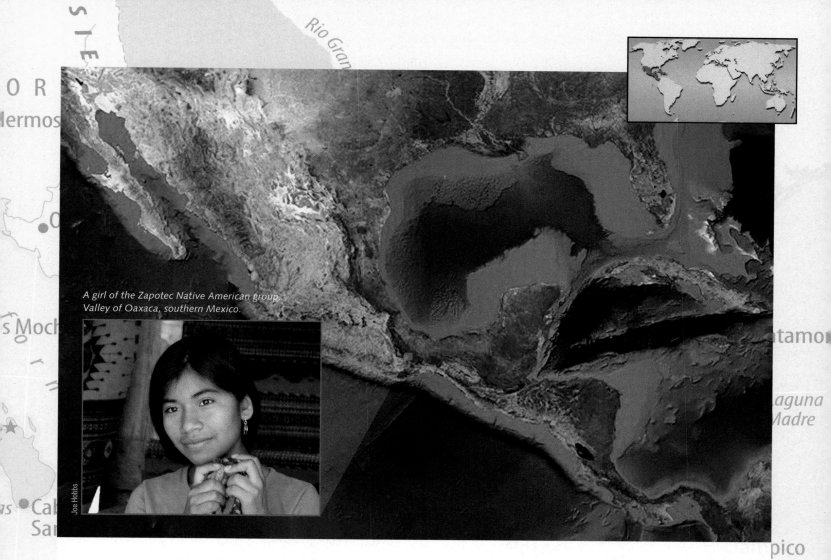

A girl of the Zapotec Native American group, Valley of Oaxaca, southern Mexico.

Joe Hobbs

MODULE 10.1
MIDDLE AMERICA: LAND OF THE SHAKING EARTH

The region most geographers prefer to call "Middle America," also known as Mesoamerica, includes Mexico (its largest country in area and population); the much smaller Central American countries of Guatemala, El Salvador, Belize, Honduras, Nicaragua, Costa Rica, and Panama; and the numerous island countries in and near the Caribbean Sea, the largest being Cuba, Haiti, the Dominican Republic, and Jamaica. The region is a vibrant patchwork of physical features, ethnic groups, cultures, political systems, population densities, and livelihoods. Most of the political units are independent, but a few are still dependencies of outside nations.

Middle America is a region of spectacular natural landscapes, from highland mountain chains and towering, isolated volcanoes to lush, seaside tropical rain forests. Its location astride major zones of collision between plates of the earth's crust (see Figure 2.1) is responsible for much of the region's natural beauty but has also brought suffering in the form of earthquakes and volcanic eruptions. This chapter is subtitled after *Sons of the Shaking Earth,* a fine

■ module
outline

■ module
objectives

This chapter should enable you to

→ Recognize how the maldistribution of resources, particularly of quality farmland, has contributed to dissent and war

→ Understand how the economic boom in China, where labor costs are very low, has taken a toll on the region's industries

→ See why fertile volcanic uplands came to be the core regions of most of the mainland countries

→ Appreciate the vital role played by tourism in many of the nations' economies

→ Understand how the limited resources of small island countries hampers their economies and leads them in search of Internet, banking, and other alternative ventures

depiction of the Guatemalans and Mexicans living in the geologically active heart of the region, written by the anthropologist Eric Wolf.

This survey begins with the demographic and territorial leader, Mexico.

10.1.1 Mexico: Higher and Further

The federal republic of Mexico—officially, *Los Estados Unidos Mexicanos* (United Mexican States)—is the largest, most complex, and most influential country in Middle America (●Figure 10.1.1). The country's great population of 106 million in 2007 makes Mexico by far the world's largest nation in which Spanish is the main language. Compared to most of the world's countries, Mexico is a spatial giant, but its large size tends to go unrecognized because it lies in the shadow of the United States. Triangular in shape, Mexico's area of 756,000 square miles (c. 2 million sq km) is nearly eight times that of the United Kingdom, and its elongated territory would stretch from the state of Washington to Florida if superimposed on the United States (see Figure 10.2, page 525). Its capital, Mexico City, stands now as the world's second-largest city in population (after Tokyo-Yokohama), with a whopping 22.7 million inhabitants. Mexico is one of the most urbanized countries of Middle America, with about 75 percent of its people living in cities. In these and many other attributes, Mexico in the Latin American context seems to live up to its motto: "Higher and Further."

Human Geographies

Mexico's modern cultural patterns are the legacy of three major eras: the long Native American era prior to the Spanish conquest, the Spanish colonial era of the 16th to early 19th centuries, and the Mexican era following independence in 1821. In the Native American era, the region was home to indigenous peoples speaking hundreds of languages and dialects. Many of these cultures were very advanced and among other things produced the remarkable architectural legacies of Teotihuacán and Chichén Itzá. Richly varied 531 indigenous cultures persist today (●Figure 10.1.2). While some of the Native American groups are racially almost identical to their pre-Columbian ancestors, the greater part of the people of Mexico are mestizos of mixed Spanish and Amerindian ancestry. Of Mexico's population, Native Americans (locally called *indios* or *indigenas*) make up 30 percent, mestizos 60 percent, and ethnic Europeans (nearly all of Spanish ancestry) only 9 percent.

The Spanish colonial era lasted nearly three centuries. Following the overthrow of the Aztec empire by Hernando Cortés in 1521, the Spanish crown established the **Viceroy-** 535 **alty of New Spain.** Ruled from Mexico City, the viceroyalty grew to encompass the area of modern Mexico, most of Central America, and about one-quarter of the territory of today's 48 conterminous United States. Spanish administrators, fortune hunters, settlers, and Catholic priests established the **Hispanic** pattern of life, reflecting Spanish language and culture, that is still prominent today. Spanish is spoken as a first language by 88 percent of the people, and 89 percent follow the Roman Catholic faith that the Spaniards introduced into the region.

The present settlement pattern was strongly influenced by the distribution of Native Americans at the time the Spanish arrived. The conquerors sought out concentrations of Native Americans who would become laborers and Christian converts. Spanish settlements thus tended to grow near existing population centers or were established as port cities with access to indigenous settlements and resources lying inland. A great number of today's towns and cities originated as Catholic missions established in the midst of indigenous populations. Spanish influences are found in architectural styles and urban layouts. Towns and cities characteristically have a Spanish-inspired rectangular grid of streets surrounding a plaza (*zócalo*) at the urban center. The plaza functioned both as the major site for periodic markets and as a setting for the dominant Catholic church in the settlement (see Figure 10.17, page 537). The Spanish hunger for precious metals greatly expanded mining, particularly of silver, and many of Mexico's larger cities were founded as silver-mining camps or regional service centers for these mining areas.

When the Mexican era began in the early 19th century, Mexico was overwhelmingly poor, rural, illiterate, village-centered, and church-oriented. In the capital, an educated elite governed and exploited the country with little benefit to the Native American and mestizo farmers in the coun-

● **Figure 10.1.1** Principal features of Mexico.

Joe Hobbs

● **Figure 10.1.2** This yarn weaving was made by a Native American of the Huichol people, who live in north-central Mexico. This art reflects the artist's visions prompted by ritual use of peyote.

tryside. The bloody **Mexican Revolution of 1810–1821** ended Spanish control and instituted a short-lived Mexican empire that expired in 1823, to be succeeded by a federal republic in 1824. Then came a long period of political instability. From the 1830s through the 1840s, huge blocks of the nation's territory were lost to neighboring countries, including the Central American part of former New Spain, as well as what are now Texas, California, New Mexico, Arizona, Nevada, Utah, and parts of other states.

During the 1850s and 1860s, a dynamic Native American reformist named Benito Juárez ruled, but then Mexico endured a long period (1876–1910) of despotic rule by President Porfirio Díaz. Under Díaz, democratic processes were suspended and the rural population sank deeper into poverty, even as heavy foreign investments were made in railroads, the oil industry, metal mining, coal mining, and manufacturing. Political discontent over national disunity and weakness, internal oppression and corruption, foreign exploitation, poverty, and cries for land reform ultimately triggered the **Mexican Revolution of 1910–1920,** led by Pancho Villa and Emiliano Zapata, which ousted Díaz and turned into a destructive struggle among personal armies of regional leaders. In 1917, a new constitution was enacted under which land reform was an urgent priority. After that, huge acreages were expropriated from *haciendas* by the government and redistributed or restored to landless farmers and farmworkers, most particularly in a program that provided *ejidos,* or communal farmlands, to mainly Indian village farmers. Large *haciendas* still exist, especially in the dry north, but they contain only a very small proportion of Mexico's cultivated land.

Land-centered revolutions may not be over. In Mexico's poorest and southernmost state, Chiapas, dissidents mainly from several Maya ethnic groups formed a revolutionary movement called the **Zapatista Army for National Liberation (EZLN)** in 1994. Demanding indigenous autonomy and greater control over land and other resources, one of their chief complaints was the enactment that year of the North American Free Trade Agreement (NAFTA). They denounced globalization, calling Mexico's export industries a symbol of the country's slavery to rich nations. Countering arguments that ecotourism development in Chiapas could help preserve their way of life and natural environments, the Zapatistas condemned government tourism schemes, saying these would turn the region into "an amusement park for foreigners." They called ecotourism advocates "fools trying to change our lives so that we will cease being what we are: indigenous peasants with our own ideas and culture."[1] Zapatista seizures of mines and other assets led to clashes with government forces that caused numerous deaths and refugee problems. Negotiations led to the 1996 **San Andres Accords,** which promised autonomy for many communities. The Mexican government has withdrawn its military from Zapatista-held areas and tensions have eased, but the Zapatistas complain that the government has not fulfilled its promises, so the situation remains volatile.

Mexico, and in turn the United States, has much at stake in addressing the problems of Chiapas and its other poor southern states (there has also been unrest in neighboring Oaxaca). These provinces are gateways for even poorer migrants from Guatemala and other countries farther south and outside the hemisphere (including Somalia and China). Many of these illegal immigrants are on their way to the United States, where the Border Patrol knows them as **OTMs ("other than Mexicans").** Others stay to fill the jobs of the southern Mexicans who have already fled to the United States. Mexico has a project called **Plan Puebla-Panama** designed create jobs locally by improving and integrating electricity and other infrastructure with Guatemala and other Central American countries. One of its unstated objectives is to ensure that leftist governments like Hugo Chávez's Venezuela do not come forward first with development aid for the region.

Agricultural Environments

A little over one-half of Mexico lies north of the Tropic of Cancer and is dominated by desert and steppe climates with hot summers and warm to cool winters, depending on elevation. South of the Tropic of Cancer, where about four-fifths of all Mexicans live, seasonal temperatures vary less, but elevation differences create conditions ranging from high temperatures in the *tierra caliente* of the lowlands along the Gulf of Mexico through moderate heat in the *tierra templada* of the low highlands to cool temperatures in the *tierra fría* of the highlands (these elevation zones are explained on page 529 in Chapter 10). Nearly all the tropical areas are humid enough to support rain-fed crops, but irrigation is often used to increase crop diversity and productivity.

Agriculture is still a very important source of livelihood and export revenue for Mexico. Leading agricultural exports are sugar, coffee, corn, citrus fruit, and cattle. Corn (maize), first domesticated in Mexico about 9,000 years ago, is by far the most important food crop. Beans, squash, and chili peppers, also an inheritance from the Native American past, are universal Mexican foods, interpretations of which have become widespread in North American fast-food and kitchen cultures. Mexico has great agricultural variety, despite its mountainous environment and the fact that only one-eighth of the country is cultivated. Good cropland exists mainly on the floors and lower slopes of mountain basins with volcanic soils and on scattered, small alluvial plains. In many places, even the steep slopes are farmed as *milpas* (slash-and-burn plots) by peasant farmers desperate for access to land (●Figure 10.1.3).

Mexico has a serious deforestation problem, losing more than 100,000 square miles (c. 260,000 sq km) of trees annually. Much of this activity is illegal commercial logging operated by organized criminals. Some clearing is also done to open land for marijuana and opium cultivation. Environmentalists are particularly concerned with deforestation of the lowland tropics of Mexico's southernmost state, Chiapas. The great biological diversity of these forests accounts largely for Mexico's ranking as the world's third most important megadiversity country, after Brazil and Indonesia. The jewel of these forests is the Montes Azules reserve of Chiapas, which the government of Mexico granted to a traditional Maya group called the **Lacandon** in 1978. The Lacandon themselves were responsible for some of the destruction by granting logging leases to government-backed companies in the 1980s. More recently, the Lacandon have been blaming illegal squatters for the forest destruction in Montes Azules. Most of the squatters are indigenous peoples of the **Tzeltal Maya** and **Chol Maya** groups, and many of them are affiliated with the antigovernment Zapatista movement. Evicting them is therefore as much a political as an environmental problem.

Mining, Manufacturing, and *Maquiladoras*

Mexico's turbulent and complex geology has produced diverse mineral resources. Mexico ranks highly among the world's producers of such commodities as metals, oil, gas, and sulfur, with oil the country's most important export. The Mexican oil industry is a government monopoly (carried on by a government corporation, Petróleos Mexica-

• **Figure 10.1.3** Known in Mexico and Central America as *milpa,* swidden or slash-and-burn cultivation requires the clearing and burning of forest (top) before crops like corn, beans, and squash can be grown. When the soils are exhausted after three to five years (bottom), the plot is abandoned and the farmer moves on to clear another.

nos, or PEMEX), and the United States is its largest foreign customer, taking in 90 percent of the country's oil exports. Mexican oil represented 15 percent of oil imported to the United States in 2007, making Mexico the largest foreign supplier after Canada.

Mexico's bonanza of oil strikes in the 1970s seemed so promising that the government borrowed heavily from foreign banks to finance a wave of other economic and welfare projects. Crude oil grew to two-thirds of all Mexican exports in value by 1983. A worldwide oil glut then caused prices and oil export values to skid. By the early 1990s, much export revenue was spent on servicing debts, contributing to a collapse of the Mexican peso's value in 1994. Recent restructuring of the economy and a steady increase in

the value of oil exports have created a more promising outlook for the near future. As of 2007, oil exports accounted for 10 percent of the value of Mexico's exports. Oil production in Mexico is expected to peak in 2010 and come to an end by 2030.

Mexico has made large gains in the development of manufacturing and component assembly industries, especially since the implementation of NAFTA. The manufacturing 544 sector is now Mexico's leading source of revenue, followed by oil and remittances from abroad. High-technology exports grew from 8 percent of total exports in 1990 to 20 percent in 2007. About one-third of all manufacturing employees in Mexico work in the thousands of factories located in metropolitan Mexico City. Most of those factories are small, and manufacturing is devoted largely to various consumer items, although some plants carry on more substantial manufacturing in metallurgy, oil refining, and heavy chemicals.

To take advantage of the vast U.S. market to the north, *maquiladora* factories are located close to the U.S. border (but they also cluster in the Yucatán Peninsula, where labor is less expensive). The *maquiladora* is designed to boost 543 exports of manufactured goods by using tax breaks, cheap labor, and other incentives to attract foreign industries. The factories produce jeans, cars, computers, and other goods, mainly from components that are imported duty-free because they are assembled for export only (•Figure 10.1.4). The *maquiladoras* have surged in growth and diversification since NAFTA went into effect; about 2,700 of these factories, employing nearly 1 million Mexicans, are now operating in the U.S. border region alone. There has also been growth in industries supporting the *maquiladoras,* such as small-scale retail and transportation services. Among the most recent developments are **energy *maquiladoras.*** These are power generation plants—built by American firms just inside Mexico, where pollution standards are much lower than in the United States—that provide electricity to the U.S. grid. Their power source, natural gas, comes from Texas, and Mexico supplies their coolant waters.

In the global marketplace, Mexico's *maquiladoras* are evolving. As the U.S. economy slowed while China's surged between 2000 and 2002, Mexico lost the competitive edge represented by its inexpensive labor. Hundreds of *maquiladoras* closed, and hundreds of thousands of their workers lost jobs. Mexico's clothing, textile, and other low-tech industries could not match Chinese rock-bottom labor costs. Mexico itself became flooded with Chinese imports, even of pottery imitating Mexico's traditional wares. While conceding the loss of low-tech, low-skill industrial advantage to China (and also nearby El Salvador, where labor is also cheaper), Mexican manufacturers hung on to Mexico's edge in exports of flat-screen televisions, computers, aircraft parts, and other high-tech and heavier goods that are more expensive to ship from China. The *maquiladoras* began a slow and steady recovery and are now thriving. They have also become begun to merge into specialized clusters, for

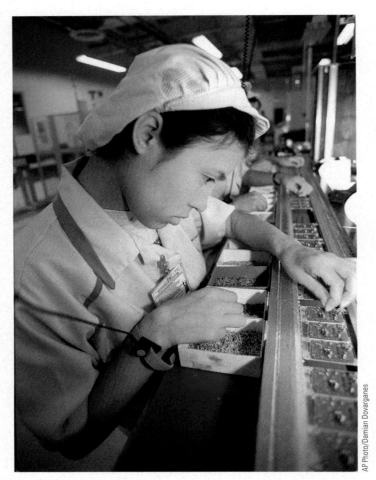

AP Photo/Damian Dovarganes

● **Figure 10.1.4** Globalization in a nutshell. Electronic devices for Samsung, a Korean firm, are assembled in a *maquiladora* in the Mexican border town of Tijuana for sale in the United States.

example, with Tijuana concentrating on flat-screen televisions and Ciudad Juárez on car parts.

One of Mexico's outstanding development problems is that relatively few of its young people are gaining multilingual and multicultural educations abroad, while Chinese and South Koreans, for example, generally see study abroad as an important investment in their national economies.

Economic relations with the United States, so well symbolized by the border *maquiladora*, are crucial to Mexico. Trade between Mexico and the United States increased over 100 percent between 1994 and 2007. Exports from Mexico to the United States have more than tripled. The trade is in Mexico's favor, with imports from Mexico exceeding exports to Mexico by 29 percent. The United States is Mexico's leading trade partner, with about 77 percent of its exports going to the United States. Mexico is the United States' second-largest trading partner (after Canada), with 13 percent of U.S. exports bound for Mexico. Only Canada and China export more goods to the United States than Mexico does.

There are both inseparable ties and divisive issues between Mexico and the United States. Huge disparities in wealth and power come between them. Mexicans have not

forgotten that Mexico lost more than half its national territory to the *gringos* in the **Mexican War** (known to Mexicans as the **American Invasion**) in 1847. Mexicans have enormous national and cultural pride and often feel misunderstood and underappreciated by their northern neighbors.

Regional Geography of Mexico

Mexico is strongly regionalized (for the country's physical geography, see Figure 10.4, page 528). Its major subregions are Central Mexico, the country's core; Northern Mexico—dry, mountainous, and intertwined with the United States through migration and *maquiladoras*; the Gulf Tropics, an area of hot, wet coastal lowlands along the Gulf of Mexico, together with an inland fringe of *tierra templada* in low highlands; and the Pacific Tropics, a poorly developed southern outlier of humid mountains and narrow coastal plains.

Central Mexico: Volcanic Highland Core Region

Many Latin American countries have a sharply defined core region where national life is centered, and Mexico is a classic example of this cultural and demographic pattern. About half of the population inhabits contiguous highland basins clustered along an axis from Guadalajara (population 4.1 million) at the northwest to Puebla (population 2.8 million) at the southeast. Toward the eastern end of the axis is massive Mexico City. The basins comprise Mexico's core region, often referred to as Central Mexico, the Central Plateau, or the Mexican Plateau. The basin floors vary in elevation from about 5,000 feet (c. 1,500 m) to 9,000 feet (c. 2,700 m), with each basin separated from its neighbors by a hilly or mountainous rim. Ash ejected from the region's many volcanoes has added fertility to the soils. The centers of the basins contain flat and sometimes marshy land. This swampiness has led populations to cluster on higher sloping land, and population pressure has led to cultivation of steep slopes, causing much soil erosion. The prevailing *tierra fría* environment limits the growth of crops to those less susceptible to frost damage, such as corn, sorghum, wheat, and potatoes. Beef, dairy cattle, and poultry fare well in this environment.

In the mid-14th century, the Aztecs came to the Basin of Mexico from the north and built their capital, Tenochtitlán, on an island in the shallow, salty Lake Texcoco. Canals and dikes were constructed to control flooding and protect water resources for the city, which was connected to the shore by causeways. This was the city that Cortés conquered in 1521. Mexico City was later founded in this same locale. Most of the lake was drained, and city structures have been built on the drained land over the past four and a half centuries. Gradual subsidence (settling of land and buildings) continues today as the lake sediments contract (●Figure 10.1.5). In recent times, overpumping of groundwater to support a metropolitan region of 22 million people has aggravated this problem. But despite the unstable subsurface

522

529

533

and the city's presence atop an active geologic fault zone, Mexico City sports imposing office towers. Fortunately, most have been built with foundations designed to withstand both subsidence and earthquake shocks. These withstood a strong test in the form of a magnitude 8.1 earthquake that rocked the city in 1985, killing 10,000 people and rendering 250,000 homeless.

The city has other planning achievements to its credit, including a modern subway system and a system to pump water to the city through the encircling mountain walls of the Basin of Mexico. Overall, though, explosive urbanism has outrun the efforts of planners to cope with the capital city's growth. Mexico City's inhabitants must cope with the all too familiar difficulties of developing world megacities, including dreadful air pollution, compounded in this case by its topographical setting. Being situated low in a valley with mountains surrounding, Mexico City (like Los Angeles) often experiences a **thermal inversion.** This is a circumstance in which air that is cooled at night sinks down the mountainsides surrounding the city and settles underneath a layer of warmer air (●Figure 10.1.6). The cooler air below is trapped by the warm air above, and without air

● **Figure 10.1.5** Mexico City's subsidence is dramatically evident at the Shrine of Our Lady of Guadalupe, where the massive cathedral (right) has sunken sharply inward toward the plaza.

● **Figure 10.1.6** Model of the thermal inversion responsible for smog in Mexico City.

circulation, the pollution from factories and millions of (often gridlocked) cars rises to dangerous levels.

Rugged Northern Mexico and the Borderlands

North of the Tropic of Cancer, Mexico is characterized by ruggedness, aridity, and a ranching economy that supports a generally sparse population. There are some widely separated spots of more intense activity based on irrigated agriculture, mining, heavy industry, and diversified border industries adjacent to the United States. Most of Northern Mexico lies above 2,000 feet (600 m) in elevation, with lowland strips along the Gulf of Mexico, the Gulf of California, and the Pacific. The climate is desert or steppe, except where elevations are high enough to yield more rainfall, and there are coniferous forests. Most of the forested land is in the Sierra Madre Occidental—the westernmost of the region's two mountain ranges—which serves as the country's main source of timber.

Northern Mexico supplies the country's greatest output of metals, coal, and irrigated crops. Iron ore, zinc, lead, copper, manganese, silver, and gold are the leading metals. Ores are smelted with coal mined in this region. Some coal is made into coke for the manufacture of iron and steel, particularly around the city of Monterrey (population 3.8 million), Mexico's leading heavy industry center and Northern Mexico's most important business and transportation hub.

Irrigated agriculture in the dry north has expanded with the building of dams to impound waters flowing from the Sierra Madre. Most irrigated farms are in the northwest and near the Rio Grande. Cotton is a major product of Northern Mexico's irrigation, and varied fruits and vegetables (notably tomatoes and cucumbers) for shipment to U.S. markets are also grown. In some irrigated areas, the Mexican government has helped develop agricultural colonies of small landowning farmers who have migrated from more crowded parts of the country.

With NAFTA came a boom in manufacturing in the **border towns** adjacent to U.S. cities. The largest are Tijuana, adjoining but fenced off from San Diego, California, and Ciudad Juárez, across the Rio Grande from El Paso, Texas. Much of the traffic in illegal drugs from South America and Mexico itself makes it into the United States through these border towns. The border towns are sin cities for U.S. citizens interested in indulging their pleasures in ways they would not dare at home: Prostitution is legal, the drinking age is 18, prescription drugs are sold without prescriptions, and illegal drugs are easy to obtain. They are also notoriously violent cities, with a high death toll related to the cross-border drugs trade.

The border towns are collection points for Mexican and Central American migrants attempting to enter the United States illegally on the countries' common 2,000-mile (3,200-km) border. In a remarkable variant of the balloon effect described on page 548 in the context of the war on drugs, the U.S. crackdown on crossings at the major border towns has led to the mushrooming of new, more remote border communities that serve as springboards for crossing weaker points in the border defenses. An estimated one-half to two-thirds of those who try actually succeed in crossing, despite the efforts of the overburdened U.S. Border Patrol. Periodic amnesties have pardoned illegal migrants who could prove long-term residence in the United States, and more measures are being considered to deal with the illicit flow (see pages 600–601 in Module 11.2). The draw of *El Norte*—the North—can be easily appreciated by the ratio of 4 to 1 in the per capita GNI PPP figures for the United States and Mexico and the fact that nearly half of Mexico's people are classified as poor.

The Gulf and Pacific Tropics: Oil, Mountains, Beaches, and Tourists

On the Gulf of Mexico side, Mexico has tropical lowlands that have a much lower population density than Central Mexico but are very important to the national economy. Their most valuable product is oil, discovered in huge quantities in the mid-1970s. Just as in coastal Texas and Louisiana, some fields are onshore and others lie underneath the gulf.

The Gulf Tropics are Mexico's main producer of tropical plantation crops. Cacao, sugarcane, and rubber are leading export crops in the lowlands, and coffee exports come from a strip of *tierra templada* along the border between the lowlands and Mexico's Central Plateau. The greater part of Mexico's Gulf Tropics lies in the large Yucatán Peninsula, where the Maya civilization flourished in pre-Columbian 531 times. The peninsula's east coast, fronting the Caribbean, is being developed and promoted as the "Riviera Maya." From Cancún southward stretches a swath of tourist resorts of varying degrees of harmony with the landscapes and seascapes. The sultry and exotic environment is an irresistible draw for U.S. tourists, who make up 80 percent of Mexico's foreign visitors and whose dollars provide Mexico with much needed revenue. Tourism overall is Mexico's fourth-largest source of foreign currency earnings (after *maquiladora* manufacturing, oil, and remittances from Mexican workers abroad), with about 20 million visitors in 2007 generating an estimated $10 billion in income for the country.

South of the Tropic of Cancer, rugged mountains with a humid climate lie between the densely populated highland basins of Central Mexico and the Pacific Ocean. This difficult terrain is far more thinly populated than Central Mexico. Native Americans carry on a traditional agricultural livelihood in this area of otherwise little economic activity. There are two exceptions to the comparative underdevelopment. One is a string of coastal resorts, including Mazatlán, Puerto Vallarta, and Acapulco, which are international

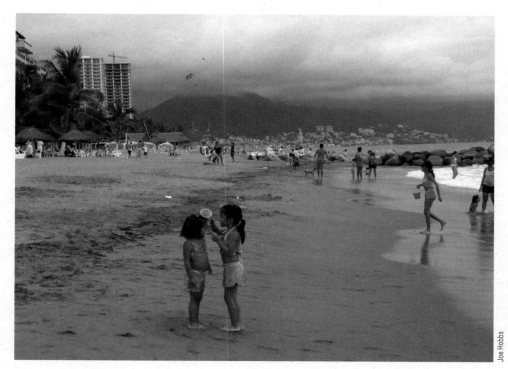

• **Figure 10.1.7** Mexico's coasts are magnets of domestic and international tourism. This is the popular Pacific resort beach of Puerto Vallarta.

tourist destinations (•Figure 10.1.7). There is also a major iron and steel complex near La Unión.

10.1.2 Central America: Beyond Banana Republics

Between Mexico and South America, the North American continent tapers southward through an isthmus containing seven less developed countries known collectively as Central America. Five of these countries—Guatemala, Honduras, Nicaragua, Costa Rica, and Panama—have seacoasts on both the Pacific and the Atlantic Oceans (via the Caribbean Sea; •Figure 10.1.8). The other two have coasts on one body of water only: El Salvador on the Pacific and Belize on the Caribbean. If the seven countries were one, they would have an area only about one-fourth that of Mexico (see Table 10.1), with a total population about 40 percent that of Mexico.

Environmentally, the region is both very rich and threatened. With about 8 percent of the world's biodiversity, Central America suffers from habitat destruction rooted in poverty and profiteering. In "debt-for-nature swaps," the United States has recently forgiven portions of the foreign debts of Guatemala, Belize, and Panama, in exchange for promises by the those countries to establish forest preserves. Preservation of the moisture-retaining tropical forests in the watershed of the Panama Canal is critical for proper

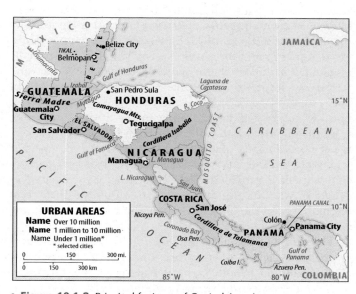

• **Figure 10.1.8** Principal features of Central America.

operation of the waterway, so investment in that resource is especially beneficial to Panama.

Fragmentation and Turbulence

Geographic fragmentation, both physically and politically, is a major characteristic of Central America. Five countries—Guatemala, El Salvador, Honduras, Nicaragua, and Costa

64

Rica—were originally governed together as a part of New Spain in a unit called the **Captaincy-General of Guatemala.** Composed of many small settlement nodes isolated from each other by mountains, empty backlands, and poor transportation, the unit never established strong geopolitical cohesion. Separate feelings of nationality became strong enough to fracture overall unity after the end of Spanish imperial rule in 1821. The area gained independence from Mexico in 1825 as the **Central American Federation,** but in 1838–1839, this loose association fragmented into the five republics of today. In recent decades, there were serious international tensions affecting the five countries, associated in part with antigovernment guerrilla warfare in various countries. However, by the mid-1990s, tensions had lessened, and the Central American area appeared to enter a new era of greater political stability.

In Guatemala, El Salvador, Honduras, and Nicaragua, the population is composed of groups that differ sharply in wealth, social standing, ethnicity, culture, and worldview. 539 Wealthy owners of large *latifundia* (estates) formed a social aristocracy that has lasted since the Spanish crown made the first land grants in early colonial times. Members of this group tend to be ethnically European (Spanish). They have, in general, monopolized wealth, owned the most and best agricultural lands, defended the status quo, and exercised great political power. Occupying a middle position in society are small landholding farmers, largely mestizos, who own and work their own land. At the bottom of 537 the scale are millions of landless mestizo tenant farmers, hired workers, and the many Native Americans (especially in Guatemala) who live essentially outside of white and mestizo society. Also low in the economic ranks are blacks and mulattoes descended from the black African slaves brought to the region beginning in the 16th century and from the migrants who came later from Jamaica to work on the region's banana plantations. Poverty, unemployment, a young population, and weapons left over from the region's civil wars have provided ideal conditions for the emergence of gangs, with an estimated total of 25,000 members, that wreak havoc in the big cities of Honduras, El Salvador, and Guatemala.

There is widespread sentiment in this region that government officials look after their own welfare but not that of the common people. Central America has a history of revolutionary movements and popular uprisings, with the United States typically backing the seated powers. In Nicaragua in 1979, revolutionaries known as **Sandinistas** threw out the governing, U.S.-supported Somoza family. After a very contentious 11-year rule characterized by continual battles with the military-backed **Contras** (also supported by the United States), the Sandinistas were replaced by an elected president, and by the early 1990s, a reasonably stable government was established, with some of the earlier Sandinista reforms institutionalized. Daniel Ortega, leader of the Sandinistas during the turbulent 1980s, was elected the country's president in 2006. His challenges include improving the welfare of a Latin American country whose poverty is surpassed only by Haiti and Bolivia. Nicaragua has meanwhile become a magnet for wealthy Americans seeking vacation and retirement properties.

In El Salvador, a reformist-minded government attempted in 1980 to redistribute some of the country's better farmlands from the wealthy elite to rural peasant cooperatives. 542 A violent civil war ensued between the right-wing landowners, supported by the military, and the left-wing peasants. An estimated 75,000 people were killed and 1 million made homeless in the 12-year conflict. Since the war ended, El Salvador has become a stable democracy with an economy based largely on exports of commercial crops and clothing. Like many other textile-exporting countries, however, El Salvador lost many jobs in the clothing industry (mainly to China) after tariffs on the global textile trade ended in 2005.   299

Guatemala fared the worst. A long civil war there, costing more than 200,000 lives, finally ended in 1996. The bloodletting began in 1954 when the U.S. Central Intelligence Agency undertook a coup to oust the country's president. President Guzman had resisted the powerful influence of the **United Fruit Company (UFC)** in Guatemala, confiscating 400,000 acres of land from UFC with an aim to subdivide it among the landless poor. U.S. President Eisenhower, interpreting this action against a U.S. company as a sign that Guatemala might drift toward communism, ordered the coup. In a scorched-earth campaign against leftist rebels, government forces carried out most of the killings in the subsequent civil conflict. The army leveled more than 400 villages and massacred tens of thousands of civilians, mainly from among the country's poor, disfranchised, Native American Maya. The war over, Guatemala still suffers from banditry and from underdevelopment; 8 in 10 Guatemalans are poor.

Costa Rica, by far the most prosperous of these small states, has had a different historical development from the other four countries of the old Central American Federation. Distant from the center of the Central American Federation in Guatemala, it was left largely to its own devices in developing its economy and politics. Early Spanish settlers eliminated the smaller numbers of indigenous people there, and the ethnic pattern that developed was dominated overwhelmingly by whites of Spanish descent. More than 85 percent of the present Costa Rican population is white, in stark contrast to the predominance of mestizos or Native Americans in El Salvador, Honduras, Nicaragua, and Guatemala. No precious metals of significance were found in Costa Rica, and no large landed aristocracy developed. The basic population of small landowning European immigrant farmers and their descendants nevertheless managed to create one of Latin America's most economically productive and democratic societies, despite some episodes of dictatorial rule.

Costa Rica is renowned for its minimal military expenditures and its strong efforts to safeguard its natural heritage. National parks and other protected areas make up nearly one-quarter of the country, and ecotourism is the leading source of foreign currency. Otherwise placid Costa Rica has a border dispute on its northwestern frontier with Nicaragua, where a separatist group would like to establish an independent country called Airrecú that would thrive on ecotourism income.

Panama and Belize were not part of the Central American Federation and achieved independence much later. Panama was governed from Bogotá as a part of Colombia until 1903, when it broke away as an independent republic. The United States was deeply involved in the political maneuvering that led to Panamanian independence. In 1904, the United States was granted control over the Panama Canal Zone and undertook the construction of the transoceanic 546 Panama Canal, which opened in 1914 (see Regional Perspective, page 564).

Belize, on the Caribbean side of Central America, was a possession of the United Kingdom during the colonial era under the name of British Honduras. Remote, poor, underdeveloped, and sparsely inhabited, the colony was used mainly as a source of timber, with African slaves brought in to do the woodcutting. Guatemala also claimed Belize as part of its national territory (and the two countries still have a disputed border). Belize became independent in 1981 as a parliamentary state governed by a prime minister and bicameral legislature. The British monarch, represented by a governor-general, is the chief of state, and English is the official language. Ecotourism is a growing industry. Both the pristine tropical rain forests of the country and the laid-back culture of its islands, called *cayes* (pronounced like *keys*), draw visitors from North America and Europe (•Figure 10.1.9). Belize has modest oil exports and produces hydroelectricity. There are concerns that the Chalillo Dam, a hydropower project recently built on the pristine Macal River, will diminish water quality and deter ecotourists. Dozens of similar dams are proposed or under construction across energy-starved Central America.

Environments, Settlements, and Economic Activities

The distribution of settlements and economic life in Central America relate strongly to three parallel physical belts: the volcanic highlands, the Caribbean lowlands, and the Pacific lowlands (see Figure 10.4, page 528).

All Central American countries except Belize share the volcanic highlands that reach south from Mexico. Here lie the core regions, including the political capitals, of Guatemala, El Salvador, Honduras, Nicaragua, and Costa Rica. The only cities with over a million people in their metropolitan areas are the capitals of those countries, respectively: 111 Guatemala City, San Salvador, Tegucigalpa, Managua, and San José. Each of these is a primate city serving as the demographic, economic, and political heavyweight of the na-

tion. Costa Rica, Panama, Nicaragua, and El Salvador are the most urbanized Central American populations, with roughly 60 percent of their peoples living in cities, while Guatemala is the least urban with about 40 percent.

The highlands are dotted with majestic volcanic cones and scenic lakes (•Figure 10.1.10). Ongoing volcanism periodically deposits ash into the highland basins, and this weathers into fertile soils that grow corn, beans, and squash for local food and the coffee that is Central America's largest export crop. Climatically, these basins are classed mainly as *tierra templada*. Only a small fraction of the land is still 529 forested. Between the highlands and the Pacific Ocean lies a coastal strip of seasonally dry lowlands where cattle and cotton are raised, mainly for export. The Atlantic side of Central America is lower, hotter, rainier, and much less populous than the Pacific side. Here, on large corporate-owned plantations, are grown most of the bananas that are Central America's agricultural trademark and second most valuable export, after coffee.

Commercial production of coffee, bananas, cotton, sugarcane, beef, and other commodities provides the majority of Central American exports. While not a major shift toward manufacturing, that is considerable diversification beyond the region's "banana republic" days. As seen earlier in the case of Guatemala, in the early 1900s, the United Fruit Company had a virtual monopoly of Central American 562 banana production. The company established railways to transport the produce from growing areas to coastal ports. The UFC acquired an enormous influence over national political systems and came to be seen by Central Americans as an extension of official U.S. interests. Banana exports continue to be very important in the economies of Honduras, Panama, and Costa Rica. A trade dispute known as the "banana war" broke out in the late 1990s between the United States and the European Union over U.S. corporate banana production from this area as the United States pressed to have Europe open its protected markets to Central American bananas. 89

All the countries except overcrowded El Salvador still have **pioneer zones,** especially on the Caribbean lowlands, where new agricultural settlement is taking place. Settlers generally come from overcrowded highlands, often aided by government colonization plans featuring land grants and the building of roads to get products to market and bring in supplies. Many peasants still practice the ancient *milpa* form of slash-and-burn agriculture.

There is some manufacturing in the few large cities, but the products are generally simple consumer goods for domestic markets. A few factories assemble foreign-made cars, but industries comparable to the *maquiladoras* of Mexico are few. "Cut-and-sew" apparel exports are crucial to El Salvador and Honduras, and baseballs hand-stitched from U.S. components are made in a free-trade zone in Costa Rica. Characteristically for the region, a large share of the profits from industrial enterprises makes its way into very few pockets.

REGIONAL PERSPECTIVE

The Panama Canal

The idea of a shortcut between the Atlantic and Pacific Oceans had been around for a long time before the Panamanian site was chosen. In the early 1800s, Nicaragua was the first choice because of natural landscape features that appeared better able to accommodate a crossing without massive construction (Nicaragua is still considering building this canal, which unlike even the upgraded Panama Canal would accommodate the world's megaships). In 1846, however, Colombia (which then controlled Panama) and the United States signed a treaty supporting a project to cut a canal across the much narrower Isthmus of Panama.

Interest in the project picked up substantially in 1849 with the discovery of gold in California. At that time, prospectors and speculators had to sail to the port of Colón on the east side of Panama, make the 50 miles of land crossing in any way possible, and then reboard a ship in Balboa and sail to San Francisco. By 1855, this pattern was modified by the construction—funded by New York business interests—of a railroad from port to port. This method of crossing, requiring a break of bulk of goods and passengers, would finally be superseded by the canal.

In 1878, Colombia granted permission to a French firm to build a canal across the isthmus near the railroad line. Ferdinand de Lesseps, who had seen the Suez Canal to completion in 1869, designed the project. The French quickly bought control of the railroad line and began digging in 1882, with a plan to build a sea-level transit requiring no **locks** (the devices that allow a ship to move from one water level to another). This venture went bankrupt in 1889 and was followed by another French attempt, which also failed. Construction difficulties and a high death toll due to yellow fever and other diseases were to blame. By the beginning of the 20th century, the French had left. The United States had just participated in the 1898 Spanish-American War and had more geopolitical interest than ever before in this prospective waterway.

In 1903, Panama broke away from Colombia, and within weeks, the United States recognized the new government. Shortly thereafter, the two governments struck an agreement to build the canal, which ultimately required three sets of locks. Construction began in 1904 and, at a cost of $380 million and many lives lost to yellow fever, bubonic plague, and malaria, was completed in 1914. The first crossing was made under a banner that read: "THE LAND DIVIDED; THE WORLD UNITED."

Work on the canal has never ended (●Figure 10.1.A). Reflecting especially the surging exports of China's manufactured goods, the canal's traffic grew by about one-third between 1999 and 2007. To meet anticipated demand, the canal is now undergoing the largest expansion ever. A project set for completion in 2014 aims to overcome problems that now delay and prohibit the passage of many ships. To double capacity and accommodate larger cargo vessels up to the size known in the shipping trade as "post-Panamax," new channels with locks that are longer and deeper will be added at each end of the canal. These and other measures are projected to double the canal's capacity. The country's citizens approved the expensive project in 2006 in the expectation that it will quadruple income from transit fees. Thanks to growing trade, expanding ports, and residential and commercial development, Panama's economy has boomed in recent years; it has invited comparison with that of Dubai, another place that has capitalized on its strategic crossroads location.

1 Deepening and widening of the Atlantic entrance channel
2 New approach channel for the Atlantic Post-Panamax locks
3 Atlantic Post-Panamax locks with 3 water-saving basins per lock chamber
4 Raise the maximum Gatún Lake operating water level
5 Widening and deepening of the navigational channel of Gatún Lake and Galliard Cut
6 New approach channel for the Pacific Post-Panamax locks
7 Pacific Post-Panamax locks with 3 water-saving basins per lock chamber
8 Deepening and widening of the Pacific entrance channel

● **Figure 10.1.A** Profile and plan of the Panama Canal.

● **Figure 10.1.9** Large parts of Belize have not been cleared of their tropical rain forests, and the country is capitalizing on its potential to draw nature-loving tourists. This is the Macal River, downstream from where the controversial Chalillo Dam was built.

10.1.3 The Caribbean Islands: From Rastafari and Reggae to Baseball and Communism

The islands of Caribbean America, also known as the West Indies, are very diverse (●Figure 10.1.11). They encompass a wide range of sizes and physical types, ethnic populations, and political and economic systems. Altogether there are 13 independent countries and 12 possessions of other powers.

General Characteristics

Cuba, the largest island, is mostly lowland, with low mountains at the eastern and western ends. The next-largest islands—Hispaniola (on which Haiti and the Dominican Republic are located), Jamaica, and Puerto Rico—are steeply mountainous or hilly. These four large islands are known together as the Greater Antilles.

Most of the remaining islands are rather small and make up the group known as the Lesser Antilles. They fall into two broad physical types: low, flat limestone islands rimmed by coral reefs, including the Bahamas (northeast of Cuba) and a few others, and volcanic islands, such as the Virgin Islands, Leeward Islands (roughly the northern half of the Lesser Antilles), and Windward Islands (roughly the southern half of the Lesser Antilles), stretching along the eastern margin of the Caribbean from Puerto Rico toward Trinidad and Tobago. Each of the volcanic islands consists of one or more volcanic cones (most of which are extinct or inactive), with limited amounts of cultivable land on lower slopes and small plains (●Figure 10.1.12). Barbados, east of the Windward Islands, is a limestone island with elevated rolling surfaces. The country of Trinidad and Tobago, just off the northeast corner of South America, includes a low mountain range in northern Trinidad that is an offshore continuation of the Andes and is level to hilly elsewhere.

The islands are warm all year. Precipitation varies, with windward northeastern slopes of mountains often receiving heavy rain but leeward southwestern slopes and low islands little rain. The natural vegetation varies from luxuriant tropical rain forests in wet areas to sparse woodland in dry areas. Most areas have two rainy seasons and two dry seasons each year. In summer and autumn, hurricanes, approaching from the east and then curving northward, are a scourge throughout the region and northward all the way to the Middle Atlantic region of the eastern United States. 531, 662

Tourism and plantation agriculture are the islands' economic mainstays, overall. Tourists, primarily from the United States, are a major source of revenue on many islands, most notably the Bahamas, Puerto Rico, the Dominican Republic, and Jamaica. Despite an official ban, increasing numbers of U.S. tourists are also visiting Cuba, already a favored destination for European tourists. Tourism has become Cuba's leading source of foreign exchange. Sugarcane, the crop around which the island plantation economies were originally built, is still the main export crop in Cuba, Barbados, and a few other places. Other commercial crops include coffee, bananas, spices, citrus fruits, and coconuts. Production comes from large plantations or estates worked by tenant farmers or hired laborers, from small farms worked by their owners, and in Cuba, from state-owned farms. In pre-Columbian times, important crops in this region included tobacco, indigo, and cacao.

Mineral production is absent or unimportant on most islands, the most important exceptions being Jamaica (bauxite), Trinidad and Tobago (petroleum and natural gas), and Cuba (metals). Manufacturing has increased and now provides the largest exports from some islands, including chemicals from Barbados and clothing from St. Kitts and Nevis. Dominica and some of the other smaller island countries are making money through Internet gambling and offshore banking, enterprises also undertaken by

● **Figure 10.1.10** The volcanic highlands of Central America have stunning landscapes and are home to most of the region's people. This is Lake Atitlan, in highland Guatemala.

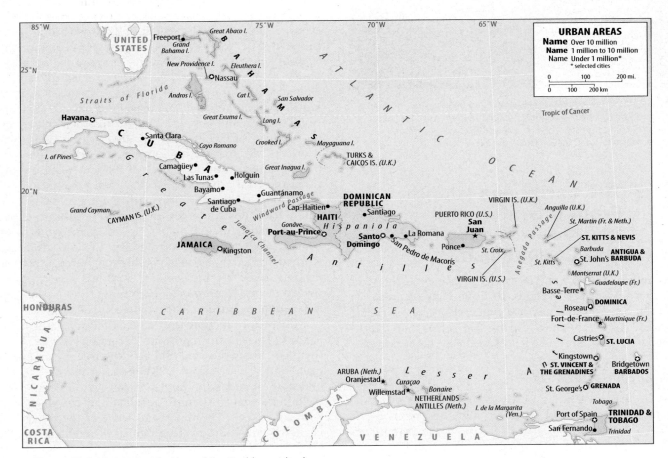

● **Figure 10.1.11** Principal features of the Caribbean Islands.

428

many small, resource-poor islands in the Pacific. Jamaica and Curaçao (in the Netherlands Antilles) are developing their communications infrastructure with the aim of luring information technology (IT) services. Dominica and Grenada get an economic lift from medical schools established on the islands to serve medical students who fail to pass the qualifying exams for U.S. schools.

Overall, the region is quite poor, with a collective average per capita GNI PPP of just $5,630. The wealthiest independent entity is Trinidad and Tobago; the poorest is Haiti, whose indexes of quality of life are among the lowest in the world, comparable to many countries in sub-Saharan Africa. Caribbean poverty has long stimulated emigration. Millions of West Indians now live in the United States, the United Kingdom, Canada, France, and the Netherlands, mainly in cities. The remittances they send back to relatives are a major source of island income.

Sometimes stopping at other islands along the way, Haitian and Cuban immigrants reach (or try to reach) U.S. shores in periodic waves that correspond generally to political tumult in their home countries (●Figure 10.1.13). The United States receives them differently. With powerful political support from Cuban Americans, Cubans are welcomed as asylum seekers, serving as symbols of American defiance of Communist Cuba. Haitians, on the other hand, are turned away because—like so many other Latin Americans—they are regarded as economic refugees. They are not afforded even temporary protected status (as citizens of Honduras, Nicaragua, and El Salvador were until 2006), because Haitians do not meet the criterion for asylum, which is granted only in the aftermath of a war or natural disaster.

547, 615

The Caribbean region has the unfortunate distinction of having the highest HIV/AIDS infection rates outside

● **Figure 10.1.12** A volcanic landscape of hot springs and fumaroles on Dominica, in the Windward Islands.

● **Figure 10.1.13** Haitian refugees seem to bring to life the overcrowded lifeboat metaphor of Garrett Hardin (see page 62).

sub-Saharan Africa. As in Africa, the disease is most widespread among heterosexuals, who transmit the virus through unprotected sex. To date, the Caribbean countries have mustered few of their own resources to fight this scourge, in part because they are fearful that attention to AIDS will deter international tourism. The travel industry has been fearful too; for some time, Royal Caribbean Cruise Lines promoted one of its resort destinations as being in Hispaniola to disguise the fact it was in Haiti. Haiti is one of just two countries in the hemisphere receiving major funds from the U.S. to fight HIV (the other is Guyana). 464

The Greater Antilles

Of the five present political units in the Greater Antilles, Cuba, the Dominican Republic, and Puerto Rico were Spanish colonies; Haiti was French; and Jamaica was British. These units have developed and maintained distinct geographic personalities shaped in part by their colonial past, which for many was not long ago.

By any standard, Cuba is the most important Caribbean island. Centrality of location, large size, a favorable environment for growing sugarcane, some useful minerals, and proximity to the United States contribute to Cuba's leading role in the region. The island would stretch from New York to Chicago if superimposed on the United States and is nearly as large in area as the other Caribbean islands combined. Cuba lies only 90 miles (145 km) across the Strait of Florida from the United States (measured from the southernmost island of the Florida Keys).

At the western end of the island, Havana (population 2.6 million; •Figure 10.1.14) developed on a fine harbor as a major colonial Spanish stronghold, and it serves today as Cuba's main industrial center and its political capital, largest city, and main seaport. Gentle relief, adequate rainfall, warm temperatures, and fertile soils create very favorable

• **Figure 10.1.14** The Malecon, Havana's seashore drive. This is a recent photo; Cubans lovingly maintain American cars from the pre-Castro 1950s.

growing conditions on the island. Agriculture has always been the core of the Cuban economy, with sugarcane as the dominant crop. Cane sugar still accounts for the largest share of Cuban exports by value.

The 1959 Communist revolution, led by Fidel Castro, expropriated sugar and other estates from large private farms and shifted their ownership to the state. About one-quarter of all farmland is now under state control (down from almost 100 percent prior to 1993) and is worked by wage laborers. As happened in most Communist regimes around the world, state-controlled agriculture led to significant decreases in crop production in Cuba. The collapse of the Soviet Union in 1991 eliminated a pillar of foreign support and a vast export market for Cuba's economy. The vital sugar industry lost its Soviet subsidies, forcing the closure of many sugar mills and the conversion of sugarcane fields to other crops. 140, 167

A decades-old U.S. **trade embargo** against Communist Cuba has made it impossible for the country to enjoy a trading relationship that could help make up for the loss of Cuba's partnership with the Soviet Union. Venezuela under Hugo Chávez has stepped into the breach by providing desperately needed oil supplies to Cuba. Even with its economic troubles, Cuba's government has been successful in raising the adequacy of medical care, housing, and education over the years, particularly in comparison to the levels of such services during the 25 years of the Batista regime that Castro replaced in 1959. Cuban doctors have a remarkable history of voluntary service to needy patients throughout the world's poorer countries. Since the 1990s, growing numbers of foreigners have been coming to Cuba for medical procedures that are too costly at home. 547

Southeast of Cuba, the mountainous island of Hispaniola is divided into two different countries, cultures, and economies: the French-speaking Republic of Haiti and the Spanish-speaking Dominican Republic. In Haiti (population 9 million; capital, Port au Prince, population 2.3 million), the black population forms an overwhelming majority, and a small mulatto elite—perhaps 1 percent of the population—controls a major share of the wealth. Population density is extreme and infrastructure is poor in this country where poverty rules and where foreign aid pays two-thirds of the national budget. Development prospects are dampened by Haiti's brain drain; more than 80 percent of skilled Haitians live abroad. The overwhelmingly agricultural economy is based mainly on subsistence rather than export. 59

The French colonial regime was driven out of Haiti at the beginning of the 19th century. Independent since 1804, the Republic of Haiti has a strong West African flavor in both culture and landscape. Unfortunately, deforestation, erosion, and poor transportation are major problems. Deforestation took place on a large scale as French colonists stripped the countryside of millions of its valuable mahogany trees. Commercial and subsistence exploitation continued after independence, and more than 90 percent of the country has been deforested. Aside from subsistence farming, Haiti's small economy concentrates on light manufacturing of clothing (aided by low-wage labor) and coffee growing.

The country has had a turbulent political history. Its first democratically elected president, Jean-Bertrand Aristide, was ousted by a military coup in 1991. U.S. military intervention restored him to power in 1994. Aristide was subsequently implicated in the trafficking of Colombian cocaine to the United States, and he fell out of favor with Washington. When rebels angry with the outcome of a flawed 2000 legislative election closed in on him in 2004, the United States did not intervene on his behalf and instead flew him out of the country. With Aristide's departure, U.S. troops and then a UN peacekeeping force were deployed to Haiti. In elections the following year, René Préval succeeded Aristide as Haiti's president.

Haiti's neighbor to the east, the Dominican Republic (known to many as "the D.R."; population 9 million), bears a mainly Spanish cultural heritage. It is much more prosperous than Haiti. Santo Domingo (population 3.1 million), the capital, is the oldest continually inhabited European city in the New World (it was founded in 1496), and it is home to the first European church, school, university, and hospital. An independent country since 1844, the republic has a history of internal war, foreign domination (including periods of occupation by U.S. military forces), and misrule, although greater stability has prevailed in recent years.

The Dominican Republic's diversified economy relies on tourism, mining, oil refining, and agricultural products such as sugar, meat, coffee, tobacco, and cacao. *Maquiladora*-style export production, patterned after Mexico's success with these assembly plants, is playing an increasing role in the republic's economy. Baseball is dear to the Dominican Republic. The country has sent some of the best major league baseball players to the United States—Pedro Martinez, Manny Ramirez, David Ortiz, and the Alou brothers, to mention a few—and produces baseballs for the U.S. market. Reflecting its stronger economy, urbanization (64 percent) is more pronounced than in Haiti (36 percent).

Puerto Rico (population 3.9 million; capital, San Juan, population 2 million), the smallest and easternmost of the Greater Antilles, is a mountainous island with coastal strips of lowland. Spain ceded the island to the United States after its defeat in the Spanish-American War of 1898, and Congress administered Puerto Rico as a federal territory until 1952. It then became a self-governing commonwealth voluntarily associated with the United States. Puerto Ricans have U.S. citizenship but do not pay taxes, vote in U.S. elections, or receive federal U.S. benefits. The U.S. dollar is the official currency, and all of the trappings of American life prevail on the island, together with a strong population of tourists and expatriates from the U.S. mainland. In referenda, a narrow majority of Puerto Ricans have rejected U.S. statehood in favor of continued commonwealth status. However, pressure for statehood continues on the island, and a small minority wants complete independence.

Puerto Rico's pre–World War II economy was agriculturally based, with an emphasis on sugar. After the war, a determined effort by Puerto Ricans to raise living standards brought great changes. With much U.S. financing, hydroelectric production, agriculture, manufacturing, and tourism grew. Tax incentives and low land prices attracted American businesses and industries. Manufacturing came to replace sugar as the island's main source of revenue, with petrochemicals, pharmaceuticals, electrical equipment, food products, clothing, and textiles among the major products. Copper and nickel deposits were found and developed. But poverty is still a problem for Puerto Rico and has long motivated emigration to the U.S. mainland, especially to New York City and, more recently, Florida and the American Southwest.

Jamaica, independent from Great Britain since 1962, is in the Caribbean Sea about 100 miles (160 km) west of Haiti and 90 miles (145 km) south of Cuba. Its population of 2.7 million (as of 2007) is largely descended from black African slaves transported there by the Spanish and (after 1655) the British. When slavery was abolished in 1838, most blacks left the sugar plantations on the coastal lowlands and migrated to the mountainous interior. Tourism, bauxite, and agriculture have been Jamaica's economic base for decades. Jamaican culture in the form of reggae music has become known worldwide, especially through the works of the late Bob Marley. He sang of the poverty and the humanity of Trenchtown, a poor section of the Jamaican capital, Kingston (population 900,000). He also introduced knowledge of the Rastafari faith to people around the world.

505, 537

The Lesser Antilles

More than 3 million people live on the Lesser Antilles islands scattered in a 1,400-mile (2,250-km) arc from the Virgin Islands (east of Puerto Rico) to near Venezuela. This string of small islands on the eastern edge of the Caribbean Sea has a variety of political entities. The Netherlands Antilles and Aruba are Dutch dependencies evolving toward greater self-government. Guadeloupe and Martinique are French overseas departments, and Montserrat (an island racked by volcanic eruptions in 1995 and 1997) and Anguilla are still British dependencies. Britain began its withdrawal from the islands with independence for Jamaica and Trinidad and Tobago in 1962. Since then, Barbados and six smaller units have gained independence from Britain. Britain also granted independence to the Bahamas in 1973 but retains as colonies the British Virgin Islands, the Cayman Islands, and the Turks and Caicos Islands. All the former colonies are still members of the Commonwealth of Nations.

The island economies suffer from many serious problems, including unstable agricultural exports, limited natural resources, unfavorable trade balances, a shortage of industries, and chronic unemployment. A shared trait of the islands is their history of European colonialism and plantation slavery. Such generalizations contribute to some understanding of the islands but give little hint of the individuality of these small units. Just one example—Trinidad and Tobago—is discussed here.

Trinidad and Tobago (a single political entity) is the largest, most resource-rich, and most ethnically varied of the eastern Caribbean island countries. Trinidad, the larger of the two islands, has offshore oil and natural gas deposits and is the Western Hemisphere's largest supplier of liquefied natural gas. Some 40 percent of the country's population is black. Most blacks live in the urban areas and favor work in the fossil fuel industry. Another 40 percent descended from immigrants from the Indian subcontinent and live mainly in the intensively cultivated countryside. Indians began coming to the islands as indentured workers after slavery was abolished. Among their descendents is V. S. Naipaul, who has written fine portraits of Indian life on Trinidad and Tobago, including *The Mystic Masseur* and *The Suffrage of Elvira*. About 18 percent of the islanders are classified as "mixed," and there are small numbers of Chinese, Madeirans, Syrians, European Jews, Venezuelans, and others. This ethnic complexity is reflected in religion, architecture, language, diet, social class, dress, and politics.

From Trinidad and Tobago, the text takes a short jump in the next module to the great landmass of South America.

SUMMARY

→ Middle America includes Mexico, the Central American countries, and countries in the Greater and the Lesser Antilles islands in the Caribbean Sea. The largest islands—Cuba, Hispaniola, Puerto Rico, and Jamaica—make up the Greater Antilles, and most of the smaller islands are in the Lesser Antilles chain.

→ Mexico is the largest, most populated, and economically most developed of the Middle American nations. It is the largest Spanish-speaking country in the world.

→ The North American Free Trade Agreement (NAFTA) has played a major role in economic change in Mexico. Much of Mexico's economy is based on production in the *maquiladora* factories of the U.S. border region, where NAFTA has benefited the country. NAFTA has been less beneficial to the mainly agricultural south, where farmers have a difficult time competing with subsidized U.S. farmers.

→ Mexico's history includes Native American, Spanish, and Mexican eras. The Aztecs and Maya were the country's most technologically advanced Native American populations. The Spanish presence is very strong. The Spanish were a major force in establishing land use patterns that continue today. The Mexican era began with the 1810–1821 period of the Mexican Revolution, and the country lost much land to the United States in the following decades. The second period of revolution in Mexico was 1910 to 1920.

→ Central Mexico is a volcanic highland area in which Mexico City, the world's second-largest city, is located. The city has problems with land subsidence and air pollution. It is Mexico's leading industrial center.

→ The Gulf Tropics of Mexico are known for gas, oil, agricultural plantations, and Mayan archaeological sites. Tourism is a vital industry. Mexican oil, a nationalized resource, has been important in economic development.

→ Central America is made up of Guatemala, Belize, El Salvador, Honduras, Nicaragua, Costa Rica, and Panama. Guatemala, Honduras, and El Salvador have large Native American populations, and there is a strong European presence in the wealthiest country, Costa Rica. Panama's history over the past century has been shaped by U.S. interests in the Panama Canal, which is now being enlarged.

→ Insurgencies and civil wars were problems in most of the mainland Middle American countries in the second half of the 20th century. El Salvador, Guatemala, and Nicaragua had the costliest conflicts.

→ Plantation agriculture, especially of bananas, is important in the *tierra caliente* on both coasts of Middle America. Sugar is most prominent on the islands. Deforestation is a major problem throughout the region, and debt-for-nature swaps have exchanged protected forests for debt forgiveness in some countries.

→ Some Caribbean islands maintain strong economic and political affiliations with the European countries that colonized them.

→ Cuba is the most important Caribbean island because of its size, population, and geopolitical status. Sugar and tourism have long been central to its economy.

→ Puerto Rico is the smallest and easternmost of the Greater Antilles and is a self-governing commonwealth of the United States.

→ Other major islands in the Caribbean include Jamaica and, in the Lesser Antilles, the Netherlands Antilles, Aruba, Guadeloupe, Martinique, Barbados, the Bahamas, the Virgin Islands, and Trinidad and Tobago. These represent a variety of political affiliations. Banking, tourism, and varied resources are important in the economic profiles of these islands.

KEY TERMS + CONCEPTS

American Invasion (p. 558)
border towns (p. 560)
Captaincy-General of Guatemala (p. 562)
Central American Federation (p. 562)
Chol Maya (p. 556)
Contras (p. 562)
energy *maquiladoras* (p. 557)
Hispanic (p. 554)
Lacandon (p. 556)

lock (p. 563)
Mexican Revolution of 1810–1821
    (p. 555)
Mexican Revolution of 1910–1920
    (p. 555)
Mexican War (p. 558)
Plan Puebla-Panama (p. 556)
pioneer zone (p. 563)
San Andres Accords (p. 556)

Sandinistas (p. 562)
thermal inversion (p. 559)
trade embargo (p. 568)
Tzeltal Maya (p. 556)
United Fruit Company (UFC) (p. 562)
Viceroyalty of New Spain (p. 554)
Zapatista Army for National Liberation
    (EZLN) (p. 556)

REVIEW QUESTIONS

1. What are the major subregions of Mexico? What characteristics define the core regions of most of the Central American countries? What unique environmental problems must Mexico City deal with?

2. What are the particular attractions of the Mexican border towns for U.S. citizens? What drives the economies of these towns?

3. What are energy *maquiladoras*? What resources do they use, where is their product exported, and why are they located where they are?

4. What are the most important plantation crops of the region, and where are they grown?

5. What role has oil played in Mexico's economy? What is its future?

6. Where are the capital cities of the Central American countries located? What are the main economic activities of those countries?

7. What considerations went into the construction of the Panama Canal? What steps are being taken to improve the canal?

8. What are the major geographic subdivisions of the Caribbean islands?

9. How does the United States deal differently with emigrants from Cuba and Haiti? What accounts for these different approaches?

10. Why is Cuba as the most important Caribbean island?

11. What factors have helped make Haiti the poorest country in the Western Hemisphere?

12. What is Puerto Rico's relationship with the United States? Does it appear likely to change?

13. What are the main characteristics of the cultures and economy of Trinidad and Tobago?

NOTE

1. Quoted in Tim Weiner, "Mexican Rebels Confront Tourism in Chiapas." *New York Times*, March 9, 2003, p. TR3.

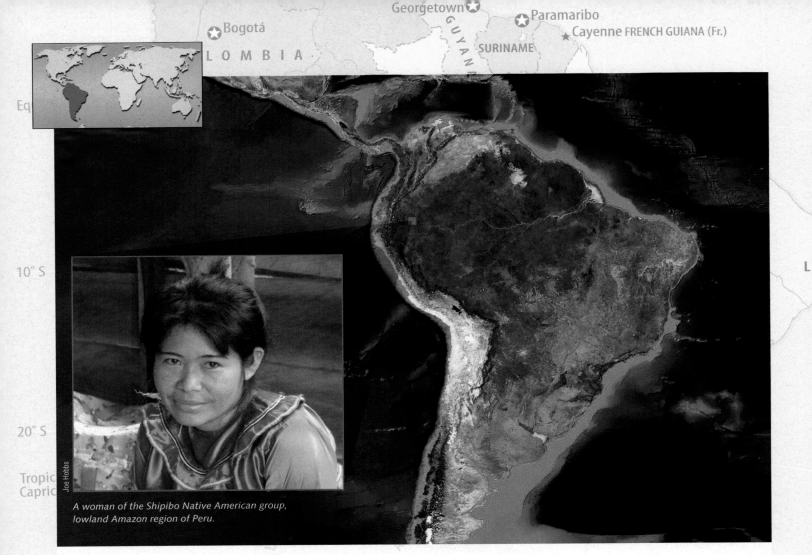

A woman of the Shipibo Native American group, lowland Amazon region of Peru.

Joe Hobbs

MODULE 10.2
SOUTH AMERICA: STIRRING GIANT

South America has the lion's share of Latin America's total land area (see Figure 10.1, page 523). It is the fourth-largest continent on earth, amounting to 6,898,400 square miles (17,866,900 sq km). It is approximately 4,500 miles (7,200 km) from north to south and extends to a maximum width of 3,200 miles (5,120 km). It includes the world's largest Portuguese-speaking nation, Brazil, which is also home to the largest remaining tropical rain forest in the world. South America has a history of Native American civilization on the west side of the Andes Mountains, which stretch from Colombia in the north all the way to Tierra del Fuego at the southern tip of the continent. The continent also has a wide range of climates, landscapes, and ecological zones, ranging from the parched dry desert of the Atacama in Chile to the lush tropical rain forests of central Brazil and eastern Peru and Ecuador. In world affairs, South America is often depicted as a "sleeping giant," a land with huge potential for economic development and political leadership. This module explores its potential and the prospects for its awakening.

■ module
objectives

This chapter should enable you to

→ Recognize the paradox of how a blessing of oil wealth can para-
   lyze national economies and political systems and be rejected by
   the poor

→ Appreciate the challenges that rebel forces have posed to national
   governments and resource wealth in the region

→ Understand the liabilities of a landlocked location as suffered in
   this region by Bolivia and Paraguay

→ Appreciate how uneven distribution of national wealth promotes
   instability and hampers development

→ Understand why "saving the rain forest" is not a simple problem
   and how various countries and national groups see the problem
   differently

→ Appreciate how excessive dependence on raw materials can cause
   an uncontrollable cycle of economic boom and bust

10.2.1 The Andean Countries: Lofty and Troubled

The South American countries of Venezuela, Colombia, Ecuador, Peru, and Bolivia—all within the tropics and traversed by the Andes Mountains—are grouped in this book as the Andean countries. Common interests among them were recognized in 1969, when five countries signed the Cartagena (Colombia) Agreement, which created a new free-trade zone called the **Andean Group.** At first, Chile was part of the group, but it later withdrew. Although the Andes Mountains run its entire length, Chile is not classified as an Andean country because its major traits are more akin to its southern neighbors. Venezuela came late to the Andean Group, which is now made up of Bolivia, Peru, Ecuador, Colombia, and Venezuela.

These countries share several traits. One is environmental zonation. The region has pronounced vertical and horizontal zonation, with extreme environmental contrasts and areas are that isolated from one another by natural barriers. Andean glaciers, like those in most other parts of the world, are retreating, with similar implications for reduced water supplies downstream. Another common trait is that all have fragmented settlement patterns, owing primarily to the broken topography of the countries; this fragmentation is especially evident in Peru. Each country contains both well-populated highlands and an extensive fringe of lightly populated interior lowlands or low uplands being settled by pioneers. And each country except Bolivia has an economically productive lowland region. (Bolivia originally had a Pacific coastal zone but lost it to Chile during the war of 1879–1883.)

The Andean countries also share some culture traits; for example, they all have significant populations of indigenous peoples. At the time of the Spanish conquest in the 16th century, large indigenous populations existed in the mountains, valleys, and basins. These old settlement areas are now major zones of dense population and strong political authority, especially in the highlands. The indigenous cultures have persisted despite the conquest. Today, Bolivia, Peru, and Ecuador have Native American majorities; the only other Latin American country with that distinction is Guatemala. In Colombia and Venezuela, Native Americans make up just 1 or 2 percent of the total populations, with mestizos forming the largest ethnic element.

Also common to the Andean countries are traditions that are predominantly Hispanic. Spanish is the official language in all five countries (although Native American languages are also official in Peru and Bolivia), and over 90 percent of each country's population is Catholic. Furthermore, all the countries in this region are relatively poor, despite oil production in most of them. Considerable money flows into several countries from the trade in illegal drugs, especially cocaine, but most of it does not find its way into the pockets of drug crop farmers, who remain poor. Poverty is evident both in the countryside and in the shantytowns of the cities.

The final common feature of the Andean countries is that in all countries except Ecuador and Venezuela, the national capital is a primate city, towering over all others in size and political and cultural importance. Let us take a closer look at each of the countries in this region.

Venezuela

The population of Venezuela (●Figure 10.2.1) is concentrated mainly in Andean valleys and basins not far from the Caribbean coast, as it was when the Spanish arrived in the 1500s. The main city and capital, Caracas (population 3.8 million), is in the highlands. At an elevation of 3,300 feet (c. 1,000 m), Caracas has a *tierra templada* climate. Around Caracas, the Spanish found gold, a large indigenous population from which to draw forced labor, and climatic conditions suitable for commercial production of sugarcane and, later, coffee. Labor needs associated with sugarcane led to the 16th-century introduction of African slaves, accounting for most ancestors of the country's 10 percent black population.

Oil resources in the coastal area around and under the large Caribbean inlet called Lake Maracaibo have been a huge blessing for Venezuela, the world's ninth-largest oil producer in 2007. Oil makes up about 60 percent of the country's exports by value. The country was one of the

● **Figure 10.2.1** Principal features of Venezuela, Guyana, Suriname, and French Guiana.

founding members of the Organization of Petroleum Exporting Countries (OPEC) in 1961 and is still a member. Oil revenues have fueled urbanization and industrialization and generated wealth. Yet Venezuela is rather like Nigeria or Angola, an oil-rich country where the vast majority of people remain poor (see Geography of Energy, page 575).

Venezuela has some important mineral assets in the south. This sparsely inhabited area of the Guiana Highlands and of lowlands along the Orinoco River and its tributaries is mostly tropical savanna. For centuries, remote interior Venezuela has been Native American country (in the highlands) or poor and lightly settled ranching country (in the lowlands), but development has been quickening. Oil and gas production and agricultural settlement projects are ongoing in the lowlands. In the Guiana Highlands, enormous iron ore reserves have been mined for export (mainly to U.S. steel mills), and huge bauxite reserves in the highlands have yet to be tapped. A major petroleum- and aluminum-based industrial center is located at Ciudad Guayana, on the Orinoco River.

Venezuela has dedicated an enormous amount of its area to protected parks and reserves, including a reserve for the Native American Yanomamö that makes up fully 9 percent of the country's territory. National planners hope that investment in protected areas will have multiple returns, especially in revenue from ecotourism. Among the most spectacular of the country's natural features are the high sandstone outcrops called *tepuis* (●Figure 10.2.2). Rising precipitously from the tropical rain forest and often protruding from fog and cloud layers, they evoke prehistoric images that were said to have inspired Arthur Conan Doyle's novel *The Lost World* (which in turn inspired Spielberg's *Jurassic Park* films). From one of the *tepuis* spills Angel Falls, the world's highest waterfall at 3,212 feet (c. 980 m).

Guyana, Suriname, and French Guiana

The three entities on the northeastern edge of South America are not classified as Andean and are culturally more similar to the Caribbean islands than the rest of the South American continent: independent Guyana (formerly British Guiana), independent Suriname (formerly Dutch Guiana), and French Guiana, which is an overseas department of France (see Figure 10.2.1) and hence is a member of the European Union. All three units (collectively called the Guianas) have populations of less than 1 million. They have a long plantation history, with sugar most important in Guyana and French Guiana and tea and coffee in Suriname. The Guianas are very ethnically diverse, with significant populations of Indians, Chinese, and other Asian groups originally brought to work on the plantations, living alongside Europeans, indigenous peoples, and the descendants of African slaves. Ethnic violence has been a problem, especially in Guyana. Christianity, Judaism, Hinduism, and Islam are all practiced in the Guianas.

The Dutch traded their settlement of New Amsterdam (later called New York City) to the British for the rights to Suriname and its coastal richness for agricultural exploitation. Separately, Guyana was initially colonized by the Dutch but was later taken over by the British. The infamous mass suicide of a religious cult of U.S. origin, known as the Jonestown massacre, took place in Guyana in 1978. French Guiana was home to Devil's Island, a notorious colonial island prison used as recently as 1954. Today, the three countries' economies are still mainly agricultural (especially with exports of sugar and timber), but revenues from bauxite mining and tourism are growing. With Brazilian investment, Guyana wants to capitalize on the current boom in sugarcane-based ethanol pro-

GEOGRAPHY OF ENERGY

Venezuela's Petroleum Politics

The most prominent member in the lineup of left-leaning Latin American leaders elected in recent years is Venezuelan President Hugo Chávez. Invoking the revered name of Simón Bolívar (1783–1830), the Venezuelan-born independence leader who spearheaded a regional quest for liberation from Spain, he has proclaimed to be conducting a **Bolivarian Revolution.** Chávez has his own ideas about liberation: he has rewritten the constitution, put cronies on the supreme court, cracked down on the media, and jailed political foes. The Chávez agenda appeals to the poorest classes by promising a redistribution of wealth like that championed by Cuba's Fidel Castro (Chávez's strongest ally in the region). Chávez's political philosophy is not communism but what he calls **21st-century socialism** (and others dub **Chavismo**), with the armed forces and the president controlling everything but a small and mainly foreign-owned private sector.

Venezuela's petroleum has long been controlled by the state-owned oil company Petróleos de Venezuela S.A. (PDVSA), and early in his presidential career, Chávez sought to bring it under his personal control. PDVSA, backed by about three-quarters of Venezuela's population, resisted. In 2002, PDVSA joined business and labor interests across the nation in a general strike, pressuring Chávez to resign or call for a new election. A petition drive by the opposition led to a 2004 recall referendum that failed to unseat Chávez.

His power consolidated, Chávez has used Venezuela's oil wealth to advance his political agenda across the Western Hemisphere. He has prominent alliances with Latin America's leftist leaders, constantly courts the center and right governments of the region to move his way, and wants to build a coalition of countries to counter U.S.-led free-trade efforts in the region. He casts himself as an advocate of the poor by offering Venezuela's oil to Cuba and to select populations in other countries at little or no profit. He embarrassed U.S. politicians by supplying cheap heating oil to hard-pressed American northeasterners in the cold winter of 2005. While shunning Chávez and his politics, Washington needs Venezuela's nearby non–Middle Eastern oil and has taken no action that would jeopardize its fourth-largest source of imported oil. Chávez for his part has periodically threatened to cut off the supply of Venezuelan oil to the United States.

duction. French Guiana is the wealthiest of the three, largely because of the subsidies it receives from France.

Colombia

Within its wide expanse of the Andes, Colombia (•Figure 10.2.3) includes many populous valleys and basins, some in the *tierra fría* and others in the *tierra templada*. The great majority of Colombia's mestizos and whites are scattered among these upland settlement clusters, which tend to be separated by sparsely settled mountain country with extremely difficult terrain. Lower valleys and basins, along with some coastal districts, were developed by plantation agriculture that employed slaves. That legacy is apparent in

Jay Dickman/Corbis

• **Figure 10.2.2** Angel Falls, the world's highest waterfall, drains one of the dramatic *tepuis* of Venezuela. These mesas are sandstone remnants of the ancient Guyana Shield.

● **Figure 10.2.3** Principal features of Colombia.

Colombia. The Bogotá area has a large output of flowers grown for air shipment to the United States.

Medellín (population 3.3 million) lies in the *tierra templada* at about 5,000 feet (1,525 m) in a tributary valley above the Cauca River. This area did not have a large indigenous population to provide estate labor and was difficult to reach, so for centuries it remained a region of shifting subsistence cultivation of corn, beans, sugarcane, and bananas. The introduction of coffee and the provision of railroad connections changed this in the early 20th century. Excellent coffee could be produced under the cover of tall trees on the steep slopes, and the Medellín region became a major producer and exporter. The city also became a textile-manufacturing center. In the 1980s and 1990s, Medellín and Cali became synonymous with the cocaine trafficking cartels that came to dominate the cities' economies until Colombian authorities cracked down on them.

Cali (population 3.1 million) is located in the *tierra templada* some 3,000 feet (c. 900 m) above sea level on a terrace overlooking the floodplain of the Cauca River. Cali had easier connections than Medellín with the outside world via the Pacific port of Buenaventura. The Spanish developed sugarcane plantations in the Cali region and also produced tobacco, cacao, and beef cattle. Native Americans and later African slaves were put to work in these enterprises. From the late 19th century onward, dams on mountain streams provided hydropower for industrial development. This asset, along with cheap labor, attracted foreign manufacturing firms, and Cali grew at a rate like that of Medellín.

Colombian Coasts and Interior Lowlands

There are sharp contrasts in environment and development between the Caribbean and Pacific coasts of Colombia. The Pacific coastal strip is the wettest area in all of Latin America, with some places receiving nearly 300 inches (760 cm) of precipitation annually. It has tropical rain forest climate and vegetation, and with adjacent sections of Panama and Ecuador, this Choco region is one of the world's biodiversity hot spots. In the north, mountains rise steeply from the sea, but there is a coastal plain farther south. The area is thinly populated, with a large proportion of the population of African ancestry. The main settlement is Buenaventura, an important port. Its principal advantage is that it can be reached from Cali via a low pass through the Andes.

The Caribbean coastal area is more populous. Three ports have long competed for the trade of the upland areas lying to the south on either side of the Magdalena and Cauca Rivers. Barranquilla and the Spanish fortress city of Cartagena (●Figure 10.2.4) are the larger and more important cities in terms of legitimate trade. But the smaller city of Santa Marta, known as a beach resort as well as a port, is probably the most important in overall trade. It is the control center for Colombia's huge illegal export trade in cocaine and marijuana.

Most of the Caribbean lowland section of Colombia is a large alluvial plain where ranching and banana and sugar-

Colombia's demographics, with 14 percent of the country's population mulatto and 4 percent black.

The Highland Metropolises: Bogotá, Medellín, Cali

Colombia's inland towns and cities have tended to grow as local centers of productive mountain basins. The three largest—Bogotá, Medellín, and Cali—are diverse in geographic character. The largest highland city in all of South America is Bogotá (population 8.1 million), Colombia's capital. It lies in the *tierra fría* at an elevation of about 8,700 feet (c. 2,650 m—much higher than Denver, Colorado).

The Spanish seized the land in this region, put the indigenous people to work on estates, and improved agricultural productivity by introducing new crops and animals—notably, wheat, barley, cattle, sheep, and horses. Exploitation and disease reduced Native American numbers so drastically that their population took a long time to recover. The Spaniards were not able to develop an export agriculture as night coolness and frosts precluded such major trading crops as sugarcane and coffee. Early transport to the coast was difficult. Mountain slopes were so challenging that a railroad connection with the Caribbean Sea was not completed until 1961. In recent decades, modern governmental functions and the provision of air, rail, and highway connections have led to explosive urban growth in upland

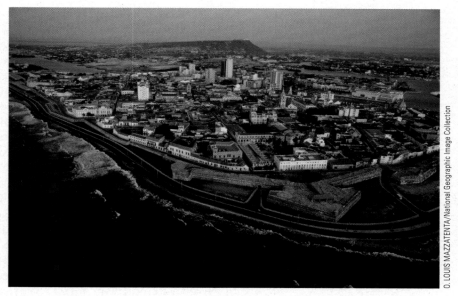

● **Figure 10.2.4** Colombia's colonial port city of Cartagena was a center of the slave trade, and was often targeted by pirates.

cane production have been carried on despite constant flood hazards. Colombia's interior lowlands east of the Andes form a resource-rich but sparsely populated region contributing little to the national economy. It is a mix of plains and hill country, with savanna grasslands in the north and rain forest in the south. The grasslands are used for ranching and the forest for shifting cultivation. Considerable mineral wealth, especially oil, awaits exploitation. As in so many parts of Latin America, and especially in South America, difficult terrain and shortages of financial resources have kept rail and road networks from reaching large portions of the countryside.

Colombia's Peril and Potential

Colombia has the resources for future economic development. Energy resources include a large hydropower capacity; coalfields in the Andes, the Caribbean coastal area, and the interior lowlands; oil in the Magdalena Valley and in the interior; and natural gas in the area near Venezuela's Lake Maracaibo. Iron ore from the Andes supplies a steel mill near Bogotá, and large Andean reserves of nickel are under development. At least 40 percent of the country is forested, mainly with the tropical rain forest. Oil plays an increasing role in the country's export economy. In 2007, petroleum and its products represented approximately 26 percent of all legal Colombian exports by value, with coal at 12 percent and coffee representing 6 percent. However, illegal cocaine and marijuana may have been the largest category of exports, possibly up to one-third by value.

Both antigovernment rebel groups like FARC (see page 547) and the paramilitary militia that oppose them have financed their operations though taxes on drugs. Widespread insecurity related to their conflicts and to the firepower and violence of the drugs trade prompted Colombians in 2002

to elect Alvaro Uribe as president; he had promised to crack down on rebellion and restore order to the country. Since assuming office, he has been eager to please both his citizens and the U.S. Congress in Washington, which wants to see some return for its huge investment in Colombia. As of 2007, he had gained considerable confidence both there and among a majority of Colombians for his efforts to improve security.

547

Ecuador

The Andes form two roughly parallel north-south ranges in Ecuador (●Figure 10.2.5). Between them lie basins filled with fertile volcanic ash. These basins are mostly in the *tierra fría* at elevations between 7,000 and 9,500 feet (c. 2,100–2,900 m). Quito (population 2 million; ●Figure 10.2.6), Ecuador's capital and second-largest city, is located almost on the equator, high up on the rim of one of the northern basins. Its elevation of 9,200 feet (2,800 m) yields daytime temperatures averaging 55°F (c. 13°C) every month of the year.

The first Spanish invaders came to Ecuador from Peru in the 1530s. They found concentrations of Native American populations in the area's Andean basins. The conquerors appropriated their land and labor and built Spanish colonial towns such as Cuenca, often on the sites of previous indigenous towns. But they found no great mineral wealth here, and both climate and isolation worked against the development of export-oriented commercial agriculture. Even today, these Andean basins support mostly subsistence agriculture.

It was from this hearth that the potato diffused into Europe and into world trade. Today, potatoes, grains, cut flowers for the U.S. market, dairy cattle, and sheep are the major crops and livestock. The isolated Ecuadorian Andes

O. LOUIS MAZZATENTA/National Geographic Image Collection

• **Figure 10.2.5** Principal features of Ecuador, Peru, and Bolivia.

• **Figure 10.2.6** Quito lies in a high, fertile valley flanked by volcanoes.

historically attracted relatively few Spaniards, except to Quito, and without plantation agriculture, there was no demand for African slave labor. As a result, the mountain population has remained predominantly Native American, with minority mestizo and white ethnicities. Growth of industries such as food processing, textiles, and chemicals not only in Quito but also in some of the smaller Andean cities has increased these minority populations outside the capital.

A marked shift of Ecuador's economy and population toward the coast took place in the 20th century. Through the 19th century, about 90 percent of Ecuador's people still lived in the highlands, but now only about half are there. The main focus of development, mainly by whites and mestizos, has been the port city of Guayaquil (population 3.3 million), at the mouth of the Guayas River, which drains a large alluvial plain where bananas are grown for export (Ecuador is the world's leading banana exporter) and rice is cultivated for domestic consumption. Coffee is grown on nearby slopes. These agricultural developments have removed almost all of the area's original forest cover.

Along the Pacific is a fishing industry centered on anchovies. This industry regularly goes bust during El Niño climatic events that warm the eastern Pacific waters. Guayaquil is the jumping-off point for the Galápagos Islands, 600 miles (960 km) to the west in the Pacific Ocean. With their extraordinary giant tortoises and other endemic species, the Galápagos are a significant resource for Ecuador because of the ecotourism revenues they generate (•Figure 10.2.7).

Oil exploitation began in the 1960s in the sparsely populated tropical rain forest of Ecuador's Amazon region. The country began exporting oil when a pipeline was built crossing the Andes to the small Pacific port of Esmeraldas. Oil products provide about 40 percent of the country's foreign-exchange earnings. Possible future mineral discoveries in the interior rain forest add urgency to a border dispute with Peru, which controls some territory claimed by Ecuador. In 1981 and again in 1994–1995, bouts of armed conflict between the two countries occurred along their undefined, mountainous eastern boundary. In 1998, the two nations signed a treaty, apparently resolving this issue. At stake for both countries is the successful future growth of fossil fuel export industries, discussed in the context of the Amazon on pages 586–587.

Peru

Peru (see Figure 10.2.5) is a poor nation with populations fragmented by great ruggedness (in the Andes), dryness (along the coast), and forested wilderness (in the Amazon lowland). More than half of the population, mainly Native Americans, lives in the Andean *altiplano,* the local name for the high intermontane plain extending from southern Peru into west central Bolivia. Some of these highlanders are descendants of the ancient civilization of the Inca, centered around Cusco (population 300,000) in Peru's southern Andes. The highlands, generally higher than in Ecuador and Colombia, are a poor environment for surplus-producing cash crop agriculture (•Figure 10.2.8). Agricultural holdings tend to be in the *tierra fría,* where cool temperatures preclude export crops such as coffee, sugarcane, and cacao. The typical household economy is focused on potatoes, grains, and livestock, especially llamas and alpacas, raised for family consumption and local sale. Farmed areas in central and southern sections of the Peruvian Andes tend to be so dry that only irrigated areas are productive. Machu Picchu and other Inca sites in the highlands are important tourist magnets, generating welcome revenue for Peru.

Growing, processing, and trafficking of coca and cocaine, 548 particularly from the eastern flank of the Andes, may represent as much as 4 percent of Peru's gross national product.

A narrow, discontinuous coastal lowland lies between the Peruvian Andes and the sea. This is an extremely arid desert owing to the cold ocean waters along the shore. Air moving landward becomes chilled over these waters. Onshore, a cool and heavy layer of surface air underlies warmer air—a thermal inversion condition similar to the one that promotes smog in Los Angeles. Air turbulence that would create rain is absent, and only fog and mist yield moisture. Most rivers on the western slopes of the Andes carry little water and disappear before reaching the sea. Where they have water, these river valleys have supported irrigated agriculture for thousands of years.

In 1535, the Spanish founded Lima (population 7.9 million; •Figure 10.2.9) a few miles inland from a usable natural harbor. The city was well located for the Spaniards to exploit the Native American communities and mineral deposits of the Andes, especially the Cerro de Pasco silver deposits that for a time led the world in silver production. Lima was designated the colonial capital of the **Viceroyalty of Peru** and became one of the main cities of Spanish America. In the late 19th century, foreign companies developed commercial agriculture in the coastal oases. These productive areas of irrigated cotton, rice, sugarcane, grapes, and olives still spot the coast. Small coastal towns linked with the oases serve as fishing ports and fish-processing centers thanks to the typically cold upwelling waters along the 532 coast, which are exceptionally rich in marine life. But this resource provides an uncertain livelihood. For a few years in the 1960s and early 1970s, Peru led the world in volume of fish caught (largely anchovies for processing into fishmeal and oil), and the port of Chimbote was the world's

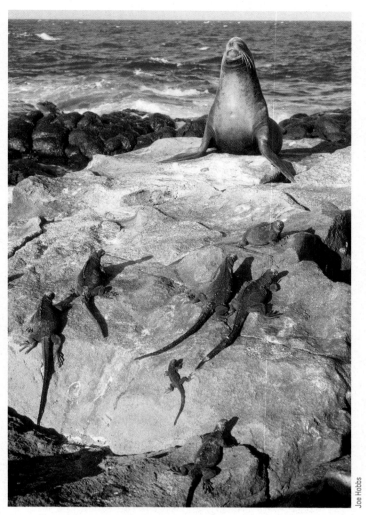

• **Figure 10.2.7** Sea lion and marine iguanas in the Galápagos Islands.

• **Figure 10.2.8** The rugged Andes rise between Peru's coast and its interior tropical rain forest. Here copper is mined, as it has been since the Inca reigned in these mountains.

● **Figure 10.2.9** A parade celebrates the Feast of St. Peter and St. Paul in Lima.

● **Figure 10.2.10** Iquitos lies on an embayment of the giant Amazon River.

leading fishing port. But since then, periodic changes in water temperature associated with El Niño have led to a drastic decline and widespread unemployment in the industry. Nevertheless, Peru typically catches more fish than every other nation except China.

Overall, Peru's agriculture is poor, and the country imports a large portion of its food. The country will depend increasingly on oil and other mineral exports to gain foreign exchange for these imports (see the discussion on page 587). At present, the overall export value of metallic ores—mainly copper, zinc, lead, and silver—exceeds that of oil, and Peru is in fact the world's largest silver producer. Metal-mining sites are divided between the Andes and the coast.

The tropical rain forest of the Amazonian plains overspread more than half of Peru. Fewer than 2 million Peruvians live in this large area. The main settlement, Iquitos (●Figure 10.2.10), was founded during the natural-rubber boom of the Amazon Basin during the 19th century. It has always been more oriented toward outside markets via the Amazon than to Andean and Pacific Peru. The Amazon River is open to oceangoing vessels for 2,200 miles (c. 3,600 km), from its mouth all the way to Iquitos (see page 585).

Peru's economic development has been constrained by gyrations between civilian and military rule. The 1980s saw widespread guerrilla activity by the Maoist group **Shining Path** (*Sendaro Luminoso*) and by the smaller **Tupac Amaru Revolutionary Movement (MRTA),** which continued its assaults on government interests well into the 1990s. Altogether an estimated 69,000 people, mostly Native American Quechuan, died, perhaps half at the hands of the Shining Path and one-third killed by Peruvian armed forces. An ethnic Japanese president named Alberto Fujimori led a largely successful counterattack against the insurgents. Following his resignation in the wake of a fraudulent election, Fujimori was succeeded in 2001 by Alejandro Toledo, Peru's first Native American president. Toledo and his successor, Alan García, have overseen a period of strong economic growth based mainly on mineral commodity exports to the United States and China. Poverty rates have fallen but remain high, and continuing inequality in the distribution of wealth led to protests against García's regime.

Landlocked Bolivia

Bolivia (see Figure 10.2.5) is a land of extreme alpine conditions. It is the poorest Andean country and one of the poorest in all of Latin America, with an annual per capita GNI PPP of only $2,890, compared to $8,900 for the Latin American average. Bolivia's unemployment rate has remained high for the past quarter century. The country's exports have grown slowly and foreign investment has been low. These were conditions ripe for the ascendancy of an advocate of the poor, Evo Morales, who became the country's first Native American president in 2006.

Like the other Andean countries, Bolivia has a sparsely populated tropical lowland east of the mountains. But unlike the others, it has no coastal lowland to supplement the production of its mountain basins and valleys. Although limited by extreme *tierra fría* conditions combined with aridity, about 10 percent of Bolivia's people are engaged in farming. La Paz (population 1.9 million) is the main city and the de facto capital. (Sucre is the legal capital, but the supreme court is the only branch of the government located there.) At an elevation of about 12,000 feet (c. 3,700 m), La Paz averages 45°F (8°C) in the daytime in its coolest month and only 53°F (12°C) in its warmest month.

Rural peoples are concentrated at a comparable elevation around Lake Titicaca (●Figure 10.2.11) on the Peruvian border. At 12,507 feet (3,812 m), this is the highest navigable lake in the world and boasts regular steamer service among lakeside ports high in the Andes. Smaller settled population clusters are scattered through Andean Bolivia, mainly along the easternmost of the two main Andean ranges and the *altiplano*. Unfavorable climate at such high elevations makes export-oriented commercial agriculture impossible.

● **Figure 10.2.11** Lake Titicaca is renowned as the world's highest navigable lake, and the hearth of this distinctive style of reed boat.

Bolivia is predominantly (more than 55 percent) Native American. Known as Upper Peru by the Spanish in the 16th and 17th centuries, Bolivia was settled by Europeans partly for the land and labor of its indigenous inhabitants and also for its minerals. In the late 16th century, a Bolivian mine in Potosí was producing about half of the world's silver output. Beginning in the late 19th century, tin was the mineral most sought in the Bolivian Andes. Tin exports are now much less valuable than the natural gas from Bolivia's eastern savanna lowlands. Since the 1920s, oil and gas discoveries have been made around the old lowland frontier town of Santa Cruz, and their control and export have become very contentious issues for Bolivia. Andean indigenous peoples, Europeans, and Japanese have settled around Santa Cruz. They want to be guaranteed a large share of revenues from the fossil fuels beneath them and worry that politicians in the poorer western region around La Paz will siphon off the money in that direction. In addition to this east-west split within Bolivia, there is contention over who should control the resources. President Morales wants to keep out foreign companies and ensure that the resources stay in Bolivian hands. Finally, Bolivia wants to be sure its historic foe, Chile, gets neither the fuels nor the transit fees Chile would charge for the fuels to be transported by pipeline across its territory. Morales is offering Chile a "gas for sea" deal, however: if Chile gives Bolivia 10 miles of coastline, Bolivia will pipe the gas to the sea and give Chile preferential gas prices.

548 Farther north, in the lowland tropical rain forest climates and in the *tierra templada* of the Andean slopes, other pioneers have move in to pan for gold, raise cattle, cut wood, and grow coffee, sugar, and coca. Bolivia's indigenous peoples have chewed coca leaves for centuries without ill effect. The country became a major producer and exporter of cocaine, a far more dangerous product of coca leaves, in the late 20th century. Most of the country's coca output comes from the Chapare region of central Bolivia. With U.S. aid, the government has promoted the substitution of bananas and pineapples for coca. Coffee has also become a replacement crop in regions of the *tierra templada* and *tierra fría*.

10.2.2 Brazil: Populous Rain-Forested Giant

Brazil is the giant of Latin America (●Figure 10.2.12). It has an area of about 3.3 million square miles (8.5 million sq km)—about 5 percent larger than the conterminous United States—and a population of 189 million as of 2007, second only to the United States in the Western Hemisphere (see Figure 10.3, page 526).

Brazil has an increasingly important role in hemispheric and world affairs—so much so that it is a candidate for a permanent seat on the United Nations Security Council. Its growth is especially notable in its overall population, in the explosive urbanism that has made São Paulo (population 19.6 million) and Rio de Janeiro (population 11.9 million; ●Figure 10.2.13) two of the world's largest cities, in the rise of manufacturing, and in the diversification of export agriculture. Brazil has the largest economy in South America. Its expanding domestic market and labor force have accelerated economic development in Brazil, but the surging population also poses serious problems; Brazil has a population growth rate of 1.4 percent. The country is hard-pressed to keep economic output ahead of population increase and to maintain adequate education and other social services.

Brazil has its millionaires, a substantial and growing middle class, and large numbers of technically skilled workers, but the benefits from development have been distributed so unevenly that 23 percent of Brazilians live in poverty. Brazil

● **Figure 10.2.12** Principal features of Brazil.

has the unfortunate distinction of having Latin America's most unequal distribution of national wealth (it ranks fourth in the world, behind three African countries). Just 1 percent of the population owns 45 percent of the country's farmlands. According to the World Bank, the poorest 10 percent of the population receive just 1 percent of the total income, and the richest 10 percent receive almost half. (For comparison, in the United States, the poorest 10 percent receive about 2 percent of the total income—making them the poorest population of all the MDCs—while the richest 10 percent gets 30 percent.) This income inequity is showing signs of shrinking, however, due in part to more generous welfare payments from the government. Brazil is also burdened with a large foreign debt owed to the International Monetary Fund and other international lenders for its ambitious and costly development projects launched after the mid-1970s. However, Brazil has many things going for its future, including abundant and varied natural resources, many allies among the world's advanced economic powers,

and remarkable internal harmony among its diverse ethnic and cultural populations.

Brazil's Environmental Regions

Brazilian has humid tropical environments that are especially challenging to the country's development. The equator crosses Brazil near the country's northern border, and the Tropic of Capricorn lies just south of Rio de Janeiro and São Paulo. Only a small southern projection extends into the midlatitude subtropics. Within the Brazilian tropics, the climate is classed as tropical rain forest or tropical savanna nearly everywhere, with small areas of *tierra templada* in some highlands and a section of tropical steppe climate near the Atlantic in the northeast. The tropical rain forest and savanna ecosystems are characterized by lateritic soils that cannot tolerate cultivation for long once the native vegetation is removed. Flooding sometimes causes health and sanitation problems, but drought also regularly challenges peoples of the savanna areas. Severe drought,

288

• **Figure 10.2.13** The statue of Christ the Redeemer, Rio de Janeiro's most famous landmark, overlooks the city's spectacular harbor.

Richard T. Nowitz/Corbis

apparently related to warmer Atlantic Ocean waters, has even visited the Amazon rain forest in recent years.

Brazil does not have much high or rugged terrain but instead a landscape of low uplands and extensive lowlands. Its three major landform divisions are the Brazilian Highlands, the Atlantic Coastal Lowlands, and the Amazon River Lowland. The country also includes a sparsely populated part of the Guiana Highlands extending across the border from Venezuela.

The Brazilian Highlands

The Brazilian Highlands is a triangular upland extending from about 200 miles (c. 325 km) south of the lower Amazon River in the north to the border with Uruguay in the south. It is a landscape of hills and river valleys, interspersed with tablelands and low mountain ranges. On the east, it is largely separated from the Atlantic Ocean by a narrow coastal plain but reaches the sea in places. Westward, the highlands stretch into parts of Paraguay and Bolivia. The northern edge runs raggedly northeastward toward the Atlantic, approaching but never reaching the Amazon River. In the south, from Salvador to Porto Alegre, the highlands descend steeply to the coastal plain and the sea.

More than half of Brazil's metropolitan cities with a million or more people are seaports along this coastal lowland: in the north are Belém, São Luís, Fortaleza, Natal, João Pessoa, Recife, Maceió, and Salvador; in the south lie Rio de Janeiro, Vitória, Santos, and Porto Alegre. Two of Brazil's three historical capitals are among them: Salvador (formerly Bahia), the colonial capital from the 16th century to 1763, and Rio de Janeiro, which was made the colonial capital in 1763 and then, with independence in 1822, the national capital until 1960. In that year, much to the chagrin of the people of Rio de Janeiro, the capital was moved to the brand-new inland city of Brasília. The shift to Bra-

sília, a frontier city at the edge of the Amazon Basin, was symbolic of the country's hopes for future development of the Amazon. Designed to be home for an estimated 500,000 people, Brasília was also supposed to be highly integrated in socioeconomic classes and ethnic populations. It has far outgrown its projected population (which currently stands at 2.3 million) and has also become one of the country's most segregated cities. As elsewhere in Brazil, the opulent homes of the well-to-do are a distressing contrast with the dense *favelas* (shantytowns). Brasília was meant to be a **growth pole** attracting new settlement, industry, and governmental activity—attempting to break the traditional centrality of Rio de Janeiro and São Paulo on the coast— but its industrial growth has been negligible.

In the northeastern Brazilian Highlands is an interior area of semiarid steppe climate. This area and the tropical savanna around it are called the Caatinga, a word referring to its natural vegetation of sparse and stunted forest. Wet enough to attract settlement at an early date, it has been a zone of recurrent disaster because of wild annual fluctuations in rainfall. Years of drought periodically witness agricultural failure, hunger, and extensive mass emigration. This region, also called the Sertão or the Northeast, is the least prosperous region of Brazil and is notorious for both drought and poverty. Brazil's government has recently approved plans to divert waters from the Amazon Basin to this thirsty region.

The Atlantic Coastal Lowlands

The narrow Atlantic Coastal Lowlands lie between the edge of the Brazilian Highlands and the Atlantic in most places, with the continuity broken occasionally by seaward extensions of the highlands. Climatically, the northern part of the discontinuous ribbon of plain is tropical savanna, the central part is tropical rain forest, and the southern part is humid subtropical. Agricultural development has removed all but about 7 percent of the area's original forest cover. For the species richness of its remaining threatened vestiges, ecologists identify Brazil's Atlantic forest as one of the earth's biodiversity hot spots. This was the first part of Brazil settled by Europeans, and it has been in agricultural use since the 1500s. Using African slave labor, the Portuguese introduced and raised sugarcane on its alluvial plains as an export crop in early transatlantic trade.

The Amazon River Lowland

Brazil's other major lowland area is the massive Amazon River Lowland, containing the world's largest continuous area of tropical rain forest, centered on what by many measures is the world's greatest river (see The World's Great Rivers, pages 586–588). This lowland extends and gradually widens westward from the Atlantic on either side of the Amazon's mouth. Far in the interior, beyond Manaus (at the strategic conjunction of the Amazon and Rio Negro Rivers; population 1.7 million), the lowland widens abruptly and extends across the frontiers of Bolivia, Peru, and Colombia to the Andes.

Tropical rain forest climate and vegetation prevail nearly everywhere, with southern fringes of tropical savanna. Manaus and the ocean port of Belém are the major cities. Large parts of the Brazilian Amazon are still undeveloped and sparsely populated. An estimated 14 million people live in this vast area, and that population is growing, partly through migration, at about 2.5 percent per year.

Boom and Bust in Brazil

In the 1494 **Treaty of Tordesillas,** Spain and Portugal agreed on a north-south line of demarcation in the New World along a meridian intersecting the South American coast just south of the mouth of the Amazon (•Figure 10.2.14). The eastward bulge of the South American continent went to Portugal. The Portuguese later expanded beyond this line into the rest of what is now Brazil, with international boundaries with Spanish-controlled areas drawn in remote and little-populated regions east of the Andes.

The colonial foundations of Brazil were laid early in the 16th century, when trade with local Indians for brazilwood (a South American tree yielding a red dye) began. To combat French and Dutch incursions, Portugal promoted settlement beyond its first small trading posts of the 1530s. The coastal settlements became increasingly valuable as they exported cane sugar to Europe and brought shiploads of slaves from Africa. Slave traders penetrated the interior in search of Native Americans. Catholic missionaries established themselves among the indigenous peoples, often bringing uncontrollable disease epidemics and slaving expeditions in their wake. In the late 1600s, sizable deposits of gold were discovered in the Brazilian Highlands north of São Paulo in what is now the state of Minas Gerais ("mines of many kinds").

In the 18th century, Brazil's center of economic gravity shifted southward in response to the expansion of gold and diamond mining in the highlands. In the mid-1700s, Brazil was producing about half the world's gold. By the time the output declined late in the century, much of the highland around the present city of Belo Horizonte had been settled. Despite the difficulty of traversing the Great Escarpment between Rio de Janeiro's magnificent harbor and the developing interior, "Rio" became the port and economic focus for the newly settled territories. Gold and diamonds were ideal high-value commodities to move by pack animals over poor roads and trails to the port.

Economic distress in the sugar industry to the north contributed to the rise of the southern areas. Sugar declined with increasing development of competing sugar production in other New World colonies, especially in the West Indies. Plantation owners and their slaves began moving to new areas in Brazil. An economy based on cattle ranching and cotton began in the Caatinga (Sertão) area inland from the sugar coast. A still-unbroken pattern of economic hardship began in Brazil's northeast, with hard-fought battles to produce sugar and cotton, regular visits by drought and floor, severe poverty, and the flight of the poor to other areas.

Brazil experienced booms in rubber and coffee production during the 19th century. This was a time of turbulent political change. Colonization by Portugal ended in 1822. Brazil was an independent empire with a strongly federal structure until 1889, then a military dictatorship until 1894, and finally a civilian-ruled republic. Only in 1888 did the country free its large slave population. The boom in wild-rubber gathering in the Amazon followed the discovery of vulcanization—the process that made rubber more flexible, durable, and valuable—by the American Charles Goodyear. At that time, Amazonia was the only home of the rubber tree. A stampede up the rivers followed, and thousands of widely scattered settlers began to tap the rubber trees and send latex downstream. Manaus grew as the interior hub of this traffic and Belém as its port. But some rubber tree seeds were smuggled to England, and seedlings transplanted to Southeast Asia became the basis of rubber plantation agriculture there. By 1910, Amazon rubber production, which in contrast depended on tapping scattered wild trees, was practically dead.

Coffee was the preeminent boom product of Brazil during the 19th century. Production exploded, especially around Rio de Janeiro and São Paulo, during the last decades of the century. This growth was related to the nearly ideal growing conditions of the Brazilian Highlands, a rising world demand for coffee, improved ocean transport, and the construction of railways to move the product. In each year of the late 1800s, over a million overseas immigrants went to work in the coffee industry. São Paulo began its rapid growth as a collecting and forwarding point and business center for the coffee industry. Aided by hydroelectric production, São Paulo also began its career as an industrial center.

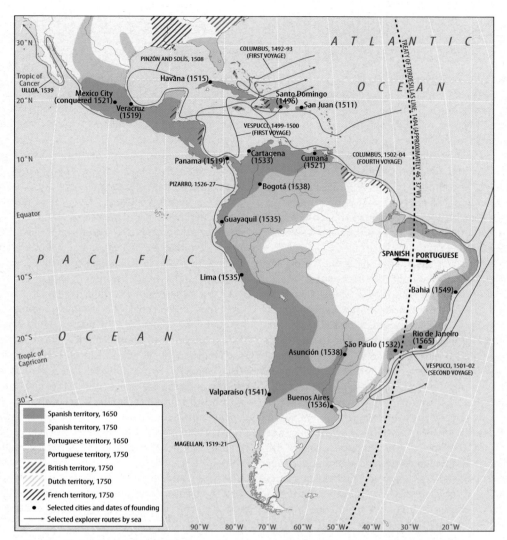

● **Figure 10.2.14** In 1494, Spain and Portugal met at Tordesillas, Spain, to divide up the non-Christian world. Spain acquired most of the lands of the New World, except for the eastern wedge of South America, which went to Portugal. Other European powers established outposts in later centuries.

But the coffee boom went bust. Brazil continued to lead the world in coffee production and export, as it does today, but increasing international competition depressed coffee prices. Industrialization still continued to grow around São Paulo in the 20th century. The railway network expanded. Japanese immigrants began arriving in 1925 and made important contributions to agriculture. There was a major expansion of irrigated rice production in the Paraíba Valley between Rio de Janeiro and São Paulo. Rice is a staple of the Brazilian diet, as are corn, beans, and manioc.

Ready for Takeoff?

The pace of development in Brazil accelerated after 1945, with periodic fluctuations. Economic growth took place under elected governments between 1945 and 1964 and occurred most rapidly in the so called "miracle years" between 1964 and 1985 under a military dictatorship with technocratic leanings and great ambitions. Unfortunately, Brazil's hoped-for economic takeoff as an industrialized power coincided with the devastating impacts of a quadrupling of oil prices in the mid-1970s due to events in the volatile Middle East. In 1985, the dictatorship relinquished power to a new elected government, which then had to face the accumulated problems of that stalled takeoff. The crushing national debt that had accumulated to finance the country's development had to be paid down, and widespread economic misery came with that effort.

Much financing went into building an effective transportation system with long-distance paved roadways, including the Trans-Amazon Highway system, and an emphasis on trucks as movers of goods. The new roads serve as ribbons of settlement, moving the flow of pioneer settlers to remote

228

THE WORLD'S GREAT RIVERS

The Amazon, Its Forest, and Its People

The Amazon is the world's great river (●Figure 10.2.A). It is not the longest; its 3,915 miles (6,264 km) are surpassed in length by the Nile. But by all other standards, the Amazon rules. It handles more volume than any other river. As much as one-fifth to one-fourth of all the world's available freshwater (excluding what is locked up in ice) is in the Amazon and its tributaries at any given moment. So much water flows from the mouth of the Amazon that it forces back the salty waters of the Atlantic as far as 200 miles (320 km) offshore. The river is so impressively large that early Portuguese explorers called it the **"River Sea."** Many creatures normally found only at sea have adapted themselves to its freshwater vastness, including dolphins, sharks, and rays.

The river drains the Amazon Basin, covering some 2.7 million square miles (4.4 million sq km), or about 10 times the size of Texas. About half of this basin is in Brazil, with the remainder sprawling into eight other countries. The basin is home to the world's largest remaining expanse of tropical rain forest and some of the world's most remote populations of indigenous peoples. It is also a region that promises, or seems to promise, prospects for economic development. The issues of pristine forest, Native American populations, and development are intertwined, often problematically. Here is a brief look at these problems.

The Amazon Basin rain forest is the world's largest storehouse of plant and animal species, the great majority of which have not yet been described by science. Ecologists argue that preservation of the rain forest's biodiversity is essential for ensuring the genetic variability nature requires for change and evolution and for helping ensure valuable future supplies of food, medicines, and other resources for people. The forest also acts as a carbon sink to mitigate possible global warming due to excess greenhouse gas emissions. The governments of Brazil and other Amazon Basin countries generally acknowledge the importance of the area's natural state but also argue that the area is so vast that portions of it can be reasonably developed to help advance their economies.

Since the 1970s, Brazil in particular has aggressively pursued development in five major areas of the Amazon: exploitation of iron ore and other minerals, resettlement of "excess" populations from Brazil's crowded southeast, farming and ranching, timber exploitation, and hydroelectric development. Each of these serves as an agent of deforestation. All are ultimately linked to Brazil's expanded road network in Amazonia. To begin its development of the Amazon, in the 1970s, Brazil began construction of the **Trans-Amazon Highway** and its feeder system. The main line runs roughly east to west from the Atlantic toward its ultimate, long-awaited destination: the Peruvian coast. Construction through 2009, part of an international project called the **Regional Initiative for the Infrastructure Integration of South America,** includes the final link crossing the Andes to give Brazilian exporters their desired easier outlet to Asian markets. Principal feeder lines include the controversial BR-364 through the Brazilian states of Acre and Rondônia. BR-163 running north-south through the heart of the region, and BR-319 now under construction between Manaus and Rondônia.

This road construction initiates a sequence of events, illustrated well by Highway BR-364. First, the main road is cut as a swath through the forest. Landless peasants, lured by cheap and sometimes free land, cut and burn the rain forest along the main road and the smaller lanes linked to it. Typical of farmers' experiences with tropical slash-and-burn agriculture elsewhere, they may be

Joe Hobbs

● **Figure 10.2.A** A view from the *middle* of the Amazon River, more than 2,000 miles upstream from its mouth.

THE WORLD'S GREAT RIVERS

The Amazon, Its Forest, and Its People *continued*

able to wrest only three to five years of corn, rice, or other crops from the soil before it is exhausted (●Figure 10.2.B). They then move on to clear and cultivate new lands—the classic pattern of shifting cultivation. If this used plot were left alone, it would return to mature tropical rain forest in a matter of decades. In Amazonia, however, the typical pattern is that large cattle operations move in to use the areas conveniently cleared for them by the farmers. As about 60 million cattle (one-third of Brazil's total cattle population) feed and tread over these Amazonian lands, the soils are exposed to excessive cycles of wetting and drying and to relentless bombardment by ultraviolet radiation. This turns the soil into the brick-like substance known as laterite, on which it is all but impossible for forest regrowth to occur. When ranchers have exhausted the land in this fashion, they move their cattle on to the next plots cleared by farmers. This **shifting ranching** follows the shifting cultivation; it is part of a cycle that is exceptionally destructive of the rain forest.

This cycle and the other uses of the forest, especially a boom in commercial soybean cultivation (see page 589), have resulted since the 1960s in the removal of about 18 percent of the tropical

rain forest cover of the Amazon Basin. Brazil lost about 8 percent of its Amazon forests between 1990 and 2007 (●Figure 10.2.C). In recent years, the annual rate of deforestation in Brazil has slowed somewhat to about 7,000 square miles (18,000 sq km), or about the size of Connecticut and Rhode Island combined.

About 40 percent of the wood cut from the Amazon is shipped overseas. The largest single destination is the United States (consuming about a third of the total), followed by China, but collectively, the European Union nations take in 40 percent. Amazon rain forest destruction has become a cause célèbre for environmentalists around the world, who denounce the development efforts of Brazil and the other countries. Most of the region's governments and people see it differently, denouncing foreigners as hypocrites who have cut down their own forests and who are intent on stunting development of the rain forest countries. In 2007, Brazil announced that it would open vast new tracts of the Amazon for large-scale logging but promised there would be strict monitoring to ensure that the resource was not overharvested and that none of the 70 percent of the forest owned by the government would end up in private hands. Brazil also shifted some of the burden back to the consuming countries, insisting that they should buy timber only from these zones.

There are alternative development schemes, such as the commercial extraction of valuable exports like rubber and brazil nuts from intact tropical forests. These take place in areas like Brazil's **extraction reserves,** but they are very limited in economic impact relative to the conventional development enterprises in Amazonia, and their long-term prospects for growth are uncertain. In the meantime, Brazil and other Amazon Basin countries have passed legislation to slow or halt illegal logging and other unlawful uses that contribute to the rain forest's destruction. Authorities in Brazil have slowed the illegal logging of valuable mahogany trees, but the problem has boomed in the Peruvian Amazon. Nearly 80 percent of the estimated 1.6 million cubic feet (45,000 cu m) of mahogany exported each year from Peru to the United States—where it becomes furniture, acoustic guitars, home decks, and coffins—is illegal. Despite U.S. support for laws prohibiting the illegal logging of mahogany, enforcement has been lax, and legal and illegal mahogany is often imported to the United States in the same shipments.

Development in the Amazon has increasingly provoked conflicts with the region's indigenous populations. There is particular concern now about how fossil fuel production will affect the Kichwa, Achuar, Schuar, and other Native Americans. After 2004, 70 percent of Peru's lowland rain forest areas, many inhabited by Native Americans, were zoned for oil and gas development, and indigenous advocates worry that exploration will introduce new diseases to the native populations. There are plans to build 800 miles (1,280 km) of pipelines that would take natural gas from lowland rain forest areas to coastal refineries and ports. Another big natural gas project is planned for a part of the Brazilian Amazon inhabited by indigenous people. In Ecuador, where oil already makes up half the country's exports by value, there are plans to

NASA/Goddard Space Flight Center Scientific Visualization Studio

NASA/Visible Earth

● **Figure 10.2.B** Satellite images of the region of BR-364 in Amazonia in 1975 (top) and 2006 (bottom).

THE WORLD'S GREAT RIVERS

The Amazon, Its Forest, and Its People *continued*

● **Figure 10.2.C** Deforestation of the Brazilian Amazon.

double production, again mainly in areas where there are significant Native American populations.

In almost all cases, the Native Americans in the areas affected believe that little, if any, benefit will come to them from these developments. They are becoming increasingly activist, using a combination of threats against the oil interests and pleas for more benefits from the projects. Some of the native groups have powerful activist allies abroad and have come to rely on the Internet to make their cause known; an example is the Kichwa people of Sarayaku (http://www.sarayacu.com), who live in one of the Ecuadorian oil blocks slated for development. The Kichwa are buoyed by the 2006 election of Ecuadorian President Rafael Correa, who as a candidate had expressed his support for their cause. Correa has appealed to the international community to use debt-for-nature swaps or other means to finance the protection of sensitive cultural and environmental sites in the rain forest. In exchange for the funding, Ecuador would prohibit oil drilling in these areas. Some Native American communities are developing their own ecotourist and cultural tourist enterprises, hoping both to demonstrate to governments that there are alternative economic uses for the rain

forest areas and to gain valuable support and publicity from international visitors (●Figure 10.2.D).

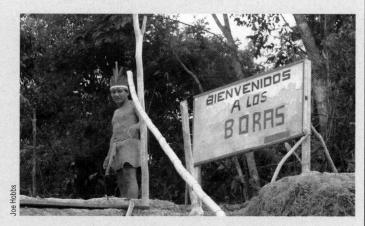

● **Figure 10.2.D** Are ecotourism and cultural tourism helping to protect indigenous people and natural environments in the Amazon? This community of Bora Native Americans promotes itself as a "genuine Indian village." Its residents wear native costume for visitors.

INSIGHTS

The Versatile Soybean

The soybean (*Glycine max*), originally domesticated in China, is the crop of the future for a number of world regions (●Figure 10.2.E). The plant has a very high percentage of protein—30 to 35 percent—and no starch. It fixes nitrogen in the soil as it grows, making it much less demanding on the soil than most. While nearly all U.S. production of soybeans is used for animal feed, the Brazilian crop is widely exported and used domestically for industrial production of chemicals, paints, ink, adhesives, insect sprays, and even a fuel blend (called *soy diesel* and akin to ethanol). The world market for soybeans is growing steadily not just because of the industrial uses of the bean but also because it can be mixed with other products to create high-protein, modestly priced foods—such as tofu (bean curd)—that are increasingly important in global efforts to diminish world hunger.

● **Figure 10.2.E** Soybeans.

Joe Hobbs

areas and connecting communities to markets for rubber, minerals, timber, fish, agricultural products, and tourism. Still, only 10 percent of the country's roads are paved, and upgrading transport infrastructure is a critical challenge in Brazil's development. Some of the huge advantages Brazil enjoys in agriculture are diminished by the high costs associated with getting produce to market and overseas shipping ports.

Major investments in agriculture have contributed to a recent rebound in the country's economy and established Brazil as one of the world's major breadbaskets. Brazil leads the world in exports of sugar, beef, poultry, coffee, orange juice, and tobacco. There is particular interest in expansion of sugar and soybeans. Brazil wants to double sugarcane production from 2007 levels by 2017, mainly to produce the biofuel **ethanol** for domestic consumption and exports. Already established as the world's leader in ethanol production, Brazil will be hard to beat; it is much more efficient and inexpensive to produce the biofuel from sugar than from corn, as the United States does. Together producing 70 percent of the world's ethanol, the two countries are exploring ways to maintain and increase that share through cooperative research and development. Brazil may be able to help the United States get the kinds of secondary uses it does from its cane processed for ethanol: the cane waste, called *bagasse,* is burned in steam boilers to produce electricity used in ethanol production and for surplus sale back to the electricity grid.

According to critics, major drawbacks to biofuel production are that it does nothing to enhance the welfare of Brazil's landless and rural poor. It puts more land in the hands of agribusiness and the elite—a classic illustration of the marginalization process described on page 46. It is also

619

grown as a monoculture, cutting down on both natural biodiversity and crop diversity. Critics levy the same accusations at Brazil's soy industry. In the 1960s, Brazil became a major soybean producer and exporter, with production centered in the subtropical southeast and in a new district in the tropical Brazilian Highlands (see Insights, above). Soybean production is now pushing into the Amazon Basin and is a major source of deforestation there.

30

This expansion is in response to growing demand for soy products in China, the world's largest soy consumer. Brazil is neck and neck with the United States as the world's leader in soybean exports and is second only to the United States in soybean production. To gain an even larger market share, Brazil has given up its long-standing opposition to genetically modified (GM) foods and is producing GM soybeans and other crops. The country is trying to maintain a delicate balance between its North American markets, which have no problem with GM foods, and its European markets, which shun them. Brazil claims it can produce GM and non-GM foods to satisfy both. American farmers look nervously at Brazil's surging agricultural exports. Having won the case it made to the World Trade Organization that U.S. farmers were too heavily subsidized, making Brazilian cotton and other exports uncompetitive, Brazil could undercut American agricultural exports on the world market if the United States abides by the WTO ruling.

89

477, 623

Brazil's manufacturing sector has experienced rapid expansion. Early-stage industrial products like textiles and shoes are still present, along with steel, machinery, chemicals, plastics, automobiles and trucks (Brazil is Latin America's leading automobile producer), ships, aircraft, and weapons. São Paulo is the industrial core, producing nearly two-thirds of Brazil's manufactured goods. Following a

300

global trend, Brazil is outsourcing an increasing share of some of its lower-tech manufacturing, for example, of shoes, to China. Brazil's metals make an important contribution to the country's industry. The Brazilian Highlands are very mineralized, with ores of iron, bauxite, manganese, tin, and tungsten. Brazil has about an eighth of the world's proven iron ore reserves and is the world's third-largest producer and exporter of iron ore. Because of these reserves, Brazil is now in the top 10 of world steel exporters, thanks to China's red-hot economy and its voracious appetite for this important product.

For decades, Brazil's greatest resource deficiency was in energy. Coal and petroleum reserves were inadequate, threatening the country's continuing industrial development. Tropical hardwoods were widely used in place of fossil fuels in some manufacturing industries. Today, however, the country is in an enviable position with respect to energy: it is virtually self-sufficient, even meeting 90 percent of its needs for oil from domestic supplies. Large natural gas and petroleum reserves have been found offshore and in the remote central Amazon. At high environmental cost, and largely with Chinese financial assistance, numerous dams have been built and are scheduled for the Amazon Basin to feed Brazil's appetite for electricity, almost 90 percent of which is produced by hydropower. Meanwhile, the country's ethanol program has boomed. All gasoline sold in Brazil is in fact **gasohol,** with a 23 percent ethanol content. "Flex cars" that can run on any combination of ethanol and gasoline now account for more than 80 percent of all new cars sold in the country.

10.2.3 The Southern Midlatitude Countries: South America's "Down Under"

Four countries of southern South America—Argentina, Chile, Uruguay, and Paraguay—differ from all other Latin American countries in being mostly midlatitude rather than low-latitude in location and environment (●Figure 10.2.15; see also Figure 10.5, page 529). Northern parts of all the countries except Uruguay extend into tropical latitudes, but the core areas lie south of the Tropic of Capricorn and are climatically subtropical. Argentina barely extends into the tropics in the extreme north, centers in subtropical plains around Buenos Aires, and reaches far southward into latitudes equivalent to those of northern Canada. Chile also extends south to these latitudes. Uruguay's small territory is entirely midlatitude and subtropical, and most Paraguayans live in the half of their country that lies poleward of the Tropic of Capricorn. Argentina, Paraguay, and Uruguay have the same humid subtropical climatic classification as the southeastern United States. Chile's midsection or core area has a Mediterranean (dry-summer subtropical) climate like that of southern California.

These four Latin American countries share certain other regional characteristics. Agriculturally, they are more similar to other midlatitude countries (in both the Northern and Southern Hemispheres) than to most other Latin American or other tropical countries. Argentina is an important export producer of such crops as wheat, corn, and soybeans and thus competes with North American farmers. Both Uruguay and Argentina are significant exporters of meat and other animal products. Paraguay exports soybeans and cotton. Chile exports fruits, vegetables, and wine.

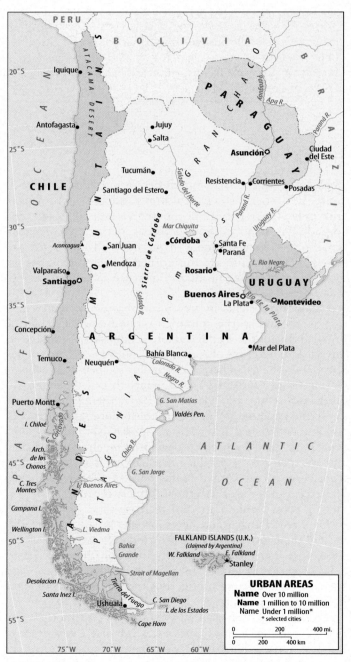

● **Figure 10.2.15** Principal features of Argentina, Chile, Uruguay, and Paraguay.

Uruguay, Argentina, and Chile are among the most economically developed countries in Latin America. They have relatively low dependence on agricultural employment, with less than 15 percent of their employed workers still in agriculture. Paraguay, with about 30 percent of its labor force still on farms, is poorer than the other countries, partly because of its landlocked isolation.

46

Because these countries never developed labor-intensive and export-oriented plantation agricultures in colonial times, they have relatively small black populations. Argentina and Uruguay have been major magnets for European immigration and have populations estimated to be 85 percent "pure" European in descent. Paraguay and Chile, in contrast, attracted fewer Europeans and are about 90 percent mestizo.

Argentina

Argentina's area of nearly 1.1 million square miles (2.8 million sq km) is second in Latin America only to that of Brazil and is about one-third the area of the 48 conterminous American states. Its population of 40 million ranks it a distant third behind Brazil and Mexico within Latin America.

Argentine Agriculture

45 Argentina, one of the "neo-Europes" endowed with quality farming and ranching lands, is one of the world's principal sources of surplus food. The leading agricultural exports are wheat, corn, soybeans, and beef, shipped mainly from the capital city of Buenos Aires (population 13.6 million). Argentine wine exports have grown significantly in recent years too. Agriculture is concentrated in the country's core region, known as the humid pampas (•Figure 10.2.16). Pampas ("plains") have a prairie vegetation similar to that of the North American Great Plains (see Figure X.X). The humid subtropical climate of the pampas is similar to that of the southeastern United States, featuring hot summers, cool to warm winters, and annual precipitation averaging 20 to 50 inches (c. 50 to 130 cm). The grasses of the original prairie provided an unusually high content of or-

28 ganic material (humus), just as in the North American prairie areas. The local beef industry is fed from the grasses and grain that flourish in this setting. Argentina is the world's third-largest exporter of beef (after Brazil and Australia), and Argentine per capita beef consumption is the second highest in the world (after Uruguay).

From the 16th century to the late 1800s, the humid pampas were sparsely settled ranching country, parts of which were still controlled by indigenous people. Then, in the late 19th and early 20th centuries, Argentina experienced a rapid shift to commercial agriculture and urbanization. Immigration from Europe skyrocketed, and the population surged. Commercial contact with Britain contributed much to Argentina's new dynamism. The British market for imported food was increasing quickly. Refrigerated shipping came into use, allowing meat from distant sources such

• **Figure 10.2.16** The Argentine pampas.

as Argentina to be imported into the United Kingdom. A British-financed rail network (the densest in Latin America) was built in Argentina to tap the countryside for trade. The network focused on Buenos Aires, which rapidly became a major port and large city. British cattle breeds, favored by British consumers, were introduced to Argentine ranches, or *estancias,* along with alfalfa as a feed crop. Soon an export agriculture developed in which wheat, corn, and other crops supplemented and then surpassed the original beef exports. European immigrant farmers assumed a major role in agriculture, generally as tenants on the large ranches that dominated the pampas economically and socially. Immigrants came mostly from Italy and Spain but also from other countries, giving present-day Argentina a complex mixture of Spanish-speaking citizens from various European backgrounds. This regional flavor is enhanced by the growing production of Argentine wines that aspire to rival those of neighboring Chile and even Mediterranean Europe.

Argentine areas outside the humid pampas are less productive and less populous. The northwestern lowland just inside the tropics, called *El Gran Chaco,* has a tropical savanna climate and supports cattle and cotton. Westward from the lower Paraná River and the southern part of the pampas, the humid plains give way to steppe and then desert, with production and population declining accordingly. Population in these western areas is concentrated in oases along the eastern foot of the Andes, where mountain streams emerging onto the eastward-sloping plains are impounded for irrigation. Vineyards and, in the hotter north, sugarcane are irrigated. Córdoba is the largest oasis service center and the second-largest city in Argentina.

To the south of the pampas lie thinly populated deserts and steppes in the Patagonian plateau, a bleak region of cold winters, cool summers, and incessant strong winds. Native American country until the early 20th century, Patagonia now has millions of sheep on large ranches but little other agricultural production. Between 1999 and 2003, mutton prices doubled and wool prices rose, leading to a boom in the Patagonian sheep industry. Several factors were responsible: a decrease in a long-standing oversupply of sheep in Australia, rising oil prices (which made synthetic wool expensive and natural wool more attractive), and the mad cow disease that reduced European appetites for beef and increased them for mutton. The clothing company Benneton owns a substantial portion of Patagonia's sheep and is Argentina's largest single landowner.

Urban and Industrial Argentina

Only about 9 percent of Argentina's labor force is directly employed in agriculture and ranching. Most Argentineans are city dwellers employed in services and manufacturing. Metropolitan Buenos Aires has about one-third of the national population. It was settled in the late 16th century as a port on the broad Paraná estuary called the Rio de la Plata. But until Argentina became reoriented toward British and then world markets, the city remained an isolated outpost of a country focused on the north and the western oases, with relatively little trade by sea. In the 19th century, Buenos Aires became the capital and grew explosively as a seaport, rail center, manufacturing city, and destination of mainly European immigrants.

Industrialization in Argentina followed a sequence found in many developing nations. The process began when the government imposed tariffs on overseas goods to help small-scale industries supply consumer items for sale within the country and to support meatpacking and other plants processing agricultural exports. Some consumer industries, such as textile manufacturing and shoemaking, were expanded with government support when the country was cut off from imported supplies during World Wars I and II. After World War II, a state-led policy of industrialization was instituted, boosting manufacturing and diversification of industrial output to include chemicals, machinery, and automobiles. Imported coal to supply power was replaced by domestic oil (mainly from fields on the coast of Patagonia), together with natural gas from the archipelago of Tierra del Fuego and the Andean piedmont. Imported ore and coal supply steel mills. Hydroelectric and nuclear power combined provide about 10 percent of Argentina's energy supply.

Argentina's industries have problems. Manufacturing has never become competitive on the world market, and trade is still dominated by agricultural and petroleum exports and industrial imports. Manufacturing contributes less to national production than should be expected of a country of Argentina's size and potential. In 1913, Argentina was actually a prosperous country due to strong exports from the pampas, much foreign (mainly British) investment in railways and other infrastructure, and immigration of capital and talent from Spain and Italy.

Growth generally stalled after that, especially because of the country's disastrous politics (described in the next section). Children and grandchildren of the hundreds of thousands of European immigrants who came to Argentina after World War II returned to their European homelands by the thousands, especially during a recession in the late 1990s and an outright collapse of the economy in 2001, when more than half of the population fell below the poverty line. Four presidents came and went in rapid succession, all unable to deal with the crisis until Eduardo Duhalde, elected in 2002, devalued Argentina's currency, instituted price controls, and encouraged use of locally made products instead of imported goods. By 2004, Argentina's economy had rebounded sharply, thanks especially to soaring prices for the country's exports of agricultural goods (especially soybeans, corn, and wheat), natural gas and oil, and new output of honey, construction materials, cosmetics, clothing, and leather goods.

Disastrous Politics

Political turmoil and government mismanagement of the Argentine economy have taken a heavy toll on the country. After initial disturbances following independence, Argentina became a constitutional democracy controlled by wealthy landowners. This situation continued for many decades until 1930, when a military coup toppled the constitutional government. Between 1930 and 1983, the country was controlled either by military juntas or by the dictatorial Juan Perón or his party, with brief intervals of semidemocratic government. Although there were variations, the military generals and the Peronists tended to be ultranationalist and semifascist. They distrusted democracy, communism, and free-enterprise capitalism and believed in authoritarianism, violent repression of opposition (which itself was often violent), and state ownership or direction of major parts of the economy. In what has come to be known as Argentina's **Dirty War**, their death squads were responsible for the disappearances and presumed deaths of an estimated 30,000 people—the so-called **disappeared generation**. They courted the support of a growing urban working class by providing it with jobs, pay, and benefits beyond what was justified by its productivity and output, and they paid for these by inflating the currency and borrowing from abroad. Inflation rates eventually reached hundreds of percentage points a year, lowering savings and leading to the flight of investment capital and an economy heavily focused on currency manipulation and speculation rather than productive investment. The end of this period also saw Argentina lose a war with Britain over the Falkland Islands (see page 452).

Chile

Like Argentina, Chile (see Figure 10.2.15) was set back by violent politics in the late 20th century. A U.S.-backed coup in 1973 resulted in the death of the country's socialist president, Salvador Allende, and swept in the brutal regime of General Augusto Pinochet (pronounced "pee-noh-*chet*"). Pinochet began the free-market reforms that were inherited and enlarged by a freely elected government in 1990. Chile after Pinochet has become one of the most prosperous countries of South America on a per capita GNI PPP basis.

It has enjoyed booming exports of copper, salmon, fruits, and wood products, particularly since concluding unilateral trade agreements with the United States in 2002, the European Union in 2003, and South Korea in 2004—and as China's demand for copper and other commodities has surged.

Chile's elongated shape conceals the country's size, which is more than double that of Germany or about that of the state of Texas. Chile stretches approximately 2,600 miles (c. 4,200 km) from its border with Peru to its southern tip at Cape Horn, but in most places, it is only about 100 to 130 miles (c. 160 to 210 km) wide (see Figure 10.2.15). The country has to struggle variously with rugged terrain, aridity, too much precipitation, and cold, and due to such extremes, only one relatively small section in the center has any significant population. The long interior boundary lies mostly near the high Andean crest, with the populous core areas of neighboring countries lying far away on the other side of the mountains. Although the Andean barrier has never been impassable, traffic across the mountains has been difficult and is still light.

The Middle Chilean Core Region

Chile's populous central region of Mediterranean (dry-summer subtropical) climate occupies lowlands between the Andes and the Pacific from about 31 degrees to 37 degrees south latitude. Mild wet winters and hot dry summers characterize this area, as they do southern California at similar latitudes on North America's West Coast (see Figure 2.10). This strip of territory has always been the heart of Chile's core area. Topographically, it consists of lower slopes in the Andes, a hilly central valley, and low coastal mountains. The area was the southern outpost of the Inca empire and was occupied in the mid-1500s by a small Spanish army. The army made use of land grants from the king of Spain to create an isolated ranching economy. Grants varied in size by the military rank of the grantees, and this fact, together with the conquered status of the indigenous population, created a society characterized by great social and economic inequality. At one end of the scale was the aristocracy, composed of the owners of enormous ranches. At the other end was a mass of landless cowboys and subsistence farmers. Men from all ranks of society married indigenous women freely, sometimes several at a time. This created the Chilean nation of today, which is over 90 percent mestizo.

Owners of the great *haciendas* dominated the country well into the 20th century, and ranching remained the primary pursuit in middle Chile. Irrigated and nonirrigated wheat, vineyards, and feed crops became agricultural specialties. During the 1960s and early 1970s, population pressure and resulting political demands led to large-scale land reform that reduced the role of many landowners.

Growing population pressure in the core region has contributed to rapid urbanization in recent decades. The national capital of Santiago, founded at the foot of the Andes in the 16th century, has a metropolitan population of 4.9 million. Metropolitan Valparaíso, the core region's main port and coastal resort, has 900,000, and the port of Concepción is slightly larger. Altogether, the area of Mediterranean climate is the home of about 80 percent of Chile's 16 million people. The country is now 87 percent urban, exceeded in South America only by Uruguay (93 percent) and Argentina (89 percent).

North and south of the core area are transition zones with lower populations. On the northern edge of the core zone, the population declines across a narrow band of steppe to the arid and almost empty wastes of the Atacama Desert, the driest region on earth. In the south, the transition zone is larger and more important economically. South of Concepción, rainfall increases, average temperatures decline, and the Mediterranean climate gives way to a wet version of the marine west coast climate, with natural vegetation of dense forests. The southernmost part of the Mediterranean climatic area and the adjacent northern fringe of the marine west coast area as far south as Puerto Montt have become part of Chile's core in the past century.

Indigenous people are still numerous in this part of Chile. They are the descendants of the indigenous Araucanian people who fought off Inca expansion and resisted most Spanish expansion for three centuries. Some eventually assimilated into Chilean society. Also distinctive is a German element descended from a few thousand immigrants who settled on this wild frontier in the mid-19th century. Today, the German community is economically and culturally very important, even putting its stamp on architectural styles. Most people in the southern transition zone, however, are descendants of mestizo settlers who came from the core of Chile in the past century as population pressure increased. Agriculture in this green landscape of woods and pastures is devoted largely to beef and dairy cattle, wheat, hay, deciduous fruits, and root crops.

Northern and Southern Chile

North of the core, Chile is desert, except high in the Andes. This part of Chile, the Atacama Desert, is one of the world's few nearly rainless areas. It is not a hot desert; mild temperatures and fog result from the almost constant winds from the cool ocean current offshore. Chile gained most of the Atacama by defeating Bolivia and Peru in the **War of the Pacific** (1879–1883). That loss deprived Bolivia of its outlet to the sea; the country had previously held a corridor across the Atacama to the sea at the port of Antofagasta, and Peru owned the part of the desert farther north. The war was fought over control of mineral resources, primarily the Atacama's world monopoly on sodium nitrate, a material much in demand at that time for use in fertilizers and smokeless gunpowder. Since then, other mineral resources of the Atacama and the adjacent Andes, especially copper, have eclipsed sodium nitrate in importance. Chile has the world's largest copper mine, Escondida, in the Atacama Desert, and has benefited from the recent surging demand for copper, prompted mainly by China's boom.

Copper mining and other enterprises require substantial inputs of energy, which in Chile is in short supply. The nation is therefore thinking of greatly expanding its hydroelectric power capacity in the rugged river-carved region of the south. South of Puerto Montt, the marine west coast climate section of Chile continues to the country's southern tip in Tierra del Fuego, a cool island region of windswept sheep pastures and wooded mountains divided between Chile and Argentina. In this thinly populated strip, the Pacific reaches the foot of the Andes along a rugged coastline penetrated by fjords. The spectacular pinnacles of the Paine Towers National Park are here (•Figure 10.2.17). Extremely high precipitation, cool temperatures, violent storms, and dense mountain forests are characteristic. Little agriculture beyond sheep grazing is possible, and population is very sparse. Environmentalists argue that the area's best economic potential is in ecotourism and that hydroelectric dams and infrastructure would destroy the area's natural attractions.

Chile has only limited coal, oil, and gas—mainly along and near the Strait of Magellan, which separates the South American mainland from Tierra del Fuego—and has come to rely on Argentine natural gas for almost half of its electricity production. The countries have invested billions of dollars since the mid-1990s on cross-border pipelines and power plants. Recently, energy shortages in Argentina have led to reductions in gas exports to Chile, leaving the country scrambling for new resources. Natural gas from Bolivia would be the obvious choice, but Bolivia's historical grudge 593 against Chile makes that prospect difficult. Meanwhile, there is talk about expanding highways to enable Argentine and Brazilian producers to transport soybeans, mineral resources, and manufactured goods to Chilean ports to promote trade around the Pacific Rim.

Uruguay: Buffer State

Little Uruguay (see Figure 10.2.15), with an area of 68,500 square miles (177,400 sq km; about the size of Missouri or half the size of Germany) and bordered by Argentina and Brazil, is a part of the humid pampas. Its independence from Argentina and Brazil resulted from its historical buffer position between them. In colonial times, there were repeated struggles between the Portuguese in Brazil and the Spanish to their south and west over possession of this territory flanking the mouth of the Rio de la Plata system. After Argentine independence from Spain, these struggles continued at the same time that a movement for political separation grew in the territory north of the Rio de la Plata. Britain eventually intervened, fearing Portuguese expansion south of the river, and the two contenders signed a treaty in 1828 recognizing Uruguay as an independent buffer state between the two larger countries.

Montevideo, Uruguay's national capital, is the only large city in the country, and its metropolitan population of 1.7 million includes about half of all Uruguayans. The city has dominated the country almost from Montevideo's

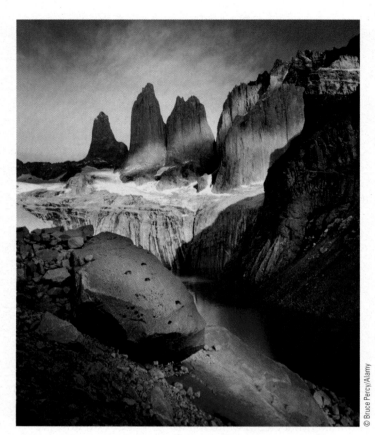

• **Figure 10.2.17** Paine Towers National Park, Chile.

inception in the 17th century, first as a Brazilian (Portuguese) and then as an Argentine (Spanish) fort. Its port facilities handle Uruguay's important waterborne trade. Montevideo directs an economy dominated by employment in government-owned services and industries. The city is the main center of Uruguay's manufacturing industries, focused on textiles, woolens, and meat processing. Agriculture and ranching, with just 12 percent of the labor force, supply the country's meat-rich diet and the greater part of all Uruguayan exports. Production of soybeans has surged to meet demand from China, as has the planting of pine and eucalyptus trees for the lumber industry.

Aside from hydropower, which supplies nearly all of the country's electricity, Uruguay has almost no natural resources to support industrialization and depends heavily on imported fuels, raw materials, and manufactured goods, which must be purchased with the proceeds from agricultural and textile exports. An extensive program is under way to privatize government-owned industries and diversify the economy.

Landlocked Paraguay

Landlocked and underdeveloped Paraguay (see Figure 10.2.15) has long been isolated from the main currents of world affairs. The Spanish founded Paraguay's capital city of Asunción (population 1.9 million) in the 1530s at a site far enough north on the Paraguay River to promise security against the Native Americans of the pampas to the south. High ground at the riverbank also offered protection from floods. Asunción became the base from which surrounding areas were occupied. But when independence came, Paraguay had no way to reach the sea except through Argentina or Brazil. The main transport route reached down the Paraguay and Paraná Rivers to Buenos Aires. But this route is much longer than it appears on maps owing to meandering of the rivers and navigation difficulties. Transport costs were so high that Paraguay could not develop an export-oriented commercial economy, and the country remained locked into a mostly self-contained economy based on its own resources.

West of the Paraguay River is the region known as the Chaco ("hunting ground"), extending into Bolivia, Argentina, and Brazil. This is a very dry area. Precipitation averages 20 to 40 inches (c. 50 to 100 cm) annually, but high temperatures cause rapid evapotranspiration, and soils are sandy. These conditions produce a xerophytic forest in the east, transitioning westward to grasses on a vast alluvial plain that is flooded during the wet season. Among the small population of Paraguayans in the western region are the Mennonites who came here in the 1920s and 1930s to farm and practice their unique lifestyle.

614

Only 6 million people lived in Paraguay in 2007. The country has experienced an outflow of population for most of its history and has never attracted any large immigration. The few whites in the country own a disproportionately high share of the land, and Native American Guaraní ethnicity is predominant in the 95 percent of the population classed officially as mestizo. Urbanization and industrialization have not gone far. Asunción is the only urban area with more than 300,000 people, and over 40 percent of the country's people are rural. Per capita GNI PPP is far below that of Argentina, Chile, and Uruguay. There are modern road and air links with the outside world and between parts of Paraguay itself. With access to outside markets, exports of soybeans and cotton have grown rapidly and have come to dominate the country's trade.

The future may see larger and more dramatic changes due to three huge new hydroelectric installations along the Paraná River on the southeastern border of the country. Paraguay is now exporting electricity to pay its share of the costs of these projects, which it undertook jointly with Argentina and Brazil. The new hydroelectric output has given Paraguay a far more favorable energy base for the future than this fuel-deficient country has had in the past.

From the far south of South America, the text turns finally to the northern end of this hemisphere.

SUMMARY

→ The Andean countries are Colombia, Venezuela, Ecuador, Peru, and Bolivia. Settlement in them is concentrated along the coasts in the *tierra caliente* and in the uplands in the *tierra templada* and *tierra fría*.

→ The Andean countries are characterized by varied environmental zonation, fragmented topographic patterns, populous highlands and sparsely settled frontiers, significant indigenous populations, dominant Hispanic traditions, poverty, and primate cities.

→ In Venezuela, major settlement is concentrated in the uplands. Discovery of oil a century ago in the coastal regions around Lake Maracaibo prompted rapid industrial growth. Iron resources in the Guiana Highlands support an active export flow via the Orinoco River.

→ Venezuelan President Hugo Chávez has led a self-proclaimed Bolivarian Revolution ostensibly aimed at using the country's oil wealth to enhance the prospects for the poor.

→ Colombia has three major highland metropolises: Bogotá, Medellín, and Cali. Coastal Colombia has less settlement than the uplands. Bananas and sugarcane have been the traditional crops.

→ Until recently, much of Colombia's territory was controlled by two rebel groups. With U.S. economic and military assistance, the government is fighting back and has wrested much control from the rebels.

→ Ecuador has major settlements in the Andean basins, with increasing populations in the lower slopes and coastal areas. Oil in both Ecuador and Peru has begun to be extracted in the tropical lowlands draining into the Amazon River. Its development is controversial because many indigenous peoples live on lands slated for oil development, and they are generally opposed to the industry.

→ Peru's population is concentrated in the high plains region called the *altiplano*. Subsistence agriculture is important for the country's large peasant population, which includes a high proportion of Native Americans. Fish are an important export but are vulnerable to the natural hazard of El Niño.

→ Bolivia is South America's poorest country, landlocked high in the Andes with its capital, La Paz, located at 12,000 feet (3,700 m). Natural gas is an important resource, and the country's Native American president has vowed to use it to benefit the poor. Bolivia has a majority Native American population.

→ Brazil is the giant of South America. It is larger than the 48 lower United States but has more than 100 million fewer people. Settlement clusters along the coast, but Brazil is promoting settlement in the interior savanna and rain forest. A new interior capital, Brasília, was built in 1960. Brazil's economy is the largest in South America, and Brazil has the region's widest disparities between wealthy and poor people.

→ Brazil has a large realm of tropical rain forest centered in the Amazon River system, which drains almost 3 million square miles (7.8 million sq km).

→ Periods of economic change in Brazil have been marked by booms and busts in sugarcane, gold, diamonds, rubber, and coffee. Soybeans, especially for the Chinese market, are increasingly important in Brazil's agricultural export economy. Beef and orange juice concentrate are also prominent exports. Sugarcane is grown mainly for Brazil's ethanol fuels, which are in increasing demand as fossil fuel costs rise. Iron ore and steel are vital nonagricultural exports.

→ The Amazon has more volume of water than any other river in the world. The Amazon Basin extends into nine countries, and about half of it lies in Brazil. Development in the Amazon region is controversial, in part because of the rich biodiversity in its rain forest. The Trans-Amazon Highway and its feeder roads are opening Amazonian wilderness to development. Shifting cultivation is often followed by shifting ranching in a very destructive land use cycle.

→ Countries of the southern midlatitudes include Argentina, Uruguay, Paraguay, and Chile. The first three all possess a humid subtropical climate like that in the southeastern United States. Chile, however, is more similar to southern California in that both regions have a Mediterranean climate.

→ Argentina is the second-largest nation in South America. It has major urban growth along the coast and in the alluvial plain of the Rio de la Plata. The agricultural productivity of the pampas grasslands define Argentina. Beef and wheat have long been major Argentine export commodities, joined recently by sheep products. Argentina was relatively prosperous early in the 20th century but suffered setbacks, largely because of political troubles. The economy rebounded after an economic collapse in 2001.

→ Chile has benefited from rising demand for its abundant copper reserves. The country is a strong advocate of globalization and has joined numerous trade pacts. Resented by gas-rich Bolivia for cutting off access to the sea and with reduced supplies from Argentina, Chile is finding it difficult to find adequate fuel. The world's driest desert, the Atacama, is in Chile.

→ Uruguay was created as a buffer state between Argentina and Brazil. The country is resource-poor, and its economy depends mainly on agricultural exports.

→ Landlocked Paraguay has few mineral resources, and many of its people have emigrated. Hydroelectric development figures prominently in future development plans.

KEY TERMS + CONCEPTS

altiplano (p. 578)
Andean Group (p. 573)
Bolivarian Revolution (p. 574)
Chavismo (p. 574)
Dirty War (p. 592)
disappeared generation (p. 592)
estancia (p. 591)
ethanol (p. 589)
extraction reserve (p. 586)

gasohol (p. 590)
growth pole (p. 583)
Regional Initiative for the Infrastructure Integration of South America (p. 585)
"River Sea" (p. 585)
shifting ranching (p. 586)
Shining Path (p. 580)
tepuis (p. 575)

Trans-Amazon Highway (p. 585)
Treaty of Tordesillas (p. 584)
Tupac Amaru Revolutionary Movement (MRTA) (p. 580)
21st-century socialism (p. 574)
Viceroyalty of Peru (p. 579)
War of the Pacific (p. 593)

REVIEW QUESTIONS

1. What are the Andean countries of South America? What traits do they share?

2. What roles do coca and cocaine play in the national economies of Colombia, Peru, and Bolivia?

3. What factors have helped make Bolivia the poorest country in South America?

4. In what ways is Brazil the giant of South America?

5. What is the Trans-Amazon Highway system? What role does it play in the region's development and deforestation?

6. What cycle of land use is particularly destructive to the Amazon rain forest?

7. Why are Native Americans of the Amazon Basin generally opposed to fossil fuel development there?

8. What have been the principal booms and busts in Brazil's economic history? What exports are particularly strong today, and why?

9. How committed is Brazil to ethanol, and what interests do other countries have in this fuel source?

10. What rising and falling economic fortunes have been experienced in Argentina, and why?

11. Why has Chile's economy been particularly strong recently?

Arizona's Monument Valley is an iconic North American landscape.

CHAPTER 11
A GEOGRAPHIC PROFILE OF THE UNITED STATES AND CANADA

The United States and Canada, together with Mexico and Central America, make up the continent of North America. Mexico and Central America are so linked with South America in culture, language, and tradition that they are best classified within the separate region of Latin America. The United States and Canada as a culture region are sometimes called Anglo America, emphasizing their British colonial origins and their contrasts with Latin America. But that term is increasingly irrelevant today, as multiculturalism has become a hallmark of both societies. This chapter and its accompanying modules depict a complex cultural mosaic overlaying an essentially British political, cultural, and economic foundation.

The island of Greenland belongs politically to Denmark but rises from the same continental crust on which Canada sits, so it is also described briefly as part of this region.

■ chapter **objectives**

This chapter should enable you to

→   Appreciate the wide range of Native American adaptations and ways of life in this region's diverse environments

→   Recognize Canada and the United States as countries shaped mainly by British influences but also by a wide range of foreign immigrant cultures

→   Understand how the United States acquired its vast land empire

→   View the prosperity of the United States and Canada in part as products of their vast natural resource wealth

→   Trace the rise of the United States to a position of global economic and political supremacy and to recent rivalries from other giants

● **Figure 11.1**   Principal features of the United States, Canada, and Greenland.

11.1 Area and Population

Canada's 3.85 million square miles (9.97 million sq km) makes it slightly larger in area than the United States' 3.72 million square miles (9.63 million sq km), but Canada is less wealthy than the United States and much less powerful on the global stage (●Figure 11.1). The region's map comparison with Europe is on page 76 and the continental United States is compared with Latin America on page 525. Canada had a population of 33 million in 2007, compared with 302 million in the United States (Table 11.1 and ●Figure 11.2). Together the two countries have 5 percent of the world's population, on 13 percent of its land surface, including Antarctica.

The United States reached the milestone of 300 million people in 2006, meaning that more than half the people who had ever lived in the country were living there that year. A snapshot of that population on a given day in 2006 would show 11,000 babies born and 3,000 immigrants arriving—both much larger figures than the number of deaths and emigrants. If the 2006 growth rate is sustained, it would put the population at 400 million—double its 1967 population—around 2043.

The majority of this region's 335 million people live in the eastern half of the region, south and east of the Saint Lawrence Lowlands and Great Lakes. The average population density of the two countries is only 81 per square mile (31/sq km) for the United States and 9 per square mile (3/sq km) for Canada, which has a tremendous expanse of sparsely settled northern lands. With its vast icecap, Greenland is even more sparsely populated, with just 57,000 people living in an area of 836,000 square miles (2.17 million sq km), or 0.1 person per square mile (0.02/sq km).

Canada and the United States share the longest international border in the world—5,527 miles (8,895 km). About 90 percent of the Canadian population lives within 100 miles (161 km) of this border, on only 12 percent of Canada's territory. In contrast, only a small percentage of the U.S. population lives within 100 miles of the Canadian border.

Canadians and Americans are overwhelmingly urban, with city dwellers accounting for 79 percent of both Canadians and Americans. Canada's population is concentrated in the cities of four main regions: the Atlantic region of peninsulas and islands at Canada's eastern edge; the culturally divided core region of maximum population and development along the lower Great Lakes and Saint Lawrence River; the Prairie region in the interior plains between the Canadian Shield on the east and the Rocky Mountains on the west; and the Vancouver region on and near the Pacific coast at Canada's southwestern corner (see Figure 11.2 and Figure 11.4).

In the United States, people are most concentrated in urban areas of the Northeast and portions of the West, with more than half that population living within 50 miles of the coast. About 54 million people, or 18 percent of the national population, live in the Northeast on 5 percent of the nation's area. The Northeast's population density is several times that of the South, Midwest, or West. The density is

TABLE 11.1 North America: Basic Data

Political Unit	Area (thousands/ sq mi)	Area (thousands/ sq km)	Estimated Population (millions)	Estimated Population Density (sq mi)	Estimated Population Density (sq km)	Annual Rate of Natural Increase (%)	Human Development Index	Urban Population (%)	Arable Land (%)	Per Capita GNI PPP ($US)
North America	**8,403.8**	**21,765.8**	**335.1**	**40.0**	**15.4**	**0.6**	**0.949**	**79**	**10**	**43,290**
Canada	3,849.7	9,970.7	32.9	8.5	3.3	0.3	0.950	79	5	34,610
Greenland (Den.)	836.3	2,166.0	0.05	0.1	0.0	0.0	N/A	80	0	N/A
United States	3,717.8	9,629.1	302.2	81.3	31.4	0.6	0.948	79	19	44,260

Sources: *World Population Data Sheet,* Population Reference Bureau, 2007; *Human Development Report,* United Nations, 2006; *World Factbook,* CIA, 2007.

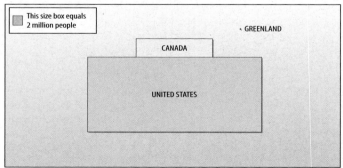

• **Figure 11.2** Population distribution (top) and population cartogram (bottom) of the United States, Canada, and Greenland.

much more extreme within the narrow urban belt stretching about 500 miles (c. 800 km) along the Atlantic coast from metropolitan Boston (in Massachusetts) through metropolitan Washington, D.C. The belt is known as the Northeastern Seaboard, Northeast Corridor, Boston-to Washington Axis, **megalopolis,** or **"Boswash."** Here, seven main metropolitan areas, including the country's largest—New York— have about 40 million people, or 1 in every 7 Americans. In the West are another 69 million people, nearly one-quarter of the nation. More than half live in California.

From the beginning of European settlement until the 1970s, this world region had a rapidly growing population. By 1800, about 200 years after the first European settlements, there were more than 5 million people in the United States and several hundred thousand in Canada, and both immigration and natural increase continued to swell their ranks. Now population growth rates (excluding migration) of the United States and Canada are low (0.6 and 0.3 percent, respectively), testifying to the passing of these countries into the final phase of the demographic transition.

Migration into North America

The United States is the only MDC in the world that is experiencing significant population growth. This is due mainly to immigration rather than natural increase, and both the United States and Canada are best appreciated as nations of immigrants and of continuing immigration. These themes are carried into the discussion of cultures later in this chapter, but an overview of modern migration flows is provided briefly here.

Each year, over a million immigrants arrive in the United States and over 200,000 in Canada (●Figure 11.3; see also

Figure 3.16, page 58). Each year, 40,000 people are granted permission to enter the United States as "guest workers." Far greater numbers do not obtain permission and enter as **illegal aliens;** most **undocumented workers** manage to obtain temporary or longer-term employment despite their illegal status.

An estimated 12 million illegal immigrants live in the United States. About 35 million people in the nation, or 12 percent of the population, are foreign-born (compared to 9.5 million, or about 5 percent of the population, in 1970). Latin America is the largest source of both legal and illegal migration into the United States, with about 470,000 Latin Americans arriving legally each year; at least that many enter illegally. Other principal source countries of immigrants, in descending order, are China, the Philippines, and India; the same three countries are the largest sources of immigration to Canada. Unfortunately, not all immigrants come to the United States and Canada of their own free will or work at the jobs they had hoped to find; their circumstances make them modern-day slaves (see Geographic Spotlight, page 601).

The status and future of illegal immigrants is a particularly contentious issue in the United States, where an estimated 8 million, or two-thirds of the total, are employed. Many U.S. citizens fear that an unstoppable tide of poor immigrants will take their jobs and bleed social and other services. Others, particularly in the business community, argue that the low-wage immigrants are vital for the American economy, generally taking jobs shunned by most Americans (but also displacing some of the mainly uneducated Americans who would take those jobs), while also contributing to the economy through their purchases. Their low wages also help keep some jobs in the country

57

544

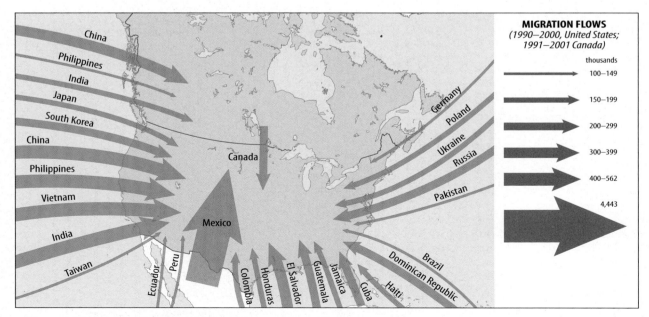

MIGRATION FLOWS
(1990–2000, United States;
1991–2001 Canada)

thousands

100–149
150–199
200–299
300–399
400–562
4,443

● **Figure 11.3** Migration flows into the United States and Canada.

GEOGRAPHIC SPOTLIGHT

The Modern-Day Slave Trade

Modern slavery is not confined to Africa. The London-based watchdog group Anti-Slavery International defines a *slave* as someone forced to work under physical or mental threat, who is completely controlled by an owner or employer, and who is bought or sold. By this definition, more than 12 million people worldwide are slaves.

Modern slaves may be found in the United States, where an estimated 50,000 people, mostly women, are brought in each year to work as sweatshop laborers, strippers, and prostitutes. The victims are struggling to escape poverty in countries such as Thailand, India, and Russia. Promising them legitimate work in a prosperous land, traffickers smuggle them into the country (particularly through Los Angeles), where they are whisked away to unfamiliar locations and put to their tawdry labors. The "master" holds the slave in submission by intimidation, such as reminding her that she is an illegal immigrant, insisting that she might be raped by police if apprehended, and threatening to have family members in the home country killed if she attempts to escape. Few of the slaves speak English, at least initially, making their escape from servitude more difficult.

Their ordeals do not necessarily end when they do find freedom; they may be deported back into the hands of traffickers in their home countries and then be trafficked back to the United States again. Recent legislation in the United States has stiffened penalties for traffickers and made it easier for the victims of trafficking to stay freely in the United States.

---

that might otherwise be offshored. While they do burden schools and others social services, most undocumented workers also pay federal taxes (a price paid for their using fake Social Security and taxpayer ID cards). The satirical film *A Day without a Mexican* depicts a U.S. economy in disarray when deprived of its undocumented workers.

Contrary to stereotypes, however, these workers are not confined to backbreaking, less-than-minimum-wage jobs. Most earn at least minimum wage, and many work outside construction and other labor-intensive industries.

Legal immigrants also add much to America's pool of talent and expertise. One in five U.S. doctors is foreign-born, as is one in five computer specialists and one in six engineers. Forty percent of the country's Ph.D.'s are foreign-born.

The typical Mexican or other Latin American migrant is poor but nevertheless manages to muster $2,000 to $10,000 to be smuggled by a "coyote" or "chicken rancher" across weak points in the U.S. border defenses. Organ Pipe Cactus National Monument of southern Arizona, along with other wild sections of the Sonoran Desert frontier, are favored crossing points. Some immigrants traveling such desert routes in the extreme conditions of summer and winter succumb to the elements. Anticipating their travails, "Good Samaritans" on the U.S. side put out water and other provisions for the migrants. Other Americans who are opposed to illegal immigration operate as vigilantes, patrolling the border to turn the travelers back.

The United States does not have a firm or effective policy for managing illegal immigration. The great majority of illegals caught within the United States are returned to the Mexican side of the border without prosecution, and an estimated half of them cross back into the United States. Many are caught yet again, perpetuating a "revolving door" of illegal immigration.

Measures are being taken to extend physical barriers that would deter new arrivals, including terrorists seeking easy entry into the United States. In the **Secure Fence Act of 2006,** the U.S. Congress passed legislation calling for lengthening existing 15-foot (4.5-m) border fences (to 700 miles or 1,120 km). Through its **Secure Border Initiative,** known as SBInet, the government is also building a so-called **virtual fence,** a multibillion-dollar, mainly high-tech surveillance system for the border region. Critics feel that determined immigrants will find their way over and around the physical barriers, which in any case, they said, will probably end up being built by illegal immigrants! Environmentalists, joined in Arizona by the Tohono O'odham Indians, who view wildlife as kindred spirits, worry about the impacts of long fences on wildlife, particularly migrating animals. The virtual fence would present no such obstacles, but its effectiveness in reducing or preventing border crossings is much debated.

U.S. lawmakers are struggling over whether and how illegal immigrants should be granted what is effectively **amnesty;** they would not be prosecuted for being in the country illegally if they met certain conditions and would be granted permanent resident status (which would eventually make them eligible for citizenship). Detractors say that amnesty programs will simply attract more illegal immigrants.

These apparently conflicting signals about immigration have led some observers to describe U.S. policy on immigration as "schizophrenic"; as an exasperated U.S. Border Patrol agent put it, "America loves illegal immigrants but hates illegal immigration."[1]

11.2 Physical Geography and Human Adaptations

The natural environments of the United States and Canada are remarkably diverse and include some of the most spectacular wild landscapes on the planet. They have presented people with a vast array of opportunities for land

NATURAL HAZARDS

Nature's Wrath in the United States

The North American continent has more natural hazards than any other, and the United States has more natural hazards than any other country in the world (●Figure 11.A). The western edge of the continent lies on a zone where two major plates of the earth's crust collide (see Figure 2.1, page 21), producing both earthquakes and volcanoes. The most consequential quake in U.S. history was the 1906 event of magnitude 8.0 on the Richter scale that produced a displacement of up to 20 feet (6 m) on the San Andreas Fault near San Francisco, California. The earthquake itself and, even more devastatingly, the fires fed by the ruptured gas lines in the city destroyed roughly 30,000 of San Francisco's buildings and killed about 3,000 people. The San Andreas is just one of the largest and best known among the numerous fault lines that trend mainly north-south through California. The pressure built up by tectonic movement is enormous and must eventually be released, so many more earthquakes are in California's future. Some of them will be catastrophic, though the effects will be mitigated somewhat by strict building codes and emergency preparedness.

A network of faults at the edge of the continent threaten Anchorage and other cities in Alaska. The coasts of Alaska, western Canada, and the western United States must also be on guard for tsunamis, the so-called tidal waves that result when offshore earthquakes displace huge volumes of water.

In the western U.S. states of Oregon and Washington are the many volcanoes of the Cascade Range. Several of them, notably Mount Rainier and Mount Hood, still have the potential for eruption and, depending on wind flow and other variables, could do serious damage to Seattle and other populated areas. The destructive potential of these mountains was demonstrated on May 18, 1980, when Washington's Mount Saint Helens blew up with an explosive force equivalent to 400 million tons of TNT. Luckily, few casualties resulted in this sparsely populated area.

Surprisingly, one of the most potentially destructive forces in the continental United States is a very ancient fault zone roughly paralleling the course of the southern Mississippi River. In 1811 and 1812, on the New Madrid fault in Missouri, three earthquakes of magnitude 7 to 8 on the Richter scale reportedly reversed the flow of the Mississippi River for some moments and rang church bells 1,000 miles (1,600 km) away in Boston, Massachusetts. The mid-Mississippi region had a small population at the time, and human losses were minimal, but a quake of that magnitude in the same region today would be very destructive to Saint Louis, Memphis, and other cities.

Many Americans of the eastern United States think Califor-

nians are insane to live in an earthquake zone, but they regularly contend with their own menacing natural hazards. Storms often pound this part of the country. It is a humid region where rainfall potential is high and where different agents trigger precipitation. The convection associated with daytime heating and the collision of air masses with different temperatures give rise to destructive thunderstorms, with heavy rain, lightning, hail, and one of the most powerful natural forces on earth, tornados. Across a swath of the Midwest known as **"Tornado Alley"** (the "high risk of tornadoes" region of Figure 11.A), these intense vortices of very low atmospheric pressure cut paths of destruction, sometimes causing large losses of life and property.

Successive days or even weeks of thunderstorm activity sometimes produce localized or widespread flooding. In the summer of 1993, weeks of heavy rains over portions of the Missouri and Mississippi River watersheds produced a flood of historic proportions. Flooding and wind damage are typical results of the hurricanes that batter the East Coast and Gulf of Mexico regions between June and October. The most catastrophic in U.S. history in terms of lives lost was a hurricane in 1900 that killed more than 8,000 people, mainly in storm surges, in Galveston, Texas. In storm damages, the most costly prior to 2005 was Hurricane Andrew, which came ashore south of Miami, Florida, in 1992, causing $27 billion in property losses. Those losses were dwarfed by the impacts of Hurricane Katrina, which roared onto the coasts of Louisiana, Mississippi, and Alabama in August 2005 (see discussion on pages 662–663). Yet another precipitation-related natural hazard is the **blizzard,** a combination punch of heavy snowfall and high winds that strikes the United States—typically in the Midwest and Northeast—an average of 11 times each year.

A more pernicious natural hazard, a kind of disaster in slow motion, is the drought that sometimes rakes the country, especially in the West. The worst was the eight-year drought that produced the **Dust Bowl** of the 1930s, ruining crops and livelihoods and causing mass migrations of broken farming families to leave Oklahoma and adjacent areas, mainly to settle in the West. This story is told superbly in John Steinbeck's novel *The Grapes of Wrath.* As this text went to press, a drought that had already lasted a decade was afflicting the western and especially southwestern United States. Climatological records suggest that such lower precipitation may be the norm for this region, which may previously have enjoyed decades of abnormally high precipitation. Human populations in the region boomed in those years, perhaps far beyond levels that can be sustained by local water supplies.

use and settlement. It is useful to consider how landforms (and what lies beneath and above them) have promoted or hindered human uses and how climates have also done the same. This wide range of environmental settings is associated with a great variety of natural hazards (see Natural Hazards, above).

Landforms and Land Uses

●Figures 11.4, 11.5, and 11.6 give details of the landforms, climates and biomes, and land uses of North America.

In the far northeast lies Greenland, the world's largest island. About 80 percent of Greenland is covered in perma-

NATURAL HAZARDS

Nature's Wrath in the United States *continued*

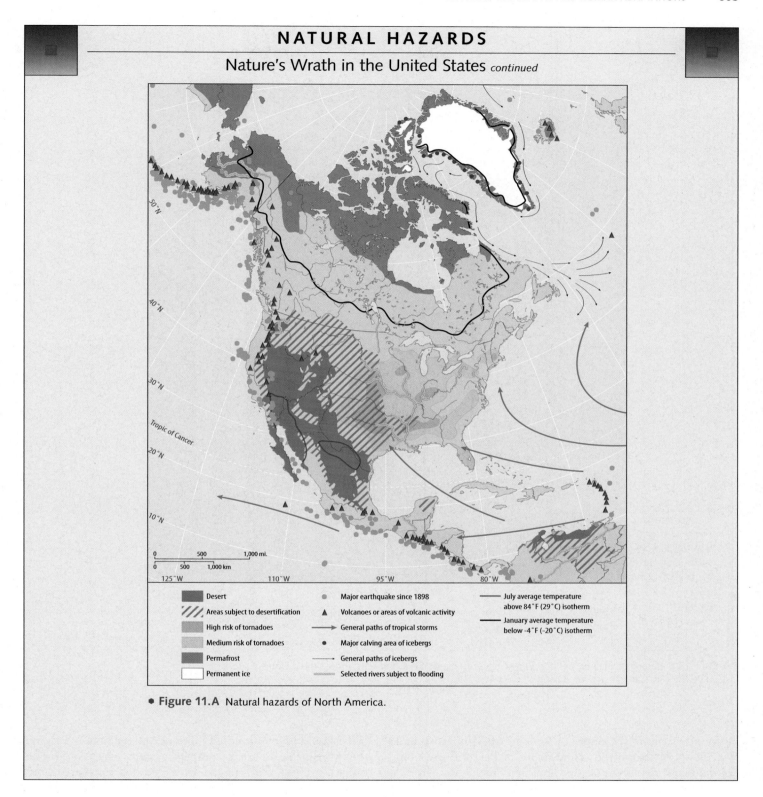

● Figure 11.A Natural hazards of North America.

nent ice. Permanent ice also covers small areas of nearby islands of the Canadian Arctic.

The ancient core of North America, with rocks up to 3 billion years old, is called the **Canadian Shield,** covering roughly the area from Nunavut south to Minnesota and northeast to Labrador. Human settlement is sparse in this vast region, and agriculture is limited by poor soils and a harsh climate similar to that of northeastern Siberia. The Canadian Shield was scoured by glaciers until about 10,000 years ago, and today the area is dotted with many large lakes and thousands of smaller ones. The surface is mostly rolling, though areas of hills and low mountains can

Figure 11.4 Reference map of the topographic features, rivers, lakes, and seas of the United States (excluding Hawaii), Canada, and Greenland.

be found, such as the Superior Upland of Minnesota and Wisconsin. Hydropower, wood, iron, nickel, and uranium are the major resources of the Canadian Shield.

Southeast of the Canadian Shield lie the Appalachian Mountains and associated highlands. The Appalachians were possibly the highest mountains in the world when they were formed 400 million years ago, but erosion has reduced the elevations to between 2,000 feet (600 m) and 6,684 feet (2,037 m). Rising in northern Alabama and Georgia and running northeast toward the Gaspé Peninsula of Quebec

(the island of Newfoundland is also Appalachian in origin), the Appalachians are a complex system where mountain ranges, ridges, and rugged dissected plateaus are interspersed with narrow valleys, isolated lowlands, and rolling uplands. The western Appalachians are known for their large coalfields. About 80 percent of the Appalachians and adjacent lowlands of New England are forested, a remarkable turnaround from the 19th century, when over two-thirds of the original forest had been removed. After 1850, farmers gravitated toward the much richer soils of the Mid-

• **Figure 11.5** Climates (left) and biomes (right) of the United States, Canada, and Greenland.

• **Figure 11.6** Land use in Canada, the United States, and Greenland.

west, and hydropower and fossil fuels replaced timber as fuel. These shifts eased pressure on the forests, and with their rebound also came increasing numbers of moose, beaver, and other wildlife.

Immediately east of the Appalachians, between Georgia and the vicinity of New York City, lies the rolling to hilly Piedmont ("foothills"), an area with a large population and good soils for farming. The Piedmont is fairly narrow, and its eastern edge is marked by the **fall line**, a natural boundary where rapids and waterfalls commonly mark the head of navigation on many rivers and have long supplied hydropower. Cities such as Columbia, South Carolina; Richmond, Virginia; Washington, D.C.; and Philadelphia, Pennsylvania, are located along the fall line.

East of the fall line is the Atlantic Coastal Plain, which is continuous with the Gulf Coastal Plain to the south. Forming a large crescent from eastern Texas up to the New Jersey shore, the coastal plains are characterized by sandy and largely infertile soils (though river valleys are more agriculturally productive), a generally flat topography, and many large river mouths and bays indenting the coast. There are many important wetland areas along the coast and inland, such as the Everglades in Florida and the Great Dismal Swamp in Virginia.

North of the Gulf Coastal Plain lies a small region of highlands comprised of the Ozark Plateau and the Ouachita Mountains, sometimes collectively referred to as the Interior Highlands. Divided by the Arkansas River Valley, this is the largest area (but not with the highest elevations) of uplands between the Appalachians to the east and the Rockies to the west.

The interior of North America is dominated by a vast, roughly triangular-shaped plain, reaching from Texas to

Ohio to northern Alberta. Millions of years ago, this land was submerged beneath a shallow sea. Today, most of this area is in the watershed of the Mississippi River and its many tributaries, including the Ohio and Missouri Rivers. The Interior Plains, as the area is collectively called, is hillier and more rolling in the east, north of the Ohio River, but becomes more level farther west, even as average elevation increases markedly from east to west. The eastern sections (the prairie), once dominated by long grasses with scattered wooded areas, have mostly been turned into built-up and agricultural land; only about 10 percent of the original ecosystem is intact. The western third of the region is called the Great Plains, a semiarid steppe dominated by short grasses that is mostly flat but higher in elevation than many of the peaks of the Appalachian Mountains (see Figure 11.4). The Great Plains, once referred to as the Great American Desert before its agricultural potential was realized, has some of the world's finest soils and most productive farmlands but is only lightly populated. A more rugged area that includes the Black Hills (site of Mount Rushmore) and the colorful terrain known as the Badlands disrupts the otherwise generally level terrain of the western plains in the western Dakotas and eastern Montana.

The Great Plains end abruptly to the west alongside the Rocky Mountains, a long chain of mountain ranges stretching from British Columbia to New Mexico. The tallest mountain in the Rockies is Mount Elbert in Colorado at 14,433 feet (4,399 m). High and rugged with few easy passes, the Rockies are the source of many major rivers, including the Rio Grande, Colorado, Missouri, Fraser, Saskatchewan, and Columbia. Recreation, especially skiing and touring preserves such as Yellowstone National Park in Wyoming and Banff National Park in Alberta (see Figures 11.1.9 and 11.2.32), is a major draw to the Rockies.

West of the Rockies is a geologically complex region of alternating basins and lower mountain ranges. The Columbia Plateau lies in the northern part of this region; south of there it includes the Great Basin to the west and the Colorado Plateau to the east and is often called the "basin and range country." Arid in the north and true desert in the south, as rainfall is blocked by tall mountains to the east and west, this is the driest area of North America. Many plateaus are carved by canyons and dry valleys; the Grand Canyon (5,000 ft/1,500 m deep) and the lowest point on the continent, Death Valley (282 ft/86 m below sea level), are located in this region. Economically, ranching, mining, and energy extraction (oil, gas and coal) are the mainstays of this region. Despite the dryness of the region, agriculture is also important here, with major dams impounding river water that is diverted toward irrigation projects and urban areas.

A series of mountain ranges parallel the Pacific Coast from Alaska south to California. These ranges—including the Alaska Range, the Coast Mountains, the Cascades, the Sierra Nevada, and the Coast Ranges—are all tall, rugged, and glaciated in places. The tallest mountains in North America (Denali or Mount McKinley, 20,320 ft/6,194 m), Canada (Mount Logan, 19,551 ft/5,959 m), and the con-tiguous United States (Mount Whitney, 14,505 ft/4,421 m) are located in these ranges. The Pacific Coast is prone to earthquake activity, and several active volcanoes are found in Alaska and the Pacific Northwest. Fishing and logging are important to the economy, as is agriculture in two very productive lowlands, the Central Valley of California and the Willamette Valley of Oregon.

Stretching across northern Alaska and into the Canadian territories lies the Arctic Coastal Plain, a flat, marshy area that has few resources except for vast quantities of oil lying offshore in the Beaufort Sea.

Climates and Land Uses

The United States has more climatic types than any other country in the world, and even Canada is quite varied. The wide range of economic opportunities afforded by climatic variety is the source of much of the economic strength of this world region.

Canada's and Alaska's tundra climate, with its long, cold winters and brief, cool summers, promotes vegetation of mosses, lichens, sedges, hardy grasses, and low bushes. The subarctic climate zone, also in Canada and Alaska, has long, cold winters and short, mild summers, with a natural vegetation of coniferous (boreal) forest resembling the Russian taiga. In 2003, an unusual coalition of energy and forestry companies, environmental groups, and Native American communities joined an agreement to preserve at least 50 percent of Canada's subarctic forests and to develop the other half in sustainable ways. Population is very sparse in the subarctic and tundra climate areas. Scattered peoples engage in trapping, hunting, fishing, mining, logging, and military activities; many live largely on welfare. The severe winter in vast sections of the tundra and subarctic zones requires unusual adaptations—like having to wait until the spring thaw to bury loved ones who died during the winter.

The humid continental climate with short summers ("humid continental cold") in Figure 11.5) is characterized by long, cold winters and short, warm summers. Agriculturally, emphasis is on dairy farming except in the Far West, where spring wheat production dominates. The humid continental long-summer climate ("humid continental warm") has cold winters and long, hot summers; agriculturally, this belt, which includes the agricultural riches of the Midwest, emphasizes dairy farming in the east and corn, soybeans, cattle, and hogs in the midwestern (interior) portion.

In the humid subtropical climate zone, winters are short and cool—with some cold snaps—and summers are long and hot. Agriculture varies, with local production of cattle, poultry, soybeans, tobacco, cotton, rice, peanuts, fruits, and vegetables.

Within each of these three humid climate regions, the pattern of natural vegetation is complex, with areas of coniferous evergreen softwoods, broadleaf deciduous hardwoods, mixed hardwoods and softwoods, and prairie grasses. Far southern Florida has a small area of tropical savanna climate, which is a major climate in adjacent Latin

America. The state of Hawaii has a tropical rain forest climate, but most of the original associated vegetation has been cleared. Along the Pacific shore of the United States and Canada, the narrow strip of marine west coast climate is associated with the barrier effect of high mountains near the sea, ocean waters that are warm in winter and cool in summer, and winds prevailing from the west. The mild, moist conditions have favored a magnificent growth of giant conifers (most notably the redwood and the Douglas fir) that provides the basis for the lumber industry. Lush pastures support dairy farming, as they do in the corresponding climatic region in northwestern Europe.

78

The Mediterranean or dry-summer subtropical climate region of central and southern California, where almost all rain comes in winter, is associated with irrigated production of cattle feeds, vineyards, vegetables, fruits, cotton, and numerous other crops. These products, together with associated livestock, dairy, and poultry, make California the leading U.S. state in total agricultural output.

In the semiarid steppe climate region, occupying an immense area between the Pacific coast of the United States and the landward margins of the humid East and extending north into Canada, temperatures range from continental in the north to subtropical in the south. The natural vegetation of short grass, bunch grass, shrubs, and stunted trees supplies forage for cattle ranching, which is the predominant form of agriculture. More moist areas are used for wheat (both winter wheat and spring wheat). Other crops are grown in scattered irrigated areas, often associated with major rivers such as the Columbia, Snake, Arkansas, and Rio Grande.

The desert climate of the U.S. Southwest is associated with scattered irrigated areas and settlements emphasizing mining, recreation, and retirement. The high, rugged mountains of the Rockies and the Sierra Nevada have undifferentiated highland climates varying with latitude, elevation, and exposure to moisture-bearing winds and to the sun.

It cannot be assumed that the long-standing distribution of biomes as mapped in Figure 11.5b will be accurate in the coming years. There is growing evidence that as global temperatures rise, agricultural and natural vegetation zones will shift poleward in the United States, as they are doing

34

elsewhere. The 2006 Arbor Foundation map of hardiness zones for typical garden plants in the United States showed that many bands are a full zone warmer, and in some places two zones warmer, than when the last map was published in 1990.

11.3 Cultural and Historical Geographies

As mentioned earlier, many geographers call the region of the United States and Canada "Anglo America," but the decision not to do so here is an acknowledgment of the rich ethnic roots and branches of this region. Native Americans began their migrations into the region as Asians, crossing what was then a land bridge between Alaska and Siberia at least 12,500 years ago. However, there is some scientific evidence, still being scrutinized, that the migrations may have begun as early as 33,000 years ago. Other controversial science holds that Polynesian sea voyagers also reached the shores of the Americas in very ancient times. Migrations across the Asian land bridge persisted until about 3,000 years ago, and a diverse pattern of indigenous ethnic groups, languages, and lifeways emerged.

423

Native American Civilizations

The indigenous cultures of Middle and South America have been described on pages 531–534 in Chapter 10. The indigenous cultures of what are now Canada and the United States, especially those of what is now the U.S. Southwest, were related to them in many ways. Some developed civilizations, the rather complex, agriculture-based ways of life associated with permanent or semipermanent settlements and stratified societies. In what are now the southwestern states of Arizona, Utah, Colorado, and New Mexico, three dominant **Native American civilizations** emerged: the **Mogollon, Hohokam,** and **Anasazi.** Despite their arid environment, these peoples developed productive agricultural systems that borrowed from and interacted with the Aztec and other cultures of Middle America. The same basic crop assemblages were found across all these cultures: corn, beans, squash, and chili peppers were the staples.

The Hohokam of what is now southern New Mexico had a very productive system of irrigated agriculture. Their culture flourished between 100 B.C.E. and 1500 C.E. The Anasazi developed a dwelling pattern based on the **pueblo,** in which interconnected mud-brick (adobe) residences and ceremonial centers, with beamed roofs of mud, sticks, and grass, were built into cliffsides or on the flat-topped mesas of the region (●Figure 11.7). The Mogollon culture that thrived between 300 B.C.E. and 1400 C.E. inherited these building traditions and was effectively absorbed by the Anasazi, whose heartland was in the Four Corners region where Colorado, New Mexico, Arizona, and Utah meet. Around 1300 C.E., the Anasazi began to abandon their pueblos and the productive agriculture associated with them. The reasons for the Anasazi decline remain unknown, but drought, invasion by hostile neighbors, or simply political decisions to relocate may have been responsible. Some Native American groups of the region today, including the Hopi, Zuni, and Tiwa peoples, carry on the architectural legacy of the Anasazi by continuing to dwell in pueblos.

Farther to the east and north, in much of the eastern half of what is now the United States, the so-called **Mound Builder civilizations** developed. There were four main culture groups among them: the **Poverty Point** and the **Mississippian** cultures of the Southeast, the **Hopewell** in the Midwest, and the **Adena** in the Northeast. The earlier Poverty Point and Adena cultures (c. 2000 B.C.E. to 200 C.E.) grew some crops but were mainly hunters and gatherers who thrived in the productive humid environments of the East.

● **Figure 11.7** Anasazi pueblo dwelling in Canyon de Chelly, Arizona.

The Hopewell (c. 200 B.C.E. to 700 C.E.) grew more crops, including corn, beans, and squash. The Mississippian (700–1700 C.E.), with an urban and suburban settlement pattern in many sites, was the greatest farming culture and had an extensive trading network. Of all the mound-building cultures, the Mississippian peoples were the greatest builders of earthen mounds. In all the cultures, these structures served as ceremonial sites, and in some, they were also tombs.

Indigenous Culture Groups and Lifeways

The Native American groups not associated with complex societies, ceremonial centers, and civilizations were varied and numerous. As many as 18 different culture regions for these groups are recognized; this text uses 11 (●Figure 11.8). The best way to categorize their essential features is to acknowledge some of the major groups within each region and their original languages. Although the past tense is used in this discussion (because these cultures flourished in pre-Columbian times), none of the languages mentioned is extinct, and speakers of all of them may still be found—although for most, English is their first or second language.

Seven Native American language families are represented in the United States, Canada, and Greenland by more than 250 languages. The **Aztec-Tanoan language family** that sprawled across most of northern Mexico was represented in the U.S. Southwest by languages including **Hopi, Comanche, Shoshone,** and **Papago.** Some of the groups that spoke these languages carried on the agricultural and pueblo-dwelling lifestyles pursued by the earlier Mogollon, Hohokam, and Anasazi civilizations, while others were nomadic hunter-gatherers who sometimes supplemented their diet by raiding the pueblo-dwelling peoples.

The larger part of what is now the contiguous United States was the hearth of the **Hokan-Siouan language family,** which includes such languages as **Iroquoian, Mohawk,** and **Lakhota.** Among its more famous culture groups were the **Sioux,** including the **Lakota** and **Dakota,** who developed their trademark subsistence pattern after the Europeans arrived in the New World. After the Spaniards and their successors introduced the horse, the Lakota and some other groups gave up what had been a village and farming way of life, migrating west to become horse-mounted hunters of the bison ("buffalo") and other game on the Great Plains.

In what is now California and Oregon, the **Penutian language family** included the **Miwok** and **Klamath-Modoc** languages, among others. Most of these peoples were not farmers but had very productive hunting and gathering economies that took advantage of the region's diverse and abundant resources. These natural bounties supported relatively dense human populations.

Farther to the north, but still on the Pacific coast, the **Mosan language family** included branches of **Chemakuan, Salish,** and **Wakashan.** The Northwest Coast peoples who spoke these languages also had rather dense populations that thrived on an abundance of sea mammals, fish, and land game. They generally developed complex societies in which the custom of the potlatch played a prominent role. **Potlatch** was a system in which rank and status were determined by the quantity and quality of material goods one could give away—an interesting variant and inversion of the "keeping up with the Joneses" tradition of later Anglo North America.

The **Algic language family** members were spread across all 10 Canadian provinces and 20 U.S. states, mainly in the northern tier. Some of its better-known languages include **Arapaho, Blackfoot, Cheyenne,** and **Cree.** The tribes that

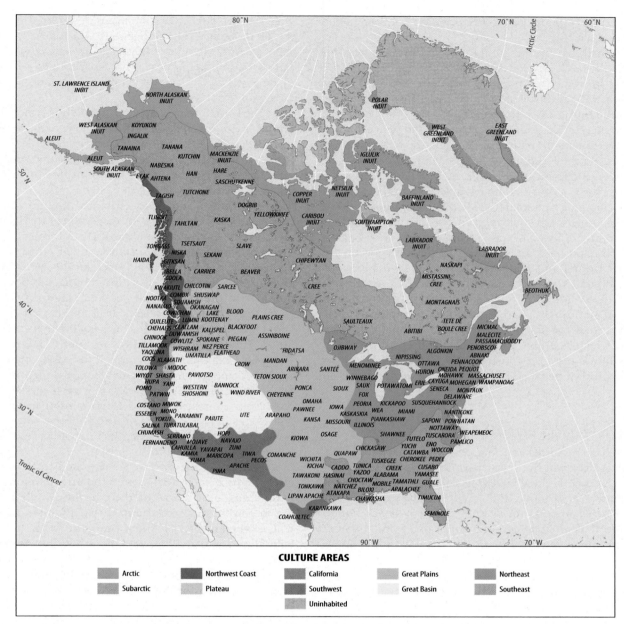

CULTURE AREAS

Arctic	Northwest Coast	California	Great Plains	Northeast
Subarctic	Plateau	Southwest	Great Basin	Southeast
		Uninhabited		

● **Figure 11.8** Native American culture areas and selected tribes of Canada, the United States, and Greenland.

spoke the Algic languages relied on the region's vast forests for game animals, fuel, shelter, and tools, complementing their hunting and gathering with fishing and farming; some groups were seminomadic and lived in villages.

To the northwest of this linguistic region, in west central Canada and southeastern Alaska, the **Na-Dene language family** was prominent. This group includes the **Athabascan subfamily** with its languages, including **Koyukon, Apache,** and in a remote outlier, the **Navajo** of the Southwest (the largest tribe in the United States, with more than 270,000 members). Most of these peoples lived either in the subarctic region of coniferous forests and wetlands or in the tun-

dra, where agriculture was impossible. They hunted caribou and other game, and their populations were limited.

Across the rest of Alaska, throughout northern Canada, and in Greenland, the **Eskimo-Aleut language family** dominated. The peoples who carried these tongues into North America were among the last to migrate from Asia, probably between 2500 and 1000 B.C.E., and they did so in boats, as the former land bridge was inundated by then. The so-called **Eskimos**—an Algonquian term meaning "raw meat eaters"—prefer to call themselves **Inuit,** meaning "the people." The Inuit and the **Aleuts** were indeed characterized by a carnivorous diet, as their livelihood was based on hunting sea mammals, caribou (known as reindeer in the Old World), and fish.

What all these diverse groups had in common was an exceptionally well-developed set of adaptations for living in the local environment. Native American economies spanned a remarkably wide continuum, from simple foraging and hunting to complex agricultural systems. These were not static adaptations but changed and generally improved over time—for example, through greater mastery of finer tools. As the Lakota and others demonstrated by giving up village agriculture in favor of a nomadic life based on horses, these cultures were capable of dramatic adjustments to changing opportunities. It was once supposed that the Native Americans always lived within the limitations imposed on them by local ecosystems, but it is now known that they transformed landscapes in significant ways by fire and other means and may even have hunted populations of large animals to extinction; this is the Pleistocene overkill hypothesis (see page 40). One trait apparently shared by most if not all the Native American groups was their deep reverence for the natural world. Animals, forests, geological features, and the people themselves were tied together intimately in animistic belief systems that emphasized the kinship between these diverse elements of life on earth.

European Impacts on Native Cultures

There are two very different narratives of what took place in North America in the centuries following 1492. For European newcomers, these were times of settlement, development, taming the frontier, and "civilizing the savages." For the Native Americans, these were times of depopulation and cultural demolition. A popular T-shirt and bumper sticker on Indian reservations in the western United States these days reads, "NATIVE AMERICANS: FIGHTING TERRORISM SINCE 1492."

How many Native Americans lived in what are now Canada, the United States, and Greenland when Columbus made landfall in 1492 is uncertain. The most common estimate is 1 million, with about three-quarters of that population in what is now the United States and almost all the rest in Canada. As in Latin America and Australia, the European contact initiated years of population losses among Native Americans through disease, famine, and warfare. In the United States, Native American populations crashed to an all-time low of fewer than 250,000 between 1890 and 1910 but are now back up to 1.8 million. Their growth has been particularly strong in the Great Plains, where there are now more Native Americans (and bison) than there were in the late 1870s. In Canada, the current population of Native Americans is about 1 million. Collectively, these peoples refer to themselves as the **First Nations** in acknowledgment of their pre-Columbian cultures and claims to the land. In Canada, this definition excludes the Inuit, who

were relative latecomers but who are referred to as among the **First Peoples.** The indigenous Greenlanders, known as the *Kalaallit,* are related linguistically and ethnically to the Inuit of Canada, Alaska, and Siberia. The official name of their island is Kalaallit Nunaat, meaning "Land of the Greenlanders."

In the United States, most of the Native American populations are in the west; in Canada, they are in the north and west (•Figure 11.9). The two countries have dealt in different ways with their indigenous populations and claims to land and other rights, with government–Native American relations historically better in Canada.

In the United States, an estimated one-third of all Native Americans today live on **reservations** that the government established as remnants of their former tribal territories or sometimes set up thousands of miles from native homelands, forcing the people to relocate there. "The Res," as Native Americans call it, are among the poorest communities of the United States, often plagued by high rates of incarceration, alcoholism, drug abuse, depression, broken families, teen suicide, and unemployment. The Lakota's Pine Ridge Reservation in South Dakota (•Figure 11.10), for example, has an annual per capita income of just $6,300 (compared to the national U.S. average of $32,000). However, it must also be noted that even these lamentable conditions represent a considerable improvement. Native Americans have growing political and economic clout, with poverty and unemployment rates dropping and education levels rising.

Gambling revenues have contributed to economic and other improvements in many Indian communities. The semiautonomous status of their territories allows casinos to circumvent many restrictions that prohibit casinos elsewhere (•Figure 11.11). About one-third of the roughly 563 federally recognized Native American tribes in the United States have developed casinos since Congress legalized gambling on their lands in 1998. This **gaming industry** has been a mixed blessing for the Native Americans, with a wide variety of successes and failures in different places. In some, gambling revenues have supported community education, health care, and cultural centers, but in others, they have lined a few pockets and failed to benefit many members of the community. Some Native Americans complain that several of the groups that have established casinos are not legitimate, historical tribes to begin with. Nonnative people often resent casinos because they view them as a windfall of special treatment for individuals who have not earned it. In all cases, casinos have introduced significant changes.

In Hawaii as in North America, incursions by Europeans—Captain Cook's visits were the first in 1778 and 1779—brought huge changes to the native people. Polynesians may have populated the island group as early as 300 C.E., making a living by growing bananas, coconuts, breadfruit, and taro and by fishing. Hawaiian society was based on a ranked caste system, and taboos governed many

• **Figure 11.9** Modern Native American reservations and other lands in the United States and Canada.

• **Figure 11.10** Lakota at the Wounded Knee Massacre Site, Pine Ridge, South Dakota.

• **Figure 11.11** Native American casino, Laguna Reservation, New Mexico.

aspects of daily life and relationships with the spiritual world. The islands were a united indigenous kingdom in 1893 when a group of American businessmen, backed by the U.S. military, overthrew the monarchy and took control (for more information on Polynesian geography and cul-

ture, see Chapter 8). The United States formally annexed Hawaii five years later. Diseases including smallpox and influenza, introduced accidentally by outsiders, nearly wiped out the indigenous population: there were an estimated 1 million ethnic Hawaiians in the late 18th century but just

22,000 in 1919. Today, the **Kanaka Maoli,** as the native Hawaiians call themselves, number 250,000. There has been much ethnic mixing among this population. The native **Hawaiian language,** a member of the Malayo-Polynesian subfamily of the Austronesian family, is undergoing a major revival following a long decline. Hawaiian became an official state language in 1978—the only indigenous language in the United States to have this distinction. Major education efforts ensued. As recently as 1983, only 50 children spoke the language, but now more than 2,000 do.

The situation for indigenous peoples in Canada is unique, even on a worldwide scale. In 1999, Canada ceded nearly one-quarter of its total land area to the Inuit peoples, who called their land **Nunavut** ("our land" in Inuktitut; see Figure 11.9). These 733,600 square miles (1.9 million sq km) were carved from Canada's Northwest Territories and generally lie north of 60°N, surrounding the northwest part of Hudson Bay and extending well within the Arctic Circle. The total population is approximately 30,000, scattered through about 30 communities. About nine-tenths of the capital required to keep the territory viable comes from Ottawa, the federal capital of Canada, making the per capita cost of maintaining Nunavut higher than for any other political unit in Canada. The Canadian government has allocated more than $1 billion to be distributed in the form of services, supplies, and subsidies until 2016 to the Inuit population, representing about $38,000 a year per person. Large subsidies also go to Native American populations in the Northwest Territories.

All across Canada, settlements of indigenous claims to ancestral lands have led to federal government financial support of the communities, a measure of self-government for the indigenous peoples, and royalty income to them from mines and fossil fuel industries. Indigenous peoples who have relocated to Canada's cities are showing major improvements in education and employment. Still, the average income of Canada's indigenous peoples is about a third lower than the population as a whole. Canada's record with its indigenous peoples is not unblemished; in the 1950s, for example, the government forcibly relocated a group of Ungava Inuit from Hudson Bay 1,200 miles (1,920 km) north to inhospitable Ellesmere Island. The move, which caused these people enormous suffering, was meant to help confirm Canadian sovereignty over the island.

European Settlers and Settlements

European settlement of what are now Canada and the United States took place in a series of waves propelled by religious persecution in Europe, colonization of new lands by European powers, and then the expansionist efforts of newly independent Canada and the United States. Canada's core region of Québec and Ontario developed with the French entry into the interior of North America. The French founded Québec City in 1608 at the point where the Saint Lawrence estuary leading to the Atlantic narrows

sharply. Their fortifications on a bold eminence allowed control of the river there.

From Québec, fur traders, missionaries, and soldier-explorers soon discovered an extensive network of river and lake routes with connecting portages reaching as far as the Great Plains and, via the Mississippi River, to the Gulf of Mexico. Montréal, founded later on an island in the Saint Lawrence River, became the fur trade's forward post for the interior wilderness. Between Québec City and Montréal, a thin line of settlement grew along the Saint Lawrence, forming the agricultural base for the colony. Population grew slowly in this northerly outpost where the winter was harsh and only French Catholics were welcome.

French Canadians numbered only about 60,000 when Britain conquered the region in 1759 (France was expelled from Canada in 1763). They did not join the English-speaking colonists of the Atlantic seaboard in the American Revolution and by their refusal laid the basis for the division of the continent of North America into two separate countries. Until the late 20th century, French Canadians increased rapidly in numbers. About 20 percent of Canada's people today are French in language and culture. They are concentrated in the lowlands along the Saint Lawrence River, where they form the majority population in the province of Québec. The French Canadians have clung tenaciously to their distinctive language and culture, with some attempting twice since 1980 to gain independence for the province of Québec. 636

A few British settlers came to Québec after the conquest, and this immigration increased rapidly during and after the American Revolution of 1775–1783. These migrations set the foundations for the British segment of Canada's core. Many of the early English-speaking immigrants were Loyalist refugees from the newly independent United States, whose rebellion French Canada had refused to join. They were soon joined by more English settlers emigrating directly from Europe. These newcomers formed a large minority in French Québec, including a British commercial elite in Montréal that persists to this day, but they also settled west of the French on the Ontario Peninsula. By 1791, the British government decided to separate the two cultural areas into different political divisions, known at that time as Lower Canada (Québec) and Upper Canada (Ontario), taking the directional terms from their positions on the Saint Lawrence River.

Farther west, the prairie region of Canada was settled much later, in the decades after 1890 (•see Figure 11.12). Settlers found some good land in an area with long and harsh winters, short and cool summers, and marginal precipitation. They could grow only hardy crops for subsistence. Later came specialization in spring wheat for export, and the prairie region became the Canadian part of the North American Spring Wheat Belt. 668

The settlers of this region were mainly English-speaking people from eastern Canada and the United States, but they included notable minorities of French Canadians, Germans,

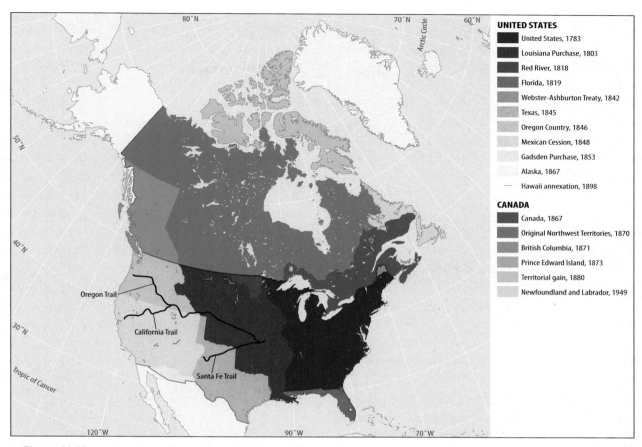

UNITED STATES

- United States, 1783
- Louisiana Purchase, 1803
- Red River, 1818
- Florida, 1819
- Webster-Ashburton Treaty, 1842
- Texas, 1845
- Oregon Country, 1846
- Mexican Cession, 1848
- Gadsden Purchase, 1853
- Alaska, 1867
- — Hawaii annexation, 1898

CANADA

- Canada, 1867
- Original Northwest Territories, 1870
- British Columbia, 1871
- Prince Edward Island, 1873
- Territorial gain, 1880
- Newfoundland and Labrador, 1949

● **Figure 11.12** Territorial acquisitions of the United States and Canada.

Ukrainians, and others. The ethnic makeup of the area still reflects them. Canada's ethnic mix today includes about 6 million people of English origin, 3 million Irish, and 4 million Scottish (see ●Figure 11.13 and Table 11.2).

Other ethnic groups would come to join Canadian society, which by law in the 1988 **Multiculturalism Act** was recognized as a multicultural society. These laws acknowledged that Canada's Native American and non-European groups would probably never become assimilated in Canadian society and that the country would probably be more stable—avoiding the kind of racial tensions that racked the United States—if traditional identities were recognized and encouraged. Most Canadians embrace this legislation, but some argue that Muslim immigrants in particular are challenging Canada's freedoms; mirroring an issue in Britain, there has been a debate over Muslim women wearing the face-cloaking veil known as the *niqab*.

The first European settlers of what is now the United States were Spanish, Dutch, English, Swedish, and French, with the majority being English. Most of their settlements were along the Eastern seaboard and were linked to Europe by the long but generally reliable sea passage across the Atlantic. Northeastern urban development began early. In the early 1600s, English settlers founded Boston, and

Dutch settlers established New Amsterdam (which the English annexed and renamed New York in 1664). The English founded Philadelphia in the late 1600s and Baltimore in the early 1700s. All four were major urban centers by the time of the American Revolution.

Development of the interior began with exploration along waterways: the British up the Chesapeake Bay and its tributaries, the Dutch up the Hudson in the early 1600s, and the French up the Saint Lawrence and Mississippi Rivers. Westward expansion proceeded very slowly; up to about 1800, there was little European settlement beyond the Appalachian region of what is now the eastern United States. The floodgates of expansion then opened, and wave after wave of immigration took place.

There were several phases of territorial acquisitions in what Americans considered their **manifest destiny**—the opening of new frontiers for their settlement all the way to the Pacific Ocean (see Figure 11.12). First, they established themselves along the Ohio and eastern Mississippi watersheds following the American Revolution in 1783. Then, in the **Louisiana Purchase** of 1803, the United States bought from France (for 3 cents per acre) about 23 percent of what are now the lower 48 states. This included much of the western watershed of the Mississippi and the lands

● **Figure 11.13** Major ethnic groups in the United States and Canada.

TABLE 11.2 Ethnic Groups of Canada and Their Origins

Major Ethnic Groups	Dates
Native peoples	pre-1600
French	1609–1755
Loyalists from the United States	1776–mid-1780s
English and Scots	mid-1600s on
Germans and Scandinavians	1830–50, c.1900, 1950s
Irish	1840s
African Americans	1850s–70s
Mennonites and Hutterites	1870s–80s
Chinese	1855, 1880s, 1990s
Jews	1890–1914
Japanese	1890–1914
East Indians	1890–1914, 1970s on
Ukrainians	1890–1914, 1940s–50s
Italians	1890–1914, 1950s–60s
Poles	1945–50
Portuguese	1950s–70s
Greeks	1955–75
Hungarians	1956–57
West Indians	1950s, 1967 on
Latin Americans	1970s on
Vietnamese	1970s on
Filipinos	1990s

Sources: E. Herberg, *Ethnic Groups in Canada: Adaptations and Transitions.* Copyright © 1989 Nelson Canada, Scarborough, Ontario; Census of Canada, 2001.

of the Missouri River watershed. The United States then picked up major lands in the south and west. It purchased Florida from Spain in 1819 and took Texas, the Southwest, and California as spoils of war with Mexico in 1845 and 1848. It established the Oregon Territory in 1846. In the 1853 **Gadsden Purchase,** the United States bought a small strip of present-day Arizona and New Mexico from Mexico. Finally, it bought Alaska from Russia for 2 cents per acre in 1867. The young nation thus acquired a vast land empire much like that built earlier by Russia, although by different means.

The **Homestead Act** of 1862 made many of these new regions almost irresistible magnets for settlement, as it allowed a pioneer family to claim up to 160 acres of farming land for a fee of about $10. Two major trails, the ancestors of today's interstate highway system, were developed in the 1840s to channel the newcomers into the frontier zones (see Figure 11.12). From a trailhead in Independence (near Kansas City), Missouri, the **Oregon Trail** took a northerly track across what are now Nebraska, Wyoming, and Idaho to the fertile Willamette Valley of the Oregon Territory, with a southern branch (the **California Trail**) leading to what is now Sacramento, California. The southern route from Independence, known as the **Santa Fe Trail,** snaked across Kansas and New Mexico. Army posts were set up along the way to facilitate and protect the westbound migrants. They also served as military garrisons in the forceful subjugation of Native American peoples, whose stories are told in broken treaties, massacres, and decimation of political leadership. Accessed in part by the Oregon and Santa Fe Trails, the U.S. Great Plains were settled rapidly between 1860 and

1890. A major impetus for settlement of the far west was the **California Gold Rush** of 1849. No Panama Canal existed at that time, so most of the prospectors, known as **Forty-Niners,** made their way either tortuously along the Oregon and California Trails, across the Isthmus of Panama by land, or by sea for a 17,000-mile (27,200-km) voyage around South America, which could take up to seven months.

Most of the homesteaders, Forty-Niners, and other settlers of the American frontier were of English, German, Scandinavian, and other western and northern European ethnicities. The United States ultimately developed into a nation in which about 99 percent of the people were either immigrants or of immigrant stock from a vast array of ethnic backgrounds. The larger groups of today include more than 45 million of German origin, over 20 million of Irish origin, and over 20 million of English origin. Although some kept their own ethnic identities—still found in ethnic neighborhoods, especially in the East, or in small rural communities in the Midwest—most were blended into the melting pot that produced what might be called the **mainstream European American culture.**[2] The ethnographer David Levinson describes that culture as one based on institutions such as the English language and British educational principles brought by the British colonists, combined with core American values that developed on U.S. soil. These include English as the national language, religious tolerance, individualism, majority rule, equality, a free-market economy, and the idealization of progress.

Added later to the mix were immigrants from eastern and southern Europe, represented today by more than 11 million Americans of Italian origin, over 6 million of Polish origin, and over 5 million Jews. Many of these immigrants settled initially in the larger cities of the U.S. Northeast, where earlier immigrants of mainly northern and western European extraction often discriminated and directed violence against them. It took several generations for these newer immigrants to become part of the mainstream ethnic fabric of America. Other large groups have had more difficult careers with the white majority—which is projected to become a minority soon after 2050—or are relatively new contributors to the American ethnic mix.

Ethnic Minorities in the United States and Canada

The George W. Bush administration's secretary of state, Condoleezza Rice, aptly described slavery as "America's birth defect."[3] The plantation economy that supported the American South was built on the exploitation of more than 500,000 black slaves brought to America from Africa between 1619 and 1807. People of mainly African ancestry, known as **blacks** and **African Americans,** occasionally as **Afro-Americans,** and formerly as **Negroes,** number about 35 million, or 12 percent of the U.S. population. The group also includes many people of mixed black and other ethnic extraction and blacks who came from regions other than Africa. Blacks were acknowledged as the largest minority in the United States until 2000, when census figures revealed that the number of people identifying themselves as Hispanic slightly exceeded the number of African Americans.

Unlike the French Canadians, African Americans are not largely confined to one part of the country. Some regions, especially in the eastern half of the United States, have a very strong black cultural presence (Figure 11.13), and African Americans are majorities in a number of large cities and particularly in the centers of those cities.

Race relations have played an important role in the country's political and social history. Conflicting attitudes toward black slavery were among the issues that nearly split the United States into two countries in the American Civil War. The **Thirteenth Amendment** to the Constitution ended slavery in 1865, but the economically and politically subordinate position of blacks has been slow to change. As recently as the 1960s, there were legal struggles to end official **segregation** between the races in parts of the United

States—for example, in the state of Alabama, where there were separate sections for whites and blacks on buses and where the races were expected to live in different neighborhoods, attend different schools, and never intermarry.

Today, while a segment of the African American population has moved into the middle class and above, a much larger portion resides in impoverished and troubled **ghettos** of central cities, where there are problems in housing, jobs, and crime (•Figure 11.14a). About 40 percent of all persons incarcerated in U.S. prisons are black. The legal battles to eliminate segregation have ended, but "hearts and minds" have been slower to change, and African Americans continue to struggle for equality. Some of their efforts parallel earlier trials by American women, who obtained the right to vote after a more than 80-year campaign in 1920 and whose wages are even today generally lower than those of men in identical occupations. The proposed **Equal Rights Amendment** to the U.S. Constitution, which would grant women official parity with men at all levels, has not made it through the legislative process.

In the 1840s, when the United States won from Mexico the great swath of the American Southwest, West, and Texas, it also became home to many Spanish speakers of European or mestizo descent. The growth rate of people of Spanish or mixed Spanish background—the Hispanic American (or **Latino**) population—has in recent years been more rapid (about 3 percent per year) than that of any other group, both domestically and through the major migration stream flowing actively from Latin America (•Figure 11.14b). About half of the growth is from natural increase and another half from immigration, both legal and illegal. About 42 million people, or 14 percent of the U.S. population, are Hispanic. The majority are of Mexican origin and are concentrated in California, Texas, New Mexico, Arizona, and Colorado. Other important Hispanic groups include people of Puerto Rican origin, most heavily represented in New York City and in the Northeast and increasingly in Florida, and people of Cuban origin, who live mainly in southern Florida. Estimates vary for numbers of other Hispanics, mainly Mexicans and Central Americans, resident in the United States as illegal immigrants, but the figure most commonly cited is 10 million. Previously, Hispanic immigrants settled overwhelmingly in the gateway states of the southwestern United States, but today, many leapfrog these states and put down roots all across the country.

Minorities make up about one-third of the U.S. population, and they are rapidly gaining momentum toward that 2050 date when non-Hispanic whites will drop below 50 percent of the population. Demographically and politically, a gap is emerging in the United States between a young, ethnically diverse population and an older, mainly white one. In 2007, there were four states with majority populations of ethnic minorities: Hawaii (75 percent), New Mexico (57 percent), California (57 percent), and Texas (52 percent). Twenty-one percent of the nation's minority populations live in California, and 12 percent live in Texas.

(a)

(b)

(c)

• **Figure 11.14** Ethnic urban landscapes of the United States: (a) a black-run business in Los Angeles; (b) a Latino neighborhood in Chicago; (c) Chinatown in San Francisco.

Asian Americans now make up about 4 percent of the U.S. population. Immigrants from China were especially important in the construction and service sectors of the West Coast economies of both Canada and the United States between 1849 and 1882. Chinese laundries and the

Chinese "coolies" who did the grueling labors of building western railways became the stuff of stereotype and legend, and there were sore points in Chinese-Anglo relations. Chinese immigrants brought an indelible cultural presence in the form of "Chinatown" landscapes in major cities, particularly in the 19th and early 20th centuries, and especially in the western regions of both countries (•Figure 11.14c). Chinese food has become even more widespread than Mexican food across the continent.

Asian Americans have long been steadily moving out of the central city and into white and mixed neighborhoods; they include among their ranks not only the Chinese but large populations of ethnic Japanese, Korean, Vietnamese, and Filipinos, among others. The closest approach to Asian minority dominance of a state is in multiethnic Hawaii, where Asian Americans, especially of Japanese descent, make up more than 40 percent of the population. Race relations between the Anglo majority and Asian groups in both the United States and Canada have generally been good in recent decades. The last major episode of tension was during World War II, when large numbers of civilian Japanese Americans, wrongly suspected of being potential collaborators with the enemy, were rounded up and interned in prison camps in the West.

Asian Americans in the United States today have the distinction of being the most highly educated of all the ethnic and religious groups except for Mormons and Jews. Among their ranks, the ethnic Japanese, Koreans, Chinese, East Indians, and Filipinos have generally prospered economically, but the ethnic Vietnamese, Lao, and Cambodians are poorer. The largest Asian populations in the United States are the Chinese and Filipino, each over 1 million. The largest in Canada are ethnic Chinese (about 1 million, more than double their numbers in 1984), East Indian (mainly of the Sikh religion), and Filipino. Recent large Asian migrant flows into Canada include Koreans and Vietnamese. Asian minorities have intermarried with other ethnic groups in both countries, and their mixed offspring call themselves **Hapas,** a Hawaiian word meaning "half" that people of mixed Pacific island and other ethnicities also use to describe themselves.

These brief sketches only scratch the surface of the ethnic mosaics of the United States and Canada. Nearly 200 non–Native American ethnolinguistic groups are represented in these countries, which are rather rare examples of nations that have generally welcomed immigrants into their social fabric.

Nonindigenous Languages and Faiths

The non–Native American religions and languages prevailing in Canada and the United States reflect both early colonial influences and more recent waves of immigration. English and French are Canada's official languages, with French usage confined largely to the traditionally French southeast (•Figure 11.15). The United States does not have an official language (despite repeated efforts to pass a constitu-

• **Figure 11.15** Nonindigenous languages of the United States, Canada and Greenland.

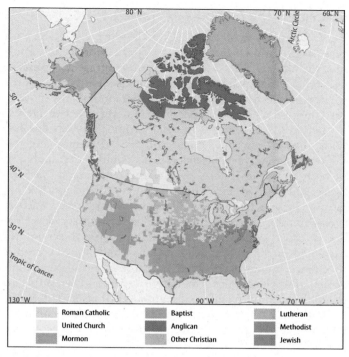

• **Figure 11.16** Religions of the United States, Canada, and Greenland.

tional amendment that would establish it as English), but 96 percent of U.S. residents speak English. The language map well illustrates the growth of Hispanic populations in the Southwest and southern Florida. **"Spanglish,"** a hybrid tongue of Spanish and English, is moving from Hispanic neighborhoods of southern California into the mainstream media and culture of the United States. Spanglish uses "code switching," substituting words or phrases from one tongue when speaking in another (for example, *"Vamos a la store para comprar milk,"* meaning "Let's go to the store to buy some milk") and also coining new words, like *perro caliente* for "hot dog."[4] In the larger cities of both the United States and Canada, there is a dizzying array of languages associated with ethnic neighborhoods.

Laws guarantee religious freedoms in both the United States and Canada, and both countries have a rich fabric of faith (•Figure 11.16). Christianity prevails in both countries, with numerous Christian churches represented. The largest single denomination in both countries is Roman Catholic (26 percent in the United States and 43 in Canada), brought to the United States by Irish, Italian, Polish, Filipino, Hispanic, and other immigrants and to Canada mainly by the French. Hispanics are replenishing the ranks of American Catholics who have left the church in recent decades and are also introducing their own forms of charismatic worship into the country. The next-largest group in the United States is Baptist (about 17 percent), then Methodist (about 7 percent) and Lutheran (about 5 percent). The other related monotheistic faiths are well represented by about 5 million Jews and by 1 million Muslims; Islam is the fastest-growing religion in the United States.

11.4 Economic Geography

The United States and Canada are very wealthy nations. Per capita GNI PPP figures for the countries are $44,260 and $34,610, respectively, compared to the world's highest, Luxembourg, at $55,970. The United States has the world's largest economy and is by far the world's largest producer and consumer of goods and services. It consumes far more than it produces and is thus the world's leading example of consumption overpopulation. With just about 5 percent of the world's population, the United States has a third of the world's wealth and each year consumes, for example, more than 25 percent of the world's energy output, 50 percent of its diamonds, 50 percent of its illegal drugs, and 75 percent of its rubber.

Generally, in the globalized economy, the United States may be seen as having built an economic structure of consumption. By contrast, China and most other major economies have built economies of production and export. The United States borrows money from other countries (it has the world's largest federal foreign debt, as well as the world's largest national debt), and it sells off assets to finance its consumption. The nation has a trade deficit (for which China alone accounts for about one-third), exporting $1 trillion worth of goods while importing $1.8 trillion annually. Once the world's largest exporter of goods, the United States lost that position to Germany in 2003. The United States' share of global gross domestic production fell from 31 to 28 percent between 2000 and 2007 (while in the same period rising by 3 percent to a combined 11 percent in the four leading emerging economies of Brazil, Russia, India and China—the so-called BRICs).

Sources of the Region's Affluence

This region, particularly the United States, is blessed with very large endowments of some of the world's most important natural assets. A number of geographic, political, and other circumstances have complemented this resource wealth in the development of these economically powerful countries. Both countries are among the five largest in the world. A wide range of environmental settings has both allowed and stimulated full use of human ingenuity in economic development. The combined population of both countries is large, yet neither country is "people-overpopulated." A large population represents both a big pool of prospective labor and talent to promote economic growth and a vast market for the goods and services produced. Both countries developed mechanized economies early and under favorable conditions. Innovations continue to replace human labor with machine labor, generally increasing efficiency and productivity (while also leading to unemployment in traditional manufacturing sectors).

Peace and stability within and between these countries have provided a good climate for investment and economic development. Since the Civil War ended in 1865, the United States has managed to keep all of its major conflicts off its own shores. Generally cooperative relations between the United States and Canada also help. Both have an overall sense of internal unity despite episodes of poor race relations in the United States and the separatist movement in Québec. Both countries have a track record of continuity in political, economic, and cultural institutions. Such stable conditions are conducive to economic growth.

The distribution of national wealth in both countries, however, is imbalanced. Large segments of the United States' population lack a "safety net" in the form of health insurance and other social services. Income distribution is skewed, with the wealthiest 1 percent of Americans taking in more than 20 percent of the country's total income and the poorest 10 percent taking in less than 2 percent of the total. This represents the greatest gap between rich and poor of all the MDCs. Canada has a more equitable distribution of income than the United States and has generally higher indexes of quality of life. Long dominated by liberal politics, Canada has a more socialistic approach to its population than the United States does and tends to be more in line with Europe than with the United States on most social and political issues. In 2006, some 12 percent of U.S. citizens lived below the poverty line (defined as an annual income of $20,650 for a family of four), while only 5 percent of Canadians were below Canada's official poverty line. Taken as a whole, however, Canada and the United States make up a region where most people enjoy the "good life."

An Abundance of Resources

The United States and Canada are the epitome of "neo-Europes," lands that European immigrants chose to settle largely because of their resemblance to the European environments and their potential for production of wheat, cattle, and other vital products. The United States has more high-quality arable land than any other country in the world. A much smaller proportion of Canada is arable, but it has more farmable land than all but a handful of countries. Such abundance has helped these two countries become the largest food-exporting region of the world.

Canada and the United States also have abundant forest resources. Canada's forests are greater in area than those of the United States but not as varied. Wood has been an essential material in the development of both countries, prompting the less developed countries of the world today, such as Brazil and Malaysia, to cry foul when American environmentalists denounce tropical deforestation. Even now, the United States cuts more wood for lumber each year, produces more wood pulp, and produces more paper than any other country in the world. The United States both exports and imports large quantities of wood, with the exports going largely to Japan. U.S. domestic industries can meet only 60 percent of their lumber needs, and their imports come mainly from Canada. Canada is the world's largest exporter of wood, with 80 percent of the products bound for the United States (•Figure 11.17).

The United States and Canada are also very rich in mineral resources. These played a major role in the earlier development of the region's economy, and mineral output is still huge. Canada is a major exporter of oil and natural gas, mainly to the United States, which in 2007 obtained 18 percent of its imported oil from Canada—more than from any other country. Although still a major mineral producer, the United States is also a major importer, especially of oil and industrial raw materials. This situation reflects the partial depletion of some American resources and the enormous demands and buying capacity of the U.S. economy.

Abundant energy resources have been essential to the region's economies. Coal was the largest source of energy

• **Figure 11.17** Canadian timber en route to East Asia.

GEOGRAPHY OF ENERGY

Energy Alternatives in the United States

The United States is the world's largest producer of nuclear power and derives 20 percent of its electricity from that source. This energy source has an uncertain future in the United States, however. In 1979, an accident occurred in the Three Mile Island nuclear power plant in Pennsylvania, releasing little radiation to the environment but spurring enormous public fear of nuclear power. The truly disastrous Chernobyl accident in 1986 in Ukraine heightened the perceived peril of nuclear energy. Due to such fears and to cost overruns and other problems, no nuclear power plants have gone online in the United States since 1978. The same concerns have troubled Canada, which gets 15 percent of its electricity from nuclear power. It has not built a new nuclear power plant since the 1970s. A 2004 blackout affecting much of southeastern Canada and the northeastern United States prompted some Canadian officials to consider building a new nuclear reactor in Ontario, but this idea was later dismissed.

The more benign sources of alternative energy—including windmills, photovoltaic panels, hydrogen fuel cells, and hydropower—meet just 6 percent of U.S. energy needs. Renewable energy sources have received only modest federal funding for research and development since the energy crisis of the 1970s. Few tax breaks or other incentives exist for the development of these alternatives by private companies, and official policy is that alternatives should develop in the free market rather than through federal funding and legislation. The United States has also not endorsed the Kyoto Protocol or other international legislation that would compel it to embrace alternative energies.

Without leadership from Washington, many states are struggling to take affairs into their own hands (12 states were so angry at the Bush administration's refusal to mandate carbon dioxide cuts that they filed lawsuits against the government in 2007). They want to cut carbon dioxide emissions through strictly mandated standards for vehicles and factories and the adoption of "greener" technologies. California leads these efforts. The state's **Global Warming Solutions Act of 2006** requires that greenhouse gas emissions be cut to 1990 levels by 2020, by which time utilities will have to generate one-third of their energy from renewable resources. Proving the maxim that "there's no such thing as a free lunch," some environmentalists in California are up in arms over some of the carbon-reducing steps proposed. They oppose Los Angeles's proposition to build a "green path" corridor for transmitting solar and geothermal power across 80 miles (c. 130 km) of southeast California's desert wildlands to the city.

Federal legislation has improved energy efficiency, especially in the vital transportation sector that accounts for about 28 percent of U.S. energy needs. In 1975, in the wake of the energy crisis, Congress enacted the **Corporate Average Fuel Economy (CAFE) standard,** requiring that the fleets of passenger cars produced by all the auto companies achieve an average of 27.5 miles per gallon. The standard for light trucks was set lower, at 20.7 miles per gallon. Sport utility vehicles (SUVs) are classified as light trucks, al-

lowing them to meet this lower standard—and allowing American consumers to pursue their love affair with these "gas guzzlers." This will change, however, as new legislation will require the CAFE standard to rise for entire fleets (including light trucks) to 35 miles per gallon by 2020.

Remarkable changes in the U.S. auto and farming sectors have accompanied the rapid embrace of ethanol fuels (●Figure 11.B). In his second term, President George W. Bush called for a 20 percent reduction of gasoline consumption within a decade. Ethanol, made primarily from corn in the United States and typically mixed with gasoline to be sold as "super unleaded" fuel, would make up the gasoline shortfall. The corn-producing states of the Midwest cheered the proposal, which brought rare good news to the farming sector. Boosted by federal ethanol subsidies of 51 cents per gallon, more and more land went into corn production (one-third

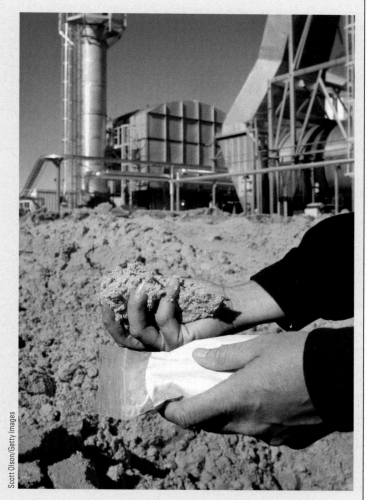

● **Figure 11.B** Pictured here is a byproduct of corn ethanol production, which can be combined with other ingredients to make high protein livestock feed.

Scott Olson/Getty Images

of the total farmed in 2007), and more corn (20 percent of the total in 2007) was turned into ethanol. With 190 ethanol plants operating or under construction, the United States has now surpassed Brazil in ethanol production. The "no free lunch" principle also applies to this apparently benign fuel source, however. Much fossil fuel–based energy and fertilizer goes into its production, raising questions about net energy gains. Diversion of the corn crop into ethanol has also raised food prices for the American (and Mexican) family; the rising price of feed corn for hogs and cattle has led to higher costs for pork and milk, among other things. The next generation of production may well focus on cellulosic ethanol produced from plant waste and from nonfood crops like switchgrass and other plants having fewer conflicting uses and requiring lower energy inputs than corn.

during the 19th-century industrialization of the United States. Today, the country leads the world in estimated coal reserves (about 25 percent of the total). In coal production, China and the United States are first and second, with China producing 35 percent and the United States 18 percent. Canada is also an important coal exporter, with reserves and production that are small on a world scale but large in terms of Canada's needs.

Environmental concerns about the impacts of this most polluting of fossil fuels raise long-term questions about what the United States can do with its generous coal endowment. As it stands, the United States gets about half of its electricity from coal-fired plants (with construction of 150 new plants planned, as of 2007), with natural gas supplying 19 percent and nuclear power 20 percent. Drawing on large uranium resources in each country, nuclear power is an important secondary source of electricity for both (see Geography of Energy, pages 619–620).

Petroleum fueled much of the development of the United States in the 20th century. For decades, the country was the world's greatest producer, having such large reserves that the question of long-term supplies received little thought. The United States is still exploiting its 21 billion barrels, an estimated 1.6 percent of the world's proven oil reserves (a figure that may increase following the recent find of a new oilfield deep under the Gulf of Mexico). But its huge economy demands 7.5 billion barrels a year—as much as the five next-largest consumers, China, Japan, Germany, Russia, and India, combined—far more oil than the country can produce. The United States does possess large reserves of petroleum that can be produced from oil shale, but only at great economic and environmental cost (see page 678). The United States therefore imports oil to meet about xx percent of its domestic need, significantly more than the 36 percent it imported before the 1973 energy crisis. Its top five suppliers, in descending order, are Canada, Mexico, Saudi Arabia, Venezuela, and Nigeria.

Canada's conventional oil reserves, situated mainly in Alberta, are small in comparison to those of the U.S., representing just 0.4 percent of the world total. However, in Alberta and neighboring Saskatchewan, Canada has the world's largest reserves of **tar sands** (also known as *oil sands*). The potential output of this unconventional oil makes Canada second only to Saudi Arabia in proven oil reserves. There is a catch, however. Tar sands are mineral deposits in which a thick liquid petroleum known as bitumen is mixed with sand and clay from an ancient ocean (•Figure 11.18). Through a water-guzzling, energy-consuming process costing about $25 per barrel, the bitumen can be converted to petroleum. When conventional oil prices are high (especially around the $70-per-barrel mark of recent years), tar sands are economically attractive to exploit. The price surge in oil beginning in 2003 led to strong growth in production from Canada's Athabascan tar sands, with most of the output going to the United States. The tar sands account for fully half of Canada's oil production.

The biggest obstacle to tar sands development for domestic consumption is Canada's 2004 ratification of the Kyoto Protocol, which requires the country to reduce its carbon dioxide emissions to 6 percent below their 1990 levels. This commitment will require a massive cutback in the use of fossil fuels, including the natural gas that must be burned to extract oil from the tar sands. Alberta's provincial government was furious at the federal government's support of the Kyoto treaty, arguing that it would cost the province billions of dollars in lost revenue and thousands of jobs. Alberta's politicians even threatened that the province might secede from Canada if the treaty were ratified. South of the border, American environmentalists pointed to Canada's ratification of the Kyoto Protocol as another example of how the United States, which refuses to ratify the treaty, is out of step with the rest of the industrialized world.

Natural gas has become an increasingly important fossil fuel to these two economies. The United States is the world's largest consumer of gas and is a major importer, especially from Canada. It ranks second in the world in output (behind Russia), producing 17 percent of the world's gas from its 3 percent of proven reserves worldwide. That equation means that the present high output cannot be sustained for long. With recent high energy prices, states like Montana and Colorado that still have reserves of natural gas are weighing whether or not to exploit them or to keep the

● **Figure 11.18** (a) Tar sands locations in Canada. (b) Oil production at the Athabasca tar sands facility.

lands above them relatively pristine for a "new economy" based on hunting, recreation, and tourism. Canada, with smaller but rapidly increasing proven reserves, produces about 6 percent of world output and is a major exporter.

The potential for future hydropower development in the United States and Canada is small. Development has been so intense in the past that the two countries rank with China and Brazil as the world leaders in hydropower capacity and output. Canada gets about 58 percent of its electricity from hydropower, and the United States 7 percent. The countries' dams are controversial because of the environmental changes they produce and also because they often result in the inundation of indigenous peoples' lands. Some dams in the United States, including several in Washington and Oregon, are scheduled for **decommissioning** (major modification or removal) primarily for environmental reasons, especially to allow a dozen endangered salmon species to reach their traditional upstream spawning grounds. Fish ladders and weirs have been constructed to aid salmon migrations, but salmon populations have nevertheless plummeted from 16 million before the dams were built to about 1 million today. A coalition of Native American groups, environmentalists, and sport fishing enthusiasts has long been demanding that dams on Washington's Columbia and Snake Rivers and the Klamath River of Oregon and California be torn down. Recently, even long-standing supporters of the dams, including farmers, have begun discussing a future without the four Snake River dams. Wind-generated electricity would replace the hydropower lost when the dams are decommissioned.

Other mineral resources in North America have presented a similar picture of abundance (Table 11.3). Many of the ores easiest to mine and richest in content are now mostly depleted, and even large mineral outputs in the United States cannot meet the huge demands of the domestic economy. In many such cases, Canada has been able to assume the role of leading foreign supplier. A major example is iron ore. The United States originally had huge deposits of high-grade ore, but by the end of World War II in 1945, this ore had been seriously depleted. New sources of high-grade ore were then developed in Québec and Labrador, mainly for the U.S. market.

Mechanization, Services, and Information Technology

Although raw materials contribute much to their wealth, the United States and Canada have become prosperous mainly because of machines and mechanical energy, complemented in recent decades by a boom in information technology (IT). From an early reliance on waterwheels powering simple machines, especially along the fall line in the Northeast, the United States increasingly exploited the power of coal-fired steam engines. Later came oil, internal-combustion engines, and electricity, all harnessed for greater and more efficient energy use.

In this energy-abundant and mechanized economy, the United States was able to take advantage of unique circumstances. Its abundant resources attracted foreign capital and stimulated domestic accumulation of capital through large-scale and often wasteful exploitation. There was also a labor shortage that attracted millions of immigrants as temporary low-wage workers but also promoted higher wages and laborsaving mechanization in the long run. A culture

TABLE 11.3   U.S. and Canadian Shares of the World's Known Natural Resource Wealth and Production

	Approximate Percent of World Total				Approximate Percent of World Total		
Item	U.S.	Canada	Total for Both	Item	U.S.	Canada	Total for Both
Area	7	7	14	**Minerals, Mining, and Manufacturing** (*continued*)			
Population	4.6	0.5	5.1	Crude oil production	10	3	13
				Crude oil refinery capacity	21	3	24
Agricultural and Fishery Resources and Production				Natural gas production	24	7	31
				Natural gas reserves	3	1	4
Arable land	13	3	16	Potential iron ore reserves	5	6	11
Meadow and permanent pasture	8	1	9	Iron ore mined	6	3	9
Production of:				Pig iron produced	8	2	10
Wheat	10	5	15	Steel produced	12	2	14
Corn (grain)	51	3	54	Aluminum produced	15	10	25
Rice	2	0	2	Copper reserves	17	5	22
Soybeans	47	2	49	Copper mined	11	5	16
Cotton	18	0	18	Lead reserves	22	13	35
Milk	16	2	18	Lead mined	15	5	20
Number of:				Zinc reserves	13	15	28
Cattle	7	1	8	Zinc mined	10	11	21
Hogs	7	1	8	Nickel mined	0	15	15
Fish (commercial catch)	5	1	6	Gold mined	14	6	20
				Phosphate rock mined	29	0	29
Forest Resources and Production				Potash produced	6	23	29
Area of forest/woodland	6	8	14	Native sulfur produced	21	0	21
Wood cut	14	5	19	Asbestos mined	1	18	19
Sawn wood production	28	17	45	Uranium reserves	9	12	21
Wood pulp production	34	16	50	Uranium produced	16	29	45
				Electricity produced	26	4	30
Minerals, Mining, and Manufacturing				Potential hydropower	4	3	7
Coal reserves	23	1	24	Production of hydropower	4	4	8
Coal production	20	2	22	Nuclear electricity produced	31	4	35
Coal exports	25	8	33	Sulfuric acid produced	26	3	29
Published proved oil reserves (tar sands and oil shale)	2	1	3				

Sources: United Nations Statistics Division, 2000; *World Population Data Sheet*, 2004.

of striving for advancement emerged. This emphasis on productivity, dubbed the **Protestant work ethic** but shared by people of many faiths and ethnicities, characterized the struggle for economic success and the **American dream** of a comfortable life. Only occasionally did national energies have to be diverted excessively into defense and war.

Most of these American conditions were duplicated in Canada. However, Canada lacked the large, unified, and growing internal market that fostered growth in its neighbor to the south. In recent decades, the advantage of the gi-

ant American domestic marketplace has been eroded by the expansion of the global market. Transportation and communication have become so relatively cheap and rapid, and so many artificial trade barriers have fallen, that the world market, rather than the domestic market, has become the essential one for major industries. This globalization of the economy has presented huge challenges to the United States, Canada, and the other MDCs, which are trying to prevent their traditional industries from being undercut by cheaper output from abroad. The United States and its industrial-

INSIGHTS

Trade Barriers and Subsidies

Many of the world's LDCs are convinced that the best way to develop is not to borrow money from the richer countries but to trade with them. After all, with their larger populations and lower labor costs, the LDCs should easily be able to churn out products at much more favorable prices than those sold by the MDCs. Buyers around the world should be snapping up those affordable products, pumping desperately needed money back into the developing countries. This exchange does not take place as freely as it could, however, because of two tools the MDCs wield as much as possible to protect their own producers: subsidies and tariffs.

Subsidies are expenses paid by a government to an economic enterprise to keep that enterprise profitable, even if the market would not. Tariffs are essentially taxes that a government imposes on foreign goods, basically for the same reason, to ensure that domestic products make it to market, even though they are not produced at competitive costs. Alone or in combination, subsidies and tariffs can make it all but impossible for a less developed country to join the world economy as a trading partner.

There are often struggles between the MDCs and LDCs over the issues of subsidies and tariffs, especially in meetings of the World Trade Organization (WTO). At these meetings, the LDCs have especially strong representation by the 21 countries that call themselves the **G-21 (Group of 21)** as a "play on numbers" against the affluent G-8.

Often bitter debate between these groups focuses on subsidies that the United States, the European Union, Japan, and other MDCs have used for several decades to prop up certain sectors of their economies, especially agriculture. The subsidies were created to help farmers and the countries get through especially difficult times, including the **Great Depression** of the 1930s in the United States, when the livelihoods of so many farmers were threatened, and post–World War II Europe and Japan, when the specter of malnutrition loomed. Over the years, the critical need for subsidies passed, but the sectors that received them were loath to part with them. Federal supporters of subsidies found political capital in maintaining them. Today, farming subsidies are something of a sacred cow in many of the MDCs, and any effort to reduce or remove them would have domestic consequences—especially at the ballot box. Altogether the MDCs pay out $1 billion a day in subsidies that support the farmers who make up less than 1 percent of their population. In the United States, subsidies are concentrated on growers of the big commodity crops of wheat, rice,

corn, soybeans, and cotton (these tend to be large, corporate-owned farms). The government views farmers of the "specialty crops" such as fruits and vegetables as self-sustainable (these tend to be smaller, family-owned farms). Under market pressure from cheaper imports, such as Chinese garlic, these farmers are now clamoring for subsidies too.

Cotton is a fine example of how the subsidies and related tariffs work domestically and what their impact is in the LDCs. The United States spends over $4 billion each year to subsidize its cotton farmers. The country produces far more cotton than it consumes, and most of the prodigious output is exported. The high level of production made possible by the subsidies drives the global price of cotton down to around 50 cents per pound. This is less than the African production cost of around 65 cents per pound. The African cotton producer not only cannot compete in the world market but loses money doing so. Cotton is, or should be, big business for Africa. If the U.S. subsidies were reduced or removed and world cotton prices rose, the potential economic gains for several African countries would be enormous. The United States, however, has been slow to reform cotton policies except in ways that clearly serve its own interest. In recent legislation, tariffs were reduced on some imported African textiles—but only those made from American cotton. The European Union, Japan, South Korea, and other countries are just as strident in protecting some of their products.

The G-21 countries, speaking for two-thirds of the world's farmers, demand that the MDCs reduce or eliminate such subsidies and tariffs or that the LDCs be compensated for the revenue lost to them. They threaten to begin erecting tariffs of their own (a proposal the MDCs say is bad for the developing economies) if the MDCs do not yield. In a familiar negotiating sequence, the MDCs give up little or no ground in such impasses, causing the G-21 delegates to walk away from trade talks in frustration.

Would removal of subsidies hurt the small American family farmer whose cause is championed by entertainers Willie Nelson and Neil Young? That would depend on how subsidy cuts were targeted. Sixty-five percent of the subsidies go the top 10 percent of America's agricultural producers, the large corporate farming industries. It is possible that the unkindest cuts could be avoided with sacrifices imposed instead on the bigger businesses, thereby perhaps helping some of the world's poorest countries escape the poverty trap on their own.

ized allies thus often find themselves in the awkward position of advocating free trade but erecting trade barriers (see Insights, above).

Mechanization in these economies has been increasingly replaced in recent decades with a new labor sector: services and information technology. The service sector, consisting of insurance, finance, medical care, education, entertainment, retail sales, and other enterprises, has become a

more significant employer than the manufacturing sector in the United States and Canada, as shoes, clothes, toys, cars, and other traditional manufactures have migrated to other countries. As telecommunications technologies have interconnected North America with the rest of the world, the United States has acquired global leadership in information technology even while many of its old heavy industries have expired in the country's Northeast—its so-called **Rust Belt**

• **Figure 11.19** The United States gained economic supremacy in the 20th century in large part because of its production of steel and other manufactured goods. It lost much of its production to other countries where labor and other costs were lower. South Korea's Posco steel mill, seen here, puts out a product that once epitomized the economy of Pittsburgh, Pennsylvania.

(•Figure 11.19). Manufacturing now only accounts for about 9 percent of the U.S economy, having fallen to the forces of globalization that moved traditional industries like textiles to countries with lower manufacturing costs, such as China. Manufactured goods make up two-thirds of U.S. exports by value, however, and the nation still has the global lead in many high-tech products.

Ironically, perhaps the best prospect for more growth in U.S. manufacturing is investment from China as that country looks to produce goods for the American market that are cheaper to produce in the United States than to export from Asia. The United States hopes to maintain its global leadership in design and innovation, profiting from a **knowledge economy** even while other countries actually make products. As one American economist put it, "We want people who can design iPods, not make them."[5] But many worry that design and innovation will also move off-shore, leaving the United States with no competitive edge.

The United States is the world leader in the production of software. In the 1990s, there was such a speculative frenzy about IT that a classic **economic bubble** developed. Tens of millions of Americans started investing in stocks for the first time, bidding up the values of companies—even companies that produced no tangible products—to lofty heights. There was talk of a "new economy" in which the old paradigms of corporate investment and growth no longer applied. However, the **tech bubble** burst in 2000. Clever or lucky investors withdrew their investments in time, but others lost fortunes that had existed on paper. An economic recession set in. The economy gradually recovered as more attention was paid to the old-fashioned brick-and-mortar industries, which actually produce tangible assets, and the tech sector was approached more cautiously. High-tech industries increasingly built what came to be known

as a **platform economy**, with companies like Dell operating headquarters in the United States but outsourcing virtually all work on its products to other countries.

U.S.-Canadian Economic Relations

Each of these countries is a vital trading partner of the other, although Canada is much more dependent on the United States than vice versa. In 2007, Canada supplied 16 percent of all U.S. imports by value and took in 22 percent of all U.S. exports, making Canada the leading country in total trade with the United States. That same year, the United States supplied 55 percent of Canada's imports and took in more than 82 percent of its exports. Except for Canadian exports of automobiles and auto parts to the United States, the main pattern of trade between the two countries is the exchange of Canadian raw and intermediate-state materials—primarily ores and metals, timber and newsprint, oil, and natural gas—for American manufactured goods.

Free-trade agreements designed to open markets and increase revenues for both countries have facilitated this exchange. Nevertheless, the countries have periodically tried to protect their industries from one another, often with poor results. In the 1990s, there was a so-called **wheat war** between the two countries as they struggled for greater market share in an environment of high yields and low prices. With agricultural prices for grains and pork dropping to the lowest levels in more than 50 years, many farmers both south and north of the U.S.-Canada border went bankrupt. U.S. producers claimed that Canada was **dumping** (selling the product for less than it cost to produce it) its durum wheat on the United States market.

There was also a **salmon war** between the two countries, marked by periodic skirmishes until a resolution was reached in 1999. Salmon hatch in Canadian and American rivers, spend most of their lives in the open seas, and return to their river birthplaces to spawn. The dispute centered on the issue of "interceptions," meaning that American fishers often intercepted fish that hatched in Canadian upstream waters but migrated into U.S. waters.

In 2003, a new dispute emerged when the United States, accusing Canada of dumping low-cost lumber in the United States, imposed heavy new tariffs of 27 percent on softwood lumber imported from Canada. An agreement reached in 2006 lowered the tariffs and required Canada to price its lumber more competitively. Other issues between the countries await resolution. There are dozens of small water disputes across the border, including one over North Dakota's plan to divert excess water into the Red River, which flows north into Manitoba's Lake Winnipeg.

Despite such volleys occasionally fired on both sides of the border, the trend has been toward more cooperation and free trade. In 1965, the two countries negotiated a free-trade agreement for the production, import, and export of automobiles. This sparked rapid expansion of the auto industry into southern Ontario, on the border with the U.S. automobile-producing state of Michigan. The two countries entered into the comprehensive **Canada-U.S. Free Trade Agreement** in 1988, and in 1994, Canada, the United States, and Mexico enacted the North American Free Trade Agreement (NAFTA). Its goals and impacts are described in detail, especially as they affect Mexico, on pages 544–545 in Chapter 10. NAFTA has brought a boom in trade between the three countries but also caused major industrial readjustments, generally characterized by the more labor-intensive firms (beginning with clothing) moving to Mexican locales from Canada's southeastern core region and from the industrial northeast of the United States.

Transportation Infrastructure

Both governments have sought to enhance national unity and economic strength with transportation networks spanning their vast landscapes (●Figure 11.20). In both countries, east-west networks have been built over long distances, against the "grain" of the land, especially the great north–south-trending ranges of the Rocky Mountains. The coasts of both countries were first tied together effectively by heavily subsidized transcontinental railroads completed during the second half of the 19th century. Later, national highway and air networks improved transportation.

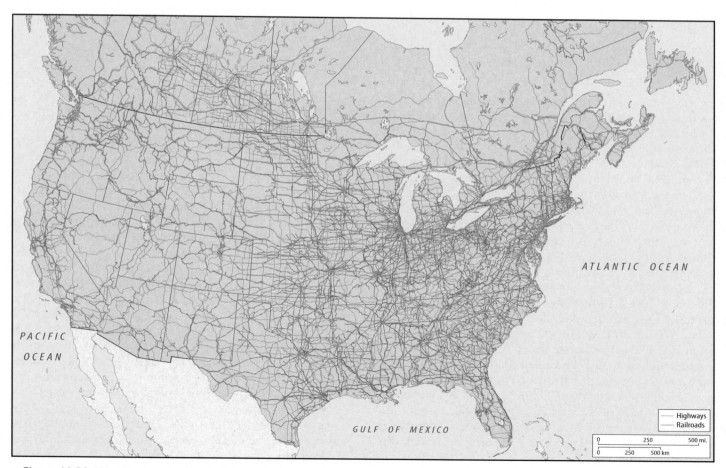

● **Figure 11.20** There is an extraordinarily dense network of highways and railways in the United States and southern Canada.

A landmark development in the United States was the growth of the **interstate highway system** beginning in the late 1950s. Today, this system serves as the primary network for the trucking of cargo across the country. Railways meet these needs in most countries. Although railways still carry the bulk of goods being transported across the country, few new lines are being built. Passenger rail traffic is particularly neglected, and once-popular routes are no longer used. In many cases, the old rails have been torn up to become bicycle and walking paths. The stagnation or decline of U.S. passenger railways is a direct reflection of the American love affair with the car. Personal prosperity (or at least the ability to take out a loan) and the freedoms associated with being on the road in one's own automobile have diminished appetites to travel by rail and bus. Public transportation is popular only in the cities, where gridlock makes it an attractive alternative.

11.5 Geopolitical Issues

Even with the meteoric economic rise of China and the collective strength of the European Union, the United States is the most important economic, military, and diplomatic power in the world. This section first examines the country's strategic relations with its northern neighbor and then with the rest of the world.

Historical Relations between the United States and Canada

For many years after the American Revolution, which secured U.S. independence from Britain, the political division of Canada and the United States—between a group of British colonial possessions to the north and the independent United States to the south—was accompanied by serious friction between the peoples and governments.

There were several sources of antagonism. The northern colonies failed to join the Revolution, and the British used those colonies as bases during the war. A large segment of the northern population had come from **Tory** stock driven from U.S. homes during the Revolution (Tories were the "friends of the king," colonists who wanted to maintain political connections with the British government). And tensions were high over the issue of who would have ultimate control of the central and western reaches of the continent.

The **War of 1812** was fought largely as a U.S. effort to conquer Canada. Even after its failure, a series of border disputes occurred, and the United States openly expressed ambitions to possess this British territory in North America. Officials in the United States threatened annexation throughout the 19th century and even into the 20th.

Canada's emergence as a unified nation came partly as a result of United States pressure. After the U.S. Civil War (1861–1865), the military power of the reunited nation seemed threatening to Canadian and British eyes. Recalling that the British had been hostile to the ultimately victorious Union (North) during the Civil War, some Americans argued that taking Canadian territory would be a just recompense. In 1867, the British Parliament passed the **British–North America Act,** bringing an independent Canada into existence by combining the separate colonies of Nova Scotia, New Brunswick, Quebec, and Ontario into one dominion named Canada. With that act, Great Britain sought to establish Canada as an independent nation capable of deterring U.S. conquest of the whole North American continent.

Although hostility between the United States and its northern neighbor did not immediately end with the establishment of an independent Canada, relations improved gradually. Canada and the United States today are strong allies with a connection only occasionally tested. But just as the United States overshadows Latin America in geopolitical affairs, it overshadows Canada as well. Canadians, like Mexicans, often feel overlooked and underappreciated by the neighbors across the border. Canada maintains a rather low profile in international affairs, sometimes disagreeing with the United States on such critical issues as the Iraq invasion of 2003 and exerting a policy of relative independence and neutrality despite membership with the United States in the North Atlantic Treaty Organization (NATO) 89 and other international agreements (Canadian soldiers have died supporting the NATO mission in Afghanistan). Some American travelers in lands where Americans are often disliked have learned to exploit Canada's relative neutrality for their own protection by sewing badges of Canada's national symbol, the maple leaf, on backpacks and jackets.

The United States' Place in the World

Historically, the United States was not the same kind of empire-building power as many European countries were. The United States did have its overseas possessions—the Philippines, Puerto Rico, and Guam, for example—and still holds a few, but its power has been and continues to be projected more through military action and trade.

The United States long enjoyed the geographic advantage of being something of an island situated far from the world's hot spots—especially Europe and the Middle East. In both world wars that ravaged Europe in the 20th century, the United States initially clung to **isolationism**—the view that those conflicts were Europe's and that the United States would do best to stay out of them. Only late in World War I, at the unpopular insistence of President Woodrow Wilson, did the United States enter the stalemated war on the side of Britain and France. The United States lost 116,516 soldiers in that war, but in helping its allies secure victory, it also gained unprecedented influence and importance on the world stage.

Again in World War II, the United States managed to stay out of the conflict for more than two years after Hitler's troops invaded Poland and ignited conflagrations across Europe. But on the infamous day of December 7, 1941, Japan forced the United States into the war with its surprise attack on Hawaii. And again, U.S. leadership and victory on both fronts of that war, which cost 405,399 American lives, helped the country achieve unprecedented strength in global affairs.

That war was succeeded by the Cold War, in which the two leading powers, the United States and the Soviet Union, faced off against one another, flanked by a host of often strong but always less powerful allies. U.S. concerns about the spread of communism from the Soviet Union and China into the newly independent countries of the postwar world—and the so-called domino theory that one after another of these countries might fall to communism—led to numerous and sometimes very costly American military interventions. The Vietnam War was the most significant of these.

The peaceful conclusion of the Cold War around 1990 was followed by a decade in which the United States sought a new role for itself on the world stage. There were military inventions to quell the conflict in disintegrating Yugoslavia and to try to deliver lawless Somalia from famine, but there was a new sense of security—and, in hindsight, complacency. There was no longer any defining framework, such as east-west relations, in international affairs. Terrorism was thought to be a problem that existed far from home.

The attacks of September 11, 2001, brought an end to the notion that geographic distance protected the United States. The events of that day represented a great watershed in U.S. geopolitical history. The United States under President George W. Bush developed a policy of **preemptive engagement:** whenever and wherever the country perceived a threat to its security, it would take military action if necessary to defuse that threat. That action would be unilateral, if necessary; in its first term, the Bush administration downplayed the need for diplomacy and partnership. The preemptive engagement policy was founded mainly on the premise that such actions would prevent potentially devastating terrorist attacks on the U.S. homeland, perhaps with

chemical, biological, or nuclear (CBN) weapons of mass destruction (WMD). Such threats could emerge not only from transnational groups like al-Qa'ida but from "rogue states," most of whom are on the United States' list of **official state sponsors of terrorism,** which includes North Korea and Iran.

The Bush administration justified the invasion of Iraq in 2003 on the grounds that Iraq was a state sponsor of terrorism (and had links with al-Qa'ida) and had WMD that it might use against the United States or its allies, especially Israel. The United States suffered a great setback in the court of world opinion when no conclusive evidence was found either of weapons of mass destruction or of links with al-Qa'ida. Worldwide opinion polls about the United States reflected the country's resulting loss of stature. Detractors also wrote of a **new American imperialism,** proposing that U.S. post-9/11 actions marked the beginning of a new era of aggressive global involvement for the country. The Bush administration denied such intentions, insisting that its actions in Iraq and elsewhere had helped protect Americans at home and strengthen democracy and the rule of law abroad.

The United States remains the world's sole **superpower,** with the strongest economy, military expenditures larger than those of the next 14 countries combined, dominance of global popular culture (from films to fast foods), the world's best universities, and headquarters to many of the world's leading international organizations (including the United Nations, the International Monetary Fund, and the World Bank). Its singular stature suggests a "unipolar" world in which its dominance is unchallenged.

A shift appears to be in the works, however. China's surging economy is the principal challenge; it is expected to be the world's largest by 2027. In his second term, President Bush began to open more diplomatic channels, recognizing, for example, that the nuclear threat from North Korea could not be confronted without China's help and the parallel threat from Iran without Russia's help.

The following modules take a closer look at Canada and the United States.

SUMMARY

→ The United States and Canada make up a region often known as Anglo America because of British influences. Both had numerous pre-European indigenous cultures and through immigration have been shaped by many others since 1492.

→ Greenland is politically a part of Denmark but physically part of this region.

→ The United States and Canada have large land areas of about the same size. Their combined population is about 335 million.

→ Physical regions of the United States and Canada are the Arctic Coastal Plains, Gulf and Atlantic Coastal Plains, Piedmont, Appalachian Highlands, Interior Highlands, Interior Plains, Rocky Mountains, Intermountain Basins and Plateaus, and Pacific Mountains and Valleys.

→ Climate types of the region are tundra, subarctic, humid continental with short and long summers, tropical savanna, Mediterranean, and undifferentiated highland, with associated patterns of vegetation and land use.

→ Major resources include agricultural land, forests, coal, petroleum, natural gas, iron ore and other minerals, and waterpower.

→ Native American settlement of the region began at least 12,500 years ago. Some groups developed civilizations based on agriculture and permanent settlement, but most were at least partly nomadic and dependent on hunting and gathering. Great linguistic and cultural diversity existed among Native American groups. All were well adapted to local environmental conditions, and all were disrupted profoundly by European contact. Native Americans are among the poorest populations in the United States but have fared better in Canada.

→ Early European settlement displaced native populations in both countries, where waves of immigration continually changed demographic, ethnic, and racial characteristics. The French are the most significant minority in Canada. Hispanics and African Americans are the largest minority groups in the United States. Asians are prominent minorities in both countries.

→ The United States built a land empire through warfare and purchases, including the Louisiana Purchase (from France) of 1803; the acquisition of Florida from Spain in 1819, the addition of Texas, the Southwest, and California as spoils of war with Mexico in 1845 and 1848; the Gadsden Purchase from Mexico of a strip of Arizona and New Mexico in 1853; and Alaska, bought from Russia in 1867.

→ Both countries industrialized in the 19th and 20th centuries. The economies have changed as manufacturing became less important while employment in service industries and information technology grew.

→ Canada and the United States are among the world's richest countries because of many factors. Both have large endowments of natural assets. Both are large in area and population. Both industrialized fairly early. Innovations have increased their efficiency and productivity. Peace and stability at home have stimulated investment and economic development. Both have strong internal unity and political stability.

→ The United States has the world's largest appetite for fossil fuel. Its energy efficiency has improved since 1973, but energy consumption and the percentage of energy imported have both increased nevertheless. Nuclear energy production has stalled because of environmental and health considerations. Production of ethanol from corn has surged in the United States, as has production of oil from tar sands in Canada.

→ The two countries have had generally good economic and political relations. Each is a major consumer of exports from the other. About 90 percent of the Canadian population lives within 100 miles of the international border.

→ The United States is the world's strongest military and economic power. Its influence in world affairs grew in the wake of the two world wars of the 20th century. The terrorist attacks of September 2001 put U.S. foreign policy on a new course that led to a more aggressively interventionist foreign policy.

KEY TERMS + CONCEPTS

African Americans (p. 615)
Afro-Americans (p. 615)
Aleuts (p. 609)
Algic language family (p. 608)
    Arapaho (p. 608)
    Blackfoot (p. 608)
    Cheyenne (p. 608)
    Cree (p. 608)
American dream (p. 622)
amnesty (p. 601)
Asian Americans (p. 616)
Aztec-Tanoan language family (p. 608)
    Comanche (p. 608)
    Hopi (p. 608)
    Papago (p. 608)
    Shoshone (p. 608)
blacks (p. 615)
blizzard (p. 602)
"Boswash" (p. 600)
British–North America Act (p. 626)
California Gold Rush (p. 614)
California Trail (p. 614)
Canada-U.S. Free Trade Agreement
    (p. 625)

Canadian Shield (p. 603)
Corporate Average Fuel Economy (CAFE)
    standard (p. 619)
Dakota (p. 608)
decommissioning (of dams) (p. 621)
dumping (p. 624)
Dust Bowl (p. 602)
economic bubble (p. 624)
Equal Rights Amendment (p. 615)
Eskimo-Aleut language family (p. 609)
Eskimos (p. 609)
fall line (p. 605)
First Nations (p. 610)
First Peoples (p. 610)
Forty-Niners (p. 614)
G-21 (Group of 21) (p. 623)
Gadsden Purchase (p. 614)
gaming industry (p. 610)
ghetto (p. 615)
Global Warming Solutions Act of 2006
    (p. 619)
Great Depression (p. 623)
Hapa (p. 616)
Hawaiian language (p. 612)

Hokan-Siouan language family (p. 608)
    Iroquoian (p. 608)
    Lakhota (p. 608)
    Mohawk (p. 608)
Homestead Act (p. 614)
illegal aliens (p. 600)
interstate highway system (p. 626)
Inuit (p. 609)
isolationism (p. 626)
Kanaka Maoli (p. 612)
knowledge economy (p. 624)
Lakota (p. 608)
Latino (p. 615)
Louisiana Purchase (p. 613)
mainstream European American culture
    (p. 615)
manifest destiny (p. 613)
megalopolis (p. 600)
Mosan language family (p. 608)
    Chemakuan (p. 608)
    Salish (p. 608)
    Wakashan (p. 608)
Multiculturalism Act (p. 613)
Na-Dene language family (p. 609)

REVIEW QUESTIONS

1. Where are populations of the United States and Canada concentrated?

2. What are the main topographic, climatic, and biotic zones and types of the region and the associated land uses?

3. What natural hazards threaten the United States?

4. What are some of the major Native American cultures of the United States and Canada? How did these people live in their diverse environments?

5. How did the French become such an important population in Canada, and how have their aspirations affected the country's politics?

6. What role has immigration had in shaping the populations of the United States and Canada?

7. What are the major nonindigenous faiths and languages of the United States and Canada?

8. What have been the major trends and developments in the economies of the United States and Canada?

9. What are the major characteristics of economic and political relations between the United States and Canada, historically and today?

10. What priorities have preoccupied U.S. foreign policy in recent years?

NOTES

1. Quoted in Eilene Zimmerman, "Border Agents Feel Betrayed by Bush Guest-Worker Plan." *Christian Science Monitor,* February 24, 2004, p. 3.

2. David Levinson, *Ethnic Groups Worldwide* (Phoenix, Ariz.: Oryx Press, 1998), p. 396.

3. Quoted in Richard W. Stevenson, "New Threats and Opportunities Redefine U.S. Interests in Africa." *New York Times,* July 7, 2003, pp. A1, A8.

4. Discussion and quotes are from Daniel Hernandez, "A Hybrid Tongue or Slanguage?" *Los Angeles Times,* December 27, 2003, pp. A1, A30.

5. Quoted in Louis Uchitelle, "Goodbye, Production (and Maybe Innovation)." *New York Times,* December 24, 2006, p. BU4.

These schoolchildren on a museum visit reflect Canada's officially multicultural society.

MODULE 11.1
CANADA: FROM SEA TO SEA

Canada is a highly developed country: affluent, industrialized, technologically advanced, and urbanized. But in some ways, it is quite unlike most developed countries. These differences are related to Canada's internal geography and to its location adjacent to the United States. This module explores the geography of Canada "From Sea to Sea" (the country's official motto): from its strongly French and English core in the southeast to the "Great White North," a land of Inuit, tundra, and ice.

■ module
outline

■ module
objectives

This module should enable you to

→ Recognize how Canada's proximity to the powerful United States has shaped the country's economic geography

→ Appreciate how resource depletion and distance from market and production centers contributed to the economic decline of the Atlantic region

→ Understand why several sectors of Canada's population—the French, Inuit, and Prairie Province inhabitants—have pushed for more autonomy and even independence from Canada

→ See how changing world energy and market conditions have altered indigenous ways of life in the North

→ Witness the scramble for resources in Arctic lands and waters that has accompanied warming temperatures in the region

→ Understand the economic rationale for Greenlanders' wanting to retain their ties with distant Denmark

11.1.1 Canada's General Traits

Canada is very urbanized, with 79 percent of its people living in cities (the same percentage as the United States). About 36 percent of Canadians live in the country's three largest metropolitan areas: Toronto (population 6.2 million), Montréal (population 3.6 million), and Vancouver (population 2.1 million) (•Figure 11.1.1). These dynamic cities appropriately convey a positive image of an advanced and prosperous country. Impressively good national statistics on health, housing, and education provide further evidence of Canada's well-being.

Canada is different in some ways from most other developed countries. It has an unusual pattern of trade. Most developed nations export mainly manufactured goods and import a mixture of manufactured and primary products. Canada is among the world's top ten MDCs in terms of manufacturing. But despite its exports of automobiles, aircraft, industrial machinery, and some other manufactured goods, Canada is primarily an exporter of raw or semifinished materials, energy, and agricultural products: wood pulp, timber, petroleum, natural gas, and hydroelectric power are among these. It is a major importer of manufactured goods, including machinery and equipment, motor vehicles, and consumer goods.

Another distinctive trait is Canada's overwhelming dependence on trade with a single partner, the United States. In 2007, the United States took in 82 percent of Canada's exports and supplied 55 percent of Canada's imports. No other developed country comes near these percentages of trade with just one other country.

Still another difference is the degree to which the Canadian economy is financed and controlled from outside the country; Canada is again overwhelmingly dependent on the United States. About two-thirds of all foreign investment in Canada is American. The extent of U.S. economic dominance in Canada has long been a sore point with many Canadians, who resent suggestions that their country is an American "economic colony." These irritations are heightened by the degree to which American mass-produced culture has permeated Canada and by how little Americans seem to know or care about Canada. Few U.S. schools offer courses on Canada, and a high percentage of American college students cannot even name Canada's capital (Ottawa). American news media give little coverage to events in Canada.

For some time, a movement has been afoot in Canada to achieve a more predominant position for Canadians within their own economy, to strengthen economic relations with industrial countries other than the United States, and to heighten cultural self-determination. But there seems to be no way for Canada to pull away from the United States without risking its own prosperity. The country's economy has long been geared to a close interchange with the gigantic economy next door. This relationship became even more deeply rooted with the North American Free Trade Agreement (NAFTA), passed in 1994.

544

Canada has a regional structure with some special problems. The country is divided between a thinly settled northern wilderness occupying most of Canada's area and a narrow, discontinuous band of more populous regions stretching from sea to sea in the extreme south (see Figure 11.2, page 599). These southern regions are very different from one other and sometimes conflict politically with each other and with the federal government in Ottawa. Following is a survey of the Canadian regions depicted in •Figure 11.1.2.

11.2 Atlantic Canada: Hardscrabble Living

The easternmost of Canada's four main populated areas has strips and pockets of population, mainly along and near the seacoast in the provinces of Newfoundland and Labrador, New Brunswick, Nova Scotia, and Prince Edward Island. These four provinces are known as the Atlantic Provinces, and three of them (excluding Newfoundland and Labrador) as the Maritime Provinces. Mountainous wilderness separates their main populated areas from the more populous and more prosperous Canadian core region in Québec and

URBAN AREAS
Name Over 10 million
Name 1 million to 10 million
Name Under 1 million*
* selected cities

KEY TO NUMBERED PROVINCES
1 Prince Edward Island
2 New Brunswick
3 Nova Scotia

• **Figure 11.1.1** Principal features of Canada and Greenland.

Ontario. Prince Edward Island, Canada's smallest and most densely populated province, is a lowland with farms and small settlements. It prides itself on taking the lead in Canada's ambitious alternative-energy future: the island wants to meet 30 percent of its energy needs from renewable sources, mainly wind power, by 2016. The main populated areas of New Brunswick, Nova Scotia, and Newfoundland are in valleys and upland coastal strips.

Politically, Newfoundland and Labrador includes the territory of Labrador on the mainland and the large Newfoundland Island opposite the mouth of the Saint Law-

rence River. Formerly a self-governing dominion within the United Kingdom, in 1949, Newfoundland (which changed its name to Newfoundland and Labrador in 2001) became the last province to join Canada. Bleak and nearly uninhabited Labrador, situated on the Canadian Shield, is physically part of Canada's North and is important for revenues from hydroelectric production, iron mining, and, most recently, nickel mining.

Isolation from markets has afflicted many major enterprises in the Atlantic region. The market in the region is small and divided into clusters of people separated by ex-

● **Figure 11.1.2** Canada's regions.

panses of sparsely populated land. The largest city in the Atlantic Provinces—Halifax, Nova Scotia—has a metropolitan population of about 400,000. Only one other city—Saint John's, Newfoundland, has more than 125,000 people. The entire four-province area has only 2.2 million people, or fewer than metropolitan Toronto or Montréal, The region is home to a large French population, especially in New Brunswick. During the 20th century, it was one of the poorest of Canada's regions and therefore experienced out-migration. Its poverty was somewhat surprising, given the region's geographic advantage of being at the mouth of a major navigable river (the Saint Lawrence Seaway) going into the heartland of a very prosperous country.

Conditions were not always like this. During the first two-thirds of the 19th century, the Maritime Provinces were a relatively prosperous area with a preindustrial economy. Local resources supported this development. There were harbors along the indented coastlines, fish for local consumption and export, timber for export and for use in building wooden sailing ships, and some agricultural land. The region had extensive trade with Britain and the West Indies. Then a period of economic decline set in. This area's ports were so far from the developing interior of the continent that most shipping bypassed them in favor of the Saint Lawrence ports or the Atlantic ports of the United States. Halifax in Nova Scotia and Saint John's in New Brunswick are the main ports of the region today, but neither is in a class with Montréal or the main U.S. ports.

The fishing industry was long the economic backbone of Canada's Atlantic region. Before the beginning of European settlement in the early 17th century, and probably before Columbus arrived in the Americas in 1492, European fishing fleets worked these waters, primarily on the **banks** along and near these shores. This area lies relatively close to Europe, and its waters were rich in fish.

But the bumper crop could not last forever, and fishing could not provide the basis for a prosperous modern economy. The United States imposed tariff barriers against Canadian fish to preserve its own domestic market. Com-

petition from newly developed fishing areas grew in other parts of the world. Fishing fleets from other nations, often with high-tech equipment, began to drive Canadian fishers from their own productive waters. The combination of expanding numbers of fishers, more sophisticated and effective fishing technology, and steady overfishing led to a major and continuing decline in the number of fish in this traditionally rich area. The cod that were once the mainstay of this region's productivity are nearly depleted, having declined more than 25 percent just between 2001 and 2006 (see Problem Landscape, page 634, and also page 31 in Chapter 2). Populations of flounder, haddock, herring, and lobster are also in calamitous decline.

Atlantic Canada is a difficult environment to farm. The four provinces lie at the northern end of the Appalachian Highlands and are mainly hilly. Under a cover of mixed forest, most soils are poor. Agriculture is concentrated in small patches of the best land (●Figure 11.1.3). Climate also makes agriculture difficult. The Maritime Provinces have a humid continental short-summer climate, and Newfoundland has a subarctic climate. The entire region is humid and windy, with much cloud cover and fog. Strong gales are frequent in the winter. Summers are cool, and colder and less hospitable conditions increase with elevation; a good part of upland Newfoundland is tundra. Although most areas in the Maritimes and Newfoundland are agriculturally marginal, there are some lowlands with better soils, especially in Prince Edward Island and the Annapolis-Cornwallis Valley of Nova Scotia.

Railroad construction also undermined agriculture in the Maritimes. The rails prompted settlement of Canada's interior and brought cheaper farm products into the Atlantic region. Today, quite a few people in the Maritime Provinces still farm (especially potatoes, dairy products, and apples), but they do so on a small-scale, part-time basis while earning their living from other jobs—a pursuit known as "hobby farming" in the United States.

Wood industries in the Maritimes suffered as overcutting depleted forests, ships began to be made of iron and steel instead of wood, and competition set in from new forest industries farther west. Pulp and paper manufacturing is making a comeback with the expansion of the international market for paper and wood products.

Having suffered economic setbacks related to natural resources, people of the Atlantic tried to develop industry. Many small cotton-textile factories were built during the 19th century, but they failed because of competition from mills in New England, central Canada, and Europe that were closer to major markets. A coal-mining industry and an iron and steel plant were developed in Nova Scotia. But in time, the plant closed and the coal mines declined.

Atlantic Canada is finally making an economic comeback in the current century. Canadian and American tourists alike are drawn to the region's stark beauty and picturesque villages. The export of electricity to adjacent provinces and the United States, especially from Labrador's large Churchill Falls hydropower installation and

PROBLEM LANDSCAPE

Overfished Waters

For many centuries, the western North Atlantic Ocean was a very rich fishery. Elevated portions of the sea bottom known as banks are located off the Atlantic coast from near Cape Cod to the Grand Banks, which lie just off southeastern Newfoundland (●Figure 11.A). Shallow depths and the mixing of waters from the cold Labrador Current and the warm Gulf Stream favor the growth of plankton, the tiny organisms on which many fish feed. These conditions helped put the Grand Banks at the heart of the fishing economy in Canada's northeastern region for nearly 500 years. Early fleets fished mainly for bottom-dwelling cod, which were cured on land before being shipped to European markets. Both the British and the French established early fishing settlements in Newfoundland, but the British drove out the French in 1763.

In 1977, worried about the effects of overfishing, the Canadian government began to enforce a 200-mile (322-km) offshore jurisdiction that prohibited foreign competition in the fishing of the Grand Banks. This delayed a crisis, but increasingly efficient fishing technology propelled this industry toward more difficult times. In the 1990s and early 2000s, overfishing of cod and other species in the banks led to a serious decline in fishing productivity. The Canadian government imposed catch quotas that were enormously unpopular with the fishing industry. After more than a decade of restrictions on fishing, the government in 2003 completely closed several cod-fishing grounds in Newfoundland waters in an effort to stop the slide. Tens of thousands of fishers and fish processors lost their jobs. The fishing tradition that was for so long a major component of the economic base for Atlantic Canada appears to have suffered a great setback and may be restored only if fish stocks can themselves be replenished. Thus far, the ban on fishing has failed to restore the fish populations, for reasons that scientists have not yet determined.

● **Figure 11.1.A** The Grand Banks.

● **Figure 11.1.3** This fertile valley in the Gaspé Peninsula is one of Atlantic Canada's rather few places for productive agriculture.

from New Brunswick, is helping. Most important, Newfoundland and Labrador has been experiencing a boom in offshore oil production and onshore refining. The offshore platforms do not require large numbers of workers, and so Atlantic Canadians are still leaving the region, relocating especially to the booming "oil patch" around Alberta's tar sands (see page 620–621). But the surge in the petroleum and nickel industries, again linked to China's commodities appetite, doubled the value of Newfoundland and Labrador's economy between 1997 and 2007. Federal subsidies also help sustain the Maritime Provinces.

11.1.3 Canada's Core Region: Ontario and Québec

The core region of Canada, with about two-thirds of Canada's total population, has developed in a narrow strip along Lakes Erie and Ontario and seaward along the Saint Lawrence River. In this area, the Interior Plains of North America extend northeastward to the Atlantic. Northern parts of the two provinces that share the core region—Ontario and Québec—include large and little-populated expanses of the Canadian Shield. Québec has a strip of Appalachian country along its border with the United States. But most of the core region lies in lowlands bordering the Great Lakes and the Saint Lawrence River from near Windsor, Ontario, to Québec City, Québec. The entire lowland area is known as the Saint Lawrence Lowlands. The Ontario Peninsula between Lakes Huron, Erie, and Ontario is often recognized as a separate section.

Ontario and Québec account for about 40 percent of products marketed from Canadian farms. Three-fourths of Canada's total manufacturing is in this core region, but in the pattern of deindustrialization also seen recently in Europe and the United States, China's output has caused factories here to shut down; household appliances, electrical equipment, plastics, rubber, and textiles have all been hit hard. The great majority of people work in the service industry, reflecting the region's status as the business and political center of the country.

The two sections of this vital region differ in their agricultural, industrial, and cultural features. The Ontario Peninsula, which has strong British roots, mirrors the U.S. Midwest's crop belts, with corn and livestock production, dairy farming, and specialty crops like tobacco. Ontario, where about 85 percent of the people speak English as a first language and only 5 percent speak French as a first language, is the most populous of the nine mainly English-speaking provinces.

In Québec, 82 percent of the people speak French as a preferred language and 8 percent speak English (•Figure 11.1.4). This is the only province in which French speakers predominate and control the provincial government, and Québec's potential devolution or even independence from Canada casts a long shadow over Canadian affairs (see

Problem Landscape, page 636). Quebec's Catholic culture historically gave the province a higher rate of population growth, producing surplus labor that attracted lower-wage industries and made the province somewhat poorer than Ontario. The French heritage is clearly evident in both rural and urban landscapes, including the French **long-lot pattern** of strip-shaped farms and the region's large Catholic churches (•Figure 11.1.5). Québec's agricultural endowment is not as great as Ontario's. It has a harsher winter climate, and dairy farming is the main agricultural livelihood.

Canada's core area has more than 20 million people, most in the country's two largest cities: Toronto, in

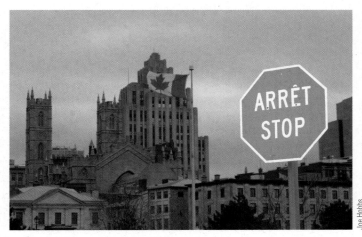

• **Figure 11.1.4** French and English multilingualism is characteristic of Canada, especially in the east.

• **Figure 11.1.5** Montréal's lovely Basilica of Notre Dame, built in the early 1800s and inspired by Paris's Sainte-Chapelle.

PROBLEM LANDSCAPE

The Québec Separatist Movement

Citizens of more developed countries are generally unaccustomed to the prospect that a major region of their own country might break off to form an independent nation. This is a real prospect for Canadians, however, if the Québec separatist movement has its way. The modern separatist movement has its roots in early Canadian history: the French were early settlers in these lands, but after a major British military victory in 1763, the broad sweep of language, government, and authority became decisively British. A cultural dichotomy developed that gave the French dominant influence in Québec but established only modest influence for them beyond the Saint Lawrence Lowlands.

After the end of World War II, French Canadian dissatisfaction spread, partly as attention to the plight of minorities grew worldwide. In Québec, this led to the formation of a separatist political party, the *Parti Québécois (PQ),* dedicated to the full independence of Québec from Canada. In 1976, the party gained control of Québec's provincial government and made French the language of commerce, requiring that all immigrants to the province be educated in French. Québec refused to ratify the country's 1982 constitution. In 1980 and again in 1985, the separatists forced the country to hold referenda on Québec's independence; both failed. In 1995, the issue was taken to provincial voters again, and those who favored staying a part of Canada again won, but by a very small margin (51 percent). In the referendum's wake, the federal government officially recognized Québec as a "distinct society" within Canada.

Although the PQ suffered a political setback in 2007, losing ground to a party favoring autonomy rather than independence for Québec, the issue of separatism is still very much alive and will likely stay on the national agenda. Separatists are frustrated by the steady influx of non–French-speaking immigrants seeking jobs and settlement in Québec City and other cities along the Saint Lawrence Seaway, who tend to vote with the pro-English bloc or otherwise have no stake in Québec's bid for independence. Québec can ill afford shutting the doors to immigrants, who now make up more than 10 percent of the province's electorate. The province's historically high birth rates have declined, and future growth in the workforce will have to come from outside.

The effects of the separatist controversy have been considerable. The Canadian government has made French an official language of Canada—along with English—all across the nation. Laws now stipulate that the children of French Canadians born outside Québec may have their children educated in French. Meanwhile, other Canadian provinces have used the separatist strategies to gain more control of their own provincial governments. As Ottawa has given French Canadians more authority in Québec in the hope that they might find the idea of full independence less attractive, other provinces, especially in the Prairie region, have come forward with their own demands for autonomy.

620

Ontario, and Montréal, in Québec—the second-largest culturally French city in the world, after Paris. Other large cities of the core include Ottawa (population 1.1 million), Québec City (700,000), and Hamilton (700,000). Ottawa, the federal capital, is just inside Ontario on the Ottawa River boundary between the two provinces. Québec City, the original center of French administration in Canada, is the capital of Québec province. Hamilton, a Lake Ontario port, is the main center of Canada's steel industry, supplying steel for the automobile and other metal-fabricating industries of Ontario.

In its evolution to industrial dominance, the Canadian core profited from a number of clear advantages over the rest of Canada: superior position with respect to transport and trade, accessibility to the largest markets, advantageous labor conditions, and access to significant resources for industry, commercial agriculture, and forestry. The region has abundant energy. On their descent from the Canadian Shield to the plains, many high-volume, steeply falling rivers drive one of the world's largest concentrations of hydroelectric plants. Major hydroelectric plants are at Niagara Falls, along the Saint Lawrence River, and at sites on

the Canadian Shield, including the James Bay plants discussed later in this module (•Figure 11.1.6). Coal requirements are met from nearby Appalachian fields in the United States, and oil comes from Atlantic or Middle Eastern ports by tanker up the Saint Lawrence. Both oil and natural gas come by pipeline from western Canada.

Among the core region's greatest geographic advantages are its lowland and water connections between the Atlantic and interior North America. Rapids on the Saint Lawrence River long barred ship traffic upstream from Montréal, but railways were built in the lowland along the river, and canals constructed in the 19th century bypassed the rapids to allow smaller ocean vessels to enter the Great Lakes. Finally, in the 1950s, the river itself was tamed by a series of dams and locks in the **Saint Lawrence Seaway Project** (•Figure 11.1.7). Since then, larger ships have been able to reach the Great Lakes via the river. The Welland Canal, which bypasses Niagara Falls between Lakes Ontario and Erie by means of a series of stair-stepped locks, predated the seaway. The canal admits shipping vessels to the four Great Lakes above the falls. Farther up the lakes, the Soo Locks and Canals allow ships to pass between Lake Hu-

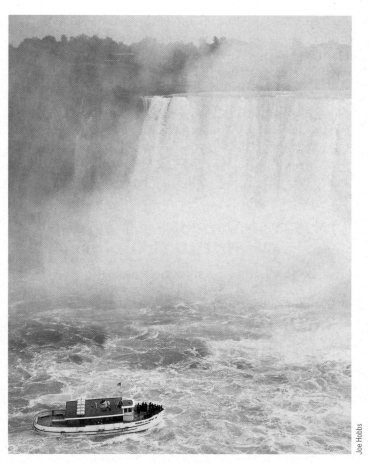

• **Figure 11.1.6** Niagara Falls, from the Canadian side.

ron and Lake Superior, and natural channels interconnect Lakes Erie, Huron, and Michigan. The total length of the Saint Lawrence Seaway and associated waterways is 2,342 miles (3,769 km). It allows cities on the Great Lakes both in Canada and the United States to serve as international ports, with wheat, iron ore, petroleum, and durable goods shipped out from about mid-April to mid-December.

The Saint Lawrence Seaway system has changed the commercial geography of North America's heartland in many ways. Through unwanted events, it has also changed the ecology of the region, particularly with the arrival of zebra mussels. The mussels, which originated in the Caspian Sea, diffused into these interior waters in the 1980s and have become a major blight on shipping and machine operations. The bivalves attach themselves to the intake pipes of irrigation, drinking-water, and power generation systems, impeding water flow and damaging machinery. They also deplete food resources for native marine life.

11.1.4 The Prairie Region: Oil, Wheat, and Wilderness

Manitoba, Saskatchewan, and Alberta make up the Prairie Provinces (see Figure 11.1.2). Isolation is one factor they have in common. Their peoples are separated from other populated areas by hundreds of miles of thinly inhabited territory. To the east, the Canadian Shield separates them from the populous parts of Ontario, and on the west, the Rocky Mountains and other highlands separate them from the Vancouver region. To the north lies subarctic wilderness, and to the south are sparsely populated areas in Montana, North Dakota, and Minnesota.

• **Figure 11.1.7** The Saint Lawrence Seaway, more than 2,300 miles (c. 3,700 km) long, connects the Atlantic Ocean with the Great Lakes in the heart of the continent. This makes interior cities like Duluth, Minnesota, and Thunder Bay, Ontario, international ports.

The prairie environment that gives the provinces their regional name exists only in some of their southern sections, within the plains between the Canadian Shield and the Rockies. There is a triangle of natural grasslands, with its southern base resting on the U.S. boundary, extending to an apex about 300 miles (c. 500 km) north of the border. On all sides except the south, vast reaches of forest border the grasslands. Most of the southern part of the triangle is a northward continuation of steppe grasslands of the U.S. Great Plains. Toward the forest edges, the soils are moister, and original settlers there found true prairies: taller grasses with scattered clumps and riverine strips of trees. Settlers were most attracted to these prairies, and these moister grasslands still form a more densely populated band near the forest edges.

Even though less than 5 percent of Canada's total area is suited to farming and although only about 3 percent of the Canadian labor force farms, Canada is a major world wheat exporter because of the Prairie region's soils and the spring wheat crop they support. The region produces about half the value of Canada's farm products. Besides wheat, there is productive farming of barley, rapeseed (for fodder and for a vegetable oil called canola), and livestock, especially cattle.

Minerals are the foundation of the recent advances in the Prairie Provinces. More than 60 percent of Canada's mineral production is here. Coal underlies the Rocky Mountain foothills and the plains in Alberta and western Saskatchewan and is exported from the region. Nickel, copper, zinc, and other metals come from the Canadian Shield of Manitoba. The Saskatchewan Shield yields uranium, and southern Saskatchewan produces major quantities of potash.

Petroleum and associated natural gas have had the most economic impact on the Prairie region. The region's oil industry began near Calgary, Alberta, in the 1940s and led to explosive growth of both Calgary and Edmonton. The industry is still mainly in Alberta, with smaller portions in Saskatchewan and southern Manitoba. An important product extracted from natural gas in Alberta is sulfur, of which Canada is a major world exporter. Most critical of all are the Athabascan tar sands of northern Alberta. Extraction of this unconventional source of energy has surged as conventional oil prices have risen to record levels, turning Alberta's Fort McMurray into a cosmopolitan boom-town of former east Canadians, Latinos, Middle Eastern-ers, South Asians, Filipinos, and Chinese.

Urbanization and urban industries have become increasingly important in this region, especially in Alberta and Manitoba. The three main metropolises are Edmonton (population 1 million) and Calgary (population 1.1 million) in Alberta and Winnipeg (population 600,000) in Manitoba. These cities have all grown near the corners of the Prairie triangle where connections to the outside world are focused. Winnipeg developed where the rail lines through the Canadian Shield crossed the Red River and entered the Prairie region, Calgary grew near a pass over the Rocky Mountains that gave a route toward Vancouver, and Edmonton was established near another Rocky Mountain pass leading to the Pacific. Both Winnipeg and Edmonton also developed as metropolitan bases for huge sections of Canada's North.

None of the three Prairie Provinces is a major manufacturing center. The environmental setting has turned this region into an important tourist destination for Canadians and Americans alike. Calgary is the eastern gateway to the spectacular Rocky Mountain region, and its annual Calgary Stampede, which celebrates the region's colorful history of farming and ranching, is one of Canada's most famous and most lucrative tourist attractions.

11.1.5 The Vancouver Region and British Columbia

Canada's transcontinental population belt (see Figure 11.2) is anchored at the Pacific end by a small region centered on British Columbia's seaport of Vancouver (•Figure 11.1.8). Metropolitan Vancouver has 2.1 million people, and another 300,000 reside in the metropolitan area of the province's capital city, Victoria. Vancouver is located on a superb natural harbor at the mouth of the Fraser River. This river valley provides a natural route for railways and highways coming from the rugged mountains and plateaus in Alberta and British Columbia. Known as the most English of Canadian cities, Victoria is located at the southern end of Vancouver Island, the southernmost and largest of a chain of islands along Canada's Pacific coast.

Victoria was the leading commercial and industrial center in British Columbia before Canada's first transcontinental railroad, the Canadian Pacific, was completed in 1886. Vancouver then became the country's main Pacific port, and it has recently become the country's largest seaport. It has profited from Canada's increasing exports of resource commodities to East Asia, especially China and Japan. This city served as a major migration destination for Hong Kong Chinese in the years preceding the 1997 British handover

• **Figure 11.1.8** The jewel of the Canadian west coast is Vancouver, British Columbia.

of Hong Kong to the People's Republic of China. Vancouver's population is about 20 percent ethnic Chinese. Tourists from Asia (mainly Japan, Korea, and China) and the United States come to Vancouver to enjoy its cosmopolitan and scenic attractions and to take cruises up the spectacular coastlines of British Columbia and Alaska. Some visitors like the liberal atmosphere; the city's tolerance of cannabis consumption has earned it the nickname "Vansterdam." Vancouver also gained renown in a recent survey as the "world's most livable city."

British Columbia, along with the Yukon Territory to its north and a narrow fringe of Alberta, is mountainous and forested (•Figure 11.1.9). The fjorded Coast Ranges along the Pacific and the Rockies of the interior are high and spectacular, and the Intermountain Plateaus between them are rugged. Only the northeastern corner of the province features a plains environment. This is Peace River country, the extreme northern outpost of the Prairie region of grain and livestock agriculture.

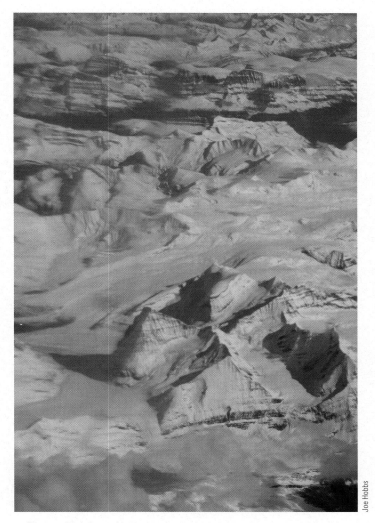

• **Figure 11.1.9** Mid-October in the Rocky Mountains northwest of Banff National Park, in spectacular British Columbia.

British Columbia is distinguished climatically within Canada by its coastal strip of marine west coast climate. Here moderate temperatures and heavy precipitation support one of the world's largest coastal temperate rain forests. The mild temperatures and spectacular scenery attract ecotourists and retirees. The forest resources are attractive to the logging industry, but a landmark agreement reached in 2006 protects a vast area of the Great Bear Rain Forest, allowing selective cutting only in the least sensitive areas. As on the Coast Ranges, the interior is largely forested, but most of its the old growth forest has been logged. The climate of the interior is subarctic in the north and varies according to elevation and exposure in the south.

Scattered small communities exploit the abundant natural resources of "B.C.," as it is sometimes affectionately known. Forest products come from both the interior and the coast. The timber industry is not as valuable as it once was, owing to competition from tropical countries like Brazil and Malaysia, which can export their hardwoods at lower cost. Salmon fishing continues along the coast as it has for more than a century. **Fish farming**, especially the raising of salmon, has become a very important industry to British Columbia and is posing a serious economic challenge to the "mom and pop" fishers of wild salmon in Canadian and Alaskan waters. With farmed salmon abundant, prices have plunged, so the traditional fisherfolk are touting the better health benefits and taste of their wild product in the hope of strengthening its market niche.

More important than forestry and fishing today is a boom in the mining of coal, natural gas, petroleum, and metals (copper, molybdenum, silver, lead, and zinc). Older metal production sites in the south are being supplemented by operations farther north. The driving factor is Chinese, Japanese, and U.S. market demand, with development being supported by foreign capital. Coal is shipped mainly to East Asia, and electricity generated from the region's abundant waters is exported to the United States. Near the coastal town of Kitimat, the firm Alcan uses hydroelectricity to produce aluminum from imported alumina (the second-stage material from bauxite ore), for exports mainly to Asia. Wages tend to be high in the small and isolated production nodes scattered over British Columbia, but so is the cost of living, and cultural amenities are few. Most new residents of British Columbia settle in cosmopolitan Vancouver or in Victoria.

11.1.6 The North: Lots of Land, Few People

About four-fifths of Canada has cold tundra and subarctic climates, with more than one-half in the glacially scoured and agriculturally sterile Canadian Shield, and a Pacific region dominated by rugged mountains (•Figure 11.1.10). These conditions create the huge expanse of nearly empty land that Canadians call "the North." More than one-half

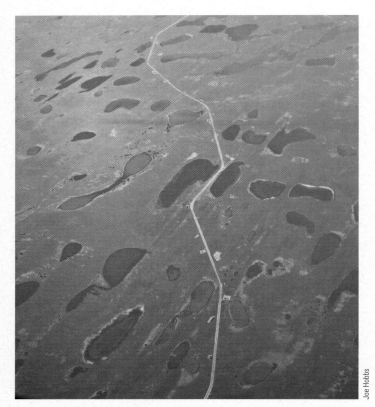

● **Figure 11.1.10** Canada's wild northland: the glaciated, lake-rich Canadian Shield of the Northwest Territories province.

of it lies in territories administered from the country's capital in Ottawa, but large parts extend into seven of the 10 provinces stretching across southern Canada. In this book, the southern edge of the subarctic climate zone is used to define most of the North's southern boundary, but the island of Newfoundland is excluded because of its Atlantic associations and denser population.

The Near North

The southern fringes of the North, often called the **Near North** (see Figure 11.1.2), have widely scattered pockets of development that are important to Canada's economy and, through trade, to the economy of the United States. Many of these are mining settlements. The Canadian Shield here has a wide variety of metallic ores, notably nickel and copper around Sudbury in Ontario, iron ore in Québec and Labrador, and uranium just north of Lake Huron and in northern Saskatchewan.

Some northern settlements manufacture pulp and paper and smelt metals. Towns producing wood pulp and paper are scattered across the southern fringes of the North. A string of towns along the Saguenay River tributary of the Saint Lawrence in Québec is supported by pulp and paper

plants, by farming, and by the manufacture of aluminum from imported raw materials brought in by ship. The other urban concentration is at Thunder Bay, Ontario, on the northwestern shore of Lake Superior. In addition to pulp and paper, Thunder Bay benefits from being at the Canadian head of Great Lakes navigation. Its port links the Prairie Provinces with the Ontario-Québec core area, the United States, and overseas ports.

Water resources are very important in the southern fringe of the North. Hydroelectricity supports local mines and mills and is sent to southern Canada and the United States. Rivers and lakes are so characteristic of the Canadian Shield that from the air it appears to be an amphibious landscape. There is a cluster of hydroelectric plants where Saint Lawrence tributaries from the North rush down through the Laurentide "Mountains"—the raised southern edge of the Shield in Québec. But increasing demand, plus increasing ability to transmit electricity for long distances, has led to major developments much farther north. These are located near James Bay (between Ontario and Québec) and in Manitoba near Hudson Bay.

Hydropower development is especially controversial in the James Bay region because of its potential impacts on indigenous people and natural environments. In the 1990s, Québec's state-owned power utility, Hydro-Québec, pursued development of the **James Bay Project** to dam and divert a number of the area's wild rivers to generate electricity. The region's Native American **Cree** concluded that the diversions, reservoirs, and other changes would be detrimental to their caribou hunting, fishing, and other forms of livelihood and would inundate much of their tribal homeland. They waged an international public relations battle against Hydro-Québec. Their campaign in the U.S. media compelled New York's governor to cancel a contract to buy some of the electricity that would have been produced by the new developments. Hydro-Québec eventually canceled plans to build a dam on the Great Whale River, which would have had the most detrimental impact.

However, in 2002, the Cree reached a settlement with the Québec government in which they have agreed to allow the building of hydropower dams on two of the region's less significant rivers. The Cree will also drop billions of dollars in environmental lawsuits against the government. For the Cree, the deal includes cash payouts of $70 million a year for 48 years. The Cree will also obtain more control over their communities and economies, more power over decisions about logging, and more jobs with Hydro-Québec.

A remarkably small number of people live in this vast area. Agricultural settlements are few, and most are declining. The port of Churchill on the western edge of Hudson Bay serves as the closest port for movement of the Prairie Provinces' wheat to the European market. It is also well known for a very specialized form of tourism, polar bear watching during the late fall season when the great predators linger on the Lake Hudson shoreline, awaiting the formation of the sea ice they will travel on through the winter.

Territorial Canada: Indigenous Landscapes

North of 60 degrees latitude lies a great part of Canada without provincial status. This wild area, known as Territorial Canada, is under the administration of the Canadian federal government and of indigenous peoples. There are three administrative territories (see Figure 11.1.1). In the northwest, the Yukon Territory is cut off from the Pacific by the panhandle of Alaska, once claimed by Canada but acquired by the United States in 1867. It is mostly rugged plateau and mountain country. The much larger Northwest Territories lie east of Yukon Territory. This area includes a fringe of the Rockies; a section of the Arctic Coastal Plain; the Mackenzie River, which flows northward to the Arctic Ocean; a vast expanse of plains and hills in the Canadian Shield; and numerous arctic islands. Two huge lakes, Great Bear Lake and Great Slave Lake, both drained by the Mackenzie River, mark the western edge of the Shield. The Yukon Territory and the western part of the Northwest Territories have mainly a subarctic climate and taiga vegetation, with tundra near the Arctic Ocean, but in the eastern part of Canada, tundra extends far to the south (see Figure 11.5). The territory known as Nunavut occupies eastern and northern lands withdrawn from the Northwest Territories in 1999 to form a semiautonomous region inhabited by Inuit peoples.

Europeans and Anglo Americans have been seeking resources in these far northern reaches since the 17th century, when England's Hudson's Bay Company first established fur-trading posts on the shore of Hudson Bay. The chief 20th-century quest was for metallic and other minerals. Exploration for oil and gas is continuing in the Mackenzie Valley and in the Beaufort Sea section of the Arctic Ocean near the Mackenzie Delta. Several diamond mines have opened in the Northwest Territories and adjacent Nunavut, and exploration for copper, zinc, gold, iron ore, and other minerals is ongoing. Future development of these resources is controversial because of potential environmental impacts. Among the issues under scrutiny is a plan to pipe natural gas from the Mackenzie Delta to Alberta, where it would provide energy for the extraction of oil from tar sands. In addition to income from the few mining enterprises, the territories also receive funding from federal government payrolls in the territorial capitals of Whitehorse (Yukon), Yellowknife (Northwest Territories), and Iqaluit (Nunavut; the town's Inuit name means "Place of Many Fish").

The population of Territorial Canada was only about 100,000 in 2007. Fewer than 2 percent of the inhabitants in each of the present territories are native First Nation peoples (indigenous groups other than the Inuit). In the Northwest Territories, about 9 percent—some 4,000—are Inuit, and in Nunavut, about 50 percent are Inuit. Most whites are government, corporate, or church employees, and they are often transient, working only until they can return south. First Nation peoples and Inuit still carry on some hunting, gathering, and fishing from homes in small fixed settlements. The trade in furs of seals, marten, beaver, fox, and other mammals was once vital to their economies, but international protests about the methods by which the animals were caught and killed led to a decline in consumer appetites for fur products and to legislation by the European Union and other governments against fur imports.

Most Inuit live in prefabricated government-built housing in widely separated villages along the Arctic Ocean and Hudson Bay (•Figure 11.1.11). Unemployment among both First Nations groups and Inuit is very high, and a large proportion live on welfare. Federal efforts to increase standards of living have included a network of schools and medical clinics across the area. Air transportation enables the seriously ill to be flown to the few hospitals in the region or to hospitals in cities such as Edmonton or Winnipeg. In Iqaluit, the capital of Nunavut, located in the southernmost part of remote Baffin Island, the school system is working to diminish the sense of isolation through an ambitious Internet linkage with schools all over the world. Education still has a big hurdle to overcome in Nunavut: instruction through fourth grade is in the Inuit language, and then students are required to switch to English, for which many are

• **Figure 11.1.11** Indigenous people still practice subsistence hunting in Canada's North. Here the hides of Arctic wolves, shot in the winter by this woman's snowmobile-mounted husband, cure in the summer sun in the village of Arviat on Hudson Bay. Canada's Inuit generally live in prefabricated houses like the one seen here. Fuel oil for household heating and cooking comes in oil drums brought by ship.

PROBLEM LANDSCAPE

Race for the Arctic

Canada has a vast arctic territory and wants to exploit it—and areas even farther north, if possible—to maximum advantage. If global warming continues on its present course—according to recent studies, the Arctic Ocean will be free of summer sea ice as soon as 2020—more resources will become readily available to Canada and other countries of the Arctic Ocean rim. The race is already on.

Explorers have long sought a **Northwest Passage** for ship navigation through the maze of Canada's arctic islands. Until recently, sea ice has made that passage impossible except under extraordinary conditions. But if arctic warming should open up a reliable route, this would be an enormous boon to shipping interests and also to Canada because it could charge shipping fees. This would be the counterpart to Russia's Northern Sea Route (see pages 184 and 195), allowing great time and cost savings: the distance traveled by ships between northeast Asia and Europe would be cut by 40 percent. The United States has rankled Canada by suggesting that the Northwest Passage should be regarded as international waters, not Canadian. Canada and the United States are unlikely to have a serious showdown over this issue. But in 2007, the Canadian prime minister stated that his country's attitude toward its arctic territory was "use it or lose it" and announced plans for construction of both more icebreakers and a deepwater Arctic Ocean port.

Canada and Russia are also talking about opening the **Arctic Bridge,** a shipping route from Canada's port of Churchill, on Hudson Bay, to the northwestern Russian port of Murmansk. That would be an eight-day voyage. Goods arriving in Churchill could be shipped quickly by rail south to ports on the Saint Lawrence Seaway that could otherwise be reached only after a 17-day sea voyage from Murmansk.

Then there are the Arctic Ocean's sea and seabed resources. As Arctic Ocean waters warm, new fishing grounds will open. One-fourth of the world's undiscovered oil and gas reserves lie in the Arctic, according to U.S. Geological Survey estimates. Five Arctic nations—Russia, Canada, Norway, Denmark, and the United States—are scrambling to stake their turf. In some cases, they are taking advantage of loopholes and legal gray areas in the Convention on the Law of the Sea (see page 407). One portion of the treaty allows a country to extend its 200-mile (320-km) exclusive economic zone (EEZ) from the edge of its undersea continental shelf rather than from its shoreline. On this basis and with the use of recent marine cartographic surveys, Russia has already laid claim to the North Pole, even using a submarine to plant a flag at the seabed spot 13,000 feet (c. 4 km) below the surface. Denmark uses Greenland's continental shelf as the basis for marking its EEZ, which also gives Denmark the North Pole. The United States is hampered by its refusal, dating to the Reagan administration of the 1980s, to ratify the Convention on the Law of the Sea (conservative lawmakers have regarded ratification as a subordination of U.S. sovereignty to a supranational entity, the United Nations). With so much at stake, negotiations and disputes among the Arctic Ocean countries will be ongoing for years to come.

● **Figure 11.1.B** Recent dramatic reductions in sea ice cover, and hopes of abundant resources, have Arctic Ocean countries staking territorial claims and drawing future trade routes.

---

unprepared. Other serious problems exist among Nunavut's overwhelmingly young population, with drug abuse, alcoholism, suicide, and family abuse all too common.

Meanwhile, across the North, the increasingly assertive First Nation and Inuit populations are pressing land claims to mineralized areas and pipeline routes, negotiating royalty arrangements with the natural gas and oil interests they once opposed, and demanding regulations to prevent further environmental disruptions by mining companies. There is a strong likelihood that resource extraction and trade in the area will increase if current climatic conditions persist (see Problem Landscape, above).

11.1.7 Greenland: A White Land

An unlikely cultural outlier of Europe, Greenland is geologically part of North America, and because it is closest to Canada, it is discussed briefly here (see Figure 11.1.1). The world's largest island (840,000 sq. mi./2,175,600 sq km), Greenland has been under Danish control since 1814 and was claimed by Norway for several centuries before that. It became a Danish province in 1953, but in 1979, the Danes granted the island self-government while retaining control over its foreign affairs. It lies more than 2,000 miles (3,200 km) west of the European nation with which it is associated. In 1985, Greenland withdrew from the European Economic Community (EEC, the precursor to the European Union) because of its frustration over EEC fishing policies and the lack of any EEC development aid. Today, however, the European Union gives Greenland almost $50 million each year in exchange for Greenland's granting of fishing quotas to EU member states in its territorial waters. Denmark subsidizes the massive island with an average support of more than $8,000 per person per year. There is increasing talk of independence among the Greenlanders, but severing the aid pipeline from Denmark and the European Union would make that a very costly proposition.

Greenland has nearly one-fourth the area of the United States, but about 80 percent of the island is covered by an icecap up to 10,000 feet (3,000 m) thick (•Figure 11.1.12).

Until recently, Greenland seemed to have bucked the worldwide trend toward melting of glacial ice, but as of 2007, it was losing as much as 80 cubic miles (c. 330 cu km) of ice each year, about three times the rate in 1996. Scientists are working to determine whether the melting is related to overall global warming and what its impacts on the "conveyer belt" of Atlantic oceanic waters and subsequent climates might be. Some models suggest that the ice sheet could shrivel completely over a period of about 1,000 years, which would raise the world's sea levels by 20 feet (6 m). With the ongoing melting, Greenland's geography is a work in progress. Maps of the coastline are outdated almost as quickly as they are drafted as new islands and inlets emerge from the thaw.

Ninety percent of Greenland's population is distributed along the southern coast. Greenland has a total population of just over 50,000, giving it a population density of less than one-tenth of a person per square mile. The capital and largest city is Nuuk (originally called Godthab; population 15,000) on the southwestern coast. The great majority of inhabitants are Inuit, with Danes making up about 15 percent of the population. The leading means of livelihood are fishing, hunting, trapping, sheep grazing, and the mining of zinc and lead. Under the auspices of NATO, joint Danish-American air bases are maintained there, and the United States has a giant radar installation at Thule in the remote northwest.

In the final module, we turn to the United States.

• **Figure 11.1.12** Greenland's spectacular glacial landscape.

SUMMARY

→ Canada is highly urbanized, with 79 percent of its population living in cities. Nearly one-third of Canadians live in Toronto, Montréal, and Vancouver.

→ Of all the world's more developed countries, Canada has the strongest economic relationship with a single country (the United States). About 82 percent of Canada's exports go to the United States. Canada has a major global role in the production of pulp and paper, hydroelectricity, commercial motor vehicles, and aluminum.

→ Ten provinces (excluding the territories and Nunavut) make up the political administration of Canada. More than one-half of Canada's area is in the North, with very low population densities. Nearly 90 percent of the Canadian population is settled within 100 miles (c. 160 km) of the U.S. border, with the largest bloc of population settled in the southeast.

→ Atlantic Canada consists of the provinces of Newfoundland and Labrador, New Brunswick, Nova Scotia, and Prince Edward Island. Cod fishing was long the mainstay of its economy, but the Grand Banks have become so overfished that government regulation has restricted the harvest, causing much unemployment. New revenues come from specialized agriculture, tourism, forestry, energy, and minerals.

→ A narrow strip across the southern parts of Ontario and Québec provinces makes up the core of Canada. Two-thirds of Canada's population live in this region. The most important economic resources are agriculture, forestry, energy generation, and bulk commodity transport.

→ Industrial development and trade moved upstream on the Saint Lawrence steadily until the completion of the Saint Lawrence Seaway in the 1950s. Québec City was outgrown by Montréal, and then Montréal was surpassed by Toronto in size and economic importance (but Vancouver on the west coast has grown to be Canada's biggest port).

→ The Ontario Peninsula is strongly British in origin and character, and Québec is strongly French, with 82 percent of the Québec population of French origin. Québec separatists have been a political presence for the past 40 years, and in 1995, the referendum for an independent Québec was defeated by less than 1 percent. But Québec has gained more provincial autonomy, and other provinces, especially in the Prairie region, have sought the same benefits by threatening to change their relationship with Ottawa.

→ The Prairie region is made up of Manitoba, Saskatchewan, and Alberta. Wheat, petroleum (including from tar sands), and coal are the major economic resources. Major urban centers include Edmonton, Calgary, and Winnipeg.

→ The Vancouver region is centered in Vancouver, British Columbia. More than half of the province's population is in the Vancouver area. It has a significant ethnic Chinese population, with trade increasingly focused on East Asia. British Columbia has become a major tourist and retirement destination.

→ The Canadian North is bounded at the south roughly by the southern limit of the region of subarctic climate. Population densities decline from south to north in this sparsely settled region. Economic resources include nickel, copper, uranium, petroleum, natural gas, forest products, and hydroelectricity. Nunavut, the Inuit homeland, is the newest political unit. The fur trade was once vital to the economies of the North's indigenous peoples, but a public backlash has all but destroyed this industry.

→ As temperatures warm in the arctic region, Canada is one of several nations capitalizing on new opportunities for trade and mineral exploitation. Several countries have claimed parts of the Arctic Ocean as sovereign territory, and some have conflicting claims to various areas, including the North Pole.

→ Greenland is geologically a part of North America but is a self-governing property of Denmark. Greenland's foreign affairs are controlled by Denmark, which, along with the European Union, spends heavily on subsidies for the 50,000 inhabitants of the world's largest island. Scientists are watchful of the potential melting of the massive icecap that covers much of the island.

KEY TERMS + CONCEPTS

Arctic Bridge (p. 642)
bank (p. 633)
Cree (p. 640)
fish farming (p. 639)

James Bay Project (p. 640)
long-lot pattern (p. 635)
Near North (p. 640)
Northwest Passage (p. 642)

Parti Québécois (PQ) (p. 636)
Saint Lawrence Seaway Project (p. 636)

REVIEW QUESTIONS

1. In what ways is Canada unique among more developed countries?

2. Why did fishing fail to provide the basis for a prosperous economy in Canada?

3. What are the major import and export patterns for Canada, and which are the most important goods and partners in the trade?

4. What environmental factors limit agriculture in Canada? Which areas are most productive agriculturally? Why is this so? What are their major agricultural products?

5. What two political units make up the core region of Canada? What are some of the major similarities and differences in their geographic characteristics?

6. What factors are behind Québec's push for independence? What are the prospects that this province might actually break away from Canada?

7. What waterways make up the Saint Lawrence Seaway system? What important roles do these waterways have in the economies of Canada and the United States?

8. What factors favored the development of Montréal?

9. What conditions have supported Canada's emergence as a major wheat producer?

10. What role has hydropower played in the development of Canada's Near North?

11. What new issues and potential conflicts have arisen as temperatures in the arctic region have increased?

12. With what nation is Greenland politically affiliated? Why do most Greenlanders want to maintain these ties?

13. What is happening to Greenland's extensive ice cover? What explains that phenomenon? What would be the result of a complete melting of that ice sheet?

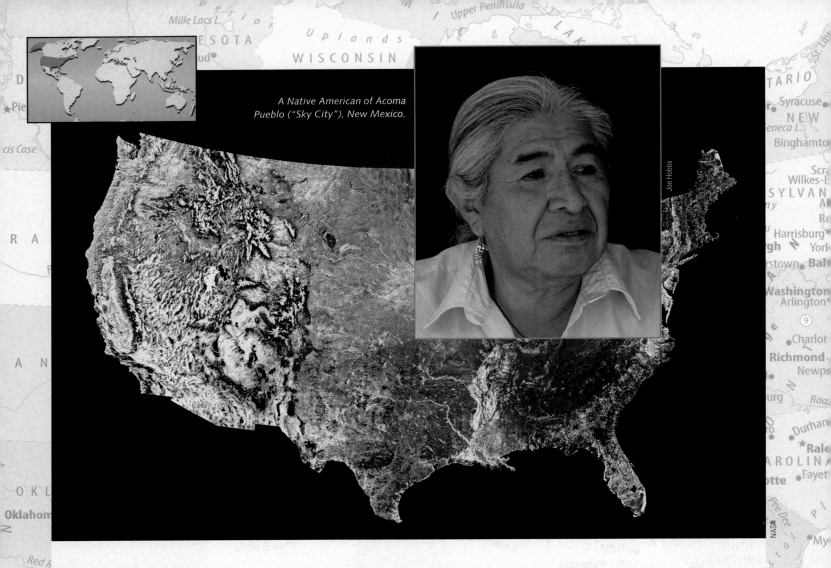

A Native American of Acoma Pueblo ("Sky City"), New Mexico.

Joe Hobbs

NASA

MODULE 11.2
THE UNITED STATES: OUT OF MANY, ONE

The book concludes with an examination of the country that is home to most of its readers (•Figure 11.2.1). This module offers Americans new insights into their home country and introduces this important nation to others. The emphasis here is on the geographic circumstances that led to the founding and growth of the main cities of the United States, and to the ways the country has been adjusting to new demographic and environmental realities at home and to fresh economic and other challenges abroad. The module is organized around five vernacular regions: the Northeast, South, Midwest, West, and Alaska and Hawaii.

■ module
outline

■ module
objectives

This module should enable you to

→ Recognize how rivers, topographic boundaries, and other geographic circumstances promoted the development of many American cities

→ Understand the environmental and political issues that complicate development of coal, oil, and natural gas in the United States but promote the use of biofuels

→ Follow the decline of traditional heavy industries in the process of deindustrialization and their replacement by high-technology and service industries

→ Evaluate the depopulation of the Great Plains, the decline of small towns, and the potential for communication technology and foreign immigration to reverse these trends

→ Gain more insight into the country's ethnic geography and immigration issues

11.2.1 The Northeast: Center of Power

Of the five regions of the United States, the Northeast is the most intensively developed, densely populated, ethnically diverse, and culturally intricate. It includes the nation's main centers of political and financial activity, and it is where most of the centers for relations with foreign nations are located (for example, the United Nations in New York City and the country's capital at Washington, D.C.; ●Figure 11.2.2). Strong traditions of intellectual, cultural, scientific, technological, political, and business leadership are rooted there. The country's political and economic systems began in the Northeast, and historically this is where most overseas immigrants arrived.

The Northeast includes six New England states (Maine, New Hampshire, Vermont, Massachusetts, Connecticut, and Rhode Island), plus five Middle Atlantic states (New York, New Jersey, Pennsylvania, Delaware, and Maryland), and the District of Columbia. Delaware and Maryland have important historical links with the states of the American South, but their character today is more like that of the Middle Atlantic states, so they are included in this region.

About 62 million people, or 20 percent of the national population, lived in the Northeast (5 percent of the nation's area) in 2007. The region's population density is many times that of the South, Midwest, or West. But the density

is much more extreme within a narrow urban belt stretching about 500 miles (c. 800 km) along the Atlantic coast from metropolitan Boston through metropolitan Washington, D.C. (see Figure 11.2, page 599). In this megalopolis, known as "Boswash" and the Northeast Corridor, seven main metropolitan areas contain about 40 million people: Boston (4.4 million), Providence (1.6 million), Hartford (1.1 million), New York (18.8 million, the country's largest city), Philadelphia (5.8 million), Baltimore (2.6 million), and Washington (5.2 million). These are major centers of political decision making, corporations, finance, sales, services, and manufacturing. 600

The Environment of the Northeast

Most of the Northeast lies in the Appalachian Mountains and its associated highlands, but small sections lie in the Atlantic Coastal Plain and the Piedmont (see Figure 11.4, page 604). Atlantic Coastal Plain areas include Cape Cod in Massachusetts, Long Island in New York State, southern New Jersey, the Delmarva Peninsula (which includes most of Delaware and Maryland's eastern shore), and parts of Maryland's western shore. The plain is low and flat to gently rolling, with sand dunes, marshes, and swamps. Soils are sandy and not very fertile. Broad, deep estuaries (low portions of rivers that have been drowned by the sea) indent this plain. The largest northeastern seaports are on the estuaries of the Hudson River (New York), the lower Delaware River (Philadelphia), and the Patapsco River (Baltimore) near the head of Chesapeake Bay.

Between the Coastal Plain and the Appalachians, the northern part of the Piedmont extends across Maryland, Pennsylvania, and New Jersey. The rolling Piedmont, with generally good soils, is higher than the Coastal Plain but lower than the Appalachians. The fall line boundary between these two physiographic regions of Piedmont and Coastal Plain (see page 605) required goods in transit by river to be offloaded and reloaded. Along with hydropower resources, this "break of bulk" requirement made the fall line a logical place to locate settlements. Along the line grew cities including Washington, Baltimore, Wilmington (Delaware), Philadelphia, Trenton (New Jersey), and the New Jersey suburbs of New York City. 376, 667

The old mountains of New England and northern New York form several ranges characterized by rough terrain, cool summers, snowy winters, and poor soils. The Adirondack Mountains of northern New York rise from lowlands as a circular mass. Heavily forested and pocked by numerous lakes, these glaciated mountains reach over 5,000 feet (1,525 m). They are bounded on the east by the Champlain Lowland (named for Lake Champlain), which extends northward into Canada as a continuation of the Hudson Valley. East of it, the low Green Mountains take up most of Vermont and run southward to become the Berkshire Hills of western Massachusetts and the Litchfield Hills of Connecticut.

● **Figure 11.2.1** Principal features of the United States.

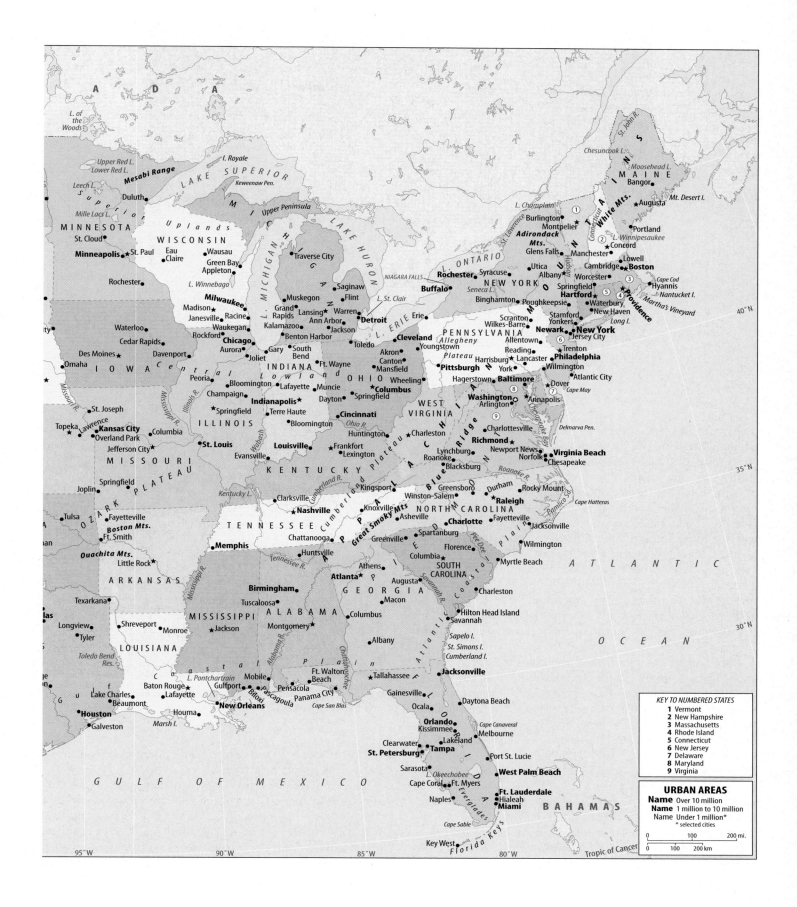

KEY TO NUMBERED STATES
1 Vermont
2 New Hampshire
3 Massachusetts
4 Rhode Island
5 Connecticut
6 New Jersey
7 Delaware
8 Maryland
9 Virginia

URBAN AREAS
Name Over 10 million
Name 1 million to 10 million
Name Under 1 million*
* selected cities

0   100   200 mi.
0   100   200 km

● **Figure 11.2.2** The National Mall in Washington D.C. , with some of America's most famous architecture: the Lincoln Memorial (in the foreground), the Washington Monument, and the U.S. Capitol.

Farther east across the narrow valley of the upper Connecticut River, the White Mountains take up northern New Hampshire and spill into Maine. Here, in the Presidential Range of New Hampshire's White Mountains, Mount Washington rises to 6,288 feet (1,917 m). This highest mountain in the Northeast is notorious for rapid weather changes. Treks to its summit have fooled climbers, sometimes fatally, because of sudden temperature drops and howling winds. Gusts of greater than 225 miles per hour (360 km/h) have been recorded on Mount Washington. There are other notable highlands besides Mount Washington; the Adirondacks and the mountains of New England are major recreation areas for the Northeast's urban populations, and their autumn colors lure tourists (●Figure 11.2.3)

Between the mountains and the sea, the hilly New England Upland is part of the Appalachians. Glaciers deposited stones across this area's soils. Early settlers cleared stones and trees to farm this area, using the stones to make New England's famous stone fences (these are like the hedgerows of Great Britain). Many have been torn down to make way for modern farming and urban development, but some fences still keep quiet guard where woodlands have overgrown abandoned farms.

The poor upland soils proved unable to support long-term commercial agriculture, and the late 18th and the 19th centuries saw a large migration of New Englanders from unproductive farms to northeastern cities and, especially with the completion of the Erie Canal in 1825, to more fertile agricultural areas to the west. This migration

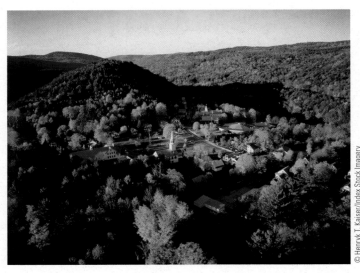

● **Figure 11.2.3** New England in the fall. Townshend, Vermont displays New England's postcard-perfect autumn beauty.

carried a New England cultural flavor across the continent, all the way to the West Coast.

The Appalachians are broken in New York by a narrow lowland corridor called the Hudson-Mohawk Trough, which is made up of the Hudson Valley, running from New York City northward to Albany, and the Mohawk Valley, which extends westward from Albany to Lake Ontario.

South of the Hudson-Mohawk Trough, the Northeastern Appalachians have three sections: the Blue Ridge in the east, the Ridge and Valley Section in the center, and the Appalachian Plateau in the north and west. They lie roughly parallel, with each trending northeast-southwest. The Blue Ridge—long, narrow, and characterized by old igneous and metamorphic rocks—has various local names. Often it forms the single ridge its name suggests, but it broadens into many ridges in the South. West of it, the Ridge and Valley Section, characterized by folded sedimentary rocks, has long, narrow, and parallel ridges trending north-south and separated by narrow valleys. Its eastern strip is essentially a single large valley, known as the Great Appalachian Valley, with a limestone floor that has decomposed into some of the better Appalachian soils.

The Appalachian Plateau is west and north of the Ridge and Valley Section. Its northern part is known as the Allegheny Plateau, while in eastern Kentucky and farther south, it is the Cumberland Plateau. Prominent east-facing escarpments—the Allegheny Front in the north and the Cumberland Front in the south—mark its eastern edge. Except in New York, this plateau is underlain by enormous, easily worked deposits of high-quality bituminous coal. These have been extremely important in the economic development of the United States, but coal's contribution to the country's energy supplies is restrained by its impacts on the
33 atmosphere and by the landscape impacts of coal mining.

Strip mining of coal throughout much of Appalachia (the Appalachian region) has long used the technique of mountaintop removal, the decapitation of whole summits by dynamite to quickly and cheaply expose thin seams of low-sulfur coal. Leftover waste is dumped into nearby valleys, creating large, unsightly, and often toxic "valley fills." Grasses are usually planted in an effort to "reclaim" strip-mined lands, but a new effort called the **Appalachian Regional Reforestation Initiative** aims to have forests—which are more compatible with the original vegetation—planted instead. The region's coal industry is nearly exhausted. Most of the big companies have moved operations to Wyoming's Powder River area, and those that remain are taking coal they can find from deep and dangerous tunnels and from dynamited mountaintops.

Early Development of the Seaboard Cities

Except for Washington, D.C., the largest metropolises of the Northeastern Seaboard began as seaports, many located at river mouths. The federal capital's location was a political compromise of 1790. It was on the border between the agrarian southern states, with their large slave populations, and the more economically diversified northern states. The site also placed the capital on the fall line. Many state capitals in the United States were located as a result of political compromises, while others were situated as centrally as possible in the states as a convenience for their citizens.

New York City was founded at the southern tip of Manhattan Island on the great natural harbor of the Upper Bay. Early on, it shipped farm produce from lands now occupied by built-up areas of metropolitan New York City and from estates farther up the Hudson River. Philadelphia (also on the fall line) was sited at the confluence of the Delaware River and its tributary Schuylkill River. Philadelphia's local hinterland was productive enough to make the city a major exporter of wheat and helped it become the largest city (population around 40,000 in 1776) in the 13 original colonies. Baltimore was founded on a Chesapeake Bay harbor at the mouth of the Patapsco River and shipped tobacco and other farm products from its local hinterland in the Piedmont and Coastal Plain.

Boston developed differently because its New England hinterland produced little agricultural surplus. Instead, colonial New England emphasized living from the sea. Its forests yielded excellent timber for shipbuilding, and both timber and ships were exported. Cod and other fish were once abundant along this coast. Whaling, based mainly on the Massachusetts island of Nantucket, was important in the 18th and 19th centuries. A "Nantucket sleigh ride" was the experience boatmen had when they harpooned a sperm whale and were dragged through Atlantic waters at speeds up to 23 miles per hour (37 kph). New Englanders also developed a wide-ranging merchant fleet, trading with far-flung partners. Such activities characterized many northeastern ports, Boston being the largest.

The Race for Midwestern Trade

As the American Midwest was settled, especially between 1800 and 1860, its agricultural surpluses and needs for manufactured goods expanded the trade of New York, Philadelphia, and Baltimore. These ports raced to establish transport connections into the interior. New York City won, especially because of its access to the Hudson-Mohawk Trough, the only continuous lowland passageway through the Appalachians. The Erie Canal was completed along this corridor in 1825 (see Figure 11.1.7, page 637). Connecting Lake Erie at Buffalo to the navigable Hudson River near Albany, the canal was a key link in an all-water route from the Great Lakes to New York City. It lowered transport costs between New York and the Midwest to a small fraction of what they had been and stimulated growth in the Midwest. In 1853, New York and the growing city of Chicago were connected by rail along this same route.

By the 1850s, New York was the busiest U.S. port, a position it has since ceded to the port of South Louisiana and the port of Houston, as measured by cargo tonnage. The competing ports of Baltimore and Philadelphia also traded with the Midwest. Major railways eventually pushed through the Appalachians into the Midwest from both ports. Boston was never a real competitor in this race, although it has remained the largest seaport in New England.

Industrialization got its first big boost during the same period that the Northeast established connections to the interior. Several factors favored industrial growth: capital from commercial profits, access to industrial techniques pioneered in Britain, skills from previous commercial operations and handicraft manufacturing, cheap labor supplied by immigrants, and the presence of energy resources and raw materials. Also favorable was a U.S. high-tariff policy restricting imports and protecting much of the nation's own emerging manufactures. The first American factory, a water-powered textile plant using pirated British weaving technology, was established on the Blackstone River at Pawtucket (near Providence), Rhode Island, in 1790. Subsequent growth of factory industries in the Northeast was rapid.

Circumstances and outcomes of industrialization varied in different parts of the Northeast. New England had suitable waterpower near its ports, with swift streams to turn waterwheels, but this area lacked coal for metalworking. It specialized in textiles, drawing wool from New England farms, cotton from the South, and labor from recent European immigrants, especially the Irish. Connecticut took a different path, however. Using small local ore deposits, it expanded a colonial specialty in metal goods to become a manufacturer of machinery, guns, and hardware. Mill towns sprang up across southern New England and up the coast into southern Maine.

In the Middle Atlantic states, early industrialization also included textiles, but there was more emphasis on metals, clothing, and chemicals. The emphasis on metals reflected local deposits of iron ore, especially in southeastern Pennsylvania. Then, in the second quarter of the 19th century, the largest deposits of anthracite coal in the United States became a major source of power. This coal lay deep under the Pennsylvania Ridge and Valley Section between Harrisburg and Scranton–Wilkes-Barre. Usable in blast furnaces without coking, anthracite fueled large-scale iron and steel-making in eastern Pennsylvania cities. Iron and steel from these early plants went into machinery, steam engines, railway equipment, and other metal goods, contributing to the Northeast's first industrial surge.

The Big Apple

The United States' largest city, New York City is centrally located on the Northeastern Seaboard's urban strip midway between Boston and Washington. An enormous harbor, an active business enterprise, and a central location within the economy of the colonies and the young United States had already made New York City the country's leading seaport before access to the Hudson-Mohawk route gave it an even more decisive advantage. Superior access to the developing Midwest then moved New York rapidly to unchallenged leadership among American cities in size, commerce, and economic impact.

New York City itself had a 2007 population of 8.2 million, and the metropolitan area (which includes much of northern and central New Jersey, Long Island east of Brooklyn and Queens, southernmost New York State just north and west of the Bronx, and southwestern Connecticut) is home to 18.8 million people. The city is centered on the

less than 24 square miles (62 sq km) of Manhattan Island (•Figure 11.2.4). The island is narrowly separated from the southern mainland of New York State on the north by the Harlem River, from Long Island on the east by the East River, and from New Jersey on the west by the Hudson River. Manhattan is one of five **boroughs** (administrative divisions) of the city proper. The others are the Bronx, on the mainland to the north; Brooklyn and Queens, on the western end of Long Island; and Staten Island, to the south across Upper New York Bay.

Expressways and mass-transit lines converge on bridges, tunnels, and ferries linking the boroughs to each other and to northern New Jersey. Port activity is concentrated in New Jersey along the Upper Bay and Newark Bay. The whole area is held together in part by an amazing 722 miles (1,155 km) of subway system, parts of which are more than a century old.

Of the city's total population today, more than 35 percent are foreign-born, with the largest contingents coming from the Dominican Republic, China, and Jamaica. Their growing numbers more than offset the out-migration of longer-resident New Yorkers, with the trends of whites and Asians moving elsewhere in the greater New York region, Asians moving to the West Coast, retiring whites to Florida, blacks whose families originated in the South moving back to the South, and some Puerto Ricans returning to their home island. Immigrants make up more than 40 percent of New York City's labor force, including more than half in restaurants and hotels, about 60 percent in construction, and nearly 70 percent in manufacturing. Most live in the outer boroughs, with two-thirds of them in Queens or Brooklyn. Whereas New York historically was an ethnic checkerboard of Irish, Italian, Jewish, Chinese, and other groups, Queens and Brooklyn are ethnic stewpots where immigrants from numerous countries live cheek by jowl. Restaurants and other services in these areas are extraordinarily diverse.

New York Mayor Michael Bloomberg promised a "sustainable city" for New Yorkers of the future. His plan, unveiled in 2006, had 10 goals, including a huge increase in affordable housing, a pledge that every New Yorker would live within 10 minutes' walk of a public park, an overhaul of the subways and other transport systems, and a target of having the cleanest air of all American cities. Not to be outdone by efforts in California, the mayor insisted in this plan that the Big Apple should reduce greenhouse gas emissions by 30 percent by 2030.

Size, financial importance, cultural complexity and variety, and national history have given New York City an uncontested position of urban prominence in the United States. The heart of this extraordinary city is the roughly the lower half of Manhattan. The island's Midtown hotel, shopping, theater, and office district includes most of the city's best-known hotels, restaurants, department stores, specialty shops, theaters, concert halls, and museums. The Midtown area also includes numerous office buildings, including the Empire State and Chrysler Buildings. At the

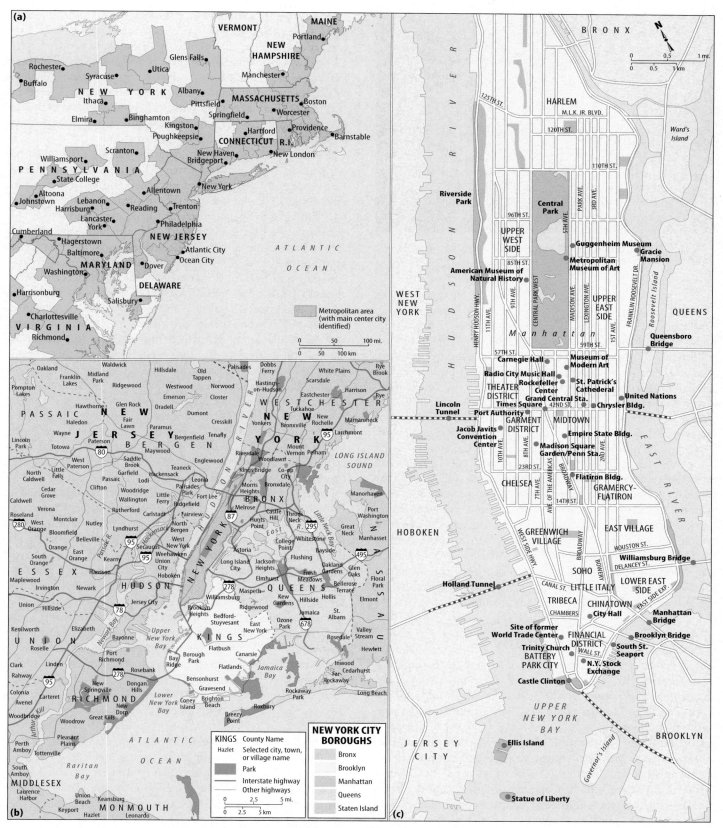

● **Figure 11.2.4** New York from three perspectives and scales. (a) New York is located at the center of a strip of highly urbanized land christened the "megalopolis" by the French geographer Jean Gottmann in 1961. Many waterfronts and other areas shown on the maps of New York City (b, c) are being redeveloped for new commercial, residential, and recreational uses. Note the complex of bridges, tunnels, and ferries tying Manhattan Island to the rest of metropolitan New York.

Joe Hobbs

• **Figure 11.2.5** "Ground Zero," former site of the twin towers of the World Trade Center, is now one of the most revered places in America.

southern end of this area are Times Square and Broadway, formerly notorious for seedy nightclubs and strip joints but spiffed up in the 1980s to be a family-friendly must-see area. Here too is the historic Garment District, also called the Fashion District to present a less industrial image. The lower Manhattan financial district includes the imposing cluster of office skyscrapers in the Wall Street area and was the home of the twin towers of the World Trade Center until 9/11. This area's critical role in the global economy made it a prime target for al-Qa'ida's September 11 attacks (and an often forgotten, more feeble attempt in 1993), and the prospect of more attacks will always loom over New York (•Figure 11.2.5). The city's economy has rebounded remarkably well from the direct and indirect costs of the 9/11 attack. Many firms lost in the rubble relocated to central Manhattan and nearby New Jersey. As the Freedom Tower, scheduled for completion in 2012, rises from the pit where the twin towers stood, the financial district is also being reborn as an affluent bedroom community.

Manufacturing in the Northeast

The Northeast rose to economic dominance through the production of 10 specialty products. First was *clothing design and manufacture*, concentrated in metropolitan New York City. The clothing industry in Manhattan was originally an outgrowth of New York's role as an importer of European-made cloth and clothing. Skilled immigrant workers of the late 19th century, among them Jewish tailors fleeing persecution in tsarist Russia, drove this industry forward. There are still clothing manufacturers in the Fashion District, but their numbers fell by about half between 1995 and 2007, and most of the vacated space has been converted to studio offices and loft apartments.

Powerful *iron and steel industries* associated with industrialization thrived from the late 19th century until after

World War II. Pittsburgh, Pennsylvania, was the center, but factories and facilities expanded east to Baltimore and west to Chicago at the beginning of the 20th century. After 1998, low-cost foreign steel began flooding American markets, sending the domestic industry into a tailspin resulting in the loss of about one-third of the producers. The recent surge in China's economy with its growing appetite for steel may help stabilize the remaining firms, which now produce about 10 percent of global output of steel.

Buffalo and Pittsburgh are good example of what has happened in the old steel towns. The opening of the Saint Lawrence Seaway in 1959 diminished Buffalo's geographic advantage of being the western terminus of the Erie Canal. This began the downtrend of the city's industries that culminated with the closing of its Bethlehem Steel plant in 1983. Free trade and outsourcing continued to eat away at manufacturing. The city's population plunged to half its 1950 number, and Buffalo became a fitting symbol for the industrial ruin of the Northeast. Images of the weary people of Buffalo regularly shoveling snowfalls measured in feet rather than inches did little to attract newcomers. But Buffalo's fortunes are changing. New businesses include a plant to process corn-based ethanol, shops to attract tourists to the renovated Erie Canal terminus, and a casino. Buffalo (population 1.1 million) wants to become a biomedical technology hub, following the 2006 inauguration of its New York State Center of Excellence in Bioinformatics and Life Sciences. There is a remarkable contrast between Buffalo's blue-collar image of the past and the white coats associated with the center's search for ways to treat or prevent serious diseases like Alzheimer's and cancer.

After peaking in 1974, steel production plummeted in Pittsburgh, and more than 75,000 workers in that industry lost their jobs. All told, almost 150,000 manufacturing jobs in the city were lost between 1975 and 2005. But in the same period, almost 175,000 jobs were added in the fields of education and medicine. Pittsburgh (population 2.3 million) has both expanded its traditional role in providing education and health care services and reinvented itself as a center of "clean industries" such as bioengineering and robotics. As these fields outgrew their space in the city center, they took advantage of what was available: Old steel mills have been converted to high-tech offices, retail shops, and housing, and Pittsburgh's cultural offerings have blossomed with new arts centers. The new buildings are generally "green"; Pittsburgh has more urban space built to national energy and sustainability standards than any other city in the country. The urban landscape of Pittsburgh is far cleaner and more appealing than in the days of steel. Yet even steel may make a turnaround in Pittsburgh and others of its historic centers, especially if China's demand for it remains strong and the cost of importing it stays high because of freight charges and a weaker dollar.

Chemical manufacturing has been diversified in the region, especially in outlying areas of metropolitan New York City. *Photographic equipment manufacturing* was long

298

637

656

dominated by Eastman Kodak of Rochester, New York, which has remained competitive in the industry only by embracing digital photographic technology and moving manufacturing offshore to Shanghai and other locations. *Advanced electronics manufacturing*, centering on computers, is widespread in the Northeast. This industry is a major specialty in eastern Massachusetts, especially along Route 128 and other beltways that ring the Boston area. Some old textile towns such as Lowell and Lawrence on the Merrimack River have largely converted to electronics or other high-tech industries. *Biotechnology* is booming in the Boston area, taking advantage of the region's top research universities, hospitals, and Ph.D.'s. Concentrated in southern New England, New York, and Pennsylvania, *electrical equipment manufacturing* is dominated by the General Electric Company, headquartered in Stamford, Connecticut, with major production facilities in Schenectady, New York.

The *manufacture of aircraft engines and helicopters* is concentrated near Hartford, Connecticut, and the *manufacture of Trident nuclear submarines* in the New London, Connecticut, area continues the long traditions of New England's shipbuilding and Connecticut's armaments making. The Northeast leads the nation in the *publishing and printing* industries, with New York and Washington the most prodigious centers.

From Rust Belt to Sun Belt

The term *Rust Belt* has come to describe parts of the Northeast and Midwest as a region of manufacturing industries that have become obsolete. This region's old **smokestack industries**—especially the steel and other heavy-metallurgy industries—experienced depressed sales and profits, high unemployment, and in many cases complete factory shutdowns. These industries have been either replaced with "greener," more high-tech industries and services (such as described here in Pittsburgh, Boston, and North Carolina)—the classic process of deindustrialization that is also typical of Europe today—or have moved elsewhere in the United States and even abroad. The so-called **Sun Belt** of the South and West has been enjoying much faster growth in population and jobs, including those in manufacturing, in part due to migration from the Rust Belt. The Sun Belt is an elastic term, but it connotes the region encompassing most of the South and the West as far north as Denver, Salt Lake City, and San Francisco.

Today, growth in the Northeast's economies focuses more on services rather than on manufacturing: banking and finance (notably Manhattan's financial district, including Wall Street), advertising (especially Manhattan's Madison Avenue), newspaper and television journalism, scientific research and education (especially in the Ivy League universities), and other white-collar industries. But despite growth in these sectors, the region has been growing in population relatively slowly, with heavy out-migration from the Middle Atlantic region. The main exceptions to this slow growth are the peripheries of some metropolitan areas into which suburban and exurban development (including into "satellite developments" or "edge cities") is expanding. Notable examples include expansion into New Hampshire from Boston, into Connecticut and New Jersey from New York City, and into Virginia from Washington. The Northeast has also recently seen strong new development in the great metropolises, both in old decaying cores and around the peripheries. Baltimore's Inner Harbor Development is a good example (see Insights, page 656).

11.2.2 The South: Dixieland

In this book, the South is defined as a bloc of 14 states: five along the Atlantic (Virginia, North Carolina, South Carolina, Georgia, and Florida); four along the Gulf of Mexico (Alabama, Mississippi, Louisiana, and Texas); and the five interior states of West Virginia, Kentucky, Tennessee, Arkansas, and Oklahoma (see Figure 11.2.1). This is of course a vernacular region, and arguments exist to exclude some states. West Virginia and Kentucky, for example, did not secede from the Union during the Civil War, and the western parts of Oklahoma and Texas are physically more typical of the West than of the South. Although Missouri was a slave state and so was part of the South before and during the Civil War, today it is more typically midwestern and is not included here.

Colloquially, since Civil War days, the South has been known as **Dixie** or **Dixieland.** Southerners also often refer to themselves as being from south of the **Mason-Dixon Line.** This was an 18th-century delimitation across the southern border of Pennsylvania and the western border of Delaware, drawn up by two surveyors (Mason and Dixon) to settle a land dispute.

The Physical Setting

Most of the South has a humid subtropical climate, with summers that are long, hot, and wet. January daytime temperatures average from the 30s Fahrenheit (c. 0°–5°C) along the northern fringe to the 50s (c. 10°–15°C) near the Gulf Coast and the 60s (c. 15°–21°C) in southern Florida and the southern tip of Texas. More than 40 inches (c. 100 cm) average annual precipitation is characteristic, rising to over 50 inches (c. 125 cm) in many Gulf Coast and Florida areas and more than 80 inches (c. 200 cm) in parts of the Great Smoky Mountains. High humidity brings discomfort in both summer and winter.

There are advantages and drawbacks to these climatic conditions. They favor rapid and abundant forest growth, and forest-based industries are important in the region. Agriculturally, they provide long growing seasons, heat, and moisture for a wide range of crops. But insects and pests flourish, and heavy rainfall leaches out soil nutrients and can bring erosion and flooding. Florida's subtropical

INSIGHTS

Adaptive Reuse and Gentrification

A characteristic urban landscape transformation, particularly in the Northeast and Midwest, is the **adaptive reuse** of older structures. In this process, abandoned commercial or factory buildings are being converted to new economic activities, generating jobs and attracting people back to the area. This shift is often associated with **gentrification,** the process by which a low-cost, run-down neighborhood undergoes physical renovation for residential or office space, resulting in increased property values and new homes and offices for more affluent urban professionals (•Figure 11.2.A). In Connecticut, for example, where the manufacturing economy shrank by nearly 25 percent between 1997 and 2007, vacated mills and other factories were reborn as art shops and antique stores.

Others were razed and replaced by big-box stores like Best Buy and discount department stores including Kohl's and Target.

Adaptive reuse takes place most often in old central business districts, but the process has also been occurring in old residential neighborhoods. Adaptive reuse and gentrification give value to the past, especially when historic buildings can be modified and used rather than destroyed. But while these processes revitalize run-down areas, attracting new wealth and revenue to them, they have a downside. People who have been living in these neighborhoods often can no longer afford to do so and are displaced to distant housing that is more affordable but far less convenient for work and services (•Figure 11.2.B).

• **Figure 11.2.A** The Baltimore Inner Harbor Development, called Harborplace, is a successful reclamation of an old warehouse and wharf area. It boasts a shopping mall with outdoor restaurants and cafés, an aquarium, power plant buildings converted into elegant townhouses and condominiums, and access to Camden Yards, the home of baseball's Baltimore Orioles.

• **Figure 11.2.B** Chicago's Pilsen area has been spiffed up in recent years. As the area becomes more expensive and attractive to urban professionals, its poorer, longtime, mainly Latino residents are driven out. With this mural, the less affluent residents condemn Pilsen's gentrification. Eastern European immigrants once lived in Pilsen, and moved out during an earlier cycle of urban change.

environment makes it a haven for hundreds of exotic species of plants and animals that have crowded out and even eliminated native species. Brazilian pepper and Australian pine are among the more prolific exotic flora. Over 200 exotic animal species, many originating as escaped or released pets, slither, crawl, fly, and swim across Florida. These include Burmese pythons, boa constrictors, green iguanas, Cuban tree frogs, vervet monkeys, macaques, capuchin monkeys, parrots, and cockatiels.

Topographically, most of the South is made up of plains with three major landform divisions: the Gulf and Atlantic Coastal Plains, the Piedmont, and the Interior Plains (see Figure 11.4, page 604). The Coastal Plain is on the seaward margin from Virginia to the southern tip of Texas, including all of Florida, Mississippi, and Louisiana and parts of the other southern states except West Virginia. It is low, with large flat areas near the sea and along the Mississippi River; in most inland sections, the surfaces are irregular.

The Interior Plains section takes up central and western Texas and all but easternmost Oklahoma. The Balcones Escarpment of Texas forms an abrupt boundary between the so-called hill country of the higher Interior Plains and the lower Coastal Plain (●Figure 11.2.6). Most of Texas's larger cities, including Dallas, Austin, and San Antonio, lie on the Coastal Plain. Irrigation, ranching of angora goats and other livestock, and dry farming of wheat are important to the rural population of this plains region.

The Piedmont extends across central Virginia, the western Carolinas, northern Georgia, and east central Alabama. Its surface is a rolling plain with some hilly areas. The natural vegetation is mixed-forest hardwood and coniferous forest with more broadleaved hardwoods, especially oak, than pines. Soils tend to be poor, with many damaged by past cultivation of tobacco, cotton, and corn on slopes subjected to heavy rains and erosion.

The region's fall line lies between the Piedmont and the Coastal Plain, with large parts on the Piedmont hilly and

● **Figure 11.2.6** The rough Texas hill country is blessed with springs, creeks, and rivers.

mountainous. Two large embayments of rugged country extend into the central and eastern South from the Northeast and Midwest. The larger one is part of the Appalachian Highlands and includes some rugged land, west of the Appalachians, known as the Unglaciated Southeastern Interior Plain. Farther west, the second large embayment is the southern part of the Interior Highlands. These three areas—Appalachians, Unglaciated Southeastern Interior Plain, and Interior Highlands—are known as the Upland South.

Southern parts of the Appalachians run from western Virginia to eastern Kentucky and southward to northern Alabama. The highlands include all the major physical subdivisions of the Middle Atlantic states. The Blue Ridge, whose highest section is the Great Smoky Mountains, extends from the Potomac River to northern Georgia (●Figure 11.2.7). Mount Mitchell in western North Carolina is the highest point in the eastern United States at 6,684 feet (2,037 m). The Ridge and Valley Section, including the Great Appalachian Valley, runs parallel to the Blue Ridge and inland from it, from western Virginia into Alabama. The southern section of the Appalachian Plateau, known as the Cumberland Plateau in parts of Kentucky and areas to the south, is wide in the north, where it occupies eastern Kentucky, most of West Virginia, and a small section of Virginia. It narrows across eastern Tennessee and then widens to occupy much of northern Alabama. As in Pennsylvania, it is bounded on the east by an abrupt escarpment, is mostly dissected into hills and low mountains, and is underlain by deposits of high-quality coal.

Central Kentucky, central Tennessee, and part of northern Alabama are in the Unglaciated Southeastern Interior Plain. This area, bounded by the Tennessee River on the west and south, was not smoothed by glaciers during the last ice age, as areas farther north were. A good part of this "plains" area is really hill country dissected by streams. Two plains areas—the Kentucky Bluegrass region around Lexington and the Nashville Basin around and south of Nashville, Tennessee—are underlain by limestone that has weathered into good soils. Both have exceptionally good soils in an otherwise poor agricultural zone.

The Interior Highlands occupy northwestern Arkansas and eastern Oklahoma. From Arkansas, they extend northward into the southern part of the midwestern state of Missouri. With their hill country, low mountains, and some plains, these highlands are split by the broad east-west-running Arkansas River Valley. North of the valley, the Boston Mountains form the southern edge of the Ozark Plateau. South of the valley in west central Arkansas and adjacent Oklahoma lie the parallel ridges and valleys of the Ouachita Mountains. Most of the Ozark Plateau ("the Ozarks") has been dissected into gentle hill country, but parts, especially the Boston Mountains, are mountainous (●Figure 11.2.8).

The Southern Interior Highlands are not resource-rich. Most soils are poor and often stony. Most of the area is forested, but it has been logged, and the present timber is

GEORGE F. MOBLEY/NGS Image Collection

● **Figure 11.2.7** The Great Smoky Mountain National Park, on the Tennessee/North Carolina border.

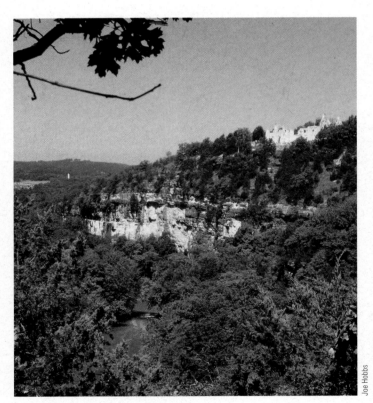

Joe Hobbs

● **Figure 11.2.8** The Missouri Ozarks.

generally poor. Most mineral deposits are not economical to exploit. The most significant resource is the lead of southern Missouri. Large artificial lakes used for recreation are major resources in both Missouri and Arkansas.

Southern Settlements and Peoples

The South's patterns of population density, urbanism, ethnicity, and income are distinct. Until recently, this was generally a region of farms, villages, and small cities. But for the past several decades, the South has been experiencing rapid population growth, second only to the pace in the West, and its farm population has plummeted. It has surpassed the Midwest in population density, although both lag behind the Northeast. Due to the swift growth of many small cities, the South has more small and medium-sized metropolitan areas than any other region of the country. It has relatively few large cities, however. The biggest, Dallas–Fort Worth, has 6 million people, followed by Houston (population 5.5 million), Miami (5.4 million), Atlanta (5.1 million), and Tampa–Saint Petersburg (2.7 million). These southern cities continue to grow rapidly. The urban patterns will be discussed in more detail shortly.

The population of the South has less ethnic variety than most other U.S. regions. The major groups are white Protestants of British ancestry, African American Protestants, and Mexican American Catholics. White Protestants of Anglo-Saxon origin are the largest contingent. The original European settlers along the Atlantic were British. Small groups of Spanish Catholics settled in Florida and Texas, and some French Catholics settled in Louisiana and elsewhere along the Gulf Coast.

The South's ethnic geography is distinguished by a high proportion of African Americans, due mainly to the trade in African slaves for plantation labor in the pre–Civil War South (see Figure 11.13, page 614). The six states with highest proportions of blacks—from about one-fourth to more than one-third—are Mississippi, Louisiana, South Carolina, Georgia, Maryland, and Alabama.

The "great migration" out of the South, mostly after 1950, brought large African American minorities to some northern states and California. New York and California have larger African American populations than any southern state, but blacks make up only 17 percent of New York's people and 7 percent of California's. African American populations outside the South are concentrated in large cities. Recently, a reverse migration of African Americans from North to South has occurred as economic and social conditions have improved in the South but declined in many northeastern cities. In this reversal of fortune, Mississippi has had the biggest uphill battle. It has the country's highest poverty rate and rates third lowest in level of education, after West Virginia and Arkansas. Hurricane Katrina hit it hard. But aircraft and auto companies are investing heavily in Mississippi, which has mounted an aggressive public relations campaign to change perceptions of the state.

Other distinctive ethnic elements in the South are Mexican Americans, Cuban Americans, others of Hispanic origin, and **Cajuns** (of French descent). In Texas, Hispanics of mainly Mexican origin make up 36 percent of the population (•Figure 11.2.9), and southern Texas is imbued with a distinctly Mexican culture. Natural increase and immigration have rapidly expanded the numbers and clout of Hispanics in Texas and elsewhere in the South in recent decades. Increasing numbers of legal and illegal Hispanic immigrants have migrated into communities where the traditional ethnic divisions were between advantaged whites and disadvantaged but economically improving blacks. The newcomers have often found themselves uneasily welcomed into this mix. Many whites express dismay at the growing need to accommodate the Spanish language and at the apparent drain of the immigrants on education and other social services. Blacks have seen many Latinos—pushed, say sociologists, by an immigrant ethic to study and work hard—take low-wage jobs and move more rapidly than

blacks into business ownership and middle-class jobs. Relations in today's South are often more testy between blacks and Hispanics than between blacks and whites.

South Florida is home base for Cuban Americans. Over 1 million refugees have fled from Communist Cuba since the revolution of 1959, and there are over 800,000 Cuban Americans and their descendents in Florida, most in metropolitan Miami. Coming mainly from the middle and upper classes of pre-Communist Cuba, Cuban Americans have been upwardly mobile in American society. They have been assimilating into the population through marriage and adoption of Anglo American culture traits while imprinting their own culture on south Florida. Cuban Americans have an important economic and political voice in the United States and take credit for the continuing poor relations between the United States and the Communist Cuba they despise.

The largest non-Hispanic minority group in the South is the French-descended Cajun culture of southern Louisiana. Small French settlements, including New Orleans, grew along the Gulf Coast in the early 1700s. But it was the arrival of the refugee **Acadians,** expelled from Nova Scotia by the British in the mid-18th century, that implanted an enduring culture of French origin. That culture mixed with American elements to create the unique Cajun culture of southern Louisiana. The region's characteristic amphibious "bayou" landscape, a distinctive dialect with strong French features, the music known as zydeco, and a culinary tradition of gumbo and hot peppers are trademarks of the Cajun region.

Agricultural Geography of the South

There is a strong agricultural heritage in southern landscapes and cultures, but change has overtaken tradition. In appearance and land use, the modern South is not overwhelmingly rural. Over large areas, less than half the land is in farms, and the region's population is only about 1 percent agricultural. Farms tend to be small in size and production. Animal products (especially beef and dairy cattle and broiler chickens) are much more important than crops in total sales. The leguminous soybean plant, which adds nitrogen to the soil and can be used for hay, is the most important southern crop. Its beans are pressed for soybean oil (for use in paints, soap, glycerin, printing ink, and many other chemical-based products), with the residue (oilcake) fed to cattle and poultry. Soybeans are use to produce ethanol. Grain sorghum, which has low moisture requirements, provides cattle feed in drier parts of Texas and Oklahoma.

Florida citrus fruits are processed mainly for frozen concentrate, with the residue fed to cattle; the main market is the large ranch industry in southern Florida. Broiler chicken production is a specialty of poorer agricultural areas, with concentrations in the Interior Highlands of Arkansas and the Appalachians of Georgia and Alabama. In Arkansas, most of the workers on the chicken farms are Mexican Americans or new immigrants from Mexico, who are bringing a distinct new culture and language to a traditionally white region.

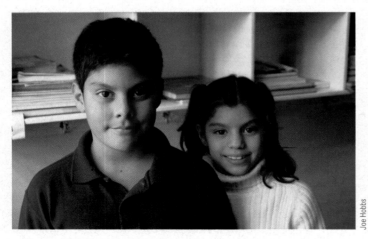

• **Figure 11.2.9** Alejandro and Julia Chavez-Lopez, Mexican-Americans.

The historical staples of tobacco and cotton are still prominent in some parts of the South. Most American tobacco is grown in the eastern and northern South, especially in the Carolinas, Kentucky, and Tennessee (•Figure 11.2.10). Nearly all the South's remaining cotton is raised on large mechanized farms in the Mississippi Alluvial Plain and in west Texas. The region's other agricultural specialties include sugarcane in Louisiana's Mississippi Lowland and near Florida's Lake Okeechobee. The building of canals for sugar production in south Florida played a major role in the 80 percent shrinkage of the Everglades wetland region, from a shallow wilderness "sheet flow" river 40 miles (64 km) wide to a much tamer marshland encircled by cane fields. Other specialties are rice in the Mississippi Lowland and the coastal prairies of Louisiana and Texas; peanuts in southern Georgia and adjoining parts of Alabama and Florida; and racehorses, especially in the Kentucky Bluegrass region around Lexington, where horse farms are showplaces of conspicuous wealth. Corn, malted barley, and rye are main ingredients of the whiskeys and bourbons of Tennessee and Kentucky. Also key is the region's fresh springwater; all the water of Jack Daniels whiskey, for example, comes from a spring issuing from a small cave in Tennessee.

Southern Industries

Most industries in the South are in the Piedmont and adjacent Coastal Plain and Appalachians. Within this Eastern South Industrial Belt, textiles long had the lead in manufacturing, with clothing, chemical products, furniture, tobacco products, and machinery also important. After the 1994 signing of the North American Free Trade Agreement (NAFTA), however, many textile and clothing manufacturers shifted to Mexico, Jamaica, and other "offshore" manufacturing sites. These were in turn undercut by lower labor costs in China and elsewhere, particularly after textile trade quotas expired on January 1, 2005. The immediate 298, 554 and cumulative impacts on textile mills in North Carolina

and other states were ruinous. More than 17,000 jobs were lost in the first three months, and the dizzying descent continued. With the imported garments costing 80 percent less than what Piedmont loomshops could produce, the factories closed.

The slide in textiles and other old manufacturing has been moderated by job creation in government, tourism, and biotechnology. And in an extraordinary turnabout of the outsourcing phenomenon, Asian manufacturers are locating new plants in the South. The Chinese company Haier opened a large refrigerator factory in South Carolina after Chinese managers concluded that making the product here for the American market would save transportation costs and ensure speedier deliveries to retailers. The 299 same economic rationale has brought a surge in auto assembly and components manufacturing to Tennessee, Kentucky, and Alabama, and ushered in a dramatic geographic shift in the industry from the Midwest to the South. Between 1996 and 2006, auto-related manufacturing jobs fell 34 percent in Michigan while rising 152 percent in Kentucky. Once viewed with wariness, the Japanese companies Toyota, Honda, and Nissan are embraced in communities like Georgetown, Kentucky, which seem a world away from the abandoned assembly lines of Flint, Michigan.

The southward migration of the auto industry is one reason why the South's economy is no longer the laggard it once was. In 1937, incomes in the South were just half the U.S. average, but in 2007, they were over 90 percent. Southern costs of living are lower than the national average, so the actual gap is probably smaller. The region's growth is due to a number of factors. After World War II, the spread of air conditioning made places like Florida renowned for winter warmth rather than notorious for insufferable summer heat. Federal defense spending grew in the region. The region had a "clean slate" for industrialization, without the North's legacy of outdated machinery and powerful labor unions.

The **Tennessee Valley Authority (TVA)** fostered much development in this region and adjacent areas in the South. The U.S. government created the TVA in 1933 as part of President Franklin Roosevelt's "New Deal" in an effort to promote economic rehabilitation of the economically depressed Tennessee River Basin. This public corporation built many dams, locks, and hydropower stations on the Tennessee River and its tributaries. These engineering works controlled floods, eased barge navigation, generated hydroelectricity, and promoted recreation around reservoirs. The demand for electricity soon outran the capacity of the hydrostations, and many thermal plants (fired by coal and nuclear energy) were built. While northern industrial development focused on the metropolises of the railway age, southern development came later, taking advantage of long-distance power transmission and high-speed truck transportation.

The southern industrial belt lies along and near the Gulf Coast from Corpus Christi, Texas, to New Orleans and Baton Rouge, Louisiana. The chemical industry, including oil refining, dominates this West Gulf Coast Industrial

• **Figure 11.2.10** Tobacco cultivation in Kentucky.

Joe Hobbs

Belt. It is based on deposits of oil, natural gas, sulfur, and salt. Including offshore fields in the Gulf of Mexico, Texas and Louisiana produce almost 30 percent of the oil, about 60 percent of the natural gas, and practically all the domestic sulfur output of the United States. Machinery and equipment for oil and gas extraction, oil refining, and chemical manufacturing are other important products. Major refineries and petrochemical plants are in several clusters, especially around metropolitan Houston and Beaumont, Texas, and Baton Rouge, Louisiana.

Other southern industries include food processing, machinery, furniture, and pulp and paper industries. Food industries are especially important in Florida. Pulp and paper plants abound near the Gulf Coast from eastern Texas to Florida and along the Atlantic Coast in Georgia. These operations draw on the pine trees that grow quickly in the wet subtropical climate. Aircraft and aerospace industries are important in the Atlanta and Dallas–Fort Worth areas.

Big Cities of the South

A distinct Texas-Oklahoma metropolitan zone includes the two largest urban clusters in the South, Dallas–Fort Worth and Houston, plus four other metropolises with populations of about 900,000 or more: San Antonio (•Figure 11.2.11) and Austin, Texas, and Oklahoma City and Tulsa, Oklahoma. Houston is a seaport by virtue of the Houston Ship Canal (or Channel). This major port serves Texas and the southern Great Plains and is the main center for the oil and gas fields and the chemical industry of the Gulf Coast. Houston is one of the country's largest ports in terms of shipping volume. Dallas–Fort Worth, San Antonio, and Austin grew along and near the western edge of the Coastal Plain in a north-south strip of unusually fertile limestone-derived soils known as the Black Land Prairie. Interstate 35, beginning at Laredo on the Mexico border and running north through San Antonio, Dallas–Fort Worth, and be-

yond, has borne a huge share of the truck traffic associated with NAFTA. A controversial, privately owned 10-lane toll road known as the Trans-Texas Corridor has been proposed to ease the strain on I-35.

Dallas–Fort Worth is the dominant business center for a huge region. San Antonio is a major military base and retirement center, and Austin is the site of the highly ranked University of Texas, many high-tech industries, the Texas state government, and a thriving music scene. All the cities are also involved in the oil and gas industries.

The Florida Peninsula is another area of strong metropolitan development. Two Florida metropolitan areas, Miami–Fort Lauderdale and Tampa–Saint Petersburg, have populations of more than 2 million each, and Orlando and Jacksonville have more than 1 million each. Proximity to the ocean and winter warmth have made Florida's cities magnets for tourists and retirees. Miami is a major intersection between the United States and Latin America in air traffic, tourist cruises, banking, and the illegal drug trade. Tampa Bay affords a harbor well placed for the trade of central Florida's citrus belt and phosphate mines. Improved transportation and communication have lured high-tech industries and regional corporate offices from the Northeast to Florida.

The third concentration of large southern cities lies along the Piedmont, the fall line, and the nearby Virginia Coastal Plain. These include Richmond, Virginia; Charlotte, Greensboro, and Raleigh, North Carolina; and Columbia and Greenville, South Carolina. Raleigh, Durham, and Chapel Hill, North Carolina, are the points of the **Research Triangle Park (RTP)**, a biotechnology and other high-tech research and development hub and a rival to Boston's thriving biotechnology business; North Carolina has the country's third-largest biotech cluster, after California and Massachusetts. Atlanta, Georgia, is the largest southeastern business and governmental center (•Figure 11.2.12). It has diverse industries and is a major inland transportation hub. Its Hartsfield-Jackson Airport is the busiest in the United

• **Figure 11.2.11** The revitalization of downtown San Antonio includes the tourist magnet of El Mercado.

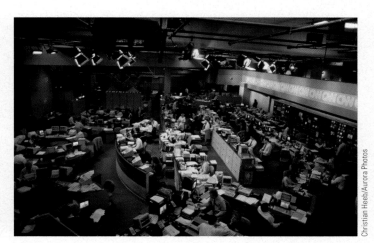

• **Figure 11.2.12** CNN television has helped to put Atlanta on the world map.

States, measured by volume of passenger traffic and operations size. CNN, the world's largest 24-hour television news service, is headquartered in Atlanta, where its founder, the tycoon Ted Turner, bases many of his business interests.

On the northern Coastal Plain is the seaport of Norfolk, Virginia. Located on the spacious Hampton Roads natural harbor at the mouth of the James River, Norfolk ships Appalachian coal by sea and is home to the world's largest naval base. The entire Upland South has only six metropolitan areas of more than 500,000 people. Louisville, Kentucky, began as a river port at the Falls of the Ohio, where goods had to be transshipped around this break-of-bulk point. It was also the closest point on the Ohio River to the state's earliest settled and most productive agricultural area, the Bluegrass region. By 1830, when a canal was built to bypass the falls, Louisville had a prominent role on the river, and it has continued to grow as a regional center.

Not far away, Nashville and Knoxville, Tennessee, also grew in locations favored by river transport and limestone soils. Nashville is located where the Cumberland River, a tributary of the Ohio, reaches farthest south into the hilly but fertile Nashville Basin of central Tennessee. It is the undisputed capital of America's country music industry. Knoxville is situated far up the Tennessee River from the Ohio, within the Ridge and Valley Section of the Appalachians. Little Rock, Arkansas, has the same advantages in transport and good soils. The main exception to this pattern of urban location in the Upland South is Birmingham, Alabama. The only large center of iron and steel industry to develop in the South, it grew near deposits of coal and iron ore in the Ridge and Valley Section of the Appalachians.

Outside the upland, southern cities with metropolitan populations of more than 500,000 include Baton Rouge, Louisiana, and Memphis, Tennessee, both Mississippi River ports, along with Charleston, South Carolina, and El Paso, Texas. Prior to its devastation by Hurricane Katrina in 2005, the historic city of New Orleans also had more than half a million people. The French founded New Orleans just upstream from the head of the "bird's foot" delta where the Mississippi divides into separate channels flowing to the Gulf. In the riverboat era before the Civil War, New Orleans was New York's main competitor as a port. After a period of decline, the port's prominence returned. Running north and south of the city, the Ports of South Louisiana and New Orleans (POSL) comprise the largest port in the nation by tonnage and the ninth-largest in the world. A substantial share of U.S. agricultural exports usually pass through POSL, as do regional international imports including oil, coal, chemicals, and fertilizer. New Orleans also grew into a major tourist center, famed for its jazz music and spring Mardi Gras festival. Then came Katrina, when New Orleans gained notoriety for the suffering of its people (see Natural Hazards, pages 664–665).

Upriver from New Orleans, Baton Rouge is accessible to medium-sized ocean tankers and is a major center of Gulf Coast oil refining and petrochemical industries. Still farther

upriver, Memphis grew to become the main business center for the greater part of the Mississippi Alluvial Plain. Its importance dates to antebellum days, when the so-called Cotton Kingdom developed in the Delta—the regional name then used for the Mississippi Alluvial Plain near Memphis in northwestern Mississippi. Capitalizing on its relatively central location on the continent, Memphis is the hub of the enormously successful FedEx mail and cargo service. One of the distinct industries of Memphis is tourism—verging on spiritual pilgrimage—to Graceland, the home and tomb of the venerated entertainer Elvis Presley, who died in 1977.

The other large metropolises lie at opposite ends of the South. Charleston was the port and commercial capital of the first major plantation area outside Virginia. It was the South's leading city at the time of the American Revolution. It languished as the plantation economy moved westward but resumed rapid growth with the rise of the modern southern economy. El Paso serves as a major railway center and tourist and business entrance to Mexico. It lies across the Rio Grande from Ciudad Juárez, which also serves as a major transportation center for northern Mexico.

11.2.3 The Midwest: Big River Country

The Midwest region as defined here includes 12 states. In the east are Ohio, Indiana, Illinois, Michigan, and Wisconsin. In the west are Minnesota, Iowa, North Dakota, South Dakota, Nebraska, Kansas, and Missouri; these correlate generally with what Americans perceive as the Heartland of the country (see Figure 11.2.1). The Mississippi River (whose Native American name means **"Father of Waters"**) separates these eastern and western groups, except for part of the Wisconsin-Minnesota boundary. All states in the eastern part of the Midwest border one or more of the Great Lakes, but only Minnesota does so in the western part of the Midwest. In the eastern Midwest, Ohio, Indiana, and Illinois are bounded on the south by the Ohio River, which carries 40 percent of the county's river traffic (•Figure 11.2.13). In the western Midwest, all states except Minnesota front the Missouri River.

The justification for the regional grouping of these 12 states as the Midwest comes from distinct spatial characteristics. They share an interior location; a distinctive plains environment; important associations with major regional lakes and rivers; historical and contemporary ethnic patterns; a productive combination of large-scale agriculture, industry, and transportation; and bountiful rural landscapes.

Environment and Agriculture in the Midwest

These vast midwestern plains produce a greater output of foods and feeds than any other area of comparable size in the world. The Midwest typically accounts for more than

• **Figure 11.2.13** The confluence of the Ohio and Mississippi Rivers at Cairo, Illinois.

• **Figure 11.2.15** A cornfield, irrigated by a center-pivot system, in the Midwest.

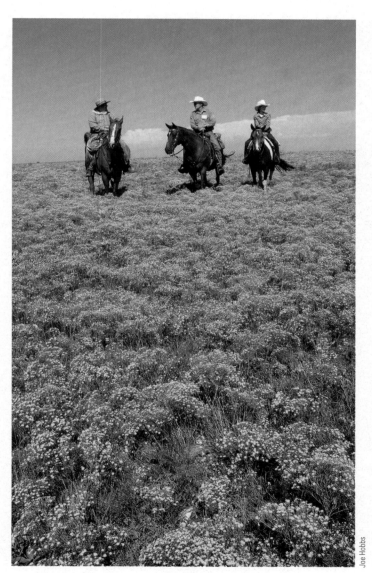

• **Figure 11.2.14** A prairie of the Flint Hills in Kansas.

40 percent of all U.S. agricultural production by value. Iowa leads in regional output and is the third-ranking state in value of farm products, exceeded only by the much larger states of California and Texas. Four other midwestern states—Nebraska, Illinois, Kansas, and Minnesota—are usually among the nation's top seven in total agricultural output. Lowest in regional output are higher-latitude Michigan, with its sandy soils, and the Dakotas, which have some dry and rough lands.

But soils are exceptional in most of the Midwest. The best soils developed under grasslands—which supplied abundant humus—in the prairies stretching from northwestern Indiana in a broadening triangle to the Dakotas, Nebraska, Kansas Oklahoma, and Texas (•Figure 11.2.14). 28, People played a role in shaping this biome; Native Americans enlarged it by using fire to clear wooded areas in order to improve hunting conditions and create pastures for game animals. Fertile soils like this are normally associated with steppe climates in the middle latitudes, where precipitation is too low to support agriculture. But in the Midwest, these soils are located in a more humid continental climatic area. Many of these soils have also been enriched by deposits of loess. These windborne soil materials accumulated in gla- 76 cial times when strong winds transported them from dry surfaces where the ice had melted.

In the eastern Midwest, the soils that formed under natural broadleaf forests are less fertile but still productive. The Midwest's poorest soils are mainly in the Ozarks and the Appalachian Plateau; in the sandy areas of coniferous forest in northern Michigan, Minnesota, and Wisconsin; and in the rougher western parts of the four westernmost states.

Corn (maize) is the leading product of midwestern farms and grows in the legendarily productive **Corn Belt** stretching from Nebraska to Illinois (•Figure 11.2.15). More than 80 percent of U.S. corn production and about 40 percent of world production comes from the Midwest. Conditions highly favorable for corn are here: fertile soils, hot and wet

NATURAL HAZARDS

Hurricane Katrina and Its Aftermath

On August 29, 2005, Hurricane Katrina made landfall as a Category 3 storm, with maximum sustained winds of 125 miles per hour (205 kph), along the vulnerable Gulf of Mexico coast from southeastern Louisiana to Alabama. New Orleans was the storm's epicenter of tragedy. Known as the "Big Easy" and the "Crescent City," New Orleans had enjoyed a reputation of carefree soulfulness, but Katrina changed that.

Hurricane Katrina was the worst natural disaster in U.S. history. Yet the city's tragedy was not unforeseeable. Geographically, New

Orleans was extremely vulnerable to the hurricane hazard. It sits in a bowl between the waters of Lake Pontchartrain to the north and the Mississippi River to the south, only 15 miles (25 km) from the waters of the Gulf of Mexico (●Figure 11.2.C). The city already lies an average of 9 feet (2.75 m) below sea level. It continues to slowly sink at a rate of 3 feet (0.9 m) each century as loose sediments are consolidated and cannot be replenished: far upstream, dams on Missouri River tributaries hold back 80 percent of the sediment that would naturally reach New Orleans, while the very

● **Figure 11.2.C** A monster storm, Hurricane Katrina overwhelmed the inadequate levees defending the vulnerable city of New Orleans.

summers, and huge areas of level land, making mechanical operations easy.

Biotechnology has enhanced the quality and productivity of this grain. Receipts from sales of beef cattle in the Midwest exceed those from corn, but here, too, corn is key because the cattle are raised mostly on corn. Soybeans are another major element in Corn Belt agriculture. Surpluses of both corn and soybeans are used in the production of ethanol (see pages 619–620).

Despite its typically impressive output, midwestern agriculture is a risky enterprise. Natural hazards including drought and plant diseases sometimes make for low har-

vests. Sharp fluctuations in world supply and demand (and therefore in market prices) and the prospect of cuts in government subsidies for key crops also hang over Midwest farming. Waves of farmers have been driven out of business at various times. Between 1930 and 2007, the number of farms in the Midwest decreased from just over 2 million to about 780,000. Mechanization has allowed farmers to work larger acreages but has also increased their costs, debts, and vulnerability to low market prices. Many who have given up farming in recent decades have been victims of debt. The machinery they borrowed money to operate has itself replaced many farmhands.

NATURAL HAZARDS

Hurricane Katrina and Its Aftermath *continued*

dikes that protect the city also keep available sediment in the Mississippi at bay.

Hurricane Katrina's storm surge penetrated the city's defenses, a 350-mile (560-km) system of elevated **levees** and 22 pumping stations. Before Katrina, the U.S. Army Corps of Engineers received less than half of the funds it had asked for to shore up this network, leaving New Orleans buffered by levees that were thought to be able to withstand a Category 3 hurricane. Katrina's **storm surge**—a powerful wave of water—put tremendous pressure on the levees, and they gave way in several places, notably along the 17th Street Canal levee near Lake Pontchartrain. Power supplies failed, and the city's pumps were unable to prevent water from inundating fully 80 percent of the city.

Preparations for the storm were inadequate. The city's mayor issued a mandatory evacuation order well in advance of the storm but did not make transportation available to tens of thousands of people who did not own automobiles. The disaster hit minorities and the disadvantaged especially hard: of the most affected, 4 in 10 were black, 2 in 10 poor, and 1 in 10 elderly.

The emergency response to Katrina was appallingly inept. Missteps and poor communication among city, state, and federal officials led to a delay of several days in deployment of security personnel and relief supplies. An estimated 25,000 people sought refuge in the giant Superdome sports stadium and an equal number in the city's Convention Center, but both sites were short on medical supplies, food, and emergency personnel. Many New Orleans police fled the city, and a few joined in the widespread looting that followed the evacuation. The appointed head of the government body responsible for disaster relief, the Federal Emergency Management Agency (FEMA), had no disaster management experience. Thousands of National Guard troops from Louisiana and neighboring states who might have been rapidly deployed to deal with the crisis were instead serving in Iraq.

New Orleans's population three months after the storm was about 280,000, or just 58 percent of its pre-Katrina level. The poorest and most vulnerable residents of the city had fled to Houston and other distant cities, and two years later, it still appeared that most would not return to New Orleans. Those who returned as

the city rebuilt were more affluent and more predominantly white, but large sections of the city remained wastelands.

The economic costs of repair and cleanup from Katrina were enormous: over $80 billion, making this by far the costliest natural disaster in U.S. history (in lives, Katrina took 1,577, well below the 8,000 lost to a hurricane in Galveston in 1900). Nearly 1 million people along the Gulf Coast were made homeless, and 300,000 jobless. Most of New Orleans's housing stock was made of wood and was irreparably damaged by floodwaters.

As the city continues to be rebuilt, questions persist about whether there should even be a new New Orleans. Not far from the sinking city, the Gulf of Mexico is eating away at the coastline: Louisiana has lost more than 1,560 square miles (4,040 sq km) of coastal land since 1930 and continues to do so at the rate of about one-tenth of a square mile (0.25 sq km) every day. There are concerns that global warming may be providing fuel for stronger hurricanes. Gulf of Mexico waters are about 1°F (0.56°C) warmer now than they were a century ago. The warmer these waters are, the greater their potential to support catastrophic hurricanes. Both Katrina and, later in the 2005 hurricane season, Hurricanes Rita and Wilma spent part of their time in the Gulf as Category 5 storms, the strongest possible.

How can the city be protected in the future? Work to repair and strengthen existing levees began in Katrina's wake, but engineers agree that these will not prevent another disaster. The U.S. Army Corps has a larger strategy, still in the planning stage, with three legs. One is to rebuild the narrow, steadily eroded barrier islands that form a punctuated chain between Louisiana's mainland and the Gulf of Mexico. Second, restored wetlands along the Gulf of Mexico coast would serve as a natural buffer against storm surge. Finally, federal and other funding would build the Morganza Project, a semipermeable levee system that would allow engineers to control the flow of both freshwater and tidal water across a wide swath of southeastern Louisiana. This "great wall of Louisiana" would not protect New Orleans directly but would reduce the losses of much of the states' coastal marshlands. These help buffer the city, serve as a nursery for Louisiana's valuable fishing industry, and are central to Cajun culture.

Following dramatic declines, the number of small farms has been relatively stable since 1990, especially as more family members commute to nonfarming jobs. But the long-term trend has been for U.S. farms to become fewer, larger, and more capitalized, with corporate farm ownership and enterprise increasing. There are about 2 million farms in the United States today, one-third the number in 1944. Farmers who have stayed on have bought or rented defunct farms, expanding their holdings. While the total amount of land farmed has decreased, the mean size of farms has increased from about 180 acres (73 hectares, or 730,000 square meters) to about 487 acres (197 ha). The

United States grows more food than it did 50 years ago on about 25 percent less land.

Industries and Cities in the Midwest

Most of the Midwest's population is urban. Most of the larger midwestern cities emerged in the early 19th century, especially on locations related to lake or river features. Even today, these cities benefit from advantageous locations for freight movement by water. When railroads came, the cities that were already established generally became rail centers as well. Then, in the mid-20th century, highways and airways also focused on the major existing cities.

One of the foundations of the region's industrialization and urbanization is its productive agriculture. Meatpacking, grain milling, and other food-processing industries have long been important. Cincinnati, dubbed **"Porkopolis,"** developed as a hog-packing center before the Civil War, and Carl Sandburg's poem *Chicago* celebrated that city's role as the "hog butcher for the world." The center of the midwestern meatpacking industry has moved westward to east central Nebraska and western Iowa. The same westward migration was followed by wheat milling, in which Minneapolis–Saint Paul now leads.

Besides processing agricultural products, midwestern industries supply many agricultural needs. A number of breakthroughs in farm technology (including Cyrus McCormick's 1840s mechanical reaper for harvesting wheat) were made in the Midwest. The world's largest farm machinery corporations (John Deere and Caterpillar) are headquartered in the region. Their activities boosted the development of other machine manufacturing. Saint Louis, for example, became second only to Detroit in the manufacture of the nation's automobiles.

The region's largest metropolitan concentration is around Chicago, Illinois. Chicago—known as the **"Windy City"** because of the boastfulness of its early promoters—is the country's third-largest city (9.5 million; ●Figure 11.2.16). The city was founded in 1833 at the mouth of the small Chicago River, which provided a harbor near the head of Great Lakes navigation. From that point the river reached up into the prime farmlands of the Midwest. A few miles upstream, a portage to tributaries of the Illinois River provided a southward-flowing connection via the Illinois to the Mississippi River near Saint Louis. A canal was built in 1848 along this portage route to connect Lake Michigan with the Illinois River. The link helped make Chicago the major shipping point for farm products from areas south and west of the city. In the 1850s, railways fanned out from Chicago into

an ever-broader hinterland and connected the city with Atlantic ports. Chicago became and still is the country's leading center of railway routes and traffic. In the 20th century, new harbors were built south of the city center, and new canals opened for barge traffic along the Illinois River. 620

Chicago's huge industrial structure rose at this major focus of transport and population growth. Major early industries were food processing, wood products made from white pine and other timber cut in the forests to the north, clothing, agricultural equipment, steel, and machinery manufacturing. Today, the most important manufactured products are machinery and steel, and Chicago is the country's leading center for both. The city's banks, corporate offices, commodity exchanges (especially the Chicago Board of Trade), and other economic institutions make it the economic capital of the Midwest. To the east, its metropolitan area reaches into northwestern Indiana, and to the north, it merges with that of Milwaukee, Wisconsin, which also has grown around a natural harbor on Lake Michigan.

The second-largest concentration of people and production in the Midwest is in metropolitan Detroit (4.4 million), together with the smaller adjoining area of Toledo, Ohio. The Detroit area extends along the water passages separating Michigan from the Ontario Peninsula in Canada and connecting Lake Huron with Lake Erie. From north to south, these passages include the Saint Clair River, Lake Saint Clair, and the Detroit River. Canada's Windsor, Ontario, lies just across the Detroit River from Detroit—somewhat surprisingly, to the city's *south*.

The French founded Detroit in 1701 as a fort and settlement on the Detroit River. The city served as a service center for the Michigan hinterland, which grew quickly after the Erie Canal opened in 1825. Explosive growth came to Detroit with the automobile boom of 20th century, when it came to be known as the **Motor City** or **Motown**. That second nickname, adopted by a local record company for its music label, became attached to the soulful music noted for its crossover appeal from black roots to a white audience starting in the late 1950s. 637

Toledo was founded on a natural harbor at the western end of Lake Erie. In the 19th century, natural gas and oil from western Ohio provided fuel for a glass industry there, with Toledo producing much of the glass for Detroit automobile production.

Several factors spurred the auto industry's growth in Detroit and nearby cities. Detroit had already built boats and horse-drawn carriages, and inventors, including Henry Ford—a Detroit native—experimented with internal-combustion engines. The availability of steel and a central location relative to markets were helpful. As the industry gathered steam early in the 20th century, managers realized that large-scale production would lower costs and increase sales. Manufacturing had to be centered in a few large plants. Huge quantities of steel were needed, as were factory locales from which the products could reach the entire national market cheaply. Detroit had a nearly optimal location. The center point for the population of the lower

● **Figure 11.2.16** Anish Kapoor's "Cloud Gate," known affectionately as "The Bean," in Chicago's Millennium Park.

Joe Hobbs

48 states was then in nearby southern Indiana (it is now in central Missouri).

From southeastern Michigan, the auto industry spread over much of the eastern Midwest: north to Flint and Lansing, east to Cleveland, south to Dayton and Indianapolis, and west to the Milwaukee area. Many midwestern cities became dependent on supplying auto parts. Assembly expanded to Saint Louis and Kansas City and eventually to other major regions of the country. The American corporate giants of the industry—General Motors, Ford, and Chrysler—still have headquarters and major plants in metropolitan Detroit—but perhaps not for long.

654 As with steelmaking, Midwest carmaking was a large industry employing many people at high wages. And as in the steel industry, employment of autoworkers began to fall in the 1980s because of decreasing plant productivity and overseas competition, especially from Japan. This industry has a history of boom and bust cycles, with the most recent trend being a bust: the three domestic automakers announced a series of factory closings and layoffs in recent years. Detroit's Big Three automakers, having continued to build gas guzzlers, found themselves unprepared for the rapid change in drivers' appetites as gasoline approached and passed the $3-per-gallon mark in 2007. Detroit employed a million autoworkers at the industry's peak in 1978 but only 266,000 in 2006, when three-quarters of them were offered incentives to give up their jobs. Where did they go? Flush with opportunities in the mining sector, Wyoming actively recruited many, while others migrated to Japanese car factories in the South.

The Lake Erie port of Cleveland, Ohio, has a metropolitan population of 2.1 million. Cleveland grew in the 19th century as a heavy-industry center at the mouth of the Cuyahoga River. Iron ore from the Lake Superior fields was offloaded there for shipment to Pittsburgh, and Appalachian coal flowed in the other direction to Great Lakes destinations. Smaller Lake Erie ports also took part in this traffic. Movement of ore one way and coal the other made it sensible to develop the iron and steel industry at Cleveland, at other Lake Erie ports, and at Youngstown, Ohio, on the route between Pittsburgh and the lake. Machinery and chemical plants and then the auto industry spread into the district. The nearby city of Akron rose as the leading center of tire manufacturing in the United States.

Northeastern Ohio was also afflicted by slow growth and closure of the older industrial enterprises. But some of the urban blight has been relieved by adaptive reuse and 656 gentrification. The environmental quality of Cleveland, a city once derided as the "mistake by the lake," is much improved. In 1969, the waters of the Cuyahoga River in Cleveland were so laden with fossil fuel waste that the river actually caught fire.

Aside from Columbus, Ohio, and Indianapolis, Indiana, the other major midwestern cities arose at strategic places on the major rivers. Although river traffic is not as important as it once was, these cities still benefit from riverside locations. Saint Louis, Missouri, known as the **"Gateway to the West,"** is the center of a metropolitan population

of about 2.8 million in Missouri and Illinois. The French founded the city in the 18th century, and Saint Louis remained the Midwest's largest city until the late 19th century. It is located at a natural crossroads of river traffic using the Mississippi to the north and south, the Missouri to the west, the Illinois toward Lake Michigan, and (via a southward connection on the Mississippi) the Ohio toward the east. Early traffic had to adjust to the fact that the Mississippi downstream from the Missouri mouth was much deeper than the upstream channel or the Missouri. Many riverboats plying the Mississippi below Saint Louis drew too much water to use the channels above the city, and Saint Louis thus became a break-of-bulk point where goods were transferred between different kinds of boats. The city has 334, suffered recently from a steep population decline, a decay 647 in its downtown area, and a ranking as one of America's most dangerous cities. City fathers have been scrambling to revitalize the downtown and restore Saint Louis's once esteemed reputation.

In Minnesota, Minneapolis–Saint Paul (population 3.1 million) developed at the head of navigation on the Mississippi River (•Figure 11.2.17). The Falls of Saint Anthony,

• **Figure 11.2.17** Minneapolis in Winter.

GEOGRAPHIC SPOTLIGHT

The Changing Geography of American Settlement

Across the United States, midsize cities with 750,000 to 2 million inhabitants—such as Cincinnati—have become more attractive communities for manufacturing and service firms. Traditional locational factors like proximity to coal or iron ore or rivers are no longer so important. Smaller cities have the infrastructure and capable workers that employers need. The Internet and other information technology allows many firms to locate almost anywhere they wish. Operating and living costs are often less than in the larger cities. Midsize cities often provide attractive incentives to new businesses, including low-cost land, reduced taxes, and subsidized transportation.

While the midsize American city has blossomed, the small town has withered. Between 1900 and the most recent census year, 2000, while the American population more than tripled, 8 of every 10 U.S. counties lost population. The decade between 1990 and 2000 saw a marked migration of people out of the rural Great Plains region in particular (●Figures 11.2.D and 11.2.E).

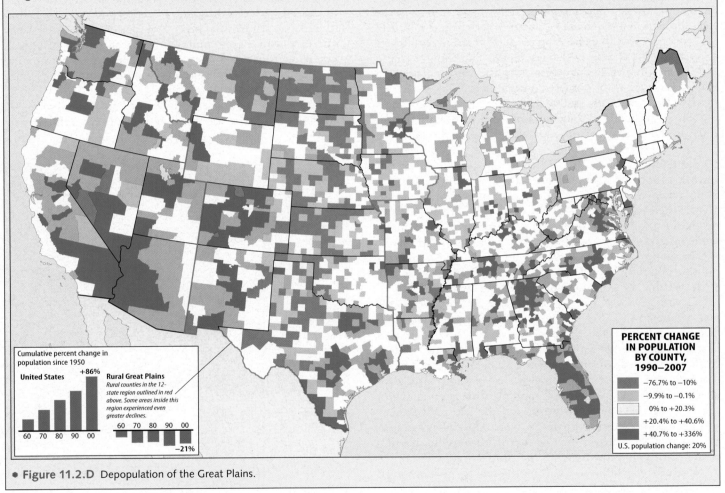

● **Figure 11.2.D** Depopulation of the Great Plains.

marking this point, supplied waterpower for early industries, especially sawmilling. Transcontinental rail lines were then built west from the **Twin Cities** to the Pacific. Business spread westward along these lines, making Minneapolis–Saint Paul the economic capital of much of the northern interior United States. From Minneapolis–Saint Paul all the way to Seattle, no center of comparable size emerged.

The business hinterland of the Twin Cities stretches across the "spring wheat belt" (where the crop matures later than at points farther south) to the foot of the Montana Rockies. Long a major focus of the grain trade and flour milling, Minneapolis–Saint Paul also has diverse industries. It boasts the nation's largest retail mall, the Mall of America. Minneapolis and other cities in Minnesota and neighboring Wis-

GEOGRAPHIC SPOTLIGHT

The Changing Geography of American Settlement *continued*

Joe Hobbs

● **Figure 11.2.E** Many small towns across the Great Plains are dying—or even dead.

A flight from small towns to larger cities was mainly responsible for this depopulation. Small towns had their heyday when they were vital clusters of merchants and townsfolk located on some avenue of transportation—highway, railway, or river—and serving a broad agricultural hinterland. The more fortunate towns became home to small industries like food processing, shoe, or textile factories.

Transportation gradually improved, and better roads led to ever-larger settlements that attracted most new businesses. By offering consistently lower prices and a wider selection of merchandise, Wal-Mart and other large retailers drove out diverse businesses in small towns and in the middle-sized and larger towns in which they were opened; this **"Wal-Mart effect"** has had a huge impact around the country. Smaller towns lost their economic significance and their residents.

The past two decades have seen some reverse migration. Some small towns in rural states from North Dakota to Kansas are actively recruiting international immigrants from countries including Vietnam, Sudan, and India to take up residence and help revive local economies. Others are enticing outsiders with offers of tax breaks, low-interest loans, and land giveaways to those who will build homes, often playing up the virtues of small-town living and the distinctive attributes of these settlements.

Another trend changing American settlement patterns has been the process of **exurbanization.** This represents a kind of reembrace or rural life, but in response to urban conditions. City dwellers who are either priced out of closer-in suburbs or drawn to the wider spaces of more distant ones move to the "exurb," an outer suburb of the city, typically so far out as to stand almost alone. The more affluent may even buy a rural "ranchette"—a plot of around 35 acres (c. 14 ha), the smallest typical subdivision—well away from the big city.

consin have become increasingly cosmopolitan in recent decades. Both states have experienced periodic culture clashes between groups. Recent examples include the refusal of Muslim Somali taxi drivers to carry airport passengers who have alcohol in their hand luggage and apparent tit-for-tat 355 killings of white and ethnic Hmong hunters.

Cincinnati (population 2.1 million), on the Ohio River, has a metropolitan area including the southwestern corner of Ohio and adjacent parts of Indiana and Kentucky. The Ohio River turns sharply southward from the city, and many early-19th-century settlers moving down the river disembarked at Cincinnati to travel overland into Indiana. The city developed a specialty in slaughtering hogs from Ohio and Indiana farms and shipping salt pork downriver to southern markets. As the meatpacking industries faded, coal mining developed in the nearby Appalachian Plateau, and Cincinnati became the economic capital for much of the mining region. It also developed an industrial base built largely on machine-working. New enterprises, including medical and scientific research and development, continued to attract professionals even as old industries died out (see Geographic Spotlight, above).

Kansas City (population 1.9 million) straddles the Missouri-Kansas border and resembles Minneapolis–Saint Paul in its situation and development. It began at the effective head of navigation westward on the Missouri River (whose nickname is the **"Big Muddy"**). Although boats could move upstream, they faced navigational hazards on a course leading sharply northwest instead of directly westward. As a result, two of the major overland trails to the West—the Santa Fe Trail and the Oregon Trail—had their eastern terminals and river connections in the Kansas 613 City area. Later, cattle drives came to Kansas City from the Great Plains, and railways reached westward from the city. Despite a major competitor to the west in Denver, Colorado, Kansas City today has a hinterland extending far to the west and is the economic capital of the "winter wheat belt," where crops mature earlier. Omaha, Nebraska (population 800,000), is a smaller riverside city resembling Kansas City and Minneapolis–Saint Paul in its functions and westward reach. Like them, Omaha lies on the western edge of the intensively farmed part of the Midwest, with drier, less productive wheat and cattle country to the west. Southwest of Kansas City, the midcontinental city of Wichita—once a major cow town on the Chisholm Trail—is a major center for aircraft manufacturing, with five producers, including Boeing and Cessna, headquartered there.

11.2.4 The West: Booming and Thirsty

Eleven states make up the West (see Figure 11.2.1). The eight interior states (New Mexico, Arizona, Colorado, Utah, Nevada, Wyoming, Montana, and Idaho) are known as the Mountain states or Mountain West. Farther west are the three Pacific Coast states of California, Oregon, and Washington. Oregon and Washington are known as the Pacific Northwest. Federal publications usually include Alaska and Hawaii in statistics for the Western states (as in Table 11.2.1), but these two states are so different that they are excluded here in generalizations about the West and are discussed later.

The West is 33 percent of the entire United States by area, or 38 percent of the conterminous United States. The federal government owns large areas of the region, and there are many military facilities on its mainly arid landscapes. There is a huge cluster of people and development in central and southern California and a handful of smaller groupings in widely dispersed locations. Otherwise, the region has a low population density. In 2007, the West had 67 million people (about 22 percent of the nation), 36 million of whom lived in California. California, the **Golden State,** is the West's economic colossus, producing almost 60 percent of the region's manufactured goods and 50 percent of its farm products (13 percent of the entire country's farm products). These states produced just 8 percent of the country's agricultural output and 4 percent of the value of its manufactures in 2007. These figures, however, conceal the fact that Nevada and Arizona have, respectively, the fastest- and second-fastest-growing populations of all U.S. states.

The Physical Setting

The overall low density of Western settlement reflects the region's dry climates, rough topography, and lack of internal water transportation. People are concentrated where there is water, so the population pattern is oasislike, with the oases separated from each other by steppe, desert, or mountains.

Topography has a big impact on western climates and settlement patterns. Precipitation is mainly orographic, with aridity due to the rain shadow effect. Prevailing dry conditions often menace the region with wildfires and droughts, punctuated ironically by floods and landslides (●Figure 11.2.18). Consensus is growing that climate change also poses a long-term hazard to the region. Since 1975, the West has warmed more rapidly than any other region of the

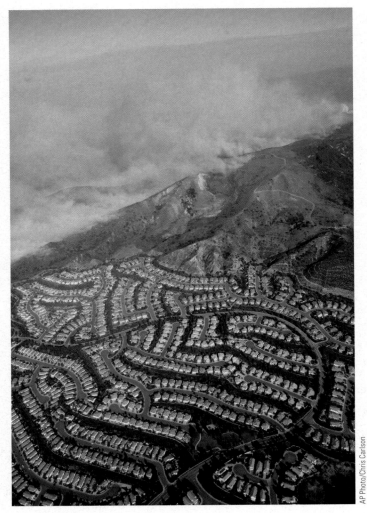

● **Figure 11.2.18** Wildfires are an increasingly common hazard in the arid American West. In October 2007, Santa Ana winds fanned flames that consumed and threatened thousands of homes in suburbs of Los Angeles.

country except Alaska. One striking result is the rapid loss of pines to an infestation of beetles that with warmer temperatures have a longer feeding season and can live higher in the mountains.

Mountains border the Pacific shore (see Figure 11.4, page 604). The relatively low Coast Ranges provide an often spectacular coastline from the Olympic Mountains in Washington to just north of Los Angeles, where the mountains swing inland and then continue south to Mexico. Inland, a line of high mountains parallels the Coast Ranges. The northernmost of these inland ranges is the Cascade Mountains, extending southward from British Columbia across Washington and Oregon to northern California. Situated on the Pacific Ring of Fire, the Cascades include many volcanic cones, with Mount Hood (near Portland) rising over 11,000 feet (3,330 m) and Mount Rainier (near Seattle) over 14,000 feet (4,267 m). In southwestern Wash-

Joe Hobbs

● **Figure 11.2.19** California's fertile Central Valley, here adjoining the Coastal Range, is a breadbasket of the United States.

ington is Mount Saint Helens, the volcano that in a major explosion in 1980 lost 1,300 feet (400 m) from its height and killed 57 people. The southern end of the Cascade Range joins the rugged Klamath Mountains, a deeply dissected plateau where the Cascades, Coast Ranges, and the Sierra Nevada almost meet. The Sierra Nevada increases in elevation southward to the High Sierra, where Mount Whitney, at 14,494 feet (4,418 m), is the highest peak in the 48 conterminous states.

Two large, hilly lowlands lie between the Coast Ranges and the inner line of Pacific mountains, and a smaller lowland lies outside the Coast Ranges in the extreme south. In Washington and Oregon, the Willamette–Puget Sound Lowland separates the Cascades from the Coast Ranges. The Oregon section, spreading from the lower Columbia River to the Klamath Mountains, is the Willamette Valley. The Washington sector, north of the Columbia River, is the

Puget Sound Lowland. The Puget Sound, a deep-water inlet of the Pacific, penetrates well inland.

The second large lowland is the Central Valley (or Great Valley) of California, enclosed by the Klamath and Coast Ranges and the Sierra Nevada. The Central Valley is a nearly flat alluvial plain, some 500 miles (800 km) long by 50 miles (80 km) wide (●Figure 11.2.19). In California's far southwest, a smaller lowland lies between the coastal mountains and the shore. The wider northern part of this lowland is the Los Angeles Basin, and a narrower coastal strip extends south to San Diego and Mexico. The Los Angeles Basin is a plain dotted by isolated high hills and overlooked by mountains. Hemmed in, it is the most densely populated urban area in the United States.

Air masses moving eastward from the Pacific Ocean bring precipitation to this region, mainly in the winter. Summer conditions are desertlike in the south, but winter brings some rain and mountain snow. The northern mountains have the most precipitation, including the deepest snows in the United States at the higher elevations. South of San Francisco, as precipitation declines, the magnificent Douglas fir and redwood forests of the north give way to sparse forests and chaparral (drought-tolerant grasses and shrubs) 29 that characterize the Mediterranean climate and biome of the southern half of California.

Lowland conditions along the Pacific strip vary from wet to arid. The Coast Ranges of Washington and Oregon are low and rainy, and the Willamette–Puget Sound Lowland is wet and forested. Incoming marine air moderates temperatures throughout the year, and this area has a marine west coast climate similar to that of Western Europe. The Coast 77 Ranges of California cast a rain shadow on the Central Valley of California. Although classed as having a Mediterranean climate—with subtropical temperatures, wet winters, and dry summers—this valley is dry enough that its natural vegetation was mainly steppe grass, grading into desert shrub in the far south. Irrigation has completely transformed the valley. The Los Angeles Basin also has a climate typically classified as Mediterranean subtropical, but with precipitation low enough to merit a steppe or desert classification.

The rain shadow effect becomes extreme to the east 25 of the high mountains. A vast area stretching east to the Rockies and around their southern end in New Mexico is mainly desert and steppe. Little moisture is left in the air masses, which shed most of their precipitation in the mountains near the Pacific. This physical region is known as the Intermountain Basins and Plateaus and includes large parts of nine western states.

The Intermountain Basins and Plateaus have three subregions. The Columbia–Snake River Plateau takes up most of eastern Washington and Oregon and southern Idaho. It is an area of mainly level terrain formed by lava flows, with scattered hills, mountains, and deep-cut river canyons. The Colorado Plateau occupies northern Arizona, northwestern New Mexico, most of eastern Utah, and western Colorado.

Here the Grand Canyon and other deep river canyons of the Colorado River and its tributaries cut rolling uplands into isolated sections (●Figure 11.2.20). The rest of the intermountain area is the Basin and Range Country, which continues beyond New Mexico into western Texas. The part between the Wasatch Range of the Rockies and the Sierra Nevada in California, called the Great Basin, has no river that reaches the sea. The whole Basin and Range Country consists of small basins separated from each other by blocklike mountain ranges. This is the driest part of the West, with connotations suited to its challenging environments. Its lore includes nuclear weapons testing, visitors from outer space, the wholesome legacy of the Mormons, and escapades of the world's oldest profession.

East of the Intermountain Basins and Plateaus, the high, rugged Rocky Mountains extend southward out of Canada through parts of Idaho, Montana, Wyoming, Utah, Colorado, and New Mexico, with a number of peaks rising above 14,000 feet (4,267 m; ●Figure 11.2.21). The Rockies are made up of a great many ranges, separated by valleys and basins. The mountainous continuity is nearly broken by the Wyoming Basin, which provides the easiest natural passageway through the Rockies. This was the route used by the first transcontinental railroad, the Central Pacific. The Rockies are a barrier to moisture-bearing air masses from either west or east, creating rain shadows for both the Great Plains at their eastern foot and the intermountain area at their western foot. The mountains themselves receive much orographic precipitation and are forested. The highest elevations rise above the tree line to mountain tundra and snowfields.

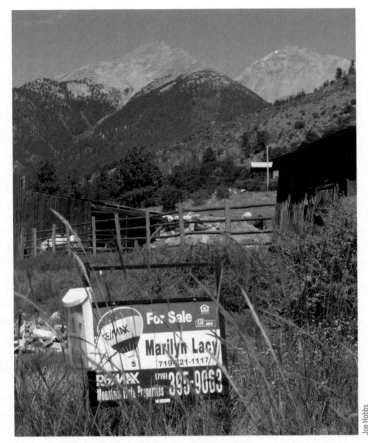

● **Figure 11.2.21** The Collegiate Peaks near Buena Vista, Colorado, include several "fourteeners," as locals call peaks over 14,000 feet high. For many, owning land with a view like this is an American Dream.

● **Figure 11.2.20** Arizona's Grand Canyon, carved deeply by the Colorado River, is but one of the many canyons that characterize the Colorado Plateau. This is the canyon's North Rim.

The higher western section of the Great Plains borders the Rockies in Montana, Wyoming, Colorado, and New Mexico. These plains are so high that Denver, at their western edge, is famed as the **Mile-High City.** The most common plains landscape here is an irregular surface punctuated by valleys. The plains region also includes hilly areas and isolated mountains such as the Black Hills of Wyoming and South Dakota. Harney Peak (7,242 ft/2,207 m) in the Black Hills, sacred to the Lakota and other Native Americans of the area, is the highest point in the United States east of the Rocky Mountains. The plains have a natural vegetation of semiarid steppe grassland, with temperatures that are continental in the north and subtropical in the south.

Water is a critical issue in the western United States. Historically, settlements were small and self-sufficient in water. But with explosive suburbanization following World War II, huge numbers of people moved west, especially to California, in search of the American dream. Populations quickly surpassed local water supplies, and new sources had to be found or redirected from elsewhere. The region's older cities, such as Denver and Phoenix, were able to grow because of federal support of giant dam, reservoir, and irrigation projects during the 20th century. But the realities of living in an arid environment have set in, and newer communities cannot meet their needs with large-scale manipulation of the environment. They must rely far more on water conservation and other ways of adapting to local conditions.

Large-scale 20th-century waterworks included canals dug to bring water from distant Mono Lake and Owens Lake, in the shadow of the Sierra Nevada, to the Los Angeles metropolitan region. These controversial projects produced decades of discord between northern and southern California. Environmentalists finally succeeded in efforts to restore these debilitated lakes with projects that have resumed water flow to both watersheds.

The region's largest water source is the Colorado River, carrying snowmelt from mountain peaks and flanks in north central Colorado and the Green River in Utah (•Figure 11.2.22). It has been highly manipulated to suit human needs. Its waters are impounded behind two major dams: the Glen Canyon in Arizona (•Figure 11.2.23) and the Hoover on the Arizona-Nevada border. Hoover Dam's Lake Mead (with the largest volume capacity of all American reservoirs) supplies a major portion of the water needs for about 25 million people in the three lower basin states of California, Arizona, and Nevada. Almost 20 million people in southern California drink these waters. Lake Powell (behind Glen Canyon Dam) and Lake Mead also generate hydroelectricity for the region. Below the dams, a canal network known as the **Central Arizona Project (CAP)** siphons water from the Colorado River to sustain large, booming, and very thirsty Phoenix and Tucson.

Colorado River water has been a subject of litigation and federal and state disputes ever since the Hoover Dam was completed in the mid-1930s. For decades, California has

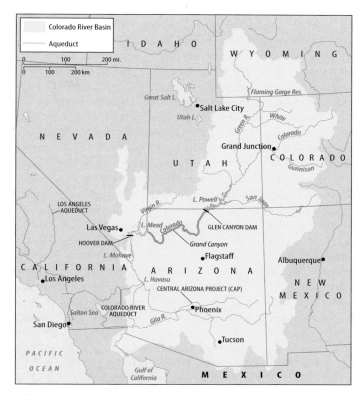

• **Figure 11.2.22** The Colorado River Basin.

• **Figure 11.2.23** The Glen Canyon Dam, just upstream from the Grand Canyon, was constructed in the 1950s. Note the low water level in this 2005 photo. A full reservoir would cover the white strip of bank that extends behind the dam.

used more water from the river than was allocated to it in a 1922 agreement known as the **Colorado River Compact,** while the six upstream river basin states (Arizona, Nevada, New Mexico, Utah, Colorado, and Wyoming) have drawn less. Meanwhile, the upstream states, especially Nevada and Arizona, have had explosive urban growth and demand for water. It took threats from the upstream states to actually cut off water to California to bring that powerful state to the negotiating table. In 2003, California agreed to reduce its use of Colorado River water over a period of 14 years to allow the six upstream basin states to use their share. There is unhappiness in California; water allocation is a classic **zero-sum game** in which any benefit to one party equals a loss to another. Meanwhile, booming cities search desperately for new water sources. Las Vegas wants to divert water by pipeline from northern Nevada—a project opposed by people there and in neighboring Utah. Saint George, Utah, a rapidly growing retirement and recreation community, wants a new pipeline from Lake Powell—much to the dismay of Nevada.

Most land east of the western coastal states is used for grazing, if at all, and the carrying capacity is so small that only low densities of animals and ranch populations are possible. Unirrigated farming (dry farming) of some steppe areas supports slightly higher population densities, notably in the wheat-growing areas of the Great Plains and the Columbia Plateau (known also as the Palouse) of Washington. There, yields from large and mechanized farms are low, and population is sparse. Most western agricultural output comes from a few large irrigated oases that grow both food crops and feeds, especially alfalfa, for dairy and beef cattle. Some oases in California and Arizona use irrigation to grow cotton, fruits, vegetables, and sugar beets.

The thirsty West is strongly urban, with most people concentrated in or near major oases. The Pacific Northwest cluster centering on Seattle and Portland might be considered an exception, but even this rain-fed area is a kind of oasis within the larger area of western drylands.

Southern California: A Sprawling Magnet

California's cities, with the unhappy distinction of being among the least affordable in the United States (at two to three times the national median home price), are concentrated in two areas with a long tradition of rivalry: a southern group focused on Los Angeles and a northern group centered on San Francisco. The southern California population cluster is by far the West's largest. This area includes the Los Angeles Basin and the narrow lowlands north to Santa Barbara and south to San Diego, along with their interior mountains.

The original Chumash and other Native American inhabitants of this area lived on wild foods and did not farm. In the 18th century, Spanish settlers from Mexico introduced small-scale irrigated agriculture and cattle ranching. By the 20th century, the Los Angeles area had become a large agricultural district. Today, sprawling urbanization has all but erased the orange groves and other agriculture that once characterized the Los Angeles Basin (•Figure 11.2.24). In 2007, about 13 million people lived in the metropolitan areas of Los Angeles—making it the country's second-largest city, after New York—and another 2.9 million in the adjacent metropolitan area of San Diego. Packed into this small corner of the West are about half of California's people, one-third of the population of the entire West, and 1 of every 15 people in the nation.

From San Diego (in 1769) to San Francisco (in 1776), Spain established Catholic missions as footholds on the northern frontier of its empire in the Americas. San Diego commands a magnificent natural harbor that is now the largest mainland base for the U.S. Navy's Pacific Fleet.

• **Figure 11.2.24** The sprawling metropolis of the Los Angeles Basin, with the skyscrapers of downtown Los Angeles in the distance.

Spain sited San Gabriel Mission and the *Pueblo* (civil settlement) of Los Angeles near the small Los Angeles River in 1781. When Mexico gained independence from Spain early in the 19th century, the California settlements became part of Mexico. The United States acquired them with the end of the Mexican-American War in 1848. Until the late 1800s, San Diego and Los Angeles were still small settlements in one of the world's most isolated corners.

The nearly continuous boom of southern California began with the arrival of railroads from the east in the 1870s and 1880s. The main lines (the Southern Pacific and the Santa Fe) crossed the area's encircling mountains via the Cajon Pass near San Bernardino. Railroads offered very low fares on westbound trains, promoting settlement. Oranges packed in refrigerated cars were shipped at high cost to far-off eastern markets, and irrigated orchards spread rapidly. Los Angeles was the main transport and business center for the citrus business. California led U.S. citrus production from the 1890s until the mid-20th century, when urban sprawl consumed the southern California groves.

The climate that favored oranges also attracted a new population seeking health, comfort, and the glamour associated with Hollywood, a small district of Los Angeles that rapidly became the world's most famous filmmaking center. In the early 1900s, pioneering moviemakers from the Northeast relocated to Los Angeles to take advantage of the area's bright sunshine, warm weather, and varied landscapes. Time was not kind to Hollywood, but like New York's Times Square, Hollywood (nicknamed "Tinseltown") has recently undergone a transformation from seedy and dangerous to chic and welcoming.

Other resources spurred development in southern California. The tar from the La Brea Tar Pits was an oil-based resource used by early settlers. In the 20th century, the expanding auto industry made the region's petroleum very important. The region still produces and refines petroleum, but environmental regulations prevent new offshore oil infrastructure from being developed. Tuna and other fishing fleets plying Pacific waters were (and still are) based in the area. In the 1920s and 1930s, the area's mild climate, open spaces, and large labor force attracted aircraft manufacturing. This industry boomed during World War II and then joined with the electronics industry in aerospace development. Up to the 1990s, the aerospace industry, heavily supported by federal military expenditures, had been southern California's largest manufacturer. Despite its glittery Hollywood reputation, Los Angeles ("L.A.") had become a major industrial city—an aerospace Detroit with palm trees.

San Francisco: The City by the Bay

About 4.1 million Californians—21 percent of them Asian Americans—lived in the San Francisco metropolis, the fifth-largest urban area in the United States in 2007. San Francisco—beloved as "The City" by its residents and neighbors—is located on a hilly peninsula between the Pacific Ocean and San Francisco Bay, but the metropolitan area includes counties on all sides of the bay (•Figure 11.2.25). The climate is unusual. Marine conditions moderate San Francisco's temperatures from an average of 48°F (9°C) in December to only 64°F (18°C) in September. Local marine and topographic circumstances give rise to San

654

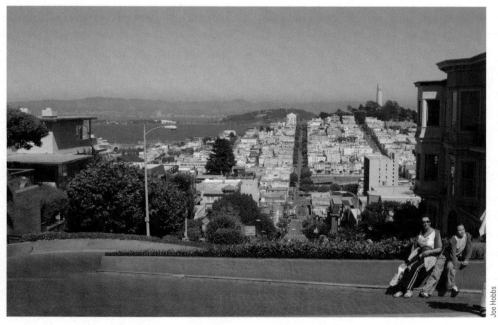

• **Figure 11.2.25** San Francisco: the top of "The Crookedest Street in the World" (Lombard Street), with Coit Tower, Angel Island and the hills of the East Bay visible in the distance.

Francisco's trademark fog, which can produce very chilly days even in the height of summer. Mark Twain is said to have quipped, "The coldest winter I ever spent was a summer in San Francisco."

The area is in the border zone between the marine west coast and Mediterranean climates. Characteristic of the Mediterranean influence, about 80 percent of the area's rainfall comes in five months from November through March. Mean annual precipitation is 20 inches (50 cm) in San Francisco, enough to produce a naturally forested landscape. There are even large groves of redwood trees within a short distance both north and south of the Golden Gate Strait. East of San Francisco Bay, warmer temperatures and lower precipitation have produced a scrubby forest (chaparral and scattered oak trees) and grassland.

San Francisco began as a Spanish fort (*presidio*) and a mission on the south side of the Golden Gate Strait. It grew as a port for the gold-mining industry and then for the expanding agricultural economy of central California. The **Gold Rush of 1849** centered on the gold-bearing gravels of Sierra Nevada streams pouring into the Central Valley north and south of Sacramento. San Francisco, with its spacious natural harbor, was the nearest port to receive immigrants and supplies for the gold camps, and the city boomed during the race for gold.

Today's San Francisco Bay Area is a major domestic and international tourist destination, famed for its scenery, wines, arts, and other cultural attractions. Manufacturing in the area is varied. In recent decades, computer-related and other electronics businesses have boomed from San Francisco south to San Jose, center of an area known as **Silicon Valley** for the silicon chips that carry microcircuits. The rise of Silicon Valley as a center of high technology is due partly to the allure of the San Francisco area as a place to live and its proximity to two outstanding universities: the University of California at Berkeley, next to Oakland, an industrial center on the east side of the bay, and Stanford University, in Palo Alto south of San Francisco. When the **dot-com bubble** (or tech bubble) popped at the turn of the 21st century, much of the software capital fled, and the region suffered a downturn. The local economy had rebounded sharply by 2005, however, in part by reaffirming its place as the country's leading biotechnology hub, with more than 800 companies in that sector. Among the most recently established is the leading center for stem cell research in the United States.

The Golden State's Rainbow

The Bay Area's fortunes reflect larger trends in the state. In the early 1990s, California's economy slumped, partly because of declining federal defense outlays. For the first time in decades, California saw more people leaving the state than entering it. By 1995, the patterns had reversed, with the economy and population growing again, only to give way in the first part of the following decade to a pronounced recession. A statewide crisis in energy services and the growth of an enormous budget deficit led in 2003 to the recall of California's governor and the election of a successor—appropriately for California, the movie megastar Arnold Schwarzenegger. California began a recovery on economic boosts from tourism and a growing trade with the booming region of East Asia. If considered apart from the United States, California would have had the world's 10th-largest economy in 2007.

California has a very diverse ethnic population. Non-Hispanic whites have recently become a minority in California for the first time in over a century, representing 43 percent of the state's population. Active in-migration by Hispanics and Asians has caused most of the demographic shift. The changes have been particularly dramatic in Los Angeles, with a population that is 46 percent Hispanic, 29 percent non-Hispanic white, 10 percent Asian, and 11 percent African American. The widely varied ethnic groups that add such a definitive flavor to California are not clustered only in homogeneous neighborhoods, such as the Koreatown and Chinatown of Los Angeles, but are scattered through the metropolitan areas. Urban California is a veritable United Nations, with a multiculturalism that is perhaps unparalleled on the planet (•Figure 11.2.26).

Not everybody is pleased with this vibrant diversity. Now in the minority, many non-Hispanic whites worry that incessant immigration will overwhelm services and opportunities in the state. Since the mid-1990s, more people have moved away from California each year than have moved into it from other parts of the United States; immigration to California from abroad, however, is vast and prevents the state's population from shrinking. In 1994, California voters passed Proposition 187, touted by proponents as "the first giant stride in ultimately ending the illegal alien invasion." The measure was designed to drive out undocumented aliens and deter future immigrants by cutting them off from medical, educational, and other public services. Governor Schwarzenegger was voted into office in part on his promise to repeal a controversial state law that allowed undocumented immigrants to obtain driver's licenses.

• **Figure 11.2.26** One face of diverse California, at a street fair in San Jose.

Seattle, Portland, and the Pacific Northwest

The Puget Sound Lowland of Washington and the Willamette Valley of Oregon have two large port cities, each located on a navigable channel extending inland through the Coast Ranges. Seattle, Washington (population 3.2 million), is located on the eastern shore of the Puget Sound and has access to the continental interior via passes over the Cascades (•Figure 11.2.27). Seattle is the manufacturing home of Boeing, the world's largest aircraft producer (with headquarters in Chicago). The abundance of the Sitka spruce (used for early wooden plane construction) and the availability of cheap electricity drew William Boeing to start the company here in 1917. Portland, Oregon (population 2.1 million), is on the navigable lower portion of the Columbia River where the Willamette tributary enters it.

Another major industry in the Pacific Northwest, also linked to pioneers in their field, is the Microsoft Corporation of Redmond, Washington. The firm was born in 1975 in Albuquerque, New Mexico, but in 1979, the two founders—Bill Gates and Paul Allen—moved their headquarters to the Seattle area. Microsoft became the largest computer software company in the world and is a major source of jobs and export revenues for the Northwest and the United States. Bill Gates, one of the world's two richest men, has directed some of his enormous financial resources toward solving the global epidemic of HIV/AIDS.

Fishing, logging, and trade built the cities of the lowland in the late 19th and early 20th centuries. But agricultural land in the Willamette Valley drew the early settlers to Oregon Country in the 1840s. Agriculture is important, even though hilly terrain and leached soils make it difficult. The leading products are milk and truck crops (vegetables grown on a large scale for shipment to distant markets), and the Oregon wine industry is growing.

Fishing, especially for salmon, is still an important occupation, but the fish catch in the Pacific Northwest is now much smaller than in the three leading states—Alaska, Massachusetts, and Louisiana. Plummeting fish stocks off

U.S. coasts have led to a dramatic overall decline in the fishing industry. As federal fishing quotas have tightened, especially since 2000, the government has stepped in to buy out the faltering fisher fleets. On the West Coast, for example, boat owners have been paid about half a million dollars each to permanently remove their vessels from fishing. 32, 634

Oregon and Washington are still the leading states in lumber production, with about 25 percent of the country's saw timber resources. Most of the trees are coniferous softwoods, and many, like the Douglas fir, are enormous. The timber industry is in a period of steep decline because of high costs and challenges from environmental advocates.

The Mountain States

The eight inland states of the Mountain West are drier than the Pacific states. They have neither the abundant irrigation waters of California nor the advantage of ocean transportation. These limiting factors kept the population of the eight states to only 18 million people in 2007, or about half of the size of California's population.

Longtime Rocky Mountain residents often express fears of what, in a barbed play on words, they call **Californication:** being bought out by wealthy "equity refugees" who cash out from their high-priced California homes, scoop up bargain properties, drive up prices, and move their quirky West Coast ways to the Rockies.

About half the region's people are in a handful of extensively irrigated metropolitan areas. The others are scattered across huge expanses of ranches and farms, Indian reservations, and small service centers and mining towns. The largest population cluster of the Mountain West lies in the Basin and Range Country of southern Arizona. Here about 75 percent of Arizona's people live in the metropolitan areas of Phoenix (population 4 million), the second-fastest-growing large city in the country (after Fort Worth, Texas), and Tucson (population 900,000). Groundwater and waters from the Salt and Colorado Rivers have supported this development in a subtropical desert.

Phoenix was originally the business center for an irrigated area along the Salt River (•Figure 11.2.28). The population grew after 1912 when the West's first federal irrigation dam in the West was completed. Today, cotton is the area's main crop, with hay grown for dairy and beef cattle. Phoenix and Tucson grew faster after the introduction of air conditioning made them attractive retirement and business centers. Each is the site of a major state university, and Phoenix is the state capital. The wealth of Sun Belt retirees, tourism, and copper mining are also basic to Arizona's economy.

Another population cluster stretches along the Great Plains at the eastern foot of the Rocky Mountains. Within this area, called the Colorado Piedmont, metropolitan Denver was home to 2.4 million people in 2007, and about 1 million more lived in other metropolitan areas along the Front Range of the Rocky Mountains, such as Boulder, Fort Collins, and Colorado Springs. These cities developed

PHIL SCHERMEISTER/NGS Image Collection

• **Figure 11.2.27** Downtown Seattle, dominated by the Space Needle

• **Figure 11.2.28** Water is lavished on golf courses and lakefront homes in the desert city of Phoenix.

The third-largest population cluster in the interior West is the Wasatch Front or Salt Lake Oasis of Utah, made up of metropolitan Salt Lake City–Ogden (population 1 million) and smaller Provo to its south. It contains about three-fourths of Utah's people, in a north-south strip between the Wasatch Range of the Rockies and the Great Salt Lake, with a southward extension along the foot of the Wasatch.

This oasis was developed as a haven for members of the **Church of Jesus Christ of the Latter-Day Saints** (popularly known as **Mormons**) when they were persecuted for their unconventional doctrines and expelled from the eastern United States in the 1840s. After a difficult trek westward, they chose the Salt Lake Basin as their "New Zion," a challenging but secure sanctuary. More than 400 other, smaller Mormon settlements dot the West, but Utah has the most, and the state's population is about 70 percent Mormon. In Salt Lake City, the Mormons built the temple that is the symbolic center of their religion and culture (•Figure 11.2.29).

on streams emerging from the Rockies. The stream valleys provided routes into the mountains, where mining booms began with a gold rush in 1859. One mineral rush followed another through the late 19th century. The Colorado Piedmont towns became supply and service bases for the mining communities, which were often short-lived.

The Colorado Rockies have recently produced molybdenum, vanadium, and tungsten, rare alloys in great demand in high-tech industries. High energy prices in the 1970s led to expanded extraction of fuels from the state's large coal reserves and scattered oil and gas deposits. So far, because of environmental concerns and shortages of water needed for processing, the region's **oil shales** have not been tapped significantly for the hydrocarbons they contain (see Figure 11.2.22). The U.S. government estimates recoverable oil shale reserves in the region at 800 billion barrels, or more than three times the proven oil reserves of Saudi Arabia. With recent oil prices reaching unprecedented highs, major oil companies are examining new ways to coax the oil out of the ground without despoiling watersheds and depleting water resources.

Irrigable land along Colorado Piedmont rivers provided food for early miners, and agricultural servicing became an important function of the towns. In time, agriculture grew into a prosperous business that was more stable than mining. Growing irrigated acreage prompted a boom in production of feed to support large cattle feedlots.

Tourism in the Rockies has grown with increasing American affluence, better transportation, and increased popularity of skiing. Denver is both a tourist destination and a major regional center for the federal government. Nearby Colorado Springs is the site of the U.S. Air Force Academy and is close to the headquarters of the **North American Aerospace Defense Command (NORAD)**. Located in a massive complex hewn inside nearby Cheyenne Mountain, NORAD monitors potential airborne threats to the United States and Canada.

231,
620

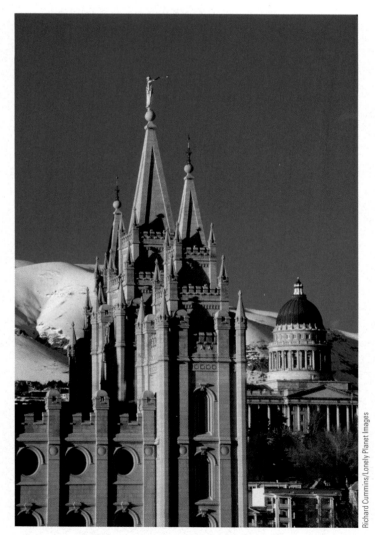

• **Figure 11.2.29** Three icons of Salt Lake City: the Mormon Temple, the Utah State Capitol building, and the snow-clad Wasatsh Range.

With no large competing metropolis within hundreds of miles, Salt Lake City is today the business capital for the huge, sparsely populated area outside the Mormon core region. It is poised to become the epicenter of a fiber-optic network for high-speed Internet and other data communication with 14 other cities in Utah. Funded as a public works project, the **Utah Telecommunication Open Infrastructure Agency (UTOPIA)** is designed to lure new high-tech firms and intellectual capital to the state.

Each of the other five states of the interior West has even fewer people than Utah. The people of New Mexico are concentrated along the north-south valley of the middle and upper Rio Grande, where Albuquerque (population 800,000; •Figure 11.2.30) is the main city, competing economically with El Paso, Texas, on the Rio Grande just south of New Mexico. New Mexico's economy is a typical western mix of irrigated agriculture, ranching, mining, tourism, some forestry in the mountains, and federally funded defense contracting. New Mexico's mining industry has encountered cultural obstacles in its quest for new resources. As natural gas prices have risen, growing pressure to expand gas production has been resisted by Navajo concerned about defilement of sacred land. Many Navajo both in New Mexico and Arizona complain that energy companies have long cheated them by paying them low royalties for coal and other mineral leases. But others have supported development of these resources for the welcome revenues they do provide.

Most people in Idaho live in oases along the Snake River, which crosses the volcanic Snake River Plains in southern Idaho. Tourism and federal money yield less income to Idaho than to New Mexico, but agriculture is more substantial. Idaho is famous for its potatoes. Idaho's capital and largest city, Boise—which prides itself on quality of life—has only about half a million people.

Except for a small section in the Sierra Nevada, the state of Nevada is desert, with little irrigation and a history of booms and busts in gold and other minerals. Metropolitan Las Vegas (population 1.7 million—in recent years the fastest-growing major city in the United States) and Reno (population 400,000) have over 80 percent of the state's population. While Las Vegas (•Figure 11.2.31) portrays itself as a family-friendly destination, it is also content to perpetuate its older reputation as "sin city"; its tourist promotion boasts "What happens in Vegas stays in Vegas." The state's laws have allowed gambling and prostitution to be Nevada's trademark industries, and gambling is an especially strong draw for tourists. Casinos and associated entertainment and convention business play a dominant role in the state's wealth.

Military interests are another leading industry for Nevada, in which almost 85 percent of the land belongs to the U.S. government. Most U.S. nuclear weapons testing has been done in Nevada. Northwest of Las Vegas is Yucca Mountain, designated to become a repository for nuclear waste from around the country. As of 2008, however, technical and environmental concerns were still preventing the underground storage site from accepting the waste.

The economies of Montana and Wyoming are based on farming, ranching, mining, and summer and some winter

• **Figure 11.2.30** Albuquerque lies in the Rio Grande Valley, at the foot of the Sandia Mountains.

● **Figure 11.2.31** The Las Vegas Strip evokes wealth and licentious-ness, making the city of Las Vegas an irresistible tourist draw.

tourism. Wyoming is the country's largest producer of coal and its associated methane gas. Recent high energy prices have been a bonanza. Natural gas production, centered on once-sleepy Pinedale, has brought a windfall to Wyoming's economy, even flooding its school system with unexpected cash. Wyoming and other western states have a history of discord between companies and politicians favoring energy production on public lands and environmentalists intent on protecting the wilderness. Montana is known as the **Big Sky Country,** famed for its wide-open spaces and the spectacular mountains for which the state was named. Neighboring Wyoming is also physically dramatic, particularly in the west, with its world-famous natural features of Yellowstone and Grand Teton National Parks (●Figure 11.2.32). Both states have seen a surge in property investment by wealthy American entertainers and businesspeople.

11.2.5 Alaska and Hawaii: The Newest States

The United States began to acquire Pacific dependencies in the mid-19th century. In 1959, two of these, Alaska and Hawaii, became the 49th and 50th states admitted to the Union. The United States bought Alaska from Russia in 1867 and annexed Hawaii in 1898. The American desire for a Pacific empire was rooted partly in military interests, which still play important roles in both states. There are large military bases in Alaska and Hawaii, and military expenditures contribute much to the economies of both. In their physical geographies, these states are quite unlike one another or the **Lower 48**—the contiguous U.S. states (see Figures 11.4 and 11.2.1).

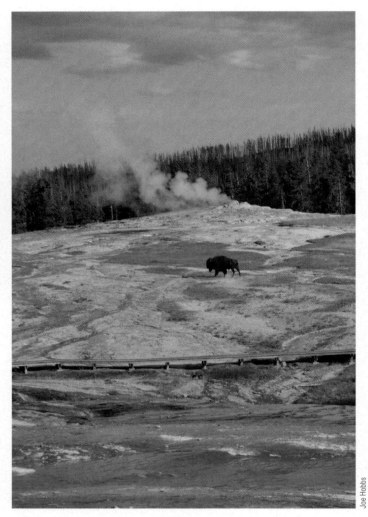

● **Figure 11.2.32** The Old Faithful Geyser Basin of Yellowstone National Park, with a steaming (but not erupting) geyser, bison, and the Firehole River. The skeletal lodgepole pines on the skyline were burned in the park's great fires of 1988.

Alaska: Land of the Midnight Sun

Alaska (from *Alyeska,* an Aleut word meaning "great land") makes up about 17 percent of the United States by area, but it is so rugged, cold, and remote that its total population was only 670,000 in 2007.

Alaska's Arctic Coastal Plain is an area of tundra along the Arctic Ocean north of the Brooks Range. Known as the North Slope, it holds the reserves of oil and natural gas that make Alaska the biggest fossil-fuel-producing state in the United States. Alaska hopes to pump natural gas from the North Slope to the Lower 48. To date, the focus has been on oil, which supplies more than 80 percent of the state's revenues. Oil income has built a hefty surplus known as the **Permanent Fund.** Interest accrued in this account pays every resident about $1,100 a year just to live in the Union's harsh but beautiful northernmost state (see Geography of Energy, page 683). The cost of living is high in this state, where virtually all foods and consumer goods must be imported.

The barren Brooks Range is sometimes considered to be the northwestern end of the Rocky Mountains, with summits ranging in elevation from about 4,000 to 9,000 feet (c. 1,200–2,800 m; •Figure 11.2.33). The Yukon River Basin lies between the Brooks Range on the north and the Alaska Range on the south. The highest mountains are in the Alaska Range, where, about 100 miles (160 km) north of Anchorage, Denali (also called Mount McKinley) rises to 20,320 feet (6,194 m). This is the highest peak in North America. The Yukon is a major river, navigable by riverboats for 1,700 miles (c. 2,700 km) from the Bering Sea to Whitehorse in Canada's Yukon Territory. In Alaska, the Yukon Basin is generally rolling or hilly, with large areas of flat and often swampy alluvium. Coniferous forest thrives in its subarctic climate.

Fairbanks (population 30,000), on the Tanana tributary of the Yukon River, is interior Alaska's main town. The Alaska Highway connects it with the Lower 48 via western Canada, while highways and the Alaska Railroad link Fairbanks with Anchorage and Seward. Fairbanks is a service center for the Trans-Alaska Pipeline from the North Slope oil area to the southern port of Valdez. At 64 degrees north latitude, this epitomizes Alaska as the American "land of the midnight sun," where the sun is above the horizon for 22 hours when summer begins—conditions that allow for some farming in the Tanana Valley. Cabbages can achieve prodigious sizes in Alaska's round-the-clock sunshine.

Alaska's largest city, Anchorage (population 400,000—over half the state's total population) lies south of the Alaska Range and near the Matanuska-Susitna Valley, one of the few places in Alaska where agriculture is viable. At the southwest, the Alaska Range merges into the lower mountains of the long, narrow Alaska Peninsula and the rough, almost treeless Aleutian Islands. The Aleutians mark the northern edge of the volcanically and seismically active Ring of Fire. Just south of Anchorage, the wildlife-abundant Kenai Peninsula lies between Cook Inlet and the Gulf of Alaska. Still farther to the southwest is Alaska's largest island, Kodiak, from which a resident subspecies of brown bear takes its name. This southwestern coastal section of the state has a subarctic climate of that is colder than that of Alaska's southeast.

Alaska's narrow southeastern portion is known as the Panhandle. The mainland ranges of the Panhandle are an extension of the Coastal Mountains of British Columbia, and there are many mountainous offshore islands. The state's capital, Juneau (population 30,000), is located in the Panhandle. Juneau is the only U.S. state capital not reachable by road, but a proposed (and locally controversial) highway to Skagway may change that.

21, 602

• **Figure 11.2.33** Alaska.

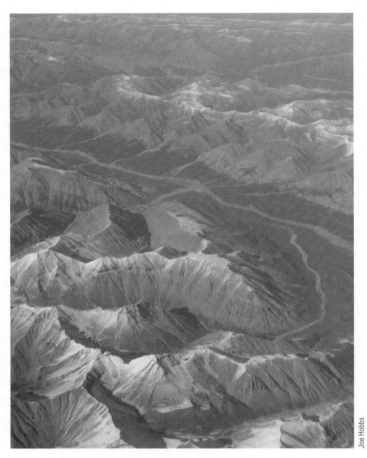

● **Figure 11.2.34** Wilderness of east-central Alaska.

The spectacular Inside Passage from Seattle and Vancouver to the Alaskan Panhandle lies between the islands and the mainland. It is a popular route for cruise ships, as some of the world's most magnificent glaciers and mountains may be seen on its shores. On the mainland, some mountain peaks rise more than 15,000 feet (4,575 m). Fjorded valleys extend their long arms of the sea inland. A cool and wet marine west coast climate prevails, favoring coniferous forests.

The general trend all over Alaska in recent decades has been toward warmer temperatures, with the average annual temperature increasing by 3.4°F (1.8°C) between 1949 and 2006. This trend has had significant impacts on the state's oil industry, cutting in half the number of days that industry vehicles are allowed to cross the tundra. There are other effects, too—for example, coniferous forests are sinking into areas where permafrost has melted, and river temperatures increased by nearly 10°F (4.5°C) between 1984 and 2007, driving salmon from their traditional habitat.

Native Alaskans are being hit particularly hard by problems attributed to global warming. Erosion and flooding in the wake of permafrost melting, late-season ice formation, and land subsidence is affecting an estimated 184 of 213 Native Alaskan villages. Many communities are being compelled to relocate, in the midst of uncertainty about who will pay for the move and what people will do in their new locales. Most villages have supported a subsistence livelihood of fishing and hunting for generations, and the sudden disruptions come as a great shock.

Alaska's Ups and Downs

After about 200 years of rule by Russia and the United States, the population of Alaska was still only 73,000 in 1940. Eighteenth-century Russian fur traders lived in small numbers along the southern and Panhandle coasts. Their legacy remains in Russian place names and Russian Orthodox churches on the landscape. In 1812, Russians established an agricultural colony on the California coast at Fort Ross, north of San Francisco Bay. They abandoned this attempt to better provision their Alaskan posts in 1841. Six years later, Russia sold remote and unprofitable Alaska to the United States at 2 cents per acre. Many Americans derided the transaction as **Seward's Folly,** after William Henry Seward, the American secretary of state who negotiated it. The value of oil pumped *every day* in Alaska now exceeds what the United States paid for its vast new territory.

By the time the United States bought Alaska, hunting had nearly wiped out the sea otters, whose costly pelts were the main incentive for Russia's Alaskan fur trade (their population has since bounced back). There was a burst of American settlement during gold rushes in the 1890s. Then came only a slow increase from military, fishing, and logging settlements. Most of the state's commercial logging is in the Tongass National Forest of the Panhandle region, but it is unprofitable and is maintained only by costly and controversial federal subsidies.

Beginning with World War II (when Japanese troops succeeded in invading the Aleutian Islands), Alaska's growth became more rapid. Growth came from military and governmental employment (the federal government spends more money in Alaska per capita than it does in any other state), fishing (for king crab and for wild salmon esteemed for not having the contaminants common in farmed salmon), tourism, and especially the boom in the oil industry after the 1973 energy crisis. Alaska has lived though a series of boom and bust cycles, with the booms always related to its natural resources. With oil production in decline, many Alaskans are asking what the next big boom might be. There is no readily apparent prospect on the horizon, and a few Alaskans are beginning to ask how the state might pursue a path of more sustainable growth.

About 15 percent of the state's residents are Alaska Natives (Inuit or Eskimos, Native Americans, and Aleuts). Long the victims of exploitation and neglect by whites, these peoples received some redress in large land grants and cash payments under the Alaska Native Claims Settlement Act passed by Congress in 1971.

Alaska's population is aging, and it does not appear that natural increase will raise the numbers of young people

164

621

228, 232

GEOGRAPHY OF ENERGY

The Arctic National Wildlife Refuge

228, 232

The 1973 energy crisis delivered a profound shock to United States and other industrialized economies. In its wake, there was a push to develop more domestic sources of energy, particularly oil. The most significant reserves were in the North Slope of Alaska, where Prudhoe Bay and nearby oilfields were established. Since 1977, their oil has been transported southward through the Trans-Alaska Pipeline, an extraordinary feat of engineering snaking 800 miles (1,300 km) across some of the most challenging environments on earth, all the way to the port of Valdez on the south coast.

Originally holding about 10 billion barrels of proven reserves (just under 4 percent of Saudi Arabia's estimated reserves in 2007), Prudhoe Bay and nearby fields now produce about 3 percent of total U.S. domestic oil production. The fields' output peaked in 1988, when 20 percent of the country's oil was produced there, and has declined steadily since then. Attention has inevitably turned eastward and southward to the estimated 3 to 8 billion barrels beneath the **Arctic National Wildlife Refuge (ANWR,** often spoken of as "Anwar"). When the 12,500 square miles (32,000 sq km) of the refuge were set aside as wilderness in 1980, the fate of the oil beneath this tundra ecosystem was left for Congress to decide at some later date.

ANWR's oil has been the subject of a bitter and polarizing debate ever since. On the one side are the oil interests, Alaskan politicians, almost all Republicans in Congress, and Republican presidents, who have argued that the United States must develop these reserves to give the country more independence from Middle Eastern oil. On the other side is a coalition of environmentalists,

Native Americans, and congressional Democrats who argue that oil production in ANWR will do irreparable damage to the coastal plain's unique environment and the indigenous people who depend on it.

The main Native American group involved is the **Gwich'in,** a 7,000-strong Athabascan tribe that derives much of its livelihood from exploiting caribou (New World reindeer). The Gwich'in live mainly south of ANWR, but the 180,000-strong population of caribou known as the Porcupine Herd spends the winter in Gwich'in territory. The Gwich'in insist that oil drilling and related activity in ANWR will disrupt caribou calving, decimate the herd, and thus endanger the Gwich'in way of life.

The indigenous perspective on ANWR is not united, however; an Inuit group called the **Inupiat** supports oil production there, especially because of the jobs it would bring them. Environmentalists argue that habitats and native cultures would be despoiled to provide just six months of U.S. oil needs, while the oil industry says that 20 years of critical U.S. supply is in ANWR. This difficult contest will likely continue in the court of public opinion and the halls of Congress for years to come.

Meanwhile attention is turning to two other potential oil bonanza sites in Alaska: the North Slope's Teshekpuk Lake, a particularly oil-rich section of the **National Petroleum Reserve,** estimated to have more reserves than ANWR—and one that successive administrations have avoided developing because of its sensitive populations of geese and other wildlife—and central Alaska's Yukon Flats National Wildlife Refuge, rich in oil and natural gas.

---

much. The quest for opportunity and adventure in a remote land has attracted many more men than women (there are 107 men for every 100 women in Alaska, compared with the national U.S. figure of 97 men for every 100 women). The single woman's prospects have been described this way: "In Alaska, the odds are good . . . but the goods are odd."

Hawaii: America in a South Seas Setting

The state of Hawaii has eight major tropical islands (see Figure 11.2.1) lying just south of the Tropic of Cancer about 2,000 miles (3,200 km) west of California. Created by the movement of the Pacific plate of the earth's surface across a

420

geologic hot spot, the islands are all volcanic in origin and largely mountainous (see Figure 8.8). They have a combined land area of about 6,400 square miles (c. 17,000 sq km), a bit larger than the combined areas of Connecticut and Rhode Island or the European nation of Slovenia. The island of Hawaii is the largest and the only one with active volcanoes. The prevailing trade winds and mountains produce extreme precipitation contrasts within short distances. Windward slopes receive heavy rainfall throughout the

year, rising to an annual average of 460 inches (1,168 cm) on Mount Waialeale on the island of Kauai. One of the wettest places on the globe, lush Kauai stands in for many tropical locations in Hollywood films. Nearby leeward and low-lying areas receive as little as 10 to 20 inches (c. 25 to 50 cm) yearly. Some areas are so dry that crops require irrigation. Temperatures are tropical except for colder spots on some mountaintops. The capital, Honolulu (on the island of Oahu; •Figure 11.2.35), averages 81°F (27°C) in August and 73°F (23°C) in January.

Hawaii's population in 2007 was 1.2 million, with about 900,000 people living in metropolitan Honolulu. Immigration from the Pacific Basin has given the state's population a unique mix by ethnic origin: 24 percent white, 17 percent Japanese, 14 percent Filipino, 6 percent native Hawaiian, 5 percent Chinese, and smaller proportions of Koreans, Samoans, and others. Although there has been tension between native islanders and immigrants from the U.S. mainland and East Asia, much ethnic mixing has taken place.

As noted in Chapter 11 (page 611), the United States overthrew the Polynesian monarchy of Hawaii in 1893. Congress formally apologized to Hawaii in 1993 for wrongfully

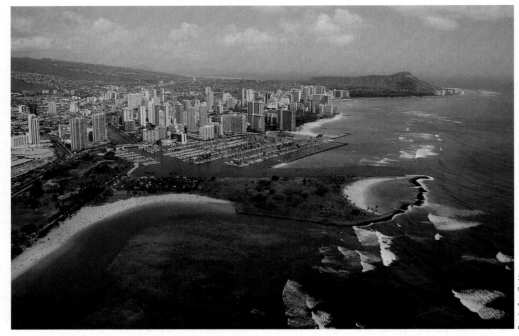

● **Figure 11.2.35** Honolulu, with the iconic range of Diamond Head in the upper right.

ejecting its indigenous rulers and promised to reconcile with the native Hawaiians. The **Native Hawaiian Government Reorganization Act,** first proposed to Congress in 2000, would go much further, extending sovereignty to native Hawaiians. This would give them the same legal standing as American Indians and native Alaskans, allowing them, among other things, to negotiate the fate of a large amount of land held in trust by the federal and state governments. A small nationalist movement is calling for the independence of the Kanaka Maoli, as the native Hawaiians call themselves.

Hawaii's economy is based largely on military expenditures, tourism, and commercial agriculture. The main defense installations, including the Pearl Harbor naval base, are on Oahu. The main agricultural products and exports—cane sugar (with a new use as an ethanol source) and pineapples—have long been grown on large estates owned by white families or corporate interests but worked mainly by Asians. New immigrants often took their first job on these estates. Agriculture now accounts for only about 1 percent of the state's economic productivity.

Propelled by better air transport and more wealth on mainland America and East Asia, tourism has been the great growth industry of recent decades. Attractions include tropical warmth, beautiful beaches, spectacular scenery, exotic cultures, and luxury resorts. Some of the "South Seas" glamour of the islands has been inevitably eroded by the relentless spread of commercial and residential development.

By some measures, Hawaii's natural environment has been altered more than that of any other state, with the highest proportion of native plant and animal species losing out to exotic competitors. Hawaii has even earned the unenviable title of being the "extinction capital of the world." 422

Hawaii has the last word in this text: *aloha!*

SUMMARY

→ This book breaks the United States into five vernacular regions: the Northeast, South, Midwest, West, and Alaska and Hawaii.

→ The Northeast includes Maine, New Hampshire, Vermont, Massachusetts, Connecticut, Rhode Island, New York, New Jersey, Pennsylvania, Delaware, Maryland, and the District of Columbia. About 62 million people, or 20 percent of the country's population, live in this subregion. There are seven major urban centers in this "Boswash" megalopolis: Boston, Providence, Hartford, New York, Philadelphia, Baltimore, and Washington, D.C.

→ The Appalachian Highlands cover a large part of the Northeast, and the other significant physical regions are the Coastal

Plain, Piedmont, and Interior Plains. Early manufacturing centers, including Baltimore, Washington, Philadelphia, and Trenton, were built on the fall line marking the Coastal Plain–Piedmont boundary.

→ The Erie Canal (opened 1825), with the Hudson-Mohawk Trough, stimulated a major migration flow to the northwestern territories, now the Midwest. By the 1850s, trade flowing through the corridor made New York the nation's leading harbor.

→ From the fall-line cities, early industrialization, beginning with steel production, spread to western Pennsylvania. The greatest center of major industry, finance, textile manufacturing, publishing, and many other enterprises was New York City. High-tech industries grew in Boston and in other former centers of heavy industry in the Northeast.

→ The South includes Virginia, North Carolina, South Carolina, Georgia, Florida, Alabama, Mississippi, Louisiana, Texas, West Virginia, Kentucky, Tennessee, Arkansas, and Oklahoma.

→ A humid subtropical climate with long, hot summers prevails in the South. Physiographic units are the Gulf and Atlantic Coastal Plains, the Piedmont, and the Interior Plains. The Upland South has hills and low mountains, and its subunits are the Appalachians, the Unglaciated Southeastern Interior Plain, and the Interior Highlands.

→ The South's agricultural significance today is minor compared with the past. Beef cattle, broiler chickens, dairy cattle, tobacco, citrus, peanuts, and racehorses come from the region.

→ Many of the South's textile mills have closed as the industry has moved offshore. Hydropower, oil, natural gas, and coal have been major energy sources.

→ Much industrial growth in the area came with the Tennessee Valley Authority projects of the 1930s. Industrial development along the Piedmont focused on small towns and midsize cities.

→ Industry also grew along and near the Gulf Coast. The Texas-Oklahoma metropolitan zone has the two largest urban centers in the South: Dallas–Fort Worth and Houston. Florida has a major metropolitan zone focusing on services and tourism rather than manufacturing. Other metropolitan zones in the South are in the Upland South and the Mississippi Delta.

→ The Midwest includes 12 states: Ohio, Indiana, Illinois, Michigan, Wisconsin, Minnesota, Iowa, North Dakota, South Dakota, Nebraska, Kansas, and Missouri. Its prolific production includes farm machinery, pork, corn, and soybeans.

→ Plains characterize the Midwest, with areas of relief at the margins of the region. Soils are very good, and in the east, rainfall supports productive farming. Rainfall diminishes westward, where irrigation is required.

→ About 40 percent of the world's corn comes from the Midwest. Corn production soared after 2006 as emphasis on ethanol production increased.

→ Wheat is very important in the Midwest, with associated sorghum, barley, and livestock production (especially of cattle and hogs).

→ Detroit was the historical center of automobile manufacturing in the United States, with Saint Louis also important. The Midwestern auto industry suffered as Japanese plants in the South produced more attractive products and absorbed many laid-off workers from the Midwest.

→ Major demographic changes have taken place in recent decades in the United States. Rural areas in the Midwest have been depopulated, while southern and western cities have exploded in population.

→ The West includes New Mexico, Arizona, Colorado, Utah, Nevada, Wyoming, Montana, Idaho, California, Oregon, and Washington. Containing about 38 percent of the conterminous United States' land area, the region is mainly arid and is home to about one-fifth of the nation's population. Much land belongs to the federal government in the form of national parks and military reserves.

→ Settlement in the West is concentrated in well-watered areas that are, in effect, oases. Topography has a strong role in the region's climates. Major mountain ranges along the West Coast have the largest population centers on their windward side, with populations diminishing eastward. The Willamette–Puget Sound Lowland in the north and California's Central Valley in the south are productive agricultural regions.

→ Grazing is the most common use of the land in the West, except where irrigation is possible. Irrigated farmlands raise alfalfa for beef cattle and dairy herds and support cotton and specialty crops, particularly in California. California's cities and farms have huge water requirements that cannot be met from local resources. California has long exceeded its allotment of Colorado River waters, and states upstream in the Colorado Basin have won court actions to increase their share and reduce California's.

→ Southern California is the West's largest population cluster and has about 5 percent of the U.S. population.

→ The San Francisco Bay Area is home to Silicon Valley, the dominant software development center in the United States. Like Los Angeles and other Californian cities, the Bay Area is very multicultural.

→ Seattle, Washington, its surrounding Puget Sound Lowland, and the Willamette Valley of Oregon have a marine west coast climate that provides abundant natural rainfall. Fishing, logging, and trade built these cities in the late 19th century, and aircraft manufacturing and software industries have boosted development.

→ In the Mountain West, water is more limited than on the coast, and population is clustered around centers that have reliable water sources. Phoenix and Tucson, Arizona; Denver, Colorado; and Salt Lake City, Utah, are the region's largest cities.

→ Alaska and Hawaii became states in 1959. Alaska stretches north of the Arctic Circle and has only about 670,000 people. Recent decades have seen a pronounced warming that has affected natural and human environments. Anchorage is the largest city, with about half the state's population. There is an ongoing dispute among Congress, the oil industry, environmentalists, and indigenous peoples about whether the oil reserves of northeastern Alaska's Arctic National Wildlife Refuge (ANWR) should be exploited.

→ Alaskan history chronicles fur trapping, gold rushes, oil exploration, military activities, tourism, and wilderness recreation. The imbalance between single men and single women is greater in Alaska than in any other state.

→ Hawaii lies about 2,000 miles (3,200 km) west of California and is comprised of eight major islands, all of volcanic origin. It has a great range of rainfall patterns. Agriculture contributes little to the state's economy today, but sugar and pineapple exports to mainland markets continue. Military expenditures and tourism, especially from the U.S. mainland and Asia, are critical to the economy.

KEY TERMS + CONCEPTS

Acadians (p. 659)
adaptive reuse (p. 656)
Appalachian Regional Reforestation Initiative (p. 651)
Arctic National Wildlife Refuge (ANWR) (p. 683)
"Big Muddy" (p. 669)
Big Sky Country (p. 680)
borough (p. 652)
Californication (p. 677)
Cajuns (p. 659)
Central Arizona Project (CAP) (p. 673)
Church of Jesus Christ of Latter-Day Saints (p. 678)
Colorado River Compact (p. 673)
Corn Belt (p. 665)
Dixie, Dixieland (p. 655)
dot-com bubble (p. 676)
exurbanization (p. 669)
"Father of Waters" (p. 664)

"Gateway to the West" (p. 667)
gentrification (p. 656)
Golden State (p. 670)
Gold Rush of 1849 (p. 676)
Gwich'in (p. 683)
Inupiat (p. 683)
levee (p. 663)
Lower 48 (p. 680)
Mason-Dixon Line (p. 655)
Mile-High City (p. 673)
Morganza Project (p. 665)
Mormons (p. 678)
Motor City (p. 666)
Motown (p. 666)
National Petroleum Reserve (p. 683)
Native Hawaiian Government Reorganization Act (p. 684)
North American Aerospace Defense Command (NORAD) (p. 678)
oil shale (p. 678)

Permanent fund (p. 680)
"Porkopolis" (p. 666)
Research Triangle Park (RTP) (p. 661)
Seward's Folly (p. 682)
Silicon Valley (p. 676)
smokestack industry (p. 655)
storm surge (p. 663)
strip mining (p. 651)
Sun Belt (p. 655)
Tennessee Valley Authority (TVA) (p. 660)
Utah Telecommunication Open Infrastructure Agency (UTOPIA) (p. 679)
Twin Cities (p. 668)
"Wal-Mart effect" (p. 669)
"Windy City" (p. 666)
zero-sum game (p. 674)

REVIEW QUESTIONS

1. What five regions (as presented in this book) make up the United States?

2. Cities generally develop at certain places to take advantage of geographic circumstances. What geographic conditions promoted the growth of Philadelphia, Saint Louis, Minneapolis, Cincinnati, San Francisco, and other cities discussed in this chapter?

3. What major topographic features characterize the U.S. Northeast? What are some of the fall-line cities? What were their roles in early industrialization and settlement?

4. When did the Erie Canal open? What route did it follow? What demographic impact did it have?

5. What are the reasons for the importance of New York City?

6. What are the topographic characteristics of the South? How do they relate to patterns of urban settlement and agriculture in the region? What are the region's most important cities?

7. Why are the soils so good in the Midwest? What are the most important agricultural products, and where are they produced?

8. Why have the rural areas and small towns of the Great Plains been losing population? What hopes are there for the revitalization of America's small towns?

9. What are the major characteristics of industrial growth and decline in the Midwest and Northeast? What cities have been most affected by these changes? What new industries emerged, and where?

10. Why is the term *oasislike* used to describe urban settlement in the American West? What major problems and controversies have emerged involving the area's water supplies?

11. What are the main topographic features of the West? What role do these features have in agriculture and urban settlement in the region?

12. What are the main climatic and settlement features of Alaska? What economic cycles has the state experienced, and what explains them?

13. What industries sustain Hawaii? What ethnic groups and concerns characterize the 50th state?

glossary

Absolute (mathematical) location Determined by the intersection of lines, such as latitude and longitude, providing an exact point expressed in degrees, minutes, and seconds.

Acid rain Precipitation that mixes with airborne industrial pollutants, causing the moisture to become highly acidic and therefore harmful to flora and bodies of water on which it falls. Sulfuric acid is the most common component of this acid precipitation.

Adaptive reuse Finding new uses for older buildings and stores, often accompanied by a shift from decline to steady renewal in an urban neighborhood.

Age of Discovery (Age of Exploration) The three to four centuries of European exploration, colonization, and global resource exploitation and trading led largely by European mercantile powers. It began with Columbus at the end of the 15th century and continued into the 19th century.

Age-structure profile (population pyramid) The graphic representation of a country's population by gender and 5-year age increments.

Agricultural (Neolithic or **New Stone Age** or **Food-Producing Revolution)** The domestication of plants and animals that began about 10,000 years ago.

Albedo The amount of the sun's energy reflected by the ground. Less vegetation cover correlates with high albedo, and vice versa.

Alfisol soils These productive soils found in steppe regions are among the world's more fertile. Also known as *chestnut soils.*

Altitudinal zonation In a highland area, the presence of distinctive climatic and associated biotic and economic zones at successively higher elevations. Ethiopia and Bolivia are examples of these kinds of zones. See also *Tierra caliente, Tierra fría, Tierra helada, Tierra templada.*

Anticyclone An atmospheric high-pressure cell. In the cell, the air is descending and becomes warmer. As it warms, its capacity to hold water vapor increases, and the result is minimal precipitation.

Anti-Semitism Anti-Jewish sentiments and activities.

Apartheid The Republic of South Africa's former official policy of "separate development of the races," designed to ensure the racial integrity and political supremacy of the white minority.

Arable Suitable for cultivation.

Archipelago A chain or group of islands.

Arid China In a climatic division of China approximately along the 20° isohyet, Arid China lies to the west of the line. It characterizes more than 50 percent of China's territory but accommodates less than 10 percent of the population.

Arms race Usually associated with the competition between the United States and the Soviet Union, it refers to rival and potential enemy powers increasing their military arsenals—each in an open-ended effort to stay ahead of the other.

Atmospheric pollution The modification of the blanket of gases that surround the earth largely through the airborne products of industrial production and the human consumption of fossil fuels.

Atoll Low islands made of coral and usually having an irregular ring shape around a lagoon.

Autonomy Self-rule, generally with reference to Palestinians' rights to run their own civil (and some security) affairs in portions of the West Bank and Gaza Strip allocated to them in the 1993–2000 peace agreements.

Balkanization The fragmentation of a political area into many smaller independent units, as in former Yugoslavia in the Balkan Peninsula.

Barrios Densely settled neighborhoods in city space that is characteristically inhabited by migrants of Latino or Hispanic origin.

Barter The exchange of goods or services in the absence of cash.

Basin irrigation The ancient Egyptian system of cultivation using fields saturated by seasonal impoundment of Nile floodwaters.

Bazaar (suq) The central market of the traditional Middle Eastern city, characterized by twisting, close-set lanes and merchant stalls.

Belief systems The set of customs that an individual or culture group has relating to religion, social contracts, and other aspects of cultural organization.

Benelux The name used to collectively refer to Belgium, the Netherlands, and Luxembourg.

Biodiversity hot spots A ranked list of places scientists believe deserve immediate attention for flora and fauna study and conservation.

Biological diversity (biodiversity) The number of plant and animal species and the variety of genetic materials these organisms contain.

Biomass The collective dried weight of organisms in an ecosystem.

Biome A terrestrial ecosystem type categorized by a dominant type of natural vegetation.

Birth rate The annual number of live births per 1,000 people in a population.

Black-earth belt An important area of crop and livestock production spanning parts of Russia, Ukraine, Moldova, and Kazakhstan. The main soils of this belt are mollisols.

Boat people Refugees who flee by sea because of conflict or political circumstances, including refugees from conflict in Indochina, Cuba, and Haiti.

Bourgeoisie In Marxist doctrine, the capitalist class.

Brain drain The exodus of educated or skilled persons from a poor to a rich country or from a poor to a rich region within a country.

Break-of-bulk point A classic geographic term describing a point in transit when bulk goods must be removed from one mode of transport and installed on another. A trainload of grain carried to a port for transshipment on cargo boats or barges is a common example.

Broadleaf deciduous forest Forests typical of middle-latitude areas with humid subtropical and humid continental climates. As cool fall temperatures set in, broadleaf trees shed their leaves and cease to grow, thus reducing water loss. They produce new foliage and grow vigorously during the hot, wet summer.

Buffer state A generally smaller political unit adjacent to a large, or between several large, political units. Such a role often enables the smaller state to maintain its independence because of its mutual use to the larger, proximate nations. Uruguay, between Brazil and Argentina, is an example.

Capital goods Goods used to produce other goods.

Carrying capacity The size of a population of any organism that an ecosystem can support.

Carter Doctrine President Jimmy Carter's declaration, following the Soviet Union's invasion of Afghanistan in 1979, that the United States would use any means necessary to defend its vital interests in the Persian Gulf Region. The "vital interests" were interpreted to mean oil, and "any means necessary" interpreted to mean that the United States would go to war with the Soviet Union if oil supplies were threatened.

Cartogram A special map in which an area's shape and size are defined by explicit characteristics of population, economy, or distribution of any stated product.

Cartography The craft of designing and making maps, the basic language of geography. In recent years, this traditional manual art has been changed profoundly through the use of computers and Geographic Information Systems (GIS) and through major improvements in machine capacity to produce detailed, colored, map products.

Cash (commercial) crops Crops produced generally for export.

Caste The hierarchy in the Hindu religion that determines a person's social rank. It is established by birth and cannot be changed.

Charney Effect Observed by an atmospheric scientist named Charney, this states that the less plant cover there is on the ground, the higher the albedo—solar energy deflected back into the atmosphere—and therefore the lower the humidity and precipitation.

Chemocline The boundary between lower, carbon dioxide-laden waters and higher, gas-free, fresh water in some African lakes. The puncture of this boundary can cause eruptions that are fatal to humans and other life around the lakes.

Chernobyl The site in the Ukraine where, in April 1986, the worst nuclear power plant accident in history occurred. It is thought that approximately 5,000 people died, and a zone with a 20-mile (32-km) radius is still virtually uninhabitable; 116,000 people were moved from the area, and cleanup continues to this day.

Chernozem A Russian term meaning "black earth." It is a grassland soil that is exceptionally thick, productive, and durable.

Chestnut soils Productive soils typical of the Russian steppe and North American Great Plains.

Chokepoint A strategic narrow passageway on land or sea that may be closed off by force or threat of force.

Choropleth maps Maps that are drawn to show the differing distribution of goods or geographic characteristics (including population) across a broad area. Such maps are good for generalizations but often mask significant local variations in the presence of the item being mapped.

Chunnel (Eurotunnel) The 16-mile (26-km) tunnel that links Britain with the European continent. It was completed in 1994 at a cost of more than $15 billion, making it the single most costly project in landscape transformation ever undertaken.

Civilization The complex culture of urban life.

Climate The average weather conditions, including temperature, precipitation, and winds, of an area over an extended period of time.

Climatology The scientific study of patterns and dynamics of climate.

Coal The residue from organic material compressed for a long period under overlying layers of the earth. Coals vary in hardness and heating value. Anthracite coal burns with a hot flame and almost no smoke. Bituminous coal is used in the largest quantities; some can be used to make coke by baking out the volatile elements. Coke is largely carbon and burns with intense heat when used in blast furnaces to smelt iron ore. Peat, which is coal in the earliest stages of formation, can be burned if dried. Concern over atmospheric pollution by the sulfur in coal smoke has recently increased the demand for low-sulfur coals.

Cold War The tense but generally peaceful political and military competition between the United States and the Soviet Union, and their respective allies, from the end of World War II until the collapse of the Soviet Union in 1991.

Collective farm A large-scale farm in the former Soviet Union that usually incorporated several villages. Workers received shares of the income after the obligations of the collective had been met.

Collectivization The process of forming collective farms in Communist countries.

Colonization The European pattern of establishing dependencies abroad to enhance economic development in the home country.

Colored A South African term for persons of mixed racial ancestry.

Command economy A centrally planned economy typical of the Soviet Union and its Communist allies, in which the government rather than free enterprise determines the production, distribution, and sale of economic goods and services.

Common Market An earlier name given to the (current) 25 countries that make up the European Union. In 1957, an initial six countries combined to form the European Economic Community (EEC), and this supranational community has grown to have considerable economic and political importance in Europe. See *European Union*.

Computer cartography Mapmaking using sophisticated software and computer hardware. It is a new, actively growing career field in geography.

Computer-controlled robots Microtechnology has allowed the creation of highly precise, robotic units that can be used in delicate manufacturing. Although expensive to create, they ultimately reduce per unit manufacturing costs because of predictable higher quality output and easier maintenance than human work organizations.

Concentric zone model A generalized model of a city. A city becomes articulated into contrasting zones arranged as concentric rings around its central business district.

Coniferous vegetation Needleleaf evergreen trees; most bear seed cones.

Consumer goods Goods that individuals acquire for short-term use.

Consumer organisms Animals that cannot produce their own food within a food chain.

Consumption overpopulation The concept that a few persons, each using a large quantity of natural resources from ecosystems across the world, add up to too many people for the environment to support.

Containerization The prepackaging of items into larger standardized containers for more efficient transport.

Continental islands Once attached to nearby continents, these are the islands north and northeast of Australia, including New Guinea.

Convectional precipitation The heavy precipitation that occurs when air is heated by intense surface radiation, then rises and cools rapidly.

Convention on the Law of the Sea A 1970s United Nations treaty permitting a sovereign power to have greater access to surrounding marine resources.

Coordinate systems A means of determining exact or absolute location. Latitude and longitude are most often used.

Cottage industry Handwork that is done in the evenings or in slack seasons in the rural world. Products created during such activity can include woven and sewn goods or piecework for local textile or manufacturing industries. This work brings cash into the rural economy, especially to those not completely involved in full-time farming activity.

Council for Mutual Economic Assistance (COMECON) An economic organization, now disbanded, consisting of the Soviet Union, Poland, East Germany, Czechoslovakia, Hungary, Romania, Bulgaria, Cuba, Mongolia, and Vietnam.

Crop calendar The dates by which farmers prepare, plant, and harvest their fields. Farmers who are involved in new agricultural patterns necessitated by greater use of chemical fertilizers and new plant strains are sometimes unable to adjust to the demands of a much more exact crop calendar than has been traditional.

Crop irrigation Bringing water to the land by artificial methods.

Crusades A series of European Christian military campaigns between the 11th and 14th centuries aimed at recapturing Jerusalem and the rest of the Holy Land from the Muslims.

Cultural diffusion One of the most important dynamics in geography, cultural diffusion is the engine of change as crops, languages, culture patterns, and ideas are transferred from one place to other places, often in the course of human migration.

Cultural geography The study of the ways in which humankind has adopted, adapted to, and modified the face of the earth, with particular attention given to cultural patterns and their associated landscapes. It also includes a culture's influence on environmental perception and assessment.

Cultural landscape The landscape modified by human transformation, thereby reflecting the cultural patterns of the resident culture.

Cultural mores The belief systems and customs of a culture group.

Culture The values, beliefs, aspirations, modes of behavior, social institutions, knowledge, and skills that are transmitted and learned within a group of people.

Culture hearth An area where innovations develop, with subsequent diffusion to other areas.

Culture System A system in which Dutch colonizers required farmers in Java to contribute land and labor for the production of export crops under Dutch supervision.

Cyclone A low-pressure cell that composes an extensive segment of the atmosphere into which different air masses are drawn.

Cyclonic (frontal) precipitation The precipitation generated in traveling low-pressure cells that bring different air masses into contact.

Death rate The annual number of deaths per 1,000 people in a population.

Debt-for-nature swap An arrangement in which a certain portion of international debt is forgiven in return for the borrower's pledge to invest that amount in nature conservation.

Deciduous trees Broadleaf trees that lose their leaves and cease to grow during the dry or the cold season and resume their foliage and grow vigorously during the hot, wet season.

Deflation A fall in prices, such as that caused by currency devaluations.

Deforestation The removal of trees by people or their livestock.

Delta A landform resulting from the deposition of great quantities of sediment when a stream empties into a larger body of water.

Demographic shift Major population redistributions as people move from the countryside to the city or from one region to another.

Demographic transition A model describing population change within a country. The country initially has a high birth rate, a high death rate, and a low rate of natural increase, then moves through a middle stage of high birth rate, low death rate, and high rate of natural increase, and ultimately reaches a third stage of low birth rate, low or medium death rate, and low or negative rate of population increase.

Dependency theory A theory arguing that the world's more developed countries continue to prosper by dominating their former colonies, the now independent less developed countries.

Desert An area too dry to support a continuous cover of trees or grass. A desert generally receives less than 10 inches (25 cm) of precipitation per year.

Desert shrub vegetation Scant, bushy plant life occurring in deserts of the middle and low latitudes where there is not enough rain for trees or grasslands. The plants are generally xerophytic.

Desertification The expansion of a desert brought about by changing environmental conditions or unwise human use.

Development A process of improvement in the material conditions of people often linked to the diffusion of knowledge and technology.

Devolution The process by which a sovereign country releases or loses more political and economic control to its constituent elements, such as states and provinces.

Diaspora The scattering of the Jews outside Palestine beginning in the Roman Era.

Digital divide The divide between the handful of countries that are the technology innovators and users and the majority of nations that have little ability to create, purchase, or use new technologies.

Dissection The carving of a landscape into erosional forms by running water.

Distributary A stream that results when a river subdivides into branches in a delta.

Domestication The controlled breeding and cultivation of plants and animals.

Donor democracy Typical of countries in Africa south of the Sahara, this is a situation in which a government makes just enough concessions on voting rights or human rights to win foreign loans and aid, without instituting any serious democratic reform.

Donor fatigue Public or official weariness of extending aid to needy people.

Double cropping Growing two crops a year on the same field.

Drought avoidance Adaptations of desert plants and animals to evade dry conditions by migrating (animals) or being active only when wet conditions occur (plants and animals).

Drought endurance Adaptations of desert plants and animals to tolerate dry conditions through water storage and heat loss mechanisms.

Dry farming Planting and harvesting according to the seasonal rainfall cycle.

Ecologically dominant species A species that competes more successfully than others for nutrition and other essentials of life.

Ecology The study of the interrelationships of organisms to one another and to the environment.

Economic shock therapy Russia's economic transformation in the early 1990s from a command economy to a free-market economy. Overseen by Boris Yeltsin, this transformation was difficult for a country accustomed to government direction in all economic matters, thus the "shock."

Ecosphere (biosphere) The vast ecosystem composed of all of Earth's ecosystems.

Ecosystem A system composed of interactions between living organisms and nonliving components of the environment.

Ecotourism Travel by people who want to both see and save the natural habitats remaining on earth.

Ejido An agricultural unit in Mexico characterized by communally farmed land or common grazing land. It is of particular importance to indigenous villages.

Endemic species A species of plant or animal found exclusively in one area.

Energy crisis The petroleum shortages and price surges sparked by the 1973 oil embargo.

Environmental assessment The process of determining the condition and value of a particular environmental setting. Used both in aspects of landscape change and in evaluating environmental perception.

Environmental determinism The concept that the physical environment has played a sovereign role in the cultural development of a people or landscape. Also known as *environmentalism.*

Environmental perception The concept that how people view the world and its landscapes and resources influences their uses of the earth and therefore the condition of the earth.

Environmental possibilism A concept that rejects environmental determinism, arguing that although environmental conditions do influence human and cultural development, people choose from various possibilities in how to live within a given environment.

Equinox On or about September 23, and again on or about March 20, Earth reaches the equinox position. Its axis does not point toward or away from the sun, so days and nights are of equal length at all latitudes on Earth.

Escarpment A steep edge marking an abrupt transition from a plateau to an area of lower elevation.

Estuary A deepened ("drowned") river mouth into which the sea has flooded.

Ethnic cleansing The relocation or killing of members of one ethnic group by another to achieve some demographic, political, or military objective.

Ethnocentrism Regarding one's own group as superior and as setting proper standards for other groups.

Euro Part of the authority of the European Union has been the institution of a new currency that has become "coin of the realm" since early 2002. Not all EU nations accepted the euro.

European Community The name that was replaced in 1993 by the term European Union. The European Community continues to serve as a governing and administrative body for the EU.

European Economic Community An economic organization designed to secure the benefits of large-scale production by pooling resources and markets. The name has been changed to Economic Union. See also *Common Market.*

European Free Trade Association (EFTA) An organization that maintains free trade among its members but allows each member to set its own tariffs in trading with the outside world. Members are Iceland, Norway, Sweden, Switzerland, Austria, and Finland.

European Union The current organization begun in the 1950s as the Common Market. It now is made up of 25 nations (see Chapter 3).

Evaporation The loss of moisture from the earth's land surfaces and its water bodies to the air through the ongoing influence of solar radiation and transpiration by plants.

Exotic species A nonnative species introduced into a new area.

Extensive land use A livelihood, such as hunting and gathering, that requires the use of large land areas.

External costs (externalities) Consequences of goods and services that are not priced into the initial cost of those goods and services.

Fall line A zone of transition in the eastern United States where rivers flow from the harder rocks of the Piedmont to the softer rocks of the Atlantic and Gulf Coastal Plain. Falls and/or rapids are characteristic features.

Fault A break in a rock mass along which movement has occurred. A break due to rock masses being pulled apart is a tensional or "normal" fault, whereas a break due to rocks being pushed together until one mass rides over the other is a compressional fault. The processes of faulting create these breaks.

Feral animals Domesticated animals that have abandoned their dependence on people to resume life in the wild.

Fertile Crescent The arc-shaped area stretching from southern Iraq through northern Iraq, southern Turkey, Syria, Lebanon, Israel, and western Jordan, where plants and animals were domesticated beginning about 10,000 years ago.

Final status issues Issues deferred to the end of the Oslo Peace Process between Israel and the PLO. Finally dealt with at Camp David in 2000, they included the status of Jerusalem, the fate of Palestinian refugees, Palestinian statehood, and borders between Israel and a new Palestinian state. The negotiations broke down over these final status issues.

Fjord A long, narrow extension of the sea into the land usually edged by steep valley walls that have been deepened by glaciation.

Folding The process creating folds (landforms resulting from an intense bending of rock layers). Upfolds are known as anticlines, and downfolds are called synclines.

Food chain The sequence through which energy, in the form of food, passes through an ecosystem.

Formal region See *Region.*

Fossil waters Virtually nonrenewable freshwater supplies, the product of ancient rainfall stored in deep natural underground aquifers, especially in the Middle East and North Africa.

Four Modernizations The effort of the Chinese after the death of Mao Zedong in 1976 to focus on economic development in agriculture, industry, science and technology, and defense.

Front A contact zone between unlike air masses. A front is named according to the air mass that is advancing (cold front or warm front).

Frontal precipitation See *Cyclonic precipitation.*

Fuelwood crisis Deforestation in the less developed countries caused by subsistence needs.

Functional region See *Region.*

Gaia hypothesis A hypothesis stating that the ecosphere is capable of restoring its equilibrium following any disturbance that is not too drastic.

Gentrification The social and physical process of change in an urban neighborhood by the return of young, often professional, populations to the urban core. These people are often attracted by the substantial nature of the original building stock of the place and the proximity to the city center, which generally continues to have major professional opportunities. While this process brings an urban landscape back into a primary role as a tax base, it does dispossess a considerable number of minority peoples.

Geographic analysis By giving attention to the spatial aspects of a distribution, geographic analysis helps to explain distribution, density, and flow of a given phenomenon.

Geographic Information Systems (GIS) The growing field of computer-assisted geographic analysis and graphic representation of spatial data. It is based on superimposing various data layers that may include everything from soils to hydrology to transportation networks to elevation. Computer software and hardware are steadily improving, enabling GIS to produce ever more detailed and exact output.

Geography The study of the spatial order and associations of things. Also defined as the study of places, the study of relationships between people and environment, and the study of spatial organization.

Geologic hot spot A small area of Earth's mantle where molten magma is relatively close to the crust. Hot spots are associated with island chains and thermal features.

Geomorphology The scientific analysis of the landforms of the earth; sometimes called *physiography*.

Geopolitics The study of geographic factors in political matters, including borders, political unity, and warfare.

Glacial deposition In the process of continental and valley glaciation, the deposition of moraines that become lateral or terminal in the act of glacial retreat. This same process also leads to glacial scouring as moving ice picks up loose rock and reshapes the landscape as the glacier moves forward or retreats.

Glacial scouring See *Glacial deposition*.

Globalization The spread of free trade, free markets, investments, and ideas across borders and the political and cultural adjustments that accompany this diffusion.

GNI PPP Per capita gross national income purchasing power parity (per capita GNI PPP) combines gross national income (GNI) and purchasing power parity (PPP) as a method of comparing the real value of output between different countries' economies.

Graben (rift valley) A landform created when a segment of Earth's crust is displaced downward between parallel tensional faults or when segments of the crust that border it ride upward along parallel compressional faults.

Great Rift Valley The result of tectonic processes, this is a broad, steep-walled trough extending from the Zambezi Valley in southern Africa northward to the Red Sea and the valley of the Jordan River in southwestern Asia.

Great Trek In what is now South Africa, a series of northward migrations in the 1830s by which groups of Boers, primarily from the eastern part of the Cape Colony, sought to find new interior grazing lands and establish new political units beyond British reach.

Green Revolution The introduction and transfer of high-yielding seeds, mechanization, irrigation, and massive application of chemical fertilizers to areas where traditional agriculture has been practiced.

Greenhouse effect The observation that increased concentrations of carbon dioxide and other gases in Earth's atmosphere cause a warmer atmosphere.

Gross domestic product (GDP) The value of goods and services produced in a country in a given year. It does not include net income earned outside the country. The value is normally given in current prices for the stated year.

Gross national product (GNP) The value of goods and services produced internally in a given country during a stated year plus the value resulting from transactions abroad. The value is normally expressed in current prices of the stated year. Such data must be used with caution in regard to developing countries because of the broad variance in patterns of data collection and the fact that many people consume a large share of what they produce.

Growth pole A new city, a resource, or some development in a heretofore undersettled area that begins to attract population growth and economic development.

Guano Seabird excrement that is also found in phosphate deposits.

Guest workers Migrants (mainly young and male) from less developed countries who are employed, sometimes illegally, in more developed countries.

Gulf Stream The strong ocean current originating in the tropical Atlantic Ocean that skirts the eastern shore of the United States, curves eastward, and reaches Europe as a part of the broader current called the North Atlantic Drift.

Gulf War The 1990–1991 confrontation between a large military coalition, spearheaded by the United States, and the Iraqi forces that had invaded Kuwait. The allies successfully drove Iraqi forces from Kuwait. Earlier, the Gulf War had referred to the 1980–1988 war between Iraq and Iran.

Hacienda A Spanish term for large rural estates owned by the aristocracy in Latin America.

Haj The pilgrimage to Mecca, the principal holy city in the Islamic religion. Every Muslim is required to make this pilgrimage at least once in a lifetime, if possible.

Han Chinese The original Chinese peoples who settled in North China on the margins of the Yellow River (Huang He) and who were central to the development of Chinese culture. Han Chinese make up more than 90 percent of China's population.

Hierarchical rule A system by which leadership for a culture group is defined by traditional and sometimes frustrating patterns of age–sex distinction. It is evident in most cultures and slowly modified by an increasing role of democratic elections and the establishment of a broader opportunity base for individual advancement.

High islands Generally the result of volcanic eruptions, these are the higher and more agriculturally productive and densely populated islands of the Pacific World.

High pressure cell (anticyclone) An air mass descending and warming because of increased pressure and weight of the air above. High pressure typically means low relative humidity and minimal precipitation.

Hinterland The region of a country that lies away from the capital and largest cities and is most often rural or even unsettled; it is often seen in the minds of economic planners as an area with development potential.

Historical geography A concern with the historical patterns of human settlement, migration, town building, and the human use of the earth. Often, this subdiscipline best blends geography and history as a perspective on human activity.

Holocaust Nazi Germany's attempted extermination of Jews, Roma (Gypsies), homosexuals, and other minorities during World War II.

Homelands Ten former territorial units in South Africa reserved for native Africans (blacks). They had elected African governments, and some were designated as "independent" republics, although they were not recognized outside South Africa. Formerly known as *Bantustans*. They were abolished in 1994. Also known as *native reserves* and *national states*.

Horizontal migration The movements of pastoral nomads over relatively flat areas to reach areas of pasture and water.

Hot money Short-term and often volatile flows of investment that can cause serious damage to the "emerging market" economies of less developed countries.

Human Development Index (HDI) A United Nations–devised ranked index of countries' development that evaluates quality of life issues (such as gender equality, literacy, and human rights) in addition to economic performance.

Humboldt (or Peru) Current The cold ocean current that flows northward along the west coast of South and North America. It plays a significant role in regional fish resources.

Humid China The eastern portion of China that is relatively well watered and where the great majority of the Chinese population has settled. An arc from Kunming in southwest China to Beijing in North China describes the approximate western margin of this zone.

Humid continental A climate with cold winters, warm to hot summers, and sufficient rainfall for agriculture, with the greater part of the precipitation in the summer half-year.

Humid pampa A level to gently rolling area of grassland centered in Argentina with a humid subtropical climate.

Humid subtropical A climate that generally occupies the southeastern margins of continents with hot summers, mild to cool winters, and ample precipitation for agriculture.

Humus Decomposed organic soil material. Grasslands characteristically provide more humus than forests do.

Hunting and gathering A mode of livelihood, based on collecting wild plants and hunting wild animals, generally practiced by preagricultural peoples. Also known as *foraging*.

Hydraulic control The ability to control water in irrigation systems, rivers, urban settlements, and for the generation of electricity. It has been central to the development of major civilizations and to culture groups throughout human history.

Ice cap A climate and biome type characterized by permanent ice cover on the ground, no vegetation (except where limited melting occurs), and a severely long, cold winter. Summers are short and cool.

Ideology A system of political and/or economic beliefs such as communism, capitalism, autocracy, or democracy.

Igneous (volcanic) rock Rock formed by the cooling and solidification of molten materials. Granite and basalt are common types. Such rocks tend to form uplands in areas where sedimentary rocks have weathered into lowlands. Certain types break down into extremely fertile soils. These rocks are often associated with metal-bearing ores.

Industrial Revolution A period beginning in mid-18th century Britain that saw rapid advances in technology and the use of inanimate power.

Inflation A rise in prices.

Information technology (IT) The Internet, wireless telephones, fiberoptics, and other technologies characteristic of more developed countries. Generally seen as beneficial for a country's economic prospects, IT is also spreading in the less developed countries.

Intensive land use A livelihood requiring use of small land areas, such as farming.

Intertillage Growing two or more crops simultaneously in alternate rows. Also known as *interplanting*.

Irredentism Movement by an ethnic group in one country to revive or reinforce kindred ethnicity in another country—often in an effort to promote succession there.

Irrigation The artificial placement of water to produce crops, generally in arid locations.

Island chain A series of islands formed by ocean crust sliding over a stationary hot spot in the earth's mantle.

Karst Landscape features associated with the dissolution of limestone or dolomite including caves, sinkholes, and towers.

Keystone species A species that affects many other organisms in an ecosystem.

Kyoto Protocol A treaty on climate change signed by 160 countries in Kyoto, Japan, in 1997, and put into force in 2005. It requires MDCs to reduce their greenhouse gas emissions by more than 5 percent below their 1990 levels by the year 2012.

Lacustrine plain The floor of a former lake where glacial meltwater accumulated and sediments washed in and settled. An extremely flat surface is characteristic.

Landscape A portion of the earth's land surface. Geographers are interested in the transformation of natural landscapes into cultural landscapes.

Landscape transformation The human process of making over the earth's surface. From initial human shelters to contemporary massive urban centers, humans have been driven to change their environmental settings.

Large-scale map A map constructed to show considerable detail in a small area.

Laterite A material found in tropical regions with highly leached soils; composed mostly of iron and aluminum oxides which harden when exposed and make cultivation difficult.

Latifundia (sing. latifundio) Large agricultural Latin American estates with strong commercial orientations.

Latitude A measurement that denotes position with respect to the equator and the poles. Latitude is measured in degrees, minutes, and seconds, which are described as parallels. Places near the equator are said to be in low latitudes; places near the poles are in high latitudes. The Tropic of Cancer and the Tropic of Capricorn, at 23.5°N and 23.5°S, respectively, and the Arctic Circle and Antarctic Circle, at 66.5°N and 66.5°S, respectively, form the most commonly recognized boundaries of the low and high latitudes. Places occupying an intermediate position with respect to the poles and the equator are said to be in the middle latitudes. There are no universally accepted definitions for the boundaries of the high, middle, and low latitudes.

Law of Return An Israeli law permitting all Jews living in Israel to have Israeli citizenship.

Law of the Sea A United Nations treaty or convention permitting coastal nations to have greater access to marine resources.

Less developed countries (LDCs) The world's poorer countries.

Lifeboat ethics Ecologist Garrett Hardin's argument that, for ecological reasons, rich countries should not assist poor countries.

Location Central to all geographic analysis is the concept of location. Where something is relates to all manner of influences, from climate to migration routes. It is a crucial component in trying to understand patterns of historic and economic development.

Loess A fine-grained material that has been picked up, transported, and deposited in its present location by wind; it forms an unusually productive soil.

Long March The 1935–1936 flight of approximately 100,000 Communist troops under Mao Zedong from south central China to the mountains of North China. Mao was fleeing attack by General Chiang Kai-shek, and the success of the Long March led to the establishment of Mao's troops as a force with the capacity to overcome the much larger and better armed troops of Chiang.

Longitude A measurement that denotes a position east or west of the prime meridian (Greenwich, England). Longitude is measured in degrees, minutes, and seconds, and meridians of longitude extend from pole to pole and intersect parallels of latitude.

Low islands Made of coral, these are the generally flat, drier, less agriculturally productive, and less densely populated islands of the Pacific World.

Malthusian scenario The model forecasting that human population growth will outpace growth in food and other resources, with a resulting population die-off.

Map projection A way to minimize distortion in one or more properties of a map (direction, distance, shape, or area).

Map scale The actual distance on the earth that is represented by a given linear unit on a map.

Maquiladoras Operations dedicated to the assembly of manufactured goods, generally in Mexico and Central America, from components initially produced in the United States or other places. With the enactment of NAFTA in 1994, there was a massive expansion of maquiladora operations.

Marginalization A process by which poor subsistence farmers are pushed onto fragile, inferior, or marginal lands that cannot support crops for long and that are degraded by cultivation.

Marine west coast A climate occupying the western sides of continents in the higher middle latitudes; it is greatly moderated by the effects of ocean currents that are warm in winter and cool in summer relative to the land.

Marshall Plan The plan designed largely by the United States after the conclusion of World War II by which U.S. aid was focused on the rebuilding of the very Germany that had been its enemy in the war just concluded. Secretary of State George Marshall (who had been Chief of Staff of U.S. Army from 1939–1945) was central to the plan's design and implementation.

Material culture The items that can be seen and associated with cultural development, such as house architecture, musical instruments, and tools. In the study of a cultural landscape, items of material culture add considerable personality to cultural identity.

Medical geography The study of patterns of disease diffusion, environmental impact on public health, and the interplay of geographic factors, migration, and population. With the increasing ease of international movement, medical geography is becoming more important as the potential for disease diffusion increases.

Medina An urban pattern typical of the Middle Eastern city before the 20th century.

Mediterranean (dry-summer subtropical) A climate that occupies an intermediate location between a marine west coast climate on the poleward side and a steppe or desert climate on the equatorward side. During the high-sun period, it is rainless; in the low-sun period, it receives precipitation of cyclonic or orographic origin.

Mediterranean scrub forest The xeroyphytic vegetation typical of hot, dry summer, Mediterranean climate regions. Local names for this vegetation type include *maquis* and *chaparral*.

Meiji Restoration The Japanese revolution of 1868 that restored the legitimate sovereignty of the emperor. *Meiji* means "Enlightened Rule." This event led to a transformation of Japan's society and economy.

Melanesia (black islands) A group of relatively large islands in the Pacific Ocean bordering Australia.

Mental maps Every individual's mind has a series of locations, access routes, physical and cultural characteristics of places, and often a general sense of the good or bad of locales. The term *mental map* is used to define such geographies.

Mercantile colonialism The historical pattern by which Europeans extracted primary products from colonies abroad, particularly in the tropics.

Meridian See *Longitude.*

Mestizo In Latin America, a person of mixed European and Native American ancestry.

Metamorphic rock Rock formed from igneous or sedimentary rock through changes occurring in the rock structure as a result of heat, pressure, or the chemical action of infiltrating water. Marble, formed of pure limestone, is a common example.

Metropolitan area A city together with suburbs, satellites, and adjacent territory with which the city is functionally interlocked.

Microcredit The lending of small sums to poor people to set up or expand small businesses.

Micronesia (tiny islands) Thousands of small and scattered islands in the central and western Pacific Ocean, mainly north of the equator.

Microstate A political entity that is tiny in area and population and is independent or semi-independent.

Middle Eastern ecological trilogy The model of mostly symbiotic relations among villagers, pastoral nomads, and urbanites in the Middle East.

Migration A temporary, periodic, or permanent move to a new location.

Milpero A Latin American farmer who engages in swidden or shifting cultivation in forest lands. The plot is often called the *milpa.*

Minifundia (sing. minifundio) Small Latin American agricultural landholdings, usually with a strong subsistence component.

Mixed forest A transitional area where both needleleaf and broadleaf trees are present and compete with each other.

Mobility The pattern of human movement, typically changing because of continual improvement in means of transportation.

Mollisols Thick, productive, and durable soils, such as the *chernozem,* whose fertility comes from abundant humus in the top layer.

Monoculture The single-species cultivation of food or tree crops, usually very economical and productive but threatening to natural diversity and change.

Monsoon A current of air blowing fairly steadily from a given direction for several weeks or months at a time. Characteristics of a monsoonal climate are a seasonal reversal of wind direction, a strong summer maximum of rainfall, and a long dry season lasting for most or all of the winter months.

Montreal Protocol A 1989 international treaty to ban chlorofluorocarbons.

Moor A rainy, deforested upland, covered with grass or heather and often underlain by water-soaked peat. Also, Muslim inhabitants of Spain.

Moraine The unsorted material deposited by a glacier during its retreat. Terminal moraines are ridges formed by long-continued deposition at the front of a stationary ice sheet.

More developed countries (MDCs) The world's wealthier countries.

Mulatto A person of mixed European and black ancestry.

Multinational companies Companies that operate, at least in part, outside their home countries.

Multiple cropping Farming patterns in which several crops are raised on the same plot of land in the course of a year or even a season. Common examples are winter wheat and summer corn or soybeans in North America or several crops of rice in the same plot in Monsoon Asia.

Nation Commonly, a nation is a term describing the citizens of a state—or that state—but it also refers to an ethnic group existing with or without a separate political entity. See *Nation-state.*

Nation-state A political situation in which high cultural and ethnic homogeneity characterizes the political unit in which people live.

Nationalism The drive to expand the identity and strength of a political unit that serves as home to a population interested in greater cohesion and often expanded political power.

Natural levee A strip of land immediately alongside a stream that has been built up by deposition of sediments during flooding.

Natural replacement rate The highest rate at which a renewable resource can be used without decreasing its potential for renewal. Also known as *sustainable yield*.

Natural resource A product of the natural environment that can be used to benefit people. Resources are human appraisals.

Near Abroad Russia's name for the now-independent former republics, other than Russia, of the USSR.

Neocolonialism The perpetuation of a colonial economic pattern in which developing countries export raw materials to, and buy finished goods from, developed countries. This relationship is more profitable for the developed countries.

Neo-Malthusians Supporters of forecasts that resources will not be able to keep pace with the needs of growing human populations.

Newly industrializing countries (NICs) The more prosperous of the world's less developed countries.

Nonblack soil zone Areas of poor soil in cool, humid portions of the Slavic Coreland, suitable for cultivation of rye.

North Atlantic Drift A warm current, originating in tropical parts of the Atlantic Ocean, that drifts north and east, moderating temperatures of Western Europe. Also known as *Gulf Stream*.

North Atlantic Treaty Organization (NATO) A military alliance formed in 1949 that included the United States, Canada, many European nations, and Turkey.

North European Plain The level to rolling lowlands that extend from the low countries on the west through Germany to Poland. These are areas rich in agricultural development, dense human settlements, and a number of major industrial centers. The plain is broken by a number of rivers flowing from the Alps and other mountain systems in central Europe into the North and Baltic Seas.

Occupied Territories The territories captured by Israel in the 1967 War: the Gaza Strip and West Bank, captured respectively from Egypt and Jordan, and the Golan Heights, captured from Syria.

Oil embargo The 1973 embargo on oil exports imposed by Arab members of the Organization of Petroleum Exporting Countries (OPEC) against the United States and the Netherlands.

Open borders Political boundaries that have minimal political or structural impediment to easy crossing between two countries.

Organization for Economic Cooperation and Development (OECD) See *Organization for European Economic Cooperation (OEEC)*.

Organization for European Economic Cooperation (OEEC) An organization created in 1948 to organize and facilitate the European response to the 1947 Marshall Plan. In the early 1960s, the OEEC became the Organization for Economic Cooperation and Development (OECD), and this served as a multinational base for continued planning in economic and social development. The Common Market came to overshadow this organization, and finally, the European Union, in 1993, became the most powerful and influential multinational organization in Europe.

Orographic precipitation The precipitation that results when moving air strikes a topographic barrier, such as a mountain, and is forced upward.

Ozone hole Areas of depletion of ozone in Earth's stratosphere, caused by chlorofluorocarbons.

Palestine Liberation Organization The Palestinian military and civilian organization created in the 1960s to resist Israel and recognized

in the 1990s as the sole legitimate organization representing official Palestinian interests internationally.

Parallel A latitude line running parallel to the equator.

Patrilineal descent system A kinship naming system based on descent through the male line.

People overpopulation The concept that many persons, each using a small quantity of natural resources to sustain life, add up to too many people for the environment to support.

Per capita For every person, or per person.

Per Capita Gross Domestic Product Purchasing Power Parity (GDP PPP) In this figure, annual per capita gross domestic product (GDP)—the total output of goods and services a country produces for home use in a year—is divided by the country's population. For comparative purposes, that figure is adjusted for purchasing power parity (PPP), which involves the use of standardized international dollar price weights that are applied to the quantities of final goods and services produced in a given economy. The resulting measure, per capita GDP PPP, provides the best available starting point for comparisons of economic strength and well-being between countries.

Perennial irrigation The year-round irrigated cultivation of crops, as in the Nile Valley following construction of barrages and dams.

Permafrost Permanently frozen subsoil.

Photosynthesis The process by which green plants use the energy of the sun to combine carbon dioxide with water to give off oxygen and produce their own food supply.

Physical geography The subdiscipline of geography most concerned with the climate, landforms, soils, and physiography of the earth's surface.

Piedmont A belt of country at an intermediate elevation along the base of a mountain range.

Pillars of Islam The five fundamental tenets of the faith of Islam.

Pivotal countries Those countries whose collapse would cause international refugee migration, war, pollution, disease epidemics, or other international security problems.

Place identity In geographic analysis of a given locale, the nature of place identity becomes a means of understanding people's response to that particular place. Determination of the environmental and cultural characteristics that are most frequently associated with a certain place helps to establish that "sense of place" for a given location.

Plain A flat to moderately sloping area, generally of slight elevation.

Plantation A large commercial farming enterprise, generally emphasizing one or two crops and utilizing hired labor. Plantations are commonly found in tropical regions.

Plate tectonics The dominant force in the creation of the continents, mountain systems, and ocean deeps. The steady, but slow, movement of these massive plates of the earth's mantle and crust has created the positions of the continents that we have and major patterns of volcanic and seismic activity. Areas where plates are being pulled under other plates are called subduction zones.

Plateau An elevated plain, usually lying 2,000 feet (610 m) above sea level. Some are known as tablelands. A dissected plateau is a hilly or mountainous area resulting from the erosion of an upraised surface.

Pleistocene overkill A hypothesis stating that hunters and gatherers of the Pleistocene Era hunted many species to extinction.

Podzol Soil with a grayish, bleached appearance when plowed, lacking in well-decomposed organic matter, poorly structured, and very low in natural fertility. Podzols are the dominant soils of the taiga.

Polder An area reclaimed from the sea and enclosed within dikes in the Netherlands and other countries. Polder soils tend to be very fertile.

Polynesia (many islands) A Pacific island region that roughly resembles a triangle with its corners at New Zealand, the Hawaiian Islands, and Easter Island.

Population change rate The birth rate minus the death rate in a population.

Population density The average number of people living in a square mile or square kilometer. It is a very handy statistic for generalized comparisons but often fails to provide a detailed sense of the real distribution of people.

Population explosion The surge in Earth's human population that has occurred since the beginning of the Industrial Revolution.

Postindustrial Description of an economy or society characterized by the transformation from manufacturing to information management, financial services, and the service sector.

Prairie An area of tall grass in the middle latitudes composed of rich soils that have been cleared for agriculture. The original lack of trees may have been due to repeated burnings or periodic drought conditions.

Primary consumers (herbivores) Consumers of green plants.

Primate city A city that dominated a country's urban scene, and usually defined as being larger than the country's second and third largest cities combined. Primate cities are generally found in developing countries, although some developed countries, such as France, have them.

Prime meridian See *Latitude*.

Producers (self-feeders) In a food chain, organisms that produce their own food (mostly, green plants).

Projection The distortion caused by the transfer of three-dimensional space on Earth's surface to the two dimensions of a flat map.

Push and pull forces of migration Emigration may be caused by so-called push factors, as when hunger or lack of land "pushes" peasants out of rural areas into cities, or by pull factors, as when an educated villager responds to a job opportunity in the city. People responding to push factors are often referred to as nonselective migrants, whereas those reacting to pull factors are called selective migrants. Both push and pull forces are behind the rural to urban migration that is characteristic of most countries.

Pyramid of biomass A diagrammatic representation of decreasing biomass in a food chain.

Pyramid of energy flow A diagrammatic representation of the loss of high quality, concentrated energy as it passes through the food chain.

Qanat A tunnel used to carry irrigation and drinking water from an underground source by gravity flow. Also known as *foggara* or *karez*.

Rain shadow A condition creating dryness in an area located on the lee side of a topographic barrier such as a mountain range.

Region A "human construct" that is often of considerable size, that has substantial internal unity or homogeneity, and that differs in significant respects from adjoining areas. Regions can be classed as formal (homogeneous), functional, or vernacular. The formal region, also known as a uniform region, has a unitary quality that derives from a homogeneous characteristic. The United States is an example of a formal region. The functional region, also called the nodal region, is a coherent structure of areal units organized into a functioning system by lines of movement or influence that converge on a central node or trunk. A major example would be the trading territory served by a large city and bound together by the flow of people, goods, and information over an organized network of transportation and communication lines. Vernacular or perceptual regions are areas that possess regional identity, such as the "Sun Belt," but share less objective criteria in the use of this regional name. General regions, such as the major world regions in this text, are recognized on the basis of overall distinctiveness.

Remote sensing Through the use of aerial and satellite imagery, geographers and other scientists have been able to get vast amounts of data describing places all over the face of the earth. Remote sensing is the science of acquiring and analyzing data without being in contact with the subject. It is used in the study of patterns of land use, seasonal change, agricultural activity, and even human movement along transport lines. This process relates closely to GIS. See also *Geographic Information Systems (GIS)*.

Renewable resource A resource, such as timber, that is grown or renewed so that a continual supply is available. A finite resource is one that, once consumed, cannot be easily used again. Petroleum products are a good example of such a resource, and because it takes too much time to go through the process of creation, they are not seen as renewable.

Resources Resources become valuable through human appraisal, and their utilization reflects levels of technology, location, and economic ambition. There are few resources that have been universally esteemed (water, land, defensible locales) throughout history, so patterns of resource utilization serve as indexes of other levels of cultural development. See also *Natural resource*.

Rift valley. See *Graben*.

Right of return, Palestinian The Palestinian Arab principle that refugees (and their descendants) displaced from Israel and the Occupied Territories in the 1948 and 1967 wars be allowed to return to the region.

Ring of Fire The long horseshoe-like chain of volcanoes that goes from the southern Andes Mountains in South America up the west coast of North America and arches over into the northeast Asian island chains of Japan, the Philippines, and Southeast Asia. This zone of seismic instability and erratic vulcanism is caused by the tension built up in plate tectonics. See also *Plate tectonics*.

Riparian A state containing or bordering a river.

Russification The effort, particularly under the Soviets, to implant Russian culture in non-Russian regions of the former Soviet Union and its Eastern European neighbors.

Sacred space (sacred place) Any locale that people hold in reverence, such as places of worship, cemeteries, and battlefields.

Salinization The deposition of salts on, and subsequent fertility loss in, soils experiencing a combination of overwatering and high evaporation.

Sand sea A virtual "ocean" of sand characteristic of parts of the Middle East, where people, plants, and animals are all but nonexistent.

Savanna A low-latitude grassland in an area with marked wet and dry seasons.

Scale The size ratio represented by a map; for example, a map with a scale of 1:12,500 is portrayed as 1/12,500 of the actual size.

Scorched earth The wartime practice of destroying one's own assets to prevent them from falling into enemy hands.

Scrub and thorn forest The low, sparse vegetation in tropical areas where rainfall is insufficient to support tropical deciduous forest.

Seamount An underwater volcanic mountain.

Second law of thermodynamics A natural law stating that high-quality, concentrated energy is increasingly degraded as it passes through the food chain.

Secondary consumers (carnivores) Consumers of primary consumers.

Sedentarization Voluntary or coerced settling down, particularly by pastoral nomads in the Middle East.

Sedimentary rock Rock formed from sediments deposited by running water, wind, or wave action either in bodies of water or on land and which have been consolidated into rock. The main classes are sandstone, shale (formed principally of clay), and limestone, including the pure limestone called chalk.

Segmentary kinship system The organization of kinship groups in concentric units of membership, as of lineage, clan, and tribe among Middle Eastern pastoral nomads.

Service sector The labor sector made up of employees in retail trade and personal services; it is the sector most likely to increase in employment significance in postindustrial society.

Settler colonization The historical pattern by which Europeans sought to create new or "neo-Europes" abroad.

Shatter belt A large, strategically located region composed of conflicting states caught between the conflicting interests of great powers.

Shi'a Islam The branch of Islam regarding male descendants of Ali, the cousin and son-in-law of the Prophet Muhammad, as the only rightful successors to the Prophet Muhammad.

Shifting cultivation A cycle of land use between crop and fallow years that is needed to work around the infertility of tropical soils. After clearing tropical forest, a farmer may get only a few years of crops before there is no fertility in the soil and must move on to clear more land. In the meantime, forest reclaims the previously farmed plots. Where new lands are not available, fallow periods on old fields are reduced or eliminated, resulting in soil deterioration. Also known as *swidden* or *slash-and-burn cultivation*.

Site and situation Site is the specific geographic location of a given place, while situation is the accessibility of that site and the nature of the economic and population characteristics of that locale.

Slavic Coreland (Fertile Triangle or **Agricultural Triangle)** The large area of the western former Soviet region containing most of the region's cities, industries, and cultivated lands.

Small-scale map A map constructed to give a highly generalized view of a large area.

Soil The earth mantle made of decomposed rock and decayed organic material.

Sovereign state A political unit that has achieved political independence and maintains itself as a separate unit.

Spatial Geographers recognize spatial distributions and patterns in Earth's physical and human characteristics. The term *spatial* comes from the noun "space," and it relates to the distribution of various phenomena on Earth's surface. Geographers portray spatial data cartographically—that is, with maps.

Special Economic Zones (SEZs) In China, the urban areas designated after the 1970s to attract investment and boost production through tax breaks and other incentives.

Spodosols Acidic soils that have a grayish, bleached appearance when plowed, lack well-decomposed organic matter, and are low in natural fertility. Also known as *podzols*.

St. Lawrence Seaway Project This seaway has a total length of 2,342 miles (3,796 km) between the Atlantic Ocean and its western terminus in Lake Superior. It allows major oceanic vessels to reach the Great Lakes, thus enabling ports as far inland as the west side of Lake Superior to serve as international ports. The seaway was completed in 1959, although Canada and the United States began to anticipate such a waterway project in the last years of the 19th century.

State A political unit over which an established government maintains sovereign control.

State farm (sovkhoz) A type of collectivized state-owned agricultural unit in the former Soviet Union; workers receive cash wages in the same manner as industrial workers.

Steppe (temperate grassland) A biome composed mainly of short grasses. It occurs in areas of steppe climate, which is a transitional zone between very arid deserts and humid areas.

Subarctic A high-latitude climate characterized by short, mild summers and long, severe winters.

Subduction zone See *Plate tectonics*.

Subsidence The settling of land by the compression of lower layers of rock and soil, usually accelerated by the removal of water from below the surface. Of particular importance in urban centers, with Mexico City having one of the most persistent problems.

Summer solstice On or about June 22, the first day of summer in the Northern Hemisphere, the northern tip of Earth's axis is inclined toward the sun at an angle of 23.5° from a line perpendicular to the plane of the ecliptic. This is the summer solstice in the Northern Hemisphere. On or about December 22, the first day of winter in the Northern Hemisphere, the southern tip of Earth's axis is inclined toward the sun at an angle of 23.5° from a line perpendicular to the plane of the ecliptic. This is the summer solstice in the Southern Hemisphere.

Sun Belt A U.S. area of indefinite extent, encompassing most of the South, plus the West at least as far north as Denver, Salt Lake City, and San Francisco. This region has shown faster growth in population and jobs than the nation as a whole during the past several decades.

Sunni Islam The branch of Islam regarding successorship to the Prophet Muhammad as a matter of consensus among religious elders.

Sustainable development (ecodevelopment) Concepts and efforts to improve the quality of human life while living within the carrying capacity of supporting ecosystems.

Sustainable yield (natural replacement rate) The highest rate at which a renewable resource can be used without decreasing its potential for renewal.

Swidden See *Shifting cultivation*.

Symbolization The representation of distinct aspects of information shown on a map, such as stars for capital cities.

Taiga A northern coniferous (needleleaf) forest.

Technocentrists (cornucopians) Supporters of forecasts that resources will keep pace with or exceed the needs of growing human populations.

Tectonic processes Processes that derive their energy from within the earth's crust and serve to create landforms by elevating, disrupting, and roughening the earth's surface.

Tertiary consumers Consumers of secondary consumers.

Theory of island biogeography A theoretical calculation of the relationship between habitat loss and natural species loss, in which a 90 percent loss of natural forest cover results in a loss of half of the resident species.

Tierra caliente A Latin American climatic zone reaching from sea level upward to approximately 3,000 feet (914 m). Crops such as rice, sugarcane, and cacao grow in this hot, wet environment.

Tierra fría A Latin American climatic zone found between 6,000 feet (1,800 m) and 12,000 feet (3,600 m) above sea level. Frost occurs and the upper limit of agriculture and tree growth is reached.

Tierra helada A Latin American climatic zone above 10,000 feet (3,048 m) that has little vegetation and frequent snow cover.

Tierra templada A Latin American climatic zone extending from approximately 3,000 to 6,000 feet (914–1,829 m) above sea level. It is a prominent zone of European-induced settlement and commercial agriculture such as coffee growing.

Tokugawa Shogunate The period 1600 to 1868 in Japan, characterized by a feudal hierarchy and Japan's isolation from the outside world.

Tradable permits A proposed mechanism for reducing total global greenhouse gases in which each country would be assigned the right to emit a certain quantity of carbon dioxide, according to its population size, setting the total at an acceptable global standard. The assigned quotas could be traded between underproducers and overproducers of carbon dioxide.

Trade winds Streams of air that originate in semipermanent high-pressure cells on the margins of the tropics and are attracted equatorward by a semipermanent low-pressure cell.

Transhumance The movement of pastoral nomads and their herd animals between low-elevation winter pastures and high-elevation summer pastures. Also known as *vertical migration*.

Triangular trade The 16th- to 19th-century trading links between West Africa, Europe, and the Americas, involving guns, alcohol, and manufactured goods from Europe to West Africa exchanged for slaves. Slaves brought to the Americans were exchanged for the gold, silver, tobacco, sugar, and rum carried back to Europe.

Trophic (feeding) levels Stages in the food chain.

Tropical deciduous forest The vegetation typical of some tropical areas with a dry season. Here, broadleaf trees lose their leaves and are dormant during the dry season and then add foliage and resume their growth during the wet season.

Tropical rain forest A low-latitude broadleaf evergreen forest found where heat and moisture are continuously, or almost continuously, available.

Tropical savanna climate A relatively moist low-latitude climate that has a pronounced dry season.

Tundra A region with a long, cold winter, when moisture is unobtainable because it is frozen, and a very short, cool summer. Vegetation includes mosses, lichens, shrubs, dwarf trees, and some grass.

Underground economy The "black market" typical of the former Soviet region and many LDCs.

Undifferentiated highland A climate that varies with latitude, altitude, and exposure to the sun and moisture-bearing winds.

Vernacular region See *Region*.

Vertical migration Movements of pastoral nomads between low winter and high summer pastures.

Virgin and idle lands (new lands) Steppe areas of Kazakhstan and Siberia brought into grain production in the 1950s.

Warsaw Pact A military alliance, now dissolved, consisting of the Soviet Union and the European countries of Poland, East Germany, Czechoslovakia, Hungary, Romania, and Bulgaria.

Waterhead In attempting to control water for the generation of electricity or a gravity flow irrigation project, it is essential to pond up water in relative quantity so that there is an adequate force—waterhead—to achieve the desired goal. Dam structures are the most common engineering responses to this requirement.

Weather The atmospheric conditions prevailing at one time and place.

Weathering The natural process that disintegrates rocks by mechanical or physical means, making soil formation possible.

West The shorthand term for the countries of the world that have been most influenced by Western civilization and that link most closely with the United States and Western Europe.

Westerly winds An airstream located in the middle latitudes that blows from west to east. Also known as *westerlies*.

Western civilization The sum of values, practices, and achievements that had roots in ancient Mesopotamia, as well as Palestine, Greece, and Rome, and that subsequently flowered in western and southern Europe.

Westernization The process whereby non-Western societies ac-quire Western traits, which are adopted with varying degrees of completeness.

Winter solstice On or about December 22, the first day of winter in the Northern Hemisphere, the southern tip of Earth's axis is inclined toward the sun at an angle of 23.5° from a line perpendicular to the plane of the ecliptic. This is the winter solstice in the Northern Hemisphere. On or about June 22, the first day of summer in the Northern Hemisphere, the northern tip of Earth's axis is inclined toward the sun at an angle of 23.5° from a line perpendicular to the plane of the ecliptic. This is the winter solstice in the Southern Hemisphere.

Xerophytic Literally, "dry plant," referring to desert shrubs having small leaves, thick bark, large root systems, and other adaptations to absorb and retain moisture.

Zero population growth (ZPG) The condition of equal birth rates and death rates in a population.

Zionist movement The political effort, beginning in the late 19th century, to establish a Jewish homeland in Palestine.

Zone of transition An area where the characteristics of one region change gradually to those of another.

pronounciation guide

The authors acknowledge with thanks the assistance of David Allen, Eastern Michigan University, in preparing an earlier pronunciation guide that the authors expanded and modified for this book. In general, names are those appearing in the text; pronunciations of other names on maps can be found in *Goode's World Atlas* (Rand McNally) or in standard dictionaries and encyclopedias. Vowel sounds are coded as: ă (uh), ā (ay), ĕ (eh), ē (ee), ī (eye), ĭ (ih), ōō (oo).

Afghanistan (af-*gann'*-ih-stann')
   Helmand R. (*hell'*-mund)
   Hindu Kush Mts. (*hinn'*-doo *koosh'*)
   Kabul (*kah'*-b'l)
   Khyber Pass (*keye'*-ber)
Albania (al-*bay'*-nee-a)
   Tirana (tih-*rahn'*-a)
Algeria (al-*jeer'*-ee-a)
   Ahaggar Mts. (uh-*hahg'*-er)
   Sahara Desert (suh-*hahr'*-a)
   Tanezrouft (*tahn'*-ez-rooft')
American Samoa
   Pago Pago (*pahng'*-oh *pahng'*-oh)
   Tutuila I. (too-too-*ee'*-lah)
Andorra (an-*dohr'*-a)
Angola (ang-*goh'*-luh)
   Benguela R.R. (ben-*gehl'*-a)
   Cabinda (exclave) (kah-*binn'*-duh)
   Luanda (loo-*an'*-duh)
   Lobito (luh-*bee'*-toh)
Anguilla (an-*gwill'*-a)
Antigua and Barbuda (an-*teeg'*-wah,
   bahr-*bood'*-a)
Argentina (ahr'-jen-*teen'*-a)
   Buenos Aires (*bwane'*-uhs *eye'*-ress)
   Córdoba (*kawrd'*-oh-bah)
   Patagonia (region) (patt'-hu-*goh'*-nee-a)
   Rio de la Plata (*ree'*-oh duh lah
     plah'-tah)
   Rosario (roh-*zahr'*-ee-oh)
   Tierra del Fuego (region)
     (tee-ehr'-uh dell *fway'*-goh)
   Tucumán (too'-koo-*mahn'*)
Armenia (ahr'-*meen'*-y-uh)
   Yerevan (yair'-uh-*vahn'*)
Aruba (ah-*roob'*-a)
Australia (aw-*strayl'*-yuh)
   Adelaide (*add'*-el-ade)
   Brisbane (*briz'*-bun)
   Canberra (*can'*-bur-a)
   Melbourne (*mell'*-bern)
   Sydney (*sid'*-nih)
   Tasmania (state) (tazz-*may'*-nih-uh)

Austria (*aw'*-stree-a)
   Graz (*grahts'*)
   Linz (*lints'*)
   Salzburg (*sawlz'*-burg)
   Vienna (vee-*enn'*-a)
Azerbaijan (ah'-zer-*beye'*-jahn)
   Baku (bah-*koo'*)

Bahamas (buh-*hahm'*-uz)
   Nassau (*nass'*-aw)
Bahrain (bah-*rayn'*)
Bangladesh (bahng'-lah-*desh'*)
   Dacca (Dhaka) (*dahk'*-a)
Barbados (bar-*bay'*-dohss)
Belarus (*bell'*-uh-roos)
   Minsk (*mensk'*)
Belgium
   Antwerp (*ant'*-werp)
   Ardennes Upland (ahr-*den'*)
   Brussels (*brus'*-elz)
   Charleroi (shahr-luh-*rwah'*)
   Ghent (*gent'*)
   Liège (lee-*ezh'*)
   Meuse R. (*merz'*)
   Sambre R. (*sohm*-bruh)
   Scheldt R. (*skelt'*)
Belize (buh-*leez'*)
Benin (beh-*neen'*)
   Cotonou (koh-toh-*noo'*)
Bhutan (boo-*tahn'*)
   Thimphu (*thim'*-poo)
Bolivia (boh-*liv'*-ee-uh)
   Altiplano (region) (al-tih-*plahn'*-oh)
   La Paz (lah *pahz'*)
   Santa Cruz (sahn-ta *krooz'*)
   Sucre (*soo'*-kray)
   Lake Titicaca (tee'-tee-*kah'*-kah)
Bosnia and Herzegovina
   (*boz'*-nee-uh herts-uh-goh-*veen'*-uh)
   Sarajevo (sahr-uh-*yay'*-voh)
Botswana (baht-*swahn'*-a)
   Gaborone (gahb'-uh-*roh'*-nee)
Brazil (bra-*zill'*)
   Amazonia (region) (am-uh-*zohn'*-ih-a)
   Belo Horizonte (*bay'*-loh haw-ruh-*zonn'*-
     tee)
   Brasília (bruh-*zeel'*-yuh)
   Caatinga (region) (kay-*teeng'*uh)
   Manaus (mah-*noose'*)
   Minas Gerais (*meen'*-us zhuh-*rice*)
   Natal (nah-*tal'*)
   Paraíba Valley (pah-rah-*ee'*-bah)
   Paraná R. (pah-rah-*nah'*)
   Pôrto Alegre (por-too' al-*egg'*-ruh)
   Recife (ruh-*see'*-fee)

Rio de Janeiro (*ree'*-oh dih juh-*nair'*-oh)
   Salvador (sal'-vah-*dor'*)
   São Paulo (sau *pau'*-loh)
Brunei (*broo'*-neye)
Bulgaria
   Sofia (*soh'*-fee-a)
Burkina Faso (bur-*kee'*-na *fah'*-soh)
   Ouagadougou (wah'-guh-*doo'*-goo)
Burma (*bur'*-ma)
Burundi (buh-*roon'*-dee) ("oo" as in
"wood")
   Bujumbura (boo-jem-*boor'*-a)
   (second "oo" as in "wood")

Cambodia
   Phnom Penh (*nom' pen'*)
Cameroon (kam'-uh-*roon'*)
   Douala (doo-*ahl'*-a)
   Yaoundé (yah-onn-*day'*)
Canada
   Montreal (mon'-tree-*awl'*)
   Nova Scotia (*noh'*-vuh *scoh'*-shuh)
   Ottawa (*aht'*-ah-wah)
   Quebec (kwih-*beck'*; Fr., kay-*beck'*)
   Saguenay R. (*sag'*-uh-nay)
Cape Verde (*verd'*)
Central African Republic
   Bangui (*bahng'*-ee)
Chile (*chill'*-ee)
   Atacama Desert (ah-tah-*kay'*-mah)
   Chuquicamata (choo-kee-*kah'*-mah-tah)
   Concepción (con-sep'-*syohn'*)
   Santiago (sahn'-tih-*ah'*-goh)
   Valparaíso (vahl'-pah-rah-*ee'*-soh)
China
   Altai Mts. (al-*teye*)
   Amur R. (ah-*moor'*) ("oo" as in
     "wood")
   Anshan (ahn-*shahn'*)
   Baotou (*bau'*-toh)
   Beijing (bay-*zhing'*)
   Canton (kan'-*tahn'*)
   Chang Jiang (chung jee-*ahng'*)
   Chengdu (chung-*doo'*)
   Chongqing (chong-*ching'*)
   Dzungarian Basin (zun-*gair'*-ree-uhn)
   Guangdong (gwahng-*dung'*)
   Guangzhou (gwahng-*joh'*)
   Hainan I. (*heye'*-nahn')
   Harbin (*hahr'*-ben)
   Huang He (*hwahng' huh'*)
   Lanzhou (lahn-*joh'*)
   Lhasa (*lah'*-suh)
   Li R. (*lee'*)
   Liao (lih-ow')

Lüda (*loo'*-dah)
Nanjing (nahn-*zhing'*)
Peking (pea-*king'*)
Shandong (shahn-*dung'*)
Shanghai (shang-*heye'*)
Shantou (shahn-*toh'*)
Shenyang (shun-*yahng'*)
Shenzhen (shun-*zún*)
Sichuan (zehch-*wahn'*)
Taklamakan Desert (*tah'*-kla-mah-*kahn'*)
Tarim Basin (*tah'-reem'*)
Tiananmen (*tyahn'*-an-men)
Tianjin (tyahn-*jeen'*)
Tien Shan (*tih'*-en *shahn'*)
Tientsin (*tinn'-tsinn'*)
Tsinling Shan (*chinn'*-leeng *shahn'*)
Ürümqi (oo-*room'*-chee)
Wuhan (*woo'-hahn'*)
Xian (shee-*ahn'*)
Xinjiang (shin-jee-*ahng'*)
Xizang (shee-*dzahng'*)
Yangtze R. (*yang'*-see)
Yunnan (yoo-*nahn'*) ("oo" as in "wood")
Colombia (koh-*lomm'*-bih-uh)
Barranquilla (bah-rahn-*keel'*-yah)
Bogotá (boh'-ga-*tah'*)
Cali (*kah'*-lee)
Cartagena (kahrt'-a-*hay'*-na)
Cauca R. (*kow'*-kah)
Magdalena R. (mahg'-thah-*lay'*-nah)
Medellín (med-deh-*yeen'*)
Comoros (*kahm'*-o-rohs')
Congo
Brazzaville (*brazz'*-uh-vill'; Fr.: brah-zah-*veel'*)
Pointe Noire (pwahnt *nwahr'*)
Costa Rica (*kohs'*-tuh *ree'*-kuh)
San José (sahn hoh-*zay'*)
Croatia (kroh-*ay'*-shuh)
Zagreb (*zah'*-grebb)
Cyprus (*seye'*-pruhs)
Nicosia (nik'-o-*see'*-uh)
Czech Republic (check)
Bratislava (bratt'-ih-*slahv'*-a, braht-)
Brno (*ber'*-noh)
Moravia (muh-*rayve'*-ih-uh)
Ostrava (*aw'*-strah-vah)
Prague (*prahg'*)

Denmark
Copenhagen (*koh'*-pen-hahg'-gen)
Djibouti (jih-*boot'*-ee)
Dominica (dahm'-i-*nee'*-ka)
Dominican Republic (duh-*min'*-i-kan)
Santo Domingo (sant'-oh d-*min'*-goh)

Ecuador (*ec'*-wa-dawr)
Guayaquil (gweye'-uh-*keel'*)
Quito (*kee'*-toh)
Egypt
Aswan (*ahs'*-wahn)
Cairo (*keye'*-roh)
Port Said (sah-*eed'*)
El Salvador (el *sal'*-va-dor)
Equatorial Guinea (*gin'*-ee)

Estonia (ess-*toh'*-nee-a)
Tallin (*tahl'*-in)
Ethiopia (ee'-thee-*oh'*-pea-a)
Addis Ababa (*ad'*-iss *ab'*-a-ba)
Asmara (az-*mahr'*-a)

Fiji (*fee'*-jee)
Suva (*soo'*-va)
Finland
Helsinki (*hel'*-sing-kee)
Karelian Isthmus (kuh-*ree'*-lih-uhn)
France
Alsace (region) (al-*sass'*)
Ardennes (region) (ahr-*den'*)
Bordeaux (bawr-*doh'*)
Boulogne (boo-*lohn'*-yuh)
Calais (kah-*lay'*)
Carcassonne (kahr-kuh-*suhn'*)
Champagne (sham-*pahn*-yuh)
Garonne R. (guh-*rahn'*)
Jura Mts. (*joo'*-ruh)
Le Havre (luh *ah'*-vruh)
Lille (*leel'*)
Loire R. (*lwahr'*)
Lorraine (loh-*rane'*)
Lyons (Fr., Lyon), both (*lee*-aw)
Marseilles (mahr-*say'*)
Massif Central (upland) (mah-*seef'* saw-*trahl*)
Meuse R. (*merz'*)
Moselle R. (moh-*zell'*)
Nancy (naw-*see'*)
Nantes (*nawnt'*)
Nice (*neece'*)
Oise R. (*wahz'*)
Pyrenees (*peer'*-uh-*neez'*)
Riviera (riv'-ih-*ehr'*-a)
Rouen (roo-*aw'*)
Saône R. (*sohn'*)
Seine R. (*senn'*)
Strasbourg (strahz-*boor'*)
Toulon (too-*law'*)
Toulouse (too-*looz'*)
Vosges Mts. (*vohzh'*)
French Guiana (gee'-*ahn*-a)
Cayenne (keye'-enn)

Gabon (gah-*baw'*)
Gambia (*gamm'*-bih-uh)
Banjul (*bahn*-jool)
Georgia
Abkhazia (ahb-kahz'-zee-uh)
Adjaria (ah-*jahr'*-ree-uh)
Caucasus Mts. (*kaw'*-kuh-suhs)
South Ossetia (oh-*see'*-shee-uh)
Tbilisi (tuh-*blee'*-see)
Germany
Aachen (*ah'*-ken)
Berlin (ber-*lin'*)
Bremen (*bray'*-men)
Chemnitz (*kemm'*-nits)
Cologne (kuh-*lohn'*)
Dresden (*drez'*-den)
Düsseldorf (*doo'*-sel-dorf)
Elbe R. (*ell'*-buh)
Erzgebirge (*ehrts'*-guh-beer-guh)
Frankfurt (*frahngk'*-foort')

Karlsruhe (*kahrls'*-roo-uh)
Leipzig (*leyep'*-sig)
Main R. (*mane'*; Ger., *mine'*)
Mannheim-Ludwïgshafen (*man'*-hime *loot'*-vikhs-hah-fen)
Munich (*myoo'*-nik)
Neisse R. (*nice'*-uh)
Oder R. (*oh'*-der)
Ruhr (region) (*roor'*)
Stuttgart (*stuht'*-gahrt)
Weser R. (*vay'*-zer)
Wiesbaden-Mainz (*vees'*-bahd-en *meynts*)
Wuppertal (*voop'*-er-tahl) ("oo" as in "wood")
Ghana (*gahn'*-a)
Accra (a-*krah'*)
Kumasi (koo-*mahs'*-ee)
Tema (*tay'*-muh)
Volta R. (*vohl'*-tuh)
Greece
Aegean Sea (uh-*jee'*-un)
Piraeus (peye-*ree'*-us)
Thessaloniki (thess'-uh-loh-*nee'*-kih)
Grenada (gruh-*nayd'*-a)
Guadeloupe (gwah'-duh-*loop'*)
Guatemala (gwah'-tuh-*mah'*-luh)
Guinea (*gin'*-ee)
Conakry (*kahn'*-a-kree)
Guinea-Bissau (biw-*ow'*)
Guyana (geye-*an'*-a; geye-*ahn*-a')

Haiti (*hayt'*-ee)
Port-au-Prince (pohrt'-oh-*prants'*)
Honduras (hahn-*dur'*-as)
Tegucigalpa (tuh-goo'-sih-*gahl'*-pah)
Hungary (*hun'*-guh-rih)
Budapest (boo-duh-*pesht'*)

Iceland
Akureyri (ah'-koor-*ray'*-ree)
Reykjavik (*ray'*-kyuh-vik')
India
Agra (*ah'*-gruh)
Ahmedabad (ah'-mud-uh-*bahd'*)
Assam (state) (uh-*sahm'*)
Bengal (region) (benn-*gahl'*)
Bihar (state) (bih-*hahr'*)
Brahmaputra R. (brah'-muh-*poo'*-truh)
Deccan (region) (*deck'*-un)
Delhi (*deh'*-lih)
Ganges R. (*gan'*-jeez)
Eastern, Western Ghats (mountains) (*gahts'*)
Himalaya Mts. (himm-*ah'*-luh-yuh, -ah-*lay*-yuh)
Jaipur (*jeye'*-poor)
Jamshedpur (juhm-*shayd'*-poor)
Kashmir (region) (*cash'*-meer)
Kerala (state) (*kay'*-ruh-luh)
Madras (muh-*drass'*, -*drahs'*)
Punjab (region) (*pun'*-jahb, -jab; pun-*jahb'*, -jab')
Rajasthan (state) (*rah'*-juh-stahn)
Uttar Pradesh (state) (*oo'*-tahr pruh-*desh'*)

Indonesia
  Bali (*bah'*-lih)
  Bandung (*bahn'*-doong)
  Celebes (Sulawesi) I.
    (*sell'*-uh-beez; sool-uh-*way'*-see)
  Irian Jaya (ihr'-ee-ahn *jeye'*-uh)
  Jakarta (yah-*kahr'*-tah)
  Java (*jah'*-vuh
  Kalimantan (kal-uh-*mann'*-tann)
  Medan (muh-*dahn'*)
  Palembang (pah-lemm-*bahng'*)
  Sumatra (suh-*mah'*-truh)
  Surabaya (soo'-ruh-*bah'*-yah)
  Ujung Pandang (oo'-jung
    pahn-*dahng'*)
Iran (ih-*rahn'*)
  Abadan (ah-buh-*dahn'*)
  Elburz Mts. (ell-*boorz'*)
  Isfahan (izz-fah-*hahn'*)
  Meshed (muh-*shed'*)
  Tabriz (tah-*breez'*)
  Tehran (teh-huh-*rahn'*)
  Zagros Mts. (*zah'*-gruhs)
Iraq (i-*rahk'*)
  Baghdad (*bag'*-dad)
  Basra (*bahs'*-ra)
  Euphrates R. (yoo-frate'-eez)
  Kirkuk (Kihr-*kook'*) ("oo" as in
    "wood")
  Mosul (moh-*sool'*)
  Shatt al Arab (R.) (*shaht'* ahl ah-*rahb'*)
  Tigris R. (*teye'*-griss)
Israel
  Gaza Strip (*gahz'*-uh)
  Eilat (*ay'*-laht)
  Haifa (*heye'*-fa)
  Tel Aviv (tel' a-*veev'*)
Italy
  Adriatic Sea (ay-drih-*at'*-ik)
  Apennines (Mts.) (*app'*-uh-nines)
  Bologna (boh-*lohn'*-ya)
  Genoa (*jen'*-o-a)
  Milan (mih-*lahn'*)
  Naples (*nay'*-pels)
  Palermo (pah-*ler'*-moh)
  Turin (*toor'*-in)
Ivory Coast (Côte d'Ivoire) (koht-d'-vwahr)
  Abidjan (abb'-ih-*jahn'*)

Jamaica (ja-*may'*-ka)
Japan
  Fukuoka (foo'-koo-*oh'*-kah)
  Hiroshima (heer'-a-*shee'*-mah)
  Hokkaido (island) (hah-*keye'*-doh)
  Honshu (island) (*honn'*-shoo)
  Kitakyushu (kee-*tah'*-*kyoo'*-shoo)
  Kobe (*koh'*-bay)
  Kyoto (kee-*oht'*-oh)
  Kyushu (island) (kee-*yoo'*-shoo)
  Nagoya (nay-*goy'*-ya)
  Osaka (oh-*sahk'*-a)
  Sapporo (sahp-*pohr'*-oh')
  Shikoku (island) (shih-*koh'*-koo)
  Tokyo (*toh'*-kee-oh)
  Yokohama (yoh'-ko-*hahm'*-a)
Jordan (*jor'*-dan)
  Amman (a-*mahn'*)

Kazakhstan (kuh-*zahk'*-stahn)
  Almaty (al-*mah'*-tee)
  Aral Sea (uh-*rahl'*)
  Karaganda (kahr'-rah-*gahn'*-dah)
  Lake Balkhash (bul-*kahsh'*)
  Syr Darya (river) (*seer'* dahr-*yah'*)
Kenya (*ken'*-ya)
  Mombasa (mahm-*bahs*-a)
  Nairobi (neye-*roh'*-bee)
Kiribati (keer'-ih-*bah'*-tih)
Korea, North
  Pyongyang (pea-ong-*yahng'*)
  Yalu River (*yah'*-loo)
Korea, South
  Inchon (*inn'*-chonn)
  Pusan (*poo'*-sahn)
  Seoul (*sohl'*)
  Taegu (teye-*goo'*)
Kuwait (koo-*wayt'*)
Kyrgyzstan (keer'-geez-*stahn'*)

Laos (*lah'*-ohs)
  Vientiane (vyen-*tyawn'*)
Latvia
  Riga (*reeg'*-a)
Lebanon (*leb'*-a-non)
  Beirut (bay-*root'*)
Lesotho (luh-*soh'*-toh)
Libya (*lib'*-yah)
  Benghazi (benn-*gah'*-zee)
  Tripoli (*tripp'*-uh-lee)
Liechtenstein (*lick'*-tun-stine')
Lithuania (lith'-oo-*ane'*-ee-uh)
  Vilnius (*vill'*-nee-us)
Luxembourg (*luk'*-sem-burg)

Madagascar
  Antananarivo (an-ta-nan'-a-*ree'*-voh)
Malawi (muh-*lah'*-wee)
  Lilongwe (lih-*lawng'*-way)
Malaysia
  Kuala Lumpur (*kwah'*-luh *loom'*-poor)
    ("oo" as in "wood")
  Sabah (*sah'*-bah)
  Sarawak (suh-*rah'*-wahk)
  Strait of Malacca (muh-*lahk'*-uh)
Maldives (*mawl'*-deevz)
Mali (*mah'*-lee)
  Bamako (*bam'*-ah-koh')
Malta (*mawl'*-tuh)
Martinique (mahr'-tih-*neek'*)
Mauritania (mahr-ih-*tay'*-nee-a)
Mauritius (muh-*rish'*-us)
Mexico
  Acapulco (a-ca-*pool*-ko)
  Ciudad Juárez (see-yoo-*dahd'*
    wahr'-ez)
  Guadalajara (gwad'-a-la-*har'*-a)
  Monterrey (mahnt-uh-*ray'*)
  Puebla (poo-*eb'*-la)
  Tijuana (tee-*hwah'*-nah)
Moldova (mohl-*doh'*-vah)
  Kishinev (kish'-in-*yeff'*)
Monaco (*mon'*-a-koh)
Mongolia (mon-*gohl'*-yuh)
  Ulan Bator (oo-*lahn'* *bah'*tor)
Montserrat (mont'-suh-*rat'*)

Morocco (moh-*rah'*-koh)
  Casablanca (kas-a-*blang'*-ka)
  Rabat (rah-*baht'*)
Mozambique (moh'-zamm-*beek'*)
  Beira (*bay'*-rah)
  Cabora Bassa Dam (kah-*bore'*-ah
    bah'-sah)
  Maputo (mah-*poot'*-oh')

Namibia (nah-*mib'*-ee-a)
  Windhoek (*Vind'*-hook)
Nauru (nah-oo-roo)
Nepal (nuh-*pawl'*)
  Katmandu (kat'-man-*doo'*)
Netherlands
  Ijsselmeer (*eye'*-sul-mahr')
  The Hague (*hayg'*)
  Utrecht (*yoo'*-trekt')
  Zuider Zee (*zeye'*-der zay')
Netherlands Antilles
  Curaçao (*koo'*-rahs-ow')
New Zealand
  Auckland (*aw'*-klund)
Nicaragua (nik'-a-*rahg'*-wah)
Niger (*neye'*-jer; Fr., nee'-*zhere'*)
  Niamey (nee-*ah'*-may)
Nigeria (neye-*jeer*-ee-a)
  Abuja (a-*boo'*-ja)
  Ibadan (ee-*bahd'*-n)
  Kano (*kahn'*-oh)
  Lagos (*lahg'*-us)
Norway
  Bergen (*berg'*-n)
  Oslo (*ahz'*-loh)
  Stavanger (stah-*vahng'*-er)
  Trondheim (*trahn'*-haym')
Oman (oh-*mahn'*)
  Muscat (*mus'*-kat')

Pakistan (pahk-i-*stahn'*)
  Hyderabad (*heyed'*-er-a-bahd')
  Islamabad (is-*lahm*-a-bahd')
  Karachi (kah-*rahch'*-ee')
  Karakoram (Mts.) (kahr'-a-*kohr*-um)
  Lahore (lah-*hohr'*)
  Peshawar (puh-*shah'*-wahr)
  Punjab (see India)
Papua New Guinea (*pap'*-yoo-uh new
gin'-ee)
  Port Moresby (*morz'*-bee)
Paraguay (*pair'*-uh-gweye')
  Asuncíon (a-*soon'*-see-*ohn'*)
Peru (puh-*roo'*)
  Callao (kah-*yow'*)
  Cerro de Pasco (*sehr'*-roh day *pahs'*-koh)
  Chimbote (chimm-*boh'*-te)
  Cuzco (*koos'*-koh)
  Iquitos (ee-*kee'*-tohs)
  Lima (*lee'*-ma)
  Machu Picchu (ma-chu *pee*-chu)
Philippines (*fil*-i-peenz')
  Luzon (island) (loo-*zahn'*)
  Mindanao (island) (minn'-dah-*nah'*-oh)
  Mt. Pinatubo (pea'-nah-*too'*-buh)
  Visayan Islands (vih-*sah'*-y'n)
Poland
  Gdansk (ga-*dahntsk'*)

Katowice (kaht-uh-*veet*'-sih)
Krakow (*krahk*'-ow)
Lodz (*looj*')
Posnan (*pohz*'-nan-ya)
Silesia (region) (sih-*lee*'-shuh)
Szczecin (*shchet*'-seen')
Wroclaw (*vrawt*'-slahf)
Portugal (*pohr*'-chi-gal)
Lisbon (*liz*'-bon)
Oporto (oh-*pohrt*'-oh)

Qatar (*kaht*'-ar)

Romania (roh-*main*'-yuh)
Bucharest (boo'-kuh-*rest*')
Carpathian Mts. (kahr-*pay*'-thih-un)
Russia
Altai Mts. (*al*'-teye)
Amur R. (ah-*moor*') ("oo" as in "wood")
Angara R. (ahn'-guh-*rah*')
Arkhangelsk (ahr-*kann*'-jelsk)
Astrakhan (as-trah-*khahn*')
Baikal, Lake (beye-*kahl*')
Bashkiria (bahsh-*keer*'-ee-uh)
Bratsk (*brahtsk*')
Buryatia (boor-*yaht*'-ee-uh)
Caucasus (*kaw*'-kah-suss)
Chechen-Ingushia (*chehch*'-yenn enn-*gooshia*')
Chelyabinsk (*chell*'-yah-binnsk)
Cherepovets (*chehr*'-yuh-puh-vyetz')
Chuvashia (choo-*vahsh*'-i-a)
Dagestan Rep. (dag'-uh-*stahn*')
Irkutsk (ihr-*kootsk*')
Ivanovo (ih-*vahn*'-uh-voh)
Izhevsk (ih-*zhehvsk*')
Kama R. (*kah*'-mah)
Karelia (kah-*reel*'-yuh)
Kazan (kuh-*zahn*')
Khabarovsk (kah-*bah*'-rawfsk)
Kola Pen. (*koh*'-lah)
Kolyma R. (kuh-*lee*'-muh)
Krasnoyarsk (kras'-nuh-*yahrsk*')
Kuril Is. (*koo*'-rill)
Kuznetsk Basin (kooz'-*netsk*')
Ladoga, Lake (*lad*'-uh-guh)
Lena R. (*lee*-nah, *lay*-nah)
Lipetsk (*lyee*'-petsk')
Magnitogorsk (mag-*nee*'-toh-gorsk')
Moscow (*mahs*'-koh or *mahs*'-kow)
Murmansk (moor'-*mahntsk*')
Nakhodka (nuh-*kawt*-kah)
Nizhniy Novgorod (*nizh*'-nee nahv-guh-*rahd*')
Nizhniy Tagil (tah-*geel*)
Novokuznetsk (naw'-voh-*kooznetsk*')
Novaya Zemlya (*naw*'-vuh-yuh zem-*lyah*')
Novosibirsk (naw-vuh-suh-*beersk*')
Oka R. (oh-*kah*')
Okhotsk, Sea of (oh-*kahtsk*')
Omsk (*ahmsk*')
Pechora R. (peh-*chore*'-a)
Rostov (rahs-*tawf*')
Sakhalin (sahk'-uh-*leen*')
Samara (suh-*mahr*'-a)

Saratov (suh-*raht*'-uff)
Sayan Mts. (sah-*yahn*')
Tatarstan (Tatar Rep.) (tah-*tahr*'-stahn')
Togliatti (tawl-*yah*'-tee)
Tuva Rep. (*too*'-va)
Ufa (oo-*fah*')
Ussuri R. (oo-*soor*'-i)
Vladivostok (vlah'-dih-vahs-*tawk*')
Volga R. (*vahl*-guh, *vohl*'-guh)
Volgograd (vahl-guh-*grahd*', vohl'-)
Voronezh (vah-*raw*'-nyesh)
Yakutsk (yah-*kootsk*')
Yekaterinburg (yeh-*kahta*'-rinn-berg)
Rwanda (roo-*ahn*'-da)

St. Kitts and Nevis (*nee*'-vis)
St. Lucia (*loo*'-sha)
St. Vincent-Grenadines (gren'-a-*deenz*')
Samoa (sa-*moh*'-a)
Apia (ah-*pea*'-uh)
São Tomé and Príncipe (sau too-*meh*', pren-*see*'-puh)
Saudi Arabia (*saud*'-ee)
Jiddah (*jid*'-a)
Riyadh (ree-*adh*')
Senegal (sen'-i-*gahl*')
Dakar (da'-*kahr*')
Seychelles (say'-*shelz*')
Sierra Leone (see-*ehr*'-a lee-*ohn*')
Slovenia (sloh-*veen*'-i-a)
Ljubljana (*lyoo*'-b'l-yah-nah)
Somalia (so-*mahl*'-ee-a)
Mogadishu (mahg'-uh-*dish*'-oo)
South Africa
Bloemfontein (*bloom*'-fonn-*tayn*')
Bophuthatswana (homeland) (boh-*poot*'-aht-*swahna*')
Drakensberg (escarpment) (*drahk*-n's-burg')
Durban (*durr*'-bun)
Johannesburg (joh-*hann*'-iss-burg)
Kwazulu (homeland) (kwah-zoo'-loo)
Natal (province) (nuh-*tahl*')
Pretoria (prih-*tohr*'-ee-uh)
Transkei (homeland) (trans-*keye*')
Transvaal (province) (trans-*vahl*')
Spain
Barcelona (bahr'-suh-*lohn*-uh)
Bilbao (bill-*bah*'-oh)
Catalonia (region) (katt'-uh-*lohn*'-ee-a)
Madrid (muh-*dridd*')
Meseta (plateau) (muh-*say*'-tuh)
Seville (suh-*vill*')
Valencia (vull-*enn*'-shee-uh)
Sri Lanka (sree *lahn*'-kah)
Sudan (*soo*'-dann)
Gezira (guh-*zeer*'-uh)
Juba (*joo*'-bah)
Khartoum (kahr'-*toom*')
Omdurman (ahm-*durr*'-man)
Wadi Halfa (*wah-di*' *hawl*'-fah)
Suriname (*soor*'-i-nahm')
Swaziland (*swah*'-zih-land')
Sweden
Göteborg (yort'-e-*bor*')
Skåne (*skohn*')
Småland (*smoh*'-lund')

Switzerland
Basel (*bah*'-z'l)
Zürich (*zoor*'-ick)
Syria (*seer*'-i-uh)
Aleppo (a-*lep*-oh')
Damascus (da-*mas*'-kus)

Tajikistan (tah-*jick*'-ih-stahn')
Tanzania (tan'-za-*nee*'-a)
Dar es Salaam (dahr' es sa-*lahm*')
Thailand (*teye*'-land')
Bangkok (*bang*'-kok')
Togo (*toh*'-goh)
Lomé (loh-*may*')
Tonga (*tawng*'-a)
Trinidad and Tobago (to-*bay*'-goh)
Tunisia (too-*nee*'-zhee-a)
Tunis (*too*'-nis)
Turkey
Anatolian Peninsula (ann'-ah-*tohl*'-yun)
Ankara (*ang*'-kah-rah)
Bosporus (strait) (*bahs*'-puhr-us)
Dardanelles (strait) (dahr'-duh-*nelz*')
Istanbul (iss'-tann-*bool*')
Izmir (izz'-*meer*')
Sea of Marmara (*mahr*'-muh-ruh)
Turkmenistan (turk'-menn-ih-*stahn*')
Ashgabat (*ash*'-guh-baht)

Ukraine (yoo'-*krane*')
Crimea (cry'-*mee*'-uh)
Dnepropetrovsk (d'nyepp-pruh-pay-*trawfsk*')
Dnieper R. (duh-*nyepp*'-er)
Donetsk (duhn-*yehtsk*')
Kharkov (*kahr*'-kawf)
Kiev (*kee*'-yeff)
Krivoy Rog (*kree*'-voi *rohg*')
Sevastopol (sih-*vass*'-tuh-*pohl*')
Yalta (*yahl*'-tuh)
Zaporozhye (zah-puh-*rawzh*'-yuh)
United Arab Emirates
Abu Dhabi (*ah*'-boo dah'-bee)
Dubai (doo-*beye*')
United Kingdom
Edinburgh (*ed*'-in-buh-ruh)
Glasgow (*glass*'-goh)
Thames R. (*timms*')
United States
Albuquerque (*al*'-buh-ker'-kih)
Aleutian Is. (uh-*loo*'-sh'n)
Allegheny R. (*al*'-ih-gay'-nih)
Appalachian Highlands (ap'-uh-*lay*'-chih-un)
Coeur d'Alene (*kurr*' duh-*layn*')
Des Moines (duh *moin*')
Grand Coulee Dam (*koo*'-lee)
Juneau (*joo*'-noh)
Mesabi Range (muh-*sahb*'-ih)
Monongahela R. (muh-*nahng*'-guh-*hee*'-luh)
Omaha (*oh*'-mah-*hah*')
Ouachita Mts. (*wosh*'-ih-taw')
Palouse (region) (pah-*loose*')
Phoenix (*fee*'-nix)
Provo (*proh*'-voh)
Rio Grande (*ree*'-oh *grahn*'-day)

San Joaquin R. (sann wah-*keen'*)
Spokane (spoh'-*kann'*)
Tucson (*too'*-sonn)
Valdez (val-*deez'*)
Uruguay (*yoor'*-uh-gweye')
  Montevideo (mahnt'-e-vi-*day'*-oh)
Uzbekistan (ooz-*beck'*-ih-stahn') ("oo" as in
  "wood")
  Amu Darya (river) (ah-moo' *dahr'*-yuh,
    dahr-*yah'*)
  Bukhara (buh-*kah'*-ruh)
  Fergana Valley (fair-*gahn'*-uh)
  Samarkand
    (*samm'*-ur-*kand'*; Rus., suh-mur-
    kahnt')
  Tashkent (tash-*kent'*, tahsh-)

Vanuatu (van'-oo-*ay'*-too)
  Port Vila (*vee'*-la)
Venezuela (ven'-ez-*way'*-la)
  Caracas (ka-*rak'*-as)

Maracaibo (mar'-a-*keye'*-boh)
Orinoco R. (oh-rih-*noh'*-koh)
Valencia (va-*len'*-see-a)
Vietnam (vee-et'-*nahm'*)
  Annam (region) (uh-*namm'*)
  Cochin China (*koh'*-chin)
  Hanoi (ha-*noi'*)
  Ho Chi Minh City (Saigon)
    (hoe chee *min'*; *seye'*-gahn)
  Mekong R. (may-*kahng'*)

Yemen
  Aden (*ah*-d'n)
  S'ana (sah-*na'*)
Yugoslavia (yoo'-goh-*slahv'*-ee-a)
  Belgrade (*bell'*-grade, -grahd)
  Kosovo (*kaw'*-suh-voh')
  Montenegro (republic)
    (mahn'-tee-*nay'*-groh)

Zaire (*zeye'*-eer')
  Kananga (kay-*nahng'*-guh)
  Kasai R. (kah-*seye'*)
  Kinshasa (*keen'*-shah-suh)
  Kisangani (kee'-sahn-*gahn'*-ih)
  Lake Kivu (*kee'*-voo)
  Lualaba R. (loo'-ah-*lahb'*-a)
  Lubumbashi (loo-boom-*bah'*-shee)
  Matadi (muh-*tah'*-dee)
  Shaba (region) (*shah'*-bah)
Zambia (*zam'*-bee-a)
  Kariba Dam (kah-*ree'*-buh)
  Lusaka (loo-*sahk'*-a)
  Zambezi R. (zamm-*bee'*-zee)
Zimbabwe (zim-*bahb'*-way)
  Bulawayo (bool'-uh-*way'*-oh)
  Harare (hah-*rahr'*-ee)
  Que Que (*kway' kway'*)
  Wankie (*wahn'*-kee)

index

credits

Figure 1.1: (right) From *Geography for Life: National Geography Standards 1994*, National Geographic Research and Exploration, Washington, DC, pp. 34–35 Table 1.1: Source: *World Population Data Sheet*, Population Reference Bureau, 2007; *U.N. Human Development Report*, United Nations, 2007; *World Factbook*, CIA, 2007. Figure 1.9: Source: *Goode's 20th edition*, published by Rand McNally. Figure 1.11: (top) After Bergman, 1995; (bottom) After Rubenstein, 1995 Table 1.3: Source: www.aag.org. Figure 2.1: Adapted from NASA, "Global Tectonic Activity Map of the Earth," DTAM-1, 2002. Figure 2.2: Adapted from Donald Hyndman and David Hyndman. 2006. *Natural Hazards and Disasters*. Belmont: Brooks Cole, p. 16. Figure 2.3: Adapted with permission from UNEP/GRID, "Mean Annual Precipitation," Geneva Data Sets-Atmosphere, GNV174 (http://www.grid.unep.ch/data). Figure 2.8: From Tom L. McKnight and Darrel Hess, *Physical Geography: A Landscape Appreciation, Animation Edition*, 7th ed., © 2004. Reproduced by permission of Pearson Education, Inc., Upper Saddle River, NJ Figure 2.10: Adapted from Rand McNally's *Classroom Atlas*, 2003. Figure 2.11: Adapted from World Wildlife Fund ecoregions data, 1999. Figure 2.13: Adapted from Conservation International, "Hotspots Explorer," 2007 (http://www.biodiversity hotspots.org/xp/Hotspots). Figure 2.16: Cornelia Dean, "Study Sees 'Global Collapse' of Fish Species." *New York Times*, November 3, 2006, p. A16. Figure 3.5: Data from World Bank, 2007. Figure 3.B: Adapted from David W. Pearce and R. Kerry Turner, *Economics of Natural Resources and the Environment* (Baltimore: Johns Hopkins University Press, 1990), p. 348. Figure 3.7: Data from Population Reference Bureau, 2007 Figure 3.8: Data from Population Reference Bureau, 2007 Figure 3.9: Data from Population Reference Bureau, 2007 Figure 3.15: Data from Population Reference Bureau, 2007 Figure 3.16: Data from Population Reference Bureau, 2007 Figure 3.21: From *Environmental Science*, 4th edition, by Miller. © 1993. Reprinted with permission of Wadsworth, a division of Thomson Learning, Inc. Table 4.1: Source: *World Population Data Sheet*, Population Reference Bureau, 2007; *U.N. Human Development Report*, United Nations, 2007; *World Factbook*, CIA, 2007. Figure 4.6: After Hoffman, 1990 Figure 4.7: (left) Adapted from Rand McNally's *Classroom Atlas*, 2003; (right) Adapted from World Wildlife Fund ecoregions data, 1999. Figure 4.8: From Patrick Wiegand, ed., *The Oxford School Atlas* (Oxford University Press, 1997). Copyright © 1997 Oxford University Press. Reprinted by permission of Oxford University Press. Figure 4.11: Adapted from Bernard Comrie et al., *The Atlas of Languages*, rev. ed. (New York: Facts on File, 2003), p. 41. Figure 4.B: Adapted from a map in *The Economist*, Sept. 20, 1997, p. 53. © 1998 The Economist Newspaper Ltd. All rights reserved. Reprinted with permission. Further reproduction prohibited. Table 4.2: From Jennifer Ehrlich and Tom Vandyck, "Belgian 'Malcolm X' Seeks Office," *The Christian Science Monitor*, May 16, 2003, p. 6. Data originates from *A Guidebook on Islam and Muslims in the Wide Contemporary Europe*, edited by B. Marechal (Louvain-la-Neuve: Academia Bruylant, 2002) Figure 4.15: Adapted from "European Union Survey," *The Economist*, May 31, 1997, p. 5. © 1997 The Economist Newspaper Ltd. All rights reserved. Reprinted with permission. Further reproduction prohibited. Figure 4.16: Adapted NATO map data from EUROSTAT (Statistical Office of the European Communities), 1996; *World Factbook*, CIA, 1995. Figure 4.1.2: Data from Global Forest Watch, 2004 (http://www.globalforestwatch.org/english/datawarehouse/index .asp). Figure 4.1.4: Based on Terry G. Jordan and Bella Bychova Jordan, *The European Culture Area: A Systematic Geography* (Lanham, Md.: Rowman and Littlefield, 2002), p. 300.

Figure 4.1.5: After Hoffman, 1990, p. 245. Figure 4.1.9: Based on Terry G. Jordan and Bella Bychova Jordan, *The European Culture Area: A Systematic Geography* (Lanham, Md.: Rowman and Littlefield, 2002), p. 326. Figure 4.2.22: Adapted from maps in John Haywood, ed., *Atlas of World History* (Barnes and Noble Books, 2004) Figure 4.2.30: Adapted from Ian Barnes and Robert Hudson, *The History Atlas of Europe* (Macmillan Inc., 1998), p. 151. Table 5.1: *World Population Data Sheet*, Population Reference Bureau, 2007; *U.N. Human Development Report*, United Nations, 2007; *World Factbook*, CIA, 2007. Figure 5.2: (top) From Patrick Wiegand, ed., *The Oxford School Atlas* (Oxford University Press, 1997). Copyright © 1997 Oxford University Press. Reprinted by permission of Oxford University Press. Figure 5.2: Data from *World Factbook*, CIA, 2007. Figure 5.4: (left) Adapted from Rand McNally's *Classroom Atlas*, 2003; (right) Adapted from World Wildlife Fund ecoregions data, 1999. Figure 5.6: From Patrick Wiegand, ed., *The Oxford School Atlas* (Oxford University Press, 1997). Copyright © 1997 Oxford University Press. Reprinted by permission of Oxford University Press. Figure 5.11: Adapted with permission from Bernard Comrie, et al., *The Atlas of Languages*, Revised Edition (New York: Facts on File, 2003). Figure 5.D: Based on Robin R. Milner-Gulland, *Cultural Atlas of Russia and the Former Soviet Union* (New York: Checkmark, 1998), map X.X. Figure 6.3: (top) From Patrick Wiegand, ed., *The Oxford School Atlas* (Oxford University Press, 1997). Copyright © 1997 Oxford University Press. Reprinted by permission of Oxford University Press. Figure 6.5: (left) Adapted from Rand McNally's *Classroom Atlas*, 2003; (right) Adapted from World Wildlife Fund ecoregions data, 1999. Figure 6.7: From Patrick Wiegand, ed., *The Oxford School Atlas* (Oxford University Press, 1997). Copyright © 1997 Oxford University Press. Reprinted by permission of Oxford University Press. Figure 6.17: Bernard Comrie, et al., *The Atlas of Languages*, Revised Edition (New York: Facts on File, 2003). Figure 6.C: Adapted from National Geographic's *Atlas of the Middle East* (National Geographic, 2003). Figure 6.1.1: Adapted from John Haywood, ed., *Atlas of World History* (Barnes and Noble Books, 1997). Figure 6.1.2: From *The Middle East and North Africa: A Political Geography*, by A. Drysdale and G. Blake. Copyright © 1985 Oxford University Press, Inc. Used by permission of Oxford University Press, Inc. Figure 6.1.3: From *The Middle East and North Africa: A Political Geography*, by A. Drysdale and G. Blake. Copyright © 1985 Oxford University Press, Inc. Used by permission of Oxford University Press, Inc. Figure 6.1.B: Zone control data from Foundation for Middle East Peace, 2002 (www .fmep.org); security wall route from IDF maps. Figure 6.1.7: Held, Colbert. 1994. *Middle East Patterns*. Figure 6.1.23: Data from CIA maps. Figure 6.1.32: Ethnicity and poppy data from CIA maps. Figure 7.3: (top) From Patrick Wiegand, ed., The Oxford School Atlas (Oxford University Press, 1997). Copyright © 1997 Oxford University Press. Reprinted by permission of Oxford University Press. Figure 7.5: (left) Adapted from Rand McNally's *Classroom Atlas*, 2003.; (right): Adapted from World Wildlife Fund ecoregions data, 1999. Figure 7.7: From Patrick Wiegand, ed., The Oxford School Atlas (Oxford University Press, 1997). Copyright © 1997 Oxford University Press. Reprinted by permission of Oxford University Press. Figure 7.11: Bernard Comrie, et al., *The Atlas of Languages*, Revised Edition (New York: Facts on File, 2003). Figure 7.F: From David E. Sanger, "China's Big Need for Oil is High on U.S. Agenda," *The New York Times*, 4-19-06, A1. Figure 7.17: From "The World in 1900," in *Atlas of World History*, ed. R. R. Palmer (Chicago: Rand McNally, 1957), pp. 169–171. Used with permission. Figure 7.1.7: A. Smith, *Atlas of Mesozoic and Cenozoic*

Coastlines (New York: Cambridge University Press, 1994). **Figure 7.2.7:** Data from Global Forest Watch, 2004 (http://www.global forestwatch.org/english/datawarehouse/index.asp). **Figure 7.2.9:** From Wayne Arnold, "A Gas Pipeline to the World Outside," *The New York Times,* 10-26-01, C1. Copyright © 2001 The New York Times Co. **Figure 7.3.E:** Sources: Erik Eckholm, "China Will Move Waters to Quench Thirst of Cities," *The New York Times,* 8-27-02, A1; "Water in China: In Deep," *The Economist,* 08-18-01, 31. **Figure 7.3.14:** *China Atlas of Population and Environment;* http://141.211.136.211/eng/macro/atlas2000/ybtableview.asp?ID=4). **Table 8.1:** *World Population Data Sheet,* Population Reference Bureau, 2007; *U.N. Human Development Report,* United Nations, 2007; *World Factbook,* CIA, 2007. **Figure 8.3:** (left) From Patrick Wiegand, ed., *The Oxford School Atlas* (Oxford University Press, 1997). Copyright © 1997 Oxford University Press. Reprinted by permission of Oxford University Press. **Figure 8.5(a):** (left) Adapted from Rand McNally's *Classroom Atlas,* 2003; (right) Adapted from World Wildlife Fund ecoregions data, 1999. **Figure 8.7:** *The Oxford School Atlas* (Oxford University Press, 1997). **Figure 8.9:** Robert E. Gabler, James F. Peterson, and L. Michael Trepasso, *Essentials of Physical Geography* (Philadelphia: Saunders, 1987), p. 562. **Figure 8.12:** Bernard Comrie et al., *The Atlas of Languages,* rev. ed (New York: Facts on File, 2003); settlement data from John Noble Wilford, "Seeking Polynesia's Beginnings in an Archipelago of Shards." *New York Times,* January 8, 2002, p. D1. **Figure 9.3:** (top) From Patrick Wiegand, ed., *The Oxford School Atlas* (Oxford University Press, 1997). Copyright © 1997 Oxford University Press. Reprinted by permission of Oxford University Press. **Figure 9.8:** (left) Adapted from Rand McNally's *Classroom Atlas,* 2003; (right) Adapted from World Wildlife Fund ecoregions data, 1999. **Figure 9.9:** From Patrick Wiegand, ed., *The Oxford School Atlas* (Oxford University Press, 1997). Copyright © 1997 Oxford University Press. Reprinted by permission of Oxford University Press. **Figure 9.A:** (left) Joint United Nations Program on H.I.V./AIDS. (right) Data from UNAIDS. **Figure 9.13:** Bernard Comrie, et al., *The Atlas of Languages,* Revised Edition (New York: Facts on File, 2003). **Figure 9.15:** Robert Stock, *Africa South of the Sahara,* 2nd ed. (New York: Guilford Press, 2004), p. 10. **Figure 9.16:** Adapted from John Haywood, ed., *Atlas of World History* (Barnes and Noble Books, 1997). **Figure 9.18:** Adapted from USGS country mineral maps. **Figure 9.1.1:** Robert Stock, *Africa South of the Sahara.* **Figure 9.1.B:** From "Linguistic Groups in Nigeria,", CIA map, 1979. **Table 10.1:** *World Population Data Sheet,* Population Reference Bureau, 2007; *U.N. Human Development Report,* United Nations, 2007; *World Factbook,* CIA, 2007. **Figure 10.3:** From Patrick Wiegand, ed., *The Oxford School Atlas* (Oxford University Press, 1997). Copyright © 1997 Oxford University Press. Reprinted by permission of Oxford University Press. **Figure 10.5(a):** (left) Adapted from Rand McNally's *Classroom Atlas,* 2003; (right) Adapted from World Wildlife Fund ecoregions data, 1999. **Figure 10.6:** David L. Clawson, *Latin America and the Caribbean* (Dubuque, Iowa: Brown, 1997), p. 46. **Figure 10.7:** From Patrick Wiegand, ed., *The Oxford School Atlas* (Oxford University Press, 1997). Copyright © 1997 Oxford University Press. Reprinted by permission of Oxford University Press. **Figures 10.A and B:** From *Living in the Environment,* 13th edition, by Miller. © 2004. Reprinted with permission of Brooks/Cole, a division of Thomson Learning, Inc. **Figure 10.11:** *The History Atlas of South America,* Edwin Early, ed., (Macmillan, 1998). **Figure 10.15:** Bernard Comrie, et al., *The Atlas of Languages,* Revised Edition (New York: Facts on File, 2003). **Table 10.2:** *World Factbook,* CIA, 2007. **Figure 10.23:** Adapted from *The Economist,* April 11, 1998, p. 125. **Figure 10.E:** Data from the U.N. Drug Control Program (UNDCP), 2007. **Figure 10.1.6:** G. Tyler Miller, *Living in the Environment,* 13th ed. © 2004. Reprinted with permission of Brooks/Cole, a division of Thomson Learning Inc. **Figure 10.2.14:** Adapted from *Atlas of World History* (Barnes and Noble Books, 1997). **Figure 10.2.C:** Data from Global Forest Watch, 2004 (http://www .globalforestwatch.org/english/interactive.maps/brazil.htm). **Table 11.1** *World Population Data Sheet,* Population Reference Bureau, 2007; *U.N. Human Development Report,* United Nations, 2007; *World Factbook,* CIA, 2007. **Figure 11.2:** (top) From Patrick Wiegand, ed., *The Oxford School Atlas* (Oxford University Press, 1997). Copyright © 1997 Oxford University Press. Reprinted by permission of Oxford University Press. **Figure 11.3:** Data from the U.S. Census Bureau and Statistics Canada. **Figure 11.5(a):** (left) Adapted from Rand McNally's *Classroom Atlas,* 2003; (right) Adapted from World Wildlife Fund ecoregions data, 1999. **Figure 11.6:** From Patrick Wiegand, ed., *The Oxford School Atlas* (Oxford University Press, 1997). Copyright © 1997 Oxford University Press. Reprinted by permission of Oxford University Press. **Figure 11.7:** Map adapted from Timothy Egan, "Vanishing Point: Amid Dying Towns of Rural Plains, One Makes a Stand," *The New York Times,* 12/01/03, p. A18. **Figure 11.8:** Carl Waldman, *Atlas of the North American Indian* (Facts on File, 1985). **Figure 11.12:** Adapted from John Haywood, ed., *Atlas of World History* (Barnes and Noble Books, 1997). **Figure 11.15:** Bernard Comrie, et al., *The Atlas of Languages,* Revised Edition (New York: Facts on File, 2003). **Figure 11.16:** *New Historical Atlas of Religion in America* (Oxford University Press, 2001). **Figure 11.18:** (left) Map from Alberta Geological Survey (http://www.ags.gov.ab.ca/GRAPHICS/CBM/oil_sands/alberta_oil_sands_map2.jpg). **Figure 11.1.7:** Getis, Arthur and Judith Getis. 1995. *The United States and Canada: The Land and Its People.* Dubuque, Iowa: Wm. C. Brown Publishers, p. 231. **Figure 11.1.B:** Map adapted from: Krauss, Clifford. October 10, 2005. "As Polar Ice Turns to Water, Dreams of Treasure Abound." *The New York Times,* pp. A1, 10, 11. **Table 11.2.1** *World Population Data Sheet,* Population Reference Bureau, 2007; *U.N. Human Development Report,* United Nations, 2007; *World Factbook,* CIA, 2007. **Endpapers:** front and rear, courtesy of National Geographic, © 2002.